ANNA A. SHER | MANUEL C. MOLLES JR.

ECOLOGY

CONCEPTS & APPLICATIONS

NINTH EDITION

ECOLOGY: CONCEPTS AND APPLICATIONS, NINTH EDITION

Published by McGraw Hill LLC, 1325 Avenue of the Americas, New York, NY 10019. Copyright © 2022 by McGraw Hill LLC. All rights reserved. Printed in the United States of America. Previous editions © 2019, 2016, and 2013. No part of this publication may be reproduced or distributed in any form or by any means, or stored in a database or retrieval system, without the prior written consent of McGraw Hill LLC, including, but not limited to, in any network or other electronic storage or transmission, or broadcast for distance learning.

Some ancillaries, including electronic and print components, may not be available to customers outside the United States.

This book is printed on acid-free paper.

1 2 3 4 5 6 7 8 9 LWI 26 25 24 23 22 21

ISBN 978-1-260-72220-8 (bound edition)
MHID 1-260-72220-1 (bound edition)
ISBN 978-1-264-36071-0 (loose-leaf edition)
MHID 1-264-36071-1 (loose-leaf edition)

Lead Product Developer: *Jodi Rhomberg*
Product Developer: *Theresa Collins, Beth Baugh*
Senior Marketing Manager: *Kelly Brown*
Content Project Managers: *Maria McGreal, Katie Reuter*
Buyer: *Laura Fuller*
Designer: *David W. Hash*
Content Licensing Specialist: *Melissa Homer*
Cover Image: *Anna A. Sher*
Compositor: *Straive*

All credits appearing on page or at the end of the book are considered to be an extension of the copyright page.

Library of Congress Cataloging-in-Publication Data

Names: Sher, Anna, author. | Molles, Manuel C., Jr., 1948- author.
Title: Ecology : concepts and applications / Anna Sher Simon, University of
    Denver, Manuel C. Molles Jr., University of New Mexico.
Description: Ninth edition. | New York : McGraw Hill LLC, [2022] | Includes
    bibliographical references and index.
Identifiers: LCCN 2021017767 (print) | LCCN 2021017768 (ebook) | ISBN
    9781260722208 (paperback) | ISBN 9781264360734 (ebook)
Subjects: LCSH: Ecology.
Classification: LCC QH541 .M553 2022 (print) | LCC QH541 (ebook) | DDC
    577—dc23
LC record available at https://lccn.loc.gov/2021017767
LC ebook record available at https://lccn.loc.gov/2021017768

The Internet addresses listed in the text were accurate at the time of publication. The inclusion of a website does not indicate an endorsement by the authors or McGraw Hill LLC, and McGraw Hill LLC does not guarantee the accuracy of the information presented at these sites.

mheducation.com/highered

Anna A. Sher is a full professor in the Department of Biological Sciences at the University of Denver, where she has been faculty since 2003. Until 2010 she held this position jointly with the Denver Botanic Gardens as the Director of Research and Conservation. As a student, she was a double major in Biology and Art at Earlham College, where she has also taught ecology, and was the co-leader of the Earlham Study Abroad Kenya Program. She received her PhD from the University of New Mexico, where she also taught botany as a visiting lecturer. As a postdoctoral researcher, Dr. Sher was awarded a Fulbright postdoctoral research fellowship to conduct research on plant interactions in Israel at Ben Gurion University's Mitrani Department of Desert Ecology, and she also studied the ecology of an invasive grass at the University of California, Davis. She has also been a visiting professor at the University of Otago, Dunedin, New Zealand.

Courtesy of Anna Sher

Dr. Sher's primary research focus has been on the ecological dynamics associated with the removal of invasive riparian plants. She is known as a leading expert in the ecology of *Tamarix,* a dominant exotic tree, and she was the lead editor of the first book exclusively on the topic. Her research interests and publications have spanned several areas within ecology, including not only restoration ecology, competition, and invasive species ecology, but also interactions between plants and soil chemistry, mycorrhizae, insect diversity and trophic cascades, ethnobotany, phenology, climate change, and rare species conservation. She is also lead author of the textbook series *An Introduction to Conservation Biology* (Oxford University Press). Dr. Sher has a particular interest in quantitative ecological methods, with her lab specializing in multivariate methods and spatial models at both individual organism and regional scales. She is currently principal investigator of a National Science Foundation award to investigate the human dimension of the restoration of damaged ecosystems, and she has been a TEDx speaker on the way ecosystems can teach us how to solve human problems.

Above all, Dr. Sher loves to teach and mentor students doing research at both undergraduate and graduate levels.

Manuel C. Molles Jr. is an emeritus Professor of Biology at the University of New Mexico, where he has been a member of the faculty and curator in the Museum of Southwestern Biology since 1975. He received his BS from Humboldt State University and his PhD from the Department of Ecology and Evolutionary Biology at the University of Arizona. Seeking to broaden his geographic perspective, he has taught and conducted ecological research in Latin America, the Caribbean, and Europe. He was awarded a Fulbright Research Fellowship to conduct research on river ecology in Portugal and has held visiting professor appointments in the Department of Zoology at the University of Coimbra, Portugal, in the Laboratory of Hydrology at the Polytechnic University of Madrid, Spain, and at the University of Montana's Flathead Lake Biological Station.

Originally trained as a marine ecologist and fisheries biologist, the author worked mainly on river and riparian ecology at the University of New Mexico. His research has covered a wide range of ecological levels, including behavioral ecology, population biology, community ecology, ecosystem ecology, biogeography of stream insects, and the influence of a large-scale climate system (El Niño) on the dynamics of southwestern river and riparian ecosystems. His current research interests focus on the influence of climate change and climatic variability on the dynamics of populations and communities along steep gradients of temperature and moisture in the mountains of the Southwest. Throughout his career, Dr. Molles has attempted to combine research, teaching, and service, involving undergraduate as well as graduate students in his ongoing projects. At the University of New Mexico, he taught a broad range of lower division, upper division, and graduate courses, including Principles of Biology, Evolution and Ecology, Stream Ecology, Limnology and Oceanography, Marine Biology, and Community and Ecosystem Ecology. He has taught courses in Global Change and River Ecology at the University of Coimbra, Portugal, and General Ecology and Groundwater and Riparian Ecology at the Flathead Lake Biological Station. Dr. Manuel Molles was named Teacher of the Year by the University of New Mexico for 1995–1996 and Potter Chair in Plant Ecology in 2000. In 2014, he received the Eugene P. Odum Award from the Ecological Society of America based on his "ability to relate basic ecological principles to human affairs through teaching, outreach and mentoring activities."

Courtesy of Manuel Molles

Design elements: Anna A. Sher

Dedication

To the Sher Lab and the whole next generation
of ecologists, who inspire me to do this work.
Also, I dedicate this edition to my co-author
and mentor, Manuel.

–Anna A. Sher

Brief Contents

1 Introduction to Ecology: Historical Foundations and Developing Frontiers   1

Section I   Natural History and Evolution   11

2 Life on Land   11
3 Life in Water   44
4 Population Genetics and Natural Selection   78

Section II   Adaptations to the Environment   101

5 Temperature Relations   101
6 Water Relations   127
7 Energy and Nutrient Relations   149
8 Social Relations   172

Section III   Population Ecology   196

9 Population Distribution and Abundance   196
10 Population Dynamics   215
11 Population Growth   237
12 Life Histories   254

Section IV   Interactions   277

13 Species Interactions and Competition   277
14 Exploitative Interactions: Predation, Herbivory, Parasitism, and Disease   299
15 Mutualism   325

Section V   Communities and Ecosystems   345

16 Species Abundance and Diversity   345
17 Species Interactions and Community Structure   365
18 Primary and Secondary Production   383
19 Nutrient Cycling and Retention   403
20 Succession and Stability   423

Section VI   Large-Scale Ecology   445

21 Landscape Ecology   445
22 Geographic Ecology   468
23 Global Ecology   490

Appendix A   Investigating the Evidence 514
Appendix B   Statistical Tables   541
Appendix C   Abbreviations Used in This Text   545
Appendix D   Global Biomes   547

Design elements: Anna A. Sher

Contents

Preface   xiii

Chapter 1   Introduction to Ecology: Historical Foundations and Developing Frontiers   1

Concepts   1

1.1   Overview of Ecology   2
Concept 1.1 Review   3

1.2   Sampling Ecological Research   3
Climatic and Ecological Change: Past and Future   7
Concept 1.2 Review   9

Applications: Ecology Can Inform Environmental Law and Policy   9

Section I
NATURAL HISTORY AND EVOLUTION

Chapter 2   Life on Land   11

Concepts   11
Terrestrial Biomes and the Importance of Plants   12

2.1   Large-Scale Patterns of Climatic Variation   14
Temperature, Atmospheric Circulation, and Precipitation   14
Climate Diagrams   16
Concept 2.1 Review   16

2.2   Other Factors That Shape Terrestrial Biomes   17
Concept 2.2 Review   19

2.3   Natural History and Geography of Biomes   19
Tropical Rain Forest   19
Tropical Dry Forest   21
Tropical Savanna   22
Desert   25
Woodland and Shrubland   27
Temperate Grassland   29
Temperate Forest   31
Boreal Forest   32
Tundra   35
Mountains: A Diversity of Biomes   37
Concept 2.3 Review   40

Applications: Finer Scale Climatic Variation over Time and Space   40

Chapter 3   Life in Water   44

Concepts   44
Aquatic Biomes and How They Differ   45

3.1   Water Cycling   47
The Hydrologic Cycle   47
The Effects of Wind and Temperature   47
Concept 3.1 Review   48

3.2   The Natural History of Aquatic Environments   49
The Oceans   49
Life in Shallow Marine Waters: Kelp Forests and Coral Gardens   54
Marine Shores: Life Between High and Low Tides   57
Transitional Environments: Estuaries, Salt Marshes, Mangrove Forests, and Freshwater Wetlands   59
Rivers and Streams: Life Blood and Pulse of the Land   64
Lakes: Small Seas   69
Concept 3.2 Review   73

Applications: Biological Integrity–Assessing the Health of Aquatic Systems   73
Number of Species and Species Composition   74
Trophic Composition   74
Fish Abundance and Condition   74
A Test   74

Chapter 4   Population Genetics and Natural Selection   78

Concepts   78

4.1   Variation Within Populations   81
Variation in a Widely Distributed Plant   81
Variation in Alpine Fish Populations   83
Concept 4.1 Review   84

4.2   Hardy-Weinberg Principle   84
Calculating Gene Frequencies   84
Concept 4.2 Review   86

4.3   The Process of Natural Selection   87
Stabilizing Selection   87
Directional Selection   87
Disruptive Selection   87
Concept 4.3 Review   88

4.4   Evolution by Natural Selection   88
Heritability: Essential for Evolution   89
Directional Selection: Adaptation by Soaperry Bugs to New Host Plants   90
Concept 4.4 Review   92

4.5 Change due to Chance 92
        Evidence of Genetic Drift in Island Crickets 93
        Genetic Diversity and Butterfly Extinctions 95
        Concept 4.5 Review 96

Applications: Evolution and Agriculture 96
        Evolution of Herbicide Resistance in Weeds 97

Section II
ADAPTATIONS TO THE ENVIRONMENT

Chapter 5   Temperature Relations   101

Concepts 101

5.1 Microclimates 102
        Altitude 103
        Aspect 103
        Vegetation 103
        Color of the Ground 103
        Presence of Boulders and Burrows 104
        Aquatic Temperatures 104
        Concept 5.1 Review 105

5.2 Evolutionary Trade-Offs 105
        The Principle of Allocation 105
        Concept 5.2 Review 106

5.3 Temperature and Performance of Organisms 106
        Extreme Temperatures and Photosynthesis 108
        Temperature and Microbial Activity 109
        Concept 5.3 Review 110

5.4 Regulating Body Temperature 110
        Balancing Heat Gain Against Heat Loss 110
        Temperature Regulation by Plants 111
        Temperature Regulation by Ectothermic Animals 113
        Temperature Regulation by Endothermic Animals 115
        Temperature Regulation by Thermogenic Plants 119
        Concept 5.4 Review 120

5.5 Surviving Extreme Temperatures 120
        Inactivity 120
        Reducing Metabolic Rate 121
        Hibernation by a Tropical Species 121
        Concept 5.5 Review 123

Applications: Local Extinction of a Land Snail in an Urban
        Heat Island 123

Chapter 6   Water Relations   127

Concepts 127

6.1 Water Availability 128
        Water Content of Air 129
        Water Movement in Aquatic Environments 130
        Water Movement Between Soils and Plants 131
        Concept 6.1 Review 132

6.2 Water Regulation on Land 133
        Water Acquisition by Animals 134
        Water Acquisition by Plants 135
        Water Conservation by Plants and Animals 136
        Dissimilar Organisms with Similar Approaches to Desert
            Life 139
        Two Arthropods with Opposite Approaches to Desert
            Life 140
        Concept 6.2 Review 141

6.3 Water and Salt Balance in Aquatic Environments 143
        Marine Fish and Invertebrates 143
        Freshwater Fish and Invertebrates 144
        Concept 6.3 Review 144

Applications: Using Stable Isotopes to Study Water Uptake by
        Plants 145
        Stable Isotope Analysis 146
        Using Stable Isotopes to Identify Plant Water Sources 146

Chapter 7   Energy and Nutrient Relations   149

Concepts 149

7.1 Photosynthetic Autotrophs 150
        The Solar-Powered Biosphere 150
        Concept 7.1 Review 153

7.2 Chemosynthetic Autotrophs 154
        Concept 7.2 Review 156

7.3 Heterotrophs 156
        Chemical Composition and Nutrient Requirements 156
        Concept 7.3 Review 162

7.4 Energy Limitation 163
        Photon Flux and Photosynthetic Response Curves 163
        Food Density and Animal Functional Response 164
        Concept 7.4 Review 165

7.5 Optimal Foraging Theory 165
        Testing Optimal Foraging Theory 166
        Optimal Foraging by Plants 167
        Concept 7.5 Review 168

Applications: Bioremediation–Using the Trophic
        Diversity of Bacteria to Solve Environmental
        Problems 168
        Leaking Underground Storage Tanks 169
        Cyanide and Nitrates in Mine Spoils 169

Chapter 8   Social Relations   172

Concepts 172

8.1 Mate Choice versus Predation 174
        Mate Choice and Sexual Selection in Guppies 175
        Concept 8.1 Review 178

8.2 Mate Choice and Resource Provisioning 178
        Concept 8.2 Review 181

8.3 Nonrandom Mating in a Plant Population 181
        Concept 8.3 Review 183

8.4 Sociality 183
        Cooperative Breeders 184
        Concept 8.4 Review 189

8.5 Eusociality 189
    Eusocial Species 189
    Evolution of Eusociality 191
    Concept 8.5 Review 193

Applications: Behavioral Ecology and Conservation 193
    Tinbergen's Framework 193
    Environmental Enrichment and Development of
        Behavior 193

Section III
POPULATION ECOLOGY

Chapter 9 Population Distribution and
Abundance 196

Concepts 196

9.1 Distribution Limits 198
    Kangaroo Distributions and Climate 198
    Distributions of Plants Along a Moisture-Temperature
        Gradient 199
    Distributions of Barnacles Along an Intertidal Exposure
        Gradient 200
    Concept 9.1 Review 202

9.2 Patterns on Small Scales 202
    Scale, Distributions, and Mechanisms 202
    Distributions of Tropical Bee Colonies 202
    Distributions of Desert Shrubs 204
    Concept 9.2 Review 205

9.3 Patterns on Large Scales 205
    Bird Populations Across North America 206
    Plant Distributions Along Moisture Gradients 207
    Concept 9.3 Review 208

9.4 Organism Size and Population Density 208
    Animal Size and Population Density 208
    Plant Size and Population Density 209
    Concept 9.4 Review 210

Applications: Rarity and Vulnerability to Extinction 210
    Seven Forms of Rarity and One of Abundance 210

Chapter 10 Population Dynamics 215

Concepts 215

10.1 Dispersal 217
    Dispersal of Expanding Populations 217
    Range Changes in Response to Climate Change 218
    Dispersal in Response to Changing Food Supply 219
    Dispersal in Rivers and Streams 220
    Concept 10.1 Review 221

10.2 Metapopulations 221
    A Metapopulation of an Alpine Butterfly 222
    Dispersal Within a Metapopulation of Lesser Kestrels 223
    Concept 10.2 Review 224

10.3 Patterns of Survival 224
    Estimating Patterns of Survival 224
    High Survival Among the Young 224
    Constant Rates of Survival 226
    High Mortality Among the Young 227
    Three Types of Survivorship Curves 227
    Concept 10.3 Review 228

10.4 Age Distribution 228
    Contrasting Tree Populations 228
    A Dynamic Population in a Variable Climate 229
    Concept 10.4 Review 230

10.5 Rates of Population Change 230
    Estimating Rates for an Annual Plant 230
    Estimating Rates When Generations Overlap 231
    Concept 10.5 Review 233

Applications: Changes in Species Distributions in Response to
    Climate Warming 233

Chapter 11 Population Growth 237

Concepts 237

11.1 Geometric and Exponential Population Growth 238
    Geometric Growth 238
    Exponential Growth 240
    Exponential Growth in Nature 241
    Concept 11.1 Review 242

11.2 Logistic Population Growth 242
    Concept 11.2 Review 245

11.3 Limits to Population Growth 245
    Environment and Birth and Death Among Darwin's
        Finches 245
    Concept 11.3 Review 248

Applications: The Human Population 248
    Distribution and Abundance 248
    Population Dynamics 249
    Population Growth 250

Chapter 12 Life Histories 254

Concepts 254

12.1 Offspring Number versus Size 255
    Egg Size and Number in Fish 256
    Seed Size and Number in Plants 258
    Seed Size and Seedling Performance 259
    Concept 12.1 Review 261

12.2 Adult Survival and Reproductive Allocation 262
    Life History Variation Among Species 262
    Life History Variation within Species 264
    Concept 12.2 Review 266

12.3 Life History Classification 266
    r and K Selection 266
    Plant Life Histories 267
    Opportunistic, Equilibrium, and Periodic Life
        Histories 268

Lifetime Reproductive Effort and Relative Offspring Size: Two Central Variables?   270

Concept 12.3 Review   272

Applications: Climate Change and Timing of Reproduction and Migration   272

Altered Plant Phenology   272

Animal Phenology   273

Section IV

INTERACTIONS

Chapter 13   Species Interactions and Competition   277

Concepts   277

Competitive Interactions Are Diverse   279

13.1   Intraspecific Competition   280

Intraspecific Competition Among Plants   280

Intraspecific Competition Among Planthoppers   281

Interference Competition Among Terrestrial Isopods   282

Concept 13.1 Review   282

13.2   Competitive Exclusion and Niches   282

The Feeding Niches of Darwin's Finches   283

Competition for Caterpillars   284

Concept 13.2 Review   285

13.3   Mathematical and Laboratory Models   285

Modeling Interspecific Competition   285

Laboratory Models of Competition   288

Concept 13.3 Review   289

13.4   Competition and Niches   289

Niches and Competition Among Plants   289

Niche Overlap and Competition Between Barnacles   290

Competition and the Niches of Small Rodents   291

Character Displacement   293

Evidence for Competition in Nature   295

Concept 13.4 Review   295

Applications: Competition Between Native and Invasive Species   296

Chapter 14   Exploitative Interactions: Predation, Herbivory, Parasitism, and Disease   299

Concepts   299

14.1   Exploitation and Abundance   300

A Herbivorous Stream Insect and Its Algal Food   300

Bats, Birds, and Herbivory in a Tropical Forest   301

A Pathogenic Parasite, a Predator, and Its Prey   303

Concept 14.1 Review   304

14.2   Dynamics   304

Cycles of Abundance in Snowshoe Hares and Their Predators   304

Experimental Test of Food and Predation Impacts   306

Population Cycles in Mathematical and Laboratory Models   307

Concept 14.2 Review   310

14.3   Refuges   310

Refuges and Host Persistence in Laboratory and Mathematical Models   310

Exploited Organisms and Their Wide Variety of "Refuges"   312

Concept 14.3 Review   314

14.4   Ratio-Dependent Models of Functional Response   314

Alternative Model for Trophic Ecology   314

Evidence for Ratio-Dependent Predation   315

Concept 14.4 Review   317

14.5   Complex Interactions   317

Parasites and Pathogens That Manipulate Host Behavior   317

The Entangling of Exploitation with Competition   320

Concept 14.5 Review   321

Applications: The Value of Pest Control by Bats: A Case Study   321

Chapter 15   Mutualism   325

Concepts   325

15.1   Plant Mutualisms   326

Plant Performance and Mycorrhizal Fungi   327

Ants and Swollen Thorn Acacias   330

A Temperate Plant Protection Mutualism   334

Concept 15.1 Review   335

15.2   Coral Mutualisms   335

Zooxanthellae and Corals   336

A Coral Protection Mutualism   336

Concept 15.2 Review   338

15.3   Evolution of Mutualism   338

Facultative Ant-Plant Protection Mutualisms   340

Concept 15.3 Review   341

Applications: Mutualism and Humans   341

Guiding Behavior   341

Section V

COMMUNITIES AND ECOSYSTEMS   345

Chapter 16   Species Abundance and Diversity   345

Concepts   345

16.1   Species Abundance   347

The Lognormal Distribution   347

Concept 16.1 Review   348

16.2   Species Diversity   348

A Quantitative Index of Species Diversity   348

Rank-Abundance Curves   350

Concept 16.2 Review   351

16.3 Environmental Complexity 351
    Forest Complexity and Bird Species Diversity 351
    Niches, Heterogeneity, and the Diversity of Algae and Plants 352
    The Niches of Algae and Terrestrial Plants 353
    Complexity in Plant Environments 354
    Soil and Topographic Heterogeneity 354
    Nutrient Enrichment Can Reduce Environmental Complexity 355
    Nitrogen Enrichment and Ectomycorrhizal Fungus Diversity 356
    Concept 16.3 Review 357

16.4 Disturbance and Diversity 357
    The Nature and Sources of Disturbance 357
    The Intermediate Disturbance Hypothesis 357
    Disturbance and Diversity in the Intertidal Zone 358
    Disturbance and Diversity in Temperate Grasslands 358
    Concept 16.4 Review 360

Applications: Disturbance by Humans 360
    Urban Diversity 361

Chapter 17 Species Interactions and Community Structure 365

Concepts 365

17.1 Community Webs 365
    Strong Interactions and Food Web Structure 367
    Concept 17.1 Review 368

17.2 Indirect Interactions 368
    Indirect Commensalism 368
    Apparent Competition 369
    Concept 17.2 Review 370

17.3 Keystone Species 371
    Food Web Structure and Species Diversity 371
    Experimental Removal of Sea Stars 373
    Snail Effects on Algal Diversity 374
    Fish as Keystone Species in River Food Webs 376
    Concept 17.3 Review 377

17.4 Mutualistic Keystones 378
    A Cleaner Fish as a Keystone Species 378
    Seed Dispersal Mutualists as Keystone Species 379
    Concept 17.4 Review 379

Applications: Human Modification of Food Webs 380
    Parasitoid Wasps: Apparent Competition and Biological Control 380

Chapter 18 Primary and Secondary Production 383

Concepts 383

18.1 Patterns of Terrestrial Primary Production 385
    Actual Evapotranspiration and Terrestrial Primary Production 385
    Soil Fertility and Terrestrial Primary Production 386
    Concept 18.1 Review 387

18.2 Patterns of Aquatic Primary Production 387
    Patterns and Models 387
    Whole-Lake Experiments on Primary Production 388
    Global Patterns of Marine Primary Production 388
    Concept 18.2 Review 389

18.3 Primary Producer Diversity 390
    Terrestrial Plant Diversity and Primary Production 390
    Algal Diversity and Aquatic Primary Production 391
    Concept 18.3 Review 391

18.4 Consumer Influences 392
    Piscivores, Planktivores, and Lake Primary Production 392
    Grazing by Large Mammals and Primary Production on the Serengeti 394
    Concept 18.4 Review 396

18.5 Secondary Production 396
    A Trophic Dynamic View of Ecosystems 397
    Top-down Versus Bottom-up Controls on Secondary Production 397
    Linking Primary Production and Secondary Production 398
    Concept 18.5 Review 399

Applications: Using Stable Isotope Analysis to Study Feeding Habits 399
    Using Stable Isotopes to Identify Sources of Energy in a Salt Marsh 400

Chapter 19 Nutrient Cycling and Retention 403

Concepts 403

19.1 Nutrient Cycles 404
    The Phosphorus Cycle 405
    The Nitrogen Cycle 406
    The Carbon Cycle 407
    Concept 19.1 Review 408

19.2 Rates of Decomposition 408
    Decomposition in Two Mediterranean Woodland Ecosystems 408
    Decomposition in Two Temperate Forest Ecosystems 409
    Decomposition in Aquatic Ecosystems 411
    Concept 19.2 Review 412

19.3 Organisms and Nutrients 412
    Nutrient Cycling in Streams and Lakes 412
    Animals and Nutrient Cycling in Terrestrial Ecosystems 415
    Plants and the Nutrient Dynamics of Ecosystems 416
    Concept 19.3 Review 417

19.4 Disturbance and Nutrients 417
    Disturbance and Nutrient Loss from Forests 417
    Flooding and Nutrient Export by Streams 418
    Concept 19.4 Review 419

Applications: Altering Aquatic and Terrestrial Ecosystems 419

Chapter 20 Succession and Stability 423

Concepts 423

20.1 Community Changes During Succession 425
Primary Succession at Glacier Bay 425
Secondary Succession in Temperate Forests 427
Succession in Rocky Intertidal Communities 427
Succession in Stream Communities 428
Concept 20.1 Review 429

20.2 Ecosystem Changes During Succession 429
Four Million Years of Ecosystem Change 429
Succession and Stream Ecosystem Properties 431
Concept 20.2 Review 432

20.3 Mechanisms of Succession 432
Facilitation 433
Tolerance 433
Inhibition 433
Successional Mechanisms in the Rocky Intertidal Zone 434
Mechanisms in Old Field Succession 435
Concept 20.3 Review 436

20.4 Community and Ecosystem Stability 436
Lessons from the Park Grass Experiment 437
Replicate Disturbances and Desert Stream Stability 438
Concept 20.4 Review 440

Applications: Ecological Succession Informing Ecological Restoration 440
Applying Succession Concepts to Restoration 440

SECTION VI
LARGE-SCALE ECOLOGY

Chapter 21 Landscape Ecology 445

Concepts 445

21.1 Landscape Structure 447
The Structure of Six Landscapes in Ohio 447
The Fractal Geometry of Landscapes 449
Concept 21.1 Review 450

21.2 Landscape Processes 450
Landscape Structure and the Dispersal of Mammals 451
Habitat Patch Size and Isolation and the Density of Butterfly Populations 452
Habitat Corridors and Movement of Organisms 453
Landscape Position and Lake Chemistry 454
Concept 21.2 Review 455

21.3 Origins of Landscape Structure and Change 455
Geological Processes, Climate, and Landscape Structure 456
Organisms and Landscape Structure 458
Fire and the Structure of a Mediterranean Landscape 462
Concept 21.3 Review 463

Applications: Landscape Approaches to Mitigating Urban Heat Islands 463

Chapter 22 Geographic Ecology 468

Concepts 468

22.1 Area, Isolation, and Species Richness 470
Island Area and Species Richness 470
Island Isolation and Species Richness 472
Concept 22.1 Review 473

22.2 The Equilibrium Model of Island Biogeography 473
Species Turnover on Islands 474
Experimental Island Biogeography 475
Colonization of New Islands by Plants 476
Manipulating Island Area 477
Island Biogeography Update 478
Concept 22.2 Review 478

22.3 Latitudinal Gradients in Species Richness 478
Latitudinal Gradient Hypotheses 478
Area and Latitudinal Gradients in Species Richness 480
Continental Area and Species Richness 481
Concept 22.3 Review 482

22.4 Historical and Regional Influences 482
Exceptional Patterns of Diversity 482
Historical and Regional Explanations 483
Concept 22.4 Review 484

Applications: Global Positioning Systems, Remote Sensing, and Geographic Information Systems 485
Global Positioning Systems 485
Remote Sensing 485
Geographic Information Systems 487

Chapter 23 Global Ecology 490

Concepts 490
The Atmospheric Envelope and the Greenhouse Earth 491

23.1 A Global System 493
The Historical Thread 493
El Niño and La Niña 494
El Niño Southern Oscillation and Marine Populations 495
El Niño and the Great Salt Lake 497
El Niño and Terrestrial Populations in Australia 498
Concept 23.1 Review 499

23.2 Human Activity and the Global Nitrogen Cycle 499
Concept 23.2 Review 500

23.3 Changes in Land Cover 500
Deforestation 500
Concept 23.3 Review 504

23.4 Human Influence on Atmospheric Composition 504
Depletion and Recovery of the Ozone Layer 507
Concept 23.4 Review 508

Applications: Impacts of Global Climate Change 508
Shifts in Biodiversity and Widespread Extinction of Species 509
Human Impacts of Climate Change 509

Appendix A   Investigating the Evidence

   1: The Scientific Method—Questions and Hypotheses   514

   2: Determining the Sample Mean   515

   3: Determining the Sample Median   516

   4: Variation in Data   517

   5: Laboratory Experiments   518

   6: Sample Size   519

   7: Scatter Plots and the Relationship Between Variables   520

   8: Estimating Heritability Using Regression Analysis   521

   9: Clumped, Random, and Regular Distributions   522

  10: Hypotheses and Statistical Significance   523

  11: Frequency of Alternative Phenotypes in a Population   524

  12: A Statistical Test for Distribution Pattern   526

  13: Field Experiments   527

  14: Standard Error of the Mean   528

  15: Confidence Intervals   530

  16: Estimating the Number of Species in Communities   531

  17: Using Confidence Intervals to Compare Populations   532

  18: Comparing Two Populations with the t-Test   533

  19: Assumptions for Statistical Tests   534

  20: Variation Around the Median   535

  21: Comparison of Two Samples Using a Rank Sum Test   537

  22: Sample Size Revisited   538

  23: Discovering What's Been Discovered   539

Appendix B   Statistical Tables   541

Appendix C   Abbreviations Used in This Text   545

Appendix D   Global Biomes   547

Glossary   548

References   558

Index   571

Design elements: Anna A. Sher

This book was written for students taking their first under-graduate course in ecology. We have assumed that students in this one-semester course have some knowledge of basic chemistry and mathematics and have had a course in general biology, which included introductions to evolution, physiology, and biological diversity.

Organization of the Book

An evolutionary perspective forms the foundation of the entire textbook, as it is needed to support understanding of major concepts. The textbook begins with a brief introduction to the nature and history of the discipline of ecology, followed by section I, which includes two chapters on earth's biomes—life on land and life in water—followed by a chapter on population genetics and natural selection. Sections II through VI build a hierarchical perspective through the traditional sub-disciplines of ecology: section II concerns adaptations to the environment; section III focuses on population ecology; section IV presents the ecology of interactions; section V summarizes community and ecosystem ecology; and finally, section VI discusses large-scale ecology, including chapters on landscape, geographic, and global ecology. These topics were first introduced in section I within its discussion of the biomes. In summary, the book begins with an overview of the biosphere, considers portions of the whole in the middle chapters, and ends with another perspective of the entire planet in the concluding chapter. The features of this textbook were carefully planned to enhance the students' comprehension of the broad discipline of ecology.

Features Designed with the Student in Mind

All chapters are based on a distinctive learning system, featuring the following key components:

Student Learning Outcomes: Educators are being asked increasingly to develop concrete student learning outcomes for courses across the curriculum. In response to this need and to help focus student progress through the content, all sections of each chapter in the ninth edition begin with a list of detailed student learning outcomes.

Introduction: The introduction to each chapter presents the student with the flavor of the subject and important background information. Some introductions include historical events related to the subject; others present an example of an ecological process. All attempt to engage students and draw them into the discussion that follows.

Concepts: The goal of this book is to build a foundation of ecological knowledge around key concepts, which are listed at the beginning of each chapter to alert the student to the major topics to follow and to provide a place where the student can find a list of the important points covered in each chapter. The sections in which concepts are discussed focus on published studies and, wherever possible, the scientists who did the research are introduced. This case-study approach supports the concepts with evidence, and introduces students to the methods and people that have created the discipline of ecology. Each concept discussion ends with a series of concept review questions to help students test their knowledge and to reinforce key points made in the discussion.

Illustrations: A great deal of effort has been put into the development of illustrations, both photographs and line art. The goal has been to create more-effective pedagogical tools through skillful design and use of color, and to rearrange the traditional presentation of information in figures and captions. Much explanatory material is located within the illustrations, providing students with key information where they need it most. The approach also provides an ongoing tutorial on graph interpretation, a skill with which many introductory students need practice.

Detailed Explanations of Mathematics: The mathematical aspects of ecology commonly challenge many students taking their first ecology course. This text carefully explains all mathematical expressions that arise to help students overcome these challenges. In some cases, mathematical expressions are dissected in illustrations designed to complement their presentation in the associated narrative.

Figure 7.16 Birds and other predators exert strong selective pressure on populations of peppered moths in Europe and elsewhere, based on the extent to which they are camouflaged. When trees were covered with soot due to industrialization, darker morphs were more abundant, whereas in the wake of environmental regulations that reduced air pollution, the bark of trees became light-colored again, resulting in dramatic shifts in the frequency of the lighter morph. Frequency data shown here were collected 37 years apart in West Kirby, Wirral. This remains one of the most commonly-used examples of natural selection, although recent research suggests that other evolutionary forces also likely played a role (data from Grant et al. 1998). Bill Coster IN/Alamy Stock Photo/H Lansdown/Alamy Stock Photo/Frank Hecker/Alamy Stock Photo

A visualization of a population bottle neck, using data from published research.

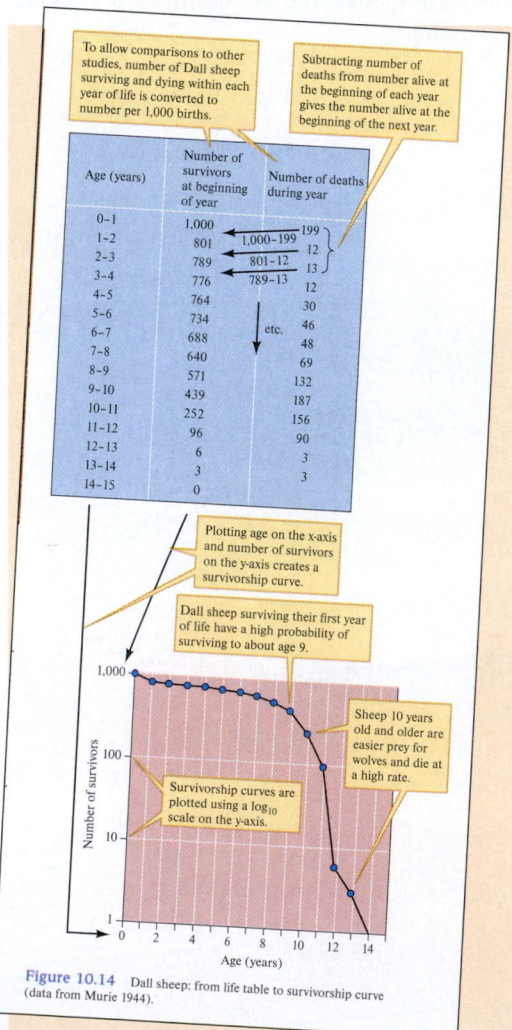

Figure 10.14 Dall sheep: from life table to survivorship curve (data from Murie 1944).

Helps students work with and interpret quantitative information, involving converting numerical information into a graph.

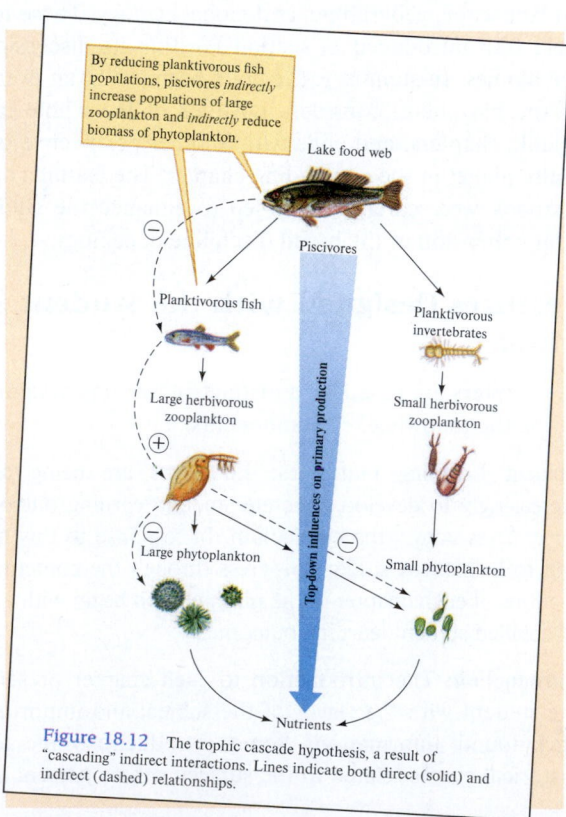

Figure 18.12 The trophic cascade hypothesis, a result of "cascading" indirect interactions. Lines indicate both direct (solid) and indirect (dashed) relationships.

Provides a visual representation of a hypothesis involving a set of complex ecological interactions.

Applications: Many students are concerned with the practical side of ecology and want to know more about how the tools of science can be applied to the environmental problems we face in the contemporary world. Including a discussion of applications at the end of each chapter can motivate students to learn more of the underlying principles of ecology. In addition, it seems that environmental problems are now so numerous and so pressing that they have erased a once easy distinction between general and applied ecology.

End-of-Chapter Material:

- *Summary*   The chapter summary reviews the main points of the content. The concepts around which each chapter is organized are boldfaced and redefined in the summary to reemphasize the main points of the chapter.
- *Key Terms*
- *Review Questions*   The review questions are designed to help students think more deeply about each concept and to reflect on alternative views. They also provide a place to fill in any remaining gaps in the information presented and take students beyond the foundation established in the main body of the chapter.

  Note: Suggested Readings are located online.

End-of-Book Material:

- *Appendixes*   Appendix A, "Investigating the Evidence," offers "mini-lessons" on the scientific method, emphasizing

statistics and study design. They are intended to present a broad outline of the process of science, while also providing step-by-step explanations. The series of features begins with an overview of the scientific method, which establishes a conceptual context for more specific material in the next 21 features. The last reading wraps up the series with a discussion of electronic literature searches. Each Investigating the Evidence ends with one or more questions, under the heading "Critiquing the Evidence." This feature is intended to stimulate critical thinking about the content. Appendix B, "Statistical Tables," is available to the student as a reference in support of the Investigating the Evidence features. Appendix C, "Abbreviations" is a handy guide to the scientific and other abbreviations used throughout the text, including units of measurement. Appendix D is a global map of the biomes.

- *Glossary*   List of all key terms and their definitions.
- *References*   References are an important part of any scientific work. However, many undergraduates are distracted by a large number of references within the text. One of the goals of a general ecology course should be to introduce these students to the primary literature without burying them in citations. The number of citations has been reduced to those necessary to support detailed discussions of particular research projects.
- *Index*

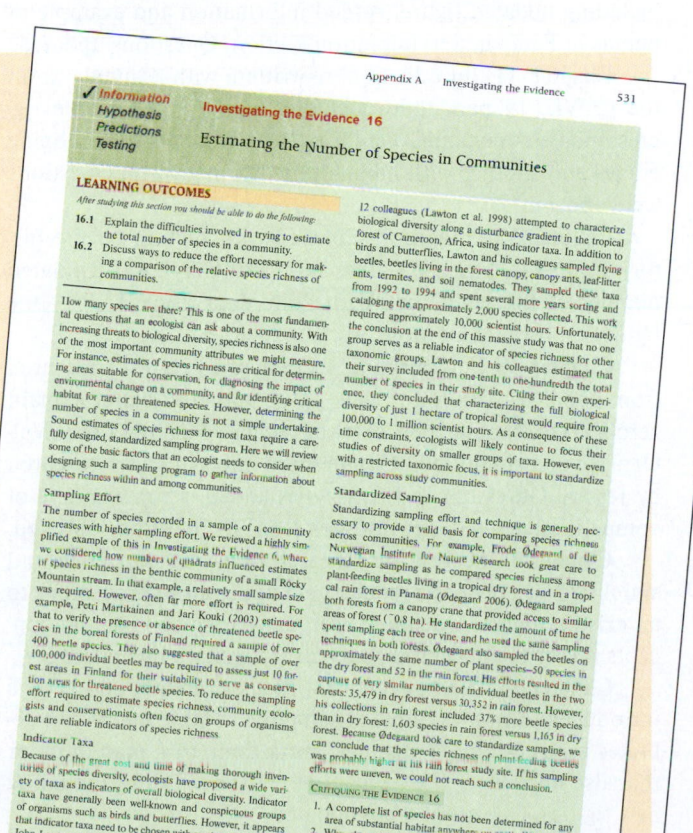

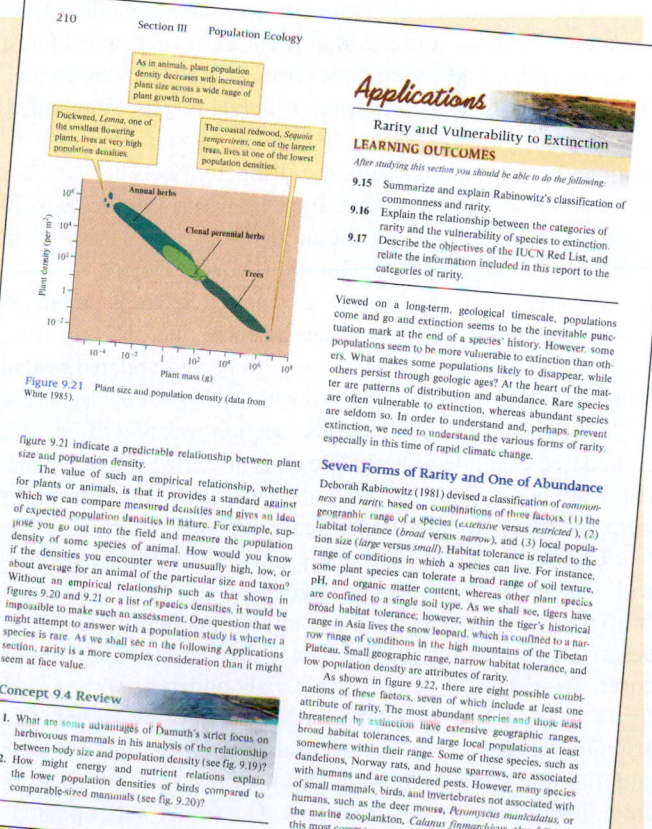

New to the Ninth Edition

Nearly every chapter has significant changes in this edition. To update content and respond to reviewers' comments, we have incorporated the research and ideas of over 140 new citations, the majority of which (73%) were authored by underrepresented scientists. A particular effort was made to cite cutting-edge ecological research by women of color. With each edition, we continue toward the goal of making this text reflect the true diversity of researchers in the field.

There are over 100 updated examples in this edition, with 42 new figures, plus improvements or updates to 20 existing figures. Dozens of new questions have been written to correspond to the new material and, in response to reviewers, many other questions have been re-written to focus more on concepts rather than specific examples. Several new terms have also been added in the text and glossary to increase student understanding and to reflect the evolving nature of the field. We have also continued to expand connections with evolution and global change in this edition.

Significant Chapter-by-Chapter Changes

Chapter 1 In response to reviewer's comments, we have created a new section that describes the different tools used by ecologists, introducing five new terms including *ex situ* and *in situ*. There are a total of nine new figures. We have revised figure 1 and added microbial ecology as an important frontier. We have added new examples from recent literature, including about evolution in alpine chipmunks. Questions were updated.

Chapter 2 Three new figures were added, including from research on habitat conversion in India. Data on tropical forest loss was updated. New examples from publications by women of color on soils and on logging of boreal forests were added. Wording in several places was clarified in response to reviewers' comments. An explanation of the distinction between weather and climate change was added. Improvements were made to 10 figures, including updating the drought data in figure 2.41 to 2020 and relating it to fires. Questions were updated.

Chapter 3 Six citations were updated. Sections added on United Nations Decade of Ocean Science, microplastics from research by Chatterjee and Sharma (2019), and updated several examples. One figure was updated with current global ice levels.

Chapter 4 The "applications" section was re-written with an updated example of herbicide resistance by Sushila Chaudhari and her colleagues, including a new figure. Questions were updated, and an existing figure improved.

Chapter 5 A total of 13 new citations, including examples with current citations were provided of how global warming is affecting ecosystems. New example and figure created to describe relationship between water temperature and canopy cover. Research on endothermic fish updated, with a new figure created and concept of RM endothermy added. Old example replaced with new section on comparisons between endothermic and ectothermic fish with research by a man of color, including another new figure. Questions were updated.

Chapter 6 Section on water-harvesting re-written with updated information and a new figure adapted from the review by Guera and Bhushan (2020). Added concept of cohesion, per reviewer request. Concepts hydrophilic and hydrophobic introduced. Water isotope section re-written to clarify per reviewer request, including a new figure to explain. Applications section was re-written with updated example from the meta-analysis by Evaristo and McDonnell (2017). Questions and one figure were updated.

Chapter 7 Information about chemosynthesis was expanded and updated with example from Naples, Italy. Peppered moth example re-written and figure replaced with one that shows actual photographs and data. Old predation examples were replaced with those using wolf spiders and coral reef fishes research from teams led by women, including new figures. Questions were updated.

Chapter 8 Opening photo replaced with a more appropriate one, six references updated. Section on nonrandom mating in plants significantly updated and clarified. Paragraph on phylogenies based on genetic analysis added. Updated number of cooperative breeding species. Updated section on lion cooperation with research by Natalia Borrego, a woman of color. Questions were updated.

Chapter 9 New example of gorillas replaces an old example, and new paragraph added based on the 2020 Living Planet Report. One new image. More information about the Breeding Bird Survey with updated references. Term endemic added, with paragraph replaced with new example of bird from Hawaii. Figure on rarity and vulnerability to extinction significantly improved in response to reviewer request. Questions were updated.

Chapter 10 Seven citations updated. Research on "killer" bees updated with genetics research led by a man of color, including updated figure. Added information and example of pumas in Patagonia to migration section. Questions updated.

Chapter 11 Introduction re-written with example from the COVID-19 pandemic, including new figures. Three figures updated, including one for current numbers of whooping cranes and another with human population growth. Questions were updated.

Chapter 12 Paragraph replaced with section on life history trade-offs, based on ideas by Anurag Agrawal. Updated number of species of fish with 2020 data from IUCN. One figure improved. Questions were updated.

Chapter 13 Self-thinning section updated with research from people of color, and new figure added to better explain zero growth isocline, per reviewer request. Existing Lotka-Volterra figure simplified. Competition meta-analysis research by Jessica Gurevitch and colleagues added. Extra example of competition deleted, per reviewer request. Questions updated.

Chapter 14 Section on research by Utida shortened and simplified per reviewer request, including an improvement to an existing figure. Questions were re-written to focus on concepts rather than specific research. Two citations updated.

Chapter 16 The concept of a species rarefaction curve is introduced. A new example of sampling benthic macroinvertebrates replaces an old example, work done by a man of color that also introduces the concept of DNA barcoding, including new figures. Questions were updated.

Chapter 17 Four examples were updated, all from research led by underrepresented scientists. This includes a new "Applications" example on hyperparasitoids with a new figure. Questions were updated.

Chapter 18 Section on primary productivity of oceans was re-written with updated environmental factors and relating this to global change. Map on marine primary productivity has been updated. Research on top-down vs. bottom-up updated with a new section and figure from meta-analysis research conducted by Mayra Vidal, a woman of color, and Shannon Murphy. Concept of tri-trophic interactions added. Eleven citations were updated, most of which from papers with under-represented lead authors. Paragraph on role of microorganisms added, per reviewer request. Questions were updated.

Chapter 20 All sections on succession at Glacier Bay section completely re-written to reflect more current research led by Brian Buma that changes interpretation of those research, including new figures. This case study becomes a more interesting story about how understanding can evolve with new information. Questions were updated.

Chapter 21 Reference to the 2020 California wildfires was added, including a short paragraph about research from UC Berkeley. Questions were updated.

Chapter 23 A total of five new figures added, including one that refers to the Australian wildfires of 2020. Figure on atmospheric CO_2 updated with current values. Section on nitrogen pollution re-written with more explanation and more current research. The forest section was re-written with forest biodiversity data from the FAO 2020 report on the State of the World's Forests and other current research. Corrections made to use of Spanish words, per reviewer request. Deforestation in Brazil was updated. There were a total of 13 new citations, 9 from underrepresented scientists. Questions were updated.

Online Materials

Available online are suggested readings and answers to concept review, chapter review, and critiquing the evidence questions.

Related Title of Interest from McGraw-Hill Education

Ecology Laboratory Manual, by Vodopich
(ISBN: 978-0-07-338318-7;
MHID: 0-07-338318-X)
Darrell Vodopich, coauthor of *Biology Laboratory Manual,* has written a new lab manual for ecology. This lab manual offers straightforward procedures that are doable in a broad range of classroom, lab, and field situations. The procedures have specific instructions that can be taught by a teaching assistant with minimal experience as well as by a professor.

Acknowledgments

First and foremost, I must thank my academic partner Dr. Eduardo González, without whose help I could have never completed this edition. I am also deeply grateful for pedagogical expertise of Julie Morris, who also went above and beyond. Thanks also go to the other members of the Sher Lab who pitched in with research and/or offered feedback on new figures for the ninth edition, including Ali Alghamdi, Violet Butler, Rhys Daniels, Alex Goetz, Annie Henry, Alex Kim, Lily Malone, Mandy Malone, and Allen Williamson. During the development of this textbook, many colleagues freely shared their expertise, reviewed sections, or offered the encouragement a project like this needs to keep it going: Anurag Agrawal, Brian Buma, Candice Galen, Diane Marshall, Scott Nichols, Mayra Vidal, and Dhaval Vyas. I am grateful to Patrick M. Burchfield and Hector Chenge Alvarez for keeping me up to date in data and photos of turtles. Special thanks to Jake Grossman for sharing his list of Ecologists of Color and Indigenous Ecologists.

We would like to especially thank Shannon Murphy for her extensive suggestions for the ninth edition, as well as for providing us with exciting new case studies to illustrate evolutionary ecology concepts. In addition, we are indebted to the many students and instructors who have helped by contacting us with questions and suggestions for improvements.

We also wish to acknowledge the skillful guidance and work throughout the publishing process given by many professionals associated with McGraw-Hill Education and Straive during this project, including Beth Baugh, Melissa Homer, Jodi Rhomberg, and Mithun Kothandath.

We gratefully acknowledge the many reviewers who, over the course of the many revisions, have given of their time and expertise to help this textbook evolve to its present ninth edition. Note that some feedback that did not make it into this edition will be incorporated into the next one. These reviewers continue our education, for which we are grateful, and we honestly could not have continued the improvement of this textbook without them.

Finally, I would like to thank my co-author Manuel Molles for entrusting me with this wonderful series, as well as my wife Fran and our son Jeremy for their support throughout the production of the ninth edition.

Reviewers for the Ninth Edition

Thomas C. Adam   *University of California, Santa Barbara*
Henry D. Adams   *Oklahoma State University*
Matthew R. Helmus   *Temple University*
Jodee Hunt   *Grand Valley State University*
Jamie Lamit   *Syracuse University, and State University of New York*
Joseph R. Milanovich   *Loyola University Chicago*
Shannon Murphy   *University of Denver*
Wiline Pangle   *Central Michigan University*
Jessica Peebles-Spencer   *Ball State University*
Jonathan Shurin   *University of California San Diego*
John F. Weishampel   *University of Central Florida*
Lan Xu   *South Dakota State University*

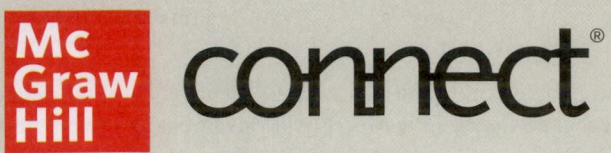

Instructors: Student Success Starts with You

Tools to enhance your unique voice

Want to build your own course? No problem. Prefer to use our turnkey, prebuilt course? Easy. Want to make changes throughout the semester? Sure. And you'll save time with Connect's auto-grading too.

65%
Less Time Grading

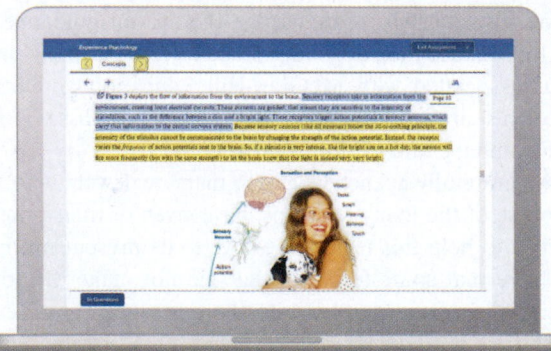

Laptop: McGraw Hill; Woman/dog: George Doyle/Getty Images

Study made personal

Incorporate adaptive study resources like SmartBook® 2.0 into your course and help your students be better prepared in less time. Learn more about the powerful personalized learning experience available in SmartBook 2.0 at **www.mheducation.com/highered/connect/smartbook**

Affordable solutions, added value

Make technology work for you with LMS integration for single sign-on access, mobile access to the digital textbook, and reports to quickly show you how each of your students is doing. And with our Inclusive Access program you can provide all these tools at a discount to your students. Ask your McGraw Hill representative for more information.

Padlock: Jobalou/Getty Images

Solutions for your challenges

A product isn't a solution. Real solutions are affordable, reliable, and come with training and ongoing support when you need it and how you want it. Visit **www. supportateverystep.com** for videos and resources both you and your students can use throughout the semester.

Checkmark: Jobalou/Getty Images

Students: Get Learning that Fits You

Effective tools for efficient studying

Connect is designed to make you more productive with simple, flexible, intuitive tools that maximize your study time and meet your individual learning needs. Get learning that works for you with Connect.

Study anytime, anywhere

Download the free ReadAnywhere app and access your online eBook or SmartBook 2.0 assignments when it's convenient, even if you're offline. And since the app automatically syncs with your eBook and SmartBook 2.0 assignments in Connect, all of your work is available every time you open it. Find out more at **www.mheducation.com/readanywhere**

"I really liked this app—it made it easy to study when you don't have your text-book in front of you."

- Jordan Cunningham,
Eastern Washington University

Calendar: owattaphotos/Getty Images

Everything you need in one place

Your Connect course has everything you need—whether reading on your digital eBook or completing assignments for class, Connect makes it easy to get your work done.

Learning for everyone

McGraw Hill works directly with Accessibility Services Departments and faculty to meet the learning needs of all students. Please contact your Accessibility Services Office and ask them to email accessibility@mheducation.com, or visit **www.mheducation.com/about/accessibility** for more information.

Top: Jenner Images/Getty Images, Left: Hero Images/Getty Images, Right: Hero Images/Getty Images

Remote Proctoring & Browser-Locking Capabilities

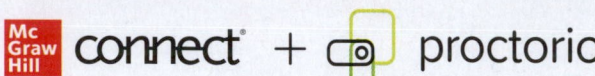

New remote proctoring and browser-locking capabilities, hosted by Proctorio within Connect, provide control of the assessment environment by enabling security options and verifying the identity of the student.

Seamlessly integrated within Connect, these services allow instructors to control students' assessment experience by restricting browser activity, recording students' activity, and verifying students are doing their own work.

Instant and detailed reporting gives instructors an at-a-glance view of potential academic integrity concerns, thereby avoiding personal bias and supporting evidence-based claims.

Writing Assignment

Available within McGraw-Hill Connect® and McGraw-Hill Connect® Master, the Writing Assignment tool delivers a learning experience to help students improve their written communication skills and conceptual understanding. As an instructor you can assign, monitor, grade, and provide feedback on writing more efficiently and effectively.

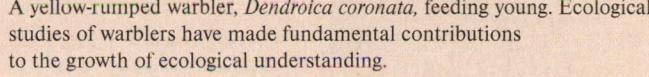

Chapter

1

Introduction to Ecology
Historical Foundations and Developing Frontiers

A yellow-rumped warbler, *Dendroica coronata,* feeding young. Ecological studies of warblers have made fundamental contributions to the growth of ecological understanding.

NPS Photo by Jeff Foott

CHAPTER CONCEPTS

1.1 Ecologists study environmental relationships ranging from those of individual organisms to factors influencing global-scale processes.   2

Concept 1.1 Review   3

1.2 Ecologists design their studies based on their research questions, the temporal and spatial scale of their studies, and available research tools.   3

Concept 1.2 Review   9

Applications:
Ecology Can Inform Environmental Law and Policy   9

Summary   10

Key Terms   10

Review Questions   10

LEARNING OUTCOME

After studying this section you should be able to do the following:

1.1 Discuss the concept of environment as it pertains to the science of ecology.

What is ecology? **Ecology** is the study of relationships among organisms and between organisms and the physical environment. These relationships influence many aspects of the natural world, including the distribution and abundance of organisms, the variety of species living together in a place, and the transformation and flow of energy in nature.

Humans are rapidly changing earth's environment, yet we do not fully understand the consequences of these changes. For instance, human activity has increased the quantity of nitrogen cycling through land and water, changed land cover across the globe, and increased the atmospheric concentration of CO_2. Changes such as these threaten the diversity of life on earth and will also endanger our life support system. Because of the rapid pace of environmental change in the early twenty-first century, it is imperative that we better understand earth's ecology.

Behind the simple definition of ecology lies a broad scientific discipline. Ecologists may study individual organisms, entire forests or lakes, or even the whole earth. The measurements made by ecologists include counts of individual organisms, rates of reproduction, and rates of processes

1

such as photosynthesis and decomposition. Ecologists often spend as much time studying nonbiological components of the environment, such as temperature and soil chemistry, as they spend studying organisms. Meanwhile, the "environment" of organisms in some ecological studies is other species. While you may think of ecologists as typically studying in the field, some of the most important conceptual advances have come from ecologists who build theoretical models or do ecological research in the laboratory. Clearly, our simple definition of *ecology* does not communicate the great breadth of the discipline or the diversity of its practitioners. To get a better idea of what ecology is, let's briefly review its scope.

1.1   Overview of Ecology

LEARNING OUTCOMES

After studying this section you should be able to do the following:

1.2   Describe the levels of ecological organization, for example, population, studied by ecologists.

1.3   Distinguish between the types of questions addressed by ecologists working at different levels of organization.

1.4   Explain how knowledge of one level of ecological organization can help guide research at another level of organization.

Ecologists study environmental relationships ranging from those of individual organisms to factors influencing global-scale processes.   This broad range of subjects can be organized by arranging them as levels in a hierarchy of ecological organization, such as that embedded in the brief table of contents and the sections of this book. Figure 1.1 attempts to display such a hierarchy graphically.

Historically, the ecology of individuals, (fig. 1.1A), has been the domain of physiological ecology and behavioral ecology. Physiological ecologists have emphasized the **evolution** (a process by which populations change over time) of physiological and anatomical mechanisms by which organisms adapt to challenges posed by physical and chemical variation in the environment. Meanwhile, behavioral ecologists have focused principally on evolution of behaviors that allow animals to survive and reproduce in the face of environmental variation.

There is a strong conceptual linkage between ecological studies of individuals and of populations particularly where they concern evolutionary processes. Population ecology is centered on the factors influencing population structure and process, where a **population** is a group of interbreeding individuals of a single species inhabiting a defined area (fig. 1.1B). The processes studied by population ecologists include adaptation, extinction, the distribution and abundance of species, population growth and regulation, and variation in the reproductive ecology of species. Population ecologists are particularly interested in how these processes

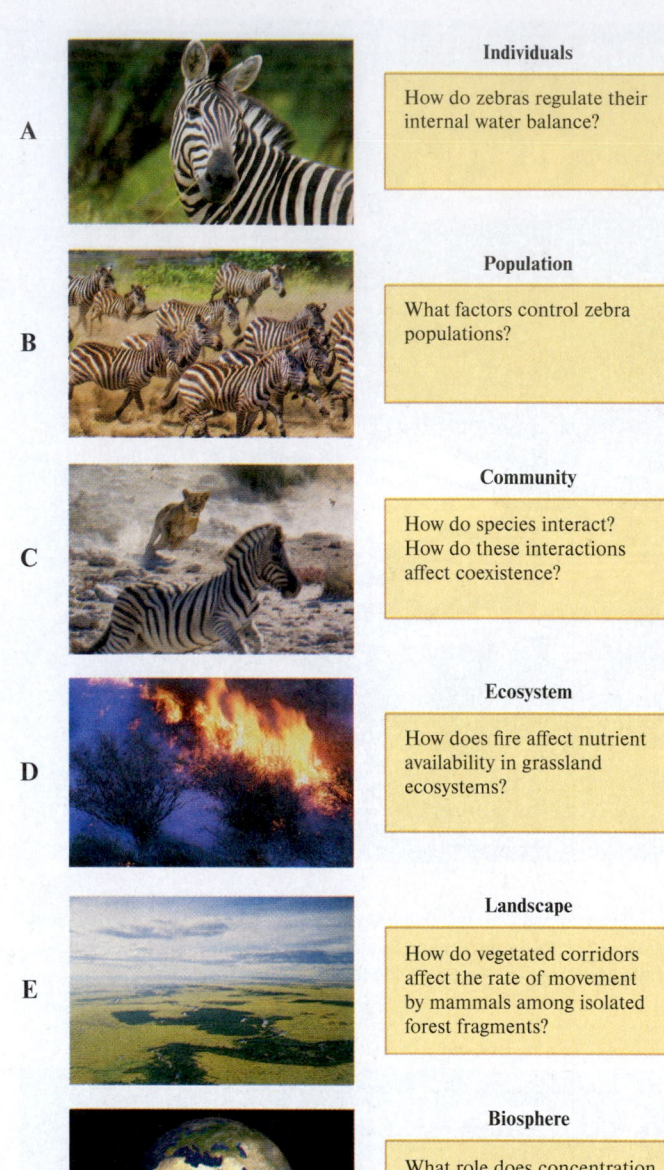

Figure 1.1   Levels of ecological organization and examples of the kinds of questions asked by ecologists working at each level. These ecological levels correspond broadly to the sections of this book.
(*A*) Glow Images; (*B*) cinoby/E+/Getty Images; (*C*) Mogens Trolle/Shutterstock; (*D*) Photo by Gary Wilson, USDA Natural Resources Conservation Service; (*E*) Comstock/PunchStock; (*F*) Calysta Images/Getty Images

are influenced by nonbiological and biological aspects of the environment.

Bringing biological components of the environment into the picture takes us to the next level of organization, the community (fig. 1.1C). A **community** is an association of interacting species. Ecologists who study interactions between species have often emphasized the evolutionary effects of the interaction on the species involved. Other approaches explore the effect of interactions on population structure or on properties of ecological communities.

The next level of organization is the **ecosystem.** An ecosystem is a biological community together with its associated physical and chemical environment. Community and **ecosystem** ecology have a great deal in common, since both are focused on multispecies systems. However, the objects of their study differ. While community ecologists concentrate on understanding environmental influences on the kinds and diversity of organisms inhabiting an area, ecosystem ecologists focus on ecological processes such as energy flow and decomposition (fig. 1.1D).

To simplify their studies, ecologists have long attempted to identify and study isolated communities and ecosystems. However, all communities and ecosystems on earth are subject to exchanges of materials, energy, and organisms with other communities and ecosystems. The study of these exchanges, especially among ecosystems, is the intellectual territory of **landscape** ecology (fig. 1.1E). Landscape ecology in turn leads us to the largest spatial scale and highest level of ecological organization—the **biosphere,** the portions of the earth that support life, including the land, waters, and atmosphere (fig. 1.1F).

While this description of ecology provides a brief preview of the material covered in this book, it is a rough sketch and highly abstract. To move beyond the abstraction represented by figure 1.1, we need to connect it to the work of the scientists who have created the discipline of ecology. To do so, let's briefly review the research of ecologists working at a broad range of ecological levels emphasizing links between historical foundations and some developing frontiers (fig. 1.2).

Concept 1.1 Review

1. How does the level of ecological organization an ecologist studies influence the questions he or she poses?
2. While an ecologist may focus on a particular level of ecological organization shown in figure 1.1, might other levels of organization also be relevant? For example, why should an ecologist studying factors limiting numbers in a population of zebras consider the influences of interactions with other species or the influences of global processes such as climate change?

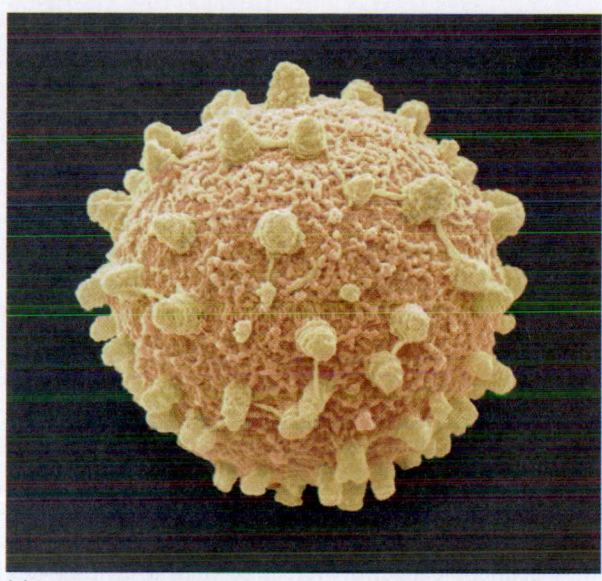

(a)

(b)

Figure 1.2   Two rapidly developing frontiers in ecology. (*a*) **Microbial ecology:** the study of the interactions among microorganisms and between them and their environment (e.g., Epps and Arnold 2019). The importance of these organisms for regulating systems from populations to ecosystems is becoming increasingly understood. (*b*) **Urban ecology:** the study of urban areas as complex, dynamic ecological systems, influenced by interconnected, biological, physical, and social components. As ecologists focus their research on the environment where most members of our species live, they have made unexpected discoveries about the ecology of urban centers such as the city of Baltimore (see chapter 19).   (*a*) STEVE GSCHMEISSNER/Science Photo Library/Getty Images; (*b*) Jon Bilous/Shutterstock

1.2   Sampling Ecological Research

LEARNING OUTCOMES

After studying this section you should be able to do the following:

1.5   List three categories of ecological research, and give an example of each.
1.6   Explain how new tools and technology can be used to advance each category of ecological research.
1.7   Describe different types of ecological models and explain how are they used.

Ecologists design their studies based on their research questions, the temporal and spatial scale of their studies, and available research tools.   Because the discipline is so broad, ecological research can draw from all the physical and biological sciences. The following section of this chapter provides a sample of ecological questions and approaches to research.

Types of Research

In the broadest sense, we can consider ecological research in three general categories: observation, experimentation, and modeling. Each of these types of research is necessary for understanding the organisms and processes at work in our world; most ecologists use at least two, if not all three approaches to answer ecological questions.

Observation

Observation refers to the collection of data in unmanipulated settings, such as counting numbers of birds in a patch of forest or describing types of fungal spores seen through a microscope. Some of this work takes place in the field, or *in situ,* meaning in the habitat where the organisms live, while other research uses specimens that have been collected in the field but are observed in a laboratory or other setting. Such specimens may have been sampled for that specific purpose or may have been collected long beforehand and stored for future study, such as those found in an herbarium or museum. For example, many researchers have been able to track the impact of climate change on plant communities using thousands of plant specimens collected over more than a century (Jones and Daehler 2018, Piao et al. 2019). Observational research can be purely descriptive or may test a hypothesis, such as to understand relationships between organisms. This may involve comparing observations over time or space.

An important, historical example of ecological observational research is that by Robert MacArthur, who observed warblers in spruce forests of northeastern North America (MacArthur 1958). Theory had predicted that two species with identical ecological requirements would compete with each other and that, as a consequence, they would not live in the same environment indefinitely. MacArthur wanted to understand how several warbler species with apparently similar ecological requirements could live together in the same forest; by observing the birds in their natural habitat, he determined that each warbler species had a distinct feeding zone (fig. 1.3). He concluded that this partitioning of the tree reduced competition among the warblers, stimulating future generations of studies of competition.

Another classic example of observational research is the work by Nalini Nadkarni (1981, 1984a, 1984b), who changed our ideas of how tropical and temperate rain forests are structured and how they function. Nadkarni was one of the first scientists to study the ecology of the unseen world of the forest canopy (fig. 1.4). Using mountain-climbing equipment, she took inventories of the distribution of nutrients in rain forests in both Costa Rica and the Pacific Northwest of the United States. She discovered through this sampling that as much as four times the nutrient content in trees leaves was found in **epiphytes**—plants such as orchids, ferns, and mosses that grow on the tree trunks and branches.

MacArthur's primary research tool was a pair of binoculars and Nadkarni's was ropes and harness; however, today there are many more means by which we can collect observational data. For example, new ways to access the forest canopy range from hot air balloons and large cranes (see Investigating the Evidence 16 in Appendix A) to unmanned aerial vehicles (UAV) such as drones. These can carry cameras and other equipment to collect data (Waite et al. 2018). Thermal sensors on UAV's have been used to survey animals in the treetops at night (Kays et al. 2019).

Another example of a new type of data being collected is stable isotope analysis (see chapter 6). Isotopes of a chemical element, such as isotopes of carbon, have different atomic masses as a result of having different numbers of neutrons. Water and nutrients from different sources can have different

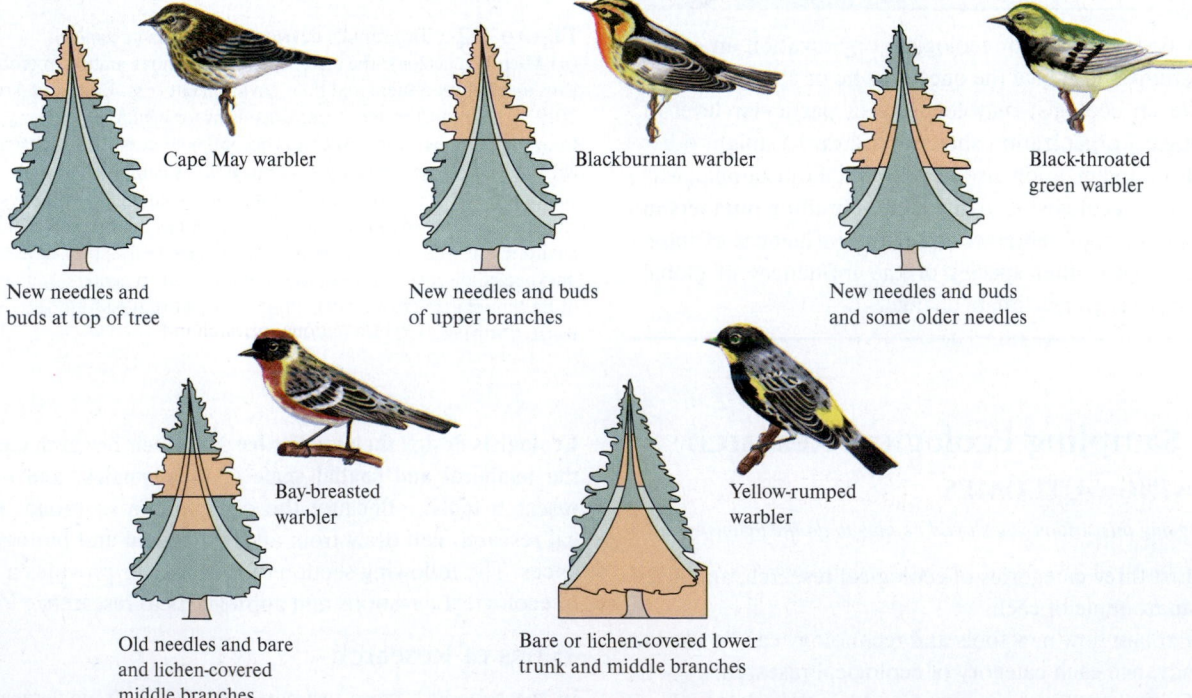

Cape May warbler

New needles and buds at top of tree

Blackburnian warbler

New needles and buds of upper branches

Black-throated green warbler

New needles and buds and some older needles

Bay-breasted warbler

Old needles and bare and lichen-covered middle branches

Yellow-rumped warbler

Bare or lichen-covered lower trunk and middle branches

Figure 1.3 Warbler feeding zones shown in beige. The several warbler species that coexist in the forests of northeastern North America feed in distinctive zones within forest trees.

Figure 1.4   Exploring the rain forest canopy. What Nalini Nadkarni discovered helped solve an ecological puzzle.   Courtesy Nalini Nadkarni, photo by Dennis Paulson

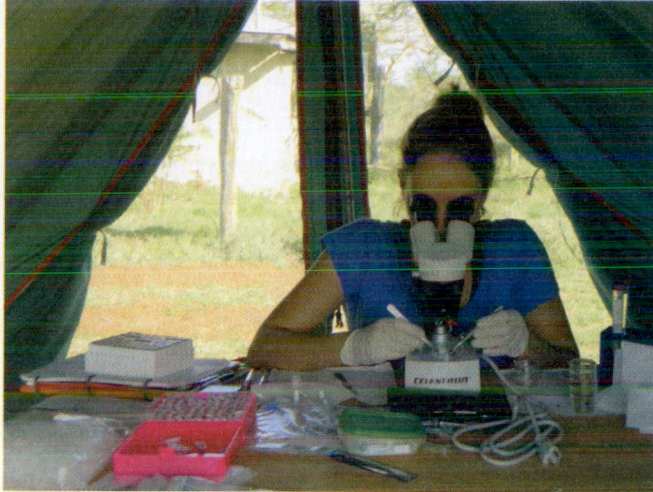

(a)

(b)

Figure 1.5   (*a*) Observational research in Kenya, East Africa, by Melissa Whittaker and colleagues on the African lycaenid butterfly, *Anthene usamba*. Since the larvae of the butterfly were found in acacia tree galls with ants (*b*), it had been hypothesized that it ate ant excretions. However, isotope research found that this was incorrect; because acacia leaves have a particular nitrogen isotope signature, researchers were able to identify it in the gut of the butterfly's larvae. (*a*) Julianne Pelaez; (*b*) Dino Martins

isotopic signatures, thus allowing us to trace them through ecosystems. In this way, stable isotope analysis provides ecologists with a new type of "lens" capable of revealing ecological relationships that would otherwise remain invisible. Melissa Whitaker and colleagues used both stable isotope and genetic analysis to identify trophic relationships for the *Anthene usamba,* a butterfly whose larval stage is found on whistling thorn acacia trees (*Vachellia drepanolobium*) (Whitaker et al. 2019). For this species, like many small organisms, there were no direct observations of feeding behavior in the field. However, because they were found on acacia trees that have a mutualism with ants, it was hypothesized that the butterfly larvae might feed upon regurgitations from ants or even eat ants themselves. By analyzing the DNA and nitrogen isotope signatures of the gut contents of butterfly larvae at field sites on Suyian Ranch in Laikipia County, Kenya (fig. 1.5*a*). Whitaker's group determined that these *A. usamba* larvae were, in fact, herbivorous, feeding almost exclusively off of the acacia tree itself (fig. 1.5*b*).

All of these examples are considered observational studies because there was no manipulation of variables in the field; data were collected on organisms as they existed in their natural environment.

Experimentation

While observational research is critically important for the field of ecology because it allows us to define patterns in nature, it is limited because observation cannot be used to definitively exclude a possible phenomenon (Tilman 1989). That is, just because we have never observed something (like a butterfly eating an ant) is not enough to say it is theoretically impossible. This is why **experiments** may also be necessary. Experimentation typically refers to research that involves manipulation of variables of interest while holding others constant in order to test a hypothesis. In her book co-authored with Elizabeth Lunbeck about scientific observation, Lorraine Daston described the relationship between observational research and experiments thus:

> "Observation, by the curiosity it inspires and the gaps that it leaves, leads to experiment; experiment returns to observation by the same curiosity that seeks to fill and close the gaps still more; thus one can regard experiment and observation as in some fashion the consequence and complement of one another" (Datson 2011)

Experiments can occur in the field or in a more controlled setting, such as in a lab, garden, greenhouse, or outdoor enclosures for animals. Ecological experimentation in the field can be difficult to interpret because of the number of factors that may influence the data being collected, including factors that may be unknown to the researcher. In contrast, *ex situ* (not in the natural environment) experimentation has the benefit of being able to exclude all but the factors of interest in the experiment; however, it has the limitation of not representing real life. Both *ex situ* and *in situ* observation and experimentation may be necessary to fully understand the ecology of an organism or system.

In a classic example of experimentation, Candace Galen tested the importance of bees for the evolution of the alpine wildflower *Polemonium viscosum* growing on Pennsylvania Mountain, Colorado (fig. 1.6a). Observational research had suggested that bumblebees prefer larger flowers, and also that more visitation by bees meant more pollen transferred and thus more seeds produced. More seeds means greater fitness. But it was also possible that some other factor such as nutrients produced both big flowers and lots of seeds in some plants. Were the bees important for the evolution of flower size? Galen was one of the first to experimentally test the hypothesis that larger flowers would have more seeds specifically because they were more attractive to bees (Galen 1989). To do this, she compared seed set in plants that were exposed to pollination by bumblebees (the treatment) versus plants that were hand-pollinated and bagged to exclude bees (the control). Both groups included plants with a range of flower sizes. She found that seed number significantly increased with flower size when pollinated by bees (fig. 1.6b), but that flower size did not predict seed number in the control group. That is, her experiment showed that not only did bees prefer larger flowers, but also that the bees were influencing the evolution of larger flowers because they caused those plants to have greater fitness.

As with observational research, new tools and technology have also advanced ecological experimentation. More than 30 years after her groundbreaking work on bumblebee pollination, Galen is still researching the ecology and evolution of these systems, most recently using a newly developed tool for analyzing audio recordings of bees' wing movements. Galen and her colleagues took advantage of a **natural experiment** created by a solar eclipse to investigate the impact of light and temperature on bees' behavior (Galen et al. 2019). They were able to

(a)

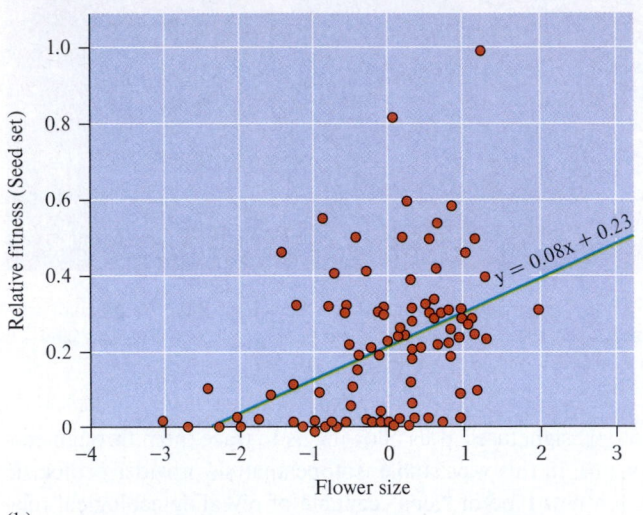

(b)

Figure 1.6   Experimental research on *Polemonium viscosum,* an alpine flower in Colorado, USA (*a*), demonstrated that increasing flower size was positively associated with fitness, as measured by seed set, in flowers pollinated by bumblebees. (*b*) We know that bumblebees are driving this relationship because no such relationship was found for flowers that were hand-pollinated (the control in the experiment) (the graph is adapted from Galen 1989).   (*a*) Candace Galen

compare behavior during darkness (the treatment created by the eclipse) to daylight (the control) within a single hour by detecting bee movements using their sound. In order to obtain a very large sample over a wide geographic area, they used another new resource: data collected by non-scientists. School children and other "**citizen scientists**" assisted in collecting recordings of bees' buzzing at 11 locations in 3 regions using tiny USB microphones dispersed among flowers (fig. 1.7a). Recordings of the bees buzzing were then digitally analyzed to document the dramatic decrease in bee movement during the eclipse (fig. 1.7b). Statistical analysis was used to determine that this decrease was due primarily to light, rather than temperature.

system works. **Conceptual models** are those which describe systems in pictures or diagrams, whereas **quantitative models** are mathematical and may involve complex equations. Like observational research, models can be purely descriptive can be designed to test a hypothesis. A line, which can be represented by the equation $y = mx + b$, is an example of a very simple quantitative model that has many applications within ecology. In Galen's early work, a line could be used to represent the relationship between flower size and fitness (fig. 1.6*b*). The line does not explain all of the variability in fitness, but it does generally define the rate at which seed set increases with increasing flower size. In this way, a model can be a tool used by ecologists to understand the data they have collected. In other cases, modeling may describe or predict patterns using data from published work by others. In chapter 6, we will learn about a modeling approach that is used to summarize findings from many different studies at once.

Models may also be used to represent a novel hypothesis that can then be tested with or compared against future observational or experimental data. In chapter 22, we will learn about McArthur and Wilson's Island Biogeography model; this is a conceptual model based on observed data to explain how size of islands and their proximity to the mainland affect species diversity. This model has been applied to many species and ecosystems, and it has even helped us understand the dynamics of protected areas as land-islands (Sher and Primack 2018).

Finally, models can be used to simulate natural systems, allowing us to test scenarios that would be too difficult, expensive, or logistically impossible to do in real life. For example, Annette Janssen and others created a computer program that can be used to predict growth of algae in lake ecosystems, based on different climate models (Janssen et al. 2019). Janssen's model includes multiple ecological feedbacks that affect algal growth in deep lakes, making it possible to quantify how the lake will respond to different nutrient levels under climate warming. The mathematical equations involved in creating the simulations can be represented by a conceptual model (fig. 1.8). Although the Janssen et al. model may seem complex, many mathematical simulations of this type are even more so in their attempt to represent the real world. Such simulations are important for understanding what has both happened in the past as well as what may happen in the future.

(a)

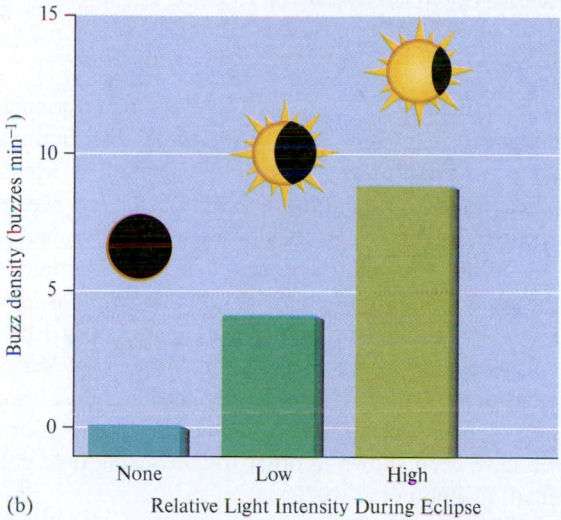

(b)

Figure 1.7 (*a*) Emilia Asante, a graduate student at the University of Missouri, assembles a microphone to record the buzzing of bees. The white, furry "jacket" screens out wind noise. (*b*) Recordings of the bees buzzing documented a dramatic decrease in bee activity as measured by buzz density during the eclipse. (*a*) Candice King; (*b*) adapted from Galen et al. 2019

Modeling

Modeling is the creation and analysis of representations of data or ideas to provide insight or make predictions. In ecology, models usually represent a hypothesis regarding how a

Climatic and Ecological Change: Past and Future

The earth and its life are always changing. However, many of the most important changes occur over such long periods of time or at such large spatial scales that they are difficult to study. Two approaches that provide insights into long-term and large-scale processes are studies of pollen preserved in lake sediments and of evolutionary change.

Margaret B. Davis (1983, 1989) carefully searched through a sample of lake sediments for pollen. The sediments had come from a lake in the Appalachian Mountains, and the pollen they contained would help her document changes in

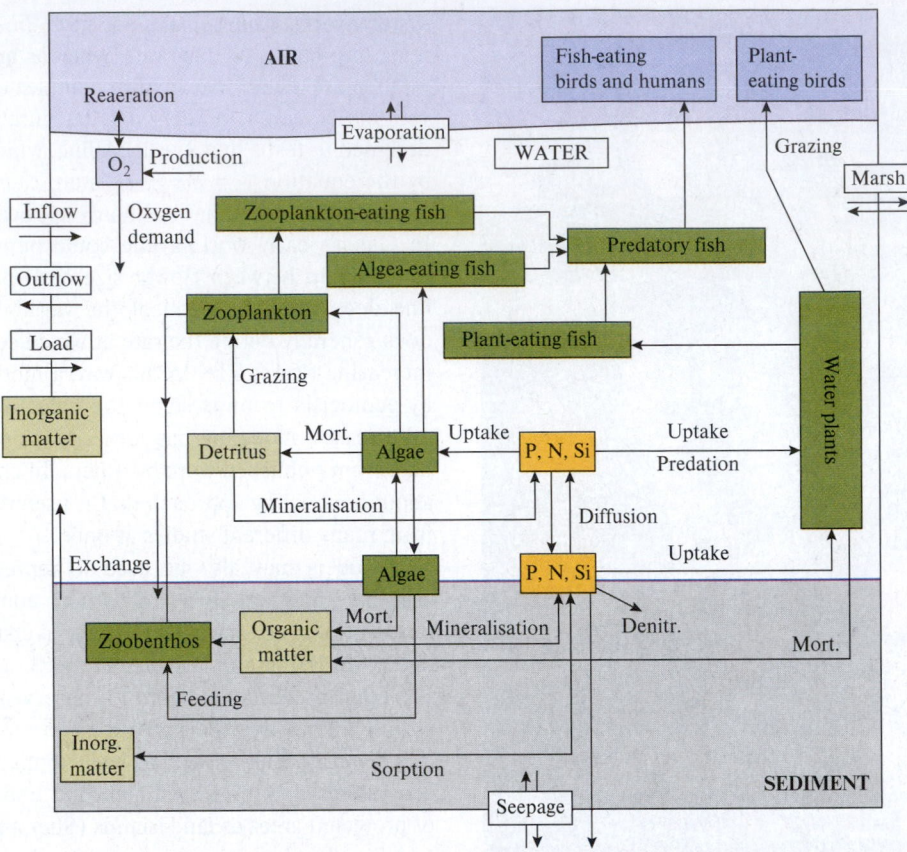

Figure 1.8   Ecological modeling research includes creating complex mathematical simulations of systems as a way of making predictions. Arrows indicate movement of energy or matter. This conceptual model or schematic of a lake ecosystem represents only a portion of a larger, a process-based mathematical model that was created to predict algal blooms under different environmental conditions (diagram based on Janssen et al. 2019).

the community of plants living near the lake during the past several thousand years. Davis is a paleoecologist trained to think at very large spatial scales and over very long periods of time. She has spent much of her professional career studying changes in the distributions of plants during the Quaternary period, particularly during the most recent 20,000 years.

Some of the pollen produced by plants that live near a lake falls on the lake surface, sinks, and becomes trapped in lake sediments. As lake sediments build up over the centuries, this pollen is preserved and forms a historical record of the kinds of plants that lived nearby. As the lakeside vegetation changes, the mix of pollen preserved in the lake's sediments also changes. In the example shown in figure 1.9, pollen from spruce trees, *Picea* spp., first appears in lake sediments about 12,000 years ago; then pollen from beech, *Fagus grandifolia,* occurs in the sediments beginning about 8,000 years ago. Chestnut pollen does not appear in the sediments until about 2,000 years ago. The pollen from all three tree species continues in the sediment record until about 1920, when chestnut blight killed most of the chestnut trees in the vicinity of the lake. Thus, the pollen preserved in the sediments of lakes can be used to reconstruct the history of vegetation in the area. Margaret B. Davis, Ruth G. Shaw, and Julie R. Etterson reviewed extensive evidence that during climate change, plants evolve, as well as disperse (Davis and Shaw 2001; Davis,

Shaw, and Etterson 2005). As climate changes, plant populations simultaneously change their geographic distributions and undergo the evolutionary process of **adaptation,** which increases their ability to live in the new climatic regime. Meanwhile, evidence of evolutionary responses to climate change has been found in many animal groups. One such example is evidence of rapid evolution in alpine chipmunks (*Tamias alpinus*) for a gene associated with high elevation stress (fig 1.10). This was discovered by researchers at the Museum of Vertebrate Zoology at the University of California using DNA from historic specimens (Bi et al. 2019). Evolution of adaptations to elevation have been shown in other rodents using field collected animals (Velotta et al. 2020).

In the remainder of this book, we will fill in the details of the sketch of ecology presented in this chapter. This brief survey has only hinted at the conceptual basis for the research described. Throughout this book we emphasize the conceptual foundations of ecology. We also explore some of the applications associated with the focal concepts of each chapter. Of course, the most important conceptual tool used by ecologists is the scientific method (see Investigating the Evidence 1 in Appendix A).

We continue our exploration of ecology in section I with natural history and evolution. Natural history is the foundation on which ecologists build modern ecology for which

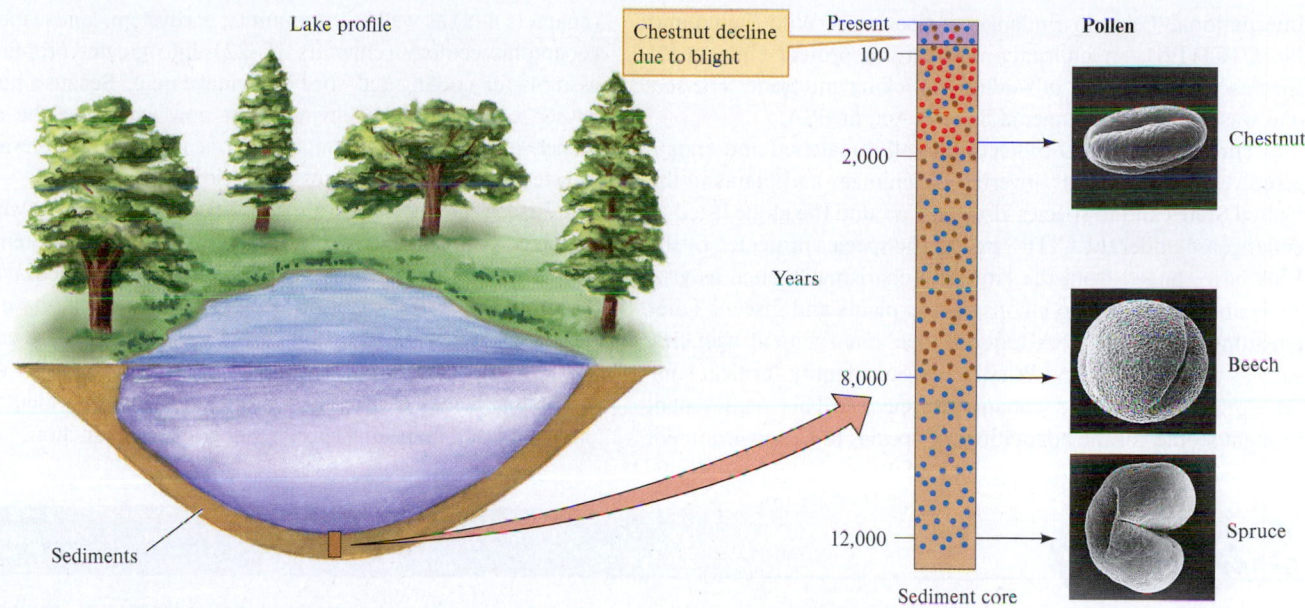

Lake profile

Chestnut decline due to blight

Present
100
2,000

Years

8,000

12,000

Pollen

Chestnut

Beech

Spruce

Sediment core

Sediments

Figure 1.9 The vegetation history of landscapes can be reconstructed using the pollen contained within the sediments of nearby lakes. (Chestnut, Beech and Spruce) Courtesy of the Gretchen and Stanley Jones Palynological Collection and the Botanical Research Institute of Texas

Figure 1.10 Studies indicate that alpine squirrels (*Tamias alpinus*) are evolving adaptations to higher elevations (Bi et al. 2019). Danita Delimont/Getty Images

evolution provides a conceptual framework. A major premise of this book is that knowledge of natural history and evolution improves our understanding of ecological relationships.

Concept 1.2 Review

1. What characterizes each of the three types of ecological research? How are they different and how might they be used together?
2. What are some of the new tools and technology being used in ecological research? Why are they valuable?
3. How are ecologists able to look backwards and forwards in time? Why is this important?

Applications

Ecology Can Inform Environmental Law and Policy

LEARNING OUTCOMES

After studying this section you should be able to do the following:

1.8 Describe the purposes of the Convention on International Trade in Endangered Species of Wild Fauna and Flora (CITES) and the U.S. Endangered Species Act (ESA).
1.9 Discuss how subject areas covered in this text are applicable to identifying and managing endangered species.

Because ecological science concerns relationships between organisms and the environment, it is natural to turn to ecology when environmental concerns arise. Consequently, ecology has contributed prominently to the development of environmental law and policy. For example, ecologists have been essential to evaluating the effects of pollution on the diversity of species in terrestrial and aquatic communities and on the functioning of ecosystems. One area where ecology has played a particularly significant role is in evaluating the status of individual species threatened by human impacts on the environment.

Ecological studies of animal and plant populations are essential to determining when species populations have declined in numbers to the point where they are in danger of extinction (see chapter 9). Reports of such declines in the 1960s eventually led to the establishment of international treaties and national laws to protect endangered species. Two prominent protections came into force in 1973. The first was the Convention on

International Trade in Endangered Species of Wild Fauna and Flora (CITES), an international treaty to protect endangered species from the threat of wildlife trafficking and trade. The second was the U.S. Endangered Species Act, or ESA.

The ESA extended protection to all threatened and endangered vertebrate animals, invertebrate animals, and plants in the United States and to species elsewhere around the globe listed as endangered under the CITES treaty. The species protected by the ESA have ranged from the large and charismatic, such as grizzly bears and whales, to inconspicuous plants and insects. Later amendments to the ESA required that management agencies, such as the U.S. Fish and Wildlife Service, identify "critical habitat" for threatened and endangered species. This requirement brought studies of the adaptations of species to the environment (chapters 4–8) as well as community, ecosystem, landscape, and geographic ecology (chapters 16–22), into greater prominence as tools for endangered species management. Because human-caused changes to the environment now extend to the entire planet, global ecology (chapter 23) is increasingly relevant to long-term endangered species protection.

Ecological studies are also essential to determining whether protected populations have recovered sufficiently to be removed from the ESA's list of endangered species, a process called *delisting*. There have been a number of high-profile species that have been delisted in recent years, including the gray whales of the eastern North Pacific Ocean and bald eagles of the contiguous 48 states. In summary, ecological science has been essential to identifying, protecting, and managing species vulnerable to extinction.

Summary

Ecologists study environmental relationships ranging from those of individual organisms to factors influencing global-scale processes. The research focus and questions posed by ecologists differ across the levels of ecological organization studied.

Ecologists use three general approaches to research: observation, experimentation, and modeling. Many scientific questions will require a combination of approaches. Historically, research of all three types used very simple approaches and tools, however advances in technology have greatly expanded our capacity to learn about our world. This includes the use of drones to collect data in tree tops, stable isotope analysis to identify the diet of a butterfly, and highly sensitive microphones that can measure bee behavior by their buzzing. Modeling as a tool to understand patterns and make predictions has also become more complex, as computer simulations have made the processing of large amounts of data easier and faster. Such simulations are particularly useful in the context of understanding processes that occur at long time scales, such as evolution and climate change. Because ecological science concerns relationships between organisms and the environment, it is often consulted when environmental concerns arise. Ecological science has been particularly important to identifying, protecting, and managing species vulnerable to extinction.

Key Terms

adaptation   8
biosphere   3
citizen scientist   6
community   2
conceptual models   7

ecology   1
ecosystem   3
epiphyte   4
evolution   2
ex situ   6

experiments   5
in situ   4
modeling   6
natural experiment   6
observation   4

population   2
quantitative models   7
urban ecology   3

Review Questions

1. Faced with the complexity of nature, ecologists have divided the field of ecology into subdisciplines, each of which focuses on one of the levels of organization pictured in figure 1.1. What is the advantage of developing such subdisciplines within ecology?

2. What are the pitfalls of subdividing nature in the way it is represented in figure 1.1? In what ways does figure 1.1 misrepresent nature?

3. What could you do to verify that the distinct feeding zones used by the warblers studied by MacArthur (see fig. 1.3) are the result of ongoing competition between the different species of warblers? How might you examine the role of competition in keeping some American redstarts out of the most productive feeding areas on their wintering grounds?

4. How has technology advanced our capacity to collect data over wider geographic and temporal scales? What other advances have benefited the scientific field of ecology?

5. What do the studies of Margaret Davis tell us about the composition of forests in the Appalachian Mountains during the past 12,000 years (see fig. 1.9)? Based on this research, what predictions might you make about the future composition of these forests?

6. Ecological models range from the very simple to the very complex. Are complex models always better? Why or why not?

Chapter

2

Life on Land

LifeJourneys/iStock/Getty Images

Despite their often barren appearance, deserts are ecosystems with a diversity of life, uniquely adapted to the temperature and moisture of those environments.

CHAPTER CONCEPTS

2.1 Uneven heating of the earth's surface by the sun and the tilt of the earth combine to produce predictable latitudinal and seasonal variation in climate.   14

Concept 2.1 Review   16

2.2 While terrestrial biome distribution is strongly associated with latitude, biomes are also influenced by microclimate and soil type.   17

Concept 2.2 Review   19

2.3 Environmental conditions shape each biome's characteristic biology.   19

Concept 2.3 Review   40

Applications: Finer Scale Climatic Variation over Time and Space   40

Summary   41

Key Terms   42

Review Questions   43

LEARNING OUTCOMES

After studying this section you should be able to do the following:

2.1 Explain why plants are the basis of life on Earth and thus define terrestrial biomes.

2.2 List the main environmental features used to differentiate the various terrestrial biomes.

If you are standing on a ground of seemingly barren rock and sand, what little vegetation being thick-leaved and/or prickly, with a few vultures overhead and a lizard eating a scorpion at your feet, where are you? Although the details differ, this general description fits areas of the southwestern United States, western China, Libya, Australia, and elsewhere; we refer to this type of ecosystem as a desert. Desert is one of several types of **biomes,** major divisions of the terrestrial environment, distinguished primarily by their predominant plants (fig. 2.1a). In figure 2.1, the boundaries between biomes appear sharp, whereas in nature these transitions generally occur gradually over long distances along gradients of environmental variation. But why do we find the same biome in such disparate locations across the globe? And conversely, why don't we find desert at the top of a mountain or at the equator? The study of how organisms in a particular area are influenced by factors such as climate, soils, predators, competitors, and evolutionary history is called **natural history.** In this chapter, we explore the natural history of different types of terrestrial biomes, including the reasons why they are distributed the way that they are.

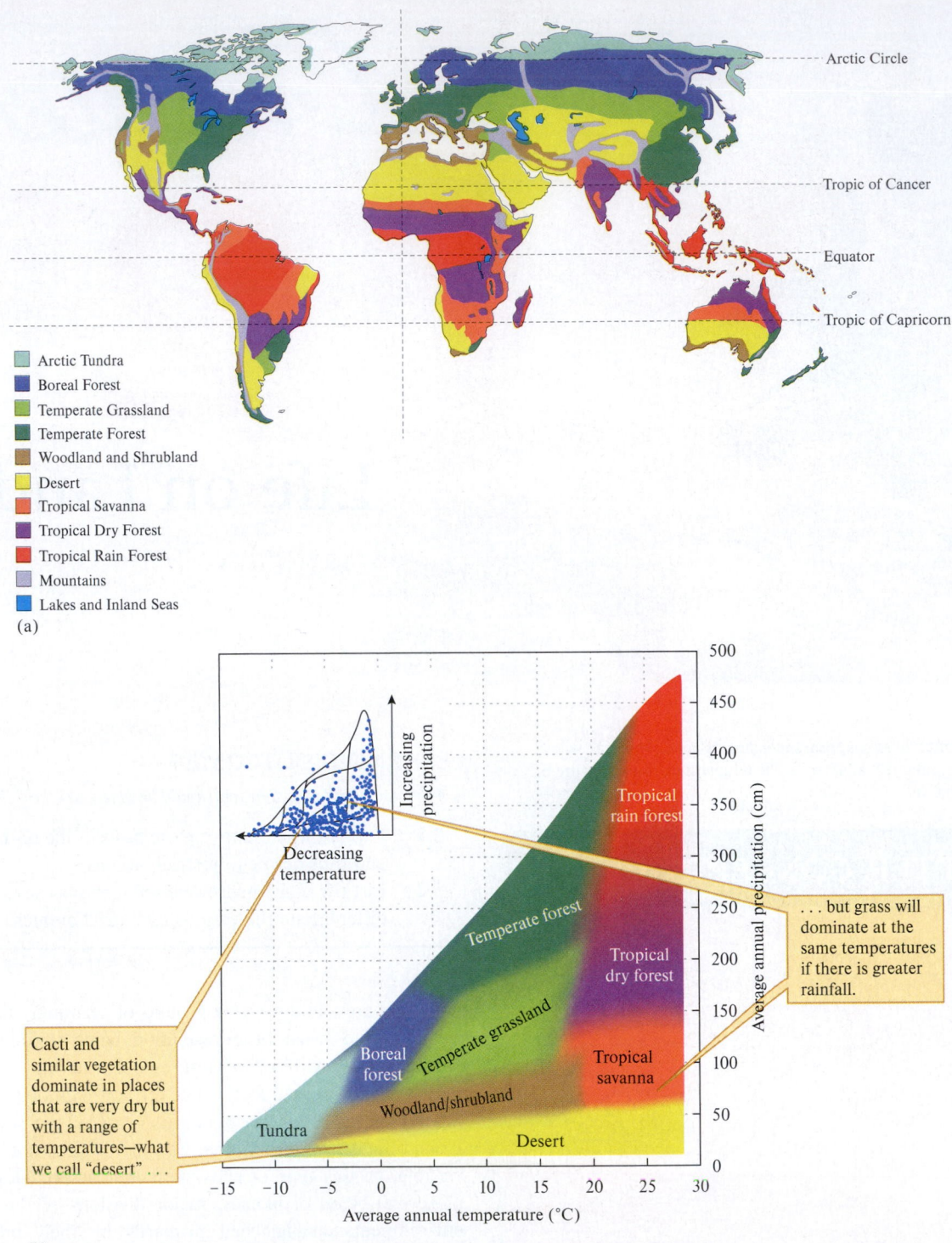

Figure 2.1 (*a*) Global distribution of biomes. (*b*) Biomes are generally defined by the average temperature (x-axis) and precipitation (rain, y-axis) of a given location. Robert Whittaker plotted the vegetation type of various locations against these two variables (inset) to determine the boundaries of the various biomes, but boundaries between these are not distinct, due to the many other influences of the environment (adapted from Whittaker 1975).

Terrestrial Biomes and the Importance of Plants

The most important factor for understanding the distribution of biomes has to do with climate. Whether an ecosystem is dominated by cacti, grass, deciduous trees, conifers, or other types of plants will primarily depend on temperature and water availability (fig. 2.1*b*). If we plot the combination of these two variables for locations across the globe by their dominant plant types, we can see clear patterns of how different natural histories can be defined by temperature and precipitation (fig. 2.1*b* inset).

But how does a particular environment result in similar plants? Returning to our desert scene described above, the dry environment will mean that regardless of which continent we are on, plants in a desert will have certain **functional traits:** particular characteristics that allow them to survive, such as waxy coatings on leaves that prevent water loss. These functional traits of plants arise through evolution via **selective pressure** by the environment, a subject explored in more detail in chapter 4. The section on deserts in this chapter contains a particularly compelling example of how selective pressure results in similar functional traits in unrelated plants on different continents.

But what about all of the animals, fungi, and other organisms in an ecosystem; why are biomes defined by plants? One reason for this is that plants form the foundation of life on this planet. On land, plants make the energy of the sun available to all other life-forms, including us, via photosynthesis (explained in more detail in chapter 7). For this reason, plants and other photosynthesizing organisms are called **primary producers** (fig. 2.2). The biomass produced by primary producers per unit time is *primary production*, which is discussed in detail in chapter 18. Animals and fungi cannot get their energy directly from the sun and so depend on this conversion by plants. For example, in the desert

we visited above, the lizard is eating a scorpion that may have eaten a grasshopper that survived by eating a succulent plant, such as a cactus. When all of these organisms die, there will be fungi and insects and bacteria that eat them. Thus, all of these **consumer** organisms can be considered **secondary producers** of energy for the organisms that eat them. We will learn more about food webs and energy flow in ecosystems in chapters 17 and 18. Selective pressure by the climate will be felt *directly* by consumers in an ecosystem, but also *indirectly,* as each experiences the selective pressure of its food source. Each consumer will have traits that make it able to capture, eat, and digest the food available in that environment. Thus, the type and diversity of plants in a given region, determined primarily by climate, will have far-reaching implications for the rest of the ecosystem.

We devote chapters 2 and 3 to an overview of general types of ecosystems and where they occur, that is, the natural history of the biosphere. In chapter 2, we examine the natural history of life on land. We will explore why natural history is primarily based on latitude—that is, why we don't find desert at the

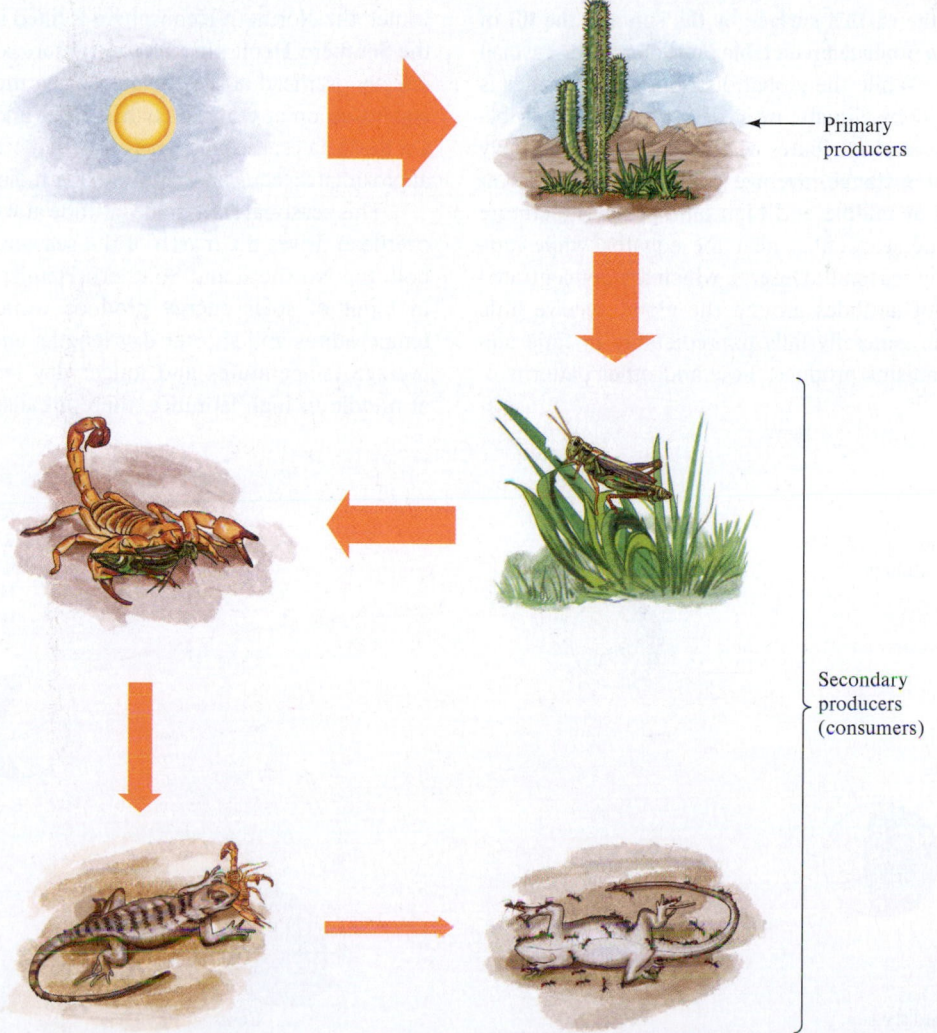

Primary producers

Secondary producers (consumers)

Figure 2.2   How energy moves through an ecosystem; size and direction of arrows indicate the movement and loss of energy through trophic groups. Plants such as cacti capture sun energy via photosynthesis; animals depend on that energy whether by eating the plants or other animals that have eaten plants, and so on. In this way, all organisms ultimately get their energy from the sun, forming the basis of all terrestrial biomes.

equator—but also how soil type and elevation can have strong influences as well, that is, why we also don't find desert on mountaintops. The main goal of chapters 2 and 3 is to take a large-scale perspective of nature before delving, in later chapters, into finer details of structure and process.

2.1   Large-Scale Patterns of Climatic Variation

LEARNING OUTCOMES

After studying this section you should be able to do the following:

2.3   Diagram the position of the sun relative to the equator and tropics of Capricorn and Cancer, during the equinoxes and solstices.

2.4   Describe how solar driven air circulation produces regional differences in precipitation.

2.5   Interpret a climate diagram.

2.6   Explain the influence of the Coriolis effect on wind direction.

Uneven heating of the earth's surface by the sun and the tilt of the earth combine to produce predictable latitudinal and seasonal variation in climate. While the global distribution of biomes is determined primarily by climate, what determines the distributions of climate? Several attributes of climate vary predictably over the earth. For instance, average temperatures are lower and more seasonal at middle and high latitudes. Temperature generally shows little seasonality near the equator, while rainfall may be markedly seasonal. Deserts, which are concentrated in a narrow band of latitudes around the globe, receive little precipitation, which generally falls unpredictably in time and space. What mechanisms produce these and other patterns of climatic variation?

Temperature, Atmospheric Circulation, and Precipitation

The uneven heating of the earth's surface results from the spherical shape of the earth and the angle at which the earth rotates on its axis as it orbits the sun. Because the earth is a sphere, the sun's rays are most concentrated where the sun is directly overhead. However, the latitude at which the sun is directly overhead changes with the seasons. This seasonal change occurs because the earth's axis of rotation is not perpendicular to its plane of orbit about the sun but is tilted approximately 23.5° away from the perpendicular (fig. 2.3).

Because this tilted angle of rotation is maintained throughout earth's orbit about the sun, the amount of solar energy received by the Northern and Southern Hemispheres changes seasonally. During the northern summer, the Northern Hemisphere is tilted toward the sun and receives more solar energy than the Southern Hemisphere. During the northern summer solstice on approximately June 21, the sun is directly overhead at the tropic of Cancer, at 23.5° N latitude. During the northern winter solstice, on approximately December 21, the sun is directly overhead at the tropic of Capricorn, at 23.5° S latitude. During the northern winter, the Northern Hemisphere is tilted away from the sun and the Southern Hemisphere receives more solar energy. The sun is directly overhead at the equator during the spring and autumnal equinoxes, on approximately March 21 and September 22 or 23. On those dates, the Northern and Southern Hemispheres receive approximately equal amounts of solar radiation.

This seasonal shift in the latitude at which the sun is directly overhead drives the march of the seasons. At high latitudes, in both the Northern and Southern Hemispheres, seasonal shifts in input of solar energy produce winters with low average temperatures and shorter day lengths and summers with high average temperatures and longer day lengths. In many areas at middle to high latitudes, there are also significant seasonal

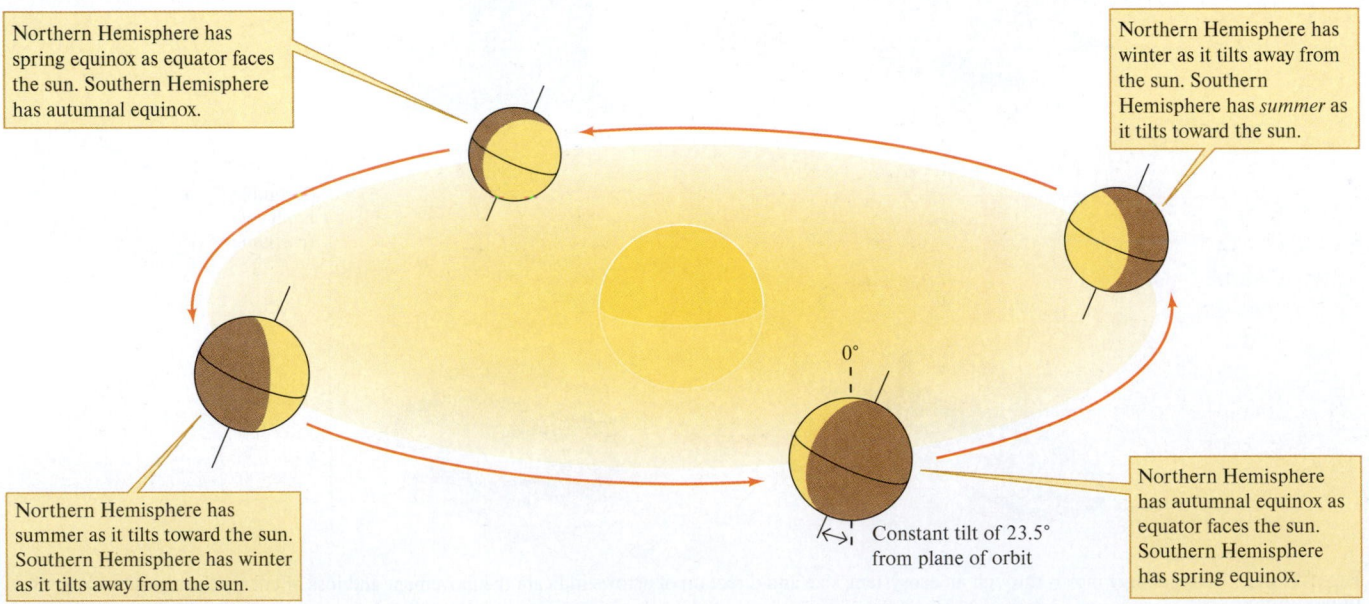

Northern Hemisphere has spring equinox as equator faces the sun. Southern Hemisphere has autumnal equinox.

Northern Hemisphere has winter as it tilts away from the sun. Southern Hemisphere has *summer* as it tilts toward the sun.

Northern Hemisphere has summer as it tilts toward the sun. Southern Hemisphere has winter as it tilts away from the sun.

0°

Constant tilt of 23.5° from plane of orbit

Northern Hemisphere has autumnal equinox as equator faces the sun. Southern Hemisphere has spring equinox.

Figure 2.3   The seasons in the Northern and Southern Hemispheres.

changes in precipitation. Meanwhile, between the tropics of Cancer and Capricorn, seasonal variations in temperature and day length are slight, while precipitation may vary a great deal. What produces spatial and temporal variation in precipitation?

Heating of the earth's surface and atmosphere drives circulation of the atmosphere and influences patterns of precipitation. As shown in figure 2.4a, the sun heats air at the equator, causing it to expand and rise. This warm, moist air cools as it rises. Since cool air holds less water vapor than warm air, the water vapor carried by this rising air mass condenses and forms clouds, which produce the heavy rainfall associated with tropical environments.

Eventually, this equatorial air mass ceases to rise and spreads north and south. This high-altitude air is dry, since the moisture it once held fell as tropical rains. As this air mass flows north and south, it cools, which increases its density. Eventually, it sinks back to the earth's surface at about 30° latitude and spreads north and south. This dry air draws moisture from the lands over which it flows and creates deserts in the process.

Air moving from 30° latitude toward the equator completes an atmospheric circulation cell at low latitudes. As figure 2.4b shows, there are three such cells on either side of the equator. Air moving from 30° latitude toward the poles is part of the atmospheric circulation cell at middle latitudes. This warm air flowing from the south rises as it meets cold polar air flowing from the north. As this air mass rises, moisture picked up at lower latitudes condenses to form the clouds that produce the abundant precipitation of temperate regions. The air rising over temperate regions spreads northward and southward at a high altitude, completing the middle- and high-latitude cells of general atmospheric circulation.

The patterns of atmospheric circulation shown in figure 2.4b suggest that air movement is directly north and south. However, this does not reflect what we observe from the earth's surface as the earth rotates from west to east. We can see how air movement changes by this simple demonstration: with your finger, trace a line from the equator to the north pole on a globe while it is slowly turning. Do it again for the south pole. Your finger will draw diagonal lines in the same direction as winds will be dragged on the surface of the earth. That is, an observer at tropical latitudes observes winds that blow from the northeast in the Northern Hemisphere and from the southeast in the Southern Hemisphere (fig. 2.5). These are the *northeast* and *southeast trades.* Someone studying winds within the temperate belt between 30° and 60° latitude would observe that winds blow mainly from the west. These are the *westerlies* of temperate latitudes. At high latitudes, our observer would find that the predominant wind direction is from the east. These are the *polar easterlies.*

Why don't winds move directly north to south? The prevailing winds do not move in a straight north–south direction because of the **Coriolis effect.** In the Northern Hemisphere, the Coriolis effect causes an apparent deflection of winds to the right of their direction of travel and to the left in the Southern Hemisphere. We say "apparent" deflection because we see this deflection only if we make our observations from the surface of the earth. To an observer in space, it would appear that winds move in approximately a straight line, while the earth rotates beneath them. However, we need to keep in mind that the perspective from the earth's surface is the ecologically relevant perspective. The biomes that we discuss in chapter 2 are as earthbound as our hypothetical observer. Their distributions across the globe are substantially influenced by global climate, particularly geographic variations in temperature and precipitation.

Geographic variation in temperature and precipitation is very complex. How can we study and represent geographic variation in these climatic variables without being overwhelmed by a mass of numbers? This practical problem is addressed by a visual device called a climate diagram.

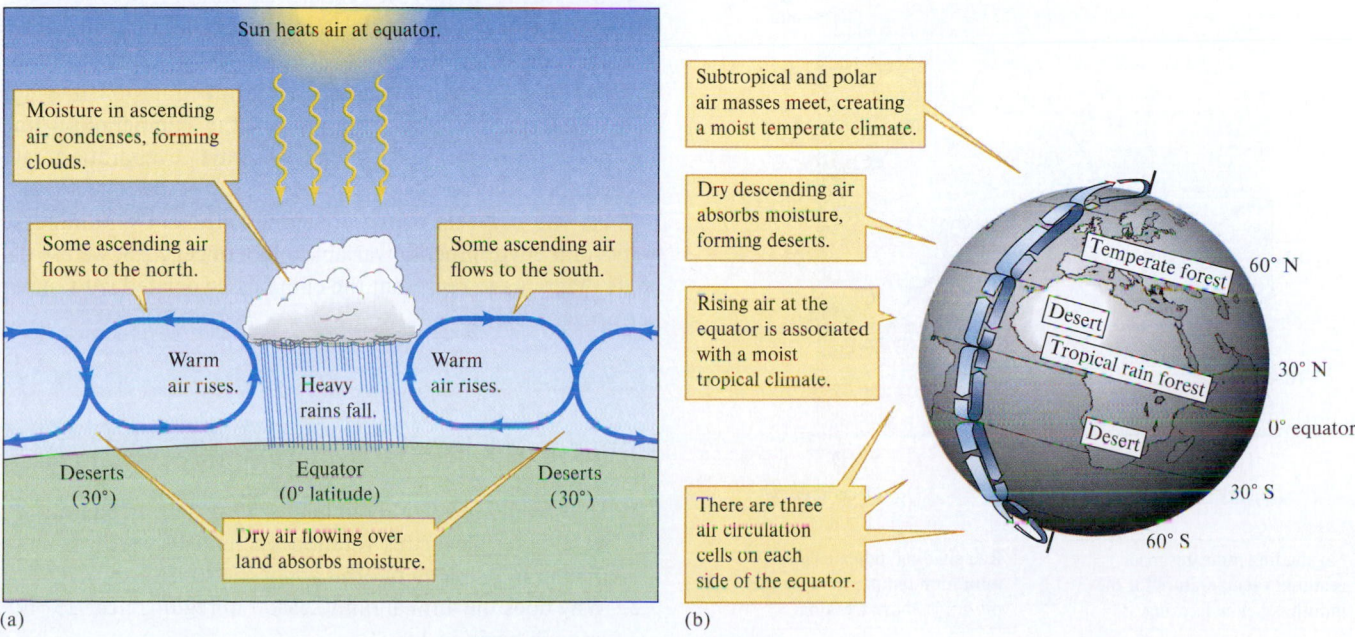

(a)   (b)

Figure 2.4   (a) Solar-driven air circulation. (b) Latitude and atmospheric circulation.

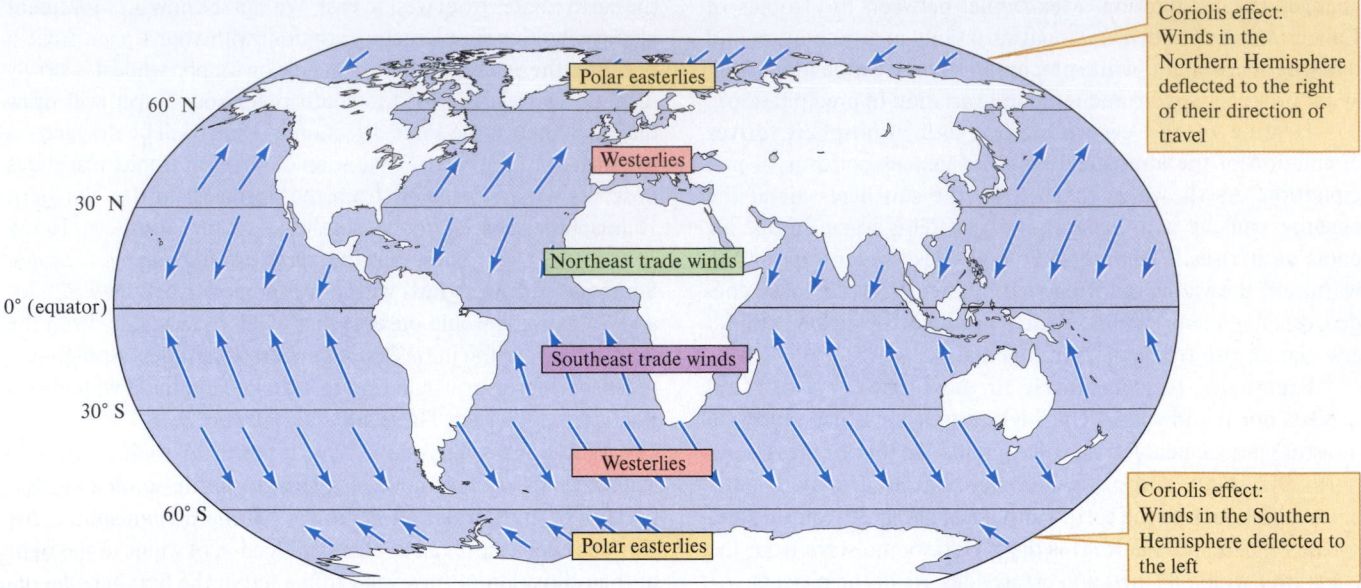

Coriolis effect: Winds in the Northern Hemisphere deflected to the right of their direction of travel

Polar easterlies

Westerlies

Northeast trade winds

Southeast trade winds

Westerlies

Polar easterlies

Coriolis effect: Winds in the Southern Hemisphere deflected to the left

60° N

30° N

0° (equator)

30° S

60° S

Figure 2.5   The Coriolis effect and wind direction.

Climate Diagrams

Climate diagrams were developed by Heinrich Walter (1985) as a tool to explore the relationship between the distribution of terrestrial vegetation and climate. Climate diagrams summarize a great deal of useful climatic information, including seasonal variation in temperature and precipitation, and the length and intensity of wet and dry seasons.

As shown in figure 2.6, climate diagrams summarize climatic information using a standardized structure. The months

of the year are plotted on the horizontal axis, beginning with January and ending with December for locations in the Northern Hemisphere and beginning with July and ending with June in the Southern Hemisphere. Temperature is plotted on the left vertical axis and precipitation on the right vertical axis. Temperature and precipitation are plotted on different scales so that 10°C is equivalent to 20 mm of precipitation. Climate diagrams for wet areas such as tropical rain forest compress the precipitation scale for precipitation above 100 mm so that 10°C is equivalent to 200 mm of precipitation. With this change in scale, rainfall data from very wet climates can be fit on a graph of convenient size.

Because the temperature and precipitation scales are constructed so that 10°C equals 20 mm of precipitation, the relative positions of the temperature and precipitation lines reflect water availability. Theoretically, adequate moisture for plant growth exists when the precipitation line lies above the temperature line. When the temperature line lies above the precipitation line, potential evaporation rate exceeds precipitation.

As you can see, climate diagrams efficiently summarize important environmental variables. In Concept 2.3, we use climate diagrams to represent the climates associated with major terrestrial biomes.

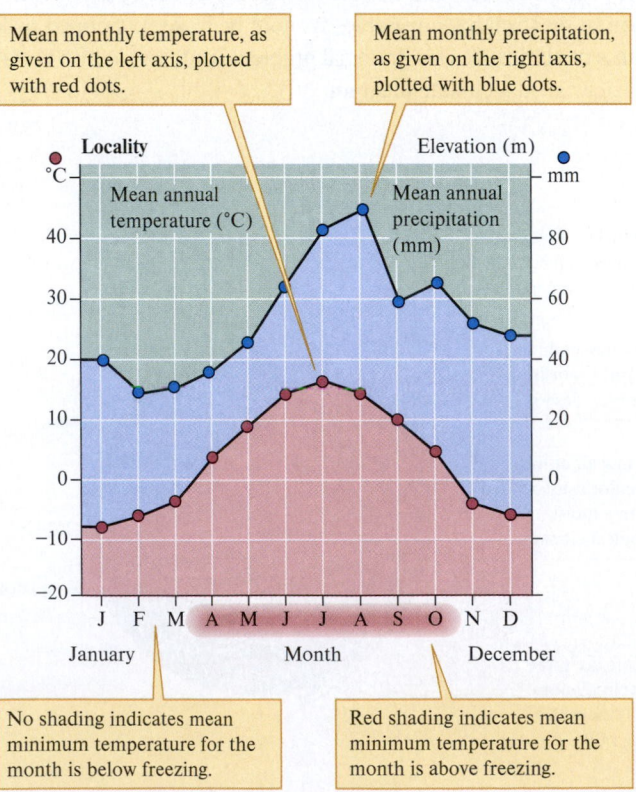

Mean monthly temperature, as given on the left axis, plotted with red dots.

Mean monthly precipitation, as given on the right axis, plotted with blue dots.

Locality   Elevation (m)

Mean annual temperature (°C)   Mean annual precipitation (mm)

No shading indicates mean minimum temperature for the month is below freezing.

Red shading indicates mean minimum temperature for the month is above freezing.

January   Month   December

Figure 2.6   Structure of climate diagrams.

Concept 2.1 Review

1. How would seasonality in temperature and precipitation be affected if earth's rotation on its axis were perpendicular to its plane of orbit about the sun?
2. Why does the annual rainy season in regions near 23° N latitude begin in June?

2.2   Other Factors That Shape Terrestrial Biomes

LEARNING OUTCOMES

After studying this section you should be able to do the following:

2.7   Explain the concept of rain shadow.

2.8   Describe the characteristics of each of the typical soil horizons.

2.9   Discuss how climate, organisms, topography, parent material, and time can influence the structure and development of soils and which organisms can survive there.

While terrestrial biome distribution is strongly associated with latitude, biomes are also influenced by microclimate and soil type.   From the previous section, we know why deserts are not found at the equator: the angle of the sun's rays on the surface of the earth results in this being a wet area of the globe. However, latitude is not the only determinant of where we find biomes; we also do not find desert on mountaintops, regardless of latitude. Also consider the fact that at 35°N we can find temperate forest in North Carolina, grassland in Oklahoma, desert in Arizona, and Mediterranean scrubland in California.

The distribution of mountains partly explains why biomes do not form perfect horizontal stripes on the earth. It is colder at higher elevations, of course, but whether it will be wetter or drier depends on which side of the mountain you are on. To understand this, we can apply what we learned in the previous section about rising air losing its moisture in the form of rain or snow. Warm, moist air from the ocean that blows toward a mountain range will have lost much of its moisture when that air reaches the leeward side (fig. 2.7). The dry climate that results is called the **rain shadow effect.** When the temperature and moisture differ from the prevailing climate, we call these local environments **microclimates,** to be discussed in greater detail in chapter 5.

Microclimates can have dramatic influences on biome distribution; the western United States is one such example (fig. 2.8). Not only do we observe particular plant communities on high-altitude mountaintops, but because of the rain shadow effect, we observe a wide range of biomes on the sides and base of a mountain, even at the same elevation and latitude. Desert, for example, is found on the eastern side of the Sierra Nevada mountains.

Biomes are not only determined by temperature and moisture, however; in addition to and interacting with climate is the effect of soils. As you will see in the biome descriptions later in this chapter, soil type can determine if a region is savanna or desert, even at the same amount of precipitation. It is partly for this reason that there are blurry, rather than crisp, boundaries between biomes in figure 2.1*b*.

Soil is a complex mixture of living and nonliving material upon which most terrestrial life depends. The nonliving component of soil comes from its underlying geology; weathering slowly breaks down parent material, often bedrock, into smaller and smaller fragments to produce sand, silt, and clay-sized particles. The size of particles and the minerals associated with them can have profound impacts on what types of plants, microorganisms, and even animals can live in the associated soils. Here we summarize the general features of soil structure and development. The biome discussions that follow include specific information about the soils associated with each. Nonetheless, it should be kept in mind that variation within a biome is often due to different types of soils occurring there.

Though soil structure usually changes gradually with depth, soil scientists generally divide soils into several discrete horizons. In the classification system used here, the soil profile is divided into O, A, B, and C horizons (fig. 2.9). The **O, or organic, horizon** lies at the top of the profile. The most superficial layer of the O horizon is made up of freshly fallen organic matter, including whole leaves, twigs, and other plant parts, which become more fragmented and decomposed

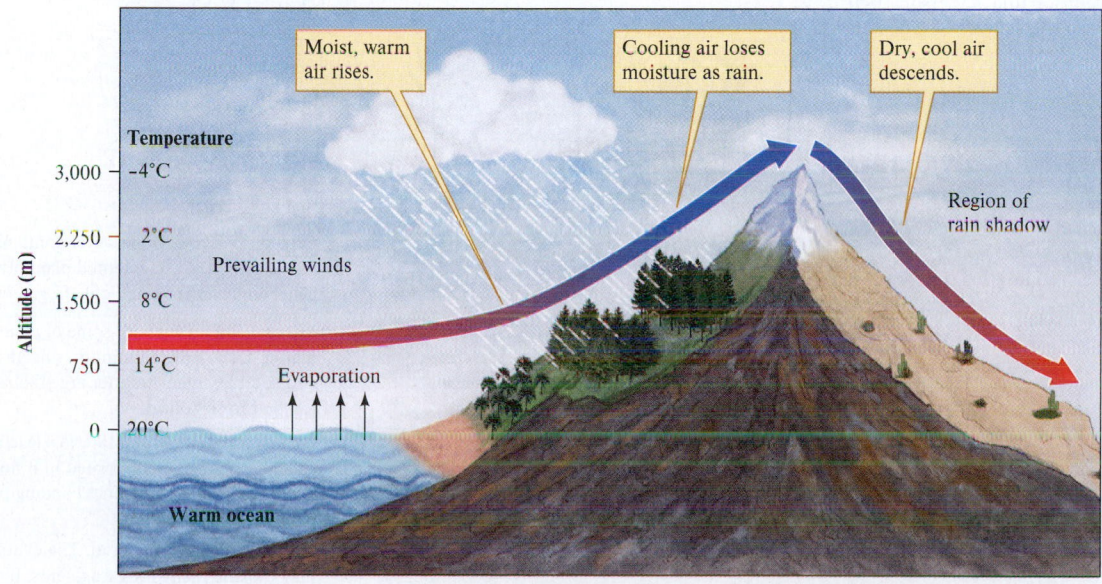

Figure 2.7   The rain shadow effect accounts for dramatic differences in climate and therefore biome on opposite sites of many mountain ranges.

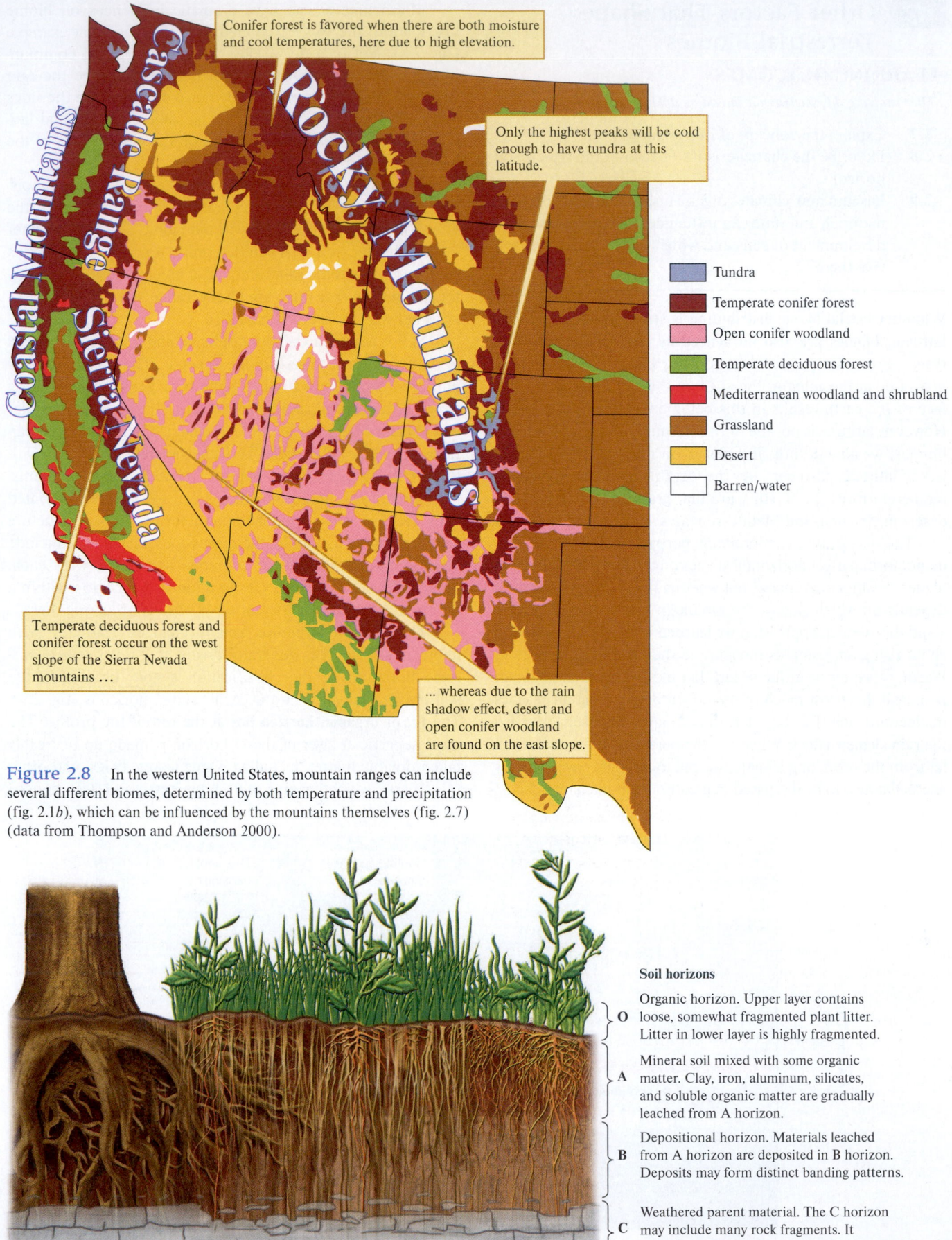

Conifer forest is favored when there are both moisture and cool temperatures, here due to high elevation.

Only the highest peaks will be cold enough to have tundra at this latitude.

Temperate deciduous forest and conifer forest occur on the west slope of the Sierra Nevada mountains …

… whereas due to the rain shadow effect, desert and open conifer woodland are found on the east slope.

Tundra

Temperate conifer forest

Open conifer woodland

Temperate deciduous forest

Mediterranean woodland and shrubland

Grassland

Desert

Barren/water

Figure 2.8   In the western United States, mountain ranges can include several different biomes, determined by both temperature and precipitation (fig. 2.1*b*), which can be influenced by the mountains themselves (fig. 2.7) (data from Thompson and Anderson 2000).

Soil horizons

O   Organic horizon. Upper layer contains loose, somewhat fragmented plant litter. Litter in lower layer is highly fragmented.

A   Mineral soil mixed with some organic matter. Clay, iron, aluminum, silicates, and soluble organic matter are gradually leached from A horizon.

B   Depositional horizon. Materials leached from A horizon are deposited in B horizon. Deposits may form distinct banding patterns.

C   Weathered parent material. The C horizon may include many rock fragments. It often lies on bedrock.

Figure 2.9   Generalized soil profile, showing O, A, B, and C horizons.

with increasing depth. Fragmentation and decomposition of the organic matter in this horizon are mainly due to the activities of soil organisms, including bacteria, fungi, and animals ranging from nematodes and mites to burrowing mammals. This horizon is usually absent in agricultural soils and deserts. At its deepest levels, the O horizon merges gradually with the A horizon.

The **A horizon** contains a mixture of mineral materials, such as clay, silt, and sand, and incorporated organic material derived from the O horizon. Burrowing animals, such as earthworms, mix organic matter from the O horizon into the A horizon. The A horizon is generally rich in mineral nutrients. It is gradually leached of clays, iron, aluminum, silicates, and humus, which is partially decomposed organic matter. These substances slowly move down through the soil profile until they are deposited in the B horizon.

The **B horizon** contains the clays, humus, and other materials that have been transported by water from the A horizon. The deposition of these materials often gives the B horizon a distinctive color and banding pattern. The B horizon gradually merges with the C horizon.

The **C horizon** is the deepest layer in our soil pit and the only one not typically dominated by plant roots. It consists of the weathered parent material, which has been worked by the actions of frost, water, and the deeper penetrating roots of plants. Because weathering is incomplete and less intense than in the A and B horizons, the C horizon may contain many rock fragments. Under the C horizon, we find unweathered parent material, which is often bedrock.

The soil profile gives us a snapshot of soil structure. However, soil structure is in a constant state of flux as a consequence of several influences. Those influences were summarized by Hans Jenny (1980) as climate, organisms, topography, parent material, and time. Climate affects the rate of weathering of parent materials, the rate of leaching of organic and inorganic substances, the rate of erosion and transport of mineral particles, and the rate of decomposition of organic matter. Living organisms, which as we know are also affected by climate, influence the quantity and quality of organic matter added to soil and the rate of soil mixing by burrowing animals. Topography affects the rates and direction of water flow and patterns of erosion. Meanwhile, parent materials, such as granite, volcanic rock, and wind- or water-transported sand, set the stage for all other influences. Last is the matter of time. Soil age influences soil structure.

As with many aspects of ecology, it is often difficult to separate organisms from their environment. The biome discussions that follow provide additional information on soils by including aspects of soil structure and chemistry characteristic of each biome.

Concept 2.2 Review

1. The organic horizon is generally absent from agricultural soils because tilling (e.g., plowing), buries organic matter. Why is an organic horizon generally absent from desert soils?

Natural History and Geography of Biomes

LEARNING OUTCOMES

After studying this section you should be able to do the following:

2.10   List the major terrestrial biomes.
2.11   Describe the climatic differences among the biomes.
2.12   Contrast the soils typical of the terrestrial biomes.
2.13   Describe the types of vegetation, animals, and other organisms characteristic of the terrestrial biomes.
2.14   Explain variation in human presence in the various terrestrial biomes.

Environmental conditions shape each biome's characteristic biology.   Early in the twentieth century, many plant ecologists studied how climate and soils influence the distribution of vegetation. Later ecologists concentrated on other aspects of plant ecology. Today, as we face the prospect of global warming (see chapter 23), ecologists are once again studying climatic influences on the distribution of vegetation. International teams of ecologists, geographers, and climatologists are exploring the influences of climate on vegetation with renewed interest and with much more powerful analytical tools. Ecologist Osvaldo Sala and others created a predictive model (for more on models, see Investigating the Evidence 1 in Appendix A) using biological and environmental data from each biome to determine where biodiversity is at the most risk. While deserts and tundra were not expected to change much over the next century, Mediterranean and grassland biomes were found to be highly sensitive to anticipated human-caused changes to the environment, including but not limited to climate change (Sala et al. 2000).

In this section, we discuss the climate, soils, and organisms of the earth's major biomes and how they have been influenced by humans.

Tropical Rain Forest

Tropical rain forest is nature's most extravagant garden (fig. 2.10). Beyond its tangled edge, a rain forest opens into a surprisingly

Figure 2.10   Tropical rain forest in Ecuador. More species live within the three-dimensional framework of tropical rain forests than in any other terrestrial biome.   Elena Kalistratova/Vetta/Getty Images

spacious interior, illuminated by dim, greenish light shining through a ceiling of leaves. The architecture of rain forests, with their vaulted ceilings and spires, has invited comparisons to cathedrals and mansions. However, this cathedral is alive from ceiling to floor, perhaps more alive than any other biome on the planet. In the rain forest, the sounds of evening and morning, the brilliant flashes of color, and rich scents carried on moist night air speak of abundant life, in seemingly endless variety.

Geography

Tropical rain forests straddle the equator in three major regions: Southeast Asia, West Africa, and South America (fig. 2.11). Most rain forest occurs within 10° of latitude north or south of the equator. Outside this equatorial band are the rain forests of Central America and Mexico, southeastern Brazil, eastern Madagascar, southern India, and northeastern Australia.

Climate

The global distribution of rain forests corresponds to areas where conditions are warm and wet year-round (see fig. 2.11). Temperatures in tropical rain forests vary little from month to month and often change as much in a day as they do over the entire year. Average temperatures are about 25°C to 27°C, lower than average maximum summer temperatures in many deserts and temperate regions. Annual rainfall ranges from about 2,000 to 4,000 mm, and some rain forests receive even more precipitation. In a rain forest, a month with less than 100 mm of rain is considered dry.

Soils

Heavy rains gradually leach nutrients from rain forest soils and rapid decomposition in the warm, moist rain forest climate keeps the quantity of soil organic matter low. Consequently, rain forest soils are often nutrient-poor, acidic, thin, and low in organic matter. In many rain forests, more nutrients are tied up in living tissue than in soil. Some rain forests, however, occur where soils are very fertile such as along rivers. Fungi, bacteria, and soil animals, such as mites and springtails, rapidly scavenge nutrients from plant litter (leaves, flowers, etc.) and animal wastes, further tightening the nutrient economy in tropical ecosystems.

Biology

Trees dominate the rain forest landscape and average about 40 m in height but some reach 80 m. These rain forest giants are often supported by well-developed buttresses. The diversity of rain forest trees is also impressive. One hectare (100 m × 100 m) of temperate forest may contain a few dozen tree species; 1 ha of tropical rain forest may contain up to 300 tree species.

Primary production in tropical rain forests is the highest of all terrestrial ecosystems, not only from the trees but also because the three-dimensional framework formed by rain forest trees is festooned with other plant growth forms. The trees are trellises for climbing vines and growing sites for epiphytes, plants that grow on other plants (fig. 2.12). This vast amount of converted energy supports a great biological richness of consumers as well. A single rain forest tree may support several thousand species of insects, many of which have not been described by scientists.

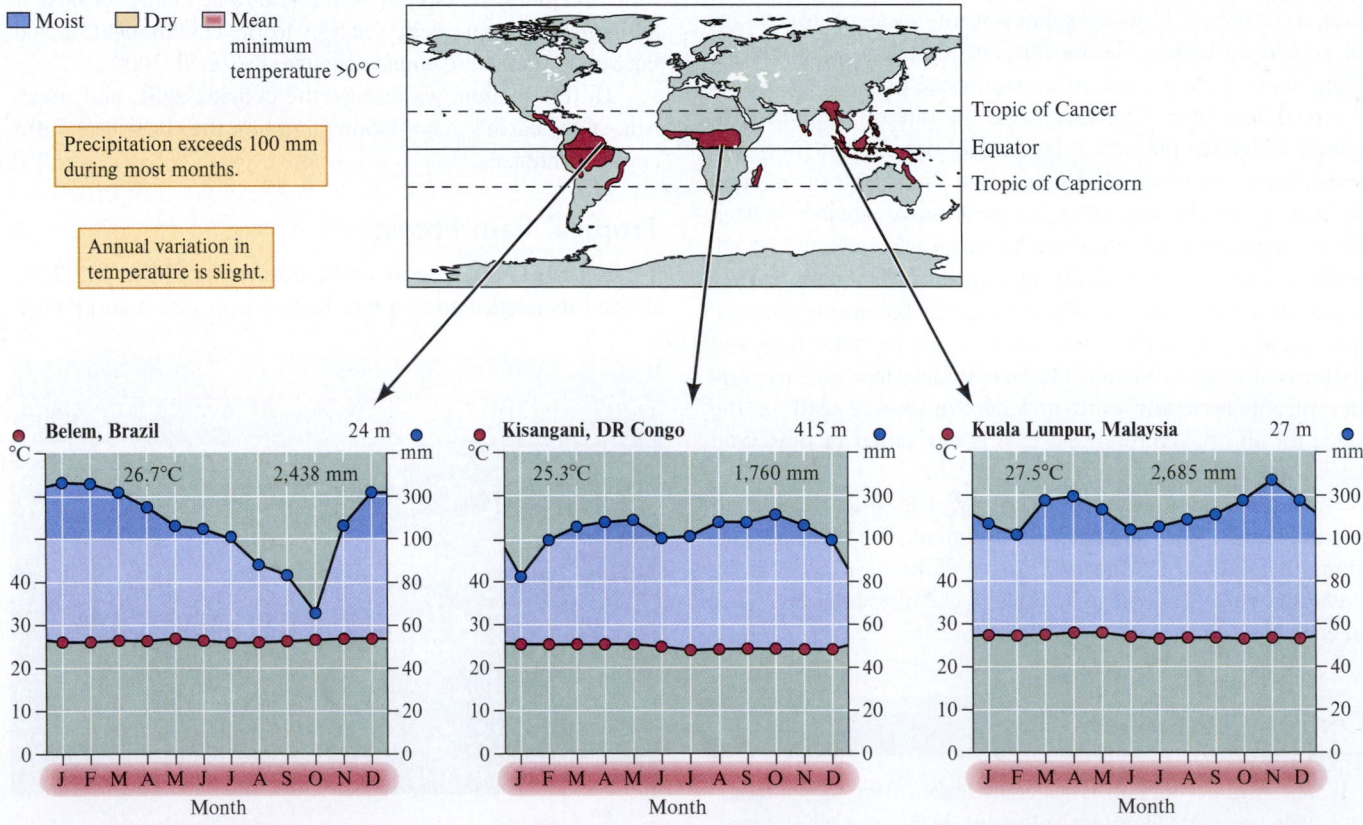

Figure 2.11   Tropical rain forest geography and climate.

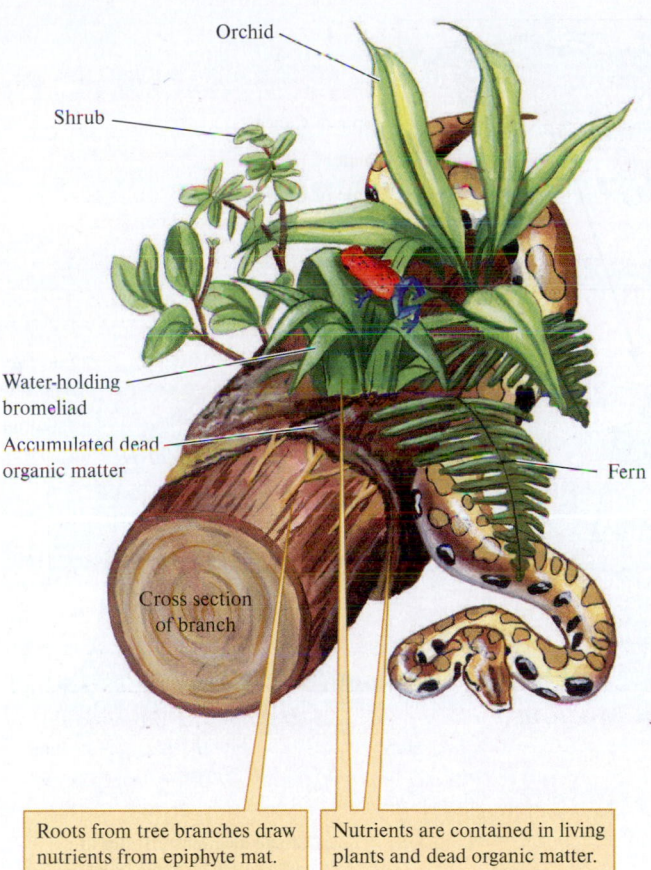

Orchid

Shrub

Water-holding
bromeliad

Accumulated dead
organic matter

Fern

Cross section
of branch

Roots from tree branches draw
nutrients from epiphyte mat.

Nutrients are contained in living
plants and dead organic matter.

Figure 2.12   An epiphyte mat in the tropical rain forest canopy.
Epiphyte mats store a substantial fraction of the nutrients in tropical rain
forests and support a high diversity of plant and animal species.

The rain forest is not, however, just a warehouse for a
large number of dissociated species. Rain forest ecology is
marked by intricate, complex relationships between species.
In the tropical rain forest, there are plants that cannot live
without particular species of fungi, which help them absorb
nutrients (called **mycorrhizae**), the hummingbirds or insects
that pollinate their flowers, and the animals that disperse
their seeds.

Human Influences

People from all over the globe owe more to the tropics than
is generally realized. Many of the world's staple foods, including maize (called corn in North America and Australia),
rice, bananas, and sugarcane, and approximately 25% of all
prescription drugs were originally derived from tropical plants.
Many more species, directly useful to humans, may await
discovery. Unfortunately, tropical rain forests are fast disappearing. According to data collected by scientists at the University
of Maryland, in 2019 an astonishing 11.9 million hectares of
tropical forest cover was lost; this is equal to a football field-
sized area every 8.25 minutes (www.globalforestwatch.org).
This loss diminishes our chances of understanding the extent
and dynamics of biological diversity.

Tropical Dry Forest

During the dry season, the **tropical dry forest** is all earth tones;
in the rainy season, it's an emerald tangle (fig. 2.13). Life in
the tropical dry forest responds to the rhythms of the annual
solar cycle, which drives the oscillation between wet and
dry seasons. During the dry season, most trees in the tropical dry forest are dormant. Then, as the rains approach, trees
flower and insects appear to pollinate them. Eventually, as the
first storms of the wet season arrive, the trees produce their
leaves and transform the landscape.

Geography

Tropical dry forests make up approximately 42% of tropical and
sub-tropical forest area (Hasnat and Hossain 2020) (fig. 2.14).
In Africa, tropical dry forests are found both north and south
of the central African rain forests. In the Americas, tropical
dry forests are the natural vegetation of extensive areas south
and north of the Amazon rain forest. Tropical dry forests
also extend up the west coast of Central America and into
North America along the west coast of Mexico. In Asia, tropical dry forests are the natural vegetation of most of India and
the Indochina peninsula. Australian tropical dry forests form
a continuous band across the northern and northeastern portions of the continent.

Figure 2.13   Tropical dry forest during the wet and dry seasons.   Created by Tomas Zrna/Getty Images; Ralph Lee Hopkins/Science Source

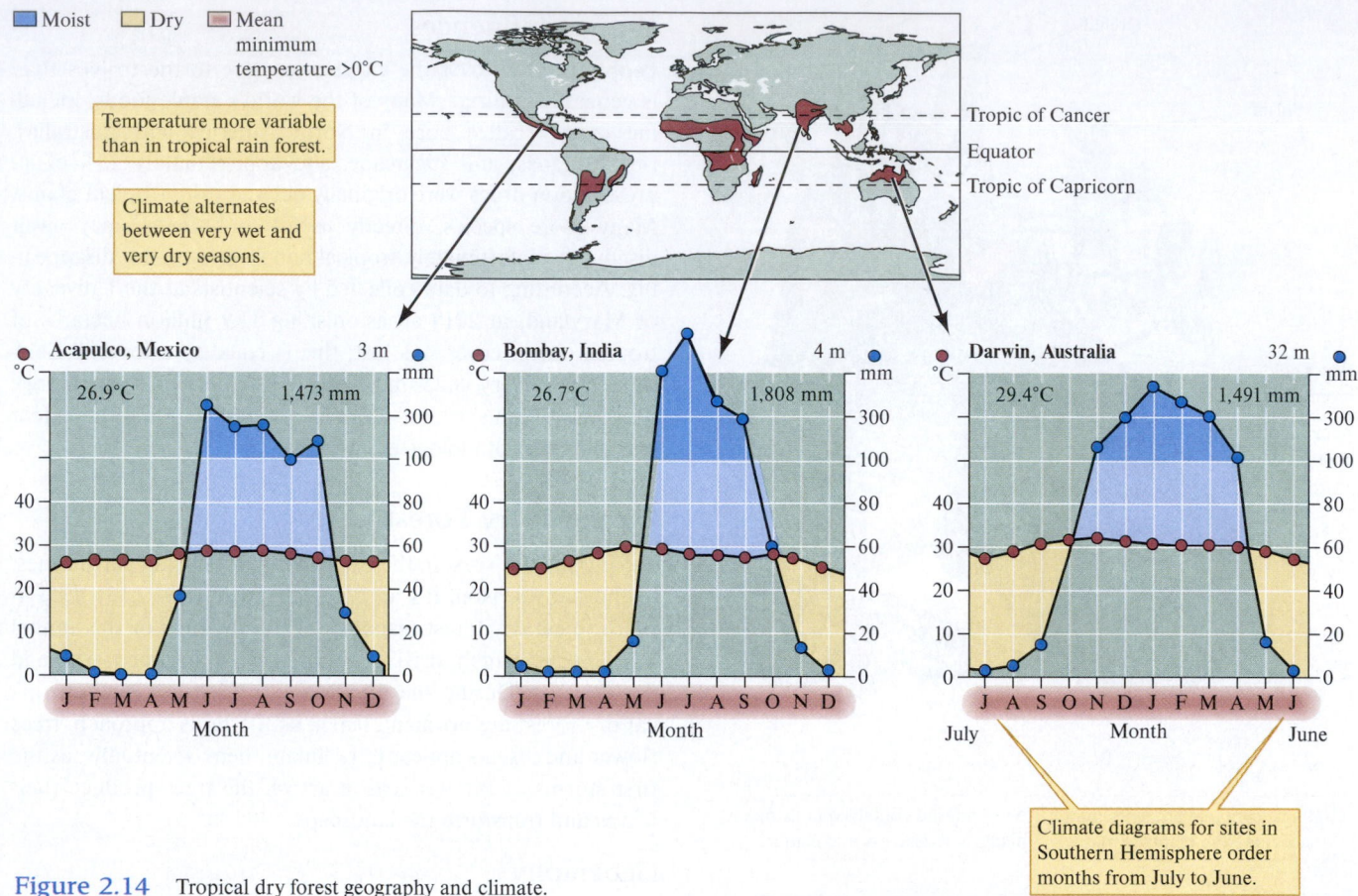

Figure 2.14   Tropical dry forest geography and climate.

Climate

The climate of tropical dry forests is more seasonal than that of tropical rain forests. The three climate diagrams shown in figure 2.14, for example, show a dry season lasting for 6–7 months, followed by a season of abundant rainfall, lasting 5–6 months. The climate diagrams also indicate more seasonal variation in temperature compared to tropical rain forest.

Soils

The soils of many tropical dry forests are of great age, particularly those in the parts of Africa, Australia, India, and Brazil that were once part of the ancient southern continent of Gondwana. The soils of tropical dry forests tend to be less acidic than those of rain forests and are generally richer in nutrients. However, the annual pulses of torrential rain make the soils of tropical dry forest highly vulnerable to erosion.

Biology

The plants of the tropical dry forest are strongly influenced by physical factors. For example, the height of the dry forest is highly correlated with average precipitation. Trees are tallest in the wettest areas. In the driest habitats, all trees drop their leaves during the dry season; in wetter areas over 50% may be evergreen. As in the tropical rain forest, many plants produce animal-dispersed seeds. However, wind-dispersed seeds are also common. Many dry forest birds, mammals, and even insects make seasonal migrations to wetter habitats along rivers or to the nearest rain forest.

Human Influences

Heavy human settlement has devastated the tropical dry forest. While the world's attention has been focused on the plight of rain forests, tropical dry forests have been quietly disappearing, including in so-called protected areas. The relatively fertile soil of tropical dry forests has attracted agricultural development, including cattle ranches, grain farms, and cotton fields. Tropical dry forests are more easily converted to agriculture compared to rain forests, since the dry season makes them more accessible and easier to burn. Using remote sensing data that spanned 40 years, Binita Kumari and colleagues found that areas of a reserve in eastern India that had more human settlements also had higher rates of deforestation (Kumari et al. 2020; fig. 2.15). In other areas, protections seem to be working; in Ghana, researchers found that rates of deforestation had decreased in recent years and new trees are colonizing in some areas (Janssen et al. 2018).

The loss of the dry forest is significant because, while rain forests may support a somewhat greater number of species, many dry forest species are found nowhere else, as many as 40% of the tree and shrub species found there are endemic (Hasnat and Hossain 2020).

Tropical Savanna

Stand in the middle of a savanna, a tropical grassland dotted with scattered trees, and your eye will be drawn to the horizon for the approach of thunderstorms or wandering herds of wildlife (fig. 2.16). The **tropical savanna** is the kingdom of

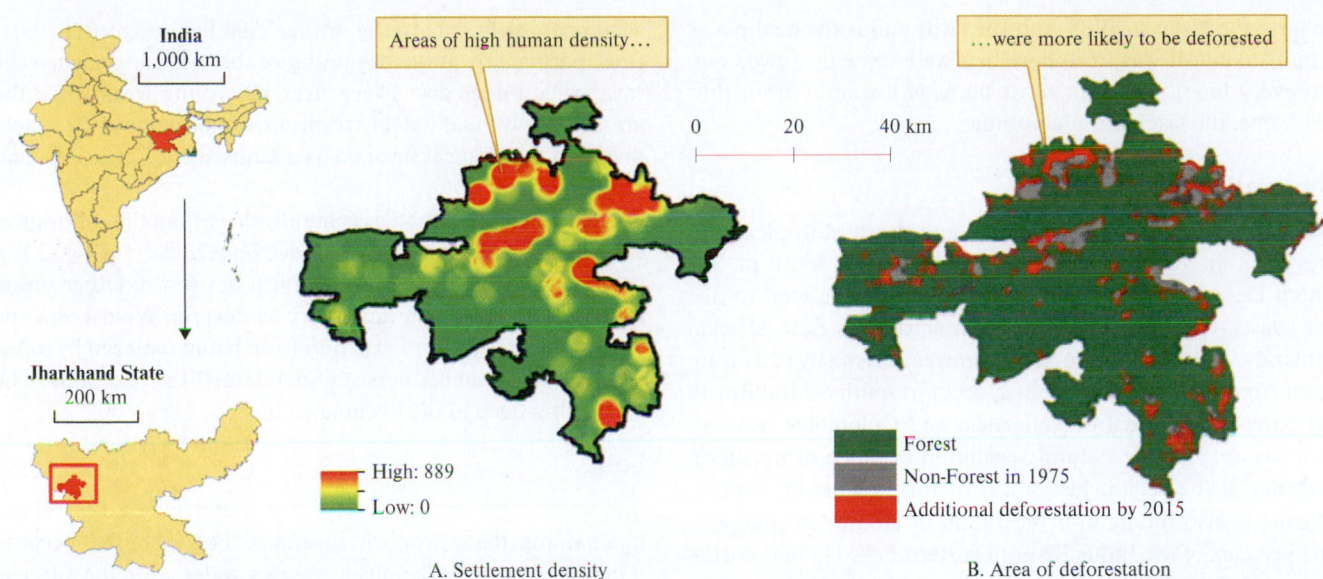

India
1,000 km

Areas of high human density...

...were more likely to be deforested

Jharkhand State
200 km

0   20   40 km

High: 889
Low: 0

Forest
Non-Forest in 1975
Additional deforestation by 2015

A. Settlement density

B. Area of deforestation

Figure 2.15   A comparison of human settlement density (A) and deforestation (B) at the Palamau Tiger Reserve, the latter using satellite imagery. Authors found that loss of forest cover over time was greatest in the central and northern regions, where population densities were highest (based on data from Kumari et al. 2020).

Figure 2.16   Tropical savanna and herbivores in East Africa. This ecosystem is characterized by a dominance of grasses with a few woody species.   Anna Sher

the farsighted, the stealthy, and the swift and is the birthplace of humankind. It was from here that we eventually moved out into every biome. Though most humans live away from this first home, the fascination continues.

Geography

Most tropical savannas occur north and south of tropical dry forests within 10–20° of the equator. In Africa south of the Sahara Desert, tropical savannas extend from the west to the east coasts, cut a north–south swath across the East African highlands, and reappear in south-central Africa (fig. 2.17). In South America, tropical savannas occur in south-central Brazil and cover a great deal of Venezuela and Colombia. Tropical savannas are also the natural vegetation of much of northern Australia in the region just south of the tropical dry forest. Savanna is also the natural vegetation of an area in southern Asia just east of the Indus River in eastern Pakistan and north-western India.

Climate

As in the tropical dry forest, life on the savanna cycles to the rhythms of alternating dry and wet seasons (see fig. 2.17). Here, however, seasonal drought combines with another important physical factor, fire. The rains come in summer and are accompanied by intense lightning. This lightning often starts fires, particularly at the beginning of the wet season when the savanna is tinder dry. These fires kill young trees while the grasses survive and quickly resprout. Consequently, fires help maintain the tropical savanna as a landscape of grassland and scattered trees.

The savanna climate is generally drier than that of tropical dry forest. However, San Fernando, Venezuela (see fig. 2.17), receives as much rainfall as a tropical dry forest. Other savannas occur in areas that are as dry as deserts. What keeps the wet savannas near San Fernando from being replaced by forest and how can savannas persist under desertlike conditions? The answer lies deep in the savanna soils.

Soils

Soil layers with low permeability to water play a key role in maintaining many tropical savannas. For instance, because a dense, impermeable subsoil retains water near the surface, savannas occur in areas of southwestern Africa that would otherwise support only desert. Impermeable soils also help savannas persist in wet areas, particularly in South America. Trees do not move onto savannas where an impermeable subsoil keeps surface soils waterlogged during the wet season. In these landscapes, scattered trees occur only where soils are well drained.

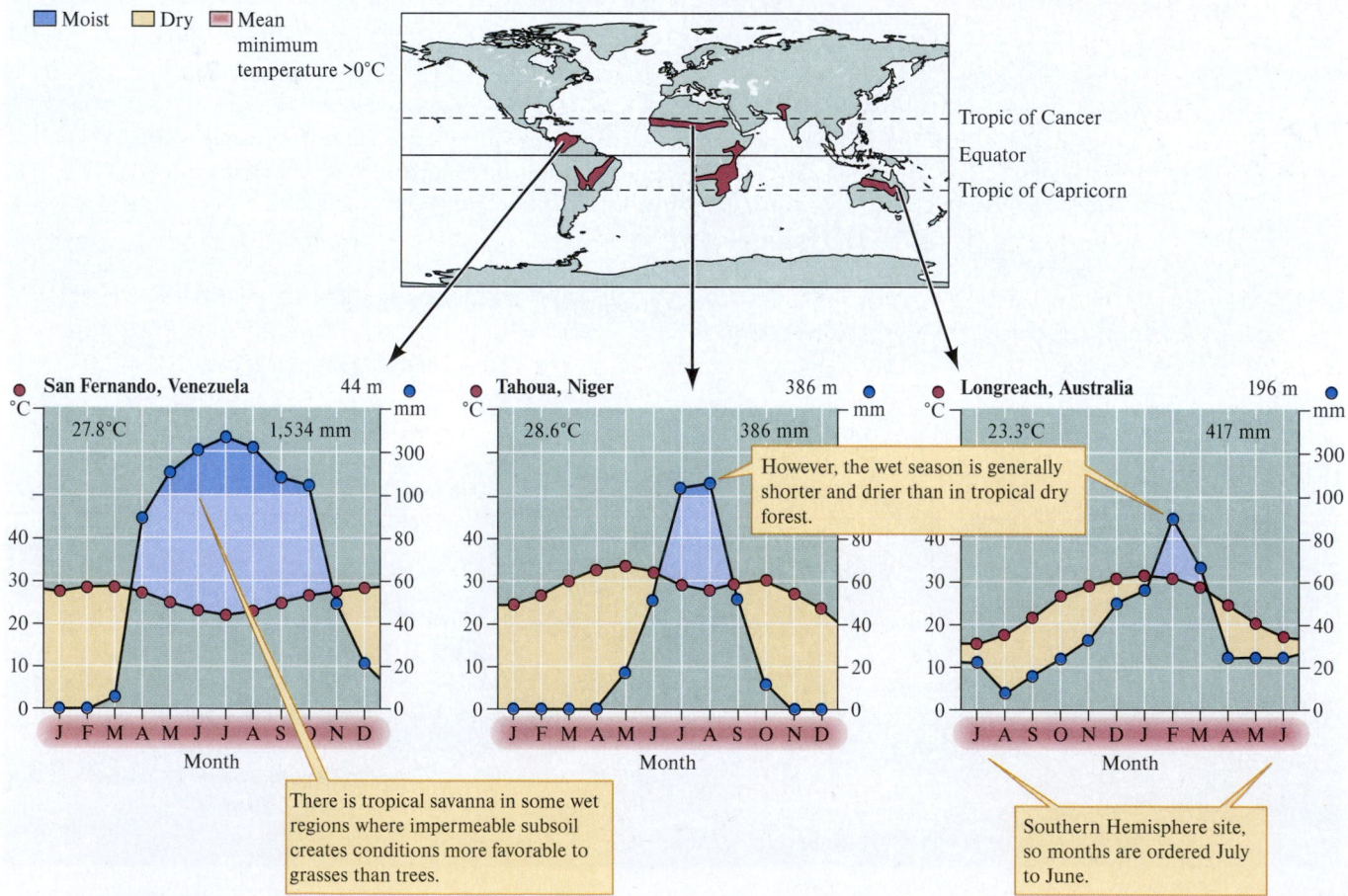

Figure 2.17 Tropical savanna geography and climate.

Biology

Even though savannas don't support many trees, their total primary production across the globe is second only to tropical rain forest; a greater proportion of the biological activity on the savanna simply takes place near ground level, primarily in grasses. Frequent fires have selected for fire resistance in the savanna flora. The few tree species on the savanna resist fire well enough to be unaffected by low-intensity fires.

The tropical savanna is populated by wandering animals that move in response to seasonal and year-to-year variations in rainfall and food availability. The wandering consumers of the Australian savannas include kangaroos, large flocks of birds, and, for about 50,000 years, humans. During droughts, some of these Australian species travel thousands of kilometers in search of suitable conditions. The African savanna is also home to a host of well-known mobile consumers, such as elephants, wildebeest, giraffes, zebras, lions, and, again, humans (see fig. 2.16).

Human Influences

Humans are, in some measure, a product of the savanna and the savanna, in turn, has been influenced by human activity. One of the factors that forged an indelible link between us and this biome is fire. Long before the appearance of hominids, fire played a role in the ecology of the tropical savanna. Later, the savanna was the classroom where early humans observed and learned to use, control, and make fire. Eventually, humans began to purposely set fire to the savanna, which, in turn, helped to maintain and spread the savanna itself. We had begun to manipulate nature on a large scale.

Originally, humans subsisted on the savanna by hunting and gathering. In time, they shifted from hunting to pastoralism, replacing wild game with domestic grazers and browsers. Today, livestock ranching is the main source of livelihood in all the savanna regions. In Africa, livestock raising has coexisted with wildlife for millennia. In modern-day sub-Saharan Africa, however, the combination of growing human populations, high density of livestock, and drought has devastated much of the region known as the Sahel (fig. 2.18).

Figure 2.18   Domestic livestock, such as these cattle on an African savanna, have had a major impact on tropical savannas around the world.   Syda Productions/Shutterstock

Desert

In the spare **desert** landscape, sculpted by wind and water, the ecologist grows to appreciate geology, hydrology, and climate as much as organisms (fig. 2.19). The often repeated description of life in the desert as "life on the edge" betrays an outsider's view. Although primary production is lower than that of other biomes, it does not follow that living conditions there are necessarily harsh. In their own way, many desert organisms flourish on meager rations of water, high temperatures, and saline soils. To understand life in the desert, the ecologist must see it from the perspective of its natural inhabitants.

Geography

Deserts occupy about 20% of the land surface of the earth. Two bands of deserts ring the globe, one at about 30° N latitude and one at about 30° S (fig. 2.20). These bands correspond to latitudes where dry subtropical air descends (see fig. 2.4), drying the landscape as it spreads north and south. Other deserts are found either deep in the interior of continents, for example, or in the rain shadow of mountains, such as the Great Basin Desert of North America as shown in figure 2.8. Still others are found along the cool western coasts of continents, for example, the Atacama of South America and the Namib of southwestern Africa, where air circulating across a cool ocean delivers a great deal of fog to the coast but little rain.

Climate

Environmental conditions vary considerably from one desert to another. Some, such as the Atacama and central Sahara, receive very little rainfall and fit the stereotype of deserts as extremely dry places. Other deserts, such as some parts of the Sonoran Desert of North America, may receive nearly 300 mm of rainfall annually. Whatever their mean annual rainfall, however, water loss in deserts due to evaporation and transpiration by plants exceeds precipitation during most of the year.

Figure 2.20 includes the climate diagrams of two hot deserts. Notice that drought conditions prevail during all months and that during some months average temperatures exceed 30°C. Shade temperatures greater than 56°C have been recorded in the deserts of North Africa and western North America. However, some deserts can be bitterly cold. For example, average winter temperatures at Dzamiin Uuded, Mongolia, in the Gobi Desert of central Asia sometimes fall to −20°C (see fig. 2.20).

Soils

Desert plants and animals can turn this landscape into a mosaic of diverse soils. Desert soils are generally so low in organic matter that they are sometimes classified as **lithosols,** which means stone or mineral soil. However, the soils under desert shrubs often contain large amounts of organic matter and form islands of fertility. Desert animals can also affect soil properties. For example, in North America, kangaroo rats change the texture and elevate the nutrient content of surface soils by burrowing and hoarding seeds. In Middle Eastern deserts, blind mole rats and isopods have been shown to strongly influence soil properties.

Figure 2.19   Life on the edge. Sparse desert vegetation stabilizes a patch of soil at the edge of a field of giant dunes in the Namib Desert in southwestern Africa.   Getty Images

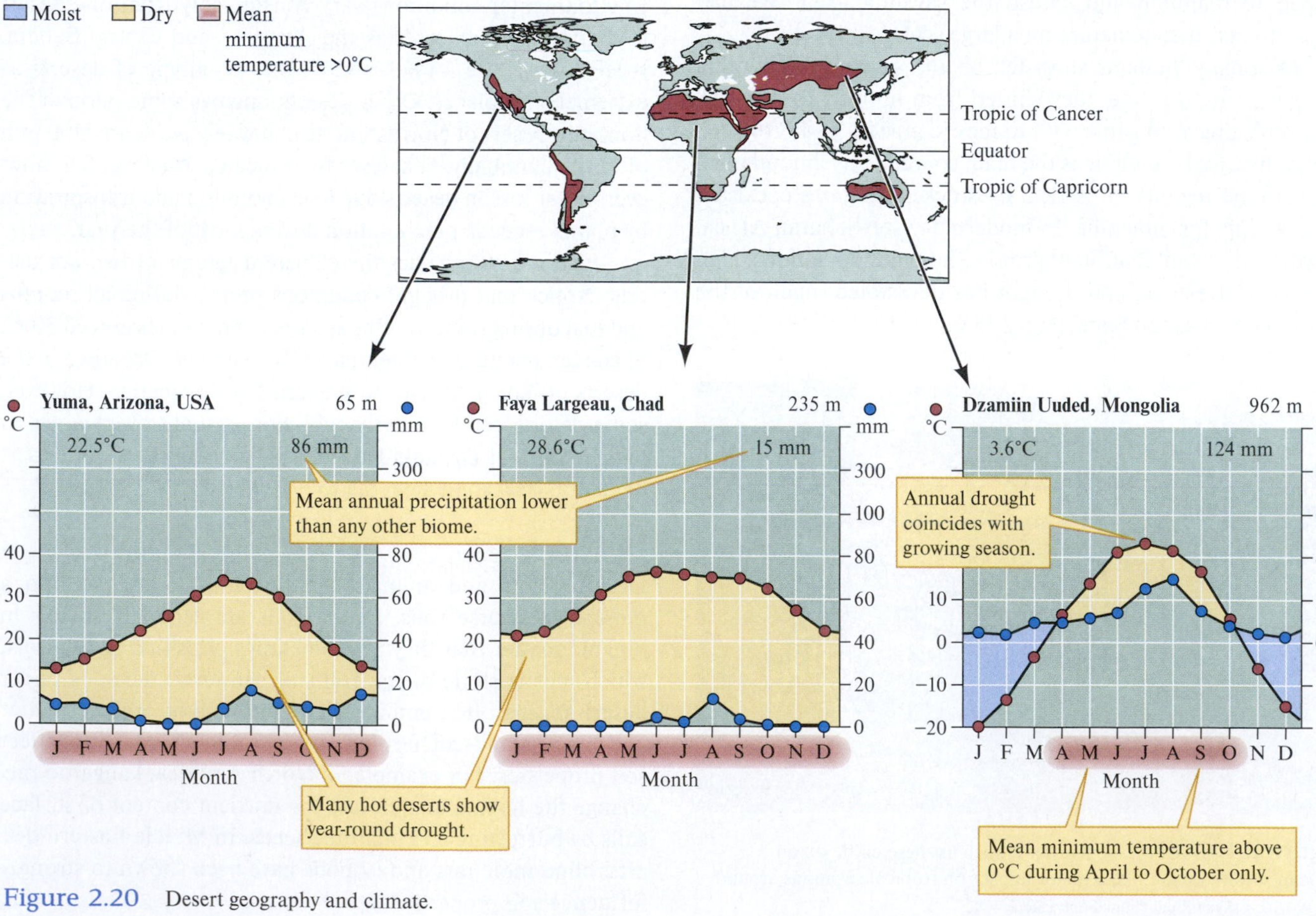

□ Moist □ Dry ▨ Mean
              minimum
              temperature >0°C

Tropic of Cancer
Equator
Tropic of Capricorn

Yuma, Arizona, USA   65 m ●
°C   mm
22.5°C   86 mm
                                          300
                                          100
40   80
30   60
20   40
10   20
0   0
  J F M A M J J A S O N D
        Month

Mean annual precipitation lower
than any other biome.

Many hot deserts show
year-round drought.

Faya Largeau, Chad   235 m ●
°C   mm
28.6°C   15 mm
                                    300
                                    100
40   80
30   60
20   40
10   20
0   0
  J F M A M J J A S O N D
        Month

Dzamiin Uuded, Mongolia   962 m ●
°C   mm
3.6°C   124 mm
                                    80
                                    60
10   40
0   20
                            0
-10
-20
  J F M A M J J A S O N D
        Month

Annual drought
coincides with
growing season.

Mean minimum temperature above
0°C during April to October only.

Figure 2.20   Desert geography and climate.

Desert soils, particularly those in poorly drained valleys and lowlands, may contain high concentrations of salts. Salts accumulate in these soils as water evaporates from the soil surface, leaving behind any salts that were dissolved in the water. Salt accumulation increases the aridity of the desert environment by making it harder for plants to extract water from the soils. As desert soils age they tend to form a calcium carbonate–rich hardpan horizon called **caliche.** The extent of caliche formation has proved a useful tool for aging these soils.

Biology

The desert landscape presents an unfamiliar face to the visitor from moist climates. Plant cover is absent from many places, exposing soils and other geologic features. Where there is plant cover, it is sparse. The plants themselves look unfamiliar. Desert vegetation often cloaks the landscape in a gray-green mantle. This is because many desert plants protect their photosynthetic surfaces from intense sunlight and reduce evaporative water losses with a dense covering of plant hairs. Other plant adaptations to drought include small leaves, producing leaves only in response to rainfall and then dropping them during intervening dry periods, or having no leaves at all (fig. 2.21). Some desert plants avoid drought almost entirely by remaining dormant in the soil as seeds that germinate and grow only during infrequent wet periods.

In deserts, animal abundance tends to be low but diversity can be high. Most desert animals use behavior to avoid environmental extremes. In summer, many avoid the heat of the day by being active at dusk and dawn or at night. In winter, the same species may be active during the day. Animals (as well as plants) use body orientation to minimize heat gain in the summer.

Human Influences

Desert peoples have flourished where nature is stingiest. Compared to true desert species, however, humans are profligate water users. Consequently, human populations in deserts concentrate at oases and in river valleys. Many desert landscapes that once supported irrigated agriculture now grow little as a result of salt accumulation in their soils (Wang et al. 2019).

The desert is the one biome that, because of human activity, is increasing in area. Humanity's challenge is to stop the spread of deserts that comes at the expense of other biomes and to establish a balanced use of deserts that safeguards their inhabitants, human and nonhuman alike.

Woodland and Shrubland

Woodlands and shrublands occur widely in temperate regions. Some are found in the interior of continents and others in coastal regions (see fig. 2.1a). Within the woodland/shrubland biome is a particular climate called Mediterranean, although it can be found in many different regions of the globe. The **Mediterranean woodland and shrubland** climate was the climate of the classical Greeks and the coastal Native

(a)

(b)

Figure 2.21   Similar environments have selected for nearly identical traits in unrelated desert plants: (*a*) cactus in North America, (*b*) *Euphorbia* in Africa.   (*a*) Lucky-photographer/Shutterstock; (*b*) Natphotos/Digital Vision/Getty Images

American tribes of Old California. The mild temperate climate experienced by these cultures was accompanied by high biological richness (fig. 2.22). The richness of the Mediterranean woodland flora is captured by a folk song from the Mediterranean region that begins: "Spring has already arrived. All the countryside will bloom; a feast of color!" To this visual feast, Mediterranean woodlands and shrublands add a chorus of birdsong and the smells of aromatic plants, including rosemary, thyme, and laurel.

Figure 2.22   A Mediterranean woodland in southern Italy.   Manuel C. Molles

Geography

Mediterranean woodlands and shrublands occur on all the continents except Antarctica (fig. 2.23). They are most extensive around the Mediterranean Sea and in North America, where they extend from California into northern Mexico. They are also found in central Chile, southern Australia, and southern Africa. Under present climatic conditions, Mediterranean woodlands and shrublands grow between about 30° and 40° latitude. This position places the majority of this biome north of the

subtropical deserts in the Northern Hemisphere, and south of them in the Southern Hemisphere. The far-flung geographic distribution of Mediterranean woodland and shrubland is reflected in the diversity of its names. In western North America, it is called *chaparral*. In Spain, the most common name for Mediterranean woodland and shrubland is *matorral*. Farther east in the Mediterranean basin the biome is referred to as *garrigue*. Meanwhile, South Africans call it *fynbos*, while Australians refer to at least one form of it as *mallee*. Although the names for this biome vary widely, its climate does not.

Climate

The Mediterranean woodland and shrubland climate is cool and moist during fall, winter, and spring, whereas summers are hot and dry (see fig. 2.23). The danger of frost varies considerably from one Mediterranean woodland and shrubland region to another. When they do occur, however, frosts are usually not severe. The combination of dry summers and dense vegetation, rich in essential oils, creates ideal conditions for frequent and intense fires.

Soils

The soils of Mediterranean woodlands and shrublands have generally low to moderate fertility and are considered fragile. Some soils, such as those of the South African fynbos, have exceptionally low fertility. Soil erosion can be severe. Fire coupled with overgrazing has stripped the soil from some

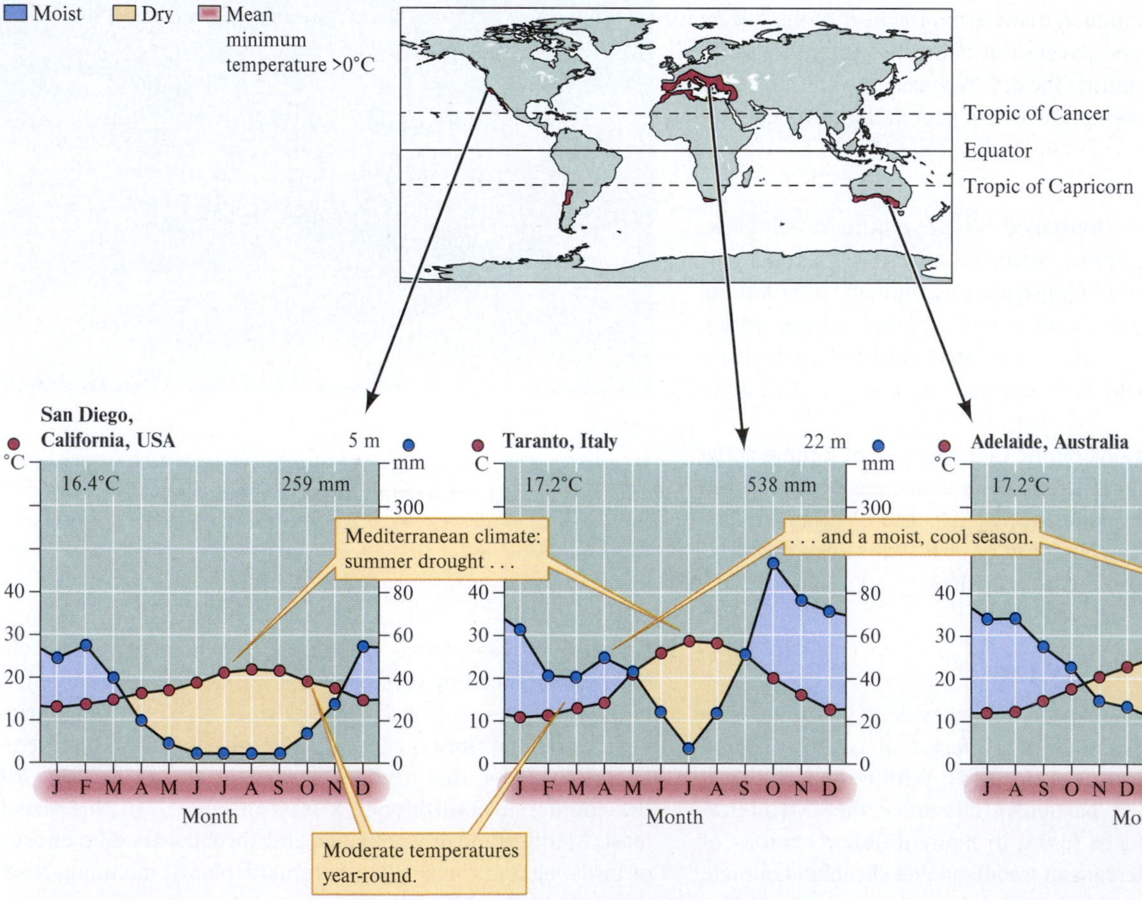

■ Moist   □ Dry   ■ Mean minimum temperature >0°C

Tropic of Cancer
Equator
Tropic of Capricorn

San Diego, California, USA   5 m
°C   16.4°C   259 mm   mm

Mediterranean climate: summer drought . . .

. . . and a moist, cool season.

Taranto, Italy   22 m
°C   17.2°C   538 mm   mm

Adelaide, Australia   6 m
°C   17.2°C   536 mm   mm

Moderate temperatures year-round.

J F M A M J J A S O N D   Month
J F M A M J J A S O N D   Month
J A S O N D J F M A M J   Month

Figure 2.23   Mediterranean woodland and shrubland geography and climate.

Mediterranean landscapes. Elsewhere, these landscapes, under careful stewardship, have maintained their integrity for thousands of years.

Biology

The plants and animals of Mediterranean woodlands and shrublands are highly diverse and, like their desert neighbors, show several adaptations to drought. Trees and shrubs are typically evergreen and have small, tough leaves, which conserve both water and nutrients (fig. 2.24). Many plants of Mediterranean woodlands and shrublands have well-developed, mutualistic relationships with microbes that fix atmospheric nitrogen.

The process of decomposition is greatly slowed during the dry summer and then started again with the coming of fall and winter rains. Curiously, this intermittent decomposition may speed the process sufficiently so that average rates of decomposition are comparable to those in temperate forests.

Fire, a common occurrence in Mediterranean woodlands and shrublands, has selected for fire-resistant plants. Many Mediterranean woodland trees have thick, fire-resistant bark. In contrast, many shrubs in Mediterranean woodlands are rich in oils and burn readily but resprout rapidly. Most herbaceous plants grow during the cool, moist season and then die back in summer, thus avoiding both drought and fire.

Human Influences

Human activity has had a substantial influence on the structure of landscapes in Mediterranean woodlands and shrublands. For example, the open oak woodlands of southern Spain and Portugal are the product of an agricultural management system that is thousands of years old. In this system, cattle graze on grasses, pigs consume acorns produced by the oaks, and cork is harvested from cork oaks as a cash crop. Selected areas are planted in wheat once every 5 to 6 years and allowed to lie fallow the remainder of the time. This system of agriculture, which

Figure 2.24   These shrubs found in South Africa's fynbos have the characteristic leaves that help prevent water loss from this Mediterranean-type climate.   Anna Sher

emphasizes low-intensity cultivation and long-term sustainability, may offer clues for sustainable agriculture in other regions.

High population densities coupled with a long history of human occupation have left an indelible mark on Mediterranean woodlands and shrublands. Early human impacts included clearing forests for agriculture, setting fires to control woody species and encourage grass, harvesting brush for fuel, and grazing and browsing by domestic livestock. Today, Mediterranean woodlands and shrublands around the world are being covered by human habitations.

Temperate Grassland

In their original state, **temperate grasslands** extended unbroken over vast areas (fig. 2.25). Standing in the middle of

Figure 2.25   Bison, native grazers of the temperate grasslands of North America.   MedioImages/PunchStock

unobstructed prairie under a dome of blue sky evokes a feeling similar to that of being on a small boat in the open ocean. It is no accident that early visitors from forested Europe and eastern North America often referred to the prairie in the American Midwest as a "sea of grass" and to the wagons that crossed them as "prairie schooners." Prairies were the home of the bison and pronghorn and of the nomadic cultures of Eurasia and North America.

Geography

Temperate grassland is the largest biome in North America, extending from 30° to 55° latitude. These grasslands are even more extensive in Eurasia (fig. 2.26). In North America, the prairies of the Great Plains extend from southern Canada to the Gulf of Mexico and from the Rocky Mountains to the deciduous forests of the east. Additional grasslands are found on the Palouse prairies of Idaho and Washington and in the central valley and surrounding foothills of California. In Eurasia, the temperate grassland biome forms a virtually unbroken band from eastern Europe all the way to eastern China. In the Southern Hemisphere, temperate grassland occurs in Argentina, Uruguay, southern Brazil, and New Zealand.

Climate

Temperate grasslands receive between 300 and 1,000 mm of precipitation annually. Though wetter than deserts, temperate grasslands do experience drought, and droughts may persist for several years. The maximum precipitation usually occurs in summer during the height of the growing season (see fig. 2.26). Winters in temperate grasslands are generally cold and summers are hot.

Soils

Temperate grassland soils are derived from a wide variety of parent materials. The best temperate grassland soils are deep, basic or neutral, and fertile and contain large quantities of organic matter. The black prairie soils of North America and Eurasia, famous for their fertility, contain the greatest amount of organic matter. The brown soils of the more arid grasslands contain less organic matter.

Biology

Temperate grassland is thoroughly dominated by herbaceous vegetation. Drought and high summer temperatures encourage fire. As in tropical savannas, fire helps exclude woody vegetation from temperate grasslands, where trees and shrubs are often limited to the margins of streams and rivers. In addition to grasses, there can be a striking diversity of other herbaceous vegetation. Spring graces temperate grasslands with showy anemones, ranunculus, iris, and other wildflowers; up to 70 species can bloom simultaneously on the species-rich North American prairie. The height of grassland vegetation varies from about 5 cm in dry, short-grass prairies to over 200 cm in the wetter, tall-grass prairies. The root systems of grasses and forbs (herbaceous plants that are not grasses) form a dense network of sod that resists invasion by both trees and the plow.

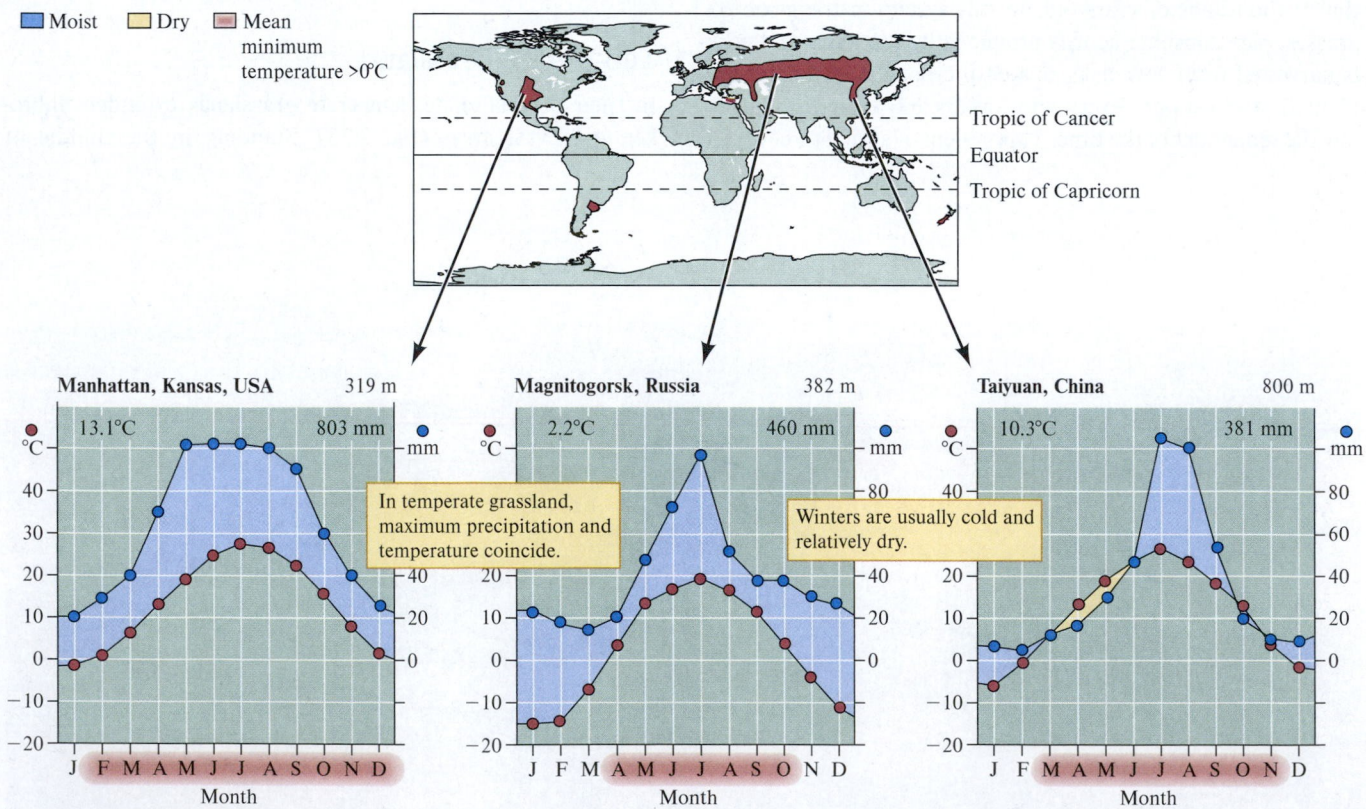

Figure 2.26   Temperate grassland geography and climate.

Temperate grasslands once supported huge herds of roving herbivores: bison and pronghorns in North America (see fig. 2.25) and wild horses and Saiga antelope in Eurasia. As in the open sea, the herbivores of the open grassland banded together in social groups, as did their attendant predators, the steppe and prairie wolves. The smaller, inconspicuous animals, such as grasshoppers and mice, were even more numerous than the large herbivores.

Human Influences

The first human populations on temperate grasslands were nomadic hunters. Next came the nomadic herders. Later, with their plows, came the farmers, who broke the sod and tapped into fertile soils built up over thousands of years. Under the plow, temperate grasslands have produced some of the most fertile farmlands on earth and fed much of the world (fig. 2.27). However, much of this primary production depends on substantial additions of inorganic fertilizers, and we are "mining" the fertility of prairie soils. Amy Molotoks and colleagues used a world soil database combined with land cover maps from satellite imagery plus data collected in the field to determine that 59% of soil organic carbon is lost when grassland is converted to cropland (Molotoks et al. 2018). In addition, the more arid grasslands, with their frequent droughts, do not appear capable of supporting sustainable farming.

Temperate Forest

For many, nothing epitomizes "nature" as do the diverse and majestic deciduous trees that characterize **temperate forest** (fig. 2.28). In the subdued light of this cool, moist realm, a world of mushrooms and decaying leaves, you can stand beside the giants of the biosphere.

Geography

Temperate forest can be found between 30° and 55° latitude. However, the majority of this biome lies between 40° and 50° (fig. 2.29). In Asia, temperate forest originally covered much of Japan, eastern China, Korea, and eastern Siberia. In western Europe, temperate forests extended from southern Scandinavia to northwestern Iberia and from the British Isles through eastern Europe. North American temperate forests are found from the Atlantic seacoast to the Great Plains and reappear on the West Coast as temperate coniferous forests that extend from northern California through southeastern Alaska. In the Southern Hemisphere, temperate forests are found in southern Chile, New Zealand, South Africa, and southern Australia.

Climate

Temperate forests, which may be either coniferous or deciduous, occur where temperatures are not extreme and where annual precipitation averages anywhere from about 650 mm to over 3,000 mm (see fig. 2.29). These forests generally receive more winter precipitation than temperate grasslands. Deciduous trees usually dominate temperate forests, where the growing season is moist and at least 4 months long. In deciduous forests, winters last from 3 to 4 months. Though snowfall may be heavy, winters in deciduous forests are relatively mild. Where winters are more severe or the summers drier, conifers are more abundant than deciduous trees. The temperate coniferous forests of the Pacific Coast of North America receive most of their precipitation during fall, winter, and spring and are subject to summer drought. Summer drought is shown clearly in the climate diagram for the H. J. Andrews Forest of Oregon (see fig. 2.29). The few deciduous trees in these coniferous forests are largely restricted to streamside environments, where water remains abundant during the drought-prone growing season.

Figure 2.28   A mixed deciduous and coniferous temperate forest in New England. This temperate forest in early autumn gives just a hint of the dramatic display of color that occurs each autumn in the New England countryside, where farms and towns occupy areas cleared of forest.   Songquan Deng/Shutterstock

Figure 2.27   Once the most extensive biome on earth, temperate grasslands have been largely converted to agriculture.
Dave Reede/Getty Images

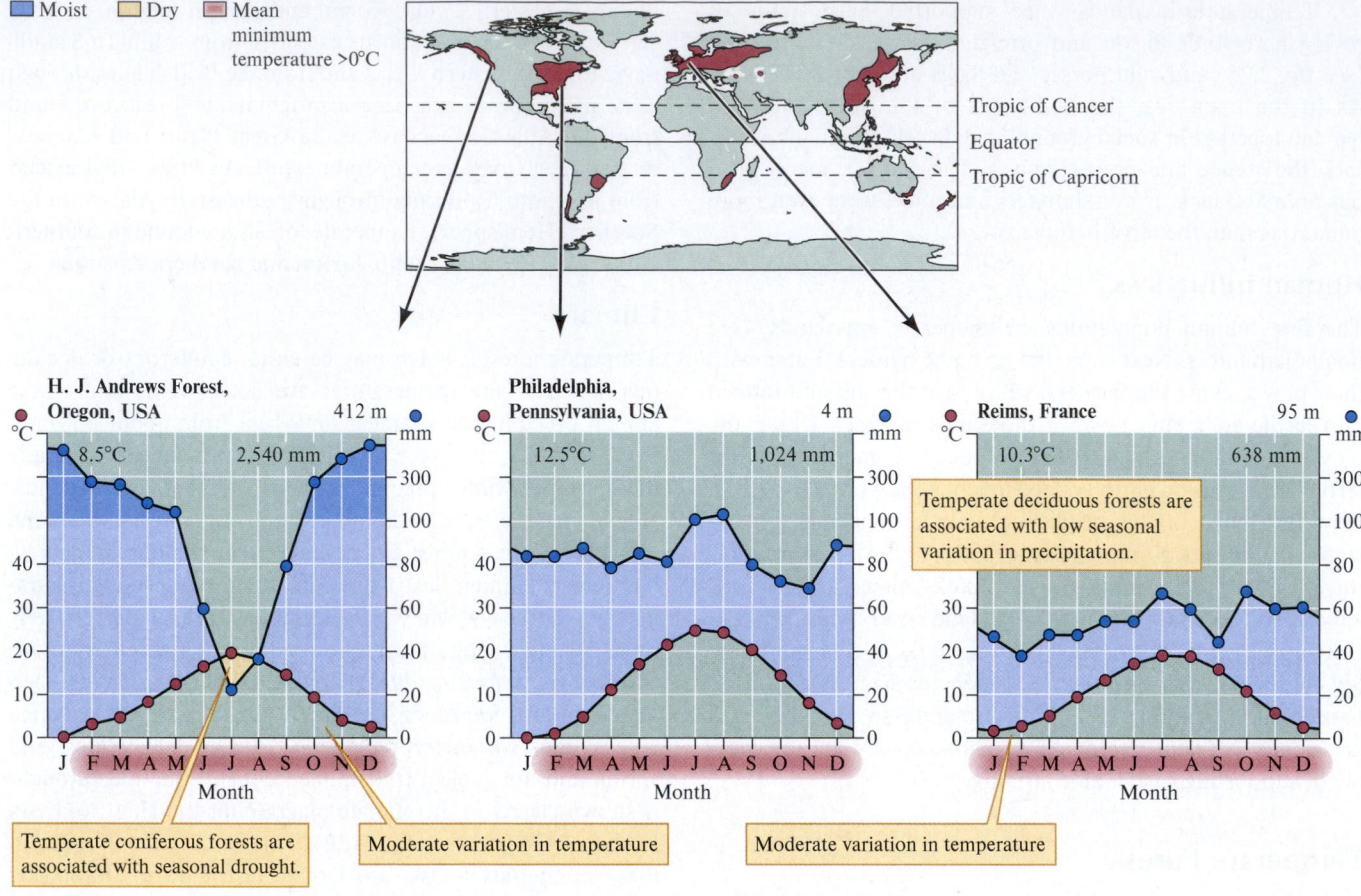

Figure 2.29   Temperate forest geography and climate.

Soils

Temperate forest soils are usually fertile. The most fertile soils in this biome develop under deciduous forests, where they are generally neutral or slightly acidic and rich in both organic matter and inorganic nutrients. Rich soils may develop under coniferous forests, but conifers are also able to grow on poorer, acidic soils. Nutrient movement between soil and vegetation tends to be slower and more conservative in coniferous forests; nutrient movement within deciduous forests is generally more dynamic.

Biology

Primary production in temperate forests is less than tropical forests, but can be very high, especially where young trees are getting established. Furthermore, while the diversity of trees found in temperate forests is lower than that of tropical forests, temperate forest biomass can be as great or greater. Like tropical rain forests, temperate forests are vertically stratified. The lowest layer of vegetation, the herb layer, is followed by a layer of shrubs, then shade-tolerant understory trees, and finally the canopy, formed by the largest trees. The height of this canopy varies from approximately 40 m to over 100 m. Birds, mammals, and insects make use of all layers of the forest from beneath the forest floor through the canopy. Some of the most important consumers are the fungi and bacteria, which, along with a diversity of microscopic invertebrate animals, consume the large quantities of wood stored on the floor of old-growth temperate forest (fig. 2.30). The activities of these organisms recycle nutrients, a process upon which the health of the entire forest depends.

Human Influences

What, besides being large cities, do Tokyo, Beijing, Moscow, Warsaw, Berlin, Paris, London, New York, Washington, D.C., Boston, Toronto, Chicago, and Seattle have in common? They are all built on lands that once supported a temperate forest. The first human settlements in temperate forests were concentrated along forest margins, usually along streams and rivers. Eventually, agriculture was practiced in these forest clearings, and animals and plant products were harvested from the surrounding forest. This was the circumstance several thousand years ago, in Europe, Asia, and North America. Since those times, most of the ancient forests have fallen to ax and saw. Few tracts of the virgin deciduous forest that once covered most of the eastern half of North America remain, and disparate interests struggle over the fate of the remaining 1% to 2% of old-growth forests in western North America.

Boreal Forest

The **boreal forest,** or **taiga,** is a world of wood and water that covers over 11% of the earth's land area (fig. 2.31). On the surface, the boreal forest is the essence of monotony. However, if you pay attention you are rewarded with plenty of variety. Forests

Figure 2.30   Key decomposers in temperate forests. The massive wood deposited on the floor of temperate forests is broken down by fungi, which are essential to the addition of organic matter to forest soils and to the cycling of nutrients in forest ecosystems.   Photo 24/Stockbyte/Brand X Pictures/Getty Images

Figure 2.31   Boreal forests, such as this one in Alaska, are dominated by a few species of conifer trees.   AlxYago/Shutterstock

of different ages, shaped by wind, fire, and other environmental forces, host diverse communities of insects, birds, rodents, and other animals. The understory may be open, with patches of fruit-bearing shrubs, or dense with young saplings. The summer forest is colored green, gray, and brown; the autumn adds brilliant splashes of yellow and red; and the long northern winter turns the boreal forest into a land of white solitude.

Geography

Boreal comes from the Greek word for north, reflecting the fact that boreal forests are confined to the Northern Hemisphere. Boreal forests extend from Scandinavia, through European

Russia, across Siberia, to central Alaska, and across central Canada in a band between 50° and 65° N latitude (fig. 2.32). These forests are bounded in the south by either temperate forests or temperate grasslands and in the north by tundra. Fingers of boreal forest follow the Rocky Mountains south along the spine of North America, and patches of boreal forest reappear on the mountain slopes of south-central Europe and Asia.

Climate

Boreal forest is found where winters are too long, usually longer than 6 months, and the summers too short to support temperate forest (see fig. 2.32). The boreal forest zone includes some fairly moderate climates, such as that at Umeå, Sweden, where the climate is moderated by the nearby Baltic Sea. However, boreal forests are also found in some of the most variable climates on earth. For instance, the temperature at Verkhoyansk, Russia, in central Siberia, ranges from about −70°C in winter to over 30°C in summer, an annual temperature range of over 100°C! Precipitation in the boreal forest is moderate, ranging from about 200 to 600 mm. Yet, because of low temperatures and long winters, evaporation rates are low, and drought is infrequent. During droughts, however, forest fires can devastate vast areas of boreal forest.

Soils

Boreal forest soils tend to be of low fertility, thin, and acidic. Low temperatures and low pH impede decomposition of plant litter and slow the rate of soil building. As a consequence, nutrients are largely tied up in a thick layer of plant litter that

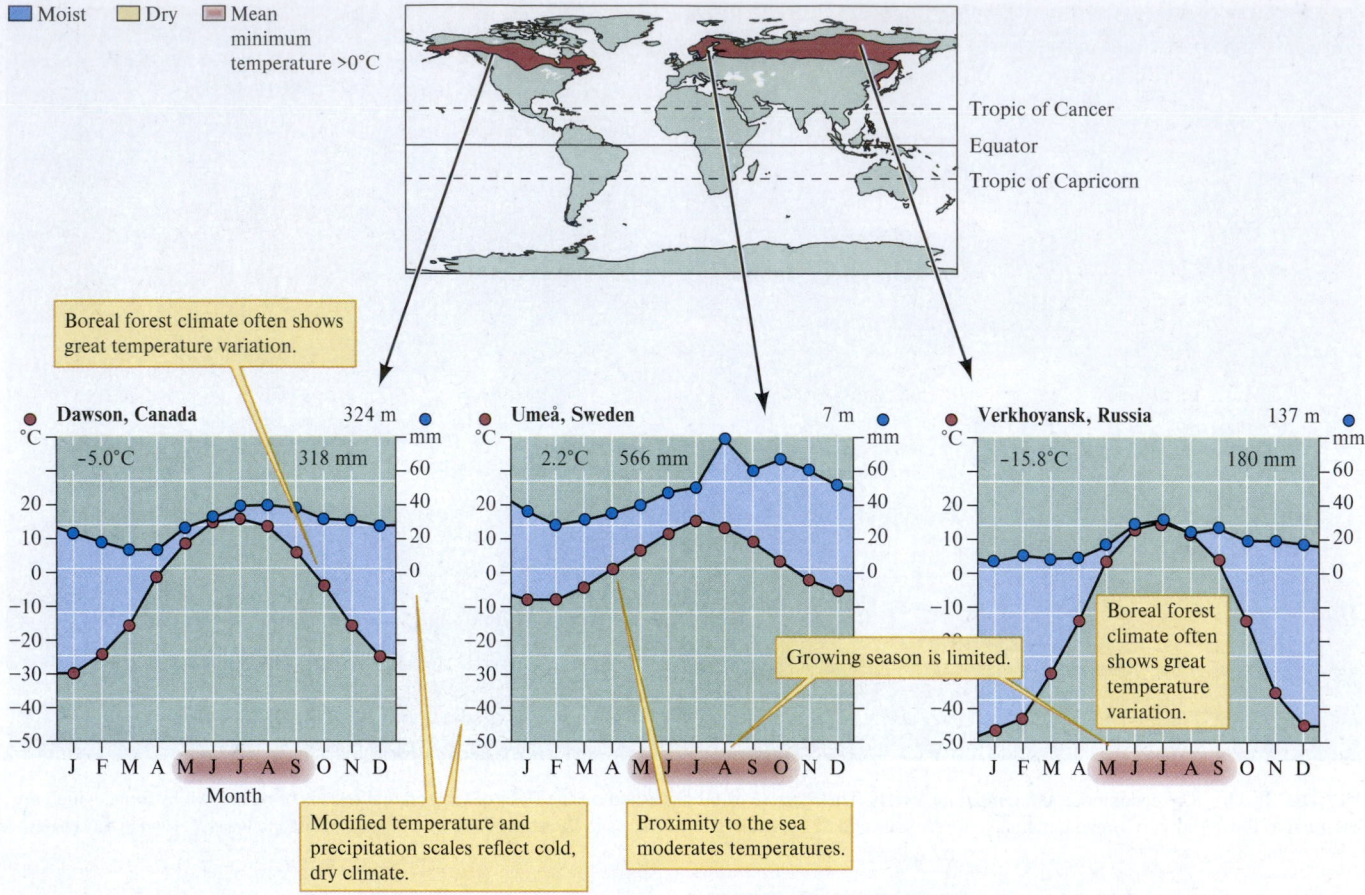

Figure 2.32   Boreal forest geography and climate.

carpets the forest floor. In turn, most trees in boreal forests have a dense network of shallow roots that, along with associated mycorrhizal fungi, tap directly into the nutrients bound up in this litter layer. The topsoil, which underlies the litter layer, is thin. In the more extreme boreal forest climates, the subsoil is permanently frozen in a layer of "permafrost" that may be several meters thick.

Biology

Boreal forest is generally dominated by evergreen conifers such as spruce, fir, and, in some places, pines. Larch, a deciduous conifer, dominates in the most extreme Siberian climates. Deciduous aspen and birch trees grow here and there in mature conifer forests and may dominate the boreal forest during the early stages of recovery following forest fires. Willows grow along the shores of rivers and lakes. There is little herbaceous vegetation under the thick forest canopy, but small shrubs such as blueberry and shrubby junipers are common.

Boreal forest is home to many animals, including migratory caribou and reindeer in winter and moose and woodland bison year-round. The wolf is the major predator of the boreal forest. This biome is also inhabited by black bears and grizzly bears in North America and the brown bear in Eurasia. A variety of smaller mammals such as lynx, wolverine, snowshoe hare, porcupines, and red squirrels also live in boreal forests. Boreal forest is the nesting habitat for many birds, such as the

American redstart (see fig. 1.4), that migrate from the tropics each spring and the year-round home of other birds such as crossbills and spruce grouse.

Our survey of the biosphere has taken us far from the rain forest, where we started. Let's reflect back on the tropical rain forest and where we've come. What has changed? Well, we're still in forest but a very different one. In the rain forest, a single hectare could contain over 300 species of trees; here, in the boreal forest, you can count the dominant trees on one hand. What about epiphytes and vines? The vines are gone and the epiphytes are limited to lichens and some mistletoe. In addition, many of the intricate relationships between species that we saw in the rain forest are absent. All the trees are wind-pollinated, and none produce fleshy fruits like bananas or papayas. Now listen to the two forests at night. Tropical rain forest echoes with a rich chorus of sounds. In contrast, the silence of the boreal forest is broken by few animal voices—the howl of a wolf, the hoot of an owl, the cry of the loon, soloists of the northern forest—accompanied by incessant wind through the trees.

Human Influences

Ancient cave paintings in southern France and northern Spain, made during the last ice age when the climate was much colder, reveal that humans have lived off boreal forest animals, for tens of thousands of years. In Eurasia, hunting of reindeer eventually gave way to domestication and herding. In northern Canada and Alaska, some Native Americans still rely on wild caribou for much of their food, and northern peoples have long harvested the berries that grow in boreal forests.

For most of history, human intrusion in the boreal forest was relatively light. More recently, however, harvesting of both animals and plants has become intense. In a review of the current literature, Tähti Pohjanmies and her colleagues found that research consistently shows that logging in boreal forests has long-reaching ecological impacts, including changing climate (Pohjanmies et al. 2017). For example, boreal forests strongly affect global climate by sequestering carbon in the soil; this carbon can be released into the atmosphere when soils are disturbed during logging (fig 2.33).

Tundra

Follow the caribou north as they leave their winter home in the boreal forest and you eventually reach an open landscape of mosses, lichens, and dwarf willows, dotted with small ponds and laced with clear streams (fig. 2.34). This is the **tundra.** If it is summer and surface soils have thawed, your progress will be cushioned by a spongy mat of lichens and mosses and punctuated by sinking into soggy accumulations of peat. The air will be filled with the cries of nesting birds that have come north to take advantage of the brief summer abundance of their plant

Figure 2.33   Deforestation in boreal forest.   Comstock Images/Alamy Stock Photo

Figure 2.34   Alaskan tundra. Tundra vegetation is mostly low-growing mosses, lichens, perennial herbaceous plants, and dwarf willows and birches.   ajliikala/Getty Images

and animal prey. After the long winter, the midnight sun signals a celebration of light and life.

Geography

Like the boreal forest, arctic tundra rings the top of the globe, covering most of the lands north of the Arctic Circle at approximately 66.5° N latitude (fig. 2.35). The tundra extends from northernmost Scandinavia, across northern European Russia, through northern Siberia, and right across northern Alaska and Canada. It reaches far south of the Arctic Circle in the Hudson Bay region of Canada and is also found in patches on the coast of Greenland and in northern Iceland.

Climate

The tundra climate is typically cold and dry. However, temperatures are not quite as extreme as in the boreal forest. Though winter temperatures are less severe, the summers are shorter (see fig. 2.35). Precipitation on the tundra varies from less than 200 mm to a little over 600 mm. Still, because average annual temperatures are so low, precipitation exceeds evaporation. As a consequence, the short summers are soggy and the tundra landscape is alive with ponds and streams.

Soils

Soil building is slow in the cold tundra climate. Because rates of decomposition are low, organic matter accumulates in deposits of peat and humus. Surface soils thaw each summer but are generally underlain by a layer of permafrost that may be many meters thick. The annual freezing and thawing of surface soil combine with the actions of water and gravity to produce a variety of surface processes that are largely limited to the tundra. One of these processes, **solifluction,** slowly moves soils down slopes. In addition, freezing and thawing bring stones to the surface of the soil, forming a netlike, or polygonal, pattern on the surface of tundra soils (fig. 2.36).

Biology

The open tundra landscape is dominated by a richly textured patchwork of perennial herbaceous plants, especially grasses, sedges, mosses, and lichens. The lichens, associations of fungi and algae, are eagerly eaten by reindeer and caribou. The woody vegetation of the tundra consists of dwarf willows and birches along with a variety of low-growing shrubs.

The tundra is one of the last biomes on earth that still support substantial numbers of large native mammals, including caribou, reindeer, musk ox, bear, and wolves. Small mammals such as arctic foxes, weasels, lemmings, and ground squirrels are also abundant. Resident birds such as the ptarmigan and snowy owl are joined each summer by a host of migratory bird species. Insects, though not as diverse as in biomes farther south, are very abundant. Each summer, swarms of mosquitoes and black flies emerge from the many tundra ponds and streams.

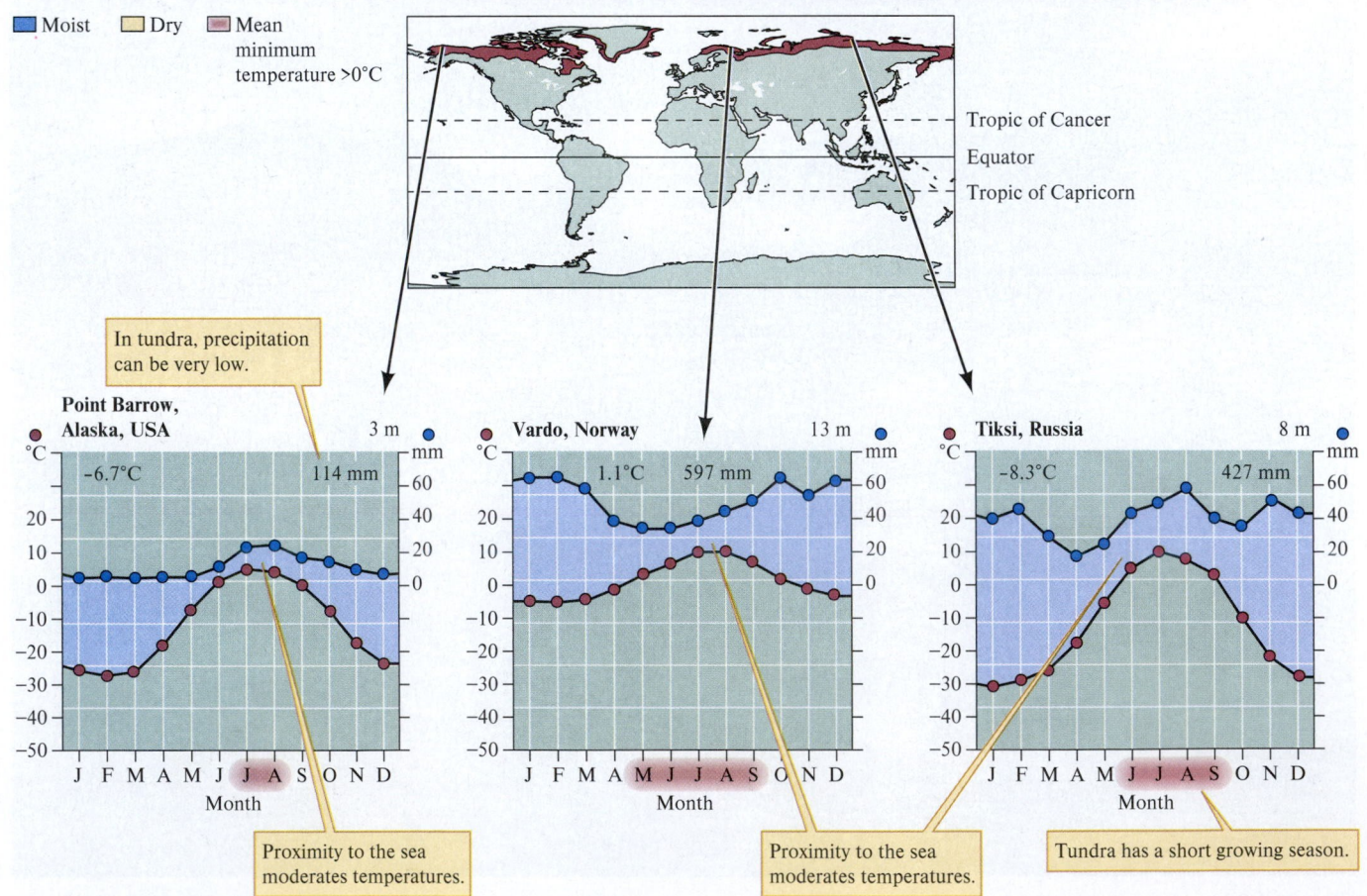

Figure 2.35   Tundra geography and climate.

Figure 2.36   Freezing and thawing form netlike polygons on the surface of the tundra as seen here in an aerial photo of Alaska.   Fletcher & Baylis/Science Source

Human Influences

Until recently, human presence in the tundra was largely limited to small populations of hunters and nomadic herders. As a consequence, the tundra has been viewed as one of the last pristine areas of the planet. Recently, however, human intrusion has increased markedly. This biome has been the focus of intense oil exploration and extraction. Airborne pesticides and radionuclides, which originate in distant human population centers, have been deposited on the tundra, sometimes with devastating results. Mercury, an element that is highly toxic to people and other animals, has increased dramatically in arctic lakes in recent years due to industrial pollution. This Hg can accumulate in fish, making it toxic to eat (Hudelson et al. 2019); this includes species that are eaten locally as well as for export. Such revelations have shattered the illusion of the tundra as an isolated biome and the last earthly refuge from human disturbance.

Mountains: A Diversity of Biomes

We now shift our attention to mountains, which are not a biome. As we explained earlier in this chapter, because of the environmental changes that occur with altitude, several biomes may be found on a single mountain, depending on elevation and which side of the mountain one is on. We include mountains here because they often introduce unique environmental conditions and organisms to regions around the globe.

Mountains capture the imagination as places of geological, biological, and climatic diversity (fig. 2.37). Mountains have long offered refuge for distinctive flora and fauna and humans alike. Like oceanic islands, they offer unique insights into evolutionary and ecological processes.

Figure 2.37   Mount Kilimanjaro, East Africa, where environmental conditions vary from tropical savanna at the base of the mountain to ice fields at its peak.   Getty Images

Geography

Mountains are built by geological processes, such as volcanism and movements of the earth's crust that elevate and fold the earth's surface. These processes operate with greater intensity in some places than others, and so mountains are concentrated in belts where these geological forces have been at work (fig. 2.38). In the Western Hemisphere, these forces have been particularly active on the western sides of both North and South America, where a chain of mountain ranges extends from northern Alaska across western North America to Tierra del Fuego at the tip of South America. Ancient low mountain ranges occupy the eastern sides of both continents. In Africa, the major mountain ranges are the Atlas Mountains of northwest Africa and the mountains of East Africa that run like beads on a string from the highlands of Ethiopia to southern Africa. In Australia, the flattest of the continents, mountains extend down the eastern side of the continent. Eurasian mountain ranges, which generally extend east to west, include the Pyrenees, the Alps, the Caucasus, and, of course, the Himalayas, the highest of them all.

Climate

On mountains, climates change from low to high altitude, but the specific changes are different at different latitudes. On mountains at middle latitudes, the climate is generally cooler and wetter at higher altitudes (fig. 2.39). In contrast, there is less precipitation at the higher elevations of polar mountains and on some tropical mountains. In other tropical regions, precipitation increases up to some middle elevation and then decreases higher up the mountain. On high tropical mountains, warm days are followed by freezing nights. The organisms on these mountains experience summer temperatures every day and winter temperatures every night. The changes in climate that occur up the sides of mountains have profound influences on the distribution of mountain organisms.

Soils

Mountain soils change with elevation and have a great deal in common with the various soils we've already discussed. However, some special features are worth noting. First, because of the steeper topography, mountain soils are generally well drained and tend to be thin and vulnerable to erosion. Second, persistent winds blowing from the lowlands deposit soil particles and organic matter on mountains, materials that can make a significant contribution to local soil building. In some locations in the southern Rocky Mountains, coniferous trees draw the bulk of their nutrition from materials carried by winds from the valleys below, not from local bedrock.

Biology

Climb any mountain that is high enough and you will notice biological and climatic changes. Whatever the vegetation at the base of a mountain, that vegetation will change as you climb and the air becomes cooler. The sequence of vegetation up the side of a mountain may remind you of the biomes we encountered on our journey from the equator to the poles. In the cool highlands of desert mountains in the southwestern United States, you can hike through spruce and fir forests much like those we encountered far to the north. However, what you see on these desert mountains differs substantially from boreal forests. These mountain populations have been isolated from the main body of the boreal forest for over 10,000 years; in the interim, some populations have become extinct, some teeter on the verge of extinction, while others have evolved sufficiently to be

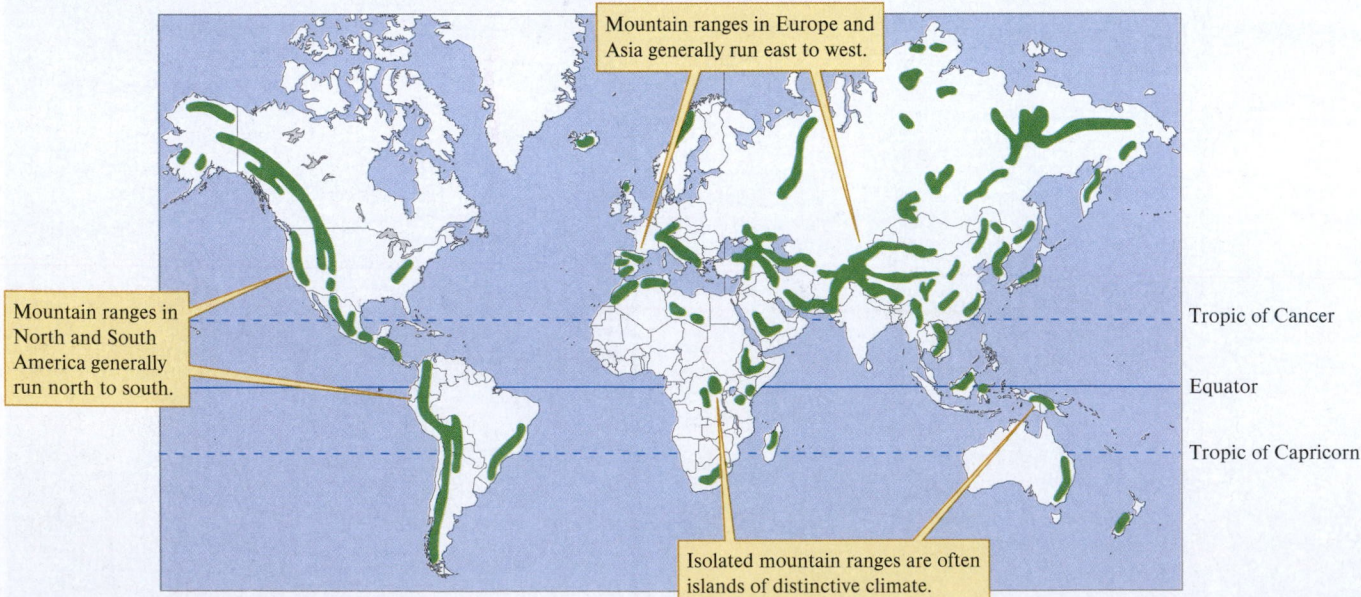

Figure 2.38   Mountain geography. Note that the shape of the continents is different in this map than in figure 2.35 and elsewhere in this chapter; that is because there is no perfect way to represent a round globe in two dimensions. Here, the Mercator map is shown, which exaggerates the Northern hemisphere so as to represent nautical distances correctly, whereas figure 2.35 is what is called the Robinson map. You can find many other types of maps on the Internet.

Moist □ **Dry** ▨ **Mean minimum temperature >0°C**

Niwot Ridge, 3,743 m
Colorado, USA High elevation
-3.7°C 930 mm

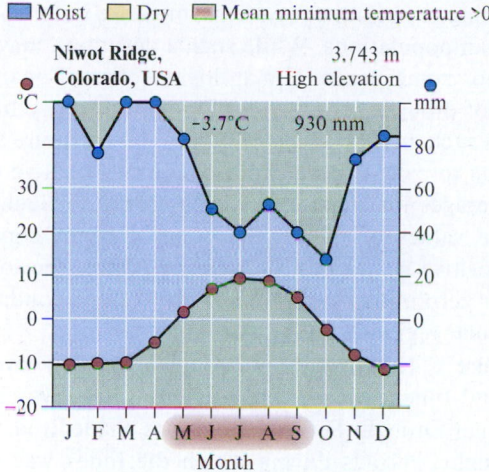

Elevation ↑ Temperature ↓ Precipitation ↑

Allenspark, 2,590 m
Colorado Middle elevation
4.7°C 520 mm

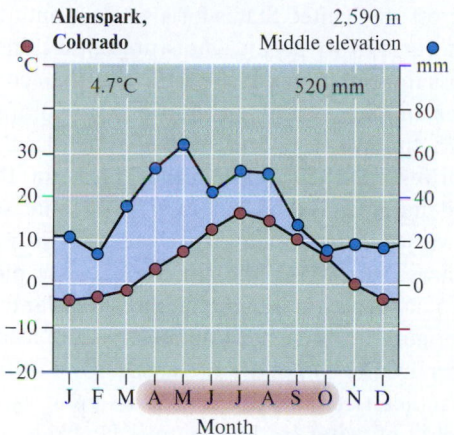

Elevation ↑ Temperature ↓ Precipitation ↑

Boulder, 1,660 m
Colorado Low elevation
11.1°C 460 mm

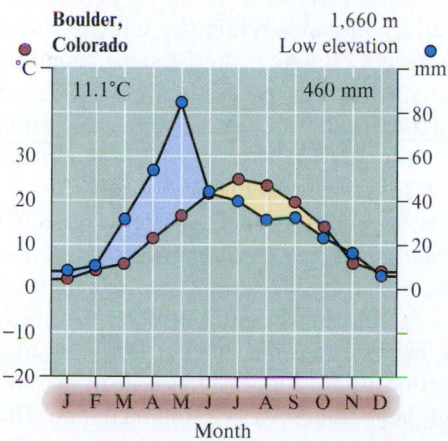

Figure 2.39 Mountain climates along an elevational gradient in the Colorado Rockies. Temperatures decrease and precipitation increases from low to high elevations in these midlatitude mountains.

(a)

(b)

Figure 2.40 Convergence among tropical alpine plants: (*a*) *Senecio* trees on Mount Kilimanjaro, Africa; (*b*) *Espeletia* in the Andes of South America. (*a*) Avatar_023/Shutterstock; (*b*) Francois Gohier/Science Source

recognized as separate species or subspecies. On these mountains, time and isolation have forged distinctive gene pools and mixes of species.

The species on high equatorial mountains are even more isolated. Think for a moment of the geography of high tropical mountains: some in Africa, some in the highlands of Asia, and the Andes of South America. The high-altitude communities of Africa, South America, and Asia share very few species. On the other hand, despite differences in species composition, there are structural similarities among the organisms on these mountains (fig. 2.40). These similarities demonstrate the power of matching evolutionary forces—such as the selective pressures of freezing nights followed by warm days.

Human Influences

Because mountains differ in climate, geology, and biota (plants and animals) from the surrounding lowlands, they have been useful as a source of raw materials such as wood, forage for animals, medicinal plants, and minerals. Some of these uses, such as livestock grazing, are highly seasonal. In temperate regions, livestock are taken to mountain pastures during the summer and back down to the lowlands in winter. Human exploitation of mountains has produced ecological degradation in many places and surprising balance in others. Increased human pressure on mountain environments has sometimes created conflict between competing economic interests, between recreation seekers and between livestock ranchers. Because of their compressed climatic gradients and biological diversity, mountains offer living laboratories for the study of ecological and evolutionary responses to climatic variation.

Concept 2.3 Review

1. Why do regions that include high mountains (in tropical, desert, or temperate biomes) tend to have greater biological diversity compared to lowland regions in the same biomes?
2. Why would the soils in tropical rain forests generally be depleted of their nutrients more rapidly compared to the nutrients in temperate forest soils?

Finer Scale Climatic Variation over Time and Space

LEARNING OUTCOMES

After studying this section you should be able to do the following:

2.15   Describe some ways that climate diagrams may oversimplify the conditions that organisms actually experience.

2.16   Interpret temporal and spatial representations of the Palmer Drought Severity Index.

In this chapter, we've used the biome concept and climate diagrams to represent the diversity of environmental selective pressure and resulting diversity of life on earth. However, both of these useful tools oversimplify what is actually occurring.

Here we explore a climatic index, the **Palmer Drought Severity Index,** which can be used to characterize climatic variation. First, what is a drought? A **drought** can be defined as an extended period of dry weather during which precipitation is reduced sufficiently to damage crops, impair the functioning of natural ecosystems, or cause water shortages for human populations. While such a definition may be sufficient for some needs, climatologists created quantitative indices of drought. The Palmer Drought Severity Index, or PDSI, is such an index. The PDSI uses temperature and precipitation to calculate moisture conditions relative to long-term averages for a particular region at a particular time. Negative values of the PDSI reflect drought conditions, while positive values indicate relatively moist periods. Values near zero indicate approximately average conditions in a particular region.

Figure 2.41 shows how drought can vary over both space and time. To ease interpretation, negative values of the Palmer Drought Severity Index are shaded red, indicating drought. Periods during which the index was positive are shaded blue, indicating moist conditions.

Figure 2.41*a* maps values of the Palmer Drought Severity Index across the United States for a single month in 2020. Notice that during this period, moisture conditions varied widely across this portion of the North American continent.

Extreme summer drought conditions in Colorado and elsewhere in the West set the stage for historical, devastating wildfires in 2020, while other areas in the East experienced hurricanes and flooding that same summer. Climate is varying not only in space but also in time. The area of Kansas from which the climate data are plotted in figure 2.41*b* falls within the temperate grassland biome. What does figure 2.41*b* suggest about moisture availability in the region around Manhattan, Kansas? One of the most apparent characteristics of this area is its great variability; the availability of water in the region is far from constant. Now compare figure 2.41*b* with the representation of climate for Manhattan, Kansas, shown in figure 2.26. How do the two figures compare? While the climate diagram and the PDSI represent climate from the same geographic location, the climate diagram, because it draws our attention to average climatic conditions, seems to suggest climatic stability. Meanwhile, the PDSI shows that the climate around Manhattan, Kansas, is in fact highly variable.

Note that variability in the short term reflects **weather,** whereas *climate* refers to conditions considered over longer time periods. For example, an analysis of temperature and precipitation in Kansas over the past century reveals that Kansas has become warmer on average, western Kansas has been getting drier, and eastern Kansas has increasing frequency of large rain events (Rahmani and Harrington 2019, https://www.epa.gov/climate-indicators/). This is an example of **climate change.** In this chapter, we have seen how climate shapes ecosystems; given this, it is not surprising that climate change can be a driver of evolution and other, significant ecological changes (see chapters 4, 5, 23, and elsewhere). Ecologists study the relationships between organisms and environment. As these examples show, in the study of those relationships both averages and variation in environmental factors need to be considered.

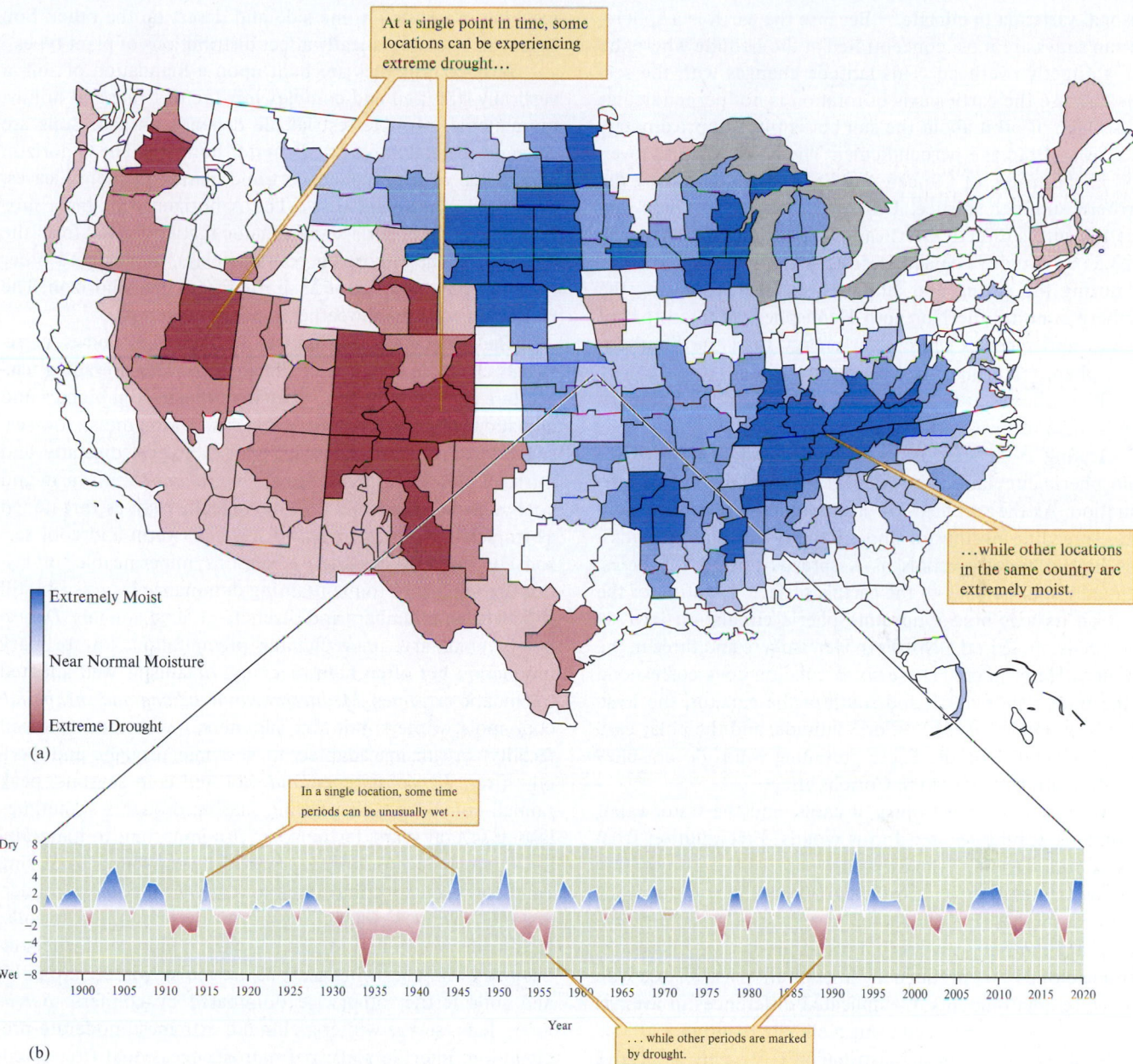

Figure 2.41 Like any feature of climate, drought can vary significantly over both space and time. (*a*) Regional variation in moisture conditions for the month of July, 2020, as indicated by the Palmer Drought Severity Index (data from NOAA 2020, www.cpc.ncep.noaa.gov). (*b*) Variation in the Palmer Drought Severity Index for Kansas region 3 near Manhattan, Kansas, plotted for the years 1895 to 2020 (data from NOAA 2020, cpc.ncep.noaa.gov).

Summary

We can understand the diversity of life on our planet in terms of the distributions of types of plant communities, called biomes. Biomes are characterized by particular functional traits of the plants, shaped by selective pressure from the environment. Because plants make sun energy available to the rest of the

ecosystem, the types of plants will have profound influences on the rest of the ecosystem. The environmental influence that determines differences between biomes is primarily climate.

Uneven heating of the earth's surface by the sun and the tilt of the earth combine to produce predictable latitudinal and

seasonal variation in climate. Because the earth is a sphere, the sun's rays are most concentrated at the latitude where the sun is directly overhead. This latitude changes with the seasons because the earth's axis of rotation is not perpendicular to its plane of orbit about the sun but is tilted approximately 23.5° away from the perpendicular. The sun is directly overhead at the tropic of Cancer, at 23.5° N latitude during the northern summer solstice. During the northern winter solstice the sun is directly overhead at the tropic of Capricorn, at 23.5° S latitude. The sun is directly overhead at the equator during the spring and autumnal equinoxes. During the northern summer the Northern Hemisphere is tilted toward the sun and receives more solar energy than the Southern Hemisphere. During the northern winter, the Northern Hemisphere is tilted away from the sun and the Southern Hemisphere receives more solar energy.

Heating of the earth's surface and atmosphere drives atmospheric circulation and influences global patterns of precipitation. As the sun heats air at the equator, it expands and rises, spreading northward and southward at high altitudes. This high-altitude air cools as it spreads toward the poles, eventually sinking back to the earth's surface. Rotation of the earth on its axis breaks up atmospheric circulation into six major cells, three in the Northern Hemisphere and three in the Southern Hemisphere. These six circulation cells correspond to the trade winds north and south of the equator, the westerlies between 30° and 60° N or S latitude, and the polar easterlies above 60° latitude. These prevailing winds do not blow directly south because of the Coriolis effect.

As air rises at the tropics, it cools, and the water vapor it contains condenses and forms clouds. Precipitation from these clouds produces the abundant rains of the tropics. Dry air blowing across the lands at about 30° latitude produces the great deserts that ring the globe. When warm, moist air flowing toward the poles meets cold, polar air, it rises and cools, forming clouds that produce the precipitation associated with temperate environments. Complicated differences in average climate can be summarized using a climate diagram.

While terrestrial biome distribution is strongly associated with latitude, biomes are also influenced by microclimate and soil type. Biomes do not exist in simple bands determined solely by latitude; this is because topography and geology also play a role. Mountain ranges create different temperature zones based on elevation as well as microclimates due to the rain shadow effect. Just as rising air in the tropics induces precipitation, so too does moist air hitting the side of a mountain, resulting in forests on one side and desert on the other. Soil types can also dramatically affect distributions of plant types.

Terrestrial biomes are built upon a foundation of soil, a vertically stratified and complex mixture of living and nonliving material. Most terrestrial life depends on soil. Soils are generally divided into O, A, B, and C horizons. The O horizon is made up of freshly fallen organic matter, including leaves, twigs, and other plant parts. The A horizon contains a mixture of mineral materials and organic matter derived from the O horizon. The B horizon contains clays, humus, and other materials that have been transported from the A horizon. The C horizon consists of weathered parent material.

The geographic distribution of terrestrial biomes corresponds closely to variation in climate, especially prevailing temperature and precipitation. The major terrestrial biomes and climatic regimes are: *tropical rain forest:* warm; moist; low seasonality; infertile soils; exceptional biological diversity and intricate biological interactions. *Tropical dry forest:* warm and cool seasons; seasonally dry; biologically rich; as threatened as tropical rain forest. *Tropical savanna:* warm and cool seasons; pronounced dry and wet seasons; impermeable soil layers; fire important to maintaining dominance by grasses; still supports high numbers and diversity of large animals. *Desert:* hot or cold; dry; unpredictable precipitation; low primary production but often high diversity; organisms well adapted to climatic extremes. *Mediterranean woodland and shrubland:* cool, moist winters; hot, dry summers; low to moderate soil fertility; organisms adapted to seasonal drought and periodic fires. *Temperate grassland:* hot and cold seasons; peak rainfall coincides with growing season; droughts sometimes lasting several years; fertile soils; fire important to maintaining dominance by grasses; historically inhabited by roving bands of herbivores and predators. *Temperate forest:* moderate, moist winters; warm, moist growing season; fertile soils; high primary production and biomass; dominated by deciduous trees where growing seasons are moist, winters are mild, and soils fertile; otherwise dominated by conifers. *Boreal forest:* long, severe winters; climatic extremes; moderate precipitation; infertile soils; permafrost; occasional fire; extensive forest biome, dominated by conifers. *Tundra:* cold; low precipitation; short, soggy summers; poorly developed soils; permafrost; dominated by low vegetation and a variety of animals adapted to long, cold winters; migratory animals, especially birds, make seasonal use. *Mountains:* temperature, precipitation, soils, and organisms shift with elevation; mountains are climatic and biological islands.

Key Terms

A horizon 19	climate change 40	functional traits 13	natural history 11
B horizon 19	climate diagram 16	lithosol 25	O (organic) horizon 17
biome 11	consumer 13	Mediterranean woodland and shrubland 27	Palmer Drought Severity Index 40
boreal forest (taiga) 32	Coriolis effect 15		
caliche 27	desert 25	microclimates 17	secondary producers 13
C horizon 19	drought 40	mycorrhizae 21	temperate forest 31

tropical rain forest   19
primary producers   13
selective pressure   13

temperate grassland   29
tropical savanna   22
rain shadow effect   17

solifluction   36
tropical dry
    forest   21

tundra   35
weather   40

Review Questions

1. Plants form the foundation of most terrestrial ecosystems. Pick a biome from this chapter and explain how the functional traits of plants from that biome could affect the evolution of other organisms in that biome.

2. Draw a typical soil profile, indicating the principal layers, or horizons. Describe the characteristics of each layer.

3. Describe global patterns of atmospheric heating and circulation. What mechanisms produce high precipitation in the tropics? What mechanisms produce high precipitation at temperate latitudes? What mechanisms produce low precipitation in the tropics?

4. Use what you know about atmospheric circulation and seasonal changes in the sun's orientation to earth to explain the highly seasonal rainfall in the tropical dry forest and tropical savanna biomes. (Hint: Why does the rainy season in these biomes come during the warmer months?)

5. We showed how the rain shadow effects biome distribution in the western United States Where else in the world can you see the impact of a rain shadow? Do you think that the height of the mountains creating it matters? Why or why not?

6. Some of the earliest studies of the geographic distribution of vegetation suggested a direct correspondence between latitudinal and altitudinal variation in climate, and our discussion in this chapter stressed the similarities in climatic changes with altitude and latitude. Now, what are some major climatic differences between high altitude at midlatitudes and high altitude at high latitudes?

7. How is the physical environment on mountains at midlatitudes similar to that in tropical alpine zones? How do these environments differ?

8. English and other European languages have terms for four seasons: spring, summer, autumn, and winter. This vocabulary summarizes much of the annual climatic variation at midlatitudes in temperate regions. Are these four seasons useful for summarizing annual climatic changes across the rest of the globe? Look back at the climate diagrams presented in this chapter. How many seasons would you propose for each of these environments? What would you call these seasons?

9. Biologists have observed much more similarity in species composition among boreal forests and among areas of tundra in Eurasia and North America than among tropical rain forests or among Mediterranean woodlands around the globe. Can you offer an explanation of this contrast based on the global distributions of these biomes shown in figures 2.11, 2.23, 2.32, and 2.35?

10. To date, which biomes have been the most heavily affected by humans? Which seem to be the most lightly affected? How would you assess human impact? How might these patterns change during this century? (You may need to consult the discussion of human population growth in the Applications section of chapter 11.)

Reinhard Dirscherl/Alamy Stock Photo

Chapter

3

Life in Water

A southern blue-ringed octopus, *Hapalochlaena maculosa,* swimming in midwater near Mabul Island, Malaysia. This small, beautiful octopus, which can defend itself with a neurotoxin delivered with a bite of its beak, is just one of the vast diversity of fascinating species inhabiting the oceans.

CHAPTER CONCEPTS

3.1 Water cycling and movement on a global scale is driven by solar energy. 47

Concept 3.1 Review 48

3.2 The biology of aquatic environments corresponds broadly to variations in physical factors such as light, temperature, and water movements and to chemical factors such as salinity and oxygen. 49

Concept 3.2 Review 73

Applications: Biological Integrity–Assessing the Health of Aquatic Systems 73

Summary 75

Key Terms 76

Review Questions 76

LEARNING OUTCOMES

After studying this section you should be able to do the following:

3.1 Summarize the major differences in the physical properties of air and water.

3.2 Describe the significance of water's physical properties to life.

3.3 Contrast the major selective forces in aquatic versus terrestrial systems.

The names that people around the world have given to our planet reveal a perspective consistent across cultures. Those names, whether in English (earth), Latin (*terra*), Greek (Γεοσ, *geos*), or Chinese (地球, *di qiu*), all refer to land or soil, revealing that cultures everywhere hold a land-centered perspective. The Hawaiians, Polynesian inhabitants of the most isolated specks of land on earth, call the planet *ka honua,* an allusion to a level landing place or dirt embankment. This universal land-centered perspective may partly explain why portraits of earth transmitted from space are so stunning. Those images challenge our sense of place by portraying our planet as a shining blue ball, as a landing place in space covered not by land but mostly by water (fig. 3.1).

Life originated in water but from our perspective as terrestrial organisms, the aquatic realm remains an alien environment governed by unfamiliar rules. New species and even whole ecosystems are still being discovered, particularly in our oceans as technology allows us to probe ever-deeper depths. In 2010, a major census of marine life that took a decade to

Figure 3.1   From space, earth shows itself as a planet covered mostly by water.   NASA Earth Observatory image by Robert Simmon with data courtesy of the NASA/NOAA GOES Project Science team

complete reported nearly 250,000 species and estimated that far more await discovery by science (COML 2010; Stępień and Błażewicz 2019). Every year we learn more about life in our oceans, rivers, and lakes.

In this chapter, we will consider aquatic ecosystems in much the same way as terrestrial systems, by beginning with an exploration of environmental forces and how they differ from terrestrial systems. As in terrestrial environments, life-forms will respond to these forces through selective pressure, resulting in the evolution of the vast aquatic diversity we see, from colorful water plants and sponges and tiny glowing protozoa, to the largest mammals on earth. We will then consider the major aquatic biomes and their particular features. Just as chapter 2 did for terrestrial systems, our goal in chapter 3 is to gain a general sense of the natural history of life in water that will prepare the way for more detailed studies of ecology.

Aquatic Biomes and How They Differ

Life originally evolved in water, and with good reason: water has many physical and chemical properties that make it the ideal medium for biotic systems (table 3.1a). Many chemical reactions necessary for life require the stability of temperature and the liquid state of water. Its buoyancy decreases the need for support structures, even while its viscosity provides resistance to movement. The much higher density of water allows the largest of animals, whales, to swim/fly through the oceans. Can you imagine animals the size of whales flying through the air? Because of the close ties of water to life as we know it, "water is life" is a common expression heard round the world. If we look closely, we find that, in fact, several of the physical

properties of water have particular significance to living systems (table 3.1b).

The differences between water and air mean different selective forces in aquatic versus terrestrial systems, resulting in very different ecosystems. Terrestrial and aquatic biomes were defined by ecologists independently, and for good reason. Characterizing aquatic biomes by their dominant photosynthesizing organisms as we did for terrestrial biomes is difficult, since in many aquatic biomes they are free-floating, single-celled organisms, such as **cyanobacteria** (formerly called "blue-green algae"). Aquatic biome classifications are more generally based on physical rather than biological traits, including salinity and water movement. Aquatic biomes are classified first by whether they are **saltwater** or **freshwater,** a chemical property. Within freshwater systems, these biomes are further defined by a physical property: whether the water is **lentic,** still as in a pond or lake; or **lotic,** moving as in a stream or river.

Water is, of course, the most important environmental feature of aquatic systems. Both physical and chemical properties of water exert selective pressure on the organisms that live there, and as we will see, the variability in these properties at different locations and depths contribute to the evolution of aquatic biodiversity. Physically, water has many properties that can affect the evolution and distribution of living things, including its viscosity (the resistance it creates to movement), its density, its translucency, and its ability to conduct heat. Water also acts as a temperature buffer and solvent for chemical reactions that take place in living cells. Chemically, water is a metabolite in many biological processes including photosynthesis; as such, water is perhaps the most important resource for living things, and its availability limits growth in terrestrial systems. However, in aquatic systems, water is not a **limiting resource** because it is abundant.

In contrast with water, O_2 is ubiquitous on land, but in aquatic systems, because of its limited solubility, O_2 becomes a limiting resource and highly variable. Like liquid water, oxygen gas is necessary for most life on earth. The concentration of O_2 in oceans, lakes, and rivers depends on other environmental factors: light from the sun, temperature, water circulation, salinity, and oxygen demand by respiring organisms.

Approximately 80% of the solar energy striking bodies of water is absorbed in the first 10 m. Most ultraviolet and infrared light is absorbed in the first few meters. Within the visible range, red, orange, yellow, and green light are absorbed more rapidly than blue light. Consequently, large bodies of water appear blue—the wavelength most likely scattered back to our eyes. In the first 10 m, the aquatic environment is bright with all the colors of the rainbow; below 50 or 60 m it is a blue twilight. When ice covers the surface, light penetration is even further compromised. The intensity and quality of light significantly affect photosynthesis and thus the availability of its products: usable energy and O_2 (see chapter 7 for more details).

Just as on land, photosynthesizing organisms convert sunlight into energy as the primary producers of aquatic systems, but the oxygen that they produce is at least as important.

Table 3.1

Life on land and water. (a) A comparison of the physical properties of air and water. (b) Some properties of water and their significance to life.

Physical Properties	Air	Water
Density	**Lower:** Support structures are needed for larger organisms (e.g., skeletons, tree-trunks)	**Higher (784×):** Water provides support for soft-bodied organisms
Buoyancy	**Lower:** Organisms must exert a great deal of energy to remain aloft (i.e., for birds, insects, or bats to fly)	**Higher:** Organisms can use lipids and air sacs to keep their integrity under pressure and to stay afloat
Viscosity	**Lower:** Moving through space takes little energy	**Higher (50×):** More energy is required to move through it
Light penetration	**Higher:** Light penetrates freely, unpolluted air is transparent; there is no restriction to where photosynthesis can occur	**Lower:** Light penetration is limited, water is translucent; photosynthesis is restricted to shallow layers
Conductor of heat	**Lower:** Easier for organisms to sustain internal temperatures that are different from temperature of surrounding air	**Higher (23×):** More difficult for organisms to sustain internal temperatures elevated above that of surrounding water
Specific heat (amount of energy it requires to increase temperature)	**Lower:** Temperatures can fluctuate widely in space and time; organisms have physiological, morphological, and behavioral adaptations to deal with heat and cold	**Higher (4.23×):** Temperatures in aquatic systems are more stable

air sac

(a)

Water is . . .	This is important for life because:
a powerful solvent	It allows and facilitates chemical reactions, making it a good medium for molecular processes necessary for life
in a liquid state at most temperatures on Earth	It is a unique property; no other common substance on the surface is liquid
highly stable: resistant to changes in temperature and resists change between states	It allows life to exist in many different temperatures and facilitates homeostasis
less dense as a solid	Ice floats, allowing bodies of water to remain liquid below surface ice
buoyant and viscose	It exerts selective pressure on aquatic organisms, resulting in reduced support systems (e.g., aquatic plants don't need strong stems), and streamlining in swimming organisms

(b)

Photosynthesizers in aquatic systems include water plants, algae, and cyanobacteria. As light becomes scarcer in aquatic systems, photosynthesis by these organisms declines. The type of photosynthesizers also changes; red algae that can use the shorter blue wavelengths are found at greater depths than green plants and green algae can tolerate. The **light compensation point** will be the depth at which oxygen is produced at the same rate as it is used; at deeper levels, photosynthesizers will require more O_2 than they produce. Oxygen and energy from photosynthesis at shallow depths is used by living things throughout the aquatic ecosystem and is transported by several mechanisms, including mixing of aquatic layers. Mixing occurs at both global and local scales, to be explained in the next section of this chapter.

Sunlight is also important because it has effects on temperature in aquatic systems, just as it has on land. For the same reason warm air rises, water that is warmed by the sun is also less dense and will float above denser cold water below. Unless mixing occurs, these warm and cold layers are separated by a **thermocline,** a layer of water through which temperature changes rapidly with depth. This layering of the water column by temperature is called *thermal stratification.* As we shall see, these differences in thermal conditions at different latitudes have far-reaching environmental and ecological consequences.

The density of air and water not only affects distributions of temperature; pressure itself can be a selective force for living things. On earth, air pressure does not vary enough to be a stressor, but deep under the ocean, hydrostatic pressures can be crushingly high. These pressures restrict the depths visited by marine birds and mammals, which have lungs, and the depth distributions of many fish species with gas-filled swim bladders. However, some large sea mammals can tolerate a range of pressures and therefore can dive to great depths; their ribs that are adapted to collapse at depths that would snap our bones. Furthermore, some marine fish make daily vertical migrations of over 1,000 m in depth, traversing 100 atmospheres of pressure as they do so. However, most marine organisms are adapted to a much narrower range of pressures and depths.

Movement of air in the form of wind, and of water in the form of currents, exerts selective pressure on living organisms that are not moving at the same velocity. For example, the trees that can tolerate the high winds of exposed mountainsides have dense wood, strong roots, and twisted growth forms that accommodate the air movement. In tide pools with crashing waves, there are barnacles, sea stars, and other hard-bodied animals that are able to cling tightly to rocks. The fish and invertebrates of swift streams are often highly streamlined or inhabit sheltered zones out of the main force of the currents. All types of living things can also evolve in ways that take advantage of the movement of air and water, particularly for dispersal.

3.1 Water Cycling

LEARNING OUTCOMES

After studying this section you should be able to do the following:

3.4 Name the major processes that move water at different scales and in different forms.

3.5 List the major reservoirs of the hydrologic cycle in order of increasing volume.

3.6 Summarize the sources and fates of water falling as precipitation on land each year.

Water cycling and movement on a global scale is driven by solar energy.   Over 71% of the earth's surface is covered by water equaling approximately 1.39 billion km^3. This water is unevenly distributed among aquatic environments; most is seawater. The oceans contain over 97% of the water in the

biosphere, and the polar ice caps and glaciers contain an additional 2%. Less than 1% is freshwater in rivers, lakes, and actively exchanged groundwater. The situation on earth is indeed as Samuel Coleridge's ancient mariner saw it: "Water, water, everywhere, nor any drop to drink" (Coleridge 1798).

The amount of water on our planet is relatively stable, with water circulating over both space and time. The same molecules of H_2O could be ice at the top of the Himalayas, melt, and flow into the Bagmati River, which empties into the Indian Ocean. Surface currents can bring this same water across the Atlantic Ocean, where it evaporates and then falls as rain on forests in Brazil, becomes part of a leaf, and then the insect that eats the leaf. In this section, we will first discuss the process by which water cycles from one form and location to another, and then how it moves by currents and mixing.

The Hydrologic Cycle

The various aquatic environments such as lakes, rivers, and oceans plus the atmosphere, and ice, can be considered as "reservoirs" within the hydrologic cycle, places where water is stored for some period of time. Figure 3.2 summarizes the dynamic movement of water among these reservoirs in a global exchange called the **hydrologic cycle.** During the hydrologic cycle, water enters each reservoir either as precipitation or as surface or subsurface flow and exits as either evaporation or flow. The hydrologic cycle is powered by solar energy, which drives the winds and evaporates water, primarily from the surface of the oceans. Water vapor cools as it rises from the ocean's surface and condenses, forming clouds. These clouds are then blown by solar-driven winds across the planet, eventually yielding rain or snow, the majority of which falls back on the oceans. The water that falls on land has several fates. Some immediately evaporates and reenters the atmosphere; some is consumed by terrestrial organisms; some percolates through the soil to become groundwater; and some ends up in lakes and ponds or in streams and rivers, which eventually find their way back to the sea.

Turnover time is the time required for the entire volume of a particular reservoir to be renewed. Because reservoirs differ in size and rates of water exchange, they turn over at vastly different rates. The water in the atmosphere turns over about every 9 days. The renewal time for river water, 12 to 20 days, is nearly as rapid. Lake renewal times are longer, ranging anywhere from days to centuries, depending on lake depth, area, and rate of drainage. But the biggest surprise is the renewal time for the largest reservoir of all, the oceans. With a renewal time of about 3,100 years, the total volume of the oceans, over 1.3 billion km^3 of water, has turned over nearly 50 times in the last 150,000 years or so, roughly since the first modern humans gazed out on an ocean.

The Effects of Wind and Temperature

Within a reservoir, wind also plays important roles for movement. At the surface, prevailing winds drive currents in oceans and lakes that transport nutrients, oxygen, and heat, as well as

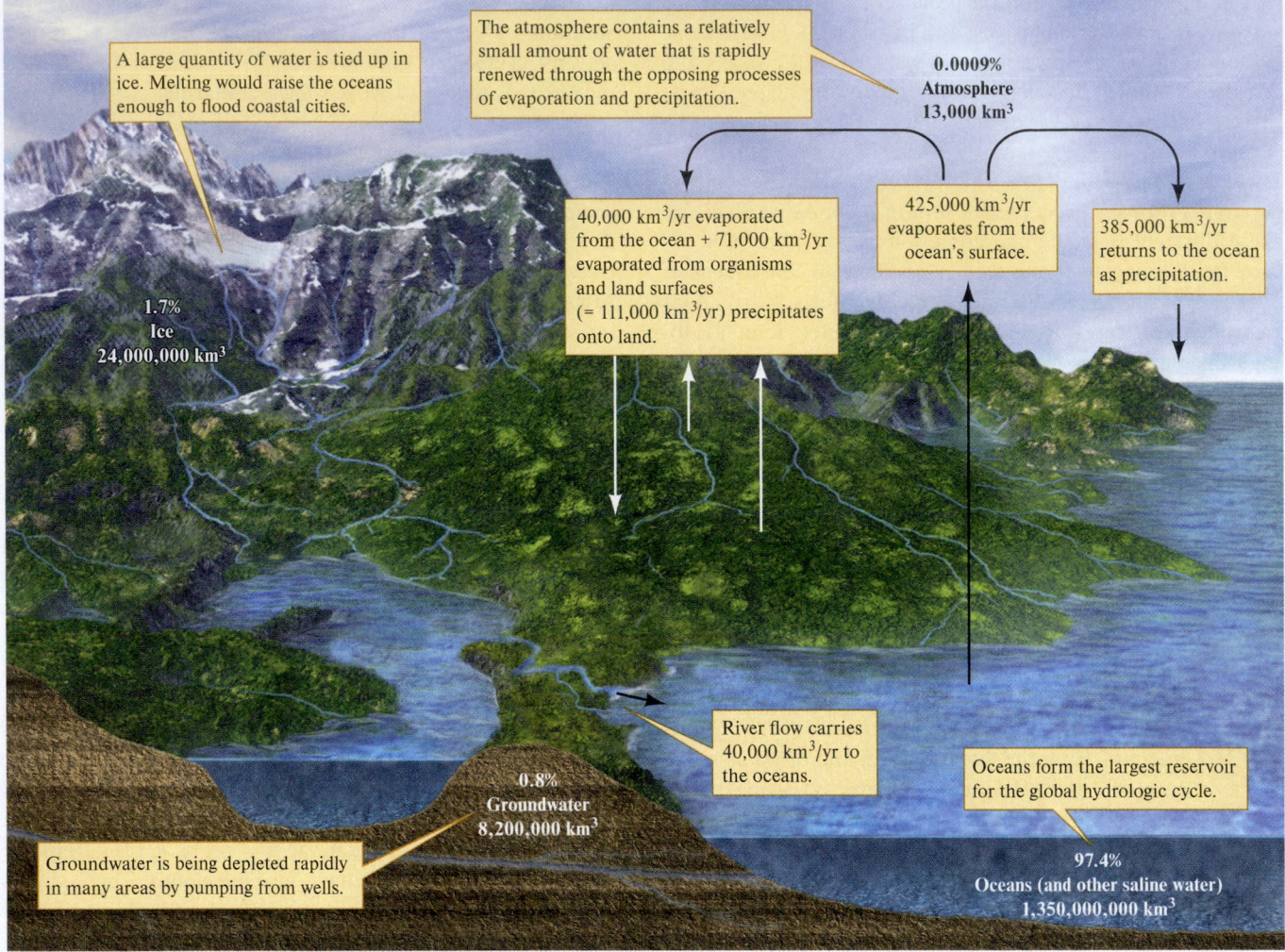

A large quantity of water is tied up in ice. Melting would raise the oceans enough to flood coastal cities.

The atmosphere contains a relatively small amount of water that is rapidly renewed through the opposing processes of evaporation and precipitation.

0.0009%
Atmosphere
13,000 km³

40,000 km³/yr evaporated from the ocean + 71,000 km³/yr evaporated from organisms and land surfaces (= 111,000 km³/yr) precipitates onto land.

425,000 km³/yr evaporates from the ocean's surface.

385,000 km³/yr returns to the ocean as precipitation.

1.7%
Ice
24,000,000 km³

River flow carries 40,000 km³/yr to the oceans.

Oceans form the largest reservoir for the global hydrologic cycle.

0.8%
Groundwater
8,200,000 km³

Groundwater is being depleted rapidly in many areas by pumping from wells.

97.4%
Oceans (and other saline water)
1,350,000,000 km³

Figure 3.2   The hydrologic cycle (data from Schlesinger 1991; USGS 2020).

organisms long distances. Currents moderate climates, fertilize the surface waters, stimulate photosynthesis, and promote gene flow among populations of marine organisms. Wind-driven surface currents sweep across vast expanses of open ocean to create great circulation systems called **gyres** that, under the influence of the Coriolis effect, move to the right in the Northern Hemisphere and to the left in the Southern Hemisphere (fig. 3.3). The great oceanic gyres transport cold water from high latitudes toward the equator and warm water from equatorial regions toward the poles, moderating climates at middle and high latitudes. A segment of one of these gyres, the Gulf Stream, moderates the climate of northwestern Europe.

Water temperature also has broad global patterns. Just as on land, surface water temperature changes with latitude, but at all latitudes, water temperatures are much more stable than terrestrial temperatures. These surface temperatures vary with season, albeit less than on land; in the temperate zone they can change 7° to 9°C while only 1°C at the equator. However, the greatest stability in water temperatures is below the surface. At 100 m depth annual variation in temperature is often less than 1°C.

Just as it drives currents, wind can also facilitate mixing of layers with different temperatures. Deep water may be moved to the surface in a process called upwelling. **Upwelling** can occur where winds blow surface water away from the shore, allowing colder water to rise to the surface. Upwelling in marine systems can also be driven by pressures created by salinity gradients, described later in this chapter. As we will see, the movement of water is critical for aquatic ecosystem functioning.

Concept 3.1 Review

1. How will global warming affect the proportion of the earth's water that resides in the oceans?
2. How do global temperature patterns in aquatic environments differ from those on land? How are they similar?

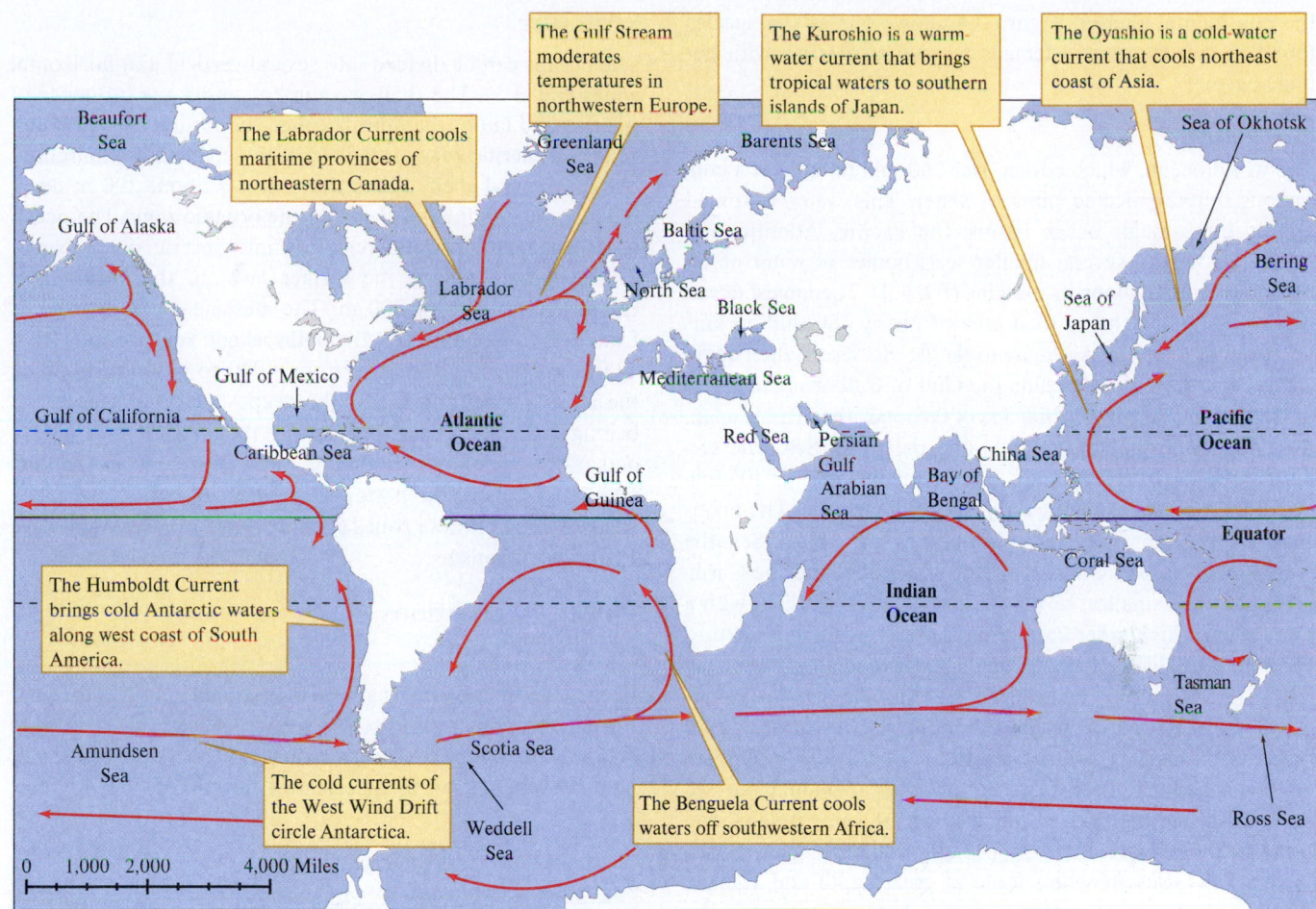

The Gulf Stream moderates temperatures in northwestern Europe.

The Kuroshio is a warm-water current that brings tropical waters to southern islands of Japan.

The Oyashio is a cold-water current that cools northeast coast of Asia.

The Labrador Current cools maritime provinces of northeastern Canada.

The Humboldt Current brings cold Antarctic waters along west coast of South America.

The cold currents of the West Wind Drift circle Antarctica.

The Benguela Current cools waters off southwestern Africa.

Figure 3.3   Oceanic circulation, which is driven mainly by the prevailing winds under the influence of the Coriolis effect, moderates earth's climate.

3.2   The Natural History of Aquatic Environments

LEARNING OUTCOMES

After studying this section you should be able to do the following:

3.7   List the major aquatic environments.

3.8   Describe the major physical characteristics of the range of aquatic environments.

3.9   Outline the chemical differences among the aquatic environments.

3.10   Discuss the differences in dominant organisms of the various aquatic environments.

3.11   Explain why human impacts on the oceans, which lagged behind our impact on terrestrial biomes for thousands of years, have rapidly increased in recent centuries.

The biology of aquatic environments corresponds broadly to variations in physical factors such as light, temperature, and water movements and to chemical factors such as salinity and oxygen.   Our discussion of the natural history of aquatic environments begins with the natural history of the oceans, the

largest aquatic environment on the planet. We continue our tour with environments found along the margins of the oceans, including kelp forests and coral reefs, the intertidal zone, and salt marshes. We then venture up rivers and streams, important avenues for exchange between terrestrial and aquatic environments. Finally, we consider lakes, inland aquatic environments that are similar in many ways to the oceans where we begin.

The Oceans

The year 2021 marks the United Nations Decade of Ocean Science for Sustainable Development, reflecting our growing awareness of the importance of this ecosystem. Preeminent marine scientists Jane Lubchenco and Steven Ganes wrote: "The ocean sustains and feeds us. It connects us. It is our past and our future. The ocean is not too big to fail, nor is it too big to fix. It is too big to ignore." (Lubchenco and Ganes 2019).

Experience with terrestrial organisms cannot prepare you for what you encounter in samples taken from the deep ocean. We dream of unknown extraterrestrial beings, some friendly and some monstrous, all with strange and shocking anatomy. We parade them through science fiction literature and films, while, unknown to most of us, creatures as odd and wonderful, some beyond imagining, live in the deep blue world beyond the continental shelves. Figure 3.4 shows one of the species found in the deep sea—a female deep-sea anglerfish with her male partner.

Geography

The world ocean, which covers over 360 million km^2, is a continuous, interconnected mass of water. This water is spread among three major ocean basins: the Pacific, Atlantic, and Indian, each with several smaller seas, bodies of water partly enclosed by land, along its margins (fig. 3.3). The largest ocean basin, the Pacific, has a total area of nearly 180 million km^2 and extends from the Antarctic to the Arctic Sea. In the Pacific Ocean, the major seas include the Gulf of California, the Gulf of Alaska, the Bering Sea, the Sea of Okhotsk, the Sea of Japan, the China Sea, the Tasman Sea, and the Coral Sea. The second largest basin, the Atlantic, has a total area of over 106 million km^2 and also extends nearly from pole to pole. The major seas of the Atlantic are the Mediterranean, the Black Sea, the North Sea, the Baltic Sea, the Gulf of Mexico, and the Caribbean Sea. The smallest of the three oceans, the Indian, with a total area of just under 75 million km^2, is mostly confined to the Southern Hemisphere. Its major seas are the Bay of Bengal, the Arabian Sea, the Persian Gulf, and the Red Sea.

The Pacific is also the deepest ocean, with an average depth of over 4,000 m. The average depths of the Atlantic and Indian Oceans are approximately equal, at just over 3,900 m. Undersea mountains stud the floor of the deep ocean, some isolated and some in long chains that run as ridges for thousands of kilometers. Undersea trenches, some of great depth and volume, rip through the seafloor. One such trench, the Marianas, in the western Pacific Ocean, is over 10,000 m deep—deep enough to engulf Mount Everest with 2 km to spare. The peak of Mauna Loa in Hawaii is a bit over 4,000 m above sea level, a modest height for a mountain. However, the base of Mauna Loa extends 6,000 m below sea level, making it, from base to peak, one of the tallest mountains on earth. What new biological discoveries might await future ecologists along this undersea slope?

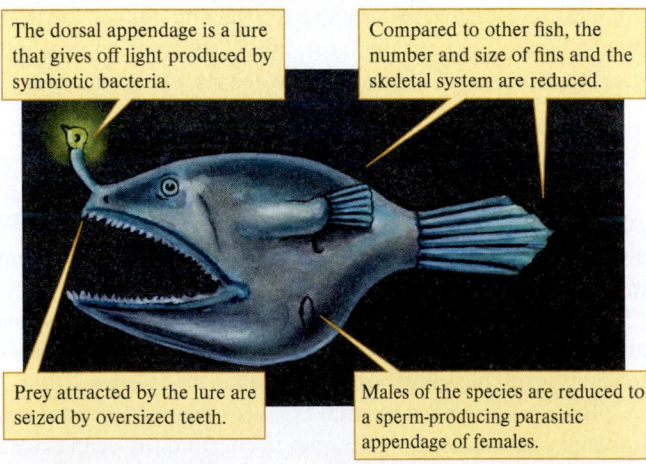

The dorsal appendage is a lure that gives off light produced by symbiotic bacteria.

Compared to other fish, the number and size of fins and the skeletal system are reduced.

Prey attracted by the lure are seized by oversized teeth.

Males of the species are reduced to a sperm-producing parasitic appendage of females.

The darkness, low food availability, and high pressures of the deep-sea environment have selected for organisms quite different from those typical of either shallow seas or the terrestrial environment. Only the females of this deep-sea anglerfish species are active predators.

Figure 3.4   Deep-sea anglerfish.

Structure

The oceans can be divided into several vertical and horizontal zones (fig. 3.5). The shallow shoreline under the influence of the rise and fall of the tides is called the **littoral**, or **intertidal, zone.** The **neritic zone** extends from the coast to the margin of the continental shelf, where the ocean is about 200 m deep. Beyond the continental shelf lies the **oceanic zone.** The ocean is also generally divided vertically into several depth zones. The **epipelagic zone** is the surface layer of the oceans that extends to a depth of 200 m. The **mesopelagic zone** extends from 200 to 1,000 m, and the **bathypelagic zone** extends from 1,000 to 4,000 m. The layer from 4,000 to 6,000 m is called the **abyssal zone,** and finally the deepest parts of the oceans belong to the **hadal zone.** Habitats on the bottom of the ocean, and other aquatic environments, are referred to as **benthic,** while those above the bottom, regardless of depth, are called **pelagic.** Each of these zones supports a distinctive assemblage of marine organisms.

Physical Conditions

Light

Because of their depth, marine environments are mostly dark. Figure 3.6 compares the colors seen by a scuba diver in deep and shallow water to demonstrate the selective absorption of light by water. Even in the clearest oceans on the brightest

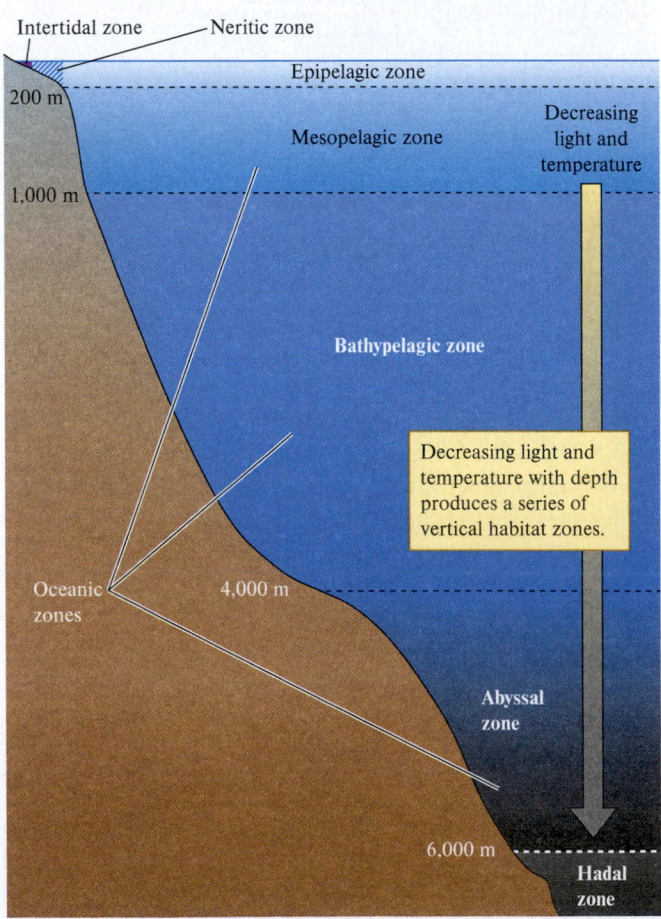

Intertidal zone   Neritic zone

Epipelagic zone

200 m

Mesopelagic zone

Decreasing light and temperature

1,000 m

Bathypelagic zone

Decreasing light and temperature with depth produces a series of vertical habitat zones.

Oceanic zones   4,000 m

Abyssal zone

6,000 m

Hadal zone

Figure 3.5   Vertical structuring of the oceans is associated with substantial variation in light and temperature with depth.

(a)

(b)

Figure 3.6   Changes in light quality with depth: (*a*) the rich colors on a shallow coral reef; (*b*) the blue of the deeper reef.
(*a*) Comstock Images/Jupiter Images; (*b*) Jody Watt/Getty Images

days, the amount of sunlight penetrating to a depth of 600 m is approximately equal to the intensity of starlight on a clear night. That leaves, on average, about 3,400 m of deep black water in which the only light is that produced by bioluminescent fishes and invertebrates.

Temperature

Oceans' depth also results in much of their volume being extremely cold, even while shallow layers may be warmed by the sun. Thermal stratification is a permanent feature of tropical oceans. Temperate oceans are stratified only during the summer, and the thermocline breaks down as surface waters cool during fall and winter. At high latitudes, thermal stratification is only weakly, if ever, developed.

The lowest average oceanic temperature, about $-1.9°C$, is around the Antarctic. The highest average surface temperatures, a bit over 27°C, occur near the equator. As we shall see, these differences in thermal conditions at different latitudes have far-reaching ecological consequences.

Water Movements

The oceans are never still. Prevailing winds drive currents that transport nutrients, oxygen, and heat, as well as organisms,

across the globe. These winds are driven by solar energy, as described in chapter 2, and shown in figure 2.5. As we have described above, oceanic gyres at the surface are even responsible for moderating terrestrial climate in some regions.

In addition to surface currents, there are deep-water currents such as those produced as cooled, high-density water sinks at the Antarctic and Arctic that move along the ocean floor. Sometimes called the ocean's "conveyor belt," **thermohaline circulation** is the movement of water caused by an interaction of temperature (thermo) and salinity (haline). At the poles, ocean water cools and freezes; when sea ice forms at the surface, it leaves behind salts (fig. 3.7). Because the surrounding ocean water is now saltier, it has a higher density and therefore sinks. This pulls in less-salty water, which then can also freeze, creating a continuous current. This displacement results in upwelling elsewhere. Upwelling also occurs along the west coasts of continents and around Antarctica, where winds blow surface water offshore, allowing colder water to rise to the surface.

Ocean life depends on water movement from both thermohaline circulation and upwelling; nutrients are eventually depleted from the shallow layers but are re-enriched by water from deeper layers. There are concerns that increased freshwater inputs due to global warming threaten thermohaline circulation by changing salinity levels (Zickfeld et al. 2007). This is just one of many threats to our oceans due to rising temperatures (see chapter 23).

Chemical Conditions

Salinity

The amount of salt dissolved in water, called **salinity,** varies with latitude and among the seas that fringe the oceans. In the open ocean, it varies from about 34 g of salt per kilogram of water (34‰, or 34 parts per thousand) to about 36.5‰. The lowest salinities occur near the equator and above 40° N and S latitudes, where precipitation exceeds evaporation. The excess of precipitation over evaporation at these latitudes is clearly shown by the climate diagrams for temperate forests, boreal forests, and tundra that we examined in chapter 2 (see figs. 2.29, 2.32, and 2.35). Highest salinities occur in the subtropics at about 20° to 30° N and S latitudes, where precipitation is low and evaporation high—precisely those latitudes where we encountered deserts (see fig. 2.20). Salinity varies a great deal more in the small, enclosed basins along the margins of the major oceans. The Baltic Sea, which is surrounded by temperate and boreal forest biomes and receives large inputs of freshwater, has local salinities of 7‰ or lower. In contrast, the Red Sea, which is surrounded by deserts, has surface salinities of over 40‰.

Despite considerable variation in total salinity, the relative proportions of the major ions (e.g., sodium [Na^+], magnesium [Mg^{2+}], and chloride [Cl^-]) remain approximately constant from one part of the ocean to another. This uniform composition, which is a consequence of continuous and vigorous mixing of the entire world ocean, underscores the physical connections across the world's oceans.

Oxygen

A liter of air contains about 200 mL of oxygen at sea level, while a liter of seawater contains a maximum of about 9 mL

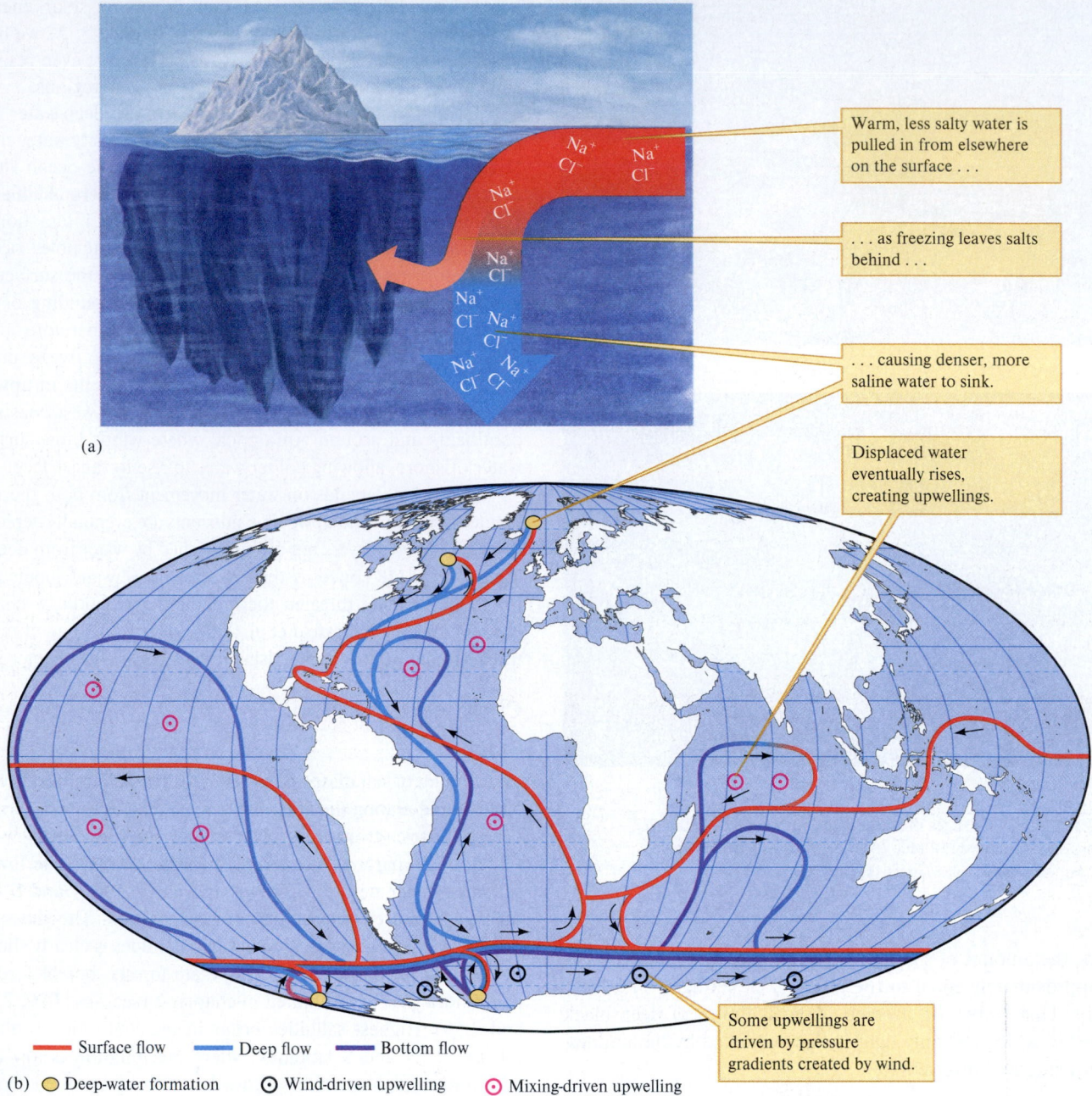

Warm, less salty water is
pulled in from elsewhere
on the surface . . .

. . . as freezing leaves salts
behind . . .

. . . causing denser, more
saline water to sink.

Displaced water
eventually rises,
creating upwellings.

Some upwellings are
driven by pressure
gradients created by wind.

—— Surface flow   —— Deep flow   —— Bottom flow

(b) ◯ Deep-water formation   ◉ Wind-driven upwelling   ◉ Mixing-driven upwelling

Figure 3.7   (*a*) Thermohaline circulation occurs when water freezes, leaving saltier and therefore denser water behind, which sinks and thereby pulls more water in. (*b*) This creates a conveyor belt of shallow and deep-water currents that span the globe. Both this process and wind patterns result in upwellings where deep water comes back up to shallow layers at several locations across the globe (data from Rahmstorf 2002; Kuhlbrodt et al. 2007).

of oxygen. Typically, oxygen concentration is highest near the ocean surface and decreases progressively with depth to some intermediate depth generally reaching a minimum at 1,000 m or less. From this minimum, oxygen concentration increases progressively to the bottom. However, some marine environments such as the deep waters in the Black Sea and in the Norwegian sill fjords are devoid of oxygen.

Biology

There is a close correspondence between physical and chemical conditions and the diversity, composition, and abundance of oceanic organisms. In oceans, the shallow upper epipelagic zone is called the *photic zone,* where microscopic organisms called **phytoplankton** (including the aforementioned cyanobacteria) drift with the currents in the open sea. The small animals that drift with these same currents are called **zooplankton.** While there is no ecologically significant photosynthesis below the photic zone, there is no absence of deep-sea organisms. Fishes, ranging from small bioluminescent forms to giant sharks, and invertebrates from tiny crustaceans to giant squid, prowl the entire water column. There is life even in the deepest trenches, below 10,000 m.

Most deep-sea organisms are nourished—whatever their place in the food chain—by organic matter fixed by photosynthesis near the surface. It was long assumed that the rain of organic matter from above was the *only* source of food for deep-sea organisms. Then, in 1977 the sea surprised everyone. There are entire biological communities on the seafloor that are nourished not by photosynthesis at the surface but by chemosynthesis on the ocean floor (see chapter 7). These oases of life are associated with undersea hot springs and harbor many life-forms (fig. 3.8).

The deep ocean shines with the blue of pure water and is often called a "biological desert." While it is true that the average rate of photosynthesis per square meter of ocean surface is similar to that of terrestrial deserts, the oceans, because they are so vast, contribute approximately one-half of the total photosynthesis in the biosphere. This oceanic primary production constitutes a substantial contribution to the global carbon and oxygen budgets.

The open ocean is home, the only home, for thousands of organisms with no counterparts on land. The terrestrial environment supports 11 animal phyla, only one of which, Onychophora, is endemic to the terrestrial environment—that is, found in no other environments. Fourteen phyla live in freshwater environments but none are endemic. Meanwhile, the marine environment supports 28 phyla, 13 of which are endemic to the marine environment (fig. 3.9).

Does the greater diversity of phyla in the marine environment shown in figure 3.9 contradict our impression of high biological diversity in biomes such as the tropical rain forest? No, it does not. The terrestrial environment is extraordinarily diverse because there are many species in a few animal and plant phyla, especially arthropods and flowering plants. Still, the number of marine species may also be very high. As mentioned above, the Census of Marine Life, which had 2,700 participating scientists, revealed thousands of new species, with undoubtedly many others yet to be discovered. In particular, the census uncovered a huge repository of diversity among microbes, which may include hundreds of millions of distinctive species. Clearly the oceans remain one of the frontiers for biological discovery.

Human Influences

Human impact on the oceans was once less than on other parts of the biosphere. For most of our history, the vastness of the oceans has been a buffer against human intrusions, but our influence is growing. The decline of large whale populations around Antarctica and elsewhere sounded a warning of what we can do to the open ocean system. The killing of whales has been curtailed, but there are plans to harvest the great whales' food supply, the small planktonic crustaceans known as *krill*. Although we may find them less engaging than their predators, the large whales, these zooplankton may be more important to the life of the open ocean. Whales are not the only marine populations that have collapsed. Overfishing has led to great declines in commercially important fish stocks, such as the cod population of Newfoundland's Grand Banks. Many marine fish populations, which once seemed inexhaustible, are now all but gone and fishing fleets sit idle in ports all over the world.

Another threat to marine life is the possibility of dumping wastes of all sorts, including nuclear and chemical wastes, into

Figure 3.8 Chemosynthesis-based community dominated by giant tube worms, mussels, and crabs on the East Pacific Rise. C. Van Dover/NOAA

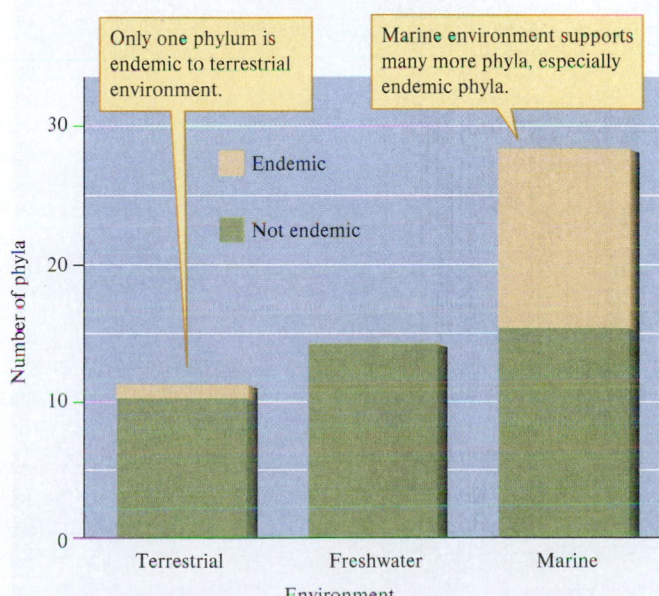

Figure 3.9 Distribution of animal phyla among terrestrial, freshwater, and marine environments (data from Grassle 1991).

the deep ocean. In recent years, chemical pollution of the sea has increased substantially, and chemical pollutants are accumulating in deep-sea sediments. In addition, plastic debris, which is resistant to decomposition, has built up in the central portions of the oceans in extensive "garbage patches." The larger pieces can entrap marine animals, while small fragments interfere with the feeding of seabirds and other marine species. The smallest fragments, called **microplastics,** have recently been found to be ubiquitous in our oceans. Their danger for marine life is amplified by their potential to absorb other chemical pollutants, thereby poisoning the food chain (Chatterjee and Sharma 2019).

Life in Shallow Marine Waters: Kelp Forests and Coral Gardens

The shallow waters along continents and around islands support marine communities of high diversity and biomass. Imagine yourself snorkeling along a marine shore, beyond the intertidal zone. If you are at temperate latitudes and over a solid bottom, you are likely to swim through groves of brown seaweed called *kelp.* Along many coasts, kelp grows so tall, over 40 m in some places, and in such densities that they resemble submarine forests (fig. 3.10).

Figure 3.10   Giant kelp forests off the California coast have structural features suggestive of terrestrial forests.   Images&Stories/Alamy

If you snorkel in the tropics, you may come across a coral reef so diverse in color and texture that it appears to be a well-tended garden (fig. 3.11). But these are forests and gardens with a difference. Here, you can soar through the canopy with fish so graceful they are called "eagle" rays or float leisurely along with "butterfly" fish.

Geography

The nearshore marine environment and its inhabitants vary with latitude. In temperate to subpolar regions, wherever there is a solid bottom and no overgrazing, there are profuse growths of kelp. As you get closer to the equator, these kelp forests are gradually replaced by coral reefs. Coral reefs are confined to low latitudes between 30° N and S (fig. 3.12).

Structure

Charles Darwin (1842b) was the first to place coral reefs into three categories: fringing reefs, barrier reefs, and atolls (fig. 3.13). **Fringing reefs** hug the shore of a continent or an island. Barrier reefs, such as the Great Barrier Reef, which stretches for nearly 2,000 km off the northeast coast of Australia, stand some distance offshore. A **barrier reef** stands between the open sea and a lagoon. Coral **atolls,** which dot the tropical Pacific and Indian Oceans, consist of coral islets that have built up from a submerged oceanic island and ring a lagoon.

Distinctive habitats associated with coral reefs include the *reef crest,* where corals grow in the surge zone created by waves coming from the open sea. The reef crest extends to a depth of about 15 m. Below the reef crest is a *buttress zone,* where coral formations alternate with sand-bottomed canyons. Behind the reef crest lies the *lagoon,* which contains numerous small coral reefs called *patch reefs* and sea grass beds.

Beds of kelp, particularly those of giant kelp, have structural features similar to those of terrestrial forests. At the water's surface is the *canopy,* which may be more than 25 m above the seafloor. The *stems,* or *stipes,* of kelp extend from the canopy to the bottom and are anchored with structures called *holdfasts.* On the stipes and fronds of kelp grow numerous species of epiphytic algae and sessile invertebrates. Other seaweed species of smaller stature usually grow along the bottom, forming an understory to the kelp forest.

Physical Conditions

Light

Both seaweeds and reef-building corals grow only in shallow waters, where there is sufficient light to support photosynthesis. The depth of light penetration sufficient to support kelp and coral varies with local conditions from a few meters to nearly 100 m.

Temperature

Temperature limits the distribution of both kelp and coral. Most kelp are limited to temperate shores, where temperatures may fall below 10°C in winter and rise to a bit above 20°C in summer. Reef-building corals are restricted to warm waters, where the minimum temperature does not fall below about 18°

Figure 3.11   Coral reefs, such as this one in the Red Sea, off the coast of Egypt, support some of the most diverse assemblages of organisms on earth.   Digital Vision/Getty Images

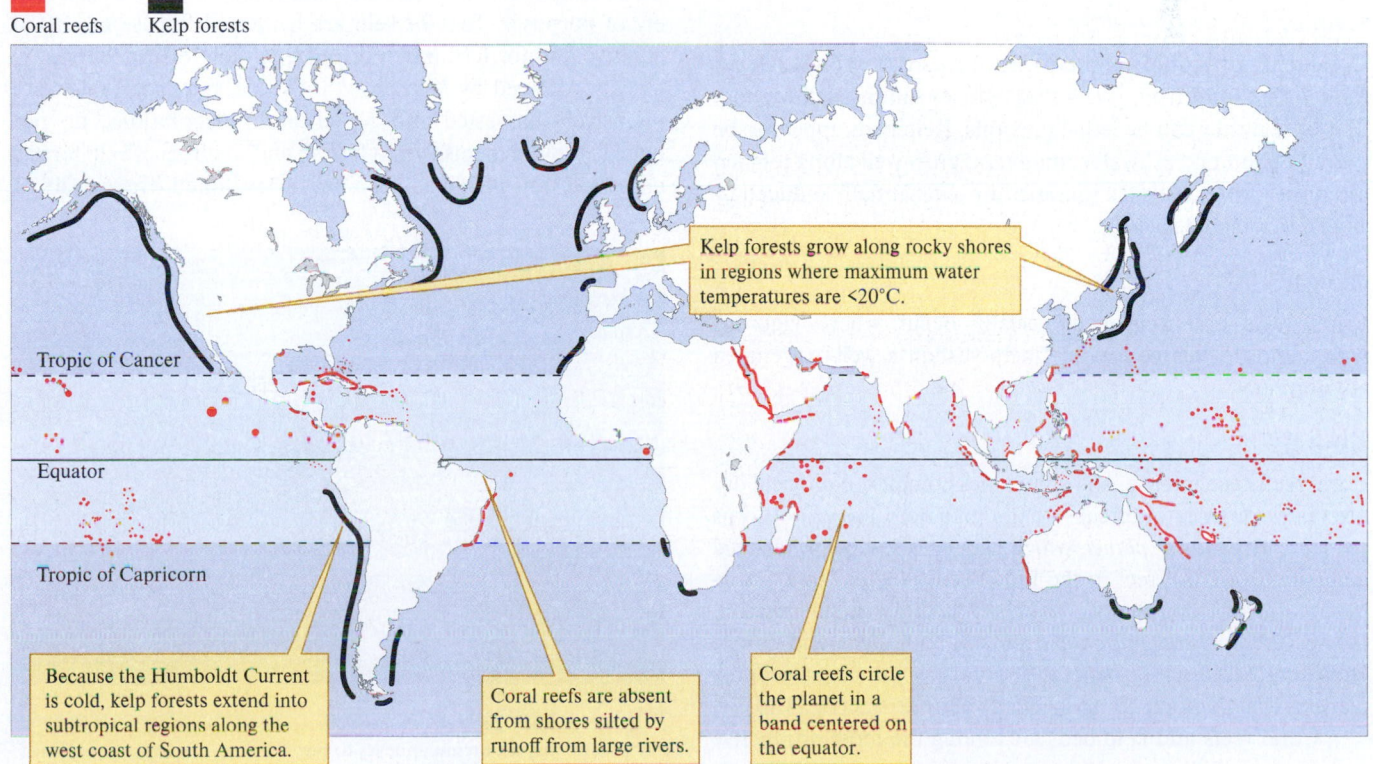

Coral reefs   Kelp forests

Kelp forests grow along rocky shores in regions where maximum water temperatures are <20°C.

Tropic of Cancer

Equator

Tropic of Capricorn

Because the Humboldt Current is cold, kelp forests extend into subtropical regions along the west coast of South America.

Coral reefs are absent from shores silted by runoff from large rivers.

Coral reefs circle the planet in a band centered on the equator.

Figure 3.12   Distribution of kelp forests and coral reefs (data from Barnes and Hughes 1988, after Schumacher 1976).

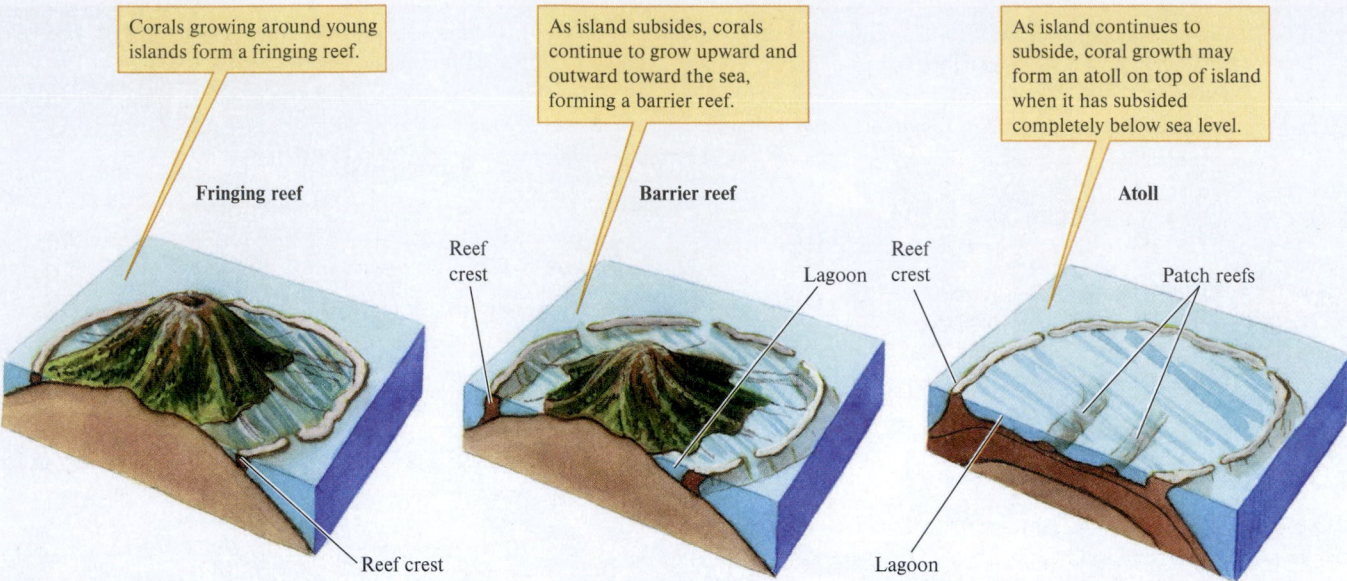

Corals growing around young islands form a fringing reef.

As island subsides, corals continue to grow upward and outward toward the sea, forming a barrier reef.

As island continues to subside, coral growth may form an atoll on top of island when it has subsided completely below sea level.

Fringing reef

Barrier reef

Atoll

Reef crest

Lagoon

Reef crest

Patch reefs

Reef crest

Lagoon

Figure 3.13   Types of coral reefs.

to 20°C and average temperatures usually vary from about 23° to 25°C. Temperatures above about 29°C are usually lethal to reef-building corals.

Water Movements

Coral reefs and kelp beds are continuously washed by oceanic currents that deliver oxygen and nutrients and remove waste products. However, extremely strong currents and wave action, as during hurricanes, can detach entire kelp forests and flatten coral reefs.

Chemical Conditions

Salinity

Coral reefs grow only in waters with fairly stable salinity. Heavy rainfall or runoff from rivers that reduces salinity below about 27% of seawater can be lethal to corals. Kelp beds appear to be more tolerant of freshwater runoff and grow well along temperate shores, where surface salinities are substantially reduced by runoff from large rivers.

Oxygen

Coral reefs and kelp beds usually occur where vigorous wave action or active currents help sustain a well-oxygenated environment.

Biology

Coral reefs face intense, and sometimes complex, biological disturbance. Periodic outbreaks of the predatory crown-of-thorns sea star, *Acanthaster planci,* which eats corals, have devastated large areas of coral reef in the Indo-Pacific region. In a Caribbean coral reef community, populations of a sea star relative, the sea urchin *Diadema antillarum,* eat both algae and corals. However, these urchins benefit the corals by reducing algal populations that compete for space with young corals (fig. 3.14).

Coral reefs and kelp beds are among the most productive and diverse of all ecological systems in the biosphere. Robert Whittaker and Gene Likens (1973) estimated that the rate of

primary production on coral reefs and algal beds exceeds that of tropical rain forests. This productivity depends on a mutually beneficial relationship between reef-building corals and algae called zooxanthellae (see chapter 15). It is the zooxanthellae that give the coral in figure 3.14 its greenish color. The center of diversity for reef-building corals is the western Pacific and eastern Indian oceans, where there are over 600 coral species and over 2,000 species of fish.

Human Influences

Coral reefs and kelp forests are increasingly exploited for a variety of purposes. Tons of kelp are harvested for use as a food additive and for fertilizer. Fortunately, most of this harvest is quickly replaced by kelp growth. Corals, however, which are intensively harvested and bleached for decorations, do not quickly replace themselves. The fish and shellfish of kelp forests and coral reefs have also been heavily exploited. Once again, it

Figure 3.14   The sea urchin *Diadema antillarum* on a coral reef. Feeding by this sea urchin appears to play a key role in the interaction between reef-building corals and benthic algae on Caribbean coral reefs.   Fine Art/Shutterstock

appears that coral reefs are more vulnerable. Some coral reefs have been so heavily fished, both for food and for the aquarium trade, that most of the larger fish are rare. Unfortunately, some especially destructive means of fishing are used on coral reefs, including dynamite and poison, with disastrous results. In the Philippines, over 60% of the area once covered by coral has been destroyed by these techniques during recent years. Pollution, especially nutrient enrichment from human population, is a growing threat to coral reefs. In addition, as the surface temperature of the oceans has warmed in recent decades, corals have increasingly expelled their zooxanthellae, resulting in a loss of their color called "coral bleaching." Bleaching episodes have been often followed by massive coral mortality.

Marine Shores: Life Between High and Low Tides

The rise and fall of the tides make the shore one of the most dynamic environments in the biosphere. The intertidal zone is a magnet for the curious naturalist and one of the most convenient places to study ecology. Where else in the biosphere does the structure of the landscape change several times each day? Where else does nature expose entire aquatic communities for leisurely exploration? Where else are environmental and biological gradients so compressed? It should be no surprise that here in the intertidal zone, immersed in tide pools, salt spray, and the distinctive smell of kelp, ecologists have found the inspiration and circumstance for some of the most elegant experiments and most enduring generalizations of ecology. The intertidal zone, the area covered by waves at high tide and exposed to air at low tides (fig. 3.15), has proved to be an illuminating window to the world.

Geography

Countless thousands of kilometers of coastline around the world have intertidal zones. From a local perspective, it is significant to distinguish between exposed and sheltered shores. Battered by the full force of ocean waves, exposed shores support very

Figure 3.15   A rocky shore at low tide. These mussels, growing on a rock outcrop, show the high population densities often seen in intertidal habitats, where competition for space is commonly intense.   Steven P. Lynch

different organisms from those found along sheltered shores on the inside of headlands or in coves and bays. A second important distinction is between rocky and sandy shores.

Structure

Although conditions vary continuously through the intertidal zone, marine biologists have divided it historically into several vertical zones (fig. 3.16). The highest zone, called the *supratidal fringe,* or *splash zone,* is seldom covered by high tides but often wetted by waves. Below this fringe is the intertidal zone proper. The upper intertidal zone is covered only during the highest tides, while the lower intertidal zone is uncovered only during the lowest tides. Between the upper and lower intertidal zones is the middle intertidal zone, which is covered and uncovered during average tides. Below the intertidal zone is the *subtidal zone,* which remains covered by water even during the lowest tides.

Physical Conditions

Light

Intertidal organisms are exposed to wide variations in light intensity. At high tide, water turbulence reduces light intensity, while at low tide, intertidal organisms are exposed to the full intensity of the sun.

Temperature

Because the intertidal zone is exposed to the air once or twice each day, intertidal temperatures are always changing. At high latitudes, tide pools, small basins that retain water at low tide, can cool to freezing temperatures during low tides, while tide pools along tropical and subtropical shores can heat to temperatures in excess of 40°C. The dynamic intertidal environment contrasts sharply with the stability of most marine environments.

Water Movements

The two most important water movements affecting the distribution and abundance of intertidal organisms are the waves that break upon the shore and the tides. The tides vary in magnitude and frequency. Most tides are *semidiurnal,* that is, there are two low tides and two high tides each day. However, in seas, such as the Gulf of Mexico and the South China Sea, there are *diurnal* tides, that is, a single high and low tide each day. The total rise and fall of the tide varies from a few centimeters along some marine shores to 15 m at the Bay of Fundy in northeastern Canada (fig. 3.17).

The sun and moon and local geography determine the magnitude and timing of tides. The main tide-producing forces are the gravitational pulls of the sun and moon on water. Of the two forces, the pull of the moon is greater because, although the sun is far more massive, the moon is much closer. Tidal fluctuations are greatest when the sun and moon are working together, that is, when the sun, moon, and earth are in alignment, which happens at full and new moons. These times of maximum tidal fluctuation are called *spring tides.* Tidal fluctuation is least when the gravitational effects of the sun and moon are working in opposition, that is, when the sun and moon, relative to earth, are at right angles to each other,

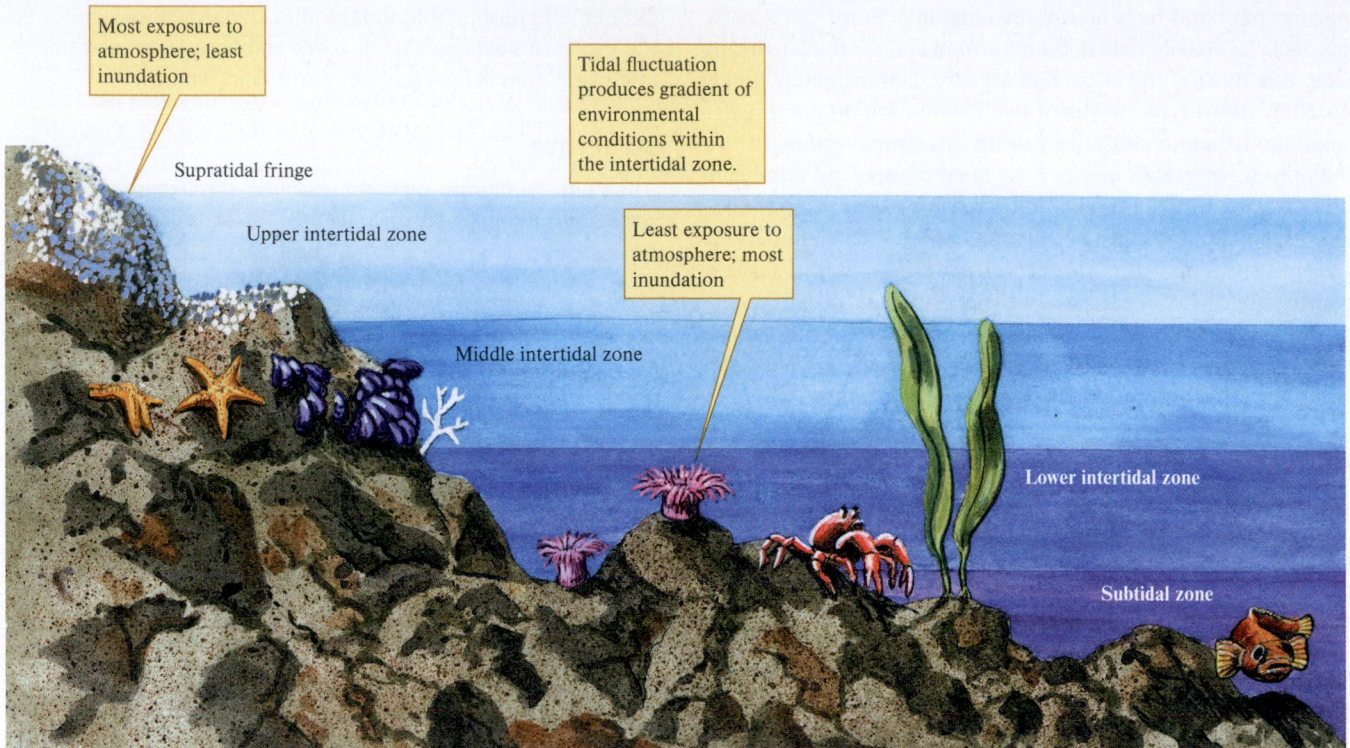

Most exposure to atmosphere; least inundation

Tidal fluctuation produces gradient of environmental conditions within the intertidal zone.

Supratidal fringe

Upper intertidal zone

Least exposure to atmosphere; most inundation

Middle intertidal zone

Lower intertidal zone

Subtidal zone

Figure 3.16   Intertidal zonation.

(a)

(b)

as they are at the first and third quarters of the moon. These times of minimum tidal fluctuation are called *neap tides*. The size and geographic position of a bay, sea, or section of coastline determine whether the influences of sun and moon are amplified or damped and are responsible for the variations in tides from place to place.

The amount of wave energy to which intertidal organisms are exposed varies considerably from one section of coast to another; this variation affects the distribution and abundance of intertidal species. Exposed headlands are hit by high waves (fig. 3.18), and they are subjected to strong currents, which are

Figure 3.17   The Bay of Fundy, a site of some of the greatest tidal fluctuations anywhere, at: (*a*) high tide and (*b*) low tide.
(*a*) Doug Sherman/Geofile; (*b*) Doug Sherman/Geofile

Figure 3.18   Storm waves such as these pounding a rocky headland have an important influence on the distribution and abundance of intertidal organisms.   Design Pics/The Irish Image Collection

at times as strong as those of swift rivers. Coves and bays are the least exposed to waves, but even the most sheltered areas may be subjected to intense wave action during storms.

Chemical Conditions

Salinity

Salinity in the intertidal zone varies much more than in the open ocean, especially within tide pools isolated at low tide. Rapid evaporation during low tide increases the salinity within tide pools along desert shores. Along rainy shores at high latitudes and in the tropics during the wet season, tide pool organisms can experience much reduced salinity.

Oxygen

Oxygen does not generally limit the distributions of intertidal organisms for two major reasons. First, intertidal species are exposed to air at each low tide. Second, the water of wave-swept shores is thoroughly mixed and therefore well oxygenated. An intertidal environment where oxygen availability may be low is in interstitial water within the sediments along sandy or muddy shores, especially in sheltered bays, where water circulation is weak.

Biology

Intertidal organisms are adapted to an amphibious existence, partly marine, partly terrestrial. All intertidal organisms are adapted to periodic exposure to air, but some species are better equipped than others to withstand that exposure. This fact produces one of the most noticeable intertidal features, **zonation of species.** Species living at the highest levels of the intertidal zone are exposed by almost all tides and remain exposed the longest. Others are exposed during the lowest tides only, perhaps once or twice per month, or even less frequently. Microtopography also influences the distribution of intertidal organisms. Tide pools support very different organisms than sections of the intertidal zone from which the water drains completely. The channels in which seawater runs, like a salty stream, during the ebb and flow of the tides offer yet another habitat.

The substratum also affects the distribution of intertidal organisms. Hard, rocky substrates support a biota different from that on sandy or muddy shores. You can see a profusion of life on rocky shores because many species are attached to rock surfaces (see fig. 3.15). The residents of the rocky intertidal zone include sea stars, barnacles, mussels, sea urchins, and seaweeds. However, most intertidal organisms are inconspicuous, since they take shelter at low tide, some among the fronds and holdfasts of kelp and others under boulders. There are even animals that burrow into and live inside rocks. As we shall see when we discuss competition in chapter 13 and predation in chapter 14, biological interactions influence the distributions of intertidal organisms.

On soft bottoms, most organisms are burrowers and shelter themselves within the sand or mud bottom. To thoroughly study the life of sandy shores, you must separate organisms from sand or mud. Perhaps this is the reason rocky shores have gotten more attention by researchers and why we know far less about the life of sandy shores. Beaches, like the open ocean, have been considered biological deserts. Careful studies, however, have shown that the intensity and diversity of life on sandy shores rivals that of any benthic aquatic community (McLachlan and Dorvlo 2005).

Human Influences

People have long sought out intertidal areas, first for food and later for recreation, education, and research. Shell middens, places where prehistoric people piled the remains of their seafood dinners, from Scandinavia to South Africa, bear mute testimony to the importance of intertidal species to human populations for over 100,000 years. Today, each low tide still finds people all over the world scouring intertidal areas for mussels, oysters, clams, and other species. But many intertidal organisms, which resist, and even thrive, in the face of twice daily exposure to air and pounding surf, are easily devastated by the trampling feet and probing hands of a few human visitors. Relentless exploitation has severely reduced many intertidal populations. Exploitation for food is not the only culprit, however. Collecting for education and research also takes its toll. The intertidal zone is also vulnerable to disruption by oil spills, which have damaged intertidal areas around the world.

Transitional Environments: Estuaries, Salt Marshes, Mangrove Forests, and Freshwater Wetlands

Estuaries are found wherever rivers meet the sea. **Salt marshes** and **mangrove forests** are concentrated along low-lying coasts with sandy shores and may, like estuaries, be associated with the mouths of rivers. **Freshwater wetlands,** such as swamps and marshes, occupy low-lying areas within landscapes and are generally inundated with water for some part of each year. All four are at the transition between one environment and another—salt marshes and mangrove forests at the transition between land and sea, estuaries at the transition between river and sea, and freshwater wetlands at the transition between land and freshwater. Because these areas are transitions between very different environments, they have a great deal in common physically, chemically, and biologically. These are particularly productive environments that teem with life. Figure 3.19 shows a rich salt marsh landscape, and figure 3.20 shows the structurally complex environment provided by dense populations of mangroves and their many prop roots.

Geography

Salt marshes, which are dominated by herbaceous vegetation, are concentrated along sandy shores from temperate to high latitudes. At tropical and subtropical latitudes, the herb-dominated salt marsh is replaced by mangrove forests (fig. 3.21). Mangroves are associated with the terrestrial climates of the tropical rain forest, tropical dry forest, savanna, and desert, due mainly to the sensitivity of mangroves to frost. Estuaries occur wherever rivers enter the sea, and freshwater marshes can form wherever runoff waters collect in low-lying areas.

Figure 3.19   The salt marsh landscape is inhabited by few species of plants, but their productivity is exceptionally high.   Index Stock/Alamy Images

Figure 3.20   The prop roots of mangroves provide a complex habitat for a high diversity of marine fish and invertebrates.   John Fedele/Blend Images

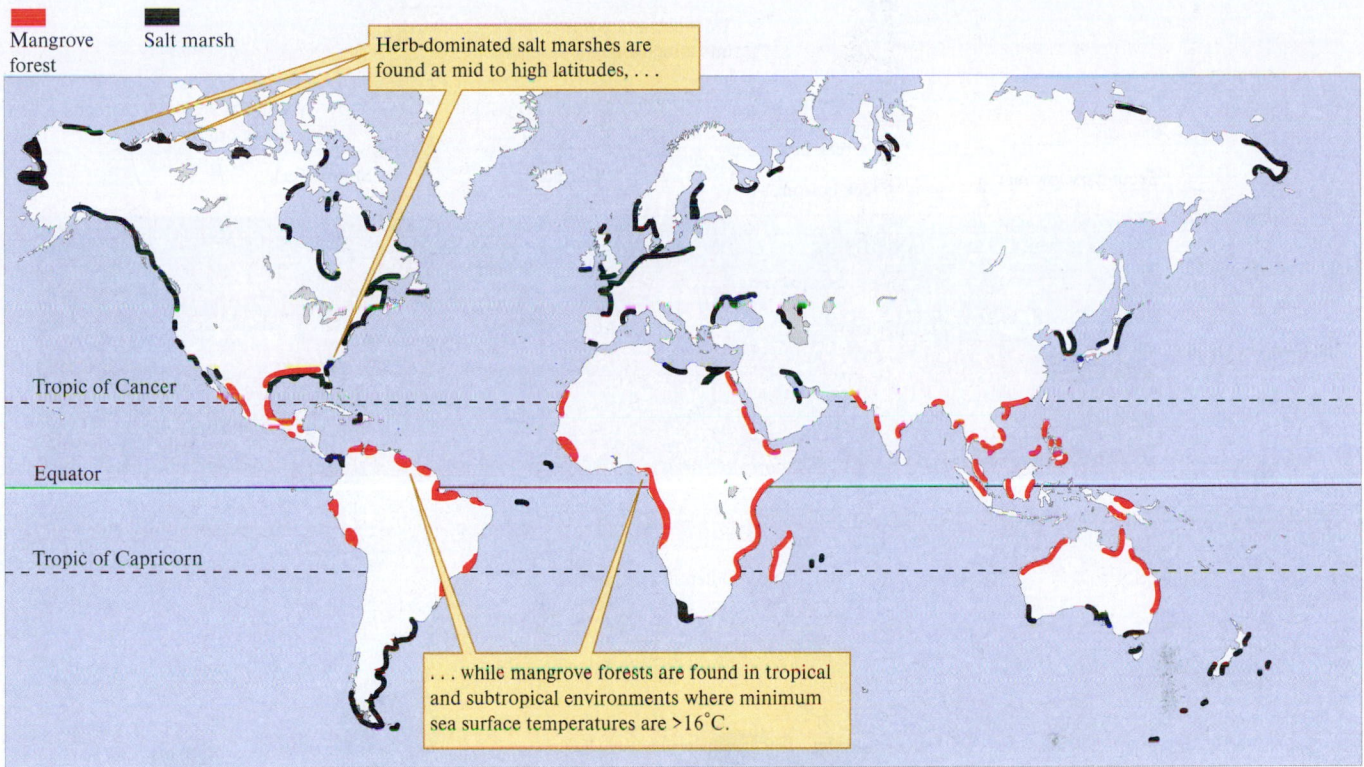

Mangrove forest

Salt marsh

Herb-dominated salt marshes are found at mid to high latitudes, . . .

Tropic of Cancer

Equator

Tropic of Capricorn

. . . while mangrove forests are found in tropical and subtropical environments where minimum sea surface temperatures are >16°C.

Figure 3.21   Salt marsh and mangrove forests (data from Chapman 1977; Long and Mason 1983).

Structure

Salt marshes generally include channels, called tidal creeks, that fill and empty with the tides. These meandering creeks can create a complex network of channels across a salt marsh (fig. 3.22). Fluctuating tides move water up and down these channels, or tidal creeks, once or twice each day, gradually sculpting the salt marsh into a gently undulating landscape (fig. 3.23). Tidal creeks are generally bordered by natural levees. Beyond the levees are marsh flats, including small basins called *salt pans* that periodically collect water that eventually evaporates, leaving a layer of salt. This entire landscape is flooded during the highest tides and drains during the lowest tides.

Figure 3.22   Viewing a salt marsh from the air reveals great structural complexity.   Stacy Pearsall/Getty Images

Mangrove trees of different species are generally distributed according to height within the intertidal zone (fig. 3.24). For instance, in mangrove forests near Rio de Janeiro, Brazil, the mangroves growing nearest the water belong to the genus *Rhizophora*. At this level in the intertidal zone, *Rhizophora* is inundated by average high tides. Above *Rhizophora* grow other mangroves such as *Avicennia*, which is flooded by the average spring tides, and *Laguncularia*, which is touched only by the highest tides.

Physical Conditions

Light

Estuaries, salt marshes, and mangrove forests experience significant fluctuations in tidal level. Consequently, the organisms in these environments are exposed to highly variable light conditions. They may be exposed to full sunlight at low tide and very little light at high tide. The waters of these environments as well as in freshwater wetlands are usually turbid because of shifting currents and agitation of their shallow waters by wind, which keep fine organic and inorganic materials in suspension.

Temperature

The temperatures of these transitional environments are highly variable. Because they are generally shallow, water temperature varies with air temperature. Since the temperatures of seawater and river water may be very different, the temperature of an estuary may change with each high and low tide. Salt marshes at high latitudes may freeze during the winter. In contrast,

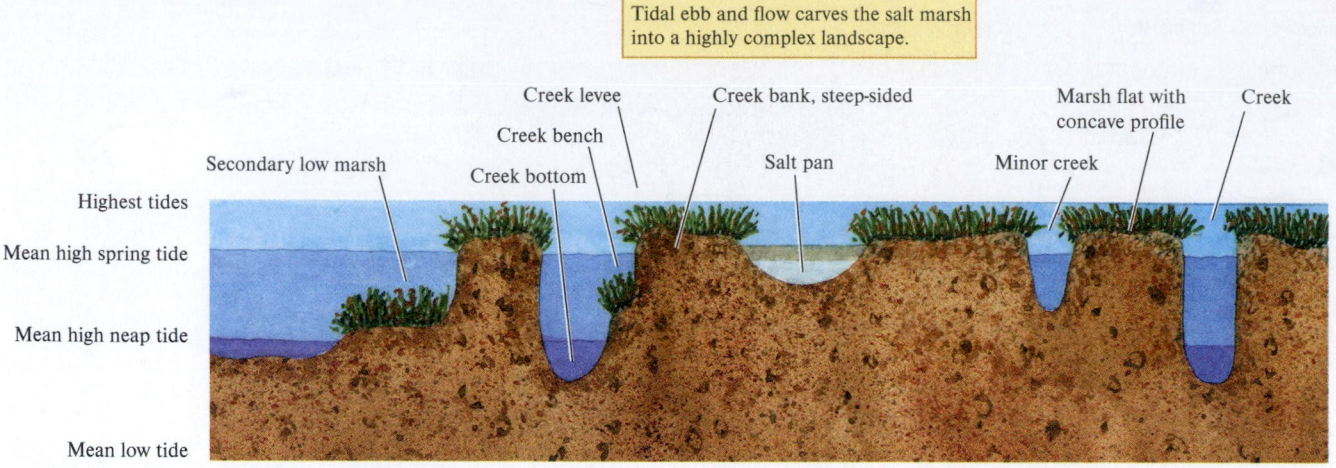

Figure 3.23   Salt marsh channels shown in cross section.

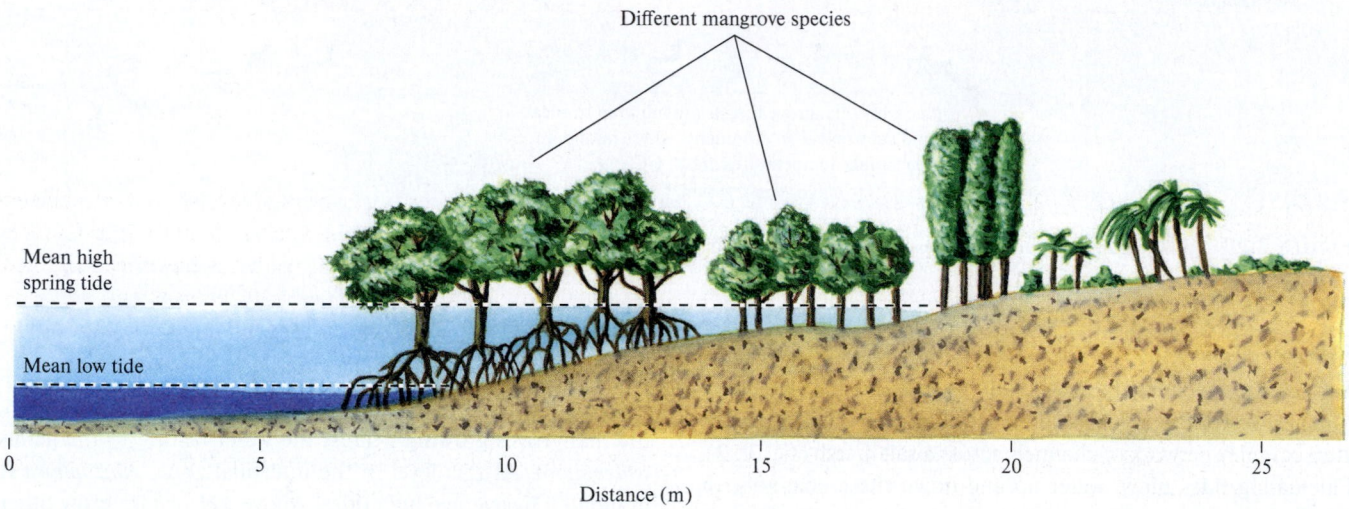

Figure 3.24   Where mangrove diversity is high, mangrove species show clear patterns of vertical zonation relative to tidal level.

mangroves grow where the minimum annual temperature is about 20°C and the shallows can heat up to over 40°C.

Water Movements

Ocean tides and river flow drive the complex currents in estuaries. These currents are at the heart of the ecological processes of the estuary because they transport organisms, renew nutrients and oxygen, and remove wastes. Tidal currents also flow in salt marshes and mangrove forests, where they are involved in these processes and fragment and transport the litter produced by salt marsh and mangrove vegetation. Once or twice a day, high tides create saltwater currents that move up the estuaries of rivers and the channels within salt marshes and mangrove forests. Low tides reverse these currents and saltwater moves seaward. Tidal height may fluctuate far from where an estuary meets the sea. For example, tidal fluctuations occur over 200 km upstream from where the Hudson River flows into the sea. Vigorous mixing, in more than one direction, makes these transitional environments some of the most physically dynamic in the biosphere. The generally limited water movements in freshwater wetlands are mainly driven by wind.

Chemical Conditions

Salinity

The salinity of estuaries, salt marshes, and mangrove forests may fluctuate widely, particularly where river and tidal flow are substantial. In such systems, the salinity of seawater can drop to nearly that of freshwater an hour after the tide turns. Because estuaries are places where rivers meet the sea, their salinity is generally lower than that of seawater. In hot, dry climates, however, evaporation often exceeds freshwater inputs and the salinity in the upper portions of estuaries may exceed that of the open ocean.

Estuarine waters are also often stratified by salinity, with lower-salinity, low-density water floating on a layer of higher-salinity water, isolating bottom water from the atmosphere. On the incoming tide, seawater coming from the ocean and river water are flowing in opposite directions. As seawater flows up the channel, it mixes progressively with river water flowing in the opposite direction. Due to this mixing, the salinity of the surface water gradually increases downriver from less than 1‰ to salinities approaching that of seawater at the river mouth (fig. 3.25).

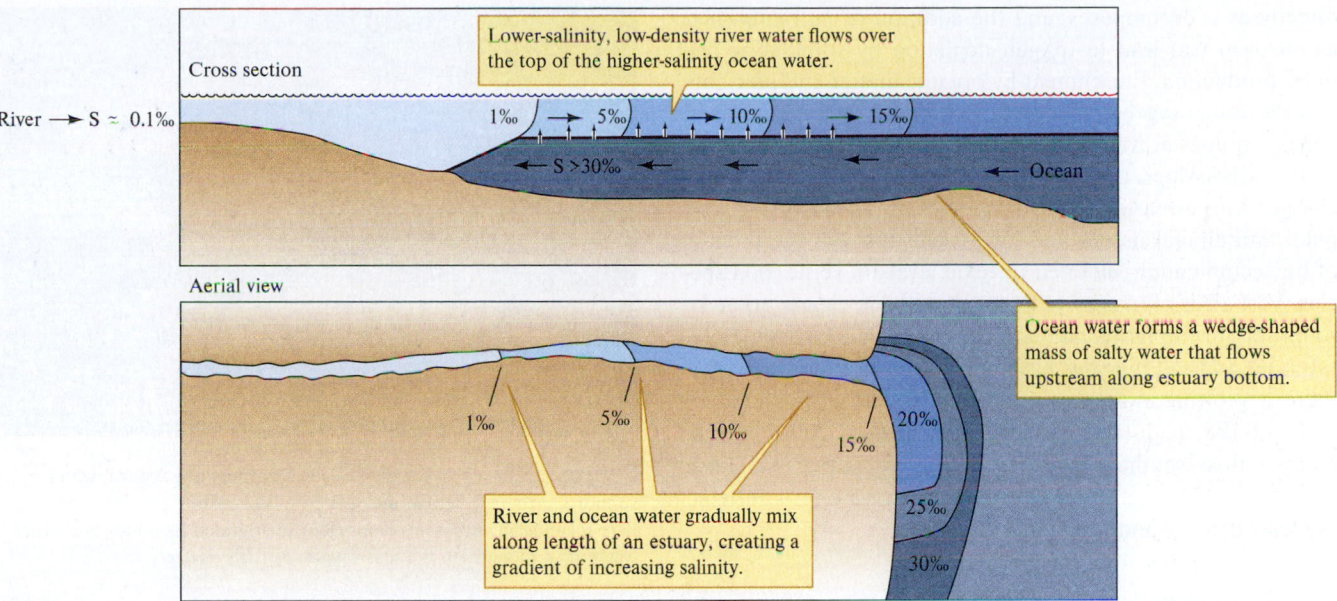

Figure 3.25   Structure of a salt wedge estuary.

Oxygen

In these transitional environments, oxygen concentration is highly variable and often reaches extreme levels. Decomposition of the large quantities of organic matter produced in these environments can deplete dissolved oxygen to very low levels, and isolation of saline bottom water from the atmosphere adds to the likelihood that oxygen will be depleted in estuaries. At the same time, however, high rates of photosynthesis can increase dissolved oxygen concentrations to supersaturated levels. Again, the oxygen concentrations to which an organism is exposed in estuaries, salt marshes, and mangrove forests can change with each turn of the tide.

Biology

The salt marshes of the world are dominated by grasses such as *Spartina* spp. and *Distichlis* spp.; by pickleweed, *Salicornia* spp.; and by rushes, *Juncus* spp. The mangrove forest is dominated by mangrove trees belonging to many genera. The species that make up the forest change from one region to another; however, within a region, there is great uniformity in species composition.

Because of their highly variable physical and chemical conditions, estuaries and salt marshes don't support a great diversity of species, but they are generally very abundant. These are places where some of the most productive fisheries occur and where aquatic and terrestrial species find nursery grounds for their young. Most of the fish and invertebrates living in estuaries evolved from marine ancestors, but estuaries also harbor a variety of insects of freshwater origin. Whatever their origins, however, the species that inhabit estuaries and salt marshes have to be physiologically tough. Estuaries and salt marshes also attract birds, especially water birds. In the mangrove forest, birds are joined by crocodiles, alligators, and, in the Indian subcontinent, by tigers. Freshwater wetlands are also among the most productive of environments.

Human Influences

Estuaries, salt marshes, mangrove forests, and freshwater wetlands are extremely vulnerable to human interference. All around the world freshwater wetlands have been drained to support agriculture. Meanwhile salt marshes and estuaries have been magnets for urban development. People want to live and work at the coast, but building sites are limited. One solution to the problem of high demand for coastal property and low supply has been to fill and dredge salt marshes, replacing wildlife habitat with human habitat (fig. 3.26). Because cities benefit from access to the sea, many, such as Boston, San Francisco, and London, have been built on estuaries. As a consequence, many estuaries have been polluted for centuries. The discharge of organic wastes into estuaries depletes oxygen

Figure 3.26   Salt marshes are vulnerable to a wide range of human-caused disturbances. Here a crew of private contractors works to clean salt marsh vegetation of oil pollution from the Deepwater Horizon oil spill of 2010.   U.S. Coast Guard photo by Pamela J. Manns

directly as it decomposes, and the addition of nutrients such as nitrogen can lead to oxygen depletion by stimulating primary production. Enrichment by organic matter and nutrients and resulting oxygen depletion have produced extensive "dead zones" in coastal waters adjacent to where large rivers, such as the Mississippi, discharge into the sea. Heavy metals discharged into estuaries and salt marshes are incorporated into plant and animal tissues and have been, through the process of bioaccumulation, elevated to toxic levels in some food species. Vast areas of mangrove forests have been cleared to make room for shrimp farms and charcoal making. The assaults on estuaries and salt marshes have been chronic and intense, but there is growing awareness of their importance. In the aftermath of the tragic Indian Ocean tsunami of 2004, governments across southern Asia have been replanting mangrove forests, since those areas with intact mangrove forests suffered the least damage and loss of human lives.

Rivers and Streams: Life Blood and Pulse of the Land

We become aware of the importance of rivers (fig. 3.27) in human history and economy as we name the major ones: Nile, Danube, Tigris, Euphrates, Yukon, Indus, Tiber, Mekong, Ganges, Rhine, Mississippi, Missouri, Yangtze, Amazon, Seine, Congo, Volga, Thames, Rio Grande. The importance of rivers, both great and small, to human history and economy is inestimable. However, river ecology has lagged behind the ecological study of lakes and oceans and is one of the youngest of the many branches of aquatic ecology. In the past few decades, however, river ecology has exploded with published research,

Figure 3.27 The Togiak River in southwestern Alaska, which supports thriving populations of five Pacific salmon species, is shown here meandering across its floodplain surrounded by ponds and other wetlands on its way to the sea. Art Wolfe/Getty Images

competing theories, controversies, and international symposia and now claims a well-earned place beside its more mature cousins.

Geography

Rivers drain most of the landscapes of the world (fig. 3.28). When rain falls on a landscape, a portion of it runs off, as either surface or subsurface flow. Some of this runoff water eventually collects in small channels, which join to form larger and larger water courses until they form a network of channels that drains the landscape. A river basin is that area of a continent or an island that is drained by a river drainage network, such as the

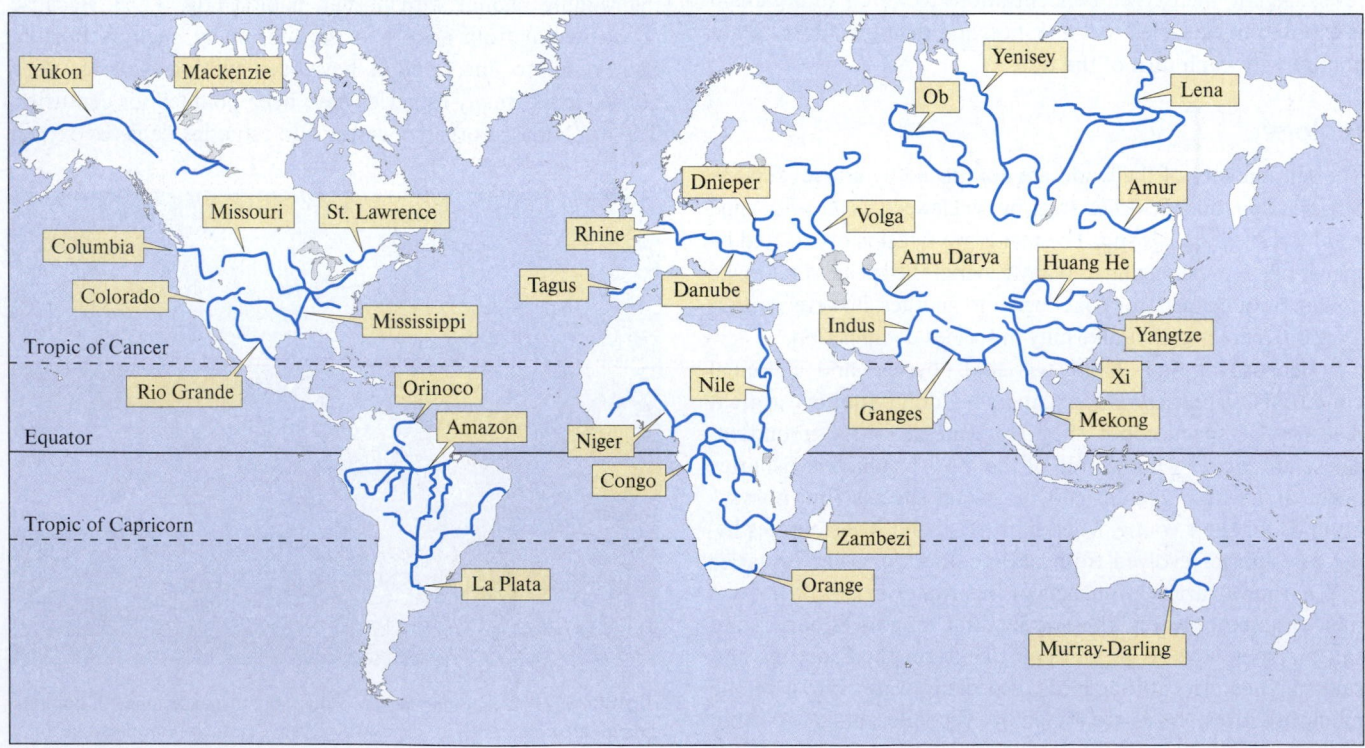

Figure 3.28 Major rivers.

Mississippi River basin in North America or the Congo River basin in Africa. Rivers eventually flow out to sea or to some interior basin like the Aral Sea or the Great Salt Lake. River basins are separated from each other by watersheds, that is, by topographic high points. For instance, the peaks of the Rocky Mountains divide runoff from melting snow. Runoff water on the east side of the peaks flows to the Atlantic Ocean, while runoff on the west side flows to the Pacific Ocean.

Structure

Rivers and streams vary along three spatial dimensions (fig. 3.29). Pools, runs, riffles, and rapids occur along their *lengths.* Because of variation in flow, rivers can also be divided across their *widths* into wetted and active channels. The wetted channel contains water even during low flow conditions, while the active channel is inundated at least annually during high flows. Outside the active channel is the **riparian zone,** a transition between the aquatic environment of the river and the upland terrestrial environment. Rivers and streams can be divided *vertically* into the water surface, the water column, and the bottom, or benthic, zone. The benthic zone includes the surface of the bottom substrate and the interior of the substrate through depths at which substantial surface water still flows. Below the benthic zone is the **hyporheic zone,** a zone of transition between areas of surface water flow and groundwater.

The area containing groundwater below the hyporheic zone is called the **phreatic zone.**

Streams and rivers can be classified by where they occur in a drainage network—that is, by **stream order.** In this system, headwater streams are first order, while a stream formed by the joining of two first-order streams is a second-order stream. A third-order stream results from the joining of two second-order streams, and so on. A lower-order stream, say a first order, joining a higher-order stream, for instance, a second-order stream, does not raise the order of the stream below the junction.

Physical Conditions

Light

Even the clearest streams are generally more turbid than clear lakes or seas. The reduced clarity of rivers, and resulting lower penetration of light, results from two main factors. First, rivers are in intimate contact with the surrounding landscape, and inorganic and organic materials continuously wash, fall, or blow into rivers. Second, river turbulence erodes bottom sediments and keeps them in suspension, particularly during floods. The headwaters of rivers are generally shaded by riparian vegetation (fig. 3.30a), which may be so dense that shading inhibits photosynthesis by aquatic primary producers. The extent of shading decreases progressively downstream as stream width increases. In arid regions, headwater streams

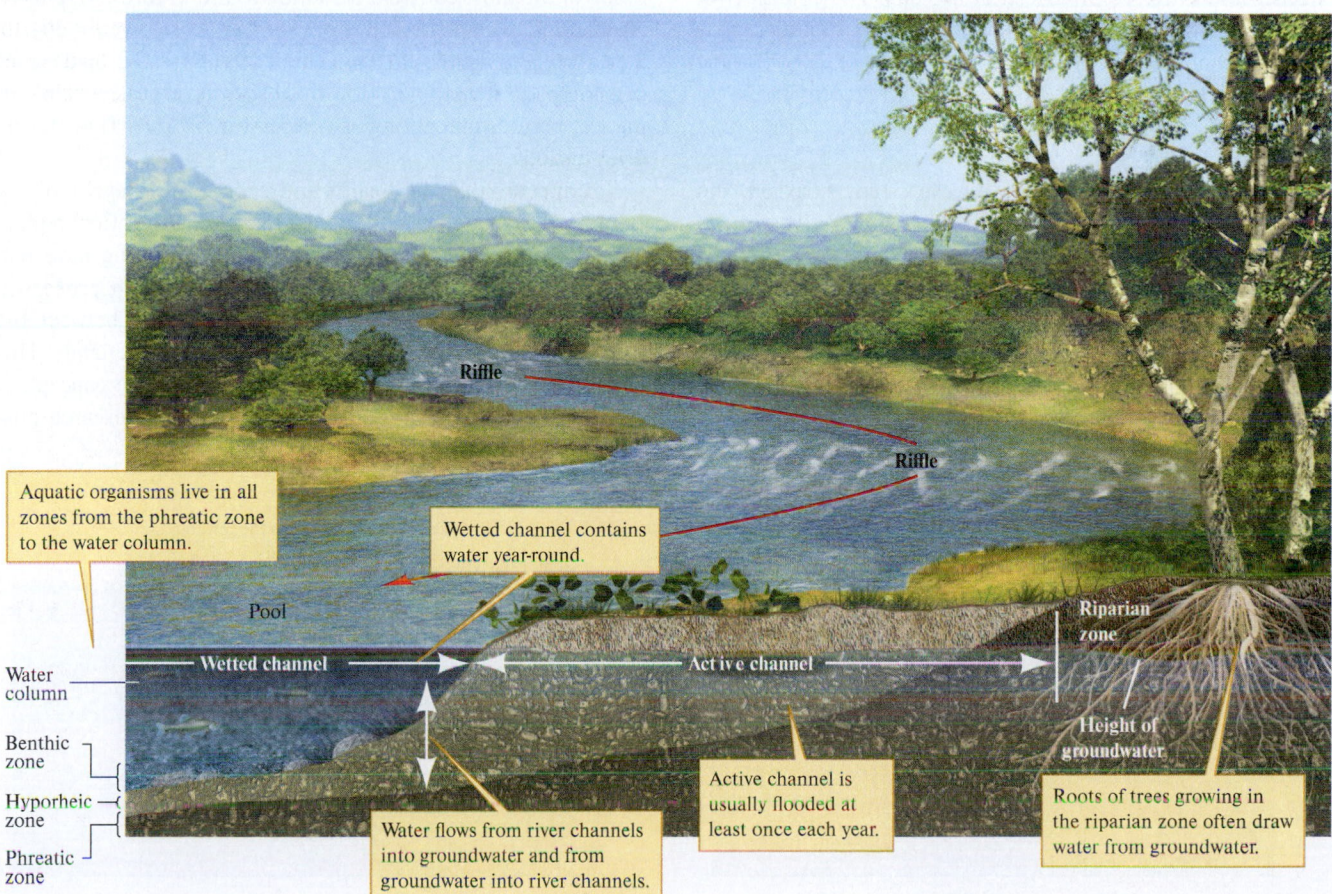

Figure 3.29   The three dimensions of stream structure.

(b)

(a)

Figure 3.30   Headwater streams in: (*a*) forested Great Smoky Mountains, Tennessee; and (*b*) arid Capitol Reef National Park, Utah. The consumers in headwater streams draining forested lands generally depend on organic matter produced by the surrounding forest. Meanwhile, arid-land streams are open to sunlight and support high levels of photosynthesis by stream algae, the main source of food energy for consumers in headwater, arid-land streams.   (*a*) Stewart Tomlinson/U.S. Geological Survey; (*b*) billperry/123RF

usually receive large amounts of solar radiation and support high levels of photosynthesis (fig. 3.30*b*).

Temperature

The temperature of rivers closely tracks air temperature but does not reach the extremes of terrestrial habitats. The coldest river temperatures, those of high altitudes and high latitudes, may drop to a minimum of 0°C. The warmest rivers are those flowing through deserts, but even desert rivers seldom exceed 30°C.

Water Movements

River currents deliver food, remove wastes, renew oxygen, and strongly affect the size, shape, and behavior of river organisms. Currents in quiet pools may flow at only a few millimeters per second, while water in the rapids of swift rivers in a flood stage may flow at 6 m per second. Contrary to popular belief, the currents of large rivers may be as swift as those in the headwaters.

The amount of water carried by rivers, which is called *river discharge,* differs a lot from one climatic regime to another

(fig. 3.31). River flows are often unpredictable and "flashy" in arid and semiarid regions, where extended droughts may be followed by torrential rains. Flow in tropical rivers also varies considerably. Many tropical rivers, which flow very little during the dry season, become torrents during the wet season. Some of the most constant flows are found in forested temperate regions, where precipitation is often fairly evenly distributed throughout the year (see fig. 2.29). Forested landscapes can damp out variation in flow by absorbing excessive rain during wet periods and acting as a reservoir for river flow during drier periods.

It appears that the health and ecological integrity of rivers and streams depend upon keeping the natural flow regime for a region intact. Historical patterns of flooding have particularly important influences on river ecosystem processes, especially on the exchange of nutrients and energy between the river channel and the floodplain and associated wetlands. This idea, which was first proposed as the **flood pulse concept,** is supported by a growing body of evidence from research conducted on rivers on virtually every continent.

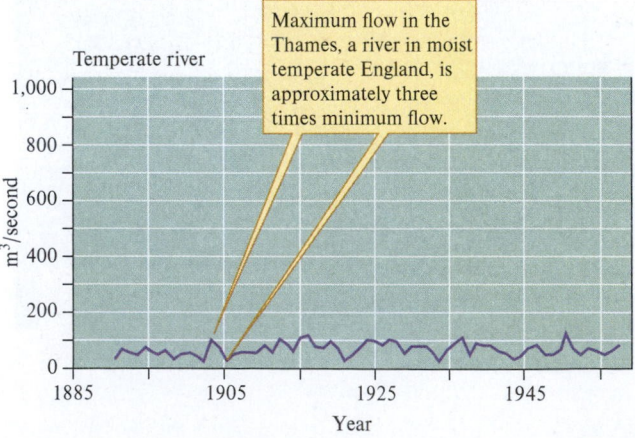

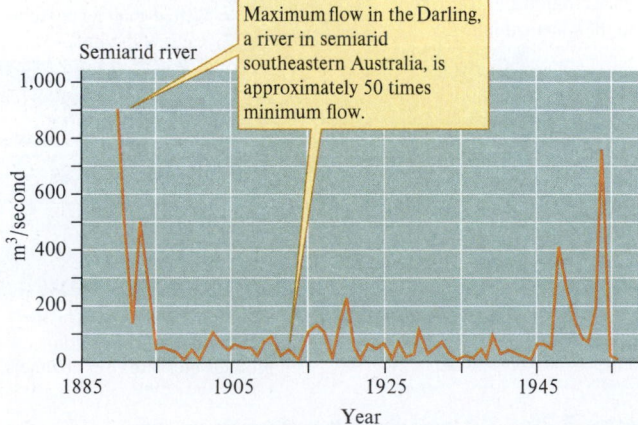

Figure 3.31   Annual flow of rivers in moist temperate and semiarid climates (data from Calow and Petts 1992).

Chemical Conditions

Salinity

Water flowing across a landscape or through soil dissolves soluble materials. The amount of salt dissolved in river water reflects the prevailing climate in its basin (fig. 3.32). As we saw in chapter 2, annual rainfall is high in tropical regions. Consequently, many tropical soils have been leached of much of their soluble materials, and it is in the tropics that the salinity of river water is often very low. Desert rivers generally have the highest salinities.

Oxygen

The oxygen content of water is inversely correlated with temperature. Oxygen supplies are generally richest in cold, thoroughly mixed headwater streams and lower in the warm, downstream sections of rivers. However, because the waters in streams and rivers are continuously mixed, oxygen is generally not limiting to the distribution of river organisms. The major exception to this generalization is in sections of streams and rivers receiving organic wastes from cities and industry. Such wastes have high **biochemical oxygen demand,** or **BOD,** a measure of organic pollution defined as the amount of dissolved oxygen required by microbes, mainly bacteria and fungi, to decompose the organic matter in a water sample.

Biology

As in the terrestrial biomes, large numbers of species inhabit tropical rivers. The number of fish species in tropical rivers is much higher than in temperate rivers. For example, the Mississippi River basin, which supports one of the most diverse temperate fish faunas, is home to about 300 fish species. By contrast, the tropical Congo River basin contains about 669 species of fish, of which over 558 are found nowhere else. The most impressive array of freshwater fish is that of the Amazon River basin, which contains over 2,000 species, approximately 10% of all the known fish species on the planet.

Most of the invertebrates of streams and rivers live on or in the sediments; that is, most are benthic. However, a great number and diversity of invertebrate animals live deep within the sediments of rivers in both the hyporheic and phreatic zones. These species can be pumped up with well water many kilometers from the nearest river.

The organisms of river systems change from headwaters to mouth. These patterns of biological variation along the courses of rivers have given rise to a variety of theories that predict downstream change in rivers and their inhabitants. One of these theories is the **river continuum concept** (Vannote et al. 1980). According to this concept, in temperate regions, leaves and other plant parts are often the major source of energy available to the stream ecosystem. Upon entering the stream, this coarse particulate organic matter (CPOM) is attacked by aquatic microbes, especially fungi. Colonization by fungi makes CPOM more nutritious for stream invertebrates. The stream invertebrates of headwater streams are usually dominated by two feeding groups: shredders, which feed on CPOM; and collectors, which feed on fine particulate organic matter (FPOM). The fishes in headwater streams are usually those, such as trout, that require high oxygen concentrations and cool temperatures.

The river continuum concept predicts that the major sources of energy in medium-sized streams will be FPOM washed down from the headwater streams and algae and aquatic plants. Algae and plants generally grow more profusely in less-shaded, medium-sized streams in which the benthic invertebrate community is dominated by collectors and grazers. The fishes of medium streams generally tolerate somewhat higher temperatures and lower oxygen concentrations than headwater fishes.

In large rivers, the major sources of energy are FPOM and, in some rivers, phytoplankton. Consequently, the benthic invertebrates of large rivers are dominated by collectors. Fish in large, temperate rivers are those, such as carp and catfish, that are more tolerant of lower oxygen concentrations and higher water temperatures, and because of the development of a plankton community, plankton-feeding fish (fig. 3.33).

Observing that river systems do not vary smoothly from headwaters to mouth, James Thorp, Martin Thoms, and Michael Delong (2006, 2008) proposed an alternative to the river continuum concept. They called their alternative model the **river ecosystem synthesis.** Thorp and his colleagues pointed out that flow conditions and geologic structure do not change continuously along the course of a river but instead have patchy distributions. For example, a river might follow a low-gradient meandering path in several sections along its length, while in other sections swift flow is constrained by steep canyon walls (fig. 3.34). The core of the river ecosystem synthesis perspective is that river sections with similar flow and geologic characteristics—for example, meandering sections—are more similar to each other ecologically than they are to sections with different flow and geologic characteristics—for example, high-gradient reaches flowing through steep-walled canyons. In other words, the river ecosystem synthesis proposes that flow conditions and geologic setting may be of greater significance in determining ecological characteristics—for example,

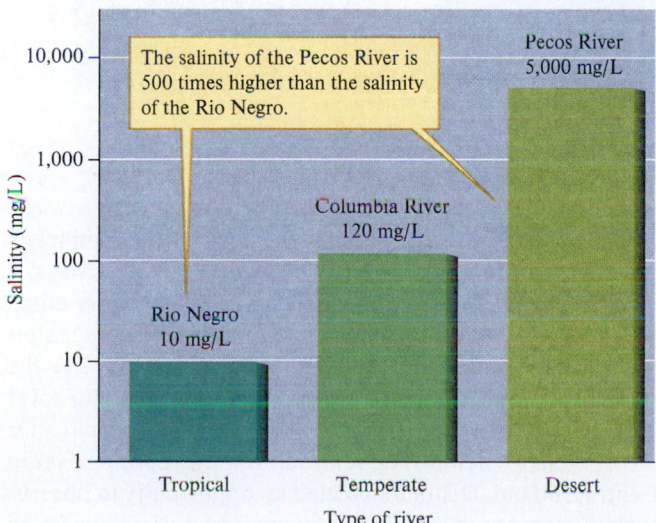

Figure 3.32 Salinities of tropical, temperate, and arid-land rivers (data from Gibbs 1970).

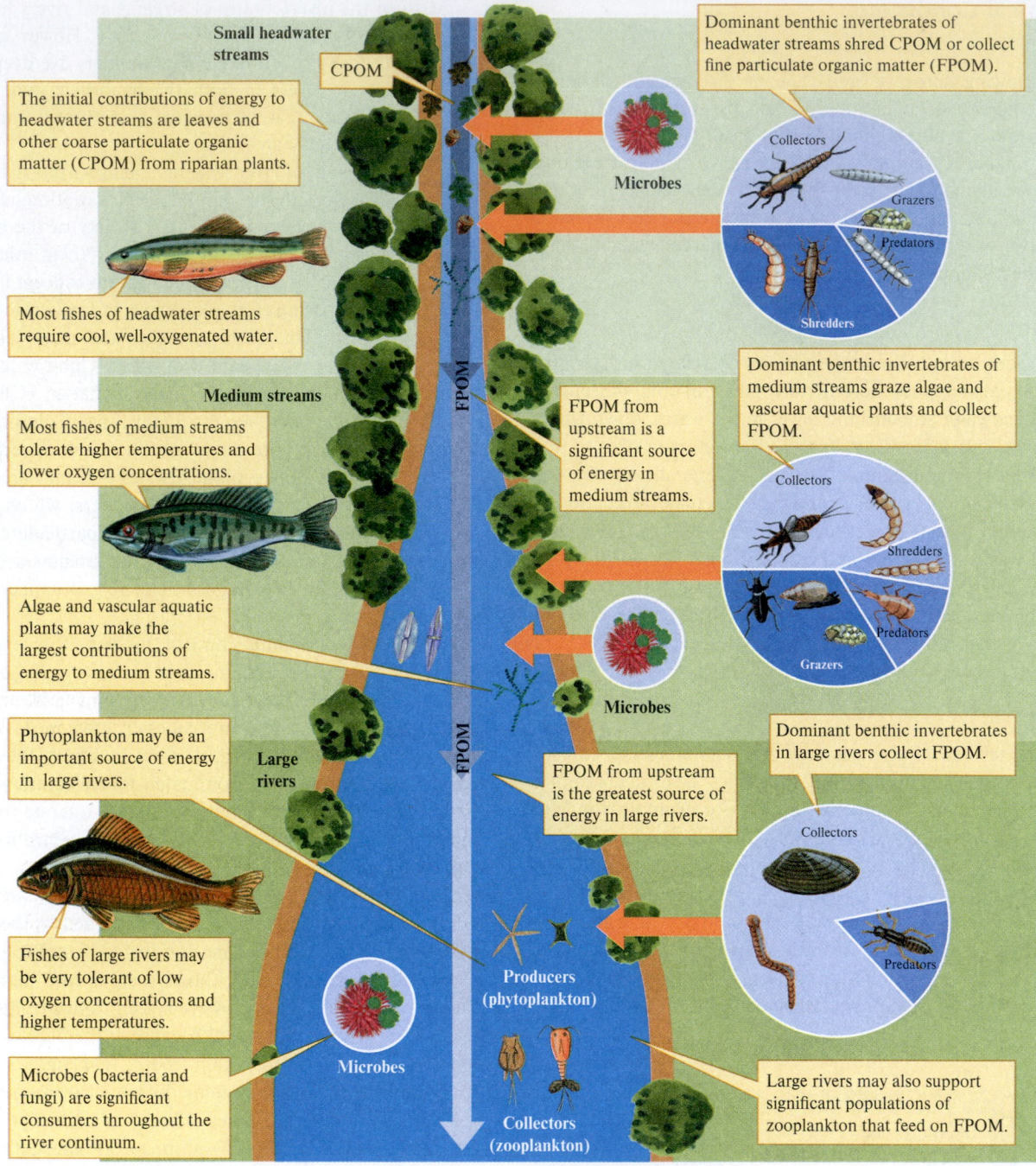

Figure 3.33 The river continuum.

the kinds of organisms living in a section of river—than is the position of a river section in a theoretical continuum. Most importantly, from a scientific perspective, the river ecosystem synthesis authors include many testable hypotheses in their theoretical framework.

Human Influences

The influence of humans on rivers has been long and intense. Rivers have been important to human populations for commerce, transportation, irrigation, and waste disposal. Because of their potential to flood, they have also been a constant threat. In the service of human populations, rivers have been channelized, poisoned, filled with sewage, dammed, filled with nonnative fish species, and completely dried. One of the most severe human impacts on river systems has been the building of reservoirs. Reservoirs eliminate the natural flow regime (including flood pulses), alter temperatures and transport of sediments, and impede the movements of migratory fish. Because of the rapid turnover of their waters, however, rivers have a great capacity for recovery and renewal. The removal of a hydropower dam in Central Jutland, Denmark created an opportunity to observe such a recovery. Kim Birnie-Gauvin and colleagues found that brown trout (*Salmo trutta*) densities increased by as

(a)   (b)

Figure 3.34   Two contrasting sections of the Yellowstone River in Yellowstone National Park. (*a*) A meandering section of the Yellowstone River flows across the broad Hayden Valley. (*b*) A few miles downriver, in another section, the river roars over Lower Yellowstone Falls before racing through the confines of the Grand Canyon of the Yellowstone River.   (*a*) NPS Photo by R. G. Johnsson; (*b*) Doug Sherman/Geofile

much as 10 times in the year following dam removal, and continued to increase in some locations in subsequent years (Birnie-Gauvin et al. 2017).

Lakes: Small Seas

In 1892, F. A. Forel defined the scientific study of lakes as the *oceanography of lakes*. On the basis of a lifetime of study, Forel concluded that lakes are much like small seas (fig. 3.35). Differences between lakes and the oceans are due, principally, to the smaller size of lakes and their relative isolation. Perhaps because they are cast on a more human scale, lakes have long captured the imagination of everyone from poets to scientists.

Figure 3.35   The historic Prince of Wales Hotel overlooking Waterton Lakes on the Canadian side of the Waterton-Glacier International Peace Park, which straddles the United States–Canadian border. The basin occupied by Waterton Lakes, which was formed by glacial action, is the deepest in the Canadian Rockies.   Alamy Stock Photo

Geography

Lakes are simply topographic depressions in the landscape that collect water. Most are found in regions worked over by the geological forces that produce such basins. These forces include shifting of the earth's crust (tectonics), volcanism, and glacial activity.

Most of the world's freshwater resides in a few large lakes. The Great Lakes of North America together cover an area of over 245,000 km^2 and contain 24,620 km^3 of water, approximately 20% of all the freshwater on the surface of the planet. An additional 20% of freshwater is contained in Lake Baikal, Siberia, the deepest lake on the planet (1,600 m), with a total volume of 23,000 km^3. Much of the remainder is contained within the rift lakes of East Africa. Lake Tanganyika, the second deepest lake (1,470 m), alone has a volume of 23,100 km^3, virtually identical to that of Lake Baikal. Still, the world contains tens of thousands of other smaller, shallow lakes, usually concentrated in "lake districts" such as northern Minnesota, much of Scandinavia, and vast regions across north-central Canada and Siberia. Figure 3.36 shows the locations of some of the larger lakes.

Structure

Lake structure parallels that of the oceans but on a much smaller scale (fig. 3.37). The shallowest waters along the lakeshore, where rooted aquatic plants may grow, is called the littoral zone. Beyond the littoral zone in the open lake is the **limnetic zone.** The **epilimnion** encompasses the surface layer of lakes. Below the epilimnion is the thermocline, or **metalimnion.** The thermocline is a zone through which temperature changes substantially with depth, generally about 1°C per meter of depth. Below the thermocline are the cold, dark waters of the **hypolimnion.**

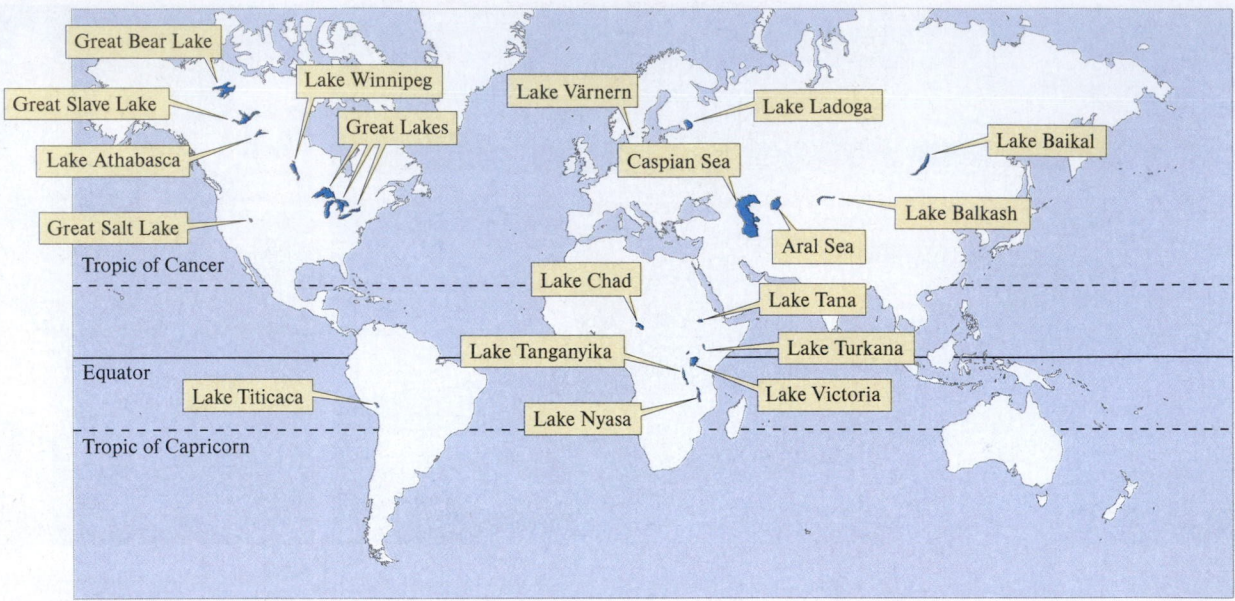

Figure 3.36 Distributions of some major lakes.

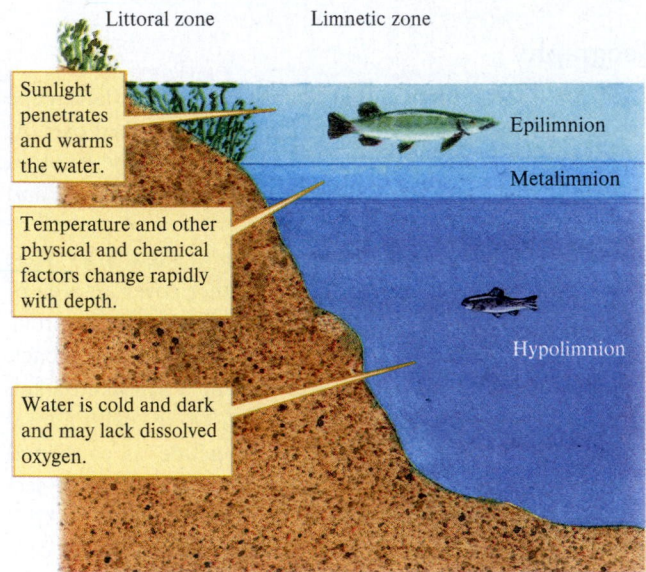

Figure 3.37 Lake structure.

Physical Conditions

Light

Lake color ranges from the deep blue of the clearest lakes to yellow, brown, or even red. Lake color is influenced by many factors but especially lake chemistry and biological activity. In lakes where the surrounding landscape delivers large quantities of nutrients, primary production is high and phytoplankton populations reduce light penetration. These highly productive lakes are usually a deep green. They are also often shallow and surrounded by cultivated lands or cities. Dissolved organic compounds, such as humic acids leached from forest soils, increase absorption of blue and green light, shifting lake color to yellow-brown. In deep lakes where the landscape delivers

low quantities of either nutrients or dissolved organic compounds, phytoplankton production is generally low and light penetrates to great depths. These lakes, such as Lake Baikal in Siberia, Lake Tahoe in California, and Crater Lake in Oregon, are nearly as blue as the open ocean.

Temperature

As in the oceans, lakes become thermally stratified as they heat. Consequently, during the warm season, they are substantially warmer at the surface than they are below the thermocline. Temperate lakes are stratified during the summer, while lowland tropical lakes are stratified year-round. As in temperate seas, thermal stratification breaks down in temperate lakes as they cool during the fall. The seasonal dynamics of thermal stratification and mixing in temperate lakes are shown in figure 3.38. In high-elevation tropical lakes, a thermocline may form every day and break down every night!

Water Movements

Wind-driven mixing of the water column is the most ecologically important water movement in lakes. As we have just seen, temperate zone lakes are thermally stratified during the summer, a condition that limits wind-driven mixing to surface waters above the thermocline. During winter on these lakes, ice forms a surface barrier that prevents mixing. During spring and fall, however, stratification breaks down and winds drive vertical currents that can mix temperate lakes from top to bottom (see fig. 3.38). These are the times when a lake renews oxygen in bottom waters and replenishes nutrients in surface waters. Like tropical seas, tropical lakes at low elevations are permanently stratified. Of Lake Tanganyika's depth of about 1,400 m, for example, only about the upper 200 m are circulated each year. Tropical lakes at high elevations heat and stratify every day and cool sufficiently to mix every night.

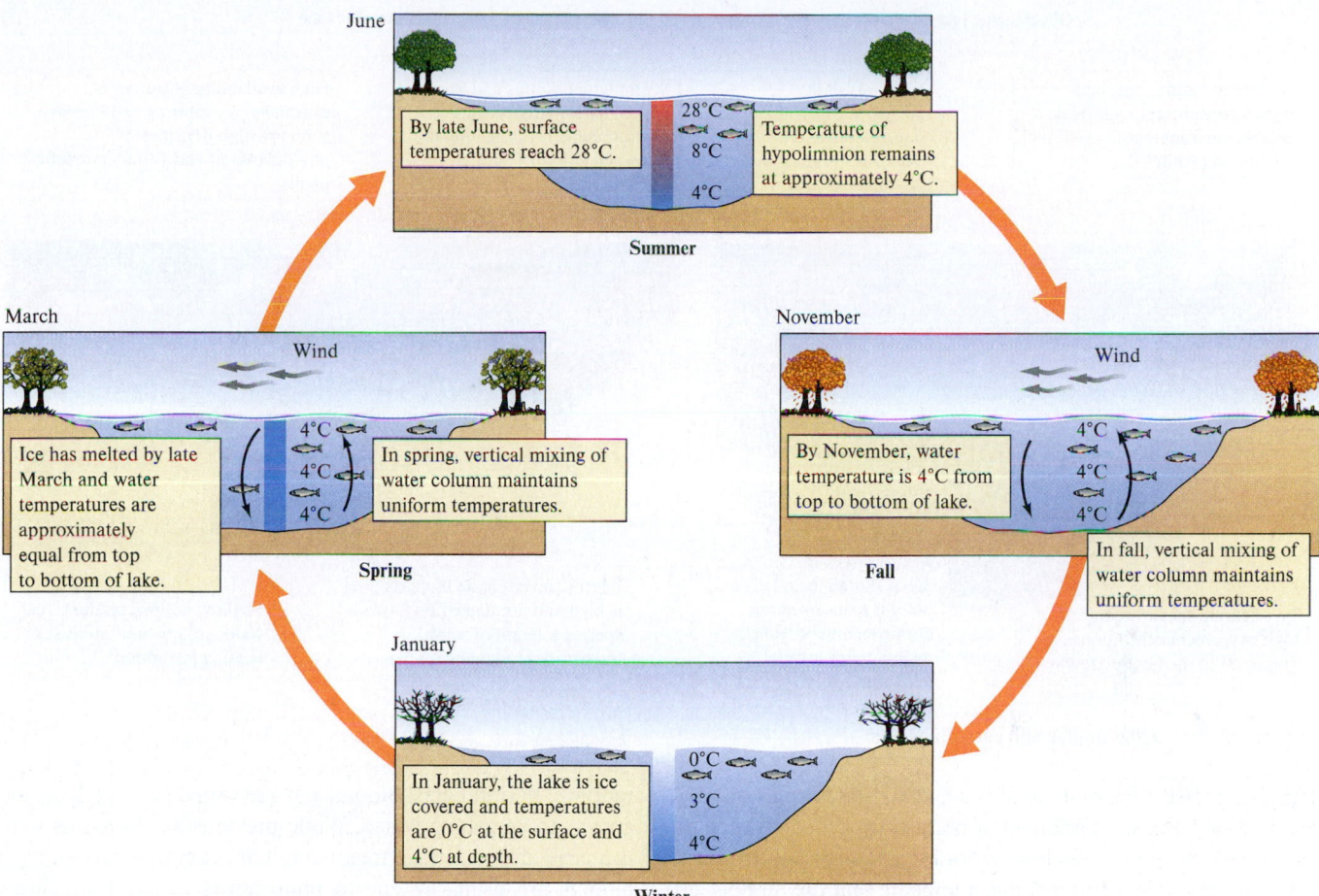

Figure 3.38   Seasonal changes in temperature in a temperate lake (data from Wetzel 1975).

Patterns of mixing have profound consequences to the chemistry and biology of lakes.

Chemical Conditions

Salinity

The salinity of lakes is much more variable than that of the open ocean. The world average salinity for freshwater, 120 mg per liter (approximately 0.120‰), is a tiny fraction of the salinity of the oceans. Lake salinity ranges from the extremely dilute waters of some alpine lakes to the salt brines of desert lakes. For instance, the Great Salt Lake in Utah sometimes has a salinity of over 200‰, which is much higher than oceanic salinity. The salinity of desert lakes may also change over time, particularly where variations in precipitation, runoff, and evaporation combine to produce wide fluctuations in lake volume.

Oxygen

Mixing and biological activities have profound effects on lake chemistry. Well-mixed lakes of low biological production, which are called **oligotrophic,** are nearly always well oxygenated. Lakes of high biological production, which are called **eutrophic,** may be depleted of oxygen. Nutrient enrichment as a consequence of human activities can accelerate the process of **eutrophication,** a process generally resulting in increased primary production, including excessive algal blooms, oxygen depletion, and reduced biodiversity. Oxygen depletion is particularly likely during periods of thermal stratification, when decomposing organic matter accumulates below the thermocline and consumes oxygen. In eutrophic lakes, oxygen concentrations may be depleted from surface waters at night as respiration continues in the absence of photosynthesis. Oxygen is also often depleted in winter, especially under the ice of productive temperate lakes. In tropical lakes, water below the euphotic zone is often permanently depleted of dissolved oxygen.

Biology

Oxygen availability can determine where fish and other organisms occur in a lake in both space and time (see fig 3.38). In addition to their differences in oxygen availability, oligotrophic and eutrophic lakes differ in factors such as availability of inorganic nutrients and temperature (fig. 3.39). Because aquatic organisms differ widely in their environmental requirements, oligotrophic and eutrophic lakes generally support distinctive biological communities.

Tropical lakes can be very productive. Also, their fish faunas may include a great number of species. Three East African lakes, Lake Victoria, Lake Malawi, and Lake Tanganyika, contain over 700 species of fish, approximately the number of

Oligotrophic lake   **Eutrophic lake**

Cool temperatures and high oxygen concentrations provide a suitable environment for fish, such as trout and whitefish.

Low availability of nutrients, especially phosphorus and nitrogen, supports low densities of phytoplankton and vascular aquatic plants.

Warm temperatures and low oxygen availability provide environments favoring tolerant fish, such as catfish and bowfins.

High availability of nutrients, especially phosphorus and nitrogen, supports high densities of phytoplankton and vascular aquatic plants.

Invertebrate species requiring high oxygen concentrations are dominant in the benthic fauna.

Steep shoreline and deep bottom reduce heating during summer and help maintain lower water temperatures.

Benthic invertebrate biomass is high and dominated by species tolerant of warm temperatures and low oxygen.

Shallow bottom reduces total water volume and increases heating in summer.

Figure 3.39   Oligotrophic and eutrophic lakes.

freshwater fish species in all of the United States and Canada; Europe and Russia combined contain only about 400 freshwater fish species. The invertebrates and algae of tropical lakes are much less studied, but it appears that the number of species may be similar to that of temperate zone lakes.

Human Influences

Human populations have had profound, and usually negative, influences on the ecology of lakes. In addition to examples of ecological degradation, however, are cases of amazing resilience and recovery—resilience in the face of fierce ecological challenge and recovery to substantial ecological integrity. Because lakes offer ready access to water for domestic and industrial uses, many human population centers have grown up around them. In both the United States and Canada, for example, large populations surround the Great Lakes. The human population around Lake Erie, one of the most altered of the Great Lakes, grew from 2.5 million in the 1880s to over 13 million in the 1980s. The primary ecological impact of these populations has been the dumping of astounding quantities of nutrients and toxic wastes. By the mid-1960s, the Detroit River alone was dumping 1.5 billion gallons of wastewater into Lake Erie each day. The Cuyahoga River, which flows through Cleveland before reaching the lake, was so fouled with oil in the 1960s that it would catch fire. In the face of such ecological challenges, much of Lake Erie, particularly the eastern end, was transformed from a healthy lake with a rich fish fauna to one that was, for a time, essentially an algal soup in which only the most tolerant fish species could live. With greater controls on waste disposal, the process of degradation began to reverse itself, and Lake Erie recovered much of its former health and vitality by the 1980s. However, after more than a decade of

reprieve, harmful algal blooms have returned to Lake Erie and appear to be getting worse. While previous algal blooms were primarily due to sewage treatment, it appears that current algal blooms are primarily due to phosphorus runoff from farms (Wilson et al. 2018). Fortunately, it appears that the lake has the capacity to recover quickly once nutrient loading decreases.

Nutrients aren't the only things that people put into lakes, however. Fish and other species are constantly moved around, either intentionally or unintentionally. As figure 3.40 shows, 139 species of fish, invertebrates, plants, and algae had been introduced to the Great Lakes by 1990.

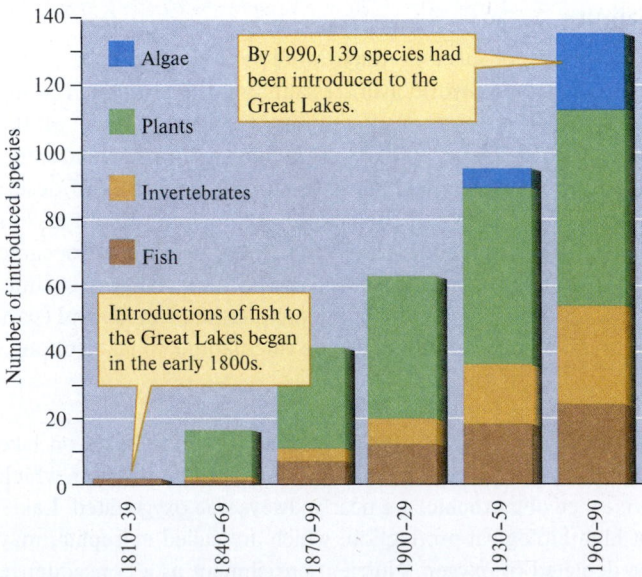

By 1990, 139 species had been introduced to the Great Lakes.

Introductions of fish to the Great Lakes began in the early 1800s.

Figure 3.40   Cumulative number of species introduced to the Great Lakes (data from Mills et al. 1994).

(a)   (b)

Figure 3.41   Two invaders of the Great Lakes: (*a*) sea lampreys, shown here attached to a lake trout; and (*b*) zebra mussels, encrusting a boat rudder. Invading species, such as these, have created ecological disasters in freshwater ecosystems around the globe.   (a) M. Gaden, Great Lakes Fishery Commission; (b) R. Griffiths/U.S. Fish & Wildlife Service

The population growth of many introduced species has been explosive and has had great ecological and economic impacts. One such introduction was that of the zebra mussel, *Dreissena polymorpha,* a bivalve mollusk native to the drainages emptying into the Aral, Caspian, and Black Seas. In 1988, zebra mussels were collected in Lake Saint Clair, which connects Lake Huron and Lake Erie. In just 3 years, zebra mussels spread to all the Great Lakes and to most of the major rivers of eastern North America.

Zebra mussels established very dense populations within the Great Lakes. Shells from dead mussels have accumulated to depths of over 30 cm along some shores. Such dense populations threaten the native mussels of the Great Lakes with extinction. Zebra mussels are also fouling water intake structures of power plants and municipal water supplies, resulting in billions of dollars in economic impact. However, the consequences of invasive species continue to unfold. Zebra mussels have been displaced from some habitats by a close relative, the quagga mussel, *Dreissena bugensis* (Ricciardi and Whoriskey 2004). Together, the two mussel species have multiple effects on Great Lakes ecology, including nutrient cycling. Quagga and zebra mussels make nitrogen and phosphorus more available, thus also contributing to algal blooms (Mohamed et al. 2019). As a consequence of introductions of zebra mussels and other species, the Great Lakes have become a laboratory for the study of human-caused biological invasions (fig. 3.41).

Concept 3.2 Review

1. After years of successful reductions in phytoplankton populations, phytoplankton blooms are on the increase in parts of Lake Erie following the introduction of zebra mussels. Why?
2. Why is the prospect of global warming considered a serious threat to coral reefs?
3. Why do physiologically tolerant rather than sensitive species inhabit estuaries and salt marshes?

Applications

Biological Integrity–Assessing the Health of Aquatic Systems

LEARNING OUTCOMES

After studying this section you should be able to do the following:

3.12   List the characteristics of fish communities included in the Index of Biological Integrity.

3.13   Explain the environmental significance of each of the elements, such as feeding biology of species, included in the calculation of an Index of Biological Integrity.

How can we put our knowledge of the natural history of aquatic life to work? A major question that biologists often face is whether a particular influence impairs the health of an aquatic system. Natural history information can play a significant role in making that judgment. Given the complex array of potential human impacts on aquatic systems, what might we use as indicators of health? An answer to this question has been proposed by James Karr and his colleagues, who suggest that we consider what they call "biological integrity," which they define as "a balanced, integrated, adaptive community of organisms having a species composition, diversity, and functional organization comparable to that of the natural habitat of the region" (Karr and Dudley 1981). These researchers proposed that a healthy aquatic community is one that is similar to the community in an undisturbed habitat in the same region. The community should be "balanced" and "integrated." Deciding what constitutes this state requires judgment based on broad knowledge of the habitats in question and their inhabitants—that is, knowledge of natural history. If we could assess the health, as defined by Karr, of a community of aquatic organisms, we would have gone a long way toward assessing the health of the aquatic ecosystem of which this community is part.

Moving beyond general definitions and broad goals, Karr developed an Index of Biological Integrity (IBI) and applied his index to fish communities which is still widely used today (Nestlerode et al. 2020). Fish communities were chosen because we know a lot about fish and their habitat requirements and they are relatively easy to sample. Karr's index has three categories for rating a stream or river:

1. number of species and species composition, which includes the number, kinds, and tolerances of fish species;
2. trophic composition, which considers the dietary habits of the fish making up the community;
3. fish abundance and condition.

Under these three categories there are 12 attributes of the fish community. The stream is assigned a score of 5, 3, or 1 for each attribute, where 5 equals best and 1 equals worst. The scores on all the attributes are added to give a total score that ranges from 12 (poor biological integrity) to 60 (excellent biological integrity). Notice that Karr has built a safeguard into his index. Judging several attributes of the fish community eliminates the bias that might creep in if assessments were made from only one or a very few attributes. In the following sections, we will examine the three community characteristics.

Number of Species and Species Composition

Heavy human impact generally reduces the number of native species in a community while increasing the number of nonnative species. The kinds of species that make up the community should also be telling, because some fish, such as trout, are intolerant of poor water quality, while others, such as carp, are highly tolerant of poor water quality. The designation of *tolerant* versus *intolerant* species must be tailored for local, or at least regional, circumstance and requires a thorough knowledge of the natural history of the waters under study, as does scoring the number and abundance of species.

Trophic Composition

The dietary habits of the fish that make up a community reflect kinds of food available in a stream as well as the quality of the environment. The attributes rated in this category are the percentage of fish such as carp that eat a wide range of food and are called **omnivores** by ecologists; the percentage of fish such as trout and bluegill that feed on insects, called **insectivores;** and the percentage of fish such as pike and largemouth bass that feed on other fish, called **piscivores.** Degradation of aquatic systems generally increases the proportion of omnivores and decreases the proportion of insectivores and piscivores in the community.

Fish Abundance and Condition

Fish are often less abundant in degraded situations and their condition is often adversely affected. Two aspects of condition are considered for the index. First, what percentage of the individuals are hybrids between different species? Second, what percentage of individuals have noticeable disease, tumors, fin damage, or skeletal deformities—all strong indicators of poor environmental quality? Figure 3.42 summarizes the process of calculating Karr's Index of Biological Integrity.

A Test

Paul Leonard and Donald Orth (1986) tested Karr's Index of Biological Integrity in seven tributary streams of the New River, which flows through the Appalachian Plateau region of West Virginia. Leonard and Orth had to adapt the index to reflect conditions in their region. In their study streams, the number of darter species, small benthic fish in

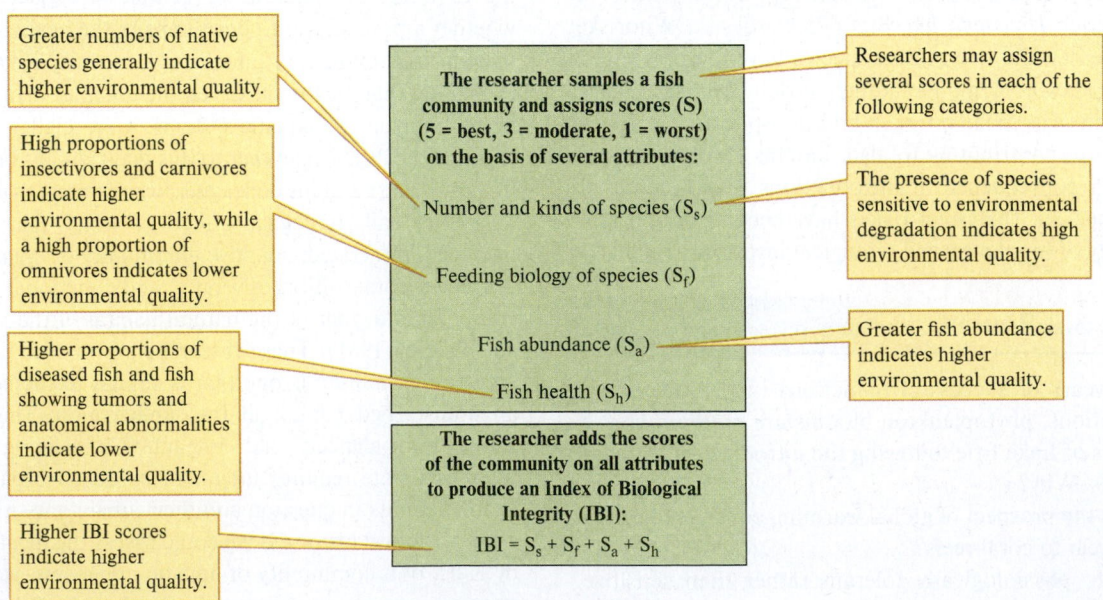

Figure 3.42   Calculating an Index of Biological Integrity.

the family Percidae, indicates high environmental quality, while increasing numbers of creek chubs indicate increasing pollution. In addition, high proportions of insectivores indicate excellent environmental conditions, while high proportions of generalist feeders, or omnivores, indicate poor conditions. High densities of fish were taken as a sign of high environmental quality, while the presence of diseased or deformed individuals indicated environmental problems.

Leonard and Orth assigned scores of 1 (worst conditions), 3 (fair conditions), or 5 (best conditions) for each of the variables they studied at each of their sampling sites in the study streams. They then summed the scores for the seven variables at each site to determine an Index of Biological Integrity. The minimum possible value was 7, poorest conditions, and the maximum possible value was 35, best conditions. They next made independent estimates of levels of pollution at each study site. Their estimates were based upon the daily discharge of municipal sewage and the local densities of septic tanks, roads, and mines. The study streams showed a wide range of environmental pollution due to sewage, mining, and urban development. Leonard and Orth found that the Index of Biological Integrity correlated well with independent estimates of pollution at each study site (fig. 3.43).

Many other investigators have tested the ability of the Index of Biological Integrity to represent the extent of environmental degradation in rivers and lakes. The index is effective in a wide range of regions and aquatic environments. The important point here is that natural history is being put to work to address important environmental problems. The foundation of natural history built in this chapter and in chapter 2 is useful now as we go forward to study ecology at levels of organization ranging from individual species through the entire biosphere.

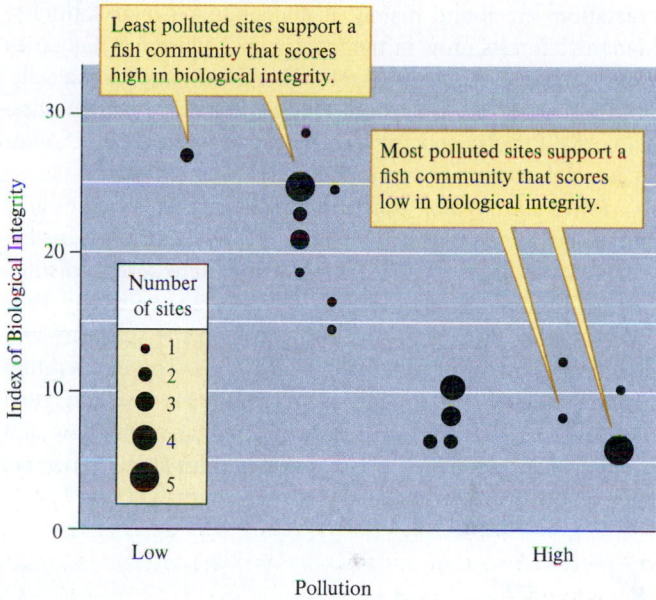

Figure 3.43   Pollution and the Index of Biological Integrity (data from Leonard and Orth 1986).

Summary

Humans everywhere hold a land-centered perspective of the planet. Consequently, there is still much to learn about aquatic ecosystems. Differences between air and water as a primary medium for life mean that selective pressures have created often wildly divergent traits in terrestrial versus aquatic systems.

Water cycling and movement on a global scale is driven by solar energy. Of the water in the biosphere, the oceans contain 97% and the polar ice caps and glaciers an additional 2%, leaving less than 1% as freshwater. The turnover of water in the various reservoirs of the hydrologic cycle ranges from only 9 days for the atmosphere to 3,100 years for the oceans.

The biology of aquatic environments corresponds broadly to variations in physical factors such as light, temperature, and water movements and to chemical factors such as salinity and oxygen. The *oceans* form the largest continuous environment on earth. An ocean is generally divided vertically into several depth zones, each with a distinctive assemblage of marine organisms. Limited light penetration restricts photosynthetic organisms to the photic, or epipelagic, zone and leads to thermal stratification. Oceanic temperatures are much more stable than terrestrial temperatures. Tropical seas are more stable physically and chemically; temperate and high-latitude seas are more productive. Highest productivity occurs along coastlines. The open ocean supports large numbers of species and is important to global carbon and oxygen budgets.

Kelp forests are found mainly at temperate latitudes. *Coral reefs* are limited to the tropics and subtropics to latitudes between 30° N and S. Coral reefs are generally one of three types: fringing reefs, barrier reefs, and atolls. Kelp beds share several structural features with terrestrial forests. Both seaweeds and reef-building corals grow only in surface waters, where there is sufficient light to support photosynthesis. Kelp forests are generally limited to areas where temperature ranges from about 10° to 20°C, while reef-building corals are limited to areas with temperatures of about 18° to 29°C. The diversity and productivity of coral reefs rival that of tropical rain forests.

The *intertidal* zone lines the coastlines of the world. It can be divided into several vertical zones: the supratidal, high intertidal, middle intertidal, and low intertidal. The magnitude and timing of the tides is determined by the interaction of the gravitational effects of the sun and moon with the configuration of coastlines and basins. Tidal fluctuation produces steep gradients of physical and chemical conditions within the intertidal zone. Exposure to waves, bottom type, height in the intertidal zone, and biological interactions determine the distribution of most organisms within this zone.

Salt marshes, mangrove forests, freshwater wetlands, and *estuaries* occur at the transitions between freshwater and marine environments and between marine and terrestrial environments. Salt marshes, which are dominated by herbaceous vegetation, are found mainly at temperate and high latitudes. Mangrove forests grow in the tropics and subtropics. Estuaries are extremely dynamic physically, chemically, and biologically. The diversity of species is not as high in estuaries, salt marshes, mangrove forests, and freshwater wetlands as in some other aquatic environments, but productivity is exceptional.

Rivers and *streams* drain most of the land area of the earth and reflect the land use in their basins. Rivers and streams are very dynamic systems and can be divided into several distinctive environments: longitudinally, laterally, and vertically. Periodic flooding has important influences on the structure and functioning of river and stream ecosystems. The temperature of rivers follows variation in air temperature but does not reach the extremes occurring in terrestrial habitats. The flow and chemical characteristics of rivers change with climatic regime.

Current speed, distance from headwaters, the nature of bottom sediments, and the geologic setting are principal determinants of the distributions of stream organisms.

Lakes are much like small seas. Most are found in regions worked over by tectonics, volcanism, and glacial activity, the geological forces that produce lake basins. A few lakes contain most of the freshwater in the biosphere. Lake structure parallels that of the oceans but on a much smaller scale. The salinity of lakes, which ranges from very dilute waters to over 200%, is much more variable than that of the oceans. Lake stratification and mixing vary with latitude. Lake flora and fauna largely reflect geographic location and nutrient content.

Potential threats to all these aquatic systems include overexploitation of populations and waste dumping. Reservoir construction and flow regulation have had major negative impacts on river ecosystems and biodiversity. Freshwater environments are particularly vulnerable to the introduction of exotic species. The nature of fish assemblages is being used to assess the "biological integrity" of freshwater communities. The application of this Index of Biological Integrity depends on detailed knowledge of the natural history of regional fish faunas.

Key Terms

abyssal zone 50	freshwater 45	limnetic zone 69	piscivore 74
atoll 54	freshwater wetland 59	littoral zone 50	riparian zone 65
barrier reef 54	fringing reef 54	lotic 45	river continuum concept 67
bathypelagic zone 50	gyre 48	mangrove forest 59	river ecosystem
benthic 50	hadal zone 50	mesopelagic zone 50	synthesis 67
biochemical oxygen demand (BOD) 67	hydrologic cycle 47	metalimnion 69	salinity 51
cyanobacteria 45	hypolimnion 69	microplastics 54	salt marsh 59
epilimnion 69	hyporheic zone 65	neritic zone 50	saltwater 45
epipelagic zone 50	insectivore 74	oceanic zone 50	stream order 65
estuary 59	intertidal zone 50	oligotrophic 71	thermocline 47
eutrophic 71	lentic 45	omnivore 74	thermohaline circulation 51
eutrophication 71	light compensation point 46	pelagic 50	upwelling 48
flood pulse concept 66	limiting resource 45	phreatic zone 65	zonation of species 59
		phytoplankton 52	zooplankton 52

Review Questions

1. Review the distribution of water among the major reservoirs of the hydrologic cycle. What are the major sources of freshwater? Explain why according to some projections availability of freshwater may limit human populations and activity.

2. The oceans cover about 360 million km^2 and have an average depth of about 4,000 m. What proportion of this aquatic system receives sufficient light to support photosynthesis? Make the liberal assumption that the photic zone extends to a depth of 200 m.

3. What environmental challenges did organisms have to overcome to move from aquatic to terrestrial systems? What advantages

were gained and lost through this change from water to air as a primary medium?

4. Darwin (1842b) was the first to propose that fringing reefs, barrier reefs, and atolls are different stages in a developmental sequence that begins with a fringing reef and ends with an atoll. Outline how this process might work. How would you test your ideas?

5. How does feeding by urchins, which prey on young corals, improve establishment by young corals? Use a diagram outlining interactions among urchins, corals, and algae to help in the development of your explanation.

6. How might a history of exposure to wide environmental fluctuation affect the physiological tolerances of intertidal species compared to close relatives in subtidal and oceanic environments? How might salinity tolerance vary among organisms living at different levels within the intertidal?

7. How might oxygen concentration of interstitial water be related to the grain size of the sand or mud sediment? How might the oxygen concentrations of tide pools in sheltered bays compare to those on the shores of exposed headlands?

8. According to the river continuum model, the organisms inhabiting headwater streams in temperate forest regions depend mainly upon organic material coming into the stream from the surrounding forests. According to the model, photosynthesis within the stream is important only in the downstream reaches of these stream systems. Explain. How would you go about testing the predictions of the river continuum model?

9. How could you test the generalization that lake primary production and the composition of the biota living in lakes are strongly influenced by the availability of nutrients such as nitrogen and phosphorus? Assume that you have unlimited resources and that you have access to several experimental lakes.

10. Biological interactions may also affect lake systems. How does the recent history of the Great Lakes suggest that the kinds of species that inhabit a lake influence the nature of the lake environment and the composition of the biological community?

Neil Holmes/Getty Images

Chapter

4

Population Genetics and Natural Selection

Flowering plants have played an important role in the discovery of the mechanisms of inheritance. There are several reasons for their importance, including the fact that plants require little more than a garden to maintain them for study and they exhibit a wide range of easily observed, genetically controlled physical variation, such as the range in colors shown by these poppies.

CHAPTER CONCEPTS

4.1 Phenotypic variation among individuals in a population results from the combined effects of genes and environment.   81
Concept 4.1 Review   84

4.2 The Hardy-Weinberg equilibrium model helps identify evolutionary forces that can change gene frequencies in populations.   84
Concept 4.2 Review   86

4.3 Natural selection is differential survival and reproduction among phenotypes.   87
Concept 4.3 Review   88

4.4 A large and rapidly growing body of research on natural populations provides robust support for the theory of evolution by natural selection.   88
Concept 4.4 Review   92

4.5 Random processes, such as genetic drift, can change gene frequencies in populations, especially in small populations.   92
Concept 4.5 Review   96

Applications: Evolution and Agriculture   96
    Summary   98
    Key Terms   99
    Review Questions   **100**

LEARNING OUTCOMES

After studying this section you should be able to do the following:

4.1 Summarize Darwin's theory of natural selection.
4.2 Explain why Mendel discovered the basic laws of genetic inheritance, while Darwin, who worked very hard to do so, did not.

Chapters 2 and 3 provide a foundation for ecology by sketching the natural history of the biosphere. We learned that the diversity of life on our planet arises in part due to variability in the **abiotic** (nonliving) environment over space and time, and that the effect of the environment can also cascade through systems due to **biotic** (living) interactions, such as who eats whom. In order to fully understand

Figure 4.1   The Galápagos Islands are like a natural library of evolution, where Charles Darwin encountered many examples of plant and animal species that differed physically from one island to another. Shown here is the chain of small islands called Las Tintoreras, located south of the much larger Isabela Island.   4FR/Getty Images

these relationships, how they arise and how they can change, we must delve into the realm of the origin of traits and how organisms' traits change over time, that is, evolution. Indeed, the theory of evolution is the centerpiece of the modern life sciences, including ecology, providing a key element in the theoretical framework of the science of ecology.

Darwin's theory of evolution by natural selection, which provided a mechanism for evolutionary change in populations, was crystallized by his observations in the Galápagos Islands. In mid-October of 1835 under a bright equatorial sun, a small boat moved slowly from the shore of a volcanic island to a waiting ship. The boat carried a young naturalist who had just completed a month of exploring the group of islands known as the Galápagos, which lie on the equator approximately 1,000 km west of the South American mainland (fig. 4.1). As the seamen rowed into the oncoming waves, the naturalist, Charles Darwin, mused over what he had found on the island. His observations had confirmed expectations built on information gathered earlier on the other islands he had visited in the archipelago. Darwin recorded his thoughts in his journal, which he later published (Darwin 1842a): "The distribution of the tenants of this archipelago would not be nearly so wonderful, if, for instance, one island had a mocking-thrush, and a second island some other quite distinct genus—if one island had its genus of lizard and a second island another distinct genus, or none whatever. [. . .] But it is the circumstance, that several of the islands possess their own species of the tortoise, mocking-thrush, finches, and numerous plants, these species having the same general habits,

occupying analogous situations, and obviously filling the same place in the natural economy of this archipelago, *that strikes me with wonder*" [emphasis added].

Darwin wondered at the sources of the differences among clearly related populations and attempted to explain the origin of these differences. He would later conclude that these populations were descended from common ancestors whose descendants had changed after reaching each of the islands. The ship to which the seamen rowed was the H.M.S. *Beagle,* halfway through a voyage around the world. The main objective of the *Beagle*'s mission, charting the coasts of southern South America, would be largely forgotten, while the thoughts of the young Charles Darwin would eventually develop into one of the most significant theories in the history of science. Darwin's wondering, carefully organized and supported by a lifetime of observation, would become the theory of evolution by natural selection, a theory that would transform the prevailing scientific view of life on earth and rebuild the foundations of biology.

Darwin left the Galápagos Islands convinced that the various populations on the islands were gradually modified from their ancestral forms. In other words, Darwin concluded that the island populations had undergone a process of **evolution,** that is, a gradual change over time. He knew that such change was possible because of those that had occurred in domesticated plants and animals as the result of selective breeding. Although Darwin was convinced the island populations had evolved, he had no mechanism to explain evolutionary changes in wild species. However, a plausible mechanism to produce evolutionary change in populations came to Darwin almost exactly 3 years after his taking leave of the Galápagos Islands. In October of 1838 while reading the essay on populations by Thomas Malthus, Darwin was convinced that during competition for limited resources, such as food or space, among individuals within populations, some individuals would have a competitive advantage. He proposed that the characteristics producing that advantage would be "preserved" and the unfavorable characteristics of other individuals would be "destroyed." As a consequence of this process of selection by the environment, populations would change over time. With this mechanism for change in hand, Darwin sketched out the first draft of his theory of natural selection in 1842. However, it would take him many years and many drafts before he honed the theory to its final form and amassed sufficient supporting information. Darwin's theory of **natural selection** can be summarized as follows:

1. Organisms beget like organisms. (Offspring appear, behave, function, and so forth like their parents.)
2. There are chance variations between individuals in a species. Some variations (differences among parents) are heritable (are passed on to offspring).
3. More offspring are produced each generation than can be supported by the environment.
4. Some individuals, because of their physical or behavioral traits, have a higher chance of surviving and reproducing than other individuals in the same population.

Darwin (1859) proposed that differential survival and reproduction of individuals would produce changes in species populations over time. That is, the environment acting on variation among individuals in populations would result in **adaptation,** an evolutionary process that changes anatomy, physiology, or behavior, resulting in an improved ability of the members of a population to live in a particular environment. He now had a mechanism to explain the differences among populations that he had observed on the Galápagos Islands. Still, Darwin was keenly aware of a major insufficiency in his theory. The theory of natural selection depended on the passage of "advantageous" characteristics from one generation to the next. The problem was that the mechanisms of inheritance were unknown in Darwin's time. In addition, the prevailing idea at the time, blending inheritance, suggested that rare traits, no matter how favorable, would be blended out of a population, preventing change as a consequence.

Darwin worked for nearly half a century to uncover the laws of inheritance. However, he did not. To do so required a facility with mathematics that Darwin had not developed. In a short autobiography, Darwin (Darwin and Darwin 1896, V.1, p. 40) remarked, "I attempted mathematics, and even went during the summer of 1828 with a private tutor . . . but I got on very slowly. The work was repugnant to me, chiefly from my not being able to see any meaning in the early steps in algebra. This impatience was very foolish, and in after years I have deeply regretted that I did not proceed far enough at least to understand something of the great leading principles of mathematics, *for men thus endowed seem to have an extra sense*" [emphasis added].

As Darwin explored the Galápagos Islands, halfway around the world in central Europe a schoolboy named Johann Mendel was developing the facility with mathematics necessary to complete Darwin's theory of natural selection. At 13, Johann was half Darwin's age, yet he had already set a course for a life of study that he followed as resolutely as did the crew of the *Beagle* on their voyage around the world. At the end of his scientific voyage, Mendel, who would be renamed Gregor Mendel when he became an Augustinian monk, would uncover the basic mechanisms of inheritance.

How did Mendel succeed, while so many others had failed? The sources of his success can be traced to his education and his own special genius. Mendel's education at the University of Vienna exposed him to some of the best minds working in the physical sciences and to an approach to science that emphasized experimentation. His introduction to the physical sciences included a solid foundation in mathematics, including probability and statistics. As a consequence, Mendel could quantify the results of his experimental research.

Mendel chose to work with plants that could be maintained in the abbey garden. His most famous and influential work was done on the garden pea, *Pisum sativum*, which has many varieties with distinct flower and seed colors, placement of flowers on the stalk, and other characteristics (fig. 4.2). Mendel identified seven traits that were passed from one generation to the next independently from each other and as morphologically distinct categories, for example, pure white versus deeply colored flowers. Although he did not understand it at the time, they were inherited independently because the characteristics

Figure 4.2   Garden pea flowers. Because its flowers are closed, the garden pea normally self-pollinates. Consequently, Mendel could keep track of and control mating in his study plants.   Shape'n'colour/Alamy

chosen by Mendel are controlled by different genes on different chromosomes. Through a great deal of hard work and perseverance breeding peas and later mice, bees, and other plants, Mendel discovered that certain traits are "dominant" whereas others are "recessive." Although humans had known for centuries that traits could be passed and mixed through breeding, it was Mendel who developed rules to predict how this occurred, by understanding that traits are passed from parent to offspring in units, called **genes,** and that these genes can have different forms, called **alleles.** It is the variety of alleles that creates variation in a population (fig 4.3).

Darwin and Mendel complemented each other perfectly, and their twin visions of the natural world revolutionized biology. The synthesis of the theory of natural selection and genetics provided a unifying conceptual foundation for modern biology and gave rise to evolutionary ecology, a very broad field of study.

Evolutionary ecology is the study of how interactions among organisms and between organisms and their environment evolve. As an example, figure 4.3 shows how differences in genes can have important ecological consequences. Krushnamegh Kunte, with colleague Marcus Kronforst and others, discovered that *Papilio polytes*, an Asian swallowtail butterfly, has a **locus**—a position on a specific chromosome—for a gene that codes for differences in wing pattern (Kunte et al. 2014). Those individuals with the dominant allele "M" at that locus have "mimetic" wings—that is, they appear similar to another, poisonous species, an evolutionary phenomenon discussed further in chapter 7. For this reason, birds avoid them. Butterflies without this allele ("mm") are much more likely to be eaten by birds and other predators that have learned to avoid the mimetic wing. In this way, natural selection acting upon the genetic diversity in the butterfly population can lead to the numerical dominance of a trait, such as a mimetic wing, in a population over time. In this chapter, we examine five major evolutionary concepts necessary for fully understanding ecology as a scientific discipline.

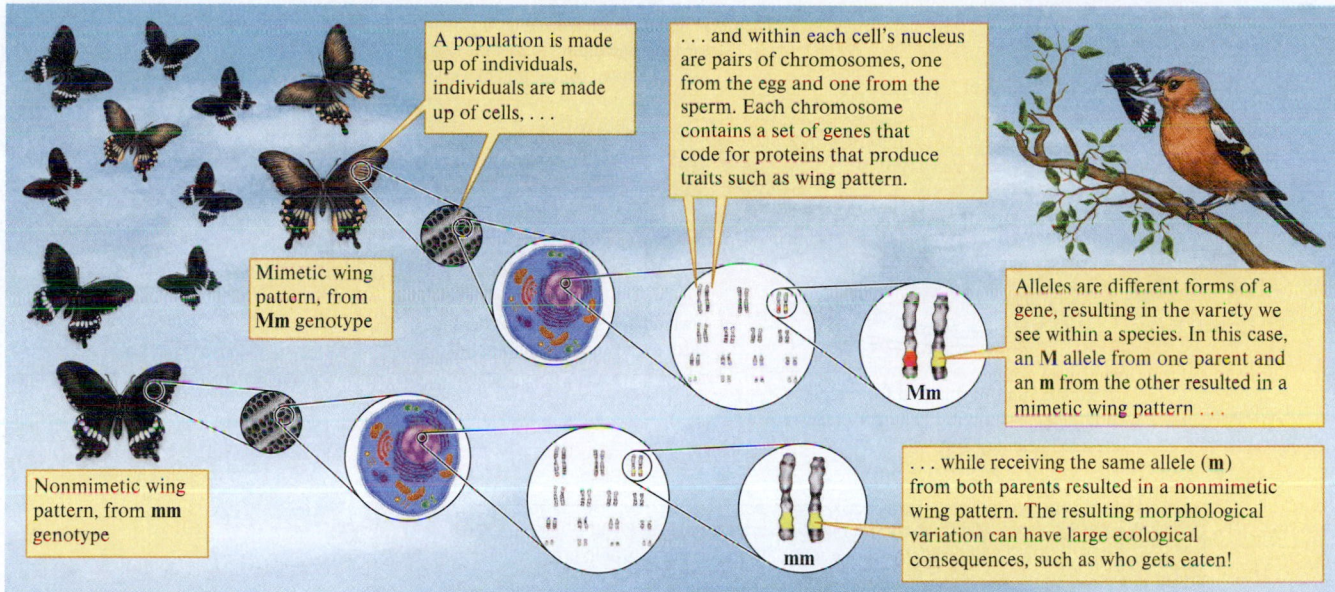

Figure 4.3 Ecologically important differences between individuals in a population can arise as the result of different genotypes. For *Papilio polytes,* an Asian swallowtail butterfly species, Marcus Kronforst and colleagues discovered that variation in wing pattern among females was due to different alleles of a single gene. When one chromosome carries the different allele, individuals appear mimetic; that is, they exhibit a similarity to another, poisonous, species of butterfly. Mimetic butterflies are much less likely to be eaten by birds and other predators than nonmimetic ones (based on Kunte et al. 2014).

4.1 Variation Within Populations

LEARNING OUTCOMES

After studying this section you should be able to do the following:

4.3 Describe phenotypic plasticity.

4.4 Explain the design of the common garden experiment used to test for genetic differences among populations of *Potentilla glandulosa.*

4.5 Contrast the methods used to study genetic variation in populations of *P. glandulosa* populations versus *Coregonus* in the Alps.

Phenotypic variation among individuals in a population results from the combined effects of genes and environment. The **phenotype** of an organism consists of its observable characteristics, which result from interactions between the genetic makeup of the individual and its environment. Because phenotypic variation among individuals is the substrate on which the environment acts during the process of natural selection, determining the extent and sources of variation within populations is one of the most fundamental considerations in evolutionary studies. Darwin's theory of natural selection sparked a revolution in thinking among biologists, who responded almost immediately by studying variation among organisms in all sorts of environments. The first of these biologists to conduct truly thorough studies of variation and to incorporate experimentation in their studies focused on plants.

Variation in a Widely Distributed Plant

Jens Clausen, David Keck, and William Hiesey, who worked at Stanford University in California, conducted some of the most widely cited studies of plant variation. Their studies provided deep insights into the extent and sources of morphological variation in plant populations, including the influence of both environment and genetics. Though this research group and its successors studied nearly 200 species, it is best known for its work on *Potentilla glandulosa,* or sticky cinquefoil (Clausen, Keck, and Hiesey 1940).

Clausen and his research team performed what is known as a **common garden experiment** in which individuals from two or more populations are transplanted and grown in the same, or "common," environment. The team worked with clones of several populations of *P. glandulosa,* which they grew in three main experimental gardens—one at Stanford near the coast at an elevation of 30 m, another in a montane environment at Mather at an elevation of 1,400 m in the Sierra Nevada, and a third garden in an alpine environment at Timberline at 3,050 m (fig. 4.4). By cloning lowland, mid-elevation, and alpine plants and growing them in experimental gardens, Clausen, Keck, and Hiesey established experimental conditions that could reveal potential genetic differences among populations. In addition, because they studied the responses of plants from all populations to environmental conditions in lowland, mid-elevation, and alpine gardens, their experiment could demonstrate adaptation by *P. glandulosa* populations to local environmental conditions.

The growth response of *P. glandulosa* to environmental conditions at the three common garden sites is summarized in figure 4.4. Plant height differed significantly among the study sites, which shows an environmental effect on plant morphology. However, the lowland, mid-elevation, and alpine plants responded differently to the three environments; while the mid-elevation and alpine plants attained their greatest height in the mid-elevation garden, the lowland plants grew the tallest in the lowland garden.

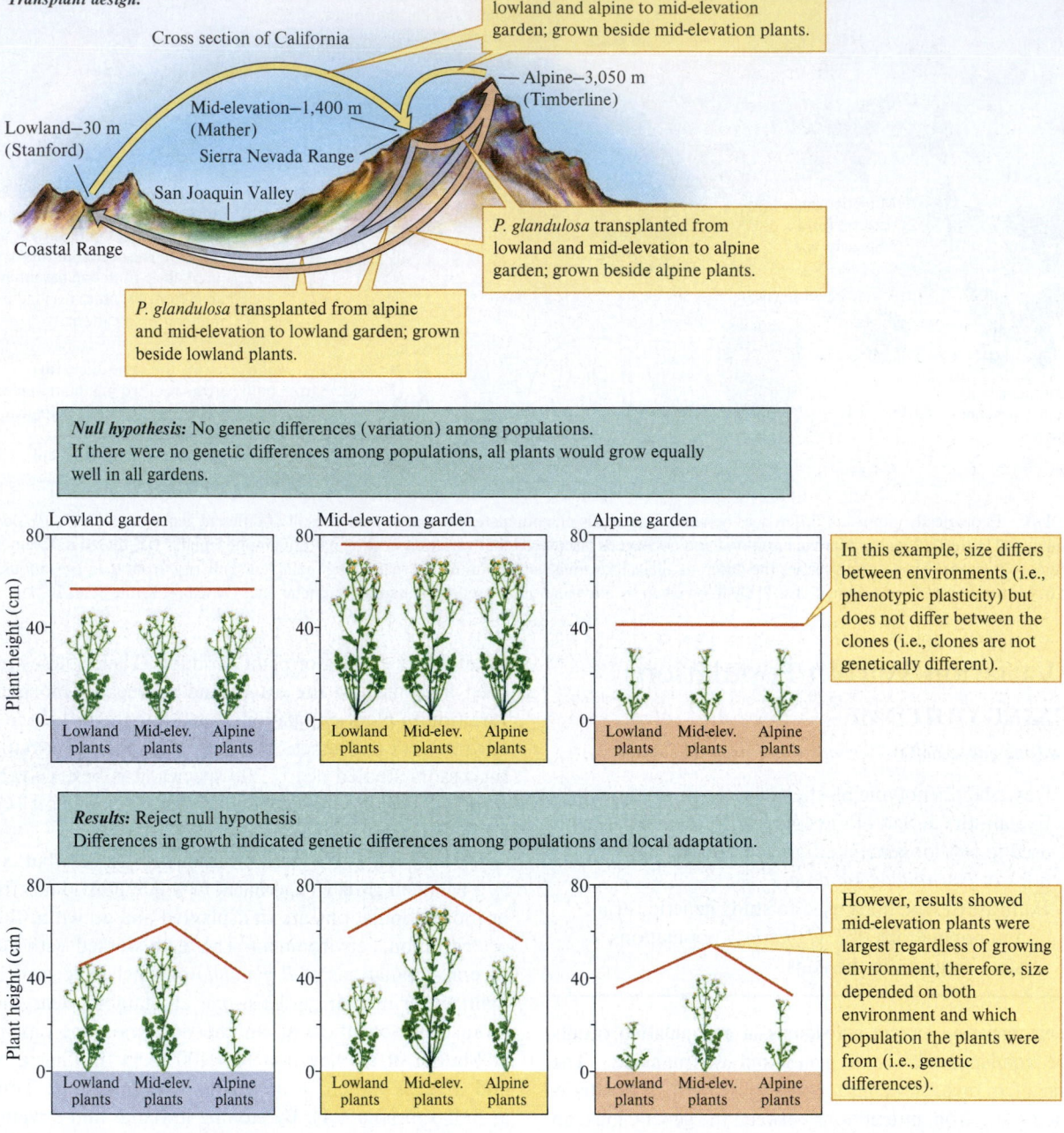

Figure 4.4   A common garden approach to studying genetic variation and phenotypic plasticity among populations of *Potentilla glandulosa* (data from Clausen, Keck, and Hiesey 1940).

The number of flowers produced by lowland, mid-elevation, and alpine plants also varied across the gardens.

Differences in response *among* and *within* clones of lowland, mid-elevation, and alpine *P. glandulosa* provide complementary information. Differences *among* clones, in growth and flower production, at the three common garden sites indicate genetic differences among lowland, mid-elevation, and alpine populations of *P. glandulosa*. Meanwhile, differences in growth and flower production *within* clones grown at the three elevations are the result of environmental differences among the common garden sites, not genetic differences. When organisms change the expression of traits in response to the environment, they are said to have **phenotypic plasticity**.

Other observations by Clausen, Keck, and Hiesey indicate that the genetic variation among the plant populations was associated with adaptation to local environments. For instance, most lowland plants died during their first winter in the alpine garden and those that survived did not produce seed. Alpine plants showed the opposite trends. They had poor survival in the lowland garden and went dormant in winter, while the lowland plants remained active. In summary, the experiments of Clausen, Keck, and Hiesey demonstrated both genetic differences among populations *and* adaptation to their natural environments. Ecologists call such locally adapted and genetically distinctive populations within a species **ecotypes**. Applying this term then, we can conclude that the lowland, mid-elevation, and alpine populations studied by Clausen, Keck, and Hiesey were ecotypes.

Using transplant and common garden approaches ecologists have learned a great deal about genetic variation among and within plant populations. These classical approaches combined with modern molecular techniques are rapidly increasing our knowledge of genetic variation in natural populations.

Variation in Alpine Fish Populations

The Alps rise out of the landscape of south-central Europe, forming a moist and cool high-elevation environment. The Alps' deep winter snows and glaciers make them the origin of four important rivers: the Danube, Rhine, Po, and Rhone Rivers. Because the headwater streams of these rivers are cool, they became refuges for cold-water aquatic organisms following the last ice age. As temperatures of the surrounding lowlands began to warm at the end of the Pleistocene, approximately 12,000 years ago, aquatic species requiring cold water migrated to the headwaters of these rivers. The movement of cold-adapted aquatic species into the headwater streams and lakes of the glacial valleys that lace the Alps created clusters of geographically isolated populations (fig. 4.5). This isolation reduced movements of individuals between populations. With reduced gene flow, populations could diverge genetically. Such genetic divergence would increase the genetic variation among populations.

Morphological differences among populations of headwater fish species in the Alps have long suggested genetic differences among them. Nowhere has morphological variation among populations been better studied and documented than among the whitefishes. Whitefish are relatives of the trout and salmon and are classified in the genus *Coregonus* (fig. 4.6). Marlis Douglas and Patrick Brunner (2002) explored the genetic and phenotypic variation among populations of *Coregonus* in the Central Alps. Douglas and Brunner pointed out that ichthyologists have described 19 indigenous *Coregonus* populations from the Central Alps. However, there has been significant disagreement over the taxonomic status of these 19 populations. The classification of these populations ranges from that of a single variable species with 19 distinctive populations to dividing the 19 populations into more than a dozen separate species.

The taxonomic status of *Coregonus* populations in the Central Alps is made more difficult by a 100-year history of intensive fisheries management. Douglas and Brunner reviewed this history, which included raising *Coregonus* in hatcheries and moving fish between lakes. One of the main purposes of the study by Douglas and Brunner was to describe the genetic variation among the present-day populations of *Coregonus* to determine if there is evidence for significant genetic differences among historically recognized populations. A second purpose was to examine the genetic similarity between introduced *Coregonus* populations and the populations from which they were drawn. Using this information, Douglas and Brunner intended to offer suggestions for the management and conservation of *Coregonus* in the Central Alps.

Figure 4.5   Lake Lucerne, Switzerland, lies nestled in the heart of the Alps, where it provides an extensive cold-water habitat for aquatic organisms, including whitefish, *Coregonus*, populations.   Glowimages/Getty Images

Figure 4.6   Whitefish, *Coregonus* sp., are adapted to cold, highly oxygenated waters like their relatives the trout and salmon. Because they are valued food fishes, whitefish have been intensively managed particularly in the Central Alps.   Tom McHugh/Science Source

Douglas and Brunner collected 907 *Coregonus* specimens from 33 populations in 17 lakes in the Central Alpine region and used a mixture of anatomical and genetic features to characterize the fish. The anatomical features were the number of rays in the dorsal, anal, pelvic, and pectoral fins, the extent of pigmentation in these fins, and the number of gill rakers on the first gill arch. The study populations were characterized genetically using **microsatellite DNA,** tandemly repetitive nuclear DNA in which a few base pairs, for example, three base pairs, are repeated up to 100 times.

Genetic analyses by Douglas and Brunner demonstrated a moderate to high level of genetic variation within all 33 study populations. They also found that genetic and morphological analyses distinguished the 19 historically recognized *Coregonus* populations of the Central Alps. Genotypic differences among populations were sufficient to correctly assign individual fish to the indigenous population from which they were sampled with approximately a 71% probability. Fin ray counts correctly assigned fish to the 19 indigenous populations with a 69% probability, while pigmentation could identify them with a 43% probability. Combining genetic and phenotypic data increased the correct assignment of specimens to the populations from which they were drawn to 79%. Genetic analyses of the introduced *Coregonus* populations revealed their genetic similarity to the populations from which they were stocked. However, these analyses also showed that the introduced populations have differentiated genetically from their source populations.

The conclusion that Douglas and Brunner drew from these results was that the *Coregonus* of the Central Alps is made up of a highly diverse set of populations that show a high level of genetic differentiation. They suggest that these populations should be considered as "evolutionarily significant units." They further conclude that the distinctiveness of local *Coregonus* populations is sufficient so that they should be managed as separate units. Douglas and Brunner recommend that *Coregonus* should not be moved from one lake basin to another.

Studies of plants and animals have repeatedly demonstrated genetic variation in populations. Such genetic variation is required for evolutionary change. However, in order to better understand how populations can evolve, we need to first understand some aspects of the genetics of populations, or **population genetics.** The theoretical foundations of population genetics were established early in the twentieth century by two investigators named Hardy and Weinberg.

Concept 4.1 Review

1. Can we be confident that differences in growth within *P. glandulosa* clones grown at different elevations were not the result of genetic differences? Why?
2. What would you expect to see in figure 4.4 if alpine, mid-elevation, and lowland populations of *P. glandulosa* were not different genetically?
3. What is a fundamental evolutionary implication of the large amounts of genetic variation commonly documented in natural populations?

4.2 Hardy–Weinberg Principle

LEARNING OUTCOMES

After studying this section you should be able to do the following:

4.6 Outline the Hardy–Weinberg principle and genetic drift.
4.7 Distinguish between allele and genotype frequencies.
4.8 Discuss the conditions required to achieve Hardy–Weinberg equilibrium in a population.

The Hardy–Weinberg equilibrium model helps identify evolutionary forces that can change gene frequencies in populations. We defined evolution as a change in a population over time. Since evolution ultimately involves changes in the frequency of heritable traits in a population, we can define evolution more precisely as a change in gene frequencies in a

population. Therefore, a thorough understanding of evolution must include some knowledge of population genetics. Mendel discovered one could mathematically predict the frequencies of genotypes and phenotypes in perfectly controlled matings. These results could be extrapolated to predict changes in genotype frequency over time, that is, evolution of the population (Mendel 1866). Though Mendel is not generally credited with studying the genetics of populations, his analysis anticipated the field of population genetics, the foundations of which would be laid 42 years later.

Calculating Gene Frequencies

Consider a population of Asian lady beetles of the species *Harmonia axyridis*. *Harmonia* populations generally include a great deal of variation in color pattern on the wing covers, or elytra, and over 200 color variants are known. Many color forms are so distinctive that early taxonomists described them as different species or even different genera. Geneticists in the first half of the twentieth century, especially Chia-Chen Tan and Ju-Chi Li (1934, 1946) and Theodosius Dobzhansky (1937), determined that the variation in color patterns shown by *Harmonia* is due to the effects of more than a dozen alternative alleles for color pattern. The phenotypic expressions of two of those alleles are shown in figure 4.7. The homozygous "19-signata" genotype of *Harmonia*, which we can represent as *SS*, has orange elytra with several black spots, while the homozygous "aulica" genotype, represented here as *AA*, has elytra with prominent black borders and a large oval area of orange. Tan and Li, who did extensive breeding experiments using *Harmonia* that they collected in southwestern China, found that crosses between 19-signata and

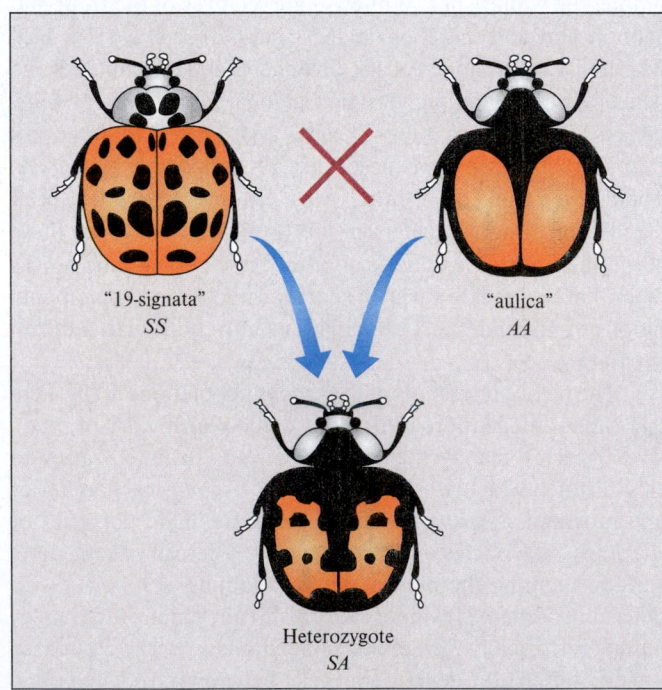

Figure 4.7 Inheritance of color patterns in the Asian lady beetle, *Harmonia axyridis.* The genetic basis of color variation in *H. axyridis* is well studied, making it a useful species for studies of population genetics and natural selection (after Dobzhansky 1937 and Tan 1946).

aulica genotypes produce heterozygous offspring, indicated here as *SA*, with a color pattern that includes elements of both the 19-signata and the aulica parental forms. One of the convenient features of knowing so much about color pattern inheritance in *Harmonia* is that color pattern can be used to determine the genotypes of many individuals. Thus, if we observed a small population of *Harmonia* in a forest in Asia that contained these three phenotypes, we would know that there are also three genotypes for wing pattern in that population: SS, SA, and AA, that is, the three possible pairwise combinations of the two alleles S and A. We can easily count the frequencies of these three genotypes in a population of 20 individuals, as shown in figure 4.8. But how many alleles are there in the population of 20 individuals? To start, because this is a species with one set of

chromosomes from each of two parents, we know there will be a total of 40 alleles, that is:

20 individuals/population × 2 alleles/individual
= 40 alleles/population

with the two alleles being either SS, SA, or AA. It is then straightforward to determine how many S alleles and how many A alleles there are in a population. For example, a group of only two individuals that are both SS will have a total of four S alleles (SS + SS). If we add a single individual that is SA, there will be five S alleles and one A allele (SS + SS + SA). In figure 4.8, you will see how we can extrapolate this approach to counting alleles for our population of 20 individuals. The number of alleles can then be converted to an allele frequency, that is, the proportion of the

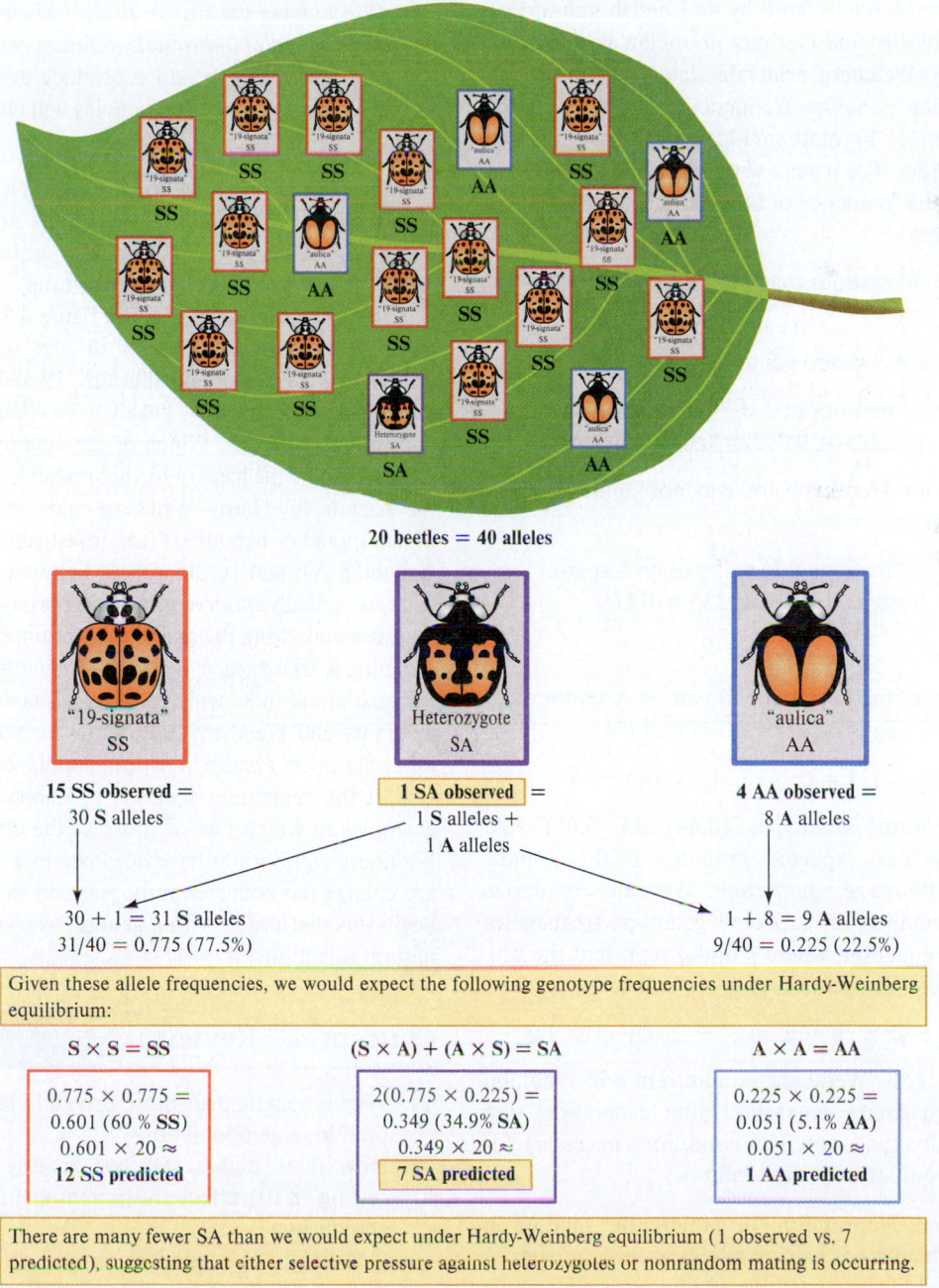

Figure 4.8   Anatomy of a Hardy–Weinberg equilibrium equation and how it is used to detect evolutionary pressures in observed populations.

alleles present in a population by dividing by the total number of alleles. For example, 31 S alleles out of 40 total alleles is:

$$31 \div 40 = 0.775$$

To convert a proportion to a percentage, multiply by 100:

$$0.775 \times 100 = 77.5\%$$

Thus, 0.775 of the total alleles are S, while the remainder, 0.225, are A alleles. These are the **allele frequencies.** The sum of all allele frequencies must equal 1 (0.775 + 0.225 = 1). This is powerful information because with it, we can determine if this population is likely experiencing forces that would cause it to evolve by using simple math.

The mathematical tool that would become one of the foundations of population genetics today was discovered shortly after Gregor Mendel's work by an English mathematician, G. H. Hardy (1908) and German physician W. Weinberg (1908). The **Hardy–Weinberg principle** states that if no evolution is taking place, genotype frequencies can be predicted from allele frequencies by mathematical formulations based on simple probabilities. For a gene with two alleles S and A, if mating is random, the frequency of SS should be the frequency of S times itself. That is:

Proportion of matings that will pair an S sperm
with an S egg is 0.775 × 0.775 = 0.601 (60.1% SS)

And similarly for the AA genotype:

Proportion of matings that will pair an A sperm
with an A egg is 0.225 × 0.225 = 0.051 (5.1% AA)

For the heterozygote SA, there are two possibilities, which must be added together:

Proportion of matings that will pair an S sperm
with an A egg is 0.775 × 0.225 = 0.174

and

Proportion of matings that will pair an A sperm
with an S egg is 0.225 × 0.775 = 0.174

0.174 + 0.174 = 0.349 (34.9% SA)

This ratio of the three genotypes (0.601 SS, 0.051 AA, and 0.349 SA) is our expected outcome if the population is at *Hardy–Weinberg equilibrium.* We can summarize Hardy–Weinberg equilibrium expected genotype frequencies with the following equation, where p and q represent the proportions of two alleles at a locus and $p + q = 1.0$:

$$(p + q)^2 = (p + q) \times (p + q) = p^2 + 2pq + q^2 = 1.0$$

A population at Hardy–Weinberg equilibrium will maintain constant allele frequencies generation after generation; that is, it won't evolve for that trait. The conditions necessary for Hardy–Weinberg equilibrium are as follows:

1. *Random mating.* Nonrandom or preferential mating, in which the probability of pairing alleles is either greater or lower than would be expected based on their frequency in the population, can change the frequency of genotypes.

2. *No mutations.* Mutations that add new alleles to the population or change an allele from one form to another have the potential to change allele frequencies in a population and therefore disrupt Hardy–Weinberg equilibrium.

3. *Large population size.* Small population size increases the probability that allele frequencies will change from one generation to the next due to chance alone. Change in allele frequencies due to chance or random events is called **genetic drift.** Genetic drift reduces genetic variation in populations over time by increasing the frequency of some alleles and reducing the frequency or eliminating other alleles.

4. *No immigration.* Immigration can introduce new alleles into a population or, because allele frequencies are different among immigrants, alter the frequency of existing alleles. In either case, immigration will disrupt Hardy–Weinberg equilibrium.

5. *All genotypes have equal fitness, where fitness is the genetic contribution of individuals to future generations.* If different genotypes survive and reproduce at different rates, then gene and genotype frequencies will change in populations.

How likely is it that all the conditions required for Hardy–Weinberg equilibrium will be present in a natural population? In places and at times the conditions appear to be present. However, it is very likely that one or more will not be met and allele frequencies will change over time.

Returning to our example in figure 4.8, we can see that our observed genotype frequencies do not match those expected under Hardy–Weinberg equilibrium. Therefore, we can conclude that the gene for this trait is likely to be changing over time in this population of beetles. Which of the assumptions above is being violated can be the basis of further research. Thus, we can see that the value of the Hardy–Weinberg equation is that it is as a type of null model or hypothesis (see Investigating the Evidence 10 in Appendix A); that is, differences between this expectation and what we actually observe give us important information. In this case, deviation from the null of equilibrium provides evidence that evolution is taking place. By carefully identifying the highly restrictive conditions under which evolution is not expected, the analysis by Hardy and Weinberg leads us to conclude that the potential for evolutionary change in natural populations is often very great.

In the remaining sections of chapter 4, we will discuss examples in which one or more of the conditions for Hardy–Weinberg equilibrium have not been met and where evolutionary change has occurred in populations as a consequence. We begin this discussion with a general overview of the process of natural selection.

Concept 4.2 Review

1. Why is genetic drift more probable in small populations than in large populations?
2. How does highly selective mating by females (e.g., see fig. 8.10) affect the potential for Hardy–Weinberg equilibrium?
3. How might immigration oppose the effects of genetic drift on genetic diversity in a small population?

4.3 The Process of Natural Selection

LEARNING OUTCOMES

After studying this section you should be able to do the following:

4.9   Discuss the concept of fitness.

4.10 Outline the processes of stabilizing, directional, and disruptive selection.

4.11 Explain how ongoing stabilizing selection can work against change in a population.

Natural selection is differential survival and reproduction among phenotypes.   In each of the examples we have seen so far, traits are controlled by a single gene, with only two possible alleles of that gene. In such cases, there are distinct categories of phenotype, as in the case of Mendel's peas. However, traits that vary along gradients, such as size or metabolic rate, may have several different alleles per gene and typically require interactions between several genes. Characteristics that are controlled by multiple genes are called **polygenic traits,** and predictions about them cannot usually be made using the simple population genetics models we have employed for wing patterns in butterflies and ladybird beetles. The mathematical treatment of continuously varying traits and how they respond to natural selection is known as **quantitative genetics.** Breeders of animals and plants have a long history of using quantitative genetics principles to develop traits of interest to humans, such as milk production in cows and sugar content in fruit. The evolution of traits we observe in nature is often similarly produced by selective pressures in the environment.

Darwin was one of the first to point out that the phenotypic characteristics of some individuals, for instance, larger or smaller size or higher or lower metabolic rate, would result in higher rates of reproduction and survival compared to other individuals with other phenotypic characteristics. In other words, some individuals in a population, because of their phenotypic characteristics, produce more offspring, which then will also eventually reproduce.

While the basic concept of natural selection is easy enough to grasp, natural selection does not take the same form everywhere and at all times. Rather, natural selection can act against different segments of the population under different circumstances and can produce quite different results. Natural selection can lead to change in populations but it can also serve as a conservative force, impeding change in a population. Natural selection can increase diversity within a population or decrease diversity. Let's begin our discussion of natural selection with a process that conserves population characteristics.

Stabilizing Selection

One of the conclusions that we might draw from the discussion of the Hardy–Weinberg equilibrium model is that most populations have a high potential for evolutionary change. However, our observations of the natural world suggest that species can remain little changed generation after generation. If the potential for evolutionary change is high in populations, why does it not always lead to obvious evolutionary change? There are many reasons for apparent absence of change in populations. For example, one form of natural selection, called **stabilizing selection,** can act to impede changes in populations.

Stabilizing selection acts against extreme phenotypes and as a consequence favors the average phenotype. One common characteristic of polygenic traits is that their variation often follows the familiar bell-shaped or "normal" curve. Figure 4.9*a* pictures stabilizing selection, using a normal distribution (see Investigating the Evidence 19 in Appendix A) of body size. Under the influence of stabilizing selection, individuals of average size have higher survival and reproductive rates, while the largest and smallest individuals in the population have lower rates of survival and reproduction. In other words, under conditions of stabilizing selection, average individuals have higher Darwinian or evolutionary **fitness** compared to individuals with extreme phenotypes. Fitness can be defined as the number of offspring, or genes, contributed by an individual to future generations. As a consequence of stabilizing selection, a population tends to sustain the same phenotype over time. Stabilizing selection occurs where average individuals in a population are best adapted to a given set of environmental conditions. If a population is well adapted to a given set of environmental circumstances, stabilizing selection may maintain the match between prevailing environmental conditions and the average phenotype within a population. However, stabilizing selection for a particular trait can be challenged by environmental change. In the face of environmental change, the dominant form of selection may be directional.

Directional Selection

If we examine the fossil record or trace the history of well-studied populations over time, we can find many examples of how populations have changed over time. For instance, there have been remarkable changes in body size or body proportions in many evolutionary lineages. Such changes may be the result of **directional selection.**

Directional selection favors an extreme phenotype over other phenotypes in the population. Figure 4.9*b* presents an example of directional selection, again, using a normal distribution of body size. In this hypothetical situation, larger individuals in the population realize higher rates of survival and reproduction, while average and small individuals have lower rates of survival and reproduction. As a consequence of these differences in survival and reproduction, the average phenotype changes over time. In the example shown in figure 4.9*b*, average body size increases with time. Directional selection occurs where one extreme phenotype has an advantage over all other phenotypes.

Disruptive Selection

Unlike in the case of directional selection, there are situations in which there are more than one extreme phenotype that have

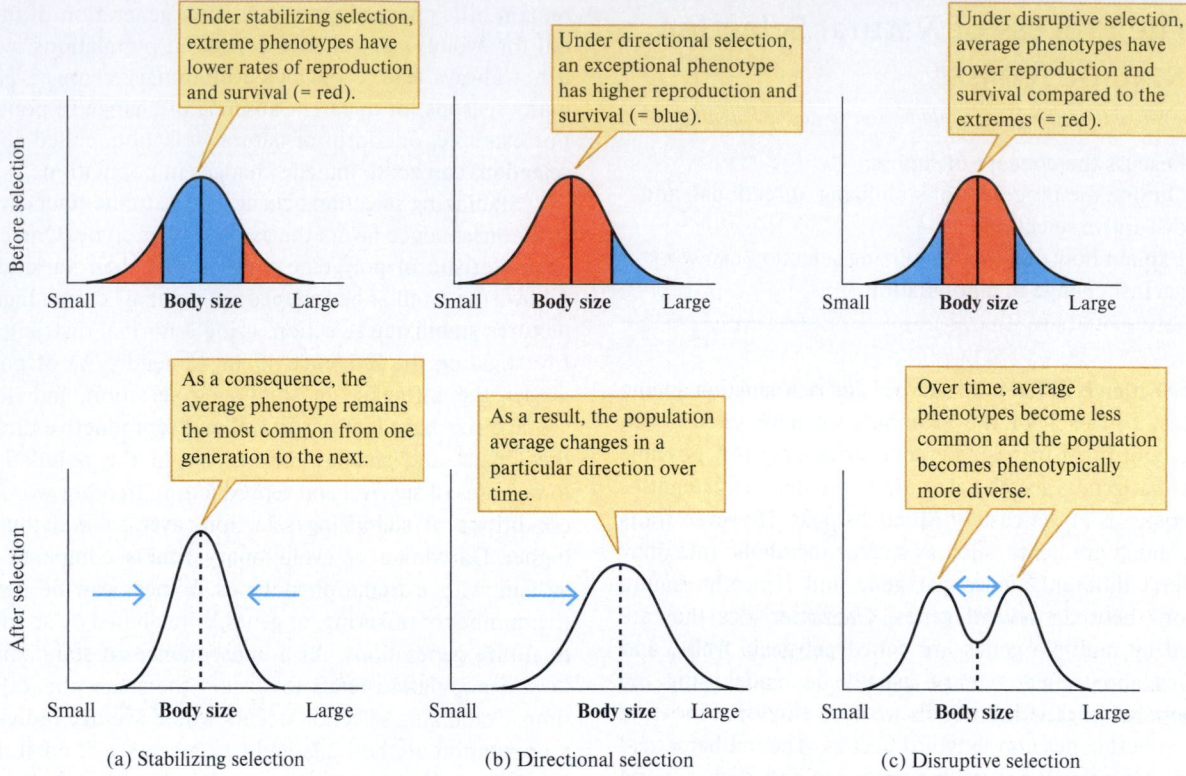

Figure 4.9   Three principal forms of natural selection: (*a*) stabilizing selection, (*b*) directional selection, and (*c*) disruptive selection.

an advantage over the average phenotype. Such a circumstance can lead to **disruptive selection.** This type of selection will lead to a trait distribution that is not normal. In a normal distribution such as those depicted in figures 4.9*a* and 4.9*b,* there is a single peak, which coincides with the population mean. That is, the average phenotype in the population is the most common and all other phenotypes are less common. However, in some populations there may be two or more common phenotypes. In many animal species, for example, males may be of two or more discrete sizes. For example, it appears that in some animal populations small and large males have higher reproductive success than males of intermediate body size. In such populations, natural selection seems to have produced a diversity of male sizes.

Disruptive selection favors two or more extreme phenotypes over the average phenotype in a population. In figure 4.9*c,* individuals of average body size have lower rates of survival and reproduction than individuals of either larger or smaller body. As a consequence, both smaller and larger individuals increase in frequency in the population over time. The result is a distribution of body sizes among males in the population with two peaks. That is, the population has many large males and many small males but few of intermediate body size.

Figures 4.9*b* and 4.9*c* indicate change in the frequencies of phenotypes in the two hypothetical populations after a period of natural selection. This change depends on the extent to which genes determine the phenotype upon which natural

selection acts. This dependence is the focus of the discussion of Concept 4.4.

Concept 4.3 Review

1. How do we know that there are traits that aren't controlled by a single gene, as flower color had been in Mendel's peas?
2. Why is rapid, human-induced environmental change a threat to natural populations?

4.4 Evolution by Natural Selection

LEARNING OUTCOMES

After studying this section you should be able to do the following:

4.12   Define heritability.
4.13   Discuss examples of stabilizing, directional, and disruptive selection that have been documented in natural populations.
4.14   Explain why a trait that is not heritable cannot evolve.

A large and rapidly growing body of research on natural populations provides robust support for the theory of evolution by natural selection.   The most general postulate of the theory of natural selection is that the environment determines the evolution of the anatomy, physiology, and behavior of organisms.

This is what Darwin surmised as he studied variation among populations and species in different environments. Darwin was keenly aware, however, that the only way natural selection can produce evolutionary change in a population is if the phenotypic traits upon which natural selection acts can be passed from generation to generation. In other words, evolution by natural selection depends on the heritability of traits.

Heritability: Essential for Evolution

We can define **heritability** of a trait—usually symbolized as h^2—in a broad sense as the proportion of total phenotypic variation in a trait, such as body size or pigmentation, that is attributable to genetic variance. In equation form, heritability can be expressed as:

$$h^2 = V_G / V_P$$

Here, V_G represents genetic variance and V_P represents phenotypic variance. (We review how to calculate variance in Investigating the Evidence 4 in Appendix A.) Many different factors contribute to the amount of phenotypic variance in a population. We will subdivide phenotypic variance into only two components: variance in phenotype due to genetic effects, V_G, and variance in phenotype due to environmental effects on the phenotype, V_E. Subdividing V_P in the heritability equation given above produces the following:

$$h^2 = V_G / (V_G + V_E)$$

Since this highly simplified expression for heritability has important implications, let's examine it. First, consider environmental variance, V_E. Environment has substantial effects on many aspects of the phenotype of organisms. For instance, the quality of food eaten by an animal can contribute significantly to its growth rate and eventual size. Similarly, the amount of light, nutrients, temperature, and so forth affect the growth form and size of plants; we saw an example of this in how growing environment affected size of *Potentilla glandulosa* in figure 4.4. If we recall, the extent that the environment can influence variation in phenotype is called *phenotypic plasticity*. So, when we consider a population of plants or animals, some of the phenotype that we might measure will be the result of environmental effects, that is, V_E. However, we are just as familiar with the influence of genes on phenotype. For example, some of the variation in stature that we see in a population of animals or plants will generally result from genetic variation among individuals in the population, that is, V_G. In the *P. glandulosa* example above, we observed that size was not only affected by the environment; it was also determined genetically.

What our equation says is that the heritability of a particular trait depends on the relative sizes of genetic versus environmental variance. Heritability increases with increased V_G and decreases with increased V_E. Imagine a situation in which all phenotypic variation is the result of genetic differences between individuals and none results from environmental effects. In such a situation, V_E is zero and $h^2 = V_G / (V_G + V_E)$ is equal to $h^2 = V_G / V_G$ (since $V_E = 0$), which equals 1.0. In this case, since all phenotypic variation is due to genetic effects, the trait is perfectly heritable. We can also imagine the opposite circumstance in which none of the phenotypic variation that we observe is due to genetic effects. In this case, V_G is zero and so the expression $h^2 = V_G / (V_G + V_E)$ also equals zero. Because all of the phenotypic variation we observe in this population is due to environmental effects, natural selection cannot produce evolutionary change in the population. Generally, heritability of traits falls somewhere in between these extremes in the very broad region where both environment and genes contribute to the phenotypic variance shown by a population. For instance, a study of morphological variation in the water lily leaf beetle, a team of Dutch scientists (Pappers et al. 2002) found that body length and mandible width had heritabilities of between 0.53 and 0.83. Now that we have established the requirement of heritable variation in a trait for its evolution, let's review studies that have explored evolution by natural selection in nature.

Stabilizing Selection for Egg Size Among Ural Owls

Egg size, which affects offspring development and survival, influences successful reproduction by organisms ranging from sea urchins, lizards, and fishes to ostriches. Egg size can be highly variable within populations. For example, in a population of birds, the largest eggs produced can be over twice the size of the smallest. Pekka Kontiainen, Jon Brommer, Patrik Karell, and Hannu Pietiäinen, of the Bird Ecology Unit at the University of Helsinki, Finland, studied heritability, phenotypic plasticity, and evolution of egg size in the Ural owl, *Strix uralensis* (Kontiainen et al. 2008). One of their key questions was how much of the variation in egg size in their study population is the result of genetic differences among females (heritability) and how much is the result of environmental influences (phenotypic plasticity).

The study team conducted their research on Ural owls in a 1,500 km^2 area in southern Finland from 1981 to 2005. The main prey species for the owls were field voles, *Microtus agrestis,* and bank voles, *Clethrionomys glareolus,* which undergo regular population cycles in the study area, fluctuating in population density up to 50-fold over time. Because population fluctuations are not synchronized across large regions, some owl species in Finland lead nomadic lives, moving to areas where vole populations are on the increase and leaving areas where they are crashing (see chapter 10, Section 10.1). In contrast, Ural owls, which are monogamous, do not move in response to changing prey populations but stay in the pair's territory. Kontiainen and his colleagues describe them as "site tenacious."

During the course of their study, the research team made repeated size measurements of eggs laid by 344 female Ural owls in 878 clutches containing a total of nearly 3,000 eggs. The great variation in the owl's prey population combined with its site tenacity provided the research group with the opportunity to study the effects of the environment versus genetics on reproduction, including egg size. Based on measurements of eggs laid by 59 females in three phases of a vole population cycle (low, increasing, and decreasing) Kontiainen and his colleagues estimated that egg size is highly heritable in Ural owls, $h^2 = 0.60$. The finding that egg size is highly heritable was a critical element in their study, since it demonstrated that the trait is potentially subject to evolution by natural selection.

In other parts of the study, the research team explored the relationship of egg size to a variety of variables, which led to the discovery that egg size in the study population is undergoing stabilizing selection (fig. 4.10). Their results indicated two main selective factors: variation in hatching success and production of fledgling owls over a female's lifetime. The results showed that very small and very large eggs hatch at a lower rate compared to intermediate-sized eggs. Kontiainen and his colleagues also found that females that produced extremely small or large eggs produced fewer fledglings over the course of their lives, mainly because females producing eggs at the extremes of the size distribution had shorter reproductive lives. The result of these combined effects is stabilizing selection for egg size in this population of Ural owls. Elsewhere, ecologists have demonstrated directional selection in populations.

Directional Selection: Adaptation by Soapberry Bugs to New Host Plants

As will be discussed in chapter 7, herbivores must overcome a wide variety of physical and chemical defenses evolved by plants. As a consequence, plants exert strong selection on herbivore physiology, behavior, and anatomy. While herbivore adaptation to plant defenses is generally inferred from the juxtaposition of plant defenses and herbivore characteristics, few studies have documented the process of herbivore adaptation. A notable exception is provided by studies of the soapberry bug and its evolution on new host plants.

The soapberry bug, *Jadera haematoloma*, feeds on seeds produced by plants of the family Sapindaceae. Soapberry bugs use their slender beaks to pierce the walls of the fruits of their host plants. To allow the bug to feed on the seeds within the fruit, the beak must be long enough to reach from the exterior of the fruit to the seeds. The distance from the outside of the fruit wall to the seeds varies widely among potential host species. Thus, beak length should be under strong selection for appropriate length.

Scott Carroll and Christin Boyd (1992) reviewed the history and biogeography of the colonization of new host plants by soapberry bugs. Historically, soapberry bugs fed on three main host plants in the family Sapindaceae: the soapberry tree, *Sapindus saponaria* v. *drummondii*, in the south-central region of the United States; the serjania vine, *Serjania brachycarpa*, in southern Texas; and the balloon vine, *Cardiospermum corindum*, in southern Florida. During the second half of the twentieth century, three additional species of the plant family Sapindaceae were introduced to the southern United States. The round-podded golden rain tree, *Koelreuteria paniculata* from East Asia, and the flat-podded golden rain tree, *K. elegans* from Southeast Asia, are both planted as ornamentals, while the subtropical heartseed vine, *Cardiospermum halicacabum*, has invaded Louisiana and Mississippi. At some point after the introduction of these plants, some soapberry bugs shifted from their native host plants and began feeding on these introduced species.

Carroll and Boyd were particularly interested in determining whether the beak length had changed in the soapberry bugs that shifted from native to introduced host plants. Figure 4.11 contrasts the fruit radius of native and introduced host plants in Florida and the south-central United States. In Florida the fruit of the native host plant *C. corindum* has a much larger radius than the fruit of the introduced *K. elegans* (11.92 mm vs. 2.82 mm). In the south-central United States soapberry bugs shifting to introduced host plants faced the opposite situation. There, the fruit of the native *S. saponaria* has a smaller radius (6.05 mm) than the fruits of the introduced *K. paniculata* (7.09 mm) and *C. halicacabum* (8.54 mm).

Carroll and Boyd reasoned that if beak length was under natural selection to match the radius of host plant fruits, bugs shifting to the introduced plants in Florida should be selected for reduced beak length, while those shifting to introduced hosts in the south-central United States should be selected for longer beaks. Figure 4.12 shows the relationship between soapberry beak length and the radius of fruits of their host plants. As you can see, there is a close correlation between fruit radius and beak length.

At this point, we should ask whether the differences in beak length observed by Carroll and Boyd might be developmental responses to the different host plants. In other words, are the differences in beak length due to genetic differences among populations of soapberry bugs or are they the result of phenotypic plasticity? Fortunately, Carroll reared juvenile bugs from the various populations on alternative host plants, so we can answer this question. As it turns out, the differences in beak length observed in the field among bugs feeding on the various native and introduced host plants were retained in bugs that developed on alternative hosts. Here we have evidence for a genetic basis for interpopulational differences among soapberry bugs. Consequently, we can conclude that the differences in beak length documented by Carroll and Boyd were likely the result of natural selection for increased or

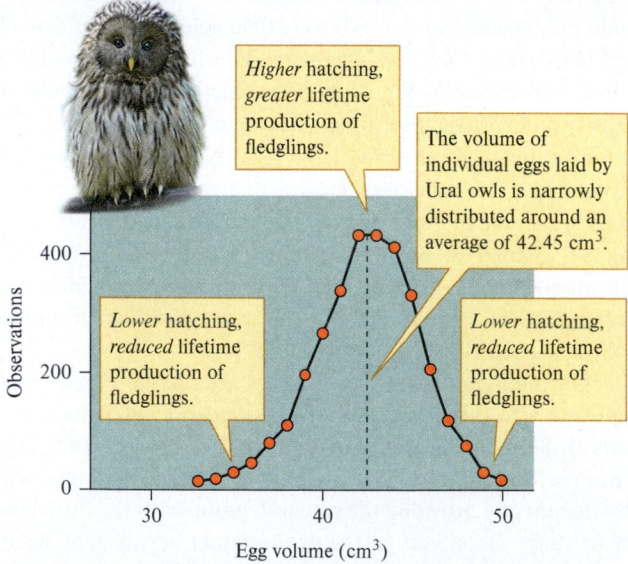

Figure 4.10 Stabilizing selection for egg volume in the Ural owl, *Strix uralensis*. Lower hatching rates by very small and very large eggs combined with reduced lifetime production of fledglings by female owls laying very small or very large eggs sustain stabilizing selection for egg volume in this population of Ural owls (data from Kontiainen et al. 2008).

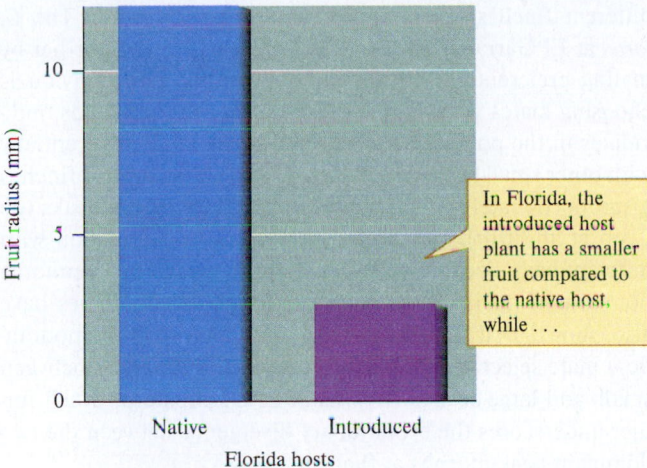

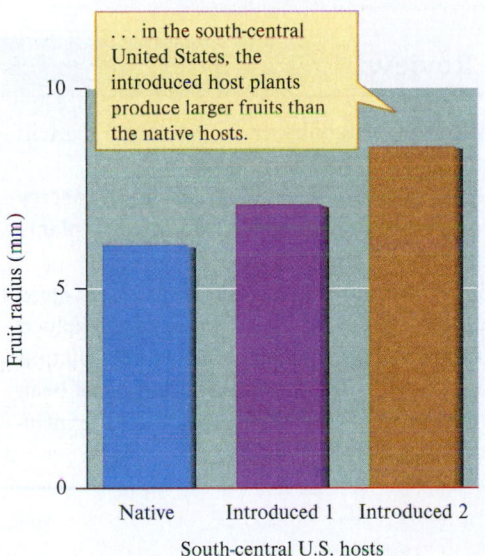

Figure 4.11 Comparison of the radius of fruits produced by native and introduced species of Sapindaceae (data from Carroll and Boyd 1992).

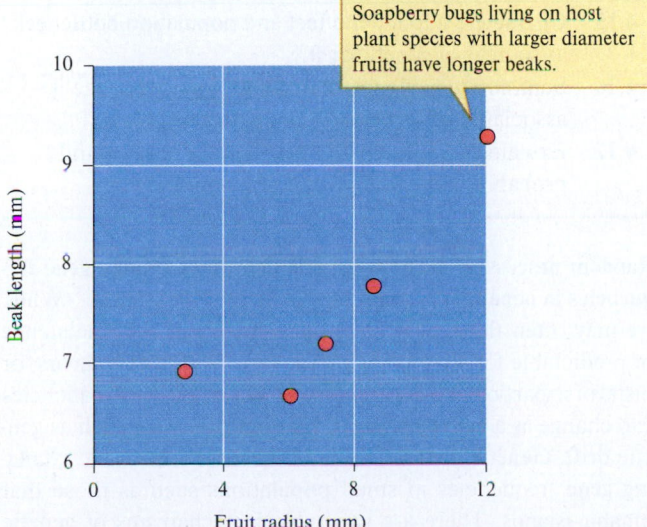

Figure 4.12 Relationship between fruit radius and beak length in populations of soapberry bugs living on native and introduced species of Sapindaceae (data from Carroll and Boyd 1992).

decreased beak length. The basis for this evolutionary change was confirmed when Carroll and colleagues (2001) demonstrated that beak length is highly heritable, with h^2 averaging approximately 0.60.

Scott Carroll, Stephen Klassen, and Hugh Dingle (1997, 1998) have done extensive additional studies of soapberry bugs that document substantial genetic differences between populations living on native versus introduced plants in the family Sapindaceae. Significantly, from the perspective of natural selection, the differences between these populations of soapberry bugs are great enough that both show reduced reproduction and survival when forced to live on the alternative host plants. That is, when soapberry bugs that normally live on native host plants are moved to introduced plants, their survival and reproductive rates decrease. Similarly, when soapberry bugs that now live on introduced plants are moved to native plants, which their ancestors fed on only 30 to 100 years ago, their reproductive and survival rates also decrease. Clearly, soapberry bugs have undergone natural selection for traits that favor their survival and reproduction on their specific plant hosts, whether native or introduced. Such directional selection has similarly been found among soapberry bugs that live on introduced, invasive Sapinaceae in South Africa as well (Foster et al. 2019).

Disruptive Selection in a Population of Darwin's Finches

As we have seen, the Galápagos Islands and their inhabitants played a key role in the development of Darwin's theory of evolution by natural selection. Darwin was particularly impressed by the variation in a group of 14 bird species now most commonly known as "Darwin's finches," or, less commonly, as "Galápagos finches." In the second edition of his journal recording his voyage on the *Beagle* (Darwin 1842a), Darwin suggests the influence of these birds on his thinking, "The most curious fact is the perfect gradation in the size of the beaks of the different species of *Geospiza* [a genus of Darwin's finches]—Seeing this gradation and diversity of structure in one small, intimately related group of birds, one might really fancy that, from an original paucity of birds in this archipelago, one species had been taken and modified for different ends."

Darwin's musing anticipated the great contribution made by studies of his finches to our understanding of the evolutionary process. For instance, studies of the variation in beak size that caught Darwin's attention have revealed the importance of this trait to the feeding ecology of Darwin's finches. Among species in the genus *Geospiza*, those with larger beaks are able to crack and feed on larger seeds (see chapter 13, Section 13.2). Studies of variation in beak size in populations of the medium ground finch, *Geospiza fortis*, have produced many notable discoveries. In a pioneering study of *G. fortis*, Peter Boag and Peter Grant (1978) showed that beak length, depth, and width are highly heritable, with heritability values of 0.62, 0.82, and 0.95, respectively. With this demonstration of heritability, Boag and Grant established a foundation for future evolutionary studies of beak size and form among Darwin's finches.

A more recent study of *G. fortis* provides one of the clearest and most complete examples of disruptive selection in a

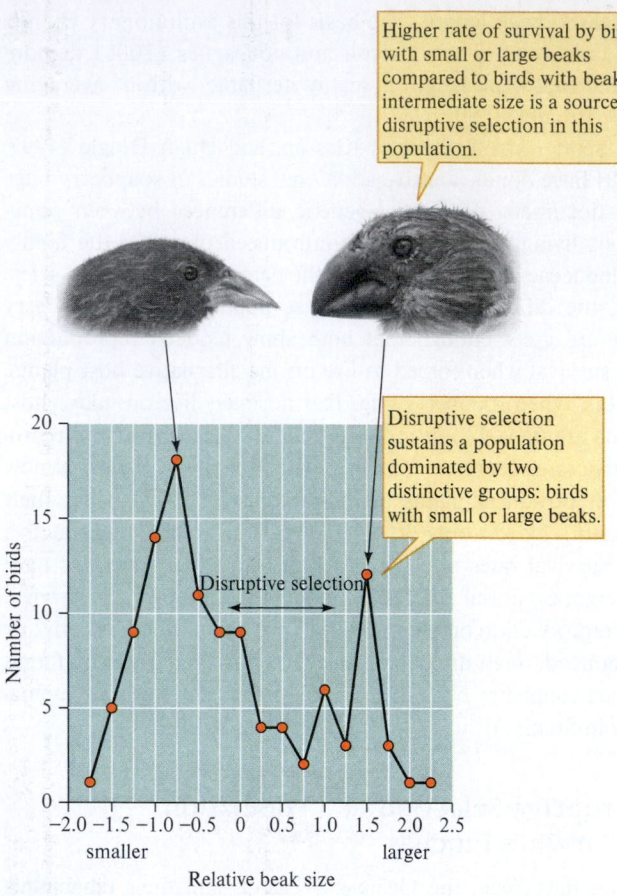

Figure 4.13   Disruptive selection in a population of medium ground finches, *Geospiza fortis,* at El Garrapatero, Santa Cruz Island, Galápagos (data from Hendry et al. 2009).

natural population. Andrew Hendry, Sarah Huber, Luis de León, Anthony Herrel, and Jeffrey Podos (2009) discovered that the *G. fortis* population at El Garrapatero, Santa Cruz Island, Galápagos, is dominated by two distinctive groups of individuals: those with small beaks and those with large beaks (fig. 4.13). Meanwhile, *G. fortis* with intermediate-sized beaks are relatively uncommon at El Garrapatero. Hendry and his colleagues uncovered the influences of disruptive selection in producing this distribution of beak sizes, when their studies revealed higher mortality, or possibly higher emigration, of birds with mid-sized beaks in this population. The researchers proposed that this higher mortality/emigration might be the result of a lack of appropriate food for these birds and/or competition with the more abundant small- and large-beaked individuals in the population.

Ongoing studies have revealed important biological details, including the finding that disruptive selection on beak size at El Garrapatero is reinforced by nonrandom patterns of mate choice in the population. Nonrandom mating itself can be a source of evolutionary change in populations (see also chapter 8) because it violates one of the conditions for Hardy–Weinberg equilibrium (see p. 86). Darwin's finches choose mates at least partly on the basis of beak size and mating song. Since different species of Darwin's finches have beaks of different size and shape and sing different songs, individuals of

different finch species rarely mate with each other. The *G. fortis* at El Garrapatero take this isolation one step further by mating preferentially within the population, with individuals choosing mates with similar-sized beaks. In other words, individuals in the population with small beaks mate preferentially with other small-beaked individuals, while large-beaked finches disproportionately choose mates that also have large beaks (de León et al. 2010). In addition, males in the population with different beak sizes sing distinctive songs, which may reinforce nonrandom mating (Podos 2010). These recent studies have also shown that disruptive selection reinforced by nonrandom mate selection has produced genetic differences between small- and large-beaked *G. fortis* at El Garrapatero, which further underscores the evolutionary divergence between the two dominant beak morphs at that site.

Concept 4.4 Review

1. Can a trait with no heritability, $h^2 = 0$, evolve? Explain your answer.
2. What must have been true for beak length in soapberry bug populations before new species of soapberry plants were introduced to the United States?
3. There is genetic evidence that mating between *Geospiza magnirostris* and *G. fortis* (see fig. 13.8) may have helped establish sufficient genetic variation in the population of *G. fortis* at El Garrapatero for the distribution of beak sizes at that site (see fig. 4.13) to emerge under the influence of disruptive selection. Explain.

4.5 Change Due to Chance

LEARNING OUTCOMES

After studying this section you should be able to do the following:

4.15 Characterize founder effect and population bottleneck as sources of genetic drift.
4.16 Summarize evidence that small population size is associated with loss of genetic diversity.
4.17 Explain the relationship between inbreeding and probability of extinction in populations.

Random processes, such as genetic drift, can change gene frequencies in populations, especially in small populations. While we may often think of evolutionary change as a consequence of predictable forces such as natural selection which favors, or disfavors, particular genotypes over others, allele frequencies can change as a consequence of random processes such as genetic drift. Genetic drift is theoretically most effective at changing gene frequencies in small populations such as those that inhabit islands. There are two typical mechanisms of genetic drift: founder effects and population bottlenecks (fig. 4.14).

Population bottlenecks can occur when large portions of a population are killed by a chance event unrelated to genotype, leaving behind reduced genetic diversity and/or different

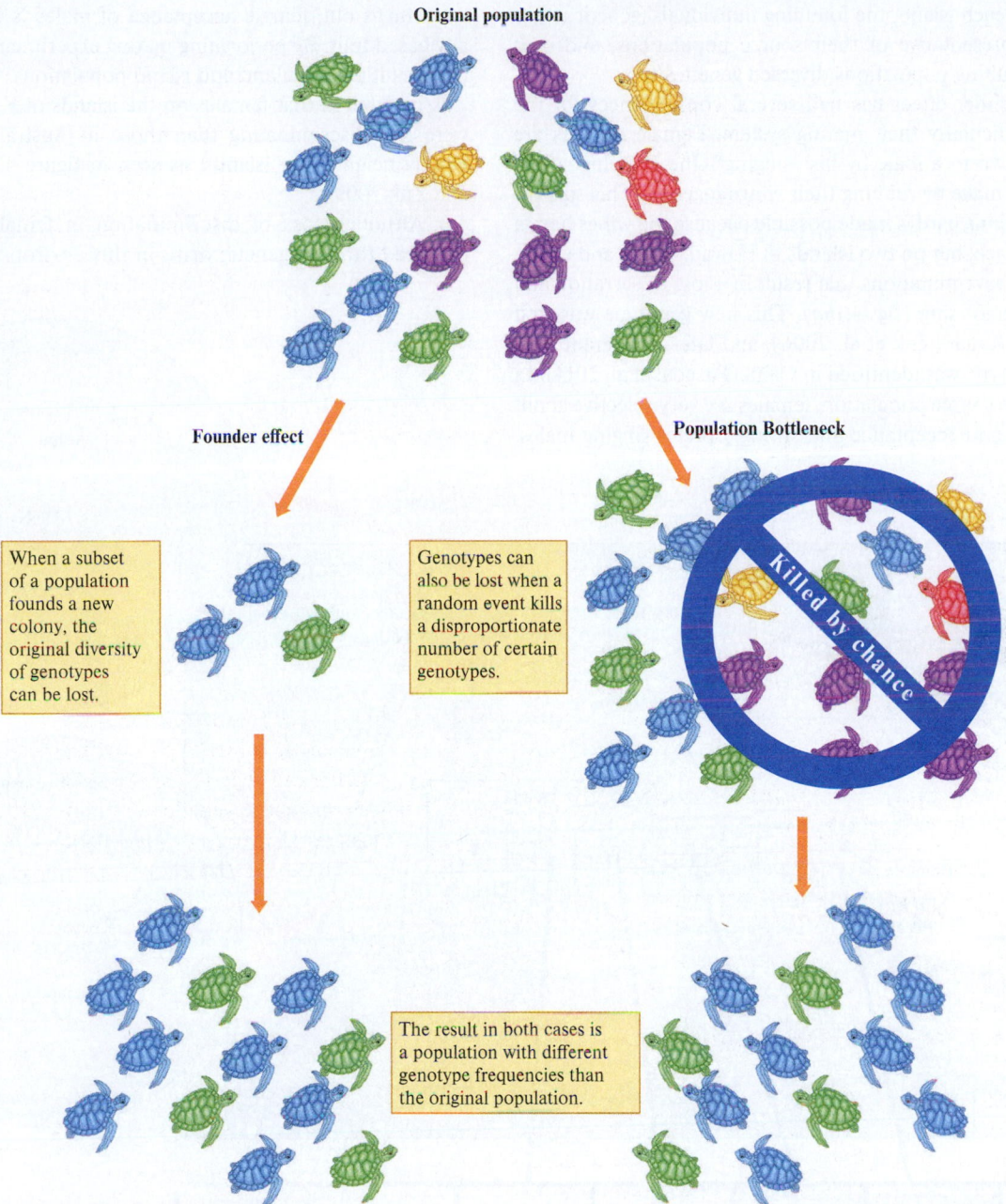

Original population

Founder effect

When a subset of a population founds a new colony, the original diversity of genotypes can be lost.

Population Bottleneck

Genotypes can also be lost when a random event kills a disproportionate number of certain genotypes.

Killed by chance

The result in both cases is a population with different genotype frequencies than the original population.

Figure 4.14   The two mechanisms of genetic drift.

frequencies of alleles than existed before the event took place. The cheetah (*Acinonyx jubatus*) has experienced at least two and possibly three significant bottlenecks over the last 12,000 years, the most recent due to humans extirpating them from 75% of their original range (Dobrynin et al. 2015). As a result, cheetahs have high rates of **inbreeding** and associated high rates of reproductive defects, birth defects, and disease. Inbreeding, which is mating between close relatives, is more likely in small populations.

Founder effects take place when a subset of the original population becomes a new population; just as with population bottlenecks, the genetic composition of this smaller group is unlikely to represent the full diversity and frequencies of the alleles in the original population. In the following examples, we consider the effects of genetic drift in both island populations and fragmented habitats.

Evidence of Genetic Drift in Island Crickets

Marlene Zuk, Robin Tinghitella, and colleagues discovered evidence of genetic drift in populations of crickets on islands in the Pacific Ocean, with important ecological implications. The field cricket, *Teleogryllus oceanicus*, is native to Australia but founded new populations on distant outlying islands, likely aided through the movement of Polynesian settlers in the region more than 3,000 years ago (Tinghitella et al. 2011). The most remote islands colonized were those in Hawaii. Through genetic analysis, Tinghitella and her colleagues determined that genetic diversity decreases with distance from the Australian mainland, and that cricket populations in the Hawaiian Islands are especially different genetically, not just from Australia, but even from the nearest populations in the Marquesas, 4,000 km away (fig. 4.15). By random chance and as a function of the small number of founders on each island, the founding individuals' genetics were not fully representative of their source populations, and over time, the resulting populations diverged genetically.

This founder effect has had several consequences for the crickets, particularly their mating system. Female crickets are typically drawn to a male by his "singing"—the chirping sound that crickets make by rubbing their wings together. This surprisingly loud calling card is made possible because the wings have a serrated surface, but on two islands in Hawaii, Kauai and Oahu, some males have mutations that result in a loss of serration and therefore cannot sing (fig. 4.16a). This new genotype was first observed in Kauai (Zuk et al. 2006), and later a different non-singing genotype was identified in Oahu (Pascoal et al. 2014). In Australia, the source population, females are very selective about mate choice and acceptance and strongly prefer singing males.

As it turns out, female acceptance of males is itself a genetically based trait. By performing mating experiments with females from multiple mainland and island populations, Tinghitella and Zuk discovered that females on the islands of Kauai and Oahu were less discriminating than those in Australia or its more closely neighboring islands, as seen in figure 4.16 (Tinghitella and Zuk 2009).

Although loss of discrimination in females could have occurred through genetic drift, in this environment it is more

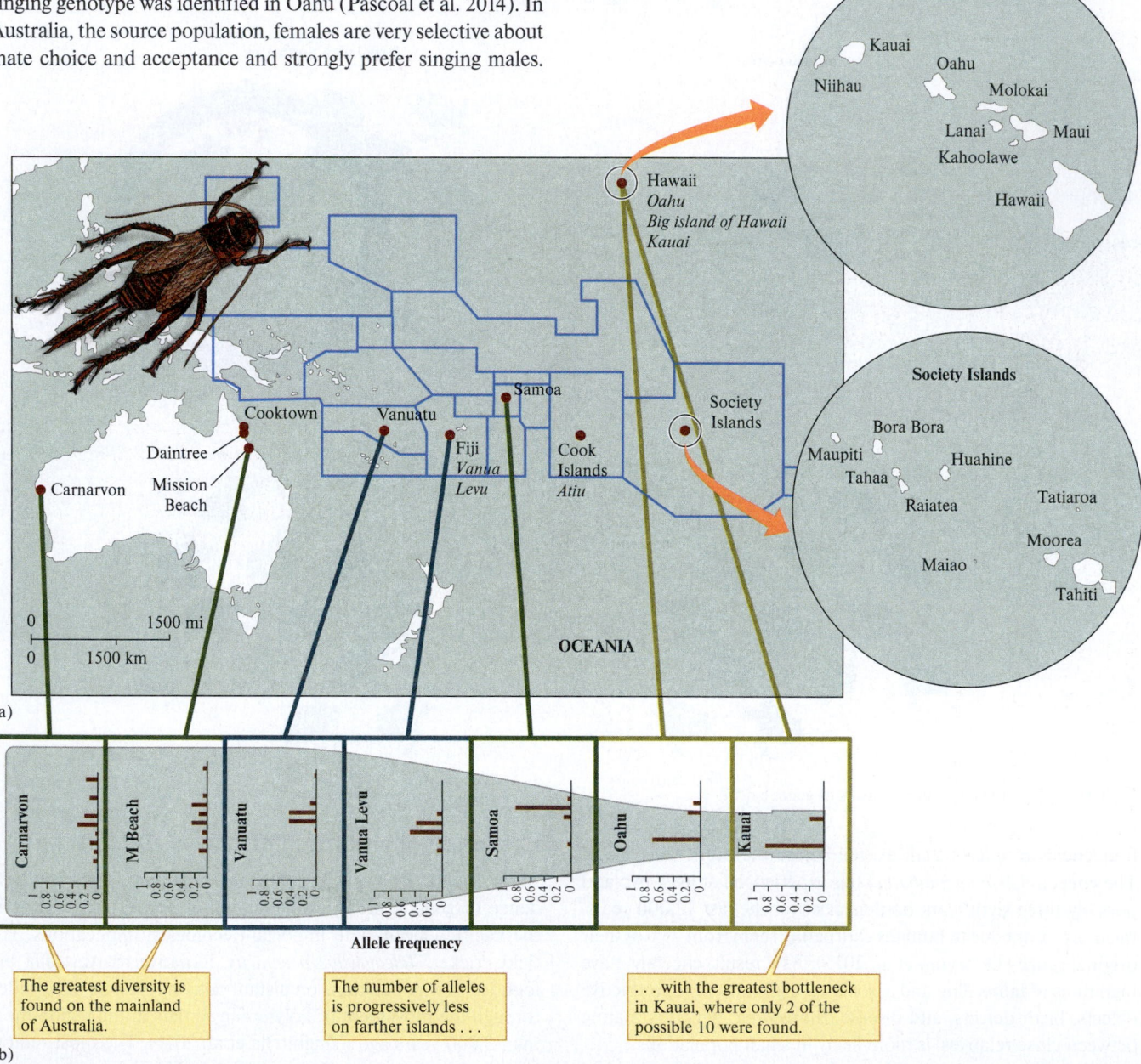

The greatest diversity is found on the mainland of Australia.

The number of alleles is progressively less on farther islands . . .

. . . with the greatest bottleneck in Kauai, where only 2 of the possible 10 were found.

Figure 4.15   (a) Map of Oceania, with those locations indicated where the genetic diversity of field crickets, *Teleogryllus oceanicus,* was measured. These include the mainland Australian cities where field crickets are native and ancestral, as well as a subset of islands where they dispersed later, some through human movement. (b) Frequencies of alleles that were sampled through microsatellite analysis for some of the locations that were sampled by Tinghitella and colleagues (2011). Each bar represents a different allele, the height of which reflecting how common it is. Frequencies sum to 1; thus, there are either many alleles at low frequencies (as in Australia) or few at high frequencies (as in Hawaii). The bottleneck in genetic diversity observed with distance from the mainland is likely due to founder effect, a mechanism of genetic drift (data from Tinghitella et al. 2011).

likely that this is an adaptive trait that arose through natural selection (Tinghitella and Zuk 2009). In the Hawaiian Islands, if females were faced with much less diversity in the quality of males (due to founder effect), more discriminating females may not have been able to find mates to their standards. Thus, through directional selection, females likely evolved to become more accepting, and thus more likely to mate with those males that could not sing when that mutation appeared.

But the ecological evolution story does not stop there. On three of the islands in Hawaii, including Kauai and Oahu, the crickets were dealing with a new predator from North America: the fly *Ormia ochracea*. This fly is attracted to the male crickets by their songs and then lays eggs on the body of the cricket; the hatching fly larvae then eat the host, killing it (learn more about parasitoids in chapter 14). From the time Zuk began observing them in 1991, crickets seemed to be disappearing from islands with flies (Zuk and Kolluru 1998). However, when the mutation arose for a non-singing phenotype in the late 1990s, those crickets had an immediate advantage: their silence allowed them to avoid parasitism by the fly (Zuk et al. 2006). The non-singing males, which already had better mating chances on islands versus the mainland due to less-choosy females, were also more likely to survive. Through this strong combination of ecological factors, non-singing males have become the dominant phenotype on the island of Kauai, and it appears that the same may be happening in Oahu (Pascoal et al. 2014). Even though the inability to sing had a different genetic origin on these two islands, similar environmental selection is resulting in the same outcome: a novel phenotype becoming common. This story reminds us of the cases of *convergent evolution* we learned about in chapter 2: plants that grow in the desert or at high altitude with similar adaptive traits but with different origins genetically (see figs. 2.21 and 2.40).

The story of the crickets' song demonstrates how mutation, genetic drift, and natural selection can interact, resulting in dramatic changes for a species. The difference between the singing and non-singing males is likely the result of just a few mutations within one or a few genes associated with wing development and song production; these mutations occurred by chance and occurred in certain island populations by chance. But because of the strong selective pressures of the environment on Kauai and Oahu, this novel genotype became common on those islands in as few as just 15 to 20 generations (Tinghitella 2008; Pascoal et al. 2014).

Genetic Diversity and Butterfly Extinctions

The landscape of Åland in southwestern Finland is a patchwork of lakes, wetlands, cultivated fields, pastures, meadows, and forest (see fig. 21.12). Here and there in this well-watered landscape you can find dry meadows that support populations of plants, *Plantago lanceolata* and *Veronica spicata,* that act as hosts for the Glanville fritillary butterfly, *Melitaea cinxia* (fig. 4.17). As discussed in chapter 21, the meadows where *Melitaea* lives vary greatly in size, and *Melitaea* population size increases directly with the size of meadows (see fig. 21.13). Careful studies of these populations by Ilkka Hanski, Mikko Kuussaari, and Marko Nieminen (1994) showed that small populations of *Melitaea* living in small meadows were most likely to go extinct.

Several factors likely influence the greater vulnerability of small populations to extinction. However, what role might genetic factors, especially reduced genetic variation, play in the vulnerability of small populations to extinction? Richard

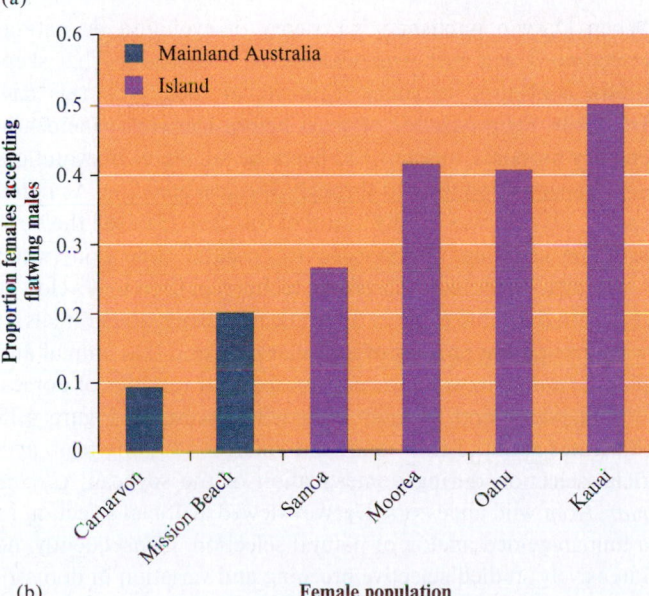

Figure 4.16   Research on cricket mating behavior shows the cascading effect of genetic drift. (*a*) The male with normal wings that are capable of song versus the mutant that lacks the ridges necessary for singing. The latter is found on the Hawaiian Islands of Oahu and Kauai. (*b*) The choosiness of female crickets as measured by rejection of non-singing males in the source populations in Australia, versus those that have experienced genetic drift on islands. (from Zuk et al. 2006; data from Tinghitella and Zuk 2009).

Figure 4.17   Long-term studies of the Glanville fritillary butterfly, *Melitaea cinxia,* have provided exceptional insights into the relationship between population size and genetic diversity.   Ken Pilsbury/Natural Visions

Frankham and Katherine Ralls (1998) point out that one of the contributors to higher extinction rates in small populations may be inbreeding, just as it has affected cheetahs in our earlier example.

Ilik Saccheri and five coauthors (1998) reported one of the first studies giving direct evidence that inbreeding contributes to extinctions in wild populations. Saccheri and his colleagues studied 1,600 dry meadows and found *Melitaea* in 524, 401, 384, and 320 of the meadows in 1993, 1994, 1995, and 1996, respectively. Over this period, they documented an average of 200 extinctions and 114 colonizations of meadows annually. As you can see, these populations are highly dynamic. In order to determine the extent that genetic factors, especially inbreeding, may contribute to extinctions, Saccheri and his colleagues conducted genetic studies on populations of *Melitaea* in 42 of the meadows. They estimated heterozygosity, an indicator of genetic variability, with respect to seven enzyme systems and one locus of nuclear microsatellite DNA. The researchers used the level of heterozygosity within each meadow population as an indicator of inbreeding, with low heterozygosity indicating high levels of inbreeding.

The results of the study indicated that influence of inbreeding on the probability of extinction was very significant. It turned out that the populations with the highest levels of inbreeding (lowest heterozygosity) had the highest probabilities of extinction. Saccheri and his colleagues found a connection between heterozygosity and extinction through effects on larval survival, adult longevity, and egg hatching. Females with low levels of heterozygosity produced smaller larvae, fewer of which survived to the winter dormancy period. Pupae of mothers with low heterozygosity also spent more time in the pupal stage, exposing them to greater attack by parasites. In addition, adult females with low heterozygosity had lower survival and laid eggs with a 24% to 46% lower rate of hatching. These effects have the potential to reduce the viability of local populations of *Melitaea* that are made up of individuals of low heterozygosity (low genetic variation) and increase their risk of local extinction.

We have seen how the small population size and isolation can influence the genetic structure of populations of many kinds of organisms, including field crickets living on Pacific islands and the Glanville fritillary, *Melitaea,* in the dry meadow environments of southwestern Finland. In situations like these, chance plays a significant role in determining the genetic structure of populations.

Concept 4.5 Review

1. Why do the managers of captive breeding and reintroduction programs for endangered species try to maintain high levels of genetic diversity?
2. Why is it significant that Tinghitella and her colleagues found that female crickets were less choosy in all island populations, not just those with selection by the parasitic fly?

Applications

Evolution and Agriculture

LEARNING OUTCOMES

After studying this section you should be able to do the following:

4.18   Contrast natural and artificial selection.
4.19   Describe what genetic engineering adds to the traditional process of selective breeding for developing new crop varieties.
4.20   Discuss research on the evolution of herbicide resistance in Palmer amaranth, and explain the evolutionary and economic implications of this work.

When Darwin published his theory of evolution by natural selection, it was met with widespread skepticism. That skepticism continues, in some segments of society, to this day. Ironically, for thousands of years before Darwin's nineteenth-century proposal, humans had been the architects of evolutionary change in populations through selective breeding. As noted earlier, Darwin was keenly aware of this fact. He used the term **artificial selection,** as opposed to "natural" selection, when referring to the selective breeding techniques used to develop or maintain desirable traits in domesticated plants and animals. It was through the process of artificial selection that animal and plant breeders have produced thousands of varieties of domesticated plants and animals from wild ancestors. Figure 4.18 illustrates some of the significant changes resulting from artificial selection during domestication of the soybean, *Glycine max,* from wild ancestors. Darwin viewed artificial selection as a human-guided analog of natural selection. Consequently, he intensively studied selective breeding and variation in domesticated plants and animals as he developed his theory of natural selection. He even wrote a book on the subject: *The Variation of Animals and Plants Under Domestication* (Darwin 1868).

The production of new varieties of plants and animals continues and may be more intense than ever. Today, the traditional selective breeding techniques of Darwin's time are combined with **genetic engineering,** the alteration of the genetic makeup of an organism through the introduction or deletion

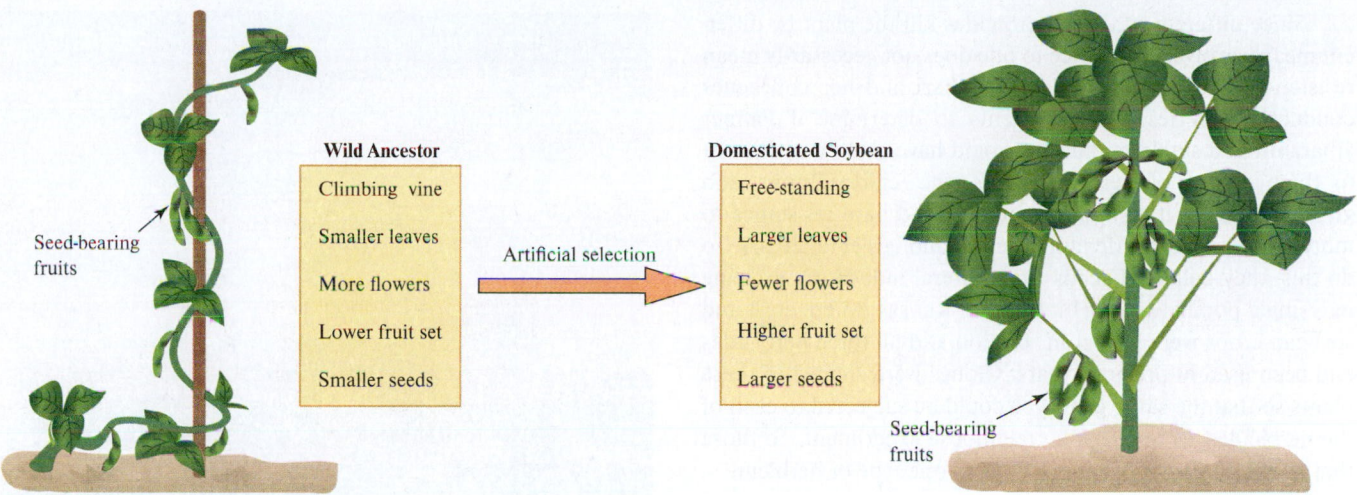

Figure 4.18   Soybeans were first cultivated in China approximately 3,000 years ago. Since that time, artificial selection has produced substantial divergence of domesticated soybeans from their wild ancestors.

of genes—for example, the introduction of bacterial genes into crop plants to give them more resistance to insect pests. Such organisms are called **genetically modified organisms,** or **GMOs.** However, **agriculture,** the growing of crops and livestock for human consumption, also induces unintended evolutionary change in wild organisms.

Evolution of Herbicide Resistance in Weeds

As farmers create environmental conditions suitable for the cultivation of particular crops, they simultaneously create selection pressures on populations of wild plants and animals that live in or near their agricultural fields (fig. 4.19). Some of the most powerful of those selective pressures occur when farmers attempt to increase crop yields by applying chemical poisons to control insect pests or weeds.

The poisons used to control weed populations are called *herbicides.* The use of herbicides is increasing as farmers shift from mechanical control of weeds, by tilling the soil, to chemical control. This shift has been facilitated by the use of genetic engineering to develop herbicide-resistant varieties of crops, including herbicide-resistant soybeans. The weeds in fields planted with genetically modified, herbicide-resistant soybeans can be controlled with herbicides, in a system of cultivation called "no-till" agriculture. No-till agriculture includes several benefits to the farmer, including reduced production costs, higher crop yields, reduced soil erosion, and better weed control. However, improved weed control through the use of herbicides has, in some cases, proved temporary.

The control of weeds with herbicides can be rendered ineffective by the evolution of herbicide resistance in weed populations.

Palmer amaranth (*Amaranthus palmeri*) is one of the most economically damaging weeds for soybean and other crops across the United States. Members of this genus are also commonly known as pigweed and are eaten as leaf vegetables not only by pigs, but by many human groups as well. Despite this, the rapid spread and growth rate of Palmer amaranth makes it a significant pest; as said by a Purdue University Extension Weed

Figure 4.19   A soybean field. Agriculture, through cultivation, planting large concentrations of a single crop, the application of pesticides, and so forth, creates environmental conditions quite different from those in surrounding natural ecosystems. Over time wild populations, including insect pests and weeds, adapt to these conditions through the process of natural selection.   sima/123RF

Specialist, "It's the only weed I've seen that can drive a farmer out of business." (https://www.ag.ndsu.edu/palmeramaranth) Because of this, herbicides are regularly applied to Palmer amaranth; however, it has evolved resistance to at least three different types, which kill the plant by inhibiting different physiological pathways. The most popular of these is glyphosate, also known by the trade name Roundup™, which works by inhibiting an important plant enzyme. In 2005 the first glyphosate-resistant Palmer amaranth was documented in the state of Georgia, and by 2016 was seen in populations in 27 states.

Since different types of herbicides kill the plant by different means, evolved resistance to one does not necessarily mean resistance to others. Sushilla Chaudhari and her colleagues conducted a series of experiments to determine if Palmer amaranth in a single population could have evolved resistance to three classes of herbicides (atrazine, chlorsulfuron, and glyphosate), and if a single genotype could have resistance to more than one herbicide at a time (Chaudhari et al. 2020). To do this, they collected seeds from several individuals growing in a single population in Hutchinson, Kansas, where corn and soybean crops were grown in rotation and all three herbicides had been used in previous years. Clones were created of these plants so that the same genotype could be subjected to each of the herbicides, as tested in a greenhouse experiment. To those that survived, that is, were resistant to one type of herbicide, a second, different herbicide was applied.

Chaudhari's group found that within the population of Palmer amaranth tested, half of the genotypes had evolved resistance to either atrazine or chlorsulfuron, and that some of those that were resistant to one of the herbicides were also resistant to a second one, including glyphosate (fig. 4.20). Thus, this work demonstrated that not only has this species

Figure 4.20   (*a*) The agricultural weed Palmer amaranth (*Amaranthus palmeri*). (*b*) Survival rate of herbicide-resistant Palmer amaranth clones subjected to commonly used herbicides (atrazine, chlorsulfuron, and glyphosate) in a greenhouse experiment. Those that survived the application of the first herbicide (approximately 50%, left) were then subjected to a second herbicide (right). Approximately 14% of all of the tested clones proved to be resistant to two herbicides (data from Chaudhari et al. 2020).   Design Pics Inc/Alamy Stock Photo

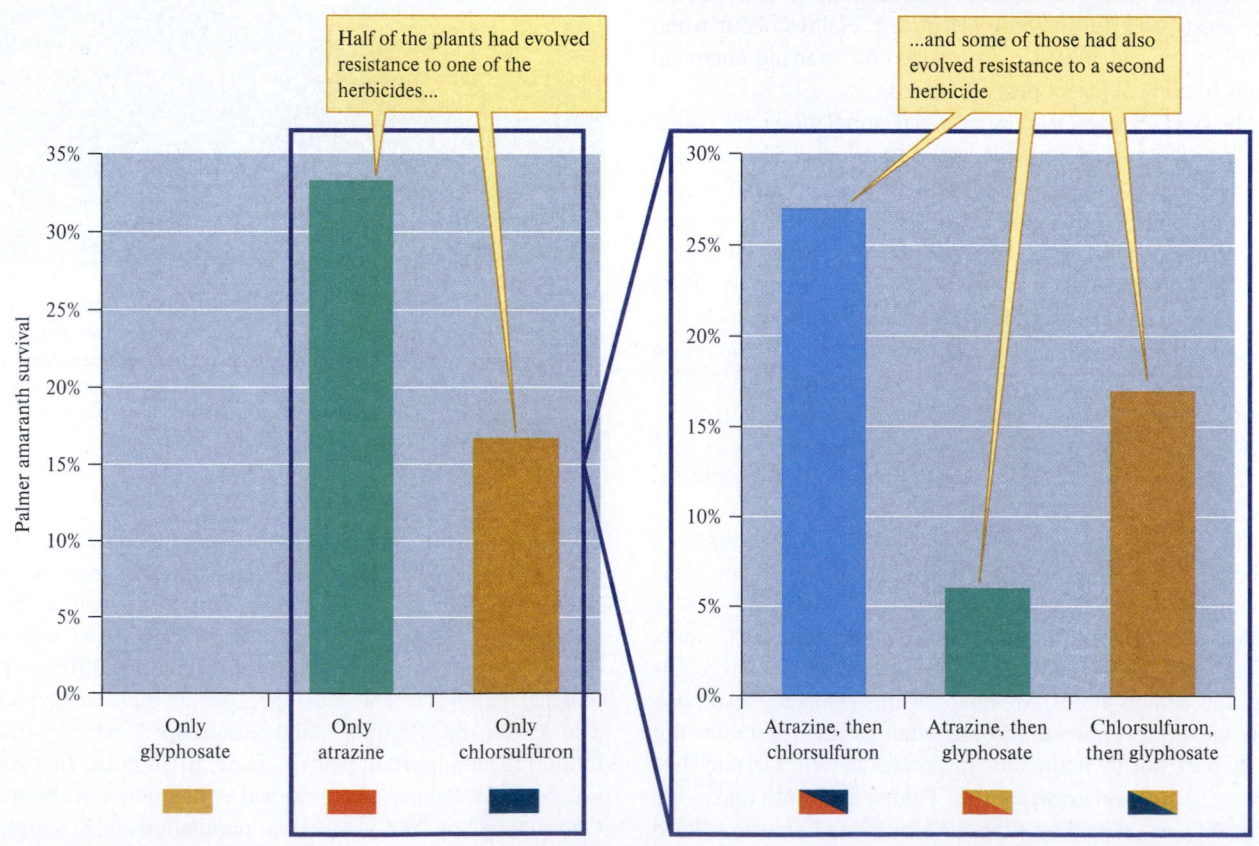

evolved in response to herbicide application, but that resistance to multiple herbicides was occurring. Further, there was diversity within this Palmer amaranth population, such that it could survive future selection pressure of the application of one, or even two, of these herbicides. The genetic diversity in this population is likely due to gene flow between populations in the area, which will facilitate future adaptation. This research has significant economic implications for weed control of Palmer amaranth, as well as being a prime example of rapid evolution.

Palmer amaranth is not the only important weed species to evolve herbicide resistance. As of 2020, there were 514 unique cases of herbicide resistance documented representing 262 species that have been documented (International Survey of Herbicide Resistant Weeds, weedscience.org). New species and populations are evolving resistance over time, and different populations of the same species can acquire it at different time. For example, Julia M. Kreiner and colleagues used genetic analysis to determine that glyphosate resistance has evolved multiple times in *Amaranthus hypochondriacus*, a significant agricultural weed. The evolution of insecticide resistance among insect species that are significant crop pests is also well documented. Clearly, sustaining agricultural production worldwide requires new approaches to pest control, especially a firm understanding of the great potential of pest populations to adapt through natural selection. In section II, we will explore adaptation of organisms to physical, chemical, and biological aspects of the environment.

Summary

Darwin and Mendel complemented each other well, and their twin visions of the natural world revolutionized biology. The synthesis of the theory of natural selection and genetics gave rise to modern evolutionary ecology. Here we examine five major concepts within the area of population genetics and natural selection.

Phenotypic variation among individuals in a population results from the combined effects of genes and environment. The first biologists to conduct thorough studies of phenotypic and genotypic variation and to incorporate experiments in their studies, focused on plants. Clausen, Keck, and Hiesey explored the extent and sources of morphological variation in plant populations, including both the influences of environment and genetics. Molecular genetic studies, such as those conducted by Douglas and Brunner on whitefish populations in the Alps, offer a powerful way of assessing the genetic variation in populations.

The Hardy–Weinberg equilibrium model helps identify evolutionary forces that can change gene frequencies in populations. Because evolution involves changes in gene frequencies in a population, a thorough understanding of evolution must include the area of genetics known as population genetics. One of the most fundamental concepts in population genetics, the Hardy–Weinberg principle, states that in a population mating at random in the absence of evolutionary forces, allele frequencies will remain constant. For a population in Hardy–Weinberg equilibrium in a situation where there are only two alleles at a particular locus, $p + q = 1.0$. The frequency of genotypes in a population in Hardy–Weinberg equilibrium can be calculated as $(p + q)^2 = (p + q) \times (p + q) = p^2 + 2pq + q^2 = 1.0$. The conditions necessary to maintain constant allele frequencies in a population are (1) random mating, (2) no mutations, (3) large population size, (4) no immigration, and (5) equal survival and reproductive rates for all genotypes. When a population is not in Hardy–Weinberg equilibrium, the Hardy–Weinberg principle helps us to identify the evolutionary forces that may be in play.

Natural selection is differential survival and reproduction among phenotypes. Natural selection can lead to change in populations but it can also serve as a conservative force, impeding change in a population. Stabilizing selection acts against extreme phenotypes and as a consequence, favors the average phenotype. By favoring the average phenotype, stabilizing selection decreases phenotypic diversity in populations. Directional selection favors an extreme phenotype over other phenotypes in the population. Under directional selection, the average of the trait under selection can change over time. Disruptive selection favors two or more extreme phenotypes over the average phenotype in a population, leading to an increase in phenotypic diversity in the population.

A large and rapidly growing body of research on natural populations provides robust support for the theory of evolution by natural selection. The most general postulate of the theory of natural selection is that the environment determines the evolution of the anatomy, physiology, and behavior of organisms. For such evolutionary change to occur, however, the variation in traits under selection must be heritable. A research team in Finland discovered that the size of eggs produced by a population of Ural owls is maintained by stabilizing selection. Their results indicated two main selective factors: variation in hatching success and production of fledgling owls over a female's lifetime. Studies by Carroll and several colleagues show that soapberry bug populations living on native and introduced host plants have undergone natural selection for traits that favor their survival and reproduction on particular host plant species. Studies of the medium ground finch, *Geospiza fortis,* provide one of the clearest and most complete examples of disruptive selection in a natural population. Researchers uncovered the influences of disruptive selection on beak sizes in a population at El Garrapatero, Santa Cruz Island, Galápagos, when their research revealed higher mortality, or possibly higher emigration, of birds with mid-sized beaks in the population. Disruptive selection on beak size at El Garrapatero is reinforced by nonrandom patterns of mate choice in the population. Hundreds of other examples of natural selection have been brought to light during the century and a half since Darwin published his theory.

Random processes, such as genetic drift, can change gene frequencies in populations, especially in small populations. Genetic drift is theoretically most effective at changing gene frequencies in small populations such as those that inhabit islands. Work by

Tinghitella and her colleagues showed that genetic diversity in crickets decreased with increasing distance from the source population, and that this reduced diversity was likely responsible for females becoming less discriminating of mates on these islands. This adaptation plus natural selection by a novel predator ended up leading to the dominance of a genotype that was extremely rare in the source population. This example thus shows us how genetic shifts based on chance can lead to large ecological differences between populations. One of the greatest concerns associated with fragmentation of natural ecosystems due to human land use is that reducing habitat availability will decrease the size of animal and plant populations to the point where genetic drift will reduce the genetic diversity within natural populations. Saccheri and his colleagues found that higher heterozygosity

(genetic diversity) was associated with lower rates of population extinction through the effects of heterozygosity on larval survival, adult longevity, and egg hatching in populations of the Glanville fritillary butterfly, *Melitaea cinxia*.

Animal and plant breeders have used selective breeding, a process Darwin called artificial selection, to produce the thousands of varieties of domesticated plants and animals from wild ancestors. Domestication of soybeans provides a good example of artificial selection in action. Today, the traditional selective breeding techniques of Darwin's time are combined with genetic engineering to alter the genetics of agricultural crops. The rapid evolution of herbicide resistance in Palmer amaranth is an example of evolution by natural selection to the modern agricultural environment.

Key Terms

abiotic   78

adaptation   80

agriculture   97

allele   80

allele frequencies   86

artificial selection   96

biotic   78

common garden
   experiment   81

directional selection   87

disruptive selection   88

ecotype   82

evolution   79

fitness   87

gene   80

genetic drift   86

genetic engineering   96

genetically modified
   organisms (GMOs)   97

Hardy–Weinberg
   principle   86

heritability   89

inbreeding   93

locus   80

microsatellite DNA   83

natural selection   79

phenotype   81

phenotypic plasticity   82

polygenic traits   87

population genetics   84

quantitative genetics   87

stabilizing selection   87

Review Questions

1. Contrast the approaches of Charles Darwin and Gregor Mendel to the study of populations. What were Darwin's main discoveries? What were Mendel's main discoveries? How did the studies of Darwin and Mendel prepare the way for the later studies reviewed in chapter 4?

2. Why is it important to conduct common garden experiments when one is interested in the evolution of a particular trait that varies in nature? If such an experiment shows that a trait is always expressed the same way, regardless of environment, is the trait plastic?

3. What is the Hardy–Weinberg principle? What is Hardy–Weinberg equilibrium? What conditions are required for Hardy–Weinberg equilibrium?

4. Review the Hardy–Weinberg equilibrium equation. What parts of the equation represent allele frequencies? What elements represent genotype frequencies and phenotype frequencies? Are genotype and phenotype frequencies always the same? Use a hypothetical population to specify alleles and allelic frequencies as you develop your presentation.

5. What is genetic drift? Under what circumstances do you expect genetic drift to occur? Under what circumstances is genetic drift

unlikely to be important? Does genetic drift increase or decrease genetic variation in populations?

6. Suppose you are a director of a captive breeding program for a rare species of animal, such as Siberian tigers, which are found in many zoos around the world but are increasingly rare in the wild. Design a breeding program that will reduce the possibility of genetic drift in captive populations.

7. How might the distribution of beak sizes in the population differ from that shown in figure 4.13, if mate choice in the population was random with respect to beak size?

8. How did the studies of Marlene Zuk, Robin Tinghitella, and their colleagues demonstrate rapid evolutionary adaptation to introduced parasitic flies? What role did genetic drift have in this rapid evolution?

9. How do classical approaches to genetic studies, such as common garden experiments, and modern molecular techniques, such as DNA sequencing, complement each other? What are the advantages and disadvantages of each?

Chapter 5

Temperature Relations

©Digital Vision/Getty Images RF

Japanese macaques, *Macaca fuscata,* huddle together, conserving their body heat in the midst of driving snow. The capacity to regulate body temperature, using behavioral, anatomical, and physiological adaptations, enables these monkeys to live through the cold winters in Nagano, Japan, site of the 1998 Winter Olympics.

CHAPTER CONCEPTS

5.1 Macroclimate interacts with the local landscape to produce microclimatic variation in temperature.   102

Concept 5.1 Review   105

5.2 Adapting to one set of environmental conditions generally reduces a population's fitness in other environments.   105

Concept 5.2 Review   106

5.3 Most species perform best in a fairly narrow range of temperatures.   107

Concept 5.3 Review   110

5.4 Many organisms have evolved ways to compensate for variations in environmental temperature by regulating body temperature.   110

Concept 5.4 Review   120

5.5 Many organisms survive extreme temperatures by entering a resting stage.   120

Concept 5.5 Review   123

Applications: Local Extinction of a Land Snail in an Urban Heat Island   123

Summary   124

Key Terms   125

Review Questions   125

LEARNING OUTCOMES

After studying this section you should be able to do the following:

5.1 Distinguish between temperature and heat.
5.2 Explain the ecological significance of environmental temperatures.

The thermometer was one of the first instruments to appear in the scientific tool kit, and we have been measuring and reporting temperatures ever since. However, what do thermometers actually quantify? **Temperature** is a measure of the average kinetic energy, or energy of motion, of the molecules, in a mass of a substance. For example, water

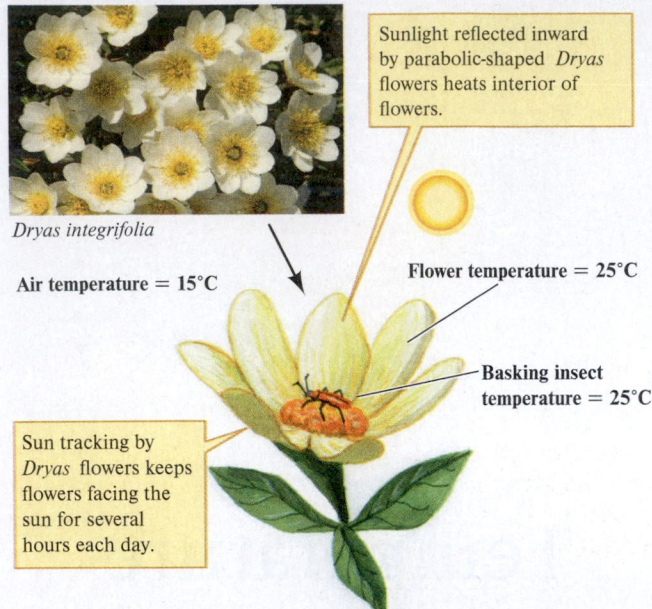

Dryas integrifolia

Sunlight reflected inward by parabolic-shaped *Dryas* flowers heats interior of flowers.

Air temperature = 15°C

Flower temperature = 25°C

Basking insect temperature = 25°C

Sun tracking by *Dryas* flowers keeps flowers facing the sun for several hours each day.

Figure 5.1   Sun-tracking behavior of the arctic plant *Dryas integrifolia* heats the reproductive parts of its flowers, making them attractive to pollinating insects.   ©All Canada Photos/Alamy

temperature is determined by the average kinetic energy of the water molecules in a mass of water. The kinetic energy in a mass of a substance is generally referred to as heat energy or simply as **heat.** Temperature is one of the most ecologically significant environmental factors. Consequently, many organisms have evolved varied mechanisms for regulating the temperature of their bodies or the temperature of parts of their anatomy. For example, at least one plant of the arctic tundra regulates the temperature of its reproductive structures. Peter Kevan went to Ellesmere Island, which lies at about 82° N latitude in the Northwest Territories of Canada, to study sun-tracking behavior by arctic flowers. It was summer, there was little wind, and the sun stayed above the horizon 24 hours each day. As the sun's position in the arctic sky changed, the flowers of one of the common tundra plants, *Dryas integrifolia,* like the sunflowers of lower latitudes, followed.

Kevan discovered that the flowers act like small solar reflectors; their parabolic shape reflects and concentrates solar energy on the reproductive structures (Kevan 1975). He also observed that many species of small insects, attracted by their warmth, basked in the sun-tracking *Dryas* flowers, elevating their body temperatures as a consequence (fig. 5.1). *Dryas* depends on these insects to pollinate its flowers. Since that time, Kevan and others have determined that there are several different ways that flowers can increase their temperature. In addition to the shape, color, and movement to follow the sun, patterns of opening and closing can protect against extreme weather, and some flowers have tiny hairs (called **pubescence**) that increase heat retention (van der Kooi et al. 2019).

Environmental temperatures are also of fundamental importance to humans. Why? The fundamental importance of temperature to organisms, including humans, is a consequence of temperature's influences on rates of chemical reactions, including those reactions that control life's essential processes, for example, photosynthesis and respiration. In addition, as we will see, all organisms are best adapted to a fairly narrow range of temperatures. For us and all other species, the impact of extreme temperatures can range from discomfort, at a minimum, to extinction. Long-term changes in temperature have set entire floras and faunas marching across continents, some species thriving, some holding on in small refuges, and others becoming extinct. Areas now supporting temperate species were at times tropical and at other times the frigid homes of reindeer and woolly mammoths.

The dynamic environmental history of the earth has become more significant as we face the reality of rapidly rising global temperatures (see chapter 23) and their potential ecological impact. We are currently observing many changes in species traits (e.g., Fernández-Pascual et al. 2019), distributions (e.g., Schivo et al. 2019), growth (Dorado-Liñán et al. 2020), and fertility (Walsh et al. 2019), just to name a few effects, in response to warming temperatures across the globe. Such changes as these have implications for species relationships and whole ecosystems (e.g., Thierry et al. 2019; Wang et al. 2019). Some specific ways species have changed include earlier spring migrations (e.g., Horton et al. 2020), earlier reproduction (e.g., Munson and Sher 2015), extensions of geographic ranges toward the poles (e.g., McBride 2019), and altitudinal ranges up mountainsides (e.g., Malakoutikhah et al. 2020). Some of these responses can be attributed to phenotypic and/or behavioral plasticity (see chapter 4). However, many studies indicate rapid adaptation to the warming climate through natural selection.

We defined ecology as the study of the relationships between organisms and their environments. In chapter 5, we examine the relationship between organisms and temperature, one of the most important environmental factors and one of the most relevant in the face of today's rapidly changing climate.

5.1   Microclimates

LEARNING OUTCOMES

After studying this section you should be able to do the following:

5.3   Describe the difference between microclimate and macroclimate.

5.4   List the major factors contributing to microclimatic differences.

Macroclimate interacts with the local landscape to produce microclimatic variation in temperature.   **Macroclimate** is the prevailing weather conditions in a region over a long period of time, which we represented with climate diagrams in chapter 2. **Microclimate** is climatic variation on a scale of a few kilometers, meters, or even centimeters, usually measured over short periods of time. You acknowledge microclimate when you choose to stand in the shade on a summer's day or in the sun on a winter's day. Many organisms live out their lives in very small areas during periods of time ranging from days to a few months. For these organisms, macroclimate may be less important than microclimate. Microclimate is influenced by landscape features such as altitude, aspect, vegetation, color of the ground, and presence of boulders. The physical nature of water and soil reduces temperature variation in aquatic environments and in animal burrows.

Altitude

As we saw in chapter 2 (see fig. 2.39), temperatures are generally lower at high elevations. Along the elevational gradient presented in figure 2.39, average annual temperature is 11.1°C at 1,660 m compared to −3.7°C at 3,743 m. Several factors contribute to lower average temperatures at higher elevations. First, because atmospheric pressure decreases with elevation, air rising up the side of a mountain expands. The energy of motion (kinetic energy) required to sustain the greater movement of air molecules in the expanding air mass is drawn from the surroundings, which cool as a result. A second reason that temperatures are generally lower at higher elevations is that there is less atmosphere above sites at higher elevations to trap and radiate heat back to the ground.

Aspect

Topographic features such as hills and mountains create microclimates. Mountains and hillsides create these microclimates by shading parts of the land, while exposing other parts of the landscape to more direct sunlight. In the Northern Hemisphere, the shaded areas are on the north-facing sides, or *northern aspects,* of hills and mountains, which face away from the equator. In the Southern Hemisphere, the *southern aspect* faces away from the equator and so receives less direct sunlight.

Because of their contrasting microclimates, the northern and southern aspects of mountains may support very different types of vegetation (fig. 5.2). The greater density of conifer trees on the north-facing slope shown in figure 5.2 is paralleled in miniature on north- and south-facing dune slopes in the Negev Desert, where north-facing slopes support a higher density of crust-forming mosses. G. Kidron, E. Barzilay, and E. Sachs, earth scientists from Hebrew University Jerusalem, documented a possible physical basis for the differences in moss cover (Kidron, Barzilay, and Sachs 2000). They found

Figure 5.2   The north-facing slope at this site supports a dense conifer forest while the vegetation on the south-facing slope is mainly grassland.   ©muha04/Getty Images RF

that north-facing dune slopes are cooler: 7.8° to 9.2°C cooler at midday in winter and 1.8° to 2.5°C cooler at midday in summer. Following rainfall these north-facing slopes remain moist approximately 2.5 times longer than south-facing slopes as a result of lower evaporation rates.

Vegetation

Because they shade the landscape, plants create microclimates. For instance, trees, shrubs, and plant litter (fallen leaves, twigs, and branches) produce ecologically important microclimates in deserts. The desert landscape, which often consists of a mosaic of vegetation and bare ground, is also a patchwork of sharply contrasting thermal environments (Zhuang et al. 2020). Such a patchwork is apparent near Kemmerer, Wyoming, where microclimates separated by only a few meters can differ in temperature by 27°C (fig. 5.3).

Color of the Ground

Another factor that can significantly affect temperatures is the color of the ground. This statement may sound a bit odd if you

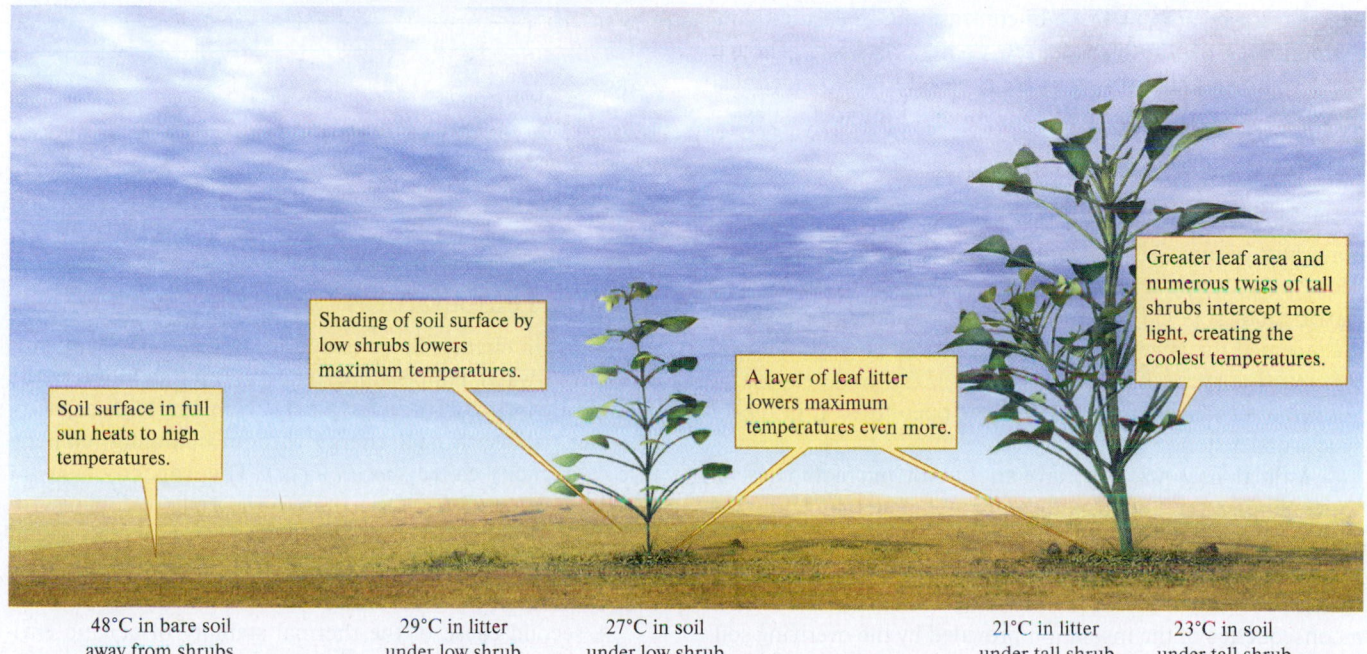

Soil surface in full sun heats to high temperatures.

Shading of soil surface by low shrubs lowers maximum temperatures.

A layer of leaf litter lowers maximum temperatures even more.

Greater leaf area and numerous twigs of tall shrubs intercept more light, creating the coolest temperatures.

| 48°C in bare soil away from shrubs | 29°C in litter under low shrub | 27°C in soil under low shrub | 21°C in litter under tall shrub | 23°C in soil under tall shrub |

Figure 5.3   Desert shrubs create distinctive thermal microclimates in the desert landscape (data from Parmenter, Parmenter, and Cheney 1989).

(a)

(b)

Figure 5.4   Ground color influences microclimate. For example, (*a*) white sand reflects most wavelengths of visible light, creating a much cooler microclimate compared to (*b*) black sand, which absorbs most wavelengths of visible light.   (*a*) ©Marc Shandro/Getty Images RF; (*b*) ©Medioimages/PunchStock RF

are from a moist climate, either temperate or tropical, where vegetation usually covers the ground. In contrast, much of an arid or semiarid landscape is bare ground, which can vary widely in color. Colors have been used to name deserts around the world, such as the central Asian deserts called Kara Kum, which means "black sand" in Turkish, and Kyzyl Kum, or "red sand," and White Sands, New Mexico (fig. 5.4*a*).

Bare ground is also the dominant environment offered by beaches. Neil Hadley and his colleagues (1992) studied the beaches of New Zealand, which range in color from white to black and offer a wide range of microclimates to beach organisms. These beaches heat up under the summer sun, but black beaches heat up faster and to higher temperatures. The black beaches heat up more because they absorb more visible light than do the white beaches (fig. 5.4*b*). When air temperatures at both beaches hovered around 30°C, Hadley and his colleagues found that the temperature of the sand on the white beach averaged around 45°C. In contrast, they measured sand temperatures on the black beach as high as 65°C. Though these white and black beaches are exposed to nearly identical macroclimates, they have radically different microclimates.

Presence of Boulders and Burrows

Many children soon discover that the undersides of stones harbor a host of organisms seldom seen in the open. This is partly because the stones create distinctive microclimates. E. B. Edney's classic studies (1953) of the seashore isopod *Ligia oceanica* documented the effect of stones on microclimate. Edney found that over the space of a few centimeters, *Ligia* could choose air temperatures ranging from 20°C in the open to 30°C in the air spaces under stones.

Animal burrows also have their own microclimates, in which temperatures are usually more moderate than at the soil surface. For example, while daily temperatures under a shrub in the Chihuahuan Desert ranged from 17.5° to 32°C, temperatures in a nearby mammal burrow ranged from 26° to 28°C, as a consequence of the insulation provided by the overlying soil. Figure 5.5 captures the insulating properties of soil at a site in the Chihuahuan Desert in central New Mexico.

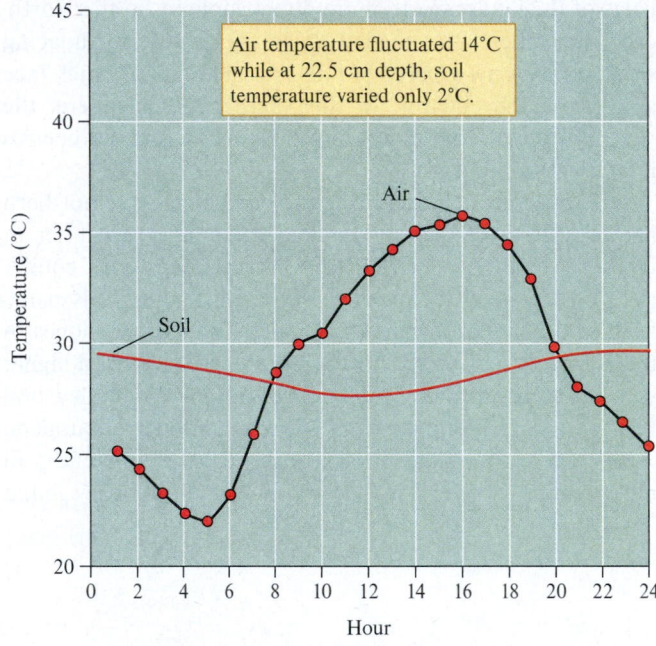

Figure 5.5   Air temperature and soil temperature at 22.5 cm below the soil surface on July 3, 2005, at the Sevilleta National Wildlife Refuge in central New Mexico (data courtesy of the Sevilleta LTER, University of New Mexico).

Aquatic Temperatures

As we saw in chapter 3, air temperature generally fluctuates more than water temperature. The thermal stability of the aquatic environment derives partly from the high capacity of water to absorb heat energy without changing temperature (a capacity called *specific heat*). This capacity is about 3,000 times higher for water than for an equal volume of air. It takes approximately 1 calorie of energy to heat 1 cm^3 of water 1°C. For an equal volume of air, this temperature rise requires only about 0.0003 calorie.

A second cause of the thermal stability of aquatic environments is the large amount of heat absorbed by water as it evaporates (which is called the *latent heat of vaporization*).

This amounts to about 584 calories per gram of water at 22°C and 580 calories per gram of water at 35°C. So, 1 g of water evaporating from the surface of a desert stream, a lake, or a tide pool at 35°C draws 580 calories of heat from its surroundings. From the definition of a calorie, this is enough energy to cool 580 g of water 1°C.

A third cause of the greater thermal stability of aquatic environments is the heat energy that water gives up to its environment as it freezes (the *latent heat of fusion*). Water gives up approximately 80 calories as 1 g of water freezes and the energy of motion of water molecules decreases as they leave the liquid state and become incorporated into the crystalline latticework of ice. So, as 1 g of pond water freezes, it gives off sufficient energy to heat 80 g of water 1°C, thus retarding further cooling.

The aquatic environments with greatest thermal stability are generally large ones, such as the open sea. These are environments that store large quantities of heat energy and where daily fluctuations are often less than 1°C. Even the temperatures of small streams, however, usually fluctuate less than the temperatures of nearby terrestrial habitats.

Other factors besides the physics of water can affect the temperature of aquatic environments. **Riparian vegetation,** that is, vegetation that grows along rivers and streams, influences the temperature in streams in the same way that vegetation modifies the temperature of desert soils—by providing shade. This can be seen in the modeling of temperatures in **Girnock Burn, a tributary of the River Dee in Aberdeenshire, Scotland** by Grace Garner and colleagues (2017). They found that mean water temperature decreased with canopy density, but that temperature was also more variable throughout the day for intermediate canopy density, relative to either high or low density (fig. 5.6). This was because over the course of a day, the movement of the sun meant that there could be both periods when the water would be in the shade or in the sun (fig. 5.6).

Concept 5.1 Review

1. Why does the temperature of streams vary throughout the day, and how is this affected by stream side vegetation?
2. Why is evaporative cooling by various animal species so effective?
3. Contrast the microclimates of the aboveground parts of desert plants to that of their roots.

5.2 Evolutionary Trade-Offs

LEARNING OUTCOMES

After studying this section you should be able to do the following:

5.5 Define the principle of allocation.
5.6 Discuss experimental evidence supporting the principle of allocation.
5.7 Explain the significance of the principle of allocation to the evolutionary process of adaptation.

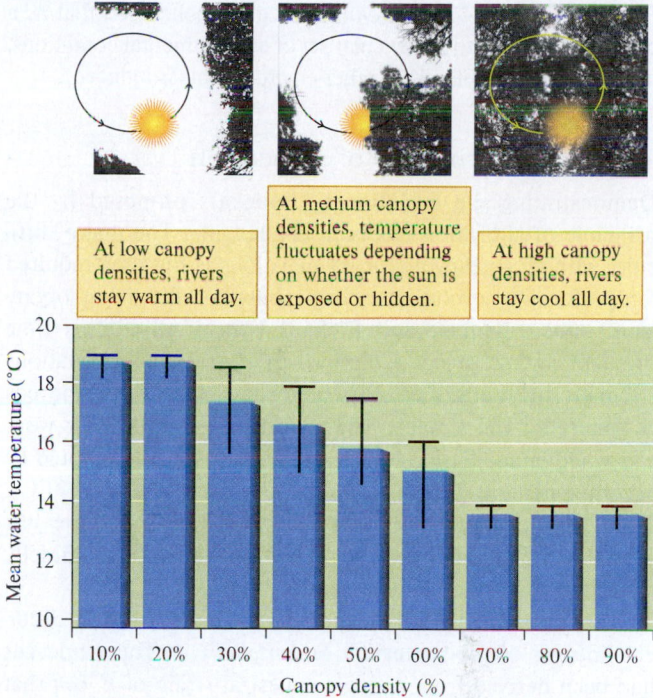

Figure 5.6 The relationship between degree of tree canopy cover over a stream and water temperature, modeled from data collected along a tributary of the River Dee in Aberdeenshire, Scotland by Grace Garner and colleagues. Column height shows mean water temperature in the stream for each of nine degrees of canopy cover, with error bars showing 90% ranges of values throughout the span of a day. Above the graph are the types of images taken by the researchers to quantify canopy cover; the curved arrows show the path of the sun over the course of a particular day. Variance of temperature is due to degree of sun exposure (Garner et al. 2017).

Adapting to one set of environmental conditions generally reduces a population's fitness in other environments. We might imagine an organism that not only is capable of living in any environment but also thrives in all environments. In the language of evolution such an organism would have high fitness across all environmental conditions. In everyday language we might refer to such a life-form as a "super" organism. Whatever we might call them, however, such life-forms do not, as far as we know, exist. All known organisms are adapted to a limited range of environmental conditions, at least partially as a consequence of energy limitation.

The Principle of Allocation

All organisms have access to limited energy supplies. We introduce the concept of energy limitation here and will examine it in detail later (see chapter 7) because it comes with significant evolutionary consequences. One of those consequences is that energy allocated to one of life's functions, such as reproduction, defense against disease, or growth, will reduce the amount of energy available for other functions. Darwin appreciated the evolutionary implications of energy limitation and included it in his writings. However, Richard Levins was the first to use a mathematical approach to analyze the evolutionary consequences of such trade-offs, which he referred to as the **principle of allocation** (Levins 1968, p. 15). In his classic book,

Evolution in Changing Environments, Levins concluded that as a population adapts to a particular set of environmental conditions, its fitness (see chapter 4) in other environments is reduced.

Testing the Principle of Allocation

Demonstrating the evolutionary trade-offs proposed by the principle of allocation has been challenging. The major difficulty with all such evolutionary questions is the time required for performing evolutionary experiments with living organisms. Albert Bennett and Richard Lenski solved this time problem by studying evolution by microbial populations (Bennett and Lenski 2007), which can go through hundreds of generations in a week. The central question of their work was whether adaptation to a low temperature (20°C) would be accompanied by a loss of fitness at a high temperature (40°C). Their working hypothesis was that they would observe just such a trade-off in fitness, a prediction that follows directly from Levins's principle of allocation.

Bennett and Lenski's experiments focused on 24 different lineages of the bacterium *Escherichia coli.* These lineages had been derived from a single ancestral strain of *E. coli* that had been grown at 37°C (human body temperature) for 2,000 generations. Bennett and Lenski used this ancestral strain to establish six replicate populations at four temperature regimes: constant 32°, 37°, and 42°C, and daily alternation between 32° and 42°C. They maintained these 24 populations at these temperatures for 2,000 generations, sufficient time for each population to adapt to its particular temperature regime. (Evolutionary research involving so many generations is feasible only with microbial populations, since they have short generation times. For example, the generation time for *E. coli* growing at 40°C is approximately 20 minutes!) Bennett and Lenski next used bacterial cells from each of their 24 populations to establish 24 new populations, which were all grown at 20°C for 2,000 generations, theoretically adapting to this relatively low temperature in the process.

To address their original question (Will adaptation to a low temperature [20°C] be accompanied by a loss of fitness at a high temperature [40°C]?), Bennett and Lenski compared the fitness of the low-temperature-selected line with the fitness of the ancestral line at 20°C and at 40°C. Their measure of fitness was the rate of population doubling of a selected line of *E. coli* compared to that of its ancestral line, when with the two lines were grown together. Two major results stand out. First, the lines grown at 20°C had *higher* (positive) *fitness* at 20°C temperature compared to their immediate ancestors. In other words, Bennett and Lenski showed that their lines grown at 20°C for 2,000 generations had indeed adapted to that lower temperature. However, the lines that had adapted to 20°C had, on average, *lower* (negative) *fitness* compared to their immediate ancestor, when grown at 40°C. Therefore, as predicted by the principle of allocation, selection for higher fitness at 20°C had been accompanied by an average loss in fitness at higher temperatures (fig. 5.7).

Bennett and Lenski's results provide the first direct experimental evidence in support of Levins's principle of allocation.

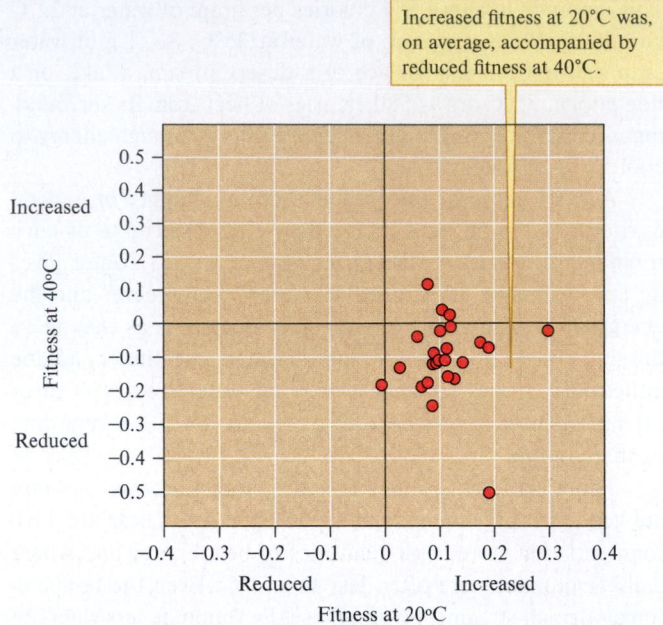

Figure 5.7   *Escherichia coli* grown at 20°C for 2,000 generations showed increased fitness at that temperature compared to ancestral lines, which were adapted to higher temperatures. However, they had reduced fitness at 40°C compared to ancestral lines (data from Bennett and Lenski 2007).

This principle, in turn, offers an explanation for the observation that most organisms perform best under a limited range of environmental conditions, including thermal conditions.

Concept 5.2 Review

1. If growing lines of *Escherichia coli* at 20°C for 2,000 generations increased their fitness at 20°C without reducing their fitness at 40°C, how would the distribution of points in figure 5.7 change?
2. If your research team obtained the hypothetical results described in question 1, what could you conclude about the principle of allocation?

5.3 Temperature and Performance of Organisms

LEARNING OUTCOMES

After studying this section you should be able to do the following:

5.8   List several measures of animal, plant, and microbial performance.

5.9   Write a summary equation for the process of photosynthesis.

5.10 Explain why the process of acclimation to temperature is not an evolutionary process.

5.11 Discuss how evidence that organisms are adapted to a restricted range of temperatures supports the principle of allocation (see section 5.2).

Most species perform best in a fairly narrow range of temperatures. Ecologists concerned with the ecology of individual organisms study how environmental factors, such as temperature, water, and light, affect the physiology and behavior of organisms: how fast they grow; how many offspring they produce; how fast they run, fly, or swim; how well they avoid predators; and so on. We can group these phenomena and say that ecologists study how environment affects the "performance" of organisms.

Whether in response to variations in temperature, moisture, light, or nutrient availability, most species do not perform equally well across the full range of environmental conditions to which they are exposed; most perform best under a narrower range of conditions.

Let's begin our discussion of temperature and animal performance by reviewing the influence of temperature on enzyme function. Enzymes usually work best in some intermediate range of temperatures, neither too hot nor too cold, where they retain both proper shape, to bind with the substrate, and sufficient flexibility, to perform their particular function. In other words, there is usually some optimal range of temperatures for most enzymes. How might you determine the optimal temperature for an enzyme? One way that molecular biologists assess the optimal conditions for enzyme performance is to determine the concentration of substrate required for an enzyme to work at a particular rate. If this concentration is low, the enzyme is performing well at low concentrations of the substrate; that is, the enzyme has a high affinity for the substrate. The affinity of an enzyme for its substrate is one measure of its performance.

John Baldwin and P. W. Hochachka (1970) studied the influence of temperature on the activity of acetylcholinesterase, an enzyme produced at the synapse between neurons. This enzyme promotes the breakdown of the neurotransmitter acetylcholine to acetic acid and choline and so turns off neurons, a process critical for proper neural function. The researchers found that rainbow trout, *Oncorhynchus mykiss,* produce two forms of acetylcholinesterase. One form has highest affinity for acetylcholine at 2°C, that is, at winter temperatures. However, the affinity of this enzyme for acetylcholine declines rapidly above 10°C. The second form of acetylcholinesterase shows highest affinity for acetylcholine at 17°C, at summer temperatures. However, the affinity of this second form of acetylcholinesterase falls off rapidly at both higher and lower temperatures. In other words, the optimal temperatures for the two forms of acetylcholinesterase are 2° and 17°C (fig. 5.8).

This influence of temperature on the performance of acetylcholinesterase makes sense if you consider the temperatures of the rainbow trout's native environment. Rainbow trout are native to the cool, clear streams and rivers of western North America. During winter, the temperatures of these streams hover between 0° and 4°C, while summer temperatures approach 20°C. These environmental temperatures are similar to the temperatures at which the acetylcholinesterase of rainbow trout performs optimally.

Studies of reptiles, especially lizards and snakes, are offering additional valuable insights into the influence of temperature on animal performance. Widely distributed species often offer the opportunity for studies of local variation in ecological relationships, including the influence of temperature on performance. For example, the eastern fence lizard, *Sceloporus undulatus,* is found across approximately two-thirds of the United States, living in a broad diversity of climatic zones (fig. 5.9). Taking advantage of this wide range of environmental conditions, Michael Angilletta (2001) studied the temperature relations of *S. undulatus* over a portion of its range. In one of his studies, Angilletta determined how temperature influences metabolizable energy intake, or MEI. He measured MEI as the amount of energy consumed (C) minus energy lost in feces (F) and uric acid (U), the nitrogen waste product produced by lizards. We can summarize MEI in equation form as:

$$MEI = C - F - U$$

Angilletta studied a population from New Jersey and one from South Carolina, regions with substantially different climates. He collected a sample of lizards from both populations and maintained lizards from both populations at 30°, 33°, and 36°C. Angilletta kept his study lizards in separate enclosures and fed them crickets that he had weighed to the nearest 0.1 mg. Since he had determined the energy content of an average cricket, Angilletta was able to determine the energy intake by each lizard by counting the number of crickets it ate and calculating the energy content of that number. He determined the energy lost as feces (F) and uric acid (U) by collecting all the feces and uric acid produced by each lizard and then drying and weighing this material. He estimated the average energy content of feces and uric acid using a bomb calorimeter.

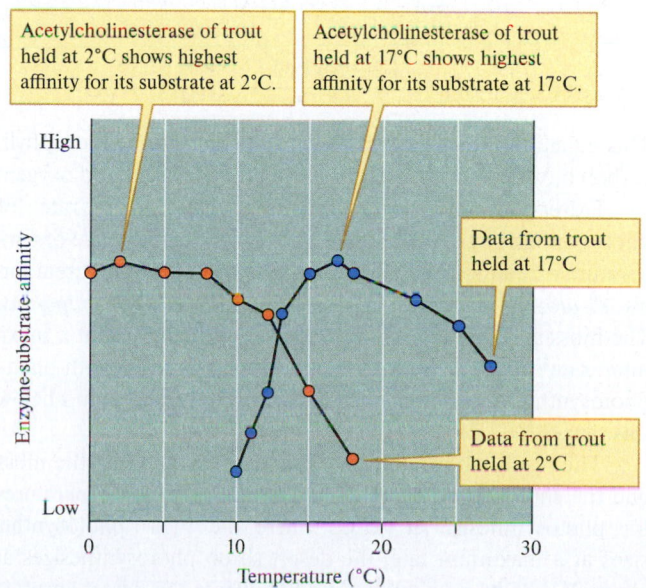

Figure 5.8 Enzyme activity is affected substantially by temperature (data from Baldwin and Hochachka 1970).

Figure 5.9 *Sceloporus undulatus,* the eastern fence lizard, is one of the most widely distributed lizard species in North America.   ©Suzanne L. and Joseph T. Collins/Science Source

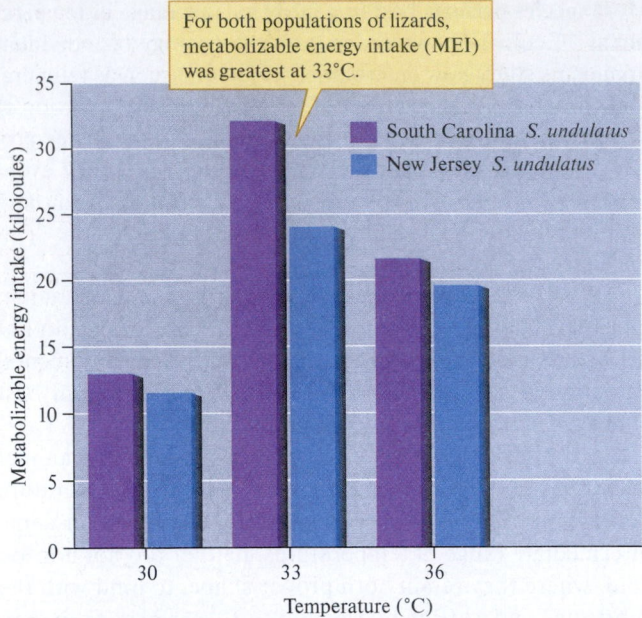

For both populations of lizards, metabolizable energy intake (MEI) was greatest at 33°C.

■ South Carolina *S. undulatus*
■ New Jersey *S. undulatus*

Figure 5.10   The rate of metabolizable energy intake by two populations of the eastern fence lizard, *Sceloporus undulatus,* peaks at the same temperature (data from Angilletta 2001).

The results of Angilletta's experiment, which are shown in figure 5.10, show clearly that MEI is highest in both populations of lizards at the intermediate temperature of 33°C. Note, however, that the South Carolina lizards took in energy at a much higher rate at 33°C. Despite this difference, lizards from both populations performed best in a fairly narrow range of temperatures. Analogous influences of temperature on performance have also been well documented in plants.

Extreme Temperatures and Photosynthesis

One of the most fundamental characteristics of plants is their ability to photosynthesize. **Photosynthesis,** the conversion of light energy to the chemical energy of organic molecules, is the basis for the life of plants—their growth, reproduction, and so on—and the ultimate source of energy for most consumer organisms.

Photosynthesis can be summarized by the following equation:

$$6CO_2 + 6H_2O \xrightarrow[\text{Chlorophyll}]{\text{Light}} C_6H_{12}O_6 + 6O_2$$

This equation indicates that as light interacts with chlorophyll, carbon dioxide and water combine to produce sugar and oxygen.

Extreme temperatures generally reduce the rate of photosynthesis by plants. Figure 5.11 shows the influence of temperature on rate of photosynthesis by a moss from a boreal forest, *Pleurozium schreberi,* and a desert shrub, *Atriplex lentiformis.* The moss and the desert shrub both photosynthesize at a maximum rate over some narrow range of temperatures. Both plants photosynthesize at lower rates at temperatures above and below this range.

The results shown in figure 5.11 demonstrate that the moss and the shrub have substantially different optimal temperatures for photosynthesis. At 15°C, where the moss photosynthesizes at a maximum rate, the desert shrub photosynthesizes at about 25% of its maximum. At 44°C, where the desert shrub is

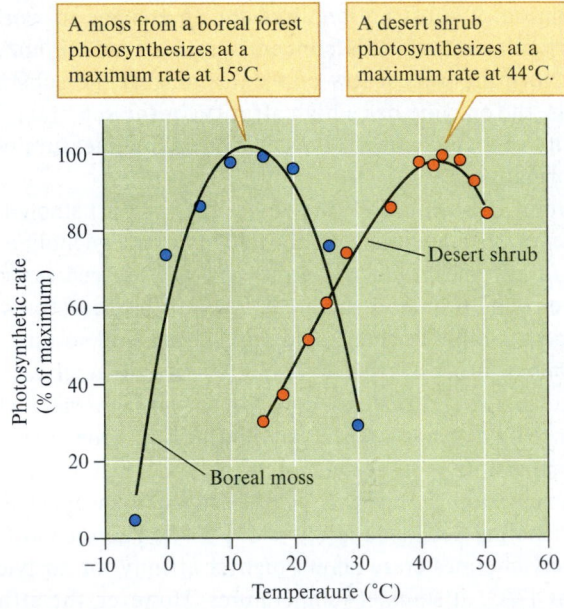

A moss from a boreal forest photosynthesizes at a maximum rate at 15°C.

A desert shrub photosynthesizes at a maximum rate at 44°C.

Desert shrub

Boreal moss

Figure 5.11   The optimal temperatures for photosynthesis by a boreal forest moss and a desert shrub differ substantially (data from Kallio and Kärenlampi 1975; Pearcy and Harrison 1974).

photosynthesizing at its maximum rate, the moss would probably die. These physiological differences clearly reflect differences in the environments where these species live and to which they are adapted. While the moss lives in the cool boreal forests of Finland, the study population of the desert shrub, *A. lentiformis,* lives near Thermal, California, in one of the hottest deserts on earth.

Plant responses to temperature, as well as those of animals, can also reflect the short-term physiological adjustments called **acclimation.** Acclimation involves physiological, not genetic, changes in response to temperature; acclimation is generally reversible with changes in environmental conditions. Studies of *A. lentiformis* by Robert Pearcy (1977) clearly demonstrate the effect of acclimation on photosynthesis. Pearcy located a population of this desert shrub in Death Valley, California, and grew plants for his experiments from cuttings. By propagating plants from cuttings, he was able to conduct his experiments on genetically identical clones. The clones from the Death Valley plants were grown under two temperature regimes: one set in "hot" conditions of 43°C during the day and 30°C at night; the other set under cool conditions of 23°C during the day and 18°C at night.

Pearcy then measured the photosynthetic rates of the two sets of plants. The plants grown in a cool environment photosynthesized at a maximum rate at about 32°C. Those grown in a hot environment photosynthesized at a maximum rate at 40°C, a difference in the optimal temperature for photosynthesis of 8°C. Figure 5.12 summarizes the results of Pearcy's experiment.

The physiological adjustments made by *A. lentiformis* correspond to what these plants do during an annual cycle. The plant is evergreen and photosynthesizes throughout the year, in the cool of winter and in the heat of summer. The physiological adjustments suggest that acclimation by *A. lentiformis* may shift its optimal temperature for photosynthesis to match seasonal changes in environmental temperature.

Temperature and Microbial Activity

Microbes appear to have adapted to all temperatures at which there is liquid water, from the frigid waters around the Antarctic to boiling hot springs. However, while each of these environments harbors one or more species of microbes, no known species thrives in all these conditions. All microbes that have been studied, like the plants and animals discussed in this section, perform best over a fairly narrow range of temperatures. Let's look at two microbes that live in environments at opposite extremes of the aquatic temperature spectrum.

In chapter 3, we saw that most of the oceanic environment, the largest continuous environment on the earth, lies below the well-lighted surface waters. The organisms that live in the deep oceans live in darkness. Their environment is also cold, generally below 5°C. This cold-water environment extends to the surface in the Arctic and Antarctic. A wide variety of organisms live in these cold waters. How do you think the performance of these organisms is affected by temperature?

Richard Morita (1975) studied the effect of temperature on population growth among cold-loving, or **psychrophilic,** marine bacteria that live in the waters around Antarctica. He isolated and cultured one of those bacteria, *Vibrio* sp., in a temperature-gradient incubator for 80 hours. During the experiment, the temperature gradient within the incubator ranged from about −2°C to just over 9°C. The results of the experiment show that this *Vibrio* sp. grows fastest at about 4°C. At temperatures above and below this, its population growth rate decreases. As figure 5.13 shows, Morita recorded some growth in the *Vibrio* population at temperatures approaching −2°C;

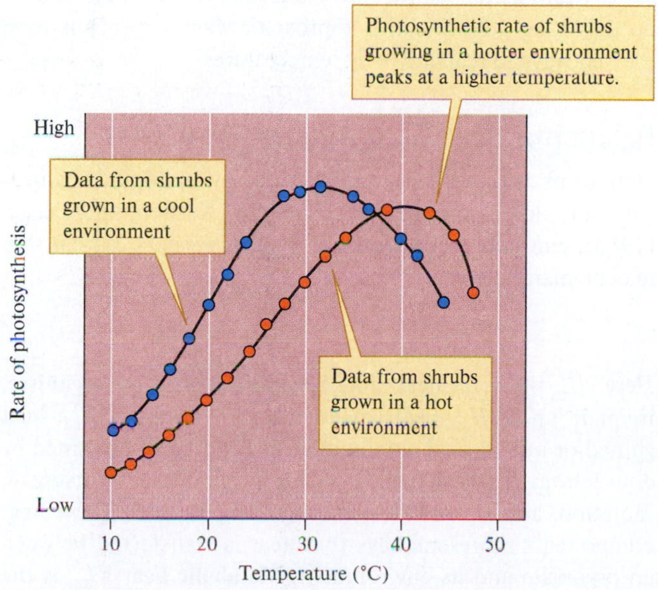

Figure 5.12 Growing the same species of shrub in cool versus hot environments altered their optimal temperature for photosynthesis. This change was a short-term physiological adjustment due to acclimation (data from Berry and Björkman 1980, after Pearcy 1977).

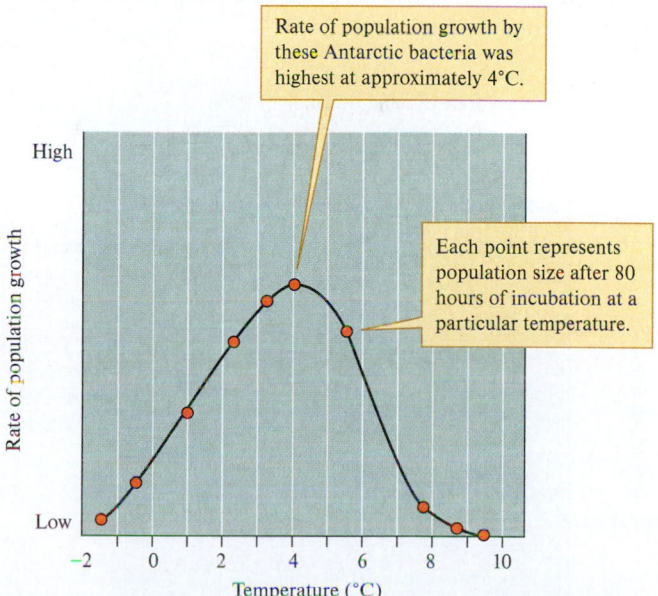

Figure 5.13 Antarctic bacteria have a very low optimal temperature for population growth (data from Morita 1975).

however, populations did not grow at temperatures above 9°C. Morita has recorded population growth among some cold-loving bacteria at temperatures as low as −5.5°C.

Some microbes can live at very high temperatures. Microbes have been found living in all of the hot springs that have been studied. Some of these heat-loving, or **thermophilic,** microbes grow at temperatures above 40°C in a variety of environments. The most heat-loving microbes are the hyperthermophiles, which have temperature optima above 80°C. Some hyperthermophiles grow best at 110°C! Some of the most intensive studies of thermophilic and hyperthermophilic microbes have been carried out in Yellowstone National Park by Thomas Brock (1978) and his students and colleagues. One of the genera they have studied is *Sulfolobus,* a member of the microbial domain Archaea, which obtains energy by oxidizing elemental sulfur. Jerry Mosser and colleagues (1974) used the rate at which *Sulfolobus* oxidizes sulfur as an index of its metabolic activity. They studied the microbes from a series of hot springs in Yellowstone National Park that ranged in temperature from 63° to 92°C. The temperature optimum for the *Sulfolobus* populations ranged from 63° to 80°C and was related to the temperature of the particular spring from which the microbes came. For instance, one strain isolated from a 59°C spring oxidized sulfur at a maximum rate at 63°C. This *Sulfolobus* population oxidizes sulfur at a high rate within a temperature range of about 10°C (fig. 5.14). Outside of this temperature range, its rate of sulfur oxidation is much lower.

We have reviewed how temperature can affect microbial activity, plant photosynthesis, and animal performance. These examples demonstrate that most organisms perform best over a fairly narrow range of temperatures. In the next Concept, we review how some organisms regulate their body temperatures in response to spatial and temporal variation in environmental temperatures.

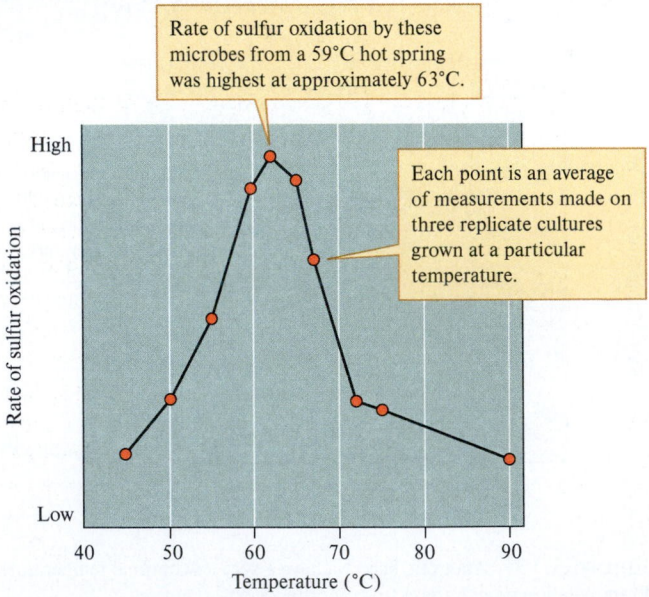

Figure 5.14 Hot spring microbes have a very high optimal temperature for population growth (data from Mosser, Mosser, and Brock 1974).

Concept 5.3 Review

1. Signs of thermal stress in fish include swimming on their sides and swimming in spirals. Using what you know about temperature and acetylcholinesterase, explain.
2. How can we be sure that the two distinctive responses to temperature shown by *Atriplex lentiformis* were due to acclimation and not the result of genetic differences (see fig. 5.12)?

5.4 Regulating Body Temperature

LEARNING OUTCOMES

After studying this section you should be able to do the following:

5.12 Define the terms *poikilotherm, ectotherm, endotherm,* and *homeotherm.*

5.13 Outline the process of thermoregulation using an equation that includes all major sources of heat gain and loss.

5.14 Compare thermoregulation by endotherms and ectotherms.

5.15 Explain the difficulty of being an endotherm in an aquatic environment.

Many organisms have evolved ways to compensate for variations in environmental temperature by regulating body temperature. So, how do organisms respond to the juxtaposition of thermal heterogeneity in the environment and their own fairly narrow thermal requirements? Do they sit passively and let environmental temperatures affect them as they will, or do they take a more active approach? Many organisms have evolved ways to regulate body temperatures.

Balancing Heat Gain Against Heat Loss

Organisms regulate body temperature by manipulating heat gain and loss. An equation, used by K. Schmidt-Nielsen (1983), can help us understand the components of heat that may be manipulated:

$$H_s = H_m \pm H_{cd} \pm H_{cv} \pm H_r - H_e$$

Here, H_s, the total heat stored in the body of an organism, is made up of H_m, heat gained from metabolism; H_{cd}, heat gained or lost through conduction; H_{cv}, heat lost or gained by convection; H_r, heat gained or lost through electromagnetic radiation; and H_e, heat lost through evaporation. These heat components represent ways that heat is transferred between an organism and its environment. **Metabolic heat,** H_m, is the energy released within an organism during the process of cellular respiration. **Conduction** is the transfer of heat between objects in direct physical contact, as occurs when you sit on a stone bench on a cold winter's day; **convection** is the process of heat flow between a solid body and a moving fluid, such as between you and wind on a cold day. During the process of conduction or convection, H_{cd} and H_{cv}, the direction of heat flow is always from the warmer region to the colder.

Heat may also be transferred through electromagnetic radiation. This transfer of heat, H_r, is often called simply **radiation.** All objects above absolute 0, above $-273°C$, give off electromagnetic radiation, but the most obvious source in our environment is the sun. Curiously, we are blind to most of this heat flux, because at sea level over half of the energy content of sunlight falls outside our visible range. Much of this radiation that we cannot see is in the infrared part of the spectrum. The electromagnetic radiation emitted by most objects in our environment, including our own bodies, is also infrared light. Infrared light is responsible for most of the warmth you feel when standing in front of a fire or that you feel radiating from the sunny side of a building on a winter's day. The chilling effect of standing outdoors under a clear, cold night sky with no wind is also mainly due to radiative heat flux, in this case from your body to the surroundings, including the night sky.

Heat may be lost by an organism through **evaporation,** H_e. In general, we need only consider the heat lost as water evaporates from the surface of an organism. The ability of water to absorb a large amount of heat as it evaporates makes cooling systems based on the evaporation of water very effective. Figure 5.15 summarizes the potential pathways by which heat can be transferred between an organism and the environment.

So, how can organisms regulate body temperature? First of all, many organisms don't. The body temperature of these organisms, called **poikilotherms,** varies directly with environmental temperatures. Of the organisms that regulate body temperature, most use external sources of energy and a combination of anatomy and behavior to manipulate H_c, H_r, and H_e. Animals that rely mainly on external sources of energy for regulating body temperature are called **ectotherms.** Organisms

that rely heavily on internally derived metabolic heat energy, H_m, are called **endotherms.** Among endotherms, birds and mammals use metabolic energy to heat most of their bodies. Other endothermic animals, including certain fish and insects, use metabolic energy to selectively heat critical organs. Endotherms that use metabolic energy to maintain a relatively constant body temperature are called **homeotherms.** The only known homeothermic organisms are birds, mammals, and some deep-sea fish such as the opah, *Lampris guttatus.*

Temperature regulation presents both plants and ectothermic animals with a similar problem. Both groups of organisms rely primarily on external sources of energy. Despite the much greater mobility of most ectothermic animals, the ways in which plants and ectothermic animals solve these problems are similar.

Temperature Regulation by Plants

What sorts of environments are best for studying temperature regulation by plants? Plant ecologists have typically concentrated their studies in extreme environments, such as the desert and tundra, where the challenges of the physical environment are greater and where ecologists believed they would find the most dramatic adaptations.

Desert Plants

The desert environment challenges plants to avoid overheating; that is, plants are challenged to reduce their heat storage, H_s. How do desert plants meet this challenge? They, like plants from other environments, use morphology and behavior to alter heat exchange with the environment. Evaporative cooling of leaves, which would increase heat loss, H_e, is not a workable option because desert plants usually have inadequate supplies of water. Also, for most plants, we can ignore H_m. Most produce only a small quantity of heat by metabolism. So, for a plant in a hot desert environment, our equation for heat balance reduces to:

$$H_s = H_{cd} \pm H_{cv} \pm H_r$$

To avoid heating, plants in hot deserts have three main options: decreasing heating by conduction, H_{cd}; increasing rates of convective cooling, H_{cv}; and reducing rates of radiative heating, H_r. Many desert plants place their foliage far enough above the ground to reduce heat gain by conduction. Many desert plants have also evolved very small leaves and an open growth form, adaptations that give high rates of convective cooling because they increase the ratio of leaf surface area to volume and the movement of air around the plant's stems and foliage. Some desert plants have low rates of radiative heat gain, H_r, because they have evolved reflective surfaces. Many desert plants cover their leaves with a dense coating of white plant hairs. These hairs reduce H_r gain by reflecting visible light, which constitutes nearly half the energy content of sunlight.

We can see how natural selection has adapted plants to different temperature regimes by comparing species in the genus *Encelia*, which are distributed along a temperature and moisture gradient from the coast of California to Death Valley. James Ehleringer (1980) showed that the leaves of the coastal

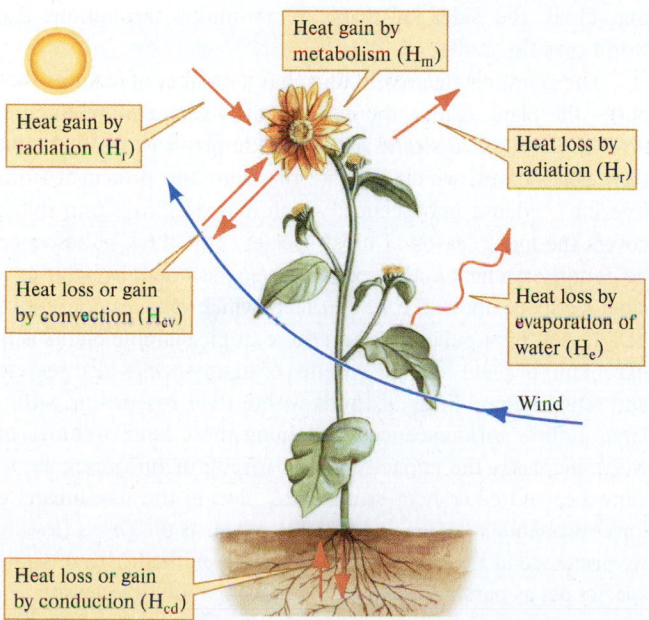

Heat gain by metabolism (H_m)

Heat gain by radiation (H_r)

Heat loss by radiation (H_r)

Heat loss or gain by convection (H_{cv})

Heat loss by evaporation of water (H_e)

Wind

Heat loss or gain by conduction (H_{cd})

Figure 5.15   The multiple pathways for heat exchange between organisms and the environment.

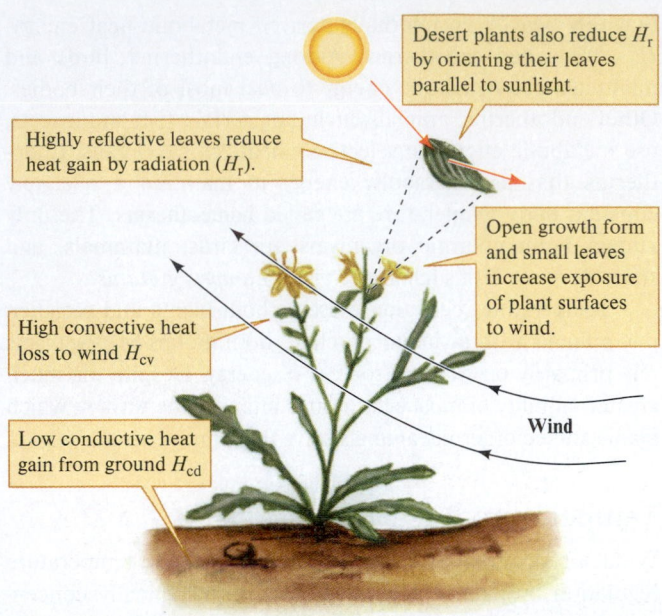

Desert plants also reduce H_r by orienting their leaves parallel to sunlight.

Highly reflective leaves reduce heat gain by radiation (H_r).

High convective heat loss to wind H_{cv}

Low conductive heat gain from ground H_{cd}

Open growth form and small leaves increase exposure of plant surfaces to wind.

Wind

Figure 5.16   The form and orientation of desert plants reduce heat gain from the environment and facilitate cooling.

species, *Encelia californica,* lack hairs entirely and reflect only about 15% of visible light. He also found that two other species that grow partway between the cool coast regions and Death Valley produce leaves that are somewhat pubescent and reflect about 26% of visible light. The desert species, *Encelia farinosa,* produces two sets of leaves, one set in the summer and another when it's cooler. The summer leaves are highly pubescent (hairy) and reflect more than 40% of solar radiation while the cool season leaves are not.

Plants can also modify radiative heat gain, H_r, by changing the orientation of leaves and stems. Many desert plants reduce heating by orienting their leaves parallel to the rays of the sun or by folding them at midday, when sunlight is most intense. Figure 5.16 portrays the main processes involved in heat balance in desert plants.

Arctic and Alpine Plants

As you would probably predict, temperature regulation by plants in cold regions, in most cases, contrasts sharply with temperature regulation by desert plants. However, we can model temperature regulation by plants from cold environments using the same equation we used for heat regulation in desert plants: $H_s = H_{cd} \pm H_{cv} \pm H_r$.

The problem here, though, is staying warm, and arctic and alpine plants have two main options: increase their rate of radiative heating, H_r, and/or decrease their rate of convective cooling, H_{cv}. It appears that many plants have evolved to do both and, as a result, can heat up to temperatures far above air temperature. So, while favoring desert plants that reflect light, natural selection has favored arctic and alpine plants that absorb light with dark pigments. These dark pigments increase radiative heat gain, H_r. Arctic and alpine plants, such as the *Dryas integrifolia* (see fig. 5.1), also increase their H_r gain by

orienting their leaves and flowers perpendicular to the sun's rays. In addition, many plants increase their H_r gain from the surroundings by assuming a "cushion" growth form that "hugs" the ground. The ground often warms to temperatures exceeding that of the overlying air and radiates infrared light, which can be absorbed by cushion plants. Cushion plants can also gain heat from warm substrate through conduction, H_{cd}.

The cushion growth form also reduces convective heat loss, H_{cv}, in two main ways. First, growing close to the ground gives them some shelter from the wind. Second, the compact, hemispherical growth form of cushion plants reduces the ratio of surface area to volume, which slows the movement of air through the interior of the plant. Reduced surface area also reduces the rate of radiative heat loss.

Figure 5.17 summarizes the processes involved in thermal regulation by a cushion plant. As a consequence of these processes, cushion plants are often warmer than the surrounding air and compared to plants with other growth forms. Y. Gauslaa (1984), who studied the heat budgets of a variety of Scandinavian plants, documented the thermal consequences of the cushion growth form. He found that while the temperature of plants with an open growth form closely matches air temperature, the temperature of cushion plants can be over 10°C higher than air temperature. The results of one of Gauslaa's comparisons are shown in figure 5.18.

Tropical Alpine Plants

Some of the most amazing examples of thermoregulation occur among the plants that inhabit the far-flung world of the tropical alpine zone. This zone is a unique environment with little annual variation in temperature but with so much daily fluctuation that freezing temperatures at night are often followed by "summer" temperatures during the day. In this environment, natural selection has produced one of the most remarkable examples of convergence, the giant rosette plants that cloak the sides of tropical mountains throughout the world (see fig. 2.40).

The giant rosette growth form has a number of features that buffer the plant against the extreme daily temperature fluctuations of the tropical alpine zone. Rosette plants generally retain their dead leaves, which insulate the stem and protect it from freezing. A dense pubescence, which may be 2 to 3 mm thick, covers the living leaves of most species. This thick pubescence helps conserve heat in the cool alpine environment by creating a dead air space above the leaf surface, which reduces convective heat losses. Leaf pubescence on these tropical alpine plants acts like a kind of plant fur. The rosettes of many species also secrete and retain several liters of fluids within their rosettes or within large, hollow inflorescences. Retaining these large volumes of water increases the capacity of the rosette or inflorescence to store heat. Greater heat storage, H_s, during the day means a lower probability of freezing at night. Much as the *Dryas* flowers we discussed at the beginning of this chapter, the leaves of some species act as parabolic mirrors, increasing radiative heating, H_r, of the apical bud and expanding leaves. The rosettes of some tropical alpine plants even close over the apical bud at night, which protects the bud from freezing.

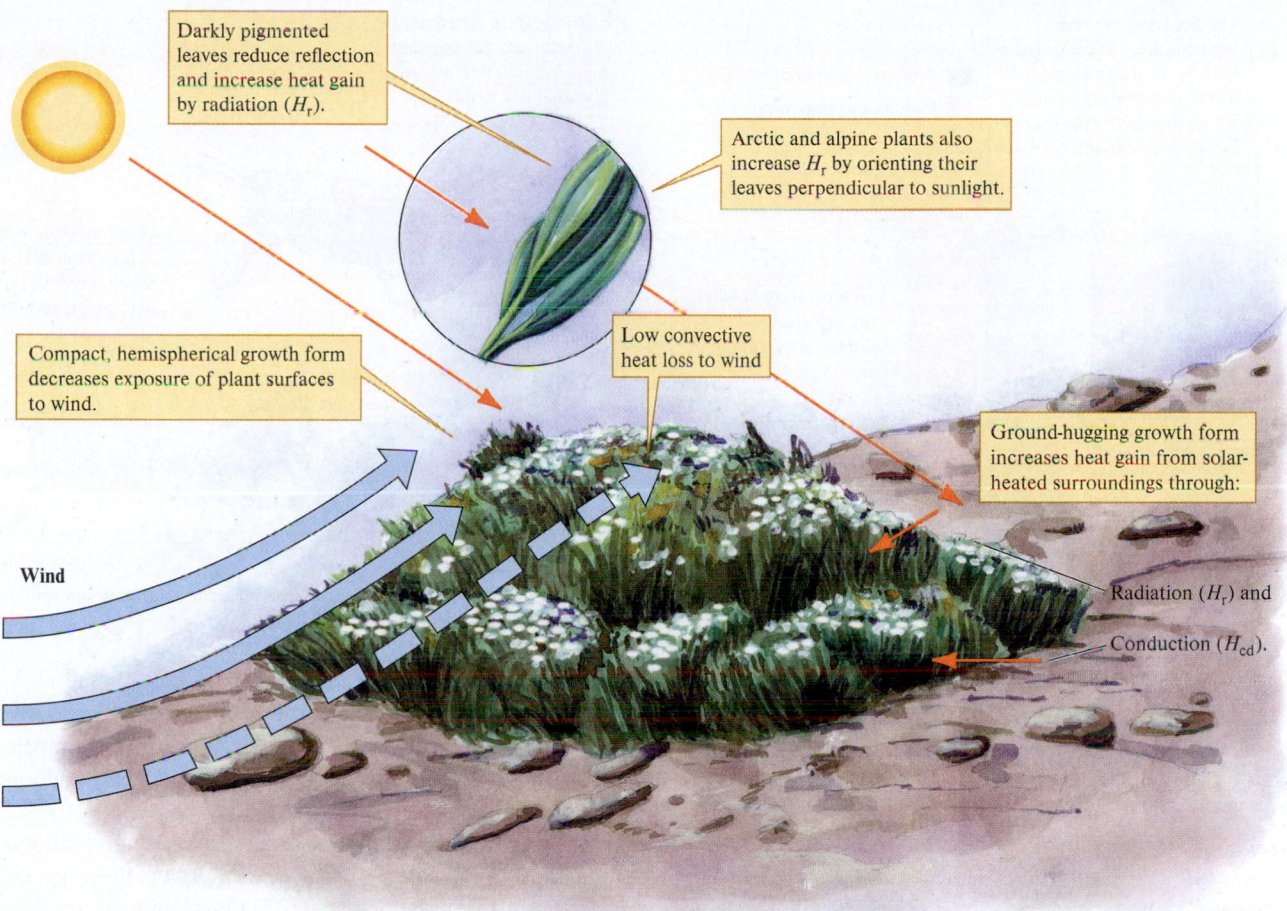

Darkly pigmented leaves reduce reflection and increase heat gain by radiation (H_r).

Arctic and alpine plants also increase H_r by orienting their leaves perpendicular to sunlight.

Compact, hemispherical growth form decreases exposure of plant surfaces to wind.

Low convective heat loss to wind

Ground-hugging growth form increases heat gain from solar-heated surroundings through:

Wind

Radiation (H_r) and

Conduction (H_{cd}).

Figure 5.17   The growth form and orientation of the cushion plants in arctic and alpine environments increase heat gain from sunlight and the surrounding landscape and conserve any heat gained.

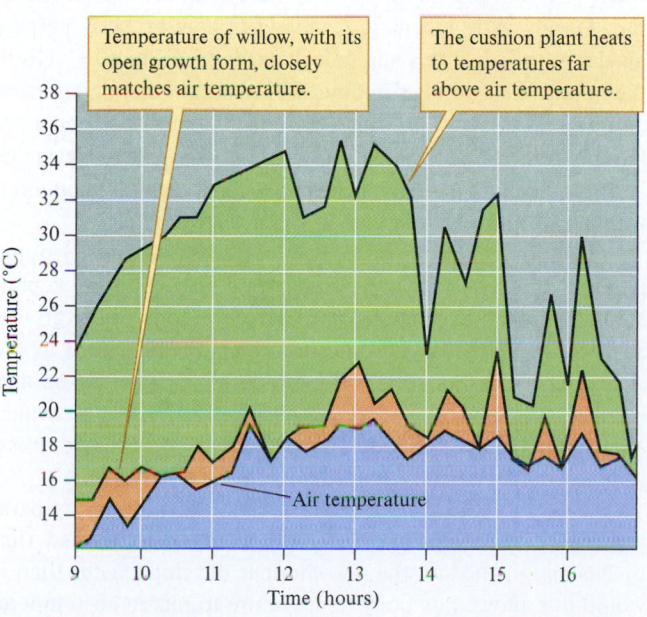

Temperature of willow, with its open growth form, closely matches air temperature.

The cushion plant heats to temperatures far above air temperature.

Air temperature

Figure 5.18   An arctic cushion plant maintains significantly higher temperatures compared to plants with a more open growth form, such as willows (data from Fitter and Hay 1987, after Gauslaa 1984).

Temperature Regulation by Ectothermic Animals

Like plants, the vast majority of animals, including fish, amphibians, reptiles, and invertebrates of all sorts, use external sources of energy to regulate body temperature. These ectothermic animals use means analogous to those used by plants, including variations in body size, shape, and pigmentation. The obvious difference between plants and ectothermic animals is that the animals have more options for using behavior to thermoregulate. Yet, as we shall see, the difference between the behavior of these animals and that of plants is more a matter of degree than of kind.

Lizards

The eastern fence lizard, *Sceloporus undulatus* (see fig. 5.9), is an ectotherm that regulates its body temperature by behaviors such as basking in the sun to warm its body or seeking shade to cool. Research by Michael Angilletta showed that the rate of metabolizable energy intake is maximized at a temperature of 33°C (see fig. 5.10). Knowing this, he studied the relationship between this optimal temperature and the preferred temperature of *S. undulatus*. Angilletta (2001) explored this

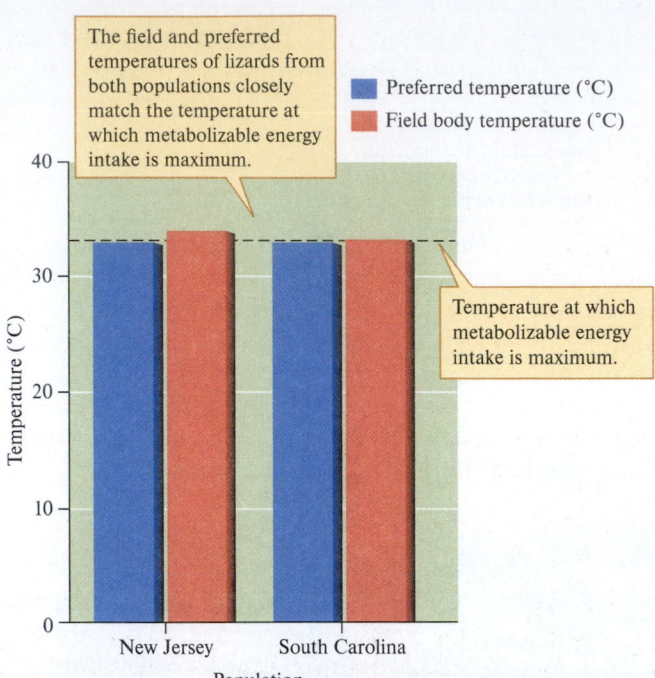

The field and preferred temperatures of lizards from both populations closely match the temperature at which metabolizable energy intake is maximum.

■ Preferred temperature (°C)
■ Field body temperature (°C)

Temperature at which metabolizable energy intake is maximum.

Figure 5.19   Two populations of the eastern fence lizard, *Sceloporus undulatus,* both regulate their body temperatures to match closely the temperature of maximum metabolizable energy intake (data from Angilletta 2001).

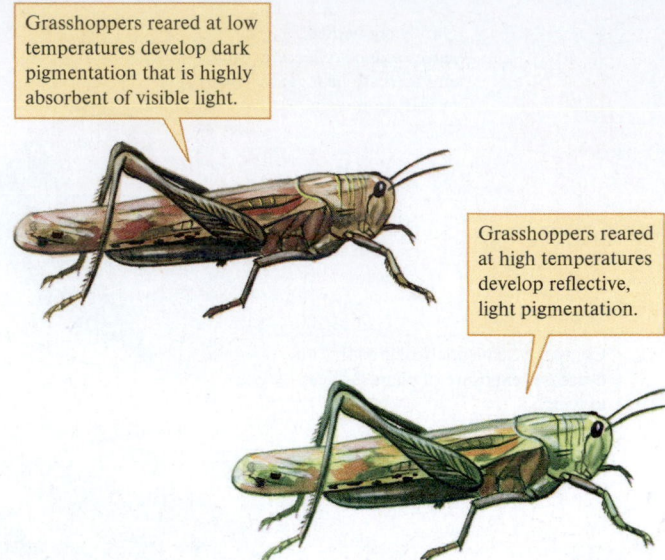

Grasshoppers reared at low temperatures develop dark pigmentation that is highly absorbent of visible light.

Grasshoppers reared at high temperatures develop reflective, light pigmentation.

Figure 5.20   Rearing temperatures influence the pigmentation of the clear-winged grasshopper.

relationship in the laboratory and the field. In the laboratory, he placed *S. undulatus* from New Jersey and South Carolina in a temperature gradient that ranged from 26°C at one end to 38°C at the other end. He determined preferred temperature early each morning by quickly measuring the body temperature of each lizard. Body temperature would indicate where each lizard had been in the temperature gradient, that is, its "preferred" temperature. Angilletta examined thermoregulation by measuring the body temperatures of active individuals in the field.

The results of Angilletta's study provide strong evidence for a correspondence among preferred temperature, thermoregulation, and optimal temperatures in *S. undulatus* (fig. 5.19). Lizards from New Jersey and South Carolina had virtually identical preferred temperatures: 32.8° versus 32.9°C, respectively. The body temperatures found by Angilletta in the field were also very similar. The body temperatures of *S. undulatus* measured in the field in New Jersey averaged 34.0°C, while the body temperatures of *S. undulatus* taken in South Carolina averaged 33.1°C. As shown in figure 5.19, both preferred temperatures determined in the laboratory and the body temperatures of *S. undulatus* measured in the field are very close to the temperature that maximizes metabolizable energy intake by these lizards. The following example shows that effective thermoregulation by ectotherms is not limited to lizards.

Grasshoppers: Some Like It Hot

Many grasshoppers also bask in the sun, elevating their body temperature to 40°C or even higher. R. I. Carruthers and his colleagues (1992) described how some species of grasshoppers even adjust their capacity for radiative heating, H_r, by varying

the intensity of their pigmentation during development. When reared at low temperatures, these species appear to compensate by developing dark pigmentation, while at higher developmental temperatures, they produce less pigmentation (fig. 5.20). How would changing pigmentation in response to developmental temperatures affect thermoregulation by these grasshoppers? Because grasshoppers reared at low temperatures develop darker pigmentation, they increase their potential for H_r gain. Because those reared at high temperatures develop lighter pigmentation, they reduce their potential for H_r gain.

The clear-winged grasshopper, *Camnula pellucida,* inhabits subalpine grasslands in the White Mountains of eastern Arizona, where the cool mornings warm up quickly under the mountain sun. During early morning, *Camnula* orients its body perpendicular to the sun's rays and quickly heats to 30° to 40°C. Given the opportunity, young *Camnula* will maintain a body temperature around 38° to 40°C, very close to its optimal temperature for development. In the laboratory, *Camnula* is able to elevate its body temperature to 12°C above air temperature and maintain it within a very narrow range (± 2°C) for many hours.

Carruthers and his colleagues divided a laboratory population of *Camnula* into two groups, both of which were kept at an air temperature of about 18°C. However, one of the groups also had access to light, while the other was restricted to the shade. The grasshoppers that had access to light basked and elevated their body temperatures about 10°C above air temperature. Meanwhile, the body temperatures of the shaded grasshoppers remained close to air temperature (fig. 5.21).

Why does *Camnula* bask and maintain a body temperature above air temperature? The researchers estimated that by basking in the sun the grasshopper develops faster than it would if it allowed its body temperature to match air temperature. What other benefits might *Camnula* gain by maintaining a high body temperature? The grasshopper may raise its body temperature to 38° to 40°C to control *Entomophaga grylli,* a fungus that infects and kills grasshoppers.

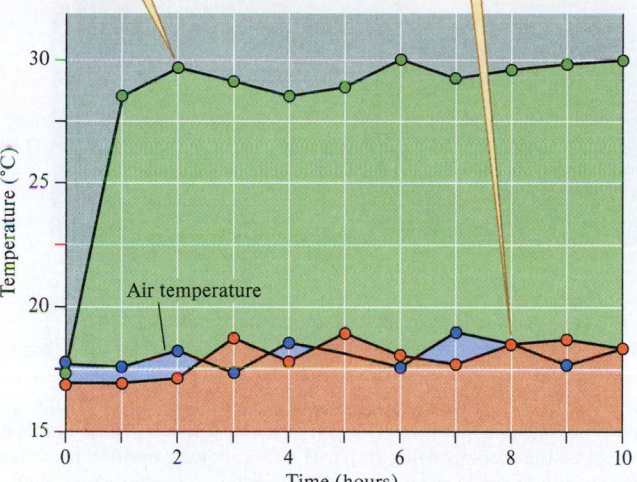

Grasshoppers with access to light bask, raising their body temperatures about 10°C above air temperature.

Body temperatures of grasshoppers confined to the shade nearly match air temperature.

Air temperature

Figure 5.21   Basking allows the clear-winged grasshopper to elevate its body temperature significantly (data from Carruthers et al. 1992).

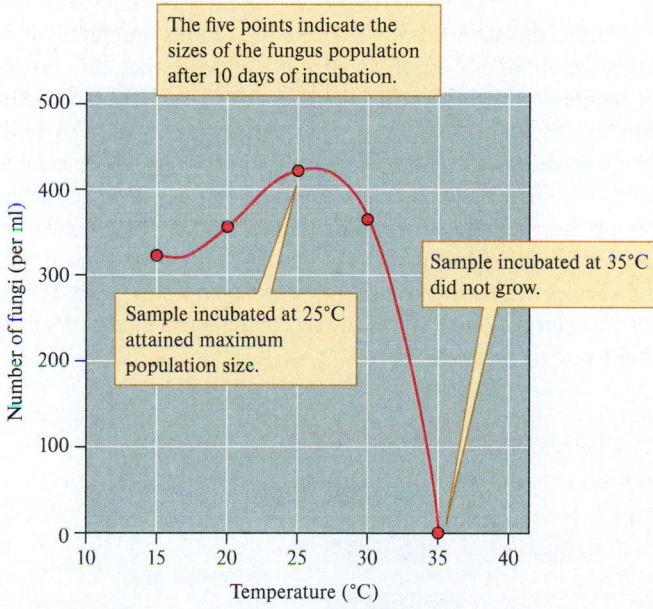

The five points indicate the sizes of the fungus population after 10 days of incubation.

Sample incubated at 25°C attained maximum population size.

Sample incubated at 35°C did not grow.

Figure 5.22   High temperatures inhibit growth by *Entomophaga grylli* (data from Carruthers et al. 1992).

The idea that high temperatures could control *Entomophaga* was tested by growing the fungus in artificial media at 15°, 20°, 25°, 30°, 35°, and 45°C. The populations grew fastest at 25°C; above and below 25°C the fungus populations grew at a slower rate; they did not grow at 35°C and were killed at 45°C (fig. 5.22).

After studying the growth of the fungus in artificial media, the researchers studied how temperature influences mortality among grasshoppers infected with the fungus. They found that exposure to 40°C temperatures for as few as 4 hours each day significantly reduced the number of grasshoppers dying of *Entomophaga* infections. The results of these experiments support the hypothesis that by maintaining body temperatures of 38° to 40°C, clear-winged grasshoppers create an environment unsuitable for one of their most serious pathogens.

Temperature Regulation by Endothermic Animals

Do endothermic animals thermoregulate differently than the other organisms we've discussed? Endotherms employ all the mechanisms used by ectothermic organisms to influence heat exchange with the environment. So, our basic equation for temperature regulation, $H_s = H_m \pm H_{cd} \pm H_{cv} \pm H_r - H_e$, still applies but with some changes in the relative importance of the terms. Most significantly, endotherms rely a great deal more on metabolic heat, H_m, to maintain internal temperatures.

Environmental Temperature and Metabolic Rates of Homeotherms

P. F. Scholander and his colleagues (1950) studied thermoregulation in several mammal species by monitoring metabolic rate while exposing them to a range of temperatures. The range of environmental temperatures over which the metabolic rate of a homeothermic animal does not change is called its **thermal neutral zone.** When environmental temperatures are within the thermal neutral zone of an inactive endothermic animal, its metabolic rate stays steady at resting metabolism. However, if the environmental temperature falls below or rises above the thermal neutral zone, a homeotherm's metabolic rate will rapidly increase to two or even three times resting metabolism.

What causes metabolic rates to rise when environmental temperatures are outside the thermal neutral zone? We can use humans as a model for the responses of endotherms generally. At low temperatures, we start shivering, which generates heat by muscle contractions. We also release hormones that increase our metabolic rate, the rate at which we metabolize our energy stores, which are mainly fats. Increasing metabolic rate increases the rate at which we generate metabolic heat, H_m. At high temperatures, heart rate and blood flow to the skin increase. This increased blood flow transports heat from the body core to the skin, where heat energy is radiated to the surrounding environment. Radiative heat loss can be increased by specialized structures. For example, Brazilian free-tailed bats have specialized regions along the sides of their bodies that radiate excess metabolic heat generated as they fly (fig. 5.23).

An evaporative cooling system based on sweating can accelerate unloading of heat to the external environment. Many large homeotherms, including humans, horses, and camels, also cool by sweating. Other homeotherms do not sweat but evaporatively cool by other means: dogs and birds pant and marsupials and rodents moisten their body surfaces by salivating and licking. Thus, in these and other species temperature regulation and water balance are closely related.

Scholander and his colleagues found that the breadth of the thermal neutral zone varies a great deal among mammal species. Based on their results, they suggested that differences in the width of the thermal neutral zone defines two groups of organisms: tropical species, with narrow thermal neutral

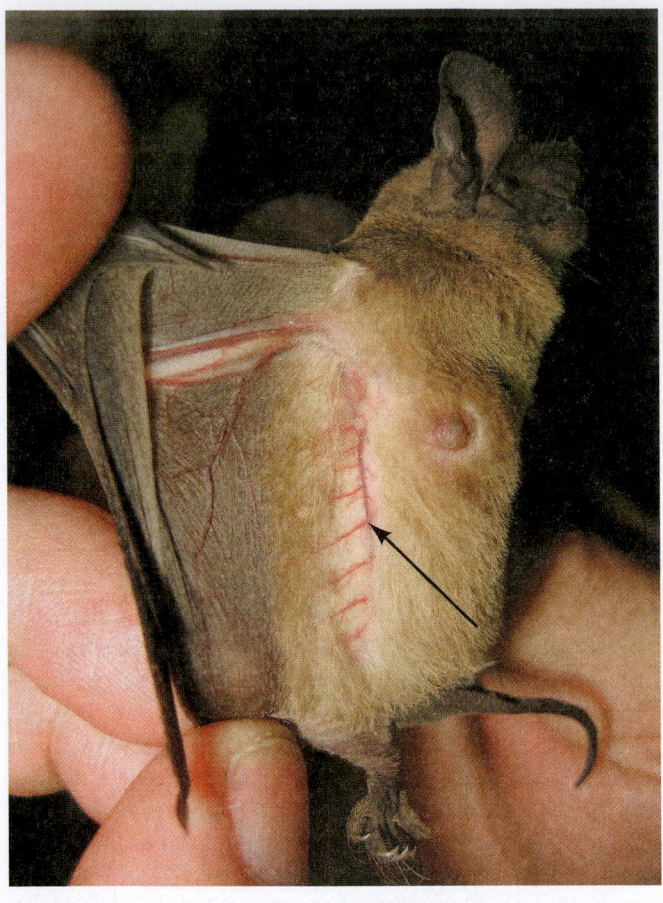

(a)

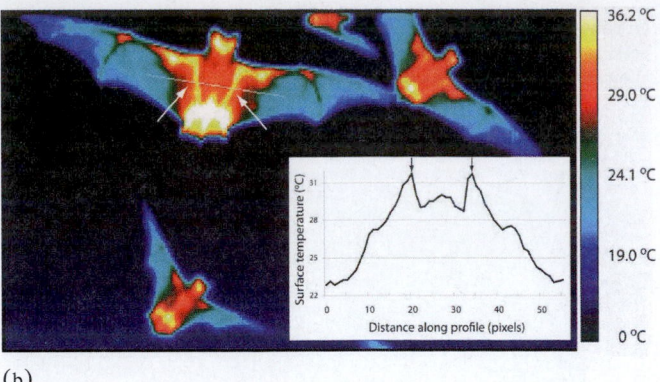

(b)

Figure 5.23   (*a*) Heat radiator on the right side of a female Brazilian free-tailed bat indicated by arrow. The radiator area is naturally hairless, with arterial and venous blood vessels passing close to the skin surface. (*b*) A thermal image of a Brazilian free-tailed bat. The white line across the bat marks the location of the surface temperature profile shown on the accompanying graph. The area showing maximum surface temperature in the thermal image is the location of the bat's heat radiator shown in *a*. The black arrows on the graph point to the bat's surface temperature at the places indicated by white arrows on the bat thermal image (after Reichard et al. 2010).   (*a, b*) ©Jonathan D. Reichard and Thomas H. Kunz/Boston University

zones, and arctic species, with broad thermal neutral zones. The researchers pointed out that the narrow thermal neutral zone of *Homo sapiens* is similar to that of several species of rain forest mammals and birds. Meanwhile, arctic species, such as the arctic fox, have impressively broad thermal neutral zones.

Since the normal body temperature of most mammals varies from about 35° to 40°C, it is no surprise that this range of temperatures falls within the thermal neutral zone of both tropical and arctic species. What distinguishes tropical and arctic species is the great tolerance that arctic species have for cold. For instance, the arctic fox can tolerate environmental temperatures down to −30°C without showing any increase in metabolic rate. Meanwhile, the metabolic rate of some tropical species begins to increase when air temperature falls below 29°C. Figure 5.24 contrasts the thermal neutral zones of some arctic and tropical species.

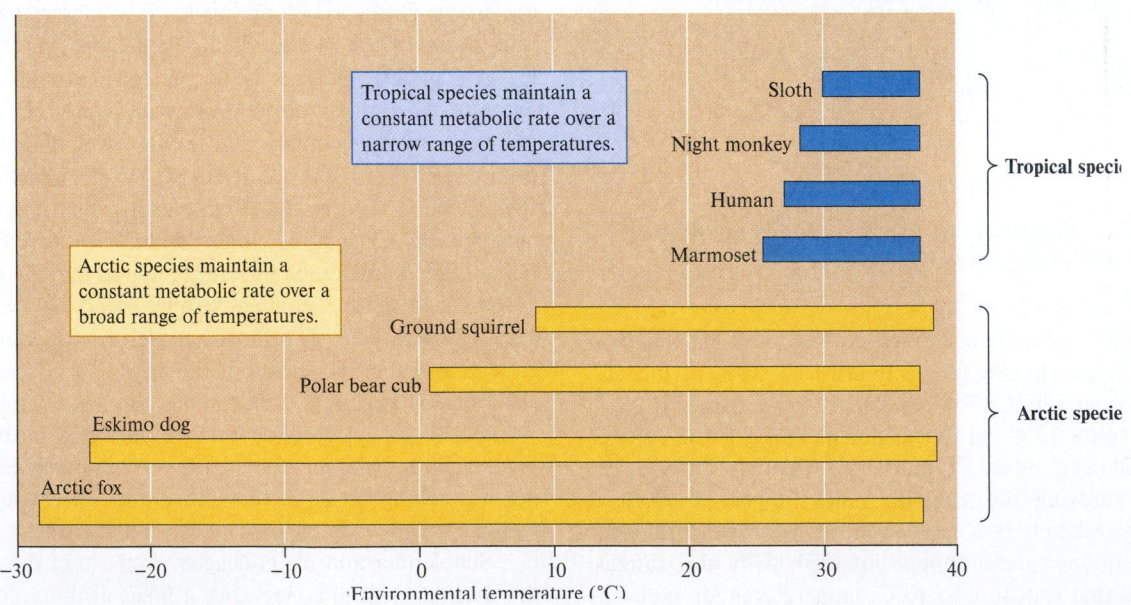

Figure 5.24   Temperature and the thermal neutral zone of arctic and tropical mammals. Bars indicate range of temperatures over which metabolic rate does not change for each species (data from Scholander et al. 1950).

From evolutionary and ecological perspectives, the important point of this discussion is that thermoregulation outside the thermal neutral zone costs energy that could be otherwise directed toward reproduction. How might such energetic costs affect the distribution and abundance of organisms in nature?

Endothermic Aquatic Animals

Now let's turn to thermoregulation by aquatic endotherms, where the aquatic environment limits the possible ways organisms can regulate their body temperatures. Why is that? First, as we have seen, the capacity of water to absorb heat energy without changing temperature is about 3,000 times that of air. Second, conductive and convective heat losses to water are much more rapid than to air, over 20 times faster in still water and up to 100 times faster in moving water. Thus, the aquatic organism is surrounded by a vast heat sink. The potential for heat loss to this heat sink is very great, particularly for gill-breathing species that must expose a large respiratory surface in order to extract sufficient oxygen from water. In the face of these environmental difficulties, only a few aquatic species are truly endothermic.

Aquatic birds and mammals, such as penguins, seals, and whales, can be homeothermic in an aquatic environment for two major reasons: First, they are all air breathers and do not expose a large respiratory surface to the surrounding water. Second, many endothermic aquatic animals, including penguins, seals, and whales, are well insulated from the heat-sapping external environment by a thick layer of fat while others, such as the sea otter, are insulated by a layer of fur that traps air. The parts of these animals that are not well insulated, principally appendages, are outfitted with *countercurrent heat exchangers,* vascular structures that reduce the rate of heat loss to the surrounding aquatic environment. Figure 5.25 diagrams the structure and functioning of a countercurrent heat exchanger in the flipper of a dolphin.

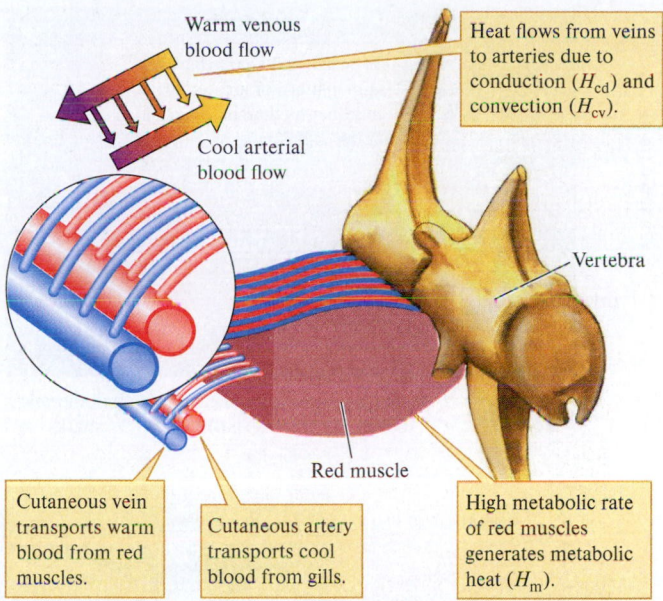

Figure 5.26   Countercurrent heat exchange in RM endothermic fish.

Tunas and some sharks, are also able to be endothermic by being well supplied with blood vessels that function as countercurrent heat exchangers. Heat is produced by the high metabolic rate of the lateral swimming muscles, which is transferred to blood in the arteries. This heat can then be transported to the rest of the body through cutaneous veins (figure 5.26). This type of endothermy, seen in at least 14 species of tuna and five species of shark, has been called **RM endothermy**, because the warming mechanism is slow-twitch, aerobic red muscle. Since tuna and sharks diverged some 450 million years ago, this shared trait is a remarkable example of convergent evolution.

The higher temperature of endotherms relative to ectotherms is energetically expensive to maintain, and so was initially somewhat of an evolutionary mystery. Such an elaborate adaptation would have to result in a significant energetic benefit for it to be selected for in natural systems. To find out what these benefits were, Yuuki Y. Watanabe of the National Institute of Polar Research, Japan, and his colleagues used data from animal-tracking studies to compare movements of endothermic versus ectothermic fish (Watanabe et al. 2015). In addition to using published data from other researchers, they collected additional data by attaching sensors to sharks in Alaska, the Bahamas, and the central Pacific to measure swimming speeds and movements.

Wantabe's team found that endothermic fish swam faster (fig. 5.27a) and typically had a greater migration range than ectothermic fish, more similar to marine mammals (fig. 5.27b). Taken together, these differences from ectotherms give endotherms significantly more access to prey. The improved hunting efficiency proves enough of a fitness benefit to have allowed RM endothermy to evolve not once but multiple times.

In more recent years, Wantabe and colleagues determined that both endothermic and ectothermic fish show a positive relationship between body size and swimming speed; however

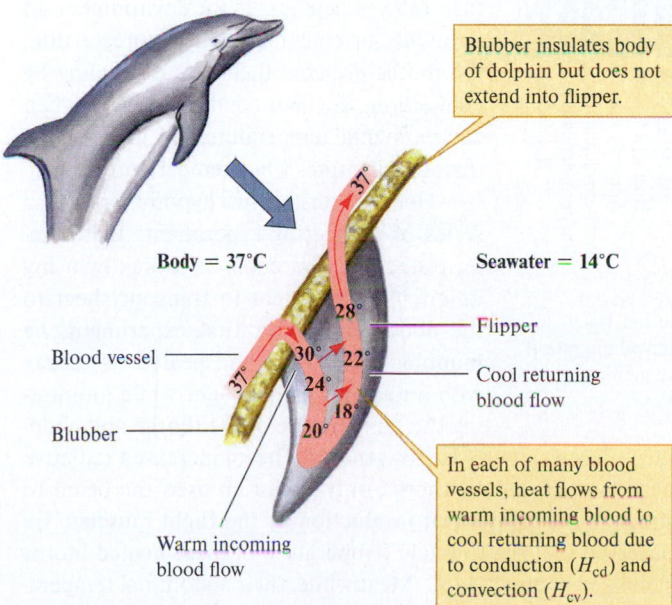

Figure 5.25   Countercurrent heat exchange in dolphin flippers promotes conservation of body heat.

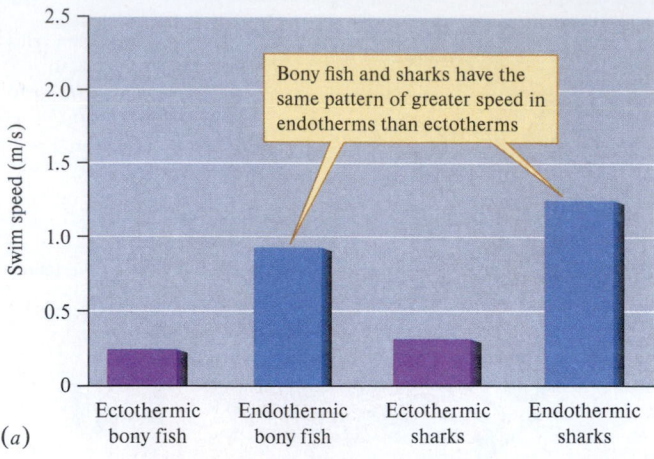

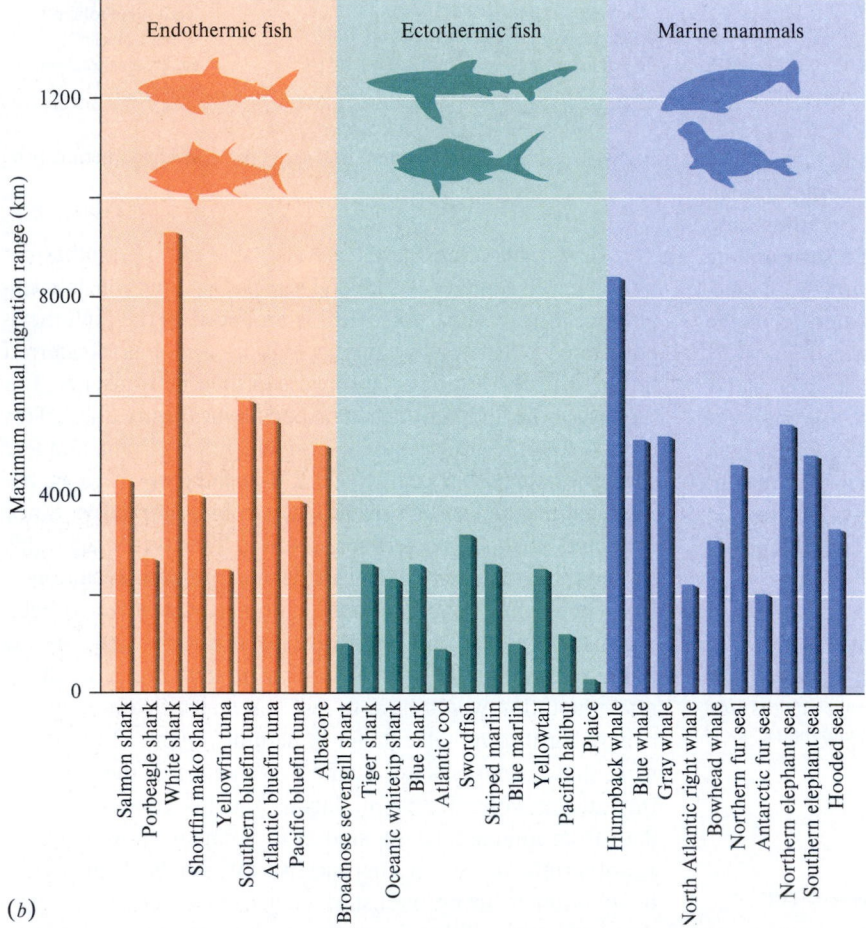

(a)

(b)

Figure 5.27 (a) The mean swimming speed of ectothermic versus endothermic bony fishes and sharks, based on data from 13 different species. (b) The maximum annual migration range of endothermic fish (9 species), ectothermic fish (11 species), and marine mammals (10 species) (data from Watanabe et al. 2015).

even small endothermic fish swim faster than much larger ectotherms (Wantabe et al. 2019).

Warming Insect Flight Muscles

Have you ever gone outside on a cool fall or spring morning when few insects were active, and yet met with bumblebees visiting flowers? Were you surprised? While you may have taken the meeting for granted, these early morning forays by bumblebees require some impressive physiology. Most insects use external sources of energy to heat their bodies, but there are some notable exceptions.

Studies of temperature regulation by moths began in the early 1800s. Many of these studies were focused on moths of the family Sphingidae, the sphinx moths. Sphinx moths are convenient insects for study because many reach impressive sizes, large enough to be sometimes mistaken for hummingbirds. Since the nineteenth century, researchers have been aware that active sphinx moths have elevated thoracic temperatures. These early researchers also knew that temperature increases within the thorax were due to activity of the flight muscles contained within the thorax that vibrated the wings. Once warmed up and actively flying, sphinx moths maintained a relatively constant thoracic temperature over a broad range of environmental temperatures.

However, a significant problem remained. No one knew *how* they did it. Phillip Adams and James Heath (1964) proposed that the moths thermoregulate by changing their metabolic rate in response to changing environmental temperatures. In terms of our equation for thermoregulation, Adams and Heath proposed that the moths increased H_m when environmental temperatures fell and decreased H_m when environmental temperatures rose.

Several observations led Bernd Heinrich (1993) to propose an alternative hypothesis. He proposed that active sphinx moths have a fairly constant metabolic rate and, therefore, generate metabolic heat, H_m, at a constant rate. Heinrich also proposed that sphinx moths thermoregulate by changing their rates of heat loss to the environment. In terms of our equation for thermoregulation, the moths *decrease* their rate of cooling by convection, H_{cv}, and conduction, H_{cd}, when environmental temperatures fall, and *increase* their cooling rates when temperatures rise.

Heinrich tested his hypothesis with a series of pioneering experiments that demonstrated *M. sexta* cools its thorax by using its circulatory system to transport heat to the abdomen. In his first experiment, he immobilized a moth and heated its thorax with a narrow beam of light while monitoring the temperature of the thorax and abdomen. Because it was narrow, the light beam increased radiative heat gain, H_r, of the thorax only. Heinrich used the beam to simulate metabolic heat production by the flight muscles. He observed that the thoracic temperature of these heated moths stabilized at about 44°C. Meanwhile, their abdominal temperatures gradually increased.

These results indicated that heat within the thorax was transferred to the abdomen. Heinrich proposed that blood

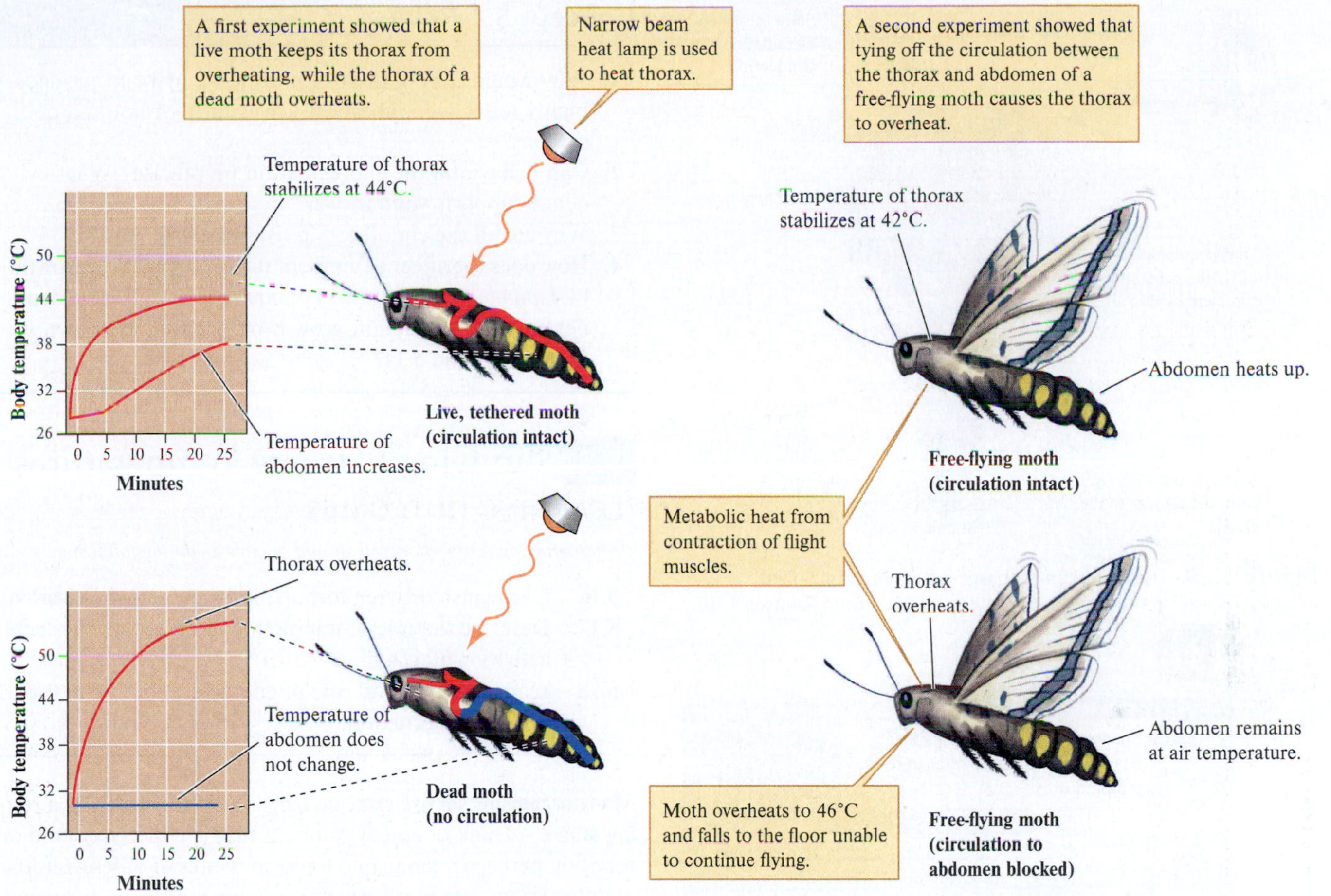

A first experiment showed that a live moth keeps its thorax from overheating, while the thorax of a dead moth overheats.

Narrow beam heat lamp is used to heat thorax.

A second experiment showed that tying off the circulation between the thorax and abdomen of a free-flying moth causes the thorax to overheat.

Temperature of thorax stabilizes at 44°C.

Temperature of abdomen increases.

Live, tethered moth (circulation intact)

Temperature of thorax stabilizes at 42°C.

Abdomen heats up.

Free-flying moth (circulation intact)

Metabolic heat from contraction of flight muscles.

Thorax overheats.

Thorax overheats.

Temperature of abdomen does not change.

Dead moth (no circulation)

Moth overheats to 46°C and falls to the floor unable to continue flying.

Abdomen remains at air temperature.

Free-flying moth (circulation to abdomen blocked)

Figure 5.28   The circulatory system plays a central role in thermoregulation by the moth *Manduca sexta* (data from Heinrich 1993).

flowing from the thorax to the abdomen was the means of heat transfer. To confirm this, he conducted a second experiment. He tied off blood flow to the thorax using a fine human hair. With this blood flow stopped, flying moths overheated and stopped flying. Instead of stabilizing at 44°C, the thoracic temperatures approached the lethal limit of 46°C. An interesting debate between two groups of researchers with competing hypotheses was decided by two decisive experiments, which are summarized by figure 5.28.

Endothermic insects were a surprise to many biologists. The existence of endothermic plants was even more surprising.

Temperature Regulation by Thermogenic Plants

Roger Knutson (1974, 1979) visited a marsh on a cold February day in northeast Iowa, where he saw eastern skunk cabbage, *Symplocarpus foetidus,* emerging from the frozen landscape. Because the skunk cabbages were encircled by bare ground, it seemed that they had generated enough heat to melt their way through the snow. Knutson returned the next day with a thermometer and so began a research project that produced some surprising observations of thermoregulation by plants.

Almost all plants are poikilothermic ectotherms. However, plants in the family Araceae have the unusual habit of using metabolic energy to heat their flowers. Some of the

temperate species in this mostly tropical family use this ability to protect their inflorescences from freezing and to attract pollinators. One of the most studied of these temperate species is the eastern skunk cabbage, which lives in the deciduous forests of eastern North America. This skunk cabbage blooms from February to March, when air temperatures vary between −15° and 15°C. During this period, the inflorescence of the plant, which weighs from 2 to 9 g, maintains a temperature 15° to 35°C above air temperature. As Knutson observed, this temperature is warm enough so that *S. foetidus* can melt its way through snow. The plant's inflorescences can maintain these elevated temperatures for up to 14 days. During this period, it functions as an endothermic organism.

How does the skunk cabbage fuel the heating of its inflorescence? It has a large root in which it stores a great quantity of starch. Some of this starch is translocated to the inflorescence, where it is metabolized at a high rate, generating heat in the process. This heat, besides keeping the inflorescence from freezing, may help attract pollinators. Various pollinators are attracted to both the warmth and the sweetish scent given off by the plant. Some of the biology of this interesting plant is summarized in figure 5.29.

The inflorescence of the skunk cabbage maintains a high respiratory rate, equivalent to that of a small mammal of similar size. However, its metabolic rate is not constant. The plant increases its metabolic rate as temperatures fall, which increases

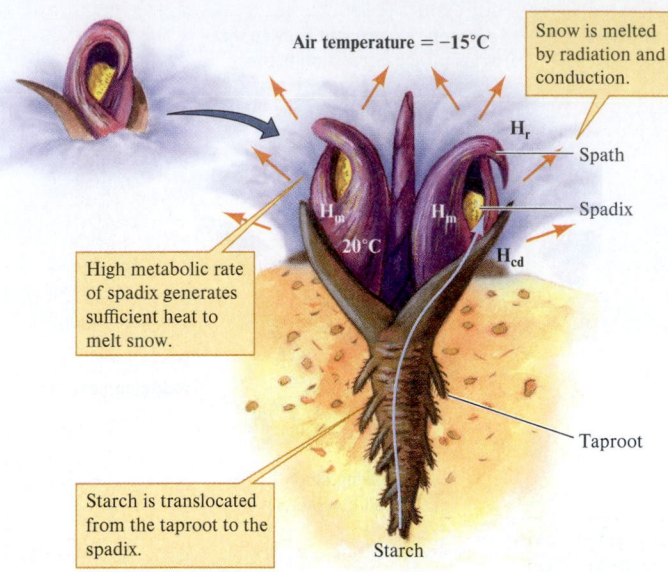

Air temperature = −15°C

Snow is melted by radiation and conduction.

High metabolic rate of spadix generates sufficient heat to melt snow.

H$_r$

Spath

H$_{vp}$ H$_{jm}$

20°C

Spadix

H$_{cd}$

Taproot

Starch is translocated from the taproot to the spadix.

Starch

Figure 5.29 Eastern skunk cabbage, an endothermic plant, can melt its way up through spring snow cover (data from Knutson 1974).

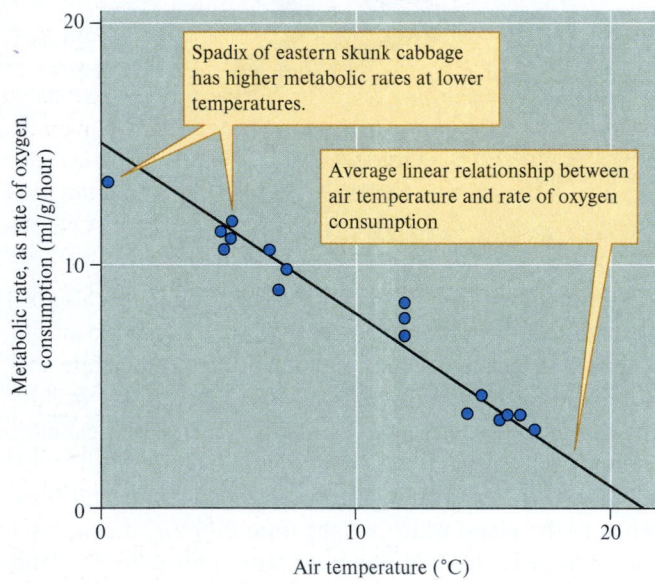

Spadix of eastern skunk cabbage has higher metabolic rates at lower temperatures.

Average linear relationship between air temperature and rate of oxygen consumption

Metabolic rate, as rate of oxygen consumption (ml/g/hour)

Air temperature (°C)

Figure 5.30 Air temperature has a clear influence on the metabolic rate of eastern skunk cabbage (data from Knutson 1974).

the rate of metabolic heat production. By adjusting its metabolic rate, the plant can maintain its inflorescence at a similar temperature despite substantial variation in environmental temperature (fig. 5.30).

In this section, we have considered how various organisms regulate their body temperatures by using external sources of energy, internal sources of energy, or both. Thermoregulation is possible where organisms face temperatures within their range of tolerance. However, in the face of extreme temperatures, thermoregulation may not be a viable option. In such circumstances, organisms may turn to other survival strategies.

Concept 5.4 Review

1. Why would it be a disadvantage for a plant to produce highly reflective, pubescent leaves in both hot and cool seasons?
2. Can behavioral thermoregulation be precise? What evidence supports your answer?
3. Why are all the endothermic fish relatively large?
4. How does the effect of temperature on population growth by *Entomophaga* (fig. 5.22) compare to the effect of temperature on population growth by bacteria as shown in figures 5.13 and 5.14?

5.5 Surviving Extreme Temperatures

LEARNING OUTCOMES

After studying this section you should be able to do the following:

5.16 Distinguish between torpor, hibernation, and estivation.
5.17 Describe the role of microclimate in tiger beetle daily activity patterns.
5.18 Explain the central role of energy availability in torpor and hibernation.

Many organisms survive extreme temperatures by entering a resting stage. Think of an environment that is either very cold or very hot, perhaps a temperate forest in winter or a desert in the middle of a hot summer's day. If you have been in such an environment, you may have noticed less obvious biological activity than at other times of the year. Many plants are dormant, few birds are active, and insects may not be evident. Many organisms have evolved ways to avoid extreme environmental temperatures by entering a resting stage. This stage may be as simple as resting in a sheltered spot during the heat of the day or may involve elaborate physiological and behavioral adjustments. Let's examine some of the ways organisms avoid extreme temperatures.

Inactivity

A simple way to avoid extreme environmental temperatures is to seek shelter during the hottest or coldest times of the day. Let's consider some beetles that take shelter during the middle of the day, when environmental temperatures are too high.

Many organisms live on the beaches of New Zealand, including small, predatory tiger beetles. One species of black tiger beetle, *Neocicindela perhispida campbelli,* lives on black sand beaches. As we saw earlier in the chapter, these black sand beaches heat up rapidly in the morning sun and reach higher temperatures than nearby white sand beaches (see fig. 5.4). The black beetles that live on these beaches also heat up quickly. By basking in the morning sun, they warm enough to become active early in the day. However, later in the day, they must work very hard to avoid overheating.

The beetles maintain their body temperature at about 36.4°C by shuttling between sun and shade and by facing into the sun to orient themselves parallel to the sun's rays. The beetles

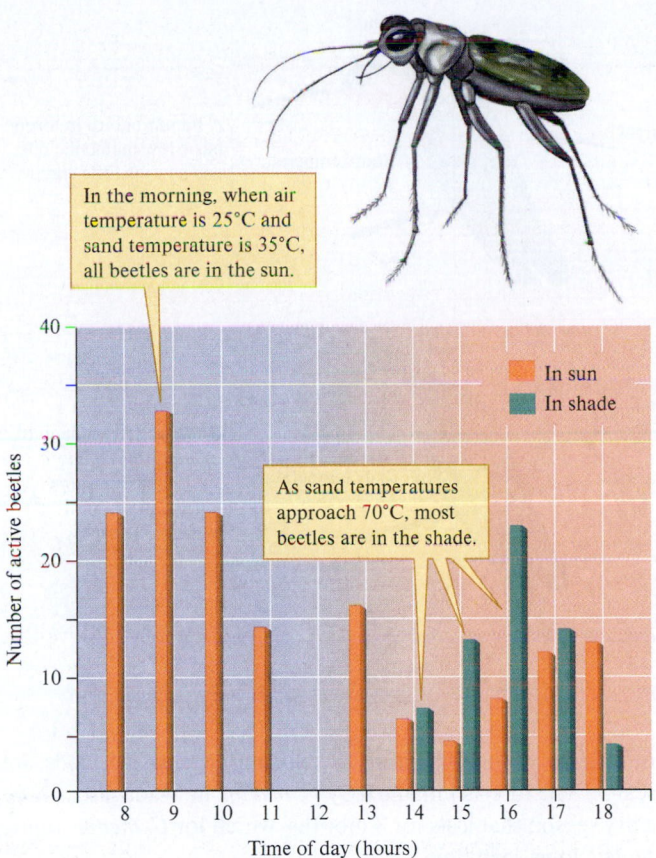

In the morning, when air temperature is 25°C and sand temperature is 35°C, all beetles are in the sun.

As sand temperatures approach 70°C, most beetles are in the shade.

Figure 5.31   Tiger beetles' avoidance of high environmental temperatures helps them avoid elevating their body temperatures above acceptable levels (data from Hadley, Savill, and Schultz 1992).

also reduce heating by increasing their rate of convective cooling. They do this by "stilting," standing on the tips of their feet and extending their legs, to get themselves a bit higher into the air. With this combination of behaviors, the beetle maintains its body temperature substantially below the temperature of the sand.

Thermoregulation becomes difficult by midday, however, when sand temperatures may reach 70°C. Most of the beetles simply avoid these high temperatures by leaving. As shown in figure 5.31, those beetles that remain active are mostly in the shade.

Now let's look at hummingbirds that live in environments, where maintaining an elevated body temperature during cold nights requires a great deal of energy.

Reducing Metabolic Rate

Hummingbirds are small birds that depend on a diet of nectar and insects to maintain a high metabolic rate and a body temperature of about 39°C. When food is abundant, they maintain these high rates throughout the day and night. However, when food is scarce and night temperatures are low, hummingbirds may enter torpor. **Torpor** is a state of low metabolic rate and lowered body temperature. During torpor, a hummingbird's body temperature is about 12° to 17°C, quite a reduction from 39°C. Because this lower body temperature is a direct consequence of a lower metabolic rate, a hummingbird in torpor saves a lot of energy. How much energy is saved? F. L.

Carpenter and colleagues (1993) estimated that rufus hummingbirds that maintained full body temperature all night lost (metabolized) about 0.24 g of fat. In contrast, birds in a state of torpor lost only 0.02 g of fat, resulting in an energy savings of over 90%.

This much was known when William Calder (1994) began a study to discover the circumstances under which hummingbirds use torpor. When he began his work, there were two major hypotheses. The "routine" hypothesis proposed that hummingbirds go into torpor regularly, perhaps every night. The "emergency-only" hypothesis proposed that hummingbirds go into torpor only when food supplies are inadequate.

Since wild hummingbirds are difficult to follow, how could Calder determine whether they go into torpor at night? By weighing hummingbirds just before they went to their night roosts and then again first thing in the morning, he could estimate the amount of fat metabolized during the night. A hummingbird in torpor would lose much less weight. To weigh hummingbirds, Calder either captured them in mist nets or rigged the perch on a hummingbird feeder with an electronic balance.

Calder's observations of broad-tailed hummingbirds did not support the routine hypothesis. He found that on most nights the hummingbirds lost 15 times the amount of weight, as a result of burning fat, than they would have if they had gone into torpor. Clearly, hummingbirds usually have elevated metabolic rates during the night and go into torpor only occasionally. Calder found that hummingbirds use torpor under two main circumstances: (1) when they arrive at breeding or wintering sites before flowers are abundant and (2) when their food intake is reduced by decreased nectar production by flowers or by storms that interfere with their feeding. As with Heinrich's work on moth thermoregulation, Calder's observations resolved a debate between two competing hypotheses (fig. 5.32).

While hummingbirds may go into torpor for several hours each night, other animals can go into a state of reduced metabolism that may last several months. If this state occurs mainly in winter, it is called **hibernation**. If it occurs in summer, it is called **estivation**. During hibernation, the body temperature of arctic ground squirrels may drop to 2°C. The metabolic rates of hibernating marmots may fall to 3% of levels during active periods. During estivation, the metabolic rate of long-neck turtles may fall to 28% of their normal metabolic rate. Such reductions in metabolic rate allow these animals to survive long arctic and alpine winters or hot, dry periods in the desert, during which they must rely entirely on stored energy reserves. Under some conditions, even tropical species may hibernate.

Hibernation by a Tropical Species

While most studies of hibernation have focused on temperate and arctic species, there are tropical animals that hibernate. One of those species is a primate called the fat-tailed dwarf lemur, *Cheirogaleus medius* (fig. 5.33). *C. medius* lives in the tropical dry forests of western Madagascar, where it is active during 5 months and hibernates for 7 months. As its common name implies, *C. medius* is a small primate, with a body length of about 20 cm

Day

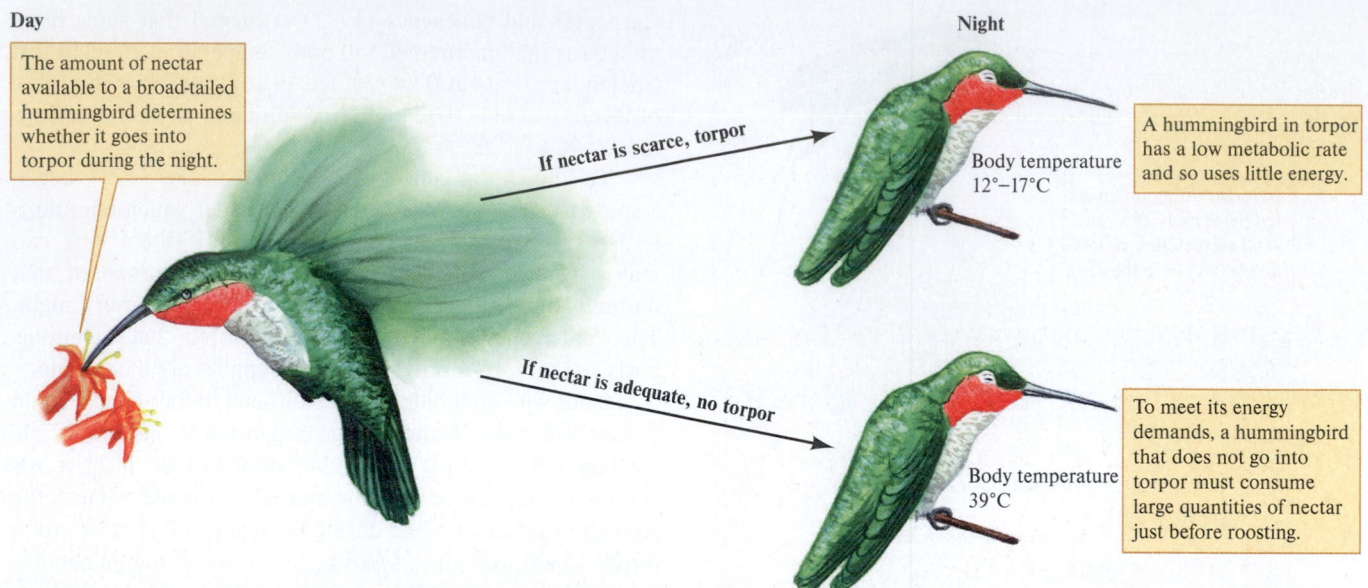

The amount of nectar available to a broad-tailed hummingbird determines whether it goes into torpor during the night.

Night

If nectar is scarce, torpor

Body temperature 12°–17°C

A hummingbird in torpor has a low metabolic rate and so uses little energy.

If nectar is adequate, no torpor

Body temperature 39°C

To meet its energy demands, a hummingbird that does not go into torpor must consume large quantities of nectar just before roosting.

Figure 5.32   The availability of nectar affects whether broad-tailed hummingbirds enter torpor at night.

Figure 5.33   The fat-tailed dwarf lemur, *Cheirogaleus medius*, inhabits tropical dry forests in western Madagascar, where it hibernates through most of the dry season.   imageBROKER/Alamy Stock Photo

and a tail of about the same length. As adults they weigh about 140 g. The tail of *C. medius* is a primary site for storing the fat that the species uses for energy during its long hibernation. *C. medius* lives in trees, sleeping in tree cavities during the day in groups of up to five individuals and foraging in the canopy at night. In contrast to its daytime roosting behavior, *C. medius* is entirely solitary while foraging at night. The main foods of *C. medius* are fruits and flowers, but it also eats some insects and small vertebrate animals such as chameleons.

Why do these tropical primates hibernate? Photos of tropical dry forest landscape during wet and dry seasons and climate diagrams of tropical dry forest provide a suggestion (see figs. 2.13 and 2.14). During the wet season, tropical dry forests are very productive of the fruits and flowers eaten by

C. medius. However, during the dry months, these foods are scarce. The forests inhabited by *C. medius* in Madagascar have a dry season that lasts for 8 months, which for *C. medius* represents a long, lean time.

The physiology of hibernation by free-ranging *C. medius* was studied by Joanna Fietz, Frieda Tataruch, Kathrin Dausmann, and Jörg Ganzhorn (2003) in the Kirindy forest of western Madagascar. This team of researchers from universities in Germany and Austria found that during hibernation, the body temperature of *C. medius* varies from about 18° to 31°C. By lowering its body temperature, *C. medius* saves energy during the food-scarce dry season. As a result of its lower energy requirement, *C. medius* is able to live off the fat it stores in its tail and in other parts of the body. These fat stores are gradually depleted as *C. medius* passes its annual 7 months of hibernation. Fietz and her colleagues found that during hibernation the body mass of *C. medius* decreases by approximately 34%, while the volume of the tail is reduced by nearly 58%. Studies of hibernation among tropical species, such as this one, broaden our understanding of hibernation generally and underscore the need of homeothermic animals to access adequate supplies of energy for maintaining their relatively constant body temperatures. Hibernation, and its associated energy savings, has selective advantage when energy supplies are inadequate for supplying these metabolic needs. This appears to be the case in cold as well as tropical regions.

The temperature relations of organisms, a fundamental aspect of ecology, is attracting increased attention. This interest is fueled by concerns about the ecological consequences of global warming. Though we discuss this issue in detail in chapter 23, in the Applications section of this chapter we look at how studies of temperature relations and climatic warming are helping explain local extinction of a species.

Concept 5.5 Review

1. Why don't hummingbirds save energy by going into torpor at night even when food supplies are abundant? In other words, what would be a possible disadvantage of routine, nightly torpor?
2. Why might the frequency of torpor and hibernation be more common among animals in tropical dry forests compared to those living in tropical rain forests?

Applications

Local Extinction of a Land Snail in an Urban Heat Island

LEARNING OUTCOMES

After studying this section you should be able to do the following:

5.19 Outline changes in the distribution of the snail *Arianta arbustorum* around Basel, Switzerland, between 1900 and 1990.

5.20 Explain how urbanization generally creates a "heat island."

5.21 Review the evidence that temperature changes around the city of Basel are responsible for local extinctions of the snail *A. arbustorum*.

Between 1906 and 1908, a PhD candidate named G. Bollinger (1909) studied land snails in the vicinity of Basel, Switzerland. Eighty-five years later, Bruno Baur and Anette Baur (1993) carefully resurveyed Bollinger's study sites near Basel for the presence of land snails. In the process, they found that at least one snail species, *Arianta arbustorum,* had disappeared from several of the sites. This discovery led the Baurs to explore the mechanisms that may have produced extinction of these local populations.

A. arbustorum is a common land snail in meadows, forests, and other moist, vegetated habitats in northwestern and central Europe. The species lives at altitudes up to 2,700 m in the Alps. The Baurs report that the snail is sexually mature at 2 to 4 years and may live up to 14 years. Adult snails have shell diameters of 16 to 20 mm. The species is hermaphroditic. Though individuals generally mate with other *A. arbustorum,* they can fertilize their own eggs. Adults produce one to three batches of 20 to 80 eggs each year. They deposit their eggs in moss, under plant litter, or in the soil. Eggs generally hatch in 2 to 4 weeks, depending upon temperature. The egg is an especially sensitive stage in the life cycle of land snails. *A. arbustorum* often lives alongside *Cepea nemoralis,* a land snail with a broader geographic distribution that extends from southern Scandinavia to the Iberian Peninsula.

How did the Baurs document local extinctions of *A. arbustorum?* If you think about it a bit, you will probably realize that it is usually easier to determine the presence of a species

than its absence. If you do not encounter a species during a survey, it may be that you just didn't look hard enough. Fortunately, the Baurs had over 13 years of experience doing fieldwork on *A. arbustorum* and knew its natural history well. For instance, they knew that it is best to search for the snails after rainstorms, when up to 70% of the adult population is active. Consequently, the Baurs searched Bollinger's study sites after heavy rains. They concluded that the snail was absent at a site only after two 2-hour surveys failed to turn up either a living individual or an empty shell of the species.

The Baurs found *A. arbustorum* still living at 13 of the 29 sites surveyed by Bollinger near Basel. Eleven of these remaining populations lived in deciduous forests and the other two lived on grassy riverbanks. However, the Baurs could not find the snail at 16 sites. Eight of these sites had been urbanized, which made the habitat unsuitable for any land snails because natural vegetation had been removed. Between 1900 and 1990 the urbanized area of Basel had increased by 500%. However, the eight other sites where *A. arbustorum* had disappeared were still covered by vegetation that appeared suitable. Four of these sites were covered by deciduous forest, three were on riverbanks, and one was on a railway embankment. These vegetated sites also supported populations of five other land snail species, including *C. nemoralis*.

What caused the extinction of *A. arbustorum* at sites that still supported other snails? The Baurs compared the characteristics of these sites with those of the sites where *A. arbustorum* had persisted. They found no difference between these two groups of sites in regard to slope, percent plant cover, height of vegetation, distance from water, or number of other land snail species present. The first major difference the Baurs uncovered was in altitude. The sites where *A. arbustorum* was extinct had an average altitude of 274 m. The places where it survived had an average altitude of 420 m. The places where the snail had survived were also cooler.

A thermal image of the landscape taken from a satellite showed that surface temperatures in summer around Basel ranged from about 17° to 32.5°C. Surface temperatures where *A. arbustorum* had survived averaged approximately 22°C, while the sites where the species had gone extinct had surface temperatures that averaged approximately 25°C. The sites where the snail was extinct were also much closer to very hot areas with temperatures greater than 29°C. Figure 5.34 is based on the Baurs' thermal image of the area around Basel and shows where the snail was extinct and where it persisted.

The Baurs attributed the higher temperatures at the eight sites where the snail is extinct to heating by thermal radiation from the urbanized areas of the city. Buildings and pavement store more heat than vegetation. In addition, the cooling effect of evaporation from vegetation is lost when an area is built over. Increased heat storage and reduced cooling make urbanized landscapes thermal islands. Heat energy stored in urban centers is transferred to the surrounding landscape through thermal radiation, H_r.

The Baurs documented higher temperatures at the sites near Basel where *A. arbustorum* is extinct and identified a well-studied mechanism that could produce the higher temperatures

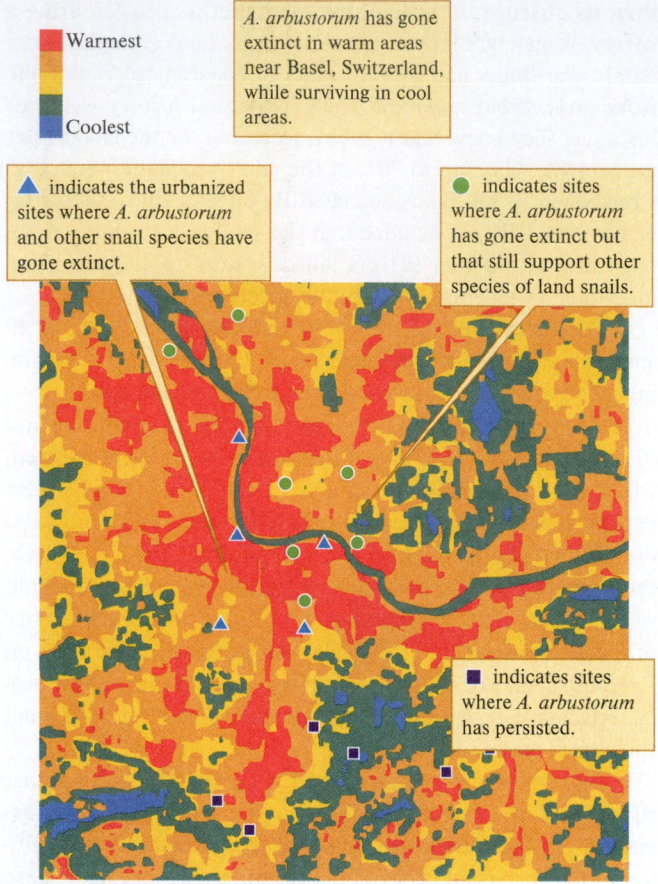

Warmest

Coolest

A. arbustorum has gone extinct in warm areas near Basel, Switzerland, while surviving in cool areas.

▲ indicates the urbanized sites where *A. arbustorum* and other snail species have gone extinct.

● indicates sites where *A. arbustorum* has gone extinct but that still support other species of land snails.

■ indicates sites where *A. arbustorum* has persisted.

Figure 5.34   Relative surface temperatures and patterns of extinction and persistence by the snail *Arianta arbustorum* around Basel, Switzerland (data from Baur and Baur 1993).

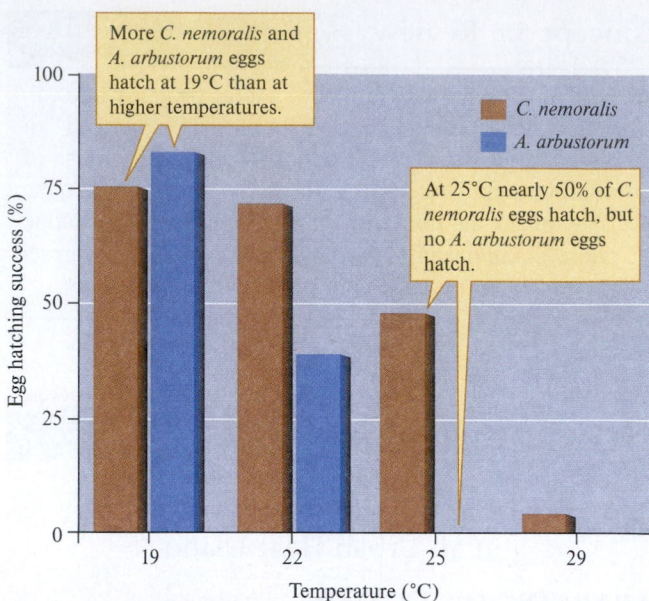

More *C. nemoralis* and *A. arbustorum* eggs hatch at 19°C than at higher temperatures.

At 25°C nearly 50% of *C. nemoralis* eggs hatch, but no *A. arbustorum* eggs hatch.

C. nemoralis
A. arbustorum

Figure 5.35   Temperature and hatching success of two snail species; the eggs of *Arianta arbustorum* are sensitive to high temperatures (data from Baur and Baur 1993).

of these sites. However, are the temperature differences they observed sufficient to exclude *A. arbustorum* from the warmer sites? The researchers compared the temperature relations of *A. arbustorum* and *C. nemoralis* to find some clues. They concentrated their studies on the influence of temperature on reproduction by these two snail species.

The eggs of each species were incubated at four temperatures—19°, 22°, 25°, and 29°C. Notice that these temperatures fall within the range measured by the satellite image (see fig. 5.34). The eggs of both species hatched at a high rate at 19°C. However, at higher temperatures, their eggs hatched at

significantly lower rates. At 22°C, less than 50% of *A. arbustorum* eggs hatched, while the eggs of *C. nemoralis* continued to hatch at a high rate. At 25°C, no *A. arbustorum* eggs hatched, while approximately 50% of the *C. nemoralis* eggs hatched. At 29°C, the hatching of *C. nemoralis* eggs was also greatly reduced. Figure 5.35 summarizes the results of this hatching experiment.

The results of this study show that the eggs of *A. arbustorum* are more sensitive to higher temperatures than are the eggs of *C. nemoralis*. This greater thermal sensitivity can explain why *A. arbustorum* is extinct at some sites, while *C. nemoralis* survived. In a recent United Nations report, at least 1 million species of plants and animals are threatened with extinction due to climate change and human activity (data from Intergovernmental Science-Policy Platform on Biodiversity and Ecosystem Services [IPBES], https://ipbes.net/news/Media-Release-Global-Assessment#_ftn1). As we face the prospect of warming on a global scale, studies of temperature relations will assume greater importance. In chapter 6, we look at a related topic, water relations.

Summary

Macroclimate interacts with the local landscape to produce microclimatic variation in temperature.   The sun's uneven heating of the earth's surface and earth's permanent tilt on its axis produce macroclimate. Macroclimate interacts with the local landscape—mainly altitude, aspect, vegetation, color of the ground, and small-scale structural features such as boulders

and burrows—to produce microclimates. For the individual organism, macroclimate may be less significant than microclimate. The physical nature of water limits temperature variation in aquatic environments.

Adapting to one set of environmental conditions generally reduces a population's fitness in other environments.   The principle of allocation, which is supported by research on bacterial populations, proposes that evolutionary trade-offs are inevitable, since organisms have access to limited amounts of energy.

Most species perform best in a fairly narrow range of temperatures.   The influence of temperature on the performance of organisms begins at the molecular level, where extreme temperatures impair the functioning of enzymes. Rates of photosynthesis and microbial activity generally peak in a narrow range of temperatures and are much lower outside of this optimal temperature range. How temperature affects the performance of organisms often corresponds to the current distributions of species and their evolutionary histories.

Many organisms have evolved ways to compensate for variations in environmental temperature by regulating body temperature.   Temperature regulation balances heat gain against heat loss. Plants and ectothermic animals use morphology and behavior to modify rates of heat exchange with the environment. Birds and mammals rely heavily on metabolic energy to regulate body temperature. The physical nature of the aquatic environment reduces the possibilities for temperature regulation by aquatic organisms. Most endothermic aquatic species are air breathers. Some organisms, mainly flying insects and some large marine fish, improve performance by selectively heating parts of their anatomy. The energetic requirements of thermoregulation may influence the geographic distribution of species.

Many organisms survive extreme temperatures by entering a resting stage.   This stage may be as simple as resting in a sheltered spot during the heat of the day or may involve elaborate physiological adjustments. Hummingbirds may enter a state of torpor, a state of low metabolic rate and lowered body temperature, when food is scarce and night temperatures cold. Other animals can go into a state of reduced metabolism that may last several months. If this state occurs mainly in winter, it is called hibernation. If it occurs in summer, it is called estivation. Such reductions in metabolic rate allow these animals to survive extreme environmental conditions during which they must rely entirely on stored energy reserves.

Long-term studies of populations of land snails around Basel, Switzerland, have documented local extinctions of these land snails. These extinctions are attributable to habitat destruction and climatic warming. The results of these studies suggest that climatic warming can lead to the local extinction of species. As we face the prospect of climatic warming at a global scale, studies of temperature relations will assume greater importance.

Key Terms

acclimation 109	heat 102	poikilotherm 111	temperature 101
conduction 110	hibernation 121	principle of allocation 105	thermal neutral zone 115
convection 110	homeotherm 111	psychrophilic 109	thermophilic 110
ectotherm 111	macroclimate 102	pubescence 102	torpor 121
endotherm 111	metabolic heat 110	radiation 111	
estivation 121	microclimate 102	riparian vegetation 105	
evaporation 111	photosynthesis 108	RM endothermy 117	

Review Questions

1. Many species of plants and animals that are associated with boreal forests also occur on mountains far to the south of the boreal forests. Using what you have learned about microclimates, predict how aspect and elevation would influence their distributions on these southern mountains.

2. Imagine a desert beetle that uses behavior to regulate its body temperature above 35°C. How might this beetle's use of microclimates created by shrubs, burrows, and bare ground change with the season?

3. Given that endothermy is relatively rare in gill-breathing marine species, what is remarkable about the fact that it is found in both sharks and tuna? Using our understanding of the potential costs and benefits of endothermy in this environment, what can we infer about the evolutionary processes that may have led to this case?

4. Figure 5.8 shows how temperature influences the activity of acetylcholinesterase in rainbow trout. Assuming that the other enzymes of rainbow trout show similar responses to temperature, how would trout swimming speed change as environmental temperature increased above 20°C?

5. The Applications section reviews how the studies of Bruno Baur and Anette Baur (1993) have documented the local extinction of the land snail *Arianta arbustorum*. Their research also shows that these extinctions may be due to reduced egg hatching at higher temperatures. Do these results show conclusively that the direct effect of higher temperatures on hatching success is responsible for the local extinctions of *A. arbustorum*? Propose and justify alternative hypotheses. Be sure you take into account all of the observations of the Baurs.

6. Butterflies, which are ectothermic and diurnal, are found from the tropical rain forest to the Arctic. They can elevate their body temperatures by basking in sunlight. How would the percentage of time butterflies spend basking versus flying change with latitude? Would the amount of time butterflies spend basking change with daily changes in temperature?

7. When we reviewed how some organisms use torpor, hibernation, and estivation to avoid extreme temperatures, we discussed the idea of energy savings. However, organisms do not always behave in a way that saves energy. For instance, when food is abundant hummingbirds do not go into torpor at night. This suggests that there may be some disadvantages associated with torpor. What are some of those potential disadvantages?

8. The section on avoiding temperature extremes focused mainly on animals. What are some of the ways in which plants avoid temperature extremes? Bring cold and hot environments into your discussion. Some of the natural history included in chapter 2 might be useful as you formulate an answer.

9. Some plants and grasshoppers in hot environments have reflective body surfaces, which make their radiative heat gain, H_r, less than it would be otherwise. The beetles on the black sand beaches of New Zealand are black, and the beetles on the white sand beaches are white. What do the matches between the color of these beetles and their beaches tell us about the relative roles of thermoregulation and predation pressure in determining beetle color?

10. Natural selection does not always produce an optimal, or even a good, fit of organisms to their environments. What are some of the reasons for a mismatch between organisms and environments? Develop your explanation using the environment, the characteristics of organisms, and the nature of natural selection.

©Digital Vision Ltd.

Chapter 6

Water Relations

A black-backed jackal, *Canis mesomelas,* takes a drink at a water hole in Botswana. As a relatively small predator, the black-backed jackal takes a risk each time it visits a water hole, where it may become prey for the larger predators, such as leopards, with which it shares the southern African landscape. It is a risk that must be taken, however, since the jackal cannot live without water.

CHAPTER CONCEPTS

6.1 Concentration gradients influence the movement of water between an organism and its environment. **129**
Concept 6.1 Review 132

6.2 Terrestrial plants and animals regulate their internal water by balancing water acquisition against water loss. **133**
Concept 6.2 Review 143

6.3 Marine and freshwater organisms use complementary mechanisms for water and salt regulation. **143**
Concept 6.3 Review 146

Applications: Using Stable Isotopes to Study Water Uptake by Plants **146**
Summary **147**
Key Terms **148**
Review Questions **148**

LEARNING OUTCOMES

After studying this section you should be able to do the following:

6.1 Explain why extreme environments such as deserts are useful to ecologists studying water relations.

6.2 Describe some mechanisms of desert plants and animals that allow them to capture water in arid environments.

Water plays a central role in the lives of all organisms. However, water acquisition and conservation are particularly critical for desert organisms. As a consequence, many ecologists studying water relations have focused their attention on desert species. It is a fact that all living things need water (see chapter 3), so how is it that places with no rivers or lakes and without any rain for months at a time have life? As it turns out, even the driest deserts are not entirely without water; sources include oases, ephemeral rivers, groundwater, moisture produced through metabolism (for animals), and moisture in the air.

Understanding the adaptations of organisms that have evolved in arid environments can help humans live in these places as well. Dev Gurera and Bharat Bhushan at the Nanoprobe Laboratory for Bio- & Nanotechnology and Biomimetics (NLBB) were interested in the mechanisms whereby plants and animals in desert environments harvest water, as a way of informing development of water collection technologies for human use (Gurera and Bhushan 2020).

Gurera and Bhushan (2020) considered dozens of arid-land species that use physical structures to harvest water, including from fog and water vapor at night. These structures facilitate condensation, promote movement of water droplets to where they can be used, and/or store water until it can be used (fig. 6.1). For example, the Texas horned lizard (*Phrynosoma cornutum*) has channels between its scales that allow water that has condensed on its skin to flow toward its mouth (fig. 6.1*a*), while many grasses (such as *Setaria viridis,* green bristlegrass or foxtail) have grooves that direct water droplets to the base of the plant and thus its roots for absorption (fig. 6.1*b*). The male sandgrouse (*Syrrhaptes* sp.) is known to transport water to its young by holding water it has collected from puddles or other water sources in its feathers (fig. 6.1*c*), whereas pubescence on leaves of the creosote bush (*Larrea tridentata*) promote condensation and protect against evaporation (fig. 6.1*d*). In the previous chapter, we learned how pubescence helps to regulate temperature as well; indeed, strategies to manage heat will often be linked to those that protect against water loss because one can lead to the other.

In this chapter, we will also learn about mechanisms used by desert insects to get the water that they need, such as hydrophobic surfaces. Such structures are found across many species across the globe. It should be no surprise that the need for water would be a strong selective force; the similarity of these structures on such different species may be considered examples of convergent evolution (chapter 2).

Water and life on earth are closely linked. The high water content of most organisms, which ranges from about 50% to 90%, reflects life's aquatic origins. Life on earth originated in salty aquatic environments and is built around biochemistry within an aquatic medium. To survive and reproduce, organisms must maintain appropriate internal concentrations of water and dissolved substances. To maintain these internal concentrations, organisms must balance water losses to the environment with water intake. How organisms maintain this water balance is called their water relations.

In some environments, organisms face the problem of water loss. Elsewhere, water streams in from the environment. The problem of maintaining proper water balance is especially strong for those organisms that live in arid terrestrial environments. A parallel challenge faces organisms that live in aquatic environments with a high salinity. In these extreme environments, the water relations of organisms stand out in bold relief. However, most organisms must expend energy to maintain their internal pool of water. In the study of relationships between organisms and the environment, which we call ecology, the study of water relations is fundamental.

6.1   Water Availability

LEARNING OUTCOMES

After studying this section you should be able to do the following:

6.3   Define relative humidity, water vapor pressure, saturation water vapor pressure, and vapor pressure deficit.

6.4   Diagram the movements of salts and water between the surrounding environment and aquatic organisms that are isosmotic, hyperosmotic, and hypoosmotic.

6.5   Using gradients in water potential, diagram and explain the movement of water from the soil, through a plant, and to the atmosphere.

Concentration gradients influence the movement of water between an organism and its environment.   The tendency of water to move down concentration gradients and the magnitude of those gradients from an organism to its environment determine whether an organism tends to lose or gain water from the environment. To understand the water relations of

(a)

Grooves and channels that allow transport of moisture collected from the air to places where it can be used by the organism can be found in both desert animals and plants.

(b)

(c)

Fine "hairs", such as those found in feathers and on pubescent leaves, can collect and even store water for short periods.

(d)

Figure 6.1   Many desert plants and animals have evolved similar structures to harvest water. Here are examples of (*a*) Texas horned lizard (*Phrynosoma cornutum*), which uses channels in the skin to direct water toward its mouth and (*b*) foxtail (*Setaria viridis*) that uses channels in the leaves to direct droplets to the roots. Other organisms use fine hairs to capture water such as (*c*) the feathers of the sandgrouse (*Syrrhaptes* sp.) or (*d*) the pubescence of the creosote bush (*Larrea tridentata*) (adapted from Gurera and Bhushan 2020).

organisms, we first review the basic physical behavior of water in terrestrial and aquatic environments.

In chapter 2, we saw that water availability on land varies tremendously, from the tropical rain forest with abundant moisture throughout the year (see fig. 2.11) to hot deserts with year-round drought (see fig. 2.20). In chapter 3, we reviewed the considerable variation in salinity among aquatic environments, ranging from the dilute waters of tropical rivers draining highly weathered landscapes to hypersaline lakes. Salinity, as we shall see, reflects the relative "aridity" of aquatic environments.

These preliminary descriptions in chapters 2 and 3 do not include the situations faced by individual organisms within their microclimates—microclimates such as those experienced by a desert animal that lives at an oasis, where it has access to abundant moisture, or a rain forest plant that lives in the forest canopy, where it is exposed to full tropical sun and drying winds. As with temperature, to understand the water relations of an organism we must consider its microclimate, including the amount of water in the environment.

Water Content of Air

As we saw when we reviewed the hydrologic cycle in chapter 3, water vapor is continuously added to air as water evaporates from the surfaces of oceans, lakes, and rivers. On land, evaporation also accounts for much of the water lost by organisms. The potential for such evaporative water loss depends on the temperature and water content of the air around the organisms. As the amount of water vapor in the surrounding air increases, the water concentration gradient from organisms to the air is reduced and the rate at which organisms lose water to the atmosphere decreases. This is the reason that evaporative air coolers work poorly in humid climates, where the water content of air is high. These mechanical systems work best in arid climates, where there is a steep gradient of water concentration from the evaporative cooler to the air. A steep water concentration gradient is conducive to a high rate of evaporation.

How is the water content of air measured? The most familiar measure of the water content of air is **relative humidity,** the percent water content relative to a potential maximum:

$$\text{Relative humidity} = \frac{\text{Water vapor density}}{\text{Saturation water vapor density}} \times 100$$

The actual amount of water in air is measured directly as the mass of water vapor per unit volume of air. This quantity, the *water vapor density,* is given either as milligrams of water per liter of air (mg H_2O/L) or as grams of water per cubic meter of air (g H_2O/m^3). The maximum quantity of water vapor that air at a particular temperature can contain is its *saturation water vapor density,* the denominator in the relative humidity equation. Saturation water vapor density increases with temperature, as you can see from the red curve in figure 6.2.

One of the most useful ways of expressing the quantity of water in air is in terms of the pressure it exerts. If we express the water content of air in terms of pressure, we can use similar units to consider the water relations of organisms in air, soil, and water. Using pressure as a common currency to represent

water relations in very different environments helps us unify our understanding of this very important area of ecology. We usually think in terms of *total atmospheric pressure,* the pressure exerted by all the gases in air, but you can also calculate the partial pressures due to individual atmospheric gases such as oxygen, nitrogen, or water vapor. We call this last quantity **water vapor pressure.** At sea level, atmospheric pressure averages approximately 760 mm of mercury, the height of a column of mercury supported by the combined force (pressure) of all the gas molecules in the atmosphere. The international convention for representing water vapor pressure, however, is in terms of the pascal (Pa), where 1 Pa is 1 newton of force per square meter. Using this convention, 760 mm of mercury, or one atmosphere of pressure, equals approximately 101,300 Pa, 101.3 kilopascals (kPa), or 0.101 megapascal (MPa = 10^6 Pa).

The pressure exerted by the water vapor in air that is saturated with water is called **saturation water vapor pressure.** As the black curve in figure 6.2 shows, this pressure increases with temperature and closely parallels the increase in saturation water vapor density shown by the red curve.

We can also use water vapor pressure to represent the *relative* saturation of air with water. You calculate this measure, called the **vapor pressure deficit,** as the difference between the actual water vapor pressure and the saturation water vapor pressure at a particular temperature. In terrestrial environments, water flows from

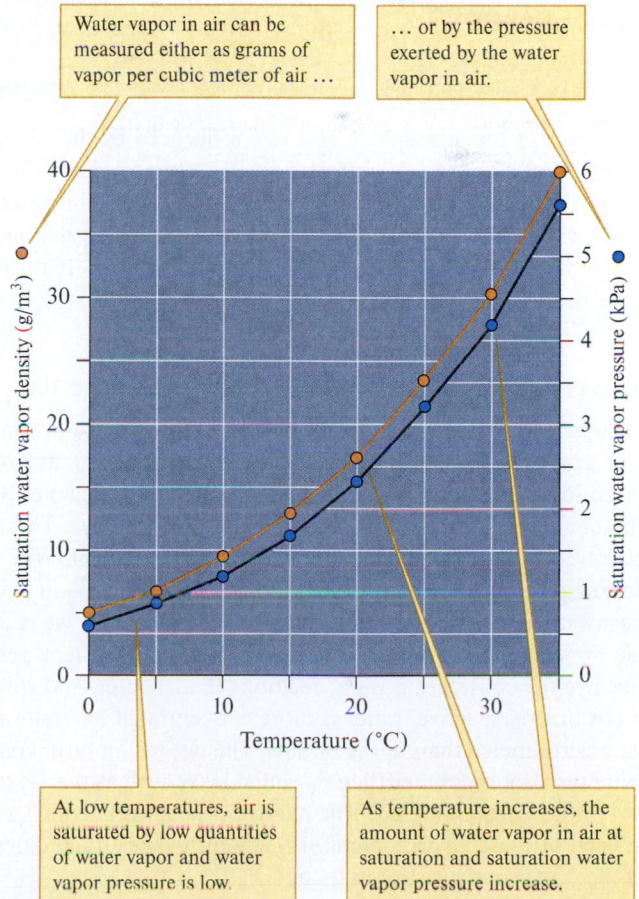

Figure 6.2   The relationship between air temperature and two measures of water vapor saturation of air.

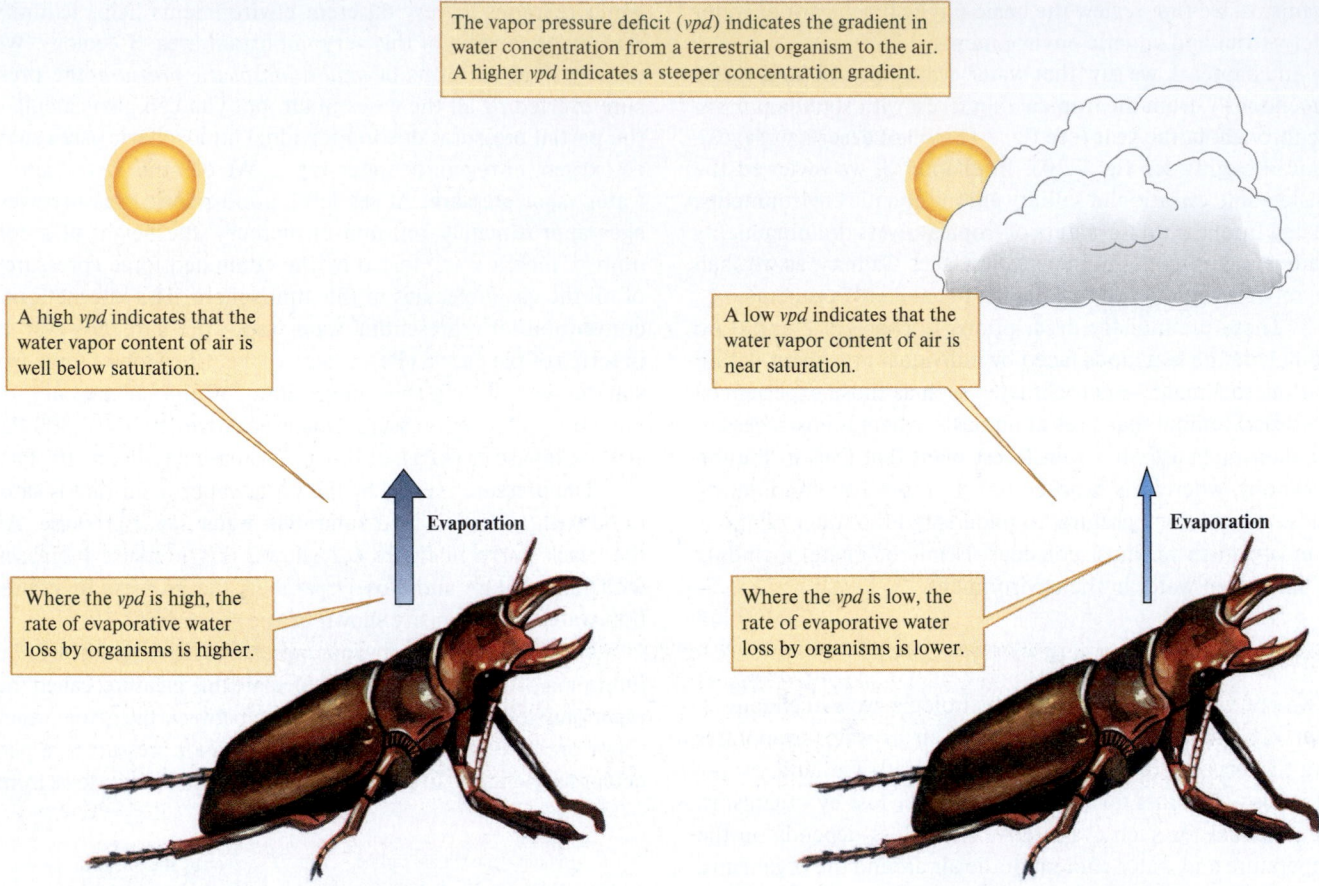

The vapor pressure deficit (*vpd*) indicates the gradient in water concentration from a terrestrial organism to the air. A higher *vpd* indicates a steeper concentration gradient.

A high *vpd* indicates that the water vapor content of air is well below saturation.

A low *vpd* indicates that the water vapor content of air is near saturation.

Evaporation

Evaporation

Where the *vpd* is high, the rate of evaporative water loss by organisms is higher.

Where the *vpd* is low, the rate of evaporative water loss by organisms is lower.

Figure 6.3   The potential for evaporative water loss by terrestrial organisms increases as vapor pressure deficit increases.

organisms to the atmosphere at a rate influenced by the vapor pressure deficit of the air surrounding the organism. Figure 6.3 shows the relative rates of water loss by an organism exposed to air with a low versus high vapor pressure deficit. Again, one of the most useful features of vapor pressure deficit is that it is expressed in units of pressure, generally kilopascals.

Water Movement in Aquatic Environments

In aquatic environments, water moves down its concentration gradient. It may sound silly to speak of the amount of water in an aquatic environment but, as we saw in chapter 3, all aquatic environments contain dissolved substances. These dissolved substances, however slightly, dilute the water. While oceanographers and limnologists (those who study bodies of freshwater) generally focus on salt content, or salinity, we take the opposite point of view in order to build a consistent perspective for considering water relations in air, water, and soil. From this perspective, water is more concentrated in freshwater environments than in the oceans. The oceans, in turn, contain more water per liter than do saline lakes such as the Dead Sea or the Great Salt Lake. The relative concentration of water in each of these environments strongly influences the biology of the organisms that live in them.

The body fluids of all organisms contain water and solutes, for example, inorganic salts and amino acids. We can think of aquatic organisms and the environment that surrounds them as two aqueous

solutions separated by a selectively permeable membrane. If the internal environment of the organism and the external environment differ in concentrations of water and salts, these substances will tend to move down their concentration gradients. This movement is the process of **diffusion.** We give the diffusion of water across a semipermeable membrane a special name, however: **osmosis.**

In the aquatic environment, water moving down its concentration gradient produces osmotic pressure. Osmotic pressure, like vapor pressure, can be expressed in pascals. The strength of the osmotic pressure across a semipermeable membrane, such as the gills of a fish, depends on the difference in water concentration across the membrane. Larger differences, between organism and environment, generate higher osmotic pressures.

Aquatic organisms generally live in one of three circumstances. Those with body fluids containing the same concentration of water and solutes as the external environment are **isosmotic.** Organisms with body fluids with a higher concentration of water and lower solute concentration than the environment are **hypoosmotic** and tend to lose water to the environment. Those with body fluids with a lower concentration of water and higher solute concentration than the external medium are **hyperosmotic** and are subject to water flooding inward. In the face of these osmotic pressures, aquatic organisms expend energy to maintain a proper internal environment. How much energy the organism must expend depends on the magnitude of the osmotic pressure between them and the environment and the permeability of their body surfaces. Figure 6.4

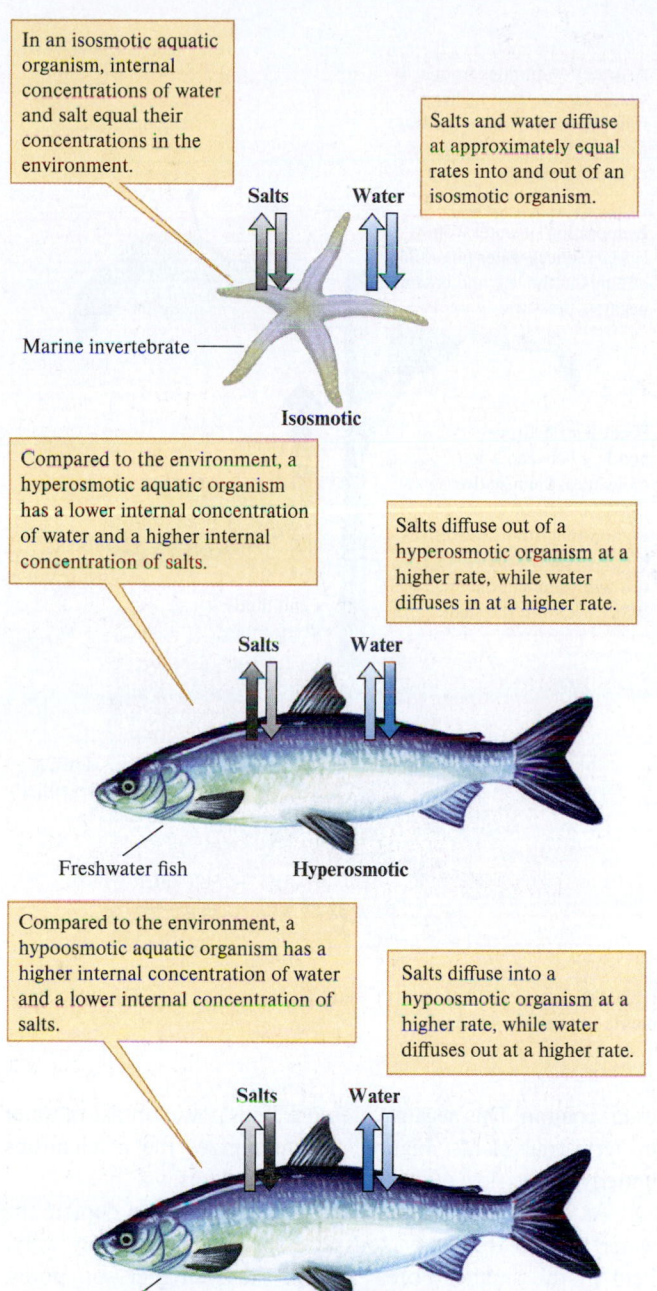

In an isosmotic aquatic organism, internal concentrations of water and salt equal their concentrations in the environment.

Salts and water diffuse at approximately equal rates into and out of an isosmotic organism.

Salts   Water

Marine invertebrate

Isosmotic

Compared to the environment, a hyperosmotic aquatic organism has a lower internal concentration of water and a higher internal concentration of salts.

Salts diffuse out of a hyperosmotic organism at a higher rate, while water diffuses in at a higher rate.

Salts   Water

Freshwater fish   **Hyperosmotic**

Compared to the environment, a hypoosmotic aquatic organism has a higher internal concentration of water and a lower internal concentration of salts.

Salts diffuse into a hypoosmotic organism at a higher rate, while water diffuses out at a higher rate.

Salts   Water

Marine fish   **Hypoosmotic**

Figure 6.4   Water and salt movements between the environment and isosmotic, hyperosmotic, and hypoosmotic aquatic organisms.

summarizes the movement of water and salts into and out of isosmotic, hyperosmotic, and hypoosmotic organisms.

Water Movement Between Soils and Plants

In the aquatic environment, water may flow either to or from the organism, depending on the relative concentrations of water and solutes in body fluids and the surrounding medium. On land, water also flows down a concentration gradient. This gradient can be created by concentrations of solutes in tissues and soil, by gravity, and by vapor pressure deficits. We say that in all of these cases, water is flowing down a gradient of **water potential**, which is the capacity of water to do work. This is obvious in the

case of flowing water, which can turn the water wheel of an old-fashioned water mill or the turbines of a hydroelectric plant. The capacity of water to do work depends on its free energy content. Water flows from positions of higher to lower free energy. Under the influence of gravity, water flows downhill from a position of higher free energy, at the top of the hill, to a position of lower free energy, at the bottom of the hill.

In the section "Water Movement in Aquatic Environments," we saw that water flows down a concentration gradient from locations of higher water concentration (*hypo*osmotic) to locations of lower water concentration (*hyper*osmotic). The measurable "osmotic pressure" generated by water flowing down these concentration gradients shows that water flowing in response to osmotic gradients has the capacity to do work. We measure water potential, like vapor pressure deficit and osmotic pressure, in pascals, usually megapascals ($MPa = Pa \times 10^6$). By convention, water potential is represented by the symbol ψ (psi), and the water potential of pure water is set at 0. In nature, water potentials are generally negative. Figure 6.5 shows that water is flowing down a gradient of water potential that goes from a slightly negative water potential in the soil through the moderately negative water potentials of the plant to the highly negative water potential of dry air.

Now let's look at some of the mechanisms involved in producing a gradient of water potential such as that shown in figure 6.5. We can express the water potential of a solution as:

$$\psi = \psi_{solutes}$$

$\psi_{solutes}$ is the *reduction* in water potential due to dissolved substances, which is a negative number.

Within small spaces, such as the interior of a plant cell or the pore spaces within soil, other forces, called **matric forces,** are also significant. Water is not only sticky to itself, but also to other particles. Matric forces are a consequence of water's tendency to adhere to the walls of containers such as cell walls or the soil particles lining a soil pore. Matric forces lower water potential. The water potential for fluids within plant cells is approximately:

$$\psi_{plant} = \psi_{solutes} + \psi_{matric}$$

In this expression, ψ_{matric} is the *reduction* in water potential due to matric forces within plant cells. At the level of the whole plant, another force is generated as water evaporates from the air spaces within leaves into the atmosphere. Evaporation of water from leaves generates a negative pressure, or tension, on the column of water that extends from the leaf through the plant all the way down to its roots. This means that water can move counter to gravity. A property of water that makes its upward movement possible is **cohesion.** That is, water molecules stick to each other, making it possible to suck water up in a straw. Similarly, land plants can draw water up from its roots simply by losing it through their leaves—no pumping structure is necessary, even in very tall trees.

So, the water potential of plant fluids is affected by solutes, matric forces, and the negative pressures exerted by evaporation. Consequently, we can represent the water potential of plant fluids as:

$$\psi_{plant} = \psi_{solutes} + \psi_{matric} + \psi_{pressure}$$

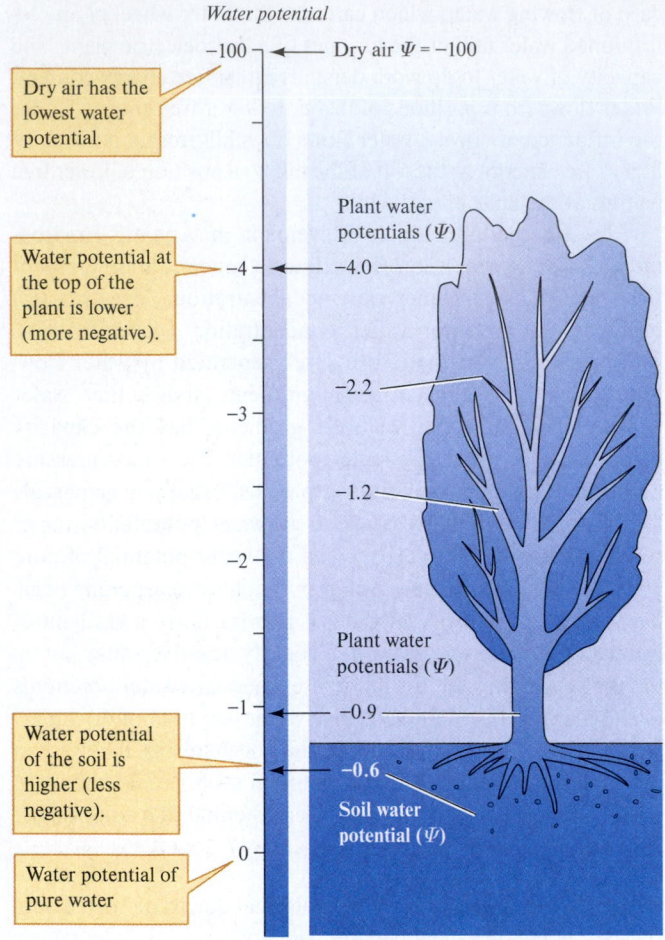

Water potential

Dry air $\Psi = -100$

−100

Dry air has the lowest water potential.

Plant water potentials (Ψ)

−4   −4.0

Water potential at the top of the plant is lower (more negative).

−3   −2.2

−1.2

−2

Plant water potentials (Ψ)

−1   −0.9

Water potential of the soil is higher (less negative).

−0.6

Soil water potential (Ψ)

Water potential of pure water

0

Figure 6.5   Water potentials decrease (become more negative) from soil to plant to air (data from Wiebe et al. 1970).

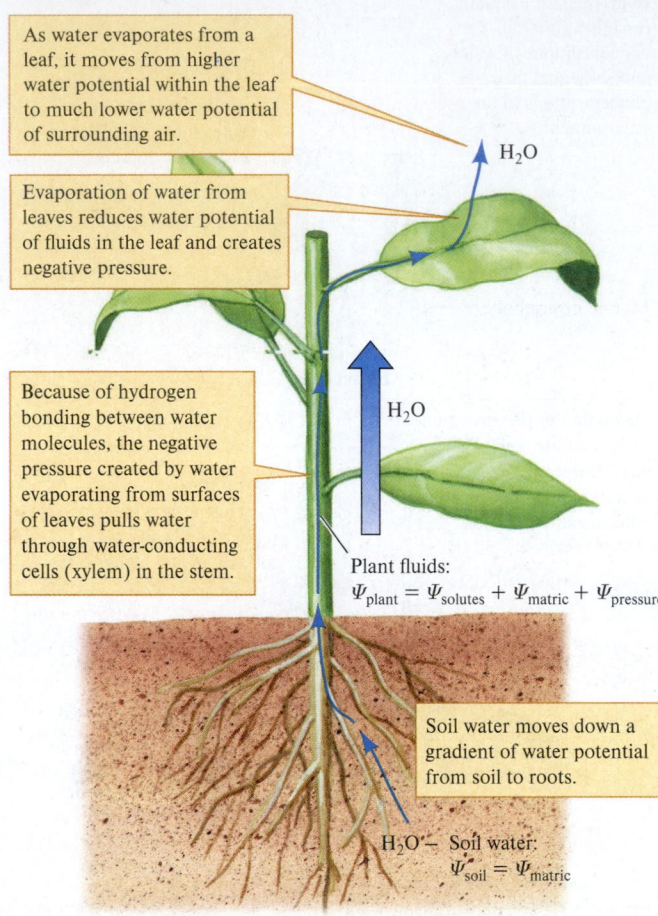

As water evaporates from a leaf, it moves from higher water potential within the leaf to much lower water potential of surrounding air.

H_2O

Evaporation of water from leaves reduces water potential of fluids in the leaf and creates negative pressure.

Because of hydrogen bonding between water molecules, the negative pressure created by water evaporating from surfaces of leaves pulls water through water-conducting cells (xylem) in the stem.

H_2O

Plant fluids:
$\Psi_{plant} = \Psi_{solutes} + \Psi_{matric} + \Psi_{pressure}$

Soil water moves down a gradient of water potential from soil to roots.

H_2O − Soil water:
$\Psi_{soil} = \Psi_{matric}$

Figure 6.6   Mechanisms of water movement from soil through plants to the atmosphere.

Here again, $\psi_{pressure}$ is the *reduction* in water potential due to negative pressure created by water evaporating from leaves.

Meanwhile, the solute content of soil water is often so low that soil matric forces account for most of soil water potential:

$$\psi_{soil} > \psi_{matric}$$

Matric forces vary considerably from one soil to another, depending primarily upon soil texture and pore size. Coarser soils, such as sands and loams, with larger pore sizes exert lower matric forces, while fine clay soils, with smaller pore sizes, exert higher matric forces. So, while clay soils can hold a higher quantity of water compared to sandy soils, the higher matric forces within clay soils bind that water more tightly. As long as the water potential of plant tissues is less than the water potential of the soil, $\psi_{plant} < \psi_{soil}$, water flows from the soil to the plant.

The higher water potential of soil water compared to the water potential of roots induces water to flow from the soil into plant roots. As water enters roots from the surrounding soil, it joins a column of water that extends from the roots through the water-conducting cells, or xylem, of the stem to the leaves. Due to the cohesion of water molecules, as water molecules at the upper end of this column evaporate into the air at the surfaces of leaves, they exert tension, or negative pressure, on the entire

water column. This negative pressure helps power uptake of water by terrestrial plants. Figure 6.6 summarizes the mechanisms underlying the flow of water from soil to plants.

As plants draw water from the soil, they soon deplete the water held in the larger soil pore spaces, leaving only water held in the smaller pores. Within these smaller soil pores, matric forces are greater than in the larger pores. Consequently, as soil dries, soil water potential decreases and the remaining water becomes harder and harder to extract.

This section has given us a basis for considering the availability of water to organisms living in terrestrial and aquatic environments. Let's use the foundation we have built here to explore the water relations of organisms on land and in water. In the face of variation in water availability, organisms have been selected to regulate their internal water.

Concept 6.1 Review

1. Why are the two curves shown in figure 6.2 so similar?
2. Which has a higher free energy content, pure water or seawater?
3. Why are water potentials in nature generally negative?

6.2   Water Regulation on Land

LEARNING OUTCOMES

After studying this section you should be able to do the following:

6.6   List the major avenues for water gain and loss in terrestrial plants and animals.

6.7   Discuss the response of plant roots to differences in water availability.

6.8   Compare water conservation by animals from different environments.

6.9   Explain how the Sonoran Desert cicada can remain active when environmental temperatures exceed its lethal maximum temperature.

Terrestrial plants and animals regulate their internal water by balancing water acquisition against water loss.   When organisms moved into the terrestrial environment, they faced two major environmental challenges: potentially massive losses of water to the environment through evaporation and reduced access to replacement water. Terrestrial organisms evolved by natural selection to meet these challenges, eventually acquiring the capacity to regulate their internal water content on land. We can summarize water regulation by terrestrial animals as:

$$W_{ia} = W_d + W_f + W_a - W_e - W_s$$

This says simply that the internal water of an animal (W_{ia}) results from a balance between water acquisition and water loss. The major sources of water are:

$$W_d = \text{water taken by drinking}$$
$$W_f = \text{water taken in with food}$$
$$W_a = \text{water absorbed from the air}$$

The avenues of water loss are:

$$W_e = \text{water lost by evaporation}$$
$$W_s = \text{water lost with various secretions and excretions including urine, mucus, and feces}$$

We can summarize water regulation by terrestrial plants in a similar way:

$$W_{ip} = W_r + W_a - W_t - W_s$$

The internal water concentration of a plant (W_{ip}) results from a balance between gains and losses, where the major sources of water for plants are:

$$W_r = \text{water taken from soil by roots}$$
$$W_a = \text{water absorbed from the air}$$

The major ways that plants lose water are:

$$W_t = \text{water lost by transpiration}$$
$$W_s = \text{water lost with various secretions and reproductive structures, including nectar, fruit, and seeds}$$

The main avenues of water gain and loss by terrestrial plants and animals are summarized in figure 6.7. The figure presents

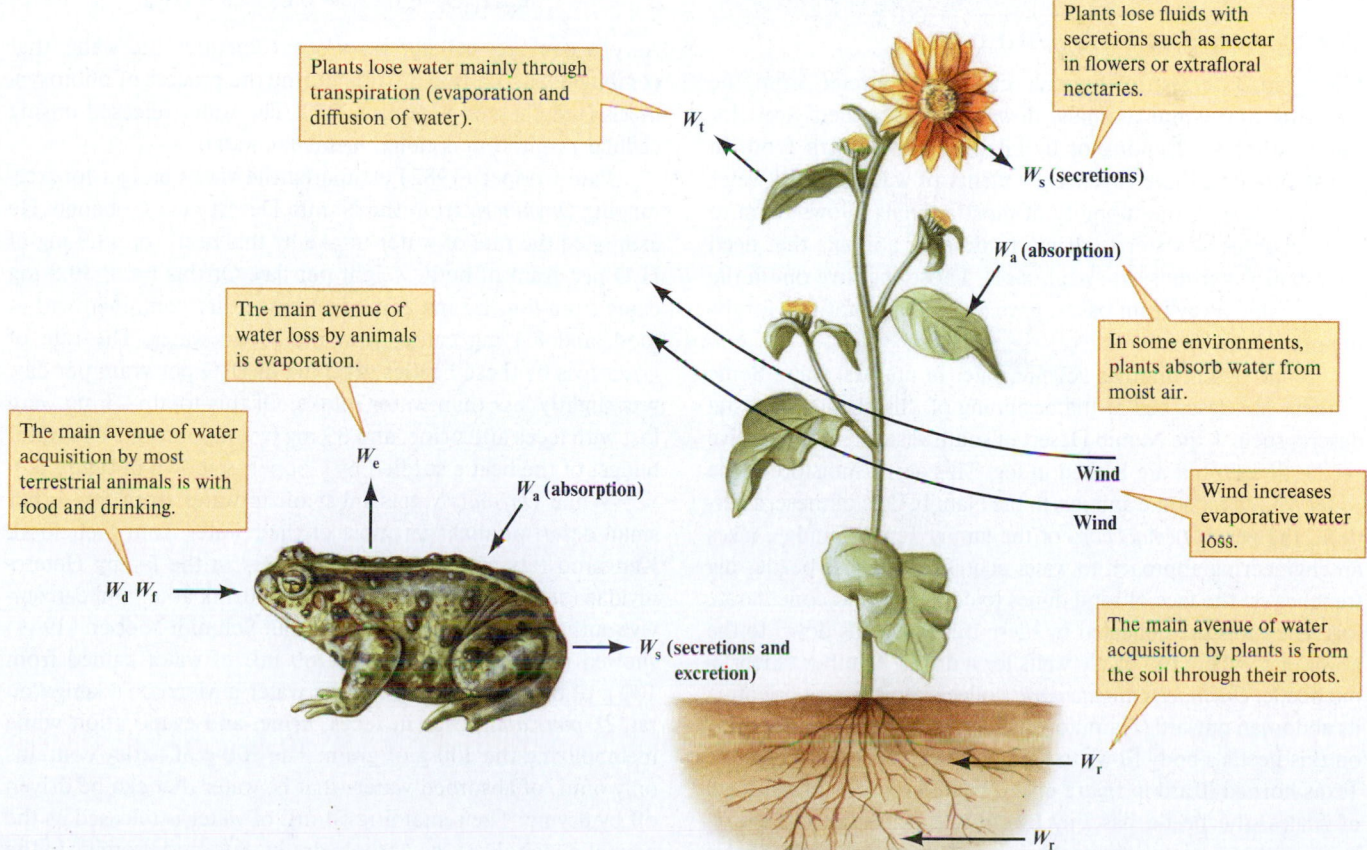

Plants lose water mainly through transpiration (evaporation and diffusion of water).

W_t

Plants lose fluids with secretions such as nectar in flowers or extrafloral nectaries.

W_s (secretions)

W_a (absorption)

In some environments, plants absorb water from moist air.

The main avenue of water loss by animals is evaporation.

W_e

The main avenue of water acquisition by most terrestrial animals is with food and drinking.

W_a (absorption)

W_d W_f

Wind

Wind

Wind increases evaporative water loss.

W_s (secretions and excretion)

The main avenue of water acquisition by plants is from the soil through their roots.

W_r

W_r

Figure 6.7   The water relations of terrestrial plants and animals can be summarized by analogous pathways for water gain and loss.

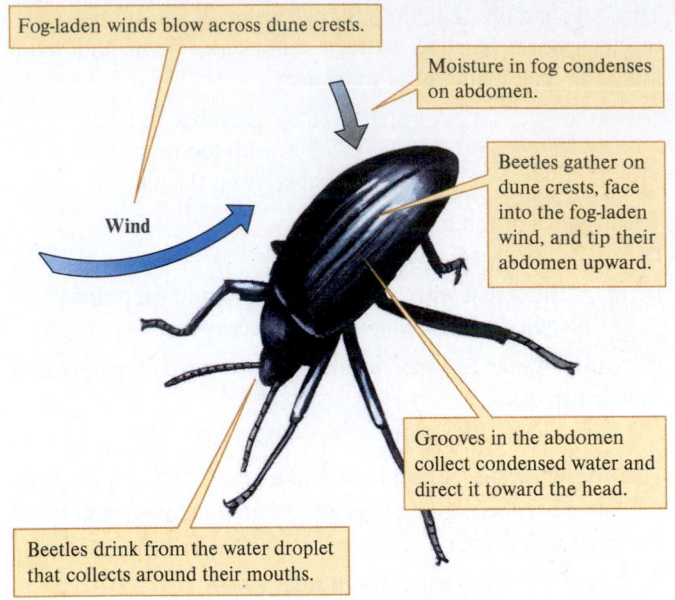

Fog-laden winds blow across dune crests.

Moisture in fog condenses on abdomen.

Wind

Beetles gather on dune crests, face into the fog-laden wind, and tip their abdomen upward.

Grooves in the abdomen collect condensed water and direct it toward the head.

Beetles drink from the water droplet that collects around their mouths.

Figure 6.8   Some beetles of the Namib Desert can harvest sufficient moisture from fog to meet their needs for water.

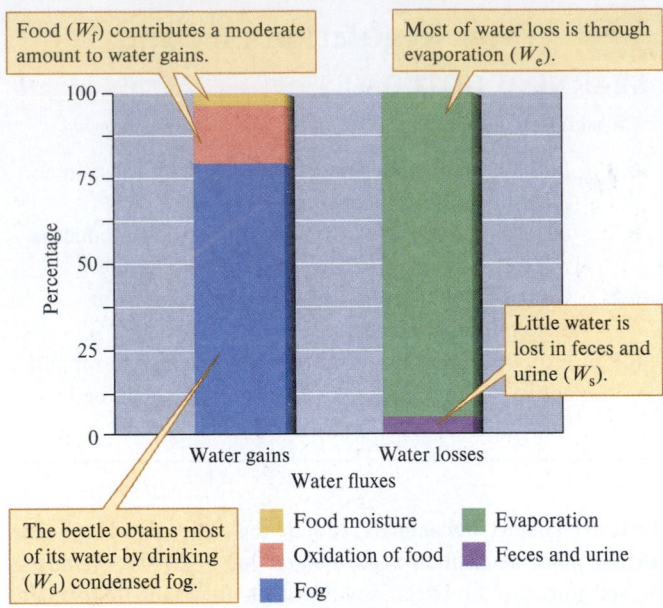

Food (W_f) contributes a moderate amount to water gains.

Most of water loss is through evaporation (W_e).

Little water is lost in feces and urine (W_s).

The beetle obtains most of its water by drinking (W_d) condensed fog.

Food moisture   Evaporation
Oxidation of food   Feces and urine
Fog

Figure 6.9   Water budget of the desert beetle, *Onymacris unguicularis* (data from Cooper 1982).

a generalized picture of the water relations of terrestrial organisms. However, organisms in different environments face different environmental challenges to which they have evolved a wide variety of responses.

Water Acquisition by Animals

Many small terrestrial animals can absorb water from the air. Most terrestrial animals, however, satisfy their need for water either by drinking or by taking in water with food. In moist climates, there is generally plenty of water, and, if water becomes scarce, the mobility of most animals allows them to go to sources of water to drink. In deserts, animals that need abundant water must live near oases. Those that live out in the desert itself, away from oases, have evolved adaptations for living in arid environments.

Some desert animals acquire water in unusual ways. Some of these are described at the beginning of this chapter. Coastal deserts such as the Namib Desert of southwestern Africa receive very little rain but are bathed in fog. This aerial moisture is the water source for some animals in the Namib. One of these, a beetle in the genus *Lepidochora* of the family Tenebrionidae, takes an engineering approach to water acquisition. These beetles dig trenches on the face of sand dunes to condense and concentrate fog. The moisture collected by these trenches runs down to the lower end, where the beetle waits for a drink. Another tenebrionid beetle, *Onymacris unguicularis,* collects moisture by orienting its abdomen upward (Hamilton and Seely 1976). Fog condensing on this beetle's body flows to its mouth (fig. 6.8) not unlike the Texas horned lizard in figure 6.1a. The difference is that instead of scales, the beetle has tiny **hydrophilic** (moisture attracting) bumps that create channels in an otherwise **hydrophobic** (moisture repelling), waxy coating on the beetle's back. *Onymacris* also takes in water with its food. Some of this water is absorbed

within the tissues of the food. The remaining water is produced when the beetle metabolizes the carbohydrates, proteins, and fats contained in its food. We can see the source of this water if we look at an equation for oxidation of glucose:

$$C_6H_{12}O_6 + 6\ O_2 \rightarrow 6\ CO_2 + \mathbf{6\ H_2O}$$

As you can see, cellular respiration liberates the water that combined with carbon dioxide during the process of photosynthesis (see chapter 5, section 5.3). The water released during cellular respiration is called **metabolic water.**

Paul Cooper (1982) estimated the water budget for free-ranging *Onymacris* from the Namib Desert near Gobabeb. He estimated the rate of water intake by this beetle at 49.9 mg of H_2O per gram of body weight per day. Of this total, 39.8 mg came from fog, 1.7 mg came from moisture contained within food, and 8.4 mg came from metabolic water. The rate of water loss by these beetles, 41.3 mg of H_2O per gram per day, was slightly less than water intake. Of this total, 2.3 mg were lost with feces and urine, and 39 mg by evaporation. The water budget of the beetle studied by Cooper is shown in figure 6.9.

While *Onymacris* gets most of its water from fog, other small desert animals get most of their water from their food. Kangaroo rats of the genus *Dipodomys* in the family Heteromyidae (see fig. 13.22a) don't have to drink at all and can survive entirely on metabolic water. Knut Schmidt-Nielsen (1964) showed that the approximately 60 mL of water gained from 100 g of barley makes up for the water a Merriam's kangaroo rat, *D. merriami,* loses in feces, urine, and evaporation while metabolizing the 100 g of grain. The 100 g of barley contains only 6 mL of absorbed water—that is, water that can be driven off by drying. The remaining 54 mL of water is released as the animal metabolizes the carbohydrates, fats, and proteins in the grain. The importance of metabolic water in the water budget of Merriam's kangaroo rat is pictured in figure 6.10.

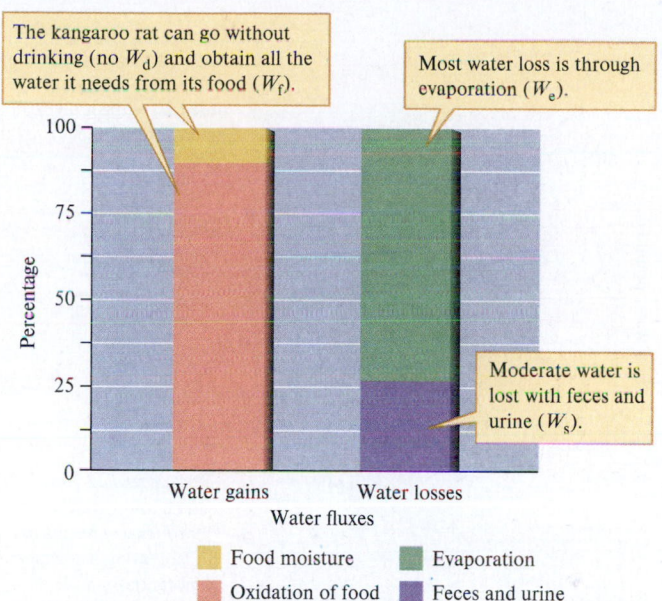

The kangaroo rat can go without drinking (no W_d) and obtain all the water it needs from its food (W_f).

Most water loss is through evaporation (W_e).

Moderate water is lost with feces and urine (W_s).

Food moisture   Evaporation
Oxidation of food   Feces and urine

Figure 6.10   Water budget of Merriam's kangaroo rat, *Dipodomys merriami* (data from Schmidt-Nielsen 1964).

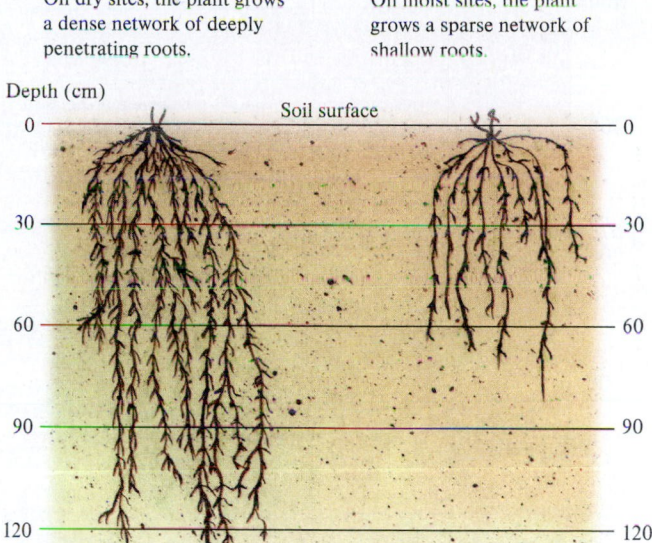

On dry sites, the plant grows a dense network of deeply penetrating roots.

On moist sites, the plant grows a sparse network of shallow roots.

Figure 6.11   Soil moisture influences the extent of root development by *Artemesia frigida* (data from Coupland and Johnson 1965).

While animals generally obtain most of their water by drinking or with their food, most plants get the bulk of their water from the soil through their roots.

Water Acquisition by Plants

The extent of root development by plants often reflects differences in water availability. Studies of root systems in different climates show that plants in dry climates grow more roots than do plants in moist climates. In dry climates, plant roots tend to grow deeper in the soil and to constitute a greater proportion of plant biomass. The taproots of some desert shrubs can extend 9 or even 30 m down into the soil, giving them access to deep groundwater. Roots may account for up to 90% of total plant biomass in deserts and semiarid grasslands. In coniferous forests, roots constitute only about 25% of total plant biomass.

You don't have to compare forests and deserts, however, to observe differences in root development. R. Coupland and R. Johnson (1965) compared the rooting characteristics of plants growing in the temperate grasslands of western Canada. During their study, they carefully excavated the roots of over 850 individual plants, digging over 3 m deep to trace some roots. They found that microclimate affects root development in many grassland species. For instance, the roots of fringed sage, *Artemesia frigida*, penetrate over 120 cm into the soil on dry sites; on moist sites, its roots grow only to a depth of about 60 cm (fig. 6.11).

Deeper roots often help plants from dry environments extract water from deep within the soil profile. This generalization is supported by studies of two common grasses that grow in Japan, *Digitaria adscendens* and *Eleusine indica*. The grasses overlap broadly in their distributions in Japan; however, only *Digitaria* grows on coastal sand dunes, which are among the most drought-prone habitats in Japan.

Y.-M. Park (1990) was interested in understanding the mechanisms allowing *Digitaria* to grow on coastal dunes

where *Eleusine* could not. Because of the potential for drought in coastal dunes, Park studied the responses of the two grasses to water stress. He grew both species from seeds collected at the Botanical Gardens at the University of Tokyo. Seeds were germinated in moist sand, and the seedlings were later transplanted into 10 cm by 90 cm polyvinyl chloride (PVC) tubes filled with sand from a coastal dune. Park planted two seedlings of *Digitaria* in each of 36 tubes and two of *Eleusine* in 36 other tubes. He watered all 72 tubes with a nutrient solution every 10 days for 40 days. At the end of the 40 days, Park divided the 36 tubes of each species into two groups of 18. One group of each species was kept well watered for the next 19 days, while the other group remained unwatered.

Unwatered *Digitaria* and *Eleusine* responded differently. The root mass of *Digitaria* increased almost sevenfold over the 19 days of no watering, while the root mass of *Eleusine* increased about threefold. In addition, the roots of *Digitaria* were still growing at the end of the experiment, while those of *Eleusine* stopped growing about 4 days before the end of the experiment. Figure 6.12 summarizes these results.

Park found that the differences in root growth were greatest in the deeper soil layers. Below 60 cm in the growing tubes, the unwatered group of *Eleusine* showed suppressed root growth, while *Digitaria* did not. With its greater mass of more deeply penetrating roots, *Digitaria* maintained high leaf water potential throughout the 19 days of no watering. During this same period, *Eleusine* showed a substantial decline in leaf water potential. The leaf water potentials of *Digitaria* and *Eleusine* over the 19 days are shown in figure 6.13.

Park's results suggest that *Digitaria* can be successful in the drier dune habitat because it grows longer roots, which exploit deeper soil moisture. With these deeper roots, *Digitaria* can keep the water potential of its tissues high even in relatively dry soils, where *Eleusine* suffers lowered water potential.

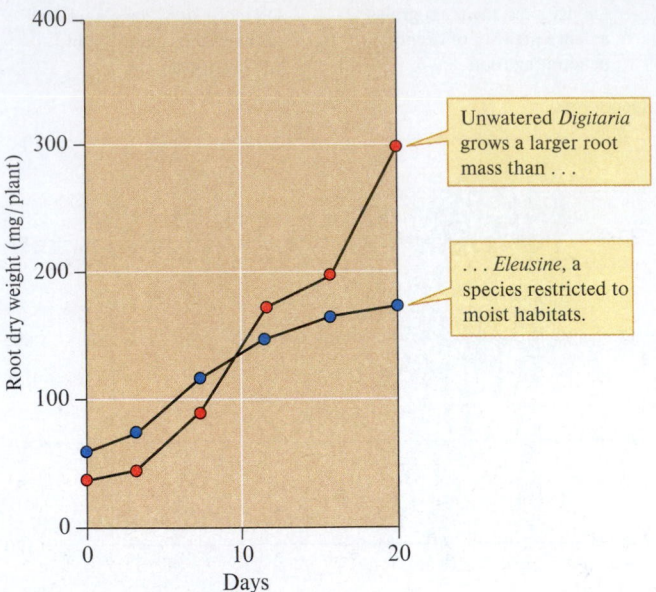

Figure 6.12   A grass species from a dry habitat responded to a simulated drought by greater root growth compared to a grass species from a moist habitat (data from Park 1990).

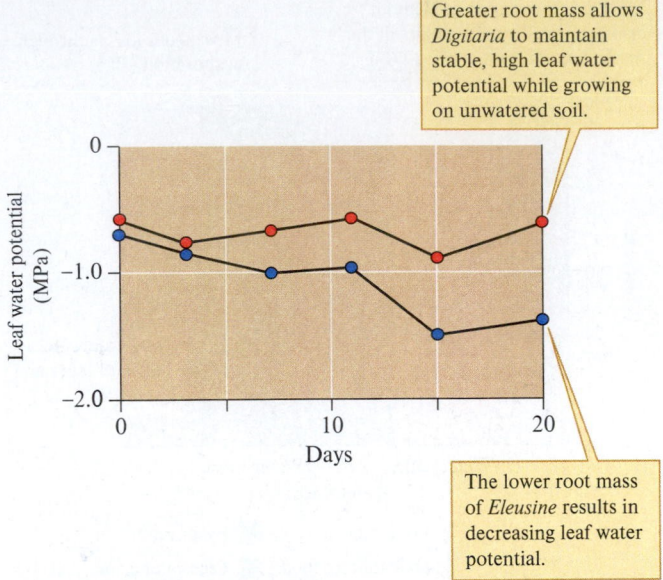

Figure 6.13   A grass species from a dry habitat maintained a higher water potential during a simulated drought compared to a grass species from a moist habitat (data from Park 1990).

In other words, *Digitaria* maintains higher leaf water potentials because its greater root development maintains a higher rate of water intake—higher W_r.

The examples we've just reviewed concern rooting by individual plant species either in the field or under experimental conditions. An important question that we might ask is whether there have been enough root studies to make tentative generalizations about the rooting biology of plants. Jochen Schenk and Robert Jackson (2002) conducted an analysis of 475 root profile studies from 209 geographic localities from around the world. In over 90% of the 475 root profiles, at least 50% of roots were in the top 0.3 m of the soil and at least 95% of roots were in the upper 2 m. However, there were pronounced geographic differences in rooting depth. Schenk and Jackson found that rooting depth increases from 80° to 30° latitude—that is, from Arctic tundra to Mediterranean woodlands and shrublands and deserts. However, there were no clear trends in rooting depth in the tropics. Consistent with our present discussion, deeper rooting depths occur mainly in water-limited ecosystems.

Water Conservation by Plants and Animals

Another way to balance a water budget is by reducing water losses. One of the most common adaptations to arid environments is waterproofing to reduce evaporative water loss. Many terrestrial plants and animals cover themselves with a fairly waterproof "hide" impregnated with a variety of waterproofing waxes. However, some organisms are more waterproof than others, and rates of evaporative water loss vary greatly from one animal or plant species to another.

Why do the water loss rates of organisms differ? One reason is that species have evolved in environments that differ greatly in water availability. As a consequence, selection for water conservation has been more intense in some environments than others. Species that evolved in warm deserts are generally much more resistant to desiccation than relatives that evolved in moist tropical or temperate habitats. In general, populations that evolved in drier environments lose water at a slower rate. For instance, turtles from wet and moist habitats lose water at a much higher rate than do desert tortoises (fig. 6.14). As the following example shows, however, the water loss rates of even closely related species can differ substantially.

Neil Hadley and Thomas Schultz (1987) studied two species of tiger beetles in Arizona that occupy different microclimates. *Cicindela oregona* lives along the moist shoreline of streams and is active in fall and spring. In contrast, *Cicindela obsoleta* lives in the semiarid grasslands of central and southeastern Arizona and is active in summer. The researchers suspected that these differences in microclimate select for differences in waterproofing of the two tiger beetles.

Hadley and Schultz studied the waterproofing of the tiger beetles by comparing the amount of water each species lost while held in an experimental chamber. They pumped dry air through the chamber at a constant rate and maintained its temperature at 30°C. They weighed each beetle at the beginning of an experiment and then again after 3 hours in the chamber. The difference between initial and final weights gave them an estimate of the water loss rate of each beetle. By determining water loss for several individuals of each species, they estimated the average water loss rates for *C. oregona* and *C. obsoleta*. Hadley and Schultz found that *C. oregona* loses water two times as fast as *C. obsoleta* (fig. 6.15). In other words, the species from the drier microclimate, *C. obsoleta*, appears to be more waterproofed.

Waterproofing of terrestrial insect cuticles is usually provided by hydrocarbons, which include organic compounds, such as lipids and waxes. When Hadley and Schultz analyzed

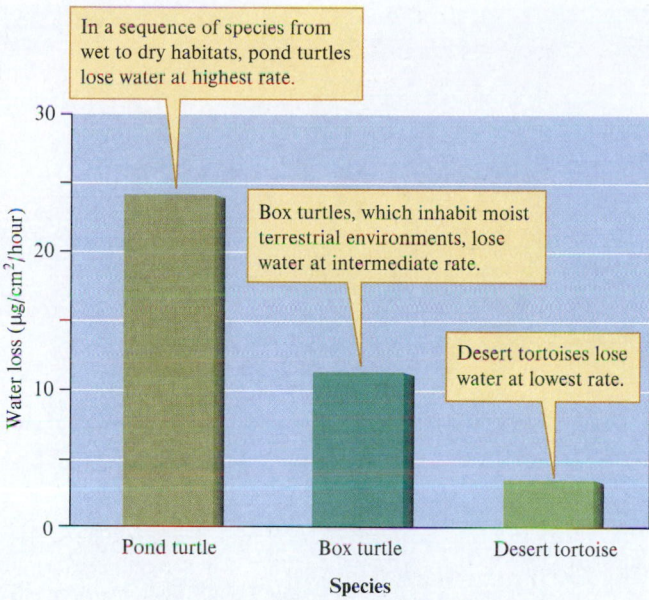

Figure 6.14 Rates of water loss by two turtles and a tortoise indicate an inverse relationship between the dryness of the habitat and water loss rates (data from Schmidt-Nielsen 1969).

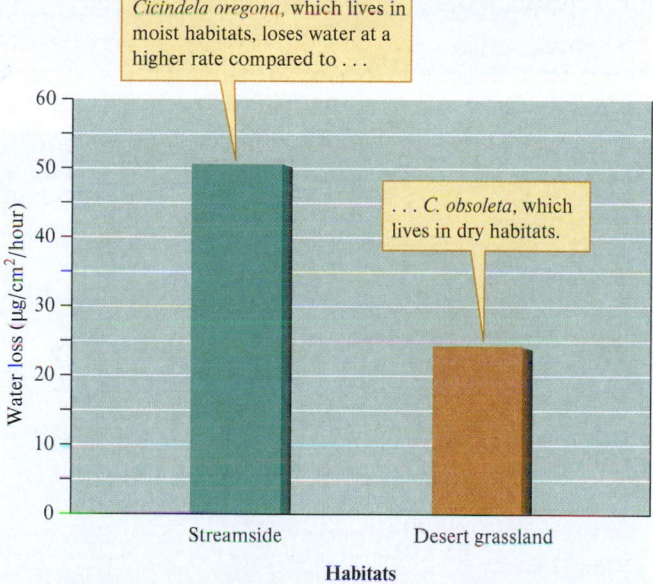

Figure 6.15 A tiger beetle species from a moist habitat lost water at a higher rate than one from a dry habitat (data from Schultz and Hadley 1987).

the cuticles of the two species of tiger beetles, they found that the concentration of hydrocarbons in the cuticle of *C. obsoleta* is 50% higher than in the cuticle of *C. oregona* (fig. 6.16). In addition, the two species differ in the percentages of cuticular hydrocarbons that are saturated with hydrogen. Fully saturated hydrocarbons are much more effective at waterproofing. One hundred percent of the hydrocarbons in the cuticle of *C. obsoleta* are saturated. In contrast, only 50% of the cuticular hydrocarbons of *C. oregona* are saturated. These results support the hypothesis that *C. obsoleta* loses water at a lower rate because its cuticle contains a higher concentration of waterproofing hydrocarbons.

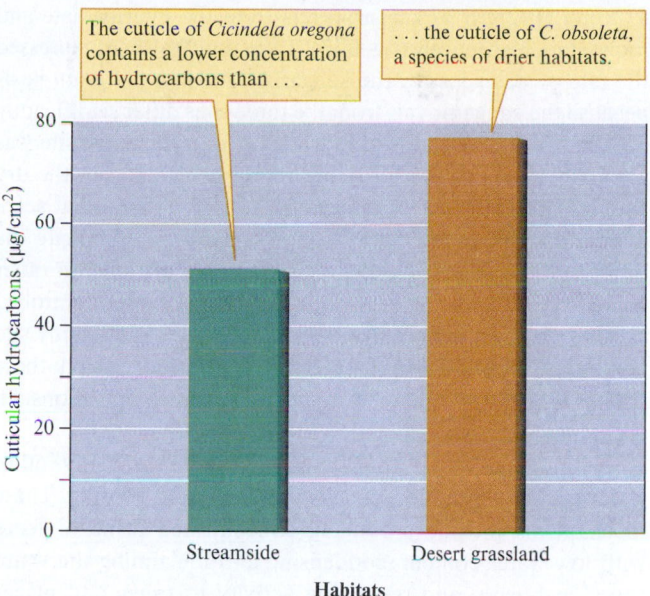

Figure 6.16 The cuticles of tiger beetles from dry habitats tend to contain a higher concentration of waterproofing hydrocarbons compared to those of tiger beetles from moist habitats (data from Schultz and Hadley 1987).

Merriam's kangaroo rats conserve water sufficiently that they can live entirely on the moisture contained within their food and on metabolic water (see fig. 6.10). This capacity is assumed to be an adaptation to desert living. Over long periods of time as the American Southwest became increasingly arid, the ancestors of today's Merriam's kangaroo rats were subject to natural selection that favored a range of adaptations to dry environments, including water conservation. However, Merriam's kangaroo rat is a widespread species that lives from 21° N latitude in Mexico to 42° N latitude in northern Nevada. Over this large geographic range, Merriam's kangaroo rat populations are exposed to a very broad range of environmental conditions.

Intrigued by their large geographic range and exceptional adaptation to desert living, Randall Tracy and Glenn Walsberg studied three populations of Merriam's kangaroo rats across a climatic gradient. Their main objective was to determine if different populations of the species vary in their degree of adaptation to dry environments (Tracy and Walsberg 2000, 2001, 2002). The three populations studied live in southwestern Arizona near Yuma, central Arizona, and north-central Arizona, at elevations of 150 m, 400 m, and 1,200 m, respectively. Mean annual maximum temperatures at the study sites are 31.5°, 29.1°, and 23.5°C, while mean annual precipitation at the three sites is 106 mm, 336 mm, and 436 mm. Climatic differences at the three study sites are reflected in the vegetation. The habitat at the driest site consists of sand dunes with scattered shrubs; the intermediate site is a desert shrubland; and the vegetation at the moist site consists of temperate, pinyon-juniper woodland.

The results of Tracy and Walsberg's study showed clear differences among the study populations. The mean rate of evaporative water loss by kangaroo rats from the dry site was 0.69 mg of water per gram per hour, compared to 1 mg H_2O/g/h and

1.08 mg H_2O/g/h by kangaroo rats from the intermediate and moist sites, respectively (fig. 6.17). Tracy and Walsberg expressed the rate of water loss by the kangaroo rats on a per gram basis because the kangaroo rats from the three sites differ significantly in size. The average mass of individuals from the moist site was approximately 33% greater than the mass of rats from the dry site. In additional studies, Tracy and Walsberg found that acclimating animals to laboratory conditions did not eliminate the differences in water conservation among populations. In other words, even after being kept in the laboratory under controlled conditions, Merriam's kangaroo rats from the driest study site continued to lose water at a lower rate. The evidence from these studies supports the conclusion that these three populations differ in their degree of adaptation to desert living.

Animals adapted to dry conditions have many other water conservation mechanisms besides waterproofing. These mechanisms include producing concentrated urine or feces with low water content, condensing and reclaiming the water vapor in breath, and restricting activity to times and places that decrease water loss.

Plants have also evolved a wide variety of ways to conserve water. Those with less leaf surface per length of root lose less water and, not surprisingly, arid land plants generally have less leaf area per unit area of root compared to plants from moist climates. Many plants reduce leaf area over the short term by dropping leaves in response to drought. Some desert plants produce leaves only in response to soaking rains and then shed them when the desert dries out again. These plants reduce leaf area to zero in times of drought. Figure 6.18 shows one of these plants, the ocotillo of the Sonoran Desert of North America.

Other plant adaptations that conserve water include thick leaves, which have less transpiring leaf surface area per unit volume of photosynthesizing tissue than thin leaves

(a)

(b)

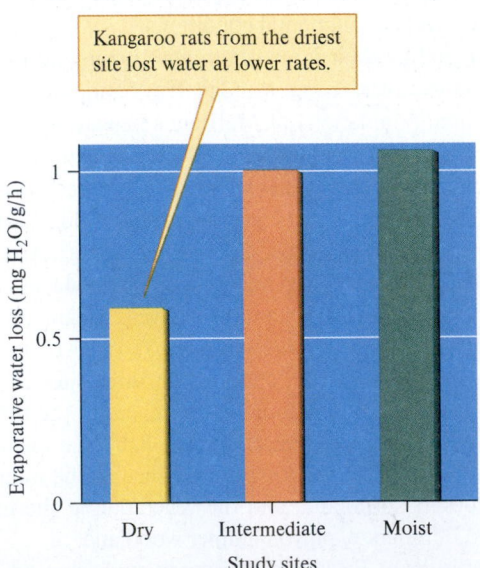

Figure 6.17   Water loss rates by Merriam's kangaroo rats from across a moisture gradient suggest adaptation to local climate by each of the populations (data from Tracy and Walsberg 2001).

Figure 6.18   Changing leaf area: (*a*) following rainfall, ocotillo plants of the Sonoran Desert develop leaves; (*b*) during dry periods, they lose their leaves.   (*a*) natureguy/123RF; (*b*) mikelane45/123RF

do; few stomata on leaves rather than many; structures on the stomata that impede the movement of water; dormancy during times when moisture is unavailable; and alternative, water-conserving pathways for photosynthesis (C_4 and CAM). (We discuss these alternative pathways for photosynthesis in chapter 7.)

We should remember that plants and animals in terrestrial environments other than deserts also show evidence of selection for water conservation. For instance, Nona Chiariello and her colleagues (1987) discovered an intriguing example of adjusting leaf area in the moist tropics. *Piper auritum,* a large-leafed, umbrella-shaped plant, grows in clearings of the rain forest. Because it grows in clearings, the plant often faces drying conditions during midday. However, it reduces the leaf area it exposes to the midday sun by wilting. Wilting at midday reduces leaf area exposed to direct solar radiation by about 55% and leaf temperature by up to 4° to 5°C. These reductions decrease the rate of transpiration by 30% to 50%, which is a substantial water savings. The behavior of this tropical rain forest plant reminds us that even the rain forest has its relatively dry microclimates, such as the forest clearings where *P. auritum* grows. The rapidity of the wilting response by *P. auritum* is shown in figure 6.19.

Organisms balance their water budgets in numerous ways. Some rely mainly on water conservation. Others depend on water acquisition. However, every biologist who studies organisms in their natural environment knows that nature is marked by diversity and contrast. To sample nature's variety, let's review the variety of approaches to desert living.

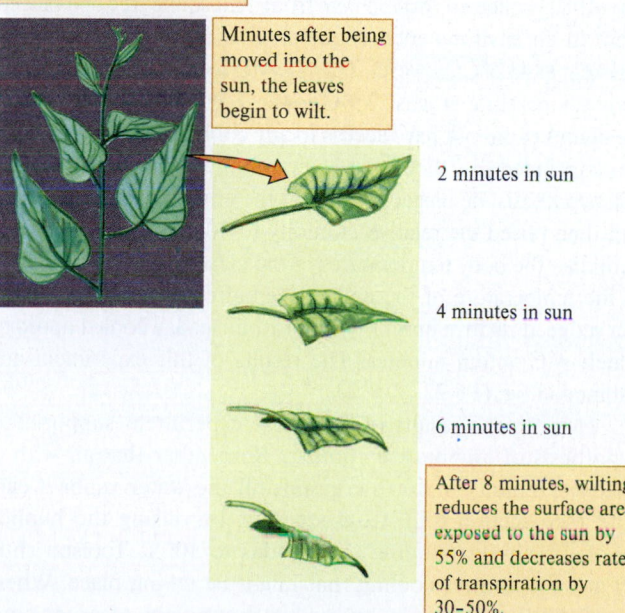

In a shaded portion of a greenhouse, the leaves of the rain forest plant are unwilted and fully exposed to incoming light.

Minutes after being moved into the sun, the leaves begin to wilt.

2 minutes in sun

4 minutes in sun

6 minutes in sun

After 8 minutes, wilting reduces the surface area exposed to the sun by 55% and decreases rate of transpiration by 30–50%.

Figure 6.19   Temporary wilting by this rain forest plant decreases rates of water loss (data from Chiariello, Field, and Mooney 1987).

Dissimilar Organisms with Similar Approaches to Desert Life

On the surface, camels and saguaro cactus appear entirely different. If you look deeper into their biology, however, you find that they take very similar approaches to balancing their water budgets. Both the camel and the saguaro cactus acquire massive amounts of water when water is available, store water, and conserve water.

The camel can go for long periods in intense desert heat without drinking, up to 6 to 8 days in conditions that would kill a person within a day. During this time, the animal survives on the water stored in its tissues and the metabolic water released when it metabolizes the fat stored in its hump. In addition, a camel can withstand water losses of up to 20% of its body weight without harm. For humans, a loss of about 10% to 12% is near the fatal limit. When the camel has the opportunity, it can drink and store prodigious quantities of water, up to one-third of its body weight at a time.

Between opportunities to drink, the camel is a master of water conservation. One way it conserves body water is by reducing its rate of heat gain. Like overheating tiger beetles (see chapter 5, section 5.5), the camel faces into the sun, reducing the body surface it exposes to direct sunlight. In addition, its thick hair insulates it from the intense desert sun, and rather than sweating sufficiently to keep its body temperature down, the camel allows its body temperature to rise by up to 7°C. This reduces the temperature difference between the camel and the environment and so decreases the rate of additional heating, which decreases water loss by evaporation.

The saguaro cactus takes a similar approach. The trunk and arms of the plant act as organs in which the cactus can store large quantities of water. During droughts, the saguaro draws on these stored reserves and, as a result, can endure long periods without water. When it rains, the saguaro, like a camel at an oasis, can ingest great quantities of water, but instead of drinking, the saguaro gets its water through its dense network of shallow roots. These roots extend out in a roughly circular pattern to a distance approximately equal to the height of the cactus. For a 15 m tall saguaro, this means a root coverage of over 700 m² of soil.

The saguaro also reduces its rate of evaporative water loss in several ways. First, like other cactus, it keeps its stomata closed during the day when transpiration losses would be highest. In the absence of transpiration, in full sun, the internal temperature of the saguaro can rise to over 50°C, which is among the highest temperatures recorded in plants. However, as we noted for the camel, higher body temperature can be an advantage because it reduces the rate of additional heating. The saguaro's rate of heating is also reduced by the shape and orientation of its trunk and arms. At midday, when the potential for heating is greatest, the saguaro exposes mainly the tips of its arms and trunk to direct sunlight. In addition, the tips of the saguaro's arms and trunk are insulated by a layer of plant hairs and a thick tangle of spines, which reflect sunlight and shade the growing tips of the cactus.

The parallel approaches to desert living seen in saguaro cactus and camels are outlined in figure 6.20. Now let's examine two organisms that live in the same desert but have very different water relations.

The saguaro reduces heat gain by exposing only tops of its trunk and branches to the midday sun.

The trunk and branch tips are shaded and insulated with a high density of spines, which reduces heat gain.

The camel stores fat in its hump, a source of metabolic water.

The camel reduces heat gain by facing into the sun.

The camel is covered with dense hair which reduces heat gain.

Water is stored in the massive trunk and arms.

The saguaro reduces water loss by transpiration by keeping stomata closed and allowing its temperature to rise.

When water is available, both the saguaro and camel take in massive quantities.

The camel reduces evaporative water loss by not sweating and allowing body temperature to rise.

Figure 6.20   Dissimilar organisms with similar approaches to desert living. The improbable pairing of a dromedary camel, native to southwestern Asia and North Africa, with a saguaro cactus, from western North America, actually occurred during the mid-nineteenth century, when the U.S. Army imported camels for use as pack animals.

Two Arthropods with Opposite Approaches to Desert Life

Although cicadas and scorpions may live within a few meters of each other, they take sharply contrasting approaches to living in the desert (fig. 6.21). The scorpion's approach is to slow down, conserve, and stay out of the sun. Scorpions are relatively large and long-lived arthropods with very low metabolic rates. A low rate of metabolism means that they can subsist on low rations of food and lose little water during respiration. In addition, scorpions conserve water by spending most of their time in their burrows, where the humidity is higher than at the surface. They come out to feed and find mates only at night, when it's cooler. In addition, desert scorpions are well waterproofed; hydrocarbons in their cuticles seal in moisture. With this combination of water-conserving characteristics, scorpions can easily satisfy their need for water by consuming the moisture contained in the bodies of their arthropod prey.

The Sonoran Desert cicada, *Diceroprocta apache,* is active on the hottest days, when air temperature is near its lethal limit. How can *Diceroprocta* do this and not die?

A series of investigations by Eric Toolson and Neil Hadley showed conclusively that cicadas, including *Diceroprocta,* are capable of evaporative cooling. In one of these studies, Eric Toolson (1987) collected *Diceroprocta* from a mesquite tree and placed them in an environmental chamber. The chamber temperature was kept at 45.5°C; however, *Diceroprocta* was able to maintain its body temperature at least 2.9°C lower. Since the cicadas within the chamber did not have access to any cool microclimates, Toolson concluded that they must be evaporatively cooling. To verify this hypothesis, he placed cicadas in the environmental chamber and then raised the relative humidity to 100%. At 100% relative humidity, the body temperatures of the cicadas quickly increased to the temperature of the environmental chamber. When Toolson reduced relative humidity to 0%, the cicadas cooled approximately 4°C within minutes. The results of this experiment are outlined in figure 6.22.

How do the results of Toolson's experiment support the hypothesis of evaporative cooling? Remember that air with a relative humidity of 100% contains all the water vapor it can hold (see section 6.1). Consequently, by raising the humidity of the air surrounding the cicadas to 100%, Toolson shut off any evaporative cooling that might be taking place. When he reintroduced dry air, he created a gradient of water concentration from the cicada to the air and evaporative cooling resumed. This experiment by Toolson was analogous to

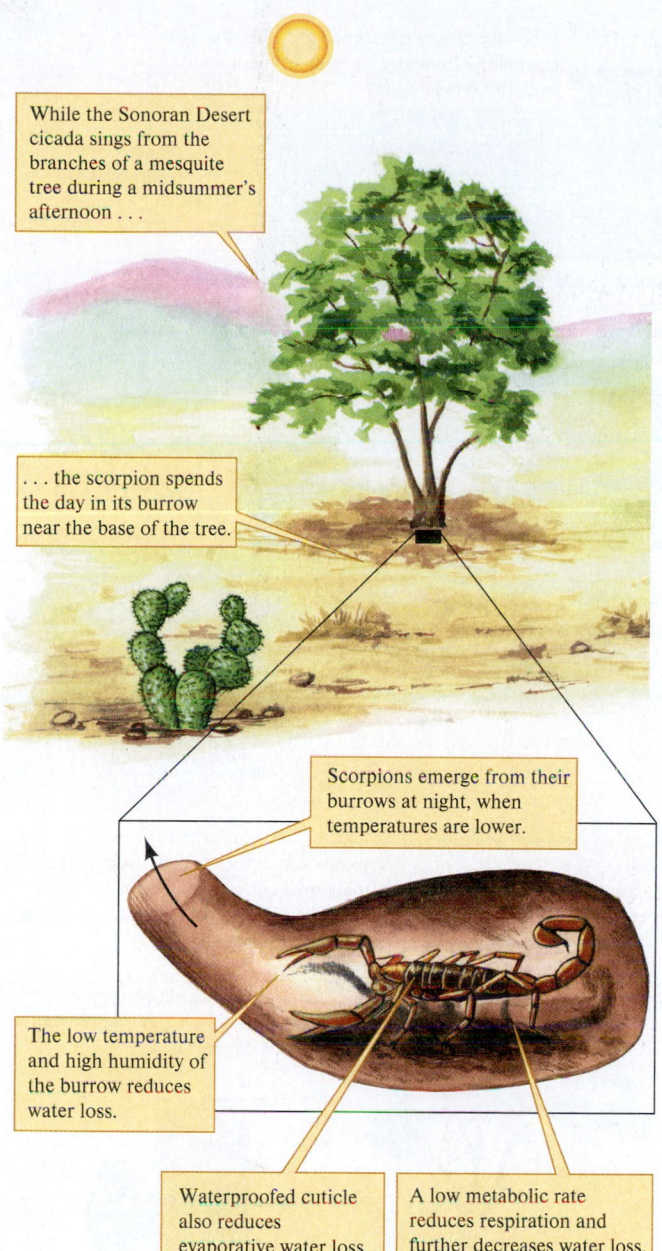

While the Sonoran Desert cicada sings from the branches of a mesquite tree during a midsummer's afternoon . . .

. . . the scorpion spends the day in its burrow near the base of the tree.

Scorpions emerge from their burrows at night, when temperatures are lower.

The low temperature and high humidity of the burrow reduces water loss.

Waterproofed cuticle also reduces evaporative water loss.

A low metabolic rate reduces respiration and further decreases water loss.

Figure 6.21   These two desert arthropods, a scorpion and a cicada, have evolved very different approaches to living in the desert.

Heinrich's tying off the circulatory system of a sphinx moth to determine the role of the circulatory system in thermoregulation (see chapter 5, section 5.4).

Toolson's results are consistent with the hypothesis that *Diceroprocta* evaporatively cools but does not demonstrate that capacity directly. Consequently, Toolson and Hadley (1987) conducted observations to make a direct demonstration. First, they placed a live *Diceroprocta* in an environmental chamber with a humidity sensor just above its cuticle. If *Diceroprocta* evaporatively cools, then this sensor would detect higher humidity as the temperature of the environment was increased. This is exactly what occurred.

As the temperature was increased from 30° to 43°C, the rate of water movement across the cicada's cuticle increased in three steps. When Toolson and Hadley increased the temperature from 37° to 39°C, water loss increased from 5.7 to 9.4 mg H_2O per square centimeter per hour. At 41°C water loss increased from 9.4 to 36.1 mg H_2O per square centimeter per hour, and at 43°C water loss increased from 36.1 to 61.4 mg H_2O per square centimeter per hour. These results are graphed in figure 6.23.

The rate of water loss by *Diceroprocta* is among the highest ever reported for a terrestrial insect. How does water cross the cuticle of this cicada at such a high rate? Toolson and Hadley searched the cuticle of *Diceroprocta* for avenues of water movement. They found three areas on the dorsal surface with large pores that might be involved in evaporative cooling. When they plugged these pores, *Diceroprocta* could no longer cool itself. In summary, Toolson and Hadley verified a previously unknown phenomenon, evaporative cooling by cicadas, and carefully demonstrated the underlying mechanisms.

So, it turns out that these cicadas can sing in the hottest hours of the desert day because they sweat! *Diceroprocta* is able to maintain this seemingly impossible lifestyle because it has tapped into a rich supply of water. Cicadas are members of the order Homoptera and distant relatives of the aphids. Like aphids, cicadas feed on plant fluids. So, though the cicada lives in the same macroclimate as the scorpion, it has tapped into a totally different microclimate. The cicada's scope for water acquisition is extended up to 30 m deep into the soil by the taproots of its mesquite host plant, *Prosopis juliflora*. *Diceroprocta* can sustain high rates of water loss through evaporation, high W_e, because it is able to balance these losses with a high rate of water acquisition, high W_d. Figure 6.24 illustrates how *Diceroprocta* uses mesquite trees to get access to deep soil moisture.

Sometimes, similar organisms employ radically different approaches to balancing their water budgets. Sometimes, organisms of very different evolutionary lineages employ functionally similar approaches. In short, the means by which terrestrial organisms balance water acquisition against water loss are almost as varied as the organisms themselves. Similar variation occurs among aquatic organisms.

Concept 6.2 Review

1. The tiger beetle *Cicindela oregona* (see figs. 6.15 and 6.16) has a distribution that extends from Arizona through the temperate rain forests of Alaska. Why should the amounts of cuticular hydrocarbons vary geographically among populations of *C. oregona*?
2. During severe droughts, some of the branches of shrubs and trees die, while others survive. How might losing some branches increase the probability that an individual plant will survive a drought?
3. How are water and temperature regulation related in many terrestrial organisms?

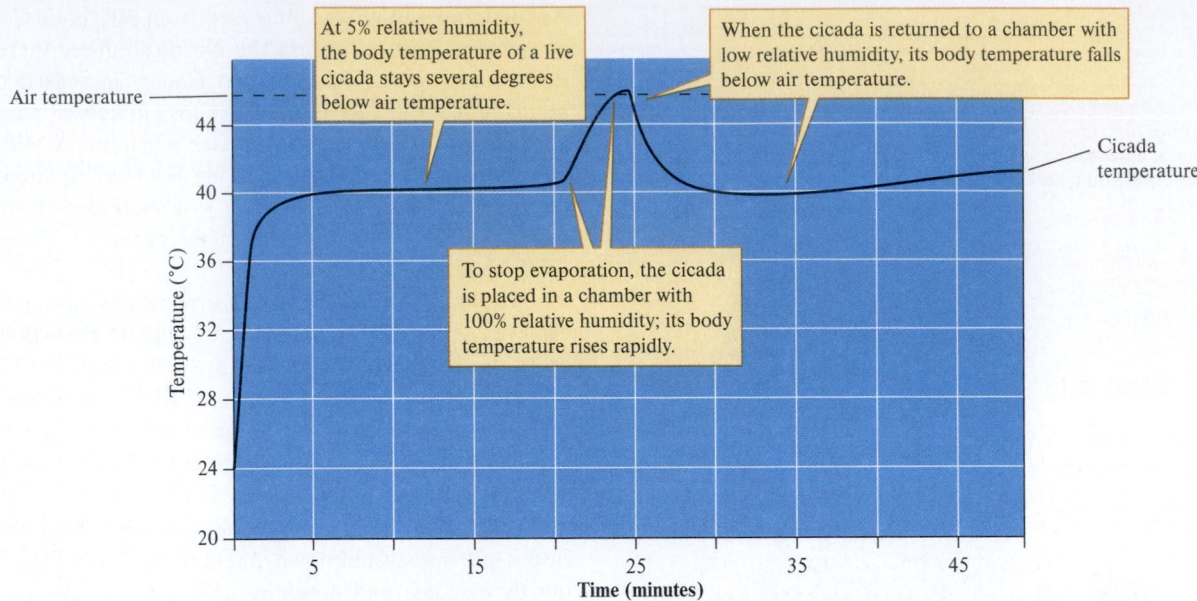

At 5% relative humidity, the body temperature of a live cicada stays several degrees below air temperature.
When the cicada is returned to a chamber with low relative humidity, its body temperature falls below air temperature.
Air temperature
Cicada temperature
To stop evaporation, the cicada is placed in a chamber with 100% relative humidity; its body temperature rises rapidly.
Temperature (°C)
44
40
36
32
28
24
20
5
15
25
35
45
Time (minutes)

Figure 6.22 A laboratory experiment verified evaporative cooling by the cicada, *Diceroprocta apache* (data from Toolson 1987).

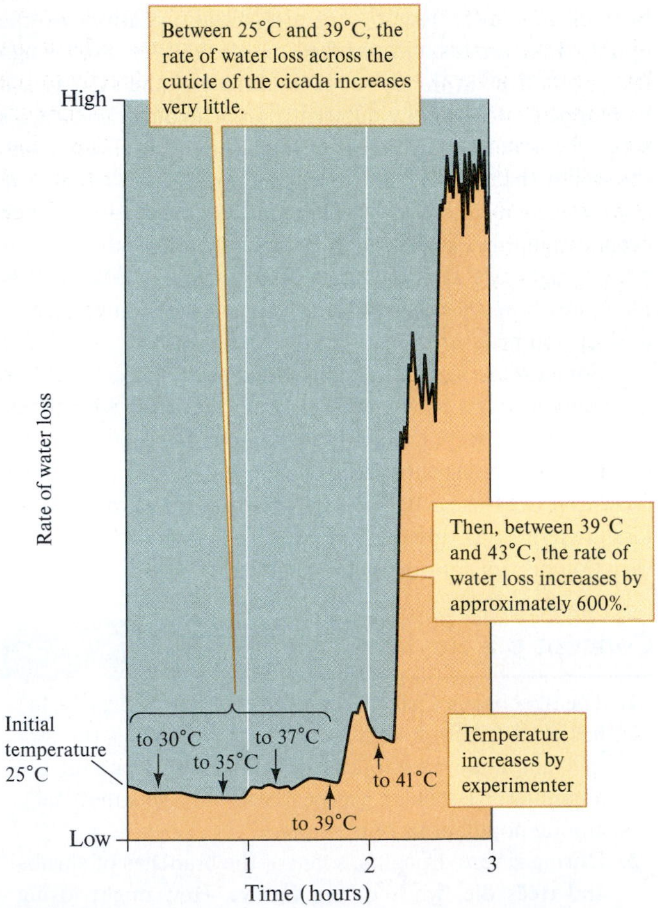

Between 25°C and 39°C, the rate of water loss across the cuticle of the cicada increases very little.
High
Low
Rate of water loss
Then, between 39°C and 43°C, the rate of water loss increases by approximately 600%.
Initial temperature 25°C
to 30°C
to 35°C
to 37°C
to 39°C
to 41°C
Temperature increases by experimenter
1
2
3
Time (hours)

Figure 6.23 High temperatures induce massive rates of water loss by the cicada, *Diceroprocta apache* (data from Toolson and Hadley 1987).

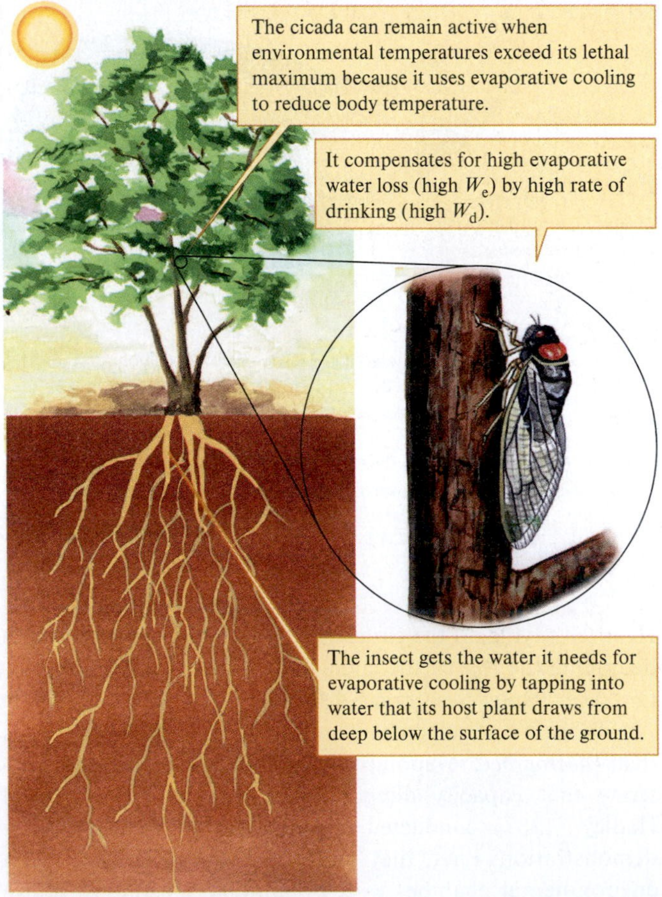

The cicada can remain active when environmental temperatures exceed its lethal maximum because it uses evaporative cooling to reduce body temperature.
It compensates for high evaporative water loss (high W_e) by high rate of drinking (high W_d).
The insect gets the water it needs for evaporative cooling by tapping into water that its host plant draws from deep below the surface of the ground.

Figure 6.24 An ecological puzzle solved.

6.3 Water and Salt Balance in Aquatic Environments

LEARNING OUTCOMES

After studying this section you should be able to do the following:

6.10 Define osmoregulation.
6.11 Contrast osmoregulation by sharks versus marine bony fish.
6.12 Compare osmoregulation by freshwater bony fish and freshwater mosquitoes.

Marine and freshwater organisms use complementary mechanisms for water and salt regulation. Aquatic organisms, like terrestrial species, regulate internal water, W_i, by balancing water gain against water loss. We can represent water regulation in aquatic environments by modifying our equation for terrestrial water balance to:

$$W_i = W_d - W_s \pm W_o$$

Drinking, W_d, is a ready source of water for aquatic organisms. Secretion of water with urine, W_s, is an avenue of water loss. By osmosis, W_o, an aquatic organism may either gain or lose water, depending on the organism and the environment.

Marine Fish and Invertebrates

Most marine invertebrates maintain an internal concentration of solutes equivalent to that in the seawater around them. What does the animal gain by remaining isosmotic with the external environment? The isosmotic animal does not have to expend energy overcoming an osmotic gradient. This strategy is not without costs, however. Although the total concentration of solutes is the same inside and outside the animal, there are still differences in the concentrations of some individual solutes. These concentration differentials can only be maintained by active transport, which consumes some energy.

Sharks, skates, and rays generally elevate the concentration of solutes in their blood to levels slightly hyperosmotic to seawater. However, inorganic ions constitute only about one-third of the solute in shark's blood; the remainder consists of the organic molecules urea and trimethylamine oxide, or TMAO. As a consequence of being slightly hyperosmotic, sharks slowly gain water through osmosis; that is, W_o is slightly positive. The water that diffuses into the shark, mainly across the gills, is pumped out by the kidneys and exits as urine. Sodium, because it is maintained at approximately two-thirds its concentration in seawater, diffuses into sharks from seawater across the gill membranes and some sodium enters with food. Sharks excrete excess sodium mainly through a specialized gland associated with the rectum called the salt gland. The main point here is that sharks and their relatives reduce the costs of **osmoregulation,** regulation of internal salt and water concentrations, by decreasing the osmotic gradient between themselves and the external environment (fig. 6.25).

In contrast to most marine invertebrates and sharks, marine bony fish have body fluids that are strongly hypoosmotic to the surrounding medium. As a consequence, they lose water to the surrounding seawater, mostly across their gills. Marine bony fish make up these water losses by drinking seawater. However, drinking seawater increases salt gain. The fish rid themselves of excess salts in two ways. Specialized "chloride" cells at the base of their gills secrete sodium and chloride directly to the surrounding seawater, while the kidneys excrete magnesium and sulfate ions, which are expelled with urine. The urine, because it is hypoosmotic to the body fluids of the fish, represents a loss of water. However, the loss of water through the kidneys is low because the volume of urine is low.

The larvae of some mosquitoes in the genus *Aedes* live in saltwater. These larvae meet the challenge of a high-salinity environment in ways analogous to those used by marine bony fish. Like marine bony fish, saltwater mosquitoes are hypoosmotic to the surrounding environment, to which they

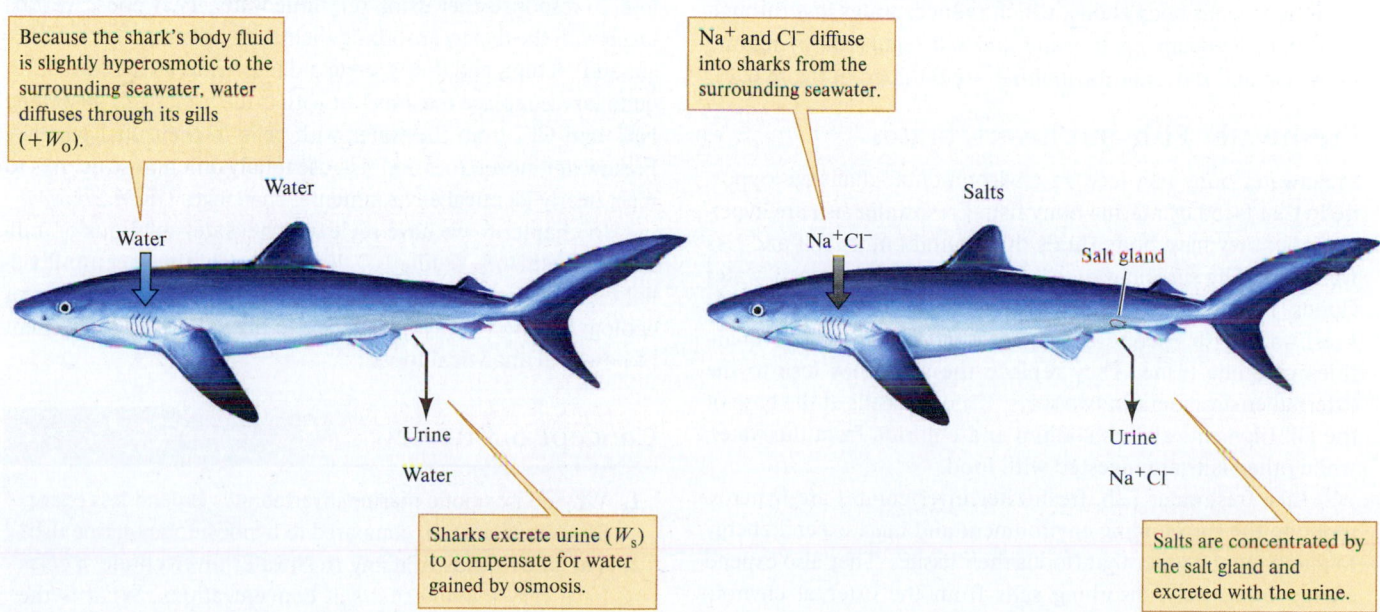

Because the shark's body fluid is slightly hyperosmotic to the surrounding seawater, water diffuses through its gills ($+W_o$).

Water

Water

Urine

Water

Sharks excrete urine (W_s) to compensate for water gained by osmosis.

Na^+ and Cl^- diffuse into sharks from the surrounding seawater.

Salts

Na^+Cl^-

Salt gland

Urine

Na^+Cl^-

Salts are concentrated by the salt gland and excreted with the urine.

Figure 6.25 Osmoregulation by sharks.

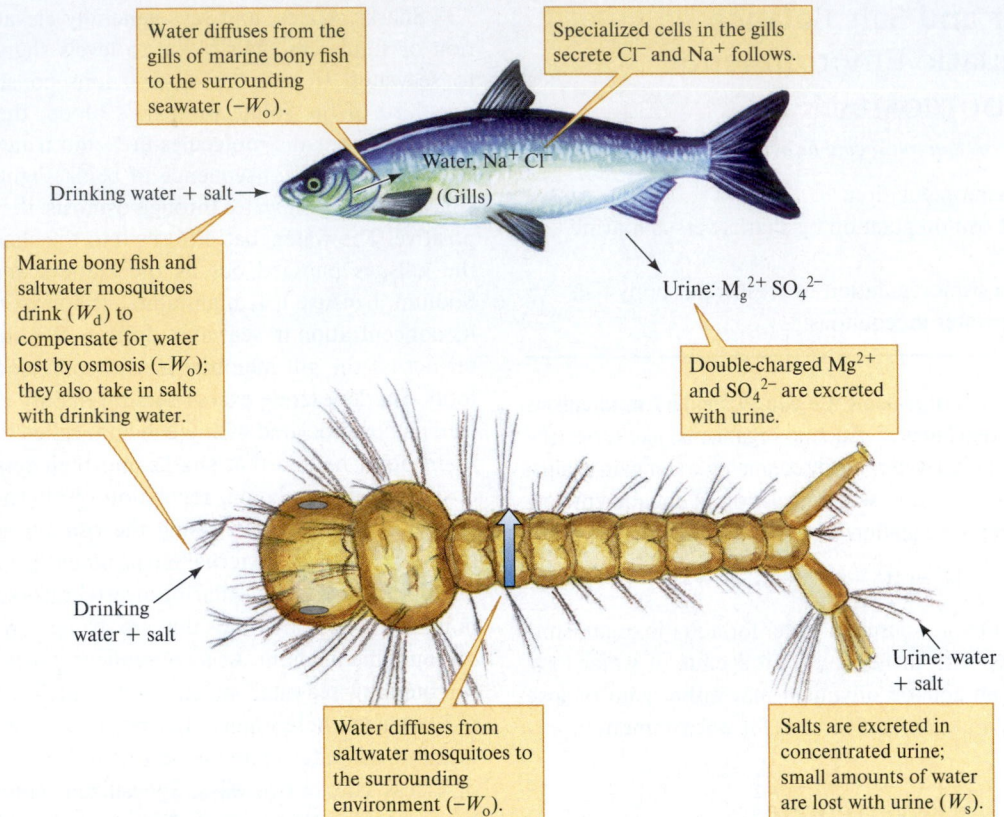

Water diffuses from the gills of marine bony fish to the surrounding seawater ($-W_o$).

Specialized cells in the gills secrete Cl^- and Na^+ follows.

Water, Na^+ Cl^-
(Gills)

Drinking water + salt →

Urine: Mg^{2+} SO_4^{2-}

Marine bony fish and saltwater mosquitoes drink (W_d) to compensate for water lost by osmosis ($-W_o$); they also take in salts with drinking water.

Double-charged Mg^{2+} and SO_4^{2-} are excreted with urine.

Drinking water + salt

Urine: water + salt

Water diffuses from saltwater mosquitoes to the surrounding environment ($-W_o$).

Salts are excreted in concentrated urine; small amounts of water are lost with urine (W_s).

Figure 6.26   Main avenues of osmoregulation by hypoosmotic marine fish and saltwater mosquitoes.

lose water. Saltwater mosquitoes also make up this water loss by drinking large amounts of seawater, up to 130% to 240% of body volume per day! While this prodigious drinking solves the problem of water loss, it imports another: large quantities of salts that must be eliminated. Saltwater mosquitoes secrete these salts into the urine using specialized cells that line the posterior rectum. Here, saltwater mosquitoes do something that marine bony fish cannot. They excrete a urine that is hyperosmotic to their body fluids, which reduces water loss through the urine. The parallels in water and salt regulation by marine bony fish and saltwater mosquitoes are outlined in figure 6.26.

Freshwater Fish and Invertebrates

Freshwater bony fish face an environmental challenge opposite to that faced by marine bony fish. Freshwater fish are hyperosmotic; they have body fluids that contain more salt and less water than the surrounding medium. As a consequence, water floods inward and salts diffuse outward across their gills. Freshwater fish excrete excess internal water as large quantities of dilute urine. They replace the salts they lose to the external environment in two ways. Chloride cells at the base of the gill filaments absorb sodium and chloride from the water, while other salts are ingested with food.

Like freshwater fish, freshwater invertebrates are hyperosmotic to the surrounding environment and must expend energy to pump out the water that floods their tissues. They also expend energy by actively absorbing salts from the external environment. However, the concentration of solutes in the body fluids of freshwater invertebrates ranges from between about one-half

and one-tenth that of their marine relatives. This lower internal concentration of solutes reduces the osmotic gradient between freshwater and the outside environment and so reduces the energy freshwater invertebrates must expend to osmoregulate.

The larvae of approximately 95% of mosquito species live in freshwater, where they face osmotic challenges very similar to those presented to freshwater fish. Like freshwater fish, mosquito larvae must solve the twin problems of water gain and ion loss. In response, they drink very little water. They conserve ions taken with the diet by absorbing them with cells that line the midgut and rectum, and they secrete a dilute urine. Freshwater mosquito larvae replace the ions lost with urine by actively absorbing Na^+ and Cl^- from the water with cells in their anal papillae. Freshwater mosquitoes and fish use totally different structures to meet nearly identical environmental challenges (fig. 6.27).

In chapter 6, we have reviewed the water relations of individual organisms. Studies of the relationship between individual organisms and the environment, a fundamental aspect of ecology, are now being advanced rapidly with the development of powerful analytical tools.

Concept 6.3 Review

1. Why do isosmotic marine invertebrates expend less energy for osmoregulation compared to hypoosmotic marine fish?
2. The body fluids of many freshwater invertebrate species have very low internal salt concentrations. What is the benefit of such dilute internal fluids?

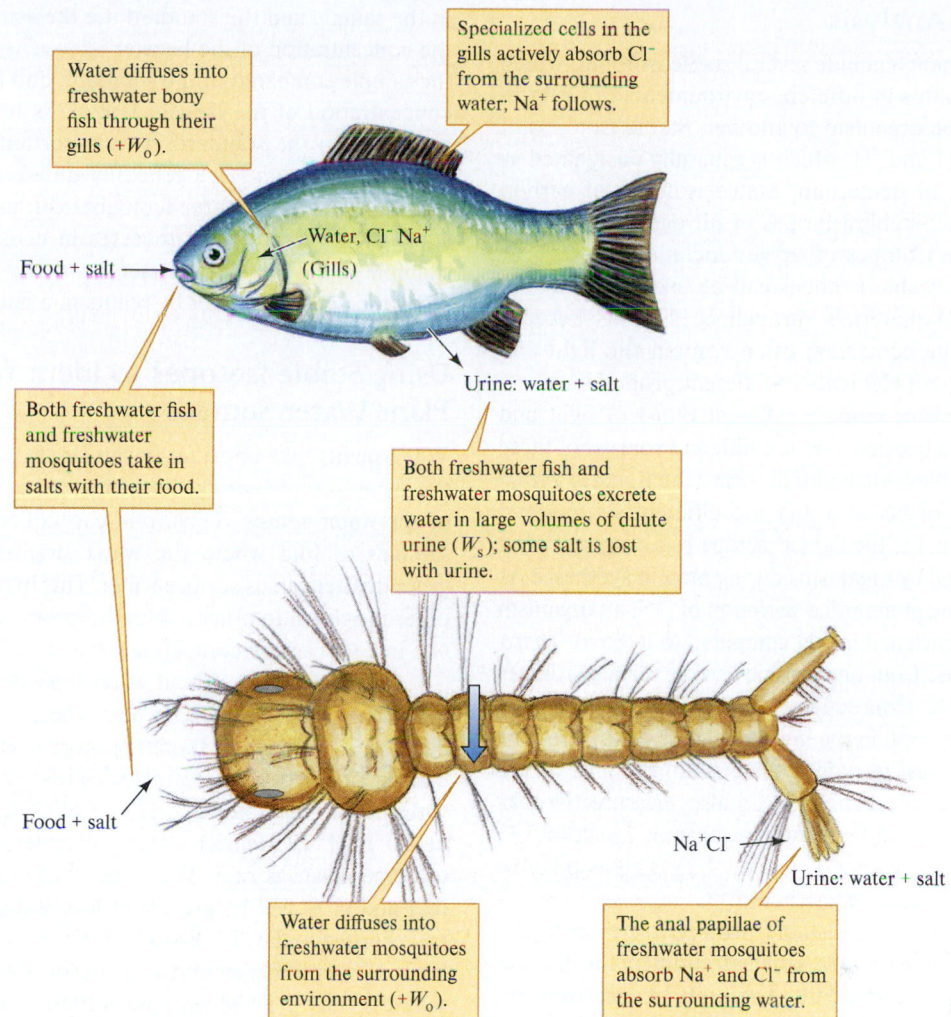

Water diffuses into freshwater bony fish through their gills ($+W_o$).

Specialized cells in the gills actively absorb Cl⁻ from the surrounding water; Na⁺ follows.

Water, Cl⁻ Na⁺
(Gills)

Food + salt →

Both freshwater fish and freshwater mosquitoes take in salts with their food.

Urine: water + salt

Both freshwater fish and freshwater mosquitoes excrete water in large volumes of dilute urine (W_s); some salt is lost with urine.

Food + salt

Na⁺Cl⁻

Urine: water + salt

Water diffuses into freshwater mosquitoes from the surrounding environment ($+W_o$).

The anal papillae of freshwater mosquitoes absorb Na⁺ and Cl⁻ from the surrounding water.

Figure 6.27   Main avenues of osmoregulation by hyperosmotic freshwater fish and mosquitoes.

Applications

Using Stable Isotopes to Study Water Uptake by Plants

LEARNING OUTCOMES

After studying this section you should be able to do the following:

6.13   Describe stable isotope analysis.

6.14   List some major stable isotopes that have proved useful in ecological studies.

6.15   Describe examples of how stable isotopes have been used in ecological studies.

6.16   Explain how $\delta\,^2H$ and $\delta\,^{18}O$ can be used to identify what water sources plants are using.

In order to fully understand the ecology of an individual plant or the dynamics of an entire landscape, ecologists need information about what happens below the earth's surface,

as well as about surface structure and processes. However, ecologists have produced much more information about the surface realm than about the subsurface, the domain of soil microbes, burrowing animals, and roots. While many ecologists have worked very hard to fill this gap in our knowledge, their work on belowground ecology has been historically slow. Fortunately, progress has accelerated in recent years. A major contributor to recent progress in belowground ecology has been the development of new tools. One of the most important of those tools is **stable isotope analysis,** which involves the analysis of the relative proportions of stable isotopes, such as the oxygen include ^{13}O and ^{12}O, in materials. Stable isotope analysis is increasingly used in ecology (Dawson et al. 2002). As we saw in chapter 1, stable isotope analysis is proving useful for understanding the diets of animals whose eating patterns are difficult to observe directly. It is also a very powerful tool in studies of water uptake by plants. To understand the application of this analytical tool, we need to know a little about the isotopes themselves and about their behavior in ecosystems.

Stable Isotope Analysis

Most chemical elements include several stable isotopes, which occur in different ratios in different environments or differ in their ratios from one organism to another. Stable isotopes of hydrogen include ^1H and ^2H, which is generally designated as D, an abbreviation of deuterium. Stable isotopes of carbon include ^{13}C and ^{12}C; stable isotopes of nitrogen include ^{15}N and ^{14}N; and stable isotopes of oxygen include ^{16}O and ^{18}O. The ratios of these stable isotopes can be used to study the flow of energy and materials through ecosystems because different parts of the ecosystem often contain the light and heavy isotopes of these elements in different proportions.

Different organisms contain different ratios of light and heavy stable isotopes because they use different sources of these elements, because they preferentially use (fractionate) different stable isotopes, or because they use different sources and fractionate. For instance, the lighter isotope of nitrogen, ^{14}N, is preferentially excreted by organisms during protein synthesis. As a consequence of this preferential excretion of ^{14}N, an organism becomes relatively enriched in ^{15}N compared to its food. Therefore, as materials pass from one trophic level to the next, tissues become richer in ^{15}N. Consequently, the highest trophic levels within an ecosystem contain the highest relative concentrations of ^{15}N, while the lowest trophic levels contain the lowest concentrations. Stable isotope analysis can also measure the relative contribution of C_3 and C_4 plants (see chapter 7, section 7.1) to a species' diet (see chapter 1, section 1.2). This is possible because C_4 plants are relatively richer in ^{13}C.

The concentrations of stable isotopes are generally expressed as differences in the concentration of the heavier isotope relative to some standard. The units of measurement are differences ($\pm$) in parts per thousand ($\pm\ ^0/_{00}$). These differences are calculated as:

$$\delta X = \left[\left(\frac{R_{\text{sample}}}{R_{\text{standard}}}\right) - 1\right] \times 10^3$$

where:

$\delta = \pm$

X = the relative concentration of the heavier isotope, for example, D, ^{13}C, ^{15}N, or ^{18}O in $^0/_{00}$

R_{sample} = the isotopic ratio in the sample, for example, D:^1H, ^{13}C:^{12}C, or ^{15}N:^{14}N

R_{standard} = the isotopic ratio in the standard, for example, D:^1H, ^{13}C:^{12}C, or ^{15}N:^{14}N

The reference materials used as standards in the isotopic analyses of hydrogen, nitrogen, carbon, and oxygen are the D:^1H and ^{18}O:^{16}O ratios in Standard Mean Ocean Water, the ^{15}N:^{14}N ratio in atmospheric nitrogen, and the ^{13}C:^{12}C ratio in PeeDee limestone.

The ecologist measures the ratio of stable isotopes in a sample and then expresses that ratio as a difference relative to some standard. If $\delta X = 0$, then the ratios of the isotopes in the sample and the standard are the same; if $\delta X = -X^0/_{00}$, the concentration of the heavier isotope is lower (e.g., ^{15}N) in the sample compared to the standard, and if $\delta X = +X^0/_{00}$, the concentration of the heavier isotope is higher in the sample compared to the standard. The important point here is that these isotopic ratios are generally different in different parts of ecosystems. Therefore, ecologists can use isotopic ratios to study the structure and processes in ecosystems. Here is an example of how hydrogen isotope ratios have been used to study the uptake of water by plants in a natural ecosystem.

Using Stable Isotopes to Identify Plant Water Sources

When plants take up water in the xylem, that water maintains the same isotopic composition (i.e., percent of δ^2H and $\delta^{18}O$) of the water source. Thus, one can collect water from plant samples to find where the water originated, such as from ground water versus surface water. This in turn can then give us other insights into other ecological processes, such as whether two tree species are competing for water (Sher et al. 2010).

The reason why ground water typically has a different isotopic signature than surface water is because a normal $^1H^1H^{16}O$ molecule is lighter and therefore more likely to be lost to evaporation than those water molecules that have any combination with the heavier isotopes δ^2H and $\delta^{18}O$, the heaviest of which being $^2H^2H^{18}O$. Heavier water molecules are also more likely to come down as rain. Thus, the concentration of deuterium δ^2H and $\delta^{18}O$ will be greater in lake water that is fed by rain and exposed to the air, versus water protected from evaporation because it is underground (fig. 6.28). Because the reference, seawater, has the highest concentrations of heavy molecules, measurements of other water isotopes found in nature are represented in terms of negative values. The most negative values are found in ground water, which is almost entirely made up of the lighter molecule $^1H^1H^{16}O$.

Isotope analysis is now so broadly used that we can use patterns in the published literature to discover generalizations across locations or taxa. One way that this is done is through **meta-analysis**—a way to statistically test hypotheses using many, separate studies. Jaivime Evaristo and Jeffrey J. McDonnell (2017) conducted a meta-analysis of 138 studies that used water isotope analysis to quantify the prevalence of ground water use across plant species and regions. These studies represented 414 different species and 7,367 xylem water measurements. Of these, 37% represented isotope signatures that indicated groundwater use. Perhaps unsurprisingly, this percentage was positively related to aridity; dry sites were more likely to have plants that used groundwater. Such a generalization was only possible because of the wide range of individual studies that used this technique in the past.

Stable isotope analysis opens a window to the water relations of plants that would not be accessible without this innovative tool. We will explore the use of stable isotope analysis further in chapter 18, where we discuss its use in studies of energy flow in ecosystems.

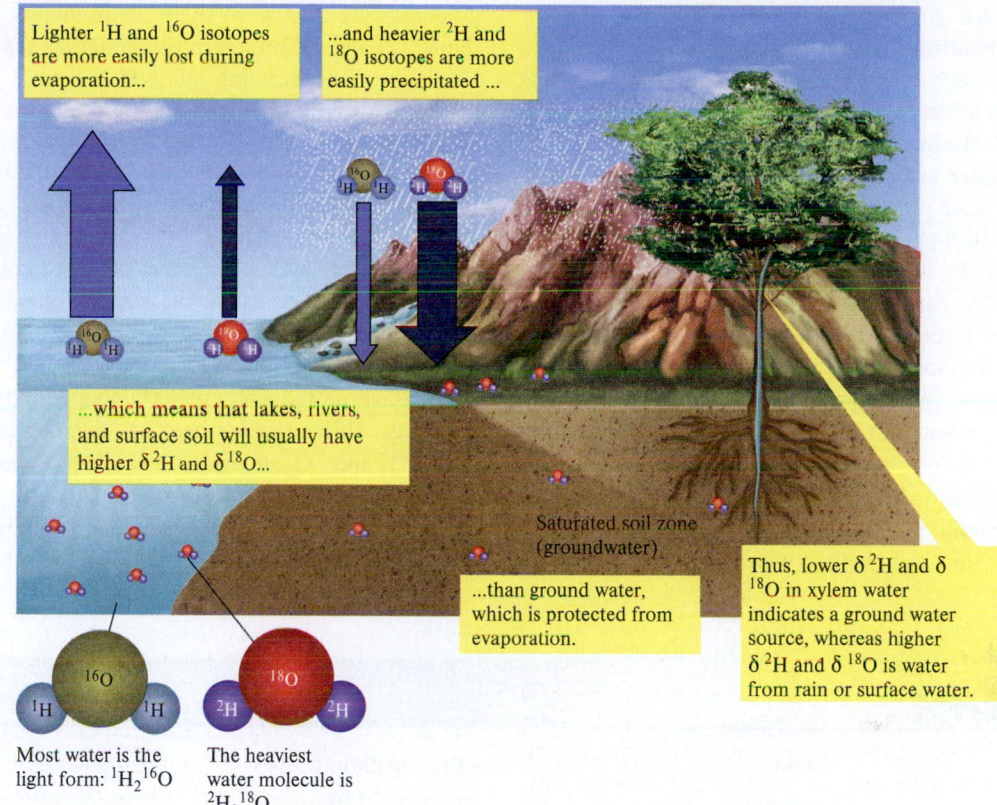

Figure 6.28   Water from different parts of an ecosystem will have different isotopic signatures because of the behavior of different types of water molecules based on their weight. Most water is the light form, which has the isotopes ^{1}H and ^{16}O, shown in light blue and yellow, but forms of the water molecule that have ^{2}H and/or ^{18}O have a heavier molecular weight. The heaviest water molecule is ^{2}H$_2$^{18}O, shown here in red and dark blue. Our ability to measure the concentration of the heavier isotopes makes it possible to identify where a plant is taking its water. This is done by sampling the xylem water and matching its isotopic signature (δ^{18}O:^{16}O) to a source.

Summary

Concentration gradients influence the movement of water between an organism and its environment.   The most familiar relative measure of the water content of air is relative humidity, defined as water vapor density divided by saturation water vapor density multiplied by 100. On land, the tendency of water to move from organisms to the atmosphere can be approximated by the vapor pressure deficit of the air. Vapor pressure deficit is calculated as the difference between the actual water vapor pressure and the saturation water vapor pressure.

In the aquatic environment, water moves down its concentration gradient, from solutions of higher water concentration and lower salt content (hypoosmotic) to solutions of lower water concentration and higher salt content (hyperosmotic). This movement of water creates osmotic pressure. Larger osmotic differences, between organism and environment, generate higher osmotic pressures.

Water flows from areas of higher water potential to areas of lower water potential. The water potential of pure water, which by convention is set at zero, is reduced by adding solute and by matric forces, the tendency of water to cling to soil particles and to plant cell walls. Typically, the water potential of plant fluids is determined by a combination of solute concentrations and matric forces, while the water potential of soils is determined mainly by matric forces. In saline soils, solutes may also influence soil water potential. Water potential, osmotic pressure, and vapor pressure deficit can all be measured in pascals (newtons/m^2), a common currency for considering the water relations of diverse organisms in very different environments.

Terrestrial plants and animals regulate their internal water by balancing water acquisition against water loss.   Water regulation by terrestrial animals is summarized by $W_{ia} = W_d + W_f + W_a - W_e - W_s$, where W_d = drinking, W_f = taken in with food, W_a = absorption from the air, W_e = evaporation, and W_s = secretions and excretions. Water regulation by terrestrial plants is summarized by $W_{ip} = W_r + W_a - W_t - W_s$, where W_r = uptake by roots, W_a = absorption from the air, W_t = transpiration, and W_s = secretions and reproductive structures. Some very different terrestrial plants and animals, such as the camel and saguaro cactus, use similar mechanisms to survive in arid climates. Some organisms, such as scorpions and cicadas, use radically different mechanisms. Comparisons such as these suggest that natural selection is opportunistic.

Marine and freshwater organisms use complementary mechanisms for water and salt regulation. Marine and freshwater organisms face exactly opposite osmotic challenges. Water regulation in aquatic environments is summarized by: $W_i = W_d - W_s \pm W_o$, where W_d = drinking, W_s = secretions and excretions, W_o = osmosis. An aquatic organism may either gain or lose water through osmosis, depending on the organism and the environment. Many marine invertebrates reduce their water regulation problems by being isosmotic with seawater. Some freshwater invertebrates also reduce the osmotic gradient between themselves and their environment. Sharks, skates, and rays elevate the urea and TMAO content of their body fluids to the point where they are slightly hyperosmotic to seawater. Marine bony fish and saltwater mosquito larvae are hypoosmotic relative to their environments, while freshwater bony fish and freshwater mosquito larvae are hyperosmotic.

While the strength of environmental challenge varies from one environment to another, and the details of water regulation vary from one organism to another, all organisms in all environments expend energy to maintain their internal pool of water and dissolved substances.

Stable isotope analysis, an important tool in ecology, involves the analysis of the ratios of stable isotopes in materials. Examples of stable isotopes include the stable isotopes of hydrogen ^2H (which is usually symbolized by D, referring to deuterium) and ^1H, and the stable isotopes of carbon, ^{13}C and ^{12}C. Stable isotope analysis has proved a very powerful tool in studies of water uptake by plants. This is possible because the more unusual isotopes for water, ^2H and ^{18}O, are heavier and therefore less likely to be evaporated and more likely to be precipitated by rain. Thus, surface water has a different isotopic ratio than groundwater. By sampling water in the plant, it is possible to identify which sources it is using.

Key Terms

cohesion 131	hypoosmotic 130	osmoregulation 143	stable isotope analysis 145
diffusion 130	isosmotic 131	osmosis 130	vapor pressure deficit 129
hydrophillic 133	matric forces 131	relative humidity 129	water potential 131
hydrophobic 133	meta-analysis 146	saturation water vapor	water vapor pressure 129
hyperosmotic 130	metabolic water 133	pressure 129	

Review Questions

1. The body temperature of the seashore isopod *Ligia oceanica* is 30°C under stones, where the relative humidity is 100%, but 26°C on the surface, where it is exposed to full sun and the relative humidity is 70%. Explain why evaporative cooling by this isopod would be effective in the open air but nearly impossible under stones.

2. Distinguish among vapor pressure deficit, osmotic pressure, and water potential. How can all three phenomena be expressed in the same units of measure: pascals?

3. Leaf water potential is typically highest just before dawn and then decreases progressively through midday. Assuming no change in soil water potential during the day, will lower leaf water potentials at midday increase or decrease the rate of water movement from soil to a plant?

4. Compare the water budgets of the tenebrionid beetle, *Onymacris*, and the kangaroo rat, *Dipodomys*, shown in figures 6.9 and 6.10. Which of these two species obtains most of its water from metabolic water? Which relies most on condensation of fog as a water source? In which species do you see greater losses of water through the urine?

5. In this chapter, we discussed water relations of tenebrionid beetles from the Namib Desert. However, members of this family also occur in moist temperate environments. How should water loss rates vary among species of tenebrionids from different environments? On what assumptions do you base your prediction? How would you test your prediction?

6. How would traits such as those shown in figure 6.1 evolve? Explain why it makes sense that the structures used to capture water found in both plants and animals are very similar. How might this fact be used to help humans living in arid environments?

7. Many desert species are well waterproofed. Evolution cannot, however, eliminate all evaporative water loss. Why not? (Hint: Think of the kinds of exchanges that an organism must maintain with its environment.)

8. Review water and salt regulation by marine and freshwater bony fish. Which of the two is hypoosmotic relative to its environment? Which of the two is hyperosmotic relative to its environment? Some sharks live in freshwater. How should the kidneys of marine and freshwater sharks function?

9. Given what you have learned about the isotopes found in water, why might two trees growing on the same river have different δ^2H and $\delta^{18}O$ in their xylem water? What implications does this have for competition between the two trees?

Steven P. Lynch

Chapter
7

Energy and Nutrient Relations

A painted caterpillar, *Vanessa cardui,* feeding on a leaf. The energy and nutrients the caterpillar assimilates from its food plants on southern wintering grounds will be used to build a colorful painted lady butterfly, well stocked with fat stores to fuel its long migration north.

CHAPTER CONCEPTS

7.1 Photosynthetic autotrophs synthesize organic molecules using CO_2 as a source of carbon and light as an energy source.   150
Concept 7.1 Review   154

7.2 Chemosynthetic autotrophs synthesize organic molecules using CO_2 as a carbon source and inorganic molecules as an energy source.   154
Concept 7.2 Review   156

7.3 Heterotrophic organisms use organic molecules both as a source of carbon and as an energy source.   156
Concept 7.3 Review   162

7.4 The rate at which organisms can take in energy is limited.   162
Concept 7.4 Review   165

7.5 Optimal foraging theory models feeding behavior as an optimizing process.   165
Concept 7.5 Review   168

Applications: Bioremediation–Using the Trophic Diversity of Bacteria to Solve Environmental Problems   168
Summary   170
Key Terms   170
Review Questions   171

LEARNING OUTCOMES

After studying this section you should be able to do the following:

7.1 Define the major forms of trophic biology.
7.2 Compare the trophic diversity found among prokaryotes, protists, plants, fungi, and animals.

Whether on coral reef, rain forest, or abandoned urban lot, organisms engage in an active search for energy and nutrients (fig. 7.1). For most organisms, life boils down to converting energy and nutrients into descendants. Nutrients are the raw materials an organism must acquire from the environment to live. The energy used by

149

Figure 7.1   The moray eel meets its energy needs by being an effective predator.   Westend61 GmbH/Alamy Stock Photo

different organisms to power life's processes comes in the form of light, organic molecules, or inorganic molecules.

We generally group organisms on the basis of shared evolutionary histories, creating taxa such as vertebrate animals, insects, coniferous trees, and orchids. However, we can also classify them by their **trophic (feeding) biology.** Organisms that use inorganic sources of both carbon and energy are called **autotrophs** ("self-feeders"). **Photosynthetic autotrophs** use carbon dioxide (CO_2) as a source of carbon and light as a source of energy to synthesize **organic compounds,** molecules that contain carbon, such as sugars, amino acids, and fats. **Chemosynthetic autotrophs** synthesize organic molecules using CO_2 as a carbon source and inorganic chemicals, such as hydrogen sulfide (H_2S), as their source of energy. **Heterotrophs** ("other-feeders") are consumers that use organic molecules as a source of both carbon and energy.

Prokaryotes, which include heterotrophic, photosynthetic, and chemosynthetic species, show more trophic diversity than do other groups of organisms (fig. 7.2). Prokaryotes, which have cells with no membrane-bound nucleus or organelles, include the bacteria and the **archaea.** The archaea are prokaryotes distinguished from bacteria on the basis of structural, physiological, and other biological features. Though first discovered in association with extreme environments, the archaea are now known to be widely spread in the biosphere. The protists are either photosynthetic or heterotrophic, most plants are photosynthetic, and all fungi and animals are heterotrophic.

Some of the most ecologically significant discoveries of prokaryotic trophic diversity have come from studies of marine prokaryotes. For instance, Oded Béjà and Edward Delong and their research team from the Monterey Bay Aquarium Research Institute in California and the University of Texas Medical School discovered that energy production from light involving bacterial **rhodopsin** is widely distributed in the oceans (Béjà et al. 2000, 2001). Rhodopsins are light-absorbing pigments found in animal eyes and in the bacteria and archaea. The rhodopsin in bacteria and archaea performs a variety of functions, including that of a proton pump involved in ATP synthesis—that is, in the production of energy-rich molecules. A particularly intriguing discovery is that the light sensitivity of bacterial rhodopsin appears adapted to local variations in light quality. For instance,

bacterial rhodopsin from deep, clear waters absorbs light most strongly within the blue range of the visible spectrum, whereas that from shallow coastal waters absorbs most strongly in the green range. Discoveries such as these and others (e.g., Pushkarev et al. 2018, Inoue et al. 2020, Chen et al. 2019) are rapidly revolutionizing our understanding of how the biosphere works.

7.1   Photosynthetic Autotrophs

LEARNING OUTCOMES

After studying this section you should be able to do the following:

7.3   Define photosynthetically active radiation, photon flux density, rubisco, photorespiration, and bundle sheath.

7.4   Describe C_3, C_4, and CAM photosynthesis.

7.5   Explain the environmental conditions favorable to C_3, C_4, and CAM photosynthesis.

Photosynthetic autotrophs synthesize organic molecules using CO_2 as a source of carbon and light as an energy source. Because photosynthetic organisms use light as a source of energy, we now review the physical nature of light.

The Solar-Powered Biosphere

As we saw in chapters 2 and 3, solar energy powers the winds and ocean currents, and annual variation in sunlight intensity drives the seasons. In chapter 5, we also discussed how organisms use sunlight to regulate body temperature. Here, building on those discussions, we look at light as a source of energy for photosynthesis.

Light propagates through space as a wave, with all the properties of waves such as frequency and wavelength. When light interacts with matter, however, it acts not as a wave but as a particle.

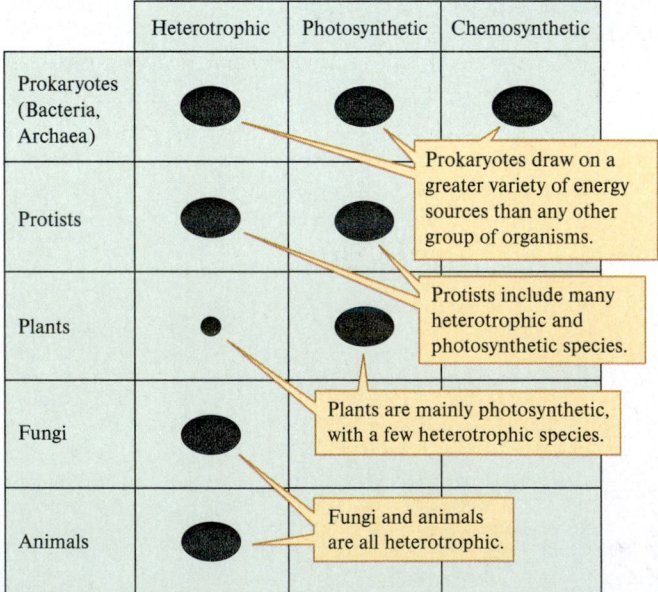

Figure 7.2   A plot of trophic diversity across the major groups of organisms shows highest trophic diversity among the prokaryotic bacteria and archaea.

Particles of light, called *photons,* bear a finite quantity of energy. Longer wavelengths, such as *infrared light,* carry less energy than shorter wavelengths, such as *visible* and *ultraviolet light.*

Infrared light, as we saw in chapter 5 (see section 5.4), is very important for temperature regulation by organisms. This is because its main effect on matter is to increase the motion of whole molecules, which we measure as increased temperature. However, infrared light does not carry enough energy to drive photosynthesis. At the other end of the solar spectrum, ultraviolet light carries so much energy that it breaks the covalent bonds of many organic molecules. Consequently, ultraviolet light can destroy the complex biochemical machinery of photosynthesis. Between these extremes is the light we can see, so-called visible light, which is also called **photosynthetically active radiation,** or **PAR.** PAR, with wavelengths between about 400 and 700 nm, carries sufficient energy to drive the light-dependent reactions of photosynthesis but not so much as to destroy organic molecules. PAR makes up about 42% of the total energy content of the solar spectrum at sea level. Because the pigments involved in photosynthesis absorb unevenly across the PAR wavelengths, however, the energy actually available for photosynthesis amounts to about 26% of the total (Agrawal 2010). Meanwhile, infrared light accounts for about 46% and ultraviolet light for most of the remainder.

Measuring PAR

Ecologists quantify PAR as photon flux density. **Photon flux density** is the number of photons striking a square meter surface each second. The number of photons is expressed as micromoles (μmol), where 1 mole is Avogadro's number of photons, 6.023×10^{23}. To give you a point of reference, a photon flux density of about 4.6 μmol photons $m^{-2}\,s^{-1}$ (μmol per square meter per second) equals about 1 watt per square meter. Measuring light as photosynthetic photon flux density makes sense ecologically because chlorophyll absorbs light as photons.

Light changes in quantity and quality with latitude, with the seasons, with the weather, and with the time of day. In addition, landscapes, water, and even organisms themselves change the amount and quality of light. For example, in aquatic environments (see chapter 3), only the superficial euphotic zone receives sufficient light to support photosynthetic organisms. In addition, light changes in quality, as well as quantity, within the euphotic zone, which ranges in depth from a few meters to about 100 m (see fig. 3.6).

As in the sea, sunlight changes as it shines through the canopy of a forest. A mature temperate or tropical forest can reduce the total quantity of light reaching the forest floor to about 1% to 2% of the amount shining on the forest canopy (fig. 7.3). However, forests also change the quality of sunlight. Within the range of photosynthetically active radiation, leaves absorb mainly blue and red light and transmit mostly green light with a wavelength of about 550 nm. As in the deep sea, the organisms on the forest floor live in a kind of twilight. Only here, the twilight is green.

Alternative Photosynthetic Pathways

During photosynthesis, the photosynthetic pigments of plants, algae, and bacteria absorb photons of light and transfer their

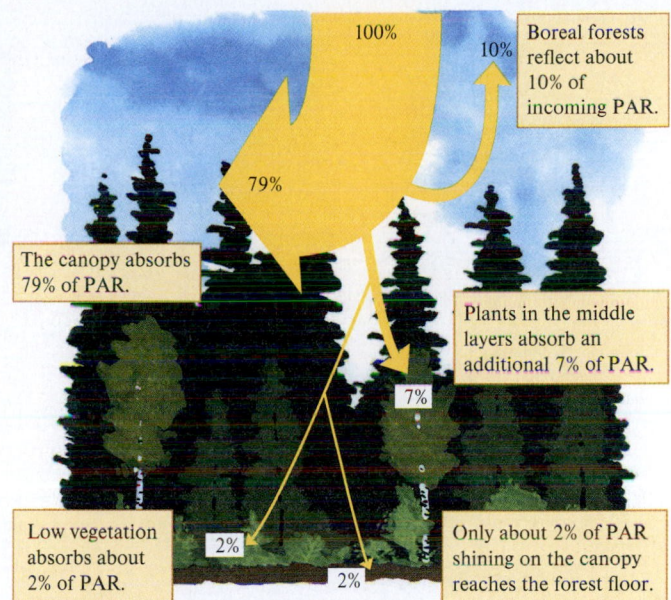

Figure 7.3   Photosynthetically active radiation (PAR) diminishes substantially with passage through the canopy of a boreal forest (data from Larcher 1995, after Kairiukstis 1967).

energy to electrons. Subsequently, the energy carried by these electrons is used to synthesize ATP and NADPH. These molecules, in turn, serve as donors of electrons and energy for the synthesis of sugars. In this way, photosynthetic organisms convert the electromagnetic energy of sunlight into energy-rich organic molecules, the fuel that feeds most of the biosphere. Within photosynthetic organisms, specific biochemical pathways carry out this energy conversion; three different biochemical pathways are known: C_3 photosynthesis, C_4 photosynthesis, and CAM photosynthesis.

C_3 Photosynthesis

Biologists often speak of photosynthesis as "carbon fixation," which refers to the reactions in which CO_2 becomes incorporated into a carbon-containing acid. In the photosynthetic pathway used by most plants and all algae, the CO_2 first combines with a five-carbon compound called *ribulose bisphosphate,* or *RuBP.* The product of this initial reaction, which is catalyzed by the enzyme RuBP carboxylase/oxygenase, or **rubisco,** is *phosphoglyceric acid,* or *PGA,* a three-carbon acid. Therefore, this photosynthetic pathway is called C_3 **photosynthesis** and the plants that employ it, including crop plants such as rice, wheat, and soybeans, are called C_3 plants (fig. 7.4).

To fix carbon, plants must open their stomata, usually located on the underside of leaves, to let CO_2 in. However, as CO_2 enters, diffusing down its concentration gradient from the surrounding air to the leaf interior, water exits. Water vapor flows out faster than CO_2 flows in. The movement of water is more rapid because the gradient in water concentration from the leaf to the atmosphere is generally much steeper than the gradient in CO_2 concentration from the atmosphere to the leaf, particularly in arid climates. Relatively high rates of water loss are not a problem for plants that live in moist conditions but in hot, dry climates, high rates of water loss can close the stomata and shut down photosynthesis.

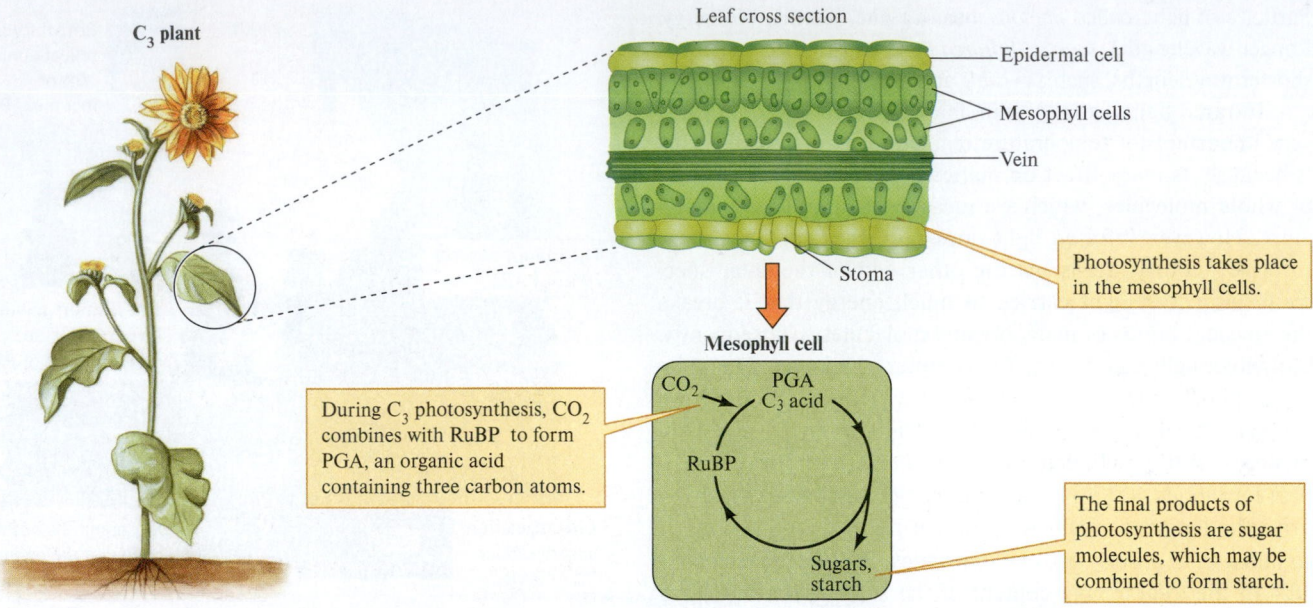

Figure 7.4 C_3 photosynthesis.

Rubisco, the photosynthetic enzyme that catalyzes the reaction between CO_2 and RuBP, catalyzes another reaction that combines O_2 with RuBP. This is the first reaction in a process called **photorespiration,** which occurs in the light, consumes energy, and produces CO_2. Since it consumes energy, the net effect of photorespiration is to reduce the efficiency of photosynthesis. Rubisco is capable of catalyzing the initial reaction of photorespiration because it has an affinity for O_2, but rubisco's affinity for CO_2 is higher. Therefore, photorespiration is most significant energetically when CO_2 concentrations are reduced, which occurs mainly when a plant conserves water, during hot, dry periods, by closing stomata. Under these conditions, photosynthesis reduces CO_2 and increases O_2 concentrations within the air spaces of a leaf. In addition, higher temperatures reduce rubisco's affinity for CO_2 relative to O_2, which further favors photorespiration at higher temperatures. However, alternative photosynthetic pathways that reduce the problem of photorespiration have evolved in some groups of plants. These plants avoid high levels of photorespiration mainly through mechanisms that concentrate CO_2. One of those alternative pathways is C_4 photosynthesis.

C_4 Photosynthesis

Two alternative photosynthetic pathways that fix and store CO_2 in acids containing four carbon atoms have evolved. Light plays no part in carbon fixation, but the reactions that follow depend on light. Both alternative pathways separate the initial fixation of carbon from the light-dependent reactions.

One of these alternative pathways, **C_4 photosynthesis,** separates carbon fixation and the light-dependent reactions of photosynthesis into separate cells (fig. 7.5). C_4 plants fix CO_2 in mesophyll cells by combining it with *phosphoenol pyruvate,* or *PEP,* to produce a four-carbon acid, the source of the name "C_4" photosynthesis. This initial reaction, which is catalyzed by PEP carboxylase, concentrates CO_2. Because PEP

carboxylase is specialized for fixing CO_2, for which it has a high affinity C_4 plants can reduce their internal CO_2 concentrations to very low levels. Low internal concentration of CO_2 increases the gradient of CO_2 from atmosphere to leaf, which in turn increases the rate of diffusion of CO_2 inward. Consequently, compared to C_3 plants, C_4 plants need to open fewer stomata to deliver sufficient CO_2 to photosynthesizing cells. By having fewer stomata open, C_4 plants conserve water.

In C_4 plants, the acids produced during carbon fixation diffuse to specialized cells comprising a structure called the **bundle sheath.** There, deeper in the leaf, the four-carbon acids are broken down to a three-carbon acid and CO_2. In this way, C_4 plants can build up the CO_2 concentration in bundle sheath cells to high levels, increasing the efficiency with which RuBP carboxylase combines CO_2 with RuBP to produce PGA. C_4 plants do better than C_3 plants under conditions of high temperature, high light intensity, low CO_2, and limited water.

C_4 photosynthesis has a very complex evolutionary history, ecology, and geography (Ehleringer, Cerling, and Helliker 1997, Sage 1999). This pathway, which has evolved independently over 30 times, has been documented in approximately 8,000 to 10,000 plant species belonging to 18 plant families. Currently, about half the grass species employ C_4 photosynthesis, which enables them to thrive in hot, semiarid environments. The expansion 8 to 5 million years ago of tropical and subtropical grasslands, which are dominated by C_4 grasses, was associated with the evolution of a high diversity of grazing mammals and their predators. Though the vast majority of plants employ C_3 photosynthesis, C_4 plants are estimated to contribute up to 20% of global terrestrial primary production. They are also of great economic significance, since many important crops, such as corn, and many noxious weeds are C_4 plants. Because the C_4 pathway is more efficient than the C_3 pathway under conditions of low atmospheric CO_2, ecologists have predicted that rising levels of atmospheric CO_2 (see chapter 23,

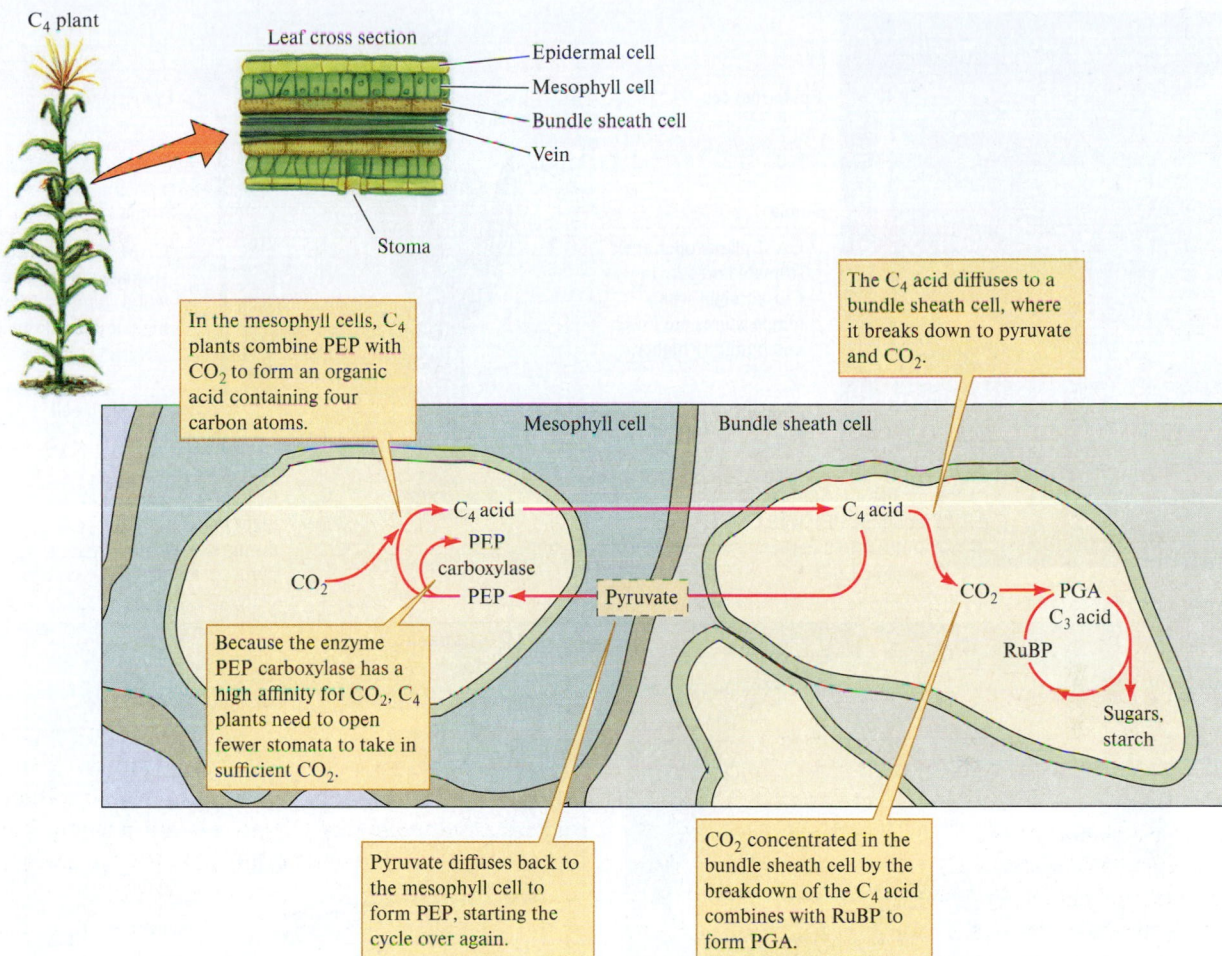

C₄ plant

Leaf cross section
— Epidermal cell
— Mesophyll cell
— Bundle sheath cell
— Vein

Stoma

In the mesophyll cells, C₄ plants combine PEP with CO_2 to form an organic acid containing four carbon atoms.

The C₄ acid diffuses to a bundle sheath cell, where it breaks down to pyruvate and CO_2.

Mesophyll cell   Bundle sheath cell

C₄ acid   C₄ acid

PEP carboxylase

CO_2 PEP ← Pyruvate CO_2 → PGA
                                            C₃ acid
                                    RuBP

                                            Sugars, starch

Because the enzyme PEP carboxylase has a high affinity for CO_2, C₄ plants need to open fewer stomata to take in sufficient CO_2.

Pyruvate diffuses back to the mesophyll cell to form PEP, starting the cycle over again.

CO_2 concentrated in the bundle sheath cell by the breakdown of the C₄ acid combines with RuBP to form PGA.

Figure 7.5   C₄ photosynthesis.

fig. 23.21) will favor C₃ plants. However, we now know that any future changes in the relative abundance of C₃ and C₄ plants will be mediated by a complex interaction between levels of atmospheric CO_2 and the influences of temperature, moisture, and nutrient availability (Feggestad et al. 2004).

CAM Photosynthesis

CAM (crassulacean acid metabolism) **photosynthesis** is largely found in succulent plants in arid and semiarid environments and among epiphytes growing in the canopies of tropical forests, where they are subjected to intense sunlight and drying winds. In this pathway, carbon fixation takes place at night, when lower temperatures reduce the rate of water loss during CO_2 uptake. CAM plants fix carbon by combining CO_2 with PEP to form four-carbon acids. As in C₄ plants, this reaction is catalyzed by PEP carboxylase. These acids are stored until daylight, when they are broken down into pyruvate and CO_2, which then combines with RuBP to form PGA (fig. 7.6). In CAM plants, all these reactions take place in the same cells. While CAM plants do not normally show very high rates of photosynthesis, their water use efficiency, as estimated by the mass of CO_2 fixed per kilogram of water used, is higher than that of either C₃ or C₄ plants.

Separating initial carbon fixation from the other reactions reduces water losses during photosynthesis: C₃ plants lose

from about 380 to 900 g of water for every gram (dry weight) of tissue produced. C₄ plants lose from about 250 to 350 g of water per gram of tissue produced, while CAM plants lose approximately 50 g of water per gram of new tissue. The differences in these numbers give us one of the reasons C₄ and CAM plants do well in hot, dry environments.

Whether the pathway of carbon fixation is CAM, C₃, or C₄, plants and photosynthetic algae and bacteria capture energy from sunlight and carbon from CO_2. Now let's turn from photosynthetic autotrophs to autotrophs that obtain their energy from inorganic molecules, the chemosynthetic autotrophs. Although it may be less familiar to most of us, chemosynthesis may be the oldest way of making a living.

Concept 7.1 Review

1. What environmental conditions favor plants with C₃ photosynthesis? Why?
2. How are C₄ and CAM photosynthesis similar? How are they different?
3. Why does the ongoing increase in atmospheric CO_2 (see chapter 23, fig. 23.21) not give guaranteed advantage to C₃ plants over C₄ plants?

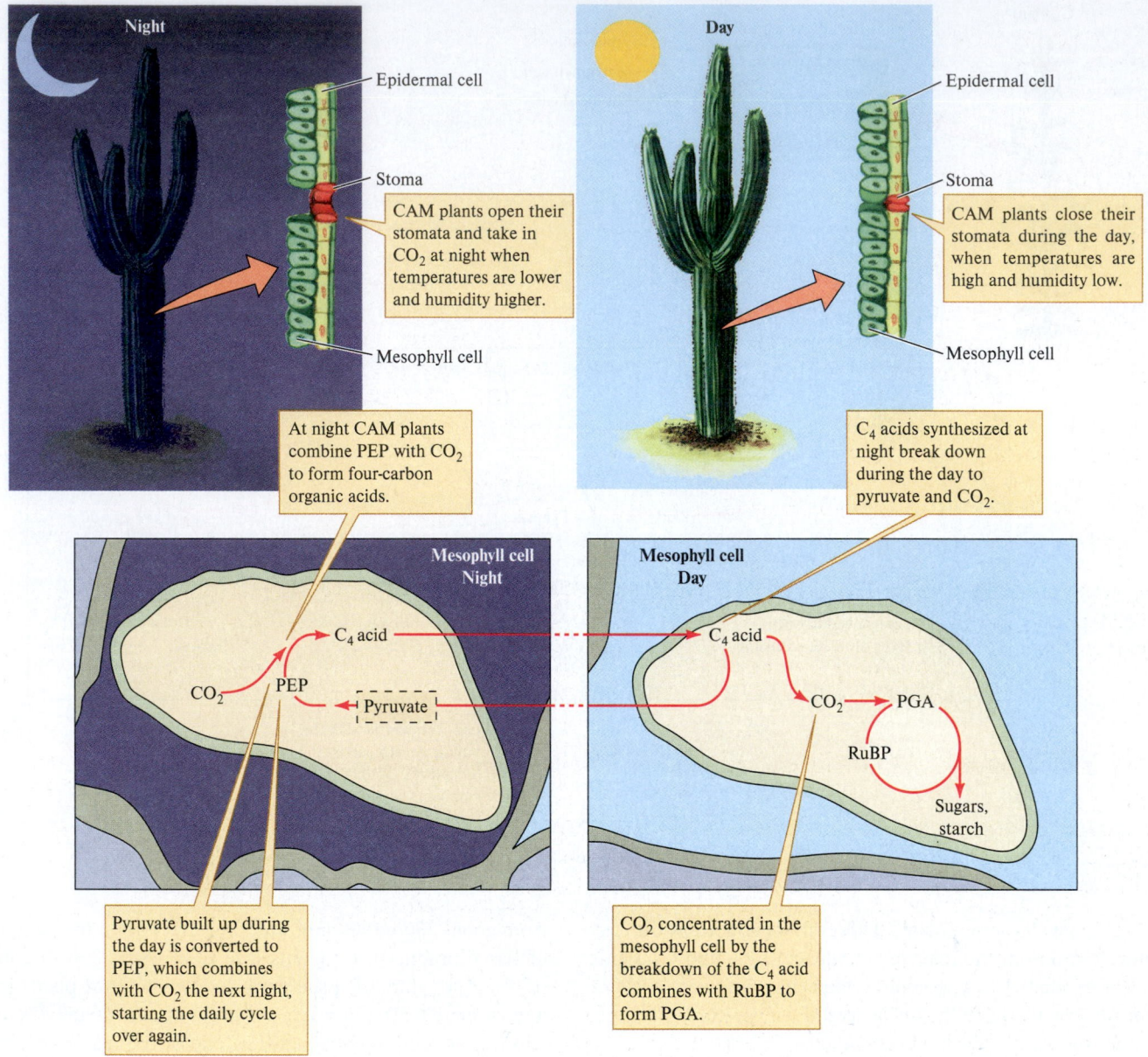

Figure 7.6   Crassulacean acid metabolism, or CAM, photosynthesis.

7.2   Chemosynthetic Autotrophs

LEARNING OUTCOMES

After studying this section you should be able to do the following:

7.6   List several types of chemosynthetic autotrophs.

7.7   Discuss the ecological significance of chemosynthetic autotrophy.

Chemosynthetic autotrophs synthesize organic molecules using CO_2 as a carbon source and inorganic molecules as an energy source.   In 1977, a routine dive by a small submersible carried scientists exploring the Galápagos rift to a grand discovery. Their discovery changed our view of how a biosphere can be structured (Jannasch and Wirsen 1979). Ecologists had long

assumed that photosynthesis provides the energy for nearly all life in the sea. However, it had long before been proposed that there was another process by which organisms could produce organic matter (Ackert 2006). In 1890, Sergei Vinogradskii discovered that some microbes could live off of inorganic matter, including sulfur, iron, and nitrogen bacteria. However, it wasn't until the late 1970's that an entire ecosystem based on this energy source was discovered. These unsuspecting scientists came across a world based on energy captured by chemosynthesis. The world they discovered was inhabited by giant worms up to 4 m long with no digestive tracts, by filter-feeding clams, and by carnivorous crabs tumbling over each other in tangled abundance (see fig. 3.8). These organisms lived on nutrients discharged by deep-sea volcanic activity through an oceanic rift, a crack in the seafloor. Interconnected systems of rifts extend tens of thousands of kilometers along the seafloor. Subsequent explorations have confirmed that chemosynthetic communities exist at many points of volcanic discharge along the seafloor (Dick 2019).

The autotrophs on which these submarine oases depend are chemosynthetic bacteria. Some of the most common are sulfur oxidizers, bacteria that use CO_2 as a source of carbon and get their energy by oxidizing elemental sulfur, hydrogen sulfide, or thiosulfite. The submarine volcanic vents with which these organisms are associated discharge large quantities of sulfide-rich warm water. The sulfur-oxidizing bacteria that exploit this resource around the vents are of two types: free-living forms and those that live within the tissues of a variety of invertebrate animals, including the giant tube worms (fig. 7.7). Other communities dependent on sulfur-oxidizing bacteria have been discovered in thermal vents in deep freshwater lakes, in surface hot springs, and in caves. For example, Laure Bellec and colleagues described a community of chemosynthetic microbes in relatively shallow (<200 m deep) hydrothermal vents near Naples, Italy (Bellec et al. 2020). Using electron microscopy and genetic analysis, they discovered an entire community of sulfur-oxidizing bacteria in the gut of a marine nematode. These shallow hydrothermal vents are globally distributed and, despite being more accessible than deep-sea vents, represent ecosystems that are relatively unexplored.

Other chemosynthetic bacteria oxidize ammonium (NH_4^+), nitrite (NO_2^-), iron (Fe^{2+}), hydrogen (H_2), or carbon monoxide (CO). Of these, the nitrifying bacteria, which oxidize ammonium to nitrite and nitrite to nitrate, are undoubtedly among the most ecologically important organisms in the biosphere. Figure 7.8 summarizes one of the energy-yielding reactions exploited by nitrifying bacteria. The importance of these bacteria is due to their role in cycling nitrogen. As we will see in section 7.3, nitrogen is a key element in the chemical makeup of individual organisms. It also plays a central role in the economy of the entire biosphere (see chapter 19). Nitrogen will frequently enter our discussions in later chapters. In the Applications section of this chapter, we will see how nitrifying bacteria have contributed to a pollution problem associated with an old gold mine and how other bacteria and fungi can be stimulated to solve the problem.

The chemosynthetic and photosynthetic autotrophs opened the way for the evolution of organisms that could meet their energy and carbon needs using organic molecules. And this way of making a living did indeed evolve. We call these organisms heterotrophs.

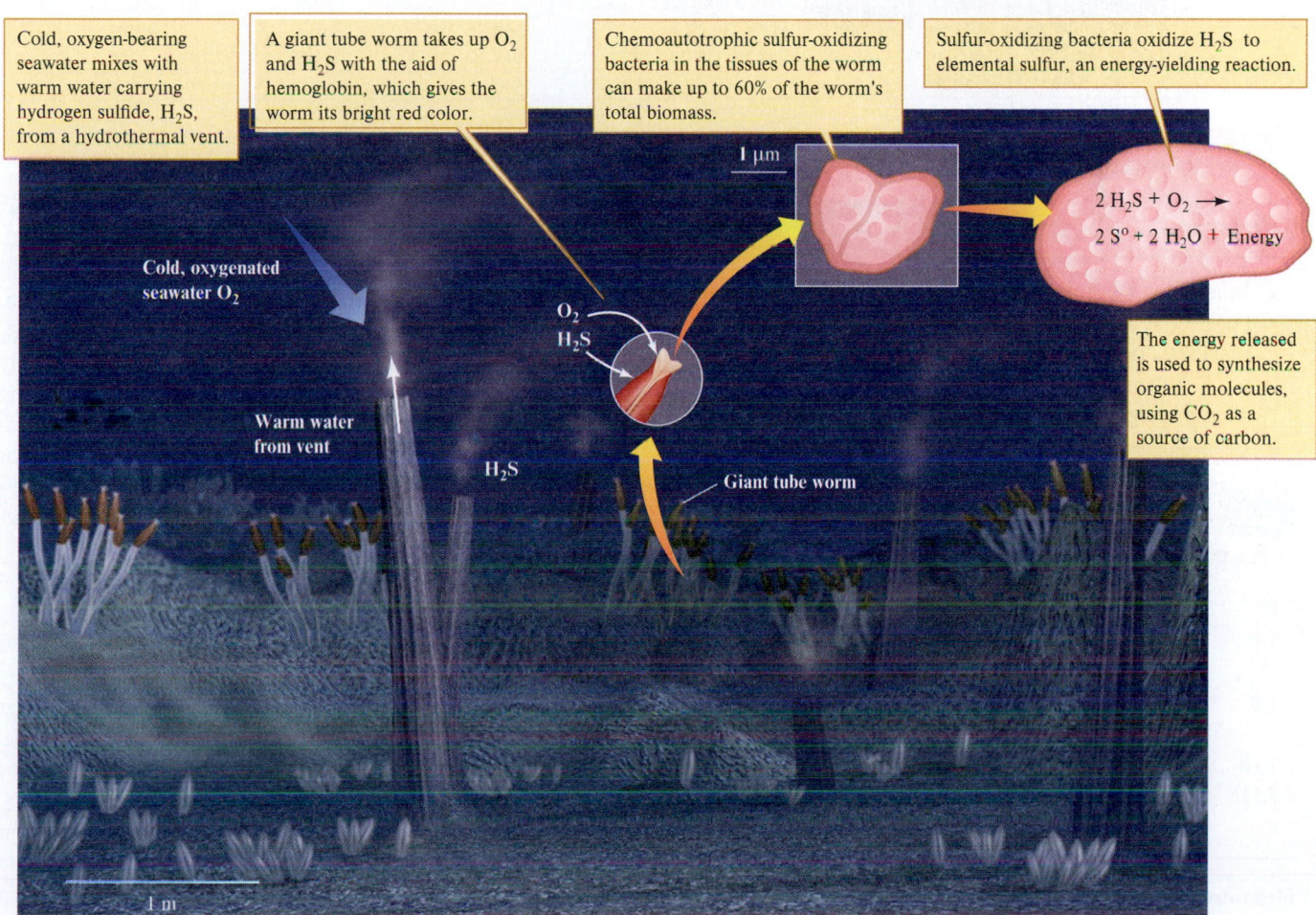

Figure 7.7 Hydrogen sulfide as an energy source for chemoautotrophic bacteria in the deep sea.

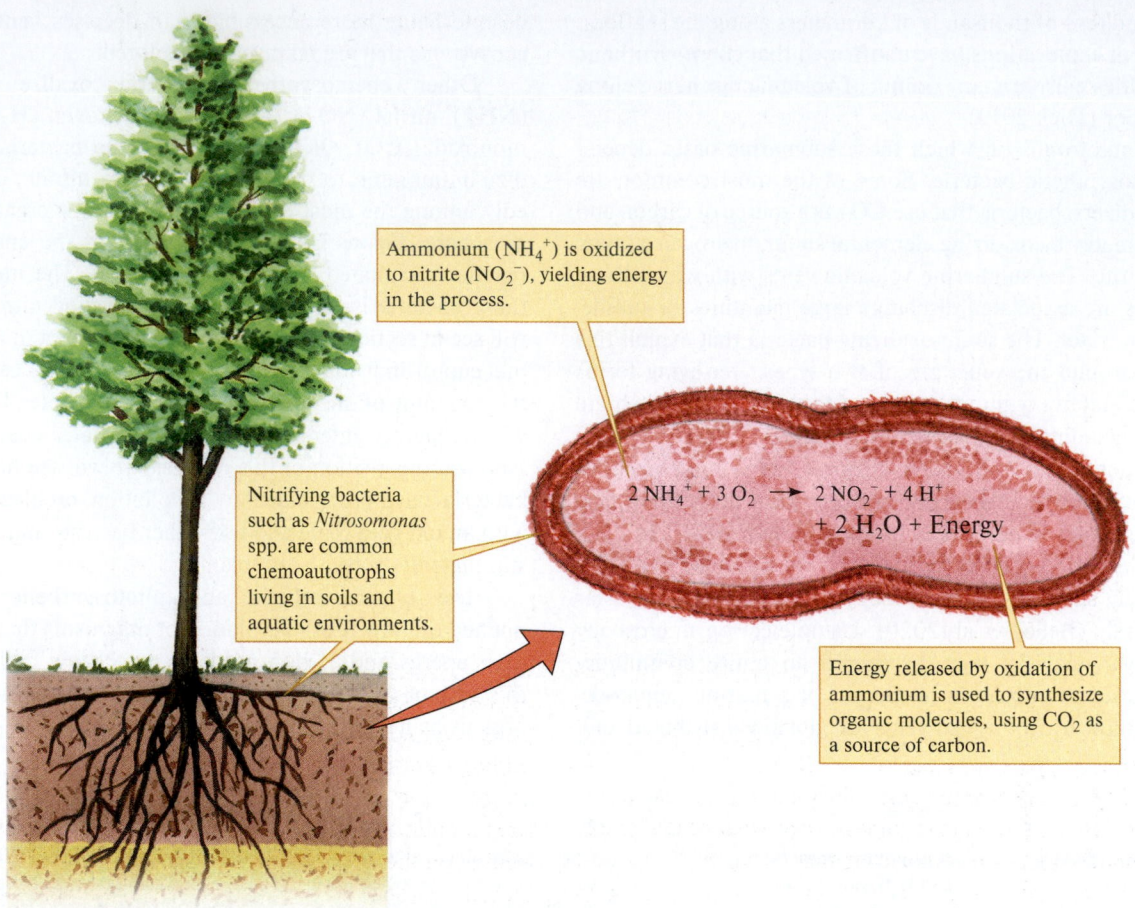

Ammonium (NH_4^+) is oxidized to nitrite (NO_2^-), yielding energy in the process.

Nitrifying bacteria such as *Nitrosomonas* spp. are common chemoautotrophs living in soils and aquatic environments.

$$2\,NH_4^+ + 3\,O_2 \longrightarrow 2\,NO_2^- + 4\,H^+ + 2\,H_2O + Energy$$

Energy released by oxidation of ammonium is used to synthesize organic molecules, using CO_2 as a source of carbon.

Figure 7.8   Ammonium as an energy source for chemoautotrophic bacteria in soil.

Concept 7.2 Review

1. In general, what must be true of the chemical energy of the products of chemosynthesis compared to that of the reactants, for instance, the chemical energy of the product S^0 (elemental sulfur) versus that of the reactant H_2S shown in figure 7.7?

7.3   Heterotrophs

LEARNING OUTCOMES

After studying this section you should be able to do the following:

7.8   Distinguish between herbivores, carnivores, and detritivores.
7.9   Define ecological stoichiometry, and Batesian and Müllerian mimicry.
7.10   Describe size-selective predation.
7.11   Compare the trophic challenges faced by herbivores, carnivores, and detritivores.

Heterotrophic organisms use organic molecules both as a source of carbon and as an energy source.   They depend, ultimately, on the carbon and energy fixed by autotrophs. Heterotrophs have evolved numerous ways of feeding. This trophic variety has inspired numerous terms to describe the ways heterotrophs feed. A full list of the trophic categories proposed by ecologists would be impossibly long and not especially useful to this discussion. So, we will concentrate on three major categories: **herbivores,** organisms that eat plants; **carnivores,** organisms that mainly eat animals; and **detritivores,** organisms that feed on nonliving organic matter, usually the remains of plants. While these categories do not capture all the trophic diversity in nature, they are not arbitrary. Herbivores, carnivores, and detritivores have evolved to live off fundamentally different sources of energy and nutrients.

Chemical Composition and Nutrient Requirements

We can get some idea of the nutrient requirements of organisms by examining their chemical composition. Biologists have found that the chemical composition of organisms is very similar. Just five elements (carbon [C], oxygen [O], hydrogen [H], nitrogen [N], and phosphorus [P]) make up 93% to 97% of the biomass of plants, animals, fungi, and bacteria. But organisms allocate elements in different ways. Of these four groups, plants have the most variable nutrient content. Plant tissues generally contain about 45% carbon and much lower concentrations of nitrogen and phosphorus. Plant leaves contain around 2% nitrogen and less than 0.3% phosphorus.

In contrast, invertebrates, bacteria, and fungi average 5% to 10% nitrogen and about 1% phosphorus. Vertebrate animals have an even higher requirement for phosphorus to support the growth and maintenance of a mineral-rich internal skeleton. Fast-growing organisms require relatively more nutrient-rich resources to support their higher allocation to tissue building compared to those with slower growth rates. Thus, how an organism allocates elements has consequences for its elemental requirements.

Ecologists use the principles of stoichiometry, which means literally "measuring chemical elements," to study the relationships between ratios of elements in a food resource to those ratios in an organism eating that food resource. Such studies are the subject of **ecological stoichiometry,** which concerns the balance of multiple chemical elements in ecological interactions, for example, the balance of multiple chemical elements between plants and the herbivores that consume them. In this instance, the ratio of carbon to nitrogen in plants is much higher than in the herbivores eating them. When such a large difference, or imbalance, in elemental composition exists, the consumer must eat more food to obtain the limiting nutrient, in this case nitrogen. Differences in elemental ratios among tissues or among organisms significantly influence what organisms eat, how rapidly consumers reproduce, and how rapidly organisms decompose (see chapter 19, section 19.2).

If carbon, oxygen, hydrogen, nitrogen, and phosphorus make up 93% to 97% of living biomass, then what accounts for the remainder? Dozens of other elements occur in the tissues of organisms. Essential plant nutrients include potassium (K), calcium (Ca), magnesium (Mg), sulfur (S), chlorine (Cl), iron (Fe), manganese (Mn), boron (B), zinc (Zn), copper (Cu), and molybdenum (Mo). Most of these nutrients are also essential for other organisms. Some organisms require additional nutrients. For instance, animals also require sodium (Na) and iodine (I).

Plants obtain carbon from the air through their stomata. They obtain other essential nutrients from the soil through their roots. For the most part, animals obtain both the energy they require and essential nutrients with their food. Let's now turn to the energy and nutrient relations of herbivorous, detritivorous, and carnivorous animals.

Herbivores

While a band of zebra grazing on the plains of Africa or a sea turtle feeding on sea grass in a tropical lagoon may suggest a life of ease, this image is not accurate. Herbivores face substantial problems that begin at the level of nutritional chemistry.

Herbivores must compensate for large differences between the nutrient content of their food and their own requirements for growth and metabolism. How ubiquitous is the nutrient imbalance between herbivores and their food? James Elser and colleagues (2000) sought to answer this question and compiled a large data set for C:N and C:P ratios of plants and herbivorous insects. They found that herbivorous insects must overcome an average five- to tenfold difference in the ratio of carbon to nutrients they require (fig. 7.9). To compensate for these differences in elemental ratios, insects must consume large amounts of high C:N and C:P plant tissue to meet their requirements for nitrogen and phosphorus.

Herbivores must also overcome the physical and chemical defenses of plants. Some physical defenses are obvious, such as thorns that deter some herbivores entirely and slow the rate of feeding of others (fig. 7.10). However, plants also often deploy a variety of more subtle physical defenses. Grasses incorporate large amounts of abrasive silica into their tissues, which makes feeding on them difficult and which has apparently selected for specialized dentition among grazing mammals. Many plants toughen their tissues with large quantities of cellulose and lignin, producing leaves that are fibrous and difficult to chew.

The use of cellulose and lignin to strengthen tissues may also provide plants with a kind of chemical defense. Increasing the cellulose and lignin content of tissues increases their C:N ratios. An increased C:N ratio decreases the nutritional value of plant tissues. Some plant tissues have C:N ratios that are far higher than the average values we saw in figure 7.9. For instance, the tree trunks that make up most of the plant biomass

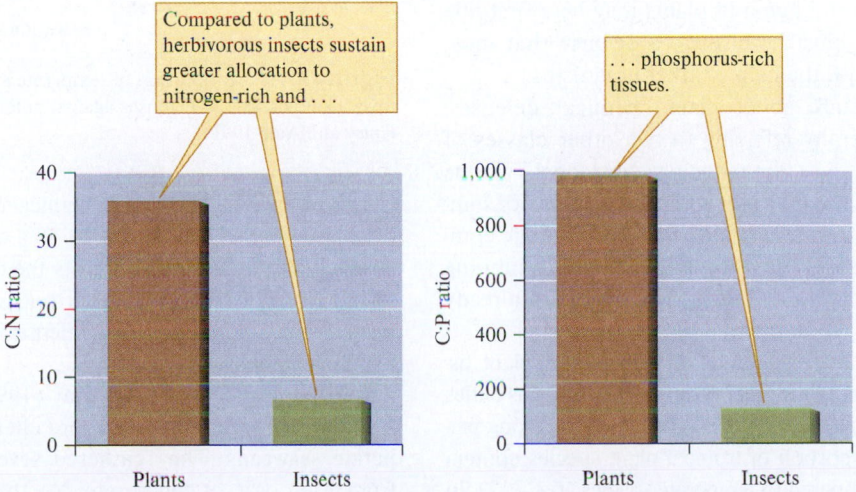

Figure 7.9   A comparison of carbon:nitrogen (C:N) and carbon:phosphorus (C:P) ratios in plants and herbivorous insects. Plants are relatively carbon-rich and nutrient-poor compared to the herbivorous insects that feed on them (data from Elser et al. 2000).

Figure 7.10 Herbivores must overcome the wide variety of defenses evolved by plants, including physical defenses, such as the spines on these acacia branches that a giraffe must delicately negotiate as it feeds, and, often, an arsenal of potent chemical defenses.
Chestertonjp/iStock/Getty Images

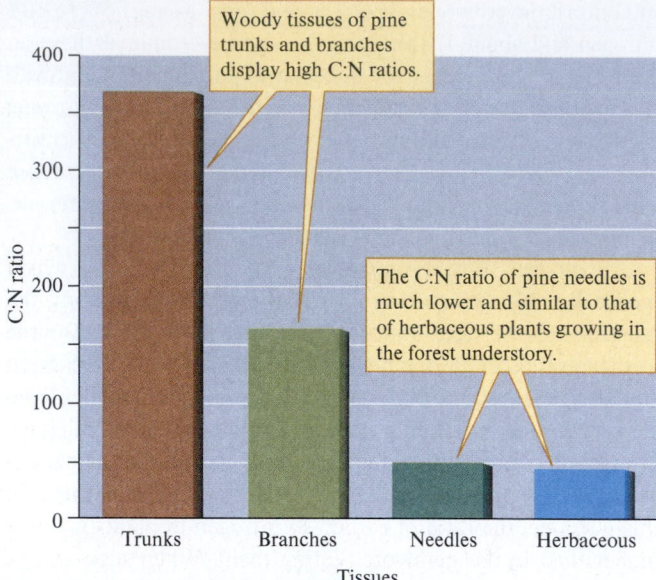

Figure 7.11 C:N ratios differ a great deal among the tissues of pines and between the woody tissues of pines and those of herbaceous plants on the forest floor (data from Klemmedson 1975).

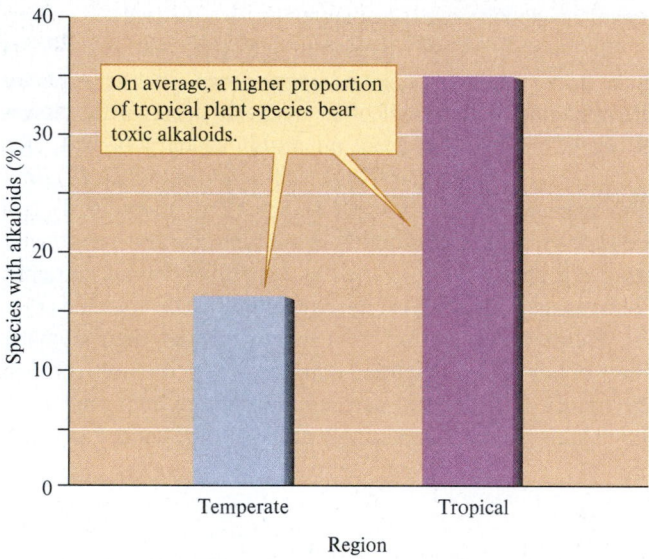

Figure 7.12 Proportion of temperate and tropical plants bearing toxic alkaloids, potent defenses against potential herbivores (data from Coley and Aide 1991).

in a pine forest have a C:N ratio of over 300:1 (fig. 7.11). This ratio is much higher than that of either branches or needles. The living needles of pine trees have C:N ratios very similar to those of understory herbs living on the forest floor.

In addition, most animals cannot digest either cellulose or lignin. Those that can, generally do so with the help of bacteria, fungi, or protists that live in their digestive tracts. This suggests that the cellulose and lignin in plants may be a first line of chemical defense against herbivores, a defense that most herbivores overcome with the help of other organisms.

When ecologists talk about plant chemical defenses, however, they are generally referring to two other classes of chemicals: toxins and digestion-reducing substances. Toxins are chemicals that kill, impair, or repel most would-be consumers. Digestion-reducing substances are generally phenolic compounds such as tannins that bind to plant proteins, inhibiting their breakdown by enzymes and further reducing the already low availability of nitrogen in plant tissues.

Chemists have isolated thousands of toxins from plant tissues, and the list is growing. The great variety of plant toxins defies easy description and generalization. However, one interesting pattern is that a higher proportion of tropical plant species contain toxic alkaloids when compared to temperate species (fig. 7.12). In addition, on average, the alkaloids produced by tropical plants are more toxic than those produced by their temperate counterparts.

Despite these higher levels of chemical defense, herbivores appear to remove approximately 11% to 48% of leaf biomass in tropical forests, while in temperate forests they remove about 7%. These higher levels of herbivore attack on tropical plants suggest that natural selection for chemical defense is more intense in tropical plant populations.

Robin Bolser and Mark Hay (1996) tested the hypothesis that tropical seaweeds have more chemical defenses than temperate seaweeds. They gathered several species of seaweeds from the coast of temperate North Carolina and from the tropical Bahama Islands. Bolser and Hay were careful to pick the same species of seaweed in the two study sites or at least to pick species that belonged to the same genus. They tested the relative palatability of temperate and tropical seaweeds using temperate and tropical sea urchins.

The researchers were careful to preserve any chemical defenses their study algae might contain. They cleaned and then froze the seaweeds in a freezer (−20°C) on the research ship. On shore, the seaweeds were transferred to a colder freezer (−70°C) to minimize chemical changes.

To remove the potential confounding effect of various physical factors, Bolser and Hay created artificial algae to test their hypothesis. They did this by freeze-drying samples of each seaweed and grinding them in a coffee mill. The powdered algae was then mixed with agar at a concentration of 0.1 g alga per milliliter of agar. The warm alga and agar mixture was poured into a mold set on screening. As the mixture gelled, it attached to the screening. The result was strips of artificial seaweed that could be cut up into equal-sized squares and presented to sea urchins in equal numbers. This method of presentation also provided an easy means of quantifying the actual amount of seaweed eaten.

The results of this study showed a clear preference for temperate species of seaweed (fig. 7.13). When given a choice the urchins removed approximately twice as much of the available temperate seaweed. In addition, both temperate and tropical urchins showed a similar preference for temperate seaweeds. What caused the lower palatability of the tropical seaweeds? In additional tests, Bolser and Hay showed that the tropical seaweeds have more potent chemical defenses. So, we see that this study produced a pattern in the sea that parallels the better-known pattern in tropical and temperate forests. Tropical plants and algae appear to possess stronger chemical defenses.

No defense is perfect; the defenses of most plants work against some herbivores, but not all. The tobacco plant uses nicotine, a toxic alkaloid, to repel herbivorous insects, most of which die suddenly after ingesting nicotine. However, several insects specialize in eating tobacco plants and manage to avoid the toxic effects of nicotine. Some simply excrete nicotine, while others convert it to nontoxic molecules. Similarly, toxins and repellents produced by plants in the cucumber family repel most herbivorous insects but attract the spotted cucumber beetle. This beetle is a specialist that feeds mainly on members of the cucumber family. Some specialized herbivores go even further by using plant toxins as a source of nutrition! The world may appear green to us, but to herbivores only some shades of green are edible.

Detritivores

The problems faced by herbivores in their search for energy and nutrients are related to those faced by detritivores, which feed on dead plant material. Because they play key roles in the cycling of nutrients (see chapter 19), detritivores are essential to sustaining life on earth. These organisms consume food that is rich in carbon and energy but very poor in nitrogen. In fact, plant tissues, already relatively low in nitrogen when living (see figs. 7.9 and 7.11), are even lower in nitrogen content when cast off by plants as detritus. Keith Killingbeck and Walt Whitford (1996) averaged the nitrogen contents of living and dead leaves of many plant species of environments from tropical rain forests through deserts and temperate forests. Their results show that in all these environments, living leaves contain about twice the nitrogen as dead leaves (fig. 7.14).

Carnivores

Carnivores consume animal prey, to which they are stoichiometrically similar; that is, both predators and prey have low C:N and C:P ratios. However, carnivores cannot go out into their environment and choose their nutritionally rich prey at

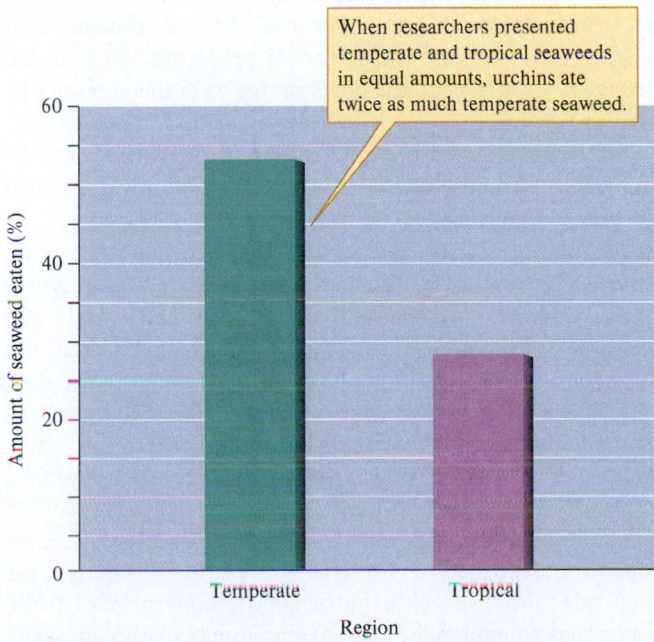

When researchers presented temperate and tropical seaweeds in equal amounts, urchins ate twice as much temperate seaweed.

Figure 7.13 Sea urchin preference for temperate versus tropical seaweeds suggests that tropical seaweeds are better defended against attack by herbivores (data from Bolser and Hay 1996).

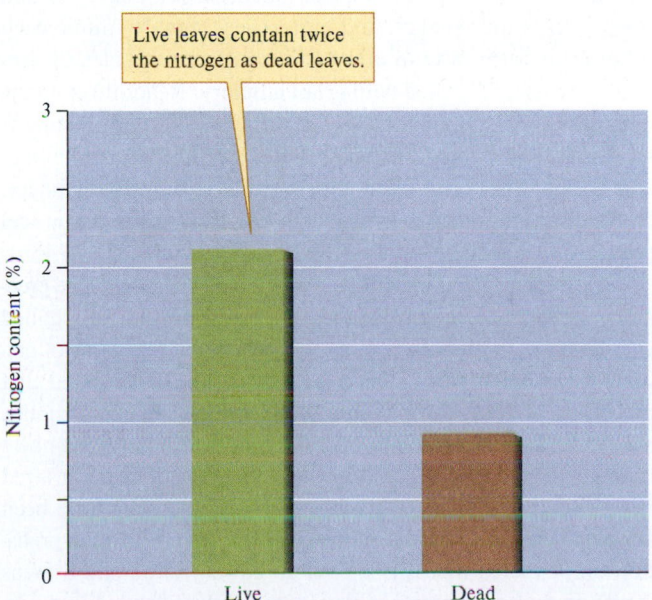

Live leaves contain twice the nitrogen as dead leaves.

Figure 7.14 Nitrogen content of live and dead leaves (data from Killingbeck and Whitford 1996).

(a)

(b)

Figure 7.15   (*a*) Poisonous Müllerian (honeybee) and (*b*) nonpoisonous Batesian (hoverfly) mimic.   (*a*) Alan and Sandy Carey/Getty Images; (*b*) IT Stock/age fotostock

will. Most prey species are masters of defense. One of the most basic prey defenses is camouflage. Predators cannot eat prey they cannot find. Other prey defenses include anatomical and chemical features such as spines, shells, repellents, and poisons, and behavioral defenses such as flight, taking refuge in burrows, banding together in groups, playing dead, fighting, flashing bright colors, spitting, hissing, and screaming at predators. It's enough to spoil your appetite!

Prey that carry a threat to predators often advertise that fact, usually by being brightly colored or conspicuous in some other way. The conspicuous, or *aposematic,* colors of many distasteful or toxic butterflies, snakes, and nudibranchs warn predators that "feeding on me may be hazardous to your health." Aposematic, or warning, coloration generally consists of sharply contrasting patches of orange or yellow and black. Many noxious organisms, such as stinging bees and wasps, venomous snakes, and butterflies, seem to mimic each other. This form of comimicry among several species of noxious organisms is called **Müllerian mimicry.** In addition, many harmless species appear to mimic noxious ones. For instance, king snakes mimic coral snakes, and syrphid flies mimic bees and wasps. This form of mimicry is called **Batesian mimicry.** In Batesian mimicry, the noxious species serves as the model (fig. 7.15*a*) and the harmless species is the mimic (fig. 7.15*b*).

How have prey populations evolved their defenses? The predators themselves are usually the agents of selection for refined prey defense. In one of the most thoroughly studied cases of natural selection for prey defense, H. Kettlewell (1959) found that predation by birds favors camouflage among peppered moths, *Biston betularia.* Birds eliminate the more conspicuous members of the peppered moth population, leaving the better camouflaged (fig. 7.16). Kettlewell's research methods and results have been strongly supported by the more recent research on bird predation on peppered moths (Cook et al. 2012, Walton and Stevens 2018). In general, predators eliminate poorly defended individuals and leave the well defended. Consequently, the average level of defense in the prey population increases with time.

As a consequence of prey defenses, the rate of prey capture by predators is highly variable. Stephanie Green and colleagues compared predation between Caribbean coral reef fishes on different types of prey by creating feeding trials in large tanks (Green et al. 2019). Specifically, they were interested in whether lionfish, which are newly introduced and invasive in the Caribbean, had different success rates relative to the native fish. The lionfish is a stalking predator—a different hunting strategy from the resident native fish. Green and colleagues discovered that the lionfish had capture rates that were much higher than the native species tested. Among attempts (or "strikes"), capture rates by the native fish were as low as 20%, whereas the lion fish had nearly 100% success on every type of prey fish tested (fig. 7.17). This is likely because the prey species had evolved defense strategies that were effective against the historical predators, whereas the prey were vulnerable to the novel strategies of the more newly-arrived lionfish.

Though elusive, the prey of carnivores are generally similar in nutrient content. Consequently, carnivores, which are often widely distributed geographically, can vary their diets from one region to another. The Eurasian river otter, *Lutra lutra,* which is distributed from Europe and North Africa through northern and central Asia, changes its diet based on the local availability of prey. Manuel Graça and F. X. Ferrand de Almeida (1983) compared Eurasian river otter diets along a gradient from northern to southern Europe. On the Shetland Islands, the diets of the river otters were over 91% fish, with the remainder consisting almost entirely of crabs. To this staple diet of fish, the river otters of England added frogs, mammals, birds, and crustaceans. Meanwhile, the diets of river otters in central Portugal are less than one-third fish, with the remainder consisting of frogs, water snakes, mammals, and aquatic insects. Although seemingly very different, these three diets are all fairly similar in terms of their carbon, nitrogen, and phosphorus content; they are just packaged differently.

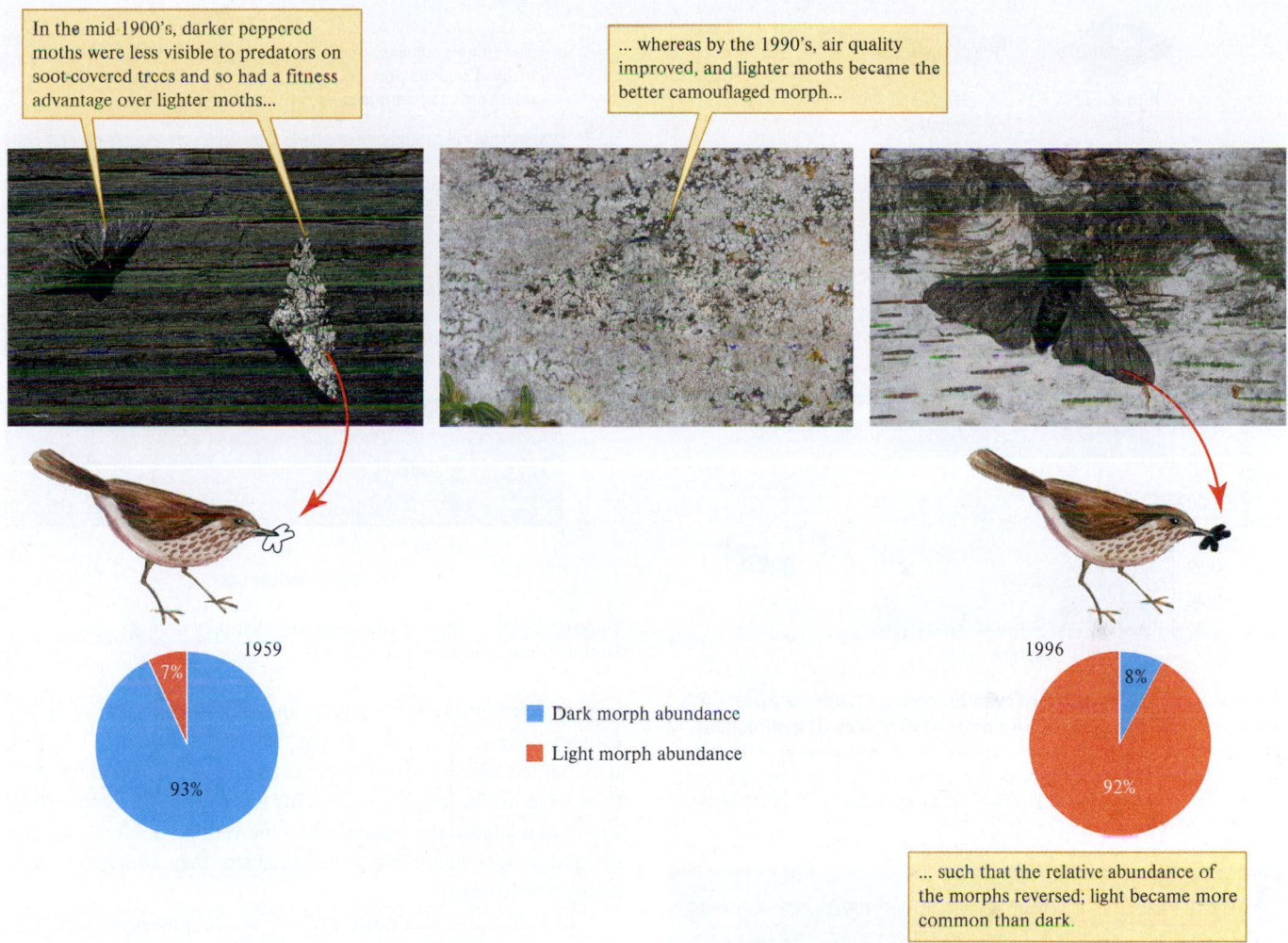

In the mid 1900's, darker peppered moths were less visible to predators on soot-covered trees and so had a fitness advantage over lighter moths...

... whereas by the 1990's, air quality improved, and lighter moths became the better camouflaged morph...

1959

7%

Dark morph abundance

Light morph abundance

93%

1996

8%

92%

... such that the relative abundance of the morphs reversed; light became more common than dark.

Figure 7.16 Birds and other predators exert strong selective pressure on populations of peppered moths in Europe and elsewhere, based on the extent to which they are camouflaged. When trees were covered with soot due to industrialization, darker morphs were more abundant, whereas in the wake of environmental regulations that reduced air pollution, the bark of trees became light-colored again, resulting in dramatic shifts in the frequency of the lighter morph. Frequency data shown here were collected 37 years apart in West Kirby, Wirral. This remains one of the most commonly used examples of natural selection, although recent research suggests that other evolutionary forces also likely played a role (data from Grant et al. 1998). Bill Coster IN/Alamy Stock Photo/H Lansdown/Alamy Stock Photo/Frank Hecker/Alamy Stock Photo

Because predators must catch and subdue their prey, they often select prey by size, a behavior that ecologists call **size-selective predation.** Because of this behavior, prey size is often significantly correlated with predator size, especially among solitary predators. One such solitary predator, the puma, or mountain lion, *Puma concolor,* ranges from the Canadian Yukon to the tip of South America (fig. 7.18). Puma size changes substantially along this latitudinal gradient. Augustin Iriarte and his colleagues (1990) found that as pumas increase in size, the average size of their prey also increases (fig. 7.19). Mammals make up over 90% of the puma's diet, and large mammals, especially deer, are its main prey in the northern part of its range in North America, where pumas are larger. In the tropics where pumas are smaller, they feed mainly on medium and small prey, especially rodents. Why should different-sized pumas feed on different-sized prey? One reason is that large prey may be difficult to subdue and may even injure the predator, while small prey may be

difficult to find or catch. As we shall see later in chapter 7, size-selective predation may also have an energetic basis.

Size of the predator can affect prey selection in other ways. Shannon Murphy, Danny Lewis, and Gina Wimp studied the impacts of size by asking if the diets of wolf spiders (*Pardosa littoralis*) in salt marshes changed as they aged (Murphy et al. 2020). Older spiders are as much as 30 times larger than newly hatched spiders; however, in theory small and large spiders could have the same diet because they digest their prey externally and so don't have to be larger than their prey. Using nitrogen isotope analysis (see chapter 6) of field-collected spiders, they were able to determine that diet did change: young spiders primarily hunted detritovores, which occur on the soil surface under the thatch, whereas the larger, older wolf spiders were hunting herbivores out in the open. However, a few small spiders had isotope signatures similar to large spiders, confirming that there was nothing physiological that kept them from eating the same herbivore diet. Murphy and her colleagues concluded that the mechanism for the difference

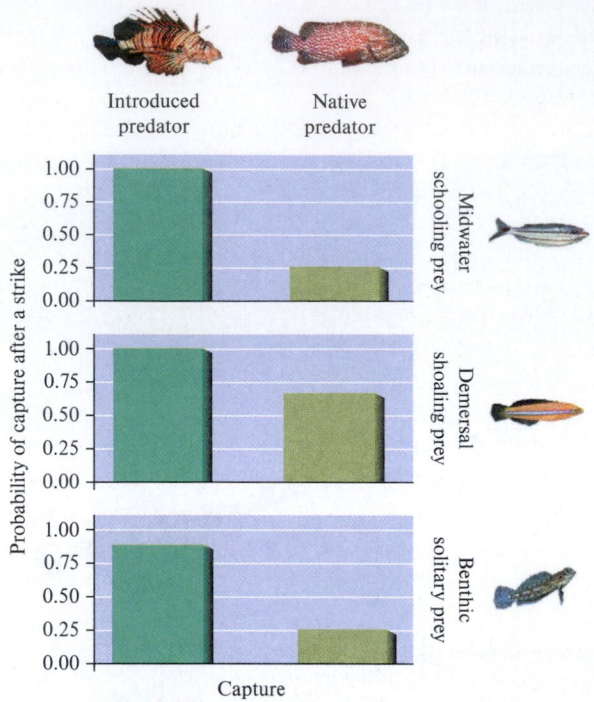

Introduced predator   Native predator

Midwater schooling prey

Demersal shoaling prey

Benthic solitary prey

Probability of capture after a strike

Capture

Figure 7.17   Capture rates by introduced and native predatory fish from the Caribbean coral reefs for three types of prey. The introduced lionfish are stalking predators, whereas the native species shown here are ambush predators. The prey do not appear to have evolved defenses yet against the newly arrived lionfish, resulting in higher capture rates. (data from Green et al. 2019).

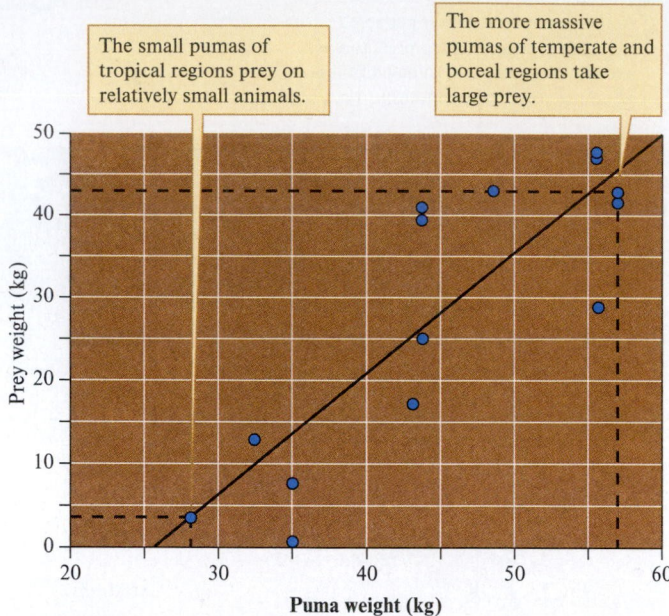

The small pumas of tropical regions prey on relatively small animals.

The more massive pumas of temperate and boreal regions take large prey.

Figure 7.19   The size of pumas and their prey changes with latitude (data from Iriarte et al. 1990).

Figure 7.18   A puma, *Puma concolor*, walking in snow. Pumas are noted for their stealth and are one of the most efficient predators in the Americas.   Jerry & Barb Jividen/Moment/Getty Images

in prey selection between young and old spiders was likely the ecology of fear: When the younger spiders were newly hatched, they hunted where they were concealed to avoid becoming prey themselves to the larger spiders. This was supported by the observation that later in the season, when the younger spiders were bigger and so less vulnerable to cannibalism, they switched to also eating herbivores.

In summary, predators are subject to pressures not only from the need to capture prey, but also often from the need to avoid becoming prey themselves. In the co-evolutionary race between predators and prey, predators eliminate poorly defended individuals in the population and average prey defenses improve. As average prey defenses improve, the poorer hunters go hungry and leave fewer offspring. Consequently, improved hunting skills evolve in the predator population, which exerts further selection on the prey population. Such selection was central to chapter 4, which focused on population genetics and natural selection.

As we have seen, the trophic diversity among organisms is great. However, at least one ecological characteristic is shared by all organisms, regardless of the trophic group to which they belong—all organisms take in energy at a limited rate.

Concept 7.3 Review

1. Why do pumas face fewer challenges from the perspective of stoichiometry compared to herbivores, such as deer, on which they prey?
2. Compare the dietary challenges associated with being a detritivore versus an herbivore. Consider figure 7.14.
3. Explain how a Batesian mimic, such as the hoverfly in figure 7.15b, could evolve, through natural selection, from a nonaposematic ancestor.

7.4   Energy Limitation

LEARNING OUTCOMES

After studying this section you should be able to do the following:

7.12   Define P_{max}, irradiance, I_{sat}, and net photosynthesis.

7.13   Distinguish between type 1, 2, and 3 functional response curves.

7.14   Compare plant photosynthetic response curves and animal functional response.

The rate at which organisms can take in energy is limited. As children, many of us imagined that if we had free access to a candy or ice cream shop, we would consume an infinite quantity of goodies. But even if we were given a chance to do this, our rate of intake would be limited, not by supply but by the rate at which we could process what we ate. In reality, the rate of intake of candy or ice cream for most children is limited, at least in the short term, by how much is available. The same is true in nature. However, if organisms are not limited by the availability of energy in the environment, their energy intake is limited by internal constraints. Limits on the potential rate of energy intake by animals have been demonstrated by studying how feeding rate increases as the availability of food increases. Limits on rates of energy intake by plants have been demonstrated by studying how photosynthetic rate responds to photon flux density.

Photon Flux and Photosynthetic Response Curves

Plant physiologists generally test the photosynthetic potential of plants in environments that are ideal for particular species being studied. These environments have abundant nutrients and water, normal concentrations of oxygen and carbon dioxide, ideal temperatures, and high humidity. If you gradually increase the quantity of light shining on plants growing under these conditions—that is, if you increase the photon flux density—the plants' rates of photosynthesis gradually increase and then level off. At low light intensities, photosynthesis increases linearly with photon flux density. At intermediate light levels, photosynthetic rate rises more slowly. Finally, at higher light levels, but well below that of full sunlight, photosynthesis levels off. Organisms that show this type of photosynthetic response curve include terrestrial plants, lichens, planktonic algae, and benthic algae.

The photosynthetic response curves of different plant species generally level off at different maximum rates of photosynthesis. This rate in figure 7.20 is indicated as P_{max}. A second difference among photosynthetic response curves is the photon flux density, or **irradiance,** required to produce the maximum rate of photosynthesis. The irradiance required to saturate photosynthesis is shown in figure 7.20 as I_{sat}.

Differences in photosynthetic response curves have been used to divide plants into "sun" and "shade" species. The response curves of plants from shady habitats suggest selection for efficiency at low light levels, that is, low irradiance. The photosynthetic rate of shade plants levels off at lower irradiance, and they are often damaged by high irradiance. However, at very low light levels, shade plants usually have higher photosynthetic rates than sun plants. Park Nobel (1977) determined the photosynthetic response curve for the maidenhair fern, *Adiantum decorum.* This plant generally grows at low light levels in forests. In one of Nobel's trials, the maximum rate of **net photosynthesis** by *A. decorum,* P_{max}, was approximately 9 $\mu mol\ CO_2\ m^{-2}\ s^{-1}$. Net photosynthesis can be measured as total, or gross, CO_2 uptake during photosynthesis minus the CO_2 produced by the plant's own respiration. The amount of light required to achieve this maximum rate of photosynthesis, I_{sat}, was a PAR photon flux density of about 300 μmol per square meter per second (fig. 7.21). The values of P_{max} and I_{sat} shown by *A. decorum* are much lower than those observed in plants that have evolved in sunny environments.

Herbs and short-lived perennial shrubs that have evolved in sunny environments show high maximum rates of photosynthesis, P_{max}, at relatively high irradiance, I_{sat}. One such plant, *Encelia farinosa,* grows in the hot deserts of North America. James Ehleringer and his colleagues (1976) found that *E. farinosa* has a high P_{max}, more than four times that of *A. decorum.* In addition, *E. farinosa* reaches these maximum rates of photosynthesis at a photon flux density of about 2,000 μmol photons $m^{-2}\ s^{-1}$ (see fig. 7.21). This combination of high P_{max} and I_{sat} allows *E. farinosa* to fix energy at a high rate during the infrequent times when water is plentiful in its desert environment.

Whether of shade or sun plant, photosynthetic response curves eventually level off. In other words, the rate at which photosynthetic organisms can take in energy is limited. Animals also take in energy at a limited rate.

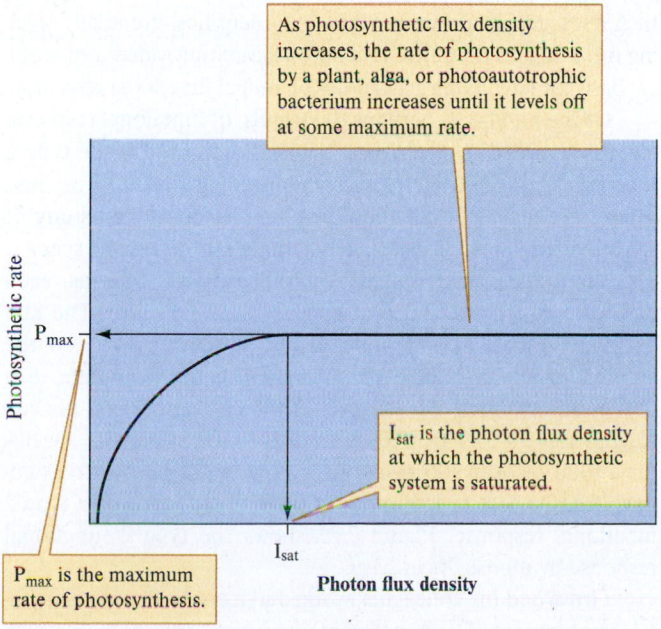

Figure 7.20   A theoretical photosynthetic response curve.

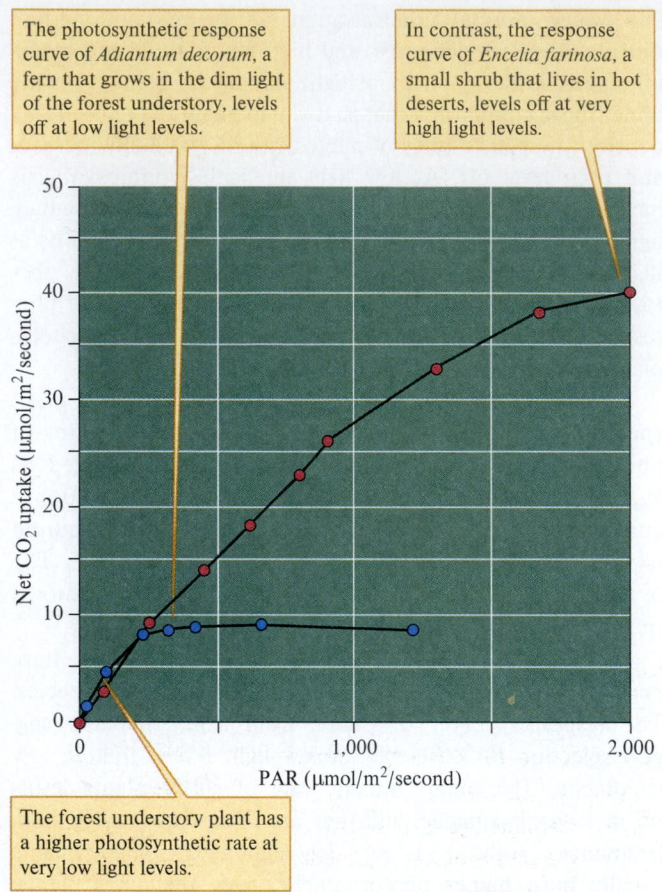

The photosynthetic response curve of *Adiantum decorum*, a fern that grows in the dim light of the forest understory, levels off at low light levels.

In contrast, the response curve of *Encelia farinosa*, a small shrub that lives in hot deserts, levels off at very high light levels.

The forest understory plant has a higher photosynthetic rate at very low light levels.

Figure 7.21   Contrasting photosynthetic response curves indicate adaptation to very different environmental light conditions (data from Ehleringer, Björkman, and Mooney 1976, after Nobel 1977).

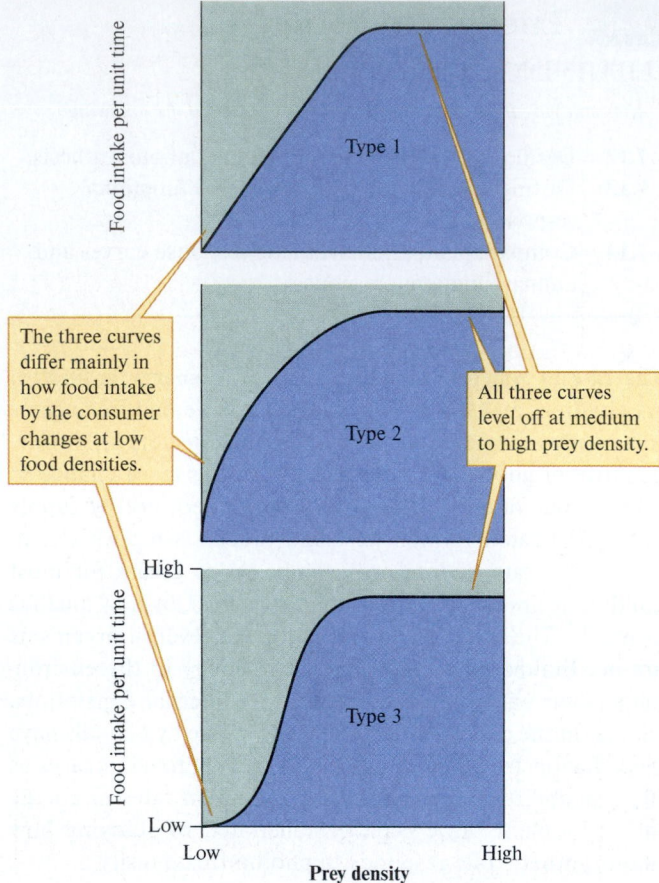

The three curves differ mainly in how food intake by the consumer changes at low food densities.

All three curves level off at medium to high prey density.

Figure 7.22   Three theoretical functional response curves.

Food Density and Animal Functional Response

If you gradually increase the amount of food available to a hungry animal, its rate of feeding increases and then levels off. This relationship is called the **functional response.** Ecologists use graphs to describe functional responses. C. S. Holling (1959a) described three types of functional responses, all of which level off at a maximum feeding rate (fig. 7.22).

Type 1 functional responses are those in which feeding rate increases linearly (as a straight line) as food density increases and then levels off abruptly at some maximum feeding rate. The only animals that have type 1 functional responses are consumers that require little or no time to process their food—for example, some filter-feeding aquatic animals that feed on small prey.

In a type 2 functional response, feeding rate at first rises linearly at low food density, rises more slowly at intermediate food density, and then levels off at high densities. At low food densities, feeding rate appears limited by how long it takes the animal to find food. At intermediate food densities, the animal's feeding rate is partly limited by the time spent searching for food and partly by the time spent handling food. "Handling" refers to such activities as cracking the shells of nuts or snails, removing distasteful scent glands from prey, and chasing down elusive prey. At high food densities, an animal does not have to search for food at all and feeding rate is determined almost entirely by

how fast the animal can handle its food. At these very high densities, the animal, in effect, has "all the food it can handle."

The type 3 functional response is S-shaped. At low food densities, type 3 functional response curves increase more slowly than during either type 1 or type 2 functional response. Food intake then rises steeply at intermediate food densities, eventually leveling off at higher densities. Holling's research provided a theoretical basis for later empirical studies of animal functional response.

Of the hundreds, perhaps thousands, of functional response curves described by ecologists, the most common is the type 2 functional response. Here are some examples. John Gross and several colleagues (1993) conducted a well-controlled study of the functional responses of 13 mammalian herbivore species. The researchers manipulated food density by offering each herbivore various densities of alfalfa, *Medicago sativa*. The rate of food intake was measured as the difference between the amount of alfalfa offered to an animal at the beginning of a trial and how much was left over at the end. Gross and his colleagues ran 36 to 125 feeding trials for each herbivore species for a total of over 900 trials. Every species of herbivore examined, from moose to lemmings to prairie dogs, showed a type 2 functional response. Figure 7.23 shows the type 2 functional response by moose, *Alces alces*.

Gross and his colleagues worked in a controlled experimental environment. Do consumers in natural environments also show a type 2 functional response? To answer this question, let's examine the functional response of wolves, *Canis lupus,* feeding on moose. François Messier (1994) examined the interactions between moose and wolves in North America. He focused on areas where moose are the dominant large prey species eaten by wolves. When moose density in various regions was plotted against the rate at which they were killed by wolves, the result was a clear type 2 functional response (fig. 7.24).

Type 2 functional responses are remarkably similar to the photosynthetic response curves shown by plants (see fig. 7.21) and have the same implications. Even if you provide an animal with unlimited food, its energy intake eventually levels off at some maximum rate. This is the rate at which energy intake is limited by internal rather than external constraints. As we shall see in the next Concept, limited energy intake is a fundamental assumption of optimal foraging theory.

Concept 7.4 Review

1. In type 3 functional response, what mechanisms may be responsible for low rates of food intake—compared to type 1 and type 2 functional response—at low food densities?
2. Why are plants such as mosses living in the understory of a dense forest, which show higher rates of photosynthesis at low irradiance, unable to live in environments where they are exposed to full sun for long periods of time?
3. What conclusion can we draw from the parallel between photosynthetic response curves in plants and functional response curves of animals?

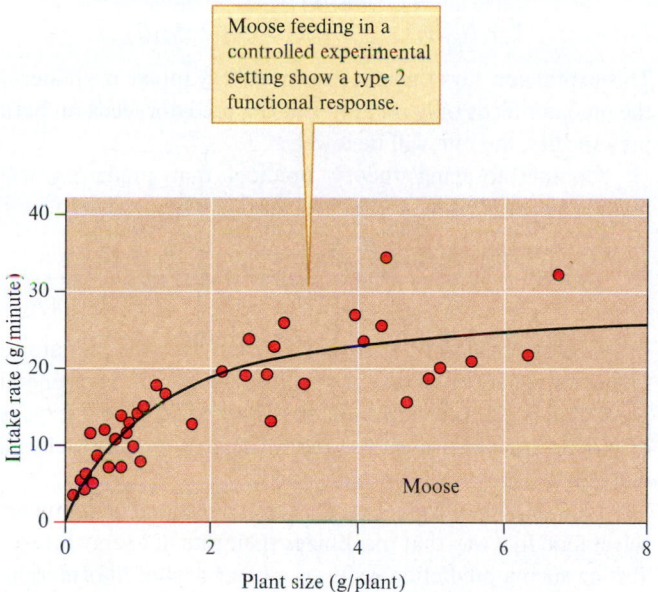

Moose feeding in a controlled experimental setting show a type 2 functional response.

Figure 7.23   A functional response by moose (data from Gross et al. 1993).

7.5   Optimal Foraging Theory

LEARNING OUTCOMES

After studying this section you should be able to do the following:

7.15   Define optimal foraging theory and optimization.

7.16   Critique the match between a model for predicting the diet breadth of a predator and the actual diet of bluegill sunfish.

7.17   Compare optimal foraging by plants and animals.

Optimal foraging theory models feeding behavior as an optimizing process.   Evolutionary ecologists predict that if organisms have limited access to energy, natural selection is likely to favor individuals within a population that are more effective at acquiring energy. This prediction spawned an area of ecological inquiry called optimal foraging theory. **Optimal foraging theory** assumes that if energy supplies are limited, organisms cannot simultaneously maximize all of life's functions; for example, allocation of energy to one function, such as growth or reproduction, reduces the amount of energy available to other functions, such as defense. As a consequence, there must be compromises between competing demands. This seemingly inevitable conflict between energy allocations is called the **principle of allocation,** which we introduced in chapter 5 (see section 5.2).

Optimal foraging theory attempts to model how organisms feed as an optimizing process, a process that maximizes or minimizes some quantity. In some situations, the environment may favor individuals that assimilate energy or nutrients at a high rate (e.g., some filter-feeding zooplankton and short-lived weedy annual plants growing in disturbed habitats). In other situations, selection for minimum water loss appears much stronger (e.g., cactus and scorpions in the desert). Optimal foraging theory attempts to predict what consumers will eat, and when and where they will feed. Early work in this area concentrated on animal behavior. More recently the acquisition of energy and nutrients by plants has been modeled, using ideas borrowed from economic theory.

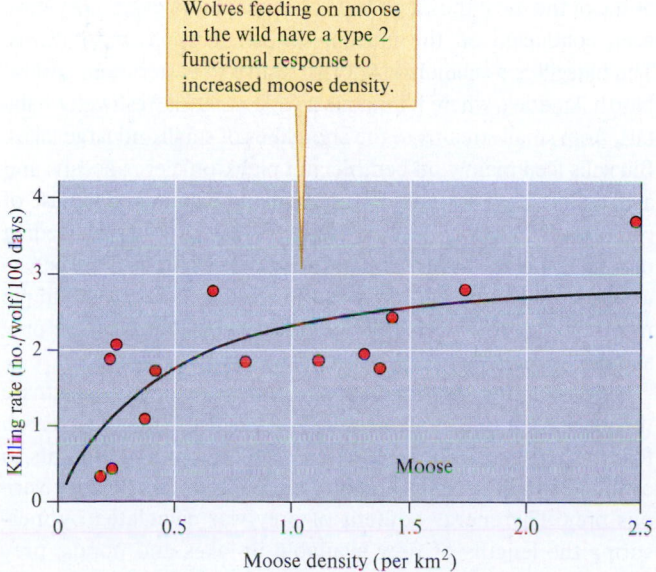

Wolves feeding on moose in the wild have a type 2 functional response to increased moose density.

Figure 7.24   Wolf functional response (data from Messier 1994).

Testing Optimal Foraging Theory

How can you test optimal foraging theory? Unfortunately, you cannot test this theory, or any other complex theory, directly in one grand experiment. Consequently, researchers chip away at the problem by testing specific predictions of the theory. One of the most productive avenues of research has been to use optimal foraging theory to predict the composition of animal diets.

When ecologists consider potential prey for a consumer, they try to identify the prey attributes that may affect the rate of energy intake by the predator. One of the most important factors is the abundance of a potential food item. All things being equal, a more abundant prey item yields a larger energy return than an uncommon prey. In optimal foraging studies, prey abundance is generally expressed as the number of the prey encountered by the predator per unit of time, N_e. Another prey attribute is the amount of energy, or costs, expended by the predator while searching for prey, C_s. A third characteristic of potential prey that could affect the energy return to the predator is the time spent processing prey in activities such as cracking shells, fighting, removing noxious scent glands, and so forth. Time spent in activities such as these are summarized as handling time, H. Ecologists ask, given the searching and handling capabilities of an animal and a certain array of available prey, do animals select their diet in a way that yields the maximum rate of energy intake? We can rephrase this question mathematically by incorporating the terms for prey encounter rate, N_e, searching costs, C_s, and handling time, H, into a model.

A Model for Diet Breadth

One of the most basic questions that we might ask about feeding by a predator concerns the number of prey items that should be included in its diet. Put another way, what mix of prey will maximize energy intake by a predator feeding under a particular set of circumstances? Early theoretical work on this question was published by MacArthur and Pianka (1966), Charnov (1973), and several others. We can represent the rate of energy intake of a predator as E/T, where E is energy and T is time. Earl Werner and Gary Mittelbach (1981) modeled the rate of energy intake for a predator feeding on a single prey species as follows:

$$\frac{E}{T} = \frac{N_{e1}E_1 - C_s}{1 + N_{e1}H_1}$$

In this equation, N_{e1} is the number of prey 1 encountered per unit of time. E_1 is the net energy gained by feeding on an individual of prey 1. C_s is the cost of searching for the prey. H_1 is the time required for "handling" an individual of prey 1. Once again, this equation expresses the net rate at which a predator takes in energy when it feeds on a particular prey species.

What would be the rate of energy intake if the predator fed on two types of prey? That rate is calculated as:

$$\frac{E}{T} = \frac{(N_{e1}E_1 - C_s) + (N_{e2}E_2 - C_s)}{1 + N_{e1}H_1 + N_{e2}H_2}$$

This is an extension of the first equation. Here, we've added encounter rates for prey 2, N_{e2}, the energetic return from feeding on prey 2, E_2, and the handling time for prey 2, H_2. The searching costs, C_s, are assumed to be the same for prey 1 and prey 2.

The rate of energy intake by a predator feeding on several prey can be represented as:

$$\frac{E}{T} = \frac{\sum_{i=1}^{n} N_{ei}E_i - C_s}{1 + \sum_{i=1}^{n} N_{ei}H_i}$$

Here, $\sum$ means "the sum of" and i equals 1, 2, 3, etc., to n, where n is the total number of prey. Remember that this equation gives an estimate of the rate of energy intake. The question that optimal foraging theory asks is whether organisms feed in a way that maximizes the rate of energy intake, E/T.

Optimal foraging theory predicts that a predator will feed exclusively on prey 1, ignoring other available prey, when:

$$\frac{N_{e1}E_1 - C_s}{1 + N_{e1}H_1} > \frac{(N_{e1}E_1 - C_s) + (N_{e2}E_2 - C_s)}{1 + N_{e1}H_1 + N_{e2}H_2}$$

This expression says that the rate of energy intake is greater if the predator feeds only on prey 1. If the predator feeds on both prey species, the rate will be lower.

Optimal foraging theory predicts that predators will include a second prey species in their diet when:

$$\frac{(N_{e1}E_1 - C_s) + (N_{e2}E_2 - C_s)}{1 + N_{e1}H_1 + N_{e2}H_2} > \frac{N_{e1}E_1 - C_s}{1 + N_{e1}H_1}$$

In this case, feeding on two prey species gives the predator a higher rate of energy intake than if it feeds on one. The general prediction is that predators will continue to add different types of prey to their diet until the rate of energy intake reaches a maximum. This is called **optimization.**

Now let's get back to our basic question: Do animals select food in a way that maximizes their rate of energy intake? Testing such a prediction requires a great deal of information. Fortunately, mathematical models such as this one help focus experiments and observations on a few key variables.

Foraging by Bluegill Sunfish

Some of the most thorough tests of optimal foraging theory have been conducted on the bluegill sunfish, *Lepomis macrochirus*. The bluegill is a medium-sized fish native to eastern and central North America, where it inhabits a wide range of freshwater habitats, from small streams to the shorelines of small and large lakes. Bluegills feed mainly on benthic and planktonic crustaceans and aquatic insects, prey that differ in size and habitat and in ease of capture and handling. Bluegills often choose prey by size, feeding on organisms of certain sizes and ignoring others. This behavior is convenient because it gives the ecologist a relatively simple measure, size, to describe the composition of the available prey and the composition of the theoretically optimal diet.

Werner and Mittelbach used published studies to estimate the amount of energy expended by bluegills while they search for (C_s) and handle prey. They used laboratory experiments to estimate handling times (H) and encounter rates (N_e) for various prey. The energy content of prey was calculated by measuring the lengths of prey available in lakes and ponds; prey length was converted to mass, and then mass was converted to energy content using published values.

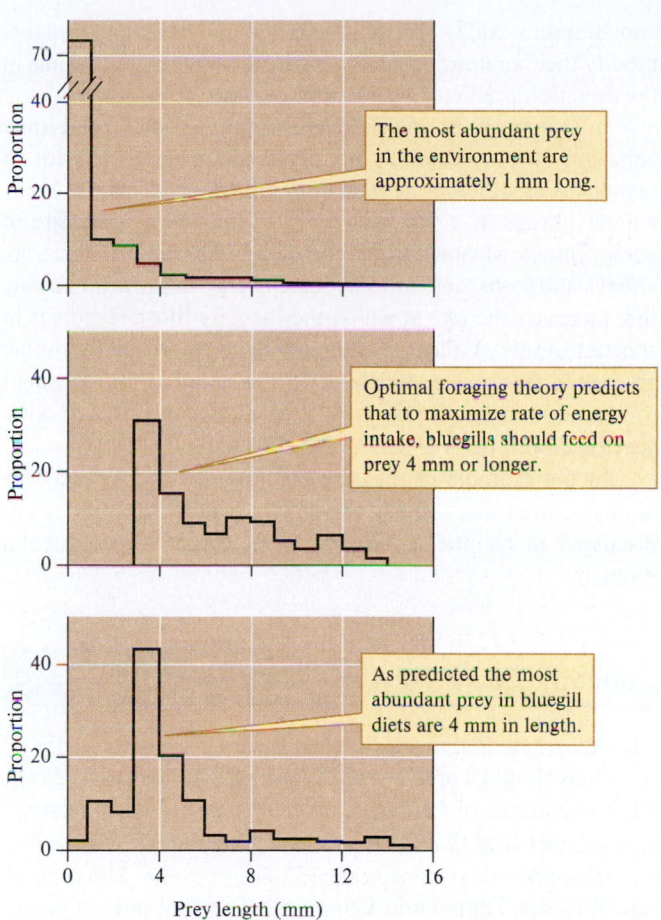

Figure 7.25 Optimal foraging theory predicts composition of bluegill sunfish diets (data from Werner and Mittelbach 1981).

The upper graph in figure 7.25 shows the size distribution of potential prey for bluegills in vegetation in Lawrence Lake, Michigan. The middle graph shows the composition of the optimal diet as predicted by the optimal foraging model just presented. Finally, the bottom graph shows the actual composition of bluegill diets in Lawrence Lake. Bluegills feeding in vegetation selected prey that were uncommon and larger than average. The optimal diet matches well with the actual diets of bluegills in Lawrence Lake. A similar match was obtained for bluegills feeding on zooplankton in open water.

Werner and Mittelbach found that optimal foraging theory provides reasonable predictions of prey selection by natural populations of bluegills. Ecologists studying plants have developed an analogous predictive framework for foraging by plants.

Optimal Foraging by Plants

How do plants "forage"? What animals do with behavior, plants do with growth. Plants forage by growing and orienting structures that capture either energy or nutrients. They grow leaves or other green surfaces to capture light and roots to capture nutrients. Terrestrial plants harvest energy from sunlight aboveground and nutrients and water from soil. Because of the structure of their environment and the distribution of their resources, plants forage in two directions at once. Like

animals, however, plants face limited supplies of energy and nutrients and so face the prospect of compromises between competing demands for energy. Allocation of energy to leaves and stems reduces the amount of energy available for root growth. Increased allocation to root growth reduces energy available for leaves and stems.

In some environments, such as deserts, plants have access to an abundance of light but face shortages of water. In other environments, such as in the understory of temperate forests, there is little light but the soil may be rich in moisture and nutrients. In the face of such environmental heterogeneity, how do plants invest their energy? Using economic theory, Arnold Bloom and his colleagues (1985) suggested that plants adjust their allocation of energy to growth in such a way that all resources are equally limited. They predicted that plants in environments with abundant nutrients but little light, would invest more energy in the growth of stems and leaves and less in roots to match their supply of energy to the supply of nutrients. They predicted that in environments rich in light but poor in nutrients, plants would invest more in roots.

The predictions of this economically based model have been supported by numerous studies showing that plants in light-poor environments invest more aboveground, while plants from nutrient-poor environments invest more belowground. So, it appears that plants allocate energy for growth to those structures that gather the resources that most limit growth in a particular environment. These patterns are consistent with concepts developed by Justus Liebig in the midnineteenth century (see chapter 18, section 18.1). Some of the most revealing tests of these predictions come from studies of plants growing along gradients of nutrient availability.

Experimental studies show that the same plant species grown in nutrient-poor soils often develop a higher ratio of root biomass to shoot biomass, the so-called root:shoot ratio, than when grown on nutrient-rich soils. For example, when H. Setälä and V. Huhta (1991) grew birch tree seedlings in boreal forest soils of low and high nitrogen content, those in the nitrogen-poor soils developed higher root:shoot ratios (fig. 7.26).

David Tilman and M. Cowan (1989) obtained similar results when they grew four species of grass and four species of forbs on soils of different nitrogen content. They created a nitrogen gradient by mixing three different soils in different proportions. The soils were a subsoil (B horizon, see fig. 2.9) containing approximately 25 mg of nitrogen per kilogram of soil, a topsoil (A horizon) with 350 mg of nitrogen per kilogram of soil, and a black loam topsoil (A horizon) from a nearby site containing 5,000 mg of nitrogen per kilogram of soil. These soils were mixed to produce seven levels of soil nitrogen ranging from about 125 to 1,800 mg N per kilogram of soil. Several other nutrients were added to the experimental soils so that other nutrients would not limit growth of the experimental plants.

Tilman and Cowan conducted their growth experiments in 504 flowerpots that were 30 cm wide by 30 cm deep. They filled several pots with each of their seven soil mixtures and grew each of the eight study species from seed at high densities, 100 plants per pot, and low densities, 7 plants per pot. In each of the soil types, six pots of each species were planted at low density and three pots of each species at high density.

In general, the study species had lower root:shoot ratios when grown with more nitrogen. Figure 7.27 shows the pattern for the grass *Sorghastrum nutans* when grown at high density. Like the other species, *S. nutans* reduced its root biomass and increased its shoot biomass in the presence of higher nitrogen availability. In a study of plants of the Great Plains of North America, Nichole Levang-Brilz and Mario Biondini found that 62% of the 55 plant species in their study increased root:shoot ratios in the face of reduced nitrogen availability (Levang-Brilz

and Biondini 2002). These results indicate that many plants modify their anatomy to match environmental circumstance in the direction predicted by economic theory.

In summary, ecologists, using simple models of feeding behavior, have successfully predicted the feeding behavior of organisms as different as predatory animals and plants. Many animals forage in a way that tends to maximize their rate of energy intake. Meanwhile, plants, from birch trees to temperate grasses and forbs, appear to allocate energy to growth in a way that increases the rate at which they acquire those resources in shortest supply. While the optimal foraging approach to trophic ecology is far from complete, it is a substantial improvement over a body of knowledge consisting of a long list of species-specific descriptions of diet and feeding habits.

In the example in the Applications section, we see how biologists are using some of the patterns and concepts we have discussed in chapter 7 to address important environmental problems.

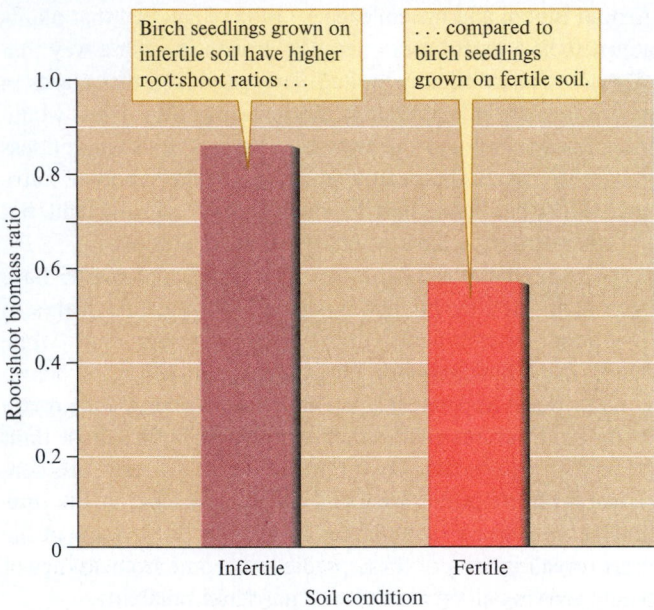

Figure 7.26   Soil fertility and ratio of root biomass to shoot biomass (data from Setälä and Huhta 1991).

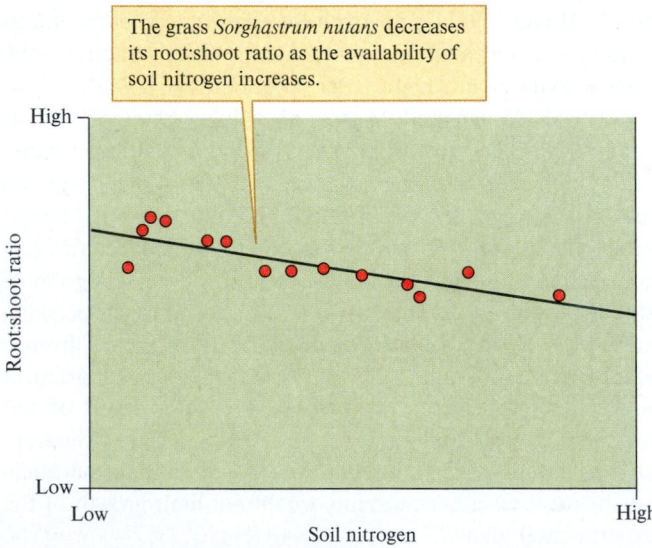

Figure 7.27   The root:shoot ratios of a grass species change along a gradient of nitrogen availability (data from Tilman and Cowan 1989).

Concept 7.5 Review

1. According to optimal foraging theory, under what conditions should a predator add a new prey species to its diet?
2. Do patterns of feeding by bluegills (see fig. 7.25) include any evidence that these consumers ignore certain potential prey?
3. Why did Tilman and Cowan plant several pots of each species in each of their growing conditions?

Applications

Bioremediation–Using the Trophic Diversity of Bacteria to Solve Environmental Problems

LEARNING OUTCOMES

After studying this section you should be able to do the following:

7.18   Describe research demonstrating the usefulness of bacteria for cleaning up pollutants.

7.19   Explain how ecological stoichiometry can be useful to bioremediation.

Imagine yourself in the center of a densely populated region with thousands of leaky gasoline tanks or complex mine wastes contaminating the groundwater. How would you solve these environmental problems? Where would you turn for help? Increasingly, we are turning to nature's own cleanup crew, the bacteria. Environmental managers are taking advantage of the exceptional trophic diversity of bacteria to perform a host of environmental chores.

Leaking Underground Storage Tanks

Gasoline and other petroleum derivatives are stored in underground storage tanks all over the planet. Those that leak are a serious source of pollution. Maribeth Watwood and Cliff Dahm (1992) explored the possibility of using bacteria to clean up soils and aquifers contaminated by leaking storage tanks. The first step in their work was to determine if there are naturally occurring populations of bacteria that can break down complex petroleum derivatives such as benzene.

Watwood and Dahm collected sediments from a shallow aquifer that contained approximately 8.5×10^8 bacterial cells per gram of wet sediment. Of these, 6.55×10^4 bacterial cells per milliliter were capable of living on benzene as their only source of carbon and energy. By exposing sediments from the aquifer to benzene for 6 months, the researchers increased the populations of benzene-degrading bacteria approximately 100 times.

How rapidly can these bacteria break down benzene? Watwood and Dahm found that with no prior exposure, bacterial populations could break down 90% of the benzene in their test flasks within 40 days (fig. 7.28). Exposing sediments to benzene prior to their tests increased the rate of breakdown.

Briefly, this study demonstrated that naturally occurring populations of bacteria can rapidly break down benzene leaking from underground storage tanks. This study suggests that these bacteria will eventually clean up the organic contaminants from leaking gasoline storage tanks without manipulation of the environment. However, in the next example, environmental managers found that they had to manipulate the environment to stimulate the desired bacterial cleanup of a contaminant.

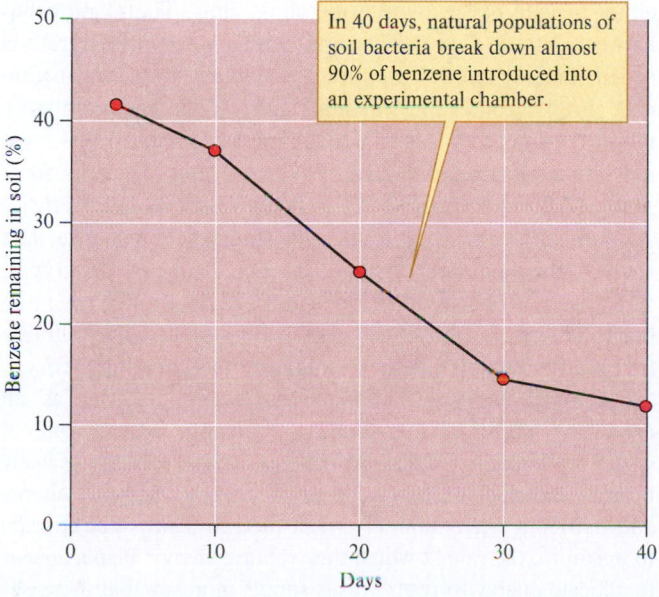

In 40 days, natural populations of soil bacteria break down almost 90% of benzene introduced into an experimental chamber.

Figure 7.28   Benzene breakdown by soil bacteria (data from Watwood and Dahm 1992).

Cyanide and Nitrates in Mine Spoils

Many gold mines were abandoned when they could not be mined profitably with the mining technology of the nineteenth and early twentieth centuries. Then, in the 1970s, techniques were developed to economically extract gold from low-grade ores. One of the main extraction techniques was to leach ore with cyanide (CN^-). Dissolved CN^- forms chemical complexes with gold and other metals. The solution containing gold-bearing CN^- can be collected and the gold and CN^- removed by filtering the solution with activated charcoal.

This new method of mining solved a technical problem but contaminated soils and groundwater. When the leaching process is finished, the leached ore is stored in piles; however, much CN^- remains. Several kinds of bacteria can break down CN^- and produce NH_3. This NH_3 can, in turn, be used by nitrifying bacteria as an energy source, producing NO_3^- (see "The Nitrogen Cycle" in chapter 19, section 19.1). Thus, leaching gold-bearing ores and subsequent microbial activity can contaminate soil and groundwater with CN^-, a deadly poison, and with nitrate, another contaminant.

Carleton White and James Markwiese (1994) studied a gold mine that had been worked with the CN^- leaching process. The leached ores from the mine were gradually releasing CN^- and NO_3^- into the environment. The researchers looked to bacteria to solve this environmental problem. They first documented the presence of CN^- degraders by looking for bacterial growth in a diagnostic medium. This medium contained CN^- as the only source of carbon and nitrogen. Using this growth medium, White and Markwiese estimated that each gram of ore contained approximately 10^3 to 10^5 cells of organisms capable of growing on, and breaking down, CN^-.

The leached ores presented bacteria with a rich source of nitrogen in the form of CN^- and NO_3^- but the ores contained little organic carbon. White and Markwiese predicted that adding a source of carbon to the residual ores would increase the rate at which bacteria break down CN^- and reduce the concentration of NO_3^- in the environment. Why should adding organic molecules rich in carbon increase bacterial use of nitrogen in the environment? Bacteria have a carbon:nitrogen ratio of about 5:1. In other words, growth and reproduction by bacteria require about five carbon atoms for each nitrogen atom.

White and Markwiese tested their ideas in the laboratory. In one experiment, they added enough sucrose to produce a C:N ratio of 10:1 within leached ores. This experiment included two controls, both of which contained leached ores without sucrose. One of the controls was sterilized to kill any bacteria. The other control was left unsterilized.

Bacteria in the treatments containing sucrose broke down all the CN^- within the leached ore in 13 days. Meanwhile, only a small amount of CN^- was broken down in the unsterilized control and no CN^- was broken down in the sterilized control (fig. 7.29). Why did the researchers include a sterilized control? The sterilized control demonstrated that nonbiological processes were not responsible for the observed breakdown of CN^-.

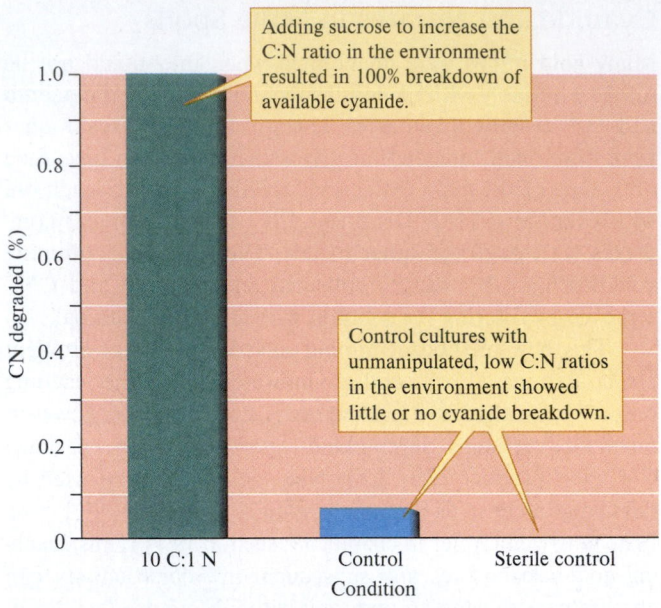

Figure 7.29 Manipulating C:N ratios to stimulate breakdown of cyanide (CN^-) (data from White and Markwiese 1994).

Figure 7.29 shows that adding sucrose to the residual ore stimulates the breakdown of CN^-. However, remember that this process ultimately leads to the production of NO_3^-. Does adding sucrose to eliminate CN^- lead to the buildup of NO_3^-, trading one pollution problem for another? No, it does not. In another experiment, White and Markwiese showed that adding sucrose also stimulates uptake of NO_3^- by heterotrophic bacteria and fungi. These organisms use organic molecules, in this case sucrose, as a source of energy and carbon and NO_3^- as a source of nitrogen. The nitrogen taken up by bacteria and fungi becomes incorporated in biomass as complex organic molecules. Nitrogen in this form is recycled within the microbial community and is not a source of environmental pollution.

White and Markwiese recommended that sucrose be added to leached gold-mining ores to stimulate breakdown of CN^- and uptake of NO_3^- by bacteria. This environmental cleanup project was successful because the researchers were thoroughly familiar with the energy and nutrient relations of bacteria and fungi. Another key to the project's success was the great trophic diversity of bacteria. Bacteria will likely continue to play a great role as we address some of our most vexing environmental problems.

Summary

Photosynthetic autotrophs synthesize organic molecules using CO_2 as a source of carbon and light as an energy source. Photosynthetic plants and algae use CO_2 as a source of carbon, and light, of wavelengths between 400 and 700 nm, as a source of energy. Light within this band, which is called photosynthetically active radiation, or PAR, accounts for about 45% of the total energy content of the solar spectrum at sea level. PAR can be quantified as photosynthetic photon flux density, generally reported as μmol per square meter per second. Among plants, there are three major alternative photosynthetic pathways, C_3, C_4, and CAM. C_4 and CAM plants are more efficient in their use of water than are C_3 plants.

Chemosynthetic autotrophs synthesize organic molecules using CO_2 as a carbon source and inorganic molecules as an energy source. They consist of a highly diverse group of chemosynthetic bacteria and archaea. Prokaryotes are the most trophically diverse organisms in the biosphere.

Heterotrophic organisms use organic molecules both as a source of carbon and as an energy source. Herbivores, carnivores, and detritivores face fundamentally different trophic problems. Herbivores feed on plant tissues, which often contain a great deal of carbon but little nitrogen. Herbivores must also overcome the physical and chemical defenses of plants. Detritivores feed on dead plant material, which is even lower in nitrogen than living plant tissues. The study of the balance of multiple chemical elements in ecological interactions, for example, in trophic interactions, is the subject of the field of ecology known as ecological stoichiometry. Carnivores consume prey that are nutritionally rich but also energetically expensive to

obtain. In general, carnivores eliminate poorly defended individuals, leaving well-defended individuals to reproduce and pass on their traits. This leads to the evolution of improved prey defenses over time.

The rate at which organisms can take in energy is limited, either by external or by internal constraints. The relationship between photon flux density and plant photosynthetic rate is called photosynthetic response. Herbs and short-lived perennial shrubs from sunny habitats have high maximum photosynthetic rates that level off at high irradiance. The lowest maximum rates of photosynthesis occur among plants from shady environments. The relationship between food density and animal feeding rate is called the functional response. The shape of the functional response is generally one of three types. The forms of photosynthetic response curves and type 2 animal functional responses are remarkably similar. Energy limitation is a fundamental assumption of optimal foraging theory.

Optimal foraging theory models feeding behavior as an optimizing process. Evolutionary ecologists predict that if organisms have limited access to energy, natural selection is likely to favor individuals that are more effective at acquiring energy and nutrients. Many animals select food in a way that appears to maximize the rate at which they capture energy. Plants appear to allocate energy to roots versus shoots in a way that increases their rate of intake of the resources that limit their growth. Plants in environments with abundant nutrients but little light tend to invest more energy in the growth of stems and leaves and less in roots. In environments rich in light but poor in nutrients, plants tend to invest more energy in the growth of roots.

The trophic diversity of bacteria, which is critical to the health of the biosphere, can also be used as a tool to address some of our most challenging waste disposal problems. Bacteria can be used to clean up soils and aquifers polluted by petroleum products such as benzene, and to eliminate the pollution caused by some kinds of mine waste. The success of these projects requires that ecologists understand the energy and nutrient relations of bacteria. Bacteria will likely continue to play a great role as we address some of our most vexing environmental problems.

Key Terms

archaea   150
autotroph   150
Batesian mimicry   160
bundle sheath   152
C_3 photosynthesis   151
C_4 photosynthesis   152
CAM photosynthesis   153
carnivore   156

chemosynthetic
    autotroph   150
detritivore   156
ecological
    stoichiometry   157
functional response   164
herbivore   156
heterotroph   150
irradiance   163
I_{sat} 163

Müllerian mimicry   160
net photosynthesis   163
optimal foraging theory   165
optimization   166
organic compounds   150
photon flux density   151
photorespiration   152
photosynthetic
    autotroph   150

photosynthetically active
    radiation (PAR)   151
P_{max} 163
principle of allocation   165
prokaryote   150
rhodopsin   150
rubisco   151
size-selective predation   161
trophic (feeding)
    biology   150

Review Questions

1. Why don't plants use highly energetic ultraviolet light for photosynthesis? Would it be impossible to evolve a photosynthetic system that uses ultraviolet light? Does the fact that many insects see ultraviolet light change your mind? Would it be possible to use infrared light for photosynthesis? (Hint: Photosynthetic bacteria tap into the near infrared range.)

2. In what kinds of environments would you expect to find the greatest predominance of C_3, C_4, or CAM plants? How can you explain the co-occurrence of two, or even all three, of these types of plants in one area? (Hint: Think about the variations in microclimate that we considered in chapters 5 and 6.)

3. In chapter 7, we emphasized how the C_4 photosynthetic pathway saves water, but some researchers suggest that the greatest advantage of C_4 over C_3 plants occurs when CO_2 concentrations are low. What is the advantage of the C_4 pathway when CO_2 concentrations are low?

4. What are the relative advantages and disadvantages of being an herbivore, a detritivore, or a carnivore? What kinds of organisms were left out of our discussions of herbivores, detritivores, and carnivores? Where do parasites fit? Where does *Homo sapiens* fit?

5. What advantage does advertising give to noxious prey? How would convergence in aposematic coloration among several species of Müllerian mimics contribute to the fitness of *individuals* in each species? In the case of Batesian mimicry, what are the costs and benefits of mimicry to the model and to the mimic?

6. Design a planetary ecosystem based entirely on chemosynthesis. You might choose an undiscovered planet of some distant star or one of the planets in our own solar system, either today or at some distant time in the past or future.

7. What kinds of animals would you expect to have type 1, 2, or 3 functional responses? How should natural selection for better prey defense affect the height of functional response curves? How should natural selection for more effective predators affect the height of the curves?

8. The rivers of central Portugal have been invaded, and densely populated by the Louisiana crayfish, *Procambarus clarki*, which looks like a freshwater lobster about 12 to 14 cm long. The otters of these rivers, which were studied by Graça and Ferrand de Almeida (1983), can easily catch and subdue these crayfish. Using the model for prey choice:

$$\frac{E}{T} = \frac{\sum_{i=1}^{n} N_{ei}E_i - C_s}{1 + \sum_{i=1}^{n} N_{ei}H_i}$$

explain under what conditions the diets of the otters of central Portugal would shift from the highly diverse menu shown in figure 7.17, which included fish, frogs, water snakes, birds, and insects, to a diet dominated by crayfish.

9. How is plant allocation to roots versus shoots similar to plant regulation of temperature and water? (We discussed these topics in chapters 5 and 6.) Consider discussing these processes under the more general heading of homeostasis. (Hint: Homeostasis is the maintenance of a relatively constant internal environment.)

Design elements: (Yellowstone thermal pool):©flickrRF/Getty Images

Natalia Kuzmina/Alamy Stock Photo

Chapter 8

Social Relations

Why do bull elk, *Cervus elaphus,* grow large antlers each year whereas cow elk do not? Darwin puzzled over the evolution of male ornaments, such as antlers, concluding that the behaviors associated with selection of mates by females and male competition for mates were responsible.

CHAPTER CONCEPTS

8.1 The effects of female mate choice on evolution of ornamentation in males can be reduced by other sources of natural selection.   174

Concept 8.1 Review   178

8.2 Females of some species select mates based on the male's ability to provide important resources.   178

Concept 8.2 Review   181

8.3 Mating in wild plant populations can be nonrandom.   181

Concept 8.3 Review   183

8.4 Evolution of sociality in many species appears driven by the need for group defense of high-quality territories and/or defense of mates and young.   183

Concept 8.4 Review   188

8.5 Kin selection and ecological constraints may have played key roles in the evolution of eusociality.   189

Concept 8.5 Review   192

Applications: Behavioral Ecology and Conservation   193

Summary   194

Key Terms   195

Review Questions   195

LEARNING OUTCOMES

After studying this section you should be able to do the following:

8.1 Define behavioral ecology, sociobiology, female, male, and hermaphrodite.

8.2 Describe some of the variety of social systems found on a coral reef.

8.3 Explain how social relations can influence evolutionary fitness.

Nowhere are the social interactions and other behaviors of animals easier to observe than on tropical reefs. In early evening as the sun's rays shine obliquely through the clear waters over a coral reef, the activity of some of its inhabitants quickens. As if activated by some remote switch, a vast school of fish that had remained in the lagoon all day begins to move steadily toward an opening in the reef. The school is leaving the lagoon's protection and going out to the open ocean for a night of feeding. Living in a school appears to have favored uniformity among its members. Approached underwater, the edge of the school looks like a giant, translucent curtain stamped with the silhouettes of thousands of identical fish. Their coloration, countershaded dark above and silvery below, their similar size, highly coordinated movements, and great numbers give the fish within the school some protection from predators. Though seabirds and predaceous fish ambush the school as it makes its way, the schooling fish are so numerous and their individual movements so difficult to follow that only a small proportion of them are eaten. Gradually the school, moving like a gigantic, shape-shifting organism, passes through the channel connecting the lagoon to the open ocean. The school of fish will be back by daybreak only to repeat its seaward journey next evening in a cycle of comings and goings that helps mark the rhythm of life on the reef.

Meanwhile, along the reef, damselfish are distributed singly on territories. The damselfish retain exclusive possession of their territorial patches of coral rubble, living coral, and sand by patrolling the boundaries and driving off any fish attempting to intrude, especially other damselfish that would take their territory or other fish that would prey on eggs or consume food within the territory. Each day at this time, however, some territory-holding males are joined by females. For the space of time that they court and deposit eggs and sperm on the nest site prepared by the male, the territory contains two fish. Once mating is complete, however, the male is again alone on the territory, guarding the food and shelter contained within its boundaries as well as the newly deposited eggs that he fertilized minutes before.

Higher along the reef face a male bluehead wrasse mates with a member of the harem of females that live within his territory (fig. 8.1). In contrast to the male with his blue head, black bars, and green body, the female is mostly yellow with a large, black spot on her dorsal fin. As the male bluehead extrudes sperm to fertilize the eggs laid by the female, small males, similar in color to the female, streak by the mating pair, discharging a

Figure 8.1   Bluehead wrasse males with yellow females of the species. If the bluehead male is removed from a territory, the largest female in the territory can change to a fully functional bluehead male within days.
Gregory G. Dimijian, M.D./Science Source

cloud of sperm as they do. Some of the female's eggs will be fertilized by the large territorial bluehead male, while others will be fertilized by the sperm discharged by the smaller yellow streakers. In addition to differences in color and courtship behavior, bluehead and yellow males have distinctive histories. While the yellow males began their lives as males, the bluehead male began life as a female and transformed to a male only when the local bluehead male was eaten by a predator or met some other end. At that point, because she was the largest yellow phase among the local females and males, she was in line to become the dominant local male and so changed from the yellow to the bluehead form of the species. Within a week the former female was producing sperm and fertilizing the eggs produced by the females in the territory.

While male bluehead wrasses patrol their individual mating territories and male damselfish fight with each other at the boundaries of theirs, elsewhere on the reef groups of snapping shrimp live cooperatively in colonies that may contain over 300 individuals. Most of the individuals in the colonies are juveniles or males along with a single reproductive female. The female snapping shrimp, which plays a role much like the queen ant in an ant colony, breeds continuously and so is easily identified by her ripe ovaries or by the eggs she carries. Meanwhile, the males of the colony, most of which will probably never mate, vigorously defend the nest site, with its "queen" shrimp and numerous juveniles, against intruders. In this shrimp society most males serve the colony and its queen by protecting her offspring and the sponge where they live. While the queen reproduces profusely, the chance to reproduce is probably rare for an individual male. The colony thrives but reproduction is restricted to a few individuals in the population.

The study of social relations is the territory of **behavioral ecology,** which concentrates on relationships between organisms and environment that are mediated by behavior. In the case of social relations, other individuals of a species are the part of the environment of particular interest. A branch of biology concerned with the study of social relations is **sociobiology.** Social relations, from dominance relationships and reproductive interactions to cooperative behaviors, are important, since they often directly impact the reproductive contribution of individuals to future generations, a key component of Darwinian or evolutionary fitness. Fitness, which can be defined as the number of offspring, or genes, contributed by an individual to future generations, can be substantially influenced by social relations within a population.

One of the most fundamental social interactions between individuals takes place during sexual reproduction. The timing of those interactions and their nature are strongly influenced by the reproductive system of a species. The behavioral ecologist considers several factors. Does the population engage in sexual reproduction? Are the sexes separate? How are the sexes distributed among individuals? Are there several forms of one sex or the other? Questions such as these have drawn the attention of biologists since Darwin (1862), who wrote, "We do not even in the least know the final cause of sexuality; why new beings should be produced by the union of the two sexual elements, instead of by a process of parthenogenesis [production of offspring from unfertilized eggs] . . . The whole subject is as yet hidden in darkness." As you will see, behavioral and evolutionary ecologists have learned a great deal about the evolution and ecology of reproduction in the nearly one and a half centuries since Darwin published this statement. However, much remains to be discovered.

Since mammals and birds reproduce sexually, from a human perspective sexual reproduction may appear the norm. However, asexual reproduction is common among many groups of organisms such as bacteria, protozoans, plants, and some vertebrates. But, most described species of plants and animals include male and female functions, sometimes in separate individuals or within the same individual. This brings us to a fundamental question in biology: What is female and what is male? From a biological perspective, the answer is simple. **Females** produce larger, more energetically costly gametes (eggs or ova), whereas **males** produce smaller, less-costly gametes (sperm or pollen). Because of the greater energetic cost of producing their gametes, female reproduction is thought to be generally limited by access to the necessary resources. In contrast, male reproduction is generally limited by access to female mates. Biologists long ago proposed that this difference in investment in gametes has usually led to a fundamental dichotomy between actively courting males and highly selective females.

Despite the basic differences between males and females, distinguishing the two sexes in nature is sometimes difficult. While it is easy to distinguish between males and females in species where males and females differ substantially in external morphology, the males and females of other species appear very similar and are very difficult to distinguish using only external anatomy. Still other species are **hermaphrodites,** organisms that combine male and female function in the same individual (fig. 8.2). The most familiar examples of hermaphrodites are plants, among which the vast majority of species produce flowers that have both male and female parts.

Clearly, the way populations are divided between the sexes will influence social relations, which will in turn affect the fitness of individuals, particularly through influences on their reproductive rates. Studies of social relations provide numerous examples of the complex relationships between social interactions, such as how individuals choose mates, and fitness.

(a)

(b)

Figure 8.2   Male and female function: (*a*) male and female Canada geese, a species in which males and females have very similar external anatomy (i.e., are monomorphic); (*b*) a "perfect" flower, which includes both male (stamens) and female (pistil) parts and function.
(*a*) dgm photography/Alamy; (*b*) Steven P. Lynch

8.1 Mate Choice versus Predation

LEARNING OUTCOMES

After studying this section you should be able to do the following:

8.4   Describe sexual selection.

8.5   Contrast intrasexual selection and intersexual selection.

8.6   Outline field and laboratory experiments designed to test the influences of mate choice and predation in determining colorfulness of male guppies.

8.7   Discuss the roles of mate choice and predation in determining colorfulness of male guppies.

The effects of female mate choice on evolution of ornamentation in males can be reduced by other sources of natural selection.   Darwin (1871) proposed that the social environment, particularly the mating environment, could exert significant influence on the characteristics of organisms. He was particularly intrigued by the existence of what he called "secondary sexual characteristics," the origins of which he could not explain except by the advantages they gave to individuals during competition for mates. Darwin used the term *secondary sexual characteristics* to mean characteristics of males or females not directly involved in the process of reproduction. Some of the traits that Darwin had in mind were "gaudy colors and various ornaments . . . the power of song and other such characters." How do we explain the existence of characteristics such as the antlers of male deer, the bright peacock's tail, or the gigantic size and large nose of the male elephant seal? In order to explain the existence of such secondary sexual characteristics, Darwin proposed a process that he called sexual selection. **Sexual selection** results from differences in reproductive rates among individuals as a result of differences in their mating success.

Sexual selection is thought to be important under two circumstances. The first is where individuals of one sex compete among themselves for mates, which results in a process called **intrasexual selection.** For instance, when male mountain sheep or elephant seals fight among themselves for dominance or mating territories, the largest and strongest generally win such contests. In such situations, the result is often selection for larger body size and more effective weapons such as horns or teeth. Since this selection is the result of contests within one sex, it is called intrasexual selection.

Sexual selection can also occur when members of one sex consistently choose mates from among members of the opposite sex on the basis of some particular trait. Because two sexes are involved, this form is called **intersexual selection.** Examples of traits used for mate selection include female birds choosing among potential male mates based on the brightness of their feather colors or on the quality of their songs. Darwin proposed that once individuals of one sex begin to choose mates on the basis of some anatomical or behavioral trait, sexual selection would favor elaboration of the trait. For instance, the plumage of male birds' color might become brighter over time or their songs more elaborate, or both (fig. 8.3).

However, how much can sexual selection elaborate a trait before males in the population begin to suffer higher mortality due to other sources of natural selection? Darwin proposed that sexual selection will continue to elaborate a trait until balanced by other sources of natural selection, such as predation. Since Darwin's early work on the subject, research has revealed a great deal about how organisms choose mates and the basis of sexual selection. An excellent model for such studies is the guppy, *Poecilia reticulata.*

Mate Choice and Sexual Selection in Guppies

It would be difficult for experimental ecologists interested in mate choice and sexual selection to design a better experimental animal than the guppy (fig. 8.4). Guppies are native to the streams and rivers of Trinidad and Tobago, islands in the southeastern Caribbean, and in the rivers draining nearby parts of the South American mainland. The waters inhabited by guppies range from small, clear mountain streams to murky, lowland

rivers. Along this gradient of physical conditions, guppies also encounter a broad range of biological situations. In the headwaters of streams above waterfalls, guppies live in the absence of predaceous fish or with the killifish *Rivulus hartii,* which preys mainly on juveniles and is not a very effective predator on adult guppies. In contrast, guppies in lowland rivers live with a wide variety of predaceous fish, including the pike cichlid, *Crenicichla alta,* a very effective visual predator of adult guppies.

Male guppies show a broad range of coloration both within and among populations. What factors may produce this range of variation? It turns out that female guppies, if given a choice, will mate with more brightly colored and behaviorally dominant males (Kodric-Brown 1993, Houde 1997). Why would a female guppy prefer to mate with such males? The characteristics associated with male mating success among guppies, brighter coloration and behavioral dominance, reflect a male's genetics, health, and nutritional state. A female choosing to mate with such males may increase her chances of producing offspring with a more robust immune system and better at finding food and at competing with other members of the population. In addition, since we also know that coloration in guppies has a strong genetic basis, the sons of females mating with colorful males would be more likely to be colorful themselves and, therefore, attractive to females. In other words, a female guppy choosing to mate with colorful, socially dominant males would be likely to produce offspring with higher fitness.

However, brightly colored males are attacked more frequently by visual predators. This trade-off between higher mating success by bright males but greater vulnerability to predators provides a mechanistic explanation for variation in male coloration among different habitats. The most brightly colored male guppies are found in populations exposed to few predators, while those exposed to predators, such as the pike cichlid, are less colorful (Endler 1995). Thus, the coloration of male guppies in local populations may be determined by a dynamic interplay between natural selection exerted by predators and by female mate choice.

Figure 8.3 A peacock in full courtship display. Colorful male birds of many different species, such as this peacock, are testimony to the potential of sexual selection by females, choosing mates from among available males, to produce elaborate ornamentation.
Carole_ R/Flickr Flash/Getty Images

Figure 8.4 A colorful aquarium variety of male guppy courting a female: How do mate selection by female guppies and natural selection by predators influence male ornamentation?
Dr. Paul Zahl/Science Source

Experimental conditions

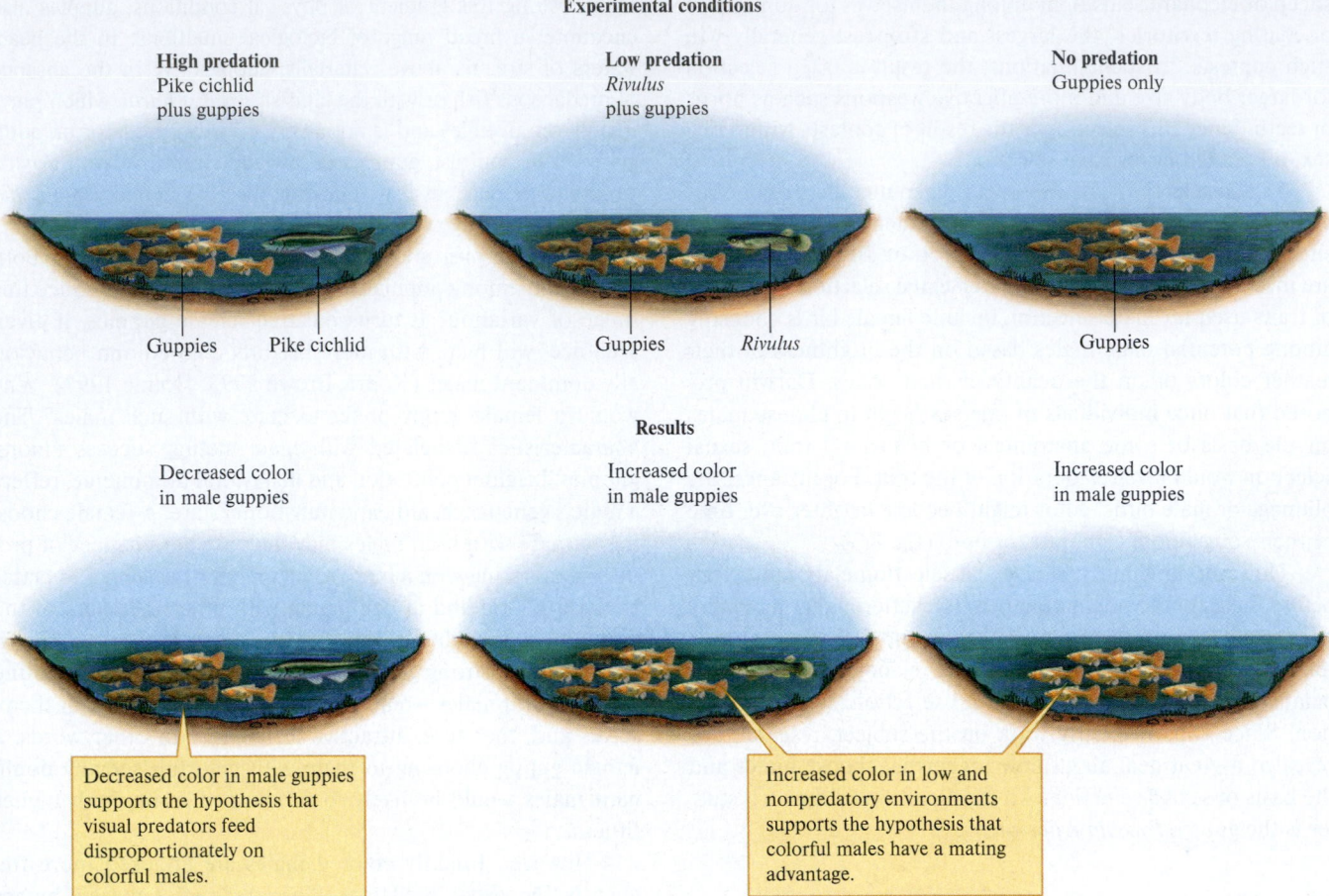

High predation
Pike cichlid
plus guppies

Low predation
Rivulus
plus guppies

No predation
Guppies only

Guppies   Pike cichlid

Guppies   *Rivulus*

Guppies

Results

Decreased color
in male guppies

Increased color
in male guppies

Increased color
in male guppies

Decreased color in male guppies
supports the hypothesis that
visual predators feed
disproportionately on
colorful males.

Increased color in low and
nonpredatory environments
supports the hypothesis that
colorful males have a mating
advantage.

Figure 8.5   Summary of greenhouse experimental design and results (information from Endler 1980).

While field observations are consistent with a trade-off between sexual selection due to mate choice and natural selection due to predation, the evidence would be more convincing with an experimental test. John Endler (1980) performed such a test.

Experimental Tests

Endler did two experiments, one in artificial ponds in a greenhouse at Princeton University (fig. 8.5) and one in the field. For the greenhouse experiments, Endler constructed 10 ponds designed to approximate pools in the streams of the Northern Range in Trinidad. Four of the ponds were of a size (2.4 m × 1.2 m × 40 cm) typical of the pools inhabited by a single pike cichlid in smaller streams. During the final phase of the experiment, Endler placed a single pike cichlid in each of these ponds. The six other ponds were the size (2.4 m × 1.2 m × 15 cm) of headwater stream pools that support an average of six *Rivulus*. Endler eventually placed six *Rivulus* in four of these ponds and maintained the other two ponds with no predators as controls. What did Endler create with this series of pools and predator combinations? These three groups of ponds represented three levels of predation: high predation (pike cichlid), low predation (*Rivulus*), and no predation.

First, Endler established similar physical environments in the pools. He lined all ponds with commercially available dyed gravel, taking care to put the same proportions of gravel colors

in each. The gravel consisted of 31.4% black, 34.2% white, 25.7% green, plus 2.9% each of blue, red, and yellow.

Endler stocked each pond with 200 guppies descended from 18 different populations in Trinidad and Venezuela. By drawing guppies from so many populations, Endler ensured that the experimental populations would include substantial genetic variation for color, an essential requirement for evolutionary change in the study populations.

Endler conducted his second experiment within the drainage network of the Aripo River (fig. 8.6), where he encountered three distinctive situations. Within the mainstream of the Aripo River, guppies coexisted with a wide variety of predators, including pike cichlids, which provided a "high predation" site. Upstream from the high predation site, Endler discovered a small tributary, which flowed over a series of waterfalls near its junction with the mainstream. Because the waterfalls prevented most fish from swimming upstream, this tributary was entirely free of guppies but supported a population of the relatively ineffective predator *Rivulus*. This potential "low predation" site provided an ideal situation for following the evolution of male color. The third site, which was a bit farther upstream, was a small tributary that supported guppies along with *Rivulus*. This third site gave Endler a low predation reference site for his study. Endler captured 200 guppies in the high predation environment, measured the coloration of these

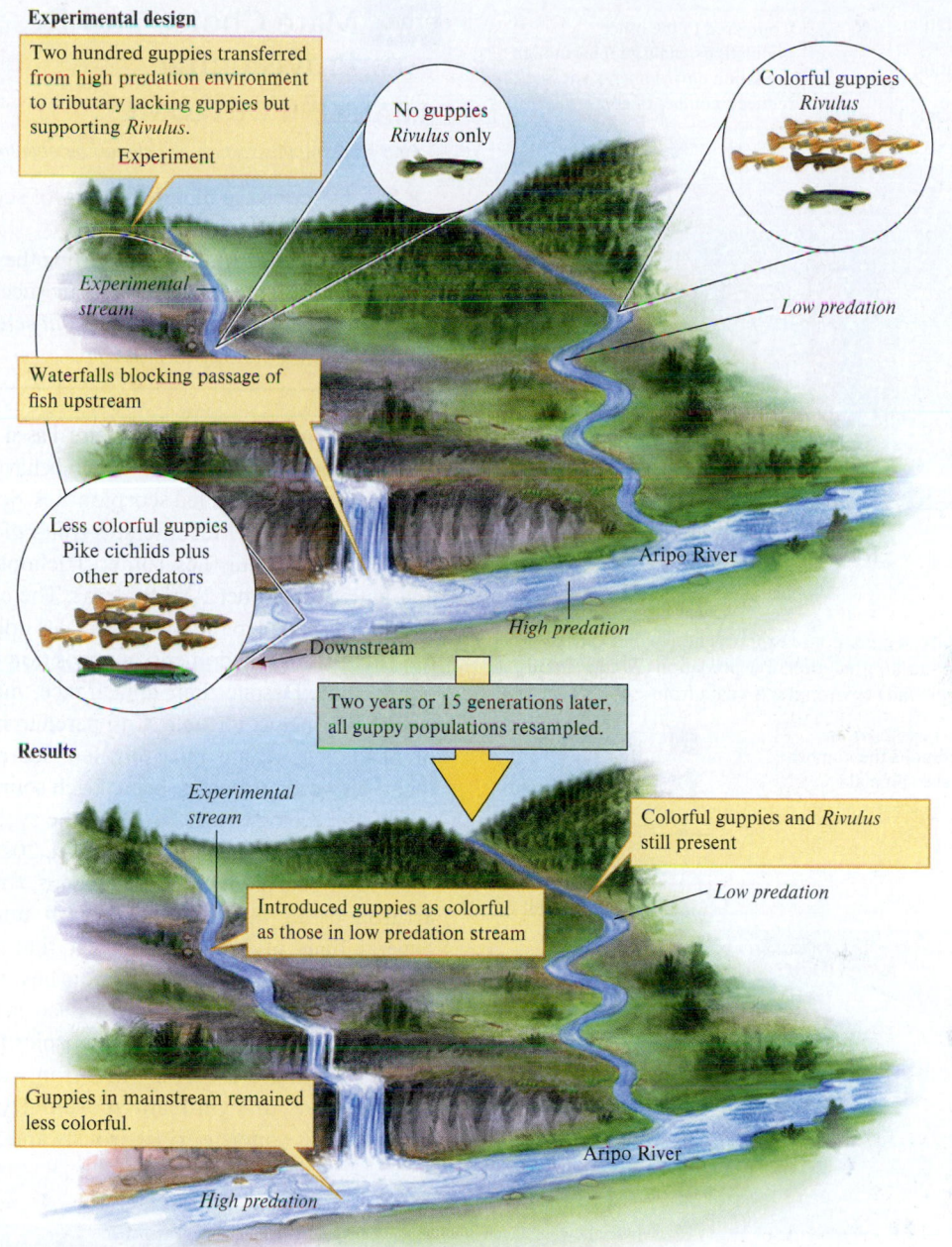

Figure 8.6   Field experiment on effects of predation on male guppy coloration (information from Endler 1980).

guppies, and then introduced them to the site lacking guppies. Six months later the introduced guppies and their offspring had spread throughout the previously guppy-free tributary. Finally, 2 years, or about 15 guppy generations after the introduction, Endler returned and sampled the guppies at all three study sites.

The results of the greenhouse and field experiments supported each other. As shown in figure 8.7, the number of colored spots on male guppies increased in the greenhouse ponds with no predators and with *Rivulus* but decreased in the ponds containing pike cichlids. Figure 8.8, which summarizes the results of Endler's field experiment, compares the number of spots on males in high predation and low predation stream environments with guppies transferred from the high predation environment to a low predation environment. Notice that the transplanted population converged with the males at the low predation reference site during the experiment. In other words, when freed from predation, male guppies in the population had an increase in the average number of spots. This result, along with those of the greenhouse experiment, support the hypothesis that predation reduces male showiness in guppy populations and thereby moderates the effects of intersexual selection by female guppies.

The main characteristics influencing female mate choice in Endler's studies, color and numbers of spots, are anatomical. Let's look now at a mating system where male attractiveness is dependent on complex behaviors that provide females with an important resource: food.

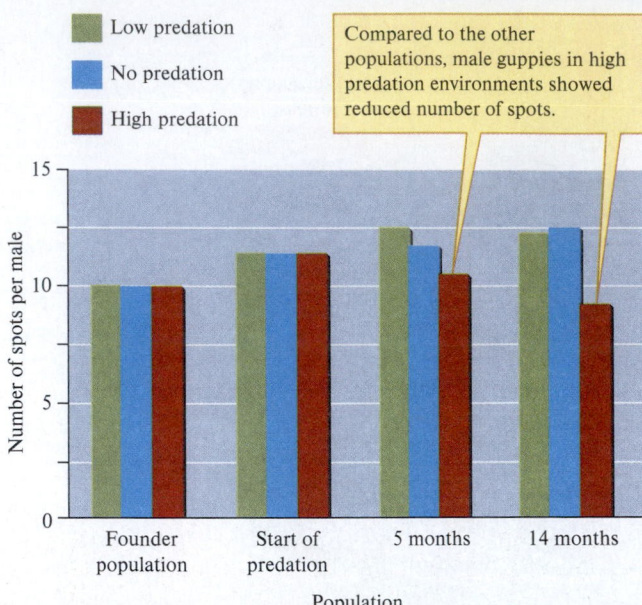

Compared to the other populations, male guppies in high predation environments showed reduced number of spots.

Figure 8.7   Results of greenhouse experiment, which exposed populations of guppies to no predation, low predation (*Rivulus*), and high predation (pike cichlid) environments (data from Endler 1980).

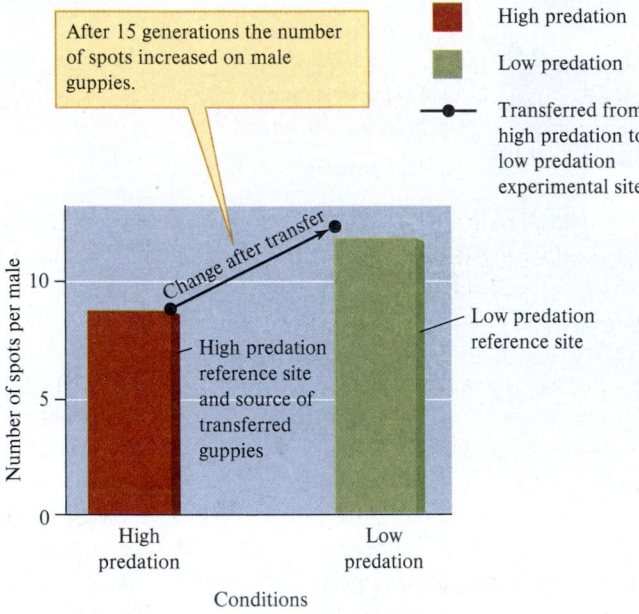

After 15 generations the number of spots increased on male guppies.

Figure 8.8   Results of field experiment involving transfer of guppies from high predation site to site with *Rivulus*, a fairly ineffective predator (data from Endler 1980).

Concept 8.1 Review

1. Why did John Endler take great care to put the same colors of gravel in the same proportions into all of his greenhouse ponds (see fig. 8.5)?
2. In Endler's field experiment (see fig. 8.6), why did male colorfulness increase in the absence of effective predators and not remain unchanged?
3. What do female guppies potentially gain by mating with colorful males?

8.2 Mate Choice and Resource Provisioning

LEARNING OUTCOMES

After studying this section you should be able to do the following:

8.8   Describe the mating biology of scorpionflies in the genus *Panorpa*.

8.9   Assess the evidence supporting the influence of *intrasexual* selection in mating success by male *Panorpa*.

8.10   Discuss the role of *intersexual* selection in the mating biology of *Panorpa*.

Females of some species select mates based on the male's ability to provide important resources.   Such behavior has been observed in a group of insects called scorpionflies. Scorpionflies (fig. 8.9) belong to the order Mecoptera, a group of insects most closely related to the caddisflies (order Trichoptera) and the moths and butterflies (order Lepidoptera). The common name "scorpionfly" is related to the way that males hold their genitalia over the back of their abdomens in a position that suggests a scorpion's sting. Despite their appearance, male scorpionflies are entirely harmless to people. Compared to insects such as moths or beetles, there are relatively few scorpionfly species alive today. However, they have been a rich source of information on behavioral ecology, particularly on the evolution and ecology of mating systems. (e.g., Tinghitella et al. 2020, Dore et al. 2020).

Adult scorpionflies in the genus *Panorpa* feed on dead arthropods in the shrub and herb understory of forests. Several lines of evidence suggest that the supply of dead arthropods available to scorpionflies is limited and that the intensity of competition for dead arthropods is intense, especially among males. Male scorpionflies fight over dead arthropods and even steal them from spider webs, a behavior that leads to significant scorpionfly mortality. Why do scorpionflies compete so vigorously and risk death over dead

Figure 8.9   Male scorpionflies such as this one compete vigorously with each other for female scorpionflies.   Richard Parker/Science Source

arthropods? One reason they fight is that male *Panorpa* use dead arthropods to attract females. If a male finds a dead arthropod and can successfully defend it from other males, he will stand next to the arthropod and secrete a pheromone, which can attract females from several meters away. A female attracted by the pheromone will usually feed on the arthropod while the male mates with her. However, if an arthropod is not available as a nuptial offering, males will secrete a mass of saliva from their enlarged salivary glands and use that to attract females.

In a series of experimental studies, Randy Thornhill explored the details of alternative male mating strategies and the ecological conditions associated with each. In one study, he asked whether there is a difference in mating success among males using different mating strategies. Thornhill created an enclosed environment where he could control the availability of dead arthropods and the number of male *Panorpa* competing for them. He set up 12 replicate environments in 10-gallon terrariums. He included six dead crickets—two large, two medium, and two small—in each terrarium and added 12 male *Panorpa latipennis* to each. Male aggression over crickets, which began soon after they were introduced, was finished after about 3 hours. At that time each of the crickets had been won by a single male, which stood near their respective prizes and secreted pheromone. The majority of the remaining six males secreted a mass of saliva, which they guarded, while secreting pheromones. Finally, some males had no nuptial offerings.

Once the competition among male *Panorpa* for possession of the dead crickets had been decided, Thornhill introduced 12 females and recorded mating activity once per hour for 3 hours. Across the 12 terrariums, there were 144 male *Panorpa* and 144 females. Of the males, 72 males took possession of crickets, 45 had secreted salivary masses, and 27 had no nuptial offerings. How did mating success differ among these groups of males? Figure 8.10 shows that males with a medium or large cricket as a nuptial offering had a clear advantage over those that offered females a small cricket, a salivary mass, or no nuptial offering.

What benefit do females gain by mating with males that offer larger arthropods? One of the clearest benefits is that females feeding on the arthropods offered by males do not have to forage for their own and avoid the risk of being eaten by a spider or other predator as they fly through the forest understory. In addition, feeding on these larger nuptial offerings gives females a reproductive advantage. Thornhill documented that rate of egg laying is higher among females mating with males that provide arthropod prey compared to females that mate with males offering saliva only. Meanwhile, females mated to males with no offering lay very few eggs. What produces this contrast among females mated to males with different nuptial gifts? Thornhill's results likely reflect the greater nutritional benefit of arthropod prey versus saliva and the lack of a nutritional contribution by males without gifts.

Next, Thornhill studied the factors that determine whether males compete successfully for arthropod offerings or resort to the alternative strategies. One of the most basic questions that one could ask is whether males are fixed in particular

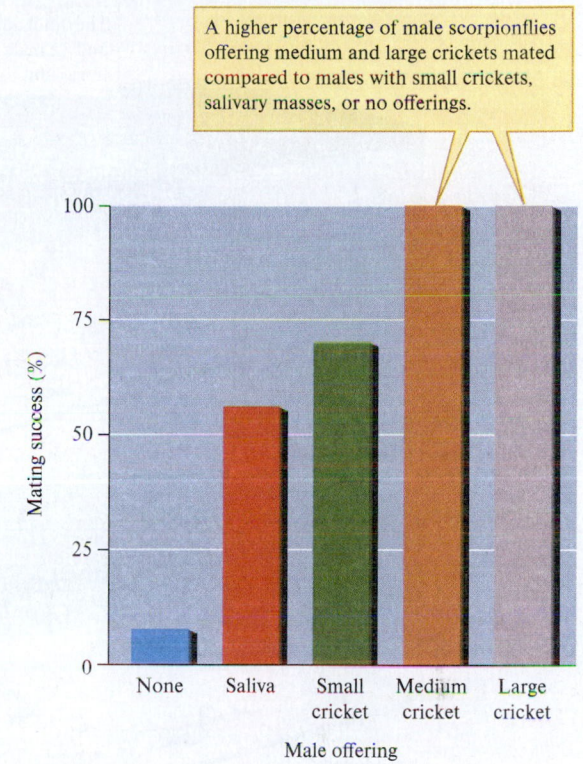

Figure 8.10   Influence of alternative nuptial offerings on mating success by male scorpionflies, *Panorpa latipennis* (data from Thornhill 1981).

behaviors. That is, if males that have not competed successfully for possession of a dead arthropod are given access to one, will they take possession of it and advertise their possession by secreting pheromone? Thornhill addressed this question with a series of controlled experiments with enclosures. Again, he placed six crickets, all medium, in each of 12 terrariums and added 12 male *Panorpa* to each. As in previous experiments, six males took possession of the dead crickets in each terrarium, leaving six males without arthropods. Again, the males without arthropods secreted a salivary mass, which they stood beside as they secreted pheromone. At this point, Thornhill removed all the males possessing crickets in all the terrariums. Within half an hour, almost all the remaining males moved from their salivary masses to the available crickets and secreted pheromone (fig. 8.11). It therefore appears that given the opportunity, male *Panorpa* will take possession of and guard dead arthropods.

What factors determine whether male *Panorpa* will be able to successfully claim a dead arthropod in a competitive environment? Males contesting over a dead arthropod will usually first display to each other. However, visual displays often quickly escalate to head butting and lashing at each other with the scorpion-like genital bulb and its pair of sharp claspers. The claspers of male scorpionflies are capable of tearing wings or other body parts of an opponent. As a consequence, these battles over bugs can be dangerous to both opponents. Because male body size varies widely within populations and male aggression over dead arthropods often

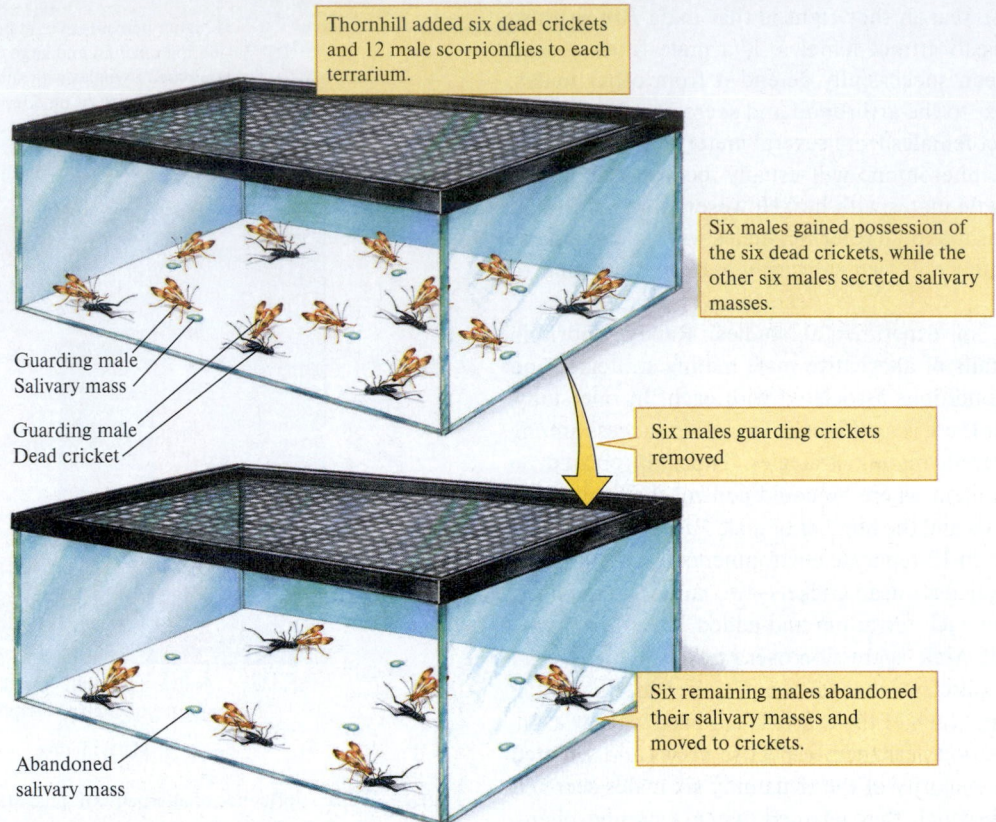

Thornhill added six dead crickets and 12 male scorpionflies to each terrarium.

Six males gained possession of the six dead crickets, while the other six males secreted salivary masses.

Six males guarding crickets removed

Six remaining males abandoned their salivary masses and moved to crickets.

Guarding male
Salivary mass

Guarding male
Dead cricket

Abandoned
salivary mass

Figure 8.11   Experimental test of the influence of nuptial offerings on mating success by male scorpionflies (information from Thornhill 1981).

involves direct combat, Thornhill predicted that larger males would be most successful as competitors over arthropods.

Thornhill tested the relationship between male size and ability to compete for and retain possession of arthropod prey in another experiment. This time he conducted his experiment in 14 larger, 3′ × 3′ × 3′ screen enclosures set out on the forest floor of his study area. Because the enclosures had no bottom panel, they just enclosed a 9-square-foot area of scorpionfly habitat. Thornhill placed four crickets in seven of the enclosures and two in the other seven. He then added 10 female and 10 male scorpionflies to each of the enclosures, which were similar to natural population densities. The males in each enclosure consisted of three large males (55–64 mg), four medium males (42–53 mg), and three small males (33–41 mg). Because scorpionflies are nocturnal, Thornhill monitored the scorpionflies from sunset to sunrise with night vision equipment. Observations continued every night for a week during which Thornhill periodically added fresh dead crickets and replaced any female or male scorpionflies that died with new individuals to keep population densities constant.

The results of the field experiment clearly support the hypothesis that during competition for dead arthropods, larger males have an advantage over small males. Figure 8.12 compares the nuptial offerings of small, medium, and large males in the enclosures with two crickets. While most small males either had no offerings or had a salivary mass, medium

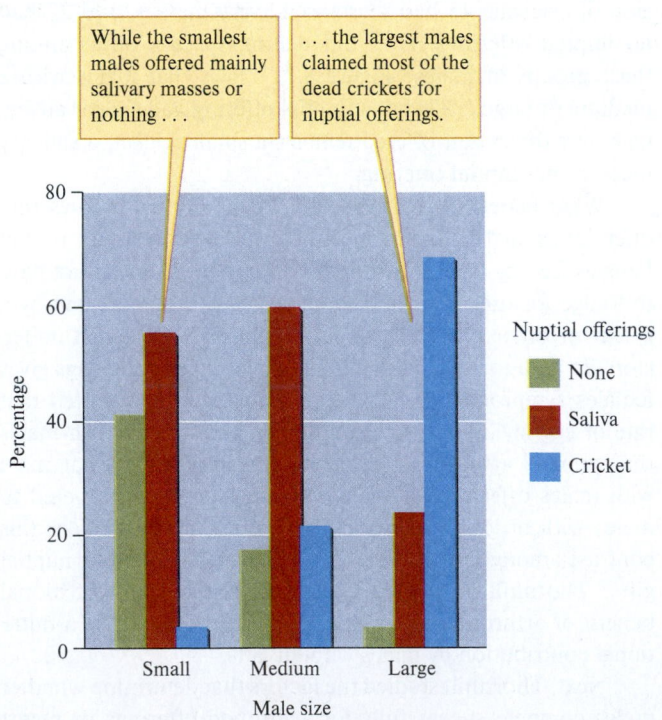

While the smallest males offered mainly salivary masses or nothing . . .

. . . the largest males claimed most of the dead crickets for nuptial offerings.

Nuptial offerings

None

Saliva

Cricket

Percentage

Male size

Small   Medium   Large

Figure 8.12   Type of nuptial offering has a significant influence on mating success among male scorpionflies (data from Thornhill 1981).

males generally offered salivary masses and occasionally competed successfully for a cricket. In contrast, large males generally offered females a cricket and only occasionally offered saliva or had no nuptial offering.

Thornhill's study revealed that larger males are more likely to successfully defend the available arthropod offerings due to their advantages in aggressive encounters. Now, does this difference in offerings translate into different mating success among males? The answer is given in figure 8.13, which shows the percentage of matings observed by Thornhill in cages with two crickets. Large males were involved in 60% of the matings observed, compared to 27% for medium males and 13% for small males. Clearly, the ability of large males to defend higher-quality nuptial offerings translates directly into higher mating success.

Why would female scorpionflies mate preferentially with males that provide them with food? As we saw previously, Thornhill found that larger and higher-quality food offerings by males translate directly into higher egg production by females. In addition, there are likely other benefits to females choosing such males. The ability to provide food in an environment where potential food items are in scarce supply and competition for them is intense certainly reflects a male's competitive ability. To the extent that those competitive traits are heritable, they have the potential to increase the fitness of a female's offspring.

Let us now consider plants. Though we know much less about the mating behavior of plants, it appears that their reproductive ecology also includes the potential for mate choice and sexual selection.

Concept 8.2 Review

1. What evidence is there that the availability of dead insects for scorpionfly feeding is limited in nature?
2. What led Thornhill to conclude that mating success by male scorpionflies is tied to the quality of nuptial offerings presented by males?
3. Which results clearly show the influence of intrasexual selection on male scorpionfly mating success?

8.3 Nonrandom Mating in a Plant Population

LEARNING OUTCOMES

After studying this section you should be able to do the following:

8.11 Explain the possible mechanisms of nonrandom mating and mate choice in flowering plants.
8.12 Describe the evidence indicating nonrandom mating among wild radish plants.

Plants can exhibit nonrandom mating and sexual selection. Sexual selection is usually straightforward to observe in animals and has been understood as a driver of animal evolution for a long time. In contrast, until somewhat recently, plants were assumed to be incapable of sexual selection. Like many others who would follow him, Darwin could not imagine that an organism with "imperfect senses and low mental powers" could engage in competition between males or mate choice (Kaul and Raina 2020). However, even though plants cannot move on their own, there is a growing body of research demonstrating that both do occur in plants (Christopher et al. 2020, Kaul and Raina 2020).

A requirement for sexual selection to occur is that there is nonrandom mating. In flowering plants, mating occurs when pollen is carried by bees or other animals or by wind to flowers that contain the female component, the ovules or eggs. The pollen contains the male component, sperm, which travels through pollen tubes to fertilize the ovules, which if successful, produce seeds. Although this process may appear to be random, males can compete both by having traits that make them more likely to reach an ovule-bearing flower, and by being more likely to fertilize, or "sire", the ovules once the pollen has arrived. Meanwhile, the ovule-bearing, maternal flowers are hardly passive recipients of pollen, given that they provide the resources that allow the pollen tubes to grow and for the seeds to develop. Thus, female mate choice can occur by differentially allowing certain pollen donors to fertilize eggs or by favoring some seeds over others.

The consequence of these selective processes will be nonrandom mating. We can observe this when there is a pattern of some individuals siring more seeds with their pollen, seeds they sire are larger, or even if they are siring seeds in a nonrandom position within a fruit. Position within a fruit can be important if it is associated with more investment of resources from the maternal plant.

One of the best-studied mating systems in plants where nonrandom mating has been documented is that of the wild

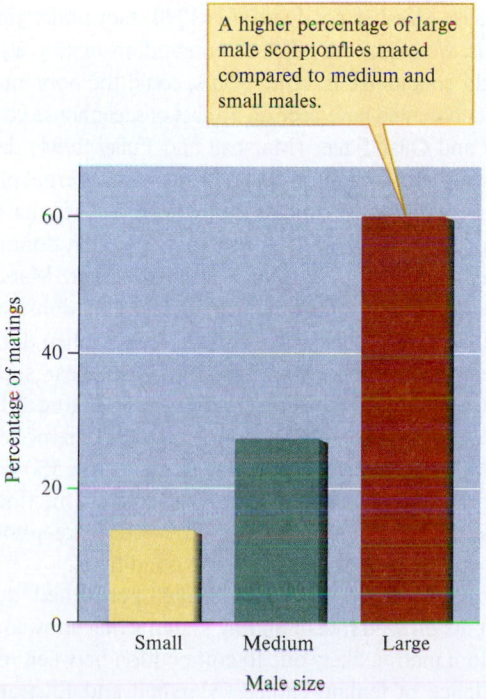

A higher percentage of large male scorpionflies mated compared to medium and small males.

Figure 8.13 Male size has a significant influence on mating success among male scorpionflies (data from Thornhill 1981).

radish, *Raphanus sativus.* Wild radish grows as an annual weed in California where it can be commonly seen along roadways and in abandoned fields (fig. 8.14). During their flowering season, wild radishes are pollinated by a wide variety of insects, including honeybees, hoverflies (see fig. 7.15), and butterflies. The insects that pollinate wild radish generally arrive at flowers carrying pollen from several different plants, and as a consequence a wild radish plant typically has about seven mates. Wild radish flowers have both male (**stamens**) and female (**pistils**) parts and produce both pollen and ovules. However, a wild radish plant cannot pollinate itself, a condition called **self-incompatibility.** Because they must mate with other plants, a researcher working on wild radish can more easily control matings between plants.

One such researcher, Diane Marshall, has spent decades exploring the question of whether siring of offspring in wild radish is a random process. In other words, do the seven mates of a typical wild radish plant have an equal probability of fertilizing the available ovules? The alternative, nonrandom mating, would suggest the potential for mate choice and sexual selection. What mechanisms might produce nonrandom mating among wild radish? Nonrandom mating could result from

maternal control over the fertilization process, competition among pollen, or a combination of the two processes. If it does occur in plants, nonrandom mating establishes the conditions necessary for sexual selection in plants.

Marshall and her colleagues have repeatedly demonstrated nonrandom mating in wild radish. For instance, Marshall (1990) carried out greenhouse experiments that showed nonrandom mating among three maternal plants and six pollen donors. In this experiment, Marshall mated three seed parents or maternal plants with six pollen donors, the plants that would act as sources of pollen to pollinate the flowers of the seed plants.

Marshall used the 6 pollen donors to make 63 kinds of crosses, 6 single donor crosses plus 57 mixed donor crosses, on each maternal plant. Her crosses included all possible mixtures of pollen from 1 to 6 donors. Plants were pollinated in the greenhouse by hand. All pollinations were performed on freshly opened flowers in the morning. Pollen was collected by tapping flowers lightly on the bottom of small petri dishes from an equal number of flowers of each pollen donor. Pollen was then mixed and applied to the stigmas of flowers on the maternal plant using forceps wrapped in tissue. Sufficient pollen was applied to cover each stigma. Because each cross was replicated from 2 to 20 times depending on the type of cross, the total number of pollinations performed on each plant was 300.

One of the ways that Marshall assessed the possibility of nonrandom mating was through performance of pollen donors. She estimated pollen donor performance in three ways: (1) number of seeds sired in mixed pollinations, (2) positions of seeds sired, and (3) weight of seeds sired. The results of this analysis, which are shown in figure 8.15, indicate clearly that pollen donors vary widely in their performance. In other words, mating in this experiment was nonrandom.

Because Marshall conducted her 1990 study under greenhouse conditions, we might ask whether nonrandom mating also occurs under field conditions. In other words, could the nonrandom mating she documented have been an artifact of greenhouse conditions? Marshall and Ollar Fuller (Marshall and Fuller 1994) designed a study to address this question. They chose four maternal plants and grew their offspring in a field setting. Three other maternal lineages (A, B, and C) were chosen to act as pollen donors. Using the forceps and tissue method described earlier, Marshall and Fuller performed several kinds of hand-pollinations, including mixed pollinations using pollen from all three pollen donors.

The result of this experiment provided clear support for nonrandom mating in the field population. Figure 8.16 shows that during the mixed pollen donor pollinations, pollen donor C1 sired a much greater proportion of seeds (56.5%) compared to pollen donors A1 (24.8%) and B1 (18.7%). This finding suggests that the nonrandom matings observed in greenhouse studies were not an artifact of greenhouse conditions.

Over the ensuing years, Marshall performed dozens of experiments on wild radish mating systems that showed not only nonrandom mating likely due to competition between males, but also evidence of female choice (Marshall and Ellstrand 1986, 1988; Marshall and Diggle 2001). She found that a maternal plant could effectively choose which paternal line sired their

Figure 8.14   The wild radish, *Raphanus sativus,* has become a model for studying the mating behavior of plants.   Larry West/Science Source

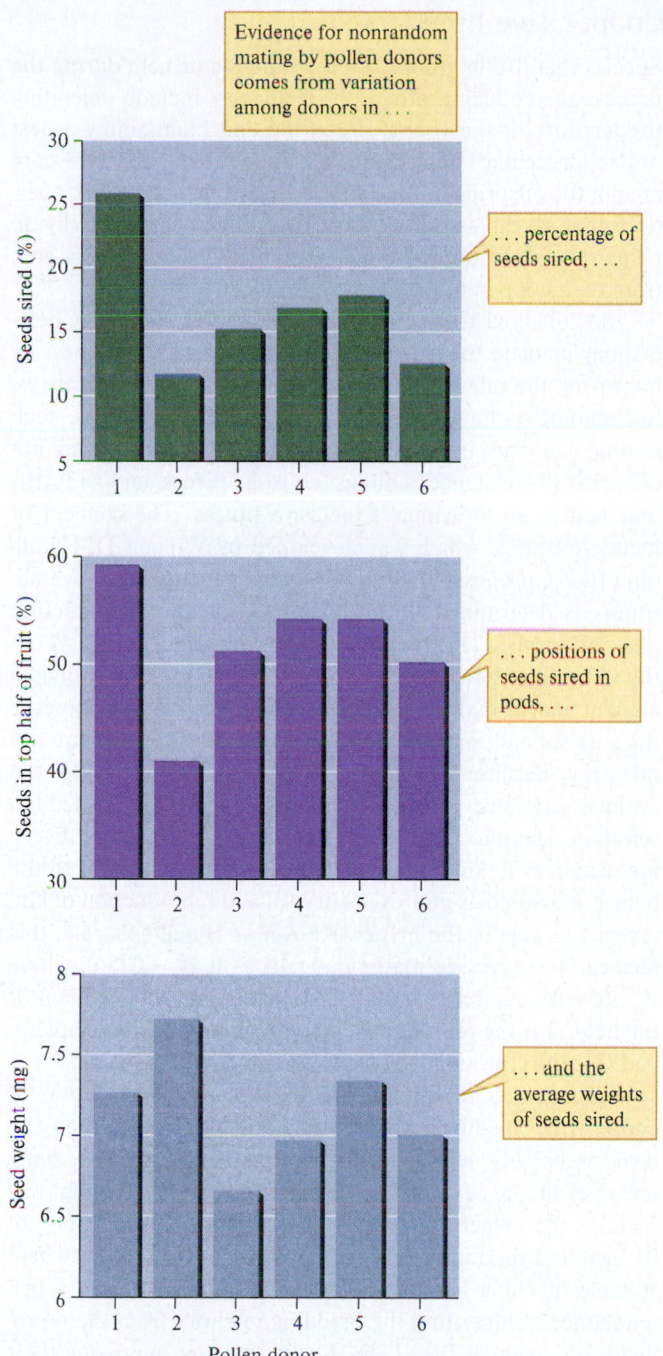

Evidence for nonrandom mating by pollen donors comes from variation among donors in . . .

. . . percentage of seeds sired, . . .

. . . positions of seeds sired in pods, . . .

. . . and the average weights of seeds sired.

Figure 8.15   Evidence for unequal mating success among wild radish pollen donors in a greenhouse environment (data from Marshall 1990).

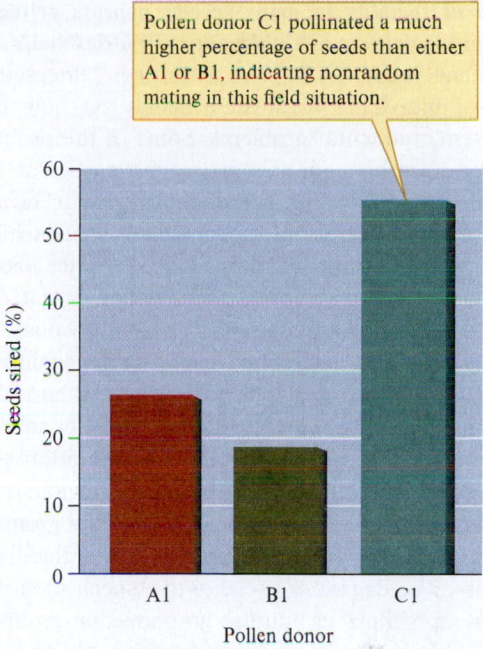

Pollen donor C1 pollinated a much higher percentage of seeds than either A1 or B1, indicating nonrandom mating in this field situation.

Figure 8.16   Variation in wild radish pollen donor mating success in a field environment (data from Marshall and Fuller 1994).

In sections 8.1 to 8.3, we have seen how organisms as different as fish, insects, and plants compete for and select mates. While competition for mates may be intense, the vast majority of mature females in most populations mate and a large proportion of males may also mate. In populations that have evolved a high degree of sociality, however, the opportunities for mating are often restricted to relatively few individuals in the population.

Concept 8.3 Review

1. What types of experiments and evidence are necessary to demonstrate nonrandom mating and sexual selection in plants?
2. How would you expect Marshall and Fuller's data to be different if mating in the wild radish populations that they studied was truly random?

8.4   Sociality

LEARNING OUTCOMES

After studying this section you should be able to do the following:

8.13   Distinguish between eusociality and simpler forms of sociality.

8.14   Define inclusive fitness, kin selection, Hamilton's rule, natal territory, philopatry, and lifetime reproductive success.

8.15   Discuss the evidence indicating a role for kin selection in the evolution of cooperative behavior among woodhoopoes and African lions.

seeds by selective abortion; that is, withholding resources to some seeds so that they did not develop. Female mate choice in plants has now been documented many times and in different species (Christopher et al. 2020).

However, demonstrating that mating is nonrandom is not sufficient evidence for sexual selection. The traits that allow the nonrandom mating to occur must be heritable. Many years after the aforementioned experiments, Diane Marshall and her colleague Ann Evans conducted greenhouse experiments which showed that siring a specific fruit position could be, in fact, selected for in wild radish (Marshall and Evans 2016).

Evolution of sociality in many species appears driven by the need for group defense of high-quality territories and/or defense of mates and young.   Chapters 5 through 7 focused on the ecology of individual organisms, mainly on how individuals solve environmental problems. Some of the problems we considered were how animals maintain a particular range of body temperatures in the face of much greater variation in environmental temperatures or how plants sustain high rates of photosynthesis while avoiding excessive water loss. In the preceding parts of chapter 8 we've also concentrated on the ecology of individuals, examining how individuals choose mates. However, a fundamental change in relationships among individuals within a population takes place when individuals begin living in groups, such as colonies, herds, or schools, and begin to cooperate with each other. Cooperation generally involves exchanges of resources between individuals or various forms of assistance, such as defense of the group against predators. Group living and cooperation signal the beginnings of **sociality.** The degree of sociality in a social species ranges from acts as simple as mutual grooming or group protection of young to highly complex, stratified societies such as those found in colonies of ants or termites. This more complex level of social behavior, which is considered to be the pinnacle of social evolution, is called **eusociality.** Eusociality is generally thought to include three major characteristics: (1) individuals of more than one generation living together, (2) cooperative care of young, and (3) division of individuals into sterile, or nonreproductive, and reproductive castes.

Because individuals in social species often appear to have fewer opportunities to reproduce compared to individuals in nonsocial species, the evolution of sociality has drawn a great deal of attention from behavioral ecologists. The apparent restriction of reproductive opportunities that comes with sociality appears to challenge the idea that the fitness of an individual is determined by the number of offspring it produces. How does sociality challenge this concept of fitness? The challenge emerges from the observation that in social species, many individuals do not reproduce themselves, while helping others in the population to do so. How can we explain such behavior that on first glance appears to be self-sacrificing? It can be argued that such behavior should be quickly eliminated from populations. However, since eusocial species such as bees and ants have survived for millions of years, behavioral ecologists have assumed that in some circumstances, the benefits of sociality must outweigh the costs.

Behavioral ecologists have assumed that the key to understanding the evolution of sociality will result from careful assessment of its costs and benefits. The ultimate goal of sociobiology has been a comprehensive theory capable of explaining the evolution of the various forms of sociality, particularly its most specialized form, eusociality. However, in our quest for such a theory, where should we begin the accounting of costs and benefits? David Ligon (1999) pointed the way when he wrote, "Most, if not all, of the important issues relevant to cooperative breeding systems are . . . related to the costs and benefits of sociality." Following Ligon's suggestion, this Concept section focuses on cooperative breeders.

Cooperative Breeders

Species that live in groups often cooperate or help during the process of producing offspring. Help may include defending the territory or the young, preparing and maintaining a nest or den, or feeding young. Since the young that receive the care are not the offspring of the helpers, one of the most basic questions that we can ask about these breeding systems is: Why do helpers help? In other words, what benefits do helpers gain from their cooperation?

Sociobiologists have offered two main reasons. First, helpers may increase their own evolutionary, or genetic, fitness by improving the rates of survival and reproduction of relatives. Sociobiologists have suggested that investing resources, such as time or energy, in genetically related individuals that are not offspring (for instance, siblings, cousins, nieces, and so forth) may add to an individual's **inclusive fitness.** The concept of inclusive fitness, which was developed by William D. Hamilton (1964), proposes that an individual's inclusive, or overall, fitness is determined by its own survival and reproduction plus the survival and reproduction of individuals with whom the individual shares genes. Under some conditions, individuals can increase their inclusive fitness by helping increase the survival and reproduction of genetic relatives that are not offspring. Because this help is given to relatives, or kin, the evolutionary force favoring such helping behavior is called **kin selection.** Hamilton proposed that selection will favor diverting resources to kin under conditions where its benefit to the helper, measured as improved survival and reproduction of kin, exceeds its cost to the helper. Known as **Hamilton's rule,** this idea can be expressed mathematically as $R_gB - C > 0$, where R_g is the genetic relatedness of the helper and the recipient of the help, B is the reproductive benefit gained by the recipient, and C is the reproductive cost to the helper of giving aid.

The second reason offered to explain the evolution of cooperative breeding is that helping may improve the helper's own probability of successful reproduction. Because helping gives the helper experience in raising young, helping may increase the helper's chances of successfully raising young of its own and recruiting helpers of its own. In addition, where suitable breeding habitat is limited, helpers may have a better chance of inheriting the breeding territory from the reproductive individuals they help. Again, they are improving their chances of eventually raising their own young.

What sorts of species engage in cooperative breeding? By some estimates, as many as 900 species of birds are cooperative breeders (Cockburn 2006). In addition, several species of mammals such as wolves, wild dogs, African lions, and meerkats engage in cooperative breeding (fig. 8.17). Let's review two intensively studied species where several benefits of cooperative breeding have been demonstrated.

Green Woodhoopoes

We know a great deal about the cooperative breeding and general ecology of green woodhoopoes due to the pioneering, long-term studies of J. David Ligon and Sandra Ligon (Ligon and Ligon 1978, 1982, 1989, 1991). Adult green woodhoopoes,

Phoeniculus purpureus, have reddish-orange bills and feet and black feathers with a metallic green and blue-purple sheen (fig. 8.18). Meanwhile, juvenile green woodhoopoes have black bills and feet, which allowed the Ligons to distinguish between mature and immature individuals in the field. Of the eight species of woodhoopoes, all of which are restricted to sub-Saharan Africa, the green woodhoopoe is the most common and widespread. Green woodhoopoes live in a wide variety of habitats at elevations from sea level to over 2,000 m. However, their most common habitat is open woodlands with trees large enough to provide cavities for nesting and roosting. For instance, the Ligons' long-term study site, near Lake Naivasha in the Central Rift Valley of Kenya, was located in a woodland dominated by yellow-barked acacia.

Tree cavities keep the birds warm at night and provide some protection from predators. The habit of cavity roosting also makes green woodhoopoes ideal for field studies. To place unique color bands on green woodhoopoes in their study area, all the Ligons had to do was plug the opening to a roosting cavity after dark and then place a clear plastic bag over the opening in the morning to catch the woodhoopoes as they left the roost. Using this technique, they placed unique color bands on 386 green woodhoopoes. By closely studying the movements and interactions of banded individuals over a long period of time, the Ligons learned a great deal of the social relations of green woodhoopoes. For instance, they eventually knew the parentage of over 93% of the birds in their study area, the number and fates of offspring produced by each flock, and the identity of all breeders and non-breeders in each flock. The results of this long-term study provide clues to the costs and benefits of cooperative breeding.

The Ligons found that green woodhoopoes live in territories that are occupied and defended by flocks of 2 to 16 individuals. Average flock size varied from approximately 4 to 6 over the course of their studies. Within a group, only one pair breeds, while the remainder act as helpers. Males, which are approximately 20% larger than females, are particularly vigorous in their defense of breeding territories. The Ligons (1989) suggested that the larger body size of males is related to their intense competition with other males over territories and females. Territory defense is very important because territories appear to vary widely in quality.

One of the most obvious differences among territories is the quality of the cavities they contain. Cavity characteristics are important, since predation while the birds are in their roosting or nesting cavities is a major source of mortality. The Ligons documented annual mortality rates of 30% for females and 40% for males, most of which was due to predation. Predators on nestlings include driver ants, hawks, and owls. Roosting adults are attacked at night by driver ants and large-spotted genets—small, slender predators related to the mongoose. The vulnerability of nestlings and roosting adults to these predators depends on the characteristics of the cavity, especially its depth, the size of the opening, and the soundness of the wood.

Green woodhoopoes stay very close to their **natal territories,** the territories where they were raised. Out of 38 females that the Ligons banded as nestlings or fledglings and later observed breeding, 18 bred on their natal territory, 14 bred on an adjacent territory, and 6 only two or three territories away from their natal territory. Male dispersal is also limited. In other words, this population of green woodhoopoes shows a great deal of **philopatry.** Philopatry, which means literally "love of place," is a term that behavioral ecologists use to describe the tendency of some organisms to remain in the same area throughout their lives.

Why do green woodhoopoes stay at home and help raise young, which are close relatives, rather than disperse to produce their own offspring? The Ligons suggested that the major factor producing this high degree of philopatry is that roost cavities on which green woodhoopoes depend in the highlands of Kenya are scarce. By staying home, a young green woodhoopoe gets a warm and relatively safe place to roost at night and may eventually inherit the territory and its cavities.

Over the course of their study, the Ligons found that 91% of females and 89% of males died without leaving any descendants. However, they also documented very high reproductive success among some woodhoopoes. Variation in reproductive success

Figure 8.17   Meerkats, *Suricata suricatta,* are native to southern Africa, where they live in cooperatively breeding groups of up to 30 individuals. Within a group of meerkats, commonly called a "mob," a dominant female and a dominant male produce most of the young, with other adults sharing in guarding and defending the mob and caring for pups. ©Digital Vision/Getty Images

Figure 8.18   Studies of African green woodhoopoes have made major contributions to our understanding of the evolution of cooperative breeding among vertebrate animals.   nomis_g/Getty Images

within the study area seemed to have two major sources, spatial and temporal variation. Year-to-year variation in breeding success was associated with variation in rainfall, which appears to influence the woodhoopoes' food supply. The main food that woodhoopoes give to nestlings is moth larvae that pupate in the soil and are sensitive to soil moisture. In general, rainfall during the dry season correlates with high mortality of these pupae and reproductive failure among the woodhoopoes (fig. 8.19).

The second source of variation in reproductive success appears to be differences in territory quality. The Ligons found that territories fell into two clearly distinctive groups, which they called high-quality and low-quality territories. Territory quality appeared to be mainly determined by the availability of roosting cavities capable of protecting the birds from predators. Figure 8.20 compares the average number of young produced

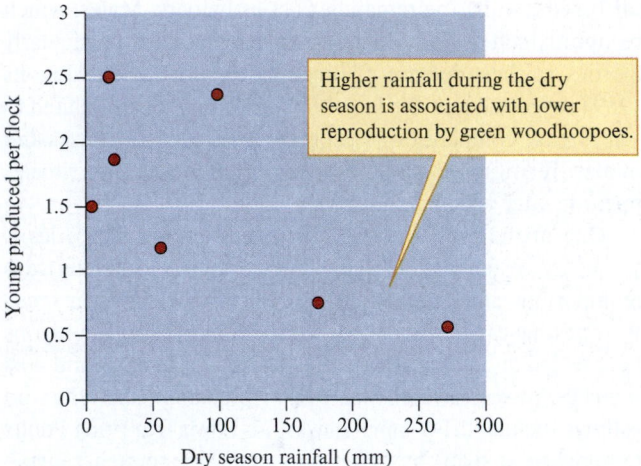

Figure 8.19   Rainfall is correlated with reproductive rate by green woodhoopoes (data from Ligon and Ligon 1989).

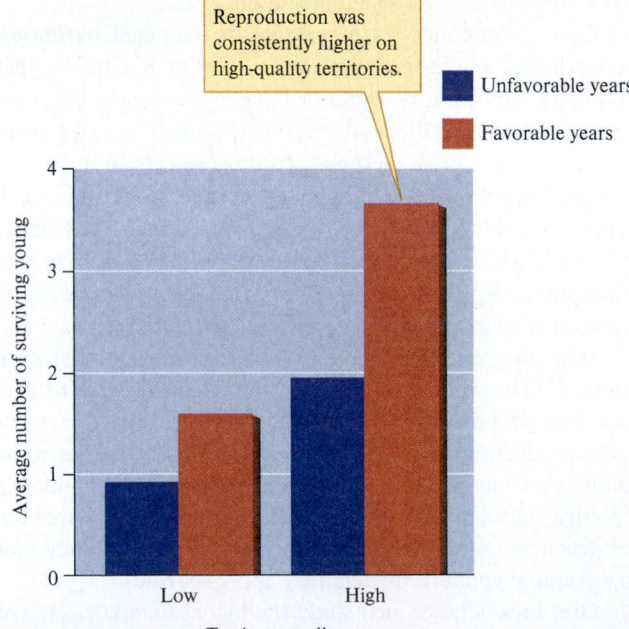

Figure 8.20   Relationship between territory quality and reproduction by green woodhoopoe flocks (data from Ligon and Ligon 1989).

per year on low-quality and high-quality territories. As you can see, the number of offspring produced on high-quality territories is approximately twice as high as on low-quality territories during both favorable and unfavorable years.

While the birds can do nothing about the chances of rainfall during the dry season, they can and do compete for territories. Those flocks that successfully compete for the best territories have a clear reproductive advantage. So, returning to our original question, why do green woodhoopoes stay home and help? The first reason seems to be that by helping to raise and protect close relatives, the helpers may increase their inclusive fitness (Hawn, Radford, and du Plessis 2007). The Ligons found that the bulk of the young tended by helpers ranged from half siblings to full siblings. We should keep in mind that a full sibling, on average, would share as many genes (50%) with the helper as its own son or daughter. The second and perhaps clearest potential benefit to a helper is that since high-quality territories are limited in number, the chance of inheriting the natal territory and advancing to breeding status may be greater than finding another suitable territory elsewhere.

Long-term studies have shown that delaying reproduction by green woodhoopoes, as they help tend nests, has opposite effects on males and females. Amanda Hawn, Andrew Radford, and Morné du Plessis (2007) used a 24-year study of green woodhoopoes in South Africa to explore the influence of delayed reproduction on **lifetime reproductive success**—the total number of offspring produced over the course of a lifetime. The pattern that they uncovered for males was what you might expect. Males that began reproducing at a younger age produced more offspring that successfully fledged over the course of their lives (fig. 8.21a). In contrast, female green woodhoopoes that delay reproduction until they are older have higher lifetime reproductive success (fig. 8.21b). The main reason for the positive relationship between female age at first reproduction and lifetime reproductive success is that females that started reproducing at 1 to 3 years of age had higher mortality compared to those that delayed reproducing until they were 4 to 6 years old. Over 60% of young females died 1 to 2 years after beginning to reproduce, whereas only 30% of older females died within their first 2 reproductive years. Hawn, Radford, and du Plessis also point out that females that delay reproduction also spend more years as helpers raising close relatives, which adds significantly to their inclusive fitness.

What might we learn about the evolution of cooperative breeding from other species? Several cooperative species live in sub-Saharan Africa. For example, the African lion, a species that shares the same landscape with green woodhoopoes, also seems to be forced by a variety of environmental circumstances into a cooperative social system.

African Lions

At about the same time that the Ligons were studying cooperative breeding among green woodhoopoes, Craig Packer and Anne E. Pusey were studying cooperation among African lions in the Serengeti (Packer and Pusey 1982, 1983, 1997, Packer et al. 1991). Their studies have revealed a great deal of complexity in lion societies. Female lions live in groups of related

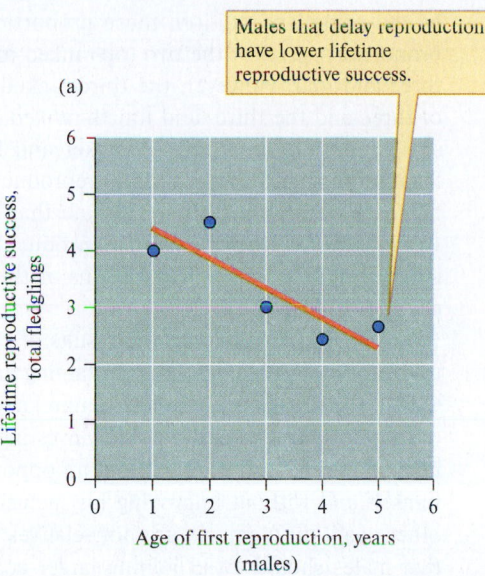
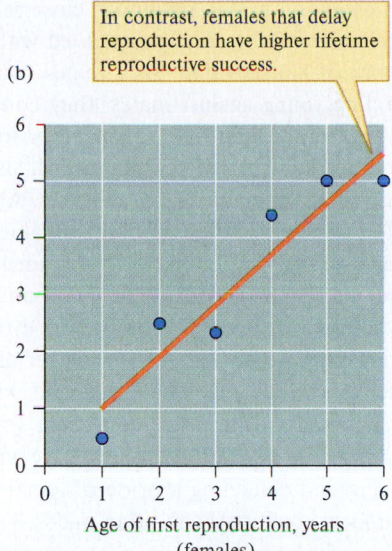

Males that delay reproduction have lower lifetime reproductive success.

In contrast, females that delay reproduction have higher lifetime reproductive success.

Figure 8.21   The relationship between age at first reproduction and lifetime reproductive success by male (*a*) and female (*b*) green woodhoopoes (data from Hawn, Radford, and du Plessis 2007).

individuals called prides (fig. 8.22). Prides of female lions generally include 3 to 6 adults but may contain as many as 18 or as few as 1. In addition to adult females, prides also include their dependent offspring and a coalition of adult males (fig. 8.23). Male coalitions may be made up of closely related individuals or of unrelated individuals.

Within lion society one can observe many forms of cooperation. Female lions nurse each other's cubs. They also cooperate when hunting large, difficult-to-kill game such as zebra and buffalo. In addition, females cooperatively defend their territory against encroaching females. However, the most critical form of cooperation among females is their group defense of the young against infanticidal males. These attacks on the young generally take place as a male coalition is displaced by another invading coalition. While a single female lion has little chance in a fight against a male lion, which are nearly 50% larger, cooperating females are often successful at repelling attacking males. Males, in turn, cooperate in defending the territory against invading males, which threaten the young they have sired, and against threats from other predators such as

hyenas. Research by Natalia Borrego on both captive and wild lions has revealed that males can cooperate on a variety of tasks, and that working together allows them to solve problems more readily then when working alone (Borrego 2020). The challenge for the behavioral ecologist has been to determine whether these various forms of cooperation can be reconciled with evolutionary theory.

Figure 8.23   The mane of adult male African lions has evolved under the complex influences of sexual and natural selection. Female choice (intersexual selection) and competition between males (intrasexual selection) favor long, dark manes, while thermal regulation under the hot equatorial sun favors short, light-colored manes (West and Packer 2002). Christina Krutz/Radius Images/Getty Images

Figure 8.22   A pride of female African lions. Female lions cooperate in their hunts, in the raising of cubs, and in defense of the pride.   Robert Muckley/Moment/Getty Images

Since the females in lion prides are always close relatives, their cooperative behavior can be readily explained within the conceptual framework of kin selection. As females cooperate in nursing or defending young against males, they contribute to the growth and survival of their own offspring or to those of close kin. Cooperative hunting and sharing the kill also contribute to the welfare of offspring and close relatives. All these contributions add to the inclusive fitness of individual females.

In contrast, because male coalitions are sometimes made up of close relatives and sometimes not, cooperation within coalitions has represented a greater challenge to evolutionary theory. However, on close consideration, Packer and colleagues (1991) discovered that the rules associated with the formation and behavior of coalitions are consistent with predictions of evolutionary theory. Single males have virtually no chance of claiming and defending a pride of female lions; therefore, they must form coalitions with other males. This represents a type of ecological constraint on viable choices open to males. If males form a coalition with brothers and cousins, cooperative behavior that increases the production and survival of offspring of the coalition will increase an individual male's inclusive fitness. However, theoretically, a male within a coalition with unrelated males must produce some offspring of his own or he is merely increasing the fitness of others at the expense of his own fitness.

The first question we should ask is: Do all males within a coalition have an equal opportunity to reproduce? If all males within a coalition have an equal probability of reproducing, then forming coalitions with unrelated males is easier to reconcile with evolutionary theory. However, if there is significant variation in reproductive opportunities within coalitions, then cooperating with unrelated males is more difficult to reconcile with theories predicting that individuals will attempt to maximize their inclusive fitness. It turns out that the probability of a male siring young depends on his rank within a coalition and on coalition size. As shown in figure 8.24, males in coalitions of two sire relatively similar proportions of the young produced

by the pride. In addition, these proportions are close to the proportions sired by the two top-ranked males in coalitions of three and four. However, the third-ranked males in coalitions of three and the third- and fourth-ranked in coalitions of four sire almost no young lions. Packer and his team concluded from these data that variation in reproductive success is much higher in coalitions of three and four than in coalitions of two. In other words, the chance of reproducing is less evenly distributed among males in coalitions of three and four than in coalitions of two.

What implications do the results of Packer's studies have to the formation of coalitions containing unrelated individuals? One of the implications is that an unrelated male in a coalition of three or more runs the risk of investing time and energy in helping maintain a pride without an opportunity to reproduce himself and without improving his inclusive fitness, since the other coalition members are not relatives. This result suggests that males should avoid joining larger coalitions of unrelated males, and this is just what Packer and his colleagues found. Figure 8.25 shows the percentage of males with unrelated partners in coalitions of different sizes. These patterns show clearly that males that team up with unrelated individuals mostly do so in coalitions of two or three. Larger coalitions of four to nine individuals are almost entirely made up of relatives.

In summary, cooperation among green woodhoopoes and African lions appears to be a response to environmental conditions that require cooperation for success. In the case of green woodhoopoes, the scarcity of high-quality territories and intense competition between flocks for those territories create conditions that favor staying in the natal territory and helping raise related young and perhaps inheriting the territory at a later date. Packer and Pusey (1997) captured the situation facing African lions in a fascinating article titled "Divided We Fall: Cooperation Among Lions." To survive, reproduce, and successfully raise offspring to maturity, African lions must work in cooperative groups. The lone lion has no chance of meeting the ecological challenges presented by living on the

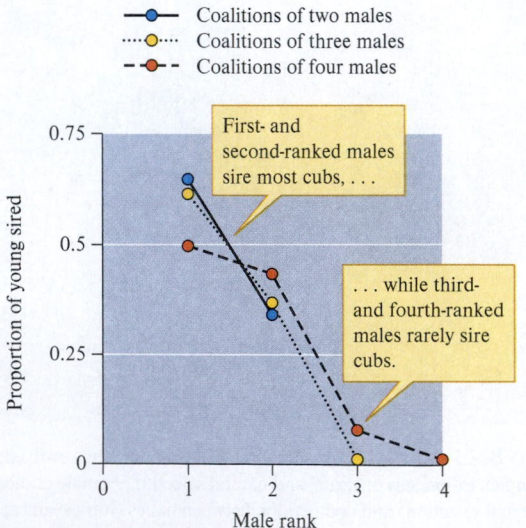

Figure 8.24   Male lion rank and proportion of cubs sired in male coalitions of different sizes (data from Packer et al. 1991).

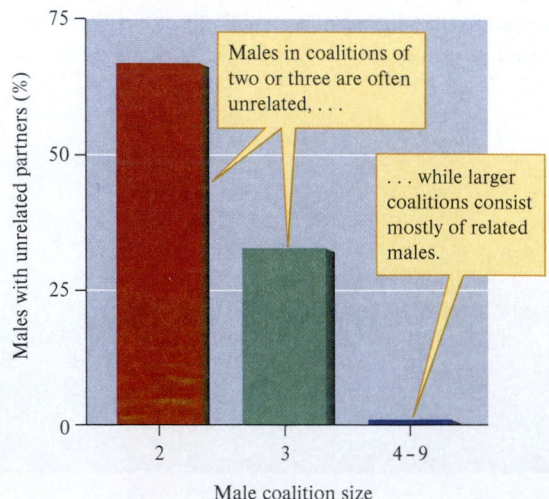

Figure 8.25   Relatedness and size of male coalitions among African lions (data from Packer et al. 1991).

Serengeti in lion society with its aggressive prides and invasive and infanticidal male coalitions. However, as we have seen, within the constraints set by their environments, both green woodhoopoes and African lions appear to behave in a way that contributes positively to their overall fitness.

While the complexities of African lion and green woodhoopoe societies have taken decades to uncover, they pale beside the intricacies of life among eusocial species, such as bees, termites, and ants. Let's explore eusociality in some animal populations to get some insights into the evolution of these complex social systems and to introduce the comparative method, one of the most valuable tools in evolutionary ecology.

Concept 8.4 Review

1. The Ligons found that most of the young woodhoopoes tended by helpers were either full siblings or half siblings of the helpers. If full siblings were genetically related to the helpers by an average of 50%, what was the genetic relationship between helpers and half siblings?
2. According to Hamilton's rule, would helpers derive greater benefit through kin selection by contributing the same amount of help toward raising a full sibling or a half sibling?
3. What are the evolutionary implications of the fact that larger coalitions of male lions consist almost entirely of close relatives (see fig. 8.25)?

8.5   Eusociality

LEARNING OUTCOMES

After studying this section you should be able to do the following:

8.16   Describe the comparative method.
8.17   Compare the social organization of colonies of leaf-cutter ants and naked mole rats.
8.18   Describe the evidence for and against the influence of haplodiploidy on the evolution of eusociality.

Kin selection and ecological constraints may have played key roles in the evolution of eusociality.   Behavioral ecologists are concerned both with how particular social systems work and with determining the mechanisms responsible for their evolution and maintenance. In most cases, however, the evolutionary origins of biological traits lie deep in the past and biologists cannot observe their evolution directly. So, how do scientists construct evolutionary hypotheses, test them, and eventually construct evolutionary theories? Many tools are used in such a process. We have already employed one of those tools, in a rudimentary way, without giving it a name. As we explored mate choice and sexual selection using guppies, scorpionflies, and wild radish, we were employing one of the most valuable tools available to evolutionary biologists. That tool is the **comparative method.** The comparative method involves comparisons of the characteristics of different species or populations

of organisms in a way that attempts to isolate a particular variable or characteristic of interest, such as sociality.

Let's review a remarkable case of convergence in social organization between a eusocial insect and a eusocial mammal. Such comparisons, the foundation of the comparative method, if quantified and replicated across many species, can help disentangle the evolution of complex characteristics, including the evolution of social systems.

Eusocial Species

Probably the most thoroughly studied of eusocial species are the ants. Scientists have described over 12,000 species of ants, all belonging to the family Formicidae, which along with their relatives the wasps and bees, are members of the insect order Hymenoptera. Hölldobler and Wilson (1990) wrote a monumental summary of what was known about ants near the end of the twentieth century in a book titled simply *The Ants.* However, despite that book and the thousands of studies done on ants since its publication, much is left to learn about this group of insects.

One of the most socially complex groups of ants are the leaf-cutters (fig. 8.26). The 39 described species of leaf-cutter ants, which belong to two genera, are found only in the Americas, from the southern United States to Argentina. Leaf-cutter ants make their living by cutting and transporting leaf fragments to their nest, where the leaf material is fragmented and used as a substrate upon which to grow fungi. The fungi provide the primary food source for leaf-cutter ants.

Among the various species of leaf-cutter ants, some of the most thoroughly studied are species belonging to the genus *Atta.* *Atta* species live mainly in tropical Central and South America. However, at least two species reach as far north as Arizona and Louisiana in the United States. Leaf-cutter ants are important consumers in the tropical ecosystems, where they move large amounts of soil and process large quantities of leaf material in their nests. The nests of leaf-cutter ants can attain great size. For instance, the nests of *A. sexdens* can include over 1,000 entrance holes and nearly 2,000 occupied and abandoned chambers. In one excavation of an *A. sexdens* nest (cited in Hölldobler and Wilson 1990), researchers estimated that the ants had moved

Figure 8.26   Leaf-cutter ants carrying leaf fragments back to the nest, where they will be processed to create a substrate for growing the fungi that the ants eat.   Alamy Stock Photo

more than 22 m³ of soil, which weighed over 40,000 kg. Within this nest, the occupants had stored nearly 6,000 kg of leaves. Mature nests of *A. sexdens* contain a queen, various numbers of winged males and females, which disperse to mate and found colonies elsewhere, and up to 5 to 8 million workers.

Though involving far fewer individuals, there are striking analogies between the organization of ant colonies and colonies of naked mole rats, *Heterocephalus glaber,* one of the few species of eusocial mammals (fig. 8.27). Despite their common name, naked mole rats are not completely naked and they are neither moles nor rats. Like moles, naked mole rats live underground but they are rodents, not moles. However, the family of rodents to which they belong is more closely related to porcupines and chinchillas than to rats.

Naked mole rats live in underground colonies in the arid regions of Kenya, Somalia, and Ethiopia. Colonies often include 70 to 80 individuals but can sometimes contain as many as 250 individuals. The burrow system of a single colony of naked mole rats is extensive and can cover up to approximately 100,000 m², or about 20 football fields. Most of the digging required to maintain their large burrow systems is done with the naked mole rats' teeth and massive jaws. It turns out that the jaw muscles of naked mole rats make up about 25% of their entire muscle mass. This would be approximately equivalent to having muscles the size of those in your legs powering your jaws!

Both naked mole rats and leaf-cutter ants live in social groups in which individuals are divided among castes that engage very different activities. A **caste** is a group of physically distinctive individuals that engage in specialized behavior within the colony. E. O. Wilson (1980) studied how labor is divided among castes of ants in a laboratory colony of *A. sexdens* that he established and studied over a period of 8 years. During this period, Wilson carefully cataloged the behaviors of individual colony members. Because the colony lived in a closed series of clear plastic containers, their behavior could be studied easily. In addition to recording behaviors, Wilson also estimated the sizes of individuals engaging in each behavior by measuring their head widths to the nearest 0.2 mm. He made his estimates visually by comparing an ant to a standard array of preserved *A. sexdens* specimens of known size.

When Wilson compared the leaf-cutter ant *A. sexdens* with three non-leaf-cutter ant species, he found that the leaf-cutter ants included a larger number of castes and engaged in a wider variety of behaviors (fig. 8.28). Wilson identified a total of 29 distinctive tasks performed by the leaf-cutter ants compared to an average of 17.7 tasks performed by the three other species. He found that the division of labor within the *A. sexdens* colony was mainly based on size. Possibly because of the large number of specialized tasks that need to be performed by leaf-cutter ants, they have one of the most complex social structures and one of the greatest size ranges found among the ants. Within *A. sexdens* colonies, the head width of the largest individuals (5.2 mm) is nearly nine times the head width of the smallest individuals (0.6 mm). On the basis of size, Wilson identified four castes within his leaf-cutter colony. However, because the tasks performed by some of the size classes change as they age, Wilson discovered three additional temporal or developmental castes for

Figure 8.27   Naked mole rats live in colonies of closely related individuals ruled by a single dominant female, or queen.
Gregory G. Dimijian, M.D./Science Source

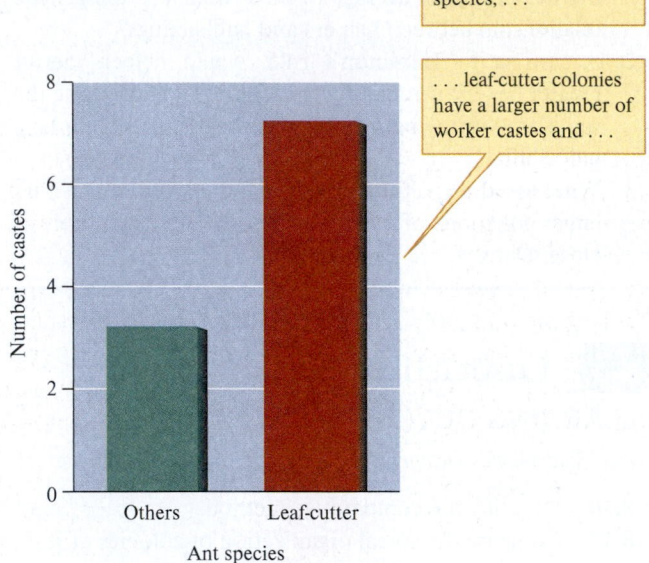

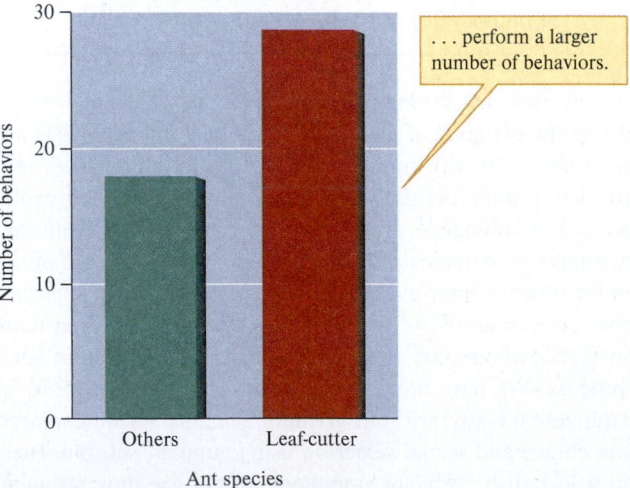

Figure 8.28   Comparison of the number of castes and number of behaviors in a colony of leaf-cutter ants, *Atta sexdens,* and in colonies of three other ant species (data from Wilson 1980).

a total of seven castes within the colony, compared to an average of three castes in the non-leaf-cutter ant species he studied.

As a consequence of this great variation, someone watching a trail of leaf-cutter ants bring freshly cut leaf fragments back to their nest is treated to a rich display of size and behavioral diversity. While medium-sized ants carry the leaf fragments above their heads, the largest ants line the trail like sentries, guarding against ground attacks on the column of ants carrying leaf fragments. Very small ants ride on many of the leaf fragments, protecting the ant carrying a leaf fragment from aerial attacks by parasitoid flies. Meanwhile, other size classes of leaf-cutters performing behaviors associated with processing leaves, tending larvae, and maintaining fungal gardens remain hidden in the nest. It was the activity of these smaller individuals that Wilson's laboratory colony was able to reveal so clearly.

Careful study has revealed some remarkable parallels in the structures of naked mole rat and leaf-cutter ant societies. The social behavior of naked mole rats was first reported by Jennifer Jarvis, professor at the University of Cape Town, South Africa, in a paper in the journal *Science* (Jarvis 1981). Her published study was based on more than 6 years of observation and experimentation with colonies of naked mole rats that she had established in the laboratory. Jarvis dug up a number of colonies and relocated them to a laboratory habitat analogous to that used by Wilson in his study of leaf-cutter ants. She waited approximately a year after bringing the naked mole rats to the laboratory before attempting to quantify their behaviors. Once this period of acclimation was over, Jarvis spent approximately 100 hours detailing how the members of her laboratory population of naked mole rats spent their time.

The picture of naked mole rat society that emerged from Jarvis's study was immediately intriguing to behavioral ecologists. The social organization of the colony appeared more similar to an ant colony than to any other mammal population known. Jarvis's paper in *Science* stimulated dozens of studies of naked mole rats and of related species. The results provide interesting insights into the evolution of social behavior. Within a colony of naked mole rats, one female and a few males breed. This group of reproductive individuals functions basically as a queen and her mates, while the rest of the colony is nonreproductive. Behavioral ecologists have found that life in a naked mole rat colony centers on the queen and her offspring, and the queen's behavior appears to maintain this focus. She is the most active member of the colony and literally pushes her way around the colony. By physically pushing individuals, she appears to call them to action when there is work to be done or when the colony is threatened and needs defending. The aggressiveness of the queen also appears to maintain her dominance over other females in the colony and prevents them from coming into breeding condition. If the queen dies or is removed from the colony, one of the other females in the colony will assume the role of queen. If two or more females compete for the position of queen, they may fight to the death during the process of establishing the new social hierarchy.

In contrast to leaf-cutter ant colonies, where all workers are females, both males and females work in naked mole rat colonies. Jarvis found that work is divided among colony members, as in leaf-cutter ant colonies, according to size. However, in contrast to leaf-cutter ants, colonies of naked mole rats include only two worker size classes, small and large. Small workers are the most active. Small workers excavate tunnels, build the nest, which is deeper than most of the passage ways, and line the nest with plant materials for bedding. In addition, small workers also harvest food, mainly roots and tubers, and deliver it to other colony members, including the queen for feeding. Since they spend most of their time sleeping, the role of large nonbreeders was unclear for some time. However, eventually researchers working in the field were able to observe these large nonbreeders in action. It turns out that the large workers, as in ant colonies, are a caste specializing in defense. If the tunnel system is breached by members of another colony, the large nonbreeders move out quickly from their resting places to defend the colony from the invaders, literally throwing themselves into the breach. Eventually the large nonbreeders push up enough soil to wall off the intruders. However, they may be most important in defending against snakes, the most dangerous predators of naked mole rats. When confronted with a snake, the large nonbreeders will try to kill the snake or spray it with soil until it is driven off or buried.

Evolution of Eusociality

The studies of Wilson and Jarvis suggest interesting parallels in the organizations of leaf-cutter ants and naked mole rat colonies, despite their distinctive evolutionary histories and other biological differences (fig. 8.29). Similarities include division of labor within colonies based on size, with smaller workers specializing in foraging, nest maintenance, and excavation of extensive burrow systems. Meanwhile, larger workers in both species specialize in defense. In addition, reproduction in both species is limited to a single queen and her mates. These areas of convergence in social organization between such different organisms may help shed light on the forces responsible for the evolution of eusociality. Such comparisons form the basis of the comparative method.

While the comparative method is still used in this way, it is now more often applied in the context of phylogenies, that is, hypothesized relationships between species reflecting their evolution (Huey et al. 2019). Phylogenies are now usually created from genetic analysis, which can tell us the relative sequence of trait evolution and how closely related one group is to another. Solomon Tin Chi Chak and colleagues used the phylogeny of a genus of sponge-dwelling shrimps, *Synalpheus* spp., to explore whether eusociality had evolved from species that rear young communally (Chak et al. 2017). *Synalpheus* spp. were ideal to investigate the evolution of reproductive systems because they include eusocial, pair-forming, and communal species. The researchers found that pair-forming is the ancestral condition, whereas eusociality and communal reproduction each likely evolved independently from pair-forming ancestors, rather than eusociality having evolved from communal species.

What factors may have been important in the evolution and maintenance of naked mole rat and leaf-cutter ant sociality? Kin selection may play a role. Leaf-cutter ants along with other Hymenoptera, such as bees and wasps, have an inheritance system called **haplodiploidy.** The term *haplodiploid* refers to the number of chromosome sets possessed by males and

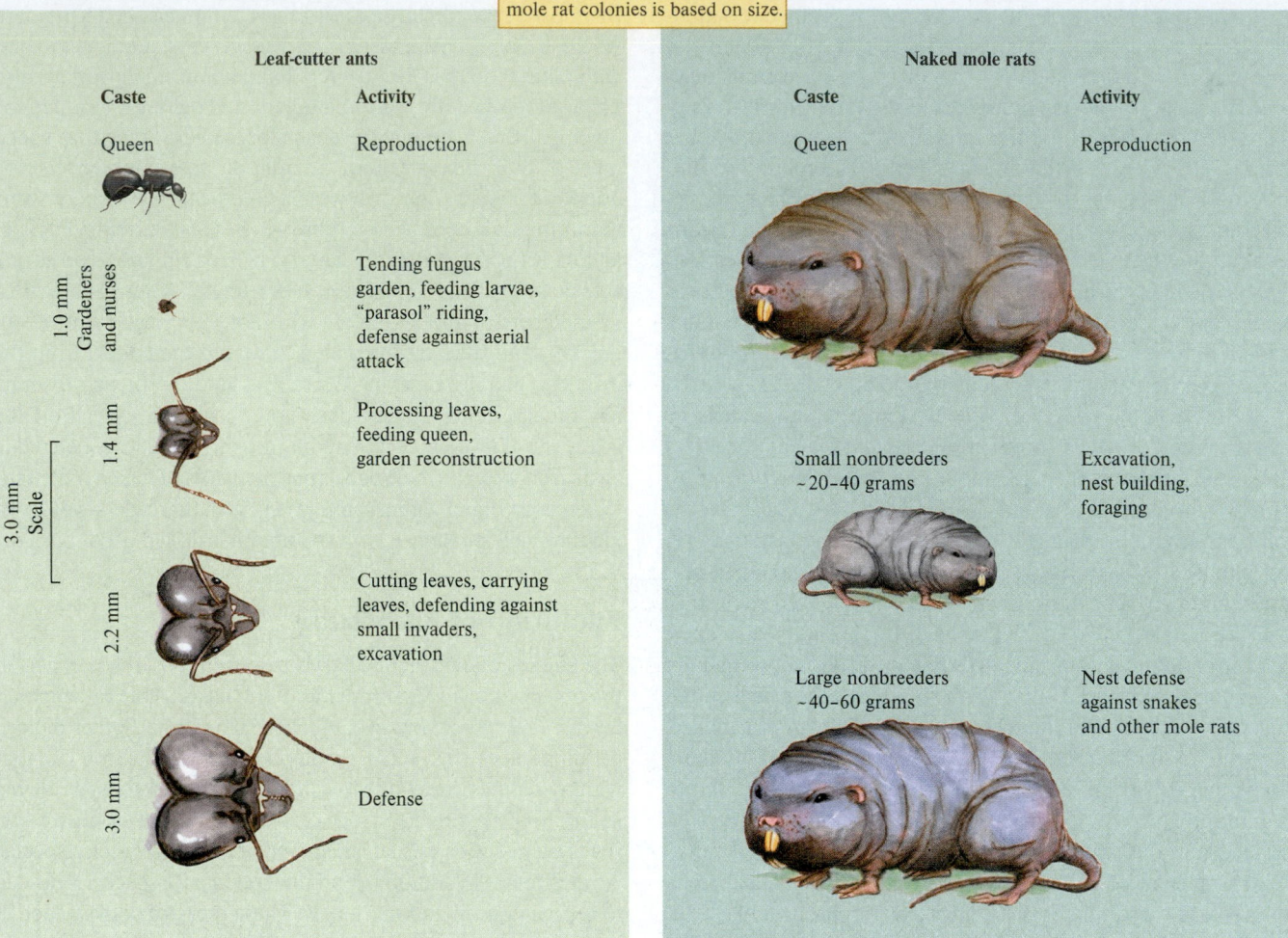

Division of labor in both leaf-cutter ant colonies and naked mole rat colonies is based on size.

Figure 8.29   Division of labor among castes of leaf-cutter ants, *Atta sexdens,* and naked mole rats, *Heterocephalus glaber.* Ant sizes are head widths of workers typically engaged in each activity (data from Wilson 1980, Jarvis 1981, Sherman, Jarvis, and Braude 1992).

females. In haplodiploid systems, males develop from unfertilized eggs and are haploid, whereas females develop from fertilized eggs and are diploid. One of the consequences of haplodiploidy is that worker ants within a colony can be very similar genetically. In an ant colony where there is a single queen that mated with a single male, the workers will be more related to each other than they would be to their own offspring. W. D. Hamilton (1964) was the first to point out that under these conditions the average genetic similarity among workers would be 75%, while their relationship with any offspring they might produce would be 50%.

What is the source of this high degree of relatedness? The queen mates only during her mating flight and stores the sperm she receives to fertilize all the eggs she lays to produce daughters. If she mates with a single male, since he is haploid, all her daughters will receive the same genetic information from their male parent. As a consequence, the 50% of the genetic makeup that workers receive from their male parent will be identical. In addition, workers will share an average of 25% of their genes through those that they receive from the queen, yielding an average genetic relatedness of 50% + 25% = 75%. Of course, where queens mate

with more than one male, relatedness among colony members will be reduced. And, queens in many species of the highly social Hymenoptera, such as honeybees and leaf-cutter ants, do generally mate with multiple males (Strassmann 2001). This discovery weakens kin selection as an explanation for the evolution of eusociality in these species. The important role of haplodiploidy in the evolution of eusociality had been long challenged by the existence of eusociality in termites, which are not haplodiploid, and the discovery of eusociality in other non-haplodiploid organisms, including shrimp, beetles, and spiders along with the naked mole rats discussed here (Gadagkar 2010).

A high degree of relatedness can, however, be maintained in populations that are not haplodiploid. For instance, because naked mole rat colonies are relatively closed to outsiders, the individuals within each colony, like the workers within leaf-cutter ant colonies, are also very similar genetically. Paul Sherman, Jennifer Jarvis, and Stanton Braude (1992) reported that approximately 85% of matings within a colony of naked mole rats are between parents and offspring or between siblings. As a consequence of these matings between close relatives, the relatedness between individuals within a colony is about 81%, suggesting that kin selection may be involved in the maintenance of nonreproductive helpers in colonies of naked mole rats.

What factors other than kin selection may have contributed to the evolution of eusociality? Many factors have been implicated. While researchers working on ants and other social Hymenoptera have emphasized the potential importance of kin selection, studies of cooperative-breeding vertebrate species have emphasized ecological constraints. What sorts of ecological common constraints are faced by leaf-cutter ants and naked mole rats? One of the most obvious is the work associated with the creation, maintenance, and defense of extensive burrow systems against invasion by competing colonies or predators. The more social organisms are studied, the less likely it has become that one or a few simple mechanisms will be adequate to explain their evolution. The relative roles of kin selection, ecological constraints, and other factors in the evolution of eusociality is an area of much current debate and research (Quiñones and Pen 2017, Arsenault et la. 2019, Field and Toyoizumi 2020).

Concept 8.5 Review

1. How would a queen ant, or other eusocial hymenopteran queen, mating with several males affect the relatedness of workers within a colony? If common in social Hymenoptera, how would queens mating with several males affect the potential of kin selection to account for the evolution of eusociality in the Hymenoptera?

2. What are two major ecological challenges favoring colony living that are shared by leaf-cutter ants and naked mole rats?

3. What evidence does colony structure offer in support of the idea that both leaf-cutter ants and naked mole rats must vigorously defend their colonies from predators and invaders?

Behavioral ecology has yielded many intriguing insights into the evolution of animal behavior, including some of the most complex of behavioral phenomena, such as eusociality. However, behavioral ecology also has much to offer of direct practical value, particularly in the area of conservation.

Applications

Behavioral Ecology and Conservation

LEARNING OUTCOMES

After studying this section you should be able to do the following:

8.19   Outline the questions that Tinbergen proposed for understanding a behavior.

8.20   Discuss evidence indicating that environmental enrichment can improve survival by captive-bred animals when they are released into the wild.

Ecologists are increasingly conscious of the essential role of behavioral ecology in conservation. However, Wayne Linklater (2004) points out that while behavioral ecology contributes a great deal to conservation, it has the potential to contribute much more. To achieve its potential, Linklater suggests more balance across the conceptual framework for behavioral ecology proposed by one of its founders, Niko Tinbergen (1963).

Tinbergen's Framework

Tinbergen proposed that a thorough understanding of any behavior requires answers to four fundamental questions:

1. What are the mechanistic causes of the behavior?
2. How does development, including learning, influence the behavior?
3. What is the evolutionary history of the behavior?
4. How does the behavior contribute to fitness?

Reflecting contemporary emphasis within behavioral ecology, the central concepts of this chapter have focused principally on the last question, the adaptive significance of behavior. However, Linklater suggests that information critical to conservation requires more attention to Tinbergen's three other questions. The following section discusses how the development of behavior can be influenced by the environment in which animals are reared.

Environmental Enrichment and Development of Behavior

Programs to conserve many species often combine captive rearing and reintroduction to the wild. However, Fiona Mathews and her colleagues at the University of Oxford's Conservation Research Unit (Mathews et al. 2005) point out that most reintroductions of captive-bred animals fail. These researchers suggest that one reason for failure may be that captive-bred animals behave in ways that reduce their chance of survival. For instance, Mathews and colleagues found that captive-bred bank voles, *Clethrionomys glareolus*, were unable to process a key natural food (hazelnuts) and were less dominant compared to wild-caught bank voles. These researchers suggest that increasing the complexity of the captive environment, an approach called **environmental enrichment,** may help captive animals retain behaviors critical to their survival in the wild. Their suggestion is supported by the results of a study of captive-bred fish.

Victoria Braithwaite and Anne Salvanes (2005), of the Universities of Edinburgh and Bergen, respectively, found that environmental enrichment increases the behavioral flexibility of captive-bred Atlantic cod, *Gadus morhua* (fig. 8.30). Based on the results of their study, they suggest that enriching the hatchery environment may improve the success of fish stocking programs. Hatchery rearing and stocking, a common practice in fisheries management, is intended to counter the effects of overfishing and environmental degradation. However, survival of captive-bred fish in the wild is often poor. Braithwaite and Salvanes suggest that poor survival among captive-bred fish may be due to behaviors developed in response to the typically uniform hatchery environment. To address this

Figure 8.30 The Atlantic cod, *Gadus morhua*, is one of the most valued commercial fishes and is the focus of intensive management. The species is associated with deep-sea bottom environments, which can be very complex spatially, across the North Atlantic.   Matthew Lawrence, Stellwagen Bank NMS, NOS, NOAA

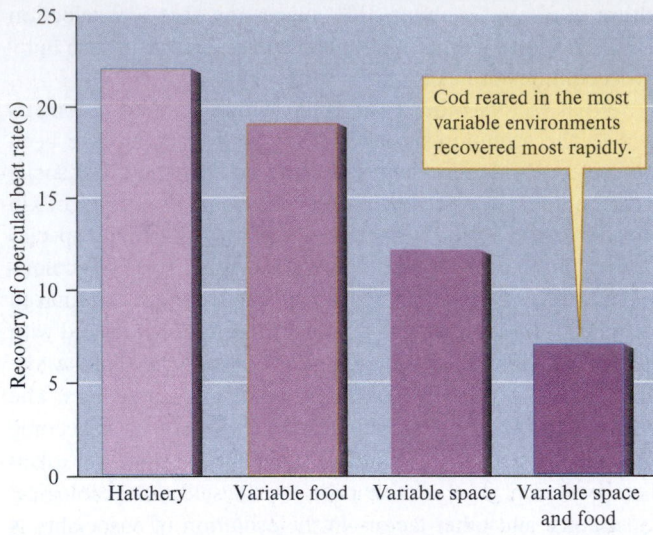

Cod reared in the most variable environments recovered most rapidly.

Figure 8.31   Influence of rearing environment on time required by hatchery-reared North Sea cod to recover from a simulated encounter with a predator; recovery measured as the time required for opercular beat rate to return to resting level (data from Braithwaite and Salvanes 2005).

question, they raised juvenile cod under four environmental circumstances: (1) hatchery—a conventional bare hatchery tank with food available continuously in small amounts from a single point; (2) variable food—a bare tank in which the availability of food varied in space and time; (3) variable space—a tank with a natural stony bottom and seaweeds but with constant food available; and (4) variable food and space—a tank with a natural bottom and a variable food supply.

Braithwaite and Salvanes raised 100 juvenile cod in each condition. Cod raised under typical hatchery conditions grew faster than any other experimental groups. However, cod exposed to variable environments were quicker to join another cod when given the opportunity, were quicker to recover physiologically from a simulated predator attack (fig. 8.31), responded to live prey more rapidly, and made a more rapid transition from pelleted, artificial food to live prey. Clearly, environmental variability affects the development of several behaviors in cod that would likely influence their chances of surviving in the wild. There is great potential for further work on cod and for similar studies on other species. Such behavioral studies will undoubtedly yield a great deal of information helpful to conservation.

Summary

Social relations are important, since they often directly impact a key component of evolutionary fitness, the number of offspring, or genes, contributed by an individual to future generations. One of the most fundamental social interactions between individuals takes place during sexual reproduction.

The effects of female mate choice on evolution of ornamentation in males can be reduced by other sources of natural selection.   Sexual selection results from differences in reproductive rates among individuals as a result of differences in their mating success. Sexual selection is thought to work either through intrasexual selection, where individuals of one sex compete with each other for mates, or intersexual selection, when members of one sex choose mates from among members of the opposite sex on the basis of some particular trait. Experimental evidence supports the hypothesis that the coloration of male guppies in local populations is determined by a dynamic interplay between natural selection exerted by predators, under which less-colorful males have higher survival, and by female mate choice, which results in higher mating success by more-colorful males.

Females of some species select mates based on the male's ability to provide important resources.   Among scorpionflies, larger males are more likely to successfully defend available arthropod offerings due to their advantages in aggressive encounters and consequently mate more frequently than smaller males without arthropod offerings.

Mating in wild plant populations can be nonrandom. Studies of mating in the wild radish, *Raphanus sativus,* in greenhouse and field experiments indicate nonrandom mating and suggest competition among pollen from different pollen donors.

Evolution of sociality in many species appears driven by the need for group defense of high-quality territories and/or defense of mates and young.   The degree of sociality in a social species ranges from acts as simple as mutual grooming or group protection of young to highly complex, stratified societies such as those found in colonies of ants or termites. This more complex level of social behavior, which is considered to be the pinnacle of social evolution, is called eusociality. Eusociality is generally thought to include three major characteristics: (1) individuals of more than one generation living together, (2) cooperative care of young, and (3) division of individuals into sterile, or nonreproductive, and reproductive castes.

Cooperation among green woodhoopoes and African lions appears to be a response to environmental conditions that require cooperation for success. For green woodhoopoes, the scarcity of high-quality territories and intense competition between flocks for those territories create conditions that favor staying in the natal territory and helping raise related young and perhaps inheriting the territory at a later date. To survive, reproduce, and successfully raise offspring to maturity, African lions must work in cooperative groups of females, which are called prides, and of males, which are called coalitions.

Kin selection and ecological constraints may have played key roles in the evolution of eusociality. The comparative method has been used to study the evolution of eusociality among a wide variety of animal species, including leaf-cutter ants and naked mole rats, both of which live in social groups in which individuals are divided among castes that engage in very different activities. Compared to other ant species, leaf-cutter ant colonies have a larger number of castes that engage in a wider variety of behaviors. In contrast to leaf-cutter ant colonies, where all workers are females, both males and females work in naked mole rat colonies. However, as in leaf-cutter ant colonies, work in naked mole rat colonies is divided among members according to their size. Many factors have likely contributed to the evolution of eusociality in leaf-cutter ants and naked mole rats, including kin selection and ecological constraints.

Behavioral ecology has much to offer of direct practical value to conservation biology, such as the influence of environment on the development of behavior.

Key Terms

behavioral ecology 173	Hamilton's rule 184	kin selection 184	self-incompatibility 181
caste 190	haplodiploidy 191	lifetime reproductive	sexual selection 174
comparative method 189	hermaphrodite 174	success 186	sociality 184
environmental	inclusive fitness 184	male 174	sociobiology 173
enrichment 193	intersexual	natal territory 185	stamen 181
eusociality 184	selection 175	philopatry 185	
female 174	intrasexual selection 174	pistil 181	

Review Questions

1. The introduction to chapter 8 included sketches of the behavior and social systems of several fish species. Using the concepts that you have learned in this chapter, revisit those examples and predict the forms of sexual selection occurring in each species.

2. One of the basic assumptions of the material presented in chapter 8 is that the form of reproduction will exert substantial influence on social interactions within a species. How should having several forms of one sex, for example, large and small males, influence the diversity of behavioral interactions within the population?

3. Endler (1980) pointed out that though field observations are consistent with the hypothesis that predators may exert natural selection on guppy coloration, some other factors in the environment could be affecting variation in male color patterns among guppy populations. What other factors, especially physical and chemical factors, might influence male color?

4. Endler set up two experiments, one in the greenhouse and one in the field. What were the advantages of the greenhouse experiments? What were the shortcomings of the greenhouse experiments? Endler also set up field experiments along the Aripo River. What were the advantages of the field experiments and what were their shortcomings?

5. Discuss the scorpionfly mating system. Pay particular attention to the potential roles of intersexual and intrasexual selection in scorpionflies.

6. The results of numerous studies indicate nonrandom mating among plants. These results lead to questions concerning the biological mechanisms that produce these nonrandom matings. How might the maternal plant control or at least influence the paternity of her seeds? What role might competition between pollen determine in the nonrandom patterns observed?

7. The details of experimental design are critical for determining the success or failure of both field and laboratory experiments. Results often depend on some small details. For instance, Jennifer Jarvis waited 1 year after establishing her laboratory colony of naked mole rats before attempting to quantify the behavior of the laboratory population. What might have been the consequence of beginning to quantify the behavior of the colony soon after it was established?

8. Behavioral ecologists have argued that naked mole rats are eusocial. What are the major characteristics of eusociality and which of those characteristics are shared by naked mole rats?

9. Choose a problem in the ecology of social relations, formulate a hypothesis, and design a study to test your ideas. Take two approaches. In one approach use field and laboratory experiments to test your ideas. In the second design, develop a study that will employ the comparative method.

Chapter

9

Population Distribution and Abundance

A population of saguaro cactus, *Carnegiea gigantea,* at Saguaro National Park near Tucson, Arizona. The natural distribution of the saguaro cactus is restricted to the Sonoran Desert, extending from central Arizona through northwestern Mexico. Because saguaro cactus are very sensitive to freezing, the northern limit and highest elevations occupied by the species appear limited by low temperatures.

Concept 9.4 Review 210

Applications: Rarity and Vulnerability
   to Extinction 210
   Summary 213
   Key Terms 213
   Review Questions 213

CHAPTER CONCEPTS

9.1 Environment limits the geographic distribution of species. 198
   Concept 9.1 Review 201

9.2 On small scales, individuals within populations are distributed in patterns that may be random, regular, or clumped. 201
   Concept 9.2 Review 205

9.3 On large scales, individuals within a population are clumped. 205

   Concept 9.3 Review 208

9.4 Population density declines with increasing organism size. 208

LEARNING OUTCOMES

After studying this section you should be able to do the following:

9.1 Define population, density, and abundance.
9.2 Compare the characteristics of gray whale, monarch butterfly, and Monterey pine populations.
9.3 Outline some of the reasons ecologists study populations.

The distributions and dynamics of populations vary widely among species. While some populations are small and have highly restricted distributions, other populations number in the millions and may range over vast areas of the planet. Standing on a headland in central California overlooking the Pacific Ocean, a small class of students spots a group of gray whales, *Eschrichtius robustus,* rising to the

(a)

(b)

Figure 9.1   (*a*) A young gray whale, *Eschrichtius robustus*. During their annual migration, gray whales migrate from subtropical waters off Baja California to the Arctic and back again, passing along the coast of California as they do so. (*b*) Along that same coast, monarch butterflies, *Danaus plexippus*, winter in huge numbers, some having flown thousands of kilometers from the Rocky Mountains to reach the trees where they roost in winter. In contrast, the entire natural population of the Monterey pine, *Pinus radiata*, is restricted to five small areas along the California coast.   (*a*) Getty Images; (*b*) Sexto Sol/Photodisc/Getty Images

surface and spouting water as they swim northward (fig. 9.1*a*). The whales are rounding the point of land on their way to feeding grounds off the coasts of Alaska and Siberia. This particular group is made up of females and calves. The calves were born during the previous winter along the coast of Baja California, the gray whale's wintering grounds. Over the course of the spring, the entire population of over 20,000 gray whales will round this same headland on their way to the Bering and Chukchi Seas. Gray whales travel from one end of their range to the other twice each year, a distance of about 18,000 km. Home to the gray whale encompasses a swath of seacoast extending from southern Baja California to the coast of northeastern Asia.

The grove of pine trees on the headland where the students stand gazing at the whales is winter home to another long-distance traveler: monarch butterflies, *Danaus plexippus* (fig. 9.1*b*). The lazy flying of the bright orange and black monarch butterflies gives no hint of their capacity to migrate. Some

of the butterflies flew to the grove of pines the previous autumn from as far away as the Rocky Mountains of southern Canada. As the students watch the whales, the male monarch butterflies pursue and mate with the female butterflies. After mating, the males die, while the females begin a migration that leads inland and north. The females stop to lay eggs on milkweeds they encounter along the way and eventually die; however, their offspring continue the migration. Monarch caterpillars grow quickly on their diet of milkweed and then transform to a pupa or chrysalis. The monarch butterflies that emerge from the chrysalises (or chrysalides) mate and, like the previous generation, fly northward and inland. By moving farther north and inland each generation, some of the monarch butterflies eventually reach the Rocky Mountains of southern Canada, far from where their ancestors fluttered around the group of students on the pine-covered coastal headland.

Then, as the autumn days grow shorter, the monarch butterflies begin their long flight back to the coastal grove of pines. This autumn generation, which numbers in the millions, flies southwest to its wintering grounds on the coast of central and southern California. Some of the monarchs might fly over 3,000 km. Those that survive the trip to the pine grove overwinter, hanging from particular roost trees in the thousands. They mate in the following spring and start the cycle all over again.

Gray whales and monarch butterflies, as different as they may appear, lead parallel lives. The Monterey pines, *Pinus radiata*, covering the headland where the monarch butterflies overwinter and by which the gray whales pass twice each year, are quite different. The Monterey pine population does not migrate each generation and has a highly restricted distribution. Its current natural range is limited to a few sites on the coast of central and northern California and to two islands off the coast of western Mexico. These scattered populations are the remnants of a large, continuous population that extended for over 800 km along the California coast during the cooler climate of the last glacial period. This history of the Monterey pine underscores a very important fact about species distributions: They are highly dynamic, especially over long periods of time.

With these three examples, we begin to consider the ecology of populations. Ecologists usually define a **population** as a group of individuals of a single species inhabiting a specific area. A population of plants or animals might occupy a mountaintop, a river basin, a coastal marsh, or an island, all areas defined by natural boundaries. Just as often, the populations studied by biologists occupy artificially defined areas such as a country, county, or national park. The areas inhabited by populations range in size from the few cubic centimeters occupied by the bacteria in a rotting apple to the millions of square kilometers occupied by a population of migratory whales. A population studied by ecologists may consist of a highly localized group of individuals representing a fraction of the total population of a species, or it may consist of all the individuals of a species across its entire range.

Ecologists study populations for many reasons. Population studies hold the key to saving endangered species, controlling pest populations, and managing fish and game populations. They also offer clues to understanding and controlling disease epidemics. Finally, the greatest environmental challenge to biological diversity and the integrity of the entire biosphere is at its heart a population problem—the growth of the human population.

All populations share several characteristics. The first is their **distribution.** The distribution of a population includes the size, shape, and location of the area it occupies. A population also has a characteristic pattern of spacing of the individuals within it. It is also characterized by the number of individuals within it and their **density,** which is the number of individuals per unit area. Additional characteristics of populations—their age distributions, birth and death rates, immigration and emigration rates, and rates of growth—are the subject of chapters 10 and 11. In chapter 9 we focus on two population characteristics: distribution and **abundance,** the total number of individuals, or biomass, of a species in a specified area.

9.1 Distribution Limits

After studying this section you should be able to do the following:

9.4   Define niche and distinguish between the fundamental niche and the realized niche.

9.5   Compare the perspectives of Grinnell, Elton, and Hutchinson on the nature of the niche.

9.6   Discuss the factors limiting the distributions of plants and animals on geographic and local spatial scales.

Environment limits the geographic distribution of species. A major theme in chapters 5, 6, and 7 is that populations have evolved physiological, anatomical, and behavioral characteristics that compensate for environmental variation. Organisms compensate for temporal and spatial variation in the environment by regulating body temperature and water content and by foraging in ways that maintain energy intake at relatively high levels. However, there are limits on how much organisms can compensate for environmental variation.

While there are few environments on earth without life, no single species can tolerate the full range of earth's environments. For each species, some environments are too warm, too cold, too saline, or unsuitable in other ways. As we saw in chapter 7, organisms take in energy at a limited rate. It appears that at some point, the metabolic costs of compensating for environmental variation may take up too much of an organism's energy budget. Partly because of these energy constraints, the physical environment places limits on the distributions of populations. The environmental limits of a species are related to its **niche.** The word *niche* has been in use a long time. Its earliest and most basic meaning was that of a recessed place in a wall where one could set or display items. For about a century, however, ecologists have given a broader meaning to the word. To the ecologist, the niche summarizes the environmental factors that influence the growth, survival, and reproduction of a species. In other words, a species' niche consists of all the factors necessary for its existence—approximately when, where, and how a species makes its living.

The niche concept was developed independently by Joseph Grinnell (1917, 1924) and Charles Elton (1927), who used the term *niche* in slightly different ways. In his early writings, Grinnell's ideas of the niche centered around the influences of the physical environment, while Elton's earliest concept included biological interactions as well as abiotic factors. However their thinking and emphasis may have differed, it is clear that the views of these two researchers had much in common and that our present concept of the niche rests squarely on their pioneering work.

A single paper authored by G. Evelyn Hutchinson (1957) crystallized the niche concept. In this seminal paper titled simply "Concluding Remarks," Hutchinson defined the niche as an *n-dimensional hypervolume,* where *n* equals the number of environmental factors important to the survival and reproduction by a species. Hutchinson called this hypervolume, which specifies the values of the *n* environmental factors permitting a species to survive and reproduce, as the **fundamental niche** of the species. The fundamental niche defines the physical conditions under which a species might live, in the absence of interactions with other species. However, Hutchinson recognized that interactions such as competition may restrict the environments in which a species may live. He referred to these more restricted conditions as the **realized niche,** which is the actual niche of a species whose distribution is limited by biotic interactions such as competition, predation, disease, and parasitism. In a single word, *niche* captures much of what we discussed in section II and now turn to in section III. In this section, we consider how environment affects the growth, survival, reproduction, distribution, and abundance of species, a particularly timely topic as we face the ecological consequences of global warming.

Kangaroo Distributions and Climate

The family Macropodidae includes the kangaroos and wallabies, which are some of the best known of the Australian animals. However, this group of large-footed mammals includes many less-familiar species, including rat kangaroos and tree kangaroos. While some species of macropods can be found in nearly every part of Australia, no single species ranges across the entire continent. All are confined to a limited number of climatic zones and biomes.

G. Caughley and his colleagues (1987) found a close relationship between climate and the distributions of the three largest kangaroos in Australia (fig. 9.2). The eastern grey kangaroo, *Macropus giganteus,* is confined to the eastern third of the continent. This portion of Australia includes several biomes (see chapter 2). Temperate forest grows in the southeast and tropical forests in the north. Mountains, with their varied climates, occupy the central part of the eastern grey kangaroo's range (see figs. 2.14, 2.29, and 2.38). The climatic factor that distinguishes these varied biomes is little seasonal variation in precipitation or dominance by summer precipitation. The western grey kangaroo, *M. fuliginosus,* lives mainly in the southern and western regions of Australia, which coincides largely with the distribution of the Mediterranean woodland and shrubland biome in Australia. The climatically distinctive feature of this biome is a predominance of winter rainfall

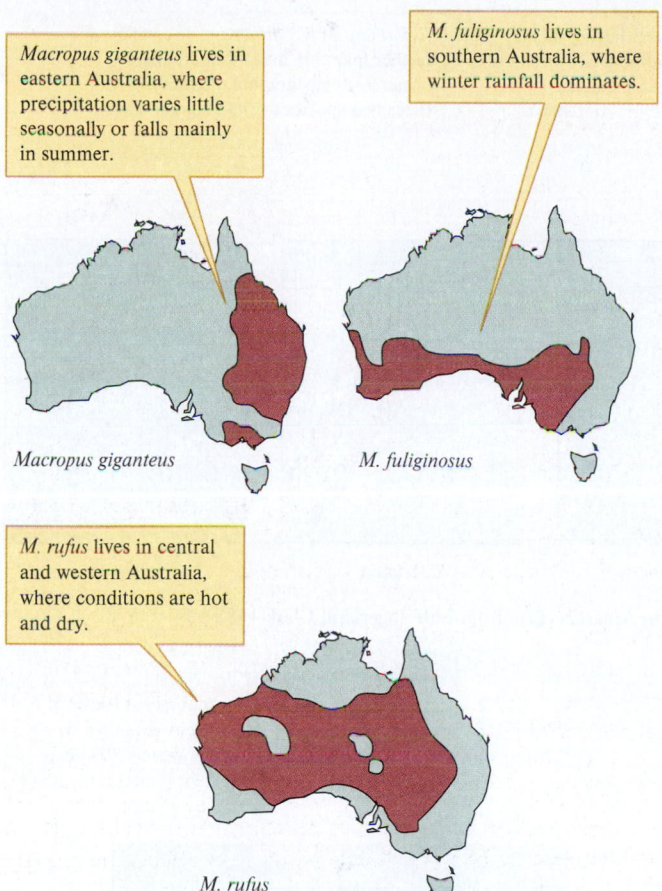

Macropus giganteus lives in eastern Australia, where precipitation varies little seasonally or falls mainly in summer.

M. fuliginosus lives in southern Australia, where winter rainfall dominates.

Macropus giganteus   *M. fuliginosus*

M. rufus lives in central and western Australia, where conditions are hot and dry.

M. rufus

Figure 9.2   Climate and the distributions of three kangaroo species (data from Caughley et al. 1987).

Figure 9.3   The red kangaroo (*Macropus rufus*) occupies the hottest and driest areas.   Jami Tarris/Getty Images

(a)   (b)

(c)   (d)

Figure 9.4   Four *Encelia* species that have adapted to different temperatures and have different distributions across North America: (a) *E. californica,* (b) *E. actoni,* (c) *E. farinosa,* (d) *E. fructens.*
(*a*) Wirestock Creators/Shutterstock; (*b* and *d*) Jared Quentin/iStock/Getty Images; (*c*) Christopher Bellette/Alamy Stock Photo

(see fig. 2.23). Meanwhile, the red kangaroo, *M. rufus,* wanders the arid and semiarid interior of Australia, areas dominated by savanna and desert (see figs. 2.17 and 2.20). Of the three species of large kangaroos, the red kangaroo occupies the hottest and driest areas (fig. 9.3).

The distributions of these three large kangaroo species cover most of Australia. However, as you can see in figure 9.2, none of these species lives in the northernmost region of Australia. Caughley and his colleagues explain that these northern areas are probably too hot for the eastern grey kangaroo, too wet for the red kangaroo, and too hot in summer and too dry in winter for the western grey kangaroo. However, they are also careful to point out that these limited distributions may not be determined by climate directly. Instead, they suggest that climate often influences species distributions through factors such as food production, water supply, and habitat. Climate also affects the incidence of parasites, pathogens, and competitors.

Regardless of how the influences of climate are played out, the relationship between climate and the distributions of species can be stable over long periods of time. The distributions of the eastern grey, western grey, and red kangaroos have been stable for at least a century.

Now, let's consider how the physical environment may limit the distribution of plants. Our example is drawn from the arid and semiarid regions of the American Southwest.

Distributions of Plants Along a Moisture-Temperature Gradient

In chapter 5, we discussed the influence of pubescence on leaf temperature in plants of the genus *Encelia.* Variation in leaf pubescence among *Encelia* species appears to correspond directly to the distributions of these species along a moisture-temperature gradient from the California coast eastward (Ehleringer and Clark 1988). *Encelia californica,* the species with the least pubescent leaves, occupies a narrow coastal zone that extends from southern California to northern Baja California (fig. 9.5). Inland, *E. californica* is replaced by *E. actoni,* which has leaves that are slightly more pubescent. Still farther to the east, *E. actoni* is in turn replaced by *E. frutescens* and *E. farinosa.*

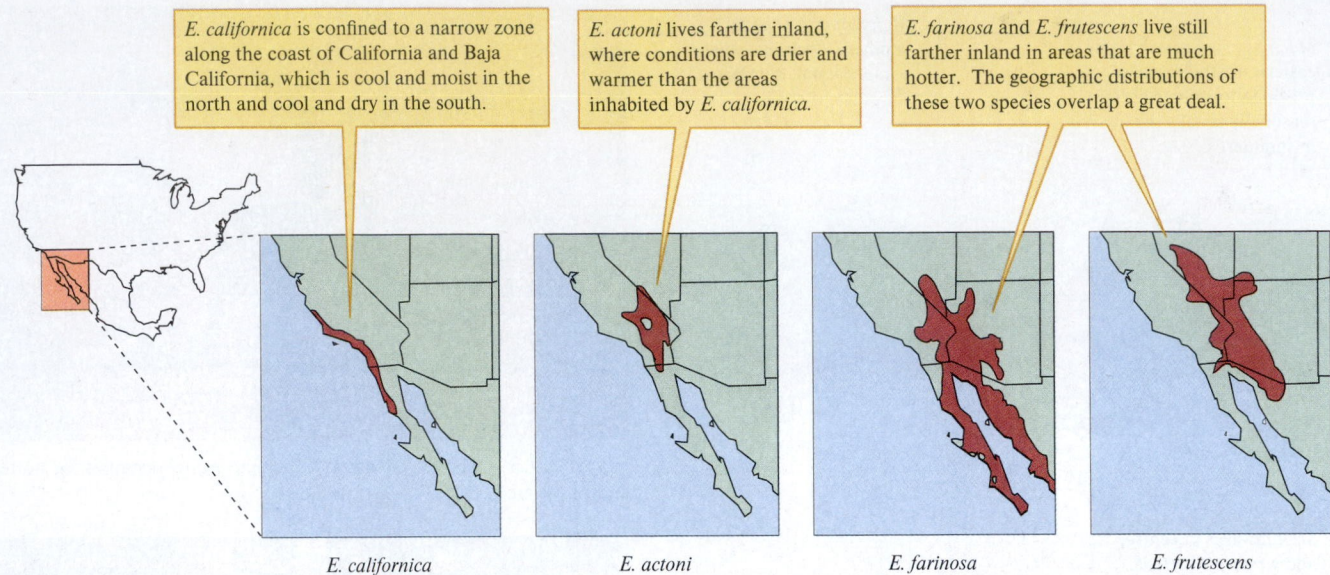

E. californica is confined to a narrow zone along the coast of California and Baja California, which is cool and moist in the north and cool and dry in the south.

E. actoni lives farther inland, where conditions are drier and warmer than the areas inhabited by E. californica.

E. farinosa and E. frutescens live still farther inland in areas that are much hotter. The geographic distributions of these two species overlap a great deal.

E. californica   E. actoni   E. farinosa   E. frutescens

Figure 9.5   The distributions of four *Encelia* species in southwestern North America (data from Ehleringer and Clark 1988).

These geographic limits to these species' distributions correspond to variations in temperature and precipitation. The coastal environments where *E. californica* lives are all relatively cool. However, average annual precipitation differs a great deal across the distribution of this species. Annual precipitation ranges from about 100 mm in the southern part of its distribution to well over 400 mm in the northern part. By comparison, *E. actoni* occupies environments that are only slightly warmer but considerably drier. The rainfall in areas occupied by *E. frutescens* and *E. farinosa* is similar to the amount that falls in the areas occupied by *E. actoni* and *E. californica*. However, the environments of *E. frutescens* and *E. farinosa* are much hotter.

Variation in leaf pubescence does not correspond entirely to the macroclimates inhabited by *Encelia* species. The leaves of *E. frutescens* are nearly as free of pubescence as the coastal species, *E. californica*. However, *E. frutescens* grows side by side with *E. farinosa* in some of the hottest deserts in the world. Because they are sparsely pubescent, the leaves of *E. frutescens* absorb a great deal more radiant energy than the leaves of *E. farinosa* (fig. 9.6). Under similar conditions, however, leaf temperatures of the two species are nearly identical. How does *E. frutescens* avoid overheating? The leaves do not overheat because they transpire at a high rate and are evaporatively cooled as a consequence.

Evaporative cooling solves one ecological puzzle but appears to create another. Remember that these two shrubs live in some of the hottest and driest deserts in the world. Where does *E. frutescens* get enough water to evaporatively cool its leaves? Although the distributions of *E. frutescens* and *E. farinosa* overlap a great deal on a geographic scale, these two species occupy distinctive microenvironments. As shown in figure 9.7, *E. farinosa* grows mainly on upland slopes, while *E. frutescens* is largely confined to ephemeral stream channels, or desert washes. Along washes, runoff infiltrates into the deep soils increasing the availability of soil moisture to *E. frutescens*. This example reminds us of a principle that we first considered

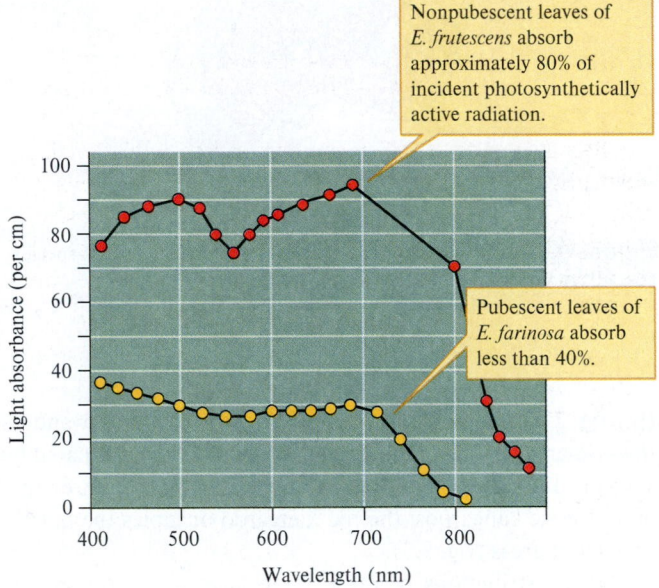

Nonpubescent leaves of *E. frutescens* absorb approximately 80% of incident photosynthetically active radiation.

Pubescent leaves of *E. farinosa* absorb less than 40%.

Figure 9.6   Light absorption by leaves of *Encelia frutescens* and *E. farinosa* (data from Ehleringer and Clark 1988).

in chapter 5: Organisms living in the same macroclimate may, because of slight differences in local distribution, experience substantially different microclimates. This is certainly true of the two barnacle species we consider in the following example.

Distributions of Barnacles Along an Intertidal Exposure Gradient

The marine intertidal zone presents a steep gradient of physical conditions from the shore seaward. As we saw in chapter 3, the organisms high in the intertidal zone are exposed by virtually every tide while the organisms that live at lower levels in the intertidal zone are exposed by the lowest tides only. Exposure to air differs at different levels within the intertidal

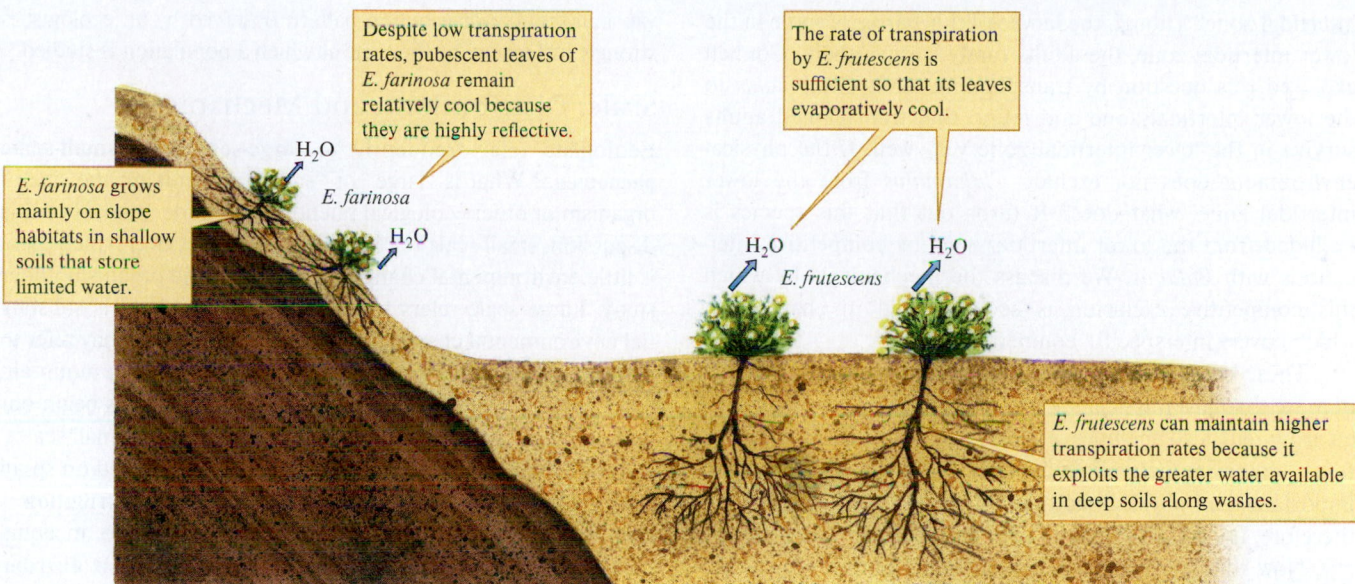

E. farinosa grows mainly on slope habitats in shallow soils that store limited water.

Despite low transpiration rates, pubescent leaves of *E. farinosa* remain relatively cool because they are highly reflective.

The rate of transpiration by *E. frutescen*s is sufficient so that its leaves evaporatively cool.

E. frutescens can maintain higher transpiration rates because it exploits the greater water available in deep soils along washes.

H_2O *E. farinosa* H_2O H_2O *E. frutescens* H_2O

Figure 9.7   Temperature regulation and distributions of *Encelia farinosa* and *E. frutescens* across microenvironments.

zone. Organisms that live in the intertidal zone have evolved different degrees of resistance to drying, a major factor contributing to zonation among intertidal organisms (see fig. 3.16).

Barnacles, one of the most common intertidal organisms, show distinctive patterns of zonation within the intertidal zone. For example, Joseph Connell (1961a, 1961b) described how along the coast of Scotland, adult *Chthamalus stellatus* are restricted to the upper levels of the intertidal zone, while adult *Balanus balanoides* are limited to the middle and lower levels (fig. 9.8). What role does resistance to drying play in the intertidal zonation of these two species? Unusually, calm and warm weather combined with very low tides gave Connell

some insights into this question. In the spring of 1955, warm weather coincided with calm seas and very low tides. As a consequence, no water reached the upper intertidal zone occupied by both species of barnacles. During this period, *Balanus* in the upper intertidal zone suffered much higher mortality than *Chthamalus* (fig. 9.9). Meanwhile, *Balanus* in the lower intertidal zone showed normal rates of mortality. Of the two species, *Balanus* appears to be more vulnerable to desiccation. Higher rates of desiccation may exclude this species of barnacle from the upper intertidal zone.

Vulnerability to desiccation, however, does not completely explain the pattern of intertidal zonation shown by *Balanus* and *Chthamalus*. What excludes *Chthamalus* from the lower

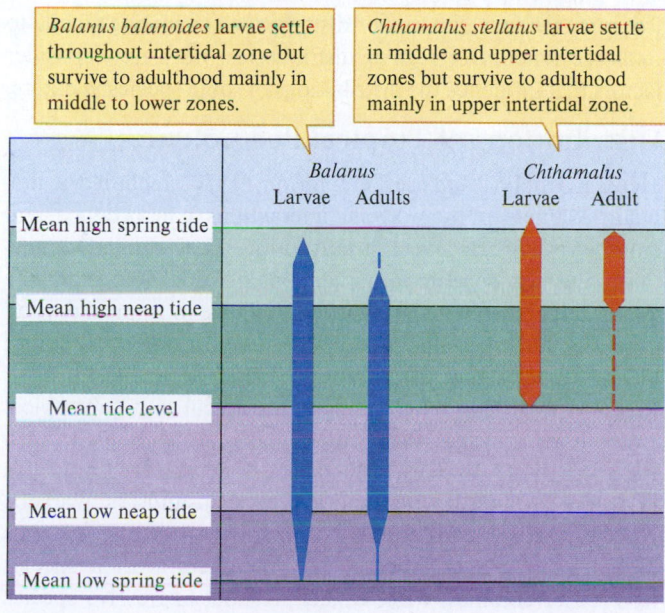

Balanus balanoides larvae settle throughout intertidal zone but survive to adulthood mainly in middle to lower zones.

Chthamalus stellatus larvae settle in middle and upper intertidal zones but survive to adulthood mainly in upper intertidal zone.

Balanus   Larvae   Adults   *Chthamalus*   Larvae   Adult

Mean high spring tide
Mean high neap tide
Mean tide level
Mean low neap tide
Mean low spring tide

Figure 9.8   Distributions of two barnacle species within the intertidal zone (data from Connell 1961b).

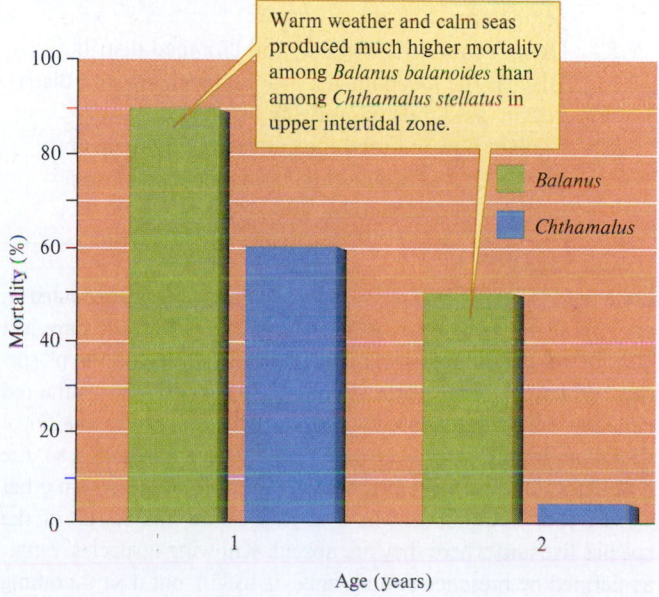

Warm weather and calm seas produced much higher mortality among *Balanus balanoides* than among *Chthamalus stellatus* in upper intertidal zone.

Mortality (%)

Balanus
Chthamalus

Age (years)

Figure 9.9   Barnacle mortality in the upper intertidal zone (data from Connell 1961b).

intertidal zone? Though the larvae of this barnacle settle in the lower intertidal zone, the adults rarely survive there. Connell explored this question by transplanting adult *Chthamalus* to the lower intertidal zone and found that transplanted adults survive in the lower intertidal zone very well. If the physical environment does not exclude *Chthamalus* from the lower intertidal zone, what does? It turns out that this species is excluded from the lower intertidal zone by competitive interactions with *Balanus.* We discuss the mechanisms by which this competitive exclusion is accomplished in chapter 13, which covers interspecific competition.

These barnacles remind us that the environment consists of more than just physical and chemical factors. An organism's environment also includes biological factors. In many situations, biological factors may be as important as or even more important than physical factors in determining the niche, and therefore, the distribution and abundance of a species.

Now that we have considered factors limiting the distributions of populations, let's consider patterns of distribution of individuals within their habitat. Let's begin by considering three basic patterns of distribution.

Concept 9.1 Review

1. How may a species respond to climate change?
2. How might biological and physical aspects of the environment interact to influence a species' geographic distribution?

9.2   Patterns on Small Scales

LEARNING OUTCOMES

After studying this section you should be able to do the following:

9.7   Define small and large scale from an ecological perspective.

9.8   Describe random, regular, and clumped distributions.

9.9   Discuss the mechanisms producing changes in distribution, as stands of creosote bush age.

9.10   Explain how interactions between individuals in a population are thought to influence distribution patterns in populations.

On small scales, individuals within populations are distributed in patterns that may be random, regular, or clumped.   We have just considered how the environment limits the distributions of species. When you map the distribution of a species such as the red kangaroo in Australia (see fig. 9.2), or the zoned distribution of *Chthamalus* and *Balanus* in the intertidal zone (see fig. 9.8), the boundaries on your map indicate the range of the species. In other words, your map shows where at least some individuals of the species live and where they are absent. Knowing a species' range, as defined by presence and absence, is useful, but it says nothing about how the individuals that make up the population are distributed in the areas where they are present. Are individuals randomly distributed across the range? Are they regularly distributed? As

we shall see, the distribution pattern observed by an ecologist is strongly influenced by the scale at which a population is studied.

Scale, Distributions, and Mechanisms

Ecologists refer frequently to **large-scale** and **small-scale phenomena.** What is "large" or "small" depends on the size of organism or other ecological phenomenon under study. For this discussion, small scale refers to small distances over which there is little environmental change significant to the organism under study. Large scale refers to areas over which there is substantial environmental change. In this sense, large scale may refer to patterns over an entire continent or patterns along a mountain slope, where environmental gradients are steep. Let's begin our discussion with patterns of distribution observed at small scales.

Three basic patterns of distribution are observed on small scales: random, regular, or clumped. A **random distribution** is one in which individuals within a population have an equal chance of living anywhere within an area. A **regular distribution** is one in which individuals are uniformly spaced. In a **clumped distribution,** individuals have a much higher probability of being found in some areas than in others (fig. 9.10).

These three basic patterns of distribution are produced by the kinds of interactions that take place between individuals within a population, by the structure of the physical environment, or by a combination of interactions and environmental structure. Individuals within a population may *attract* each other, *repel* each other, or *ignore* each other. Mutual attraction creates clumped, or aggregated, patterns of distribution. Regular patterns of distribution are produced when individuals avoid each other or claim exclusive use of a patch of landscape. Neutral responses contribute to random distributions.

The patterns created by social interactions may be reinforced or damped by the structure of the environment. An environment with patchy distributions of nutrients, nesting sites, water, and so forth fosters clumped distribution patterns. An environment with a fairly uniform distribution of resources and frequent, random patterns of disturbance (or mixing) tends to reinforce random or regular distributions. Let's now consider factors that influence the distributions of some species in nature.

Distributions of Tropical Bee Colonies

Stephen Hubbell and Leslie Johnson (1977) recorded a dramatic example of how social interactions can produce and enforce regular spacing in a population. They studied competition and nest spacing in populations of stingless bees in the family Trigonidae. The bees they studied live in tropical dry forest in Costa Rica. Though these bees do not sting, rival colonies of some species fight fiercely over potential nesting sites.

Stingless bees are abundant in tropical and subtropical environments, where they gather nectar and pollen from a wide variety of flowers. They generally nest in trees and live in colonies made up of hundreds to thousands of workers. Hubbell and Johnson observed that some species of stingless bees are highly aggressive to other members of their species from other colonies, while others are not. Aggressive species usually forage in groups and feed mainly on flowers that occur in high-density clumps. The nonaggressive species feed singly or in small groups and on more widely distributed flowers.

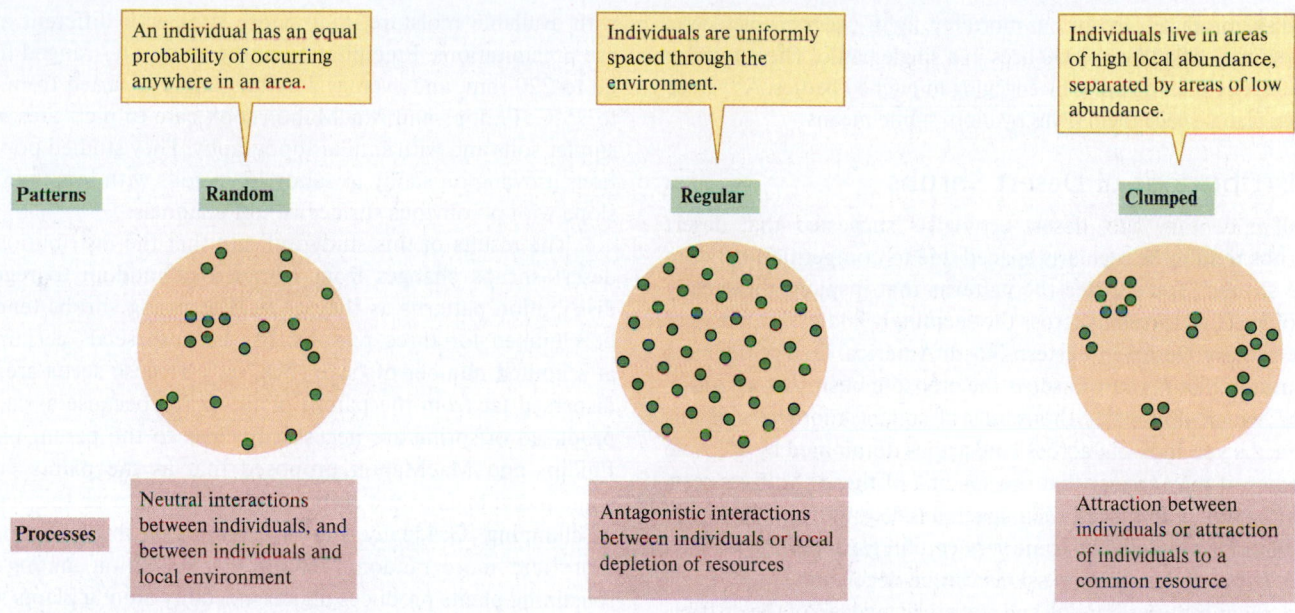

Figure 9.10   Random, regular, and clumped distributions.

Hubbell and Johnson studied several species of stingless bees to determine whether there is a relationship between aggressiveness and patterns of colony distribution. They predicted that the colonies of aggressive species would show regular distributions while those of nonaggressive species would show random or clumped distributions. They concentrated their studies on a 13 ha tract of tropical dry forest that contained numerous nests of nine species of stingless bees.

Though Hubbell and Johnson were interested in how bee behavior might affect colony distributions, they recognized that the availability of potential nest sites for colonies could also affect distributions. So, in one of the first steps in their study, they mapped the distributions of trees suitable for nesting. They found that potential nest trees were distributed randomly through the study area and that the number of potential nest sites was much greater than the number of bee colonies.

Hubbell and Johnson were able to map the nests of five of the nine species of stingless bees accurately. The nests of four of these species were distributed regularly. As they had predicted, all four species with regular nest distributions were highly aggressive to bees from other colonies of their own species. The fifth species, *Trigona dorsalis,* was not aggressive and its nests were randomly distributed over the study area. Figure 9.11 contrasts the random distribution of *T. dorsalis* with the regular distribution of one of the aggressive species, *T. fulviventris.*

The researchers also studied the process by which the aggressive species establish new colonies. In the process, they made observations that provide insights into the mechanisms that establish and maintain the regular nest distributions of species such as *T. fulviventris.* This species and the other aggressive species apparently mark prospective nest sites with a pheromone. **Pheromones** are chemical substances secreted by some animals for communication with other members of their species. The pheromone secreted by these stingless bees attracts and aggregates members of their colony to the prospective nest site; however, it also attracts workers from other nests.

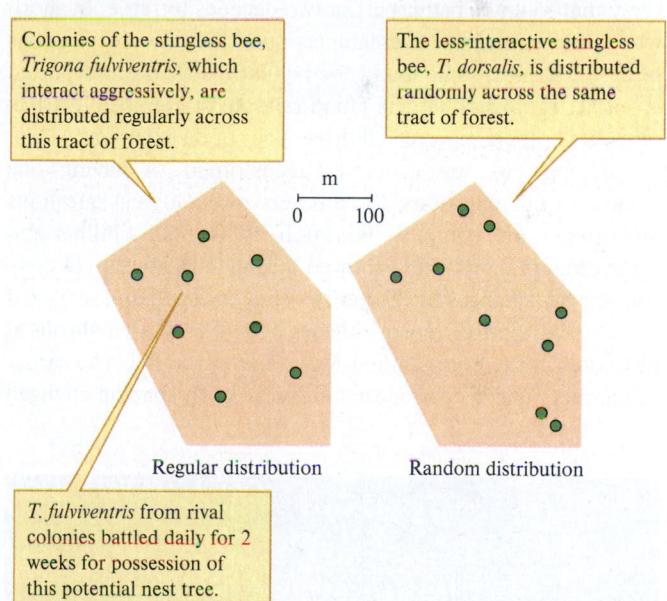

Figure 9.11   Regular and random distributions of stingless bee colonies in the tropical dry forest are related to levels of aggression (data from Hubbell and Johnson 1977).

If workers from two different colonies arrive at the prospective nest, they may fight for possession. Fights may escalate into protracted battles. Hubbell and Johnson observed battles over a nest tree that lasted for 2 weeks. Each dawn, 15 to 30 workers from two rival colonies arrived at the contested nest site. The workers from the two rival colonies faced off in two swarms and displayed and fought with each other. In the displays, pairs of bees faced each other, slowly flew vertically to a height of about 3 m, and then grappled each other to the ground. When the two bees hit the ground, they separated, faced off, and performed another aerial display. Bees did not appear to be injured in these fights, which were apparently ritualized. The two swarms abandoned the battle at about 8 or 9 A.M. each day, only to re-form and begin again the next day just after dawn. While this contest over an unoccupied nest site produced no obvious mortality, fights over occupied nests sometimes killed over 1,000 bees in a single battle. These tropical bees space their colonies by engaging in pitched battles. As we see next, plants space themselves by more subtle means.

Distributions of Desert Shrubs

Half a century ago, desert ecologists suggested that desert shrubs tend to be regularly spaced due to competition between the shrubs. You can see the patterns that inspired these early ecologists by traveling across the seemingly endless expanses of the Mojave Desert in western North America. One of the most common plants you will see is the creosote bush, *Larrea tridentata,* which dominates thousands of square kilometers of this area. As you look out across landscapes dominated by creosote bushes, it may appear that the spacing of these shrubs is regular (fig. 9.12). In places, their spacing is so uniform they appear to have been planted by some very careful gardener. As we shall see, however, visual impressions can be deceiving.

Quantitative sampling and statistical analysis of the distributions of creosote bushes and other desert shrubs led to a controversy that took the better part of two decades to settle. In short, when different teams of researchers quantified the distributions of desert shrubs, some found the regular distributions reported by earlier ecologists. Others found random or clumped distributions. Still others reported all three types of distributions.

Though we are generally accustomed to having one answer to our questions, the answers to ecological questions are often more complex. Research by Donald Phillips and James MacMahon (1981) showed that the distribution of creosote bushes changes as they grow. They mapped and analyzed the distributions of creosote bushes and several other shrubs at nine sites in the Sonoran and Mojave Deserts. Because earlier researchers had suggested that creosote bush spacing changed

Figure 9.12 Observing the distribution of individuals in local populations of creosote bush, *Larrea tridentata,* ecologists suggested that their spacing is regular. Notice on the left side of this photo how the spacing of creosote bushes in the distance suggest that they have been planted in regularly spaced rows. This view is similar to what you would see if you looked between two rows of trees in a fruit orchard.   Charlie Ott/Science Source

with available moisture, they chose sites with different average precipitations. Precipitation at the study sites ranged from 80 to 220 mm, and average July temperature varied from 27° to 35°C. Phillips and MacMahon took care to pick sites with similar soils and with similar topography. They studied populations growing on sandy to sandy loam soils with less than 2% slope with no obvious surface runoff channels.

The results of this study indicate that the distribution of desert shrubs changes from clumped to random to regular distribution patterns as they grow. The young shrubs tend to be clumped for three reasons: (1) because seeds germinate at a limited number of "safe sites," (2) because seeds are not dispersed far from the parent plant, or (3) because asexually produced offspring are necessarily close to the parent plant. Phillips and MacMahon proposed that as the plants grow, some individuals in the clumps die, which reduces the degree of clumping. Gradually, the distribution of shrubs becomes more and more random. However, competition among the remaining plants produces higher mortality among plants with nearby neighbors, which thins the stand of shrubs still further and eventually creates a regular distribution of shrubs. This hypothetical process is summarized in figure 9.13.

Phillips and MacMahon and other ecologists proposed that desert shrubs compete for water and nutrients, a competition that takes place belowground. How can we study these belowground interactions? Work by Jacques Brisson and James Reynolds (1994) provides a quantitative picture of the belowground side of creosote bush distributions. These researchers carefully excavated and mapped the distributions of 32 creosote bushes in the Chihuahuan Desert. They proposed that if creosote bushes compete, their roots should grow in a way that reduces overlap with the roots of nearby individuals.

The 32 excavated creosote bushes occupied a 4 m by 5 m area on the Jornada Long Term Ecological Research site near Las Cruces, New Mexico. Creosote bush was the only shrub within the study plot. Their roots penetrated to only 30 to 50 cm, the depth of a hardpan calcium carbonate deposition layer. Because they did not have to excavate to great depths, Brisson and Reynolds were able to map more root systems than previous researchers. Still, their excavation and mapping of roots required 2 months of intense labor.

The complex pattern of root distributions uncovered confirmed the researchers' hypothesis: Creosote bush roots grow in a pattern that reduces overlap between the roots of adjacent plants (fig. 9.14). Notice that the root systems of creosote bushes overlap much less than they would if they had circular distributions. Brisson and Reynolds concluded that competitive interactions with neighboring shrubs influence the distribution of creosote bush roots. Their work suggests that creosote bushes compete for belowground resources.

After more than two decades of work on this single species of plant, desert ecologists have a much clearer understanding of the factors that influence the distribution of individuals on a small scale. On small scales, the creosote bush may have clumped, random, or regular distributions. Hubbell and Johnson (1977) showed that stingless bee colonies may also show different patterns of distribution, depending on the level of aggression

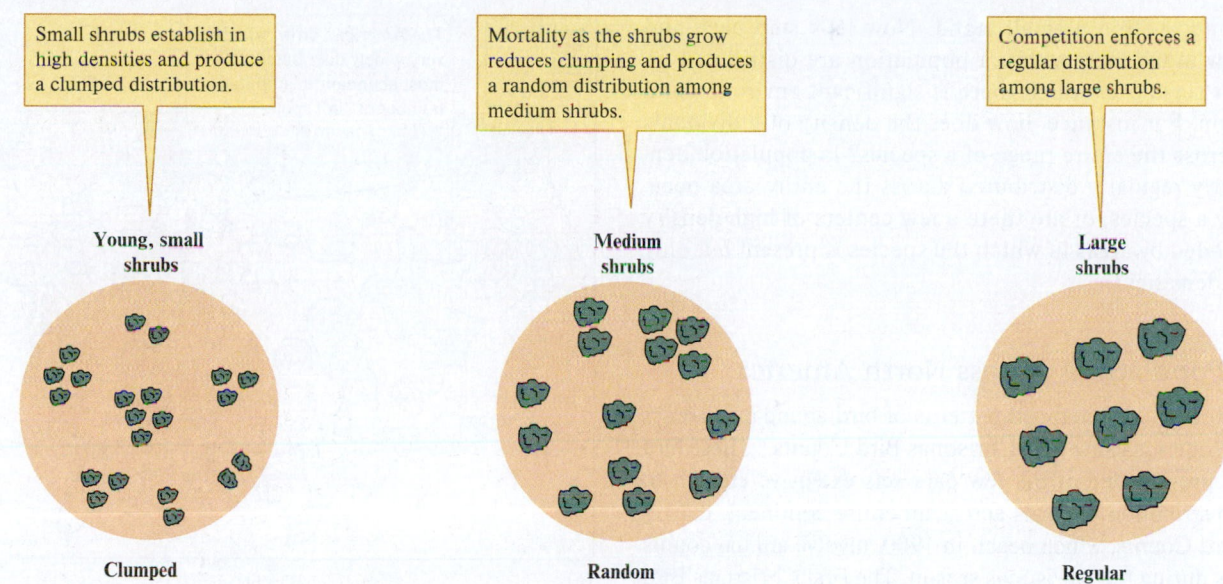

Small shrubs establish in high densities and produce a clumped distribution.

Mortality as the shrubs grow reduces clumping and produces a random distribution among medium shrubs.

Competition enforces a regular distribution among large shrubs.

Young, small shrubs

Medium shrubs

Large shrubs

Clumped

Random

Regular

Figure 9.13   Change in creosote bush distributions with increasing shrub size.

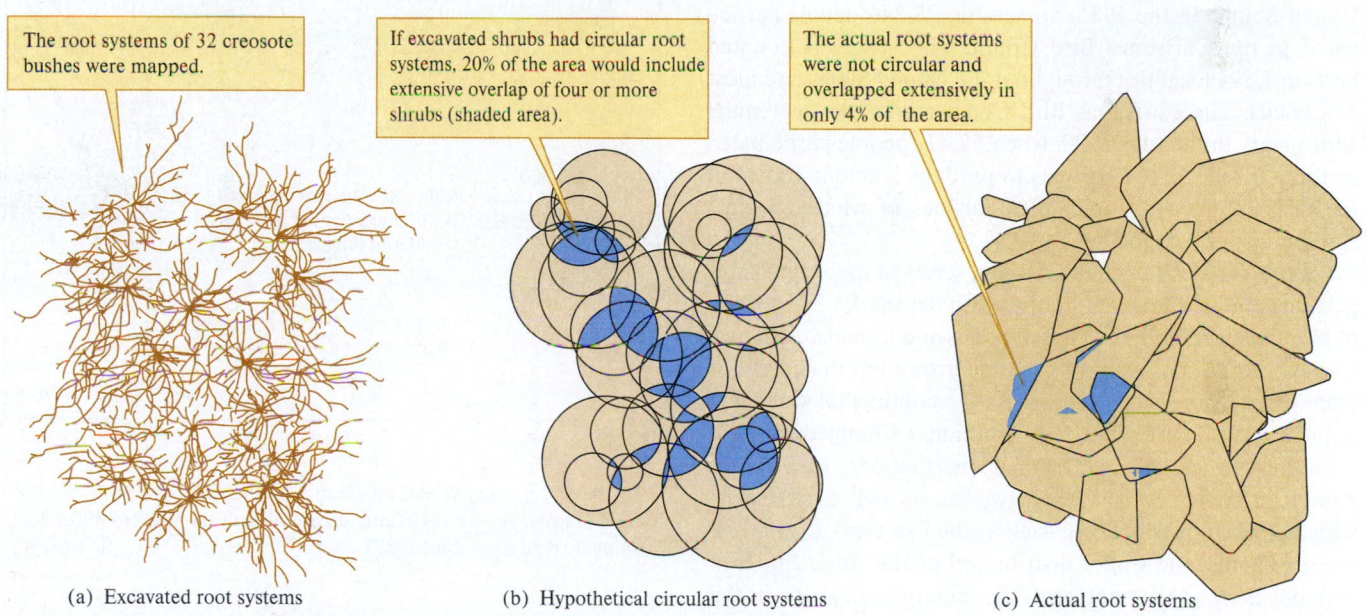

The root systems of 32 creosote bushes were mapped.

If excavated shrubs had circular root systems, 20% of the area would include extensive overlap of four or more shrubs (shaded area).

The actual root systems were not circular and overlapped extensively in only 4% of the area.

(a) Excavated root systems

(b) Hypothetical circular root systems

(c) Actual root systems

Figure 9.14   Creosote bush root distributions: hypothetical versus actual root overlap (data from Brisson and Reynolds 1994).

between colonies. As we shall see in the following section, however, on larger scales, individuals have clumped distributions.

Concept 9.2 Review

1. Are the concepts of "small" versus "large" scale the same for all organisms?
2. How could you test the hypothesis that low overlap in root systems in creosote bush populations (see fig. 9.14) is the result of ongoing competition?
3. In the study of the distribution of stingless bee colonies (see section 9.2), why were measurements of the number and distribution of potential nest trees necessary?

9.3 Patterns on Large Scales

LEARNING OUTCOMES

After studying this section you should be able to do the following:

9.11   Review evidence that wintering and breeding birds are clumped at large scales.
9.12   Explain the clumped distributions of trees along moisture gradients in terms of niches (Concept 9.1).

On large scales, individuals within a population are clumped. We have considered how individuals within a population are distributed on a small scale: How bee colonies are distributed within a few acres of forest and how shrubs are distributed within a small stand. Now let's step back and ask how individuals within a population are distributed on a larger scale over which there is significant environmental variation. For instance, how does the density of individuals vary across the entire range of a species? Is population density fairly regularly distributed across the entire area occupied by a species, or are there a few centers of high density surrounded by areas in which the species is present but only in low densities?

Bird Populations Across North America

Terry Root (1988) mapped patterns of bird abundance across North America using the "Christmas Bird Counts." These bird counts provide one of the few data sets extensive enough to study distribution patterns across an entire continent. Christmas Bird Counts, which began in 1900, involve annual counts of birds during the Christmas season. The first Christmas Bird Count was attended by 27 observers, who counted birds in 26 localities—2 in Canada and the remainder in 13 states of the United States. In the 1985–86 season, 38,346 people participated in the Christmas Bird Count. The observers counted birds in 1,504 localities throughout the United States and most of Canada. The Christmas Bird Count marked its centennial anniversary in the year 2000, when 52,471 people participated at 1,823 localities. It continues to produce a unique record of the distribution and population densities of wintering birds across most of a continent.

Root's analysis centers around a series of maps that show patterns of distribution and population density for 346 species of birds that winter in the United States and Canada. Although species as different as swans and sparrows are included, the maps show a consistent pattern. At the continental scale, bird populations show clumped distributions. Clumped patterns occur in species with widespread distributions, such as the American crow, *Corvus brachyrhynchos,* as well as in species with restricted distributions, such as the fish crow, *Corvus ossifragus.* Though the winter distribution of the American crow includes most of the continent, the bulk of individuals in this population are concentrated in a few areas. These areas of high density, or "hot spots," appear as red patches in figure 9.15a. For the American crow population, hot spots are concentrated along river valleys, especially the Cumberland, Mississippi, Arkansas, Snake, and Rio Grande. Away from these hot spots the winter abundance of American crows diminishes rapidly.

The fish crow population, though much more restricted than that of the American crow, is also concentrated in a few areas (fig. 9.15b). Fish crows are restricted to areas of open water near the coast of the Gulf of Mexico and along the southern half of the Atlantic coast of the United States. Within this restricted range, however, most fish crows are concentrated in a few hot spots—one on the Mississippi Delta, another on Lake Seminole west of Tallahassee, Florida, and a third in the everglades of southern Florida. Like the more widely distributed American crow, the abundance of fish crows diminishes rapidly away from these centers of high density.

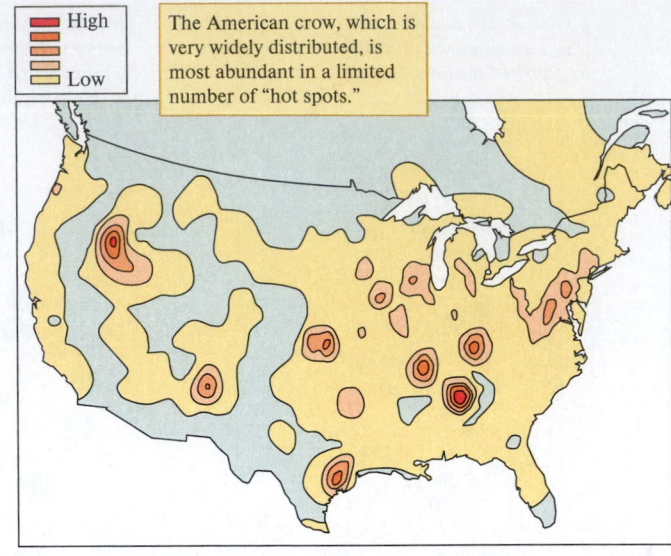

(a)

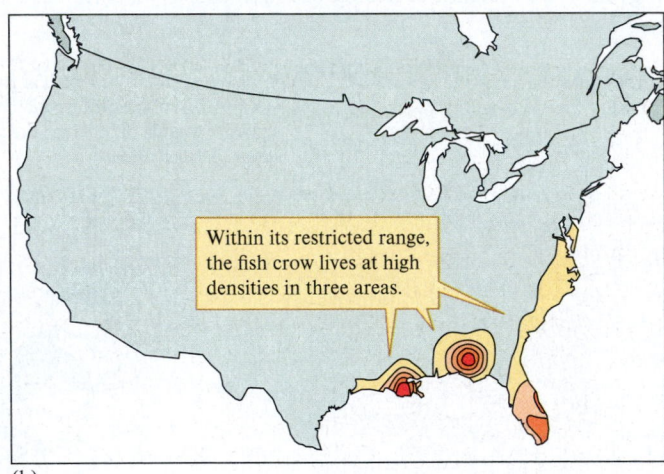

(b)

Figure 9.15   (a) Winter distribution of the American crow, *Corvus brachyrhynchos.* (b) Winter distribution of the fish crow, *Corvus ossifragus* (data from Root 1988).

Might bird populations have clumped distributions only on the wintering grounds? James H. Brown, David Mehlman, and George Stevens (1995) analyzed large-scale patterns of abundance among birds across North America during the breeding season, the opposite season from that studied by Root. In their study these researchers used data from the Breeding Bird Survey, which consists of standardized counts by amateur ornithologists conducted each June at approximately 2,000 sites across the United States and Canada under the supervision of the Fish and Wildlife Services of the United States and Canada. For their analyses, they chose species of birds whose geographic ranges fall mainly or completely within the eastern and central regions of the United States, which are well covered by study sites of the Breeding Bird Survey. This powerful data set includes abundances for over 500 species and is still widely used for ecological studies and management more than 50 years after its inception (Hudson et al. 2017, Edwards and Smith 2020).

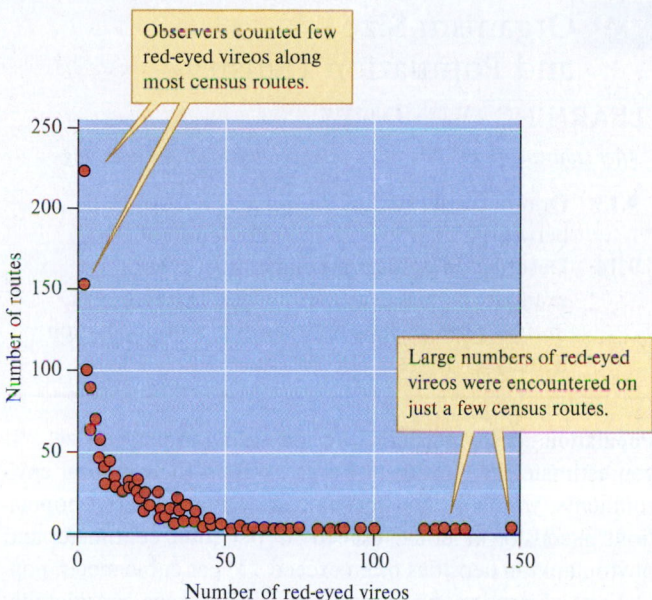

Figure 9.16   Red-eyed vireos, *Vireo olivaceus,* counted along census routes of the Breeding Bird Survey (data from Brown, Mehlman, and Stevens 1995).

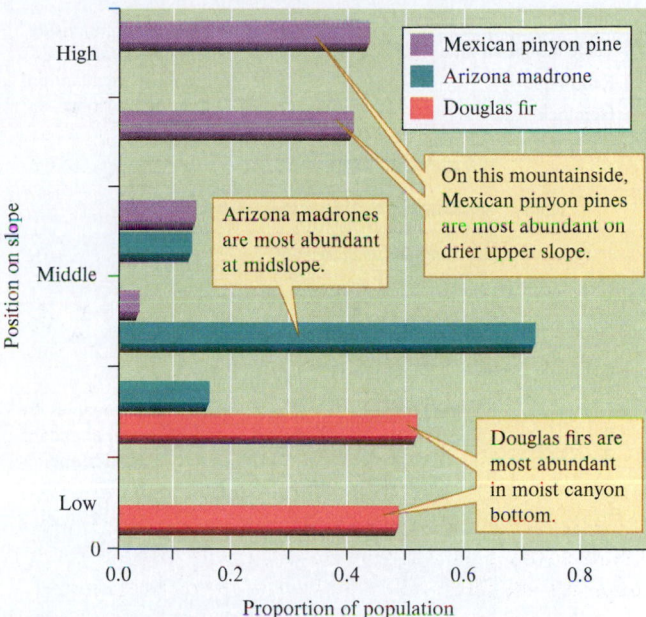

Figure 9.17   Abundances of three tree species on a moisture gradient in the Santa Catalina Mountains, Arizona (data from Whittaker and Niering 1965).

Like Root, Brown and his colleagues found that a relatively small proportion of study sites yielded most of the records of each bird species. That is, most individuals were concentrated in a fairly small number of hot spots. For instance, the densities of red-eyed vireos are low in most places (fig. 9.16). Clumped distributions were documented repeatedly. When the numbers of birds across their ranges were totaled, generally about 25% of the locations sampled supported over half of each population. By combining the results of Root and Brown and his colleagues, we can say confidently that at larger scales, bird populations in North America show clumped patterns of distribution. In other words, most individuals within a bird species live in a few hot spots, areas of unusually high population density.

Brown and his colleagues propose that these distributions are clumped because the environment varies and individuals aggregate in areas where the environment is favorable. What might be the patterns of distribution for populations distributed along a known environmental gradient? Studies of plant populations provide interesting insights.

Plant Distributions Along Moisture Gradients

Decades ago, Robert Whittaker gathered information on the distributions of woody plants along moisture gradients in several mountain ranges across North America. As we saw in chapter 2 (see fig. 2.39), environmental conditions on mountainsides change substantially with elevation. These steep environmental gradients provide a compressed analog of the continental-scale gradients to which the birds studied by Root and Brown and his colleagues were presumably responding.

Let's look at the distributions of some tree species along moisture gradients in two of the mountain ranges studied by

Whittaker. Robert Whittaker and William Niering (1965) studied the distribution of plants along moisture and elevation gradients in the Santa Catalina Mountains of southern Arizona. This mountain range rises out of the Sonoran Desert near Tucson, Arizona, like a green island in a tan desert sea. Vegetation typical of the Sonoran Desert, including the saguaro cactus and creosote bush, grows in the surrounding desert and on the lower slopes of the mountains. However, the summit of the mountains is topped by a mixed conifer forest. Forests also extend down the flanks of the Santa Catalinas in moist, shady canyons.

There is a moisture gradient from the moist canyon bottoms up the dry southwest-facing slopes. Whittaker and Niering found that along this gradient the Mexican pinyon pine, *Pinus cembroides,* is at its peak abundance on the uppermost and driest part of the southwest-facing slope (fig. 9.17). Along the same slope, Arizona madrone, *Arbutus arizonica,* reaches its peak abundance at middle elevations. Finally, Douglas firs, *Pseudotsuga menziesii,* are restricted to the moist canyon bottom. Mexican pinyon pines, Arizona madrone, and Douglas fir are all clumped along this moisture gradient, but each reaches peak abundance at different positions on the slope. These positions appear to reflect the different environmental requirements of each species.

Whittaker (1956) recorded similar tree distributions along moisture gradients in the Great Smoky Mountains of eastern North America. Again, the gradient was from a moist valley bottom to a drier southwest-facing slope. Along this moisture gradient, hemlock, *Tsuga canadensis,* was concentrated in the moist valley bottom and its density decreased rapidly upslope (fig. 9.18). Meanwhile, red maple, *Acer rubrum,* grew at highest densities in the middle section of the slope, while table mountain pine, *Pinus pungens,* was concentrated on the driest upper sections. As in the Santa Catalina Mountains of Arizona,

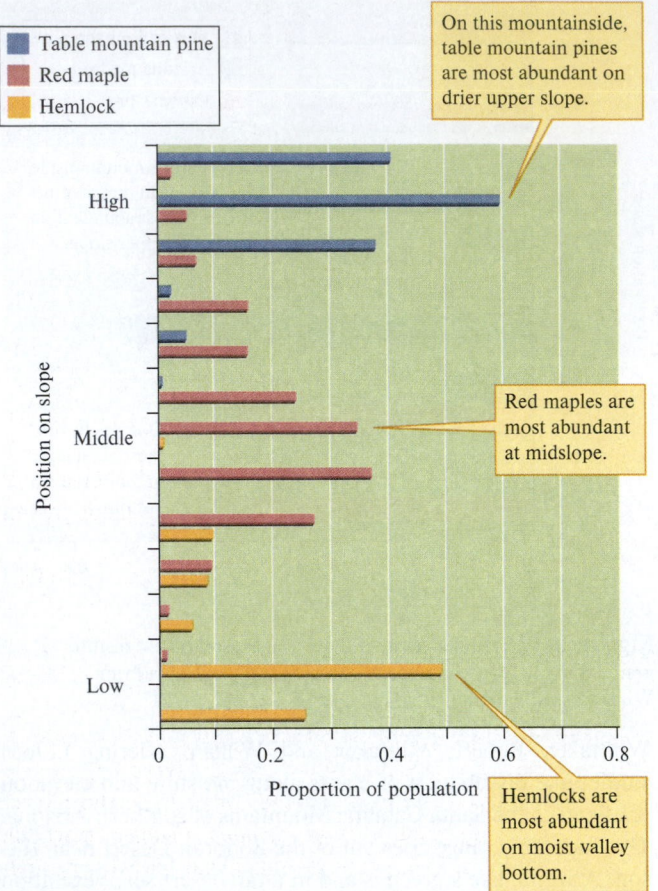

Figure 9.18   Abundance of three tree species on a moisture gradient in the Great Smoky Mountains, Tennessee (data from Whittaker 1956).

these tree distributions in the Great Smoky Mountains reflect the moisture requirements of each tree species.

The distribution of trees along moisture gradients seems to resemble the clumped distributional patterns of birds across the North American continent but on a smaller scale. All species of trees discussed here showed a highly clumped distribution along moisture gradients, and their densities decreased substantially toward the edges of their distributions. In other words, like birds, tree populations are concentrated in hot spots. As we shall see in the next Concept, population density is influenced by organism size.

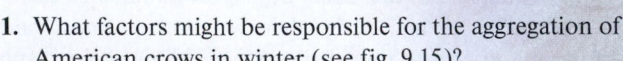

Concept 9.3 Review

1. What factors might be responsible for the aggregation of American crows in winter (see fig. 9.15)?
2. Why might the winter aggregations of crows occur mainly along river valleys?
3. What does the position of pines along moisture gradients in both the Santa Catalina Mountains of Arizona (see fig. 9.17) and the Great Smoky Mountains of Tennessee (see fig. 9.18) suggest about pine-water relations?

9.4 Organism Size and Population Density

LEARNING OUTCOMES

After studying this section you should be able to do the following:

9.13   Draw a scatter plot of the general relationship between organism size and population density.

9.14   Describe differences among animal groups, for example, mammals versus aquatic invertebrates, in their relationships between size and population density.

Population density declines with increasing organism size.   If you estimate the densities of organisms in their natural environments, you will find great ranges. While bacterial populations in soils or water can exceed 10^9 per cubic centimeter and phytoplankton densities often exceed 10^6 per cubic meter, populations of large mammals and birds can average considerably less than one individual per square kilometer. What factors produce this variation in population density? The densities of a wide variety of organisms are highly correlated with body size. In general, densities of animal and plant populations decrease with increasing size.

While it makes common sense that small animals and plants generally live at higher population densities than larger ones, quantifying the relationship between body size and population density provides valuable information. First, quantification translates a general qualitative notion into a more precise quantitative relationship. For example, you might want to know how much population density declines with increased body size. Second, measuring the relationship between body size and population density for a wide variety of species reveals contrasting size-density relationships for different groups of organisms, suggesting divergent environmental requirements.

Animal Size and Population Density

John Damuth (1981) produced one of the first clear demonstrations of the relationship between body size and population density. He focused his analysis on herbivorous mammals, ranging in size from small rodents, with a mass of about 10 g, to large herbivores such as rhinoceros, with a mass well over 10^6 g. Meanwhile, average population density varied from 1 individual (10^{-1}) per 10 km^2 to 10,000 (10^4) per 1 km^2, which spans approximately five "orders of magnitude," or powers of 10, in population density. As figure 9.19 shows, Damuth found that the population density of 307 species of herbivorous mammals decreases, from species to species, with increased body size. The regression line (Investigating the Evidence 8 in Appendix A) in the graph shows the average decrease in population density with increased body size.

Building on Damuth's analysis, Robert Peters and Karen Wassenberg (1983) explored the relationship between body size and average population density for a wider variety of animals. Their analysis included terrestrial invertebrates, aquatic invertebrates, mammals, birds, and poikilothermic

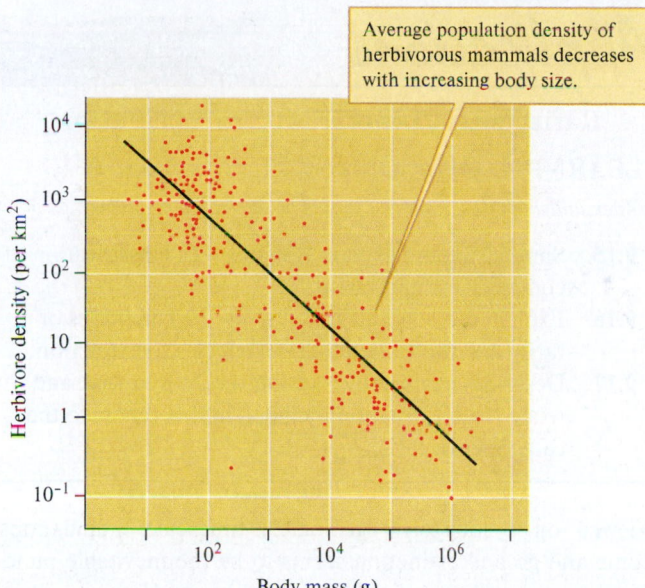

Average population density of herbivorous mammals decreases with increasing body size.

Figure 9.19 Body size and population density of herbivorous mammals (data from Damuth 1981).

of similar size. Peters and Wassenberg suggest that it may be appropriate to analyze aquatic invertebrates and birds separately from the other groups of animals.

The general relationship between animal size and population density has held up under careful scrutiny and reanalysis. Plant ecologists have found a qualitatively similar relationship in plant populations, as we see next.

Plant Size and Population Density

James White (1985) pointed out that plant ecologists have been studying the relationship between plant size and population density since early in the twentieth century. He suggests that the relationship between size and density is one of the most fundamental aspects of population biology. White summarized the relationship between size and density for a large number of plant species spanning a wide range of plant growth forms (fig. 9.21).

The pattern in figure 9.21 illustrates that as in animals, plant population density decreases with increasing plant size. However, the biological details underlying the size-density relationship shown by plants are quite different from those underlying the patterns shown by animals. The points in figures 9.19 and 9.20 represent different species of animals. A single species of tree, however, can span a very large range of sizes and densities during its life cycle. Even the largest trees, such as the giant sequoia, *Sequoia gigantea*, start life as small seedlings. These tiny seedlings can live at very high densities. As the trees grow, density declines progressively until the mature trees live at low densities. We discuss this process, which is called *self-thinning*, in chapter 13. Thus, the size-density relationship changes dynamically within plant populations and differs significantly between populations of plants that reach different sizes at maturity. Despite differences in the underlying processes, the data summarized in

vertebrates, representing a great range in size and population density. Animal mass ranged from 10^{-11} to about $10^{2.3}$ kg, while population density varied from less than 1 per square kilometer to nearly 10^{12} per square kilometer. When Peters and Wassenberg plotted animal mass against average density, they, like Damuth, found that population density decreased with increased body size.

If you look closely at the data in figure 9.20, however, it is clear that there are differences among the animal groups. First, aquatic invertebrates of a given body size tend to have higher population densities, usually one or two orders of magnitude higher, than terrestrial invertebrates of similar size. Second, mammals tend to have higher population densities than birds

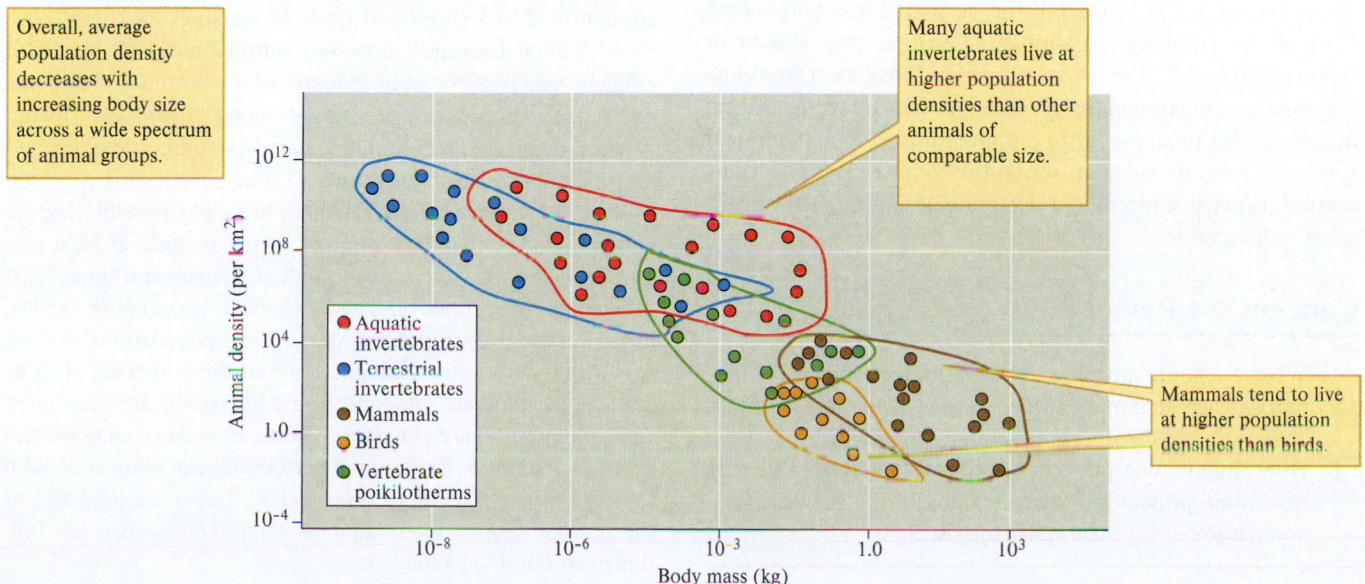

Overall, average population density decreases with increasing body size across a wide spectrum of animal groups.

Many aquatic invertebrates live at higher population densities than other animals of comparable size.

Mammals tend to live at higher population densities than birds.

Figure 9.20 Animal size and population density (data from Peters and Wassenberg 1983).

As in animals, plant population density decreases with increasing plant size across a wide range of plant growth forms.

Duckweed, *Lemna*, one of the smallest flowering plants, lives at very high population densities.

The coastal redwood, *Sequoia sempervirens*, one of the largest trees, lives at one of the lowest population densities.

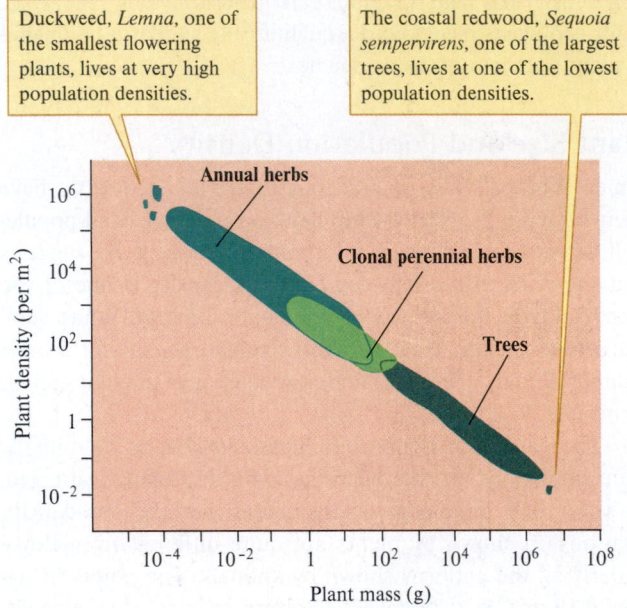

Figure 9.21   Plant size and population density (data from White 1985).

figure 9.21 indicate a predictable relationship between plant size and population density.

The value of such an empirical relationship, whether for plants or animals, is that it provides a standard against which we can compare measured densities and gives an idea of expected population densities in nature. For example, suppose you go out into the field and measure the population density of some species of animal. How would you know if the densities you encounter were unusually high, low, or about average for an animal of the particular size and taxon? Without an empirical relationship such as that shown in figures 9.20 and 9.21 or a list of species densities, it would be impossible to make such an assessment. One question that we might attempt to answer with a population study is whether a species is rare. As we shall see in the following Applications section, rarity is a more complex consideration than it might seem at face value.

Concept 9.4 Review

1. What are some advantages of Damuth's strict focus on herbivorous mammals in his analysis of the relationship between body size and population density (see fig. 9.19)?
2. How might energy and nutrient relations explain the lower population densities of birds compared to comparable-sized mammals (see fig. 9.20)?

Applications

Rarity and Vulnerability to Extinction

LEARNING OUTCOMES

After studying this section you should be able to do the following:

9.15   Summarize and explain Rabinowitz's classification of commonness and rarity.

9.16   Explain the relationship between the categories of rarity and the vulnerability of species to extinction.

9.17   Describe the objectives of the IUCN Red List, and relate the information included in this report to the categories of rarity.

Viewed on a long-term, geological timescale, populations come and go and extinction seems to be the inevitable punctuation mark at the end of a species' history. However, some populations seem to be more vulnerable to extinction than others. What makes some populations likely to disappear, while others persist through geologic ages? At the heart of the matter are patterns of distribution and abundance. Rare species are often vulnerable to extinction, whereas abundant species are seldom so. In order to understand and, perhaps, prevent extinction, we need to understand the various forms of rarity, especially in this time of rapid climate change.

Seven Forms of Rarity and One of Abundance

Deborah Rabinowitz (1981) devised a classification of *commonness* and *rarity,* based on combinations of three factors: (1) the geographic range of a species (*extensive* versus *restricted*), (2) habitat tolerance (*broad* versus *narrow*), and (3) local population size (*large* versus *small*). Habitat tolerance is related to the range of conditions in which a species can live. For instance, some plant species can tolerate a broad range of soil texture, pH, and organic matter content, whereas other plant species are confined to a single soil type. As we shall see, tigers have broad habitat tolerance; however, within the tiger's historical range in Asia lives the snow leopard, which is confined to a narrow range of conditions in the high mountains of the Tibetan Plateau. Small geographic range, narrow habitat tolerance, and low population size are attributes of rarity.

As shown in figure 9.22, there are eight possible combinations of these factors, seven of which include at least one attribute of rarity. The most abundant species and those least threatened by extinction have extensive geographic ranges, broad habitat tolerances, and large local populations at least somewhere within their range. Some of these species, such as dandelions, Norway rats, and house sparrows, are associated with humans and are considered pests. However, many species of small mammals, birds, and invertebrates not associated with humans, such as the deer mouse, *Peromyscus maniculatus,* or the marine zooplankton, *Calanus finmarchicus,* also fall into this most common category.

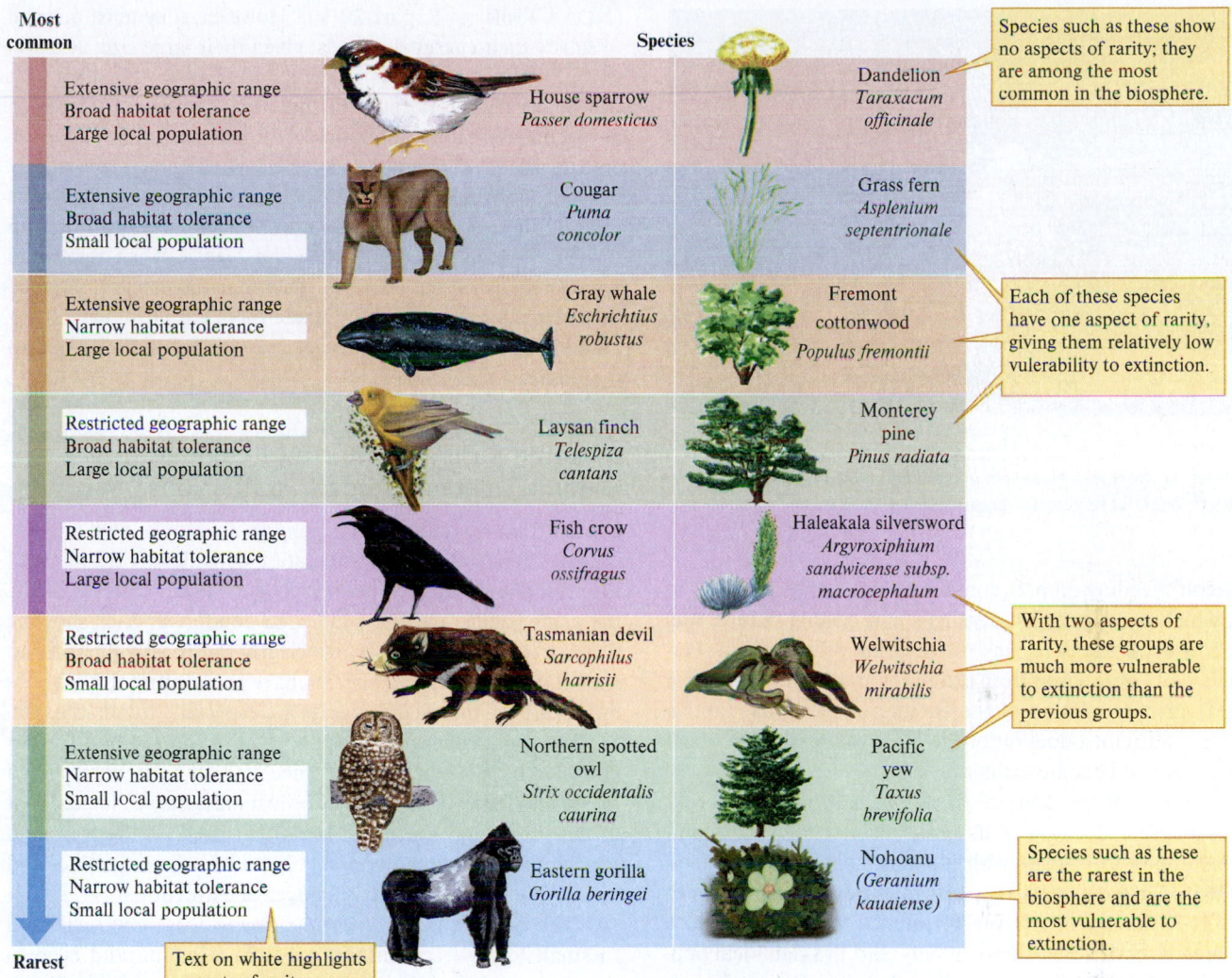

Most common

Species

Extensive geographic range Broad habitat tolerance Large local population	House sparrow *Passer domesticus*	Dandelion *Taraxacum officinale*	Species such as these show no aspects of rarity; they are among the most common in the biosphere.
Extensive geographic range Broad habitat tolerance Small local population	Cougar *Puma concolor*	Grass fern *Asplenium septentrionale*	
Extensive geographic range Narrow habitat tolerance Large local population	Gray whale *Eschrichtius robustus*	Fremont cottonwood *Populus fremontii*	Each of these species have one aspect of rarity, giving them relatively low vulerability to extinction.
Restricted geographic range Broad habitat tolerance Large local population	Laysan finch *Telespiza cantans*	Monterey pine *Pinus radiata*	
Restricted geographic range Narrow habitat tolerance Large local population	Fish crow *Corvus ossifragus*	Haleakala silversword *Argyroxiphium sandwicense subsp. macrocephalum*	
Restricted geographic range Broad habitat tolerance Small local population	Tasmanian devil *Sarcophilus harrisii*	Welwitschia *Welwitschia mirabilis*	With two aspects of rarity, these groups are much more vulnerable to extinction than the previous groups.
Extensive geographic range Narrow habitat tolerance Small local population	Northern spotted owl *Strix occidentalis caurina*	Pacific yew *Taxus brevifolia*	
Restricted geographic range Narrow habitat tolerance Small local population	Eastern gorilla *Gorilla beringei*	Nohoanu (*Geranium kauaiense*)	Species such as these are the rarest in the biosphere and are the most vulnerable to extinction.

Text on white highlights aspects of rarity.

Rarest

Figure 9.22   Rarity and vulnerability to extinction.

Ecologists exploring the relationship between size of geographic range and population size have found that they are not independent. Instead, there is a strong positive correlation between the two variables for most groups of organisms. In other words, species abundant in the places where they occur are more likely to be widely distributed within a region, continent, or ocean, whereas species living at low population densities generally have small, restricted distributions (Hanski 1982, Brown 1984). Species with distributions limited to only one region are called **endemic**. The Laysan finch (*Telespiza cantans*) is endemic to the Hawaiian Islands. There are thousands of the birds in the Hawaiian island of Laysan, and as a generalist feeder it has a broad niche, but this species is vulnerable because it exists almost nowhere else in the world. In the early 1900's, the introduction of rabbits to Laysan almost drove the birds to extinction, but the birds were able to rapidly recover when the rabbits were removed (iucnredlist.org).

Most species are uncommon; seven combinations of range, tolerance, and population size each create a kind of rarity. As a consequence, Rabinowitz referred to "seven forms

of rarity." Let's look at species that represent the two extremes of Rabinowitz's seven forms of rarity. The first two discussions concern species that are rare according to only one attribute. These are species that, before they become extinct, may seem fairly secure. The final discussion concerns the rarest species, which show all three attributes of rarity. Though these rarest species are the most vulnerable to extinction, rarity in any form appears to increase vulnerability to extinction.

Rarity I: Extensive Range, Broad Habitat Tolerance, Small Local Populations

It is easy to understand how people were drawn to the original practice of falconry. The sight and sound of a peregrine falcon, *Falco peregrinus,* in full dive at over 200 km per hour must have been one of the great experiences of a lifetime (fig. 9.23). The peregrine, which has a geographic range that circles the Northern Hemisphere and broad habitat tolerance, is uncommon throughout its range. Apparently, this one attribute of rarity was enough to make the peregrine vulnerable to extinction.

Figure 9.23 The peregrine falcon, *Falco peregrinus,* is found throughout the Northern Hemisphere but lives at low population densities throughout its range.   Brand X Pictures/PunchStock

The falcon's feeding on prey containing high concentrations of DDT, which produced thin eggshells and nesting failure, was enough to drive the peregrine to the brink of extinction. Peregrine falcons were saved from extinction by restricting the use of DDT, strict regulation of the capture of the birds, captive breeding, and reintroduction of the birds to areas where local populations had become extinct.

The range of the cougar, *Puma concolor*, extends across North America. Because of its broad habitat tolerances and uncanny ability to co-exist with humans, cougars (also called puma) are more common that one might expect and are not considered endangered or even threatened. However, because they are large carnivores, they can only exist in small local populations. Furthermore, cougars are often the target of hunters, who consider them a risk to livestock and competition for hunting deer and other game. Because of this, we must keep monitoring their populations, even if they are currently considered at low risk of extinction.

Rarity II: Extensive Range, Large Populations, Narrow Habitat Tolerance

When Europeans arrived in North America, they encountered one of the most numerous birds on earth, the passenger pigeon. The range of the passenger pigeon extended from the eastern shores of the present-day United States to the Midwest, and its population size numbered in the billions. However, the bird had one attribute of rarity: It had a narrow requirement for its nesting sites. The passenger pigeon nested in huge aggregations in virgin forests. As virgin forests were cut, its range diminished and market hunters easily located and exploited its remaining nesting sites, finishing off the remainder of the population. By 1914, the last passenger pigeon died in captivity. Extensive range and high population density, alone, do not guarantee immunity from extinction.

The grey whale (*Eschrichtius robustus*) can be found in the coastal waters of Asia and North America; the largest population inhabiting the Eastern North Pacific Ocean. That population has been estimated to be nearly 27,000 individuals

(NOAA Fisheries Report 2019). However, they must migrate to satisfy their energy demands, given their large size: at maturity, 13–15m long and as heavy as 40 tonnes (the same as a 14-wheeler truck). Because of this, they can be considered to have a narrow habitat tolerance. Although they are not considered at risk of extinction as a species, whaling has significantly reduced the Western Pacific Ocean populations (currently less than 290 individuals), and no grey whales have been seen in the North Atlantic Ocean since the 1700's, when they were completely fished out (IUCN). Their narrow habitat tolerance that kept them close to shore made them easier to hunt. International agreements that protect grey whales and have led to their current stable status.

Extreme Rarity: Restricted Range, Narrow Habitat Tolerance, Small Populations

Species that combine small geographic ranges with narrow habitat tolerances and low population densities are the rarest of the rare. This group includes species such as the eastern gorilla, the giant panda, and the California condor. Species showing this extreme form of rarity are clearly the most vulnerable to extinction. Many island species have these attributes, so it is not surprising that island species are especially vulnerable. Stuart Pimm and colleagues reported that of the 154 bird species known to have become extinct since 1500, most were restricted to islands (Pimm et al. 2006). For instance, they estimated that of the 125 to 145 species of birds that lived on the Hawaiian Islands prior to human contact, 70 to 90 are now extinct and most of the remainder are in danger of extinction.

The eastern gorilla, *Gorilla beringei*, is now restricted to small forest fragments in Rwanda, Uganda and eastern Democratic Republic of Congo (Canington 2018). There are believed to be less than 3,000 in the wild. The eastern gorilla is considered to be critically endangered, although one of the subspecies, the mountain gorilla, *Gorilla beringei berengei*, has been increasing in number due to intensive conservation.

In this era of anthropogenic change, increasing numbers of species are rare; according to the Living Planet Report of the World Wildlife Fund (WWF), there has been a 68% average decrease in abundance in monitored populations of mammals, birds, amphibians, reptiles, and fish since they began tracking in 1970 (WWF 2020). A more exhaustive evaluation of species, including plants, fungi, and invertebrate animals, is compiled by the International Union for Conservation of Nature (IUCN). The IUCN is an organization comprised of over 17,000 scientists and other experts, plus government agencies, NGO's, and indigenous people's groups. The IUCN uses carefully collected data from its members to assess the conservation status of 120,000 species, published as the Red List (https://www.iucnredlist.org/). According to the Red List, more than 32,000 species (approximately 27% of all evaluated species) are threatened with extinction, with rarity being a leading risk factor. In nearly all cases, the key to a species' survival is increased distribution and abundance, which is a key goal of conservation programs.

Summary

Ecologists define a population as a group of individuals of a single species inhabiting an area delimited by natural or human-imposed boundaries. Population studies hold the key to solving practical problems such as saving endangered species, controlling pest populations, and managing fish and game populations. All populations share a number of characteristics. Chapter 9 focused on two population characteristics: distribution and abundance.

While there are few environments on earth without life, no single species can tolerate the full range of earth's environments. Because all species find some environments too warm, too cold, too saline, and so forth, **environment limits the geographic distribution of species.** The environmental limits of a species are related to its niche. For instance, there is a close relationship between climate and the distributions of the three largest kangaroos in Australia. Large- and small-scale variation in temperature and moisture limits the distributions of certain desert plants, such as shrubs in the genus *Encelia*. However, differences in the physical environment only partially explain the distributions of barnacles within the marine intertidal zone, a reminder that biological factors constitute an important part of an organism's environment.

On small scales, individuals within populations are distributed in patterns that may be random, regular, or clumped. Patterns of distribution can be produced by the social interactions within populations, by the structure of the physical environment, or by a combination of the two. Social organisms tend to be clumped; territorial organisms tend to be regularly spaced. An environment in which resources are patchy also fosters clumped distributions. Aggressive species of stingless bees live in regularly distributed colonies, whereas the colonies of nonaggressive species are randomly distributed. The distribution of creosote bushes changes as they grow.

On large scales, individuals within a population are clumped. In North America, populations of both wintering and breeding birds are concentrated in a few hot spots of high population density. Clumped distributions are also shown by plant populations living along steep environmental gradients on mountainsides.

Population density declines with increasing organism size. This negative relationship holds for animals as varied as terrestrial invertebrates, aquatic invertebrates, birds, poikilothermic vertebrates, and herbivorous mammals. Plant population density also decreases with increasing plant size. However, the biological details underlying the size-density relationship shown by plants are quite different from those underlying the size-density patterns shown by animals. A single species of tree can span a very large range of sizes and densities during its life cycle. The largest trees start life as small seedlings that can live at very high population densities. As trees grow, their population density declines progressively until the mature trees live at low densities.

Rare species are more vulnerable to extinction than are common species. Rarity of species can be expressed as a combination of extensive versus restricted geographic range, broad versus narrow habitat tolerance, and large versus small population size. The most abundant species and those least threatened by extinction combine large geographic ranges, wide habitat tolerance, and high local population density. All other combinations of geographic range, habitat tolerance, and population size include one or more attributes of rarity. Populations that combine restricted geographic range with narrow habitat tolerance and small population size are the rarest of the rare and are usually the organisms most vulnerable to extinction.

Key Terms

abundance 198	endemic 211	pheromone 203	regular distribution 202
clumped distribution 202	fundamental niche 198	population 197	small-scale phenomena 202
density 198	large-scale phenomena 202	random distribution 202	
distribution 198	niche 198	realized niche 198	

Review Questions

1. What adaptation permits *Encelia farinosa* to live on the dry upland slopes in the very hot Mojave Desert? How does *E. frutescens* keep from overheating in the same desert?

2. Spruce trees, members of the genus *Picea*, occur throughout the boreal forest and on mountains farther south. For example, spruce grow in the Rocky Mountains south from the heart of boreal forest all the way to the deserts of the southern United States and Mexico. How do you think they would be distributed in the mountains that rise from the southern deserts?

3. What kinds of interactions within an animal population lead to clumped distributions? What kinds of interactions foster a regular

distribution? What kinds of interactions would you expect to find within an animal population distributed in a random pattern?

4. How might the structure of the environment—for example, the distributions of different soil types and soil moisture—affect the patterns of distribution in plant populations?

5. Suppose one plant reproduces almost entirely from seeds, and that its seeds are dispersed by wind, and a second plant reproduces asexually, mainly by budding from runners. How should these two different reproductive modes affect local patterns of distribution seen in populations of the two species?

6. Suppose that in the near future, the fish crow population (see fig. 9.15b) in North America declines because of habitat destruction. Devise a conservation plan for the species that includes establishing protected refuges for the species. Where would you locate the refuges? How many refuges would you recommend?

7. For a given body size, which generally has the higher population density, birds or mammals? On average, which lives at lower population densities, terrestrial or aquatic invertebrates?

8. Outline Rabinowitz's classification (1981) of rarity, which she based on size of geographic range, breadth of habitat tolerance, and population size. In her scheme, which combination of attributes makes a species least vulnerable to extinction? Which combination makes a species the most vulnerable?

9. Can the analyses by Damuth (1981) and by Peters and Wassenberg (1983) be combined with that of Rabinowitz (1981) to make predictions about the relationship of animal size to its relative rarity (see figs. 9.19 and 9.20)? What two attributes of rarity, as defined by Rabinowitz, are not included in the analyses by Damuth and by Peters and Wassenberg?

Pixtal/AGEfotostock

Chapter 10

Population Dynamics

A swarm of honeybees, *Apis melifera*. Honeybees disperse as a swarm, in which a mass of worker bees accompanies a queen as she leaves an established colony in search for a place to establish a new colony. Dispersal by individuals or groups is an important facet of the dynamics of populations.

CHAPTER CONCEPTS

10.1 Dispersal can increase or decrease local population densities.   217
Concept 10.1 Review   221

10.2 Ongoing dispersal can join numerous subpopulations to form a metapopulation.   221
Concept 10.2 Review   224

10.3 A survivorship curve summarizes the pattern of survival in a population.   224
Concept 10.3 Review   228

10.4 The age distribution of a population reflects its history of survival, reproduction, and potential for future growth.   228
Concept 10.4 Review   230

10.5 A life table combined with a fecundity schedule can be used to estimate net reproductive rate (R_0), geometric rate of increase (λ), generation time (T), and per capita rate of increase (r).   230
Concept 10.5 Review   233

Applications: Changes in Species Distributions in Response to Climate Warming   233
Summary   235
Key Terms   236
Review Questions   236

LEARNING OUTCOMES

After studying this section you should be able to do the following:

10.1 Describe population dynamics.
10.2 Express population size as a balance between the opposing processes of birth, death, immigration, and emigration.

To explore population distribution and abundance in chapter 9, we froze the populations we examined at a particular instant in time. In nature, however, populations are

215

in continuous flux and their patterns of distribution and abundance result from a dynamic balance between factors that add individuals to populations, births and immigration, and factors that remove individuals from populations, deaths and emigration.

These processes that contribute to population size can be summarized by a simple equation:

$$N_t = N_{t-1} + B + I - D - E$$

where N_t is the number of individuals in a population at some time, t; N_{t-1} is the number of individuals in the population at some previous time, $t-1$; B is the number of births that have occurred during the interval between $t-1$ and t; I is the number of immigrants to the population during that time interval; D is the number of deaths; and E is the number of individuals that have emigrated, or left, the population.

The dynamic population processes underlying distribution and abundance are the subject of chapter 10. We call this area of ecology **population dynamics,** which is concerned with the factors influencing the expansion, decline, and maintenance of populations. This is one of the most important areas of ecology, since it holds the key to understanding and, hopefully, preventing the decline and extinction of endangered species, the control of noxious pest species; including parasites and pathogens of humans; and the maintenance of economically or culturally important animal or plant populations.

Though generally unnoticed, the dynamic nature of populations becomes apparent when a conspicuous species expands its range. For instance, the expansion of Africanized honeybees through South and North America is well documented (fig. 10.1). The legendary aggressiveness of these bees ensures that their dispersal into a new area does not escape notice for long.

Honeybees, *Apis melifera,* evolved in Africa and Europe, where their native range extends from tropical to cold, temperate environments. Across this extensive environmental range, this species has differentiated into a number of locally adapted subspecies. In an attempt to improve the adaptability of managed honeybees to their tropical climate, Brazilian scientists imported queens of the African subspecies *A. melifera scutellata* in 1956. These queens mated with the European honeybees used by Brazilian beekeepers, producing what we now call Africanized honeybees, abbreviated as AHB.

AHB differ in several ways from European honeybees. Temperate and tropical environments have apparently selected for markedly different behavior and population dynamics. Natural selection by a high diversity and abundance of nest predators, including humans, has probably produced the greater aggressiveness shown by AHB. The warmer climate and greater stability of nectar sources eliminate the advantages of storing large quantities of honey and maintaining large colonies for survival through the winter. Most important to this discussion of dispersal, Africanized honeybees produce swarms that disperse to form new colonies at a much higher rate than do European honeybees.

High rates of colony formation and dispersal have caused a rapid expansion of AHB through South and North America. Their rate of dispersal has ranged from 300 to 500 km per year. Within 30 years, AHB occupied most of South America, all of Central America, and most of Mexico. The estimated number of

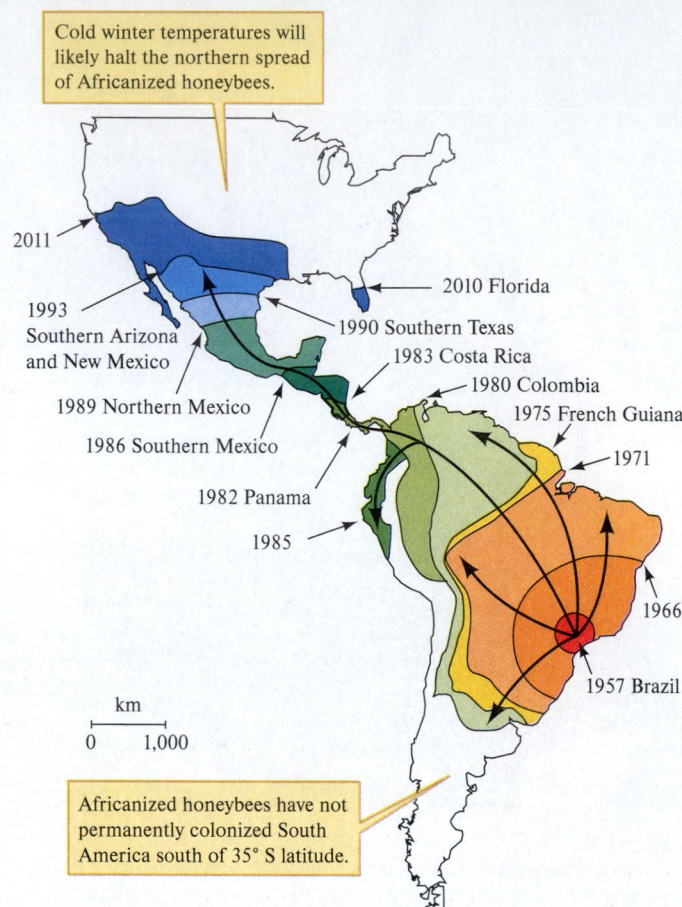

Cold winter temperatures will likely halt the northern spread of Africanized honeybees.

2011

1993 Southern Arizona and New Mexico

1989 Northern Mexico

1986 Southern Mexico

1982 Panama

1985

2010 Florida

1990 Southern Texas

1983 Costa Rica

1980 Colombia

1975 French Guiana

1971

1966

1957 Brazil

km
0   1,000

Africanized honeybees have not permanently colonized South America south of 35° S latitude.

Figure 10.1   The expansion of Africanized honeybees from South America through Central and North America, 1956 to 2010 (data from Winston 1992, USDA Agricultural Research Service 2011, Porrini et al. 2020)

wild colonies of these bees in South America alone is 50 to 100 million. Recent analysis of the mtDNA and wing morphology of colonies in Argentina suggests that the AHB genotype has not colonized south of 35°S latitude, and likely will not, due to intolerance to cold winters (Porrini et al. 2020).

Although at one time dubbed "killer bees" due to their defensive behavior, AHB are no longer considered a serious threat to humans. However, they can have ecological impacts, due to the fact that they have spread so fast and that they build hives in a wide range of habitats. As just one example, research in Brazil by Erica C. Pacífico and colleagues found that this species of bee was competing with the endangered Lear's macaw (*Anodorhynchus leari*) for nesting sites in the cavities of trees. Removal of the bees' nests resulted in a 71% increase in the breeding population of the endangered bird (Pacifico et al. 2020). Understanding the population dynamics of AHB is thus important in a conservation context.

Our discussion will be organized around the dynamic aspects of populations in space and time. We begin with spatial dynamics by considering the influence of dispersal on distribution and abundance, which began with the introductory example of Africanized honeybees. We then move to temporal dynamics, where we focus on patterns of survival and reproduction in populations, forming a transition to chapter 11, which introduces population growth.

10.1 Dispersal

LEARNING OUTCOMES

After studying this section you should be able to do the following:

10.3 Compare rates of dispersal by various animal and plant species.

10.4 Discuss the dispersal of North American trees in response to climate change at the end of the last ice age.

10.5 Describe numerical responses by owls and kestrels in response to changing prey densities.

10.6 Explain how the colonization cycle sustains populations of stream invertebrates.

Dispersal can increase or decrease local population densities. Dispersal is an important aspect of population dynamics. The seeds of plants disperse with wind or water or may be transported by a variety of mammals, insects, or birds. Adult barnacles may spend their lives attached to rocks, but their larvae travel the high seas on far-ranging ocean currents. A host of other sessile marine invertebrates, algae, and many highly sedentary reef fishes also disperse widely as larvae. Some young spiders spin a small net that catches winds and carries them for distances up to hundreds of kilometers. Young mammals and birds often disperse from the area where they were born and may join other local populations. As a consequence of movements such as these (fig. 10.2), the population ecologist studying population density must consider dispersal *into* (immigration) and *out of* (emigration) local populations.

Despite its importance, dispersal is one of the least-studied aspects of population dynamics. Its study is clearly a difficult undertaking. But dispersal is worth studying; the health and survival of many local populations may depend on this under-appreciated aspect of population dynamics. One of the richest sources of information on dispersal and some of the clearest examples come from studies of expanding populations.

Dispersal of Expanding Populations

Expanding populations are in the process of increasing their geographic range. Why should this type of population provide us with some of the best records of species dispersal? The appearance of a new species in an area is commonly quickly noted and recorded, especially if the species impacts the local economy or human health or safety, or attracts significant attention, as would a new bird species.

Eurasian Collared Doves

Birds provide some of the best examples of rapid population expansion. European starlings and house sparrows, which were purposely introduced into North America, spread across the continent in less than a century. Eurasian collared doves, *Streptopelia decaocto,* began to spread from Turkey into Europe after 1900. The expansion of Eurasian collared doves into Europe was notable in a number of ways. First, the spread began suddenly and, once begun, was relentless. By the 1980s, the doves were found in every country of western and eastern Europe (fig. 10.3). Eurasian collared doves have also rapidly colonized North America. A population of the species was established in the Bahamas in the 1970s by birds released during a burglary of a pet shop in Nassau. By the early 1980s, Eurasian collared doves colonized nearby Florida, subsequently spreading rapidly across North America and reaching California by the mid-1990s.

Another notable feature of the Eurasian collared dove expansion across Europe is that we know a great deal about the underlying population dynamics. The expansion took place in small jumps. Adult Eurasian collared doves are highly sedentary, and dispersal is limited to young doves. Most dispersing young stay within a few kilometers of their parent's nest, but some disperse hundreds of kilometers (fig. 10.4). Once they have chosen a mate, the young birds nest and become sedentary like their parents. These pulses of dispersal by young birds spread the Eurasian collared dove population across Europe at a rate of about 45 km per year.

How does this rate of expansion by Eurasian collared doves compare to rates of expansion by other populations? Compared with the dispersal rate of Africanized honeybees across the Americas, 45 km per year is a modest rate. However, compared with dispersal rates for most other animals that have been studied, 45 km per year is rapid. Figure 10.5, which summarizes rates of dispersal by several species of mammals and birds, shows that rates of dispersal differ by three orders of magnitude. While some species such as Africanized honeybees and Eurasian collared doves spread at rates of tens or hundreds of kilometers per year, others disperse only a few hundred meters per year. This is about the same rate at which North American trees expanded their distributions at the end of the last ice age.

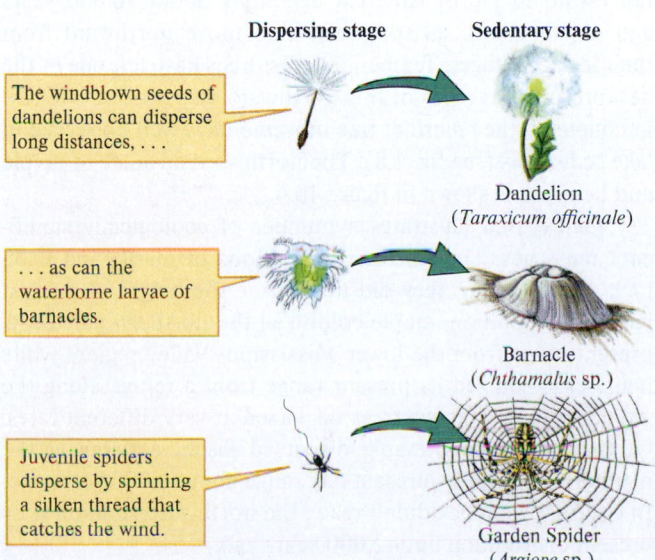

Dispersing stage   **Sedentary stage**

The windblown seeds of dandelions can disperse long distances, . . .

Dandelion
(*Taraxicum officinale*)

. . . as can the waterborne larvae of barnacles.

Barnacle
(*Chthamalus* sp.)

Juvenile spiders disperse by spinning a silken thread that catches the wind.

Garden Spider
(*Argiope* sp.)

Figure 10.2   Dispersing and sedentary stages of organisms.

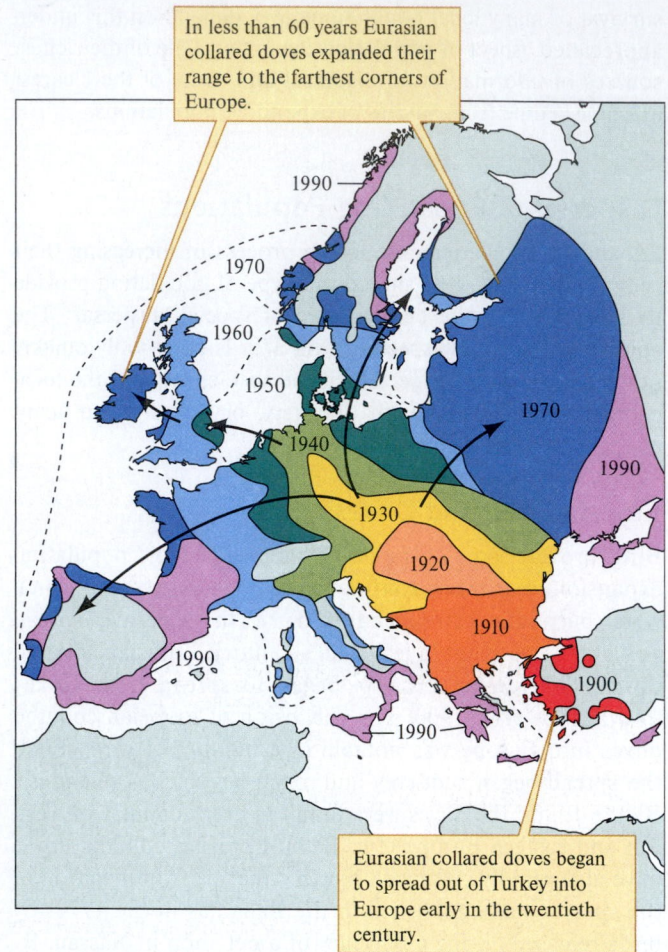

Figure 10.3   The expansion of Eurasian collared doves, *Streptopelia decaocto,* across Europe (data from Hengeveld 1988, Baptista, Trail, and Horblit 1997).

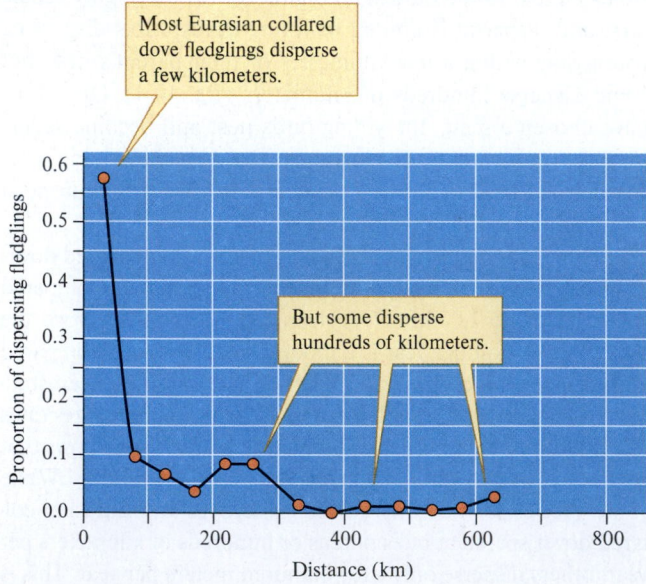

Figure 10.4   Dispersal distances by Eurasian collared dove fledglings (data from Hengeveld 1989).

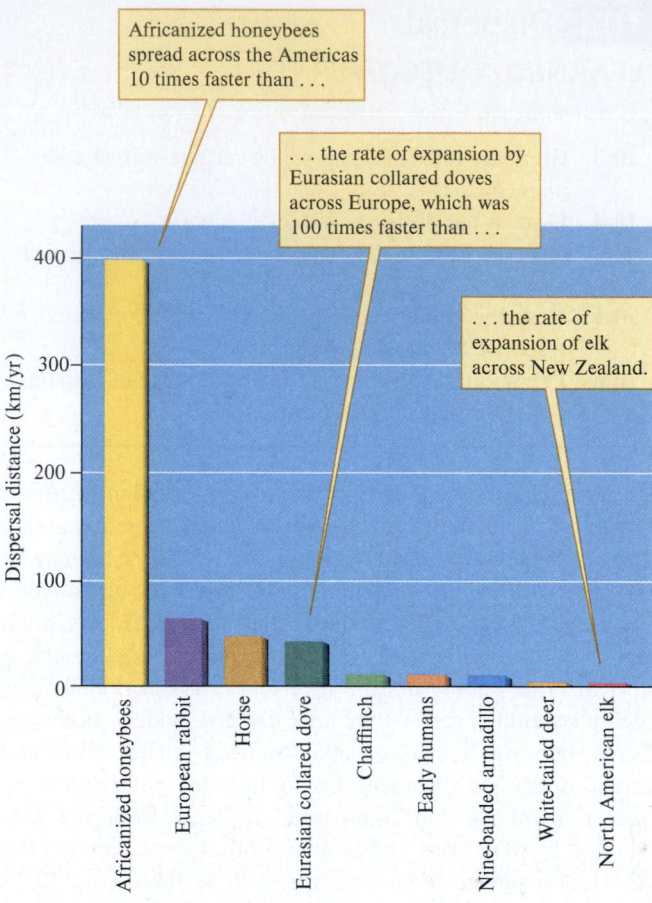

Figure 10.5   Rates of expansion by animal populations (data from Caughley 1977, Hengeveld 1988, Winston 1992).

Range Changes in Response to Climate Change

In response to climate change following retreat of the glaciers northward in North America beginning about 16,000 years ago, organisms of all sorts began to move northward from their ice age refuges. Temperate forest trees have left one of the best-preserved records of this northward dispersal. As we saw in chapter 1, the record of tree movements is well preserved in lake sediments (see fig. 1.8). The northward advance of maple and hemlock is shown in figure 10.6.

Figure 10.6 illustrates a number of ecologically significant messages. Though the distributions of maple and hemlock overlap today, they did not during the height of the last ice age. In addition, maple colonized the northern part of its present range from the lower Mississippi Valley region, while hemlock colonized its present range from a refuge along the Atlantic coast. The two trees dispersed at very different rates. Of the two species, maple dispersed faster, arriving at the northern limits of its present-day range about 6,000 years ago. In contrast, hemlock didn't reach the northwestern limit of its present distribution until 2,000 years ago.

The pollen preserved in lake sediments indicates that forest trees in eastern North America spread northward

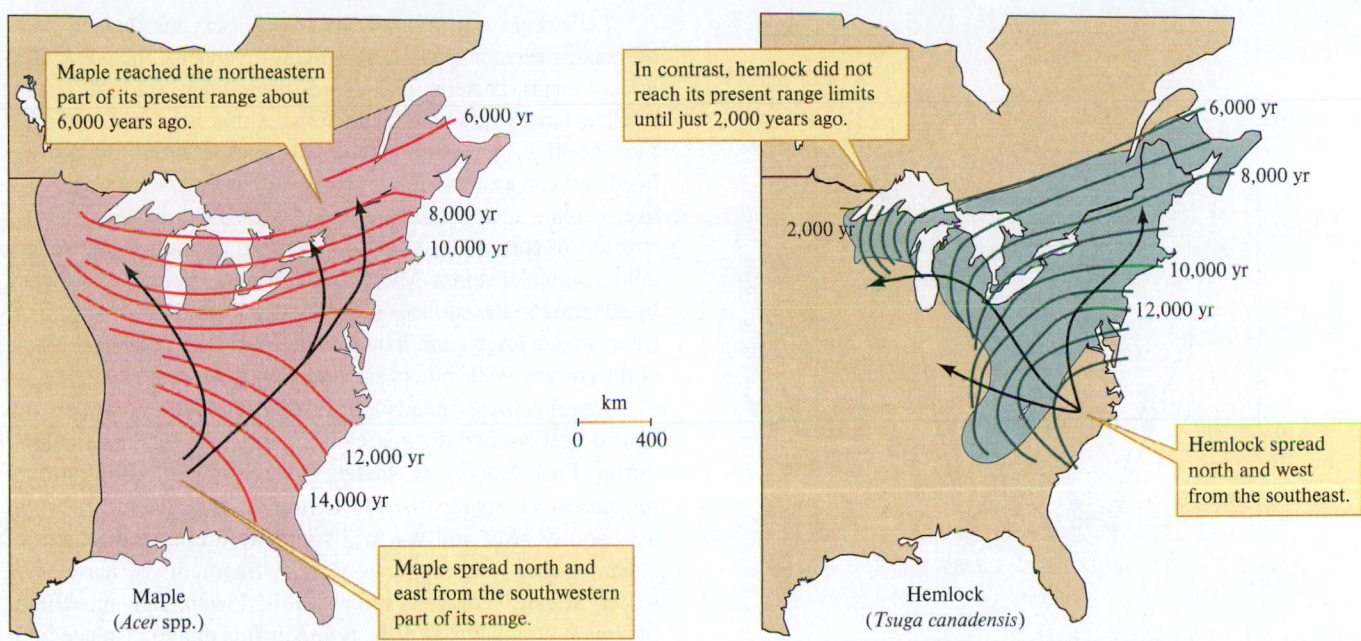

Figure 10.6   The northward expansion of two tree species in North America following glacial retreat (data from Davis 1981).

following the retreat of the glaciers at the rate of 100 to 400 m (0.1–0.4 km) per year. This rate of dispersal is similar to that of some large mammals such as the North American elk. However, it is 1/100 the rate of dispersal shown by Eurasian collared doves in Europe and 1/1,000 the dispersal rate of Africanized honeybees across South, Central, and North America. The range shifts by species in response to historical climate change help us to understand what is happening now in response to modern climatic warming. Ecologists from all over the planet are documenting rapid changes in the ranges, in both elevation and latitude, by organisms varying from aquatic bugs and birds to trees (e.g., Peterson et al. 2019, Liao et al. 2020, Mohammadi et al. 2020).

The previous examples concern dispersal by populations in the process of changing their ranges. Significant dispersal also takes place within established populations whose ranges are not changing. Movements, within established ranges, can be an important aspect of local population dynamics. We will consider two examples.

Dispersal in Response to Changing Food Supply

Predators show several kinds of responses to variation in prey density. In addition to the functional response we discussed in chapter 7, section 7.4, C. S. Holling (1959a) also observed **numerical responses** to increased prey availability. Numerical responses are changes in the density of predator populations in response to increased prey density. Holling studied populations of mice and shrews preying on insect cocoons and attributed the numerical responses he observed to increased reproductive rates. He commented that, "because the reproductive rate of small mammals is so high, there was an almost immediate increase in density with increase in food."

However, some other predators, with much lower reproductive rates, also show strong numerical responses. These numerical responses to prey density are almost entirely due to dispersal.

In some years, northern landscapes are alive with small rodents called voles, *Microtus* spp. Go to the same place during other years and it may be difficult to find any voles. In northern latitudes, vole populations usually reach high densities every 3 to 4 years. Between these peak times, population densities crash. Population cycles in different areas are not synchronized, however. In other words, while vole population density is very low in one area, it is high elsewhere.

Erkki Korpimäki and Kai Norrdahl (1991) conducted a 10-year study of voles and their predators. The study began in 1977 during a peak in vole densities of about 1,800 per square kilometer and continued through two more peaks in 1982 (960/km^2) and 1985–86 (1,980 and 1,710/km^2). The researchers estimated that between these population peaks vole densities per square kilometer fell to as low as 70 in 1980 and 40 in 1984. During this period, the densities of the European kestrel, *Falco tinnunculus*, short-eared owls, *Asio flammeus*, and long-eared owls, *Asio otus*, closely tracked vole densities (fig. 10.7).

What mechanisms produce the numerical responses by kestrels and owls to changing vole densities? Look at figure 10.7 for a clue. The peaks in raptor densities in 1977, 1982, and 1986 match the peaks in vole densities almost perfectly. If reproduction were the source of numerical response by kestrels and owls, there would have been more of a delay, or time lag, in kestrel and owl numerical response. From this close match in numbers, Korpimäki and Norrdahl proposed that kestrels and owls must move from place to place in response to local increases in vole populations.

Is there any supporting evidence for high rates of movement by kestrels and owls? Korpimäki (1988) captured

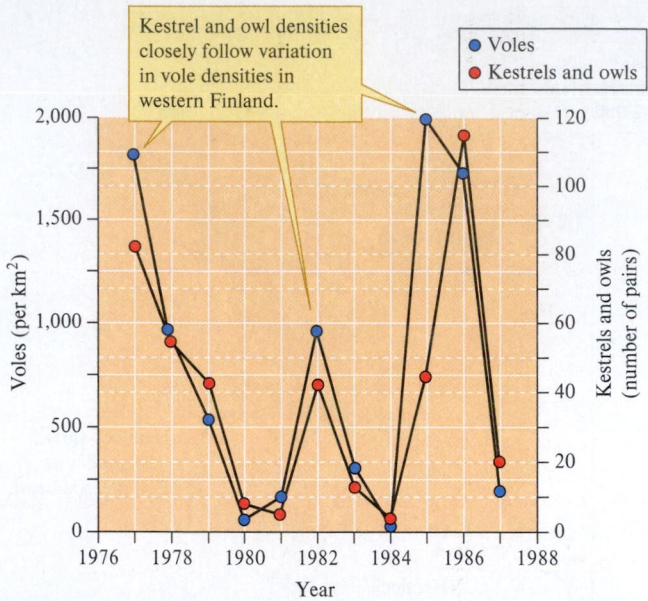

Kestrel and owl densities closely follow variation in vole densities in western Finland.

Figure 10.7   Dispersal and numerical response by predators (data from Korpimäki and Norrdahl 1991).

217 kestrels from 1983 to 1987. Because European kestrel populations have an annual survival rate of 48% to 66%, he predicted a high rate of recapture of marked birds. However, only 3% of the female and 13% of the male kestrels were recaptured. These very low rates of recapture indicated that kestrels were moving into and out of the study area. From their data, Korpimäki and Norrdahl concluded that the kestrels and owls in western Finland are nomadic, moving from place to place in response to changing vole densities. In other cases, we find predators staying put while their primary prey migrates, such as pumas in Patagonia (Gelin et al. 2017) and Eurasian lynx in Norway (Walton et al. 2017). These predators' densities do not fluctuate as much with a specific prey because they switch to other prey when necessary, and/or are able to survive on the few individuals who do not disperse.

These studies documented the contribution of dispersal to local populations of kestrels and owls. Earlier in this section, we saw how studies of expanding populations have shed light on the contribution of dispersal to local population density and dynamics. Many other local populations are strongly influenced by dispersal. One of the environments in which dispersal has a major influence on local populations is in streams and rivers.

Dispersal in Rivers and Streams

One of the most distinctive features of the stream and river environment is *current,* the downstream flow of water. What effect does current have on the lives of stream organisms? As you may recall from chapter 3 (section 3.2), the effects of current are substantial and influence everything from the amount of oxygen in the water to the size, shape, and behavior of stream organisms. In this section, we stop and consider how stream populations are affected by current.

Let's begin with a question: Why doesn't the flowing water of streams eventually wash all stream organisms, including fish, insects, snails, bacteria, algae, and fungi, out to sea? All stream dwellers have a variety of characteristics that help them maintain their position in streams. Some fish, such as trout, are streamlined and can easily swim against swift currents, while other fish like sculpins and loaches are well designed for avoiding the full strength of currents by living on the bottom and seeking shelter among or under stones. Microorganisms resist being washed away by adhering to the surfaces of stones, wood, and other substrates. Many stream insects are flattened and so stay out of the main force of the current, while others are streamlined and fast-swimming.

Despite these means of staying in place, stream organisms do get washed downstream in large numbers, particularly during flash floods, or **spates.** To observe this downstream movement of organisms, put a fine or medium mesh net in a stream or river and you will soon capture large numbers of stream insects and algae along with fragments of leaves and wood. Stream ecologists refer to this downstream movement of stream organisms as **drift.** Some drift is due to displacement of organisms during flash floods. However, some is due to the active movement of organisms downstream.

Whatever its cause, stream organisms drift downstream in large numbers. Why doesn't drift eventually eliminate organisms from the upstream sections of streams? Karl Müller (1954, 1974) hypothesized that drift would eventually wash entire populations out of streams unless organisms actively moved upstream to compensate for drift. He proposed that stream populations are maintained through a dynamic interplay between downstream and upstream dispersal that he called the **colonization cycle.** The colonization cycle is a dynamic view of stream populations in which upstream and downstream dispersal, as well as reproduction, have major influences on stream populations (fig. 10.8).

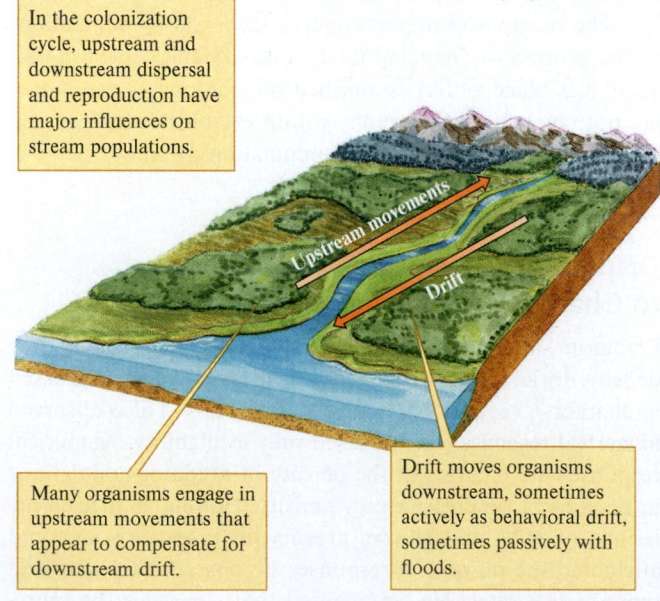

In the colonization cycle, upstream and downstream dispersal and reproduction have major influences on stream populations.

Many organisms engage in upstream movements that appear to compensate for downstream drift.

Drift moves organisms downstream, sometimes actively as behavioral drift, sometimes passively with floods.

Figure 10.8   The colonization cycle of stream invertebrates.

Many studies support Müller's hypothesized colonization cycle, especially among aquatic insects. Aquatic insect larvae disperse upstream as well as downstream by swimming, crawling, and drifting, while most adult aquatic insects disperse by flying. Because of continuous dispersal, which reshuffles stream populations, new substrates put into streams are quickly colonized by a wide variety of stream invertebrates, algae, and bacteria. Most of these dynamics are difficult to observe because they occur too quickly, within the substratum, or at night, or they involve microorganisms impossible to observe directly without the aid of a microscope. However, snails living in clear tropical streams allow researchers to observe the colonization cycle directly (Blanco and Scatena 2005, 2006, 2007).

The Rio Claro flows approximately 30 km through tropical forest on the Osa Peninsula of Costa Rica before flowing into the Pacific Ocean. One of the most easily observed inhabitants of the Rio Claro is the snail *Neritina sp.* which occupies the lower 5 km of the river. The eggs of *Neritina* hatch to produce free-living planktonic larvae that drift down to the Pacific Ocean. After the larvae metamorphose into small snails, they reenter the Rio Claro and begin moving upstream in huge migratory aggregations of up to 500,000 individuals (fig. 10.9). These aggregations move slowly and may take up to 1 year to reach the upstream limit of the population.

Daniel Schneider and John Lyons (1993) discovered that the population of *Neritina* in the Rio Claro consists of a mixture of migrating and stationary subpopulations, with exchange between them. Individual snails migrate upstream for some distance and then leave the migrating wave and enter a local subpopulation. At the same time, individuals from the local subpopulation enter the migratory wave and move upstream. Thus, individuals move upstream in steps and immigration continuously adds to local subpopulations, while emigration removes individuals. Because an organism that is visible to the naked eye does all this in a clear stream, and does it at a snail's pace, we are provided with a unique opportunity to observe that dispersal can strongly influence local population density.

Concept 10.1 Review

1. Why might a species, such as the Eurasian collared dove, be less threatened by rapid climate change than hemlock or maple trees?
2. Ecologists who have used clear plastic sheets coated with adhesive to trap the adults of aquatic insects flying over rivers have found that often the side of the sheets facing downstream traps more of the flying adults than the upstream-facing side. Explain.

(a)

(b)

Figure 10.9 The colonization cycle in action. (*a*) A wave of migrating snails, *Neritina latissima,* in the Rio Claro, Costa Rica; (*b*) a close-up of the migrating snails. (*a, b*) Daniel W. Schneider and John Lyons

10.2 Metapopulations

LEARNING OUTCOMES

After studying this section you should be able to do the following:

10.7 Describe the role of dispersal in sustaining a metapopulation.
10.8 Discuss the similarities and differences in metapopulations of Rocky Mountain Parnassian butterflies and lesser kestrels.

Ongoing dispersal can join numerous subpopulations to form a metapopulation. Populations of many species occur not as a single continuously distributed population but as spatially separated subpopulations. A **subpopulation** is a part of a larger population, with which it sustains a limited exchange of individuals through immigration and emigration. A group of subpopulations living on habitat patches connected by exchange of individuals among patches make up a **metapopulation**. The population of Glanville fritillary butterflies, *Melitaea cinxia,* which lives in dry meadows scattered through the landscape of southern Finland (see chapter 4, section 4.5), is a metapopulation. In the discussion of this butterfly metapopulation, we reviewed how the exchange of individuals among subpopulations has been well documented (Saccheri et al. 1998). However, the metapopulation of *Melitaea* in southern Finland is only one of many that are well known. Here is another example of a butterfly metapopulation.

A Metapopulation of an Alpine Butterfly

Once population biologists began to include the concept of metapopulations in their thinking, they found them everywhere. Butterflies have been well represented in studies of metapopulations. One of these butterflies is the Rocky Mountain Parnassian butterfly, *Parnassius smintheus* (fig. 10.10). The range of *P. smintheus* extends from northern New Mexico along the Rocky Mountains to southwestern Alaska. Along this range *P. smintheus* caterpillars feed mainly on the leaves and flowers of stonecrop, *Sedum* sp., in areas of open forest and meadows. Because of their tie to a narrow range of host plants, *P. smintheus* populations are often distributed among the habitat patches occupied by their host plant, appearing to form metapopulations.

One such metapopulation was studied by Jens Roland, Nusha Keyghobadi, and Sherri Fownes of the University of Alberta in Edmonton, Canada (2000). These researchers focused their attention on a series of 20 alpine meadows on ridges in the Kananaskis region of the Canadian Rocky Mountains. The study meadows ranged in area from about 0.8 ha to 20 ha. While some meadows were adjacent to each other, others were separated by up to 200 m of coniferous forest. The host plant of *P. smintheus* in the study meadows was the lanceleaf stonecrop, *Sedum lanceolatum*.

A combination of fire suppression by forest managers and global warming appears to be decreasing the size of alpine meadows and increasing their isolation from each other by intervening forest. In 1952, the study meadows averaged approximately 36 ha in area. By 1993, the average area of these meadows had declined to approximately 8 ha, a decrease in area of approximately 77%. These changes motivated the research team of Roland, Keyghobadi, and Fownes to study the influences of meadow size and isolation on movements of *P. smintheus* during the summers of 1995 and 1996.

The research team marked and recaptured butterflies to estimate population size in each meadow and to follow butterfly movements. Butterflies were hand netted and marked on the hind wing with a three-letter identification code, using a fine-tipped permanent marker. The team recorded the sex of *P. smintheus* captured and its location within 20 m. Upon recapture, dispersal distance of an individual was estimated as the straight line distance from its last point of capture.

The research team marked 1,574 *P. smintheus* in 1995 and 1,200 in 1996. Of these marked individuals, they recaptured 726 in 1995 and 445 in 1996. Over the course of the study, the size of *P. smintheus* populations in the 20 study meadows ranged from 0 to 230. The average movement distance by males and females in 1995 was approximately 131 m. In 1996, the average movement distances of males and females was 162 m and 118 m, respectively. The maximum dispersal distance for a butterfly in 1995 was 1,729 m and in 1996 the maximum dispersal distance was 1,636 m. Most of the movements determined by recaptures were the result of dispersal within meadows. In 1995 only 5.8% of documented dispersal movements were from one meadow to another, and in 1996 dispersal between meadows accounted for 15.2% of total recaptures.

One of the questions posed by Roland, Keyghobadi, and Fownes was how meadow size and population size might affect dispersal by *P. smintheus*. As shown in figure 10.11, average butterfly population size increased with meadow area. It turned out that butterflies are more likely to leave small populations than large populations. Butterflies leaving small populations generally emigrate to larger populations. The researchers pointed out that this pattern of movement, from small to large populations, has been observed in several other butterfly species, including the Glanville fritillary. Ongoing research is aimed at discovering the reasons for this pattern of emigration. The results of this study suggest that as alpine meadows in the Rocky Mountains decline in area, populations of *P. smintheus* will become progressively more compressed into fewer and fewer small meadows, perhaps disappearing entirely in parts of their range. These population bottlenecks have significant impacts on its genetic structure (Jangjoo et al. 2020).

Some of the patterns of dispersal within this metapopulation of alpine butterflies have been observed in a study of dispersal in a metapopulation of a small falcon.

Forest   Meadow

1 km

Movements of butterflies among patches of meadow separated by forest unite the local populations of this butterfly into a metapopulation.

Figure 10.10   A metapopulation of the Rocky Mountain Parnassian butterfly, *Parnassius smintheus*. Because it is tied to meadows where its larval host plants grow, *P. smintheus* lives in scattered subpopulations living in patches of meadows surrounded by forest. Butterflies dispersing between meadows connect these subpopulations, sustaining a metapopulation (after Matter et al. 2009). Alamy Stock Photo

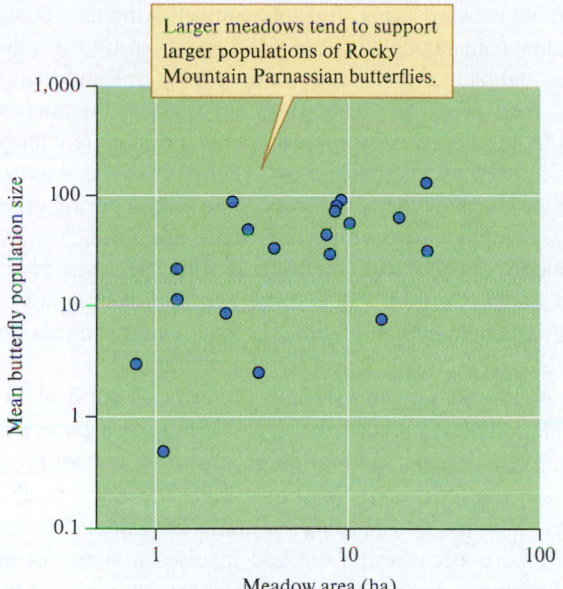

Larger meadows tend to support larger populations of Rocky Mountain Parnassian butterflies.

Figure 10.11   The relationship between meadow area and the size of Rocky Mountain Parnassian butterfly, *Parnassius smintheus,* populations. With forest encroachment into alpine meadows in the Rocky Mountains, populations of *P. smintheus* will likely decline (data from Roland, Keyghobadi, and Fownes 2000).

Dispersal Within a Metapopulation of Lesser Kestrels

The lesser kestrel, *Falco naumanni,* is a small migratory falcon that breeds in colonies of monogamous pairs in Eurasia and spends its winters in sub-Saharan Africa (fig. 10.12). Lesser

Figure 10.12   The lesser kestrel, *Falco naumanni,* breeds in scattered colonies that collectively form metapopulations. Most populations of this species have declined dramatically with modernization of agriculture. Gregory G. Dimijian/Science Source

kestrels have suffered a high rate of decline across their range and are listed as a globally threatened species. In contrast to its global circumstance, the lesser kestrel population of the Ebro River valley of northeastern Spain has grown dramatically in 7 years. David Serrano and Jose Tella, two ecologists who have conducted long-term studies of this population (Serrano and Tella 2003), documented growth in this population from 224 pairs distributed among 4 subpopulations in 1993 to 787 pairs living in 14 subpopulations in 2000. Serrano and Tella attribute this regional growth to sustained traditional farming practices in the Ebro River valley. However, they warn that plans to modernize farming practices in the Ebro River valley may lead to the population declines seen elsewhere (Grande et al. 2018).

Serrano and Tella used numbered, color leg bands to mark and track individual kestrels in their study population. From 1993 to 1999, they banded 4,901 fledgling kestrels and 640 adults. Because lesser kestrels breed within colonies, they are easier to track during the breeding season. Once locating a colony, Serrano and Tella would observe the colony members from a blind, record the numbers of pairs within the colony, and, using a telescope, read the numbers on the leg band of any banded adult birds in the colony. Serrano and Tella were able to obtain accurate counts of the entire breeding population of lesser kestrels within the Ebro River valley each year. They could also use their observations to plot the movements of any banded birds they saw. Within colonies, the percentage of banded adults of known age ranged from 60% to 90%.

The data gathered indicate that a substantial percentage of birds leave the breeding colony where they hatched to join other subpopulations in their first year of breeding. However, females in this species are more likely to move than males. The rate of emigration by first-breeding females is approximately 30% versus 22% for first-breeding males. In contrast, less than 4% of older adults emigrate from a colony on any given year. Though some lesser kestrels in the study population dispersed more than 100 km, Serrano and Tella found a negative correlation between distance between colonies and the frequency of dispersal between them.

The lesser kestrels of the Ebro River valley and the Rocky Mountain Parnassian butterflies of southern Alberta, Canada, interact with their environments on greatly different scales. In addition, the butterfly population appears to be contracting spatially; the lesser kestrel population is expanding. However, these populations also have a number of features in common. First, they both are spatially organized into metapopulations. Another feature that the two populations share is the influence of local population size on the tendency to disperse and the direction of dispersal. Like *P. smintheus,* lesser kestrels in smaller subpopulations are more likely to emigrate than are individuals in larger subpopulations. Second, lesser kestrels are more likely to disperse from small colonies to larger colonies.

In the last two Concept discussions, we have reviewed the influences of dispersal on populations. Now we compare patterns of survival in populations, a major contributor to the dynamic nature of populations.

Concept 10.2 Review

1. Figure 10.11 and the upper portion of figure 21.13 show the relationship between meadow size and population size in two butterfly species. How are the patterns shown by the two graphs similar? How do they differ? (1 ha = 10,000 m^2)

2. The Rocky Mountain Parnassian butterfly tends to disperse from small to large meadows. Why is this direction of movement more advantageous than the reverse? (*Hint:* See Glanville fritillary studies, chapter 4.)

3. Contrast human influences on metapopulations of the Rocky Mountain Parnassian butterfly versus those of the lesser kestrels of the Ebro River valley.

10.3 Patterns of Survival

LEARNING OUTCOMES

After studying this section you should be able to do the following:

10.9 Compare cohort and static life tables.

10.10 Define age distribution.

10.11 Construct a survivorship curve, using data from a life table.

10.12 Identify type I, II, and III survivorship curves.

10.13 Explain the population processes underlying type I, II, and III survivorship curves.

A survivorship curve summarizes the pattern of survival in a population. Patterns of survival vary a great deal from one species to another and, depending on environmental circumstances, can vary substantially even within a single species. Some species produce young by the millions, which, in turn, die at a high rate. Other species produce a few young and invest heavily in their care. The young of species that have evolved this pattern survive at a high rate. Still other species show intermediate patterns of reproductive rate, parental care, and juvenile survival. In response to practical challenges of discerning patterns of survival, population biologists have invented bookkeeping devices called **life tables** that list both the survivorship and the deaths, or *mortality,* in populations.

Estimating Patterns of Survival

There are three main ways of estimating patterns of survival within a population. The first and most reliable way is to identify a large number of individuals that are born at about the same time and keep records on them from birth to death. A group born during the same time period—for example, the same year—is called a **cohort.** A life table made from data collected in this way is called a **cohort life table.** The cohort studied might be a group of plant seedlings that germinated at the same time or all the lambs born into a population of mountain sheep in a particular year.

While understanding and interpreting a cohort life table may be relatively easy, obtaining the data upon which a cohort

life table is based is not. Imagine yourself lying face down in a meadow painstakingly counting thousands of tiny seedlings of an annual plant. You must mark their locations and then come back every week for 6 months until the last member of the population dies. Or, if you are studying a moderately long-lived species, such as a barnacle or a perennial herb like a buttercup, imagine checking the cohort repeatedly over a period of several years. If your study organism is a mobile animal such as a whale or falcon, the problems multiply. If your species is very long-lived, such as a giant sequoia, such an approach is impossible within a single human lifetime. In such circumstances population biologists usually resort to other techniques.

A second way to estimate patterns of survival in wild populations is to record the age at death of a large number of individuals. This method differs from the cohort approach because the individuals in your sample are born at different times. This method produces a **static life table.** The table is called *static* because the method involves a snapshot of survival within a population during a short interval of time. To produce a static life table, the biologist often needs to estimate the age at which individuals die. This can be done by tagging individuals when they are born and then recovering the tags after death and recording the age at death. An alternative procedure is to somehow estimate the age of dead individuals. The age of many species can be determined reasonably well. For instance, mountain sheep can be aged by counting the growth rings on their horns. There are also growth rings on the carapaces of turtles, in the trunks of trees, and in the "stems" of soft or hard corals.

A third way of determining patterns of survival is from the **age distribution.** An age distribution consists of the proportion of individuals of different ages within a population. You can use an age distribution to estimate survival by calculating the difference in proportion of individuals in succeeding age classes. This method, which also produces a static life table, assumes that the difference in numbers of individuals in one age class and the next is the result of mortality. What are some other major assumptions underlying the use of age distributions to estimate patterns of survival? This method requires that a population is neither growing nor declining and that it is not receiving new members from the outside or losing members because they migrate away. Since most of these assumptions are often violated in natural populations, a life table constructed from this type of data tends to be less accurate than a cohort life table. Static life tables are often useful, however, since they may be the only information available.

High Survival Among the Young

Adolph Murie (1944) studied patterns of survival among Dall sheep (fig. 10.13) in what is now Mount Denali National Park, Alaska. Murie estimated survival patterns by collecting the skulls of 608 sheep that had died from various causes. He determined the age at which each sheep in his sample died by counting the growth rings on their horns and by studying tooth wear. The major assumption of this study was that the proportion of skulls in each age class represented the typical

Figure 10.13   Dall sheep, *Ovis dalli,* a mountain sheep of far northern North America, was the subject of one of the classic studies of survivorship. Getty Images

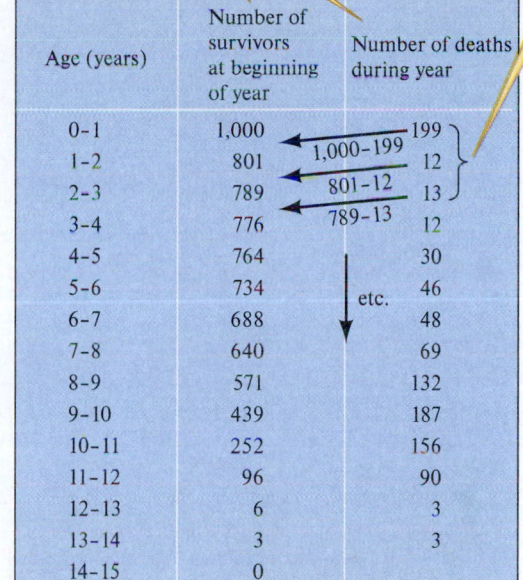

To allow comparisons to other studies, number of Dall sheep surviving and dying within each year of life is converted to number per 1,000 births.

Subtracting number of deaths from number alive at the beginning of each year gives the number alive at the beginning of the next year.

Age (years)	Number of survivors at beginning of year	Number of deaths during year
0–1	1,000	199
1–2	801	12
2–3	789	13
3–4	776	12
4–5	764	30
5–6	734	46
6–7	688	48
7–8	640	69
8–9	571	132
9–10	439	187
10–11	252	156
11–12	96	90
12–13	6	3
13–14	3	3
14–15	0	

1,000–199
801–12
789–13
etc.

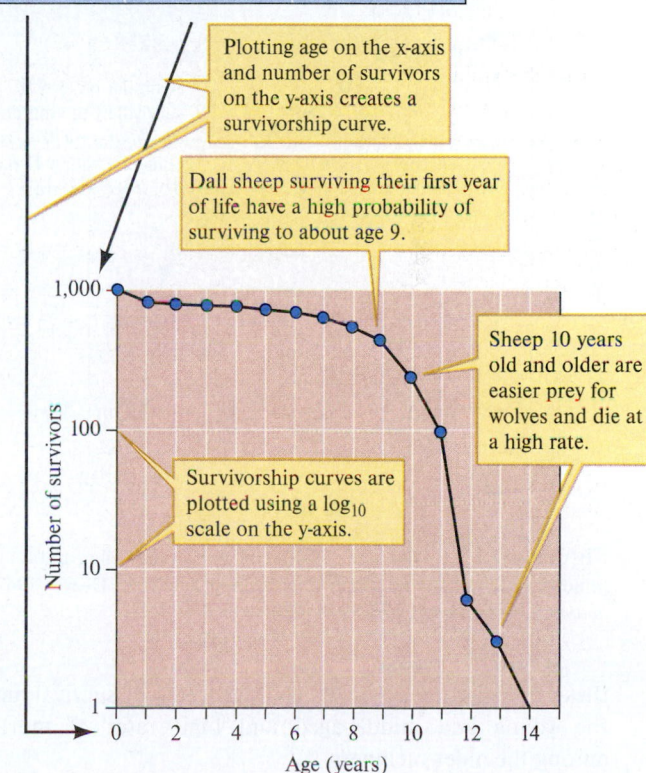

Plotting age on the x-axis and number of survivors on the y-axis creates a survivorship curve.

Dall sheep surviving their first year of life have a high probability of surviving to about age 9.

Sheep 10 years old and older are easier prey for wolves and die at a high rate.

Survivorship curves are plotted using a $\log_{10}$ scale on the y-axis.

Figure 10.14   Dall sheep: from life table to survivorship curve (data from Murie 1944).

proportion of individuals dying at that age. For example, the proportion of sheep in the sample that died before the age of 1 year represents the proportion that generally dies during the first year of life. While this assumption is not likely to be strictly true, the pattern of survival that emerges probably gives a reasonable picture of survival in the population, particularly when the sample is as large as Murie's.

Figure 10.14 summarizes the survival patterns for Dall sheep based on Murie's sample of skulls. The upper portion of the figure shows the static life table that Murie constructed. The first column lists the ages of the sheep, the second column lists the number surviving in each age class, and the third column lists the number dying in each age class. Notice that although Murie studied only 608 skulls, the numbers in the table are expressed as numbers per 1,000 individuals. This adjustment is made to ease comparisons with other populations.

The upper portion of figure 10.14 also shows how to translate numbers of deaths into numbers of survivors. Plotting number of survivors per 1,000 births against age produces the **survivorship curve** shown in the lower portion of the figure. A survivorship curve shows patterns of life and death within a population. Notice that in this population of Dall sheep, there are two periods when mortality rates are higher: during the first year and during the period between 9 and 13 years. Juvenile mortality and mortality of the aged are higher in this population, while mortality in the middle years is lower. The overall pattern of survival and mortality among Dall sheep is much like that for a variety of other large vertebrates, including red deer, *Cervus elaphus,* Columbian black-tailed deer, *Odocoileus hemionus columbianus,* East African buffalo, *Syncerus caffer,* and humans. The key characteristics of survival among

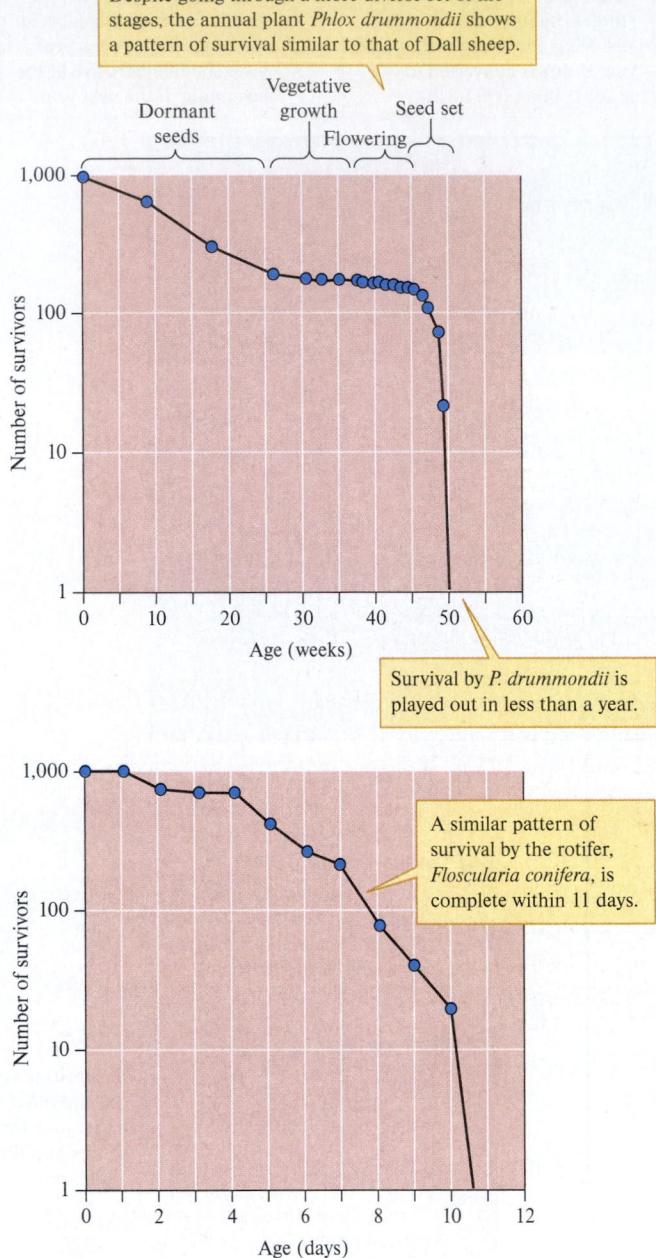

Figure 10.15   High rates of survival among the young and middle-aged in plant and rotifer populations (data from Deevey 1947, *bottom;* Leverich and Levin 1979, *top*).

in less than 11 days. These survivorship curves are based on cohort life tables.

Survival patterns can be quite different in other species. In the next example, mortality is not delayed until old age but occurs at approximately equal rates throughout life.

Constant Rates of Survival

The survivorship curves of many species are nearly straight lines. In these populations, individuals are equally likely to die at any age. This pattern of survival has been commonly observed in birds, such as the American robin, *Turdus migratorius,* and the white-crowned sparrow, *Zonotrichia leucophrys*

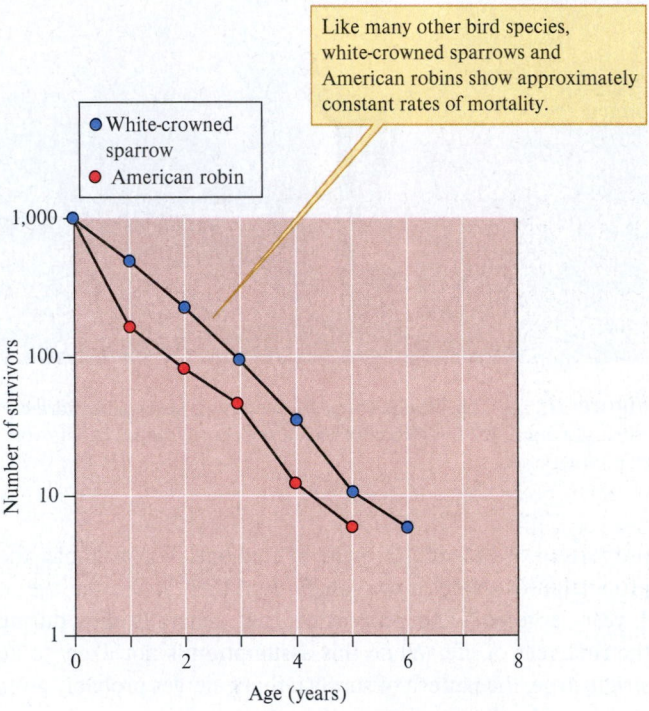

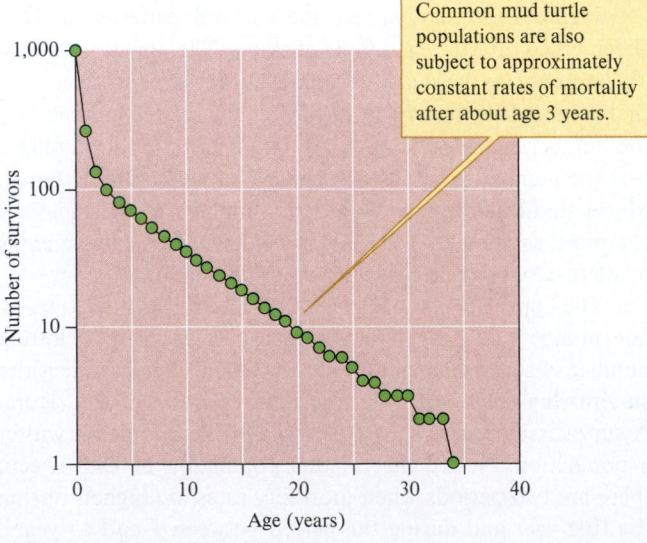

Figure 10.16   Constant rates of survival (data from Deevey 1947, Baker, Mewaldt, and Stewart 1981, Frazer, Gibbons, and Greene 1991).

these populations are relatively high rates of survival among the young and middle-aged and high rates of mortality among the older members.

This pattern of survival has also been observed in populations of annual plants and small invertebrate animals. Notice in figure 10.15 that patterns of survival in a population of a plant, *Phlox drummondii,* and a rotifer, *Floscularia conifera,* are remarkably similar to that of Dall mountain sheep. Following an initial period of higher juvenile mortality, mortality is relatively low for a period, and then is high among older individuals. In the *Phlox* population, however, this pattern of survival is played out in less than 1 year and in the rotifer population *nuttalli* (fig. 10.16). Life expectancy remains relatively constant over the whole period a cohort is in existence. While birds are the best known for showing a linear pattern of survival, many other species do as well. For instance, figure 10.16 also shows a similar pattern of survival for a population of the common mud turtle, *Kinosternon subrubrum.* Though the mud turtle has a high rate of mortality during the first year of life, thereafter, survival follows a straight line.

As we shall see next, some organisms die at a much higher rate as juveniles than we have seen in any of the populations we have considered to this point.

High Mortality Among the Young

Some organisms produce large numbers of young with very high rates of mortality. The eggs produced by marine fish such as the mackerel, *Scomber scombrus,* may number in the millions. Out of 1 million eggs laid by a mackerel, more than 999,990 die during the first 70 days of life as eggs, larvae, or juveniles. Survival rates are similar in populations of the prawn *Leander squilla* off the coast of Sweden. For each 1 million eggs laid by *Leander,* only about 2,000 individuals survive the first year of life. This period of high mortality among young prawns is followed by a fairly constant mortality over the remainder of the life span.

Similar patterns of survival are shown by other marine invertebrates and fish and by plants that produce immense numbers of seeds. One of these plants is *Cleome droserifolia,* a desert shrub studied by Ahmad Hegazy (1990). Hegazy estimated that a local population of approximately 2,000 plants produce almost 20 million seeds each year. Of these, approximately 12,500 seeds germinate and produce seedlings. Only 800 seedlings survive to become juvenile plants. Figure 10.17 traces this pattern of survival by *Cleome* expressed as

survivors per million seeds. Hegazy estimated that for each 1 million seeds produced, about 39 survive to the age of 1 year, a survival rate of only 0.0039%. Survival in this desert plant population contrasts sharply with that seen in Dall sheep. The striking difference in patterns of survival between populations such as *Cleome,* birds such as the American robin, and large mammals such as Dall sheep led early population biologists to propose a classification of survivorship curves.

Three Types of Survivorship Curves

Based on studies of survival by a wide variety of organisms, population ecologists have proposed that most survivorship curves fall into three major categories (fig. 10.18). A relatively high rate of survival among young and middle-aged individuals followed by a high rate of mortality among the aged is known as a **type I survivorship curve.** This is the pattern of survival we saw in populations of Dall sheep, *P. drummondii,* and rotifers (see figs. 10.14 and 10.15). Constant rates of survival throughout life produce the straight-line pattern of survival known as a **type II survivorship curve.** American robins, white-crowned sparrows, and common mud turtles show this pattern of survival (see fig. 10.16). A **type III survivorship curve** is one in which a period of extremely high rates of mortality among the young is followed by a relatively high rate of survival. The desert plant *Cleome* provides an excellent example of a type III survivorship curve (see fig. 10.17).

How well does this classification of survivorship represent natural populations? Most populations do not conform perfectly to any one of the three basic types of survivorship but show virtually every sort of intermediate form of survivorship between the curves. If survivorship can be so variable within species, what good are these idealized, theoretical survivorship curves? Their most important value, like most theoretical constructs, is that they set boundaries that mark

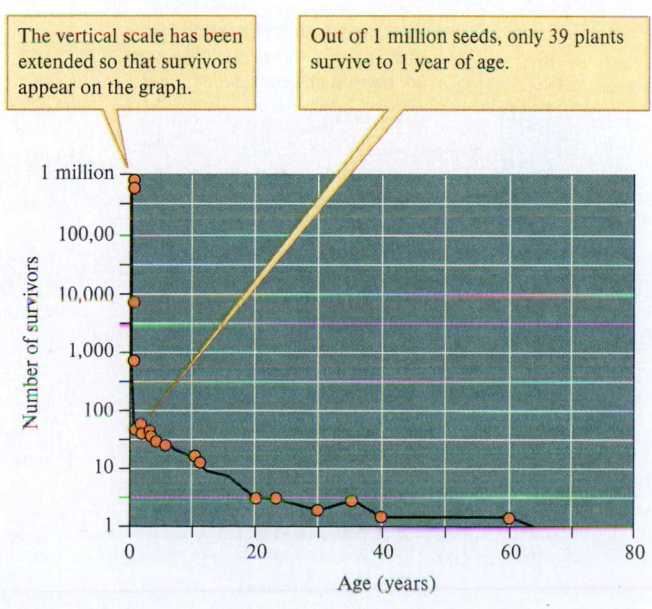

Figure 10.17   A high rate of mortality among the young of a perennial plant, *Cleome droserifolia* (data from Hegazy 1990).

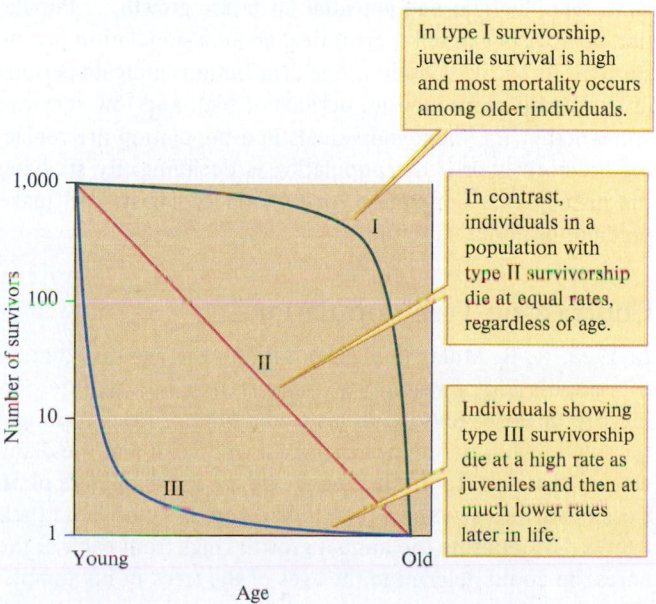

Figure 10.18   Three types of survivorship curves.

what is possible within populations. Regardless of how closely actual survivorship curves approximate the theoretical curves, they serve excellent summaries of survival patterns within populations.

We now turn to the age distributions of populations, a topic closely related to survivorship. As we have seen, the age distribution of a population can be used to construct a static life table from which a survivorship curve can be drawn. However, as we shall see next, a population's age distribution offers other insights into population dynamics, particularly patterns of survival and reproduction.

Concept 10.3 Review

1. How would substantial emigration and immigration affect estimates of survivorship within a population, where estimates are based on age distributions?
2. Female cottonwood trees (*Populus* species) produce millions of seeds each year. Does this information give you a sound basis for predicting their survivorship pattern?
3. How would human mortality patterns have to change for our species to shift from type I to type II survivorship?

10.4 Age Distribution

LEARNING OUTCOMES

After studying this section you should be able to do the following:

10.14 Describe how survival and reproduction contribute to a population's age distribution.
10.15 Interpret the age distributions of a variety of organisms.

The age distribution of a population reflects its history of survival, reproduction, and potential for future growth.   Population ecologists can tell a great deal about a population just by studying its age distribution. Age distributions indicate periods of successful reproduction, periods of high and low survival, and whether the older individuals in a population are replacing themselves or if the population is declining. By studying the history of a population, population ecologists can make predictions about its future.

Contrasting Tree Populations

In 1923, R. B. Miller published data on the age distribution of a population of white oak, *Quercus alba,* in a mature oak-hickory forest in Illinois. In his study, Miller first determined the relationship between the age of a white oak and the diameter of its trunk. To do this, he measured the diameters of 56 trees of various sizes and then took a core of wood from their trunks. By counting the annual growth rings from each of the cores, he could determine the ages of the trees in his sample. With the relationship between oak age and diameter in hand, Miller used diameter to estimate the ages of hundreds of trees.

Most white oaks in Miller's study forest were concentrated in the youngest age class of 1 to 50 years, with progressively fewer individuals in the older age classes (fig. 10.19). The oldest white oaks in the forest were over 300 years old. In other words, the age distribution of white oak in this forest was biased toward the young trees. What might we infer from this age distribution? The age distribution indicates that reproduction is sufficient to replace the oldest individuals in the population as they die. That is, this population of white oaks appeared to be either stable or growing.

The age distribution of this white oak population contrasts sharply with the age distributions of populations of Rio Grande cottonwoods, *Populus deltoides* subsp. *wislizenii.* The most extensive cottonwood forests remaining in the southwestern United States grow along the Middle Rio Grande in central New Mexico. However, studies of age distributions indicate that these populations are declining. Older trees, which can live to about 130 years, are not being replaced by younger trees (fig. 10.20). In contrast to the white oak population in Illinois, the Rio Grande cottonwood population is dominated by older individuals. At the study site represented by figure 10.20, there has been no reproduction for over a decade. At other sites along the Rio Grande there has been little reproduction for over three decades.

Why have Rio Grande cottonwoods failed to reproduce? Reproduction by Rio Grande cottonwoods depends on seasonal floods, which play two key roles. First, floods create areas of bare soil without a surface layer of organic matter and without competing vegetation, ideal conditions for germination and establishment of cottonwood seedlings. Floods also keep these nursery areas of bare soil moist until cottonwood seedlings can grow their roots deep enough to tap into the shallow water table. Historically, these conditions were created by

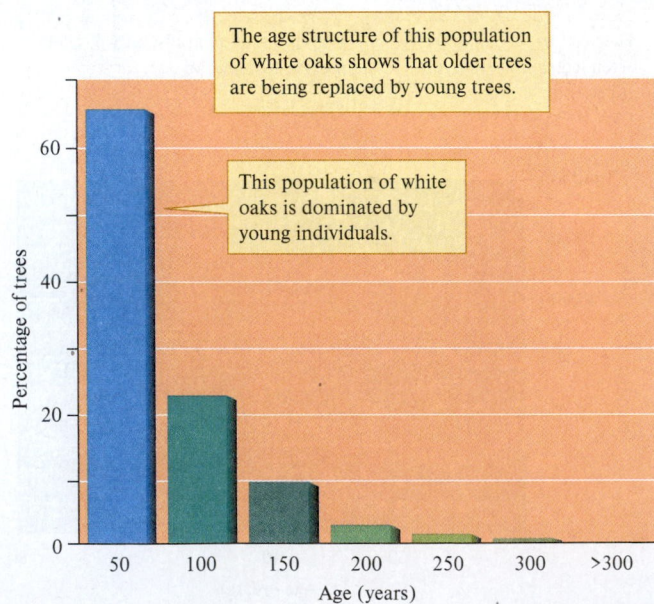

The age structure of this population of white oaks shows that older trees are being replaced by young trees.

This population of white oaks is dominated by young individuals.

Figure 10.19   The age distribution of a white oak, *Quercus alba,* population in Illinois (data from Miller 1923).

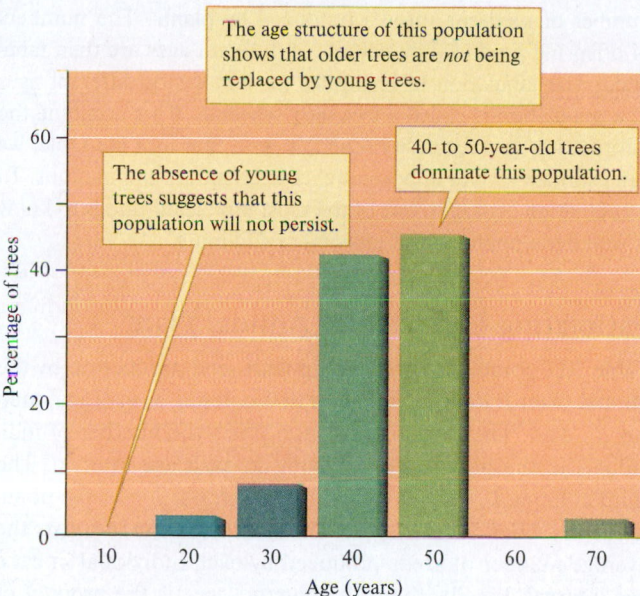

Figure 10.20   Age distribution of a population of Rio Grande cottonwoods, *Populus deltoides* subsp. *wislizenii,* near Belen, New Mexico (data from Howe and Knopf 1991).

spring floods, the timing of which coincided with dispersal of cottonwood seeds by wind. The annual rhythm of seed bed preparation and seeding has been interrupted by the construction of dams on the Rio Grande for flood control and irrigation. The tamed Rio Grande no longer floods like it once did, and though Rio Grande cottonwoods produce seeds each year, their age distribution indicates that these seeds find few suitable places to germinate.

The age distributions of tree populations change over the course of many decades or centuries. Meanwhile, other populations can change significantly on much shorter timescales. One of these dynamic populations has been thoroughly studied on the Galápagos Islands.

A Dynamic Population in a Variable Climate

Rosemary Grant and Peter Grant (1989) have spent decades studying Darwin's finch populations. One of their most thorough studies has concerned the large cactus finch, *Geospiza conirostris,* on the island of Genovesa, which lies in the northeastern portion of the Galápagos archipelago, approximately 1,000 km off the west coast of South America. The Galápagos Islands have a highly variable climate, which is reflected in the highly dynamic populations of the organisms living on the islands, including populations of the large cactus finch.

The age distributions of the large cactus finch during 1983 and 1987 show that the population can be very dynamic (fig. 10.21). The 1983 age distribution shows a fairly regular distribution of individuals among age classes. However, there were no 6-year-old individuals in the population. This gap is due to a drought in 1977, during which no finches reproduced. Now, compare the 1983 and 1987 age distributions. The distributions contrast markedly, though they are for the same population separated by only 4 years!

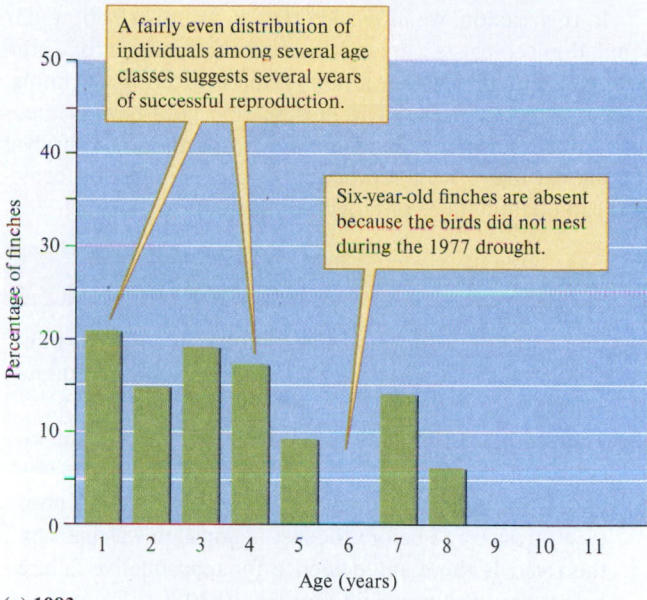

(a) **1983**

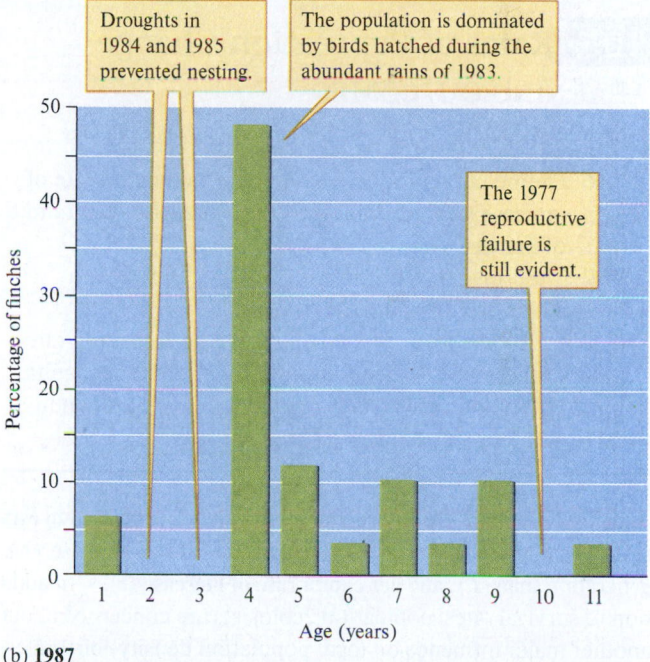

(b) **1987**

Figure 10.21   Age distribution of a population of large cactus finches, *Geospiza conirostris,* on the island of Genovesa in the Galápagos Islands during 1983 (*a*) and 1987 (*b*) (data from Grant and Grant 1989).

The 1977 gap is still present in the 1987 age distribution, and another has been added for 2- and 3-year-old finches. This second gap is the result of 2 years of reproductive failure during a drought that persisted from 1984 to 1985. Another difference is that the 1987 age distribution is dominated by 4-year-old birds that were fledged during 1983. The 1983 class dominates because wet weather that year resulted in very high production of food that the finches depend on for reproduction. This long-term study of the large cactus finch population of Genovesa Island demonstrates the responsiveness of population age structure to environmental variation.

In this section, we have seen that an age distribution tells population ecologists a great deal about the dynamics of a population, including whether a population is growing, declining, or approximately stable. The next section goes beyond these qualitative assessments. By combining information on survival and age structure with reproductive rates, population ecologists can quantify rates of growth or decline.

Concept 10.4 Review

1. Can a healthy population that is not in danger of extinction have an age structure that shows years of reproductive failure?
2. The last major natural reproduction by Rio Grande cottonwoods, which produced the large number of 40- and 50-year-old trees documented by Howe and Knopf (1991), occurred before the last major dam was built on the river. Is there any evidence for reproductive failure before that dam was built (see fig. 10.20)?

10.5 Rates of Population Change

LEARNING OUTCOMES

After studying this section you should be able to do the following:

10.16 Define net reproductive rate (R_0), geometric rate of increase (λ), generation time (T), and per capita rate of increase (r).

10.17 Describe the conditions necessary to produce a stable age distribution.

10.18 Calculate net reproductive rate (R_0), geometric rate of increase (λ), generation time (T), and per capita rate of increase (r) for a population, using fecundity and life table data.

A life table combined with a fecundity schedule can be used to estimate net reproductive rate (R_0), geometric rate of increase (λ), generation time (T), and per capita rate of increase (r). In addition to survival rates, population ecologists are concerned about another major influence on local population density—birthrates. In mammals and other live-bearing organisms, from sharks to humans, the term **birthrate** means the number of young born per female in a period of time. Population biologists also use the term *birth* more generally to refer to any other processes that produce new individuals in the population. In populations of birds, fish, and reptiles, births are usually counted as the number of eggs laid. In plants, the number of births may be the number of seeds produced or the number of shoots produced during asexual reproduction. In bacteria, the birth, or reproductive, rate is taken as the rate of cell division.

Tracking birthrates in a population is similar to tracking survival rates. In a sexually reproducing population, the population biologist needs to know the average number of births per female for each age class and the number of females in each age class. In practice, the ecologist counts the number of eggs produced by birds or reptiles, the number of fawns produced by deer, or the number of seeds or sprouts produced by plants. The numbers of offspring produced by parents of different ages are then tabulated. The tabulation of birthrates for females of different ages in a population is called a **fecundity schedule.** If we combine the information in a fecundity schedule with that in a life table, we can estimate several important characteristics of populations. To a population ecologist, one of the most important things to know is whether a population is growing or declining.

Estimating Rates for an Annual Plant

Table 10.1 combines survivorship with seed production by the annual plant *P. drummondii*. The first column, x, lists age intervals in days. The second column, n_x, lists the number of individuals in the population surviving to each age interval. The third column, l_x, lists survivorship, the proportion of the population surviving to each age x. The fourth column, m_x, lists the average number of seeds produced by each individual in each age interval. Finally, the fifth column, $l_x m_x$, is the product of columns 3 and 4.

We've already used the data in column 3, l_x, to construct the survivorship curve for this species (see fig. 10.15). Now, let's combine those survivorship data with the seed production for *P. drummondii*, m_x, to calculate the **net reproductive rate, R_0.** The calculations of reproductive rates in this section assume that l_x and m_x for each age class in a population are constant and, as a result, the population has a **stable age distribution.** In a population with a stable age distribution, the proportion of individuals in each of the age classes is constant. In general, the net reproductive rate is the average number of female offspring produced by an individual female in a population during her lifetime. In the case of the annual plant *P. drummondii,* the net reproductive rate is the average number of seeds left by an individual. We don't have to consider the sex of the *Phlox* seeds, since all germinating and surviving to maturity will produce flowers with fully functional female and male reproductive organs. You can calculate the net reproductive rate from table 10.1 by adding the values in the final column. The result is:

$$R_0 = \Sigma \, l_x m_x = 2.4177$$

To calculate the total number of seeds produced by this population during the year of study, multiply 2.4177 by 996, which was the initial number of plants in this population. The result, 2,408, is the number of seeds that this population of *P. drummondii* will begin with the next year.

Since *P. drummondii* has pulsed reproduction, we can estimate the rate at which its population is growing with a quantity known as the **geometric rate of increase λ.** The geometric rate of increase is the ratio of the population size at two points in time:

$$\lambda = \frac{N_{t+1}}{N_t}$$

In this equation, N_{t+1} is the size of the population at some future time and N_t is the size of the population at some earlier time (fig. 10.22). The time interval t may be years, days, or hours; which time interval you use to calculate the geometric

Table 10.1

Combining survivorship with seed production by *P. drummondii* to estimate net reproductive rate, R_0				
Age (days)	Number surviving to day x	Proportion surviving to day x	Average number of seeds per individual during time interval	Multiplication of l_x and m_x
x	n_x	l_x	m_x	$l_x m_x$
0–299	996	1.0000	0.0000	0.0000
299–306	158	0.1586	0.3352	0.0532
306–313	154	0.1546	0.7963	0.1231
313–320	151	0.1516	2.3995	0.3638
320–327	147	0.1476	3.1094	0.4589
327–334	136	0.1365	2.5411	0.3469
334–341	105	0.1054	3.1589	0.3330
341–348	74	0.0743	8.6625	0.6436
348–355	22	0.0221	4.3072	0.0952
355–362	0	0.0000	0.0000	0.0000

Data from Leverich and Levin 1979.

Each individual leaves an average of 2.4177 offspring.

$$R_0 = \sum l_x m_x = 2.4177$$

The value of R_0, which is greater than 1.0, indicates that this population of *P. drummondii* is growing.

Summing the final column yields R_0, the net reproductive rate per individual.

rate of increase for a population depends on the organism and the rate at which its population grows.

Let's calculate λ for the population of *P. drummondii*. What time interval should we use for our calculation? Since *P. drummondii* is an annual plant, the most meaningful time interval would be 1 year. The initial number, N_t, of *P. drummondii* in the population was 996. The number of individuals (seeds) in the population at the end of a year of study was 2,408. This is the number in the next generation, which is N_{t+1}. Therefore, the geometric rate of increase for the population over the period of this study was:

$$\lambda = \frac{2,408}{996} = 2.4177$$

This is the same value we got for R_0. But, before you jump to conclusions, you should know that R_0, which is the average number of female offspring per female per generation, does not always equal λ. In this case, λ equaled R_0 because *P. drummondii* is an annual plant with pulsed reproduction. If a species has overlapping generations and continuous reproduction, R_0 will usually not equal λ.

How long do you think this plant can continue to reproduce at the rate of λ, or $R_0 = 2.4177$? Not long, but we'll get back to this point in chapter 11. Before we do that, let's do some calculations for organisms with overlapping generations.

Estimating Rates When Generations Overlap

The population of the common mud turtle, *K. subrubrum*, whose mortality we examined in figure 10.16, contrasts with the *P. drummondii* population in various ways. Let's examine some

The geometric rate of increase, λ, is the ratio of numbers at a later time, N_{t+1}, to numbers at an earlier time, N_t.

$$\lambda = \frac{N_{t+1}}{N_t}$$

Numbers at later time, $t+1$

Numbers at earlier time, t

Numbers

N_{t+1}

N_t

Earlier time

Later time

Time

t $t+1$

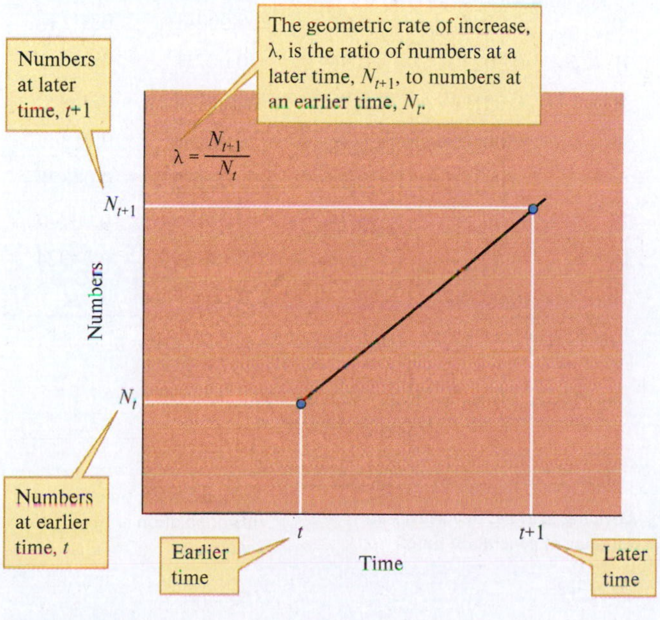

Figure 10.22   The geometric rate of increase.

of the details of this turtle's reproductive patterns in order to calculate the net reproductive rate of this population. About half (0.507) of the turtles nest each year. Of those females that do nest, most nest once during the year. However, some nest twice and a few even nest three times during the year. The average number of nests per year for the nesting turtles is 1.2, which means that 0.2, or one-fifth, of the turtles produce a second nest each year. The average **clutch size,** the number of eggs produced by a nesting female, is 3.17. So, the average number of eggs produced by nesting females each year is 3.17 eggs per nest × 1.2 nests per year = 3.8 eggs per year. However, remember that only half the females in the population nest

each year. Therefore, the number of eggs per female per year is $0.507 \times 3.8 = 1.927$ eggs per female per year. This is the average total number of eggs per female. On average, half of these eggs will develop into males and half into females. Since the sex ratio in this turtle population is 1 male:1 female, we multiply 1.927 by 0.50 to calculate the number of female eggs per adult female in the population, which equals 0.96 female egg. This is the value listed in column 3 of table 10.2.

Table 10.2 includes the life table information used to construct figure 10.16 plus the fecundity information we just calculated. As in the *Phlox* population, the sum of $l_x m_x$, $\Sigma l_x m_x$ provides an estimate of R_0, the net reproductive rate of females

Table 10.2

Calculating net reproductive rate, R_0, and generation time, T, for a population of the common mud turtle, $K.$ *subrubrum*

Age in years, x, times $l_x m_x$

x (years)	l_x	m_x	$l_x m_x$	$x l_x m_x$	x (years)	l_x	m_x	$l_x m_x$	$x l_x m_x$
0	1.0000	0	0	0	19	0.0108	0.96	0.01037	0.19703
1	0.2610	0	0	0	20	0.00945	0.96	0.00907	0.18140
2	0.1360	0	0	0	21	0.00829	0.96	0.00796	0.16716
3	0.0981	0	0	0	22	0.00725	0.96	0.00696	0.15312
4	0.0786	0.96	0.07546	0.30184	23	0.00635	0.96	0.00610	0.14030
5	0.0689	0.96	0.06614	0.33070	24	0.00557	0.96	0.00535	0.12840
6	0.0603	0.96	0.05789	0.34734	25	0.00487	0.96	0.00468	0.11700
7	0.0528	0.96	0.05069	0.40523	26	0.00427	0.96	0.00410	0.10660
8	0.0463	0.96	0.04445	0.35560	27	0.00374	0.96	0.00359	0.09693
9	0.0405	0.96	0.03888	0.34992	28	0.00328	0.96	0.00315	0.08820
10	0.0355	0.96	0.03408	0.34080	29	0.00287	0.96	0.00276	0.08004
11	0.0311	0.96	0.02986	0.32846	30	0.00251	0.96	0.00241	0.07230
12	0.0273	0.96	0.02621	0.31452	31	0.00220	0.96	0.00211	0.06541
13	0.0239	0.96	0.02294	0.29822	32	0.00193	0.96	0.00185	0.05920
14	0.0209	0.96	0.02006	0.28084	33	0.00169	0.96	0.00162	0.05346
15	0.0183	0.96	0.01757	0.26355	34	0.00148	0.96	0.00142	0.04828
16	0.0160	0.96	0.01536	0.24576	35	0.00130	0.96	0.00125	0.04375
17	0.0141	0.96	0.01354	0.23018	36	0.00114	0.96	0.00109	0.03924
18	0.0123	0.96	0.01181	0.21258	37	<0.00100	0	0	0

Data from Frazer, Gibbons, and Greene 1991.

$$R_0 = \Sigma\, l_x m_x = 0.601 \quad \Sigma\, x l_x m_x = 6.4$$

$$T = \frac{\Sigma\, x l_x m_x}{R_0} = \frac{6.4}{0.601} = 10.6$$

The value of R_0, which is less than 1.0, indicates that this population is declining.

Dividing $\Sigma\, x l_x m_x$ by R_0 gives an estimate of generation time.

The generation time for this population is 10.6 years.

in this population. In this case, $R_0 = 0.601$. We can interpret this number as the average number of daughters produced by each female in this population over the course of her lifetime. If this number is correct, the mothers in this population are not producing enough daughters to replace themselves. It appears that this population is declining. This result makes sense because during the time this study was done, the region of South Carolina where the turtle population lives was experiencing severe drought. During the drought, Ellenton Bay dried from a maximum of 10 ha of open water to about 0.05 ha.

The trend in this mud turtle population appears to reflect the declining quality of the environment. What value of R_0 would produce a stable turtle population? In a stable population, R_0 would be 1.0, which means that each female would replace only herself during her lifetime. In a growing population, such as the population of *Phlox*, R_0 would be greater than 1.0.

Population ecologists are also interested in several other characteristics of populations. One of those is the generation time, T, which is the average age of reproduction. We can use the information in table 10.2 to calculate the average generation time for the common mud turtles of Ellenton Bay:

$$T = \frac{\Sigma \, x l_x m_x}{R_0}$$

In this equation, x is age in years. To calculate T, sum the last column and divide the result by R_0. The result shows that the common mud turtles of Ellenton Bay have an average generation time of 10.6 years.

Knowing R_0 and T allows us to estimate **r**, the **per capita rate of increase** for a population:

$$r = \frac{\ln R_0}{T}$$

(ln R_0 is the natural logarithm of R_0.) We can interpret r as birthrate minus death rate: $r = b - d$. Using this method, the estimated per capita rate of increase for the common mud turtle population of Ellenton Bay is:

$$r = \frac{\ln 0.601}{10.6} = -0.05$$

The negative value of r in this case indicates that birthrates are lower than death rates and the population is declining. A value of r greater than 0 would indicate a growing population, and a value equal to 0 would indicate a stable population. While there are ways to make more accurate estimates of r, this method is accurate enough for our discussion. We will return to r in chapter 11 as we discuss population growth.

In this section we have seen how a life table combined with a fecundity schedule can be used to estimate net reproductive rate, R_0; geometric rate of increase, λ; generation time, T; and per capita rate of increase, r, which are fundamental components of population dynamics. Next, we review how the distributions of some species are changing in response to climate warming.

Concept 10.5 Review

1. Of the three populations pictured in figures 10.19, 10.20, and 10.21, which is most likely to have a stable age distribution?
2. Suppose that you are managing a population of an endangered species that has been reduced in numbers throughout its historic range and that your goal is to increase the size of the population. What values of R_0 would meet your management goals?
3. Both R_0 and r indicate that the mud turtle population in Ellenton Bay is in decline. Is there any way that this population could be maintained for many generations even with such negative indicators?

Applications

Changes in Species Distributions in Response to Climate Warming

LEARNING OUTCOMES

After studying this section you should be able to do the following:

10.19 Outline the key elements for ecological responses to climate change.
10.20 Summarize evidence of shifts in species distributions along elevation and latitude gradients during the past century.

Earlier in this chapter, we reviewed how the latitudinal distributions of maple species, *Acer* spp. and hemlock, *Tsuga canadensis*, moved northward across North America in response to climate changes, following the end of the last glacial period (see fig. 10.6). That discussion also referred briefly to ongoing studies of how species distributions are responding to today's rapidly warming climate, discussed in detail in chapter 23, in the context of the broad subject of global change. Here we review a few of the studies that have documented shifts in the elevation or latitudinal ranges of hundreds of species during the past century of environmental warming (Chen et al. 2011, Peterson et al. 2018).

Let's begin with a well-studied organism that has already entered our discussion, the land snail, *Arianta arbustorum*. In chapter 5, we reviewed a study by Bruno Baur and Anette Baur (1993) that documented local extinctions of this species around Basel, Switzerland, where the urban heat island has warmed habitats that once supported populations of the snail. The Baurs' study also provided a potential mechanism to explain those local extinctions: reduced reproduction. In experiments designed to test the influence of temperatures on reproduction, the Baurs observed reduced hatching rates of *A. arbustorum* eggs at temperatures as low as 22°C and no hatching at temperatures of 25°C and above.

How might *A. arbustorum* respond to large-scale climate warming? The Baurs explored the possibility that *A. arbustorum* may be shifting its distribution to higher elevations in Switzerland, as the landscape warms (Baur and Baur 2013). Because average air temperature is lower at higher elevations (see figure 2.39), their expectation was that the species would shift its distribution upward.

The Baurs conducted their research in the Swiss National Park, established in the Eastern Alps in 1914 to protect natural ecosystems from disturbance by humans and domestic animals. The Swiss National Park was an ideal study site. Working within a protected area was important, since many environmental factors besides climate change can potentially influence species distributions, for example, disturbance of the habitat by human activity. Another critical key element to a study, such as the one the Baurs designed, is the existence of historical information on species distributions. Because species inventories were made soon after the establishment of the Swiss National Park, the Baurs had a basis for determining whether the distribution of *A. arbustorum* has changed over time. The first inventory of the elevation range of *A. arbustorum* in the park was done in 1916–17, nearly a century before the Baurs' study. Yet another critical piece of information for such studies is a reliable long-term weather record. Fortunately, weather data have been collected at a weather station on the edge of the Swiss National Park, just 16 km from the study sites, since 1917. The weather records at the station show that during the period since the initial studies of *A. arbustorum*, the average annual temperature in the Swiss National Park has risen 1.6°C, while precipitation has not changed significantly.

The Baurs' resurvey revealed significant increases in the upper elevation limit of *A. arbustorum* along slopes in the park. Over the period since the initial population surveys, the average upper elevation limit for *A. arbustorum* along eight mountain slopes increased from an average of 2,200 m in 1916–17 to 2,361 m in 2011–12, an increase in elevation of 164 m

(fig. 10.23). In conclusion, the Baurs' research suggested that *A. arbustorum*, a species sensitive to warm temperatures, was responding to climate warming via a distributional shift.

However interesting and well documented the patterns of change in distribution of *A. arbustorum* might be, the study results are limited to a single species. How might other species in the region have responded to climate warming? The work of another research group suggests widespread movement upslope by species in the region. This study examined changes in the upper elevation limits of over 600 species of plants, insects, and birds in the Bavarian Forest National Park in southeastern Germany, located approximately 400 km northeast of the Swiss National Park, where the Baurs studied *A. arbustorum* (Bässler et al. 2013). Like the Baurs' study, researchers had access to long-term weather data at their study area. The Bavarian Forest National Park is also protected from most human disturbance and the upper elevation limits of species in the park had been documented in 1902–04, over a century before Bässler and his colleagues conducted their study. In contrast to many studies elsewhere in central and northern Europe (e.g., see Parolo and Rossi 2008), plants did not shift their elevation range in the Bavarian Forest National Park. However, the 433 species of insects in the study increased their elevation limits an average of 260 m, while 57 bird species were recorded an average of 165 m higher than previously. These results suggest that large numbers of both vertebrate and invertebrate species are extending their distributions upslope with climate warming.

Now let's turn from elevation responses to changes in the latitudinal ranges of species over the period of recent warming. As with changes in elevation, documenting latitudinal changes depends on the availability of baseline surveys of earlier distributions. One of the few places with extensive long-term records of latitudinal distributions over a large geographic area is the United Kingdom. Researchers

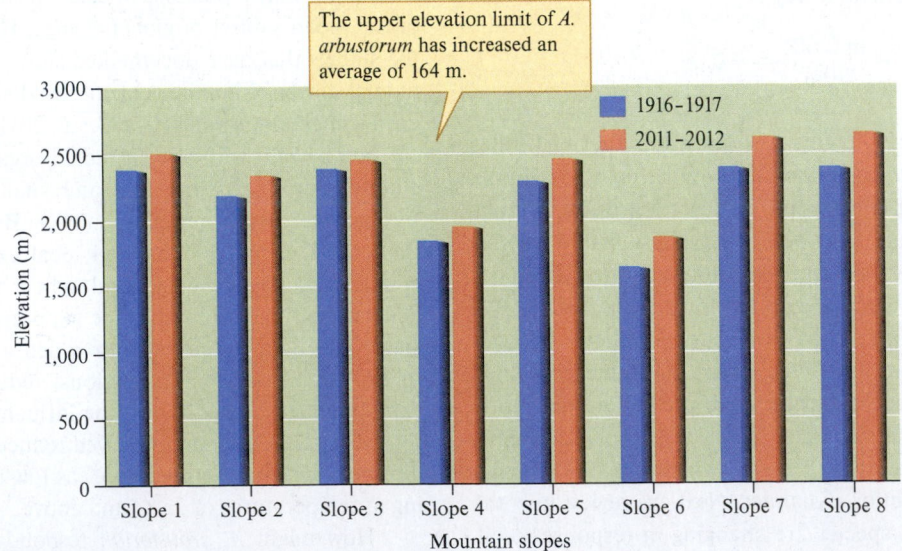

Figure 10.23 Movement upslope with climate warming: increases in the upper elevation limit of the land snail *Arianta arbustorum* on eight mountain slopes in the Swiss National Park between 1916 and 2012 (data from Baur and Baur 2013).

have taken advantage of the extensive national inventory of species available there to study whether the range limits of species have moved northward within the British Isles (Hickling et al. 2006). In this study, Rachael Hickling of the University of York and colleagues estimated the latitudinal shifts in the distributions of 16 groups of organisms ranging from birds, mammals, and butterflies to spiders, dragonflies, and ground beetles over a roughly 25-year period. Of the 329 species included in their analysis, 275, or roughly 84%, had shifted their ranges northward, 52 had moved southward, while the distributions of 2 species remained unchanged. The average shifts in distribution of species in the study ranged from 31 to 60 km. These shifts translate into average rates of movement northward of approximately 19 km per decade. Figure 10.24 shows the average latitudinal movement of four well-studied taxonomic groups—butterflies, birds, mammals, and ground beetles—during the 25 years covered by the study.

The results of the studies reviewed here and hundreds of others have demonstrated changes in population distributions in response to well-documented increases in global temperatures over the past century. The distribution of earth's biota is changing as we watch.

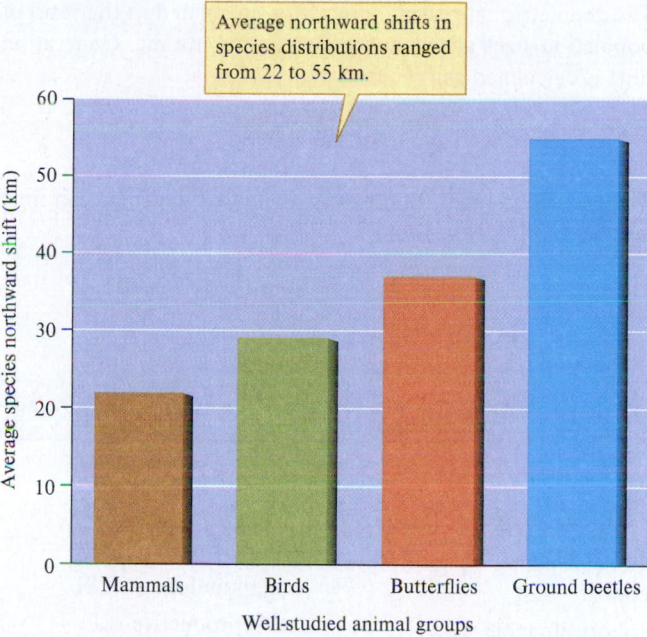

Figure 10.24   Movement northward with climate warming: average increases in the northern latitudinal limits of mammal, bird, butterfly, and ground beetle species in the United Kingdom over a 25-year period (data from Hickling et al. 2006).

Summary

Dispersal can increase or decrease local population densities. The contribution of dispersal to local population density and dynamics is demonstrated by studies of expanding populations of species such as Africanized honeybees in the Americas and Eurasian collared doves in Europe. Climate changes can induce massive changes in the ranges of species. As availability of prey changes, predators may disperse, which increases and decreases their local population densities. Stream organisms actively migrating upstream or drifting downstream increase densities of stationary and migrating populations by immigrating and decrease them by emigrating.

Ongoing dispersal can join numerous subpopulations to form a metapopulation. Populations of many species occur not as a single continuously distributed population but in spatially isolated subpopulations, with significant exchange of individuals among subpopulations. A group of subpopulations living on such patches connected by exchange of individuals among patches makes up a metapopulation. Populations of the Rocky Mountain Parnassian butterfly, *Parnassius smintheus,* in Alberta, Canada, and of the lesser kestrel, *Falco naumanni,* in Spain are examples of metapopulations. In both metapopulations individuals moved mainly from smaller subpopulations to larger subpopulations.

A survivorship curve summarizes the pattern of survival in a population. Patterns of survival can be determined either by following a cohort of individuals of similar age to produce a cohort life table or by determining the age at death of a large number of individuals or the age distribution of a population to produce a static life table. Life tables can be used to draw survivorship curves, which generally fall into one of three categories: (1) type I survivorship, in which there is low mortality among the young but high mortality among older individuals; (2) type II survivorship, in which there is a fairly constant probability of mortality throughout life; and (3) type III survivorship, in which there is high mortality among the young and low mortality among older individuals.

The age distribution of a population reflects its history of survival, reproduction, and potential for future growth. Age distributions indicate periods of successful reproduction, high and low survival, and whether the older individuals in a population are replacing themselves or if the population is declining. Population age structure may be highly complicated in variable environments, such as that of the Galápagos Islands. Populations in highly variable environments may reproduce episodically.

A life table combined with a fecundity schedule can be used to estimate net reproductive rate (R_0), geometric rate of increase (λ), generation time (T), and per capita rate of increase (r). Because these population parameters form the core of population dynamics, it is important to understand their derivation, as well as their biological meaning. Net reproductive rate, R_0, the lifetime average number of female offspring left by an individual female in a population, is calculated by multiplying age-specific survivorship rates, l_x, by age-specific birthrates, m_x, and summing the results:

$$R_0 = \Sigma \, l_x m_x$$

The geometric rate of increase, λ, is calculated as the ratio of population sizes at two successive points in time. Generation time is calculated as:

$$T = \frac{\Sigma \, x l_x m_x}{R_0}$$

The per capita rate of increase, r, is related to generation time and net reproductive rate as:

$$r = \frac{\ln R_0}{T}$$

The per capita rate of increase may be positive, zero, or negative depending on whether a population is growing, stable, or declining.

Recent studies have demonstrated shifts in the upper latitudinal and elevation limits of hundreds of species in response to the climate warming that has occurred over the past century. These range shifts provide clear evidence of ecological responses to climate change. The distribution of earth's biota is changing as we watch.

Key Terms

age distribution   224
birthrate   230
clutch size   232
cohort   224
cohort life table   224
colonization cycle   220
drift   220
fecundity schedule   230

geometric rate of
 increase (λ)   230
life table   224
metapopulation   221
net reproductive
 rate (R_0)   230
numerical
 response   219

per capita rate of
 increase (r)   233
population dynamics   216
spate   220
stable age
 distribution   230
static life table   224
subpopulation   221

survivorship curve   225
type I survivorship
 curve   227
type II survivorship
 curve   227
type III survivorship
 curve   227

Review Questions

1. Outline Müller's (1954, 1974) colonization cycle. If you were studying the colonization cycle of the freshwater snail *Neritina latissima*, how would you verify that the colonization waves exchange individuals with local populations?

2. Compare cohort and static life tables. What are the main assumptions of each? In what situations or for what organisms would it be practical to use either?

3. Of the three survivorship curves, type III has been the least documented by empirical data. Why? What makes this pattern of survivorship difficult to study?

4. Ecologists have assumed that populations of species with very high reproductive rates, those with offspring sometimes numbering in the millions per female, must have a type III survivorship curve even though very few survivorship data exist for such species. Why is this a reasonable assumption?

5. Explain how the age structure of a population with highly episodic reproduction might be misinterpreted as indicating population decline. How might population ecologists avoid such misinterpretations?

6. Concept 10.5 says that we can use the information in life tables and fecundity schedules to estimate some characteristics of populations (R_0, T, r). Why does Concept 10.5 use the word *estimate* rather than *calculate*?

7. What values of R_0 indicate that a population is growing, stable, or declining? What values of r indicate a growing, stable, or declining population?

8. From a life table and a fecundity schedule, you can estimate the geometric rate of increase, λ, the average reproductive rate, R_0, the generation time, T, and the per capita rate of increase, r. That is a lot of information about a population. What minimum information do you need to construct a life table and fecundity schedule?

9. C. S. Holling (1959a) observed predator numerical responses to changes in prey density. He attributed the numerical responses to changes in the reproductive rates of the predators. Discuss a hypothetical example of reproductive-rate numerical response by a population of predators in terms of changes in fecundity schedules and life tables. Include the terms R_0, T, and r in your discussion.

Chapter

11

Population Growth

A group of northern elephant seals, *Mirounga angustirostris,* on a breeding beach. Intense hunting of northern elephant seals for their blubber, a source of oil, drove the species to near extinction in the nineteenth century. With protection from hunting, the population has grown from as few as 100 individuals, living off the west coast of Mexico, to over 160,000, and it continues to grow as the population disperses northward.

Applications: The Human Population   247
    Summary   251
    Key Terms   252
    Review Questions   2532

CHAPTER CONCEPTS

11.1 In the presence of abundant resources, populations can grow at geometric or exponential rates.   238
    Concept 11.1 Review   241

11.2 As resources are depleted, population growth rate slows and eventually stops.   242
    Concept 11.2 Review   244

11.3 The environment influences population growth through its effects on birth and death rates.   244

    Concept 11.3 Review   247

LEARNING OUTCOME

After studying this section you should be able to do the following:

11.1 Explain why understanding population growth rate is important for human health.

11.2 Calculate the number of cases of COVID-19 that would be expected after a week if the exponential growth rate was 2 versus 3.

In early 2020, the world became aware of a new viral disease that was spreading rapidly; it was highly contagious and there was no known cure. This disease is what we all now know as COVID-19, an abbreviation of the name coronavirus disease 2019. Experts warned that unless we, as a human race, took precautions, the number of cases of this disease would grow *exponentially.* That is, they used observations of the number and growth of cases in the real world and created an equation to both describe that pattern and make predictions into the future (Kucharski et al. 2020). Many world leaders and citizens sprang to action while others did not; some blame the slow response on a lack of understanding of how fast exponential growth actually is (Lammers et al. 2020).

So how fast is it? If a single infected person infects 3 people in a day, and each of these 3 infect 3 people and so on, how many days will it take for that single person to have created over 1,000 infected people? We have described a reproductive rate of 3. You can easily calculate the spread by multiplying that first person by the reproductive rate of 3 ($1 \times 3 = 3$), then the resulting 3 people by 3 ($3 \times 3 = 9$), then 9 by 3 and so on, adding the total as you go. You may be surprised that it would only take 1 week to reach 729 new cases in a day, or 1,093 cases total. For COVID-19, the reproduction rate fluctuated between 1.6 and 4 before restrictions were implemented (Kucharski et al. 2020).

We can see from reports by the World Health Organization (WHO) that the growth of COVID-19 was indeed exponential in the first few weeks, before widespread quarantines and other policies to slow the spread were implemented (fig. 11.1a). Did these policies actually affect the growth rate? To answer this question, Solomon Hsiang and colleagues created models that predicted the growth rate of COVID-19 in the context of 1,700 types of interventions intended to slow its spread, using data from six different countries (Hsiang et al. 2020). They were able to create a population growth model that predicted significant decreases in the growth rate of COVID-19 cases; that is, the model matched observed rates (fig. 11.1b). They then modeled the rate of growth in the absence of these policies, which was constant exponential growth at a rate that would have meant that the number of COVID-19 cases would have nearly doubled every 2 days. Their findings suggest that the safety measures were likely responsible for preventing 495 million total infections and 24 million lives.

How important is it that citizens understand, even at only a basic level, growth models such as these? Ritwik Banerjee, Joydeep Bhattacharya, and Priyama Majumdar (researchers working in India and the United States) determined that individuals who had a better understanding of exponential growth were also more likely to be compliant with WHO-recommended safety measures (Banerjee et al. 2020). Taken together, the findings by Banerjee, Hsiang, and others suggest that understanding the mathematics of population growth can literally save lives.

In the field of ecology, understanding the way that a population grows, including whether it is best described by an exponential growth model or some other type, can tell us a great deal about the dynamics affecting a species. In chapter 11, we examine the factors that determine rates and patterns of population growth. We also review the environmental forces that limit population size. At this juncture in history, as our own species places increased pressure on the biosphere, and we in turn are affected by other organisms, there is no more important ecological topic.

We look at population growth in the presence of abundant resources, growth where resources are limiting, and how the environment can act to change birthrates and death rates. The concepts we review reflect the historical development of population ecology. That history shows the importance of both collecting data on populations in the real world, and then developing mathematical models to describe these populations and predict numbers into the future. Our knowledge of population growth has progressed through an interplay between modeling and observations of actual populations. Let's turn now to studies of population growth in the presence of abundant resources.

11.1 Geometric and Exponential Population Growth

LEARNING OUTCOMES

After studying this section you should be able to do the following:

11.3 Define geometric and exponential population growth.

11.4 Interpret the elements of the equations for geometric and exponential population growth.

11.5 Describe the conditions necessary for exponential population growth in natural populations.

11.6 Explain why exponential population growth by natural populations cannot be sustained for many generations.

In the presence of abundant resources, populations can grow at geometric or exponential rates.   Suppose a population had access to abundant resources, such as food, space, nutrients, and so forth. How fast could it grow? In the case of COVID-19, the "resource" was available human hosts for the virus, that is, a nearly endless supply. Imagine a plant, animal, or bacterial population reproducing at its maximum reproductive rate. What would the resulting pattern of population growth for those organisms be? Regardless of the species you choose, a population growing at its maximum rate grows slowly at first and then faster and faster. In other words, population growth accelerates.

We said earlier that COVID-19 cases grew exponentially in March 2020. We will learn in this section what that means from a mathematical perspective, and how it differs from geometric growth. Both exponential and geometric growth are models of maximum growth rates, but they differ in how they are calculated and the pattern of growth they describe.

Geometric Growth

Because it is an annual plant, populations of *Phlox drummondii* grow in discrete annual pulses and generations do not overlap. Growth by any such population without overlapping generations can be modeled as **geometric population growth,** in which successive generations differ in size by a constant ratio.

We can use the population of *Phlox* studied by Leverich and Levin (1979) to build a model of geometric population growth. In chapter 10, we calculated a geometric rate of increase, $\lambda = N_{t+1}/N_t$, for this population of 2.4177. At the end of that discussion, we asked how long the *Phlox* population could continue growing at this rate. Let's address that question here.

As we saw in chapter 10, we can compute the growth of a population of organisms whose generations do not overlap by simply multiplying λ by the size of the population at the

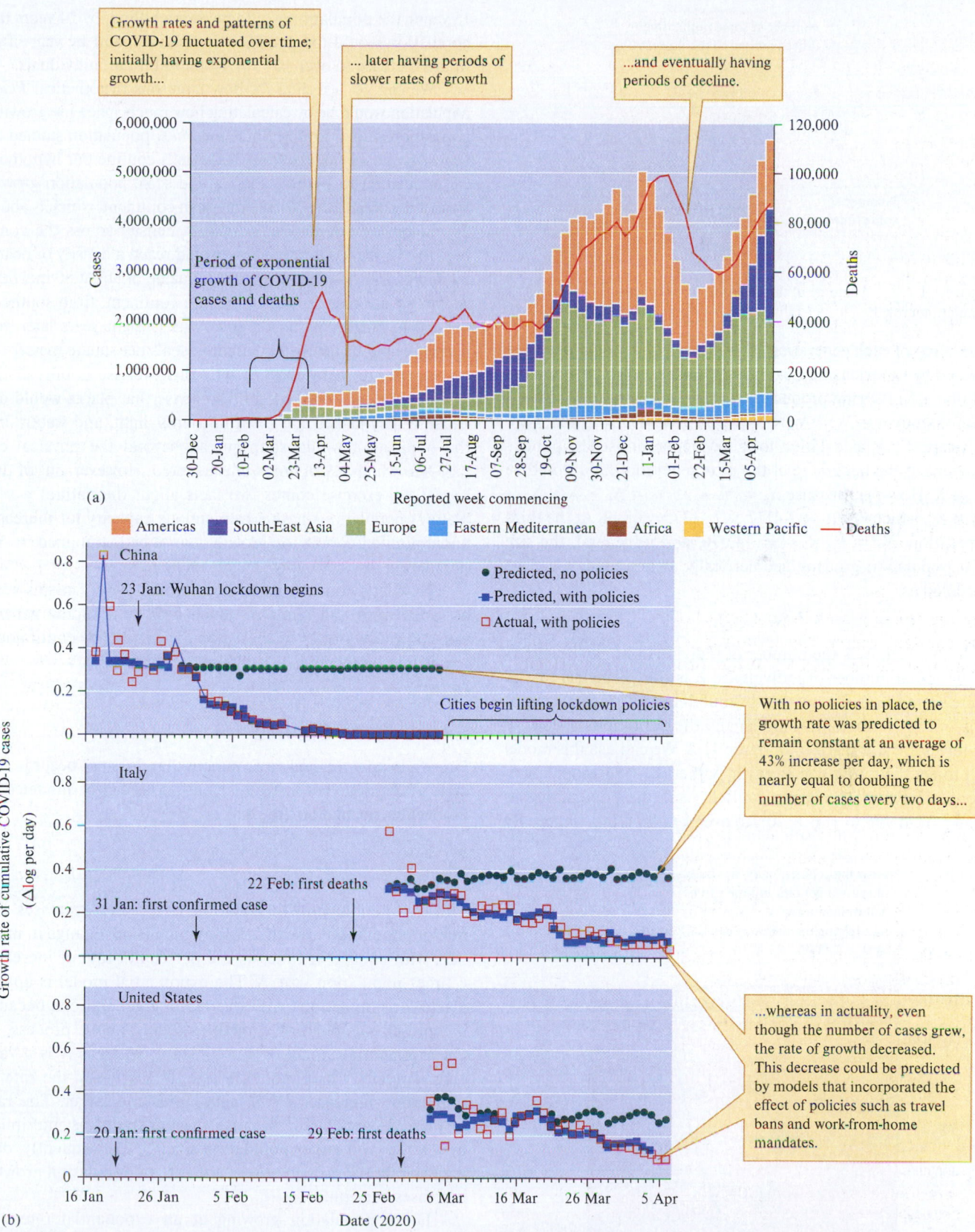

Growth rates and patterns of COVID-19 fluctuated over time; initially having exponential growth...

... later having periods of slower rates of growth

...and eventually having periods of decline.

Period of exponential growth of COVID-19 cases and deaths

Americas   South-East Asia   Europe   Eastern Mediterranean   Africa   Western Pacific   — Deaths

China

23 Jan: Wuhan lockdown begins

Predicted, no policies
Predicted, with policies
Actual, with policies

Cities begin lifting lockdown policies

With no policies in place, the growth rate was predicted to remain constant at an average of 43% increase per day, which is nearly equal to doubling the number of cases every two days...

Italy

31 Jan: first confirmed case

22 Feb: first deaths

United States

20 Jan: first confirmed case

29 Feb: first deaths

...whereas in actuality, even though the number of cases grew, the rate of growth decreased. This decrease could be predicted by models that incorporated the effect of policies such as travel bans and work-from-home mandates.

Figure 11.1   The spread of COVID-19 provides an example of how population growth can be measured and predicted. (*a*) The number of active cases and deaths over time by global region (data from WHO 2020). (*b*) Growth rate for cumulative COVID-19 cases is plotted over time for three countries. Two models are shown: growth rates predicted when quarantine and other protective policies are implemented (blue filled squares), versus predicted growth rate without those policies (green circles). Models controlled for different testing rates by country and took into account timing of policy implementation. The validity of the models (both with and without policies) is demonstrated by the close match of the predicted, with policies against the actual, with policies (red open squares). Results suggest that government policies have saved millions of lives by decreasing growth rate of COVID-19 (data from Hsiang et al. 2020).

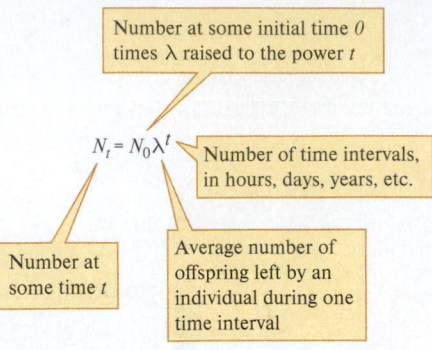

Number at some initial time *0* times λ raised to the power *t*

$$N_t = N_0\lambda^t$$

Number of time intervals, in hours, days, years, etc.

Number at some time *t*

Average number of offspring left by an individual during one time interval

Figure 11.2   Anatomy of the equation for geometric population growth.

beginning of each generation. The initial size of the population studied by Leverich and Levin was 996 (see table 10.1), and the number of offspring produced by this population during their year of study was $N_1 = N_0 \times \lambda$, or $996 \times 2.4177 = 2,408$. Now let's repeat this calculation for a few generations. The population size at the beginning of the next generation, N_2, would be $N_1 \times \lambda$. However, because $N_1 = N_0 \times \lambda$, $N_2 = N_0 \times \lambda \times \lambda$, or $N_0 \times \lambda^2$, which is $996 \times 2.4177 \times 2.4177 = 5,822$. At the third generation, $N_3 = N_0 \times \lambda^3 = 14,076$, and in general, the size of a population growing geometrically at any time, *t*, can be modeled as:

$$N_t = N_0\lambda^t$$

In this model, N_t is the number of individuals at any time *t*, N_0 is the initial number of individuals, λ is the geometric rate of increase, and *t* is the number of time intervals or generations. The interpretation of this model and the definitions of each of its terms are summarized in figure 11.2. We can use this model to project the future size of our hypothetical *Phlox* population. Notice in figure 11.3 that in only 8 years the population has grown from 996 to 1.16×10^6, to over 1 million individuals. By

16 years, the population would be over a billion, by 24 years the population would top 1 trillion individuals, and by year 40 it would increase to over 10^{18}, or 1 billion billion individuals.

We can get a feeling for how large this hypothetical *Phlox* population would be by calculating how much space the growing population would occupy. Since the *Phlox* population studied by Leverich and Levin was from Texas, let's confine our hypothetical population to North America and scale population growth against the area of the North American continent, which is about 24 million km^2. Assuming a uniform density across the continent, by 32 years our population would reach a density of nearly 80 million individuals per square kilometer, or about 80 individuals per square meter across the entire continent, from southern Mexico to northern Canada and Alaska. Eight years later, the density would be nearly 90,000 individuals per square meter!

There are many reasons why this exercise is unrealistic. The population would soon be so dense that plants would die because they lacked sufficient nutrients, light, and water; and the population would soon spread beyond the physical climates to which *P. drummondii* is adapted. However, out of this unrealistic exercise comes two facts about the natural world: Natural populations have a tremendous capacity for increase, and unlimited population growth cannot be maintained in any population for very many generations.

Now let's consider population growth by organisms such as forest trees and humans, which have overlapping generations. Because growth by these populations can be continuous, the geometric model is usually not appropriate.

Exponential Growth

Growth in an unlimited environment that does not occur in discrete generations but instead is continuous can be modeled as **exponential population growth**:

$$\frac{dN}{dt} = rN$$

The exponential growth equation (fig. 11.4) expresses the rate of population growth, dN/dt, which is the change in numbers with change in time, as the per capita rate of increase, *r*, times population size, *N*. The exponential model is appropriate for populations with overlapping generations because it represents population growth as a continuous process. In the exponential model, *r* is a constant, whereas *N* is a variable. Therefore, as population size, *N*, increases, the rate of population increase, dN/dt, gets larger and larger. The rate of increase gets larger because the constant *r* is multiplied by a larger and larger population size, *N*. Consequently, during exponential growth, dN/dt, the rate of population growth, increases over time.

For a population growing at an exponential rate, the population size at any time *t* can be calculated as:

$$N_t = N_0e^{rt}$$

In this form of the exponential growth model, N_t is the number of individuals at time *t*, N_0 is the initial number of individuals,

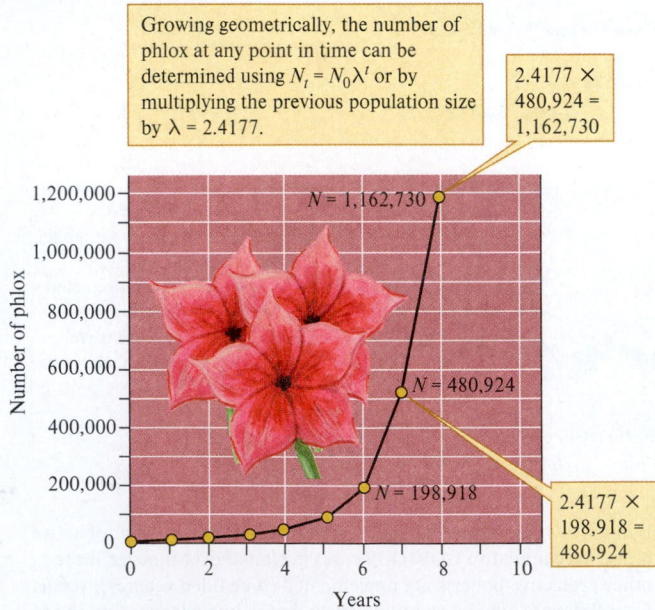

Growing geometrically, the number of phlox at any point in time can be determined using $N_t = N_0\lambda^t$ or by multiplying the previous population size by λ = 2.4177.

$2.4177 \times 480,924 = 1,162,730$

N = 1,162,730

N = 480,924

N = 198,918

$2.4177 \times 198,918 = 480,924$

Number of phlox

Years

Figure 11.3   Geometric growth by a hypothetical population of *Phlox drummondii*.

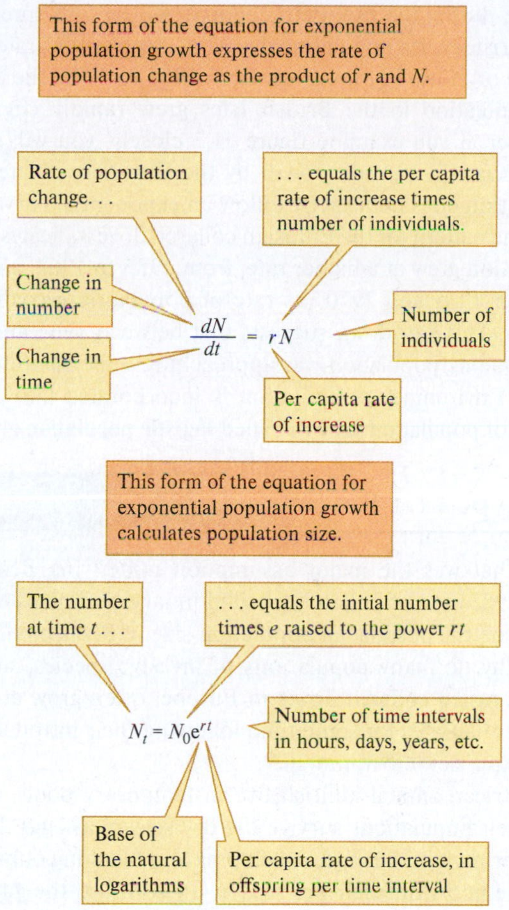

The two text boxes on the left describe the exponential growth equations.

This form of the equation for exponential population growth expresses the rate of population change as the product of r and N.

Rate of population change...

...equals the per capita rate of increase times number of individuals.

Change in number

Number of individuals

Change in time

$$\frac{dN}{dt} = rN$$

Per capita rate of increase

This form of the equation for exponential population growth calculates population size.

The number at time t...

...equals the initial number times e raised to the power rt

$$N_t = N_0 e^{rt}$$

Number of time intervals in hours, days, years, etc.

Base of the natural logarithms

Per capita rate of increase, in offspring per time interval

Figure 11.4   Anatomy of equations for exponential population growth.

e is the base of the natural logarithms, r is the per capita rate of increase, and t is the number of time intervals. The two forms of the exponential growth equation are presented and explained in figure 11.4.

Exponential Growth in Nature

Some of the assumptions of the exponential growth model, such as a constant rate of per capita increase, may seem a bit unrealistic for organisms that are not disease vectors, so it is reasonable to ask whether populations of larger organisms ever grow at an exponential rate. The answer is a qualified yes. Natural populations may grow at exponential rates for relatively short periods of time in the presence of abundant resources.

Exponential Growth by Tree Populations

As we saw in chapters 1 and 10 (see figs. 1.8 and 10.6), as the last ice age was ending, tree populations in the Northern Hemisphere followed the retreating glaciers northward. Ecologists have documented these movements by studying the sediments of lakes, where the pollen of wind-pollinated tree species is especially abundant. The appearance of pollen of a tree species in a lake sediment is a record of its establishment near the lake. The date of each establishment can be determined using ^{14}C (carbon-14) concentration to determine the age of organic matter along a sediment profile.

Pollen records have also been used to estimate the growth of several postglacial tree populations in Britain. K. Bennett (1983) estimated population sizes and growth by counting the number of pollen grains of each tree species deposited within lake sediments. By counting the number of pollen grains per square centimeter deposited each year, Bennett was able to reconstruct changes in tree population densities in the surrounding landscape. His study revealed an interesting picture of growth by postglacial tree populations in the British Isles. Populations of the tree species studied grew at exponential rates for 400 to 500 years following their initial appearance in the pollen record. Figure 11.5 shows the exponential increase in abundance of Scots pine, *Pinus sylvestris,* which first appeared in the pollen record of the study lake about 9,500 years ago.

Conditions for Exponential Growth

Natural populations of organisms as different as diatoms, whales, and trees can grow at exponential rates. However, as different as these organisms are, the circumstances in which their populations grow at exponential rates have a great deal in common. All begin their exponential growth in favorable environments at low population densities. The trees studied by Bennett began at low densities because they were invading new territory previously unoccupied by the species. Spring blooms of planktonic diatoms are the result of exponential population growth in response to seasonal increases in light in the presence of abundant nutrients.

The whooping crane provides another example of exponential growth following protection and careful management. Hunting and habitat destruction reduced the population of whooping cranes to 22 individuals by 1940. At that time, it was known that this remnant population of whooping cranes wintered on the Texas Gulf Coast but its northern breeding grounds were unknown. It was later discovered that they breed in Wood Buffalo National Park in Canada. Under the full protection and

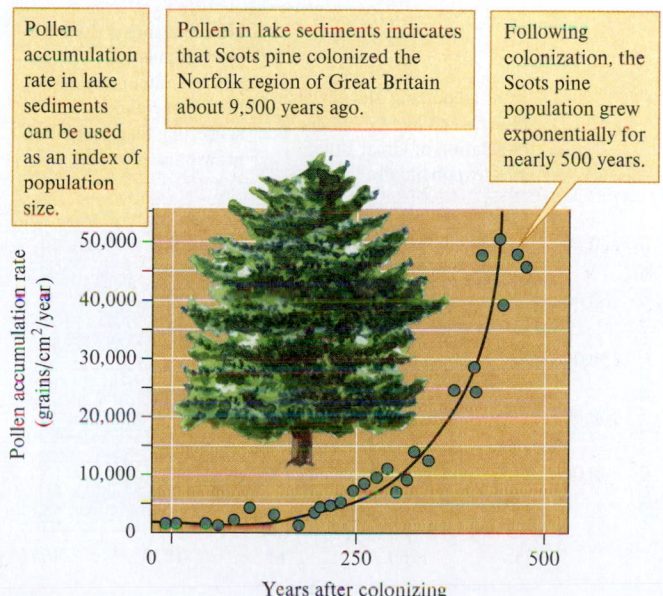

Figure 11.5   Exponential growth of a colonizing population of Scots pine, *Pinus sylvestris* (data from Bennett 1983).

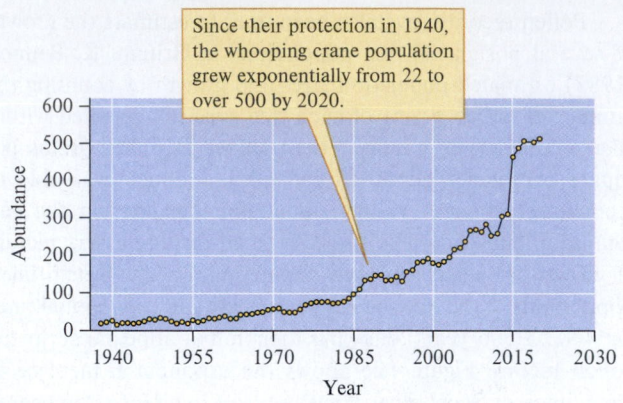

Figure 11.6   Abundance of the primary whooping cranes population, as measured in their wintering grounds in Wood Buffalo National Park, Canada. The last 10 years are estimates from surveys made via aircraft. A few additional populations have been established that have not yet exceed 22 individuals. Protection and intensive management of this main population have led to its dramatic recovery (data from USFWS 2018, USFWS 2020).

careful management in both Canada and the United States, the migratory whooping crane population grew exponentially from 22 in 1940 to 506 individuals in 2020 (fig. 11.6).

These examples suggest that exponential population growth may be very important to populations during the process of establishment in new environments, during exploitation of transient, favorable conditions, and during the process of recovery from some form of exploitation. However, as we saw with *P. drummondii,* geometric or exponential growth cannot continue indefinitely. In nature, population growth must eventually slow and population size level off.

Slowing of Exponential Growth

As we saw in chapter 10, the Eurasian collared dove, *Streptopelia decaocto,* expanded beyond its historical range into Western

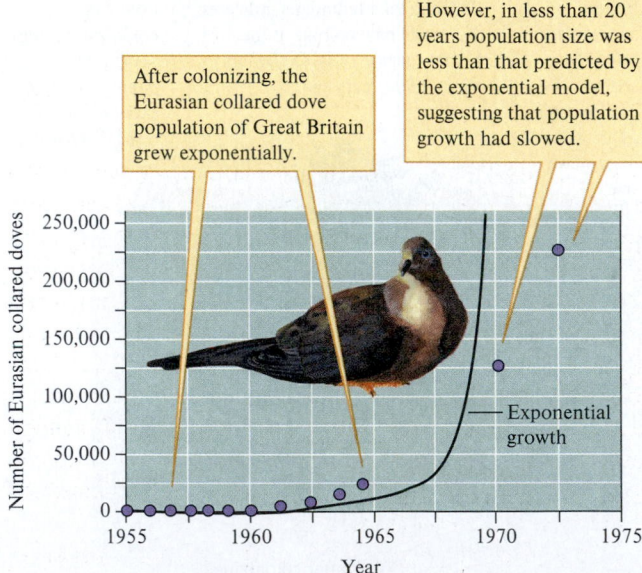

Figure 11.7   Exponential growth of the Eurasian collared dove population of Great Britain (data from Hengeveld 1988).

Europe during the twentieth century. As the bird spread into new territory, its populations grew at exponential rates for a decade or more. For instance, from 1955 to 1972, the expanding population in the British Isles grew rapidly (fig. 11.7). However, if you examine figure 11.7 closely, you will see evidence that the rate of growth by the Eurasian collared dove population does not exactly follow an exponential curve.

The pattern for the Eurasian collared dove indicates that its population grew at a higher rate, from 1955 to 1964, and then, between 1965 and 1970, its rate of population growth began to slow. This slowdown suggests that between 1965 and 1970, this invading population was approaching some environmental limits. Environmental limitation is incorporated into another model of population growth called **logistic population growth.**

Concept 11.1 Review

1. What was the major assumption underlying Bennett's (1983) use of pollen deposited in lake sediments to estimate the postglacial population size of Scots pine?
2. Why do many populations of invasive species, such as Eurasian collared doves in Europe, often grow at exponential rates for some time following their introductions into a new environment?
3. African annual killifish live in temporary pools, where their populations survive the dry season as eggs that lie dormant in the mud, developing and hatching only when the pools fill each wet season. In contrast, the guppy, a common aquarium fish, lives in populations consisting of mixed-age classes in which reproduction occurs year-round. Which model of population growth, exponential or geometric, would be most appropriate for each of these fish species?

11.2 Logistic Population Growth

LEARNING OUTCOMES

After studying this section you should be able to do the following:

11.7   Define sigmoidal population growth and carrying capacity.

11.8   Discuss examples of sigmoidal growth in laboratory and natural populations.

11.9   Interpret the elements of the equation for logistic population growth.

11.10   Describe how per capita rate of increase, *r,* is affected by population size during logistic population growth.

As resources are depleted, population growth rate slows and eventually stops.   Obviously, exponential growth cannot continue indefinitely. Eventually, populations run up against environmental limits to further increase. The effect of the environment on population growth is reflected in the shapes of population growth curves. As population size increases, growth rate eventually slows and then ceases as population size levels off. This

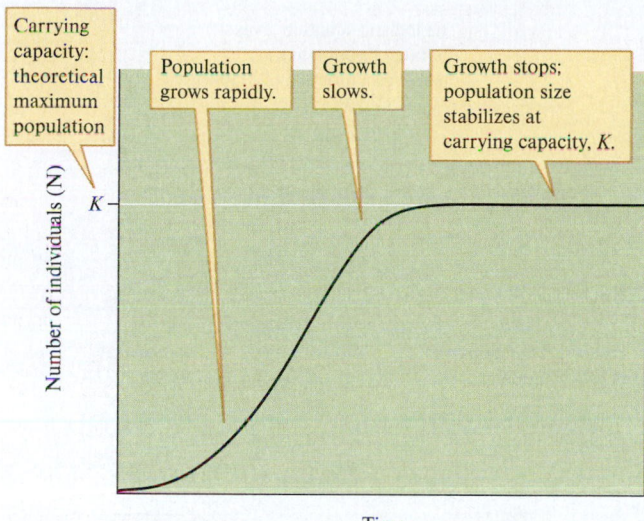

Figure 11.8 Sigmoidal population growth results from environmental limitation on population size.

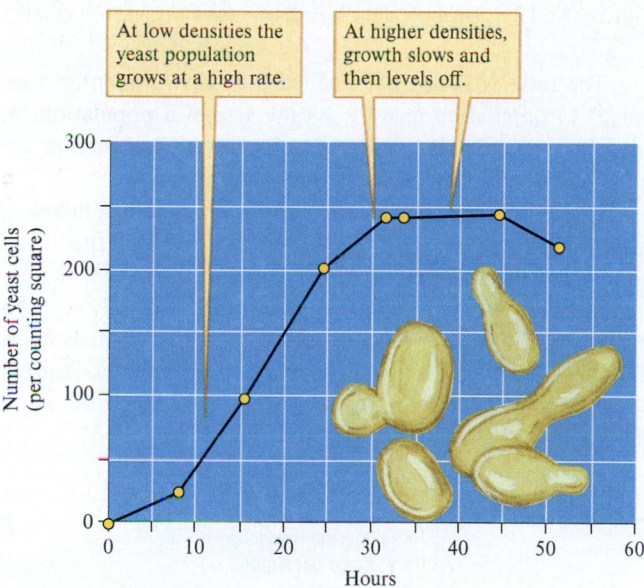

Figure 11.9 Sigmoidal growth by a population of the yeast *Saccharomyces cerevisiae* (data from Gause 1934).

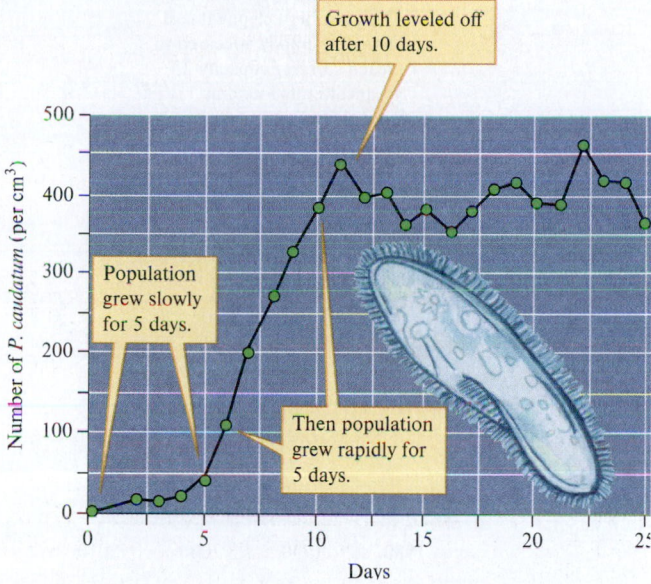

Figure 11.10 Sigmoidal growth by a population of *Paramecium caudatum* (data from Gause 1934).

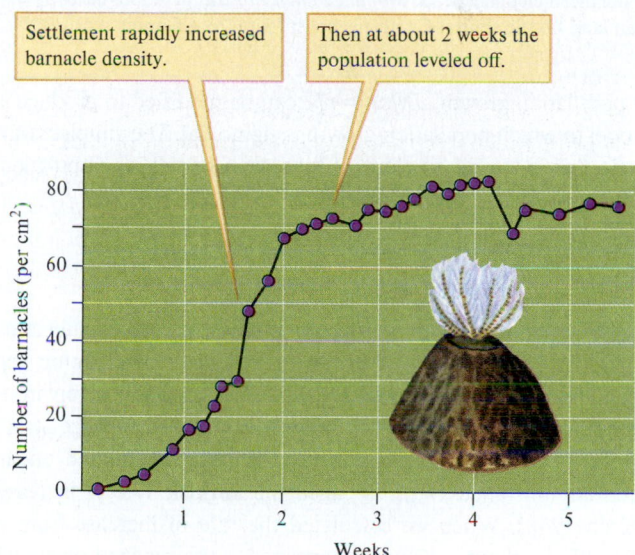

Figure 11.11 Settlement by the barnacle *Balanus balanoides* in the intertidal zone (data from Connell 1961a).

pattern of growth produces a **sigmoidal,** or S-shaped, **population growth curve** (fig. 11.8). The population size at which growth stops is generally called **carrying capacity,** or **K,** which is the number of individuals of a particular population that the environment can support. At carrying capacity, because population size is approximately constant, birthrates must equal death rates and population growth is zero.

Sigmoidal growth curves have been observed in a wide variety of populations. In the course of his laboratory experiments, G. F. Gause (1934) obtained sigmoidal growth curves for populations of several species of yeast (fig. 11.9) and protozoa (fig. 11.10). Similar patterns of population growth have been recorded for other populations, including barnacles (fig. 11.11) and northern elephant seals (fig. 11.12).

What causes these populations to slow their rates of growth and eventually stop growing at carrying capacity? The idea

behind the concept of carrying capacity is that a given environment can only support so many individuals of a particular species. For the barnacles studied by J. H. Connell (1961b), carrying capacity is largely determined by competition for the space available on rocks for attachment by barnacle larvae. Carrying capacity for females in northern elephant seal colonies also appears limited by space on the breeding beaches (Le Boeuf and Laws 1994). Yeast feed on sugars and produce alcohol. As the density of a population of yeast increases, their environment contains less and less sugar and more and more alcohol, which is toxic to them. So, yeast populations are eventually limited by competition for food and by their own waste products.

The logistic model was proposed to account for the patterns of growth shown by populations as they compete for environmental resources. Population ecologists built the logistic growth model by modifying the exponential growth model. The exponential model

Figure 11.12   Sigmoidal population growth in the breeding colony of northern elephant seals, *Mirounga angustirostris*, at Año Nuevo mainland on the central California coast (data from Condit et al. 2007).

of population growth, $dN/dt = rN$, can be modified to produce a model in which population growth is sigmoidal. The simplest way to do this is to add an element that slows growth as population size approaches carrying capacity, K:

$$\frac{dN}{dt} = r_{max}N\left(\frac{K-N}{K}\right)$$

Notice that the per capita rate of increase, r_{max}, has a subscript *max*. The subscript here indicates that this is the maximum per capita rate of increase, achieved by a species under ideal environmental conditions, where birthrates, death rates, and age structure are stable. The per capita rate of increase attained under such circumstances, r_{max}, is called the **intrinsic rate of increase** (Birch 1948). When we calculated the rate of increase from a life table in chapter 10, we determined r, the realized or actual per capita rate of increase. As we saw, realized r can be positive, zero, or negative, depending on environmental conditions. Because natural populations are usually subject to factors such as disease, competition, and so forth, the actual per capita rate of increase, realized r, is generally less than r_{max}. The inventor of this equation for sigmoidal population growth, P. F. Verhulst, called it the **logistic equation** (Verhulst and Quetelet 1838).

Rearranging the logistic equation shows more clearly the influence of population size, N, on rate of population growth:

$$\frac{dN}{dt} = r_{max}N\left(\frac{K}{K} - \frac{N}{K}\right) = r_{max}N\left(1 - \frac{N}{K}\right)$$

In the logistic equation, the rate of population growth, dN/dt, slows as population size increases because the difference, $(1 - N/K)$, becomes a smaller and smaller decimal fraction until N equals K. When N equals K, the right side of the equation becomes zero. Therefore, as population size increases, the logistic growth rate becomes a smaller and smaller fraction of the exponential growth rate and when $N = K$, population growth ceases (fig. 11.13). Logistic population growth is highest when $N = K/2$.

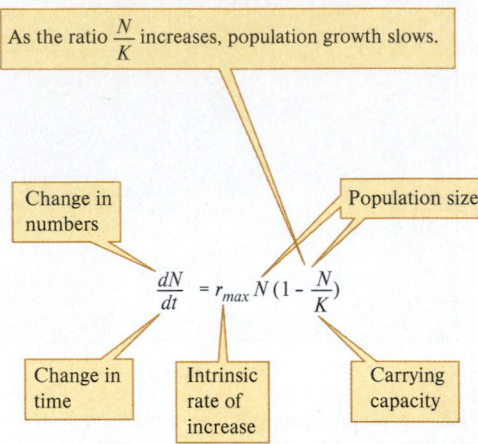

Figure 11.13   Anatomy of the logistic equation for population growth.

The ratio N/K has been called the "environmental resistance" to population growth. As the size of a population, N, gets closer and closer to carrying capacity, environmental factors increasingly impede further population growth.

In the logistic growth model, the per capita rate of increase, which is $r = r_{max}(1 - N/K) = r_{max} - r_{max}(N/K)$, depends on population size. Therefore, when population size, N, is very small, the per capita rate of increase is approximately r_{max}. As N increases, however, realized r decreases until N equals K. At that point, realized r is zero. The relationship between realized r and population size in the logistic model, which follows a straight line, is shown in figure 11.14.

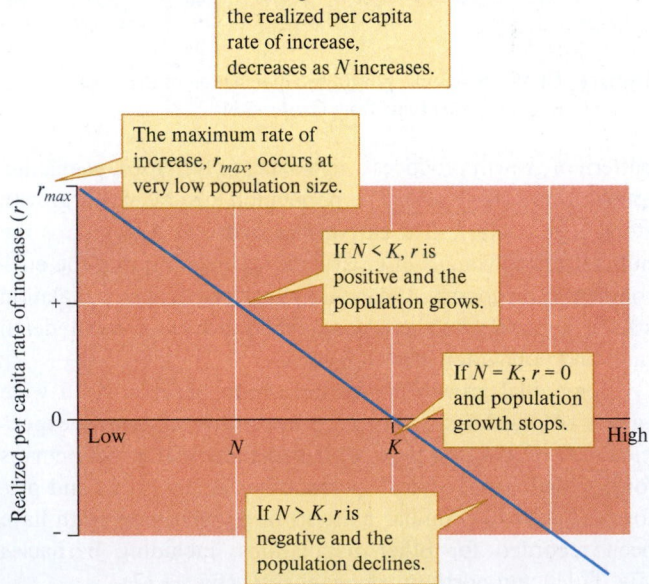

Figure 11.14   The relationship between population size, N, and realized per capita rate of increase, r, in the logistic model of population growth.

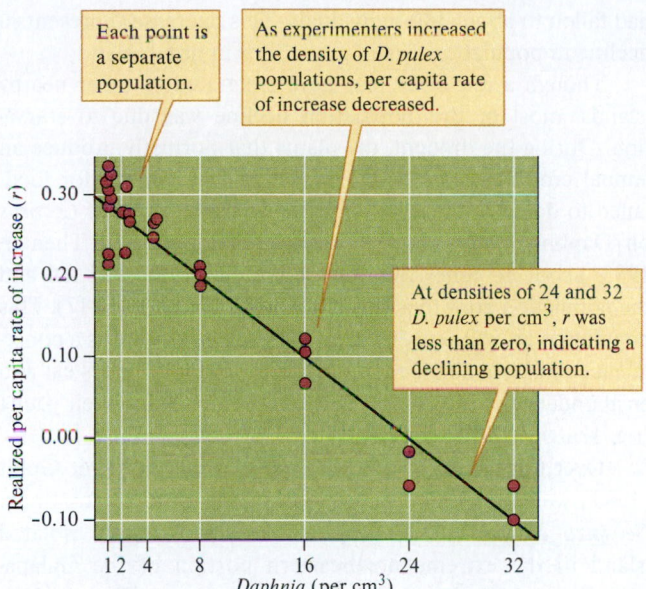

Each point is a separate population.

As experimenters increased the density of *D. pulex* populations, per capita rate of increase decreased.

At densities of 24 and 32 *D. pulex* per cm³, *r* was less than zero, indicating a declining population.

Figure 11.15 Relationship of density to per capita rate of increase in populations of *Daphnia pulex* (data from Frank, Boll, and Kelly 1957).

The response of per capita rate of increase by *Daphnia pulex* to population density closely matches the assumptions of the logistic growth model. When *D. pulex* are grown at densities ranging from 1 to 32 individuals per cubic centimeter, *r* decreases with increasing population size (fig. 11.15). As assumed by the logistic growth model, per capita rate of increase was highest at the lowest population densities. Per capita rate of increase was positive in *D. pulex* populations with densities of 16 individuals per cubic centimeter or lower. However, at densities of 24 and 32 individuals per cubic centimeter, per capita rate of increase was negative.

Ultimately, the environment limits the growth of populations by modifying birth and death rates. In the following section, we examine in detail a few examples of environmental effects on population growth.

Concept 11.2 Review

1. Interpret the pattern of population growth shown by figure 11.11 in terms of the information given in figure 11.14, and discuss the relationship between population size and *r* (realized per capita rate of increase).

2. Why is the rate of logistic population growth at very low population densities nearly equal to exponential growth?

3. Why might a manager of an exploited population, such as a commercially important fish, want to keep fish population size near one-half *K* and not much lower?

11.3 Limits to Population Growth

LEARNING OUTCOMES

After studying this section you should be able to do the following:

11.11 Distinguish between density-independent and density-dependent factors that influence populations.

11.12 Discuss how biotic and abiotic factors influence populations of Darwin's finches.

The environment influences population growth through its effects on birth and death rates. Most of us could recite an impressive list of factors affecting the size of populations. Such lists generally include food, shelter, rainfall, disease, floods, and predators—a mixture of abiotic and biotic factors. Ecologists have long been concerned with the effects of environmental factors such as these on populations. Out of this concern came a long period of debate between the champions of the importance of abiotic factors and those who argued for the importance of biotic factors. Because the effects of biotic factors, such as competition and predation, are often influenced by population density, biotic factors are often referred to as **density-dependent factors.** Meanwhile, abiotic factors, such as floods and extreme temperature, can exert their influences independently of population density and so are often called **density-independent factors.** However, many ecologists were (and are) quick to point out that abiotic factors can influence populations in a density-dependent fashion. For instance, think of the effect on mortality of an unusually cold period. At high population densities, a larger proportion of the population may inhabit less-sheltered sites and, so, mortality rate in the population is greater at high population density than at low population density. The major point of this section is that biotic and abiotic factors both act on populations by modifying birthrates and death rates. The significance of biotic and abiotic factors on populations has been well demonstrated by studies of Darwin's finches and their major food sources.

Environment and Birth and Death Among Darwin's Finches

Since Charles Darwin's visit in the 1830s, the Galápagos Islands have continued to provide scientists with a rich source of information concerning ecological and evolutionary processes. Over four decades ago, Peter Grant and B. Rosemary Grant and their students and colleagues began a long-term study of the evolution and ecology of Darwin's finches (see chapter 10, section 10.4). This long-term project has yielded significant insights into the influences of the environment on birth and death rates in natural populations.

Highly variable rainfall and responsive plant populations provided the environmental setting for these finch studies (fig. 11.16). In 1976, P. T. Boag and Peter Grant began a study of the populations of Darwin's finches inhabiting Daphne Major, an island of only 0.4 km² situated in the middle of the Galápagos Archipelago (Boag and Grant 1984a). The numerically dominant finch on Daphne Major at the beginning of the

(a)

(b)

Figure 11.16   The abundant rains of 1983 (*a*) greatly increased plant growth on the Galápagos Islands compared to (*b*) periods of lower rainfall. (*a, b*) Peter R. Grant/Princeton University

study was the medium ground finch, *Geospiza fortis,* with about 1,200 individuals. In 1977, a drought struck the Galápagos Islands and by the end of the year the population of *G. fortis*

had fallen to about 180 individuals. This decrease represents a decline in population size of about 85% in just 1 year.

Though a few birds may have emigrated to other nearby islands, most of this population decline was due to starvation. During the drought, the plants that normally produce an annual crop of seeds, on which the finches depend for food, failed to do so. From 1977 to 1982, the population of *G. fortis* on Daphne Major averaged about 300 individuals. Then in 1983, about 10 times the average amount of rainfall fell and the population grew to about 1,100 individuals (fig. 11.17). This population growth was due to an increased birthrate as a consequence of an abundance of seeds that the adult finches eat and an abundance of caterpillars that the finches feed to their young (fig. 11.18).

Over this same period, Rosemary Grant and Peter Grant (1989) were studying a population of the large cactus finch, *Geospiza conirostris,* on Genovesa, a small, highly isolated island in the extreme northeastern portion of the Galápagos Archipelago. The study continued from 1978 to 1988, long enough for the researchers to observe the effects of two droughts and two wet periods on reproductive biology. In this population of cactus finches, there was a positive correlation between the number of clutches of eggs laid by birds and the total annual rainfall (fig. 11.19). This study also showed how wet and drought cycles and cactus finches affect populations of prickly pear cactus.

Rainfall, Cactus Finches, and a Cactus Population

Darwin's finches harvest a variety of foods from several species of prickly pear cactus. Two species of finches, *Geospiza scandens* and *G. conirostris,* are well-known specialists on cacti. The Grants documented several ways in which these finches make use of cacti, including (1) opening flower buds in the dry season to eat pollen, (2) consuming nectar and pollen from mature flowers, (3) eating a seed coating called the aril, (4) eating seeds, and (5) eating insects from rotting cactus

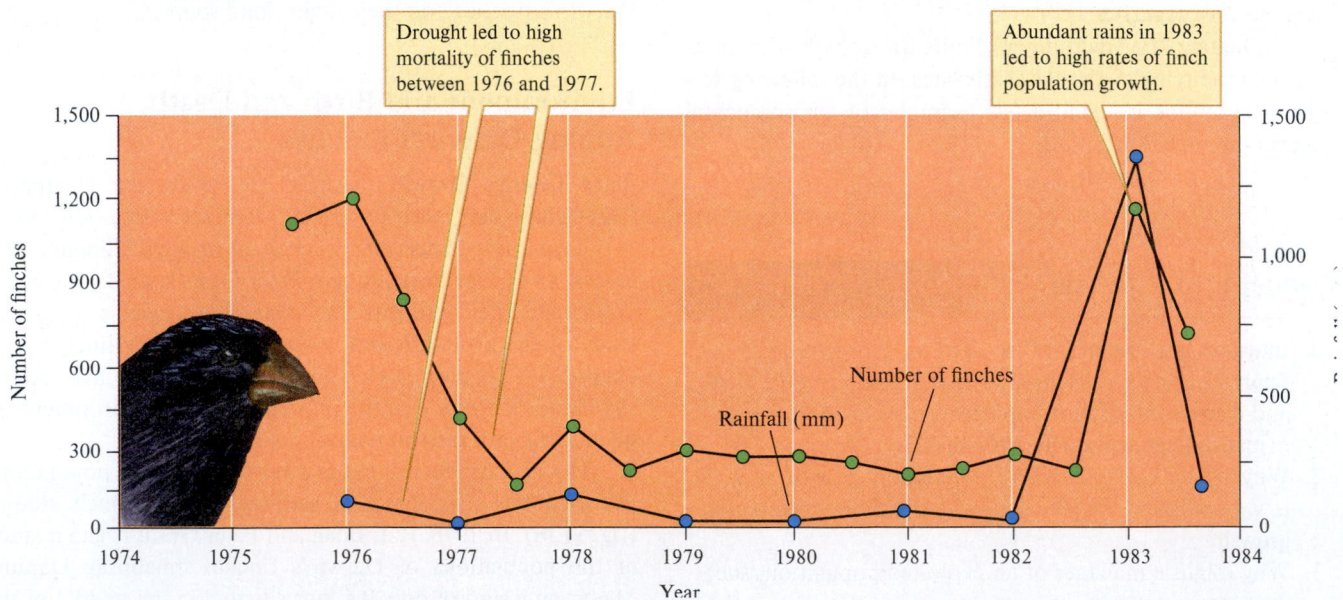

Figure 11.17   Rainfall and the medium ground finch, *Geospiza fortis,* population of Daphne Major Island (data from Gibbs and Grant 1987).

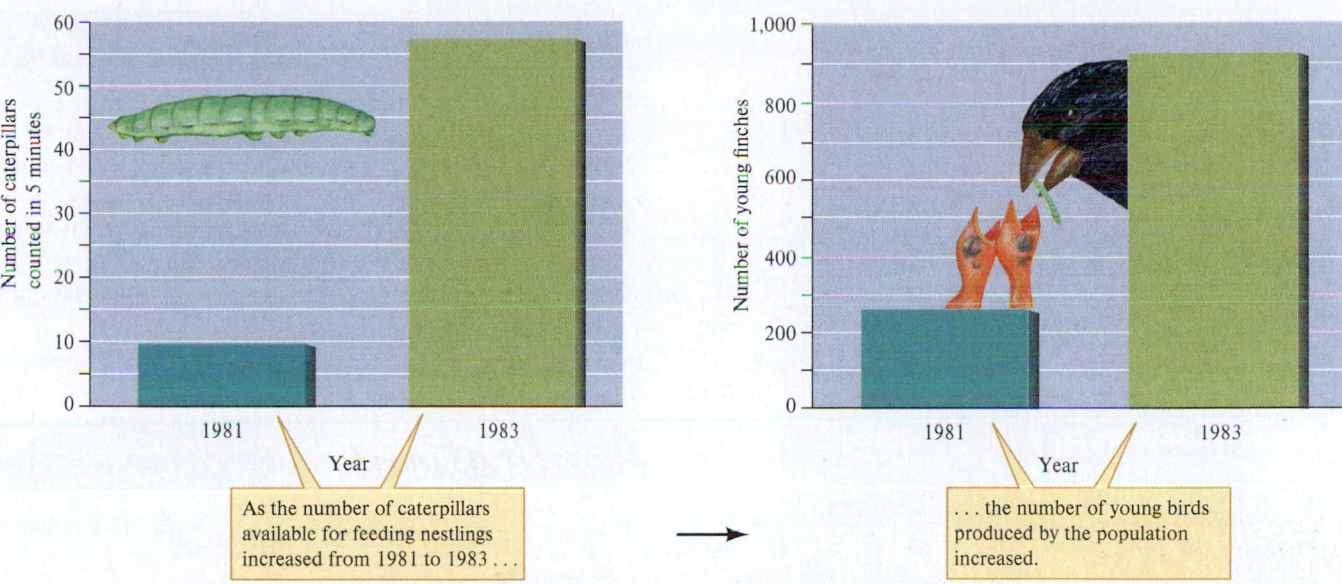

Figure 11.18   Availability of caterpillars and fledgling of young medium ground finches on Daphne Major (data from Gibbs and Grant 1987).

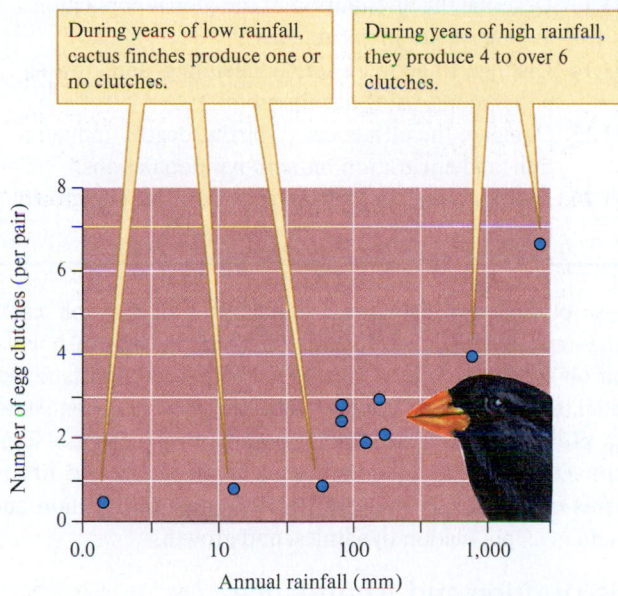

Figure 11.19   Relationship between annual rainfall and the number of egg clutches produced by large cactus finches, *Geospiza conirostris,* on Genovesa Island (data from Grant and Grant 1989).

Figure 11.20   High rainfall during the El Niño of 1983 caused increased mortality of the cactus *Opuntia helleri* on Genovesa Island. Peter R. Grant/Princeton University

pads and from underneath bark. In return, the finches disperse some cactus seeds and pollinate cactus flowers.

Finches also damage many cactus flowers, however. When they open flower buds or partially opened flowers, they snip the style and destroy the stigmas. As a consequence, the ovules of these flowers cannot be fertilized and they do not produce seeds. The Grants found that up to 78% of a population of flowers can be damaged in this way. These activities, which take place during the wet season, may reduce the seeds available to finches during the dry season.

Opuntia helleri, one of the main sources of food for cactus finches on Genovesa Island, was negatively impacted by the El Niño of 1983. This El Niño damaged the cacti in three

ways: (1) Many *O. helleri* simply absorbed so much water that their roots could no longer support them and they were blown over by wind; (2) *O. helleri* on sea cliffs were bathed in salt spray during the many storms that hit the island during 1983, which may have produced osmotic stress (see chapter 6); and (3) increased rainfall stimulated growth by a fast-growing vine that smothered many *O. helleri* (fig. 11.20). Though outright mortality of the cactus was not common, flower and fruit production was severely reduced for several years.

Reduced reproductive output by *O. helleri* was at least partly due to the activities of the cactus finches on Genovesa. The style snipping behavior of cactus finches was especially damaging during the drought years of 1984 and 1985. During normal years, stigma damage is mainly confined to the early part of the wet season from January to March. During the extremely wet 1983 season, there was

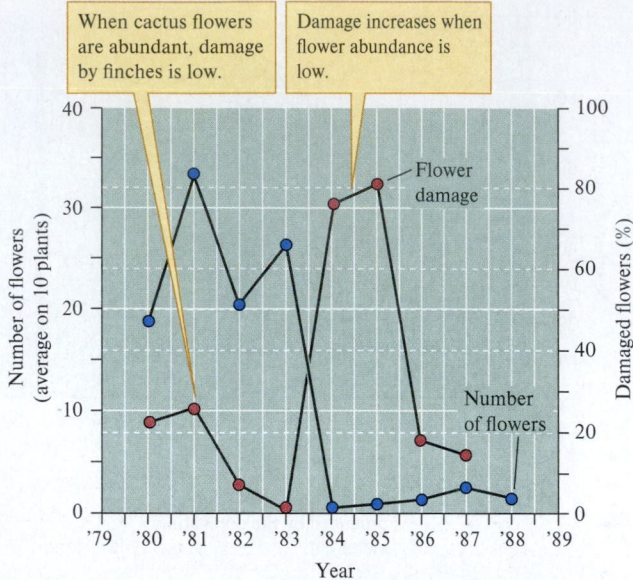

Figure 11.21   Cactus flower abundance on Genovesa Island and extent of flower damage by large cactus finches (data from Grant and Grant 1989).

very little damage to the styles of cactus flowers. However, during the drought years of 1984 and 1985, up to 95% of styles were snipped (fig. 11.21). This extensive damage to flowers helped delay recovery of flower and fruit production until 1986, when another El Niño brought heavy rains to the Galápagos Islands.

Populations of Darwin's finches and their food plants are an instructive model of how the environment can affect birth and death rates. Sometimes, as when the cacti fell because they were engorged with water during the El Niño of 1983, the effect of the physical environment is clear and direct. Sometimes, as when *G. fortis* starved in response to reduced seed supplies during the drought of 1977, the effect of the physical environment on a population is clearly mediated through a biological resource (in this case, seeds). In other cases, such as reduced fruit production by *O. helleri* on Genovesa, populations respond to a complex mixture of abiotic (drought) and biotic (damage by finches) factors that are themselves interrelated. The message to remember from these detailed studies is that both biotic and abiotic factors have important influences on birth and death rates in populations and that their effects are often tightly interconnected. In the examples just presented, environmental variation essentially changed the carrying capacity (K) of the environment for Darwin's finch populations. In section II, we focused on how various aspects of the physical environment affect the performance of organisms, including their reproductive performance. In chapters 13 to 15 of section IV, we will consider at length how biological interactions affect populations. Before we do that, however, let's use the population concepts we've reviewed to examine human populations.

Concept 11.3 Review

1. Why can we be sure that all animal and plant populations are under some form of environmental control?
2. What appears to set the carrying capacity for medium ground finches on Daphne Major Island?
3. Why might medium ground finch population responses to short-term, episodic increases in rainfall (see fig. 11.17) differ from their responses to increases in rainfall lasting for years or decades?

The Human Population

LEARNING OUTCOMES

After studying this section you should be able to do the following:

11.13   Describe the distribution of the global population among earth's major regions.
11.14   Distinguish between stable, declining, and growing populations, using age distributions.
11.15   Compare the influences of births, deaths, immigration, and emigration on national populations.
11.16   Discuss trends in global population size and growth rates.

Most of the significant environmental problems on earth today trace their origins to the effects of the human population on the environment. Therefore, it is very important that students of ecology be familiar with the history, current state, and projected growth of human populations. Let's use some of the conceptual tools we discussed in chapters 9 and 10 and in this chapter to review patterns of human distribution and abundance, population dynamics, and growth.

Distribution and Abundance

One of the most distinctive features of the human population is its distribution. Our species is virtually everywhere. We occupy all the continents—even the Antarctic includes a population of scientists and support staff—and most oceanic islands. What other species, except those dependent upon humans, is so ubiquitous? Except for the Antarctic population, the current distribution of humans did not require modern technological advances. People with stone-age technology nearly reached the present limits of our distribution over 10,000 years ago. Colonization of only the most isolated oceanic islands had to await the development of sophisticated navigational techniques by the Polynesians and Europeans.

Like other populations, human populations are highly clumped at large scales (see chapter 9). In 2020, nearly 60% of the global population, or about 4.6 billion people, was concentrated in Asia (fig. 11.22). In turn, most Asians live in

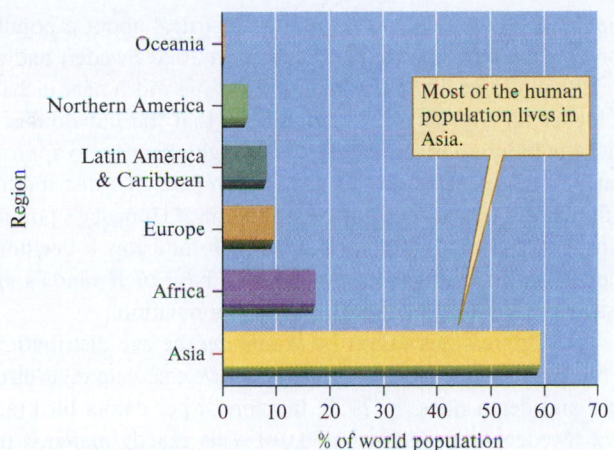

Figure 11.22 Distribution of the human population by region in 2020. Note that estimate for "Northern America" excludes Mexico, which is located in North America but is instead included in the "Latin America" estimate. (data from the Population by Regions in the World (2021) - Worldometer" https://www.worldometers.info/world-population/population-by-region/).

two countries, China and India, the most populous countries on the planet.

Within continents, human populations attain their highest densities in eastern, southeastern, and southern Asia. Other areas of high population density include western and central Europe, northern and western Africa, and eastern and western North America. The patterns shown in figure 11.23 suggest that the highest human population densities are in coastal areas and along major river valleys.

There is even more variation in human population density if viewed on a smaller scale. Within Asia, Singapore has a population density of over 8,358 persons per square kilometer, while Mongolia has a population density of only two persons per square kilometer. This is slightly less than the density on the continent of Australia, which is three persons per square kilometer. Within Europe, the Netherlands harbors nearly 508 persons per square kilometer, while Greece has a population density of about 81 per square kilometer. In North America, the United States has an average population density of about 36 per square kilometer. However, most of this population is concentrated east of the Mississippi and on the West Coast. On a state-by-state basis, population density within the United States varies from that of New Jersey, with approximately 467 people per square kilometer, to Alaska, with an average density of less than one person (0.49) per square kilometer. Canada has an average population density of four per square kilometer. Again, on a large scale, human populations are highly clumped and as a consequence, population density is highly variable. Population dynamics also vary a great deal.

Population Dynamics

Population dynamics vary widely from region to region and from country to country. Let's examine the age distributions, birthrates, and death rates of three countries that have stable, declining, and rapidly growing populations. As we saw in chapter 10,

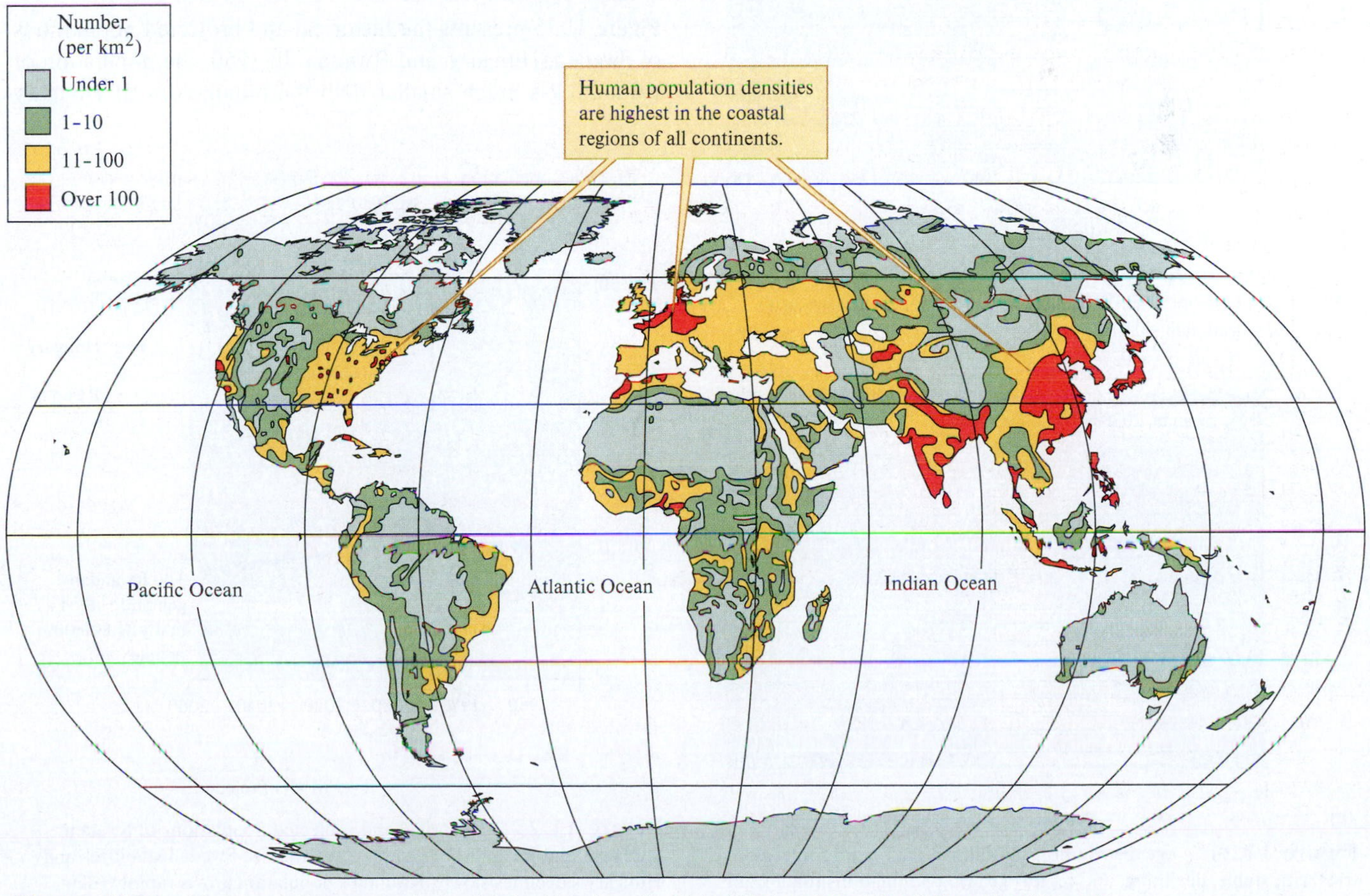

Figure 11.23 Variation in human population density (data from the United Nations Population Information Network).

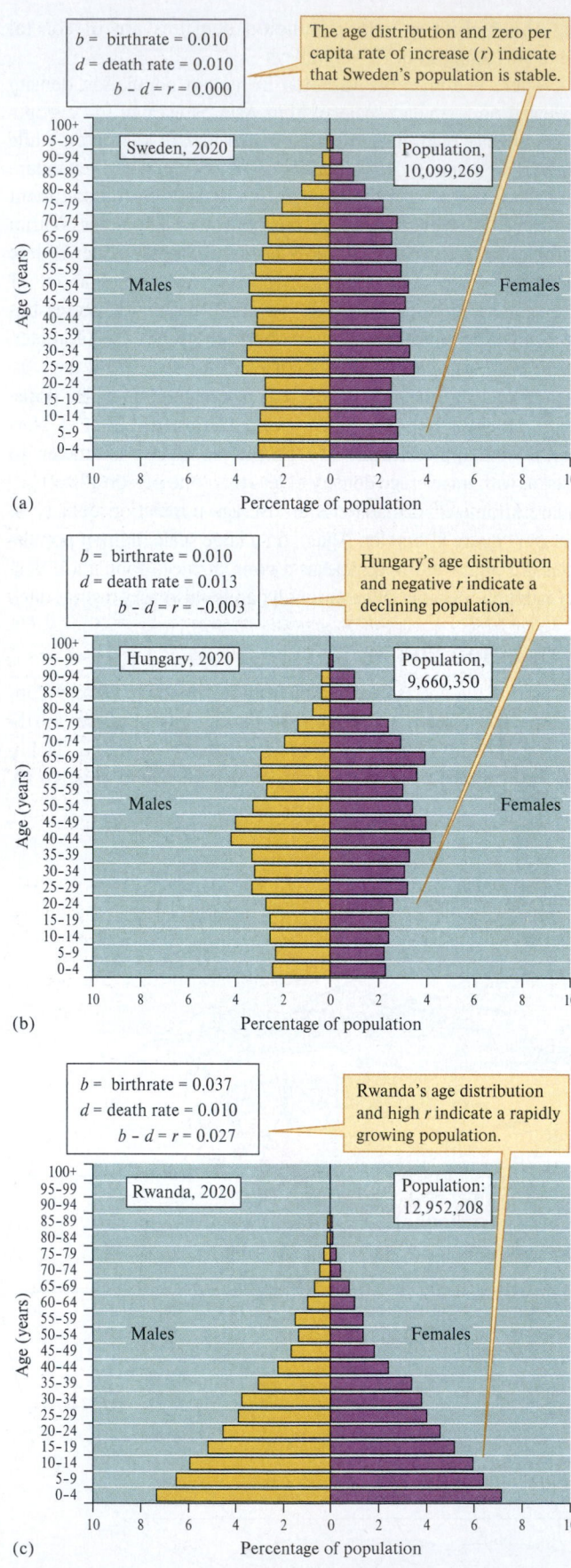

(a)

(b)

(c)

Figure 11.24 Age distributions for human populations in countries with stable, declining, and rapidly growing populations (data from the United Nations, Department of Economic and Social Affairs World Population prospects 2020).

population ecologists can surmise a great deal about a population by examining its age distribution. In 2020 Sweden had an age distribution with approximately the same width near its base as higher up (fig. 11.24). This indicates that the individuals in this population were producing just enough offspring to approximately replace losses due to death. Compare this distribution with that of Hungary. The age distribution of Hungary's population is much narrower at its base, which indicates a declining population. In contrast, the very broad base of Rwanda's age distribution indicates a rapidly growing population.

The impressions we get by examining the age distributions of these three countries are confirmed if we calculate their birthrates and death rates. In 2020, the annual per capita birthrate, b, of Sweden's population was 0.010. This exactly matched the death rate, d, in Sweden's population, which was 0.010. If we subtract Sweden's death rate from its birthrate (0.010 − 0.010), the result is a zero per capita rate of increase, r, of 0.000. In contrast, Hungary's birthrate (0.010) was lower than its death rate (0.013), which results in a per capita rate of increase, r, of −0.003. This negative value for r confirms our impression that Hungary's population is declining. At the other end of the population dynamics spectrum, Rwanda's population had a birthrate that was over two times its death rate. As a consequence, this country's annual per capita rate of increase is 0.027, which is strongly positive growth. Let's move from these estimates of rates of change to examine the longer-term population trends in these countries.

Population Growth

Figure 11.25 presents the historical and projected populations of Sweden, Hungary, and Rwanda. In 1950, the population of Rwanda was much smaller than the populations of Hungary

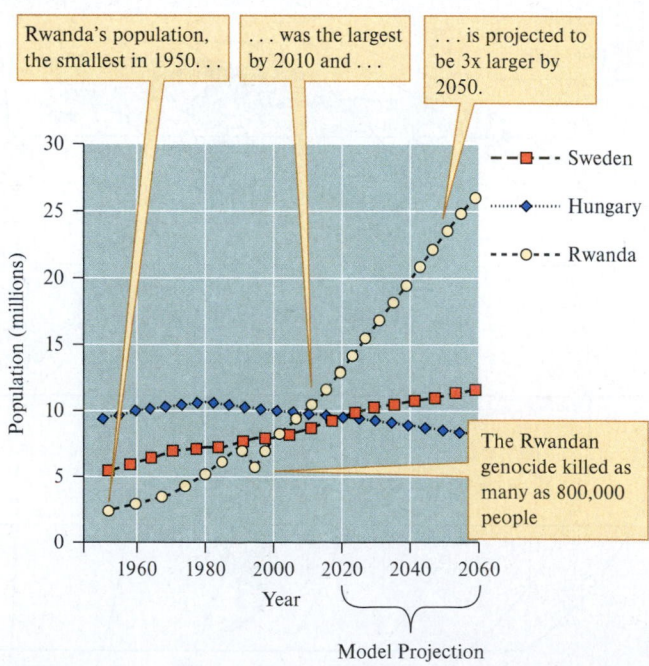

Figure 11.25 Historical and projected populations of Sweden, Hungary, and Rwanda. While the population of Sweden stabilizes and Hungary's declines slowly, Rwanda's population grows rapidly (data from 2019 United Nations DESA, Population Division https://population.un.org/wpp/Graphs/Probabilistic/POP/TOT/646).

or Sweden. Rwanda's population is, however, projected to continue growing and has exceeded that of both Hungary and Sweden. Meanwhile, Sweden's population is expected to stabilize and then decline gradually, while the population of Hungary declines at a faster rate.

If you look carefully at the population trend for Sweden, you will see that its population was still growing after 2011, when $r = 0$ (see fig. 11.24). How can a population's births and deaths balance perfectly, as in Sweden, yet still show positive growth? We encountered the answer to this question in chapter 10, where we discussed population size as the outcome of the balance between numbers of births, B, immigrants, I, deaths, D, and emigrants, E:

$$N_t = N_{t-1} + B + I - D - E$$

In short, we need to consider more than just births and deaths to understand and predict population trends. Sweden's population is growing as a result of immigration. Immigration makes a significant contribution to population trends in most developed countries but especially those of the European Union along with the United States, Canada, and Australia. Meanwhile, emigration results in losses of individuals from the populations of many less-developed nations.

How is the global population changing? While the populations of many developed countries are either stable or declining, those of most developing countries are growing, and the trend for the entire global population is continued growth. As shown in figure 11.26a, global population growth has been extremely rapid during the past 500 years. In fact,

at times global population growth has exceeded exponential rates. How can that be? Recall that in exponential growth, the per capita rate of increase, r, is constant (see fig. 11.4). However, for some time prior to the middle 1960s, the rate of growth for the global population was not constant, but increased as our population grew (fig. 11.26b)!

There are signs that global population growth is slowing. While the global population continues to grow, it is not now growing exponentially. The *rate* of global population growth has declined substantially over the past 50 years, as shown in figure 11.26b. The size of the global population is not rising as steeply as it once was and is projected to level off sometime after the middle of the twenty-first century. Figure 11.26b also displays the proximate cause of this leveling off in population size, a decline in annual growth of the global population. The rate of annual growth by the global population rose steadily from 1950 to 1957 and then took a sharp dip during a major famine in China that lasted from 1958 to 1961, resulting in the deaths of an estimated 16 to 33 million Chinese. Annual growth rate, which peaked in 1968 at 2.1%, has been decreasing in recent decades. The global growth rate is projected to decline to less than 0.5% by 2050. However, this is a projection based on current conditions and recent dynamics of the global and regional populations. Since rates of growth in human populations are currently very dynamic, projections of future global population sizes are being adjusted frequently. Nevertheless, the cost that the present human population exacts upon the global environment is already substantial (see chapter 23). However, size alone is insufficient for estimating the

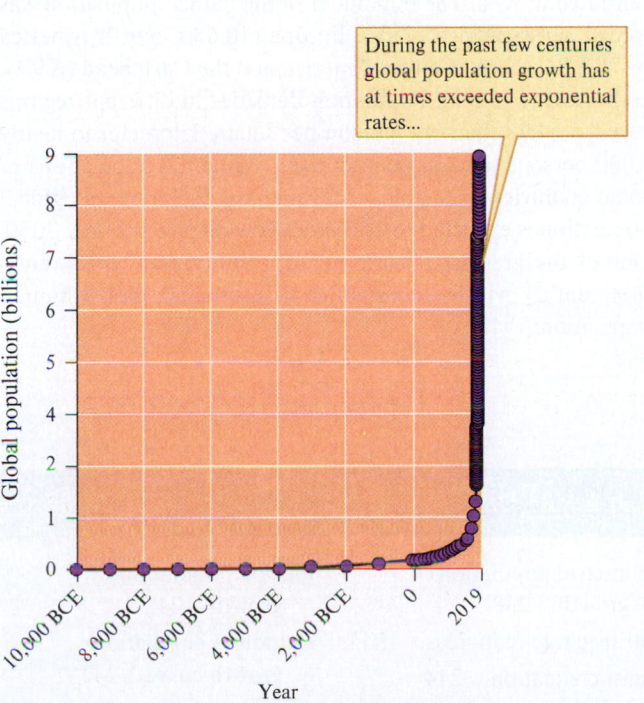

(a)

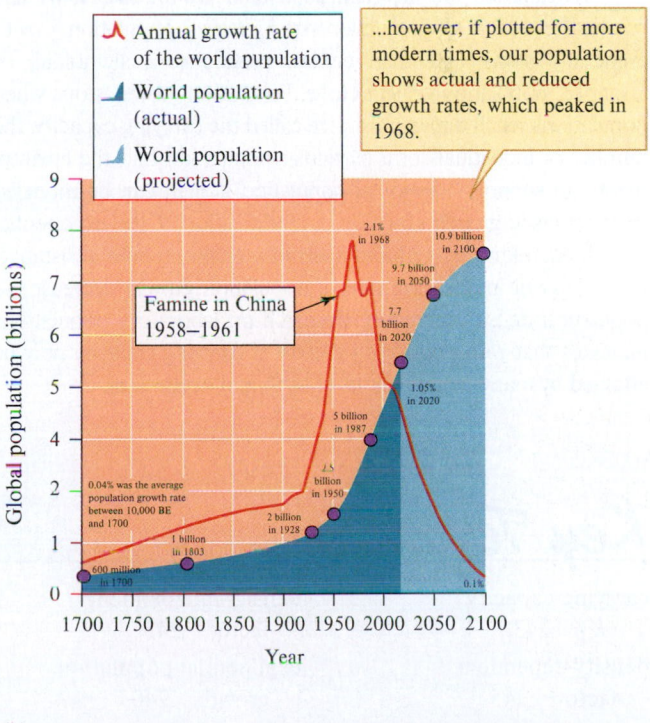

(b)

Figure 11.26   Temporal perspectives on global population growth: (*a*) very rapid growth during the past 2,000 years is evident but (*b*) the past 320 years have been a period of slowing growth by the global human population; growth is projected to continue to slow over the next half century (data from Our World in Data based on HYDE, UN and UN Population Division 2019 Revision, plus data from U.S. Census World Population Clock 2020).

environmental impact of a human population. Such impact results from a combination of population size *and* per individual use of resources. If you factor in resource use, it turns out that the populations of developed countries, on average, use natural resources at a rate eight times higher than the populations of developing countries (WWF 2006). One of the greatest environmental challenges of the twenty-first century will be to establish a sustainable global population.

Summary

Understanding population growth rates is important. Our recent experience with the spread of COVID-19 has shown us that many people do not have a reasonable grasp of how fast exponential growth is, and that a more accurate understanding is important for responding appropriately to a pandemic. It is also important to understand that a population will still be growing even if the rate of growth decreases, so long as the growth rate is greater than zero.

In the presence of abundant resources, populations can grow at geometric or exponential rates. Population growth by organisms without overlapping generations can be described by the geometric model of population growth. Population growth that occurs as a continuous process, as in human or bacterial populations, can be described by the exponential model of population growth. Examples of exponential growth from natural populations suggest that this type of growth may be very important to populations during establishment in new environments, during recovery from some form of exploitation, or during exploitation of transient, favorable conditions.

As resources are depleted, population growth rate slows and eventually stops. As population size increases, population growth eventually slows and then ceases, producing a sigmoidal, or S-shaped, population growth curve. Population growth stops when populations reach a maximum size called the carrying capacity, the number of individuals of a particular population that the environment can support. Sigmoidal population growth can be modeled by the logistic growth equation, a modification of the exponential growth equation that includes a term for environmental resistance. In the logistic model, the rate of population growth decreases as population density increases. Research on laboratory populations indicates that zero population growth at carrying capacity may be attained by many combinations of birth and death rates.

The environment influences population growth through its effects on birth and death rates. The factors affecting population size and growth include biotic factors such as food, disease, and predators and abiotic factors such as rainfall, floods, and temperature. Because the effects of biotic factors, such as disease and predation, are often influenced by population density, biotic factors are often referred to as density-dependent factors. Meanwhile, abiotic factors such as floods and extreme temperature can exert their influences independently of population density and so are often called density-independent factors. As we have already seen, both abiotic and biotic forces have important influences on populations. The significant effects of biotic and abiotic factors on populations have been well demonstrated by studies of Darwin's finches and their major food sources.

The present state of the human population can be examined using the conceptual tools of population biology discussed in chapters 9 and 10 and in chapter 11. Though humans live on every continent, their population density differs by several orders of magnitude in different regions. In 2020, nearly 60% of the global population, or about 4.6 billion people, was concentrated in Asia. The remainder of the human population was spread across Africa (15%), Europe (10.6%), North America (6.7%), South and Central America and the Caribbean (6.9%), and Oceania (0.5%). Population densities in different regions vary from less than one person per square kilometer to nearly 7,000 persons per square kilometer. While the populations of some countries are stable, and some are declining, the global population is expected to continue growing past the year 2050. One of the greatest environmental challenges of the twenty-first century will be to establish a sustainable global human population.

Key Terms

carrying capacity
  (*K*)   243
density-dependent
  factor   245

density-independent
  factor   245
exponential population
  growth   240

geometric population
  growth   238
intrinsic rate of increase   244
logistic equation   244

logistic population
  growth   242
sigmoidal population
  growth curve   243

Review Questions

1. For what types of organisms is the geometric model of population growth appropriate? For what types of organisms is the exponential model of population growth appropriate?

2. While populations of gray and blue whales have grown rapidly, the North Atlantic, right whale population remains dangerously small despite many decades of complete protection. Assuming that differences in population growth rates are not caused by external factors such as pollution, what information would you need to explain the slower growth by the right whale populations?

3. How do you build the logistic model for population growth from the exponential model? What part of the logistic growth equation produces the sigmoidal growth curve?

4. What factors in natural environments may cause sigmoidal population growth? Pick a real organism living in an environment with which you are familiar and list the things that might limit the growth of its population.

5. What is the relationship between per capita rate of increase, r, and the intrinsic rate of increase, r_{max}? In chapter 10, we estimated r from the life tables and fecundity schedules of two species. How would you estimate r_{max}?

6. Both abiotic and biotic factors influence birthrates and death rates in populations. Make a list of abiotic and biotic factors that are potentially important regulators of natural populations.

7. Can the effect of an abiotic factor on a population be somewhat dependent on population density? Explain.

8. Where on earth is human population density highest? Where is it lowest? Where on earth do no people live? Where are human populations growing the fastest? Where are they approximately stable?

9. Explain why the earth's long-term (thousands of years) carrying capacity for the human population may be much lower than the projected population size for the year 2050. Now argue the other side. Explain how the numbers projected for 2050 might be sustained over the long term.

Fuse/Getty Images

Chapter

12

Life Histories

Migrating sockeye salmon, *Oncorhynchus nerka,* leaping Brooks Falls, in Katmai National Park, Alaska. These sockeye salmon will spawn within a lake or tributary stream then die. Their offspring will spend up to 3 years in a lake before migrating to the ocean, where they will grow rapidly, maturing in 1 to 4 years before returning to the area where they hatched, completing the sockeye life cycle.

CHAPTER CONCEPTS

12.1 Because all organisms have access to limited energy and other resources, there is a trade-off between the number and size of offspring; those that produce larger offspring are constrained to produce fewer, whereas those that produce smaller offspring may produce larger numbers.   255

Concept 12.1 Review   261

12.2 Where adult survival is lower, organisms begin reproducing at an earlier age and invest a greater proportion of their energy budget into reproduction; where adult survival is higher, organisms defer reproduction to a later age and allocate a smaller proportion of their resources to reproduction.   262

Concept 12.2 Review   266

12.3 The great diversity of life histories can be classified on the basis of a few population characteristics. Examples include fecundity or number of offspring, survival, relative offspring size, and age at reproductive maturity.   266

Concept 12.3 Review   272

Applications: Climate Change and Timing of Reproduction and Migration   272
Summary   274
Key Terms   275
Review Questions   275

LEARNING OUTCOME

After studying this section you should be able to do the following:

12.1 Describe the concept of life history and some of its components.

Different species, often living side by side, reproduce at vastly different rates over lifetimes that may differ by several orders of magnitude. On a rare sunny day in temperate rain forest, a redwood tree, *Sequoia sempervirens,* shades a nearby stream. Bathed in fog all summer, soaked by rain during fall, winter, and spring, the redwood has lived through 2,000 annual cycles. The tree was well established when Rome invaded Britain and had produced seeds for 1,000 years when William the Conqueror invaded the island from across the English Channel. It was 1,800 years old when rag-tag colonials wrenched an American territorial prize from William's heirs, claiming it as their own country. Within a mere century the descendants of the colonial rebels had expanded their territory 3,000 kilometers westward and were chopping down redwood trees for lumber. Other human populations had long lived near the base of the tree and had cut some trees. However, no population had been so relentless as the newcomers who stripped vast areas of all trees. Somehow, before all the redwoods were gone the cutting stopped and the grove of the giant redwood was protected. With luck, the tree would live through several more centuries of summer fog and winter rain, during which the human order of the world would surely change many times more.

On this summer morning, other life was stirring in the nearby stream. A female mayfly along with thousands of others of her species were shedding their larval exoskeletons as they transformed from their robust crawling aquatic stage to graceful flying adults (fig. 12.1). As a larva, the mayfly had lived in the stream for a year, but her adult stage would last just this 1 day, during which she would mate, deposit her eggs in the stream, and then die. She had just this one chance to successfully complete her life cycle. For an adult mayfly, there is no tomorrow—one chance and no more. As the mayflies swarmed, some would be eaten by birds nesting in alder trees that grew along the stream, and some would be caught by bats that found roosting sites on the giant redwood. Some of the mayflies would be eaten by fish that they had successfully eluded for a year of larval life, and still others would be snared by spiders that spun their webs in azalea shrubs that grew between the alders and the redwood. However, this particular mayfly escaped all predators, mated, and laid her eggs.

Spent by the effort of depositing her eggs, the mayfly was caught by the current and washed downstream. Fifty meters from where she emerged that morning, the mayfly was taken from the surface by a trout as she floated past an old redwood log where the trout sheltered. The small splash of the feeding trout caught the attention of a man and a woman who had been

studying the stream. They knew the stream well and knew the pool where the big trout lived and they knew the trout, which they had tried to catch many times. Their grandparents' generation had cut the redwood forests. Later, their parents had worked to protect this remnant grove. The man and woman had played in the grove as children and courted there as young adults. Now that their own children were grown, they fished the stream frequently, sharing the place once again.

Redwood, mayfly, fish, and humans—lives intertwined in a web of ecological relationships but vastly different in scale and timing. All four are players in an ecological and evolutionary drama stretching into a vast past and into an unknown future. Made of the same elements and with their genetic inheritance encoded by DNA of the same basic structure, the four species have inherited vastly different lives. While the redwood has produced seeds numbering in the millions over a lifetime that has stretched for millennia, the mayfly spends a year in the stream and then emerges to lay eggs that will number in the hundreds. The trout's spawn has numbered in the thousands, deposited during the several years of her life. Meanwhile, the man and woman have produced two children during their lifetime, investing time and energy into them over a period of decades.

What are the selective forces that created and maintain this vast range of biology? Under what conditions will organisms mature at an early age and small size instead of later at a larger size? What are the costs and benefits of producing millions of tiny offspring, such as the seeds of the redwood tree, versus a few that are large and well cared for? These are the sorts of questions pondered by ecologists who study life history. **Life history** consists of the adaptations of an organism that influence aspects of its biology, such as the number of offspring it produces, its survival, and its size and age at reproductive maturity. Chapter 12 discusses some of the central concepts of life history ecology.

12.1 Offspring Number versus Size
LEARNING OUTCOMES
After studying this section you should be able to do the following:

12.2 List several plant growth forms.
12.3 Describe the general relationship between the size and number of offspring produced by an organism.
12.4 Discuss the evolutionary implications of the positive relationship between the number of eggs produced and gene flow in darter species.
12.5 Explain the reasons for the parallel relationship between the size and number of offspring in fish and plants.
12.6 Predict the size and number of seeds produced by plants associated mainly with disturbed habitats.

Figure 12.1   Adult mayflies generally live 1 day only.
NHPA/Melvin Grey

Because all organisms have access to limited energy and other resources, there is a trade-off between the number and size of offspring; those that produce larger offspring are constrained to produce fewer, whereas those that produce smaller offspring may produce larger numbers. The discussions of photosynthetic response by plants (see figs. 7.20 and 7.21) and functional response by foraging animals (see figs. 7.22 to 7.24) led us to conclude that all organisms take in energy at a limited rate. As we saw, rate of energy intake is limited either by conditions in the external environment, such as food availability, or by internal constraints, such as the rate at which the organism can process food. These constraints are fundamental to the principle of allocation, which was introduced in chapter 5 (section 5.2). The principle of allocation underscores the fact that if an organism uses energy for one function such as growth, it reduces the amount of energy available for other functions such as reproduction. This tension between competing demands for resources leads inevitably to trade-offs between functions. One of those is the trade-off between number and size of offspring. Organisms that produce many offspring are constrained, because of energy limitation, to produce smaller offspring (seeds, eggs, or live young). Viewed from the opposite perspective, organisms that produce large, well-cared-for offspring are constrained to produce fewer. Let's begin our review of examples bearing on this generalization with a survey of patterns among fish, a vertebrate group with especially large variation in life history characteristics.

Egg Size and Number in Fish

According to the IUCN Red List (see chapter 9), there are more than 35,500 species of fish (IUCN 2020). Because of their great diversity and the wide variety of environments in which they live, fish offer many opportunities for studies of life history. Kirk Winemiller (1995) pointed out that fish show more variation in many life history traits than any other group of animals. For instance, the number of offspring they produce per brood—that is, their clutch size—ranges from the one or two large live young produced by mako sharks to the 600 million eggs per clutch laid by the ocean sunfish. However, many variables other than offspring number and size change from sharks to sunfish. Therefore, more robust patterns of variation can be obtained by analyzing relationships within closely related species, such as within families or genera.

In a study of gene flow among populations of darters, small freshwater fish in the perch family, or Percidae, Tom Turner and Joel Trexler tried to determine the extent to which life history differences among species might influence gene flow between populations. Turner and Trexler (1998) pointed out that in such a study, it is best to focus on a group of closely related organisms with a shared evolutionary history. They were particularly interested in determining the relationship between egg size and egg number, or **fecundity,** and the extent of gene flow among populations. Fecundity is simply the number of eggs or seeds produced by an organism. Turner and Trexler proposed that gene flow would be higher among populations producing more numerous smaller eggs—that is, among populations with higher fecundity.

Turner and Trexler chose the darters for their studies because they are an ideal study group. Darters are small, streamlined benthic fishes that live in rivers and streams throughout eastern and central North America. Male darters are usually strikingly colored during the breeding season (fig. 12.2). The darters

Figure 12.2   Darters such as this male orangethroat darter form a diverse and distinctive subfamily of fishes within the perch family. They live only in North America.   David M. Schleser/Science Source

consist of 174 species in three genera within the family Percidae, which makes them one of the most species-rich groups of vertebrates in North America. The most diverse genus, *Etheostoma,* alone includes approximately 135 species. However, despite the fact that the darters as a whole live in similar habitats and have similar anatomy, they vary widely in their life histories. The genera most similar to the ancestors of the darters, *Crystallaria* (1 species) and *Percina* (38 species), are larger and produce more eggs than species in the genus *Etheostoma.* However, *Etheostoma* species also vary substantially in their life histories.

Turner and Trexler sampled 64 locations on streams and rivers in the Ohio, Ozark, and Ouachita Highlands regions of Ohio, Arkansas, and Missouri, the heart of freshwater fish diversity in North America, which supports one of the most diverse temperate freshwater fish faunas on earth. Of the darters they collected at these locations, they chose 15 species, 5 in the genus *Percina* and 10 *Etheostoma* species, for detailed study. Turner and Trexler chose darter species that included a wide range of variation in life history traits, especially variation in body size, number of eggs laid, and egg size.

The species in the study ranged in length from 44 to 127 mm, and the number of mature eggs that they produced ranged from 49 to 397. Meanwhile, the size of eggs produced by the study species varied from 0.9 to 2.3 mm in diameter. As they expected, Turner and Trexler found that larger darter species produce larger numbers of eggs (fig. 12.3). Their results also support the generalization that there is a trade-off between offspring size and number. On average, darters that produce larger eggs produce fewer eggs (fig. 12.4).

Turner and Trexler characterized the genetic structure of darter populations using electrophoresis of **allozymes,** different forms of an enzyme, which are gene products, produced by 21 different genes, or **loci**. They chose 21 loci out of 40 that they examined because they were polymorphic. A **polymorphic locus** is one that occurs as more than one allele. In this case, each allele synthesizes a different allozyme. Turner and Trexler assessed genetic structure using allelic frequencies. Allelic frequencies were measured as the frequencies of allozymes across the 21 different study loci. Populations with similar allelic frequencies were taken as genetically similar, while those that differed in allelic frequencies were concluded to be different

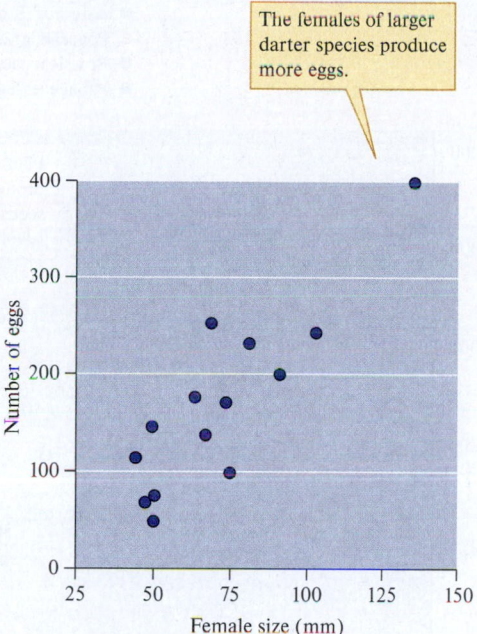

Figure 12.3   Relationship between female darter size and number of eggs. Each point represents a different darter species (data from Turner and Trexler 1998).

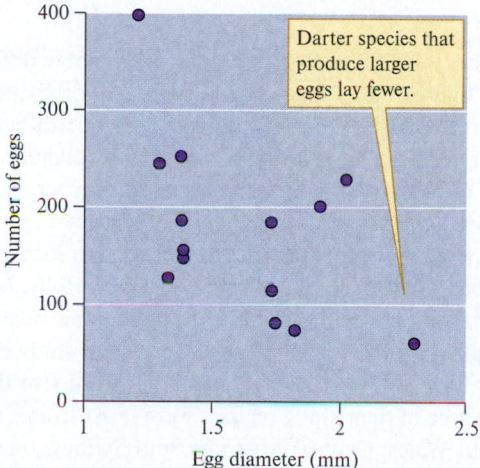

Figure 12.4   Relationship between the size of eggs laid by several darter species and the number of eggs laid (data from Turner and Trexler 1998).

genetically. Gene flow was estimated by the degree of similarity in allelic frequencies between populations.

How can the number and kinds of allozymes synthesized by a series of populations be used to determine the extent of gene flow among populations? Turner and Trexler assumed that the populations differing in allelic frequencies have lower gene flow between them than populations that have similar allelic frequencies. In other words, they assumed that genetic similarity between populations is maintained by gene flow, while genetic differences arise in the absence or restriction of gene flow.

What relationship is there between egg size and number and gene flow between populations? Turner and Trexler found a negative relationship between egg size and gene flow but a

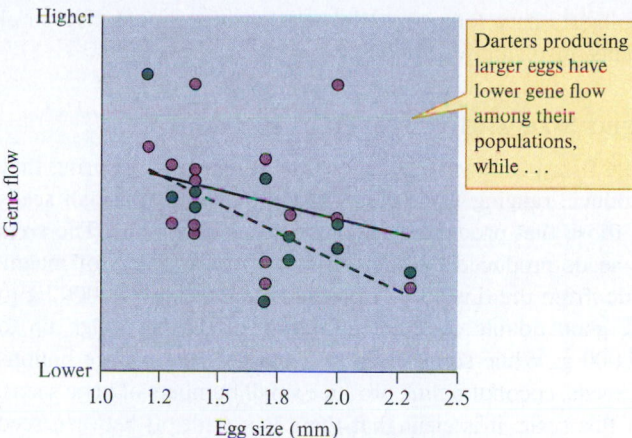

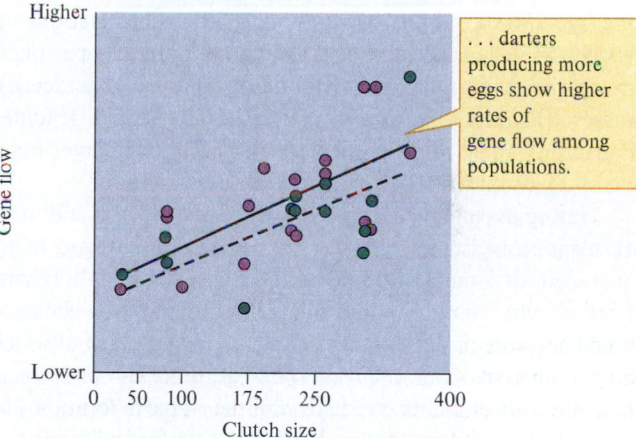

Figure 12.5   Relationship between egg size and egg number and gene flow in darter species. Solid lines and purple dots represent estimates by one genetic method and dashed lines and green dots by a second method (data from Turner and Trexler 1998).

strong positive relationship of gene flow with the number of eggs produced by females (fig. 12.5). That is, populations of darter species that produce many small eggs showed less difference in allelic frequencies across the study region than did populations that produce fewer larger eggs.

How do differences in egg size and number translate into differences in gene flow among populations? It turns out that the larvae of darters that hatch from larger eggs are larger when they hatch. These larger larvae begin feeding on prey that live on the streambed at an earlier age, and spend less time drifting with the water current. Consequently, larvae hatching from larger eggs disperse shorter distances and therefore carry their genes shorter distances. As a result, populations of species producing fewer larger eggs will be more isolated genetically from other populations. Because of their greater isolation, such populations will differentiate genetically more rapidly compared to populations of species that produce many smaller larvae that disperse longer distances.

Turner and Trexler's study not only provides data consistent with the generalization that there is a trade-off between offspring size and number, it also reveals some of the evolutionary consequences of that trade-off.

Trade-offs between offspring number and size have been found in populations of many kinds of organisms. For instance, ecologists have found parallel relationships among terrestrial plants, involving seed number and size.

Seed Size and Number in Plants

Like fish, plants vary widely in the number of offspring they produce, ranging from those that produce many small seeds to those that produce a few large seeds (fig. 12.6). The sizes of seeds produced by plants range over 10 orders of magnitude, from the tiny seeds of orchids that weigh 0.000002 g to the giant double coconut palm with seeds that weigh up to 27,000 g. While some orchids are known to produce billions of seeds, coconut palms produce small numbers of huge seeds. At this scale, it is clear that there is a trade-off between seed size and seed number, a relationship documented by botanists long ago (Stevens 1932). Figure 12.7 shows the relationship between average seed mass and the number of seeds per plant among species in four families of plants: daisies (Asteraceae), grasses (Poaceae), mustards (Brassicaceae), and legumes (Fabaceae). In all four families, species producing larger numbers of seeds on average produce fewer seeds.

Having documented a trade-off between seed size and number, plant ecologists searched for the mechanisms favoring many small seeds in some environments and few larger seeds in others. However, when venturing into the world of plants, the ecologist should be aware of the subtleties of plant biology, much of which can be inferred from their morphology. For instance, many characteristics of plants correlate with their **growth form,** or life-form, which itself constitutes an aspect of the plant life history. Therefore, comparing seed production of orchids and coconut palms, which mixes data from a species having the growth form of an epiphyte (the orchid) and another with the growth form of a tree (the palm), may not be a valid comparison. Such a comparison may not be valid, since growth form may itself influence the number and size of seeds produced by plants.

What other aspects of plant biology might influence seed size? As we saw in chapter 10, dispersal is an important facet of the population biology of all organisms, including plants. For instance, figure 10.6 shows the history of maple and hemlock dispersal northward following glacial retreat beginning

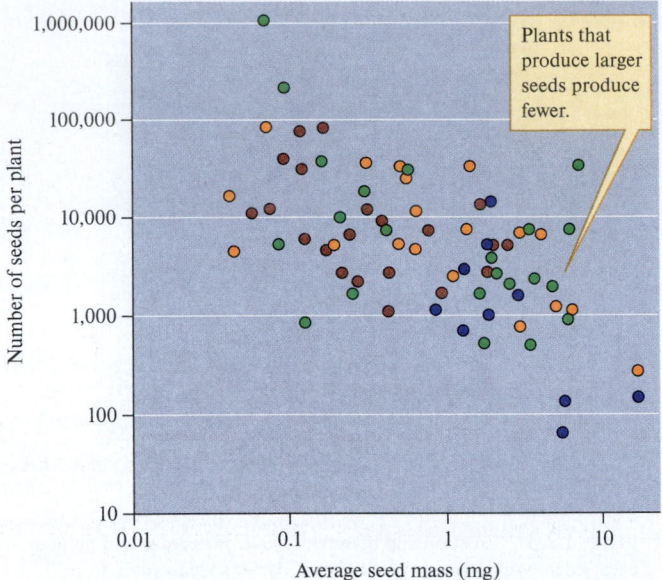

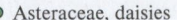
- Asteraceae, daisies
- Poaceae, grasses
- Brassicaceae, mustards
- Fabaceae, legumes

Plants that produce larger seeds produce fewer.

Figure 12.7   Relationship between seed mass and seed number (data from Stevens 1932).

approximately 14,000 years ago. One of the notable differences shown in figure 10.6 is that maple dispersed northward much faster than hemlock. What is the source of this difference in dispersal rate? Since long-distance dispersal by plants is mainly by means of seeds, we might ask whether there is a relationship between seed characteristics and mode of dispersal.

Aware of the potential influence of growth form and dispersal mode on seed characteristics, Mark Westoby, Michelle Leishman, and Janice Lord (1996) studied the relationship between plant growth form and seed size. Their study included the seeds of 196 to 641 species of plants from five different regions. Three of their study regions were in Australia; western New South Wales, Central Australia, and Sydney; one was in Europe: Sheffield, United Kingdom; and one was in North America: Indiana Dunes National Lakeshore.

Westoby, Leishman, and Lord recognized four plant growth forms. Grasses and grasslike plants, such as sedges and rushes, were classified as **graminoids.** Herbaceous plants other than graminoids were assigned to a **forb** category. Species with woody thickening of their tissues were considered as woody plants. Finally, climbing plants and vines were classified as climbers. The results showed a clear association between seed size and plant growth form (fig. 12.8*a*). In most of the floras analyzed by Westoby and his colleagues, the smallest seeds were produced by graminoid plants, followed by the seeds produced by forbs. In all five study regions, woody plants produce seeds that are far larger than those produced by either graminoids or forbs. However, the largest seeds in all regions are produced by vines. The researchers found that the seeds produced by woody plants and vines in the five floras were on average, approximately 10 times the mass of seeds produced by either graminoid plants or forbs.

Figure 12.6   A small sample of the great diversity of seed sizes and shapes.   Sabrina E. Russo/Harvard University Herbarium

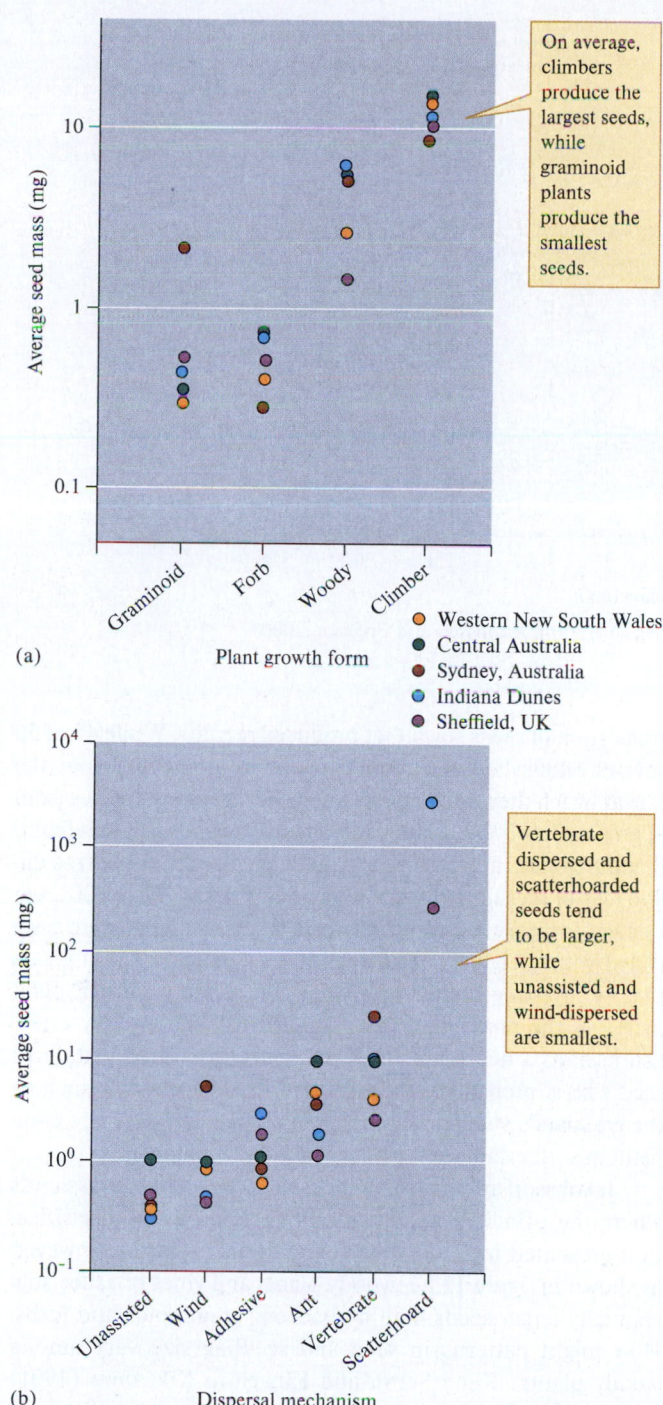

On average, climbers produce the largest seeds, while graminoid plants produce the smallest seeds.

Western New South Wales
Central Australia
Sydney, Australia
Indiana Dunes
Sheffield, UK

(a)

Vertebrate dispersed and scatterhoarded seeds tend to be larger, while unassisted and wind-dispersed are smallest.

(b)   Dispersal mechanism

Figure 12.8   Plant growth form and dispersal mechanism and seed mass (data from Westoby, Leishman, and Lord 1996).

Westoby and his coauthors recognized six dispersal strategies. They classified seeds with no specialized structures for dispersal as unassisted dispersers. Seeds with wings, hairs, or other structures that provide air resistance were assigned to a wind-dispersed category, whereas if seeds had hooks, spines, or barbs, they were classified as **adhesion-adapted.** Animal-dispersed seeds in the study included ant-dispersed, vertebrate-dispersed, and scatterhoarded. Westoby, Leishman, and Lord classified seeds with an **elaiosome,** a structure on the surface of some seeds generally containing oils attractive to ants, as

ant-dispersed. Seeds with an **aril,** a fleshy covering of some seeds that attracts birds and other vertebrates, or with flesh were classified as vertebrate-dispersed. Finally, they classified as **scatterhoarded** those seeds known to be gathered by mammals and stored in scattered caches or hoards.

Westoby, Leishman, and Lord also found that plants that disperse their seeds in different ways tend to produce seeds of different sizes (fig. 12.8b). Plants that they had classified as unassisted dispersers produced the smallest seeds, while wind-dispersed seeds were slightly larger. Adhesion-adapted seeds were of intermediate size, while animal-dispersed seeds were largest. Ant-dispersed seeds were smaller than vertebrate-dispersed seeds, with scatterhoarded seeds being the largest by far. Westoby and his team point out that between 21% and 47% of the variation in seed size in the five floras included in their study is accounted for by a combination of growth form and mode of dispersal. Angela Moles and colleagues (Moles et al. 2005a, 2005b) add support for a relationship between seed size and plant growth form through their extensive analyses, involving nearly 13,000 plant species. Their work also suggests that dispersal mode exerts a weaker, but significant, influence on seed size.

The analyses by Westoby and his colleagues show that both plant growth form and dispersal mode are associated with differences in seed size among plants. Impressively, the relationships between seed size and both growth form and dispersal mode were consistent across widely separated geographic regions. However, Westoby, Leishman, and Lord pointed out that their analysis uncovered wide variation in seed size among plants in all regions. What are the factors that maintain variation in seed size? To maintain such variation, there must be advantages and disadvantages of producing either large or small seeds. What are those advantages and disadvantages? Plants that produce small seeds can produce greater numbers of seeds. Such plants seem to have an advantage where disturbance rates are high and where plants with the capacity to colonize newly opened space appear to thrive. Though plants that produce large seeds are constrained to produce fewer, large seeds produce seedlings that survive at a higher rate in the face of environmental hazards. Those hazards include competition from established plants, shade, defoliation, nutrient shortage, deep burial in soil or litter, and drought.

Seed Size and Seedling Performance

Anna Jakobsson and Ove Eriksson (2000) of Stockholm University studied the relationships between seed size, seedling size, and seedling recruitment among herbs and grasses living in seminatural grasslands in southeastern Sweden. To estimate the influence of seed size and seedling size, Jakobsson and Eriksson germinated seeds in pots containing a standardized soil mix. The pots were maintained in a greenhouse under standardized conditions, and seedlings were harvested and weighed 3 weeks after germination. **Germination** is the process by which seeds begin to grow or develop, producing the small plant called a seedling in the process. The results of this portion of the study showed clearly that larger seeds produced larger seedlings (fig. 12.9).

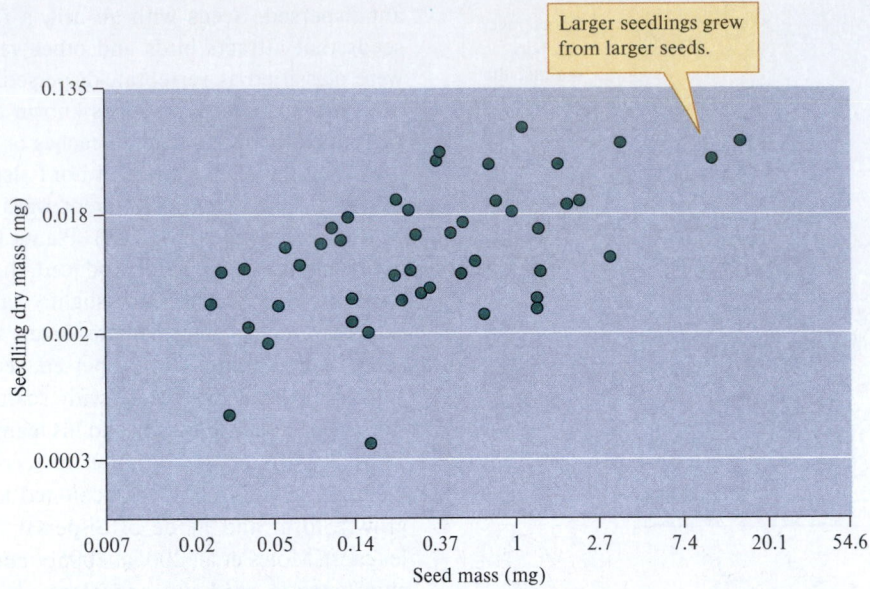

Figure 12.9   Seed mass and seedling mass among grassland plants in Sweden (data from Jakobsson and Eriksson 2000).

Jakobsson and Eriksson also investigated the relationship between seed size and recruitment among 50 plant species living in the meadows of their study region, using a field experiment. At their field sites, Jakobsson and Eriksson planted the seeds of each species in 14 small 10 × 10 cm plots. Each plot was sown with 50 to 100 seeds of the study species. They left half of the study plots undisturbed, while the other plots were disturbed before planting by scratching the soil surface and removing any accumulated litter. In addition to the 14 plots where seeds were sown, Jakobsson and Eriksson established control plots where they did not plant seeds. Again, half of these were disturbed and half left undisturbed. Why did Jakobsson and Eriksson need to establish these control plots? The control plots allowed them to estimate how much germination of each species would occur in the absence of their sowing new seeds. The seeds of many species can lie dormant in soils for long periods of time, and additional seeds of their study species might have dispersed into the study plots during the experiment. Therefore, without the control plots, Jakobsson and Eriksson would have no way of knowing if the seedlings they observed had grown from the seeds they had sown or from other seeds.

Of the 50 species of seeds planted, the seeds of 48 species germinated and those of 45 species established recruits. Jakobsson and Eriksson observed no recruitment of any of the study species on the control plots. Therefore, they could be confident that new plants recruited into their experimental plots came from seeds that they had planted. Though plants recruited to both undisturbed and disturbed plots, the number of recruits was generally higher in disturbed plots. Further, eight species of plants recruited only on disturbed plots.

What role did differences in seed size play in the rate of recruitment by different species? Jakobsson and Eriksson calculated recruitment success in various ways. One of the most basic ways was by dividing the total number of recruits by the total number of seeds of a species that they planted, giving the

proportion of seeds sown that produced recruits. While 45 of 50 species established new recruits in the experimental plots, the rate at which they established varied widely among species from approximately 5% to nearly 90%. Jakobsson and Eriksson found that differences in seed size explained much of the observed differences in recruitment success among species (fig. 12.10). On average, larger seeds, which produce larger seedlings, were associated with a higher rate of recruitment. Therefore, it appears that by investing more energy into a seed, the maternal plant increases the probability that the seed will successfully establish itself as a new plant. This advantage associated with large seed size is probably very important in environments such as the grasslands studied by Jakobsson and Eriksson, where competition with established plants is likely to be high.

Jakobsson and Eriksson focused their work on grasslands where the principal growth forms were, using the classification presented in figure 12.8a, graminoid or forbs. However, as shown in figure 12.8a, woody plants and vines produce substantially larger seeds than herbaceous graminoids and forbs. How might patterns in seed and seedling size vary among woody plants? Kenji Seiwa and Kihachiro Kikuzawa (1991) studied the relationship between seed size and seedling size among tree species native to Hokkaido, the northernmost large island of Japan. The results of their work and their interpretation of the results provide clear insights into how seed size may improve the ability of seedlings to survive environmental hazards. Seiwa and Kikuzawa were especially focused on the influences of shade on seedling establishment.

The trees studied by Seiwa and Kikuzawa were all broad-leaved deciduous trees that grow in the temperate deciduous forests of Hokkaido, either on mountain slopes between 100 and 200 m in altitude or in riparian forests. The fruits of all the study species were collected from trees growing in the arboretum of the Hokkaido Forest Experimental Station. In the laboratory the research team removed any fruit pulp

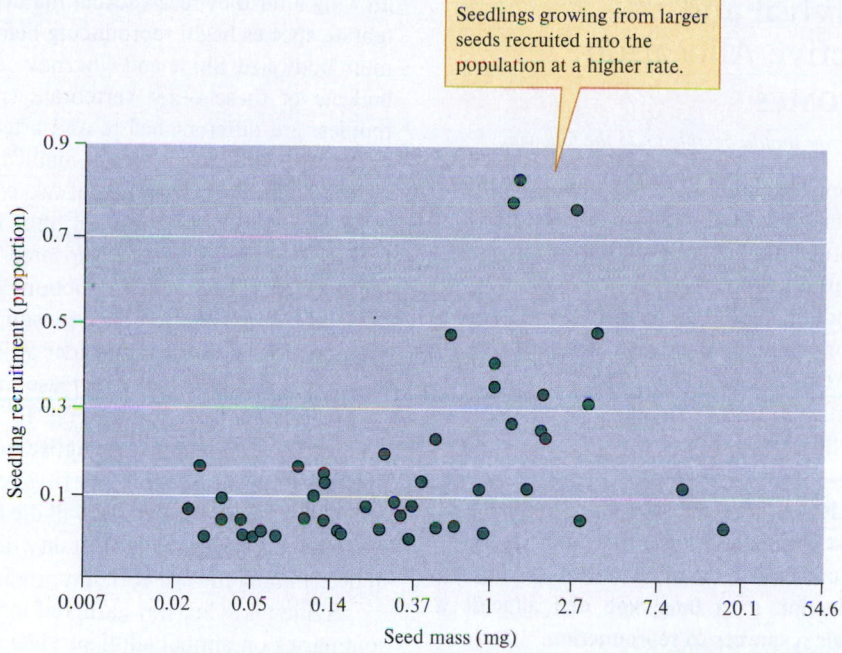

Figure 12.10 Seed mass and recruitment rates in grassland plants (data from Jakobsson and Eriksson 2000).

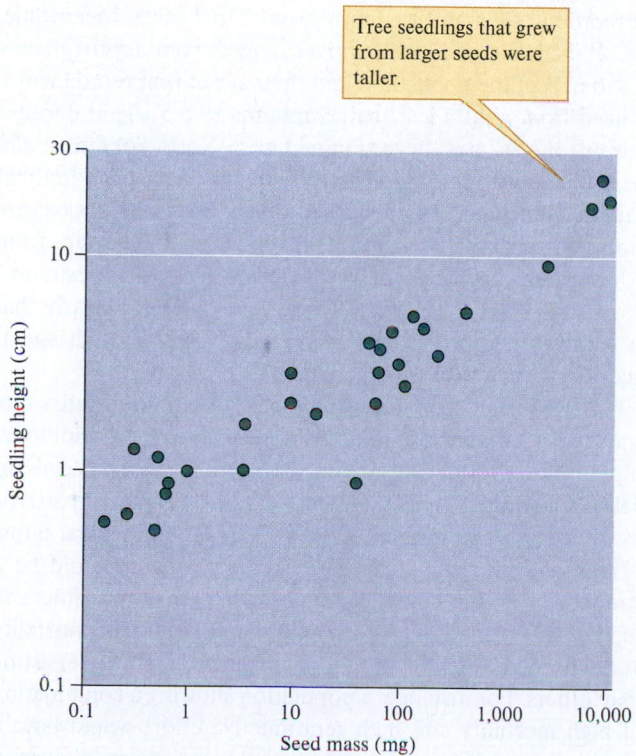

Figure 12.11 Relationship between seed mass and seedling height among trees (Seiwa and Kikuzawa 1991).

from the seeds, washed them, and then allowed them to air dry for 24 hours. Seiwa and Kikuzawa then estimated average seed mass by weighing one to five groups of 100 to 1,000 randomly chosen seeds. A week later they planted seeds at depths of 1 to 2 cm in a clay loam soil and watered, until the soil was saturated, three times a week.

Seiwa and Kikuzawa's results showed clearly that larger seeds produced taller seedlings (fig. 12.11). They explained this pattern as the result of the larger seeds providing greater energy reserves to boost initial seedling growth. Seiwa and Kikuzawa observed that seedlings from large-seeded species unfolded all of their leaves rapidly in the spring and shed all of their leaves synchronously in the autumn. They concluded that this timing allows the seedlings from large-seeded species to emerge early in the spring before the trees forming the canopy of the forest have expanded their leaves and have shaded the forest floor. Seiwa and Kikuzawa also pointed out that rapid growth would help seedlings penetrate the thick litter layer on the floor of deciduous forests and help them establish themselves as part of the forest understory.

In addition to showing variation in the number and sizes of offspring, organisms show a great deal of variation in the age at which they begin reproducing. They also differ greatly in the relative amount of energy they allocate to reproduction versus growth and maintenance.

Concept 12.1 Review

1. Why did Westoby, Leishman, and Lord (1996) include five floras on three continents in their study?
2. Why did Jakobsson and Eriksson (2000) conduct their study of the relationship between seed size and seedling size in a greenhouse?
3. Large darter species produce larger numbers of smaller eggs compared to smaller darter species (see figs. 12.3 and 12.4). Consequently, would you expect to find more genetic differences along the length of a river system among small darters or large darters? (Hint: Consider fig. 12.5.)

12.2 Adult Survival and Reproductive Allocation

LEARNING OUTCOMES

After studying this section you should be able to do the following:

12.7   Describe the connection between reproductive effort and the gonosomatic index.

12.8   Outline the relationship between adult mortality and the timing of reproduction.

12.9   Explain how finding parallel patterns in a variety of organisms, for example, lizards, snakes, and fish, strengthens our confidence in life history theory.

12.10   Discuss the contributions of studies on pumpkin-seed sunfish to life history theory.

Where adult survival is lower, organisms begin reproducing at an earlier age and invest a greater proportion of their energy budget into reproduction; where adult survival is higher, organisms defer reproduction to a later age and allocate a smaller proportion of their resources to reproduction.   Is there a relationship between the probability of an organism living from one year to the next and the age at which the organism begins reproducing? What environmental factors are responsible for variation in age at maturity and the amount of energy allocated to reproduction, which has been called **reproductive effort?** (Reproductive effort is the allocation of energy, time, and other resources to the production and care of offspring.) These are two questions central to life history ecology.

Reproductive effort generally involves trade-offs with other needs of the organism, including allocation to growth and maintenance. Because of these trade-offs, allocation to reproduction may reduce the probability that an organism will survive. However, delaying reproduction also involves risk. An individual that delays reproduction runs the risk of dying before it can reproduce. Consequently, evolutionary ecologists have predicted that variation in mortality rates among adults will be in association with variation in the age of first reproduction, or age of reproductive maturity. Specifically, they have predicted that where adult mortality is higher, natural selection will favor early reproductive maturity; and where adult mortality is low, natural selection has been expected to favor delaying reproductive maturity.

Life History Variation Among Species

The relationship among mortality, growth, and age at first reproduction or reproductive maturity has been examined in a large number of organisms. Early work, which concentrated on fish, shrimp, and sea urchins, suggested linkages between mortality or survival, growth, and reproduction. Richard Shine and Eric Charnov (1992) explored life history variation among snakes and lizards to determine whether generalizations developed through studies of fish and marine invertebrates could be extended to another group of animals living in very different environments.

Shine and Charnov began their presentation with a reminder that, in contrast to most terrestrial arthropods, birds, and mammals, including humans, many animals continue growing after they reach sexual maturity. In addition, most vertebrate species begin reproducing before they reach their maximum body size. Shine and Charnov pointed out that the energy budgets of these other vertebrate species, such as fish and reptiles, are different before and after sexual maturity. Before these organisms reach sexual maturity, energy acquired by an individual is allocated to one of two competing demands: maintenance and growth. However, after reaching sexual maturity, limited energy supplies are allocated to three functions: maintenance, growth, and reproduction. Because they have fewer demands on their limited energy supplies, individuals delaying reproduction until they are older will grow faster and reach a larger size. Because of the increase in reproductive rate associated with larger body size (see fig. 12.3), deferring reproduction would lead to a higher reproductive rate. However, again, where mortality rates are high, deferring reproduction increases the probability that an individual will die before reproducing. These relationships suggest that mortality rates will play a pivotal role in determining the age at first reproduction.

Shine and Charnov gathered information from published summaries on annual adult survival and age at which females mature for several species of snakes and lizards. The annual rate of adult survival among snakes in their data set ranged from approximately 35% to 85% of the population, while age at reproductive maturity ranged from 2 to 7 years. Meanwhile, the annual rate of lizard survival ranged from approximately 8% to 67% of the population and their age at first reproduction ranged from a little less than 8 months to 6.5 years. Because most of the species they examined were North American and were members of either one family of snakes or one family of lizards, Shine and Charnov urged that their results not be generalized to snakes and lizards generally until other groups from other regions had been analyzed. Regardless of these cautions, the results of Shine and Charnov's study showed clearly that as survival of adult lizards and snakes increases, their age at maturity also increases (fig. 12.12*a*).

Analyses of the relationship between adult mortality rate and age at maturity among fish species provide additional support for the prediction that high adult survival leads to delayed maturity. Donald Gunderson (1997) explored patterns in adult survival and reproductive effort among several populations of fish. Gunderson suggested that there should be a strong relationship between adult mortality in populations and reproductive effort because some combinations of mortality and reproductive effort have a higher probability of persisting than others. For instance, a population showing a combination of high mortality and high reproductive effort would have a higher chance of persisting than one experiencing high mortality but allocating low reproductive effort. The population with this second combination would likely go extinct in a short period of time.

The life history information Gunderson summarized in his analysis included mortality rate, estimated maximum length, age at reproductive maturity, and reproductive effort. Gunderson estimated reproductive effort as each population's **gonadosomatic index,** or **GSI.** GSI was taken as the ovary weight of each species divided by the species body weight and adjusted for the number of batches of offspring produced by each species per year. For example, because the northern anchovy spawns three times per year, the weight of its ovary was multiplied by 3 for calculating its GSI. Meanwhile, the ovary weight for dogfish sharks, which reproduce only every other year, was divided by 2. However, since most of the species included in the analysis spawn once per year, their ovary weights required no adjustment for GSI calculations.

The fish included in Gunderson's analysis ranged in size from the Puget Sound rockfish, which reaches a maximum size of approximately 15 cm, to northeast Arctic cod, which reaches a length of 130 cm. The age at maturation among these fish species ranges from 1 year in northern anchovy populations to 23 years in dogfish shark populations. Like Shine and Charnov, Gunderson gathered information about the life histories of the fish in his analysis from previously published papers and several experts on particular fish species. In his table summarizing life history information for the 28 species included in his analysis, Gunderson lists 72 references. In contrast to Shine and Charnov, Gunderson provides estimates of mortality rates rather than survival rates. In addition, his estimates are of "instantaneous" mortality rates instead of annual rates. However, like Shine and Charnov, his results show a clear relationship between adult mortality and age of reproductive maturity (fig. 12.12b). These results support the idea that natural selection has acted to adjust age at reproductive maturity to rates of mortality experienced by populations.

Gunderson's analysis also gives information on variation in reproductive effort among species. His calculations of a gonadosomatic index, or GSI, for each of the 28 species included in the analysis spanned more than a 30-fold difference from a value of 0.02 for the rougheye rockfish to 0.65 for the northern anchovy. When Gunderson plotted GSI against mortality rates (fig. 12.13), the results supported the prediction from life history theory that species with higher mortality would show higher relative reproductive effort.

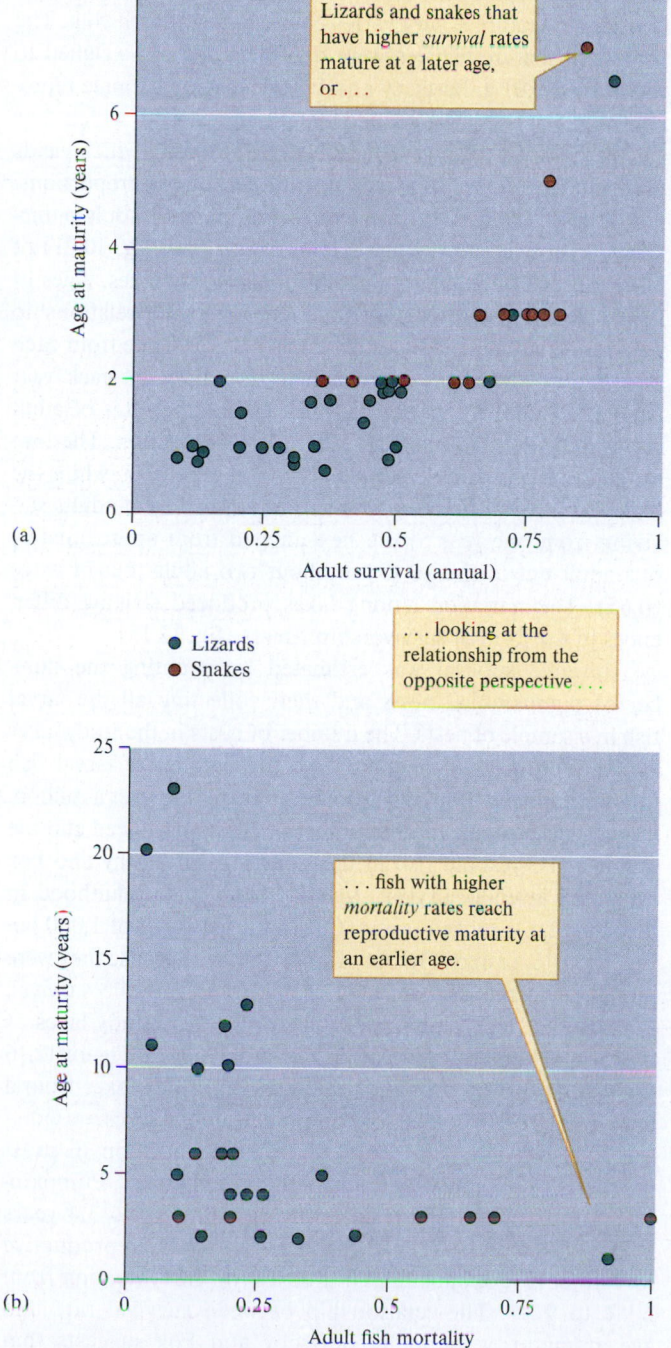

(a)

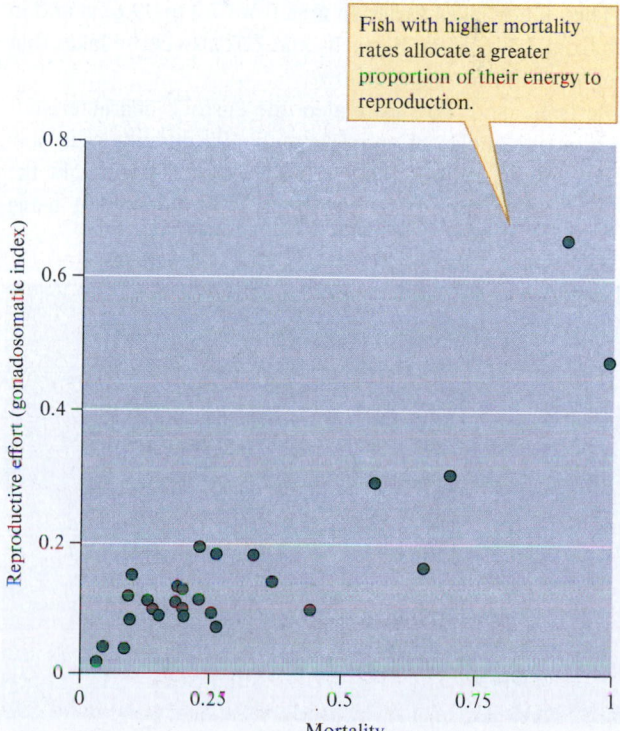

Figure 12.12   Relationship between age at maturity and (a) adult survival among lizards and snakes and (b) adult fish mortality (data from Shine and Charnov 1992; Gunderson 1997).

Figure 12.13   Relationship between adult fish mortality and reproductive effort as measured by the gonadosomatic index, or GSI (data from Gunderson 1997).

Life History Variation within Species

To this point in our discussion, we have emphasized life history differences between species, such as the lizard and snake species compared by Shine and Charnov (see fig. 12.12*a*) or the fish species compared by Gunderson (see fig. 12.12*b*). Is there evidence that life history differences will evolve within species, where different populations experience different rates of adult mortality? The data set analyzed by Shine and Charnov included nine populations of the eastern fence lizard, *Sceloporus undulatus.* Variation among those populations indicates that age at maturity within lizard populations increases with increased adult survival. Additional evidence for the evolution of such intraspecific differences comes from a comparative study of several populations of the pumpkinseed sunfish, *Lepomis gibbosus* (fig. 12.14).

Kirk Bertschy and Michael Fox (1999) studied the influence of adult survival on pumpkinseed sunfish life histories. One of the major objectives of their study was to test the prediction by life history theory that increased adult survival, relative to juvenile mortality, favors delayed maturity and reduced reproductive effort. Their study goal was to explain the evolution of life history variation within a species.

Bertschy and Fox selected five populations of pumpkinseed sunfish living in 5 lakes from a group of 27 lakes in southern Ontario, Canada. Fox had previously studied the pumpkinseed sunfish living in these lakes and so they had a considerable basis for choosing study populations. Bertschy and Fox chose lakes that were similar in area and depth and small enough that they had a reasonable chance of estimating mortality rates and variation in other life history characteristics. Their study lakes varied in area from 7.2 to 39.6 ha and in depth from 2.6 to 11 m. Bertschy and Fox also chose lakes that had no major inflows or outflows.

Bertschy and Fox estimated life history characteristics from annual samples of approximately 100 pumpkinseed sunfish taken from each of the five study lakes. They caught the fish in their shallow (0.5–2 m depth) littoral habitat using

Figure 12.14   Male pumpkinseed sunfish, *Lepomis gibbosus,* build their nests in the shallows of lakes and ponds. They guard their nests against intrusions by other males and attempt to attract females of their species to deposit eggs within them.   MattiaATH/Shutterstock.com

funnel traps and beach seine nets. Bertschy and Fox took their annual population sample in late May or early June just before the beginning or right at the beginning of the spawning season. The individuals caught were sacrificed by placing them in an ice slurry and then freezing them for later analysis. They made several measurements on each individual in their samples, including its age (by counting annual rings in scales), weight (to the nearest 0.1 g), length (in mm), sex, and reproductive status. Because female reproductive effort is largely restricted to egg production whereas male reproductive effort includes activities such as territory guarding and nest building, Bertschy and Fox studied reproductive traits in females only. A female was considered mature if her ovaries contained eggs with yolk. The ovaries of mature females were dissected out and weighed to the nearest 0.01 g. Bertschy and Fox represented female reproductive effort using the gonadosomatic index, GSI, which they calculated as 100 × (ovary mass) ÷ (body mass), which yields GSI values expressed as percentages rather than as proportions.

Bertschy and Fox estimated the number of adult pumpkinseed sunfish and the age structure (e.g., see fig. 10.21) of pumpkinseed populations in each of the study lakes. Ages of fish were estimated from their length using the relationship between length and age of individuals of known age from each population. These surveys, which were conducted each year from 1992 to 1994, gave a basis for estimating rates of adult survival for each age class in each lake's population. The lowest rate, or probability, of adult survival was 0.19, while the highest was 0.65. In other words, the proportion of adults surviving from one year to the next ranged from approximately one adult out of five (0.19) to about two adults out of three (0.65). This variation among lakes produced striking differences in the form of survivorship curves (fig. 12.15).

Juvenile survival was estimated by counting the number of pumpkinseed nests and then collecting all the larval fish in a sample of nests. The number of nests in the study lakes varied from 60 to over 1,000, and the number of larval fish produced ranged from approximately 100,000 to over a million. Using their estimate of the number of larvae produced and the number of 3-year-old fish in the same lake, Bertschy and Fox estimated juvenile survival. Juvenile survival to adulthood in the study lakes ranged from 0.004, or about 4 out of 1,000 larvae, to 0.016, or about 16 out of 1,000 larvae. Because they were interested in the relative rates of adult and juvenile survival, Bertschy and Fox represented survival in their study lakes as the ratio of adult to juvenile survival probabilities. Figure 12.16 shows that this ratio ranged widely among study lakes from a low of 10.6 to 116.8, a tenfold difference among lakes.

Bertschy and Fox found significant variation in most life history characteristics across their study lakes. Pumpkinseed sunfish matured at ages ranging from 2.4 to 3.4 years in the different study lakes, and they showed reproductive investments (gonadosomatic indexes, or GSI) ranging from 6.9% to 9.3%. The relationship between survival rate and age at maturity found by Bertschy and Fox suggests that populations with higher adult survival mature at a greater age (fig. 12.17). The correlation between survival rate and age at maturity was not high enough to be statistically

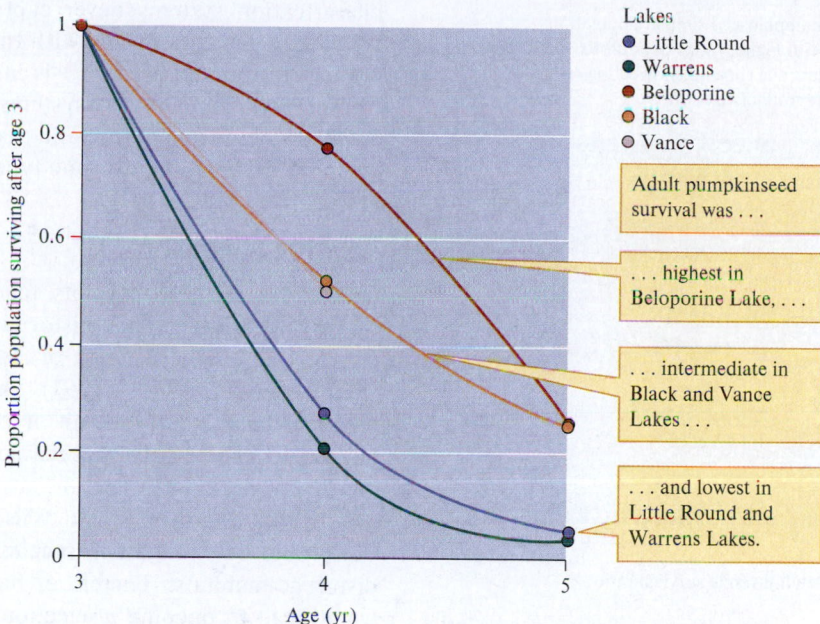

Figure 12.15   Pumpkinseed sunfish survival after age 3 years in five small lakes (data from Bertschy and Fox 1999).

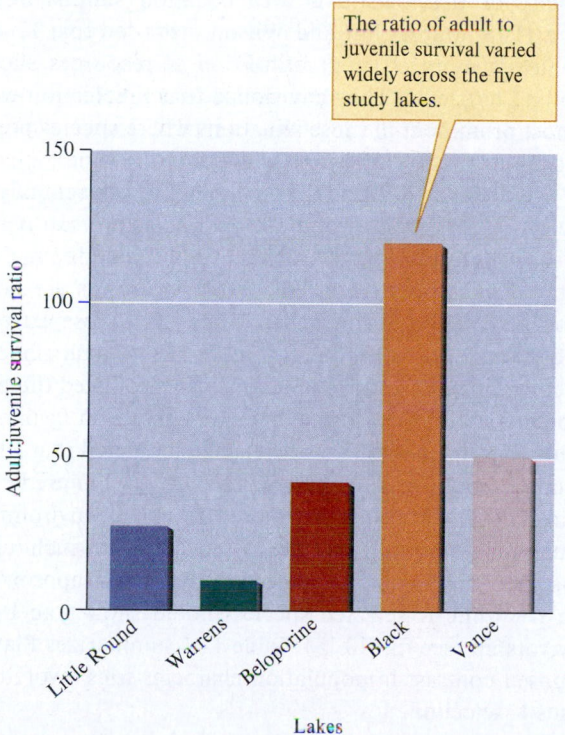

Figure 12.16   Ratio of adult to juvenile survival in pumpkinseed sunfish populations in five small lakes (data from Bertschy and Fox 1999).

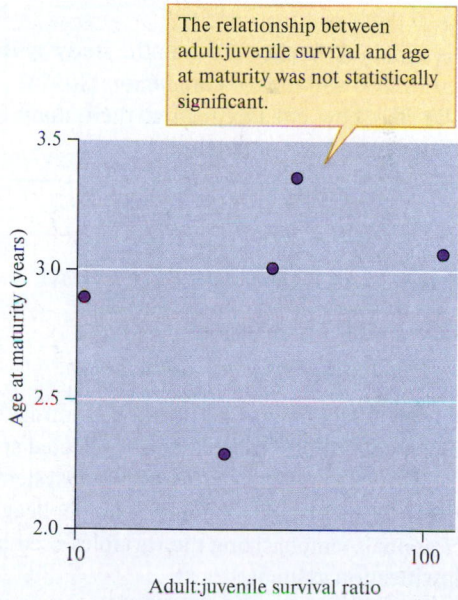

Figure 12.17   Adult:juvenile survival ratios and age at reproductive maturity in populations of pumpkinseed sunfish (data from Bertschy and Fox 1999).

significant; however, the relationship between adult survival and reproductive effort was very clear and highly significant (fig. 12.18). The patterns of life history variation across the pumpkinseed populations studied by Bertschy and Fox support the theory that, where adult survival is lower relative to juvenile survival, natural selection will favor allocating greater resources to reproduction.

As we explored the relationship between offspring size and number and the influence of mortality on the timing of maturation and reproductive effort, we've accumulated a large body of information on life histories. Let's step back now and try to organize that information to make it easier to think about life history variation in nature. Several researchers have proposed classification systems for life histories.

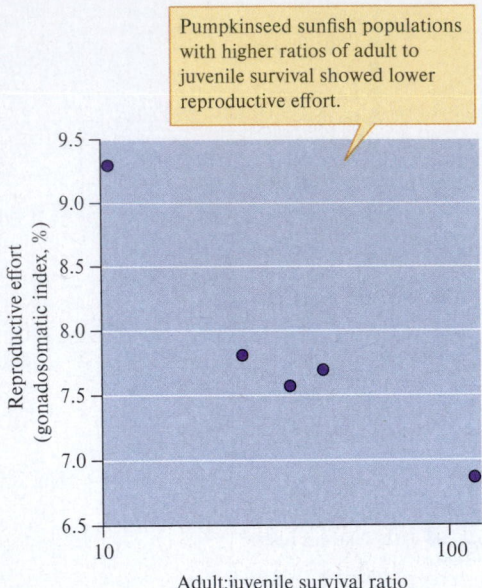

Pumpkinseed sunfish populations with higher ratios of adult to juvenile survival showed lower reproductive effort.

Figure 12.18   Adult:juvenile survival ratio and reproductive effort as measured by the gonadosomatic index, or GSI (data from Bertschy and Fox 1999).

Concept 12.2 Review

1. What do the GSI values for rougheye rockfish, 0.02, and northern anchovy, 0.65, mean in terms of the body weights of these two fish species?
2. What is a main difference between the study by Bertschy and Fox (1999) and that of Gunderson (1997)?
3. Why did Bertschy and Fox restrict their study to lakes without major inflows or outflows?

12.3 Life History Classification

LEARNING OUTCOMES

After studying this section you should be able to do the following:

12.11 Define *r* selection and *K* selection. Contrast the characteristics of *r* selected and *K* selected species.

12.12 Compare the life history classification systems of Grime for plants and Winemiller and colleagues for animals, emphasizing the variables used in the classification systems.

12.13 Explain the opportunities to life history research opened up by Charnov's life history classification, which removes the influences of size.

12.14 Describe how resource availability might influence the relationship between patterns of life history traits and evolutionary/ecological trade-offs.

The great diversity of life histories can be classified on the basis of a few population characteristics. Examples include fecundity or number of offspring, survival, relative offspring size, and age at reproductive maturity. While

classification systems never capture the full diversity of nature, they make working with the often bewildering variety of nature much easier. It is important to bear in mind when using classification systems, however, that they are an abstraction from nature and that most species fall somewhere in between the extreme types.

r and K Selection

One of the earliest attempts to organize information on the great variety of life histories that occur among species was under the heading of *r* selection and K selection (MacArthur and Wilson 1967). The term **r selection,** which refers to the per capita rate of increase, *r,* which we calculated in chapter 10, was defined by Robert MacArthur and E. O. Wilson as selection favoring a higher population growth rate. MacArthur and Wilson suggested that *r* selection would be strongest in species often colonizing new or disturbed habitats. Therefore, high levels of disturbance would lead to ongoing *r* selection. MacArthur and Wilson contrasted *r* selected species with those subjects mainly to K selection. The term **K selection** refers to the carrying capacity of the logistic growth equation summarized in figure 11.13. MacArthur and Wilson proposed that K selection favors more efficient utilization of resources such as food and nutrients. They envisioned that K selection would be most prominent in those situations where species populations are near carrying capacity much of the time.

Eric Pianka (1970, 1972) developed the concept of *r* and K selection further in two important papers. Pianka pointed out that *r* selection and K selection are the endpoints on a continuous distribution and that most organisms are subject to forms of selection somewhere in between these extremes. In addition, he correlated *r* and K selection with attributes of the environment and of populations. He also listed the population characteristics that each form of selection favors. Following MacArthur and Wilson, Pianka predicted that while *r* selection should be characteristic of variable or unpredictable environments, fairly constant or predictable environments should create conditions for K selection. In such conditions survivorship among *r* selected species will approximate type III, while K selected species should show type I or II survivorship (see fig. 10.18). Table 12.1 summarizes Pianka's proposed contrast in population characteristics favored by *r* versus K selection.

Pianka's detailed analysis clarified the sharp contrast between the two selective extremes represented by *r* and K selection by revealing biological details. The most fundamental contrasts are, of course, between intrinsic rate of increase, r_{max}, which should be highest in *r* selected species, and competitive ability, which should be highest among K selected species. In addition, according to Pianka, development should be rapid under *r* selection and relatively slow under K selection. Meanwhile, early reproduction and smaller body size will be favored by *r* selection, while K selection favors later reproduction and larger body size. Pianka predicted that reproduction under *r* selection will tend toward a single reproductive event in

Table 12.1

Characteristics favored by *r* versus K selection		
Population attribute	*r* selection	K selection
Intrinsic rate of increase, r_{max}	High	Low
Competitive ability	Not strongly favored	Highly favored
Development	Rapid	Slow
Reproduction	Early	Late
Body size	Small	Large
Reproduction	Single, semelparity	Repeated, iteroparity
Offspring	Many, small	Few, large

Source: After Pianka 1970.

Figure 12.19   The deer mouse and the African elephant represent extremes among mammals of *r* versus K selection.
(*a*) Source: James Gathany/CDC; (*b*) imagebroker/Alamy

which many small offspring are produced. This type of reproduction, which is called **semelparity,** occurs in organisms such as annual weeds and salmon. In contrast, K selection should favor repeated reproduction, or **iteroparity,** of fewer larger offspring. Iteroparity, which spaces out reproduction over several reproductive periods during an organism's lifetime, is the type of reproduction seen in most perennial plants and most vertebrate animals. Pianka's contrast may be restated as "small and fast," analogous to *r* selected species, with ones that are "large and slow," analogous to K selected species (fig. 12.19).

The ideas of *r* and K selection helped greatly as ecologists and evolutionary biologists attempted to think more systematically about life history variation and its evolution. However, ecologists who found that the dichotomy of *r* versus K did not include a great deal of known variation in life histories have proposed alternative classifications.

Plant Life Histories

J. P. Grime (1977, 1979) proposed that variation in environmental conditions has led to the development of distinctive strategies or life histories among plants. The two variables that he selected as most important in exerting selective pressure on plants were the intensity of disturbance and the intensity of stress. Grime contrasted four extreme environmental types, which he characterized by combinations of disturbance intensity and stress intensity. Four environmental extremes envisioned by Grime were (1) low disturbance–low stress, (2) low disturbance–high stress, (3) high disturbance–low stress, and (4) high disturbance–high stress. Drawing on his extensive knowledge of plant biology, Grime suggested that plants occupy three of his theoretical environments but that there is no viable strategy among plants for the fourth environmental combination, high disturbance–high stress.

Grime next described plant strategies, or life histories, that match the requirements of the remaining three environments.

His strategies were ruderal, stress-tolerant, and competitive (fig. 12.20). **Ruderals** are plants that live in highly disturbed habitats and that may depend on disturbance to persist in the face of potential competition from other plants. Grime summarized several characteristics of ruderals that allow them to persist in habitats experiencing frequent and intense **disturbance,** which he defined as any mechanisms or processes that limit plants by destroying plant biomass. One of the characteristics of ruderals is their capacity to grow rapidly and produce seeds during relatively short periods between successive disturbances. This capacity alone would favor persistence of ruderals in the face of frequent disturbance. In addition, however, ruderals also invest a large proportion of their biomass in reproduction, producing large numbers of seeds that are capable of dispersing to new habitats made available by disturbance. The term *ruderal* is sometimes used synonymously with the term *weed*. Animals associated with disturbance, like weedy plants, have high reproductive rates and are good colonists, too. Such animals are also sometimes referred to as ruderals.

Grime (1977) began his discussion of the second type of plant life history, stress-tolerant, with a definition of **stress** as "external constraints which limit the rate of dry matter production of all or part of the vegetation." In other words, stress is

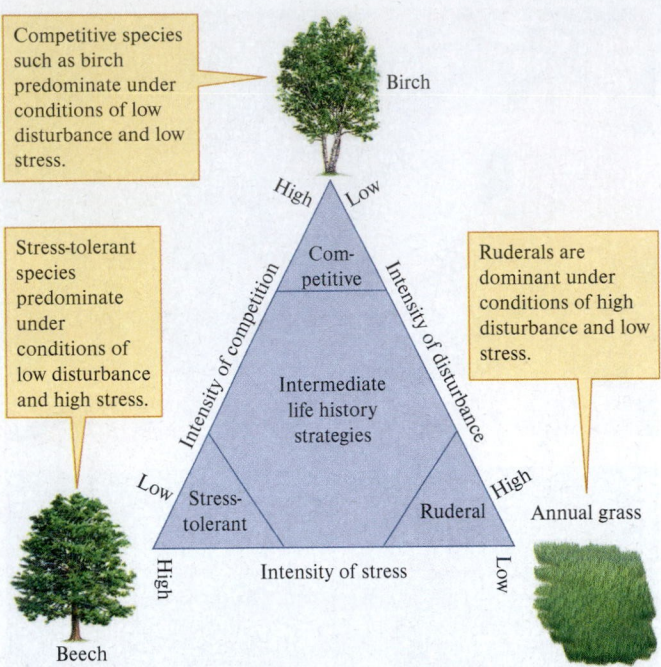

Competitive species such as birch predominate under conditions of low disturbance and low stress.

Birch

High   Low

Stress-tolerant species predominate under conditions of low disturbance and high stress.

Ruderals are dominant under conditions of high disturbance and low stress.

Intensity of competition

Intensity of disturbance

Com-
petitive

Intermediate
life history
strategies

Low   High

Stress-
tolerant   Ruderal   Annual grass

High   Intensity of stress   Low

Beech

Figure 12.20 Grime's classification of plant life history strategies (after Grime 1979).

induced by environmental conditions that limit the growth of all or part of the vegetation. What environmental conditions might create such constraints? Our discussions in chapters 5 to 7, where we considered temperature, water, and energy and nutrient relations, provide several suggestions. Stress is the result of extreme temperatures, high or low, extreme hydrologic conditions, too little or too much water, or too much or too little light or nutrients. Because different species are adapted to different environmental conditions, the absolute levels of light, water, temperature, and so forth that constitute stress will vary from species to species. In addition, conditions that induce stress will vary from biome to biome. For instance, the amount of precipitation leading to drought stress is different in rain forest and desert, or the minimum temperatures inducing thermal stress are different in tropical forest compared to boreal forest.

The important point that Grime made, however, was that in every biome, some species are more tolerant to the environmental extremes that occur. These are the species that he referred to as "stress-tolerant." **Stress-tolerant plants** are those that live under conditions of high stress but low disturbance. Grime proposed that, in general, stress-tolerant plants grow slowly; are evergreen; conserve fixed carbon, nutrients, and water; and are adept at exploiting temporary favorable conditions. In addition, stress-tolerant plants are often unpalatable to most herbivores. Because stress-tolerant species endure some of the most difficult conditions a particular environment has to offer, they are there to take advantage of infrequent favorable periods for growth and reproduction.

The third plant strategy proposed by Grime, the competitive strategy, is in many respects intermediate between the ruderal strategy and the stress-tolerant strategy. In Grime's classification, **competitive plants** occupy environments where disturbance intensity is low and the intensity of stress is also low. Under conditions of low stress and low disturbance, plants have the potential to grow well. As they do so, however, they eventually compete with each other for resources, such as light, water, nutrients, and space. Grime's model predicts that the plants living under such circumstances will be selected for strong competitive abilities.

How does Grime's system of classification compare with the r and K selection contrast proposed by MacArthur and Wilson and Pianka? Grime proposed that r selection corresponds to his ruderal strategy or life history, while K selection corresponds to the stress-tolerant end of his classification. Meanwhile, he placed the competitive life history category in a position intermediate between the extremes represented by r selection and K selection. However, while attempting this reconciliation of the two classifications, Grime suggested that a linear arrangement of life histories, with r selection and K selection occupying the extremes, fails to capture the full variation shown by organisms. He suggested that more dimensions are needed and, of course, Grime's triangular arrangement (see fig. 12.20) adds another dimension. The factors varying along the edges of Grime's triangle are intensity of disturbance, stress, and competition. Other ecologists have also recognized the need for more dimensions in representing life history diversity.

Opportunistic, Equilibrium, and Periodic Life Histories

In a review of life history patterns among fish, Kirk Winemiller and Kenneth Rose (1992) proposed a classification of life histories based on some of the aspects of population dynamics that we reviewed in chapter 10. They drew particular attention to survivorship especially among juveniles, l_x, fecundity or number of offspring produced, m_x, and generation time, T, or age at maturity, α. Table 10.2 summarized the relationship among these variables. While the analysis by Winemiller and Rose overlaps the analyses of Pianka and Grime, their system adds coherence to life history classification by its linkage to fundamental elements of population ecology, l_x, m_x, and α.

Winemiller and Rose start—as we began chapter 12—with the concept of trade-offs. Their trade-offs are among fecundity, survivorship, and age at reproductive maturity. Using variation in fish life histories as a model, Winemiller and Rose proposed that life histories should lie on a semi-triangular surface as shown in figure 12.21. They called the three endpoints on their surface "opportunistic," "equilibrium," and "periodic" life histories. The opportunistic strategy, by combining low juvenile survival, low numbers of offspring, and early reproductive maturity, maximizes colonizing ability across environments that vary unpredictably in time or space. It is important to keep in mind, however, that while

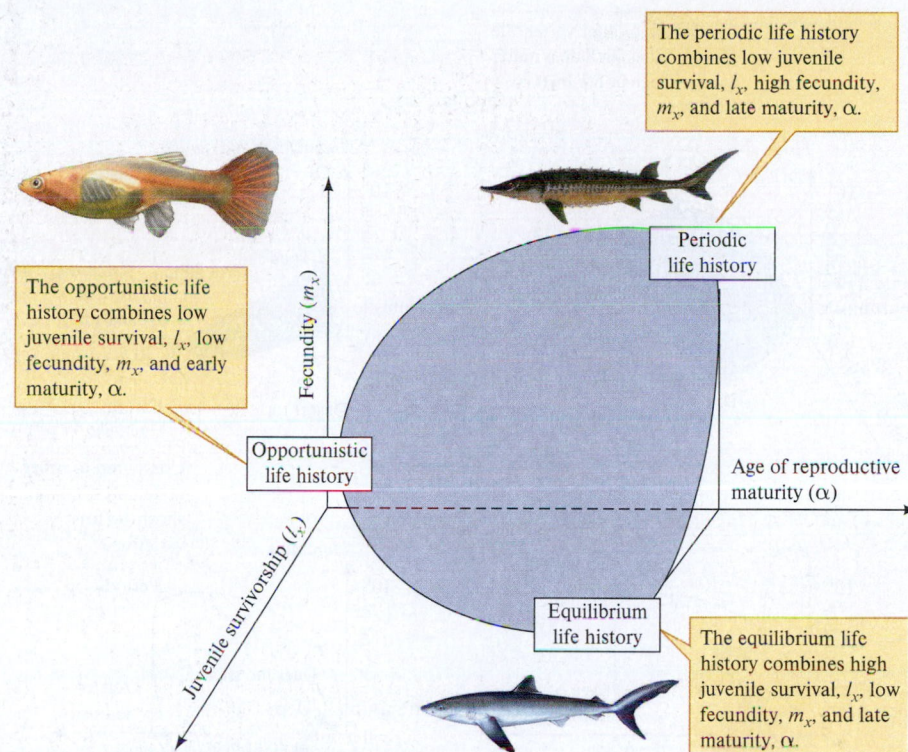

The periodic life history combines low juvenile survival, l_x, high fecundity, m_x, and late maturity, α.

The opportunistic life history combines low juvenile survival, l_x, low fecundity, m_x, and early maturity, α.

The equilibrium life history combines high juvenile survival, l_x, low fecundity, m_x, and late maturity, α.

Periodic life history

Opportunistic life history

Equilibrium life history

Fecundity (m_x)

Age of reproductive maturity (α)

Juvenile survivorship (l_x)

Figure 12.21   Classification of life histories based on juvenile survival, l_x, fecundity, m_x, and age at reproductive maturity, α (after Winemiller and Rose 1992).

the absolute reproductive output of opportunistic species may be low, the percentage of their energy budget allocated to reproduction is high. Winemiller and Rose's equilibrium strategy combines high juvenile survival, low numbers of offspring, and late reproductive maturity. Finally, the periodic strategy combines low juvenile survival, high numbers of offspring, and late maturity. Among fish, periodic species tend to be large and produce numerous small offspring. By producing large numbers of offspring over a long life span, periodic species can take advantage of infrequent periods when conditions are favorable for reproduction.

It is difficult to map the exact correspondence of Winemiller and Rose's classification of life history strategies to either the r–K continuum of MacArthur and Wilson and Pianka or the triangular classification of plant life histories developed by Grime. For instance, opportunistic species share characteristics with r selected and ruderal species. However, opportunistic species differ from the typical r selected species because they tend to produce small clutches of offspring. The equilibrium strategy, which combines production of high juvenile survival, low numbers of offspring, and late reproductive maturity, approaches the characteristics of typical K selected species. Winemiller and Rose point out, however, that many fish classified as "equilibrium" are small, while typically K selected species tend toward large body size (see table 12.1). Periodic species are not captured by the linear r to K selection gradient. Meanwhile, the periodic and equilibrium species in Winemiller

and Rose's classification share some characteristics with Grime's stress-tolerant and competitive species but differ in other characteristics.

Thus far in this review of systems for life history classification, we have focused on just three of the many that have been proposed. Even with just these three, however, translation from one classification to another is difficult. What are the sources of these differences in perspective? One of the sources is that different ecologists have worked with different groups of organisms. While MacArthur and Wilson's system was built after years of work on birds and insects, respectively, Pianka had worked mainly with lizards. Grime's classification was built on and intended for plants. Finally, the perspective of Winemiller and Rose was influenced substantially by their work with fish. Because these ecologists worked with such different groups of organisms, it is not surprising that their classifications of life histories do not overlay precisely.

However, it may be that the analysis by Winemiller and Rose has laid the foundation for a more general theory of life histories. By basing their classification system on some of the most basic aspects of population ecology, l_x, m_x, and α, Winemiller and Rose (1992) established a common currency for representing and analyzing life history information for any organism. As a model for how such a translation might be done, Winemiller (1992) plotted the distributions of life history parameters of representative animal groups on their life history classification axes (fig. 12.22). By plotting life history

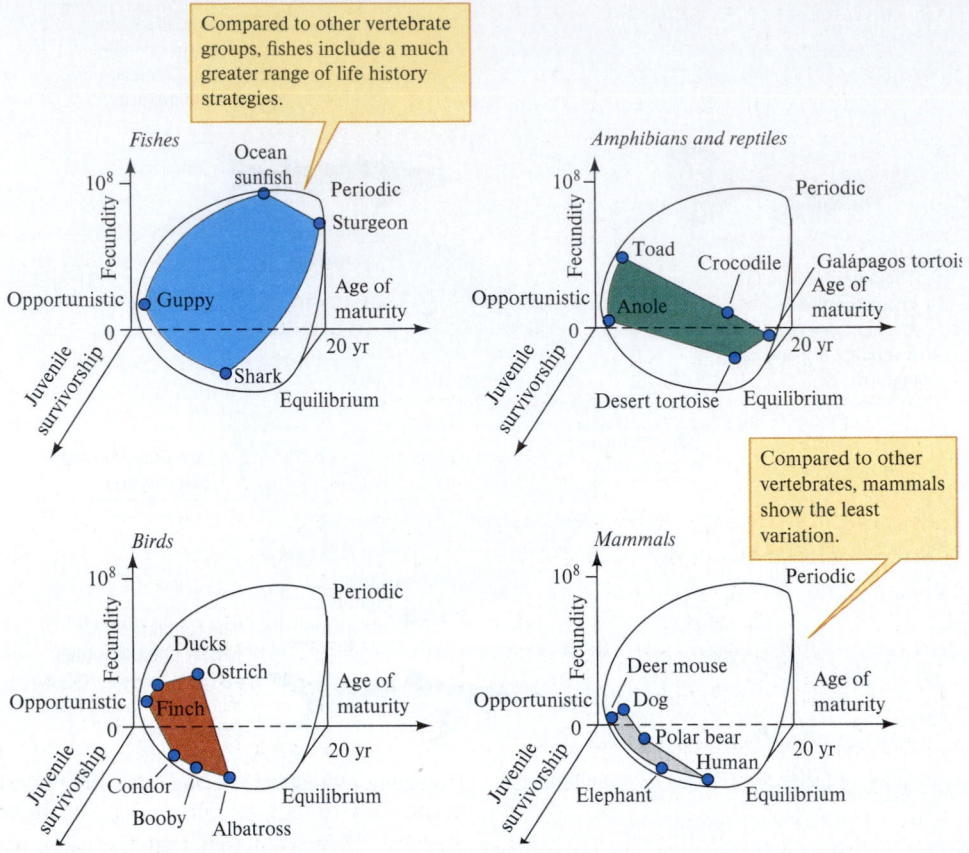

Figure 12.22   Variation in life histories within vertebrate animals (after Winemiller and Rose 1992).

variation among vertebrate groups on the same axes using the same variables, figure 12.22 demonstrates differences in the amount of life history variation between the groups. Notice that fish show the greatest variation and mammals the least, while birds and reptiles and amphibians include intermediate levels of variation.

Lifetime Reproductive Effort and Relative Offspring Size: Two Central Variables?

In response to the various attempts to classify life histories, Eric Charnov, with Robin Warne and Melanie Moses (2002, 2007), developed a new approach to life history classification. In his most recent attempts at life history classification, Charnov has focused his attention on mammals, altricial birds, and lizards. His goal has been to develop a classification free of the influences of size and time that would facilitate the exploration of life history variation within and among groups of closely related taxa. Why remove the influences of size and time? Our discussion of r and K selection underscored the relationship between size of organisms and timing of life history features (see table 12.1). The influences of size and timing are responsible for many of the obvious life history differences among species of closely related taxa, for instance, the differences among large and small mammal species, such as between a deer mouse and an African elephant (see fig. 12.19). By removing size and time effects, we may be able to more clearly detect life history differences among evolutionary lineages.

Charnov's approach was to take a few key life history features and convert them to dimensionless numbers. One of his variables was relative size of offspring. He created this dimensionless variable by dividing the mass of offspring at independence from the parent, I, by the adult mass at first reproduction, m. The result, I/m, is the size of offspring expressed as a proportion of adult body mass. While it is clear that an elephant is larger at independence than is a mouse at the same life stage, Charnov's approach allows us to determine whether one is relatively larger than the other. A young, newly independent mouse may represent as large a proportion of its parent's mass as a young elephant. The second measure was proportion of adult body mass allocated to reproduction per unit time, C, multiplied by the adult life span, E, which gives us an estimate of the fraction of adult body mass allocated to reproduction over a life span. As we have seen, higher reproductive effort is associated with shorter adult life span (see fig. 12.13), so Charnov reasoned their product might be similar for closely related taxa. Charnov chose these two dimensionless numbers (I/m and $C \cdot E$) for two particular reasons. First, R_0, the net reproductive rate (see chapter 10, section 10.5) is a measure of an individual's fitness in populations that are not growing. In addition, R_0 can be rewritten solely in terms of these two numbers and the chance of surviving to adulthood.

For his initial classification of life histories, Charnov chose three groups of well-studied organisms: mammals, lizards, and altricial birds. Altricial birds are those birds, ranging from sparrows to eagles, that are born helpless and depend entirely on parental care to mature to independence. One of the striking results of using Charnov's dimensionless analysis is that while there is little variation within mammals, lizards, or birds, there are substantial differences among these groups of animals. Figure 12.23 shows that the birds have the highest I/m (essentially 1, since they raise their young to adult size) and $C \cdot E$ values. In contrast, lizards and mammals share the *same* $C \cdot E$ value but differ a lot in I/m (0.1 vs. 0.3).

Previous classifications of life histories have revealed substantial variation within taxa, such as mammals and fish (see fig. 12.22). In contrast, Charnov's classification, by removing the influences of time and size, allows us to see both the great similarities within these groups as well as the nature of the substantial differences among them. Figure 12.24 shows the results of plotting the average values of I/m and $C \cdot E$ for mammals, lizards, and altricial birds. The striking separation

of these taxa within the two-dimensional plane of the graph suggests that mammals, lizards, and birds have life histories that are fundamentally different. But notice that lizards and mammals do not differ in the value of $C \cdot E$, only I/m.

Associations between traits are the foundation of the classification systems produced by Grime, Charnov, and others we have reviewed in this chapter. We have seen that patterns of traits among species often appear to be due to trade-offs, such as species not being able to both make many offspring and live a long time—there is only enough energy to do one or the other. However, it is important to note that there are many important exceptions. For example, the tamarisk tree (*Tamarix* spp.) is a large and long-lived organism that also produces many small offspring and is also highly stress-tolerant; in Grime's classification, it is a competitive, ruderal, and stress-tolerant species all in one (Sher 2013). How can this be?

Anurag Agrawal, a professor at Cornell and winner of the 2016 ESA MacArthur Award, pointed out that resource availability can play a very important role in the expression of life history strategies and other trait-associations, as illustrated in the "house-car tradeoff paradox" (Agrawal 2020). Because family money spent on a car cannot be also spent on a house, it would be logical to conclude that a family would own a house *or* a car, but not both. However, we know that in reality, house and car ownership are typically positively associated among families, due to differences in wealth. Agrawal points out that that the same associations can be true in natural systems: the trade-off between traits among

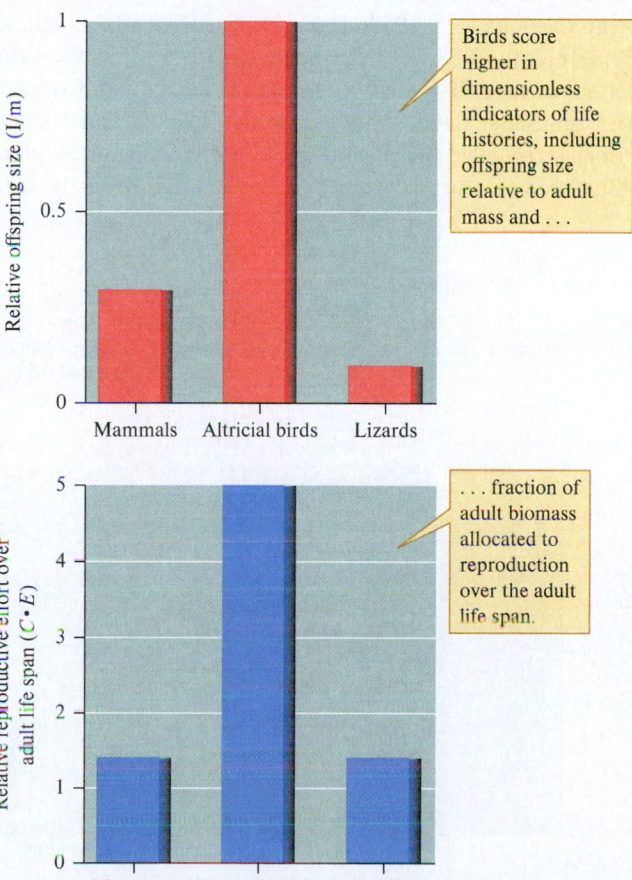

> Birds score higher in dimensionless indicators of life histories, including offspring size relative to adult mass and . . .

> . . . fraction of adult biomass allocated to reproduction over the adult life span.

Figure 12.23 Comparison of life history features of mammals, altricial birds, and lizards (data from Charnov 2002; Charnov, Warne, and Moses 2007).

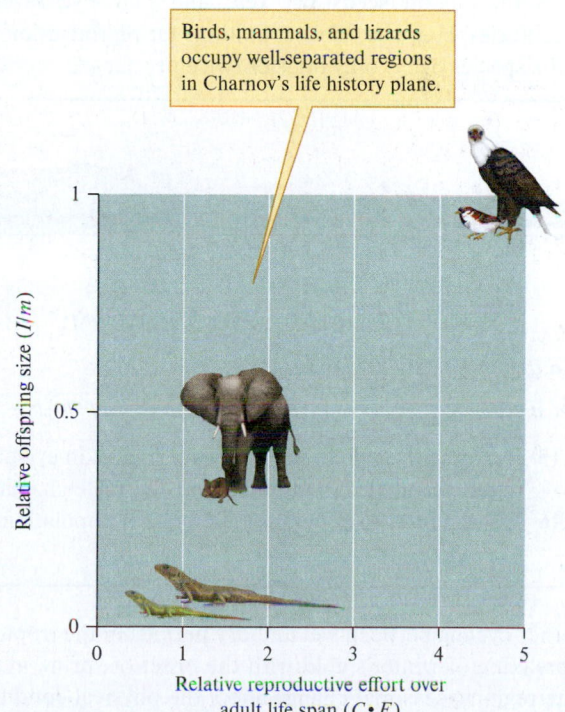

> Birds, mammals, and lizards occupy well-separated regions in Charnov's life history plane.

Figure 12.24 A planar classification of lizard, mammal, and altricial bird life histories based on two numbers, I/m and $C \cdot E$, reveals much more variation between than within taxa, although lizards and mammals share the same average $C \cdot E$ value (data from Charnov 2002; Charnov, Warne, and Moses 2007).

species will sometimes be obscured by differential access to resources. A species that is growing in a high-resource environment, or that is especially good at acquiring those resources, may express trait-combinations that are not possible under low resource environments. Thus, we can find seemingly paradoxical species like the tamarisk tree, which grows in high-resource floodplains and can rapidly utilize both light and water (Hultine and Dudley 2013).

The knowledge of species life histories revealed by the studies of life history ecologists has produced a subdiscipline of ecology rich in both theory and biological detail. In the challenges that lie ahead as we work to conserve endangered species, both theory and detailed knowledge of the life histories of individual species will be important. For instance, life history features, such as the timing of reproduction, have proved to be highly responsive to climate warming.

Concept 12.3 Review

1. If a concept, such as *r* and K selection, does not fully represent the richness of life history variation among species, can it still be valuable to science?
2. Where would you place the following plant species, in Grime's and in Winemiller and Rose's classifications of life histories (see figs. 12.20 and 12.21)? The plant species lives in an environment where it has access to plenty of water and nutrients but is subject to disturbance by flooding and wind. An average individual produces several million seeds per year and may live several centuries. However, ideal conditions for reproduction by the species occur only once or twice per decade.

Applications

Climate Change and Timing of Reproduction and Migration

LEARNING OUTCOMES

After studying this section you should be able to do the following:

12.15 Describe the role of phenological studies in evaluating the ecological consequences of climate change.
12.16 Outline responses of plant and animal populations to climate warming.

Whether cycling between wet and dry periods in the tropics or the breaking of winter's cold with the onset of spring in temperate regions, seasonal change alters the physical conditions to which organisms are exposed. Seasonal change also affects the availability of food and other resources on which organisms depend. In the face of such seasonal variation, timing is critical for successful reproduction. In chapter 10, we examined the influence of changing climate on the distributions of

organisms. Climate change is also altering the timing of life history events.

Written records show that people have been documenting seasonal events for centuries. For example, in 1736, Robert Marsham began keeping a record of the first date of flowering by selected plants, the first calls of frogs, and other phenomena, which he referred to as "indications of spring." Subsequent generations of Marsham's family continued his annual observations on their estate in eastern England for over two centuries. Although the duration of this effort is impressive, the record of cherry tree blooming dates at Kyoto, Japan, the occasion for a traditional festival, spans an incredible 1,200 years. When modern researchers (Arakawa 1956; Aono and Saito 2010) began connecting the timing of that blooming to climate, they entered the realm of **phenology.** Phenology is the study of the timing of ecological events, especially in relation to climate and weather, for example, the arrival of migratory birds on their breeding grounds or the date of flowering by a plant species, such as the cherry trees at Kyoto. The study of phenology has assumed greater significance as concern over contemporary climate change has grown.

Altered Plant Phenology

Henry David Thoreau, best known for his book *Walden,* started one of the most useful records of North American plant phenology in 1852. During the course of his studies from 1852 to 1858, Thoreau documented the first flowering dates for over 500 species of plants in the vicinity of Concord, Massachusetts. Later, Abraham Miller-Rushing and Richard Primack compiled the phenological records of Thoreau along with those of others covering the observation periods 1878,

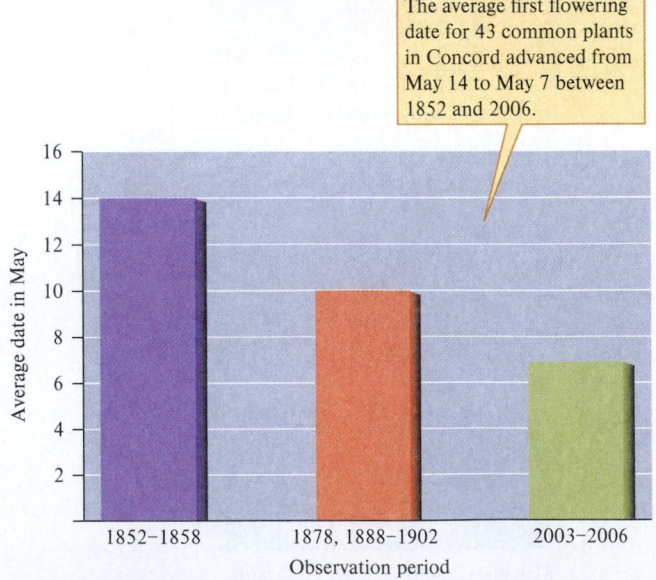

The average first flowering date for 43 common plants in Concord advanced from May 14 to May 7 between 1852 and 2006.

Figure 12.25 Advancing first flowering dates. As temperatures have warmed, plants in Concord, Massachusetts, have been blooming earlier in the year (data from Miller-Rushing and Primack 2008).

1888–1902, and 2003–06 (Miller-Rushing and Primack 2008). In all, there were 25 years of blooming records in the data set. Of the hundreds of species of plants in the record, Miller-Rushing and Primack chose 43 for which there were the most observations across the years of record. They used these data to test the hypothesis that changes in climate since Thoreau made his observations have altered the flowering phenology of plants in Concord.

Miller-Rushing and Primack found that, since 1852 when Thoreau began his observations, temperatures in Concord have increased and plants are blooming earlier in spring. Records from the Blue Hill Meteorological Observatory 33 km from Concord show that average annual temperatures have increased 2.5°C since 1852 and that the average temperatures in January, April, and May, which are highly correlated with blooming time, have increased by a similar amount: 2.3°C. Earlier blooming by the plants in Concord is correlated with this rise in temperature. The average date of first flowering by the 43 species in the study is now 7 days earlier than in 1852 (fig. 12.25). However, some of the species included in the analysis advanced their blooming much earlier in spring. For example, by 2006, highbush blueberry, *Vaccinium corymbosum,* was blooming 21 days earlier than in 1852, while yellow woodsorrel, *Oxalis stricta,* was blooming 32 days earlier.

Ecologists have combined phenological observations at single localities, such as at Concord, to test for phenological changes in plant populations across large regions. In one of the most ambitious studies, ecologists tested for phenological changes between 1971 and 2000 across 21 countries in Europe,

covering the area from the UK to Russia and from Spain to Norway. The study included over 500 plant species and over 100,000 time series of plant phenology each 15 years or more in length (Menzel et al. 2006). This massive effort showed that the average date of first flowering and leaf unfolding occurred 2.5 days earlier per decade and fruit ripening advanced nearly the same amount (fig. 12.26). In contrast, researchers documented little change in the timing of leaf coloring in autumn. Further analyses showed that the changes in plant phenology documented were clearly correlated with increases in the average temperatures of the months preceding a phenological event.

Animal Phenology

Animals are also adjusting the timing of life histories as the climate warms. For example, salmon are migrating from the ocean to freshwater earlier in the spring, and many species of butterflies and moths are increasing the number of generations they produce in a year. However, migratory birds likely offer the best-documented changes in phenology. One of the most commonly recorded migratory events has been the first arrival date of migratory birds on their breeding grounds, an event long noted by ornithologists and amateur birders.

Like first flowering date in plant populations in temperate regions, the first arrival date by migratory birds is now occurring earlier in the spring compared to historical records. Jessica Vitale and William Schlesinger analyzed a 123-year record (1885–2008) of first arrival dates for 44 species of birds migrating to Dutchess County in the Hudson River valley of

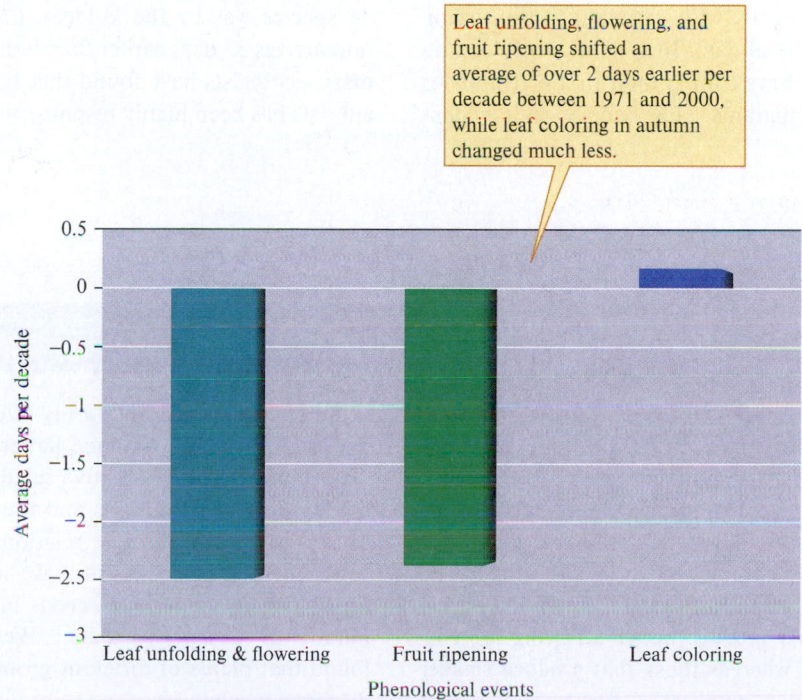

Leaf unfolding, flowering, and fruit ripening shifted an average of over 2 days earlier per decade between 1971 and 2000, while leaf coloring in autumn changed much less.

Figure 12.26   Changes in plant phenology in Europe. Climatic warming across Europe between 1971 and 2000 was accompanied by shifting of major plant phenological events to earlier in spring (data from Menzel et al. 2006).

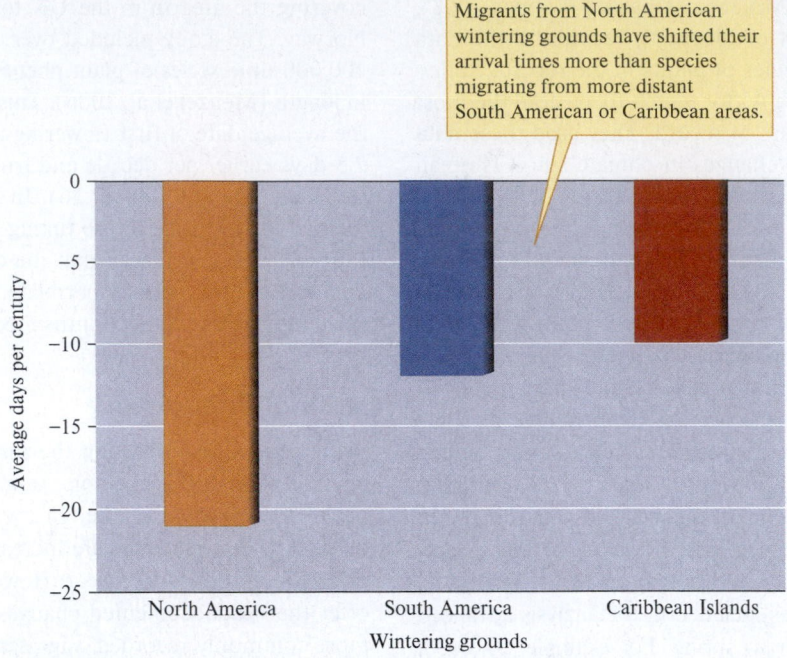

Figure 12.27 Migrating birds are arriving earlier in spring. A 123-year record of first arrival dates for birds migrating to Dutchess County, New York, shows a significant shift toward earlier average arrival dates (data from Vitale and Schlesinger 2011).

New York (Vitale and Schlesinger 2011). They calibrated shifts in migrant arrival times in days per century. Of the 44 bird species in the study, they found that 40 are now arriving significantly earlier. However, like shifts in flowering time observed in plant populations, there is significant variation among species. For example, different migration sources have been associated repeatedly with variation in the magnitude of changes in first arrival date. In general, birds migrating from greater distances have shifted their first arrival dates less than shorter-distance migrants. This association is evident in the Dutchess County bird migration record, where birds wintering in North America have shifted their first arrival dates an average of 21 days per century, while those wintering in South America or on Caribbean islands have shifted their arrivals an average of 12 and 10 days per century, respectively (fig. 12.27). The greatest shift in first arrival date among the 44 species was by the killdeer, *Charadrius vociferous,* which now arrives 53 days earlier than it did in the late 1800s. In summary, ecologists have found that the phenology of plants and animals has been highly responsive to recent climate change.

Summary

Life history consists of the adaptations of an organism that influence aspects of its biology, such as the number of offspring it produces, its survival, and its size and age at reproductive maturity. This chapter presents discussions on some of the central concepts of life history ecology.

Because all organisms have access to limited energy and other resources, there is a trade-off between the number and size of offspring; those that produce larger offspring are constrained to produce fewer, whereas those that produce smaller offspring may produce larger numbers. Turner and Trexler found that larger darter species produce larger numbers of eggs. Their results also support the generalization that there is a trade-off between offspring size and number. On average, darter species that produce larger eggs produce fewer eggs. They found a strong positive relationship between gene flow among darter populations and the number of eggs produced by females and a negative relationship between egg size and gene flow. Plant ecologists have also found a negative relationship between sizes of seeds produced by plants and the number of seeds they produce. Westoby, Leishman, and Lord found that plants of different growth form and different seed dispersal mechanisms tend to produce seeds of different sizes. Larger seeds, on average, produce larger seedlings that have a higher probability of successfully recruiting, particularly in the face of environmental challenges such as shade and competition.

Where adult survival is lower, organisms begin reproducing at an earlier age and invest a greater proportion of their energy budget into reproduction; where adult survival is higher, organisms defer reproduction to a later age and allocate a smaller proportion of their resources to reproduction. Shine and Charnov found that as survival of adult lizards and snakes increases, their age at maturity also increases. Gunderson found analogous patterns among fish. In addition, fish with higher rates of mortality allocate a greater proportion of their biomass to reproduction. In other words, they show higher reproductive effort. These generalizations are supported by comparisons both between and within species. For instance, pumpkinseed sunfish allocate greater energy, or biomass, to reproductive effort where adult pumpkinseed survival is lower.

The great diversity of life histories can be classified on the basis of a few population characteristics. Examples include fecundity or number of offspring, survival, relative offspring size, and age at reproductive maturity. One of the earliest attempts to organize information on the great variety of life histories that occur among species was under the heading of *r* selection and K selection. *r* selection refers to the per capita rate of increase, *r*, and is thought to favor higher population growth rate. *r* selection is predicted to be strongest in disturbed habitats. K selection refers to the carrying capacity in the logistic growth equation and is envisioned as a form of natural

selection favoring more efficient utilization of resources, such as food and nutrients. Grime described plant strategies, or life histories, that match the requirements of three environments: (1) low disturbance–low stress, (2) low disturbance–high stress, and (3) high disturbance–low stress. His plant strategies matching these environmental conditions were competitive, stress-tolerant, and ruderal. Based on life history patterns among fish, Kirk Winemiller and Kenneth Rose proposed a classification of life histories based on survivorship especially among juveniles, l_x, fecundity or number of offspring produced, m_x, and generation time or age at maturity, α. By basing their classification system on some of the most basic aspects of population ecology, l_x, m_x, and α, Winemiller and Rose established a common currency for representing and analyzing life history information for any organism.

Eric Charnov developed a new approach to life history classification free of the influences of size and time that facilitates the exploration of life history variation within and among groups of closely related taxa. Charnov's classification, based on relative offspring size, I/m, and fraction of adult body mass allocated to reproduction over a life span, $C \cdot E$, suggests that mammals, lizards, and altricial birds have life histories that are substantially different. Resource availability has also been shown to influence trait associations.

Shifts in first flowering dates of plants and arrival times by migratory birds on their breeding grounds in temperate regions have been highly responsive to recent climate warming.

Key Terms

adhesion-adapted 259	forb 258	K selection 266	ruderal 267
allozyme 256	germination 259	life history 255	scatterhoarded 259
aril 259	gonadosomatic	loci 256	semelparity 267
competitive plant 268	index (GSI) 262	phenology 272	stress 267
disturbance 267	graminoid 258	polymorphic locus 256	stress-tolerant plant 268
elaiosome 259	growth form 258	*r* selection 266	
fecundity 256	iteroparity 267	reproductive effort 262	

Review Questions

1. The discussion of seed size and number focused mainly on the advantages associated with large seeds. What are some of the advantages associated with producing small seeds?

2. Under what conditions should natural selection favor production of many small offspring versus the production of a few well-provisioned offspring?

3. Explain how growing from larger seeds could give an advantage to seedlings facing strong environmental challenge to their establishing.

4. The studies by Shine and Charnov (1992) and Gunderson (1997) relied heavily on data on life histories published previously by other

authors. What was it about the nature of the problems addressed by these authors that constrained them to use this approach?

5. Much of our discussion of life history variation involved variation among species within groups as broadly defined as "fish," "plants," or "reptiles." However, the work of Bertschy and Fox revealed significant variation in life history within species. In general, what should be the relative amount of variation within a species compared to that among many species?

6. Grime's classification of environments included four environments, three of which he proposed were inhabitable by plants and one of which was not. That fourth environment shows high

intensity of disturbance and high stress. What kinds of organisms can you think of that could live and perhaps thrive in such an environment?

7. Rio Grande cottonwoods can live to be well over 100 years old but their seeds experience very high rates of mortality. Female cottonwood trees produce about 25 million seeds annually. Which of the life history categories that we've discussed most closely match the life history of the Rio Grande cottonwood?

8. Using what you know about the trade-off between seed number and seed size (e.g., fig. 12.7) and patterns of variation among

plants, predict the relative number of seeds produced by the various plant growth forms and dispersal strategies listed in figure 12.8.

9. Apply Winemiller's model to plants. If you were to construct a strictly quantitative classification of plant life histories using Winemiller and Rose's approach, what information would you need about the plants included in your analysis? How many plant species would you need to have an idea of how variation in their life histories compares with those of animals (e.g., as in fig. 12.22)?

Chapter

13

Species Interactions and Competition

© image100/PunchStock

A section of the Great Barrier Reef, Australia, rich in coral species. Coral reefs are often sites of intense competition for limited space among corals, which compete using a wide variety of techniques, including overgrowing neighboring coral colonies, digesting the tissues of adjacent colonies, depositing mucus that contains stinging cells on neighboring colonies, and killing the tissues of nearby corals with elongated "sweeper" tentacles.

Applications: Competition Between Native and Invasive Species   296

Summary   297

Key Terms   298

Review Questions   298

CHAPTER CONCEPTS

13.1   Laboratory and field studies reveal intraspecific competition.   280
Concept 13.1 Review   282

13.2   The competitive exclusion principle proposes that two species with identical niches cannot coexist indefinitely, which leads to the prediction that coexisting species will have different niches.   282
Concept 13.2 Review   285

13.3   Mathematical and laboratory models provide a theoretical foundation for studying interspecific competition in nature.   285
Concept 13.3 Review   289

13.4   Competition can have significant ecological and evolutionary influences on the niches of species.   289
Concept 13.4 Review   295

LEARNING OUTCOMES

After studying this section you should be able to do the following:

13.1   List the three different outcomes when an organism interacts with another and define the six possible combinations of these.

13.2   Provide examples of commensalism and amensalism and explain why they might be rare in nature.

13.3   Define intraspecific and interspecific competition.

13.4   Distinguish between interference competition and resource, or exploitative, competition.

A monkey eats a piece of fruit, a wasp stings a bird, two trees grow roots to access a single limited store of nitrogen in the soil. These are all examples of species interacting, but with different ecological outcomes and evolutionary histories. Ecology is the study of the relationships between organisms and their environment; until now we have explored many of the ways organisms affect and are affected by their environment—in this section we turn our attention to the interaction between organisms, both within and between species.

When two organisms interact, there are three possible impacts on each: positive, negative, or neutral. If the impact is negative, it means that there is energy expended or injury incurred. This negative impact may be severe, as in the case of the consequence to a mouse of being eaten by a snake, or subtle such as when a mouse has to travel a bit farther to find food because another mouse has eaten everything nearby. Table 13.1 shows all possible combinations of the three categories of impact. While most students are somewhat familiar with the first three (which will be covered in depth in the next three chapters), interactions that result in no consequence to one or both of the participants are generally less discussed by ecologists.

Some associations between species are extremely intimate, but this does not dictate the impact on the organisms involved. Symbioses are interactions in which most or all of the life cycle of one organism occurs inside or on another; an example of this is an association in which a mycorrhizal fungus takes energy from a host plant while also increasing the host's water or nutrient uptake (+/+); mycorrhizal fungi are typically considered to be mutualists (discussed in more detail in chapter 15). However, in some cases, the host plant is harmed by the energy lost and will actively resist fungal infection by producing toxins (−/−) (Johnson et al. 1997). These counteracting impacts could theoretically cancel each other, thus conferring no net benefit or loss to either the mycorrhiza or the host plant. Although theoretically possible (Holland and DeAngelis 2009), such neutralism is rarely observed by scientists.

A range of species interactions can be seen in the context of insects that use galls, abnormal growths on twigs or leaves (fig. 13.1). Swollen thorn acacia trees, *Vachellia drepanolobium,* grow hollow galls used as homes for stinging ants that provide defense against herbivory, creating a **mutualism** (+/+; see chapter 15). In most cases, however, galls are formed by damage

Figure 13.1   Insects stimulate abnormal growths on plants, called galls, in which to create a shelter for all or part of their life cycle. Galls can reflect not only parasitic relationships between the insect and the plant, but also competition for resources between insects, and other types of species interactions.   Avalon/Photoshot License/Alamy Stock Photo

or chemical stimulation by a wasp or other gall-forming insect that forces the plant to create the abnormal plant growth where the insect spends some of its life, that is, **parasitism** (+/−; see chapter 14). Gall-forming insects can also compete (−/−; this chapter) with each other for space to form galls (Inbar et al. 1995). Galls are sometimes also used by other, often related, insects in a commensalistic relationship (+/0). **Commensalism** takes place when there is a benefit to one organism in the interaction without any impact on the other, including when one organism uses another for its residence (*inquilinism*). For example, Man-Miao Yang and her colleagues found such a relationship among psyllids (jumping plant lice); newly hatched nymphs of the iniquiline species *Pachypsylla cohabitans* feed next to nymphs of other, gall-forming psyllids on leaves, becoming enveloped in the leaf gall as it forms (Yang et al. 2001). The *P. cohabitans* has

Table 13.1

Types of species interactions, as defined by the net impact on each species or individual involved, with examples of each (adapted from Smith 1996).			
	The effect on species 1	The effect on species 2	Example
Competition	−	−	A lion and a hyena fighting over a carcass; regardless of "winner," both expend energy and risk being injured.
Mutualism	+	+	A penstemon and the bee that pollinates it; the penstemon spreads its genes and the bee gets food.
Predation/Parasitism	+	−	A fungus growing on a wild rosebush; the fungus gains energy while the rose experiences it as a disease.
Commensalism	+	0	An orchid growing on a tree; the orchid has a support structure that protects it and makes it more accessible to pollinators, with no impact on the tree.
Amensalism	−	0	A sunflower struggles to grow in the shade of a walnut tree; the tree experiences no impact from the presence of the sunflower.
Neutralism	0	0	Two species of insects live on the same plant in close proximity but because they use different parts of the plant, they have no effect on each other.

a protective home in which to grow, without any negative impact on the insect that actually formed the gall, thus making this a commensal relationship. Commensalism can also include when one organism uses another for transport, such as a seed of a water plant being dispersed on the feet or feathers of a duck (*phoresy*).

Amensalism, in contrast, occurs when the outcome of the interaction between organisms is that one is harmed with no impact on the other ($-/0$). In ecology, most research on this topic is in one of two areas: asymmetric competition, in which only one organism is negatively affected by the interaction; and when the release of a chemical by one organism harms or even kills another, called **allelopathy.** However, in both categories, true amensalism is likely to be rare because usually there will be some energetic cost to both organisms.

For example, saltcedar trees (*Tamarix* spp.) from Eurasia are adapted to be able to use saline water that they take up from the soil by exuding the salts from glands on their leaves. When these leaves fall to the ground, they consequently increase the salinity of the soil (Ohrtman and Lair 2013). When saltcedar grows in North America, native trees and most microorganisms cannot grow in this salty soil (Rowland et al. 2004; Meinhardt and Gehring 2013), leading some to call the interaction between saltcedar and native species allelopathic. But is it amensalism? Energy is expended by salt glands; therefore, there is a cost to the saltcedar, possibly an evolutionary adaptation to decrease competition ($-/-$) with native trees by preventing them from growing. Thus, amensalism generally refers to cases where the negative impact to one organism is negligible or difficult to quantify, rather than where there is actually no effect.

In general, competition will be negative for both parties ($-/-$), because even the "winner" will have expended more energy to gain that resource than it would have if there was no competition. For example, when saltcedar seedlings are planted next to native cottonwood seedlings, both grow less well than if they were planted alone (Sher et al. 2000). In this chapter, we explore the different types of competition and how they are studied; the following chapters focus on predation and mutualism, respectively.

Competitive Interactions Are Diverse

Careful observation and experimentation can reveal different types of competition between species in nature (fig. 13.2). For instance, on reefs off the north coast of Jamaica, three spot damselfish guard small territories of less than 1 m². Damselfish constantly patrol and survey the borders of their territories, vigorously attacking any intruder that presents a threat to their eggs and developing larvae, or to their food supply.

If you create a vacancy on the reef by removing one of the damselfish holding a territory, other damselfish appear within minutes to claim the vacant territory. Some of the new arrivals are threespot damselfish like the original resident, and some are cocoa damselfish, which generally live a bit higher on the reef face. These new arrivals fight fiercely for the vacated territory. The damselfish chase each other, nip each other's flanks, and slap each other with their tails. The melee ends within

(a)

(b)

Figure 13.2 Competition on a coral reef is often easily observed, as for example, the two blennies, shown here, competing intensely for space (*a*). Competition in a forest can be as intense as on a coral reef. However, much of the competition in a forest takes place underground, where the roots of plants compete for water and nutrients (*b*).
(*a*) Paulo Oliveira/Alamy Stock Photo; (*b*) Pete Ryan/Getty Images

minutes, and life among the damselfish settles back into a kind of tense tranquility.

This example demonstrates several things. First, individual damselfish maintain possession of their territories through ongoing competition with other damselfish, and this competition takes the form of **interference competition,** which involves direct interactions between individuals (fig. 13.2*a*). In the case of threespot and cocoa damselfish, interference competition takes the form of aggressive defense of territories. Other organisms engaging in interference competition produce chemical toxins that harm potential competitors. Still others interfere with competitors by growing over them, causing direct harm, or reducing the competitor's access to resources, such as food or light. Second, though it may not appear so to the casual observer, there is a limited supply of suitable space for damselfish territories, a condition that ecologists call **resource limitation.** Third, the threespot damselfish are subject to **intraspecific competition,** competition with members of their own species, as well as **interspecific competition,** competition between individuals of two species that reduces the fitness of both. The effects of competition on the two competitors may not be equal, however. The individuals of one species may suffer greatly reduced fitness while those of the second are affected very little. The observation that threespots generally win in aggressive encounters with cocoa damselfish suggests this sort of competitive asymmetry.

Competition is not always as dramatic as fighting damselfish nor is it always resolved so quickly (fig. 13.2*b*). In a mature white pine forest in New Hampshire, tree roots grow throughout the soil taking up nutrients and water as they provide support. In 1922, James Toumey designed an experiment to determine whether the high density of pine roots suppresses the growth of other plants. Researchers cut a trench, 0.92 m deep, around a plot 2.74 m by 2.74 m in the middle of the forest. In so doing, they cut 825 roots, mainly those of pines, which removed potential competition by these roots for soil resources. They also established untrenched control plots on either side of the trenched plot and then watched as the results of their experiment unfolded. The experiment continued for 8 years, with retrenching every 2 years and over 100 roots cut each time. By retrenching, the researchers maintained their experimental treatment, suppression of potential root competition.

In the end, this 8-year experiment (Toumey and Kienholz 1931) yielded results as dramatic as those with the damselfish. Vegetative cover in trenched plots of forest floor that had been released from root competition was 10 times higher than in the control plots. Apparently, the roots of white pines exert interspecific competition for limited supplies of nutrients and water that is strong enough to suppress the growth of forest floor vegetation especially growth of herbaceous plants, hemlock, and white pine seedlings. The high density of white pine seedlings in trenched plots showed that intraspecific competition was occurring as well. Competition involving the use of such limited resources is called **resource** or **exploitative competition.**

Ecologists have long thought that competition is pervasive in nature. For instance, Darwin thought that interspecific competition was an important source of natural selection. While ecologists have shown that interspecific competition substantially influences the distribution and abundance of many species, they have also questioned the assumption that competition is an all-important organizer of nature. Such questioning has stimulated more careful research and more rigorous testing of the influence of competition on populations.

13.1 Intraspecific Competition

LEARNING OUTCOMES

After studying this section you should be able to do the following:

13.5   Outline the $-3/2$ self-thinning rule.

13.6   Explain how the observation of self-thinning among plants provides evidence for intraspecific competition.

13.7   Discuss studies demonstrating intraspecific competition among planthoppers (*Prokelisia marginata*) and terrestrial isopods (*Porcellio scaber*).

Laboratory and field studies reveal intraspecific competition. In chapter 11, we saw that slowing population growth at high densities produces a sigmoidal, or S-shaped, pattern in which population size levels off at carrying capacity. Our assumption in that discussion was that intraspecific competition for limited resources plays a key role in slowing population growth at higher densities. The effect of intraspecific competition is included in the model of logistic population growth. If competition is an important and common phenomenon in nature, then we should be able to observe it among individuals of the same species, that is, individuals with identical or very similar resource requirements. Thus, we begin our discussion of competition with intraspecific competition.

Intraspecific Competition Among Plants

In chapter 7, we reviewed experiments by David Tilman and M. Cowan (1989) that showed how plants alter root:shoot ratios in response to availability of soil nitrogen (see fig. 7.26). The same experiments also included evidence for intraspecific competition. When Tilman and Cowan grew the grass *Sorghastrum nutans* at low density (7 plants per pot) and high density (100 plants per pot), those grown at low density grew to a larger size at all nitrogen concentrations (fig. 13.3). These results suggest that competition for nitrogen (resources) was more intense at the higher plant population density. Such competition for limited resources occurs in natural populations where it may lead to mortality among the competing plants.

The development of a stand of plants from the seedling stage to mature individuals suggests competition for limited resources. Each spring as the seeds of annual plants germinate, their population density often numbers in the thousands per square meter. However, as the season progresses and individual plants grow, population density declines. This same pattern occurs in the development of a stand of trees. As the stand of trees develops, more and more biomass is composed of fewer and fewer individuals. This process is called **self-thinning.**

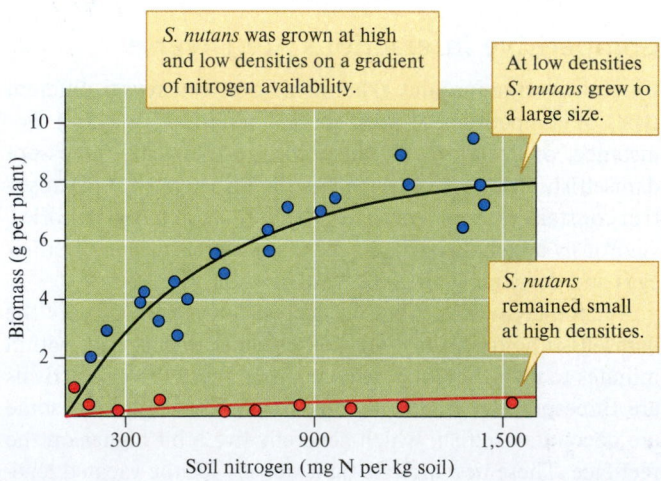

Figure 13.3   Population density, soil nitrogen, and the size attained by the grass *Sorghastrum nutans* (data from Tilman and Cowan 1989).

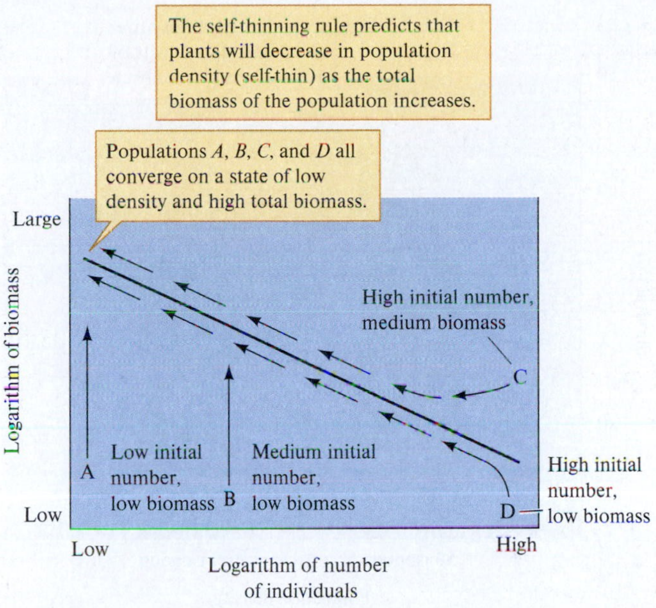

The self-thinning rule predicts that plants will decrease in population density (self-thin) as the total biomass of the population increases.

Populations *A*, *B*, *C*, and *D* all converge on a state of low density and high total biomass.

High initial number, medium biomass

Low initial number, low biomass

Medium initial number, low biomass

High initial number, low biomass

Figure 13.4 Self-thinning in plant populations (data from Westoby 1984).

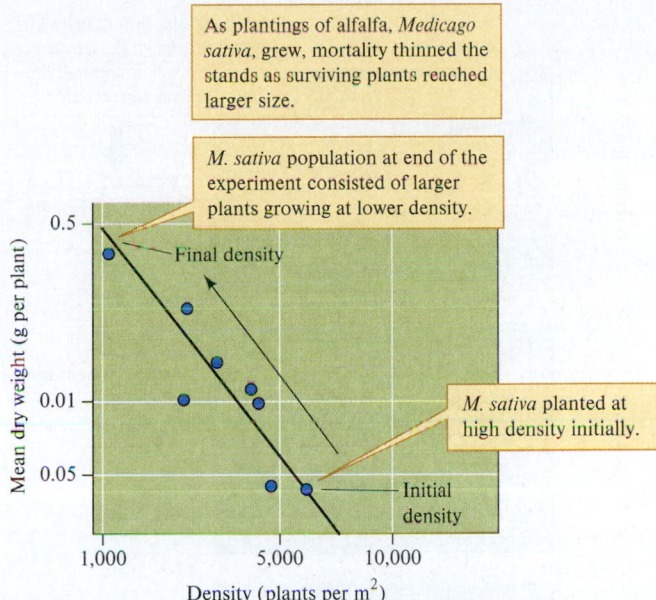

As plantings of alfalfa, *Medicago sativa*, grew, mortality thinned the stands as surviving plants reached larger size.

M. sativa population at end of the experiment consisted of larger plants growing at lower density.

Final density

M. sativa planted at high density initially.

Initial density

Figure 13.5 Self-thinning in populations of alfalfa, *Medicago sativa* (data from White and Harper 1970).

Self-thinning appears to result from intraspecific competition for limited resources. As a local population of plants develops, individual plants take up increasing quantities of nutrients, water, and space for which some individuals compete more successfully. The losers in this competition for resources die, and population density decreases, or "thins," as a consequence. Over time the population is composed of fewer and fewer large individuals.

One way to represent the self-thinning process is to plot total plant biomass against population density. If we plot the logarithm of plant biomass against the logarithm of plant density, the slope of the resulting line averages around $-\frac{1}{2}$. In other words, plant population density declines more rapidly than biomass increases (fig. 13.4).

Another way to represent the self-thinning process is to plot the average weight of individual plants in a stand against density (fig. 13.5). The slope of the line in such plots averages around $-\frac{3}{2}$. Because self-thinning by many species of plants comes close to a $-\frac{3}{2}$ relationship, this relationship has come to be called the **$-\frac{3}{2}$ self-thinning rule.** The $-\frac{3}{2}$ self-thinning rule was first proposed by K. Yoda and colleagues (1963) and amplified by White and Harper (1970), who provided many additional examples (e.g., see fig. 13.5). Subsequently, the self-thinning rule became widely accepted among ecologists.

Recent analyses have shown that self-thinning in some plant populations deviates significantly from the $-\frac{3}{2}$ (or $-\frac{1}{2}$ for biomass-numbers) slope. Although mode and intensity of competition is the most important factor for determining slope, other factors such as environmental stress can also play a role (Ma et al. 2020). For example, research by Xiongqing Zhang and colleagues found that winter minimum temperature was a primary determinant of the self-thinning slope of Chinese fir (*Cunninghamia lanceolata*) (Zhang et al. 2018). In this context, intraspecific plant competition has important implications for the forestry industry, which depends on predictions of

yield per unit area. The important point, from the perspective of our present discussion, is that self-thinning appears to be the consequence of intraspecific competition for limited resources. Resource limitation has also been demonstrated in experiments on intraspecific competition within animal populations.

Intraspecific Competition Among Planthoppers

Ecologists have often failed to demonstrate that insects, particularly herbivorous insects, compete. However, one group of insects in which competition has been repeatedly demonstrated is the Homoptera, including the leafhoppers, planthoppers, and aphids. Robert Denno and George Roderick (1992), who studied interactions among planthoppers (Homoptera, Delphacidae), attribute the prevalence of competition among the Homoptera to their habit of aggregating, to their rapid population growth, and to the mobile nature of their food supply, plant fluids.

Denno and Roderick demonstrated intraspecific competition within populations of the planthopper *Prokelisia marginata*, which lives on the salt marsh grass *Spartina alterniflora* along the Atlantic and Gulf coasts of the United States. The population density of *P. marginata* was controlled by enclosing the insects with *Spartina* seedlings at densities of 3, 11, and 40 leafhoppers per cage, densities that are within the range at which they live in nature. At the highest density, *P. marginata* showed reduced survivorship, decreased body length, and increased developmental time (fig. 13.6). These signs of intraspecific competition were probably the result of reduced food quality at high leafhopper densities, since plants heavily populated by planthoppers show reduced concentrations of protein, chlorophyll, and moisture. As demonstrated in the following example, however, intraspecific interference competition may occur in the absence of obvious resource limitation.

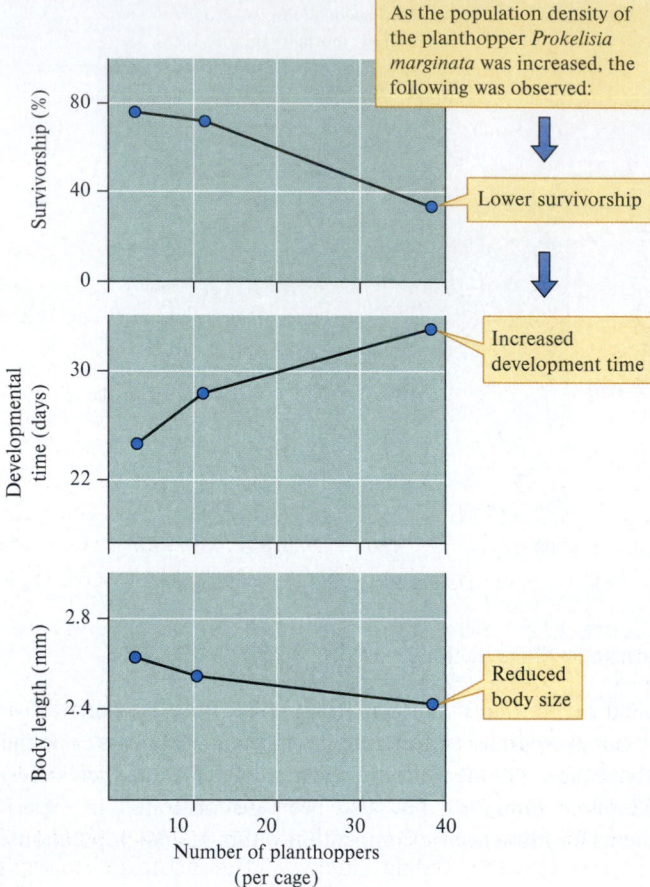

Figure 13.6 Population density and planthopper performance (data from Denno and Roderick 1992).

Interference Competition Among Terrestrial Isopods

Edwin Grosholz (1992) used a field experiment to study the effects of a wide range of biotic interactions on the population biology of the terrestrial isopod *Porcellio scaber*. This organism, which is associated with human activities such as farming and gardening and is found throughout the world, sometimes lives at densities in excess of 2,000 individuals per square meter. Such high densities suggest a strong potential for intraspecific competition.

Grosholz conducted his experiments on an outdoor grid of 48, 0.36 m² plots enclosed by aluminum flashing. To control isopod movements, he buried the flashing 12.5 cm into the soil and extended it 12.5 cm above the soil surface. Two experimental treatments were used: (1) to test for food limitation, the food within the enclosures was supplemented by adding sliced carrots and potatoes, and (2) to test for density effects, study plots were stocked with either 100 or 50 *P. scaber*. Supplementing food had no effect on survival by *P. scaber,* indicating that food was not limiting survival. However, survival was lower at the higher population density (fig. 13.7). Grosholz attributed lower survival at the higher density to cannibalism, a common occurrence in terrestrial isopods. The study offers interesting insights into the role that interference may play in intraspecific competition, even in the absence of obvious resource limitation.

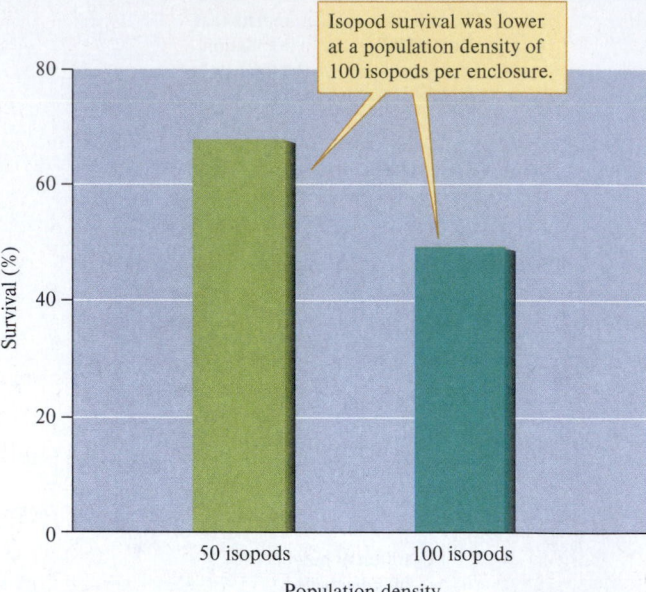

Figure 13.7 Population density and survival in populations of a terrestrial isopod, *Porcellio scaber* (data from Grosholz 1992).

As we move from discussions of intraspecific to interspecific competition, we need to back up a bit and reconsider the niche. We do this because interspecific competition usually occurs among species with similar environmental requirements, that is, among species with similar niches.

Concept 13.1 Review

1. Do you think that Grosholz might have observed food limitation if he had used higher densities of *Porcellio scaber* in his experiments?
2. How might using other indicators of competition, such as growth rate, reproductive rate, and size at maturity, have affected Grosholz's conclusions regarding lack of food limitation in his study populations?

13.2 Competitive Exclusion and Niches

LEARNING OUTCOMES

After studying this section you should be able to do the following:

13.8   Define the competitive exclusion principle.
13.9   Relate the correspondence between the niches of species and the potential for interspecific competition between them.
13.10 Describe the relationship between the beaks of Darwin's finch species and their feeding niches.
13.11 Outline the evidence that niche partitioning by wasp and fly parasitoids of slug caterpillars is based on caterpillar size.

The competitive exclusion principle proposes that two species with identical niches cannot coexist indefinitely, which leads to the prediction that coexisting species will have different niches. As we saw earlier (see chapter 9, section 9.1), the niche concept was developed over a period of several decades; however, it was within the context of interspecific competition that the importance of the niche concept was fully realized. The work of G. F. Gause (1934), whose principal interest was interspecific competition, helped ensure a prominent place for the niche concept in modern ecology. Particularly important was Gause's **competitive exclusion principle,** which states that two species with identical niches cannot coexist indefinitely. Gause experimented with competition in the laboratory and obtained results indicating that when two species compete, one will be a more effective competitor for limited resources, that is, will be more effective at converting resources into offspring. As a consequence, the more effective competitor will have higher fitness (higher reproductive success) and will eventually exclude all individuals of the second species. The competitive exclusion principle set the niche concept in a broader context. After Gause's experiments and studies, describing the niches of species was a stepping-stone to understanding interactions between species—a potential key to understanding the organization of nature. In the case of interspecific competition, the competitive exclusion principle leads to the prediction that species living together will generally have different niches, a concept we first encountered in chapter 1 in connection with Robert MacArthur's warbler studies.

Why do we revisit the niche concept here? The reason is that we, like the first ecologists to use the term, need a concept that represents all the environmental interactions of a species. The niche concept carries us beyond the details of individual species' requirements to a position where we can more easily consider the ecology of interactions between species, through differences, similarities, or complementarities of their niches. Thomas Schoener (2009) provides a comprehensive discussion of the various perspectives on the nature of the ecological niche.

Do you think it's possible to completely describe Hutchinson's *n*-dimensional hypervolume niche (see chapter 9, section 9.1) for any species? Probably not, since there are so many environmental factors that potentially influence survival and reproduction. Fortunately, it appears that niches are, in some cases, determined mostly by a few environmental factors and therefore ecologists are able to apply a simplified version of Hutchinson's comprehensive niche concept. In studies of animals, ecologists have frequently described niches in terms of their feeding biology.

The Feeding Niches of Darwin's Finches

As we saw in chapter 10, availability of suitable food significantly affects the survival and reproduction of Darwin's finches. In other words, food has a major influence on the niches of Darwin's finches. Because the kinds of food used by birds are largely reflected by the form of their beaks, David Lack (1947) linked differences in beak size and form among Darwin's finches to differences in their feeding niches. Building on Lack's earlier work, Peter Grant (1986) and his colleagues represented the feeding niches of Darwin's finches by their beak morphology. Because of selective pressure from past competition for food, these species evolved to have different feeding niches, as reflected by beak size. The large ground finch, *Geospiza magnirostris,* eats larger seeds; the medium ground finch, *G. fortis,* eats medium-sized seeds; while the small ground finch, *G. fuliginosa,* eats small seeds (fig. 13.8). Within this population, individuals with the deepest beaks fed on the hardest seeds, while individuals with the smallest beaks fed on the softest seeds.

The importance of beak size to seed use was also demonstrated by the effects of the 1977 drought on the *G. fortis* population of Daphne Major. In chapter 11, we saw how this drought caused substantial mortality in this population (see fig. 11.17). However, this mortality did not fall equally on all segments of the population. As seeds were depleted, the birds ate the smallest and softest seeds first, leaving the largest and toughest seeds (fig. 13.9). In other words, following the drought, not only were seeds in short supply, the remaining seeds were also tougher to crack. Because they could not crack the remaining seeds, mortality fell most heavily on smaller birds with smaller beaks. Consequently, at the end of the drought, the *G. fortis* population on Daphne Major was dominated by larger individuals with larger, stronger beaks that had survived by feeding on hard seeds (fig. 13.10).

These studies show that beak size provides significant insights into the feeding biology and evolution of Darwin's

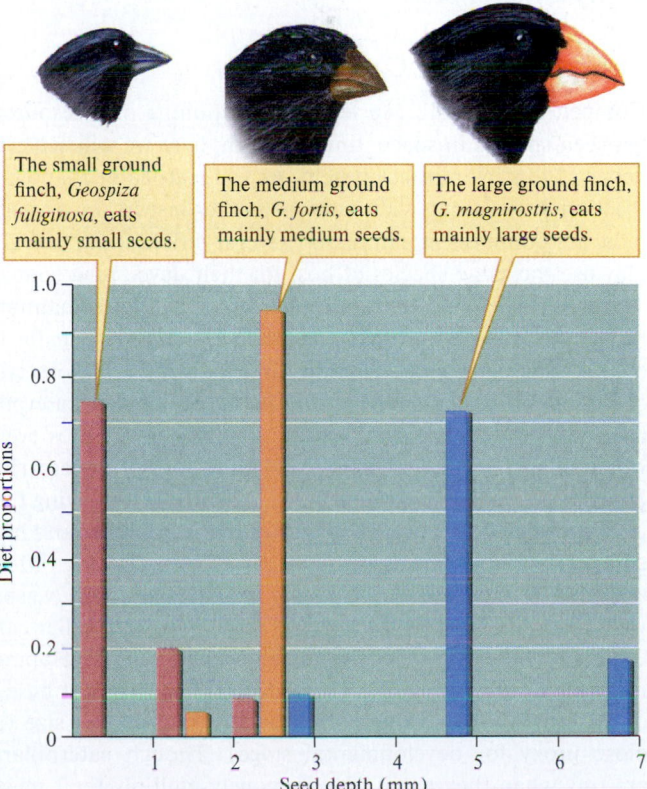

The small ground finch, *Geospiza fuliginosa,* eats mainly small seeds.

The medium ground finch, *G. fortis,* eats mainly medium seeds.

The large ground finch, *G. magnirostris,* eats mainly large seeds.

Figure 13.8 Relationship between beak size and seed size in Darwin's finch species (data from Grant 1986).

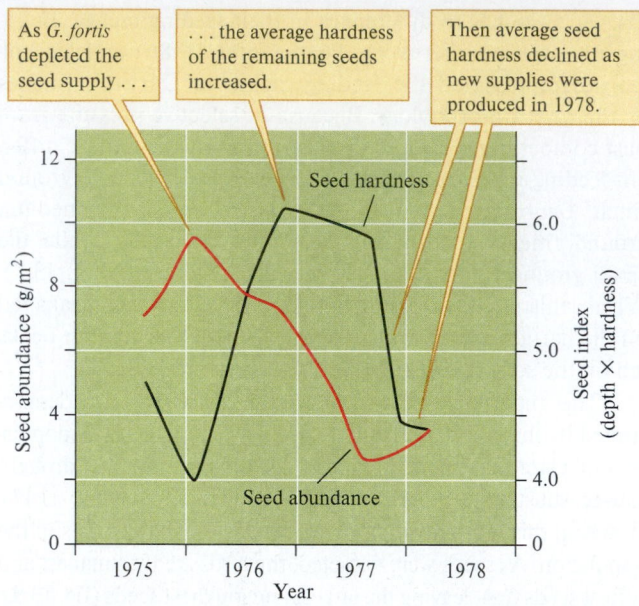

As *G. fortis* depleted the seed supply . . .

. . . the average hardness of the remaining seeds increased.

Then average seed hardness declined as new supplies were produced in 1978.

Figure 13.9 Seed depletion by the medium ground finch, *Geospiza fortis,* and average seed hardness (data from Grant 1986).

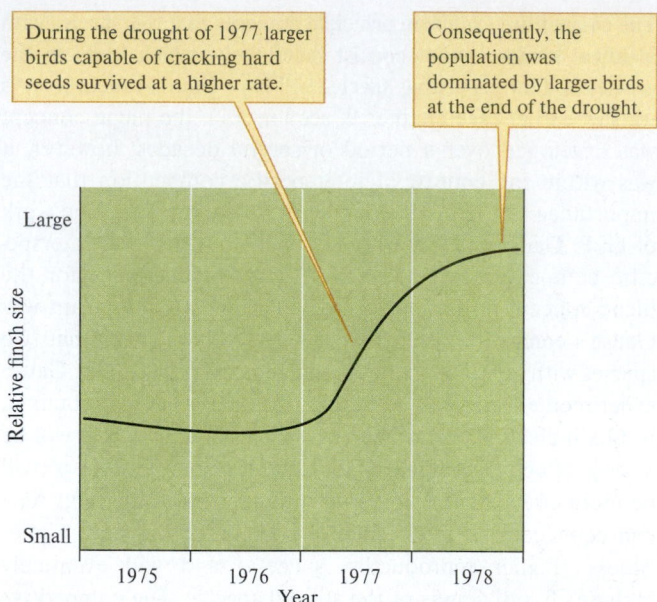

During the drought of 1977 larger birds capable of cracking hard seeds survived at a higher rate.

Consequently, the population was dominated by larger birds at the end of the drought.

Figure 13.10 Selection for larger size among medium ground finches, *Geospiza fortis,* during a drought on Daphne Major Island (data from Grant 1986).

ground finches. Since food is the major determinant of survival and reproduction among these birds, beak morphology gives us a very good picture of their niches. However, the niches of other kinds of organisms are determined by entirely different environmental factors. Let's consider the niche of caterpillars as prey to other insects.

Competition for Caterpillars

Competitive pressure can lead to partitioning of a resource between species in space, time, or both; here we will see evidence of this in an insect system. As you learned in chapter 4 (section 4.5), insects that parasitize other insects, or parasitoids, are common in nature and several different parasitoids may use the same species of host for their developing young. Teresa Stoepler and her colleagues John Lill and Shannon Murphy studied competition between fly and wasp parasitoids of slug caterpillars (family Limacodidae) in an eastern deciduous forest (Stoepler et al. 2011). Some slug caterpillars have bright warning coloration and stinging spines even when small to defend against predation but not necessarily parasitism (fig. 13.11) (Murphy et al. 2010). By observing the parasitoids of 11 species of slug caterpillars, Stoepler and her colleagues found that flies and wasps showed a strong differentiation in size of the caterpillars that they attacked: wasps were much more likely to be found developing in smaller caterpillars while flies were more likely to be found developing in larger ones (fig. 13.12). These differences were not based on caterpillar species but rather on the caterpillar's size (a close proxy for developmental stage). Though caterpillars are tiny when they hatch, they famously multiply their mass exponentially by the end of the growing season. It appeared that wasps and flies reduced competition with each other through **niche partitioning.** That is, because of the selective

Figure 13.11 Slug caterpillars (family *Limacodidae*) have several physical and chemical defenses against predators, including stinging spines, but these do not protect them from parasitoids, which compete strongly for the caterpillars as a resource.   Shannon Murphy

pressure caused by interspecific competition, fly and wasp parasitoids may have evolved to prefer different sizes of hosts in this system.

But was this difference between flies and wasps actually due to competition for caterpillar hosts? An alternate hypothesis was that the observed differentiation by size was simply a function of **phenology** (seasonal life cycle timing): perhaps wasps simply emerged and attacked earlier in the season when caterpillars were small, whereas flies were more active later when the hosts had grown to be bigger. Did flies really prefer larger caterpillar hosts or were they just tracking host availability? To test this possibility, Shannon Murphy and colleagues

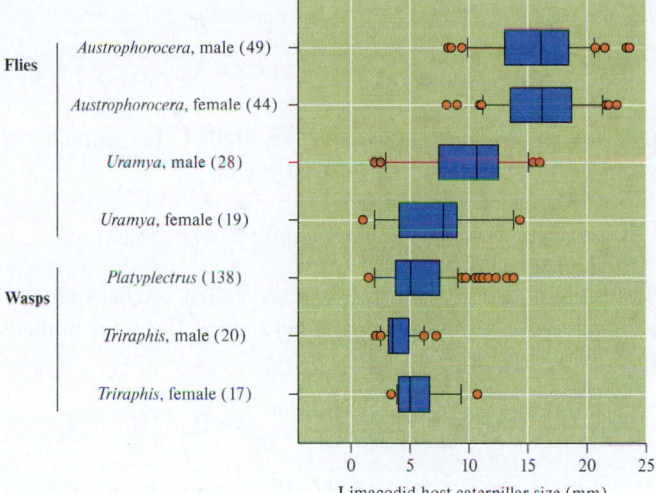

Figure 13.12   Fly and wasp parasitoids differ in the average size host caterpillar that they attack. Numbers in parentheses following parasitoid species names are number of reared parasitoids (data from Stoepler et al. 2011).

exposed 538 early (small) and late (large) developmental-stage slug caterpillars to parasitism at the same time in a "choice" experiment (Murphy et al. 2014). As in the previous study, wasps significantly preferred small caterpillars and flies significantly preferred large caterpillars.

Taken together, these studies provided strong evidence that the wasps and flies were avoiding competition for caterpillar hosts by partitioning the niche of their shared host based on size. This is yet another example of what a strong selective force competition can be.

Concept 13.2 Review

1. The competitive exclusion principle states that two species cannot occupy the same niche indefinitely. What is a fundamental assumption of this principle?
2. Do resources have to be present in limited supplies for competition to shape species niches?

13.3 Mathematical and Laboratory Models

LEARNING OUTCOMES

After studying this section you should be able to do the following:

13.12 Describe the competition coefficient.
13.13 Predict competitive exclusion or coexistence of species using graphs showing the orientations of the isoclines for zero population growth for two species involved in interspecific competition.
13.14 Explain the general biological conditions for coexistence of two species according to the Lotka-Volterra competition model.

13.15 Describe how laboratory models of competition have provided insights into the biological details involved in interspecific competition between closely related species.

Mathematical and laboratory models provide a theoretical foundation for studying interspecific competition in nature.   In the study of interspecific competition, mathematical and laboratory models have played complementary roles. Both mathematical and laboratory models are generally much simpler than the natural circumstances the ecologist wishes to understand. However, while sacrificing accuracy, this simplicity offers a degree of control that ecologists would not have in most natural settings.

D. B. Mertz (1972) began a review of four decades of research on *Tribolium* beetle populations with an astute summary of the characteristics of models in general and of the "*Tribolium* model" in particular: (1) the *Tribolium* model is an abstraction and simplification, not a facsimile, of nature; (2) except for the beetles themselves, it is a man-made construct, partly empirical and partly deductive; and (3) it is used to provide insights into natural phenomena. The predictions of these simplified models can be tested in natural systems and either supported or falsified. If falsified, a theory can be modified to accommodate the new information. Ideally, scientific understanding proceeds as a consequence of this exchange between theory and observation and experiment.

Modeling Interspecific Competition

As we saw in chapter 11, the model of logistic population growth includes a term for intraspecific competition but can be expanded to include the influence of interspecific competition on population growth. The first to do so was Vito Volterra (1926), who was interested in developing a theoretical basis for explaining changes in the composition of a marine fish community in response to reduced fishing during World War I. Alfred Lotka (1932b) independently repeated Volterra's analysis and extended it using graphics to represent changes in the population densities of competing species during competition.

Let's retrace the steps of Lotka and Volterra's modeling exercise, beginning with the logistic model for population growth discussed in chapter 11:

$$\frac{dN}{dt} = r_{max}N\left(\frac{K - N}{K}\right)$$

We can express the population growth of two species of potential competitors with the logistic equation:

$$\frac{dN_1}{dt} = r_1N_1\left(\frac{K_1 - N_1}{K_1}\right)$$

and

$$\frac{dN_2}{dt} = r_2N_2\left(\frac{K_2 - N_2}{K_2}\right)$$

where N_1 and N_2 are the population sizes of species 1 and 2, K_1 and K_2 are their carrying capacities, and r_1 and r_2 are the intrinsic rates of increase for species 1 and 2, which are expressed when their population sizes are very low and potentially limiting resources, such as food or space, are abundant.

In these models, population growth slows as N increases and the relative level of intraspecific competition is expressed as the ratio of numbers to carrying capacity, either N_1/K_1 or N_2/K_2. The assumption here is that resource supplies will diminish as population size increases due to intraspecific competition for resources. Resource levels can also be reduced by interspecific competition.

Lotka and Volterra included the effect of interspecific competition on the population growth of each species as:

$$\frac{dN_1}{dt} = r_1 N_1 \left(\frac{K_1 - N_1 - \alpha_{12}N_2}{K_1} \right)$$

and

$$\frac{dN_2}{dt} = r_2 N_2 \left(\frac{K_2 - N_2 - \alpha_{21}N_1}{K_2} \right)$$

In these models, the rate of population growth of a species is reduced both by conspecifics (individuals of the same species) and by individuals of the competing species, that is, interspecific competition. The effects of intraspecific competition ($-N_1$ and $-N_2$) are already included in the logistic models for population growth. The effect of interspecific competition is incorporated into the Lotka-Volterra model by $-\alpha_{12}N_2$ and $-\alpha_{21}N_1$. The terms α_{12} and α_{21} are called **competition coefficients** and express the competitive effects of the competing species. Specifically, α_{12} is the effect of an individual of species 2 on an individual of species 1, relative to the competitive effects of an individual of species 1. Meanwhile, α_{21} is the effect of an individual of species 1 on an individual of species 2 relative to the competitive effects of an individual of species 2. In this model, interspecific competitive effects are expressed in terms of intraspecific equivalents. If, for example, $\alpha_{12} > 1$, then the competitive effect of an individual of species 2 on the population growth of species 1 is greater than that of an individual of species 1. If, on the other hand, $\alpha_{12} < 1$, then the competitive effect of an individual of species 2 on the population growth of species 1 is less than that of an individual of species 1.

In general, the Lotka-Volterra model predicts coexistence of two species when, for both species, interspecific competition is weaker than intraspecific competition. Otherwise, one species is predicted to eventually exclude the other. These conclusions come from the following analysis.

Populations of species 1 and 2 stop growing when:

$$\frac{dN_1}{dt} = r_1 N_1 \left(\frac{K_1 - N_1 - \alpha_{12}N_2}{K_1} \right) = 0$$

and

$$\frac{dN_2}{dt} = r_2 N_2 \left(\frac{K_2 - N_2 - \alpha_{21}N_1}{K_2} \right) = 0$$

That is, when:

$$(K_1 - N_1 - \alpha_{12}N_2) = 0 \quad \text{and} \quad (K_2 - N_2 - \alpha_{21}N_1) = 0$$

Or, rearranging these equations, we predict that population growth for the two species will stop when:

$$N_1 = K_1 - \alpha_{12}N_2 \quad \text{and} \quad N_2 = K_2 - \alpha_{21}N_1$$

These are equations for straight lines, called **isoclines of zero population growth,** where everywhere along the lines population growth is stopped:

$$\frac{dN_1}{dt} = 0 \quad \text{and} \quad \frac{dN_2}{dt} = 0$$

Above an isocline of zero growth, the population of a species is decreasing; below it the population is increasing (fig. 13.13).

The isoclines of zero population growth show how the environment can be filled up or, in other words, the relative population sizes of species 1 and species 2 that will deplete the critical resources. At one extreme, for example, for species 1, the environment is completely filled by species 1 and species 2 is absent. This occurs where $N_1 = K_1$. At the other extreme, again for species 1, the environment can be saturated entirely by species 2, while species 1 is absent. This occurs where $N_2 = K_1/\alpha_{12}$. In between these extremes, the environment is saturated with a mixture of species 1 and 2. The graph of the isocline for zero population growth for species 2 can be interpreted in a similar way.

Putting the isoclines of zero population growth for the two species on the same axis allows us to predict if one species will exclude the other or whether the two species will coexist. The precise prediction depends on the relative orientation of the two isoclines. As shown in figure 13.14, there are four possibilities.

The Lotka-Volterra model predicts that one species will exclude the other when the isoclines do not cross. If the isocline for species 1 lies above that of species 2, species 1 will eventually exclude species 2. This exclusion occurs because all growth trajectories lead to the point where $N_1 = K_1$ and $N_2 = 0$ (see fig. 13.14b portrays the opposite situation in which the isocline for species 2 lies completely above that of species 1 and species 2 excludes species 1. In this case, all trajectories of population growth lead to the point where $N_2 = K_2$ and $N_1 = 0$.

Coexistence is possible only in the situations in which the isoclines cross. However, only one of these situations leads to stable coexistence. Figure 13.14c shows the situation in which coexistence is possible at the point where the isoclines of zero population growth cross but coexistence is unstable. In this situation, $K_1 > K_2/\alpha_{21}$ and $K_2 > K_1/\alpha_{12}$ and most population growth trajectories lead either to the points where $N_1 = K_1$ and $N_2 = 0$ or to where $N_2 = K_2$ and $N_1 = 0$. The populations of species 1 and 2 may arrive at the point where the lines cross, but any environmental variation that moves the populations off this point eventually leads to exclusion of one species by the other. Figure 13.14d represents the only situation that predicts stable coexistence of the two species. In this situation, $K_2/\alpha_{21} > K_1$

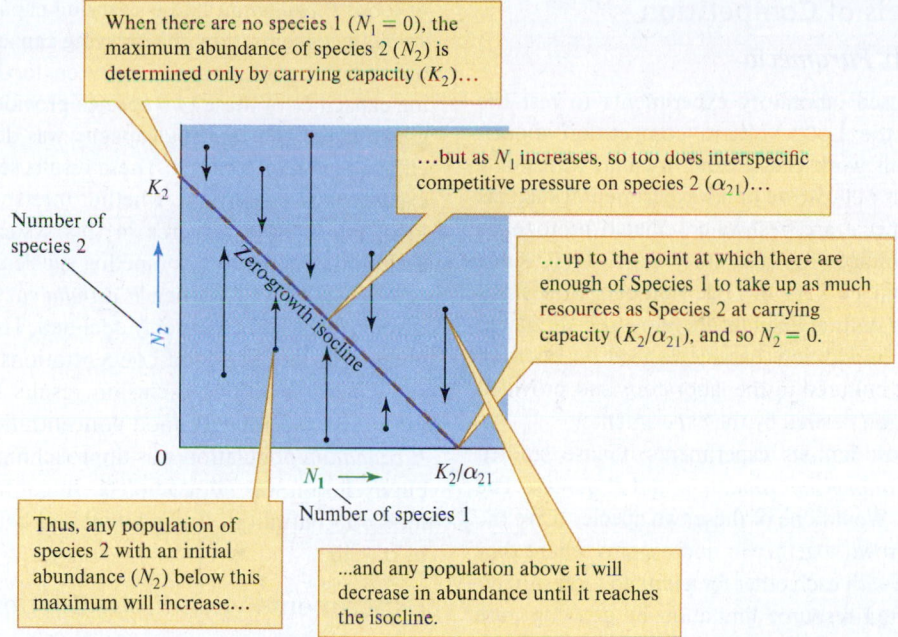

When there are no species 1 ($N_1 = 0$), the maximum abundance of species 2 (N_2) is determined only by carrying capacity (K_2)...

...but as N_1 increases, so too does interspecific competitive pressure on species 2 (α_{21})...

...up to the point at which there are enough of Species 1 to take up as much resources as Species 2 at carrying capacity (K_2/α_{21}), and so $N_2 = 0$.

Thus, any population of species 2 with an initial abundance (N_2) below this maximum will increase...

...and any population above it will decrease in abundance until it reaches the isocline.

Figure 13.13   Lotka-Volterra models make predictions of how populations will change in response to both intraspecific and interspecific competition for resources. The x-axis is the number of species 1 (N_1), and the y-axis is the number of species 2 (N_2). The zero population growth isocline for species 2 (in blue) designates the maximum population abundance that can be maintained, that is when $dN_2/dt = 0$. The slope of the line is determined by the carrying capacity (K_2, a function of intraspecific competition) and the affect of interspecific competition with the other species (a). The point K_2/α_{12} for species 2 is where no individuals of that species is present, but instead the carrying capacity has been reached with the equivalent number of species 1. The black arrows show how any given abundance of species 2 would be expected to change (increase or decrease) depending on whether it is above or below the zero-growth isocline.

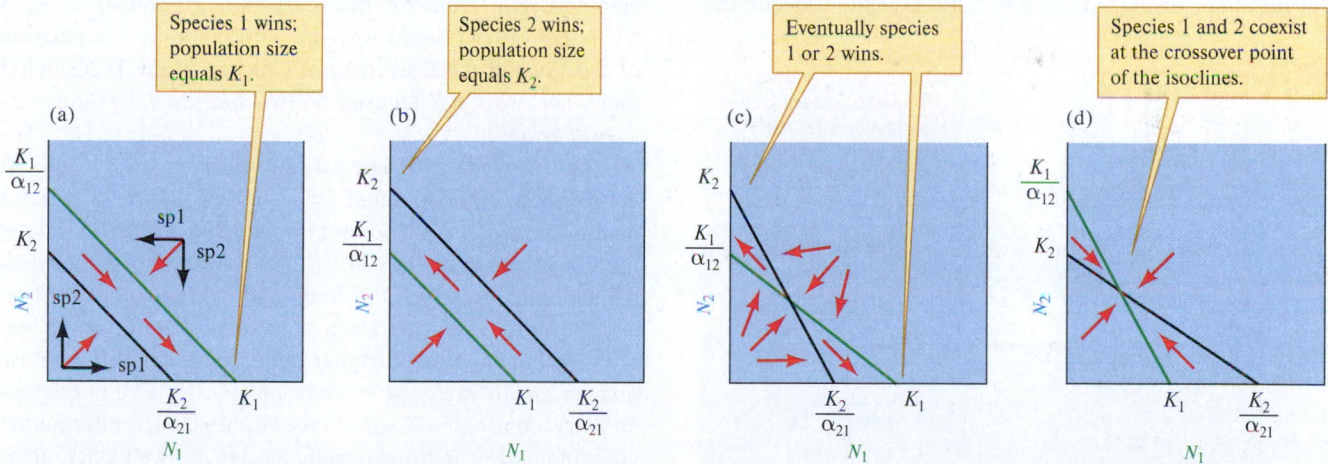

Species 1 wins; population size equals K_1.

Species 2 wins; population size equals K_2.

Eventually species 1 or 2 wins.

Species 1 and 2 coexist at the crossover point of the isoclines.

Figure 13.14   The outcomes of competition as predicted by Lotka-Volterra models, determined by the isoclines of zero population growth for species 1 (in green) and species 2 (in blue). Given a starting abundance of each species (N_1 and N_2), these graphs show us how the population is likely to change in response to competition until it reaches a stable point. Red arrows show the combined trajectories of population change predicted for both species 1 and 2. (a) Where the isocline for species 1 is greater than for species 2; (b) the opposite condition, where species 2 will dominate. (c) The situation when intraspecific competition is greater than interspecific competition for both species (i.e., $K_1 > K_2/\alpha_{21}$ and $K_2 > K_1/\alpha_{12}$); (d) the outcome when intraspecific competition is greater than interspecific competition for both species (i.e., $K_1 < K_2/\alpha_{21}$ and $K_2 < K_1/\alpha_{12}$).

and $K_1/\alpha_{12} > K_2$ and all growth trajectories lead to the point where the isoclines of zero population growth cross.

What is the biological meaning of saying that all growth trajectories lead to the point where the isoclines of zero growth cross? What this means is that the relative abundances of species 1 and 2 will eventually arrive at the point where the isoclines cross, a point where the abundances of both species are greater than

zero. In this situation, each species is limited more by members of their own species than they are by members of the other species. In other words, the Lotka-Volterra model predicts that species coexist when intraspecific competition is stronger than interspecific competition. This prediction is supported by the results of laboratory experiments on interspecific competition.

Laboratory Models of Competition

Experiments with *Paramecia*

G. F. Gause (1934) used laboratory experiments to test the major predictions of the Lotka-Volterra competition model. During the course of his work Gause experimented with many organisms, but the most well-known of his experimental subjects were paramecia. Paramecia are freshwater, ciliated protozoans that offer several advantages for laboratory work. First, since they are small, they can be kept in large numbers in a small space and some of their natural habitats are fairly well simulated by laboratory aquaria. In addition, paramecia feed on microorganisms, which can be cultured in the laboratory and provided in whatever concentration desired by the experimenter.

In one of his most famous experiments, Gause studied competition between *Paramecium caudatum* and *P. aurelia*. The question he posed was: Would one of these two species drive the other to extinction if grown together in microcosms where they were forced to compete with each other for a limited food supply?

Gause demonstrated resource limitation by growing pure populations of *P. caudatum* and *P. aurelia* in the presence of two different concentrations of their food, the bacterium *Bacillus pyocyaneus*. Gause observed sigmoidal growth with an obvious carrying capacity at both full- and half-strength concentrations of the food supply (fig. 13.15). In a full-strength concentration of food, the carrying capacity of *P. aurelia* was 195. When food availability was halved, the carrying capacity of this species was reduced to 105. *P. caudatum* showed a similar response to food concentration. In the presence of a full-strength concentration of food, *P. caudatum* had a carrying capacity of 137. At a half-strength concentration, the carrying capacity was 64. The nearly one-to-one correspondence between food level and the carrying capacities of these two species provides evidence that when grown alone, the carrying capacity was determined by intraspecific competition for food. These results set the stage for Gause's experiment to determine whether interspecific competition for food, the limiting resource in this system, would lead to the exclusion of one of the competing species.

When grown together, *P. aurelia* survived, while the population of *P. caudatum* quickly declined. The difference in results obtained at the two food concentrations supports the conclusion that competitive exclusion results from competition for food. At a full-strength food concentration, the decline in the *P. caudatum* population was approaching exclusion by 16 days but exclusion was not complete. In contrast, at a half-strength food concentration, *P. caudatum* had been entirely eliminated by day 16.

Experiments with Flour Beetles

Tribolium, beetles of the family Tenebrionidae, infest stored grains and grain products. The discovery of an infestation of these beetles in an urn of milled grain in the tomb of an Egyptian pharaoh buried about 4,500 years ago suggests that these beetles have been engaged in this occupation for some time. Their habit of attacking stored grains makes them a convenient laboratory model. Since all life stages of *Tribolium* live in finely milled flour, small containers of flour provide all the environmental requirements to sustain a population. R. N. Chapman (1928) began working with laboratory populations of *Tribolium* at the University of Chicago in the 1920s, where ever since, work has focused on two species: *T. confusum* and *T. castaneum.*

Thomas Park (1954) worked extensively on interspecific competition between these two species under six environmental conditions, including hot-wet (34°C, 70% RH, relative humidity) and cool-dry (24°C, 30% RH). In hot-wet environments, both species established healthy populations that persisted over the entire duration of the experiment (fig. 13.16*a*). However, when grown together under these conditions, *T. castaneum* usually excludes *T. confusum* (fig. 13.16*b*). In contrast, cool-dry conditions favor *T. confusum.* Under intermediate environmental conditions, each species did well when grown alone but the outcome of interspecific competition was not completely predictable. Under such intermediate conditions, the species establishing itself in greater numbers first generally wins out in competition, a phenomenon known as the **priority effect.**

How can we interpret the results of these laboratory experiments in terms of the effects of competition on these species' niches? Growing the two species separately showed that the fundamental niches of both species include a broad range of environmental conditions. However, growing the two species together suggests that interspecific competition restricts the realized niches of both species to fewer environmental conditions. We turn now to consider how competition influences the niches of species in nature.

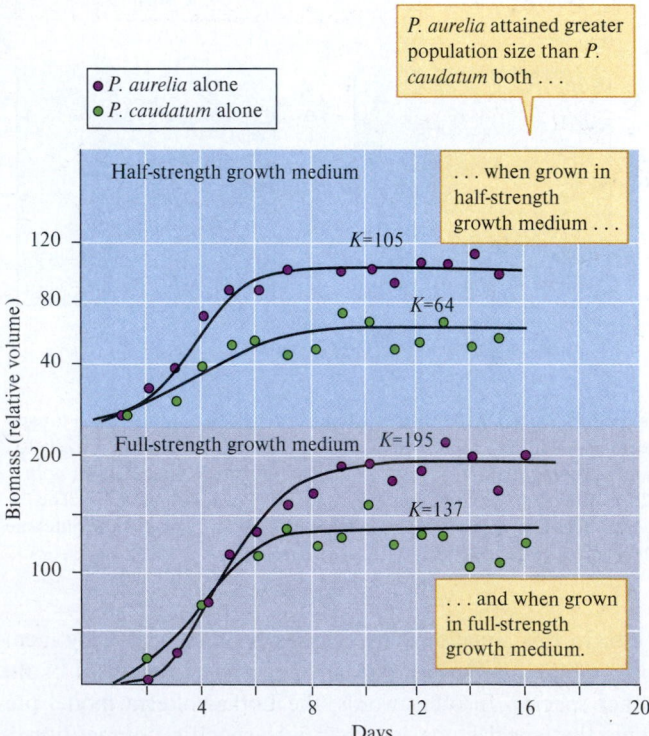

Figure 13.15 Population growth and population sizes attained by *Paramecium aurelia* and *P. caudatum* grown separately (data from Gause 1934).

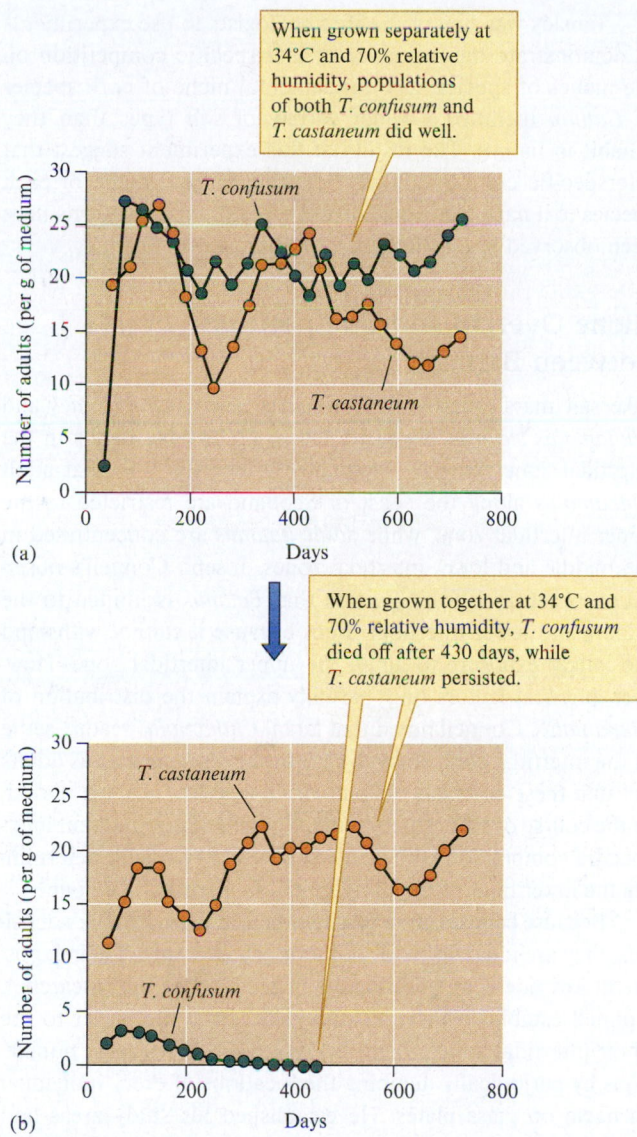

When grown separately at 34°C and 70% relative humidity, populations of both *T. confusum* and *T. castaneum* did well.

When grown together at 34°C and 70% relative humidity, *T. confusum* died off after 430 days, while *T. castaneum* persisted.

Figure 13.16 Populations of *Tribolium confusum* and *T. castaneum* grown separately (*a*) and together (*b*) under hot-wet conditions (data from Park 1954).

Concept 13.3 Review

1. *Paramecium aurelia* and *P. caudatum* coexisted for a long period when fed full-strength food compared to when they were fed half that amount. What does this contrast in the time to competitive exclusion suggest about the role of food supply on competition between these two species?

2. Can we conclude that interspecific competition commonly restricts species to realized niches in nature, based on the results of mathematical models and laboratory experiments?

3. Is there any way that predators could alter the outcome of competition as shown in figure 13.14*a*, where species 1 excludes species 2, and in figure 13.14*b*, where species 2 excludes species 1?

13.4 Competition and Niches

LEARNING OUTCOMES

After studying this section you should be able to do the following:

13.16 Describe and discuss evidence connecting interspecific competition to the difference between fundamental and realized niches.

13.17 Define the terms allopatric, sympatric, and character displacement; and discuss the relevance of these terms to our understanding of the evolutionary influence of competition on species niches.

13.18 Discuss evidence suggesting character displacement among Darwin's finches.

13.19 Summarize what decades of field research has taught us about the ecological and evolutionary importance of competition in biological communities.

Competition can have significant ecological and evolutionary influences on the niches of species. Competition can have short-term ecological effects on the niches of species by restricting them to realized niches, yet these species may retain their capacity to inhabit the fuller range of environments we call the fundamental niche. However, if competitive interactions are strong and pervasive enough, they may produce an evolutionary response in the competitor population—an evolutionary response that changes the fundamental niche. In this section, we explore the evidence for both ecological and evolutionary influences on the niches of natural populations. Field experiments show that interspecific competition may restrict the niches of populations in nature.

Niches and Competition Among Plants

Arthur Tansley (1917) conducted one of the first experiments to test whether competition was responsible for the separation of two species of plants on different soil types. In the introduction to his paper, Tansley pointed out that while the separation of closely related plants had long been attributed to mutual competitive exclusion, it was necessary to perform manipulative experiments to demonstrate that this interpretation is correct. That is exactly what Tansley did in pioneering work to account for the mutually exclusive distributions of *Galium saxatile* and *G. sylvestre* (now *G. pumilum*), two species of small perennial plants commonly called bedstraw (fig. 13.17). In the British Isles, *G. saxatile* is largely confined to acidic soils and *G. sylvestre* to basic limestone soils.

Tansley conducted his experiment at the Cambridge Botanical Garden from 1911 to 1917, where seeds of the two species of plants were sown in planting boxes of acidic and basic soils. The seeds were sowed in single-species plantings and in mixtures of the two species. Both species germinated on both soil types, in both single- and mixed-species plantings (fig. 13.18). Like the paramecia studied by Gause, both *Galium* species established healthy populations on both soil types when grown by themselves and these single-species plantings persisted to the end

Figure 13.17 These two species of bedstraw grow predominately on different soil types: *Galium saxatile* (shown here) grows mainly on acidic soils, whereas *G. sylvestre (now G. pumilum)* grows mainly on basic limestone soils.   Heather Angel/Natural Visions

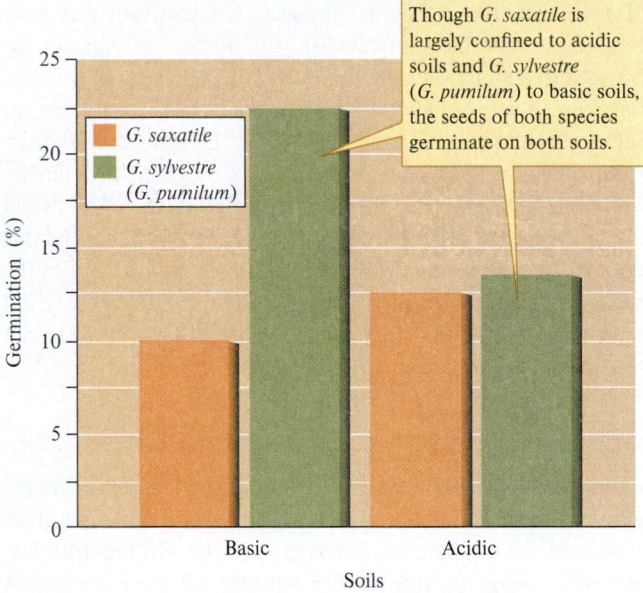

Though *G. saxatile* is largely confined to acidic soils and *G. sylvestre* (*G. pumilum*) to basic soils, the seeds of both species germinate on both soils.

Figure 13.18 Percentage seed germination by *Galium saxatile* and *G. sylvestre (G. pumilum)* in basic calcareous soils and acidic peat soil (data from Tansley 1917).

of the 6-year study. However, as the two species grew in mixed plantings, Tansley observed clear competitive dominance by each species on its normal soil type.

On limestone soils, *G. sylvestre,* the species naturally found on limestone soils, overgrew and eliminated *G. saxatile,* the acidic soil species, by the end of the first growing season. On acidic soils, the relationship was reversed and *G. saxatile* was competitively dominant but competitive exclusion was not completed. Growth by both species was so slow on the acidic soils that it took until the end of the 6-year experiment for *G. saxatile* to completely cover the planting boxes containing acidic soils, a density attained by *G. sylvestre* on limestone soils in just 1 year. However, among the abundant *G. saxatile* in the mixed plantings, Tansley found a few "quite healthy" plants of *G. sylvestre.*

Tansley was one of the first ecologists to use experiments to demonstrate the influence of interspecific competition on the niches of species. The fundamental niche of both species of *Galium* included a wider variety of soil types than they inhabit in nature. The results of this experiment suggest that interspecific competition restricts the realized niche of each species to a narrower range of soil types. Similar patterns have been observed in intertidal environments.

Niche Overlap and Competition Between Barnacles

Like salt marsh plants, the barnacles *Balanus balanoides* and *Chthamalus stellatus* are restricted to predictable bands in the intertidal zone. We saw in chapter 9 (see fig. 9.8) that adult *Chthamalus* along the coast of Scotland are restricted to the upper intertidal zone, while adult *Balanus* are concentrated in the middle and lower intertidal zones. Joseph Connell's observations (1961a, 1961b) indicate that *Balanus* is limited to the middle and lower intertidal zones because it cannot withstand the longer exposure to air in the upper intertidal zone. However, physical factors only partially explain the distribution of *Chthamalus.* Connell noted that larval *Chthamalus* readily settle in the intertidal zone below where the species persists as adults but that these colonists die out within a relatively short period. In the course of field experiments, Connell discovered that interspecific competition with *Balanus* plays a key role in determining the lower limit of *Chthamalus* within the intertidal zone.

Because barnacles are sessile, small, and grow in high densities, they are ideal for field studies of survivorship. Their exposure at low tide is an additional convenience for the researcher. Connell established several study sites from the upper to the lower intertidal zones where he kept track of barnacle populations by periodically mapping the locations of every individual barnacle on glass plates. He established his study areas and made his initial maps in March and April of 1954, before the main settlement by *Balanus* in late April. He divided each of the study areas in half and kept one of the halves free of *Balanus* by scraping them off with a knife. Connell determined which half of each study site to keep *Balanus*-free by flipping a coin.

By periodically remapping the study sites, Connell was able to monitor interactions between the two species and the fates of individual barnacles. The results showed that in the middle intertidal zone, *Chthamalus* survived at higher rates in the absence of *Balanus* (fig. 13.19). *Balanus* settled in densities up to 49 individuals per square centimeter in the middle intertidal zone and grew quickly, crowding out *Chthamalus* in the process. In the upper intertidal zone, removing *Balanus* had no effect on *Chthamalus* survivorship because the population density of *Balanus* was too low to compete seriously. Connell's results provided direct evidence that *Chthamalus* is excluded from the middle intertidal zone by interspecific competition with *Balanus.*

How does interspecific competition affect the niche of *Chthamalus*? In the absence of *Balanus,* it can live over a broad zone from the upper to the middle intertidal zones. Using the terminology of Hutchinson (1957), we can call this broad range of physical conditions the *fundamental niche* of *Chthamalus.*

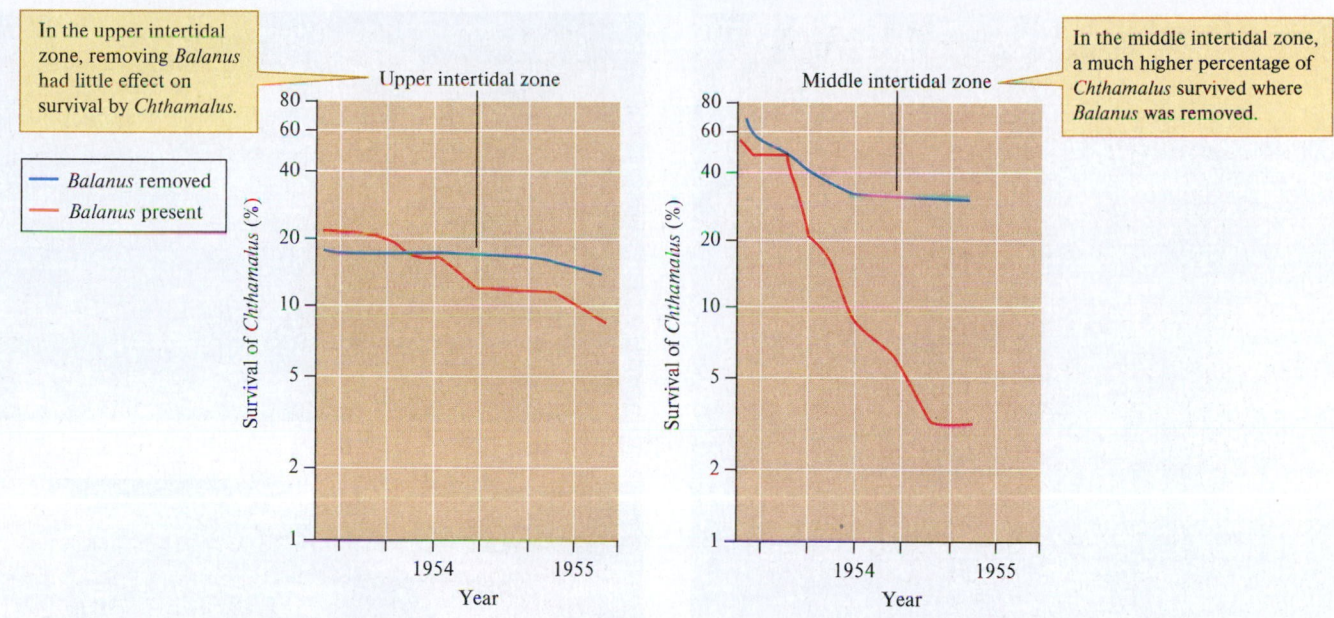

In the upper intertidal zone, removing *Balanus* had little effect on survival by *Chthamalus*.

Upper intertidal zone

—— *Balanus* removed
—— *Balanus* present

In the middle intertidal zone, a much higher percentage of *Chthamalus* survived where *Balanus* was removed.

Middle intertidal zone

Figure 13.19 A competition experiment with barnacles: removal of *Balanus* and survival by *Chthamalus* in the upper and middle intertidal zones (data from Connell 1961a, 1961b).

However, competition largely restricts *Chthamalus* to the upper intertidal zone, a more restricted range of physical conditions constituting the species' *realized niche* (fig. 13.20).

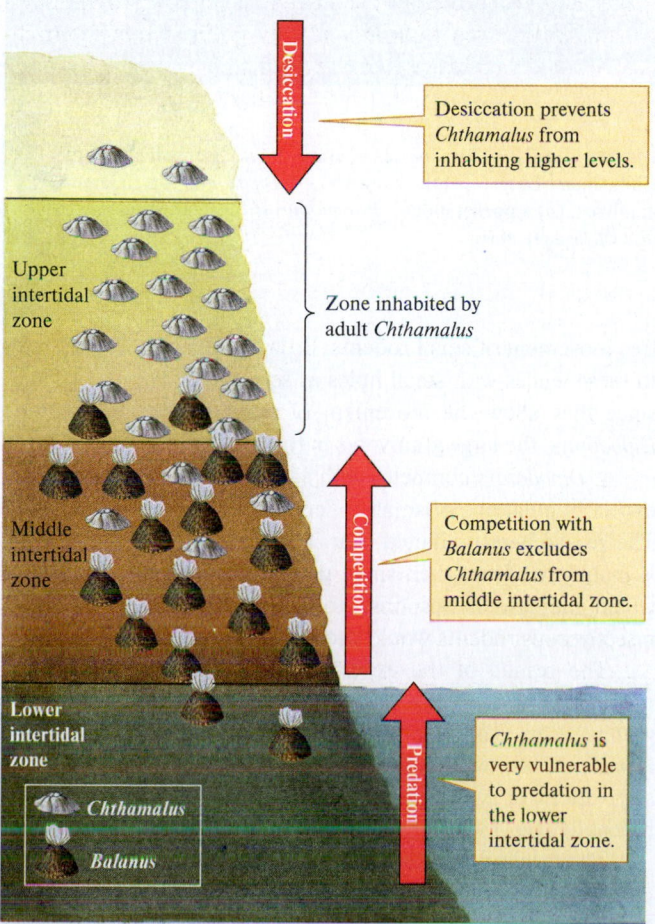

Desiccation prevents *Chthamalus* from inhabiting higher levels.

Zone inhabited by adult *Chthamalus*

Competition with *Balanus* excludes *Chthamalus* from middle intertidal zone.

Chthamalus is very vulnerable to predation in the lower intertidal zone.

Chthamalus

Balanus

Figure 13.20 Environmental factors restricting the distribution of *Chthamalus* to the upper intertidal zone.

Does variation in interspecific competition completely explain the patterns seen by Connell? At the lowest levels in the lower intertidal zone, *Chthamalus* suffered high mortality even in the absence of *Balanus* (see fig. 13.19). What other factors might contribute to high rates of mortality by *Chthamalus* in the lower intertidal zone? Experiments have shown that this species can withstand periods of submergence of nearly 2 years, so it seems that it is not excluded by physical factors. It turns out that the presence of predators in the lower intertidal zone introduces complications that we will discuss in chapter 14 when we examine the influences of predators on prey populations. Intertidal plants are also subject to competition.

Competition and the Niches of Small Rodents

One of the most ambitious and complete of the many field experiments ecologists have conducted on competition among rodents focused on desert rodents in the Chihuahuan Desert near Portal, Arizona. This experiment, conducted by James H. Brown and his students and colleagues (Munger and Brown 1981; Brown and Munger 1985), is exceptional in many ways. First, it was conducted at a large scale; the 20 ha study site includes 24 study plots, each 50 m by 50 m (fig. 13.21). Second, the experimental trials have been well replicated, both in space and in time. Third, the project has been long term; it began in 1977 and continued over three decades. These three characteristics combine to demonstrate subtle ecological relationships and phenomena that would not otherwise be apparent.

The rodent species living on the Chihuahuan Desert study site can be divided into groups based on size and feeding habits. Most species are **granivores,** rodents that feed chiefly on seeds. The large granivores consist of three species of kangaroo rats (fig. 13.22*a*) in the genus *Dipodomys*—*D. spectabilis*, 120 g; *D. ordi*, 52 g; and *D. merriami*, 45 g. In addition, the study site is home to

Figure 13.21   Aerial photo showing the placement of 24 study plots, each 50 m by 50 m, in the Chihuahuan Desert near Portal, Arizona (courtesy of J. H. Brown).   Dr. James H. Brown

(a)

(b)

Figure 13.22   Two species of granivorous rodents living in the Chihuahuan Desert: (*a*) the kangaroo rat, *Dipodomys* spp., a large granivore; (*b*) a pocket mouse, *Perognathus* sp., a small granivore. (*a,b*) Dr. James H. Brown

four species of small granivores (fig. 13.22*b*)—*Perognathus penicillatus,* 17 g; *P. flavus,* 7 g; *Peromyscus maniculatus,* 24 g; *Reithrodontomys megalotis,* 11 g—and two species of small insectivorous rodents—*Onychomys leucogaster,* 39 g, and *O. torridus,* 29 g.

In one experiment, Brown and his colleagues set out to determine whether large granivorous rodents (*Dipodomys* spp.) limit the abundance of small rodents on their Chihuahuan Desert study site. They also wanted to know whether the rodents might be competing for food. The researchers addressed their questions with a field experiment in which they enclosed 50 m by 50 m study plots with mouse-proof fences. The fences were constructed with a wire mesh with 0.64 cm openings, which were too small for any of the rodent species to crawl through. They also buried the fencing 0.2 m deep so the mice couldn't dig under it, and they topped the fences with aluminum flashing so the mice couldn't climb over it. This may sound like a lot of work, but to answer their questions, the researchers had to control the presence of rodents on the study plots.

The researchers next cut holes 6.5 cm in diameter in the sides of all the fences to allow all rodent species to move freely in and out of the study plots. With this arrangement in place, the rodents in the study plots were trapped live and marked once a month for 3 months. Following this initial monitoring period, the holes on four of eight study plots were reduced to 1.9 cm, small enough to exclude *Dipodomys* but large enough to allow

free movement of small rodents. Brown and his colleagues refer to these fences with small holes as semipermeable membranes, since they allow the movement of small rodents but exclude *Dipodomys,* the large granivores in this system.

If *Dipodomys* competes with small rodents, how would you expect populations of small rodents to respond to its removal? The researchers predicted that if competition among rodents is mainly for food, then small granivorous rodent populations would increase in response to *Dipodomys* removal, whereas insectivorous rodents would show little or no response.

The results of the experiment were consistent with the predictions. During the first 3 years of the experiment, small granivores were approximately 3.5 times more abundant on the *Dipodomys* removal plots compared to the control plots, while populations of small insectivorous rodents did not increase significantly (fig. 13.23).

The results presented in figure 13.23 support the hypothesis that *Dipodomys* spp. competitively suppress populations of small granivores. But would they do so again in response to another experimental manipulation? We cannot be certain unless we repeat the experiment. That's just what Edward Heske,

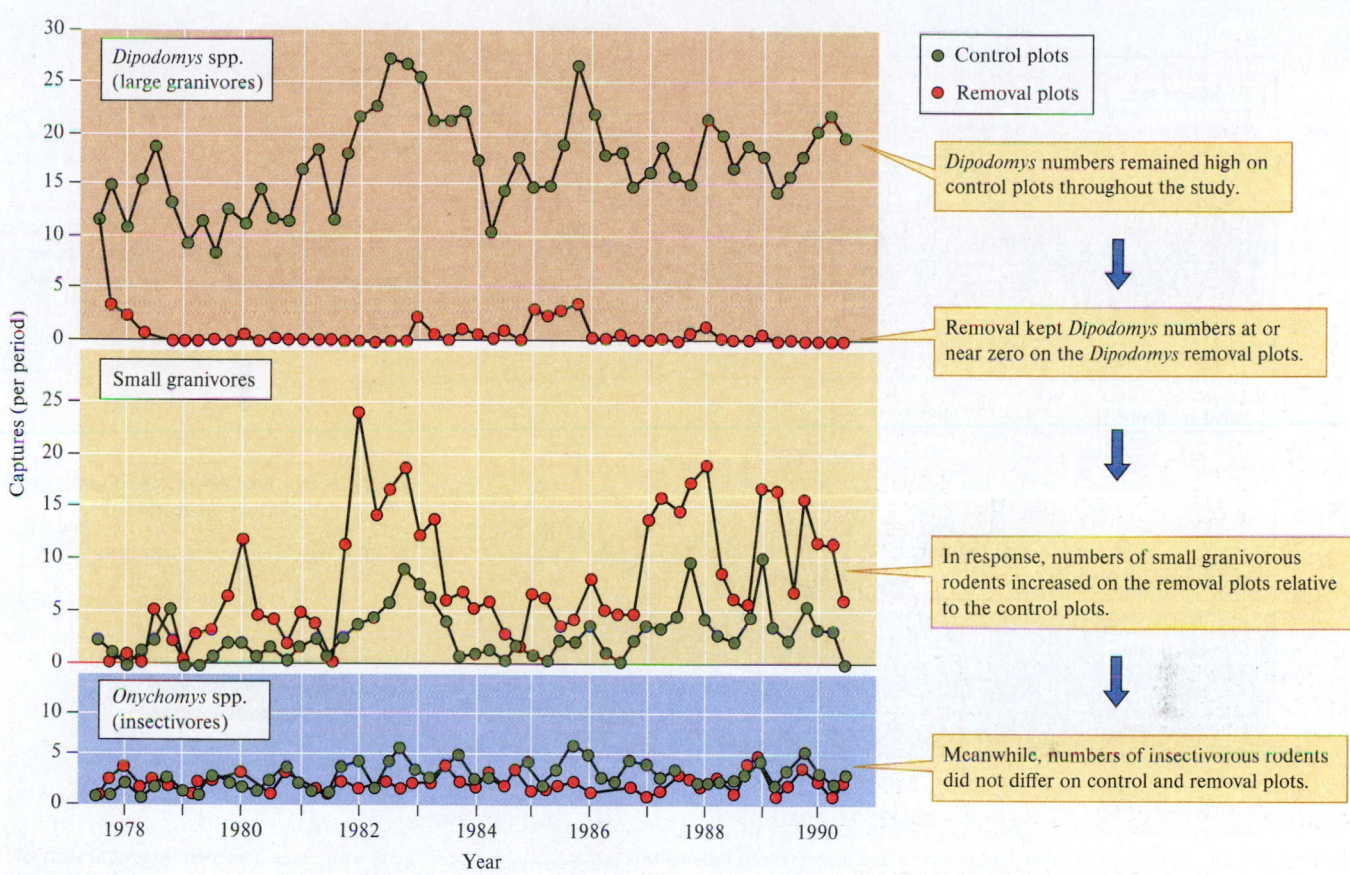

Figure 13.23   Responses by small granivorous and insectivorous rodents to removal of large granivorous *Dipodomys* species (data from Heske, Brown, and Mistry 1994).

James H. Brown, and Shahroukh Mistry (1994) did. In 1988, they selected eight other fenced study plots that they had been monitoring since 1977, installed their semipermeable barriers on four of the plots, and removed *Dipodomys* from them. The result was an almost immediate increase in small granivore populations on the removal plots (fig. 13.24). Brown and his students and colleagues have continued to monitor these study plots for more than 30 years. During this period, they have made discoveries about interactions among rodent species, the influences of a changing climate on the vegetative community, and the great impact of infrequent, catastrophic climatic events on rodent populations—discoveries that would have been impossible in a short-term field study. For instance, Katherine H. Thibault and James H. Brown (2008) report how a single, 2-hour storm "reset" the relative abundances of large and small granivores on their study plots and how the altered abundances had continued through the 8 years prior to their publication. Undoubtedly, such discoveries would continue if this monitoring of this landmark study was sustained. As we see next, long-term competition between two species can lead to evolutionary divergence in morphology.

Character Displacement

Because interspecific competition reduces the fitness of competing individuals, those individuals that compete less should have higher fitness than individuals that compete more.

Consequently, interspecific competition has been predicted to lead to **character displacement**—the circumstance in which two species differ more from each other in geographic areas where they occur together than where their distributions do not overlap.

Darwin's finches *Geospiza fortis,* the medium ground finch, and *G. fuliginosa,* the small ground finch, provide one of the most convincing cases of character displacement. These two species occur apart from each other, that is, they are **allopatric,** on Daphne Major and Los Hermanos Islands and occur together, that is, they are **sympatric,** on the island of Santa Cruz (fig. 13.25). Where the two species are allopatric, they have very similar beak sizes. However, where they are sympatric, the sizes of their beaks do not overlap. The allopatric *G. fortis* on Daphne Major have smaller beaks than those sympatric with *G. fuliginosa* on Santa Cruz, while the *G. fuliginosa* on Los Hermanos Island have beaks that are significantly larger than those sympatric with *G. fortis* on Santa Cruz. Since beak size correlates with diet in Darwin's finches, we can say that the sympatric populations of the two species on Santa Cruz have different feeding niches. Natural selection has apparently favored divergence in the beak morphology of these sympatric populations (Lack 1947; Schluter, Price, and Grant 1985; Grant 1986).

Other studies have demonstrated similar patterns of character displacement among a variety of animal species,

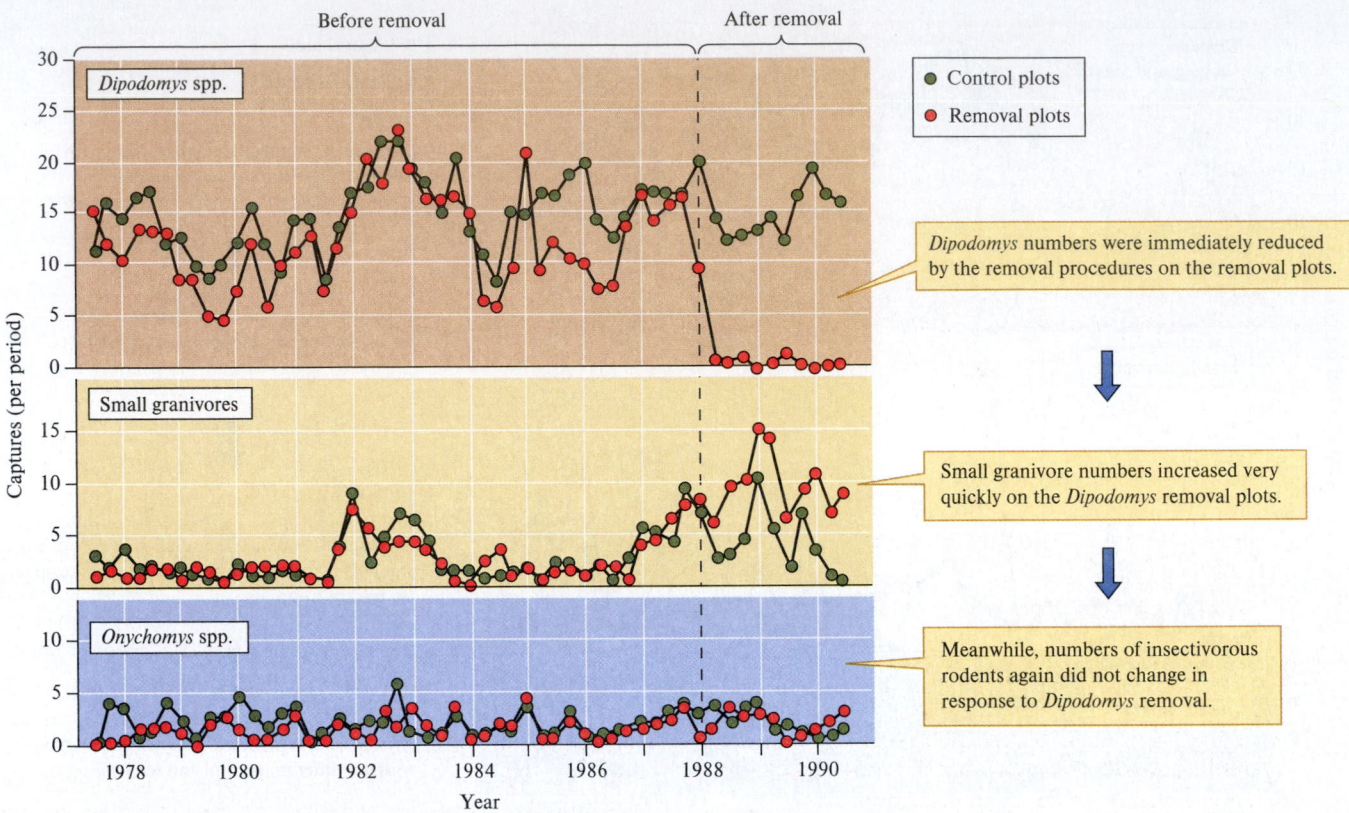

Figure 13.24 Responses of small granivorous and insectivorous rodents to a second removal experiment, which was preceded by several years of study before initiating *Dipodomys* removal (data from Heske, Brown, and Mistry 1994).

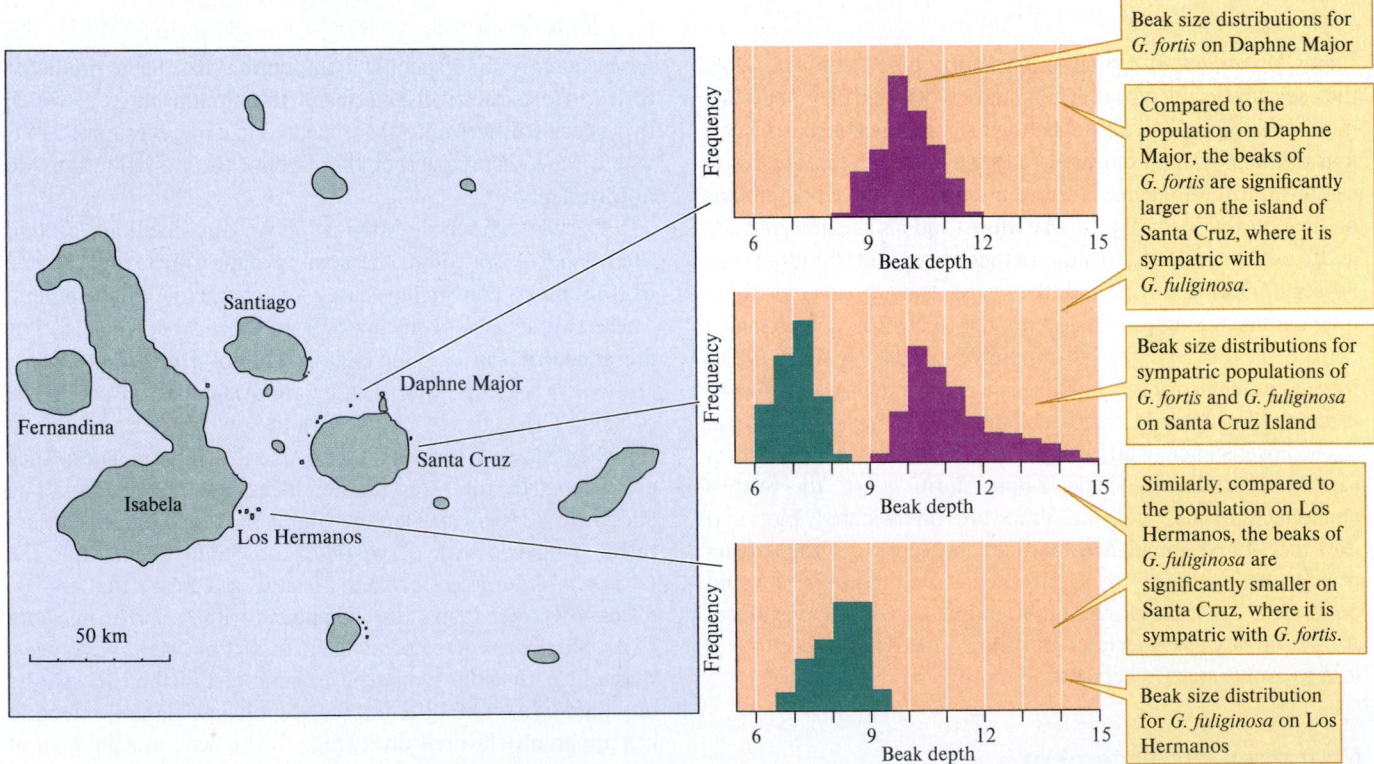

Figure 13.25 Evidence for character displacement in beak size in populations of Darwin's finches *Geospiza fortis* and *G. fuliginosa* (data from Grant 1986).

including *Cnemidophorus* lizards on islands off the coast of Baja California, *Anolis* lizards on Caribbean islands, and sticklebacks inhabiting small lakes around Vancouver Island, Canada. Character displacement has also been observed in laboratory populations of bean weevils. Many studies have provided preliminary data suggesting character displacement among populations but not establishing definitive evidence. Why is that? The main reason is that a definitive demonstration requires a great deal of evidence that is difficult to provide.

Mark Taper and Ted Case (1992) list six criteria that must be met to build a definitive case for character displacement:

1. Morphological differences between a pair of sympatric species (e.g., *G. fortis* and *G. fuliginosa* on Santa Cruz Island) are statistically greater than the differences between allopatric populations of the same species (*G. fortis* on Daphne Major and *G. fuliginosa* on Los Hermanos Island).
2. The observed differences between sympatric and allopatric populations have a genetic basis.
3. Differences between sympatric and allopatric populations must have evolved in place and they must not be due to the sympatric and allopatric populations having been derived from different founder populations already differing in the character under study (e.g., beak size).
4. Variation in the character (e.g., beak size) must have a known effect on use of resources (e.g., seed sizes).
5. There must be demonstrated competition for the resource under question (e.g., food), and competition must be directly correlated with similarity in the character (e.g., overlap in beak size).
6. Differences in the character cannot be explained by differences in the resources available to sympatric and allopatric populations (e.g., differences in the availability of seeds on one island versus another).

You can see how difficult it would be to satisfy all six of these criteria. It is fitting that one of the few studies that addresses all six criteria reasonably well is that of Darwin's finches (Grant 1986; Taper and Case 1992), in the place where Darwin started the whole discussion. Since these studies, there have been an increasing number of convincing demonstrations of character displacement (Dayan and Simberloff 2005).

Evidence for Competition in Nature

What have we learned since Darwin initiated our enduring discussion of the role of competition as an organizing force in nature? The study of competition has gone through several phases. There was an early theoretical phase, followed by work with laboratory models, which was in turn followed by intensive observation and experimentation in the field. These phases were followed by a period of vigorous questioning of the assumption that competition is an important

force in nature. One of the significant research developments during this period was the testing of supposed cases of character displacement against randomly assembled sets of species and the application of proper statistical analysis of patterns of morphological variation across communities (e.g., Simberloff and Boeklin 1981; Strong, Szyska, and Simberloff 1981). This questioning forced renewed attention to careful experimental design and stimulated a reanalysis of past research in order to weigh the existing evidence bearing on competition in nature.

In order to identify the importance of competition in nature, Jessica Gurevitch and colleagues used meta-analysis (see chapter 6) to evaluate field research on species interactions that had been published in six different journals over the previous decade (Gurevitch et al. 1992). This yielded 46 articles that reported data on 217 comparisons of competition versus no competition in plants and animals. By this point, studies of competition had become much more robust, with more replication than had been done in the early years. Gurevitch found strong effects of competition across studies, but also that the effect of competition was quite different between types of organisms. Herbivores had the greatest competitive effects on each other, with primary producers (plants) next, and carnivores showing the least effects. As expected, they also found that intraspecific competition was more intense on average than interspecific competition. More recent reviews have also confirmed this finding (e.g., Adler et al. 2018). Thus, it is clear, after decades of research that competition is an important force that contributes to the organization of nature.

What other forces besides competition may be responsible for the patterns of distribution and abundance that we observe in natural populations? We've already reviewed the influences of the physical environment (sections I, II, and III). In chapters 14 and 15 we will consider two other forms of biotic interaction: predation, herbivory, parasitism, and disease in chapter 14 and mutualism in chapter 15. But first, let's review how field experiments have been used to study competition between a native and an invasive marine snail.

Concept 13.4 Review

1. What do you think would have happened to the *Galium sylvestre* plants growing on acidic soil if Tasley had continued his experiment for a few more years? What does this experiment tell us about the influence of interspecific competition on the niches of the two *Galium* species?
2. What does the increase in small granivore populations but lack of response by populations of insectivorous rodents suggest about the nature of competition between rodents in Brown's Arizona study area (see fig. 13.23)?
3. Why did Brown and colleagues repeat their large granivore experiment (see fig. 13.24)?

Applications

Competition Between Native and Invasive Species

LEARNING OUTCOMES

After studying this section you should be able to do the following:

13.20 Describe the field experiments used to study competition between *Batillaria attramentaria,* an invasive marine snail, and *Cerithidea californica,* a snail native to coastal California and Baja California.

13.21 Explain the key difference in the biology of *Batillaria* and *Cerithidea* that indicates the mechanism underlying *Batillaria's* competitive superiority.

Some of the most significant contemporary environmental problems involve invasive species (see chapter 3, section 3.2). Because of the ecological disruption caused by invasive species, it is important to understand the mechanisms allowing them to invade communities of native species. Such studies can also help us improve our understanding of ecological relationships generally.

One of the principal mechanisms thought to allow introduced species to invade communities of native species is superior competitive ability. James Byers (2000) used field experiments to explore the ecological relationships of native and an invasive species of mud snail on the Pacific Coast of North America. The native mud snail, *Cerithidea californica,* occurs from San Ignacio Bay, Mexico, to Tomales Bay, California. Once abundant throughout this range, *Cerithidea* has declined in abundance in some bays, while an invasive mud snail, *Batillaria attramentaria,* has increased in abundance. In places along the California coast where *Batillaria* has reached very high densities—for example, up to 10,000 individuals per m^2 in Elkhorn Slough—*Cerithidea* has disappeared entirely. *Batillaria* was introduced accidentally to the California coast with the purposeful introduction of Japanese oysters early in the twentieth century. Because it does not have a planktonic larval stage, *Batillaria* has remained largely restricted to the bays where it was introduced.

Byers conducted his field studies of *Cerithidea* and *Batillaria* in one of the bays where they co-occur: Bolinas Lagoon, approximately 20 km north of San Francisco, California. In the first phase of his study, Byers determined the influence of *Cerithidea* and *Batillaria* on their main food supply, benthic diatoms. He placed each of the snails in single-species groups in 35 cm diameter cages, embedded in the sediments along a tidal channel within Bolinas Lagoon. The densities of *Cerithidea* and *Batillaria* in these cages were 0, 12, 23, 35, 46, 69, and 92 individuals. Byers set up several replicates of each of these snail densities.

When Byers compared the influences of *Cerithidea* and *Batillaria* on diatoms, he found no difference in their effect on diatom densities. Figure 13.26 shows the average density of diatoms after approximately 39 days in cages containing a low density (12 per cage) versus a higher density (92 per cage) of large snails. These results indicate that both snail species reduce diatom cover when they are present at high population densities and that both *Cerithidea* and *Batillaria* have the

potential to reduce the availability of their food supply. This part of Byers's experiment documents a potential for resource competition between the two species.

Having shown the influence of the snails on their food supply, Byers next explored the effects of snail density on their growth rates. He found that the growth rates of both *Cerithidea* and *Batillaria* decline as their population densities increase. However, at all population densities, *Batillaria* grows faster than *Cerithidea.* The relative growth rates of large individuals of these two species at densities of 12 and 46 individuals per cage are shown in figure 13.27. These results indicate that *Batillaria*

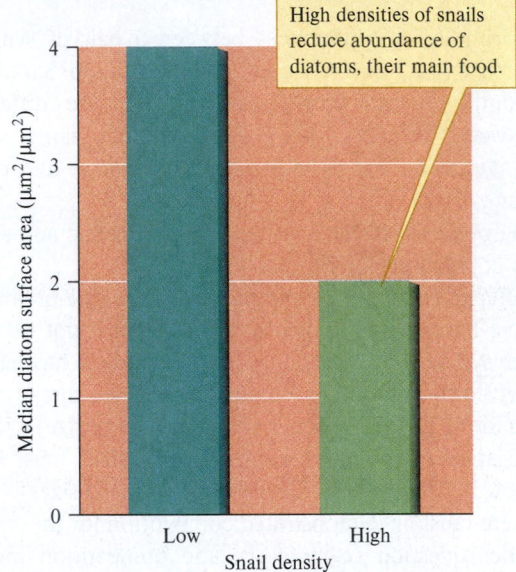

Figure 13.26 The effects of large *Batillaria attramentaria,* an invasive mud snail, and *Cerithidea californica,* a native mud snail, on the abundance of diatoms, expressed as diatom surface area per unit sediment area, within experimental enclosures after 39 days of residence (data from Byers 2000).

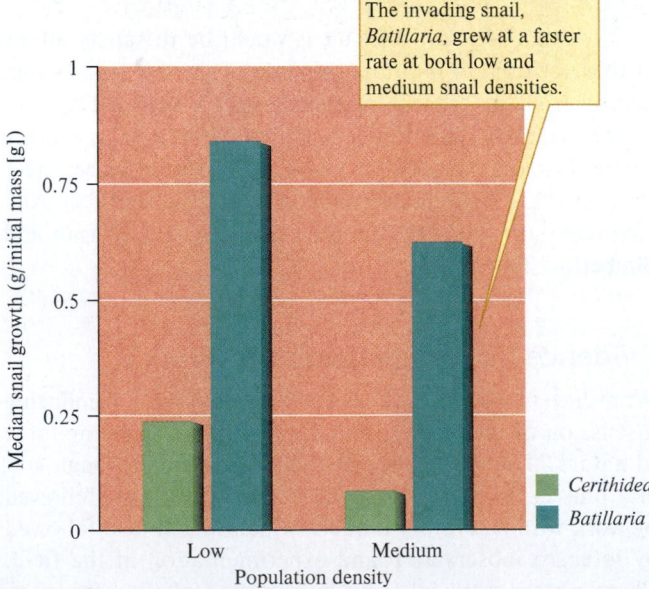

Figure 13.27 Growth, expressed as a proportion of initial mass, of large *Batillaria attramentaria,* an invasive mud snail, and *Cerithidea californica,* a native mud snail, over a period of 60 days within experimental enclosures (data from Byers 2000).

is much more efficient at converting available food into its own biomass. At high densities of 92 per cage, *Batillaria* continued to grow at a relatively high rate, while *Cerithidea* lost weight.

Byers used field experiments to show the potential for competition between native and an invasive species. His work also identified the mechanism involved, resource competition. In subsequent research, Byers and Goldwasser (2001) used the data generated by these experiments to build a computer simulation of the interaction between *Cerithidea* and *Batillaria*. That experimentally derived model predicts a time to competitive exclusion of *Cerithidea* by *Batillaria* of 55 to 70 years. This result matches the actual times to exclusion that have been recorded in the bays where *Batillaria* has been introduced. In conclusion, detailed experimental work on competitive interactions between native and invasive species has the potential to generate information that will predict the potential pathways of interaction between species and the time to competitive exclusion.

Summary

Interactions between organisms can result in positive (+), negative (−), or neutral (0) outcomes; in ecology the pairwise combinations of these are defined as competition (−/−), predation (+/−), mutualism (+/+), commensalism (+/0), amensalism (−/0), and neutralism (0/0). Careful scientific study is often necessary to determine impacts to individuals, and a single ecological phenomenon (such as plant-gall formation) can produce many different types of interactions. Seemingly neutral outcomes, as defined by commensalism, amensalism, and neutralism, will sometimes be a function of the limits to our ability to measure impacts, rather than a true lack of effect. For example, although one organism may be the winner in a contest for resources, both participants likely expend more energy than they would in the absence of the other, thus defining the interaction as competition (−/−).

Competition, is generally divided into *intraspecific competition,* competition between individuals of the same species, and *interspecific competition,* competition between individuals of different species. Competition can take the form of interference competition, involving direct interactions between individuals; or resource competition, in which individuals compete through their dependence on the same limiting resources.

Laboratory and field studies reveal intraspecific competition. Experiments with herbaceous plants show that soil nutrients may limit plant growth and that competition for nutrients increases with plant population density. Plants reflect their competition for resources, including water, light, and nutrients, through the process of self-thinning. Competition among leafhoppers also varies with population density and is reflected in reduced survivorship, smaller size, and increased developmental time at higher population densities. Experiments with terrestrial isopods show that even where there is no food shortage, intraspecific competition through interference may be substantial.

The competitive exclusion principle proposes that two species with identical niches cannot coexist indefinitely, which leads to the prediction that coexisting species will have different niches. The niche concept was developed early in the history of ecology and has had a prominent place in the study of interspecific competition because of the competitive exclusion principle: two species with identical niches cannot coexist indefinitely. While a species' niche is theoretically defined by a very large number of biotic and abiotic factors, Hutchinson's *n*-dimensional hypervolume, the most important attributes of the niche of most species, can often be summarized by a few variables. For instance, the niches of Darwin's finches are largely determined by their feeding requirements, while the niche of slug caterpillar parasitoids can be defined by the size of the caterpillar host.

Mathematical and laboratory models provide a theoretical foundation for studying interspecific competition in nature. Lotka and Volterra independently expanded the logistic model of population growth to represent interspecific competition. In the Lotka-Volterra competition model, the growth rate of a species depends on both numbers of conspecifics and numbers of the competing species. In this model, the effect of one species on another is summarized by competition coefficients. In general, the Lotka-Volterra competition model predicts coexistence of species when interspecific competition is less intense than intraspecific competition. Competitive exclusion in laboratory experiments suggests the potential for competitive exclusion in nature. Even in the laboratory, however, organisms yield results that are much less predictable than the predictions of the Lotka-Volterra competition equations.

Competition can have significant ecological and evolutionary influences on the niches of species. Field experiments involving organisms from herbaceous plants to desert rodents have demonstrated that competition can restrict the niches of species to a narrower set of conditions than they would otherwise occupy in the absence of competition. In some cases, this leads to niche partitioning. Theoretically, natural selection may lead to divergence in the morphology of competing species, a situation called character displacement. After many decades of theoretical and experimental work on competition, we can conclude that competition is a common and strong force operating in nature, but not always and not everywhere. Experimental studies of competition between native and invasive species have the potential to improve our general understanding of interspecific competition.

Key Terms

allelopathy   278

allopatric   293

amensalism   278

character
displacement   293

competition
coefficients   286

commensalism   278

competitive exclusion
principle   283

granivore   291

interference
competition   279

interspecific
competition   279

intraspecific
competition   279

isoclines of zero
population growth   286

$-^3/_2$ self-thinning rule   281

mutualism   278

niche partitioning   284

parasitism   278

phenology   284

priority effect   288

resource (or exploitative)
competition   280

resource limitation   279

self-thinning   280

sympatric   293

Review Questions

1. Design a greenhouse (glasshouse) experiment to test for intraspecific competition within a population of herbaceous plants. Specify the species of plant, the volume (or size of pot) and source of soil, the potentially limiting resource you will focus on (e.g., Tilman and Cowan [1989] studied competition for nitrogen) and how you will manipulate it, and the measures of plant performance you will make.

2. How can the results of greenhouse experiments on competition help us understand the importance of competition among natural populations? How can a researcher enhance the correspondence of results between greenhouse experiments and the field situation?

3. Explain how self-thinning in field populations of plants can be used to support the hypothesis that intraspecific competition is a common occurrence among natural plant populations.

4. Researchers have characterized the niches of Darwin's finches by beak size (which correlates with diet) and the niches of parasitoids by size of caterpillar. How would you characterize the niches of sympatric canid species such as red fox, coyote, and wolf in North America? Or felids, such as ocelots, pumas, and jaguars in South America? What characteristics or environmental features do you think would be useful for representing the niches of desert plants? Or the plants in temperate forest or prairie?

5. Explain why species that overlap a great deal in their fundamental niches have a high probability of competing. Now explain why species that overlap a great deal in their realized niches and live in the same area probably do not compete significantly.

6. Draw the four possible ways in which Lotka's (1932a) isoclines of zero population growth (see fig. 13.14) can be oriented with respect to each other. Label the axes and the points where the isoclines intersect the horizontal and vertical axes. Explain how each situation represented by the graphs leads to either competitive exclusion of one species or the other or stable or unstable coexistence.

7. How was the amount of food that Gause (1934) provided in his experiment on competition among paramecia related to carrying capacity? In Gause's experiments on competition, *P. aurelia* excluded *P. caudatum* faster when he provided half the amount of food than when he doubled the amount of food. Explain.

8. In his experiments on competition between *T. confusum* and *T. castaneum*, Park (1954) found that one species usually excluded the other species but that the outcome depended on physical conditions. In which circumstances did *T. confusum* have the competitive advantage? In which circumstances did *T. castaneum* have the competitive advantage? Could Park predict the outcomes of these experiments with complete certainty? What does this suggest about competition in nature?

9. In their experiments on caterpillar predation, Stoepler and her colleagues (2011) found evidence of niche partitioning. How might this have evolved? How could you tell if this was a genotypic versus simply plastic behavioral response to competition?

10. One of the conclusions that seems justified in light of several decades of studies of interspecific competition is that competition is a common and strong force operating in nature, but not always and not everywhere. List the environmental circumstances in which you think intraspecific and interspecific competition would be most likely to occur in nature. In what circumstances do you think competition is least likely to occur? How would you go about testing your ideas?

RonIsarin/iStock/Getty Images

Chapter 14

Exploitative Interactions

Predation, Herbivory, Parasitism, and Disease

A praying mantis, *Mantis religiosa*, feeding on a grasshopper. No relationship between species is more indicative of exploitative interactions, in which one species lives at the expense of another, than in predator–prey relations. These relationships, inherently interesting ecologically, are also a source of substantial economic benefit, as we shall see later in this chapter.

CHAPTER CONCEPTS

14.1 Predators, parasites, and pathogens influence the distribution, abundance, and structure of prey and host populations. 300
Concept 14.1 Review 304

14.2 Predator-prey, host-parasite, and host-pathogen relationships are dynamic. 304
Concept 14.2 Review 310

14.3 To persist in the face of exploitation, hosts and prey need refuges. 310
Concept 14.3 Review 314

14.4 Models incorporating the ratio of prey to predator numbers better predict predator functional responses in many ecological circumstances. 314
Concept 14.4 Review 317

14.5 Exploitative interactions weave populations into a web of relationships that defy easy generalization. 317
Concept 14.5 Review 321

Applications: The Value of Pest Control by Bats: A Case Study 321
Summary 323
Key Terms 324
Review Questions 324

LEARNING OUTCOMES

After studying this section you should be able to do the following:

14.1 Describe several forms of exploitation.
14.2 Contrast the typical effects of predators and parasites on the organisms they exploit.

In nature, the consumer eventually becomes the consumed. A moose browses intently on the twigs and buds of a small tree that barely protrudes above the deep snow of midwinter (fig. 14.1). With each mouthful it chews and swallows, the

Figure 14.1 This moose exploits the twigs and buds of woody plants for the food it needs to survive the cold northern winter. Eventually, wolves may prey upon the moose to meet their needs for food.
NPS photo by Jim Peaco

moose reduces the mass of the small tree and adds to the growing energy store in its own large and complex stomach, energy stores that the moose will need to make it through one more northern winter. Then, a familiar scent catches the moose's attention and, startled, it runs off.

Suddenly, the clearing where the moose had been feeding is a blur of bounding forms dashing headlong in the direction the moose has gone—a pack of wolves in pursuit of its own meal. A portion of the pack has already run ahead of the moose and is cutting off its retreat. This time, unlike so many times before, the old moose will not escape. After a fierce struggle, the moose is down and the wolves settle in to feed.

But the wolves are not the only organisms to benefit from this great quantity of food. Within the intestines of the wolves live several species of parasitic worms that will soon claim their share of the wolves' hard-won feast. The worms will turn some of the energy and structural compounds they absorb into the infective stages of their own kind, which after being shed into the environment may attach themselves to other hosts, who will serve as their unwitting providers.

Some of the strongest links between populations are those between herbivore and plant, between predator and prey, and between parasite or pathogen and host. The conceptual thread that links these diverse interactions between species is that the interaction enhances the fitness of one individual—the predator, the pathogen, etc.—while reducing the fitness of the exploited individual—the prey, host, etc. Because of this common thread we can group these interactions under the heading of *exploitative interactions.*

Let's consider some of the most common forms of exploitation. Herbivores consume live plant material but do not usually kill plants. **Predators** kill and consume other organisms. Typical predators are animals that feed on other animals—wolves that eat moose, snakes that eat mice, etc.

Parasites live on the tissues of their host, often reducing the fitness of the host, but not generally killing it. A **parasitoid** is an insect whose larva consumes its host and kills it in the process; parasitoids are functionally equivalent to predators. **Pathogens** induce disease, a debilitating condition, in their hosts. These diverse interactions have one thing in common: **exploitation,** that is, one organism makes its living at the expense of another.

14.1 Exploitation and Abundance

LEARNING OUTCOMES
After studying this section you should be able to do the following:

14.3 Outline the field experiment Lamberti and Resh used to determine the influence of the caddisfly *Helicopsyche* on algal biomass.

14.4 Describe the experimental manipulations and control in the studies by Kalka, Smith, and Kalko to determine the relative influences of birds and bats on foliage arthropods in tropical forest.

14.5 Discuss possible sources of uncertainty regarding the conclusion that foxes control hare populations in Sweden, which was based on hare population increases following parasite-caused fox mortality.

Predators, parasites, and pathogens influence the distribution, abundance, and structure of prey and host populations. One of the main reasons ecologists are interested in exploitative interactions between species is that these interactions have the potential to influence prey and host populations. A rapidly growing pool of studies suggests that predators, parasites, and pathogens substantially affect the abundance of the organisms they exploit.

A Herbivorous Stream Insect and Its Algal Food

Gary Lamberti and Vincent Resh (1983) studied the influence of a herbivorous stream insect on the algal and bacterial populations on which it feeds. The herbivorous insect was the larval stage of the caddisfly (order Trichoptera) *Helicopsyche borealis.* This insect inhabits streams across most of North America and is most notable for the type of portable shelter it builds as a larva. The larvae cement sand grains together to form a helical portable home that looks just like a small snail shell. In fact, the species was originally described as a freshwater snail. Larval *Helicopsyche* graze on the algae and bacteria growing on the exposed surfaces of submerged stones. This feeding habit requires that *Helicopsyche* spend considerable time out in the open, where it would be far more vulnerable to predators were it not for its case.

Lamberti and Resh found that larval *Helicopsyche* grow and develop through the summer and fall, attaining densities of over 4,000 individuals per square meter in Big Sulphur Creek, California. At this density, they make up about 25% of the total biomass of benthic animals. A consumer that reaches such high population densities clearly has the potential to reduce the density of its food supply. Lamberti and Resh got an indication of the potential of *Helicopsyche* to influence its food supply in a preliminary experiment. In this first experiment, they placed unglazed ceramic tiles (15.2 cm × 7.6 cm) on the bottom of the creek and followed colonization of these artificial substrates by algae and *Helicopsyche* over a period of 7 weeks.

Algae rapidly colonized the tiles, reaching peak density 2 weeks after the tiles were placed in Big Sulphur Creek. The *Helicopsyche* population reached its highest density 1 week later. Algal biomass decreased from week 2 to week 5 of the study and then rose again during the last 2 weeks, as *Helicopsyche* numbers declined. These results (fig. 14.2) suggest that the caddisfly larvae depleted their food supply.

In a follow-up study, the researchers used an exclusion experiment to test for the effect of *Helicopsyche* on its food supply (fig. 14.3). They placed unglazed ceramic tiles in two 3×6 grids of 18 tiles each. One grid was placed directly on the stream bottom, while the other was placed on a metal plate supported by an upside-down J-shaped metal bar. This arrangement, which raised the tiles 15 cm above the bottom but still 35 cm below the stream surface, allowed colonization of tiles by algae and most invertebrates while preventing colonization by *Helicopsyche*. *Helicopsyche* could not colonize the tiles because their heavy, snail-shaped case confines them to the stream bottom. To reach the tiles, *Helicopsyche* would

have to crawl up the J-shaped support bar, out of the water, and then back down, while most other invertebrates could colonize either by drifting downstream with the current or by swimming to the raised tiles. Lamberti and Resh coated the above-water parts of the bar with an adhesive to prevent adult *Helicopsyche* from crawling down to the tiles to deposit their eggs. The experimental arrangement excluded *Helicopsyche* while allowing other invertebrates to colonize the raised tiles. Such selective manipulations of natural populations are not easy to attain.

The results of this experiment clearly show that *Helicopsyche* reduces the abundance of its food supply. Figure 14.4 shows that the tiles without *Helicopsyche* supported higher abundances of both algae and bacteria. The large effect of *Helicopsyche* on its food supply is apparent from paired photos of the experimental and control tiles at the beginning and the end of the experiment.

The influence of exploitation on populations is often best seen when populations are released from exploitation. This is what we saw when *Helicopsyche* were excluded from experimental habitats. Similar responses have been documented in terrestrial ecosystems when herbivorous insects are shielded from their predators.

Bats, Birds, and Herbivory in a Tropical Forest

The dominant herbivores in the complex three-dimensional framework of forests are arthropods. At times, the number and biomass of herbivorous arthropods in forests can be so

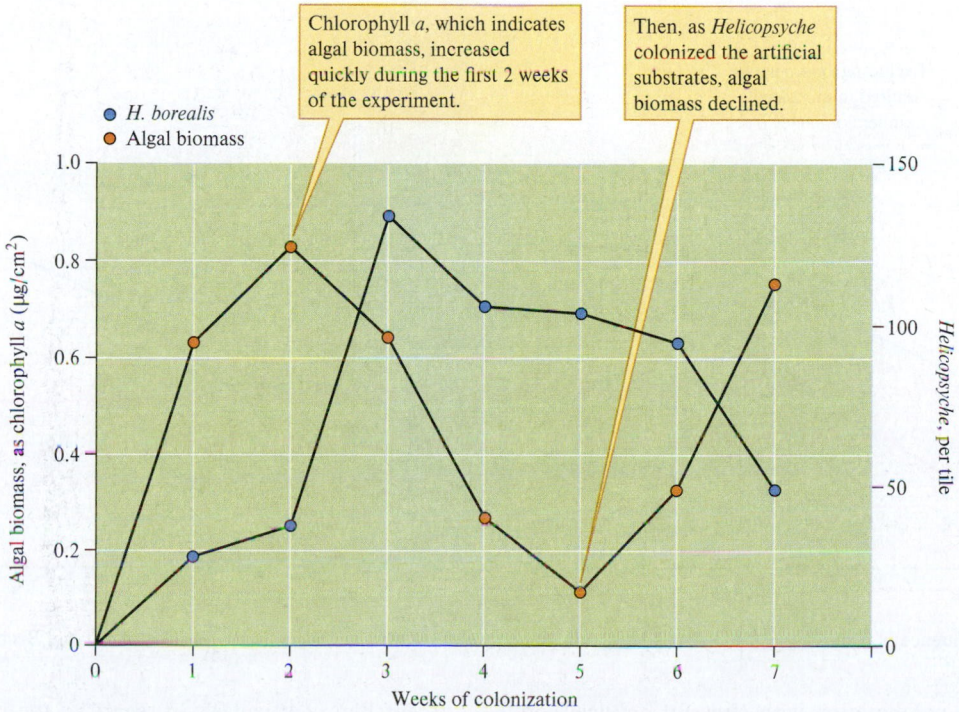

Figure 14.2   Biomass of algae and numbers of the grazing caddisfly *Helicopsyche borealis* (data from Lamberti and Resh 1983).

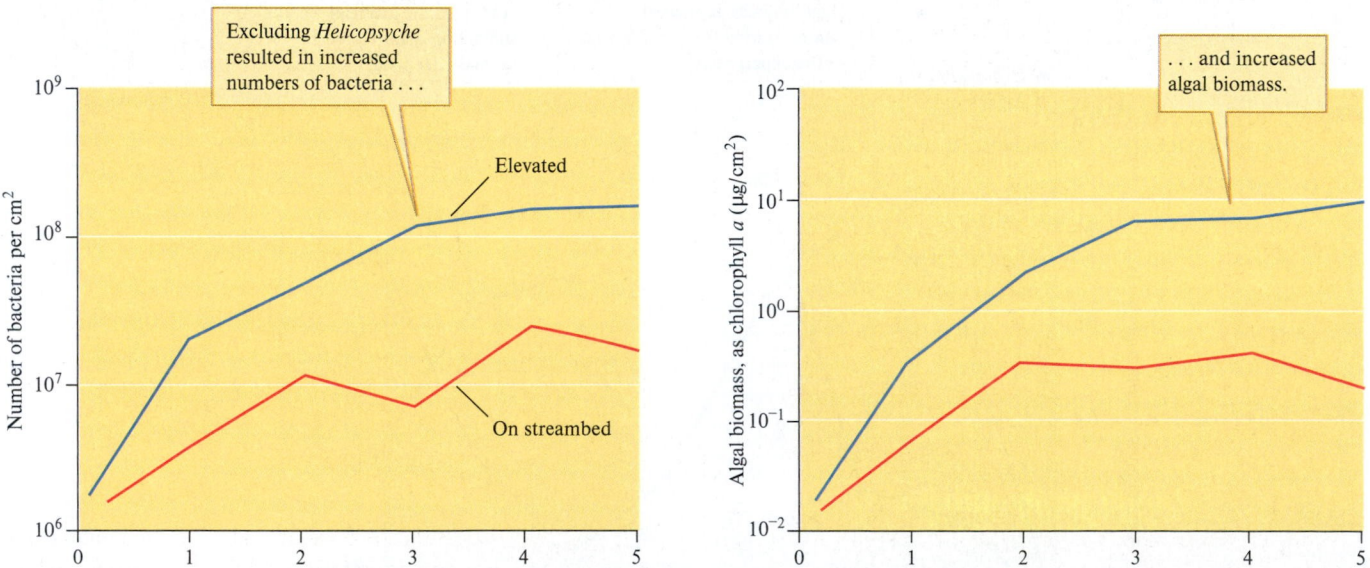

(a)   (b)

Figure 14.3   Effects of excluding *Helicopsyche borealis* on benthic algal biomass: (*a*) two sets of tiles at the beginning of the experiment; exclusion tiles in foreground; (*b*) same tiles 5 weeks into the experiment.   (*a, b*) Gary Lambertini

Figure 14.4   Influence of excluding *Helicopsyche borealis* on abundance of bacteria and algae (data from Lamberti and Resh 1983).

great that they, in combination with stressful physical conditions, such as drought, cause large-scale tree mortality (e.g., Kane and Kolb 2010). Can predators limit arthropod numbers in forests? This question has been addressed by a number of researchers, particularly in regard to the potential of birds to depress forest arthropod numbers (e.g., Holmes, Schultz, and Nothnagle 1979; Marquis and Whelan 1994; Van Bael, Brawn, and Robinson 2008).

Among the first to do so, Margareta Kalka, Adam Smith, and Elisabeth Kalko (2008) designed an experiment to distinguish between the influences of birds and bats on tropical forest arthropods. Many species of bats are known to feed on arthropods living on foliage in forests. These so called "foliage gleaning" bats are particularly common in tropical forests. Kalka, Smith, and Kalko conducted their research in a lowland tropical forest in Panama, where they, like previous researchers, covered plants with mesh to exclude arthropod-feeding vertebrates. Unlike previous researchers, however, Kalka and her colleagues covered one set of plants during the day only and another set of plants during the night only. As a result, they excluded birds from plants that they covered during the day and bats from the plants that they covered at night. Their study included 43 plants with day exclosures (− birds, + bats), 42 plants with night exclosures (−bats, +birds), and 35 uncovered control plants. Previous studies that had used exclosures had left them in place for the duration of the study, excluding birds and bats, and therefore unable to distinguish between the influences of the two groups of predators. The research team covered and uncovered their experimental plants at the appropriate time each day for 10 weeks and visually censused the arthropods on all three sets of plants.

The results of Kalka's study showed that both birds and bats reduce foliage-living arthropods significantly (fig. 14.5). However, the results also show that previous researchers had not anticipated the great effect that bats can have on arthropod densities in forests. As shown in figure 14.5a, plants from which birds had been excluded, but that continued to be visited by bats, supported arthropods at densities 65% higher than control plants. Meanwhile, those plants from which bats had been excluded and on which birds could feed during the day supported 150% higher arthropod densities compared to control plants (fig. 14.5b). With this well-designed experiment, Kalka, Smith, and Kalko disentangled the influences of birds and bats and revealed the great potential of bats to limit arthropod numbers in tropical forests, a level of influence that appears to have been unsuspected previously by most researchers working on foliage gleaning in tropical forests.

A Pathogenic Parasite, a Predator, and Its Prey

One of the great challenges of ecology is to work at large scales. Ecologists rarely have the opportunity to conduct large-scale experiments; however, nature sometimes provides such opportunities. One such opportunity arose in Sweden when a pathogen severely reduced the population of red foxes, *Vulpes vulpes.*

Erik Lindström and his colleagues (1994) at the Grimsö Wildlife Research Station in Örebro County reported that mange mites, *Sarcoptes scabiei,* were first found infesting red foxes in north-central Sweden in 1975. The researchers studied the spread of mange because mange mites are a serious external parasite of foxes that causes hair loss, skin deterioration,

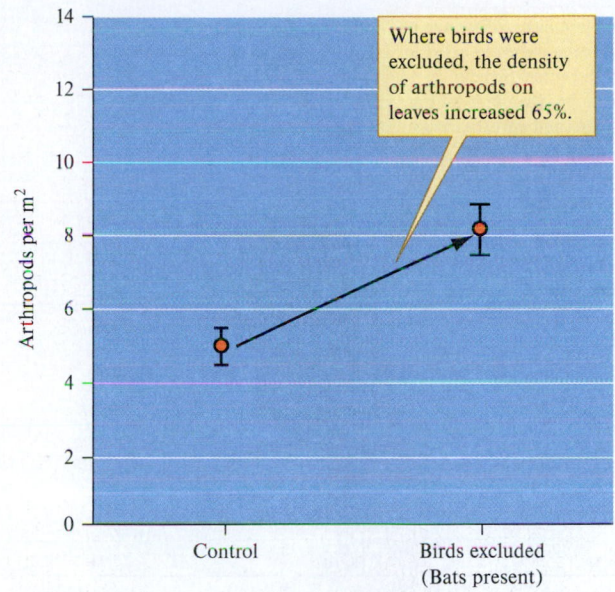

(a)

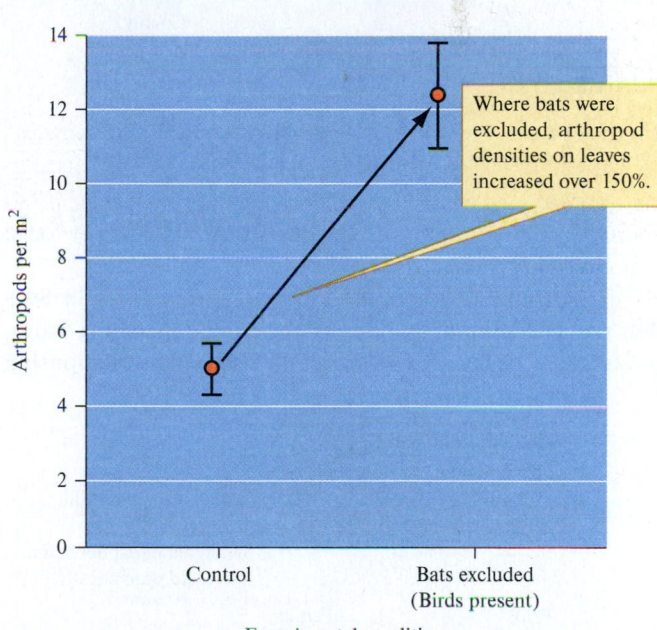

(b)

Figure 14.5   The influence of bird and bat exclusion on arthropod densities in a tropical forest: (*a*) comparison of arthropod densities (mean ± standard error) on control plants and plants from which birds have been excluded; (*b*) comparison of arthropod densities (mean ± standard error) on control plants and plants from which bats have been excluded (data from Kalka, Smith, and Kalko 2008).

and death. Within a decade of its arrival in Sweden, mange had spread over the entire country (fig. 14.6). As it spread, mange reduced the population of red foxes in Sweden by over 70%.

As wildlife ecologists, Lindström's research team was keenly interested in how the prey of red foxes would respond. Would they find evidence of population control by this predator? From 1972 to 1993, the research team studied several prey populations as well as red fox populations. They used many

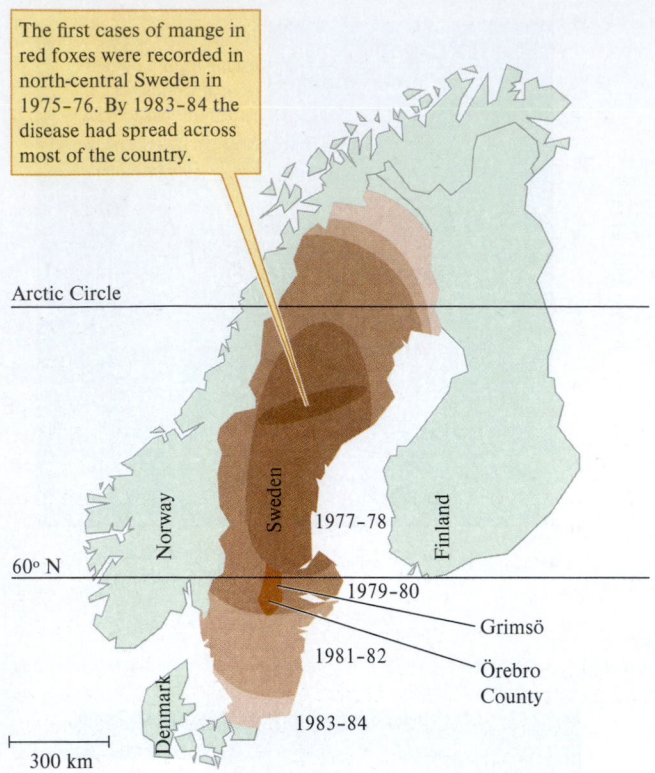

The first cases of mange in red foxes were recorded in north-central Sweden in 1975–76. By 1983–84 the disease had spread across most of the country.

Arctic Circle

Norway

Sweden

Finland

1977–78

60° N

1979–80

Grimsö

1981–82

Örebro County

Denmark

1983–84

300 km

Figure 14.6   The spread of mange in red foxes across Sweden from 1975 to 1984 (data from Lindström et al. 1994).

sources of information and conducted their studies at local, regional, and national spatial scales.

The results of the study were clear. Red foxes in Sweden reduced the populations of their prey, including hares, grouse, and roe deer fawns. Figure 14.7 shows the relationship

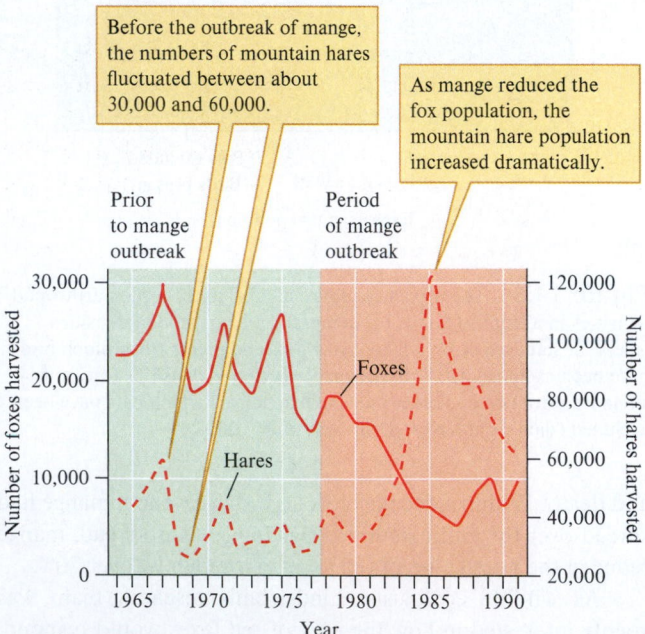

Before the outbreak of mange, the numbers of mountain hares fluctuated between about 30,000 and 60,000.

As mange reduced the fox population, the mountain hare population increased dramatically.

Prior to mange outbreak

Period of mange outbreak

Foxes

Hares

Number of foxes harvested

Number of hares harvested

Year

Figure 14.7   The numbers of foxes and mountain hares in five counties in Sweden estimated from hunters' harvest records (data from Lindström et al. 1994).

between numbers of red foxes and mountain hares, *Lepus timidus,* in Sweden. Following the reduction in the red fox population, the number of mountain hares increased two to four times. This is an especially thorough and convincing demonstration of the influence of a terrestrial vertebrate predator on its prey populations. The study also suggests that red foxes may have a significant influence on the cyclic abundance of some of their prey species. The dynamics of prey populations have been the subject of research by ecologists for some time. Studies of predation have been central to this research.

Concept 14.1 Review

1. The patterns shown in figure 14.2 suggest that *Helicopsyche* depletes its algal food supply. However, Lamberti and Resh were not certain and so conducted their second set of experiments. Why could they not reach a firm conclusion regarding the effect of *Helicopsyche* on its food supply based on their first experiment?
2. How could you test experimentally for the combined influence of bats and birds on numbers of arthropods on foliage, as well as their individual contributions?

14.2 Dynamics

LEARNING OUTCOMES

After studying this section you should be able to do the following:

14.6   List the major hypotheses proposed to explain snowshoe hare population cycles.

14.7   Discuss the roles of food supply and predators in producing snowshoe hare population cycles.

14.8   Explain how the field experiment by Krebs and his colleagues improved our understanding of the relative effects of food and predators on snowshoe hare populations, beyond what had been learned through observational studies.

14.9   Describe the unique contributions of mathematical and laboratory models to the study of population cycles.

Predator–prey, host–parasite, and host–pathogen relationships are dynamic.   In the previous section, we saw how some predators, parasites, and pathogens affect the populations they exploit. The picture that emerges from these studies is that the biology of exploitation is complex. As complex as this emerging picture of exploitation may be, it belies an even deeper underlying complexity. In this section, we add another level of complexity as we take up the topic of *temporal dynamics.* Populations of a wide variety of predators and prey are not static but cycle in abundance over periods of days to decades.

Cycles of Abundance in Snowshoe Hares and Their Predators

Population cycles are well documented for a wide variety of animals living at high latitudes, including lemmings, voles,

Figure 14.8   A lynx, *Lynx canadensis,* pursuing a snowshoe hare, *Lepus americanus,* one of the lynx's staple prey species. The large, fur-padded feet of both lynx and hare enable them to run at high speeds over snow.
Ed Cesar/Science Source

muskrats, red fox, arctic fox, ruffed grouse, and porcupines. We have already seen in chapter 10 how periodic outbreaks of voles lead to local increases in the abundance of avian predators due to numerical responses by owls and kestrels (Korpimäki and Norrdahl 1991).

One of the best-studied cases of animal population cycles is that of the snowshoe hare, *Lepus americanus,* and the lynx, *Lynx canadensis,* one of the snowshoe hare's chief predators (fig. 14.8). The population cycles of these two species are especially well documented because the Hudson Bay Company kept trapping records during most of the eighteenth, nineteenth, and twentieth centuries. Drawing on this unique historical record, ecologists were able to estimate the relative abundances of Canadian lynx and snowshoe hare over a period of about 200 years. That record, shown in figure 14.9, demonstrates a remarkable match in the cycles of the two populations.

By the 1950s, several hypotheses had been proposed to explain these and other cycles among northern populations. Charles Elton (1924) proposed that cycles of abundance in

snowshoe hare and lynx populations are driven by variation in amount of solar radiation as a consequence of sunspot cycles. He proposed that variation in intensity of solar radiation may directly affect snowshoe hares and their food supply and that lynx populations, in turn, respond to the changing abundance of the snowshoe hare, their main prey.

The sunspot hypothesis was rejected by D. MacLulich (1937) and P. Moran (1949), who showed that sunspot cycles do not match snowshoe hare population cycles. The second group of hypotheses, which Lloyd Keith (1963) referred to as "overpopulation theories," suggested that periods of high population growth are followed by (1) decimation by disease and parasitism, (2) physiological stress at high densities leading to increased mortality as a consequence of nervous disorders, and (3) starvation due to reduced quantity and quality of food at high population densities. An alternative to the overpopulation hypothesis was that cycles like that of the snowshoe hare are driven by predators. According to this hypothesis, predators increase in number in response to increasing prey availability and then eventually reduce prey populations.

Keith observed that none of these hypotheses completely accounts for population cycles in snowshoe hare and other northern populations. He went on to say that "the 10-year cycle is not likely to become better understood by further theorizing. Clearly the present need is for comprehensive long-term investigations by a diversified team of specialists." Heeding his own advice, Keith organized such studies. After three decades of research by his team and several other groups in North America and Europe, we now have a reasonable picture of the roles played by predators and food supply in producing population cycles in the far north.

The Role of Food Supply

Snowshoe hares live in the boreal forests of North America. As we saw, the boreal forest is dominated by a variety of conifers such as spruce, *Picea* spp., jackpines, *Pinus banksiana,* and tamarack, *Larix laricina,* and deciduous trees such as balsam poplar, *Populus balsamifera,* aspen, *Populus tremuloides,* and

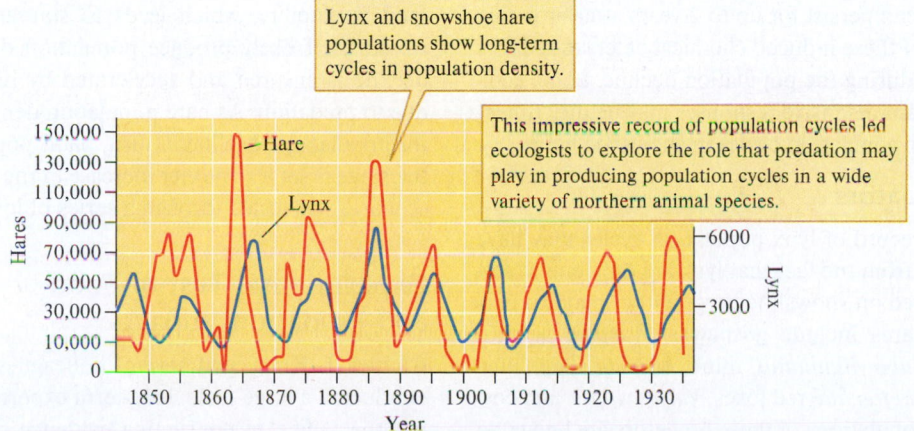

Figure 14.9   Historical fluctuations in lynx and snowshoe hare populations based on the number of pelts purchased by the Hudson Bay Company (data from MacLulich 1937).

paper birch, *Betula papyrifera*. Within the boreal forest, snow-shoe hares associate with dense growths of understory shrubs, which provide both cover and winter food, the most critical portion of the snowshoe hare's food supply.

Snowshoe hares have the potential to reduce the quantity and quality of their food supply. The hares live up to the legendary reproductive capacity of rabbits and hares. Estimated geometric rate of increase, λ (see chapter 10), during the growth phase of a hare population cycle can average as high as 2.0. In other words, snowshoe hare populations can double in size each generation. Keith and his colleagues (1984) have observed snowshoe hare population densities of up to 1,100 to 2,300 per square kilometer. However, local densities are highly dynamic. Keith cites 100-fold fluctuations in snowshoe hare densities in some areas and states that 10- to 30-fold fluctuations are common. Similar densities are sometimes observed in populations of the mountain hare, *Lepus timidus,* which shows pronounced population cycles across the Eurasian taiga (Keith 1983) and which destroys considerable vegetation at high densities.

Snowshoe hares spend the long northern winter (6–8 months) browsing on the buds and small stems of shrubs such as rose, *Rosa* spp., and willow, *Salix* spp. Where deep snow provides access, snowshoe hares browse on the saplings of trees such as spruce and aspen. The most nutritious portions of these shrubs and trees are the small stems (<4–5 mm diameter). Over the winter, each hare requires about 300 g of these stems each day. In some areas, however, snowshoe hares have been observed to remove over 1,500 g of food biomass per day, possibly wasting a great deal of potential food in the process. Feeding at these rates, one population of snowshoe hares reduced food biomass from 530 kg per hectare in late November to 160 kg per hectare by late March. Many ecologists have demonstrated food shortages during winters of peak snowshoe hare density.

Snowshoe hares also influence the quality of their food supply. Feeding by snowshoe hares induces chemical defenses in their food plants, defenses like those we discussed in chapter 7. Shoots produced after substantial browsing contain elevated concentrations of terpene and phenolic resins, defensive chemicals that repel hungry hares. Elevated concentrations of plant defensive chemicals can persist for up to 2 years after browsing by hares. The effect of these induced chemical defenses reduces *usable* food supplies during the population decline. Some ecologists suggest that plant defensive responses may be the "timer" that produces 10-year population cycles in snowshoe hares.

The Role of Predators

The long historical record of lynx population cycles may have distracted ecologists from the fact that lynx are only one of several predators that feed on snowshoe hares. Other major predators of snowshoe hares include goshawks, *Accipiter gentilis;* great horned owls, *Bubo virginianus;* mink, *Mustela vison;* long-tailed weasels, *Mustela frenata;* red foxes, *Vulpes vulpes;* and coyotes, *Canis latrans.* Populations of these predators are known to cycle synchronously with snowshoe hare populations. Though the lynx is considered to be a specialist on snowshoe hares,

the diet of a generalist predator such as the coyote may also be dominated by snowshoe hares. This is particularly true when snowshoe hare populations are at peak density. A. Todd and L. Keith (1983) report that snowshoe hares made up 67% of the coyote diets in central Alberta, Canada. Ecologists have estimated that predation can account for 60% to 90% of snowshoe hare mortality during peak densities.

Research by Mark O'Donoghue and several colleagues from Canada, Argentina, and Alaska (O'Donoghue et al. 1997, 1998) provides clear evidence of predator *functional response* (chapter 7, section 7.4) and *numerical response* (chapter 10, section 10.1) to increased hare densities. O'Donoghue and his colleagues focused their research on two of the most important predators of adult snowshoe hares: lynx and coyotes. Their study shows that coyote and lynx numbers increase six- to sevenfold, a numerical response, following increases in snowshoe hare populations. The two predators also showed functional responses to increased hare densities. However, coyote and lynx functional responses differed in their timing and form. The lynx killed more hares when hare numbers were declining, while coyotes showed higher predation rates when the hare population was increasing. O'Donoghue and colleagues discovered that lynx show a clear type 2 functional response (see fig. 7.22) to increasing hare densities, reaching a maximum number of 1.2 hares per day at medium hare densities. In contrast, coyotes preyed on up to 2.3 hares per day at the highest hare densities and their functional response showed no signs of leveling off. At high hare densities, coyote and lynx predation rates exceeded their daily energetic needs. Coyotes killed more hares early in the winter, caching many and retrieving them later in the season. In some instances, individual coyotes returned to eat hares over 4 months after they were cached. The combination of numerical and functional responses by lynx and coyotes indicate great potential for these predators to reduce snowshoe hare populations.

In summary, several decades of research provided evidence that both predation and food can make substantial contributions to snowshoe hare population cycles (Haukioja et al. 1983; Keith 1983; Keith et al. 1984). The food availability and predation hypotheses are not mutually exclusive alternatives but rather are complementary. As hare populations increase, they reduce the quantity and quality of their food supply. Reduced food availability, which leads to starvation and weight loss, would itself likely produce population decline. This potential decline is ensured and accelerated by high rates of mortality due to predation. As hare population density is reduced, predator populations decline in turn, plant populations recover, and the stage is set for another increase in the hare population. This scenario was tested through a series of long-term experiments.

Experimental Test of Food and Predation Impacts

Charles J. Krebs and several colleagues (Krebs et al. 1995) conducted a large-scale, long-term experiment designed to sort out the tangle of conflicting evidence regarding the impacts of food and predation on snowshoe hare population cycles. Over a period of 8 years, Krebs and colleagues conducted an ambitious field experiment. Their experimental plots consisted of nine 1 km^2 blocks of undisturbed boreal forest, each separated from other experimental blocks by a minimum of 1 km. Three blocks served as controls for comparison to the six other blocks where experimental treatments were applied. To test the impact of food, hares were given unlimited supplemental food on two experimental blocks during the entire period of the study. To test for the possible influences of plant tissue quality on hare numbers, the researchers applied a nitrogen-potassium-phosphorus fertilizer from the air to two of the experimental blocks. Finally, they built electric fences around two of the 1 km^2 blocks, which excluded mammalian predators but not hawks and owls. One of these predator reduction blocks received supplemental food. Krebs and his colleagues report that, due to maintenance requirements, they could not replicate the predator reduction and predator reduction + food experimental manipulations. They could not, since the fences on both predator reduction areas (8 km of fence) had to be checked every day through the winter, when temperatures would sometimes dip as low as −45°C. Krebs's research team maintained these experimental conditions through one cycle in snowshoe hare numbers.

During the 8 years of the experiment, the researchers observed an increase in hare numbers to a peak, followed by a decline on all the study plots. The application of fertilizer increased plant growth within the fertilizer treatment blocks but did not increase numbers of snowshoe hares. Meanwhile, compared to control plots, hare numbers increased substantially on food addition, predator reduction, and predator reduction + food study plots. Averaged over the peak and decline phases during the study, reducing predators doubled hare density, adding food tripled hare density, and excluding predators and adding food increased hare density to 11 times that of the controls (fig. 14.10). What factors contributed to these increased densities within treatment blocks? Krebs and his colleagues found that higher densities on experimental plots were the result of both higher survival and higher reproduction.

After many decades of research, we can conclude that the population cycle in snowshoe hares is the result of an interaction among three trophic levels: the hares, their food supply, and their predators. Krebs and colleagues (2001) point out, however, that in order to understand the controls on hare numbers, researchers have had to work with all three trophic levels simultaneously. In addition, the critical experiment had to be done on a large scale and in the field. Still, a great deal of insight into this large-scale predator–prey system has come from laboratory and mathematical studies.

Population Cycles in Mathematical and Laboratory Models

Now let's shift our focus from population cycles in the vast world of the boreal forest to population cycles in mathematical models and controlled laboratory conditions. Mathematical and laboratory models offer population ecologists the opportunity to manipulate variables that they cannot control in the field. Our question here is whether predator–prey or

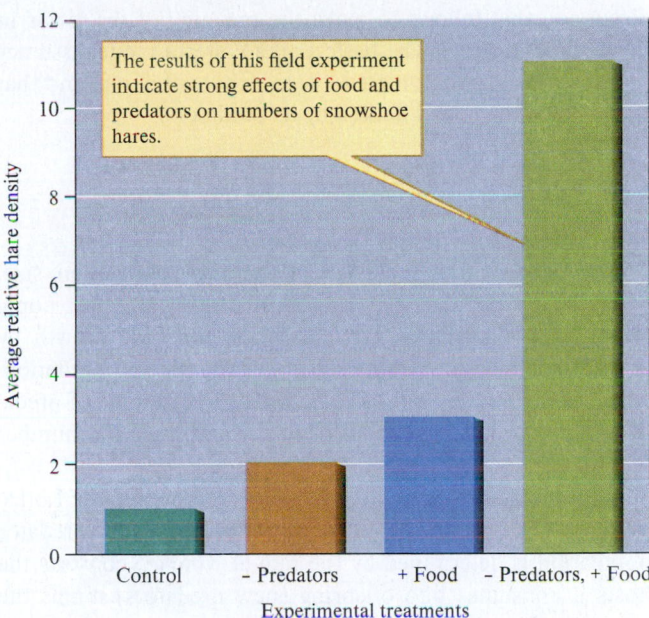

Figure 14.10 The densities of snowshoe hares averaged from the peak in hare density through the period of declining density observed during the study. Hare densities are expressed relative to the densities on the control plots where no experimental manipulation was applied (data from Krebs et al. 1995, 2001).

parasite–host cycles can be produced in mathematical and laboratory models without the complications introduced by factors, such as the effects of the prey on its food supply and uncontrolled weather cycles. In other words, can the interactions among exploited populations themselves generate population cycles of the type observed in snowshoe hares? The answer to this question is a qualified yes.

Mathematical Models

The first ecologists to model predator–prey interactions mathematically were Alfred Lotka (1925) and Vito Volterra (1926). Both researchers built their models based on observations of interactions among natural populations. Lotka was impressed by the reciprocal oscillations of populations of moth and butterfly larvae and the parasitoids that attack them. Volterra was inspired by the response of marine fish populations to cessation of fishing during World War I. Volterra observed that the responses of fish populations were uneven. Predaceous fish, particularly sharks, increased in abundance, while the populations upon which they fed decreased. This reciprocal change in numbers suggested that predators have the potential to reduce the abundance of their prey. In this single observation, Volterra somehow saw the potential for predator–prey population cycles and suggested that similar cycles should occur in parasite-host and pathogen–host systems, including those in which humans are involved. With these observations in mind, Lotka and Volterra then set out to build mathematical models that would produce the cycles that they thought occurred in nature.

The Lotka-Volterra predator–prey equations demonstrated that very simple models will produce cycling of predator and prey populations (called predators and "hosts" in the discussion that follows to make the meaning of the terms in the equations clear). The basic Lotka-Volterra model assumes that the host population grows at an exponential rate and that host population size is limited by its predators:

$$\frac{dN_h}{dt} = r_h N_h - p N_h N_p$$

This equation models change in the host population size, where $r_h N_h$ represents exponential growth by the host population. In the Lotka-Volterra model, exponential growth by the host population is opposed by deaths due to predation, which is represented by $-p N_h N_p$, where p is the rate of predation, N_h is again the number of hosts, and N_p is the number of predators.

On the other side of the predator–prey system, the Lotka-Volterra model assumes that the rate of growth by the predator population is determined by the rate at which it converts the hosts it consumes into offspring (new predators) minus the mortality rate of the predator population:

$$\frac{dN_p}{dt} = cp N_h N_p - d_p N_p$$

Here again, N_h and N_p are the numbers of hosts and predators, respectively. The rate at which the predators convert hosts into offspring is $cp N_h N_p$, which is the rate at which they destroy hosts, $p N_h N_p$, multiplied by a conversion factor, c, the rate at which hosts are converted to predator offspring. In the Lotka-Volterra equation, the growth rate of the predator population is opposed by predator deaths, $d_p N_p$. Notice that in these equations the only variables are N_h and N_p. All the other terms in the Lotka-Volterra model, p, c, d_p, and r_h, are constants. The Lotka-Volterra predator-prey model is summarized in figure 14.11.

Now let's reflect on the behavior of this model. Because the host population grows at an exponential rate, its population growth accelerates with increasing population size. However, this tendency to grow faster and faster with increasing N_h is opposed by predation. As N_h increases, the rate of predation, $p N_h N_p$, also increases. Consequently, in the Lotka-Volterra model, reproduction by the host is translated immediately into killing of hosts by the predator. In addition, increased predation, $p N_h N_p$, is translated directly and immediately into more predators by $cp N_h N_p$. Increased numbers of predators increase the rate of predation since increasing N_p increases $p N_h N_p$. Growth of the predator population eventually reduces the host population, which in turn leads to declines in the predator population. Therefore, like host success, predator success carries the seeds of its own destruction.

These reciprocal effects of prey (host) and predator produce oscillations in the two populations, which we can represent in two ways graphically. In figure 14.12a, population oscillations are presented as we looked at them in snowshoe hare and lynx populations (see fig. 14.9), while figure 14.12b gives an alternative representation. The time axis has been eliminated and the two remaining axes represent the numbers

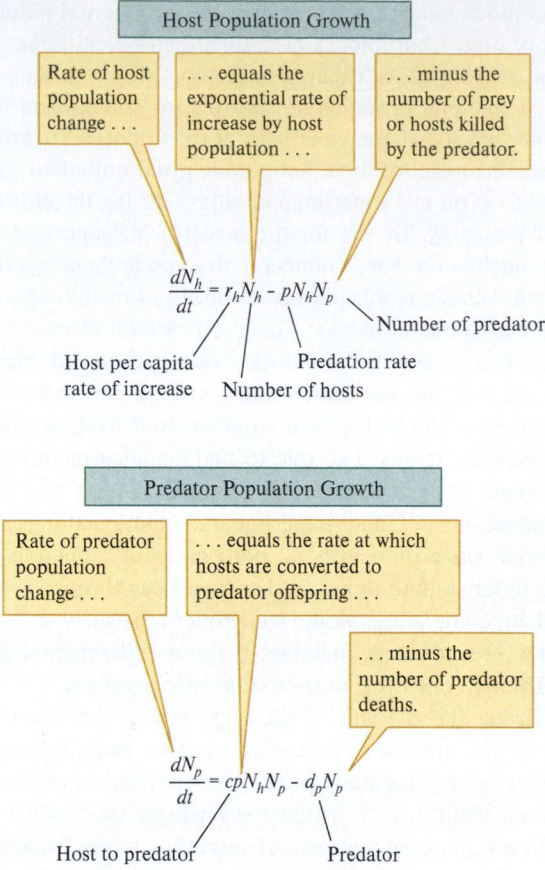

Figure 14.11 Anatomy of the Lotka-Volterra equations for predator-host population growth ("host" is used in place of "prey" to make meanings of equation terms clear).

of predators and prey. When we plot population data in this way, we see that the Lotka-Volterra model produces oscillations in predator and prey populations that follow an elliptical path, whose size depends on the initial sizes of prey and predator populations. Whatever the ellipse size, however, the prey and predator populations just go round and round on the same path forever.

The prediction of eternal oscillations on a very narrowly defined path is obviously unrealistic. Another unrealistic assumption is that neither the prey nor the predator populations are subject to carrying capacities. Another is that changes in either population are instantaneously translated into responses in the other population. Despite these unrealistic assumptions, Lotka and Volterra made valuable contributions to our understanding of predator–prey systems. They showed that simple models with a minimum of assumptions produce reciprocal cycles in populations of predator and prey analogous to those that biologists had observed in natural populations. They demonstrated that predator–prey interactions themselves can, in theory, produce population cycles without any influences from an outside force such as climatic variation. Later refinements of the Lotka-Volterra model generate predictions and behave in ways more consistent with natural

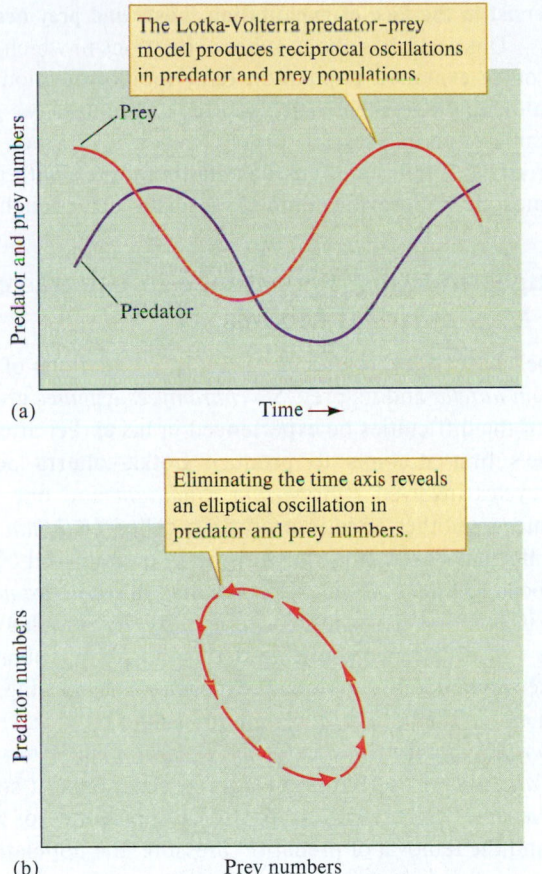

Figure 14.12   A graphical view of the Lotka-Volterra predator–prey model (data from Gause 1934).

predator–prey systems. (The Lotka-Volterra equations can also be used to model parasite–host, pathogen–host, or parasitoid–host populations.)

Laboratory Models

One of the most successful attempts to produce Lotka-Volterra-type population cycles in the laboratory was that of Syunro Utida (1957) of Kyoto University, Japan. Utida studied interactions between the adzuki bean weevil, *Callosobruchus chinensis,* and a hymenopteran parasitoid wasp,

Heterospilus prosopidis, which attacks the bean weevil. Adult weevils lay their eggs on adzuki beans, *Paseolus angularis,* and upon hatching the larvae feed on the beans until they metamorphose into pupae. When they emerge from the pupal stage, the adult weevils mate and seek out new beans on which to lay their eggs. The entire life cycle, from egg to egg, takes approximately 20 days. While the weevil works at completing its life cycle, the parasitoid wasp searches for weevil larvae and pupae, where they lay their eggs. The larvae of the wasps feed on the larvae and pupae of the weevils and, in the process, kill them. Though the details of their behavior differ, the wasps are predators of the weevils, no less than are lynx predators of snowshoe hares.

Utida's experimental populations lived in petri dishes 1.8 cm tall by 8.5 cm in diameter where temperature was maintained at a constant 30°C and relative humidity at 75%. Within the petri dishes, Utida placed 10 g of adzuki beans with a water content of 15% and added different mixtures of adult adzuki bean weevils and parasitoid wasps; some in which there were more weevils, some with equal numbers, and some with more wasps than weevils. Every 10 days, 10 g of fresh beans were added and the leavings of the old beans were placed in another dish. Any beetles moved with the spent food were recorded over a period of 20 days.

Utida followed these populations for dozens of generations and many years, carefully recording changes in population densities over time. Mistakes in handling killed some of the populations, and others simply died out. The population that he had started with 64 weevils and 8 wasps survived the longest: 112 generations over 6 years, after which the population was accidentally destroyed. It was only by following the beetle and wasp populations for so many generations that Utida was able to see the pattern we can see now (fig. 14.13). For several generations, Utida observed reciprocal fluctuations in his beetle and parasitoid populations that look very similar to those we saw for lynx and hare populations (see fig. 14.9). After an initial phase of high-magnitude oscillations, the population cycles were decreased in amplitude, remained in a situation of low-amplitude fluctuations for some time, and then increased in amplitude once again. In population A, high-amplitude cycles continued for the first 20 generations, were damped from then until generation 30,

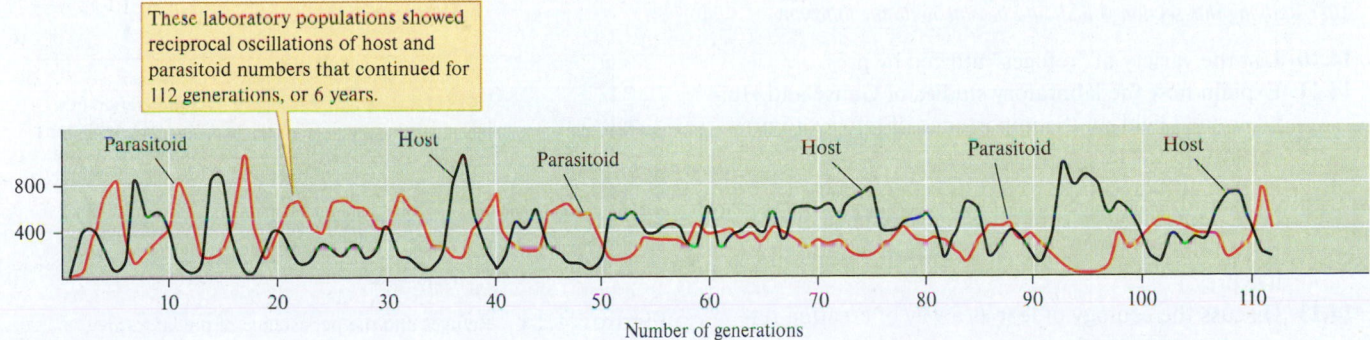

Figure 14.13   Laboratory populations of a host, the adzuki bean weevil, and a parasitoid wasp (data from Utida 1957).

then resumed high-amplitude cycling until about generation 54, when the oscillations were damped once again.

Utida's results are analogous to the patterns of reciprocal fluctuation seen in the Lotka-Volterra model along with some behavior not predicted by the mathematical model. However, despite these differences, like the Lotka-Volterra model, Utida's laboratory model shows that parasitoid–host populations can show reciprocal oscillations without significant temporal variation in the physical environment.

G. Gause (1935) produced similar results when he studied a laboratory population of *Paramecium aurelia* preying upon yeast. He followed the populations through three cycles, which took only 20 days. Though Gause's experiments were much shorter than Utida's, they also produced oscillations like those predicted by the Lotka-Volterra model.

Utida's and Gause's successes make work with laboratory models look far easier than it is. Most attempts to produce Lotka-Volterra-type oscillations in laboratory populations have failed. Most laboratory experiments have led to extinction of the predator or prey population in a fairly short period of time. To sustain oscillations even for a short period, researchers have generally had to provide the prey with refuges of some sort, which indicates another generalization about natural predator–prey systems.

Concept 14.2 Review

1. When the coupled cycling of lynx and snowshoe hare populations (see fig. 14.9) was first described, many concluded that lynx control snowshoe hare populations. Why are lynx not the primary factor controlling snowshoe hare populations even though their population cycles are highly correlated?

2. Why is it not surprising that snowshoe hare populations are controlled by a combination of factors, food and predators (see fig. 14.10), and not by a single environmental factor?

3. Both mathematical and laboratory models offer valuable insights into the dynamics of predator–prey systems. What are some advantages and limitations of each approach?

14.3 Refuges

LEARNING OUTCOMES

After studying this section you should be able to do the following:

14.10 List the variety of "refuges" utilized by prey.

14.11 Explain how the laboratory studies of Gause and Huffaker shed light on the importance of refuges to prey survival in the face of predation.

14.12 Design an experiment to test the effectiveness of large numbers as a form of refuge, emphasizing the role of predator satiation. (You do not have to use live prey.)

14.13 Discuss the ecology of fear as a way of creating refuges for some species.

To persist in the face of exploitation, hosts and prey need refuges. This section is about *refuges,* situations in which members of an exploited population have some protection from predators and parasites. When we think of refuges, we generally think of an inaccessible place. There are, however, many other kinds of refuges. Many have nothing to do with places and most do not provide complete security—just enough.

Refuges and Host Persistence in Laboratory and Mathematical Models

Gause's success at producing cycles in populations of *Paramecium aurelia* and its prey, *Saccharomyces exiguus,* gives no hint of the difficulties he experienced in his earlier attempts. Gause's first attempts to produce Lotka-Volterra population cycles involved *Paramecium caudatum* and one of its predators, another aquatic protozoan called *Didinium nasutum.* If Gause grew these organisms in a simple laboratory microcosm, *Didinium* quickly consumed all the *Paramecium* (fig. 14.14). The absence of a refuge for the prey led eventually to extinction of both predator and prey populations. Gause responded in subsequent experiments by putting some sediment on the bottom of his microcosm to provide a refuge for *Paramecium.* In this case, once *Didinium* had eaten all of the *Paramecium* not hiding in bottom sediments, it starved and became extinct. Following the disappearance of *Didinium* and the removal of predation pressure, the population of

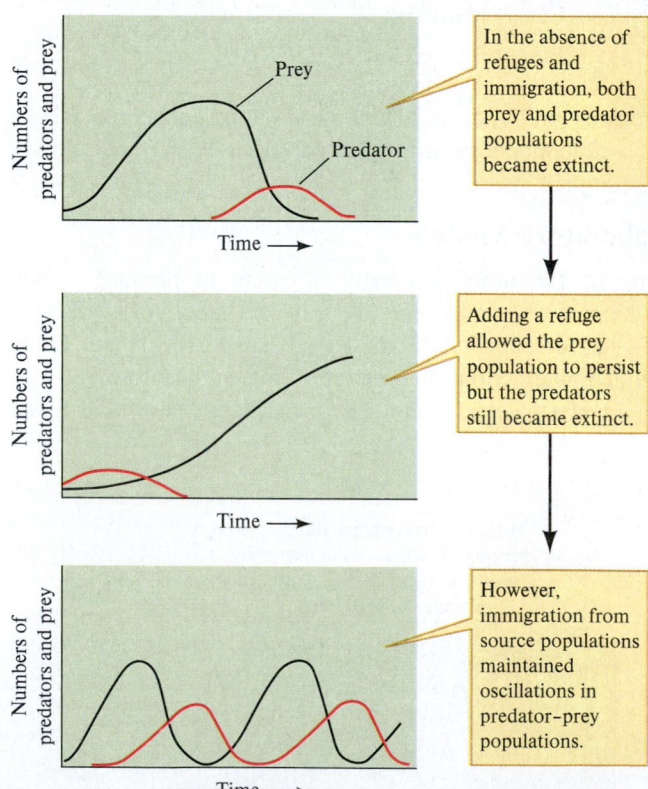

Figure 14.14 Refuges and the persistence of predator–prey oscillation in laboratory populations of prey (*Paramecium aurelia*) and predators (*Didinium nasutum*) (data from Gause 1935–v48).

Paramecium quickly increased. Here, a simple refuge for the prey population led to extinction of the predator.

Gause was able to maintain oscillations in predator–prey populations only if he periodically restocked the populations from his laboratory cultures. In this experiment, the microcosm contained no refuges for *Paramecium*, but every 3 days Gause would take one of each organism from his pure laboratory cultures and add them to the experimental microcosm. Using these periodic immigrations, he was able to produce Lotka-Volterra-type predator–prey oscillations (see fig. 14.14). For him to do so, however, the experimental system had to include a refuge for the prey and a reservoir for the predator (the laboratory cultures), and Gause had to create periodic immigrations from those populations to the experimental microcosm.

Are these experimental requirements entirely artificial, or do they correspond with anything we already know about natural populations? Actually, Gause's experimental results match many of our observations in natural populations. In chapter 9, we saw that on larger scales, populations show clumped distributions. Most species are much more common in some parts of their range than in others. Then in chapter 10, we saw how dispersal is an important contributor to population dynamics and that some local populations are maintained entirely by dispersal from other areas. Some biologists have combined observations such as these to hypothesize the existence of population sources and population sinks—local populations maintained by immigration from source populations. In Gause's experiment, the laboratory cultures were population hot spots, or sources, while the microcosms where predator and prey interacted were population sinks. The requirements of Gause's experiment are consistent with the results of later experiments.

C. Huffaker (1958) set out to test whether Gause's results could be reproduced in a situation in which the predator and prey are responsible for their own immigration and emigration among patches of suitable habitat. Huffaker chose the six-spotted mite, *Eotetranychus sexmaculatus,* a mite that feeds on oranges, as the prey and the predatory mite *Typhlodromus occidentalis,* which attacks *E. sexmaculatus,* as the predator. Huffaker's experimental setups, or "universes" as he called them, consisted of various arrangements of oranges, or combinations of oranges and rubber balls, separated by partial barriers to mite dispersal consisting of discontinuous strips of petroleum jelly.

An important point of natural history is that the predatory mite had to crawl in order to disperse from one orange to another, while the herbivorous mite can disperse either by crawling or by "ballooning," a means of aerial dispersal. A mite balloons by spinning a strand of silk that can catch wind currents. Huffaker gave the herbivorous mite the chance to balloon by providing small wooden posts that could serve as launching pads and by having a fan circulate air across his experimental setup.

While Huffaker's simpler experimental universes did not produce predator–prey oscillations, his most elaborate setup of 120 oranges did. These oscillations spanned several months

(fig. 14.15). Huffaker observed three oscillations that spanned about 6 months. They were maintained by the dispersal of predator and prey among oranges in a deadly game of hide-and-seek, in which the prey managed to keep ahead of the predator for three full oscillations. These results are similar to those obtained by Gause, but we need to remember that Huffaker did not directly manipulate dispersal. In Huffaker's experiment both predator and prey moved from patch to patch under their own power.

The importance of refuges was recognized by Lotka (1932a) and incorporated into his mathematical theory of predator–prey relations. The starting point for his discussion was the Lotka-Volterra predator–prey equations that we discussed previously:

$$\frac{dN_h}{dt} = r_h N_h - p N_h N_p \text{ and } \frac{dN_p}{dt} = c p N_h N_p - d_p N_p$$

The part of this equation that provided the starting point for Lotka's discussion was *p*, the capture or consumption rate of the predator. Lotka pointed out that while it may be reasonable to assume that *p* is a constant for a particular environment, its value should change from one environment to another if the environments differ structurally, particularly if there is a difference in the availability of refuges in the two environments. Specifically, *p* should be lower where the prey or hosts have access to more refuges. This refinement of the

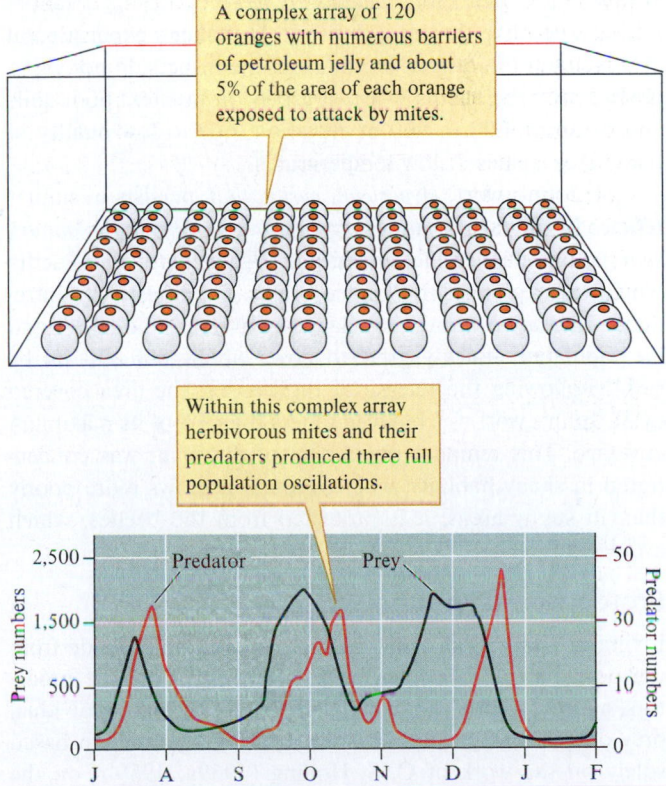

A complex array of 120 oranges with numerous barriers of petroleum jelly and about 5% of the area of each orange exposed to attack by mites.

Within this complex array herbivorous mites and their predators produced three full population oscillations.

Figure 14.15 Environmental complexity and oscillations in laboratory populations of a herbivorous mite and a predatory mite (data from Huffaker 1958).

Lotka-Volterra predator–prey model anticipated recent theoretical analysis of the role that refuges and spatial diversity, in general, play in the persistence of predator–prey and parasite-host systems.

Exploited Organisms and Their Wide Variety of "Refuges"

While his analysis concentrated on physical refuges that could shelter terrestrial prey, Lotka recognized the wide variety of forms that refuges could take. He pointed out, for instance, that flight is a refuge for birds from terrestrial predators. Large size can also serve as a type of refuge. For example, while young African elephants may be vulnerable to predation by lions, adult elephants are not. However, most of our discussion has focused on "spatial" refuges.

Space

Many forms of spatial refuge are familiar: burrows, trees, air, water (if faced with terrestrial predators), and land (if faced with aquatic predators). However, some spatial refuges differ in subtle ways from other areas.

The invasive cactus *Opuntia stricta,* which had completely covered vast areas of Australia, was eventually controlled by a combination of an introduced herbivorous insect, *Cactoblastis cactorum,* and pathogenic microbes. The introduction of the insects did not drive the cactus to extinction, however. One reason for the persistence of the cactus is that it has a number of spatial refuges. Small isolated cactus populations are difficult for *Cactoblastis* to find. This is a spatial refuge much like that designed into Huffaker's experimental arrangement of oranges. In addition, the insects do not vigorously attack the cactus where it grows on nutrient-poor soils and/or above 600 to 900 m elevation, due to low quality of the cactus tissues or low temperatures.

St. John's wort, *Hypericum perforatum,* persists in similar refuges in the face of attacks by the beetle *Chrysolina quadrigemina,* one of the chief enemies of *Hypericum* in the Pacific Northwest region of the United States. *Hypericum* was introduced into areas along the Klamath River around 1900, and its population quickly grew to cover about 800,000 ha by 1944. Following the release of the beetles, the area covered by St. John's wort was reduced to less than 1% of its maximum coverage. This remnant population of the plant was concentrated in shady habitats, where, though it grows more poorly than in sunny areas, it is protected from the beetles, which avoid shade.

Protection in Numbers

Living in a large group provides a type of refuge. Aside from the potential of social groups to intimidate would-be predators, numbers alone can reduce the probability of an individual prey or host being eaten. We can make this prediction based solely on the work of C. S. Holling (1959a, 1959b) on the responses of predators to prey density. In chapter 7, we looked at the functional responses of several predators and herbivores. Briefly, predator functional response results in increasing rate of food intake as prey density increases. Eventually, however, the predator's feeding rate levels off at some maximum rate. In chapter 10, we looked at numerical response, a second component of predator response to prey density that results in increased predator density as prey density increases. As with functional response, the numerical response eventually levels off at the point where further increases in prey density no longer produce increased predator density.

Now let's put functional response and numerical response together to predict the predator's **combined response** to increased prey density. We can combine the two responses by multiplying the number of prey eaten per predator times the number of predators per unit area:

$$\frac{\text{Prey consumed}}{\text{Predator}} \times \frac{\text{Predators}}{\text{Area}} = \frac{\text{Prey consumed}}{\text{Area}}$$

By dividing the prey consumed per unit area by the population density of the prey (prey consumed/area), we can determine the percentage of the prey population consumed by the predator. If we plot percentage of the prey consumed against prey density over a broad range of prey densities, the prediction is that the percentage of the prey population consumed will be lower at high prey densities (fig. 14.16).

Why should the percentage of the prey consumed by the predator decline at high prey densities? The answer to this question, which may not be obvious at first, lies in the predator functional and numerical responses. We see this effect because both numerical and functional responses level off at intermediate prey densities; that is, beyond a certain threshold, further increases in prey density do not lead to either higher predator densities or increased feeding rates. Meanwhile, the density of the prey population continues to increase and the proportion of the prey eaten by predators declines. This work by Holling suggests that prey can reduce their individual probability of

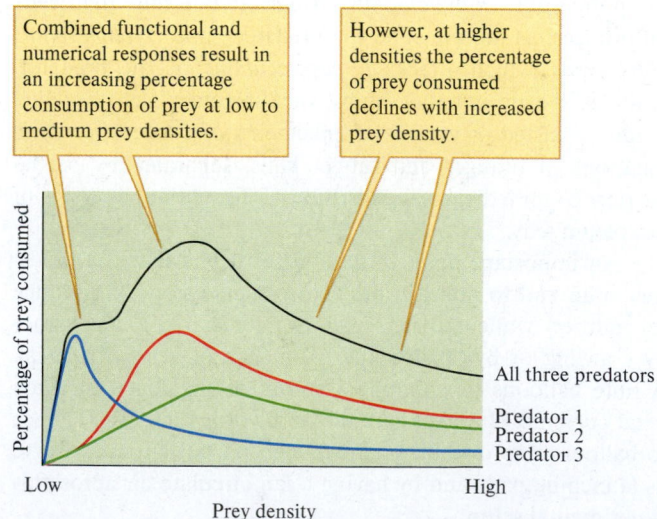

Figure 14.16 Prey density and the percentage of prey consumed due to combined functional and numerical responses (data from Holling 1959a, 1959b).

being eaten by occurring at very high densities. It appears that this defensive tactic, which is called **predator satiation,** is employed by a wide variety of organisms from insects and plants to marine invertebrates and African antelope. Nowhere is predator satiation more apparent than among the periodical cicadas of North America (fig. 14.17).

The Ecology of Fear and Refuges

When considering the effects of predators on prey populations, we generally focus on lethal effects—that is, on predators killing their prey. However, as William Ripple and Robert Beschta (2004, 2007) point out, predators can also influence prey populations by altering their behavior. Such behavioral effects of predators on prey populations, which have been broadly referred to as "the ecology of fear," are the result of prey avoiding high-risk situations. Ripple and Beschta have been studying the ecological consequences of the reintroduction of gray wolves, *Canis lupus,* to Yellowstone National Park in 1995–96 (fig. 14.18*a*) and report that as wolf populations have increased, one of their major prey species, elk, *Cervus elaphus* (fig. 14.18*b*), has altered its distribution. Most significantly, elk have reduced their use of riparian areas, where they may be more vulnerable to hunting wolf packs. As a consequence, wolves have created a refuge for some of the elk's main food sources, especially riparian tree populations (fig. 14.18*c*), particularly willow, which are now showing renewed growth in these refuges, maintained not by physical barriers, but by fear.

How much wolves have influenced the observed recovery of riparian vegetation in Yellowstone is the focus of recent debate (e.g., Kauffman, Brodie, and Jules 2010, 2013; Beschta and Ripple 2013; Beschta et al. 2018; Fleming 2019). However, the role of wolves and other predators in altering the use of space by their prey is well documented across a wide range of other ecosystems (Laundré 2010; Latombe, Fortin, and Parrott 2014).

(a)

(b)

(c)

Figure 14.17   Two periodical cicadas, *Magicicada* spp. An emergence of periodical cicadas produces a sudden flush of singing insects whose density can approach four million individuals per ha, which translates into a biomass of 1,900 to 3,700 kg of cicadas per ha, the highest biomass of a natural population of terrestrial animals ever recorded. Ogrim/Shutterstock

Figure 14.18   The ecology of fear in action. Before the introduction of (*a*) gray wolves to Yellowstone National Park, (*b*) elk focused much of their feeding in riparian areas, where they heavily browsed young willow and aspen trees; in the presence of wolves, elk now tend to avoid riverside areas, where they are more vulnerable to predation. As a consequence, riparian tree populations are beginning to recover and may eventually grow as (*c*) healthy, dense stands along Yellowstone's rivers and streams.   (*a*) NPS photo by Doug Smith; (*b*) Mark Dierker/McGraw-Hill Education; (*c*) Photodisc Collection/Getty Images

Concept 14.3 Review

1. What factors make coexistence of predators and prey less likely in a laboratory setting than in nature?
2. Why should there be strong selection on periodical cicadas for highly synchronous emergence?

14.4 Ratio-Dependent Models of Functional Response

LEARNING OUTCOMES

After studying this section you should be able to do the following:

14.14 Outline the differences between prey-dependent and ratio-dependent models of functional response.

14.15 Identify elements of the equations of prey-dependent and ratio-dependent models of functional response.

14.16 Discuss the ecological circumstances in which prey-dependent and ratio-dependent models of predator functional response are likely to be most appropriate.

Models incorporating the ratio of prey to predator numbers better predict predator functional responses in many ecological circumstances. The classical Lotka-Volterra predator–prey model (see fig. 14.11) has stimulated a vast amount of research since its publication in the early twentieth century. The utility of this model results partly from its simplicity. However, as pointed out earlier, that simplicity includes some biologically unrealistic assumptions. One of those assumptions is that the rate at which an individual predator consumes prey is determined solely by prey abundance. In this model, the size of the predator population has no effect on feeding rate by individual predators, whether the predator population is small or very large. Another assumption is that the conversion of prey to predator offspring is instantaneous, when, in fact, there can be substantial delays particularly among large predators with seasonal reproduction. Although these assumptions may be justified where predator population densities are very low or for micro-predators with fast life cycles, alternative models may be more appropriate in other systems.

Alternative Model for Trophic Ecology

Recognizing the shortcomings of the Lotka-Volterra model, Roger Arditi, now at the Sorbonne University in Paris and Lev Ginzburg, of Stony Brook University, proposed an alternative approach (Arditi and Ginzburg 1989). The core of that pioneering work focused on the functional response of predators, which we explored graphically in chapter 7 (see figs. 7.22 and 7.23). Those figures in chapter 7 project the assumption that rates of consumption by a predator are strictly a function of prey abundance:

$$g = g(N)$$

where g is the predator's functional response, that is, its rate of prey consumption. In this model, functional response is determined only by the abundance of the prey, N. As a result, Arditi

and Ginzburg referred to this type of model as **prey-dependent functional response.** A commonly used prey-dependent functional response model was proposed by C. S. Holling (1959b), which, following the notation of Arditi and Ginzburg (2012) is:

$$g(N) = \frac{aN}{1 + ahN}$$

Here, the rate at which prey are consumed by a predator, g, depends on a, the searching efficiency; h, the handling time for a particular type of prey; and N, the abundance of prey in the environment. This functional response is the result of predator feeding behavior as predators search for, capture, and handle prey. In this model, the rate of prey consumption increases with prey density, eventually leveling off at higher prey densities where consumption rate is limited by how quickly a predator can capture and handle prey (fig. 14.19a).

Arditi and Ginzburg proposed that a functional response model based entirely on predatory behavior by individual predators will not accurately predict rates of consumption in many—perhaps most—circumstances in nature. In particular, they were concerned that the number of predators in the environment can influence feeding rates in the predator population. For example, variation in predator population density may influence the likelihood of interference among predators. While predator interference would be highly unlikely at very low population densities, higher predator densities may well be accompanied by interference. Consequently, they proposed an alternative functional response model based on the per capita availability of prey:

$$g = g\left(\frac{N}{P}\right)$$

Their model of functional response, which they called "ratio-dependent," is:

$$g\left(\frac{N}{P}\right) = \frac{\alpha N/P}{1 + \alpha hN/P} = \frac{\alpha N}{P + \alpha hN}$$

In the Arditi-Ginzburg model, α is not searching time, as in the Holling model, but the rate at which prey become available to the predator population, by whatever mechanism, while h is the handling time, just as in the Holling prey-dependent model. The main distinction of this model is that the rate of prey consumption, that is, the functional response, is determined by

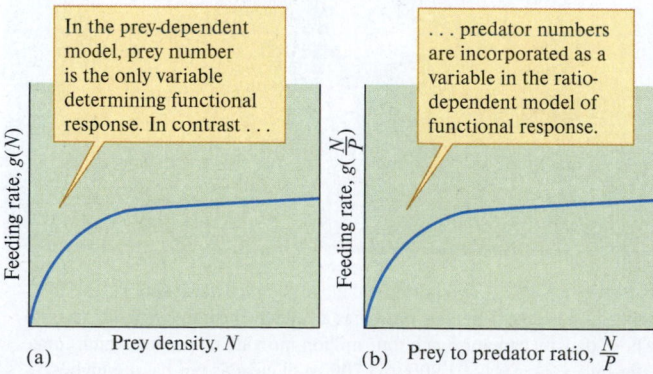

Figure 14.19 Alternative models for predator functional response: (*a*) prey-dependent functional response; (*b*) ratio-dependent functional response.

the ratio of prey numbers, N, to predator numbers, P—in other words, the per capita availability of prey (fig. 14.19b). A growing body of evidence supports the Arditi-Ginzburg model of **ratio-dependent functional response.**

Evidence for Ratio-Dependent Predation

In their 1989 paper, Arditi and Ginzburg extracted data from a study of predatory interactions between the Atlantic oyster drill, *Urosalpinx cinerea*, a predaceous snail, and its prey, the acorn barnacle, *Balanus balanoides* (now *Semibalanus balanoides*) (Katz 1985). In one set of field experiments, the author of that study, Clifford Katz of the University of Connecticut, varied both the density of the prey, acorn barnacles, and the density of the predator, Atlantic oyster drills. Consequently, the results of this experiment provide a basis for evaluating the relative effectiveness of prey-dependent versus the ratio-dependent models of functional response. When Katz analyzed the results of this experiment, he found that both prey density and predator density contributed significantly to feeding rates by the Atlantic oyster drill. In their evaluation of the same data, Arditi and Ginzburg plotted feeding rate by the predator against two different independent variables: prey population density, N, or the ratio of prey numbers to predator numbers, N/P. The result of this graphing exercise clearly supports a ratio-dependent functional response by the Atlantic oyster drill (fig. 14.20).

The three decades since the 1989 publication of the Arditi-Ginzburg ratio-dependent model have been marked by controversy and discussion between supporters of the traditional prey-dependent models and those favoring a ratio-dependent

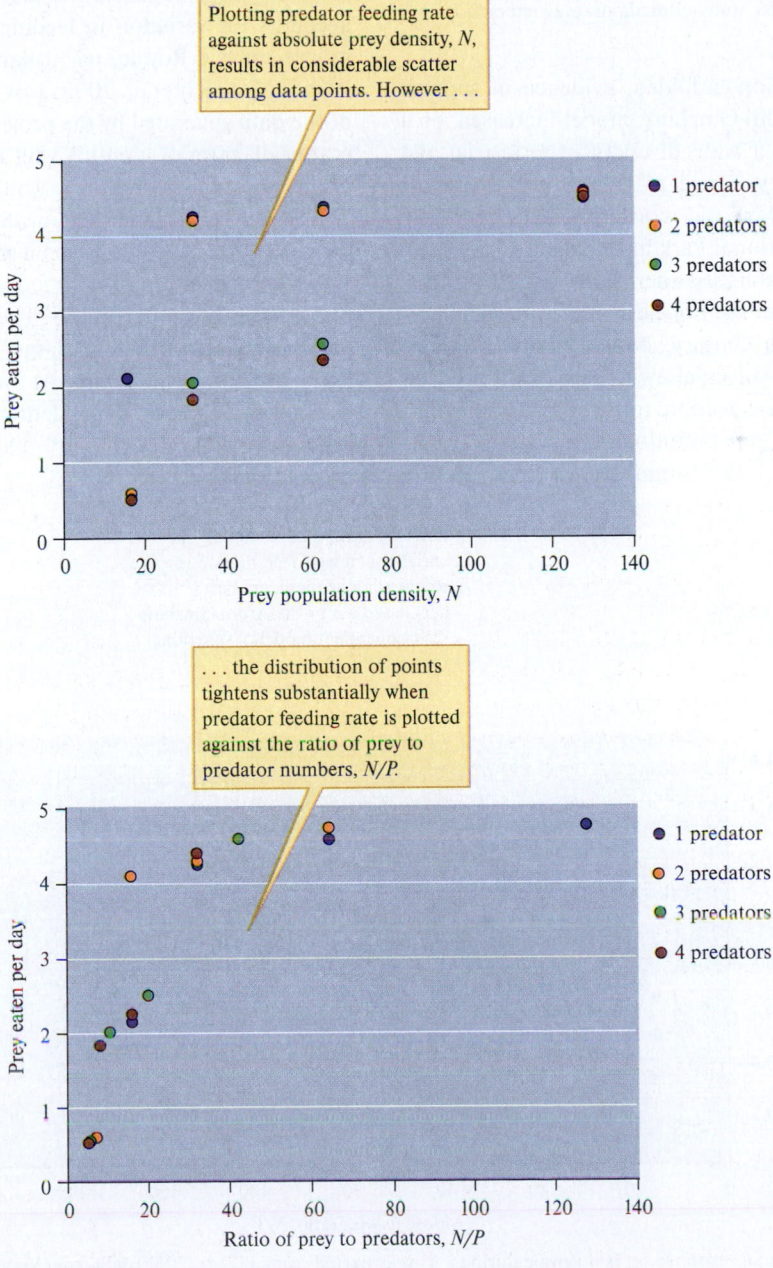

Figure 14.20 Rates of feeding by the Atlantic oyster drill, *Urosalpinx cinerea*, a predaceous marine snail, preying on the acorn barnacle, *Balanus balanoides*, lends support for ratio-dependent modeling of the functional response by this predator (data from Arditi and Ginzburg 1989).

Figure 14.21   A wolf pack pursuing a moose across the winter landscape of Isle Royale National Park. Intensively studied for more than half a century, the interactions between wolves and moose on Isle Royale have become a model case study of predator–prey interactions.
JA Vucetich and RO Peterson

approach. As this discussion unfolded, evidence in support of the ratio-dependent Arditi-Ginzburg model increased, as a result of studies involving a wide diversity of terrestrial and aquatic predators and prey. Some of the strongest support has come from studies of wolf and moose interactions on Isle Royale, a 544 km^2 U.S. National Park in Michigan (fig. 14.21). Moose colonized and established themselves on the island, which is located 24 km from the northern shore of Lake Superior, in the early twentieth century. Wolves arrived decades later in the 1940s. While wolves elsewhere generally feed on several species of prey, moose account for over 90% of the diets of Isle Royale wolves. This concentration on a single prey species combined with a near lack of immigration or emigration

by either wolves or moose, makes Isle Royale an ideal natural laboratory for studying interactions between a large mammalian predator and its prey. Since the initial colonization in the 1940s, only one additional wolf, a lone male from Canada, is known to have immigrated to Isle Royale in 1997.

Systematic study of interactions between moose and wolves on Isle Royale began in 1958, with the founding of the Isle Royale moose-wolf project. Researchers have collected mountains of data since then, including, critically for the purposes of this discussion, the number of moose and wolves on the island and the number of moose kills. Huge numbers of people, from paid professional researchers to volunteers, have been needed to sustain this long-term, large-scale research project. This project now claims to be the longest continuous study of predator–prey relationships in the world (isleroyalewolf.org). Significantly, comparative studies have verified that the Arditi-Ginzburg ratio-dependent model of functional response best accounts for variation in feeding rates by wolves preying on moose on Isle Royale, particularly at the scale of the whole island (Vucetich et al. 2002; Jost et al. 2005). A small sample of the data generated by the project is shown in figure 14.22, a scatter diagram of monthly wolf kill rates for a 41-year period against the ratio of moose to wolf numbers.

Based on increasing evidence in support of ratio-dependent models of functional response, Arditi and Ginzburg (2012) proposed that such models become the standard view of predator–prey interactions. They argue that such models are particularly appropriate for larger-scale studies of larger organisms with slower life cycles. At the same time, they are careful to point out that prey-dependent models work well for accounting for feeding rates that are mainly the result of short-term predator behavior.

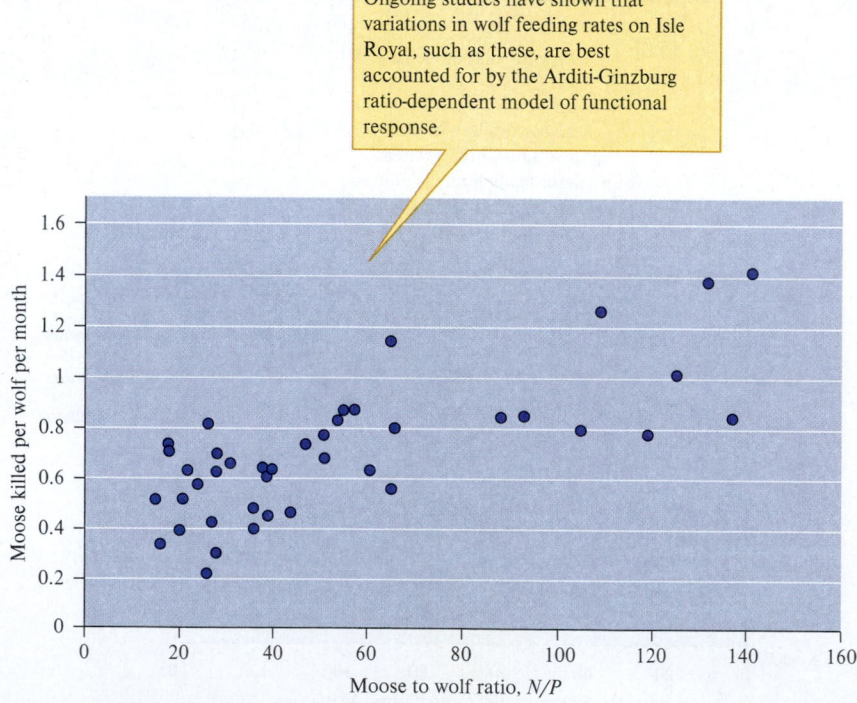

Ongoing studies have shown that variations in wolf feeding rates on Isle Royal, such as these, are best accounted for by the Arditi-Ginzburg ratio-dependent model of functional response.

Figure 14.22   Wolf functional response on Isle Royale during a 41-year period from 1971 to 2011 (data from Vucetich and Peterson 2012).

Concept 14.4 Review

1. Why do the points representing feeding rates by one predator in figure 14.20 appear in exactly the same locations in the upper and lower panels, whereas the other points representing feeding rates for two, three, and four predators do not? Do these differences suggest hypotheses for where a prey-dependent model of functional response may be appropriate?

2. What qualities of the Isle Royale moose-wolf interaction make it ideal for the study of predator functional responses in nature? What are some of the major difficulties associated with effectively studying this system?

14.5 Complex Interactions

LEARNING OUTCOMES

After studying this section you should be able to do the following:

14.17 Describe examples of how some parasites and pathogens alter host behavior.

14.18 Explain how alterations of host behavior impact the survival and reproductive success of the parasite/pathogen.

14.19 Explain how exploitative interactions (such as predation and parasitism) might impact other community-level interactions (such as interspecific competition).

Exploitative interactions weave populations into a web of relationships that defy easy generalization. By conservative estimate, the number of species in the biosphere is on the order of 10 million. As huge as this number may seem, the number of exploitative interactions between species is far greater. Why is that? Because every one of those 10 million species is food for a number of other species and is host to a variety of parasites and pathogens. In addition, most feed on other species. Exploitative interactions weave species into a tangled web of relationships. For instance, K. E. Havens (1994) estimated that the approximately 500 known species occupying Lake Okeechobee, Florida, are linked by about 25,000 exploitative interactions—50 times the number of species! Exploitation provides much of the detail in the tapestry we call nature. In this section, we try to capture some of the richness of that tapestry by discussing the natural history of a number of interactions.

Parasites and Pathogens That Manipulate Host Behavior

The most obvious form of exploitation occurs when one organism consumes part or all of another. Exploitation, however, can assume far more subtle forms. Some species alter the behavior of the species they exploit.

Parasites That Alter the Behavior of Their Hosts

A number of parasites alter the behavior of their hosts in ways that benefit transmission and reproduction of the parasite. Acanthocephalans, or spiny-headed worms, change the behavior of amphipods, small aquatic crustaceans, in ways that make it more likely that infected amphipods will be eaten by a suitable vertebrate host, especially ducks, beavers, and muskrats. Uninfected amphipods avoid the light—that is, show **negative phototaxis**. They spend most of their time near the bottoms of ponds and lakes, away from well-lighted surface waters, where the surface-feeding vertebrate hosts of acanthocephalans spend the majority of their time. In contrast, infected amphipods swim toward light—that is, show **positive phototaxis,** a behavior that places them near the pond surface in the path of feeding ducks, beavers, and muskrats (Bethel and Holmes 1977). Interestingly, amphipod behavior remains unaltered until the acanthocephalan has reached a life stage, called a *cystacanth,* that is capable of infecting the vertebrate host. If eaten earlier, the acanthocephalan would die without completing its life cycle.

Janice Moore (1983, 1984a, 1984b) studied a similar parasite–host interaction involving an acanthocephalan, *Plagiorhynchus cylindraceus,* a terrestrial isopod or pill bug, *Armadillidium vulgare,* and the European starling, *Sturnus vulgaris.* In this interaction, the pill bug serves as an intermediate host for *Plagiorhynchus,* which completes its life cycle in the starling (fig. 14.23).

At the outset of her research, Moore predicted that *Plagiorhynchus* would alter the behavior of *Armadillidium.* She based this prediction on several observations. One was the relative frequency of infection of *Armadillidium* and starlings by *Plagiorhynchus.* Field studies had demonstrated that even where *Plagiorhynchus* infects only 1% of the *Armadillidium* population, over 40% of the starlings in the area were infected. Some factor was enhancing rates of transmission to the starlings, and Moore predicted that it was altered host behavior. Moore thought that the size of *Plagiorhynchus* might also be a factor. At maturity, the cystacanth stage of *Plagiorhynchus* grows to about 3 mm, a substantial fraction of the internal environment of an 8 mm pill bug!

Moore brought *Armadillidium* into the laboratory and established two laboratory populations: an uninfected control group and an infected experimental group. She infected half the laboratory populations of *Armadillidium* by feeding them pieces of carrot coated with *Plagiorhynchus* eggs, while keeping the remaining laboratory populations free of *Plagiorhynchus.* After 3 months, the *Plagiorhynchus* in the infected populations matured to the cystacanth stage. At this point Moore mixed the infected and uninfected populations.

Because *Plagiorhynchus* does not alter the outward appearance of *Armadillidium,* Moore could not determine whether an *Armadillidium* was infected or not until she dissected it at the completion of an experiment. Consequently, all the behavioral experiments were conducted "blind"—that is, without the possibility of observer bias due to prior knowledge of the identity of experimental and control animals.

① Adult female *Plagiorhynchus* lays eggs within the intestines of infected birds. The eggs are shed with feces.

② A terrestrial isopod eats the feces of an infected bird. The eggs of *Plagiorhynchus* hatch within a few hours; they develop into a mature larva in 60–65 days.

③ The mature larvae of *Plagiorhynchus* alter isopod behavior; infected isopods leave sheltered areas and wander in the open.

④ Leaving shelter makes the isopods more conspicuous and vulnerable to predation by birds. When eaten by a bird, the mature *Plagiorhynchus* attaches to the bird's intestinal wall.

Figure 14.23   The life cycle of *Plagiorhynchus cylindraceus,* an intestinal parasite of birds.

Moore found that *Plagiorhynchus* alters the behavior of *Armadillidium* in several ways. Infected *Armadillidium* spend less time in sheltered areas and more time in low-humidity environments and on light-colored substrates. These changes in behavior would increase the time an *Armadillidium* spends in the open, where it could be easily seen and eaten by a bird.

In laboratory experiments Moore demonstrated that captive starlings consistently captured *Armadillidium* from light rather than dark substrates. She provided caged starlings access to a mixture of 10 infected and 10 uninfected *Armadillidium,* which wandered freely across the bottom of the cage, half of which was covered by black sand and half by white sand. Under these conditions, starlings ate 72% of the infected *Armadillidium* but only 44% of the uninfected individuals (fig. 14.24). The starlings took isopods mainly from the surface of white sand, so it seems that the tendency of *Armadillidium* to seek out light substrates does make them more vulnerable to predation by birds.

A critical step in this research was to determine whether the changed behavior of infected *Armadillidium* translates into their being eaten more frequently by wild birds. Moore collected the arthropods that starlings feed to their nestlings and from these collections estimated the rate at which they delivered *Armadillidium*—about one every 10 hours. Using this delivery rate and the proportion of the *Armadillidium* population infected by *Plagiorhynchus* (about 0.4%), she was able to predict the expected rate of infection among starling nestlings if the adults capture *Armadillidium* at random from the natural population. The proportion of infected nestlings was about twice the rate of infection predicted if starlings fed randomly on the *Armadillidium* population. These results support Moore's hypothesis that the altered behavior of infected *Armadillidium* increases their probability of being eaten by starlings.

Moore emphasized that *Plagiorhynchus* does not just alter *Armadillidium*'s behavior but alters its behavior in a particular

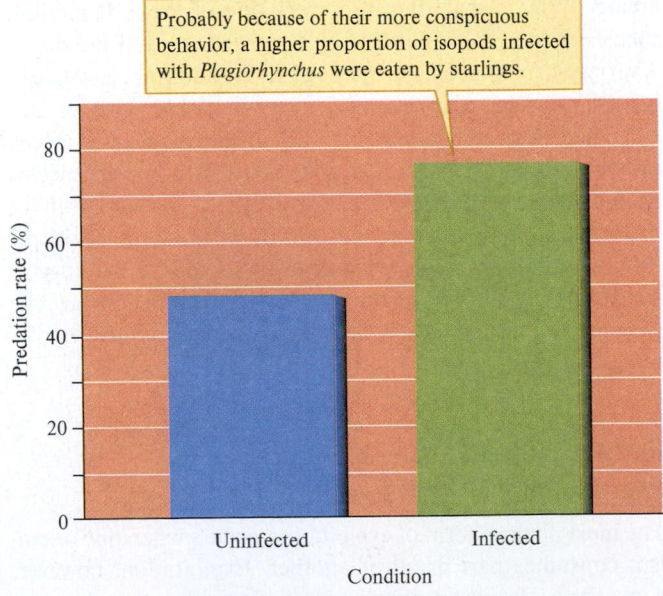

Probably because of their more conspicuous behavior, a higher proportion of isopods infected with *Plagiorhynchus* were eaten by starlings.

Figure 14.24   Starling predation on uninfected and infected *Armadillidium vulgare* (data from Moore 1984b).

way—in a way that increases the rate at which starlings, the final host of the parasite, are infected.

A Plant Pathogen That Mimics Flowers

Every year the slopes of the southern Rocky Mountains are decorated with the colorful blossoms of wildflowers. Some of these wildflowers, however, are not quite what they seem. One bright yellow and sweet-smelling "blossom" is actually produced by a pathogenic fungus that manipulates the growth of its host plant. This pathogen belongs to a group of fungi called rusts because of their rust-colored spores that appear on the surface of the infected host plant. This particular rust is *Puccinia monoica,* and its hosts are mustard plants in the genus *Arabis. Arabis* spp. are herbaceous plants that spend anywhere from a few months to several years as a rosette, a low-growing growth form with a high density of leaves. During the rosette stage, *Arabis* invests heavily in root development and storage of energy in the roots. At the end of the rosette stage, *Arabis* grows tall quickly, a process called *bolting,* and flowers (fig. 14.25*a*). Once pollinated, the flowers form seeds that mature, completing the life cycle of *Arabis.*

However, *Puccinia* completely alters the life history of *Arabis.* It attacks the rosette stage, manipulating its development to produce a growth form that promotes reproduction by the fungus and usually kills the plant. *Puccinia* infects the rosettes of *Arabis* in late summer and then invades the **meristematic tissue,** the actively dividing tissue responsible for plant growth, during the following winter. As it invades the meristematic tissue, *Puccinia* manipulates future development by the rosette. Infected rosettes elongate rapidly the following spring, maintain a high density of leaves along their entire length, and are topped by a cluster of bright yellow leaves. This cluster of yellow leaves forms a pseudoflower that looks very much like the flowers of the buttercup, *Ranunculus* spp. (fig. 14.25*b*).

The pseudoflowers of infected rosettes are produced by various fungal structures, including spermogonia containing spermatia (fungal reproductive cells), sexually receptive fungal hyphae, and secretions of sticky, sugar-containing spermatial fluid. Most rusts require outcrossing for sexual reproduction, which is accomplished by insects transferring spermatia from one fungus (thallus) to the receptive hyphae of another thallus. Barbara Roy (1993) found that the combination of yellow color and sugary fluid attracts a wide variety of flower-visiting insects, including butterflies, bees, and flies (fig. 14.26). Flies, the most common visitor to pseudoflowers at her Colorado study site, have been demonstrated to be effective carriers of rust spermatia.

Roy's studies demonstrated that *Puccinia* truncates the life cycle of *Arabis* and in the process generally kills the host plant. The *Arabis* that survive attack by *Puccinia* may go on to flower but none form seeds. Thus, destruction by *Puccinia* is total.

(a)   (b)

Figure 14.25   The effects of the fungus *Puccinia monoica* on morphology: (*a*) a normally developed mustard plant, *Arabis holboellii;* (*b*) a pseudoflower formed by *A. holboellii* infected by *Puccinia.*   (*a, b*) B.A Roy

Figure 14.26   The pseudoflowers formed by *Arabis holboellii* infected by the fungus *Puccinia monoica,* such as the one shown here, are attractive to a wide variety of pollinating insects, such as this *Polygonia* butterfly.   B.A Roy

The Entangling of Exploitation with Competition

We often arrange our thoughts about nature in neat categories, like the chapters of this book. In chapter 13, we discussed competition, in chapter 14 we discuss exploitative interactions, and in chapter 15 we examine mutualism. Nature itself is not so neatly arranged, nor are natural phenomena so easily isolated. One process is usually connected to several others. The distinction between exploitation and competition is blurred when competitors eat each other.

Predation, Parasitism, and Competition in Populations of *Tribolium*

Thomas Park and his colleagues (Park 1948; Park et al. 1965) uncovered one of the first examples of competitors eating each other during their work on competition among flour beetles. As we saw in chapter 13, the outcome of competition between *Tribolium castaneum* and *T. confusum* depended on temperature and moisture. It turns out that the presence or absence of a protozoan parasite of *Tribolium, Adelina tribolii,* also influences the competitive balance between flour beetle species. The effects of this parasite are also entangled with predation among the flour beetles and cannibalism, which we might think of as a form of intraspecific exploitation.

Park showed that various strains of *T. castaneum* and *T. confusum* differ in their rates of cannibalism. Of the two species, *T. castaneum* is the most cannibalistic, but it preys on the eggs of *T. confusum* at an even higher rate than it cannibalizes its own eggs. In the light of its predatory behavior, it's not surprising

that *T. castaneum* eliminated *T. confusum* in 84% of 76 competition experiments spanning a period of about 10 years. This predatory strategy works best, however, in the absence of *Adelina.*

Adelina invades the cells of its host and lives its life as an intracellular parasite. Beetles become infected when they ingest the oocysts of this parasite, either as they feed on flour or as they consume infected larvae, pupae, or adult beetles. Once in the gut of the beetle, the oocyst eggs rupture, liberating a life stage of *Adelina* called a *sporozoite.* The sporozoites penetrate the beetle's gut and enter the body cavity, or haemocoel. Once in the haemocoel, the sporozoites invade various cells, where they reproduce asexually and produce a second life stage called the *merozoite.* The motile merozoites invade yet other host cells eventually, producing male and female sex cells that combine to form zygotes. The zygotes eventually give rise to new sporozoites, which are encased in oocysts. Ingestion of these oocysts by another beetle completes the life cycle of *Adelina.*

Several biologists before Park had noted that *Adelina* caused "sickness" and death among *Tribolium* populations. It was Park, however, who demonstrated that *Adelina* reduces the density of *Tribolium* populations and can alter the outcome of competition between *T. confusum* and *T. castaneum. Adelina* strongly reduces the population density of *T. castaneum* populations but has little effect on *T. confusum* populations. In the absence of the parasite, *T. castaneum* won 12 of 18 competitive contests against *T. confusum.* When the parasite was included, however, *T. confusum* won 11 of 15 contests (fig. 14.27). In other words, parasitism completely reverses the outcome of competitive interactions between the two species. From insects to African lions, interference escalated to the point of predation appears to be a common occurrence among competitors. However, Park's experiments with *Tribolium* indicate that parasites may make the outcome of a predaceous competitive strategy difficult to predict.

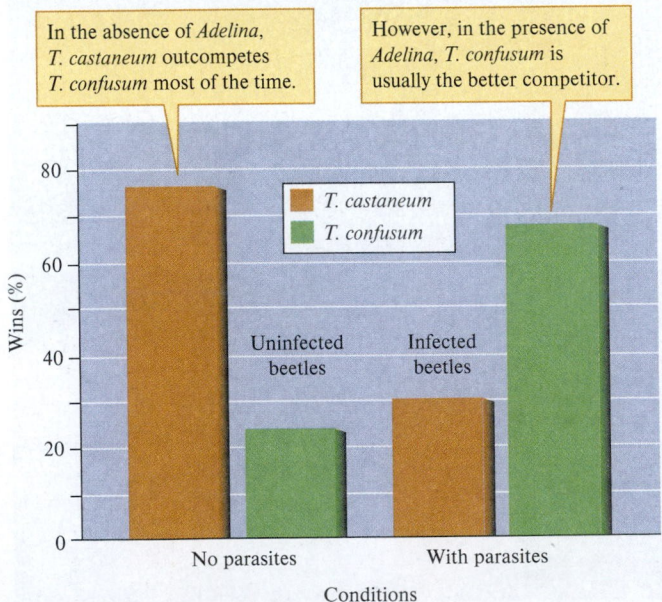

Figure 14.27   The influence of the protozoan parasite *Adelina tribolii* on competition between the flour beetles *Tribolium castaneum* and *T. confusum* (data from Park 1948).

Concept 14.5 Review

1. Why did Moore conduct "blind" behavioral observations—that is, without knowing whether individual *Armadillidium* were infected or not?
2. How did Moore's laboratory and field experiments complement each other?

The Value of Pest Control by Bats: A Case Study

LEARNING OUTCOMES

After studying this section you should be able to do the following:

14.20 Outline the methods used to study the economic benefit of Brazilian free-tailed bats in the control of cotton bollworm in southwestern Texas.

14.21 Summarize the economic benefits of bats, resulting from their eating of insects.

14.22 Explain why the economic benefit of bats to agriculture will usually vary from year to year.

We saw earlier in the chapter that bats reduce the population density of arthropods living on foliage in a lowland tropical forest (see fig 14.5). In the same study, the researchers found that plants from which bats were excluded not only supported higher densities of arthropods but also suffered over three times the amount of herbivore damage to leaves compared to plants to which bats had access (Kalka, Smith, and Kalko 2008). Arthropods of many kinds also attack crop plants and reduce agricultural yields. In defense of their crops, farmers spend billions of dollars annually to control insect pests. Can bats be valuable allies to farmers in their attempts to protect crops? Research in a tropical coffee plantation suggests so. Kimberly Williams-Guillén, Ivette Perfecto, and John Vandermeer (2008) discovered that bats limit arthropod numbers in a coffee plantation in an area of Mexico that would naturally support tropical dry forest (see chapter 2, section 2.3). Where the researchers excluded bats during the wet season, arthropod densities on coffee plants were 84% higher.

If bats provide some protection to crops from insect pests, how much might their pest control services be worth? This is the question addressed by a team of U.S. and Mexican researchers working in an eight-county region in southwestern Texas (Cleveland et al. 2006). The agricultural production of the area includes 10,000 acres of cotton with an annual value of $4.6 to $6.4 million. The area is also home to Brazilian free-tailed bats, *Tadarida brasiliensis,* which emerge from their roosts in caves and under bridges in spectacular feeding flights (fig. 14.28*a*). During a given

(a)

(b)

Figure 14.28   Brazilian free-tailed bats, *Tadarida brasiliensis,* emerge from roost sites in south-central Texas each evening to forage across the surrounding landscape. (*a*) The Brazilian free-tailed bats emerging at dusk from larger colonies pour out into the twilight in a streaming mass of agile, insect-devouring biomass. (*b*) A Brazilian free-tailed bat flying with a moth, one of the species' most frequent prey items, protruding from its mouth.   (*a, b*) Merlin D. Tuttle/Bat Conservation International (www.batcon.org)

night in the growing season, over 100 million Brazilian free-tailed bats may be feeding on flying insects over south-central Texas. These are mostly female bats living in maternity colonies, where they give birth, generally to a single pup, and raise their young. Lactating female Brazilian free-tailed bats rear their pups to essentially adult size in 6 to 7 weeks. Consequently, they face enormous energy demands and, in response, a lactating female Brazilian free-tailed bat ingests up to two-thirds of its body weight in flying insects each night (Kunz, Whitaker, and Wadanoli 1995), including an abundance of moths (fig. 14.28*b*). One of the moths commonly found in the diets of Brazilian free-tailed bats in south-central Texas is *Helicoverpa zea.* Though it feeds on a wide range of plants, the caterpillar of *H. zea* is known as the cotton, or corn, bollworm, as a result of its attacks on these two economically important crop plants. In fact, the cotton bollworm is one of the most serious agricultural pests in the Americas, causing billions of dollars in damage to crops annually (Mitter, Poole, and Matthews 1993).

With millions of bats emerging from their roost sites, flying at speeds of up to 40 km per hour, and covering distances that may exceed 100 km per night, how can an ecologist, or even a team of ecologists, keep track? One of the ways that the Cleveland team followed bats as they left and returned to their roost sites was with NEXRAD (NEXt generation RADar) Doppler radar (fig. 14.29). The radar images gave researchers an indication of the direction in which bats left their roosts to feed and the direction from which they returned. Both sets of directional data indicated that the bats were foraging over the agricultural fields in their study area. To supplement radar tracking, researchers observed bats from the ground to verify that they are feeding over agricultural fields at the time when cotton bollworm moths were emerging and flying in large numbers at elevations from 200 to 1,200 m over the fields. Based on historic estimates of numbers in the study area and recent census data, Cleveland and his colleagues estimated that 1.5 million Brazilian free-tailed bats were feeding over the agricultural fields in their study area each night.

The study team assigned a value to pest control by Brazilian free-tailed bats in the study area using an avoided-cost approach. Their analysis involved two costs. The first avoided cost was the value of the cotton crop that would have been lost without bat pest control. The second part of the calculation incorporated the cost of pesticides that were not applied due to pest control by bats. Based on a long series of calculations, Cleveland's research team estimated that each Brazilian free-tailed bat consumes an average of 1.5 female bollworm moths each night, which would have otherwise gone on to deposit eggs on a cotton plant. Although each female lays 600 to 1,000 eggs, survivorship among caterpillars is low as a result of losses to predators, such as ants, parasites, and pathogens. Taking these sources of mortality into account, the researchers estimated that an individual feeding rate by Brazilian free-tailed bats of 1.5 female moths per night would result in 5 fewer caterpillars damaging cotton plants, which would, in turn, reduce the number of damaged cotton bolls by 10 per night. By limiting the numbers of bollworms on the cotton crop, the Brazilian free-tailed bats would also delay the bollworm population reaching the density threshold at which farmers apply pesticides, resulting in the second avoided cost.

Using the avoided-cost approach, Cleveland's research team estimated that, during their study, the value of pest control by Brazilian free-tailed bats on their southwestern Texas study area was $741,000. They realized, however, that the value of this pest control would differ across years (fig. 14.30):

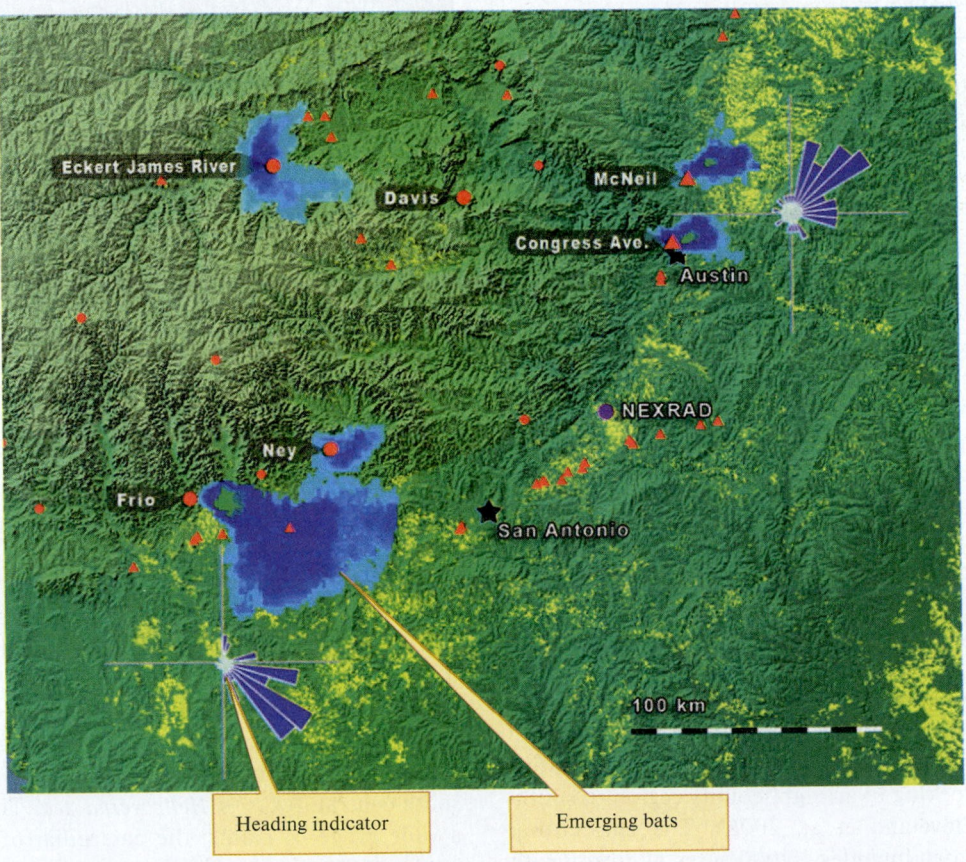

Figure 14.29   NEXRAD (NEXt generation RADar) Doppler radar is providing useful information on the foraging activities of Brazilian free-tailed bats. This image shows the mean heading (direction of travel) of Brazilian free-tailed bats emerging from roost sites near San Antonio and Austin, Texas. The emerging bats appear as irregular patches of green and blue. In the image, cave colonies are marked as circles, bridge colonies as triangles, and agricultural areas as yellow patches (after Horn and Kunz 2008).   Horn, J. W., and T. H. Kunz. 2008. Analyzing NEXRAD doppler radar images to assess nightly dispersal patterns and population trends in Brazilian free-tailed bats (*Tadarida brasiliensis*). *Integrative and Comparative Biology* 48: 24–39.

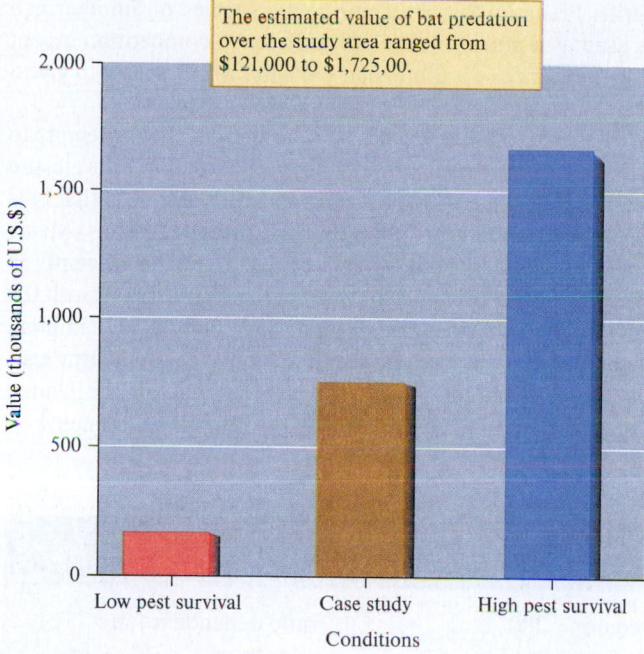

The estimated value of bat predation over the study area ranged from $121,000 to $1,725,00.

Figure 14.30 The value of cotton bollworm, *Helicoverpa zea,* control by Brazilian free-tailed bats, *Tadarida brasiliensis,* in an eight-county area in southwestern Texas. The value of this ecological service was calculated on the basis of costs avoided by farmers as a result of these bats feeding on cotton bollworm moths (data from Cleveland et al. 2006).

less valuable in years when cotton bollworm eggs and larvae had low survival, $121,000; and more valuable in high-survival years, $1,725,000. Let's put these numbers in context. Consider that the study by Cleveland and his colleagues assigned a value to a single predator, Brazilian free-tailed bats, feeding on a single agricultural pest, the cotton bollworm moths, relative to its impact on one crop, cotton, in a relatively small geographic area. How valuable are the pest control services of the other 41 species of insectivorous bats living in the United States across the entire country? Justin Boyle, Paul Cryan, Gary McCracken, and Thomas Kunz (2011) estimated the value of those services at $22.9 billion annually. However, even this estimated value is highly limited geographically and taxonomically, since it is for bats only. How valuable are the services of *all* the predators, parasites, and pathogens of agricultural pests, not just in the United States, but globally? No one knows but, undoubtedly, the value would be astronomical, a fact that is increasingly recognized.

Summary

The diversity of interactions between herbivores and plants, between predators and prey, and between parasites, parasitoids, pathogens, and hosts can be grouped under the heading of *exploitative interactions*—interactions between species that enhance the fitness of one individual at the expense of another.

Predators, parasites, and pathogens influence the distribution, abundance, and structure of prey and host populations. Herbivorous stream insects have been shown to control the density of their algal and bacterial food. An experimental study in Panama showed that both birds and bats reduce foliage-living arthropods significantly in lowland tropical forests, with bats having the larger effect. A parasitic infestation reduced the red fox population in Sweden by 70%, which in turn led to increases in the abundance of several prey species eaten by foxes. This parasitic disease revealed the influence of a predator on its prey populations.

Predator–prey, host–parasite, and host–pathogen relationships are dynamic. Populations of a wide variety of predators and prey show highly dynamic fluctuations in abundance ranging from days to decades. A particularly well-studied example of predator–prey cycles is that of snowshoe hares and their predators, which have been shown to result from the combined effects of the snowshoe hares on their food and of the predators on the snowshoe hare population. Mathematical models of predator–prey interactions by Lotka and Volterra suggest that exploitative interactions themselves can produce population cycles without any influences from outside forces such as weather. Predator–prey cycles have also been observed in a few laboratory populations under restricted circumstances.

To persist in the face of exploitation, hosts and prey need refuges. The refuges that promote the persistence of hosts and prey include secure places to which the exploiter has limited access. However, living in large groups can be considered as a kind of refuge, since it reduces the probability that an individual host or prey will be attacked. Growing to large size can also represent a kind of refuge when the prey species is faced by size-selective predators. Fear of predators can alter the distribution of herbivores creating a plant refuge as a result.

The Lotka-Volterra predator–prey model assumes that the rate at which an individual predator consumes prey is determined solely by prey abundance. In this model, the size of the predator population has no effect on feeding rate by individual predators.

Models incorporating the ratio of prey to predator numbers better predict predator functional responses in many ecological circumstances. In contrast, the Arditi-Ginzburg, ratio-dependent model proposes that both predator and prey abundance can influence predator functional response. A growing body of evidence indicates that this alternative model is particularly appropriate for larger-scale studies of larger organisms with slower life cycles.

Exploitative interactions weave populations into a web of relationships that defy easy generalization. The number of exploitative interactions between species far exceeds the number of species in the biosphere, and the nature of exploitation goes far beyond the typical consumption of one organism by another. For instance, many parasites and pathogens manipulate host behavior to enhance their own fitness at the expense of the host. Spiny-headed worms alter the behavior of a variety of crustacean hosts in a way that increases the probability that the one host species will be eaten by another. A pathogenic fungus manipulates the growth program of its host plant in a way to produce "pseudoflowers," structures aimed at promoting the reproduction of the pathogen. In the process, the pathogen usually kills the host plant and always renders it sterile. Predation by one flour beetle species on another can be used as a potent means of interference competition except in the presence of a protozoan parasite, which seems to give a competitive advantage to less predaceous species.

Bats can be valuable allies to farmers in their attempts to protect crops from arthropod pests. For example, exclosure experiments in a coffee plantation in Mexico showed that bats reduce arthropod population density on coffee plants. Meanwhile, Brazilian free-tailed bats appear to help significantly in the control of cotton bollworm in southwestern Texas, with the value of this control ranging from $121,000 to $1.7 million, depending on the level of survival by cotton bollworm eggs and larvae. Pest control by insectivorous bats in the United States was estimated in 2011 at nearly $23 billion annually.

Key Terms

combined response 312	parasite 300	predator 300	ratio-dependent functional response 315
exploitation 300	parasitoid 300	predator satiation 313	
meristematic tissue 319	pathogen 300	prey-dependent functional response 314	
negative phototaxis 317	positive phototaxis 317		

Review Questions

1. Predation is one of the processes by which one organism exploits another. Others are herbivory, parasitism, and disease. What distinguishes each of these processes, including predation, from the others?

2. How are manipulation of host behavior by spiny-headed worms and manipulation of plant growth by the rust *Puccinia monoica* the same? How are they different?

3. Predation by one flour beetle species on another can be used as a potent means of interference competition. However, the predatory strategy seems to fail consistently in the presence of the protozoan parasite *Adelina tribolii*. Explain how the predatory strategy works in one environmental circumstance and fails in another.

4. In chapter 14, we have seen how a herbivorous stream insect controls the density of its food organisms, how a herbivorous moth larva and pathogenic microbes combine to control an introduced cactus population, and how decimation of a red fox population led to increases in the populations of the foxes' prey. We do not know the specific environmental factors controlling most populations. Explain why such factors must exist. (Hint: Think back to our discussions of geometric and exponential growth in chapter 11.)

5. Early work on exploitation focused a great deal of attention on predator–prey relations. However, parasites and pathogens represent a substantial part of the discussions in chapter 14. Is this representation by parasites and pathogens just the result of biased choices by the author, or do you think that parasites and pathogens have the potential to exert significant controls on natural populations? Justify your answer.

6. Researchers have suggested that predators could actually increase the population density of a prey species heavily infected by a pathogenic parasite (Hudson, Dobson, and Newborn 1992). Explain how predation could lead to population increases in the prey population.

7. Explain the roles of food and predators in producing cycles of abundance in populations of snowshoe hare. Populations of many of the predators that feed on snowshoe hares also cycle substantially. Explain population cycles among these predator populations.

8. What contributions have laboratory and mathematical models made to our understanding of predator–prey population cycles? What are the shortcomings of these modeling approaches? What are their advantages?

9. We included spatial refuges, predator satiation, and size in our discussions of the role played by refuges in the persistence of exploited species. How could time act as a refuge? Explain how natural selection could lead to the evolution of temporal "refuges."

10. Joseph Culp and Gary Scrimgeour (1993) studied the timing of feeding by mayfly larvae in streams with and without fish. These mayflies feed by grazing on the exposed surfaces of stones, where they are vulnerable to predation by fish, which in the streams studied are size-selective feeders and feed predominantly during the day. In the study streams without fish, both small and large mayflies have a slight tendency to feed during the day but feed at all hours of the day and night. In the streams with abundant fish populations, small mayflies fed around the clock, while large mayflies fed mainly at night. Explain these patterns in terms of time as a refuge and size-selective predation.

Royalty-Free/CORBIS

Chapter 15
Mutualism

An anemonefish, *Amphiprion ocellaris,* and its mutualistic anemone partner. The anemonefish finds secure shelter among the stinging tentacles
of the anemone, where it is unharmed, while the anemone receives a
steady supply of nutrients excreted by the plankton-feeding anemonefish.

Applications: Mutualism and Humans   341
    Summary   343
    Key Terms   344
    Review Questions   344

CHAPTER CONCEPTS

15.1 Plants benefit from mutualistic
partnerships with a wide variety
of bacteria, fungi, and animals.   326

Concept 15.1 Review   335

15.2 Reef-building corals depend
on mutualistic relationships
with algae and animals, with an
exchange of benefits paralleling
those between terrestrial
mutualists.   335

Concept 15.2 Review   338

15.3 Theory predicts that mutualism will
evolve where the benefits of mutualism
exceed the costs.   338

Concept 15.3 Review   341

LEARNING OUTCOMES

After studying this section you should be able to do the following:

15.1 Distinguish between mutualism and commensalism.
15.2 Describe the ecological consequences of facultative
versus obligate mutualism.

Positive interactions between species are found throughout the biosphere. A hummingbird darts among the
red blossoms of a plant growing at the edge of a forest
glade. As it inserts its bill into a flower, hovering to sip nectar, the hummingbird head brushes up against the anthers of
the flower and picks up pollen (fig. 15.1). This pollen will be
deposited on the stigmas of other flowers as the hummingbird
goes about gathering its meal of nectar. The hummingbird disperses the plant's pollen in trade for a meal of nectar.

Belowground we encounter another partnership. The roots
of the hummingbird-pollinated plant are intimately connected
with fungi in an association called **mycorrhizae,** a mutualistic
association between fungi and the roots of plants. The hyphae
of the mycorrhizal fungi extend out from the roots, increasing

Figure 15.1   A female ruby-throated hummingbird, *Archilochus colubris,* feeding from a scarlet beebalm flower, *Monarda didyma.* Hummingbirds feeding on nectar transfer pollen from flower to flower.
Ingram Publishing/SuperStock

the capacity of the plant to harvest nutrients from the environment. In exchange for the nutrients, the plant delivers sugars and other products of photosynthesis to its fungal partner.

Meanwhile, back aboveground, a deer enters the forest glade and wanders over to the plant recently visited by the hummingbird. The deer systematically grazes it to the ground, lightly chews the plant material, and then swallows it. As the plant material enters the deer's stomach, it is attacked by a variety of protozoans and bacteria. These microorganisms break down and release energy from compounds such as cellulose, which the deer's own enzymatic machinery cannot handle. In return, the protozoans and bacteria receive a steady food supply from the feeding activities of the deer as well as a warm, moist place in which to live.

These are examples of **mutualism**—that is, interactions between individuals of different species that benefit both partners. In some cases, an interaction between two species benefits one of them, while the other is neither benefited nor harmed. We call such an interaction a **commensalism** (see chapter 13, introduction). Some species can live without their mutualistic partners and so the relationship is called **facultative mutualism.** Other species are so dependent on the mutualistic relationship that they cannot live in its absence. Such a relationship is an **obligate mutualism.** It is a curious fact that though observers of nature as early as Aristotle recognized such mutualisms, mutualistic interactions have received much less attention from ecologists than have either competition or exploitation. Does this lack of attention reflect the rarity of mutualism in nature? As you will see in this chapter, mutualism is virtually everywhere.

Mutualism may be common, but is it important? Does it contribute substantially to the ecological integrity of the biosphere? The answer to both these questions is yes. Without mutualism the biosphere would be entirely different. Let's remove some of the more prominent mutualisms from the biosphere and consider the consequences. An earth without mutualism would lack reef-building corals as we know them. Therefore, we can erase the Great Barrier Reef, the largest biological structure on earth, from our hypothetical world. We can also eliminate all the

coral atolls that dot the tropical oceans as well as all the fringing reefs. The deep oceans would have no bioluminescent fishes or invertebrates. In addition, the deep-sea oases of life associated with ocean-floor hydrothermal vents (see chapter 7, section 7.2) would be reduced to nonmutualistic microbial species.

On land, there would be no animal-pollinated plants: no orchids, no sunflowers, and no apples. The pollinators themselves would also be gone: no bumblebees, no hummingbirds, and no monarch butterflies. Gone, too, would be all the herbivores that depend on animal-pollinated plants. Without plant-animal mutualisms, tropical rain forests, the most diverse terrestrial biome on the planet, would be all but gone. Many wind-pollinated plants would remain. However, many of these species would also be significantly affected, since approximately 90% of all plants form mycorrhizae. Those plants capable of surviving without mycorrhizal fungi would likely be restricted to the most fertile soils.

Even if wind-pollinated, nonmycorrhizal plants remained on our hypothetical world there would be no vast herds of African hoofed mammals, no horses, and no elephants, camels, or even rabbits or caterpillars. There would be few herbivores to feed on the remaining plants, since herbivores and detritivores depend on microorganisms to gain access to the energy and nutrients contained in plant tissues. The carnivores would disappear along with the herbivores. And so it would go. A biosphere without mutualism would be biologically impoverished.

The impoverishment that would follow the elimination of mutualism, however, would go deeper than we might expect. Lynn Margulis and colleagues (Margulis and Fester 1991; Margulis et al. 2006) have amassed convincing evidence that all eukaryotes, both heterotrophic and autotrophic, originated as mutualistic associations between different organisms. Eukaryotes are apparently the product of mutualistic relationships so ancient that the mutualistic partners have become cellular organelles (e.g., mitochondria and chloroplasts) whose mutualistic origins long went unrecognized. Consequently, without mutualism, all the eukaryotes, from *Homo sapiens* to the protozoans, would be gone and the history of life on earth and biological richness would be set back about 1.4 billion years.

But back here in the present, let's accept that mutualism is an integral part of nature and review what is known of the ecology of mutualism. The first part of this brief review emphasizes experimental studies. Then, in the last part of chapter 15, we examine some theoretical approaches to the study of mutualism.

15.1   Plant Mutualisms

LEARNING OUTCOMES

After studying this section you should be able to do the following:

15.3   List several mutualistic relationships important to plants.

15.4   Compare the structure of arbuscular mycorrhizae and ectomycorrhizae.

15.5   Summarize the mutualistic relationship between mycorrhizal fungi and plants, emphasizing the exchanges of materials between the two partners.

15.6 Describe Johnson's experiments to determine whether a history of adding fertilizer to soils alters the mutualistic relationship between plants and mycorrhizal fungi.

15.7 Explain the functional equilibrium model of plant–mycorrhizal fungi relationships.

15.8 Review experimental evidence that plant protection mutualisms increase the health of the plants that possess them.

Plants benefit from mutualistic partnerships with a wide variety of bacteria, fungi, and animals. Plants are the center of mutualistic relationships that provide benefits ranging from nitrogen fixation and nutrient absorption to pollination and seed dispersal. It is no exaggeration to say that the structure and processes characteristic of terrestrial ecosystems depend on these plant-centered mutualisms. Here are some examples drawn from studies of mycorrhizae.

Plant Performance and Mycorrhizal Fungi

The fossil record shows that mycorrhizae arose early in the evolution of land plants, perhaps as long as 400 million years ago. Over evolutionary time, a mutualistic relationship between plants and fungi evolved in which mycorrhizal fungi provide plants with greater access to inorganic nutrients while feeding off the root exudates of plants. In 1885, Albert B. Frank was the first to correctly recognize that mycorrhizae involve a mutualistic relation between plants and fungi, but it took over half a century for his revolutionary insights to be verified and accepted (Trappe 2005). The two most common types of mycorrhizae are (1) **arbuscular mycorrhizae,** in which the mycorrhizal fungus produces **arbuscules,** sites of exchange between plant and fungus, **hyphae,** fungal filaments, and **vesicles,** fungal energy storage organs within root cortex cells, and (2) **ectomycorrhizae,** in which the fungus forms a mantle around roots and a netlike structure around root cells (fig. 15.2). Mycorrhizae are especially important in increasing plant access to phosphorus and other immobile nutrients (nutrients that do not move freely through soil), such as copper and zinc, as well as to nitrogen and water.

Mycorrhizae and the Water Balance of Plants

Mycorrhizal fungi appear to improve the ability of many plants to extract soil water. Edie Allen and Michael Allen (1986) studied how mycorrhizae affect the water relations of the grass *Agropyron smithii* by comparing the leaf water potentials of plants with and without mycorrhizal fungi. Figure 15.3 shows that *Agropyron* with mycorrhizal fungi maintained higher leaf water potentials than those without them. This means that when growing under similar conditions of soil moisture, the presence of mycorrhizal fungi helped the grass maintain a higher water potential. Does this comparison show that mycorrhizal fungi are directly responsible for the higher leaf water potential observed in the mycorrhizal grass? No, they do not. These higher water potentials may be an indirect effect of

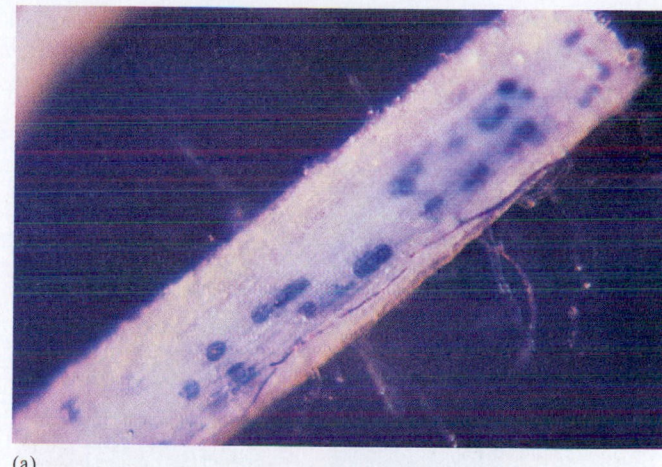
(a)

(b)

Figure 15.2 Mutualistic associations between fungi and plant roots: (*a*) arbuscular mycorrhizae stained so that fungal structures appear blue; and (*b*) ectomycorrhizae, which give a white, fuzzy appearance to these roots.   (*a*) Dr. Nancy Collins Johnson; (*b*) Dr. Jeremy Burgess/Science Source

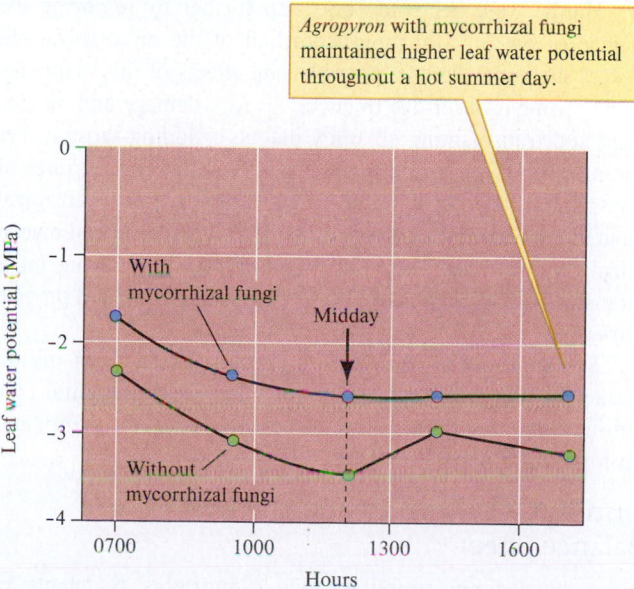

Agropyron with mycorrhizal fungi maintained higher leaf water potential throughout a hot summer day.

Figure 15.3 Influence of mycorrhizal fungi on leaf water potential of the grass *Agropyron smithii* (data from Allen and Allen 1986).

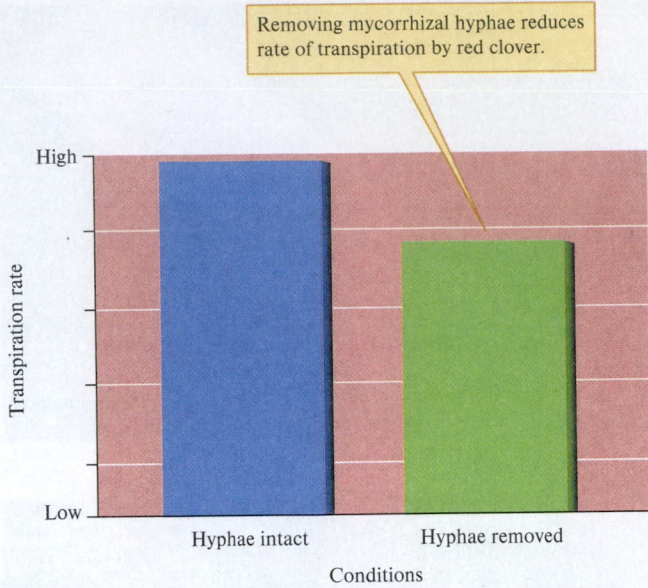

Removing mycorrhizal hyphae reduces rate of transpiration by red clover.

Figure 15.4   Effect of removing mycorrhizal hyphae on rate of transpiration by red clover (data from Hardie 1985).

greater root growth resulting from the greater access to phosphorus provided by mycorrhizal fungi.

Plants with greater access to phosphorus may develop roots that are more efficient at extracting and conducting water; mycorrhizal fungi may not be directly involved in the extraction of water from soils. Kay Hardie (1985) tested this hypothesis directly with an ingenious experimental manipulation of plant growth form and mycorrhizae. First, she grew mycorrhizal and nonmycorrhizal red clover, *Trifolium pratense*, in conditions in which their growth was not limited by nutrient availability. These conditions produced plants with similar leaf areas and root:shoot ratios. Under these carefully controlled conditions, mycorrhizal red clover showed higher rates of transpiration than nonmycorrhizal plants.

Hardie took her study one step further by removing the hyphae of mycorrhizal fungi from half of the mycorrhizal red clover. She controlled for possible side effects of this manipulation by using a tracer dye to check for root damage and by handling and transplanting all study plants, including those in her control group. Removing hyphae significantly reduced rates of transpiration (fig. 15.4), indicating a direct role of mycorrhizal fungi in the water relations of plants. Hardie suggests that mycorrhizal fungi improve water relations of plants by giving more extensive contact with moisture in the rooting zone and provide extra surface area for absorption of water.

So far, it seems that plants always benefit from mycorrhizae. That may not always be the case. Environmental conditions may change the flow of benefits between plants and mycorrhizal fungi.

Nutrient Availability and the Mutualistic Balance Sheet

Mycorrhizal fungi supply inorganic nutrients to plants in exchange for carbohydrates, but not all mycorrhizal fungi deliver nutrients to their host plants at equal rates. The relationship between fungus and plant ranges from mutualism to parasitism, depending on the environmental circumstance and mycorrhizal species or even strains within species.

Nancy Johnson (1993) performed experiments designed to determine whether fertilization can select for less mutualistic mycorrhizal fungi. Before discussing her experiments, we have to ask what would constitute a "less mutualistic" association. In general, a less mutualistic relationship would be one in which there was a greater imbalance in the benefits to the mutualistic partners. In the case of mycorrhizae, a less mutualistic mycorrhizal fungus would be one in which the fungal partner received an equal or greater quantity of photosynthetic product in trade for delivering a lower quantity of nutrients.

Johnson pointed out that there are several reasons to predict that fertilization would favor less mutualistic mycorrhizal fungi. The first is that plants vary the amount of soluble carbohydrates in root exudates as a function of nutrient availability. Plants release more soluble carbohydrates in root exudates when they grow in nutrient-poor soils and decrease the amount of carbohydrates in root exudates as soil fertility increases. Consequently, fertilization of soils should favor strains, or species, of mycorrhizal fungi capable of living in a low-carbohydrate environment. Johnson suggested that the mycorrhizal fungi capable of colonizing plants releasing low quantities of carbohydrates will probably be those that are aggressive in their acquisition of carbohydrates from their host plants, perhaps at the expense of host plant performance. She addressed this possibility using a mixture of field observations and greenhouse experiments.

In the first phase of her project, Johnson examined the influence of inorganic fertilizers on the kinds of mycorrhizal fungi found in soils. She collected soils from 12 experimental plots in a field on the Cedar Creek Natural History Area in central Minnesota that had been abandoned from agriculture for 22 years. Six of the study plots had been fertilized with inorganic fertilizers for 8 years prior to Johnson's experiment, while the other six had received no fertilizer over the same period.

Johnson's samples of mycorrhizal fungi from fertilized and unfertilized soils showed that the composition of mycorrhizal fungi differed substantially. Of the 12 mycorrhizal species occurring in the samples, unfertilized soil supported higher densities of three mycorrhizal fungi, *Gigaspora gigantea, G. margarita,* and *Scutellospora calospora,* whereas fertilized soil supported higher densities of one species, *Glomus intraradices.* Spores of *G. intraradices* accounted for over 46% of the spores recovered from fertilized soils but only 27% of the spores from unfertilized soils.

Johnson used greenhouse experiments to assess how these differences in the composition of mycorrhizal fungi might affect plant performance. She chose big bluestem grass, *Andropogon gerardii,* as a study plant for these experiments because it is native to the Cedar Creek Natural History Area and is well adapted to the nutrient-poor soils of the area. Seedlings of *Andropogon* were planted in pots containing 980 g of a 1:1 mixture of sterilized subsurface sand from the Cedar Creek Natural History Area and river-washed sand. Johnson added a composite sample of other soil microbes living in the soils of fertilized and unfertilized study plots. She prepared the composite by washing a composite soil sample from all fertilized and unfertilized study plots with deionized water and passing this water through a 25 µm screen.

To each pot, Johnson added a mycorrhizal "inoculum" of 30 g of soil of one of three types: (1) a fertilized inoculum consisting of 15 g of soil from fertilized study plots mixed with 15 g of sterilized unfertilized soil, (2) an unfertilized inoculum consisting of 15 g of soil from unfertilized study plots mixed with 15 g of sterilized fertilized soil, or (3) a nonmycorrhizal inoculum consisting of 30 g of a sterilized composite from the soils of fertilized and unfertilized study plots. The first two inocula acted as a source of mycorrhizal fungi for colonization of *Andropogon*. The design of Johnson's experiment is summarized in figure 15.5.

Pots were next assigned to one of four nutrient treatments in which Johnson added (1) no supplemental nutrients (None), (2) phosphorus only (+P), (3) nitrogen only (+N), or (4) both nitrogen and phosphorus (+N+P). The subsurface sand from the Cedar Creek Natural History Area contained a fairly low concentration of nitrogen but considerably higher concentrations of phosphorus. Nutrient additions were adjusted so that the supplemented treatments offered nitrogen and phosphorus concentrations comparable to those of the topsoil in the fertilized study plots.

Johnson harvested five replicates of each of the treatments at two points in time: at 4 weeks, when *Andropogon* was actively growing, and at 12.5 weeks, when the grass was fully grown. At each harvest she measured several aspects of plant performance: plant height, shoot mass, and root mass; and at 12.5 weeks, she also recorded the number of inflorescences per plant.

At 12.5 weeks, shoot mass was significantly influenced by nutrient supplements and by whether or not plants were mycorrhizal but not by the source of the mycorrhizal inoculum (fig. 15.6). Shoot mass was greatest in the double nutrient supplement treatment (+N+P), somewhat lower in the nitrogen supplement (+N), and very low in the other two treatments (None and +P). Figure 15.6a also indicates a definite influence

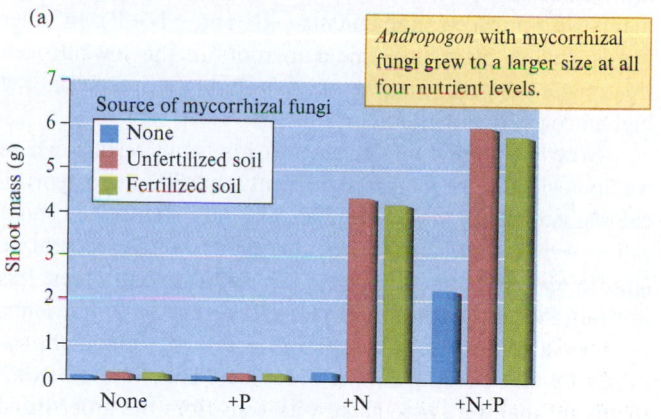

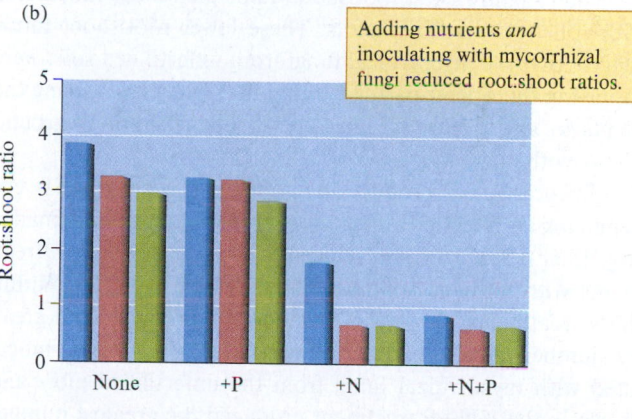

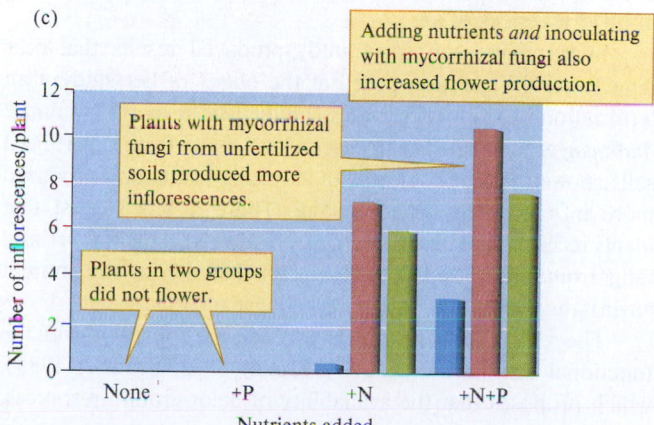

Figure 15.6   Effect of nutrient additions and mycorrhizae on the grass *Andropogon gerardii* (data from Johnson 1993).

Question: Does fertilizing soil select for less mutualistic mycorrhizal fungi?

Experimental Design

Two sources of mycorrhizal fungi

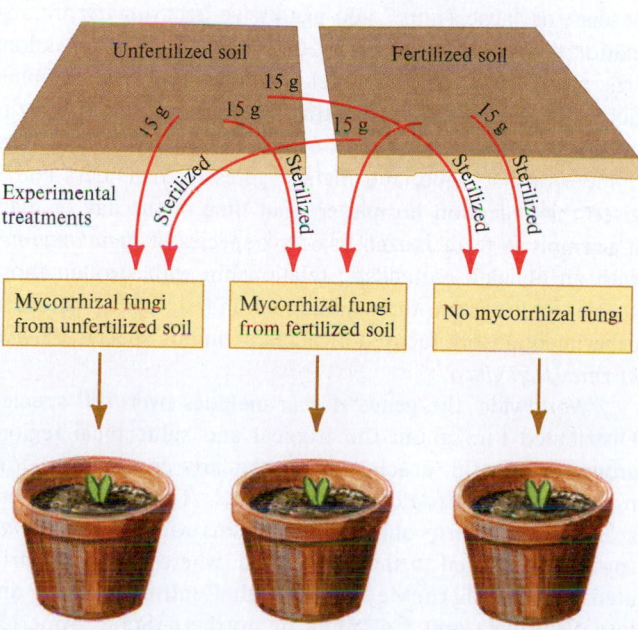

Compare: Growth, root:shoot ratios, and number of inflorescences produced by three treatments.

Figure 15.5   Testing the effects of long-term fertilizing on interactions between mycorrhizal fungi and plants on agricultural lands.

of mycorrhizae on performance. Shoot mass was significantly greater for mycorrhizal plants across all nutrient treatments.

Nutrient supplements and mycorrhizae also significantly influenced root:shoot ratios (fig. 15.6b). As we saw in chapter 7 (fig. 7.26), plants invest differentially in roots and shoots depending on nutrient and light availability. It also appears that variation in investment is aimed at increasing supplies of resources in short supply. For instance, in nutrient-poor environments many plants invest disproportionately in roots and consequently have high root:shoot ratios, which decline with increasing nutrient availability. The results of Johnson's experiments are consistent with this generalization. Root:shoot ratios were highest in the treatments without nitrogen supplements (None and +P) and lowest in the treatments with nitrogen supplements (+N and +N+P). In other words, higher plant investment in roots in the low-nitrogen treatments suggests greater nutrient limitation than in the high-nitrogen treatments.

Now let's look a bit deeper into Johnson's results, where we find evidence for increased nutrient availability to mycorrhizal plants. In both the +N and "None" treatments, root:shoot ratios were significantly lower in plants with mycorrhizae (fig. 15.6b). Mycorrhizal plants in these treatments invest less in roots, suggesting that they have greater access to nutrients. Here we also see a hint that the source of the inoculum significantly influenced plant performance. Plants in the +N+P treatment that were inoculated with soils from the unfertilized plots had slightly lower root:shoot ratios than those inoculated with soil from fertilized plots. These lower root:shoot ratios suggest that the mycorrhizal fungi from unfertilized soils were supplying their plant partners with more nutrients, freeing the plants to invest more of their energy budget in aboveground photosynthetic tissue.

Inflorescence production provides the strongest evidence for an effect of inoculum source on plant performance (fig. 15.6c). Andropogon produced inflorescences only in treatments with nitrogen supplements (+N and +N+P). Within these treatments, the mycorrhizal plants produced the greatest number of inflorescences. In addition, Andropogon inoculated with mycorrhizal fungi from the unfertilized plots and grown in the +N+P treatment produced the greatest number of inflorescences of all.

In summary, Johnson's study produced results that bear directly on the question posed at the outset of her study: Can fertilization of soil select for less mutualistic mycorrhizal fungi? Andropogon inoculated with mycorrhizal fungi from unfertilized soils showed faster shoot growth as young plants and produced more inflorescences when mature. These results suggest that plants receive more benefit from association with the mycorrhizal fungi from unfertilized soils. It appears that altering the nutrient environment does alter the mutualistic balance sheet.

The results of Johnson's experiments are consistent with the functional equilibrium model (Mooney 1972; Brouwer 1983), which proposes that the availability of belowground resources, mainly nutrients and water, and aboveground resources, mainly light, controls how plants allocate to roots, mycorrhizae, shoots, and leaves (fig. 15.7). According to the model, plants growing in

fertilized soils will allocate less energy to roots and mycorrhizae. Over the long term, this lower allocation by plants could shift the composition of the mycorrhizal fungal community toward forms that compete well for reduced supplies of carbohydrates and supply lower nutrients to the plant in exchange. However, ecologists are still far from having a complete understanding of the factors controlling the exchange of resources between plants and mycorrhizal fungi. Consequently, researchers from around the world, conducting research at local through global scales, have proposed many different models as they work toward a better understanding of this key mutualism in terrestrial ecosystems (Johnson et al. 2006).

Plants engage in a wide variety of mutualisms with many other organisms. One of those involves associations that provide protection from herbivores and competitors. Writing about the natural history of mutualism, Daniel Janzen (1985) included "plant-ant protection mutualisms" as one of his general categories of mutualism. Janzen (1966, 1967a, 1967b) himself is responsible for studying one of the best known of these mutualisms, the obligate mutualism between ants and swollen thorn acacias in Central America.

Ants and Swollen Thorn Acacias

The ants that are mutualistic with bullhorn, or swollen thorn, acacias are members of the genus Pseudomyrmex in the subfamily Pseudomyrmecinae. This subfamily of ants is dominated by genera and species that have evolved close relationships with living plants. Pseudomyrmex spp. are generally associated with trees and show several characteristics that Janzen suggested are associated with arboreal living. They are generally fast and agile runners, have good vision, and forage independently. To this list, the Pseudomyrmex spp. associated with swollen thorn acacias, or "acacia-ants," add aggressive behavior toward vegetation and animals contacting their home tree, larger colony size, and 24-hour activity outside of the nest. This combination of characteristics means that any herbivore attempting to forage on an acacia occupied by acacia-ants is met by a large number of fast, agile, and highly aggressive defenders and is given this reception no matter what time of the day or night it attempts to feed. Janzen listed six species of Pseudomyrmex with an obligate mutualistic relationship with swollen thorn acacias and refers to three additional undescribed species. His experimental work focused principally on one species, Pseudomyrmex ferruginea.

Worldwide, the genus Acacia includes over 700 species. Distributed throughout the tropical and subtropical regions around the world, acacias are particularly common in drier tropical and subtropical environments. The swollen thorn acacias, which form obligate mutualisms with Pseudomyrmex spp., are restricted to the New World, where they are distributed from southern Mexico, through Central America, and into Venezuela and Colombia in northern South America. Across this region, swollen thorn acacias occur mainly in the lowlands up to 1,500 m elevation in areas with a dry season of 1 to 6 months. Swollen thorn acacias show several characteristics related to their obligate association with ants, including

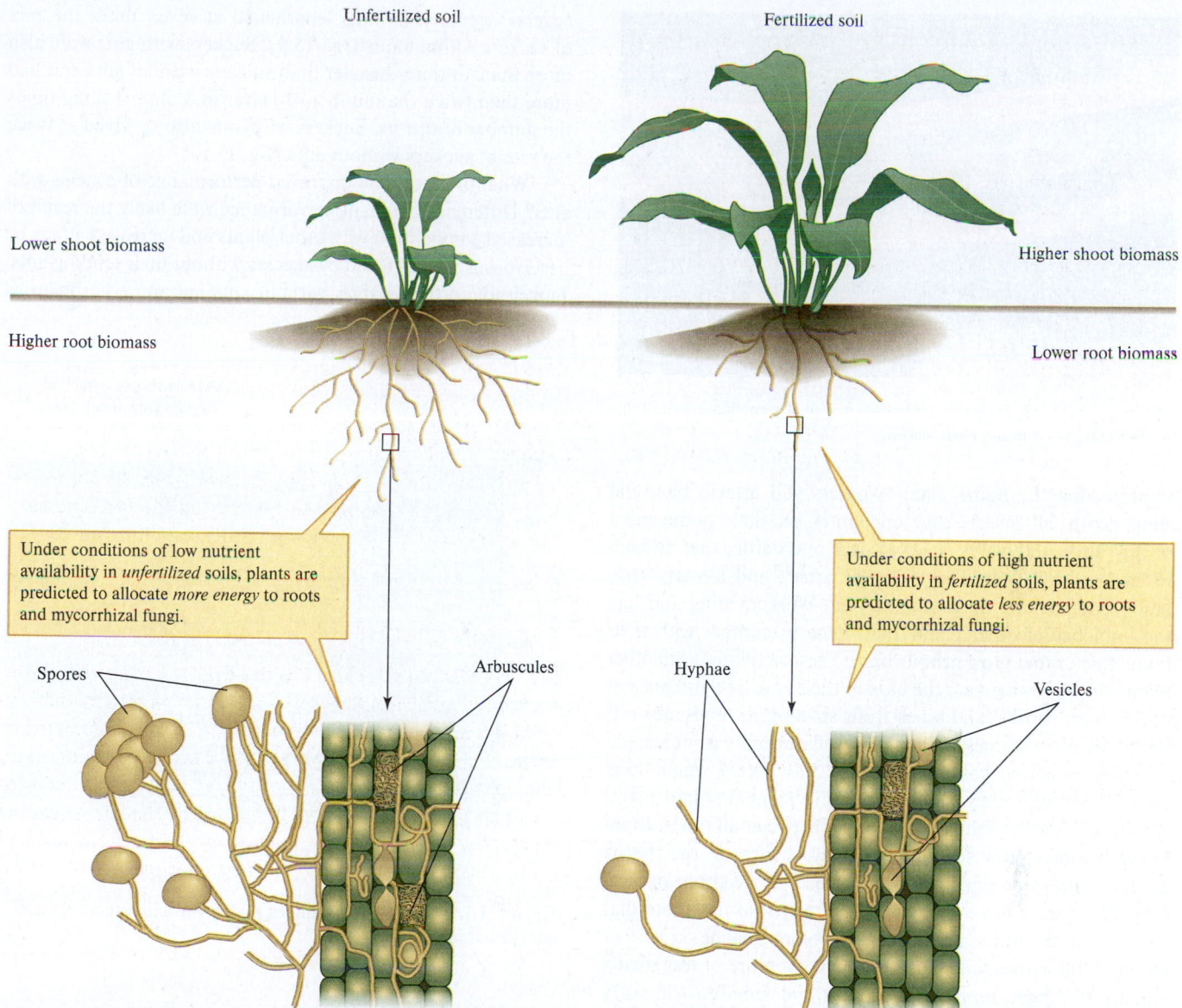

Unfertilized soil

Fertilized soil

Lower shoot biomass

Higher shoot biomass

Higher root biomass

Lower root biomass

Under conditions of low nutrient availability in *unfertilized* soils, plants are predicted to allocate *more energy* to roots and mycorrhizal fungi.

Under conditions of high nutrient availability in *fertilized* soils, plants are predicted to allocate *less energy* to roots and mycorrhizal fungi.

Spores

Arbuscules

Hyphae

Vesicles

Figure 15.7   The functional equilibrium model. This model predicts that plants will allocate energy derived from photosynthesis preferentially to acquire the most limiting resources under a particular set of environmental conditions. In unfertilized soils, nutrients often limit growth, which leads to greater allocation to roots and mycorrhizal fungi. In fertilized soils, light may limit growth, which leads to greater allocation to aboveground photosynthetic tissues (after Johnson et al. 2003).

enlarged thorns with a soft, easily excavated pith; year-round leaf production; enlarged foliar nectaries; and leaflet tips modified into concentrated food sources called Beltian bodies. The thorns provide living space, while the foliar nectaries provide a source of sugar and liquid. Beltian bodies are a source of oils and protein. Resident ants vigorously guard these resources against encroachment by nearly all encroachers, including other plants.

Janzen's detailed natural history of the interaction between bullhorn acacia, *Acacia cornigera,* and ants paints a rich picture of mutual benefits to both partners (fig. 15.8). Newly mated *Pseudomyrmex* queens fly and run through the vegetation searching for unoccupied seedlings or shoots of bullhorn acacia. When a queen finds an unoccupied acacia,

she excavates an entrance in one of the green thorns or uses one carved previously by another ant. The queen then lays her first eggs in the thorn and begins to forage on her newly acquired home plant. She gets nectar for herself and her developing larvae from the foliar nectaries and gets additional solid food from the Beltian bodies. As time passes, the number of workers in the new colony increases, and while they take up all the chores of the colony, the queen shifts to a mainly reproductive function; her abdomen enlarges and she becomes increasingly sedentary.

In exchange for food and shelter, ants protect acacias from attack by herbivores and competition from other plants. Workers have several duties, including foraging for themselves, the larvae, and the queen. One of their most important activities

Figure 15.8   Split thorn of a bullhorn acacia, revealing a nest of its ant mutualists.   Robert & Linda Mitchell

is protecting the home plant. Workers will attack, bite, and sting nearly all insects they encounter on their home plant or any large herbivores, such as deer and cattle, that attempt to feed on the plant. They will also attack and kill any vegetation encroaching on the home tree. Workers sting and bite the branches of other plants that come in contact with their home tree or that grow near its base. These activities keep other plants from growing near the base of the home tree and prevent other trees, shrubs, and vines from shading it. Consequently, the home plant's access to light and soil nutrients is increased.

Once a colony has at least 50 to 150 workers, which takes about 9 months, they patrol the home plant day and night. About one-fourth of the total colony is active at all times. Eventually colonies grow so large that they occupy all the thorns on the home tree and may even spread to neighboring acacias. The queen, however, generally remains on the shoot that she colonized originally. When the colony reaches a size of about 1,200 workers, it begins producing a more or less steady stream of winged reproductive males and females, which fly off to mate. The queens among them may eventually establish new colonies on other bullhorn acacias or one of the other Central American swollen thorn acacias. Colonies may eventually reach a total population of 30,000 workers.

Experimental Evidence for Mutualism

While much of the natural history of this mutualism was known at the time Janzen conducted his studies, no one had experimentally tested the strength of its widely supposed benefits. Janzen took his work beyond natural history to experimentally test for the importance of ants to bullhorn acacias. It was clear that the ant needs swollen thorn acacias, but do the acacias need the ants? Janzen's experiments concentrated on the influence of ants on acacia performance. He also tested the effectiveness of the ants at keeping acacias free of herbivorous insects. Janzen removed ants from acacias by clipping occupied thorns or by cutting out entire shoots with their ants. He then measured the growth rate, leaf production, mortality, and insect population density on acacias with and without ants.

Janzen's experiments demonstrated that ants significantly improve plant performance. Suckers growing from stumps of

acacias occupied by ants lengthened at seven times the rate of suckers without ants (fig. 15.9). Suckers with ants were also more than 13 times heavier than suckers without ants and had more than twice the number of leaves and almost three times the number of thorns. Suckers with ants also survived at twice the rate of suckers without ants (fig. 15.10).

What produces the improved performance of acacias with ants? Differences in plant performance were likely the result of increased competition with other plants and increased attack by herbivorous insects faced by acacias without their tending ants. Janzen found that acacias without ants had more herbivorous

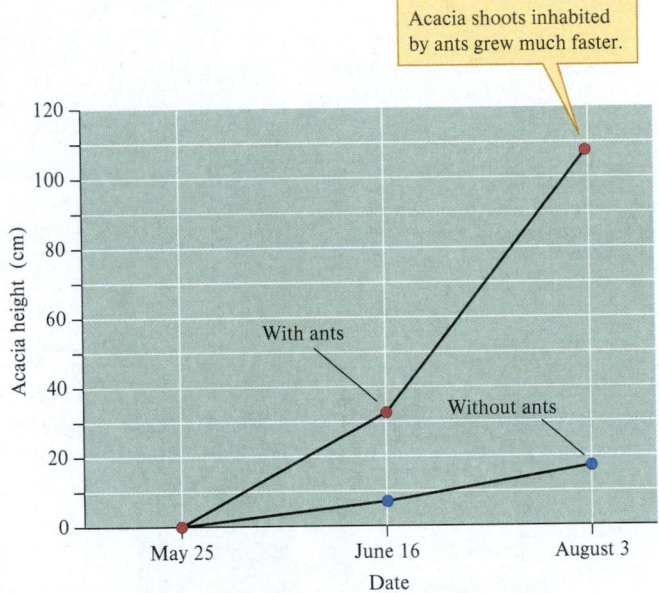

Figure 15.9   Growth by bullhorn acacia with and without resident ants (data from Janzen 1966).

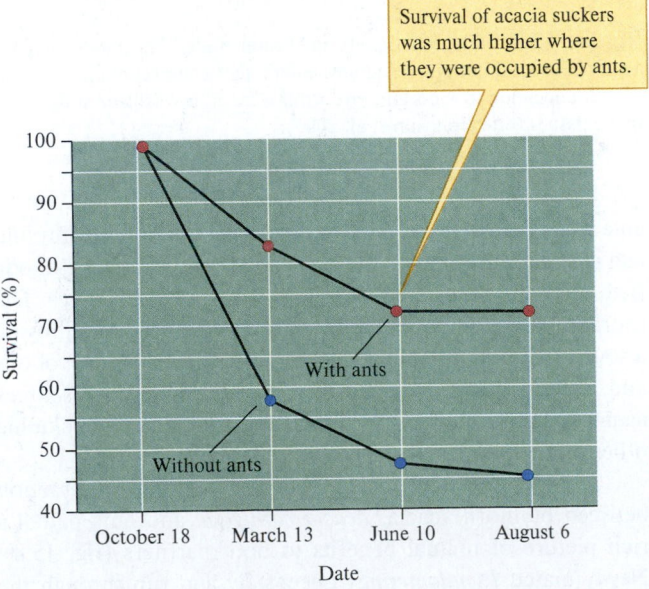

Figure 15.10   Survival of bullhorn acacia shoots with and without resident ants (data from Janzen 1966).

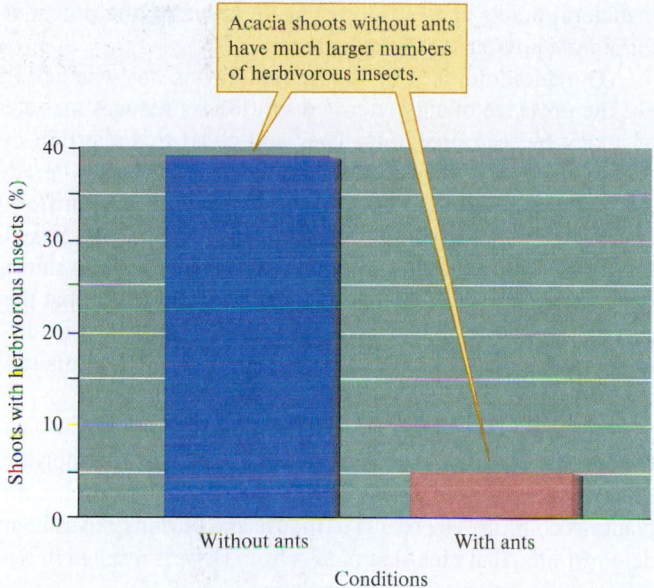

Figure 15.11   Ants and the abundance of herbivorous insects on bullhorn acacia (data from Janzen 1966).

insects on them than did acacias with ants (fig. 15.11). Janzen's experiments suggest that bullhorn acacias need ants as much as the ants need the acacia. It appears that this is a truly mutualistic situation.

Potential Conflict Between Mutualists

Most research on swollen thorn acacias has focused on their ant-protection mutualisms. However, these trees depend on many other mutualistic relationships. Belowground, their roots shelter nitrogen-fixing bacteria in nodules and harbor mycorrhizal fungi. Aboveground, besides sheltering swarms of *Pseudomyrmex,* ants that drive away herbivorous insects, the acacia's flowers depend on other insects, mainly bees, for pollination. The acacia's ant guards could come into conflict with pollinators in two ways. First, the ants could remove nectar from flowers and reduce their attractiveness to potential pollinators. Second, the ants could guard flowers, driving pollinators away.

This potential conflict between mutualists of swollen thorn acacias attracted the attention of Nigel Raine, Pat Willmer, and Graham Stone (2002), researchers from three different British universities. They conducted their research at the Chamela Biological Station of the Universidad Nacional Autónoma de México, where they studied *Acacia hindsii* and its ant protector, *Pseudomyrmex veneficus.* Raine, Willmer, and Stone first examined the distribution of ants and pollinators to see if they overlapped in space or time. They found that ant and pollinator activity overlaps in time. However, while ants and pollinators are active on *A. hindsii* at the same time, they rarely overlap spatially. The ants rarely visit acacia inflorescences. Why is that? Raine, Willmer, and Stone observed that the foliar nectaries and Beltian bodies used by the ants occur on new growth, whereas flowers are restricted to older shoots. In addition, in contrast to acacia species that do not support protective ants, the inflorescences of *A. hindsii* produce no nectar. Lack of nectar would make the flowers less attractive to

patrolling ants. Still, because new and older shoots can grow in close proximity, the researchers wondered whether some other factor might keep the ants from patrolling the inflorescences on older shoots.

Since Willmer and Stone (1997) had discovered previously that the flowers of some African acacias contained an ant repellent, they tested for the presence of a repellent in the flowers of *A. hindsii* by rubbing several acacia tissues on the bark of branches actively patrolled by ants. The tissues tested were new inflorescences, old inflorescences, leaves, and buds. Each of these tissues was rubbed within 3 cm squares marked on the bark of patrolled branches using water-based markers. As a control, Raine, Willmer, and Stone marked one set of squares but did not rub any plant tissues on them. Once experimental and control squares had been established, the researchers watched patrolling ants, noting whether they entered experimental and control squares or avoided crossing them.

Figure 15.12 shows the results of the repellent experiment. Raine, Willmer, and Stone found that new inflorescences were strongly repellent to patrolling ants and that older inflorescences, though repellent, were rejected less often. Meanwhile, leaf and bud rubbings were rejected no more frequently than were control squares. In summary, protection and pollination mutualisms do not come into conflict on *A. hindsii* because of spatial separation of inflorescences and resources used by guarding ants, because *A. hindsii* inflorescences lack a potential ant attractant (nectar), and contain a chemical repellent.

While tropical plant protection mutualisms are most often cited, there are many examples of mutualism between plants and ants in the temperate zone. A particularly well-studied interaction is that between ants and the aspen sunflower, *Helianthella quinquenervis.*

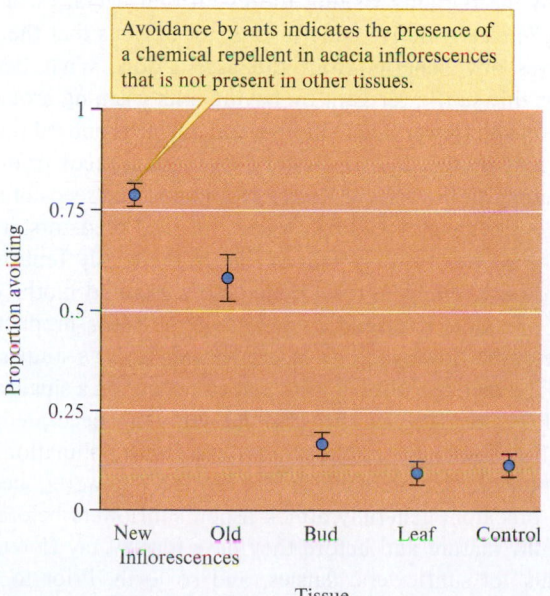

Figure 15.12   Proportion of ants avoiding control areas and areas rubbed with tissues of new and old inflorescences, buds, and leaves of the swollen thorn acacia, *Acacia hindsii;* values are means ± one standard error (data from Raine, Willmer, and Stone 2002).

A Temperate Plant Protection Mutualism

Aspen sunflowers live in wet mountain meadows of the Rocky Mountains from Chihuahua, Mexico, to southern Idaho, at elevations as low as 1,600 m in the northern part of its range to 4,000 m in the south. Ants are attracted to aspen sunflowers because they produce nectar at **extrafloral nectaries,** nectar-producing structures outside of the flowers. In the case of aspen sunflowers, the extrafloral nectaries are associated with structures called *involucral bracts,* modified leaves that first enclose the flower head prior to flowering and then surround the base of the flower after it opens. Some early researchers hypothesized that extrafloral nectaries function to attract ants, whereas others suggested that they are primarily excretory organs.

The extrafloral nectar produced by aspen sunflowers is rich in sucrose and contains high concentrations of a wide variety of amino acids. Therefore, like the swollen thorn acacias studied by Janzen, the aspen sunflower provides food to ants. In contrast to swollen thorn acacias, however, this sunflower does not provide living places. This contrast is general across temperate ant-plant mutualisms, which involve food as an attractant but no living quarters.

David Inouye and Orley Taylor (1979) recorded five species of ants on aspen sunflowers, including *Formica obscuripes, F. fusca, F. integroides planipilis, Tapinoma sessile,* and *Myrmica* sp. These ants are not obligately associated with aspen sunflowers and can be found tending aphids on other species of plants or even collecting flower nectar on some plants. However, Inouye and Taylor never observed them collecting nectar from aspen sunflower blossoms nor tending aphids on this plant. Apparently, the extrafloral nectar produced by the aspen sunflower is a sufficient attractant. Ants find the plant so attractive that Inouye and Taylor observed up to 40 ants on a single flower stalk.

While the ants visiting the extrafloral nectaries of *Helianthella* clearly derive benefit, it is not obvious that the plant receives any benefits from the association. What benefits might this sunflower gain by having ants roaming around its flowers and flower buds? Inouye and Taylor proposed that the ants may protect the sunflower's developing seeds from seed predators. In the central Rocky Mountains, the seeds of aspen sunflowers are attacked by a variety of seed predators, including the larvae of two species of flies in the family Tephritidae, a fly in the family Agromyzidae, and a phycitid moth. These seed predators damaged over 90% of the seeds produced by some of the flowers at one of Inouye and Taylor's study sites.

The high densities of ants that can occur on a single aspen sunflower certainly have the potential to deter seed predators, but these same ants might also interfere with pollination. This potential for interference is not realized, however, because seed predators generally attack aspen sunflowers before they are fully mature and before they have formed ray florets, the "petals" of sunflowers, daisies, and so forth. Prior to opening of the flower bud, when tephritid and agromyzid flies try to oviposit on the bud, ants visiting the extrafloral nectaries patrol the whole surface of the flower bud in large numbers. Later, the fully formed ray florets, which act as attractants for pollinators (mainly bumblebees), form a screen between the involucral bracts and the flower head, reducing the potential for ants to interfere with pollinators.

The question asked by Inouye and Taylor was whether or not the presence of ants on aspen sunflowers reduces the rate of attack by seed predators. They addressed this question in several ways. First, they compared rates of attack by seed predators on flowers tended by ants with rates of attack on flowers where ants were naturally absent. This comparison showed that flowers without ants suffered two to four times as much seed predation (fig. 15.13). The researchers also found that the average number of ants per flower stalk decreased with distance from an ant nest and that the plants with fewer ants suffered higher rates of seed damage by seed predators.

Next, Inouye and Taylor performed an experiment in which they prevented ants from moving onto some plants by applying a sticky barrier to the base of flower stalks. They used adjacent plants as controls. The results of Inouye and Taylor's experiment demonstrated that exclusion of ants from flowers resulted in significantly higher rates of seed predation (fig. 15.14).

As in the tropical swollen thorn acacia-ant mutualism, ants associated with aspen sunflowers provide protection while receiving substantial benefits in the form of food. Unlike the tropical system, the association between aspen sunflowers and ants incorporates a significant degree of flexibility. This flexibility may be a hallmark of many temperate mutualisms.

Why does the relationship between ants and aspen sunflowers remain facultative? In other words, why hasn't there been strong selection for an obligate relationship? Continuing studies by David Inouye provide clues. He estimated the abundance of aspen sunflowers on two study plots for more than two decades. This long-term study shows that every few years, the flower heads of aspen sunflowers are killed by late frosts. From 1974 to 1995, aspen sunflowers produced very few flower heads in 1976, 1981, 1985, 1989, and 1992 (fig. 15.15). Inouye has continued to monitor his study plots (Inouye 2008) and

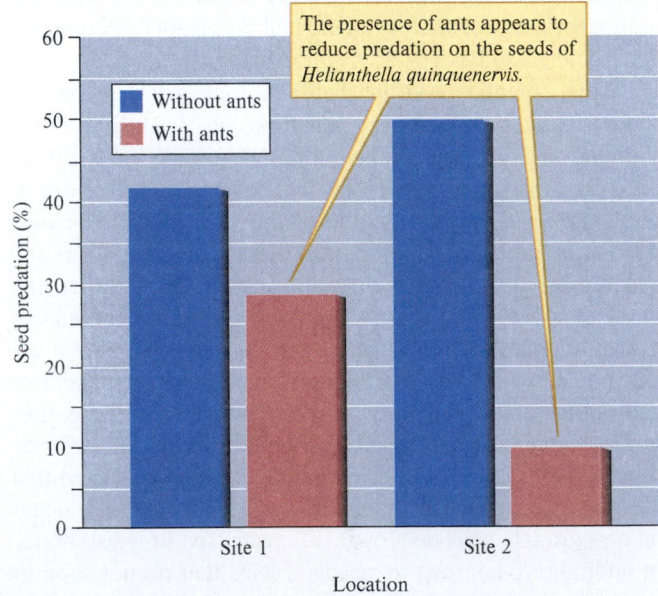

Figure 15.13   Predation on the seeds of aspen sunflower with and without ants (data from Inouye and Taylor 1979).

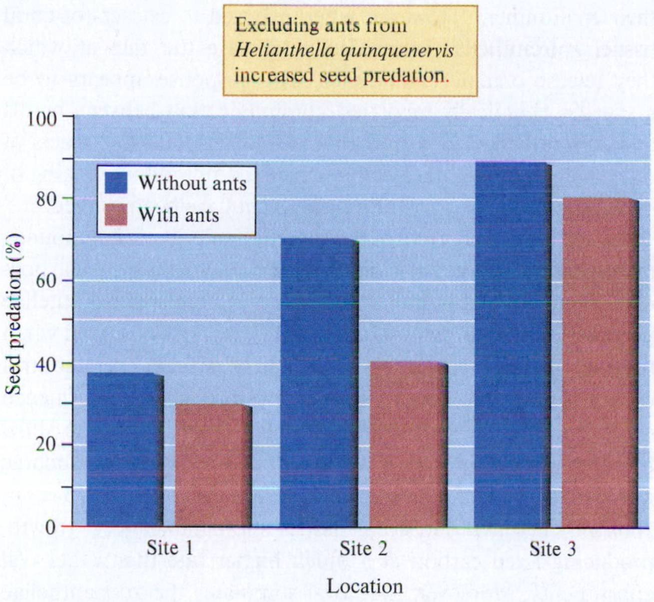

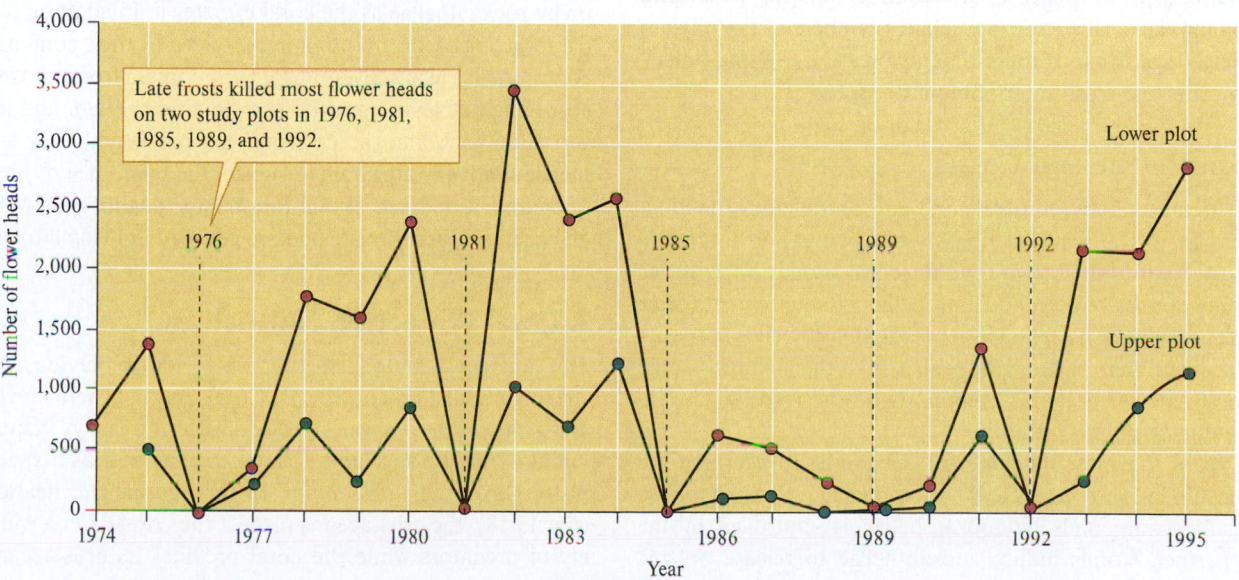

Figure 15.14 Effect of excluding ants on rates of seed predation on aspen sunflowers (data from Inouye and Taylor 1979).

has discovered that, due to earlier spring snowmelt with global warming, *Helianthella* flower buds are even more vulnerable to killing by late spring frosts (fig. 15.16). An ant species with an obligate mutualistic relationship with the aspen sunflower and that relied entirely on it as a source of nectar would not survive long. Inouye points out that, paradoxically, the frosts are beneficial to aspen sunflowers in the long run because they reduce populations of seed predators such as tephritid flies, which have no place to lay their eggs when hard frost kills the flower heads. In the coevolutionary relationships between the aspen sunflower and its predators, the physical environment plays a significant role. In temperate climates generally, the physical environment seems to play as large a role as biological relationships in determining ecological patterns and processes.

If we venture into tropical seas and probe their inhabitants, we soon uncover a wide variety of mutualistic relationships at least as rich as those we examined between terrestrial plants and their partners. The most striking marine counterparts to the mutualisms of terrestrial plants are those centered around reef-building corals.

Concept 15.1 Review

1. Why did Johnson create her inocula by mixing sterilized and unsterilized soils from the fertilized and unfertilized study areas?
2. Why did Johnson's control consist of a sterilized mixture of soils from the fertilized and unfertilized study areas?
3. In Inouye and Taylor's study, why wasn't the comparison of seed predation on plants naturally with and without ants sufficient to demonstrate the influence of ants on rates of seed predation?

15.2 Coral Mutualisms

LEARNING OUTCOMES

After studying this section you should be able to do the following:

15.9 Outline the mutualistic relationship between reef-building corals and zooxanthellae.
15.10 Compare the mutualistic relationship between reef-building corals and zooxanthellae with the mutualistic relationship between mycorrhizal fungi and plants.
15.11 Describe the mutualistic relationships between reef-building corals and crustaceans.
15.12 Discuss the parallels between coral and plant protection mutualisms.

Figure 15.15 Annual variation in numbers of flower heads produced by aspen sunflowers on two plots at the Rocky Mountain Biological Station (data courtesy of David W. Inouye).

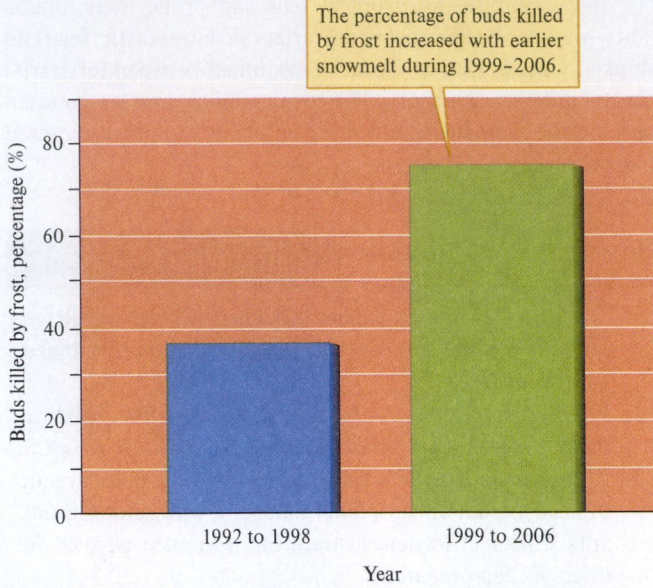

The percentage of buds killed by frost increased with earlier snowmelt during 1999–2006.

Figure 15.16   Earlier snowmelts as a consequence of warmer spring temperatures, from 1999 to 2006, made the flower buds of *Helianthella* more vulnerable to late spring frosts during occasional dips in temperature (data from Inouye 2008).

Reef-building corals depend on mutualistic relationships with algae and animals, with an exchange of benefits paralleling those between terrestrial mutualists. Because of the importance of mutualism in the lives of reef-building corals, it appears that the ecological integrity of coral reefs depends on mutualism. Coral reefs show exceptional productivity and diversity. Recent estimates put the number of species occurring on coral reefs at approximately 0.5 million, and coral reef productivity is among the highest of any natural ecosystem. Paradoxically, this overwhelming diversity and exceptional productivity occurs in an ecosystem surrounded by nutrient-poor tropical oceans. The key to explaining this paradox lies with mutualism—in this case, between reef-building corals and unicellular algae called zooxanthellae, members of the phylum Dinoflagellata. Most of these organisms are free-living unicellular marine and freshwater photoautotrophs.

Zooxanthellae and Corals

The association between corals and zooxanthellae is functionally similar to the relationship between plants and mycorrhizal fungi. Zooxanthellae live within coral tissues at densities averaging approximately 1 million cells per square centimeter of coral surface. Like plants, zooxanthellae receive nutrients from their animal partner. In return, like mycorrhizal fungi, the coral receives organic compounds synthesized by zooxanthellae during photosynthesis.

One of the most fundamental discoveries concerning the relationship between corals and zooxanthellae is that the release of organic compounds by zooxanthellae is controlled by the coral partner. Corals induce zooxanthellae to release organic compounds with "signal" compounds that alter the permeability of the zooxanthellae cell membrane. Zooxanthellae grown in isolation from corals release very little organic material into

their environment. However, when exposed to extracts of coral tissue, zooxanthellae immediately increase the rate at which they release organic compounds. This response appears to be a specific chemically mediated communication between corals and zooxanthellae. Zooxanthellae do not respond to extracts of other animal tissues, and coral extracts do not induce leaking of organic molecules by any other algae that have been studied.

Corals not only control the secretion of organic compounds by zooxanthellae, they also control the rate of zooxanthellae population growth and population density. In corals, zooxanthellae populations grow at rates 1/10 to 1/100 the rates observed when they are cultured separately from corals. Corals exert control over zooxanthellae population density through their influence on organic matter secretion. Normally, unicellular algae show **balanced growth,** growth in which all cell constituents, such as nitrogen, carbon, and DNA, increase at the same rate. However, zooxanthellae living in coral tissues show unbalanced growth, producing fixed carbon at a much higher rate than other cell constituents. Moreover, the coral stimulates the zooxanthellae to secrete 90% to 99% of this fixed carbon, which the coral uses for its own respiration. Fixed carbon secreted and diverted for use by the coral could otherwise be used to produce new zooxanthellae, which would increase population growth.

What benefits do the zooxanthellae get out of their relationship with corals? The main benefit appears to be access to higher levels of nutrients, especially nitrogen. Corals feed on zooplankton, which gives them a means of capturing nutrients, especially nitrogen and phosphorus. When corals metabolize the protein in their zooplankton prey, they excrete ammonium as a waste product. L. Muscatine and C. D'Elia (1978) showed that coral species such as *Tubastrea aurea* that do not harbor zooxanthellae continuously excrete ammonium into their environment, while corals such as *Pocillopora damicornis* do not excrete measurable amounts of ammonium (fig. 15.17). What happens to the ammonium produced by *Pocillopora* during metabolism of the protein in their zooplankton prey? Muscatine and D'Elia suggested that this ammonium is immediately taken up by zooxanthellae as the coral excretes it. In addition to internal recycling of the ammonium produced by their coral partner, zooxanthellae also actively absorb ammonium from seawater. By absorbing nutrients from the surrounding medium and leaking very little back into the environment, corals and their zooxanthellae gradually accumulate substantial quantities of nitrogen. Therefore, as in tropical rain forest, large quantities of nutrients on coral reefs accumulate and are retained in living biomass.

A Coral Protection Mutualism

The ant-acacia mutualism that we reviewed previously has a striking parallel on coral reefs. Corals in the genera *Pocillopora* and *Acropora* host a variety of crabs in the family Xanthidae, mainly *Trapezia* spp. and *Tetralia* spp. as well as a species of pistol shrimp, *Alpheus lottini*. In this mutualistic relationship (fig. 15.18), the crustaceans protect the coral from a wide variety of predators while the coral provides its crustacean partners with shelter and food.

Peter Glynn (1983) surveyed the coral-crustacean mutualism and found that the eastern, central, and western areas of

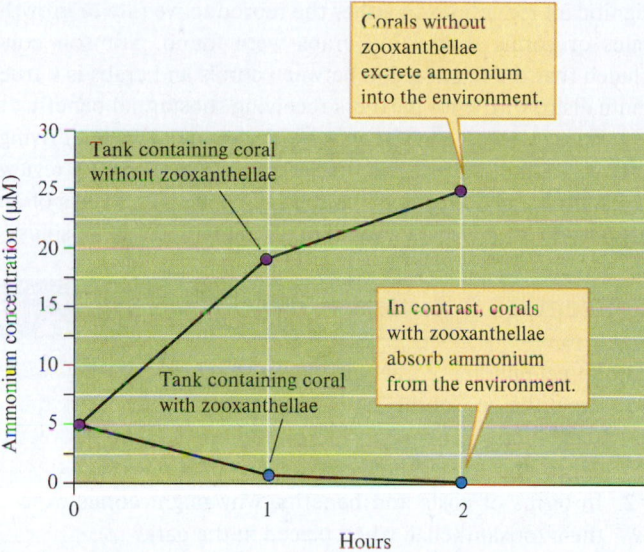

Figure 15.17   Zooxanthellae, corals, and ammonium flux (data from Muscatine and D'Elia 1978).

the Pacific Ocean contain 13 species of corals that are protected by crustacean mutualists, including 17 species of crabs and 1 species of shrimp, all of which are found only on corals in what is apparently an obligate mutualism. These crustaceans protect the corals from a variety of sea stars that prey on corals but especially from attacks by the crown-of-thorns

Figure 15.18   Pistol shrimp will defend their home coral from attacking predators.   Cigdem Sean Cooper/Shutterstock

sea star, *Acanthaster planci.* At the approach of the sea star, the crabs become highly disturbed and then attack by pinching and clipping the sea star's spines and tube feet, grasping it and jerking it up and down and resisting its retreat. The mutualistic shrimp also attacks the sea star by snipping spines and tube feet and making loud snapping sounds with an enlarged pincer specialized for the purpose. The loud popping sounds, which have given shrimp in the genus *Alpheus* the name "pistol shrimp," are so intense they stun small fish.

Glynn used field and laboratory experiments to test whether this aggression by crustaceans is effective at repelling attacks by predatory sea stars. He conducted a field experiment at 8 to 12 m depth on a reef in Guam, where he removed the crustaceans from an experimental group of corals and gave sea stars a choice between these and an equal number of corals that retained their crustacean partners. Over a period of 2 days, the sea stars attacked the unprotected corals at a much higher frequency (fig. 15.19). Glynn obtained similar results in a laboratory study of the corals and crustacean mutualists of Panama, in which sea stars attacked 85% of the unprotected colonies. These results show that the crustacean mutualists of corals substantially improve the chances that a coral will avoid attack by sea stars.

Observations by Glynn and John Stimson (1990) suggest that mutualistic crabs also protect corals from other less conspicuous attackers. Glynn observed that the presence of crabs seems to enhance the condition of coral tissues. Stimson found that when he removed crabs, corals showed tissue death in the deep axils of their branches and that these areas were soon invaded by algae, sponges, and tunicates. It appears that, in addition to protecting corals from the attacks of large predators, the activities of crabs promote the health and integrity of coral tissues. If this is a mutualistic relationship, what do the crabs receive in return for their investment?

Like swollen thorn acacias, corals provide their crustacean mutualists with shelter and food. The corals harboring crabs and

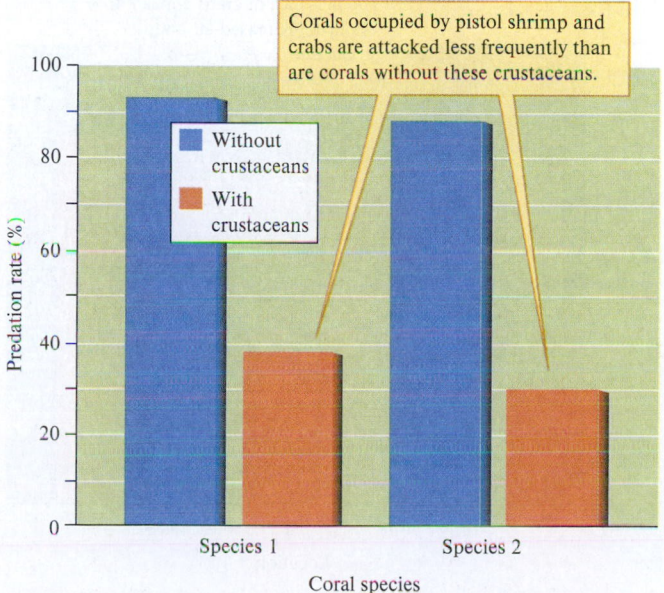

Figure 15.19   Attacks on corals with and without pistol shrimp and crabs (data from Glynn 1983).

pistol shrimp have a tightly branched growth form that offers shelter, and the crustaceans feed on the mucus produced by the corals. *Trapezia* spp., the most common crabs guarding pocilloporid corals, stimulate mucus flow from corals by inserting their legs into coral polyps, a behavior not reported for any other crabs. Corals contain large quantities of lipids that constitute 30% to 40% of the dry weight of their tissues, much of which they release with mucus. This release may constitute up to 40% of the daily photosynthetic production by zooxanthellae.

The pocilloporid corals that host crustaceans concentrate some of this lipid into fat bodies that are 300 to 500 μm in length. Glynn suggested that the fat bodies produced by pocilloporid corals hosting protective crabs may be a part of their mutualistic relationship. Stimson tested this hypothesis by determining whether commensal crabs influence the production of fat bodies by coral polyps. He conducted his experiments at the Hawaii Institute of Marine Biology on Coconut Island in Kaneohe Bay, Oahu, Hawaii. He collected colonies of *Pocillopora* 8 to 10 cm in diameter from the midbay region of Kaneohe Bay, placed them in buckets of seawater, and took them back to the marine laboratory on Coconut Island. There, he divided the corals into experimental and control groups. He then removed crabs and pistol shrimp from the experimental coral colonies by "teasing" them out with a small wire. Corals with and without crabs were then kept separately in outdoor tanks supplied with flowing seawater.

After 24 days, Stimson compared the number of fat bodies on corals with and without crabs. He also compared these experimental results with the density of fat bodies on *Pocillopora* in Kaneohe Bay that naturally hosted or lacked mutualistic crabs. The results of these experiments and field observations show clearly that *Pocillopora* increases its production of fat bodies in the presence of crabs both in the laboratory and in the field (fig. 15.20). Stimson also examined the digestive tract of crabs inhabiting corals and found that it contained large quantities of lipids. At the same time, no

significant reductions in either the reproductive rate or growth rates of corals supporting crabs were found. Stimson concluded that the relationship between corals and crabs is a true mutualism, with both partners receiving substantial benefit.

The *extent* of benefit may be the essential factor driving the evolution of mutualisms. In the following section, we review theoretical analyses of how the relative benefits and costs of an association influence the evolution of mutualistic relationships.

Concept 15.2 Review

1. If reef-building corals are placed in the dark, they will expel the zooxanthellae in their tissues. What does this suggest concerning controls on the relationship between corals and zooxanthellae?
2. In terms of costs and benefits, why might corals expel their zooxanthellae when placed in the dark?

15.3 Evolution of Mutualism

LEARNING OUTCOMES

After studying this section you should be able to do the following:

15.13 Interpret the elements of Keeler's model for the evolution of mutualism.
15.14 Use specific examples to elaborate on the conditions under which Keeler's model for the evolution of mutualism predicts that a mutualism will persist.
15.15 Explain the application of Keeler's model to the evolution of a facultative plant protection mutualism.

Theory predicts that mutualism will evolve where the benefits of mutualism exceed the costs. We have reviewed several complex mutualisms both on land and in marine environments. There are many others (fig. 15.21), every one is a fascinating example of the intricacies of nature. Ecologists not only study the present biology of those mutualisms, but also seek to understand the conditions leading to their evolution and persistence. Theoretical analyses point to the relative costs and benefits of a possible relationship as a key factor in the evolution of mutualism.

Modeling of mutualism has generally taken one of two approaches. The earliest attempts involved modifications of the Lotka-Volterra equations to represent the population dynamics of mutualism. The alternative approach has been to model mutualistic interactions using cost-benefit analysis to explore the conditions under which mutualisms can evolve and persist. In chapters 13 and 14, where we discussed models of competition and predation, we focused on the population dynamic approach to modeling species interactions. Here, we concentrate on cost-benefit analyses of mutualism.

Kathleen Keeler (1981, 1985) developed models to represent the relative costs and benefits of several types of mutualistic interactions. Among them are two of the mutualistic interactions we discussed earlier in this chapter: ant-plant protection mutualisms and mycorrhizae. Keeler's approach

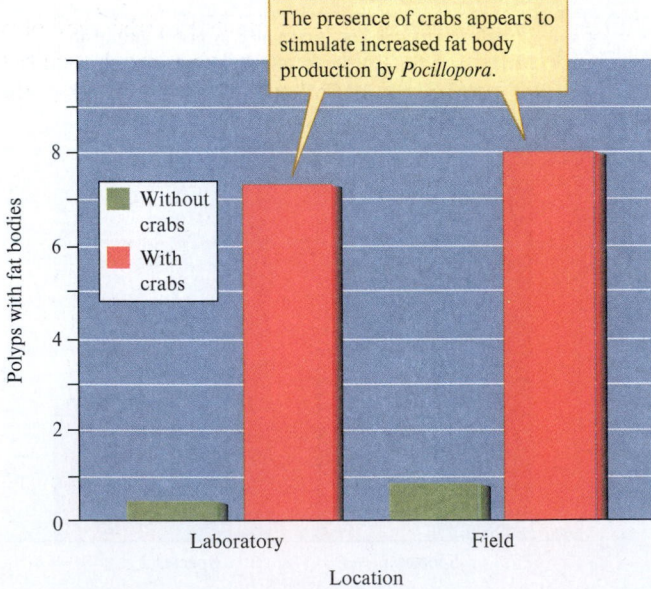

The presence of crabs appears to stimulate increased fat body production by *Pocillopora*.

- Without crabs
- With crabs

Polyps with fat bodies

Laboratory   Field

Location

Figure 15.20 Fat body production by the coral *Pocillopora damicornis* in the presence and absence of crabs (data from Stimson 1990).

(a)

(b)

(c)

(d)

Figure 15.21   A diversity of mutualisms: (*a*) cleaner wrasse and sea bass; (*b*) lichens are an association between fungi and cyanobacteria; (*c*) soybeans fix molecular nitrogen through their association with bacteria within nodules on their roots; (*d*) Cape buffaloes access the energy stored in plant tissues through the activities of a community of mutualistic microorganisms living in their gut. Meanwhile, the cattle egret gleans ticks and flies from the surface of the buffalo and feeds on any suitable prey the buffalo may flush as it forages.   (*a*) Rand McMeins/Getty Images; (*b*) Perry Mastrovito/Corbis/Getty Images; (*c*) Image Source, All rights reserved; (*d*) Franz Aberham/Stockbyte/Getty Images

requires that we consider a population polymorphic for mutualism containing three kinds of individuals: (1) *successful mutualists,* which give and receive measurable benefits to another organism; (2) *unsuccessful mutualists,* which give benefits to another organism but, for some reason, do not receive any benefit in return; and (3) *nonmutualists,* neither giving nor receiving benefit from a mutualistic partner. The bottom line in Keeler's approach is that for a population to be mutualistic, the fitness of successful mutualists must be greater than the fitness of either unsuccessful mutualists or nonmutualists. In addition, the combined fitness of successful and unsuccessful mutualists must exceed that of the fitness of nonmutualists. If these conditions are not met, Keeler proposed that natural selection will eventually eliminate the mutualistic interaction from the population.

In general, we can expect mutualism to evolve and persist in a population when and where mutualistic individuals have higher fitness than nonmutualistic individuals.

Keeler represented the fitness of nonmutualists as:

$$w_{nm} = \text{fitness of nonmutualists}$$

(Fitness has been traditionally represented by the symbol w and though it might be clearer to use another symbol, such as *f,* the traditional symbol is used here.) Keeler represents the fitness of mutualists as:

$$w_m = pw_{ms} + qw_{mu} \qquad (1)$$

where:

p = the proportion of the population consisting of successful mutualists

w_{ms} = the fitness of successful mutualists

q = the proportion of the population consisting of unsuccessful mutualists

w_{mu} = the fitness of unsuccessful mutualists

We can represent Keeler's conditions for the evolution and persistence of mutualism as:

$$pw_{ms} + qw_{mu} > w_{nm} \qquad (2)$$

or

$$w_m > w_{nm} \qquad (3)$$

Keeler predicts that mutualism will persist when the combined fitness of successful and unsuccessful mutualists exceeds the fitness of nonmutualists. Why do we have to combine the fitness of successful and unsuccessful mutualists? Remember that both confer benefit to their partner, but only the successful mutualists receive benefit in return.

The analysis is more convenient if we think of these relationships in terms of **selection coefficients,** the relative selective costs associated with being a successful mutualist, or an unsuccessful mutualist, or a nonmutualist:

$$s = 1 - w \quad \text{and} \quad w = (1 - s)$$

Using selective coefficients, Keeler expressed the selective cost of being a successful mutualist, an unsuccessful mutualist, or a nonmutualist as:

$$s_{ms} = (H)(1 - A)(1 - D) + I_A + I_D \qquad (4)$$

$$s_{mu} = (H)(1 - D) + I_A + I_D \qquad (5)$$

$$s_{nm} = H(1 - D) + I_D \qquad (6)$$

where:

H = the proportion of the plant tissue damaged in the absence of any defenses

D = the amount of protection given to the plant tissues by defenses other than ants (e.g., chemical defenses), so, $1 - D$ is the amount of tissue damage that would occur in spite of these alternative defenses

A = the amount of herbivory prevented by ants (so, again, $1 - A$ is the amount of herbivory that occurs in spite of ants)

I_A = the investment by the plant in benefits extended to the ants

I_D = investment in defenses other than ants

Using these selective coefficients, we can express Keeler's conditions for evolution and persistence of the ant-plant mutualism as:

$$p(1 + s_{ms}) + q(1 - s_{mu}) > 1 - s_{nm}$$

into which Keeler substituted the relationships given in equations (4)-(6). By simplifying the resulting equation, she produced the following expression of benefits relative to costs:

$$p[H(1 - D)A] > I_A$$

Facultative Ant-Plant Protection Mutualisms

Keeler applied her cost-benefit model to facultative mutualisms involving plants with extrafloral nectaries and ants that feed at the nectaries and provide protection to the plant in return. These are mutualisms like that between *Helianthella quinquenervis* and ants, which we discussed in Concept 15.1 on plant mutualism. Her model is not appropriate for obligate mutualisms like that between swollen thorn acacias and their mutualistic ants. In addition, Keeler wrote her model from the perspective of the plant side of the mutualism. Let's step through the general

model and connect each of the terms with the ecology of facultative plant-ant protection mutualisms.

In this model, w_{ms} is the fitness of a plant that produces extrafloral nectaries and that successfully attracts ants effective at guarding it, while w_{mu} is the fitness of a plant that produces extrafloral nectaries but that has not attracted enough ants to mount a successful defense. You may remember that Inouye and Taylor found that *Helianthella* far away from ant nests attracted few ants. These plants would correspond to Keeler's unsuccessful mutualists. In addition, Keeler includes the fitness of nonmutualistic plants, w_{nm}, which would be the fitness of individuals of a plant such as *Helianthella* that does not produce extrafloral nectaries.

Keeler's model represents potential benefits to the host plant as:

$$p[H(1 - D)A]$$

where:

p = the proportion of the plant population attracting sufficient ants to mount a defense

Keeler's model represents the plant's costs of mutualism as:

$$I_A = n[m + d(a + c + h)]$$

where:

n = the number of extrafloral nectaries per plant

m = the energy content of nectary structures

d = the period of time during which the nectaries are active

a = costs of producing amino acids in nectar

c = costs of producing the carbohydrates in nectar

h = costs of providing water for nectar

Again, Keeler's hypothesis is that for mutualism to persist, benefits must exceed costs. In terms of her model:

$$p[H(1 - D)A] > I_A$$

This model proposes that for a facultative ant-plant mutualism to evolve and persist, the proportion of the plant's energy budget that ants save from destruction by herbivores must exceed the proportion of the plant's energy budget that is invested in extrafloral nectaries and nectar.

The details of Keeler's model offer insights into what conditions may produce higher benefits than costs. First, and most obviously, I_A, the proportion of the plant's energy budget that is invested in extrafloral nectaries and nectar should be low. This means that plants living on a tight energy budget—for example, plants living in a shady forest understory—should be less likely to invest in attracting ants than those living in full sun. Higher benefits result from (1) a high probability of attracting ants—that is, high p; (2) a high potential for herbivory, H; (3) low effectiveness of alternative defenses, low D; and (4) highly effective ant defense, high A.

The task for ecologists is to determine how well these requirements of the model match values of these variables in nature.

Concept 15.3 Review

1. Suppose you discover a mutant form of *Helianthella quinquenervis* that does not produce extrafloral nectaries. What does Keller's theory predict concerning the relative fitness of these mutant plants and the typical ones that produce extrafloral nectaries?

2. According to Keller's theory, under what general conditions would the mutant *Helianthella quinquenervis*, lacking extrafloral nectaries, increase in frequency in a population and displace the typical plants that produce extrafloral nectaries?

Applications

Mutualism and Humans

LEARNING OUTCOMES

After studying this section you should be able to do the following:

15.16 Describe the guiding behavior of the greater honeyguide, particularly the information conveyed by the bird to humans, as traditionally understood by the Boran people of East Africa.

15.17 Summarize the mutual benefits between humans and the greater honeyguide.

15.18 Explain the experimental and observational evidence in support of the traditional understanding of guiding behavior by the greater honeyguide.

Mutualism has been important in the lives and livelihood of humans for thousands of years. Historically, much of agriculture has depended upon mutualistic associations between species and much of agricultural management has been aimed at enhancing mutualisms, such as nitrogen fixation, mycorrhizae, and pollination to improve crop production. Agriculture itself has been viewed as a mutualistic relationship between humans and crop and livestock species. However, there may be some qualitative differences between agriculture, as it has been generally practiced, and mutualisms among other species. How much of agriculture is pure exploitation and how much is truly mutualistic remains an open question.

There is, however, at least one human mutualism that fits comfortably among the earlier discussions in this chapter, a mutualism involving communication between humans and a wild species with clear benefit to both. This mutualism joins the traditional honey gatherers of Africa with the greater honeyguide, *Indicator indicator* (fig. 15.22). Honey gathering has long been an important aspect of African cultures, important enough that there are scenes of honey gathering in rock art painted over 20,000 years ago (Isack and Reyer 1989). No one knows how long humans have gathered honey in Africa, but it is difficult to imagine the earliest hominids resisting such sweet temptation. Whenever honey gathering began, humans have apparently had a capable and energetic partner in their searches.

Guiding Behavior

The mutualistic association between humans and honeyguides may have developed from an earlier association between the bird and the ratel, or honey badger, *Mellivora capensis*. The honey badger is a powerful animal, well equipped with strong claws and powerful muscles to rip open bees' nests, that readily follows honeyguides to bees' nests. The honey badger, though secretive, has been observed often following honeyguides while vocalizing. African honey gatherers also vocalize to attract honeyguides, and Friedmann (1955) reported that some of their vocalizations imitate the calls of honey badgers.

The most detailed and quantitative study of this mutualism to date is that of H. Isack of the National Museum of Kenya and H.-U. Reyer of the University of Zurich (Isack and Reyer 1989), who studied the details of the interaction of the greater honeyguide with the Boran people of northern Kenya. The Boran regularly follow honeyguides and have developed a penetrating whistle that they use to attract them. The whistle can be heard over 1 km away, and Isack and Reyer found that it doubles the rate at which Boran honey gatherers encounter honeyguides. If they are successful in attracting a honeyguide, the average amount of time it takes to find a bees' nest is 3.2 hours. Without the aid of a honeyguide the average search time per bees' nest is about 8.9 hours. This is an underestimate of the true time, however, since Isack and Reyer did not include days in which no bees' nests were found in their analysis. The benefit of the association to the bird seems apparent from Isack and Reyer's analysis, since they report that 96% of the nests to which the Boran were guided would have been inaccessible to the birds without human help.

Figure 15.22 The greater honeyguide, *Indicator indicator*.
Nigel Dennis/Science Source

The greater honeyguide attracts the attention of a human by flying close and calling as it does so. Following this initial attention-getting behavior, the bird will fly off in a particular direction and disappears for up to 1 minute. After reappearing, the bird again perches in a conspicuous spot and calls to the following humans. As the honey gatherers follow, they whistle, bang on wood, and talk loudly in order to "keep the bird interested." When the honey gatherers approach the perch from which the honeyguide is calling, the bird again flies off, calling and displaying its white tail feathers as it does so, only to reappear at another conspicuous perch a short time later. This sequence of leading, following, and leading is repeated until the bird and the following honey gatherers arrive at the bees' nest.

Isack, who is a Boran, interviewed Boran honey gatherers to determine what information they obtained from honeyguides. The main purpose of the study was to test assertions by the honey gatherers that the bird informs them of (1) the direction to the bees' nest, (2) the distance to the nest, and (3) when they arrive at the location of the nest. The data gathered by Isack and Reyer support all three assertions.

Honey gatherers reported that the bird indicated direction to the bees' nest on the basis of the direction of its guiding flights. One method used by Isack and Reyer to test how well flight direction indicated direction was to induce honeyguides to guide them from the same starting point to the same known bees' nest on five different occasions. Figure 15.23a shows the highly restricted area covered by these five different guiding trips. Another approach was to induce the bird to guide them to a bees' nest from seven different starting points (fig. 15.23b). The result was a consistent tendency by the bird to lead directly to the site of the bees' nest.

The Boran honey gatherers said that three variables decrease as distance to the nest decreases: (1) the time the bird stays out of sight during its first disappearance following the initial encounter, (2) the distance between stops made by the bird on the way to the bees' nest, and (3) the height of the perch on the way to the nest. Data gathered by Isack and Reyer support all three statements (fig. 15.24).

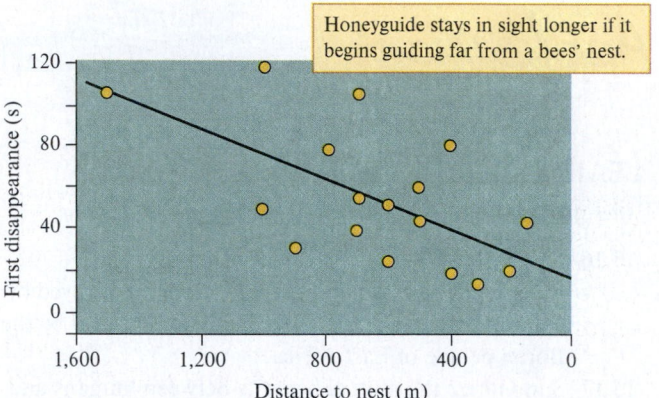

Honeyguide stays in sight longer if it begins guiding far from a bees' nest.

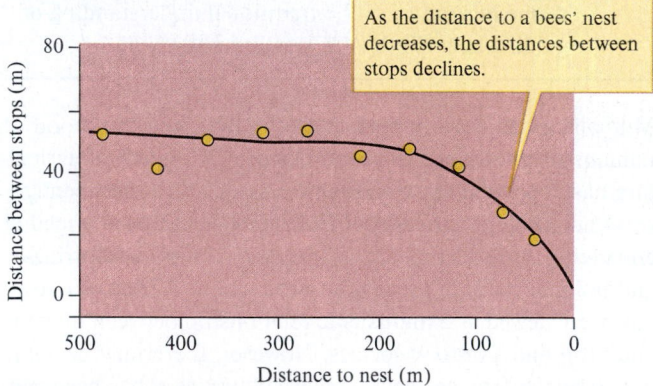

As the distance to a bees' nest decreases, the distances between stops declines.

The paths taken by a honeyguide on five separate guiding trips cover a restricted area.

Starting point

Location of bees' nest

(a)

Honeyguides lead along a nearly straight line to a bees' nest, regardless of starting point.

S_4

S_3 S_5

S_2

S_6 S_7

S_1 Location of bees' nest

(b)

Figure 15.23 Paths taken by honeyguides leading people to bees' nests (data from Isack and Reyer 1989).

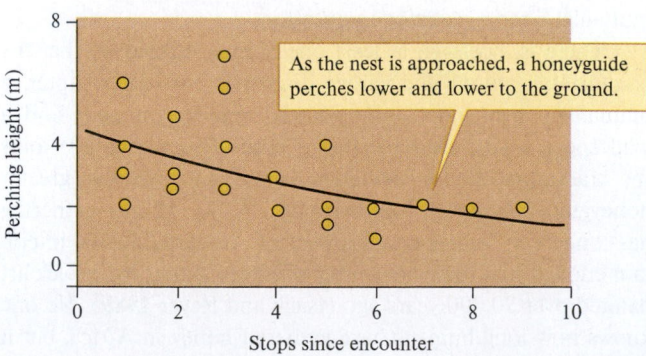

As the nest is approached, a honeyguide perches lower and lower to the ground.

Figure 15.24 Changes in behavior of the honeyguide as it nears a bees' nest (data from Isack and Reyer 1989).

On the way to a bees' nest, a honeyguide uses a particular call and responds to a human voice by increasing call frequency.

After arriving at a bees' nest, the honeyguide gives a few distinctive indication calls and then perches silently near the nest.

Figure 15.25   Vocal communication between honeyguides and humans.

The honey gatherers also report that they can determine when they arrive in the vicinity of a bees' nest by changes in the honeyguide's behavior and vocalizations (fig. 15.25). Isack and Reyer observed several of these changes. While on the path to a bees' nest, a honeyguide emits a distinctive guiding call and will answer human calls by increasing the frequency of the guiding call. On arriving at a nest, the honeyguide perches close to the nest and gives off a special "indication" call. After a few indication calls, it remains silent and does not answer to human sounds. If approached by a honey gatherer, a honeyguide flies in a circle around the nest location before perching again nearby.

Isack and Reyer observe that their data do not allow them to test other statements by the Boran honey gatherers, including that when bees' nests are very far away (over 2 km) the honeyguide will "deceive" the gatherers about the real distance to the nest by stopping at shorter intervals. Isack and Reyer add, however, that they have no reason to doubt these other statements, since all others have been supported by the data they were able to collect. What these data reveal is a rich mutualistic interaction between wild birds and humans. The results of Isack and Reyer's study caused Robert May (1989) to wonder how much important ecological knowledge may reside with the dwindling groups of native people living in the tropical regions of the world, regions about which the field of ecology has so little information.

Summary

Mutualism, interactions between individuals that benefit both partners, is a common phenomenon in nature that has apparently made important contributions to the evolutionary history of life and continues to make key contributions to the ecological integrity of the biosphere. Mutualisms can be divided into those that are *facultative,* where species can live without their mutualistic partners, and *obligate,* where species are so dependent on the mutualistic relationship that they cannot live without their mutualistic partners.

Plants benefit from mutualistic partnerships with a wide variety of bacteria, fungi, and animals.   Mutualism provides benefits to plants ranging from nitrogen fixation and enhanced nutrient and water uptake to pollination and seed dispersal. Ninety percent of terrestrial plants form mutualistic relationships with mycorrhizal fungi, which make substantial contributions to plant performance. Mycorrhizae, which are mostly either arbuscular mycorrhizae or ectomycorrhizae, are important in increasing plant access to water, nitrogen, phosphorus, and other nutrients. In return for these nutrients, mycorrhizal fungi receive energy-rich root exudates. Experiments have shown that the mutualistic balance sheet between plants and mycorrhizal fungi can be altered by the availability of nutrients. Plant-ant protection mutualisms are found in both tropical and temperate environments. In tropical environments, many plants provide ants with food and shelter in exchange for protection from a variety of natural enemies. In temperate environments, mutualistic plants provide ants with food but not shelter in trade for protection.

Reef-building corals depend on mutualistic relationships with algae and animals, with an exchange of benefits paralleling those between terrestrial mutualists.   The coral-centered mutualisms of tropical seas show striking parallels with terrestrial plant-centered mutualisms. Mutualistic algae called zooxanthellae provide reef-building corals with their principal energy source; in exchange for this energy, corals provide zooxanthellae with nutrients, especially nitrogen, a scarce resource in tropical seas. The mutualism between corals and zooxanthellae appears to be largely under the control of the coral partner, which chemically solicits the release of organic compounds from zooxanthellae and controls zooxanthellae population growth. Crabs and shrimp protect some coral species from coral predators in exchange for food and shelter.

Theory predicts that mutualism will evolve where the benefits of mutualism exceed the costs. Keeler built a cost-benefit model for the evolution and persistence of facultative plant-ant protection mutualisms in which the benefits of the mutualism to the plant are represented in terms of the proportion of the plant's energy budget that ants protect from damage by herbivores. The model assesses the costs of the mutualism to the plant in terms of the proportion of the plant's energy budget invested in extrafloral nectaries and the water, carbohydrates, and amino acids contained in the nectar. The model predicts that the mutualism will be favored where there are high densities of ants and potential herbivores and where the effectiveness of alternative defenses is low.

Humans have developed a variety of mutualistic relationships with other species, but one of the most spectacular is that between the greater honeyguide and the traditional honey gatherers of Africa. In this apparently ancient mutualism, humans and honeyguides engage in elaborate communication and cooperation with clear benefit to both partners. The mutualism offers the human side a higher rate of discovery of bees' nests, while the honeyguide gains access to nests that it could not raid without human help. Careful observations have documented that the honeyguide informs the honey gatherers of the direction and distance to bees' nests as well as of their arrival at the nest.

Key Terms

arbuscular mycorrhizae 327	ectomycorrhizae 327	hyphae 327	obligate mutualism 326
arbuscule 327	extrafloral nectary 334	mutualism 326	selection coefficient 340
balanced growth 336	facultative mutualism 326	mycorrhizae 327	vesicle 327
commensalism 326			

Review Questions

1. List and briefly describe key mutualisms that are critical to the lives of diverse organisms across the biosphere.
2. What contributions do mycorrhizal fungi make to their plant partners? What do plants contribute in return for the services of mycorrhizal fungi? How did Hardie (1985) demonstrate that mycorrhizae improve the water balance of red clover?
3. Outline the experiments of Johnson (1993), which she designed to test the possibility that artificial fertilizers may select for less mutualistic mycorrhizal fungi. What evidence does Johnson present in support of her hypothesis?
4. Explain how mycorrhizal fungi may have evolved from ancestors that were originally parasites of plant roots. Do any of Johnson's results (1993) indicate that present-day mycorrhizal fungi may act as parasites? Be specific.
5. Janzen (1985) encouraged ecologists to take a more experimental approach to the study of mutualistic relationships. Outline the details of Janzen's own experiments on the mutualistic relationship between swollen thorn acacias and ants.
6. Inouye and Taylor's study (1979) of the relationship between ants and the aspen sunflower, *Helianthella quinquenervis,* provides a reasonable representative of temperate ant-plant protection mutualisms. Compare this mutualism with that of the tropical mutualism between swollen thorn acacias and ants.
7. How are coral-centered mutualisms similar to plant-centered mutualisms? How are they different?
8. Outline the benefits and costs identified by Keeler's (1981, 1985) cost-benefit model for facultative ant-plant mutualism. Does Keeler's model view this mutualism from the perspective of plant or ant? What would be some of the costs and benefits to consider if the model were built from the perspective of the other partner?
9. How could you change the Lotka-Volterra model of competition we discussed in chapter 13 into a model of mutualism? Would the resulting model be a cost-benefit model or a population dynamic model?
10. Outline how the honeyguide-human mutualism could have evolved from an earlier mutualism between honeyguides and honey badgers.

Chapter 16

Species Abundance and Diversity

John Wang/Getty Images

Wildflowers blooming near Mt. Timpanogos in the Wasatch Mountains of Utah. Perhaps no other setting reveals species diversity so conspicuously as a diverse community of wildflowers in peak bloom.

Applications: Disturbance by Humans   360

    Summary   363

    Key Terms   363

    Review Questions   364

CHAPTER CONCEPTS

16.1 Most species are moderately abundant; few are very abundant or extremely rare.   347

    Concept 16.1 Review   348

16.2 A combination of the number of species and their relative abundance defines species diversity.   348

    Concept 16.2 Review   351

16.3 Species diversity is higher in complex environments.   351

    Concept 16.3 Review   357

16.4 Intermediate levels of disturbance promote higher diversity.   357

    Concept 16.4 Review   360

LEARNING OUTCOMES

After studying this section you should be able to do the following:

16.1 Describe the concepts of community and community structure.

16.2 Discuss the parallels between animal guild and plant life-form.

Different areas within the same region may differ substantially in the number of species they support. Vast areas of flat or gently sloping land in the hot deserts of North America are dominated by a single species of shrub, the creosote bush, *Larrea tridentata*. While grasses and forbs grow in the spaces between these shrubs, creosote bushes make up most of the plant biomass. In these areas, you can travel many kilometers and see only subtle changes in a landscape dominated by a single species of plant (fig. 16.1).

The uniformity of the creosote flats contrasts sharply with the biological diversity of other places in these hot deserts (fig. 16.2). For instance, a rich variety of plant life-forms covers

Figure 16.1 Sonoran Desert landscape dominated by the creosote bush, *Larrea tridentata.* Charlie Ott/Science Source

Figure 16.2 Species-rich Sonoran Desert landscape. Doug Sherman/ Geofile

Organ Pipe Cactus National Monument in southern Arizona. Here grow ocotillo, consisting of several slender branches 2 to 3 m tall springing from a common base; palo verde trees with green bark and tiny leaves; and mesquite, which reach the size of medium-sized trees. In addition, there are cactus such as the low-growing prickly pears and the shrublike teddy bear chollas. The most striking are the column-shaped squat barrel cactus, the organ pipe cactus, with its densely packed slender columns, and the saguaro, a massive cactus that towers over all the other plant species. Among these larger plants also grow a wide variety of small shrubs, grasses, and forbs.

The creosote flats, dominated by one species of shrub, convey an impression of great uniformity. The vegetation of Organ Pipe Cactus National Monument, consisting of a large number of species of many different growth forms, gives the impression of high diversity. The ecologist is prompted to ask what factors control this difference in diversity. Joseph McAuliffe (1994) has been able to explain much of the variation in woody plant diversity and dominance by *L. tridentata*

across Sonoran Desert landscapes by differences in soil age, frequency of disturbance by erosion, and soil depth (see chapter 21).

In chapters 13 to 15, we focused on competition, predation, and mutalism between pairs of species. We now consider patterns and processes that involve a larger number of species. With this shift in focus, we enter the realm of community ecology. A **community** is an association of interacting species inhabiting some defined area. Ecologists may study the community of plants and animals living on a mountainside or the community of invertebrate animals and algae living in an intertidal environment. The key point here is that communities generally consist of many species that potentially interact in all of the ways discussed in chapters 13 to 15.

Community ecologists seek to understand how various abiotic and biotic aspects of the environment influence the structure of communities. **Community structure** includes attributes such as the number of species, the relative abundance of species, and the kinds of species comprising a community.

Because it is difficult to study large numbers of species, most community ecologists work with restricted groups of organisms, focusing, for example, on communities of plants, mammals, or insects. Some community ecologists restrict their focus even more by studying guilds of species. A **guild** is a group of organisms that all make their living in a similar way. Examples of guilds include the seed-eating animals in an area of desert, the fruit-eating birds in a tropical rain forest, and the filter-feeding invertebrates in a stream. Some guilds consist of closely related species, whereas others are taxonomically heterogeneous. For instance, the fruit-eating birds on many South Pacific islands consist mainly of pigeons, whereas the seed-eating guild in the Sonoran Desert includes mammals, birds, and ants.

The main users of the guild concept have been animal ecologists. Similar terms used by botanists are **life-form** or growth form. The life-form of a plant is a combination of its structure and its growth dynamics. Plant life-forms have been classified in various ways and we have used an informal classification since chapter 1, where we discussed life-forms such as trees, vines, annual plants, grasses, and forbs.

Like the members of an animal guild, plants of similar life-form exploit the environment in similar ways. As a consequence, plant community ecologists have often concentrated their attention on plants of similar life-form by studying the ecology of tree, shrub, or herb communities. By studying animal guilds or plant life-forms, ecologists focus their energies on a manageable and coherent portion of the community, manageable in terms of number of species and coherent in terms of ecological requirements.

In 1959, G. Evelyn Hutchinson wrote a paper with the captivating title "Homage to Santa Rosalia or Why Are There So Many Kinds of Animals?" This paper stimulated generations of ecologists to explore biological diversity. One of the most fundamental questions that we still address today is what controls the number and relative abundance of species in communities. These two properties are the focus of chapter 16.

16.1 Species Abundance

LEARNING OUTCOMES

After studying this section you should be able to do the following:

16.3 Interpret a graph depicting the relative abundance of species in a community, paying particular attention to the labeling of the axes.

16.4 Explain the significance of the observation that most studies of the relative abundance of species have shown a lognormal distribution.

Most species are moderately abundant; few are very abundant or extremely rare. The relative abundance of species is one of the most fundamental aspects of community structure. This property is so fundamental that George Sugihara (1980) referred to it as "minimal community structure." We began our discussion of the abundance of species in chapter 9, where we explored the relationship between body size and abundance and considered the various forms of rarity. In this section, we expand our perspective by addressing the following question: What will you find if you go out into a community and quantify the abundance of species within a group of taxonomically or ecologically related organisms such as beetles, birds, shrubs, or diatoms?

It turns out that there are regularities in the relative abundance of species in communities that hold, whether you examine plants in a forest, moths in that forest, or algae inhabiting a nearby stream. If you thoroughly sample groups of organisms such as these, you will come across a few abundant species and a few that are very rare. Most species will be moderately abundant. This pattern was first quantified by Frank Preston (1948, 1962a, 1962b), who carefully studied the relative abundance of species in collections and communities. The "distribution of commonness and rarity" among species described by Preston is one of the best-documented patterns in natural communities.

The Lognormal Distribution

How do we think about the abundance of organisms? Preston suggested that we think of abundance in relative terms and say, for example, that one species is twice as abundant as another. This common way of expressing relative abundance led Preston to graph the abundance of species in collections as frequency distributions, where the classes of species abundance were intervals of 1–2, 2–4, 4–8, 8–16, etc., individuals. Preston made each interval twice the preceding one and plotted them on a $\log_2$ scale (e.g., $\log_2$ of 1 = 0, $\log_2$ of 2 = 1, $\log_2$ of 4 = 2, etc.). Preston's graphs plot $\log_2$ of species abundance against the number of species in each abundance interval. When the relative abundance of species was plotted in this way, he consistently obtained results like those shown in figure 16.3.

Figure 16.3a shows the relative abundance of desert plants. Robert Whittaker (1965) plotted these abundances using coverage rather than numbers of individuals, which correlates with our discussions in chapter 9 of how to represent the relative abundance of plants. Notice that few species were represented

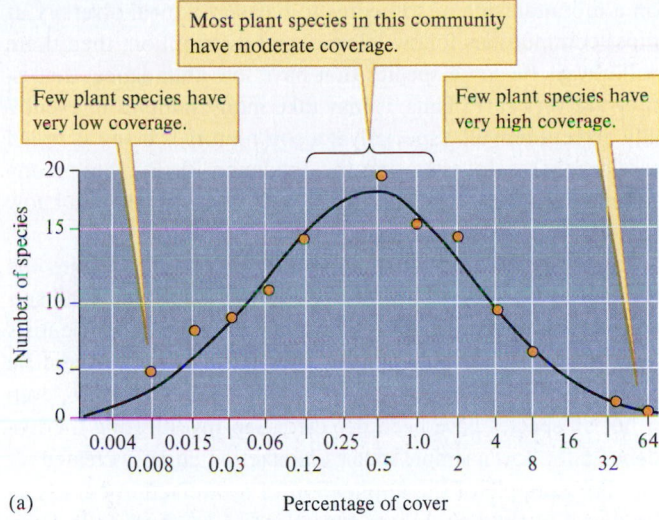

(a)

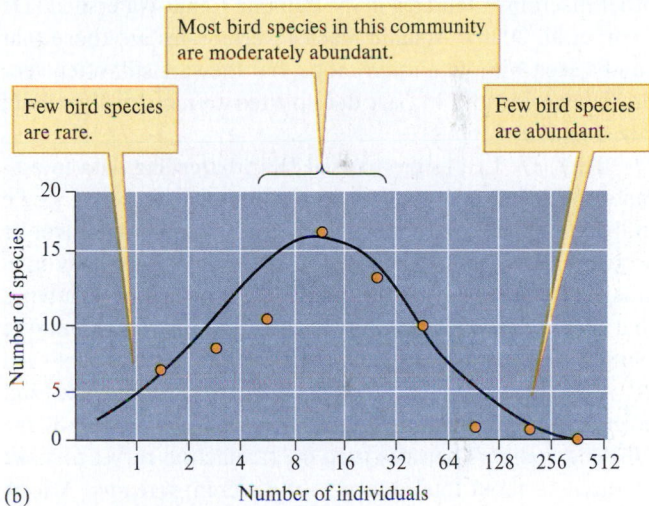

(b)

Figure 16.3 Lognormal distributions of (*a*) desert plants and (*b*) forest birds (data from Preston 1962a; Whittaker 1965).

by more than 8% cover or less than 0.15% cover. Most species had intermediate coverage. Whittaker's plot shows the most distinctive feature of Preston's distributions—that is, they are approximately "bell-shaped," or "normal." Such a distribution is called a **lognormal distribution.**

Figure 16.3*b* shows the relative abundance of 86 species of birds breeding near Westerville, Ohio, over a 10-year period (Preston 1962a). Notice that few species were represented by over 64 individuals or by a single individual. Like Whittaker's plants, most species showed intermediate levels of abundance, producing another lognormal distribution.

This principle of most species being moderately abundant has implications for our ability to document all of the species in a place. When scientists attempt to quantify species diversity in a natural community, they usually cannot survey every individual; instead, they must take samples of that community, whether that be by using quadrats to measure plant diversity on a mountainside or soil cores to measure fungal diversity. If most communities follow a lognormal distribution, then there will always be some species that have low abundance, that is, they are very rare. Thus, it may take many samples to eventually detect them all, especially if a community is being sampled randomly (see Investigating the Evidence 2). But how many sample units (e.g., quadrats, soil cores) must be taken to know that we have documented every species present?

The **species rarefaction curve** is a model that defines the number of samples needed to likely record all or almost all species at your location of interest. In this way, it instructs scientists how intense the sampling needs to be. It also can be a predictive model used to estimate what the actual species count is, even if not all species have been detected. See Investigating the Evidence 6 for how a simple species rarefaction curve is created.

An example of the application of a rarefaction curve can be found in the research of Luis F. De Léon and colleagues, who investigated the species diversity of insect larvae and other macroinvertebrates in the Panama Canal Watershed (De Léon et al. 2020). Aquatic macroinvertebrates are those that can be seen without a microscope, but they are still often very difficult to identify in part due to their tremendous diversity (fig. 16.4a, b).

Thus, De Léon's group used **DNA barcoding** data to estimate species diversity; each sample in their case was a single individual. DNA barcoding is a genetic tool that involves sequencing the same small region of DNA across many individuals. Researchers are then able to identify unique genotypes that may represent species. It is also often used to identify a sample's species by comparing the sample against published DNA sequences. In De Léon's study, they used species rarefaction curves to plot the number of species identified by the DNA barcoding. They also used the rarefaction curves to make predictions about total diversity of different streams. A look at just a few of the study locations they sampled reveals that some streams had many very rare species and high diversity, while others likely had very low diversity (fig. 16.4c).

As we can see, the distribution of species in a community can have implications for our ability to detect the diversity that is present. Analyses of thousands of ecological communities have shown that the lognormal distribution is one of several distributions that give a reasonable match to the relative abundance of species (Ulrich, Ollik, and Ugland 2010; Baldridge et al. 2016). In all these competing models, however, it remains the case that most species in well-studied communities are moderately abundant.

Concept 16.1 Review

1. What does the frequently observed lognormal distribution tell us about general patterns of species abundance and rarity?
2. Why do smaller samples result in only part of the bell-shaped curve that is characteristic of the lognormal distribution?

16.2 Species Diversity

LEARNING OUTCOMES

After studying this section you should be able to do the following:

16.5   Distinguish between species richness and species evenness.
16.6   Calculate a Shannon-Wiener diversity index, H'.
16.7   Interpret species rank-abundance curves for contrasting communities.

A combination of the number of species and their relative abundance defines species diversity.   Ecologists define **species diversity** on the basis of two factors: (1) the number of species in the community, which ecologists usually call **species richness,** and (2) the relative abundance of species in communities, or **species evenness.** The influence of species richness on community diversity is clear. A community with 20 species is obviously less diverse than one with 80 species. The effects of species evenness on diversity are more subtle but easily illustrated.

Figure 16.5 contrasts two hypothetical forest communities. Both forests contain five tree species, so they have equal levels of species richness. However, community b is more diverse than community a because its species evenness is higher. In community b, all five species are equally abundant, each comprising 20% of the tree community. In contrast, 84% of the individuals in community a belong to one species, while each of the remaining species constitutes only 4% of the community. On a walk through the two forests, you would almost certainly form an impression of higher species diversity in community b, despite equal levels of species richness in the two forests. That impression can be quantified.

A Quantitative Index of Species Diversity

Ecologists have developed many indices of species diversity, the values of which depend on levels of species richness and evenness. Table 16.1 shows how to calculate a commonly used index, the Shannon-Wiener index, H', for our two hypothetical forest communities. The contrasting values of H' for the two communities reflect the difference in species evenness that we see when we compare the two forests depicted in figure 16.5. H' for community b, the community with higher species evenness, is 1.610, while H' for community a is 0.662.

The calculations for the Shannon-Wiener index done in Table 16.1 are summarized by this equation:

$$H' = - \sum_{i=1}^{s} p_i \log_e p_i$$

where:

H' = the value of the Shannon-Wiener diversity index
p_i = the proportion of the ith species in a sample
$\log_e p_i$ = the natural logarithm of p_i
s = the number of species in the community

(a)

Figure 16.4 (*a*) Students Katherine Newman and Brittany Sprout sample benthic macroinvertebrates using a kick net, a tool that allows the tiny organisms to be collected from under rocks and in the cobble and sediment at the bottom of streams, rivers, or lakes. (*b*) Benthic macroinvertebrates are small, but "macro" because they can be seen by the naked eye, and can include insect larvae, small mollusks, worms, and other taxa, like the amphipod *Diporeia* sp. shown here. (*c*) Studies of streams of the Panama Canal Watershed used rarefaction curves to estimate benthic macroinvertebrate species diversity, based on a limited number of samples, represented by DNA barcoded individuals. The solid line represents the observed data, whereas the dashed lines are predicted diversity, based on observed data. Shading represents the degree of uncertainty associated with the estimate of diversity for each stream sampled. Stream A is Frijoles (square), stream B is Chucantı (triangle), and stream C is Blanco (circle). The blue line indicates number of species that would be detected with a sample of 100 individuals. (*a*) Wayne Armstrong; (*b*) Courtesy of NOAA Great Lakes Environmental Research Laboratory; (*c*) based on De León, L. F., Cornejo, A., Gavilán, R. G., & Aguilar, C. (2020). Hidden biodiversity in Neotropical streams: DNA barcoding uncovers high endemicity of freshwater macroinvertebrates at small spatial scales. bioRxiv.

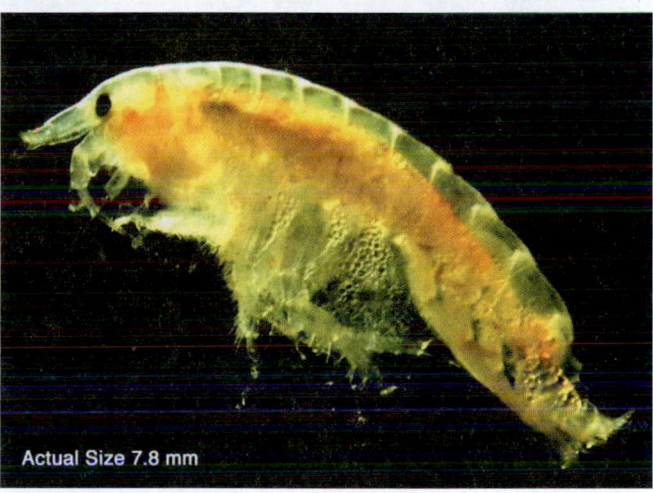

Actual Size 7.8 mm

(b)

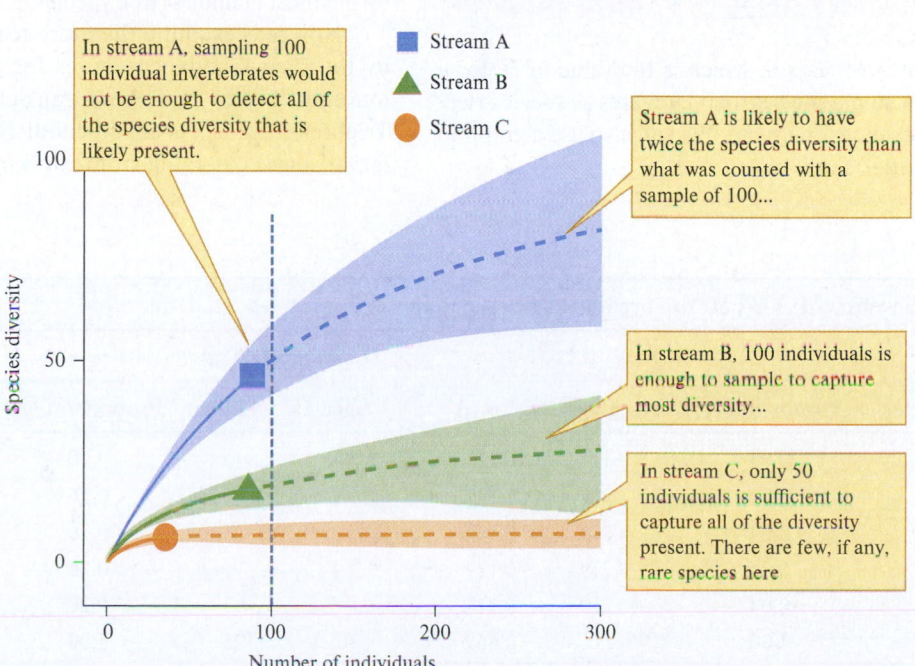

(c)

Communities *a* and *b* both contain five tree species. However, because community *b* has greater species evenness, it has higher species diversity.

Community *a* is dominated by one of its five species and so has lower species diversity than…

…community *b*, which has the same five species but in equal proportions.

(a)

Lower species evenness

(b)

Higher species evenness

Figure 16.5   Species evenness and species diversity.

To calculate H', determine the proportions of each species in a sample of the study community, p_i, and the $\log_e$ of each p_i. Next, multiply each p_i by $\log_e p_i$ and sum the results for all species from species 1 to species s, where s = the number of species in the community, that is:

$$\sum_{i=1}^{s}$$

Since this sum will be a negative number, the Shannon-Wiener index calls for taking its opposite, that is:

$$-\sum_{i=1}^{s}$$

The minimum value of H' is 0, which is the value of H' for a community with a single species, and increases as species richness and species evenness increase. We can also use a graph to contrast communities *a* and *b*.

Rank-Abundance Curves

We can also portray the relative abundance and diversity of species within a community by plotting the relative abundance of species against their rank in abundance. The resulting **rank-abundance curve** provides us with important information about a community, information accessible at a glance. Figure 16.6 plots the abundance rank of each tree species in communities *a* and *b* (see fig. 16.5) against its proportional abundance. The rank-abundance curve for community *b* shows that all five species are equally abundant, while the rank-abundance curve for community *a* shows its dominance by the most abundant tree species.

Now let's examine the more realistic differences shown by the rank-abundance curves for two actual animal communities. Figure 16.7 shows rank-abundance curves for the Trichoptera (insects called caddisflies that have an aquatic larval stage) emerging from two kinds of aquatic habitat in

Table 16.1

Calculating species diversity (H') for two hypothetical communities of forest trees									
Community a					*Community b*				
Species	Number	Proportion (p_i)	$\log_e p_i$	$p_i \log_e p_i$	Species	Number	Proportion (p_i)	$\log_e p_i$	$p_i \log_e p_i$
1	21	0.84	−0.174	−0.146	1	5	0.20	−1.609	−0.322
2	1	0.04	−3.219	−0.129	2	5	0.20	−1.609	−0.322
3	1	0.04	−3.219	−0.129	3	5	0.20	−1.609	−0.322
4	1	0.04	−3.219	−0.129	4	5	0.20	−1.609	−0.322
5	1	0.04	−3.219	−0.129	5	5	0.20	−1.609	−0.322
Total	25	1.00		−0.662	Total	25	1.00		−1.610

$$H' = -\sum_{i=1}^{s} p_i \log_e p_i = 0.662$$

$$H' = -\sum_{i=1}^{s} p_i \log_e p_i = 1.610$$

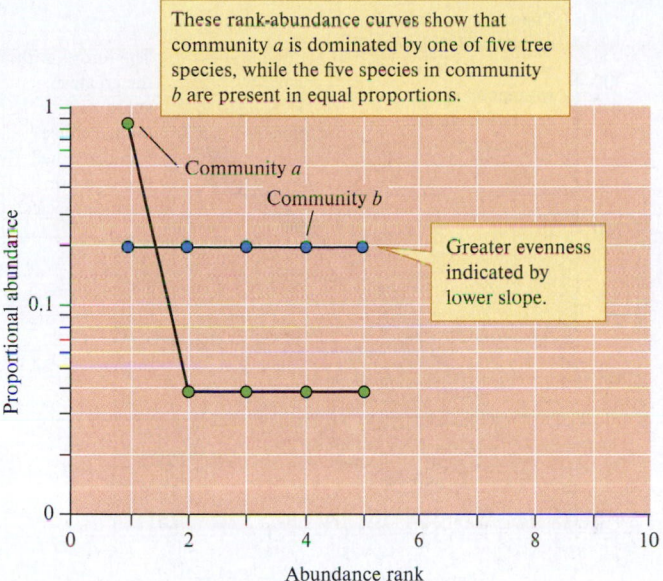

These rank-abundance curves show that community *a* is dominated by one of five tree species, while the five species in community *b* are present in equal proportions.

Community *a*

Community *b*

Greater evenness indicated by lower slope.

Figure 16.6   Rank-abundance curves for two hypothetical forests.

northern Portugal. The trichopteran community in coastal ponds at Mira contains about 18 species, while the mountain stream community at Relva includes 79 species. In addition, a few abundant species dominate the coastal pond community, while the mountain stream community shows a more even distribution of individuals among species. This difference in species evenness is shown by the much steeper rank-abundance curve for the coastal pond community.

Two reef fish communities from the Gulf of California provide a more subtle contrast in rank-abundance patterns. The reef fish communities yielded approximately similar numbers of species (52 vs. 57) but differed substantially in species evenness. The community of the central Gulf of California showed a more even distribution of individuals among species. This greater evenness is depicted in figure 16.8, which shows that after about the tenth most abundant species, the rank-abundance curve for the central Gulf of California lies above the curve for the northern Gulf. Rank-abundance curves will provide a useful representation of community structure in later discussions.

Concept 16.2 Review

1. Pollution of streams generally reduces the diversity of Trichoptera (see fig. 16.7), and several other groups of stream insects, by reducing both species richness and species evenness. Why?
2. Suppose you sample an area and find the five species of forest trees listed in table 16.1 in the following proportions: 0.35, 0.25, 0.15, 0.15, and 0.10. What is the Shannon-Wiener diversity of this community, *c*, compared to communities *a* and *b* in table 16.1?

16.3 Environmental Complexity

LEARNING OUTCOMES

After studying this section you should be able to do the following:

16.8 Explain the positive correlation generally found between foliage height diversity and bird species diversity.

16.9 Contrast the characteristics of the niches of forest birds and those of algae and plants.

16.10 Describe the significance of spatial complexity in the distribution of nutrients in lakes and soils to diversity of algae and plants.

16.11 Propose a hypothesis to explain the generally found negative relationship between soil fertility and richness of plants and mycorrhizal fungi.

16.12 Design a study to test your hypothesis to explain the generally found negative relationship between soil fertility and richness of plants and mycorrhizal fungi.

Species diversity is higher in complex environments.   How does environmental structure affect species diversity? This is one of the most fundamental questions we can ask about communities. In general, species diversity increases with environmental complexity or heterogeneity. However, an aspect of environmental structure important to one group of organisms may not have a positive influence on another group. Consequently, you must know something about the ecological requirements of species to predict how environmental structure affects their diversity. In other words, you must know something about their niches.

Forest Complexity and Bird Species Diversity

In chapter 13, we saw that competition can significantly influence the niches of species. If competition acts to produce

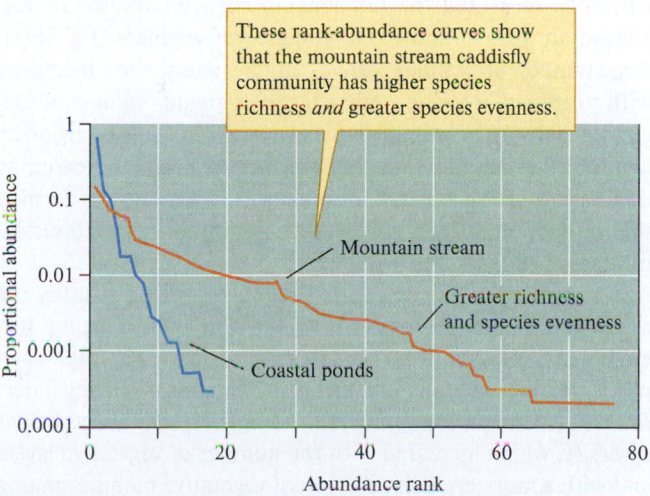

These rank-abundance curves show that the mountain stream caddisfly community has higher species richness *and* greater species evenness.

Mountain stream

Greater richness and species evenness

Coastal ponds

Figure 16.7   Rank-abundance curves for caddisflies, order Trichoptera, of two aquatic habitats in northern Portugal (data courtesy of L. S. W. Terra).

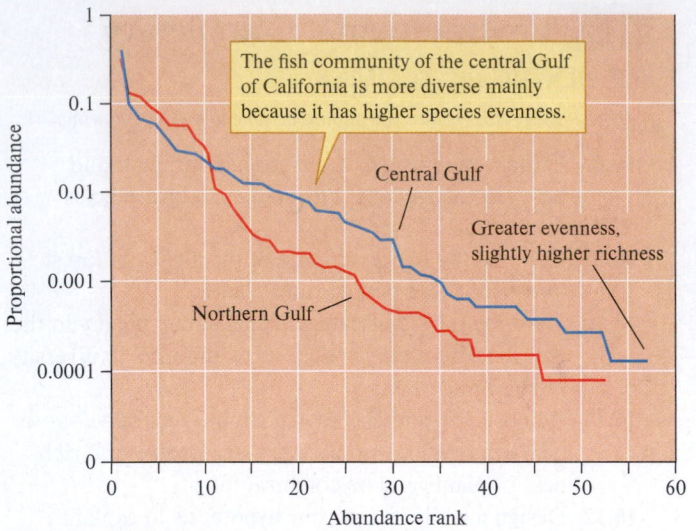

Figure 16.8 Rank-abundance curves for two reef fish communities in the Gulf of California (data from Molles 1978; Thomson and Lehner 1976; and courtesy of D. A. Thomson and C. E. Lehner 1976).

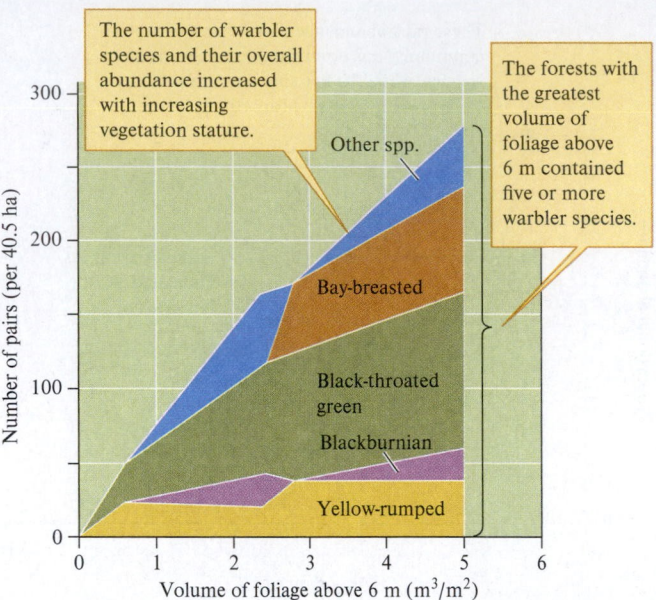

Figure 16.9 Stature of vegetation and number of warbler species (data from MacArthur 1958).

divergence in the niches of species, what would you expect to find if you characterized the niches of closely related, coexisting species? The competitive exclusion principle (see chapter 13) leads us to predict that coexisting species will have significantly different niches. As we saw in chapter 1, that is precisely what Robert MacArthur (1958) found when he examined the ecology of five species of warblers that live together in the forests of northeastern North America (see fig. 1.3).

What does MacArthur's study of warbler niches have to do with the influence of environmental complexity on species diversity? MacArthur's results suggest that since these species forage in different vegetative strata, their distributions may be influenced by variation in the vertical structure of vegetation. He explored this possibility on Mount Desert Island, Maine, where he measured the relationship between volume of vegetation above 6 m and the abundance of warblers (fig. 16.9). The number of warbler species at the study sites increased with forest stature. The study sites with greater volume of vegetation above 6 m supported more warbler species. In other words, MacArthur found that warbler diversity increased as the stature of the vegetation increased. These results formed the foundation of later studies of how foliage height diversity influences bird species diversity.

MacArthur was one of the first ecologists to quantify the relationship between species diversity and environmental heterogeneity. He quantified the diversity of species and the complexity of the environment using the Shannon-Wiener index, H'. He measured environmental complexity as foliage height diversity, which increased with the number of vegetative layers and with a more even distribution of vegetative biomass among three vertical layers, 0 to 0.6 m, 0.6 to 7.6 m, and >7.6 m. MacArthur's foliage height diversity, like species diversity, increases with richness (the number of vegetative layers) and

evenness (how evenly vegetative biomass is distributed among layers).

Robert MacArthur and John MacArthur (1961) measured foliage height diversity and bird species diversity in 13 plant communities in northeastern North America, Florida, and Panama. The vegetative communities included in their study ranged from grassland to mature deciduous forest, with foliage height diversity that ranged from 0.043 to 1.093. Plant communities with greater foliage height diversity supported more diverse bird communities (fig. 16.10). MacArthur and his colleagues went on to study the relationship between foliage height diversity and bird species diversity in a wide variety of temperate, tropical, and island settings, from North America to Australia. They again found a positive correlation between foliage height diversity and bird species diversity. The combined weight of the evidence from North and Central America and Australia suggests that the relationship is not one of chance, but instead reflects something about the way that birds in these environments subdivide space.

How is environmental complexity related to the diversity of other organisms besides birds? Ecological studies have shown positive relationships between environmental complexity and species diversity for many groups of organisms, including mammals, lizards, plankton, marine gastropods, and reef fish. Notice, however, that this list of organisms is dominated by animals. How does environmental complexity affect diversity of plants?

Niches, Heterogeneity, and the Diversity of Algae and Plants

The existence of approximately 300,000 species of terrestrial plants presents a multitude of opportunities for specialization by animals. Consequently, high plant diversity can explain much of

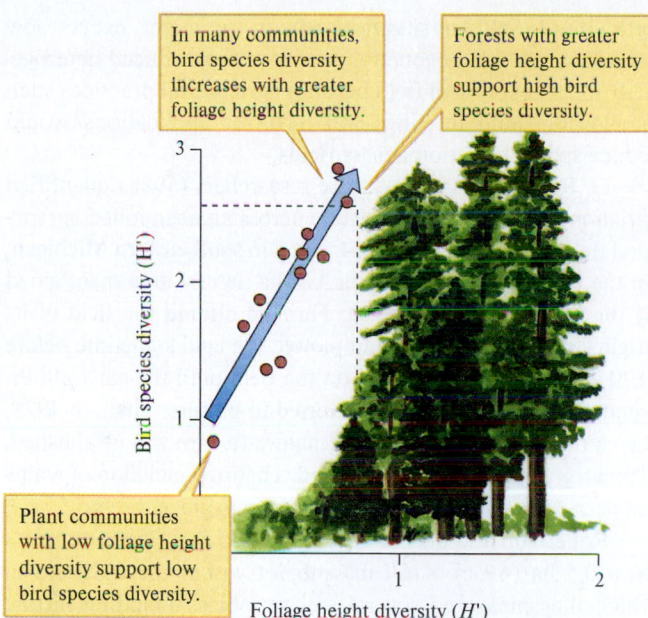

In many communities, bird species diversity increases with greater foliage height diversity.

Forests with greater foliage height diversity support high bird species diversity.

Plant communities with low foliage height diversity support low bird species diversity.

Figure 16.10   Foliage height diversity and bird species diversity (data from MacArthur and MacArthur 1961).

animal diversity. However, how do we explain the diversity of primary producers? G. Evelyn Hutchinson (1961) described what he called "the paradox of the plankton." He suggested that communities of phytoplankton present a paradox because they live in relatively simple environments (the open waters of lakes and oceans) and compete for the same nutrients (nitrogen, phosphorus, silica, etc.), yet many species can coexist without competitive exclusion. This situation seemed paradoxical because it appears to violate the competitive exclusion principle. The diversity of terrestrial plants presents a similar paradox. This paradox is sufficiently vexing that Joseph Connell (1978) proposed that environmental heterogeneity is not sufficient to account for terrestrial plant diversity, especially in tropical rain forests.

After some decades of theoretical and empirical work, however, it appears that environmental complexity can account for a significant portion of the diversity among both planktonic algae and terrestrial plants. As with animals, in order to study the influence of environment on diversity of plants and algae, we need to understand the nature of their niches.

The Niches of Algae and Terrestrial Plants

The niches of algae appear to be defined by their nutrient requirements. The importance of nutrient requirements to the niches of phytoplankton was demonstrated by David Tilman (1977). Tilman conducted experiments on competition between freshwater diatoms. His experiments were similar to those conducted by G. F. Gause (1934; also see chapter 13), on competition between paramecium. However, in addition to demonstrating competitive exclusion, Tilman's experiments also showed the conditions that allowed coexistence of diatom species. Exclusion or coexistence depended upon the ratio of two essential nutrients, silicate, SiO_4^{4-}, and phosphate, PO_4^{3-},

When Tilman grew the diatoms *Asterionella formosa* and *Cyclotella meneghiniana* by themselves, they established and maintained stable populations. However, when he grew them together, *Asterionella* sometimes excluded *Cyclotella,* and sometimes the two species coexisted. The outcome of Tilman's experiments depended on the ratio of silicate to phosphate (fig. 16.11). At high ratios *Asterionella* eventually excluded *Cyclotella,* but at lower ratios the two species coexisted. At the lowest ratio, *Cyclotella* was numerically dominant over *Asterionella.*

How can we explain Tilman's results? It turns out that *Asterionella* takes up phosphorus at a much higher rate than does *Cyclotella.* Tilman reasons that at high ratios of silicate to phosphate, *Asterionella* is able to deplete the environment of phosphorus and consequently eliminate *Cyclotella.* However, when ratios are low, silicate limits the growth rate of *Asterionella* and it cannot deplete phosphate. Consequently, when ratios are low, *Asterionella* cannot exclude *Cyclotella.* At these low ratios, silicate limits the growth rate of *Asterionella,* while phosphate limits the growth rate of *Cyclotella.* Therefore, in the presence of low ratios of silicate to phosphate, the two diatoms coexist.

What do the results of Tilman's experiments have to do with the relationship of environmental complexity to species diversity? The implication is that if the ratio of silicate to phosphate varies across a lake, then *Asterionella* will dominate some areas, while elsewhere *Cyclotella* will dominate.

Now, how might we characterize the niches of terrestrial plants? A. Tansley's experiments (1917) on competition between *Galium* species, which we discussed in chapter 13, provide insights into the niches of terrestrial plants. You may recall that Tansley studied two species: *G. saxatile,* which grows mainly on acidic soils, and *G. sylvestre,* which grows mainly on basic soils. When these two plants competed against each other in an experimental garden, each did best on the soil type that it occupies in nature. Like the diatoms *Asterionella* and *Cyclotella,* the niches of *G. saxatile* and *G. sylvestre*

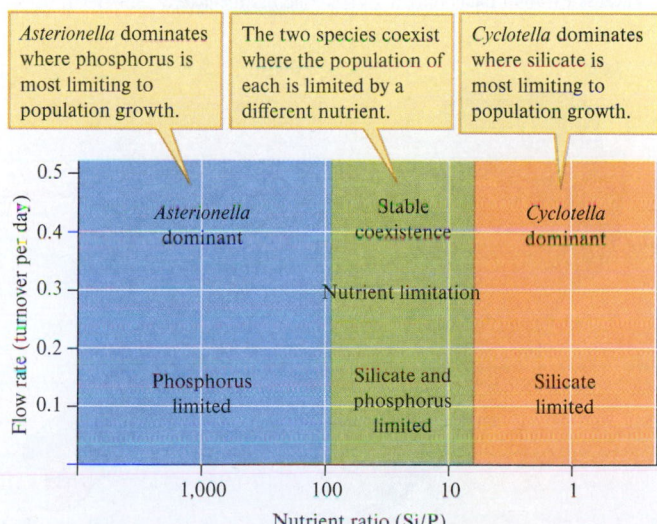

Asterionella dominates where phosphorus is most limiting to population growth.

The two species coexist where the population of each is limited by a different nutrient.

Cyclotella dominates where silicate is most limiting to population growth.

Figure 16.11   The ratio of silicate (SiO_4^{4-}) to phosphate (PO_4^{3-}) and competition between the diatoms *Asterionella formosa* and *Cyclotella meneghiniana* (data from Tilman 1977).

are significantly influenced by the chemical characteristics of the environment, in this case of the soil.

We can define the niches of algae and plants on the basis of their nutrient requirements and responses to constraining physical or chemical conditions, such as moisture and pH. Therefore, from the perspective of plants and algae, variation in the availability of limiting nutrients, such as silicate and phosphate, and variation in physical and chemical conditions, such as temperature, moisture, and pH, contribute to environmental complexity.

Complexity in Plant Environments

Let's look first at environmental heterogeneity in an aquatic environment. Martin Lebo and his colleagues (1993) studied spatial variation in nutrient and particulate concentrations in Pyramid Lake, Nevada, which has a surface area of approximately 450 km^2 and a maximum depth of 102 m.

Pyramid Lake, like other lakes, is not a uniform chemical solution. All of the nutrients studied by the researchers showed substantial variation across the lake. Figure 16.12 shows that nitrate (NO_3^-) ranged from >20 μg per liter (L^{-1}) near the inflow of the Truckee River to <5 μgL^{-1} along the western and northeastern shores. Silica (SiO_2) reached maximum concentrations of >300 μgL^{-1} at the inflow of the Truckee River and then decreased progressively northward, reaching a minimum of <200 μgL^{-1} in the north-central portion of the lake. Other nutrients also showed substantial variation across Pyramid Lake, but their pattern of variation differed from that shown by nitrate and silica. In other words, different parts of the lake offer distinctive growing conditions for phytoplankton. This environmental complexity should allow for phytoplankton diversity.

Now, let's look at variation in nutrient concentrations in a terrestrial environment. Our example concerns an abandoned

agricultural field, a situation where we might expect low environmental heterogeneity. We can expect reduced heterogeneity in an abandoned field because agricultural practices such as plowing, land leveling, and fertilizer applications would reduce spatial variation across fields.

G. Robertson and a team of researchers (1988) quantified variation in nitrogen and moisture across an abandoned agricultural field. Their study site was located in southeastern Michigan, on the E. S. George Reserve, a 490 ha natural area maintained by the University of Michigan. Farmers cleared the field of its original oak-hickory forest and plowed the land sometime before 1870. Crop raising continued on the field until the early 1900s, when most of the land was converted to pasture. Then, in 1928, the cattle were removed and the nature reserve was established. Though grazing by cattle has ceased, a dense population of white-tail deer, *Odocoileus virginianus,* continue to graze the site.

Robertson and his colleagues focused their measurements on a 0.5 ha (69 m × 69 m) subplot within the old field in which they measured several soil variables, including nitrate concentration and soil moisture, at 301 sampling points. This large number of sampling points over a small area provided sufficient data to construct a detailed map of soil properties. Figure 16.13 shows considerable patchiness in both nitrate and moisture. Both variables show at least tenfold differences across the study plot. In addition, nitrate concentration and moisture don't appear to correlate well with each other; hot spots for nitrates were not necessarily hot spots for moisture. The researchers concluded that soil conditions show sufficient spatial variability to affect the structure of plant communities.

We can see from these studies that algal and plant resources change substantially across aquatic and terrestrial environments. Now let's examine how spatial heterogeneity in these resources may affect the distribution and diversity of plants.

Soil and Topographic Heterogeneity

Carl Jordan (1985) studied the relationship between vegetation and soils in the Amazon forest. His studies led him to conclude that tropical forest diversity is organized in two ways: (1) a large number of *species* live within most tropical forest communities and (2) there are a large number of plant *communities* in a given area, with each community being distinctive in regard to species composition.

Jordan's studies showed that variation in soil characteristics influences the number of plant communities in an area. He found that slight differences in soil properties foster entirely different plant communities. Figure 16.14 shows six different plant communities that Jordan observed in a distance of just 500 m and an elevation range of less than 8 m. In the study area, the subsoil was clay weathered from a granitic bedrock. Sand, deposited on top of the clay, varied in thickness depending on local topography. Topography also determined the depth of the soil above groundwater.

Diverse mixed forests occurred on the tops of hills, where clay was close to the surface. Where the topography dipped toward streambeds and the thickness of the sand layer

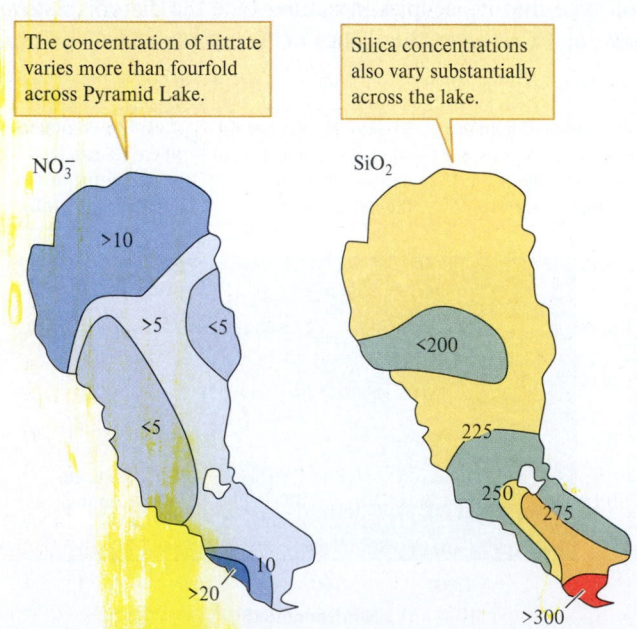

Figure 16.12 Concentrations (μg/L) of nitrate (NO_3^-) and silica (SiO_2) in the surface waters of Pyramid Lake, Nevada (data from Lebo et al. 1993).

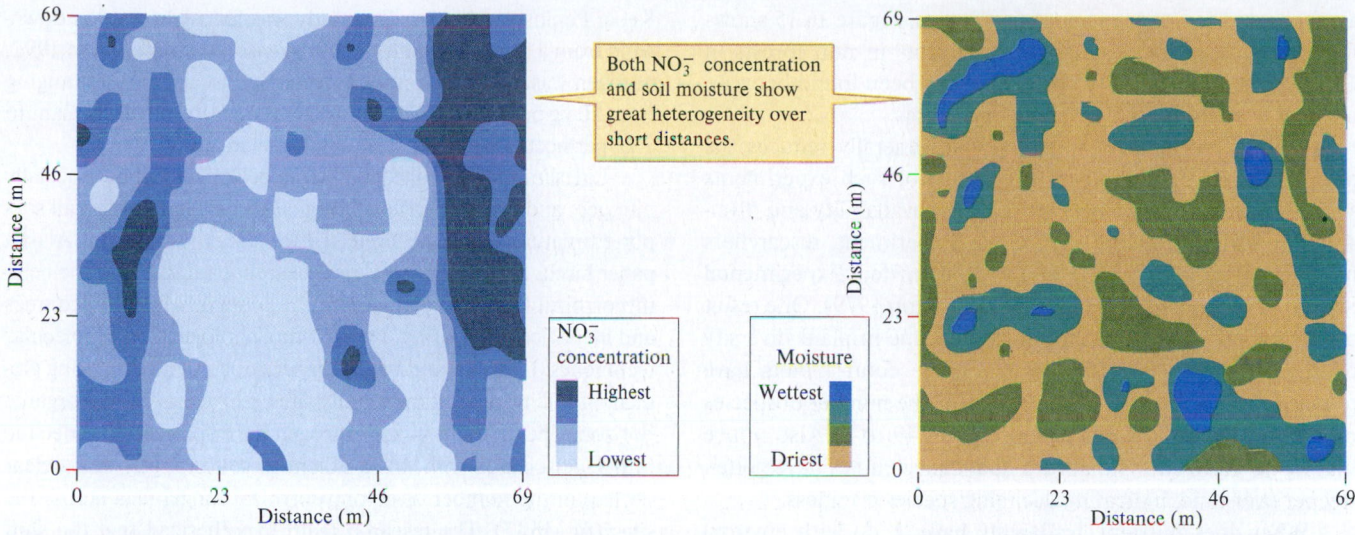

Figure 16.13   Variation in nitrate (NO_3^-) and soil moisture in a 4,761 m² area in an old agricultural field (data from Robertson et al. 1988).

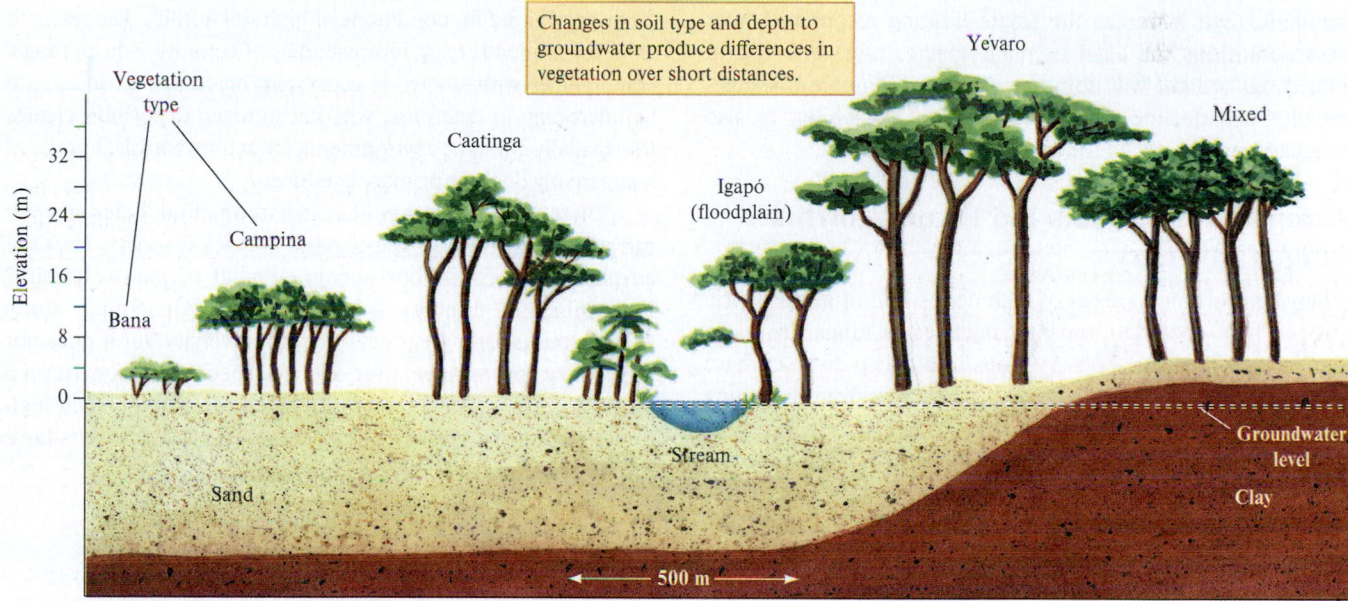

Figure 16.14   Variation in vegetation along a gradient of soil and moisture conditions (data from Jordan 1985).

increased, Jordan found a forest dominated by a tree in the legume family called yévaro, *Eperua purpurea*. Though the yévaro community was less diverse than the mixed forest, it was taller and supported higher plant growth rates. In addition, a distinctive forest called igapó grew along the edges of the streams in areas that flooded seasonally.

Away from streams on sandy soil, small changes in elevation produced substantial changes in water availability and plant communities. Near streams but above the level of seasonal flooding, there was another plant community known as caatinga. Still on sand but 1 to 2 m above the caatinga there was a low-stature forest known as campina. Finally, at elevations greater than 2 m above stream level, where water drains through the coarse sandy soil fast enough to induce water stress, Jordan found a shrub community known as bana.

He observed that while their local names vary, similar plant communities, associated with local variations in soil quality and topography, occur throughout the Amazon basin. Analogous variation in plant communities in response to differences in soil properties have also been observed in temperate regions.

Nutrient Enrichment Can Reduce Environmental Complexity

Ecologists have repeatedly observed a negative relationship between nutrient availability and algal and plant species diversity. In other words, as nutrient supplies increase, diversity of plants and algae declines. Michael Huston (1980, 1994a) reported a negative correlation between nutrient availability and plant species diversity in Costa Rican forests, a relationship also reported for African and Asian forests. Figure 16.15 shows this relationship for a series of study plots in rain forests in Ghana. A similar negative correlation has been found between nutrient availability and diversity of diatoms.

Adding nutrients to water or soils generally reduces the diversity of plants and algae. The results of such experiments suggest a causal linkage between nutrient availability and diversity. For instance, in the Park Grass Experiment, researchers have fertilized a grassland at the Rothamsted Experimental Station in Great Britain since 1856 (Kempton 1979). One result of that experiment has been a steady decline in plant diversity on the fertilized plots (fig. 16.16). While control plots have retained their diversity for over 100 years, the number of species on the fertilized plots has declined from 49 to 3. Also notice that figure 16.16 shows that rank-abundance curves have gotten steeper over time, indicating declining species evenness.

What does nutrient availability have to do with environmental complexity? Increasing nutrient availability whether due to natural variation or fertilization, reduces the number of limiting nutrients. Eventually, when or where all nutrients are abundant, light becomes the single limiting resource. Under these conditions the algal or plant species most effective at competing for light will dominate the community and species diversity will decline. Increased nutrient availability is also associated with reduced fungal diversity.

Nitrogen Enrichment and Ectomycorrhizal Fungus Diversity

Ecologists working in areas of high deposition of atmospheric nitrogen have recorded apparent declines in fungal diversity. However, most of these observations of reduced diversity have been based on declines in diversity of aboveground fruiting bodies, such as mushrooms. Such evidence may not reflect declines in fungal diversity but rather shifts from aboveground to belowground growth. To address this possibility, Erik Lilleskov, Timothy Fahey, Thomas Horton, and Gary Lovett (2002) studied the diversity and composition of ectomycorrhizal fungi along a gradient of nitrogen deposition on the

Kenai Peninsula, Alaska. The study was focused on areas downwind from a fertilizer plant that emits gaseous ammonia. In 1992, nitrogen was deposited on the forested landscape at rates ranging from 20 kg per hectare per year in areas near the fertilizer plant to 1 kg per hectare per year several kilometers away.

Lilleskov and his colleagues sampled soil nutrients, especially nitrogen, and ectomycorrhizal fungi at five sites. The sites all supported mature stands of white spruce, *Picea glauca,* and Alaska paper birch, *Betula kenaica.* The research team sampled the ectomycorrhizal fungi associated with the roots of white spruce trees and identified them using a mix of morphological and molecular techniques. Lilleskov and his colleagues documented a strong gradient in soil nitrogen at their study sites, particularly in the organic horizon. The gradient in soil nitrogen corresponded to a decline in pH. Associated with this gradient in soil nitrogen was a clear decline in the number of ectomycorrhizal fungal taxa across the sites (fig. 16.17). The research team hypothesized that the shift in diversity and composition was the result of a change from species specialized for efficient uptake of nitrogen, under conditions of low availability, to dominance by acid-tolerant ectomycorrhizal fungi, specialized for conditions of high soil fertility. The research team recommends tying future studies of ectomycorrhizal fungal communities with studies of ecosystem processes. It would also be interesting to determine whether nitrogen deposition creates less spatially complex environments for ectomycorrhizal fungi, as it apparently does for primary producers.

Therefore, it appears that environmental heterogeneity can account for a portion of plant species diversity. Can the environmental conditions account for all of plant diversity? Environmental diversity across Jordan's Amazonian study sites accounts for a great deal of plant diversity but it does not tell us how, for instance, over 300 tree species can coexist on a single hectare of Amazonian rain forest. To explain such high diversity within relatively homogeneous areas, ecologists have turned to the influences of disturbance.

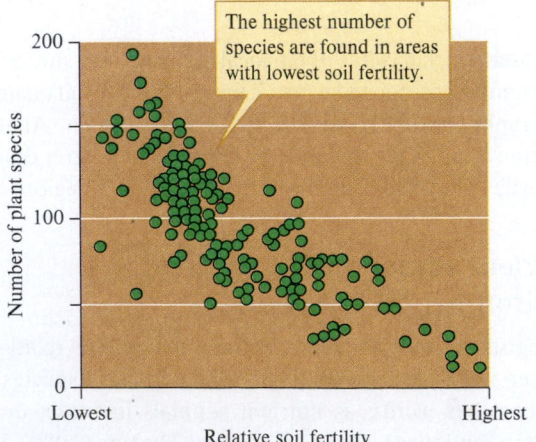

Figure 16.15   Soil fertility and the number of plant species in 0.1 ha plots of rain forest in Ghana, Africa (data from Hall and Swaine 1976).

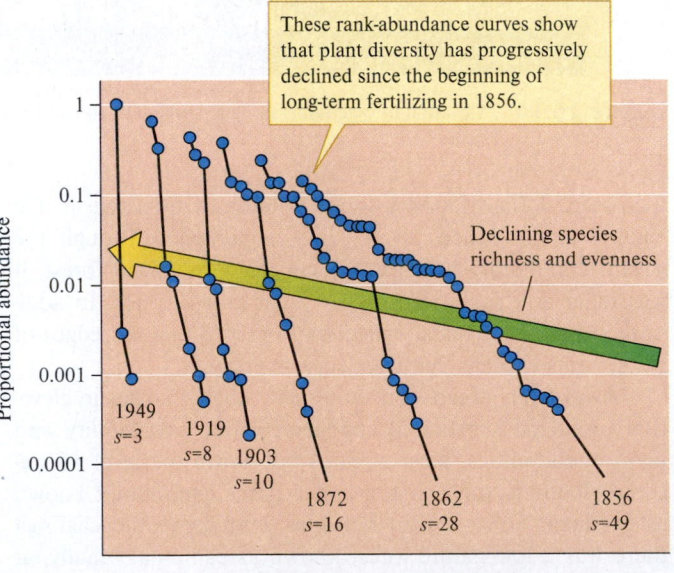

Figure 16.16   Fertilization and plant diversity at Rothamsted, England (data from Kempton 1979, after Brenchley 1958).

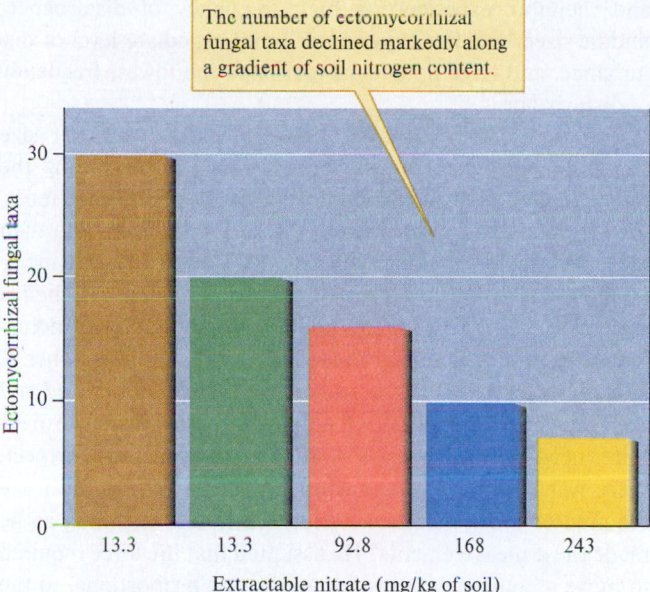

The number of ectomycorrhizal fungal taxa declined markedly along a gradient of soil nitrogen content.

Figure 16.17   Relationship between soil nitrogen (KCl-extractable) and ectomycorrhizal fungal community diversity near the Kenai Peninsula, Alaska (data from Lilleskov et al. 2002).

Concept 16.3 Review

1. Does Tilman's finding that *Asterionella* and *Cyclotella* exclude each other under certain conditions but coexist under other conditions violate the competitive exclusion principle (see chapter 13, section 13.2)?
2. Can we link increased nutrient availability during the Park Grass Experiment with decreased environmental complexity?
3. Suppose you discover that the fish species inhabiting small, isolated patches of coral reef use different vertical zones on the reef face—some species live down near the sand, some live a bit higher on the reef, and some higher still. Based on this pattern of zonation, can you predict how reef structure should affect the diversity of fish living on such reefs?

16.4 Disturbance and Diversity

LEARNING OUTCOMES

After studying this section you should be able to do the following:

16.13 Define disturbance.
16.14 Outline the intermediate disturbance hypothesis.
16.15 Discuss evidence in support of the intermediate disturbance hypothesis.

Intermediate levels of disturbance promote higher diversity. For several chapters, we have assumed that environmental conditions remain more or less stable. Ecologists refer to this state as one of **equilibrium.** In an equilibrial system, stability

is maintained by opposing forces. The Lotka-Volterra competition models (see chapter 13) and predator-prey models (see chapter 14) assumed a constant physical environment. In laboratory studies of competition, researchers have generally maintained constant environmental conditions. Even earlier in this chapter when we discussed the influences of environmental complexity on species diversity, there was an underlying assumption of a stable environmental equilibrium. However, most natural environments are subject to various forms of disturbance.

The Nature and Sources of Disturbance

What is disturbance? The answer to this question is not as simple as it may seem. What constitutes disturbance varies from one organism to another and from one environment to another. A disturbance for one organism may have little or no impact on another, and the nature of disturbance may be quite different in different environments. Wayne Sousa (1984), who examined the role of disturbance in structuring natural communities, defined disturbance as "a discrete, punctuated killing, displacement, or damaging of one or more individuals (or colonies) that directly or indirectly creates an opportunity for new individuals (or colonies) to become established." P. S. White and S. Pickett (1985) defined disturbance as "any relatively discrete event in time that disrupts ecosystem, community or population structure and changes resources, substrate availability, or the physical environment." They also caution, however, that we must be mindful of spatial and temporal scale. For instance, disturbance to bryophyte (mosses and liverworts) communities growing on boulders along the margin of a stream can occur at spatial scales of fractions of meters and annual temporal scales that are irrelevant to the surrounding forest community.

There are innumerable potential sources of disturbance to communities. White and Pickett listed 26 major sources of disturbance, roughly divided into abiotic forces, such as fire, hurricanes, ice storms, and flash floods; biotic factors, such as disease and predation; and human-caused disturbance. Regardless of the source, we can classify disturbances by a smaller set of characteristics. We will focus our discussion of disturbance on two characteristics: frequency and intensity.

The Intermediate Disturbance Hypothesis

Joseph Connell (1975, 1978) proposed that disturbance is a prevalent feature of nature that significantly influences the diversity of communities. He questioned the assumption of equilibrial conditions made by most competition-based models of diversity. Instead, he proposed that high diversity is a consequence of continually changing conditions, not of competitive accommodation at equilibrium, and proposed the **intermediate disturbance hypothesis,** which predicts that intermediate levels of disturbance promote higher levels of diversity (fig. 16.18).

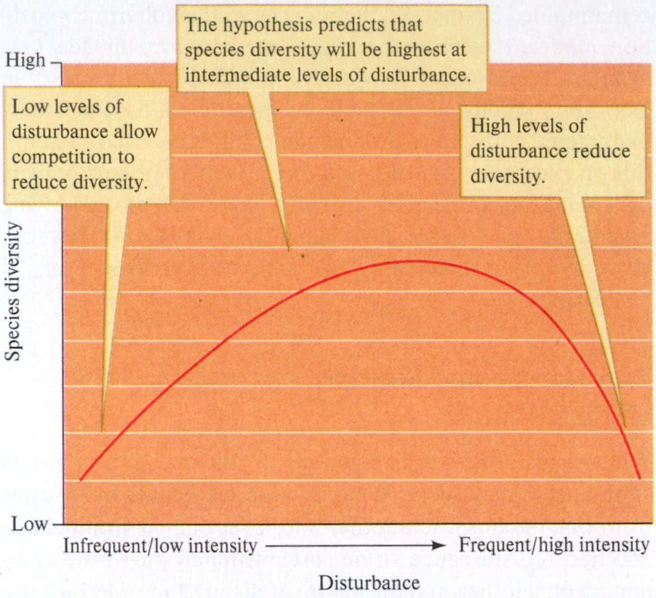

Figure 16.18 The intermediate disturbance hypothesis (data from Connell 1978).

Connell suggested that both high and low levels of disturbance would lead to reduced diversity. He reasoned that if disturbance is frequent and intense, the community will consist of those few species able to colonize and complete their life cycles between the frequent disturbances. He also predicted that diversity will decline if disturbances are infrequent and of low intensity. In the absence of significant disturbance, the community is eventually limited to the species that are the most effective competitors, effective because they are either the most efficient at using limited resources or the most effective at interference competition.

How can intermediate levels of disturbance promote higher diversity? Connell suggested that at intermediate levels of disturbance, there is sufficient time between disturbances for a wide variety of species to colonize but not enough time to allow competitive exclusion.

Disturbance and Diversity in the Intertidal Zone

Waynė Sousa (1979a) studied the effects of disturbance on the diversity of marine algae and invertebrates growing on boulders in the intertidal zone. Disturbance to this community comes mainly from ocean waves generated by winter storms. These waves, which can exceed 2.5 m in height, are large enough to overturn intertidal boulders, killing the algae and barnacles growing on their upper surfaces. Meanwhile, the newly exposed underside of the boulder is available for colonization by algae and marine invertebrates.

Because boulders of different sizes turn over at different frequencies and in response to waves of different heights, Sousa predicted that the level of disturbance experienced by the community living on boulder surfaces depends on boulder size. Smaller boulders are turned over more frequently

and therefore experience a high frequency of disturbance, middle-sized boulders experience an intermediate level of disturbance, and large boulders experience the lowest frequency of disturbance.

Sousa quantified the relationship between boulder size and probability of being moved by waves by measuring the force required to dislodge boulders of different sizes. He measured the exposed surface area of a series of boulders and then measured the force required to dislodge each. To make a measurement he wrapped a chain around a boulder, attached a spring scale to the chain, and pulled in the direction of incoming waves until the boulder moved. He recorded the number of kilograms registered on the scale when the boulder moved and then converted his measurements to force expressed in newtons (newtons [N] = kg × 9.80665). As you might expect, there was a positive relationship between size and the force required to move boulders. What was Sousa assuming as he made these measurements? He assumed that the force required to move a boulder with his apparatus was proportional to the force required for waves to move it.

Sousa verified this assumption by documenting the relationship between his force measurements and movement by waves. He established six permanent study sites and measured the force required to move the boulders in each. He next mapped the locations of boulders by photographing the study plots and then checked for boulder movements by taking additional photographs monthly for 2 years. Sousa divided the boulders in the study sites into three classes based on the force required for movement: (1) ≤49 N, (2) 50 to 294 N, and (3) >294 N. These classes translated into percentages of boulders moved per month (42% per month = frequent disturbance), intermediate movement (9% per month = intermediate disturbance), and infrequent movement (1% per month = infrequent disturbance).

The number of species living on boulders varied with frequency of disturbance (fig. 16.19). Most of the frequently disturbed boulders supported a single species, few supported five species, and none supported six or seven species. Most of the boulders experiencing a low frequency of disturbance supported one to three species, few supported six species, and none supported seven. The boulders supporting the greatest diversity of species were those subject to intermediate levels of disturbance. Most of these supported three to five species, many supported six species, and some supported seven.

Disturbance and Diversity in Temperate Grasslands

Can several species coexist where there is a single limiting resource? The Lotka-Volterra competition equations (see chapter 13) predict that under such circumstances, one species will eventually exclude all others. However, as Sousa's work shows, even where species compete for a single resource, such as space in the intertidal zone, several species may coexist if disturbance prevents competitive exclusion. David Tilman (1994) reached a similar conclusion in regard to plant diversity within North American prairies.

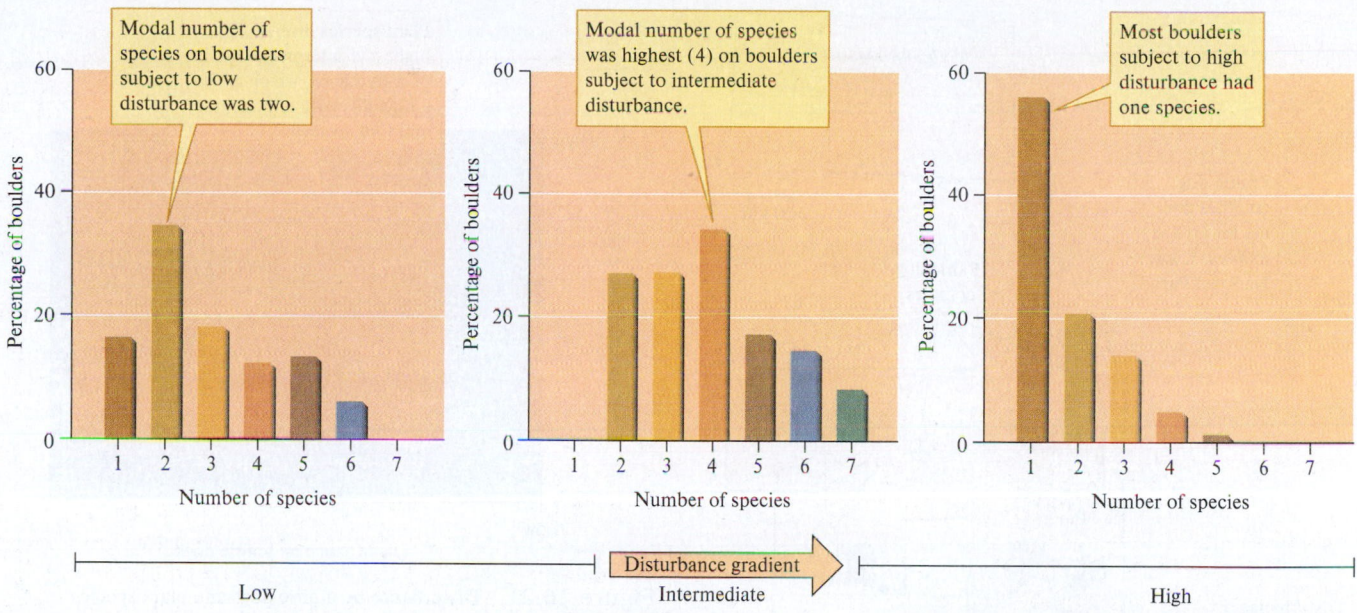

Figure 16.19   Levels of disturbance and diversity of marine algae and invertebrates on intertidal boulders (data from Sousa 1979a).

What sorts of disturbance have been important in grasslands? Historically, the magnitude of disturbance on the North American prairie ranged from trampling by bison herds and fire to the death of an individual plant. One of the most important and ubiquitous sources of disturbance to grasslands is burrowing by mammals.

April Whicker and James Detling (1988) proposed that prairie dogs (*Cynomys* spp.), which occupied about 40 million ha of North American grasslands as late as 1919, were an important source of disturbance on the North American prairies. Prairie dogs are herbivorous rodents that weigh approximately 1 kg as adults and live in colonies containing 10 to 55 individuals per hectare. Prairie dogs build extensive burrow systems that are 1 to 3 m deep and about 15 m long, with tunnel diameters of 10 to 13 cm and two entrances. To build a burrow with these dimensions, a prairie dog must excavate 200 to 225 kg of soil, which it deposits in mounds 1 to 2 m in diameter around burrow entrances.

Burrowing and grazing by prairie dogs have substantial effects on the structure of plant communities at several spatial scales. Figure 16.20 shows the areas of Wind Cave National Park in South Dakota occupied by prairie dogs. Because of the activities of these rodents, each of these areas supports vegetative communities distinctive from the surrounding landscape. Within a colony, prairie dog activities create patchiness on a smaller scale, with areas of forbs and shrubs, grass and forbs, and grass within the surrounding matrix of prairie grassland. Whicker and Detling estimate that plant species diversity is greatest in areas experiencing intermediate levels of disturbance by prairie dogs (fig. 16.21).

How does disturbance by prairie dogs foster higher diversity? The mechanisms underlying this effect are essentially the same as those operating in the intertidal boulder field studied by Sousa. By burrowing and piling earth and by grazing and clipping vegetation, prairie dogs remove vegetation from areas around their burrows. These bare patches are then open for colonization by plants. However, some plant species are more likely to colonize these open patches. Those species investing most heavily in dispersal are usually the first to arrive. But these early colonists can be displaced by better competitors that arrive later. The persistence of both good colonizers and good competitors in a plant community depends on intermediate levels of disturbance. Too much disturbance and the community is dominated by the good colonizers; too little disturbance and the better competitors dominate.

Because prairie dogs have been considered an agricultural pest, various control programs have reduced their populations dramatically across the west. The extermination also eliminated their dynamic influences on plant communities. However, other burrowing mammals remain in large numbers. One of the most important of these are the pocket gophers of the family Geomyidae. Though pocket gophers are much smaller than prairie dogs, weighing from 60 to 900 g, their effects on grassland and arid land communities are considerable. The mounds that gophers create during their burrowing may cover as much as 25% to 30% of the ground surface, which increases heterogeneity in light availability and soil nitrogen, which in turn fosters increased plant species richness.

The influences of prairie dogs and pocket gophers on plant communities are a consequence of a combination of physical disturbances due to burrowing and feeding, since both are herbivores. How might disturbance by humans affect the diversity of plant and animal communities? We are all well acquainted with examples of how severe disturbance by humans reduces biological diversity. In the following section, we consider the effects of moderate disturbance by humans on diversity.

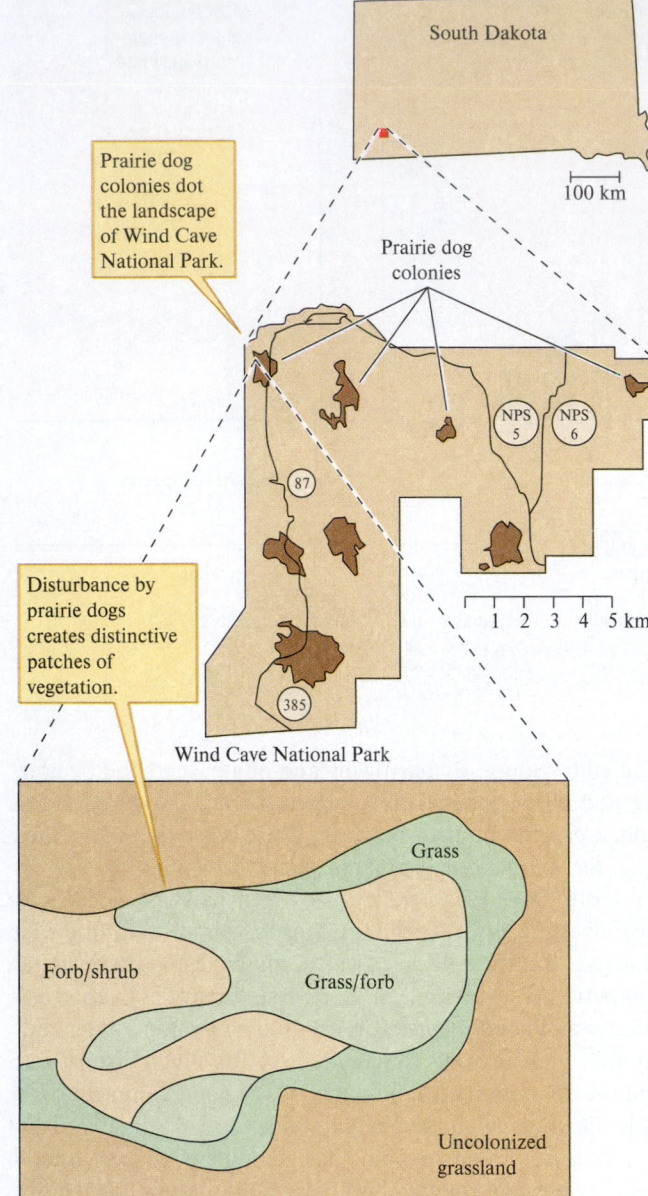

Figure 16.20   Disturbance by prairie dogs and patchiness of vegetation (data from Coppock et al. 1983; Whicker and Detling 1988).

Concept 16.4 Review

1. Could protecting forests that once burned with regular frequency, due to lightning strikes, lead to reduced plant diversity within a mountain forest landscape of 25 km² (~10 mi²)?

2. If disturbance can foster higher species diversity, why is human disturbance often (though not always) associated with reduced species diversity?

3. According to the intermediate disturbance hypothesis, could human disturbance sustain higher levels of species diversity than in the absence of human disturbance?

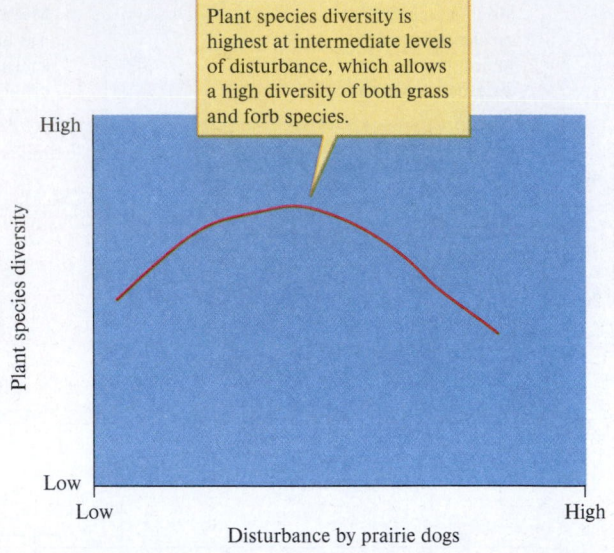

Figure 16.21   Disturbance by prairie dogs and plant species diversity (data from Whicker and Detling 1988).

Applications

Disturbance by Humans

LEARNING OUTCOMES

After studying this section you should be able to do the following:

16.16 Explain how the effects of disturbance by humans on diversity are consistent with the predictions of the intermediate disturbance hypothesis.

16.17 Describe the effects of urbanization on the diversity and composition of bird communities.

The effects of disturbance by humans are all around us. Housing developments cover the countryside as human populations continue their rapid growth. Deforestation continues at an alarming rate in both temperate and tropical regions. Industries pollute air and water. While the destructive effects of disturbance by humans may seem outside the realm of what we've discussed so far, they are not. The destruction of natural communities that we often associate with humans is a consequence of the extreme levels of disturbance of which our species is capable (fig. 16.22). These are the high levels of disturbance that the intermediate disturbance hypothesis (see fig. 16.18) predicts will lead to reduced diversity. Not surprisingly, human disturbance threatens thousands of species with extinction. The International Union for Conservation of Nature lists habitat destruction by humans as the most serious threat to endangered species worldwide (IUCN 2020). Environments disturbed by human activity are not, however, devoid of life and some, such as the chalk grasslands of Europe, support higher diversity under traditional management (e.g., Bobbink and Willems 1987, 1991). As we shall now see, even urban environments can support surprising levels of diversity.

Figure 16.22   Human populations and activity are a major source of disturbance. For example, (*a*) housing developments generally simplify natural ecosystems and feed demand for natural resources such as (*b*) the wood harvested from this clear-cut forest and (*c*) the coal from this mountaintop that is mined to supply electrical power to human populations. The growing energy demands of our population and our reliance on fossil fuels are driving global warming, disturbing habitats far beyond human population centers, for example, (*d*) by melting Arctic sea ice, which is prime hunting habitat for polar bears, *Ursus maritimus.*   (*a*) Photo by Lynn Betts, USDA Natural Resources Conservation Service; (*b*) Spaces Images/Blend Images LLC; (*c*) Stephen Reynolds; (*d*) Digital Vision/Getty Images

Urban Diversity

One of the developing ecological frontiers, urban ecology, can be found where the majority of people around the world live: in urban environments. Moreover, the proportion of people who are city dwellers is growing rapidly. In developed countries, the percentage of people living in cities stands at about 80%, a proportion that will be approached in developing countries over the next 50 years (Grimm et al. 2008). It is essential then that we better understand the ecology of cities, including their biological diversity and their functioning as ecosystems (see chapter 19, section 19.4).

While urbanization is considered a major threat to biodiversity, Stuart Pickett, director of the Baltimore Ecosystem Study Long Term Ecological Research (LTER) Program, points out that ongoing studies are revealing that the urban biota is diverse, often much more so than commonly thought (Pickett et al. 2008). Studies of urban diversity have included plants, land snails, beetles, butterflies, spiders, fish, aquatic invertebrates, mammals, and one of the most studied groups of organisms: birds.

Urbanization has been shown to reduce both bird species richness and evenness in Phoenix, Arizona, and in Baltimore, Maryland (Shochat et al. 2010), an effect observed in many other urban settings. It turns out, however, that landscapes exposed to some degree of human land use can support substantial levels of bird diversity. In a study of bird species richness along gradients of land use intensity, Robert Blair (2004) recorded the highest number of bird species in areas of intermediate use in both northern California and southwestern Ohio. Figure 16.23*a* shows that bird species richness peaked in the middle of gradients of land use intensity, peaking in both study areas in golf courses. One of the drivers of change in bird species richness along these gradients is habitat modification. Habitat changes from the nature preserve to the business district along these gradients included decreases in tree and shrub cover and increases in coverage by pavement and buildings. As might be expected, grassland and lawn coverage were highest on golf courses. As we saw in chapter 5, these urban centers would also have higher average temperatures (section 5.5), another facet of environment change associated with urbanization.

However, patterns of species richness do not tell the whole story. The composition of the bird communities also changed markedly. Most significantly, the richness of native woodland birds decreased as land use intensity increased in both California and Ohio (fig. 16.23*b*). As bird species associated

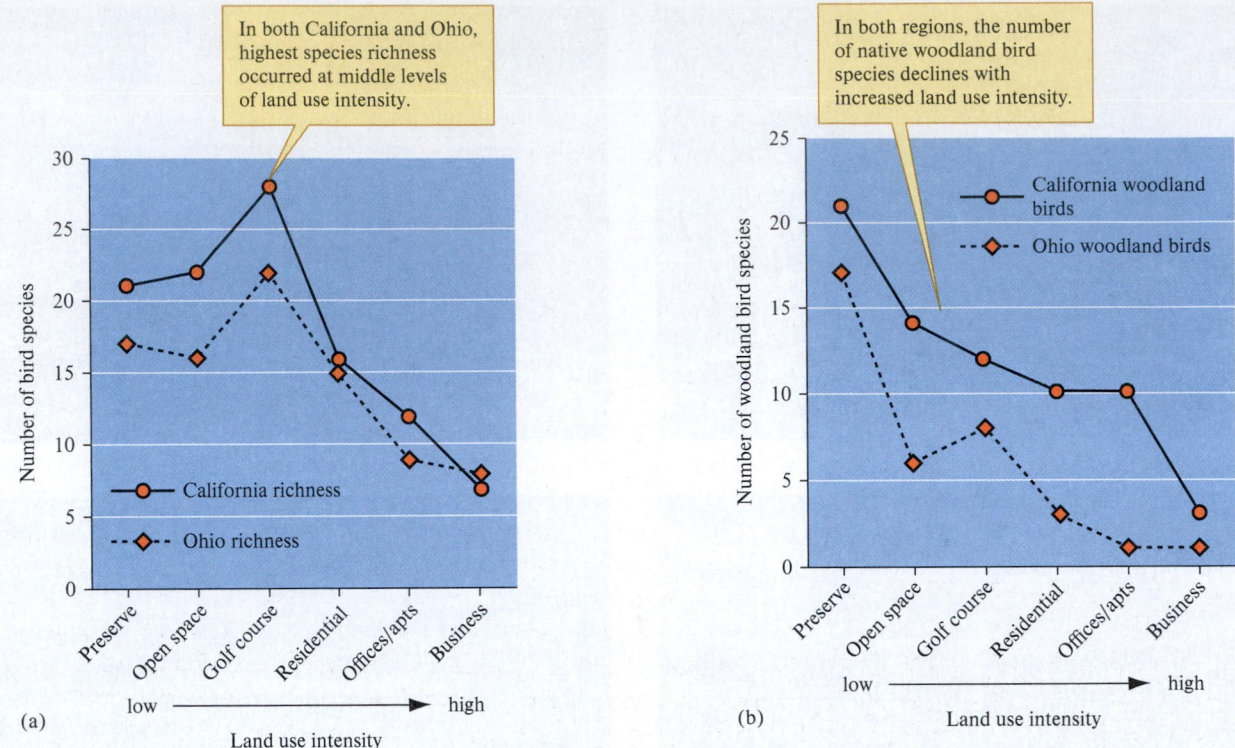

Figure 16.23 Changes in bird communities along two gradients of land use intensity. (a) Bird species richness was higher at sites of intermediate disturbance (golf courses) than at sites of lower disturbance (nature preserves and open space) or at sites of higher disturbance (residential, office/apartments, and business neighborhoods). (b) Reduced presence of native woodland bird species was observed at all levels of land use intensity (data from Blair 2004).

with natural woodland vegetation declined in representation along the gradients, they were replaced by other, generally more widespread species, such as nonnative house sparrows, *Passer domesticus,* and starlings, *Sturnus vulgaris.* As a result, the similarity between the bird communities at the California and Ohio study sites increased toward the urban core of the study gradients, where these widespread, generalist species were most common (fig. 16.24). This pattern gives some insight into the basis for concerns that an increasingly urbanized world will contribute to a homogenized global biota—that is, urbanization will reduce regional differences in biodiversity (McKinney 2002; Blair 2004). Pickett and his colleagues (2008) counter this concern by pointing out that the diversity of other groups of organisms—for example, ground beetles and plants—does not seem to be as adversely affected by urbanization. This is a rapidly growing area of ecological study; a search of peer-reviewed literature reveals that nearly 5,000 papers that mention urban ecology were published in 2020 alone. Growth in studies of urban ecology will likely soon produce the information needed to address these differences in perspectives and, perhaps, remedy remaining concerns about the ecological consequences of urbanization.

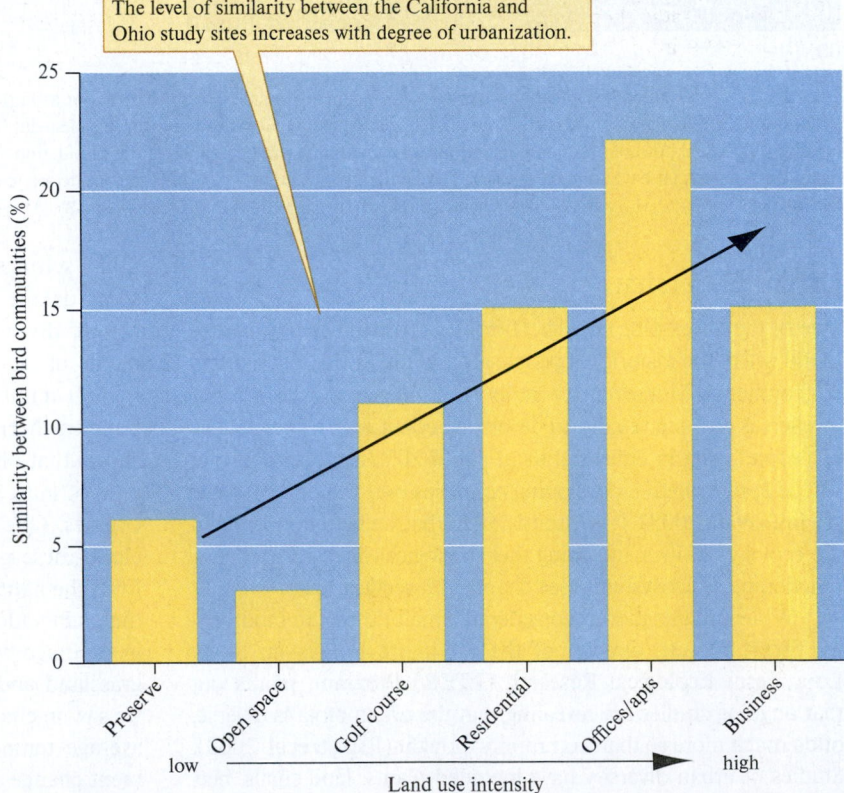

Figure 16.24 Percentage of similarity between bird communities along gradients of land use intensity in Palo Alto, California, and in Oxford, Ohio. Because the most disturbed sites support more bird species with widespread distributions and fewer native local birds, the bird communities at the widely separated California and Ohio study areas are most similar at the most disturbed urban sites (data from Blair 2004).

Summary

A *community* is an association of interacting species inhabiting some defined area. Examples of communities include the community of plants and animals on a mountainside and the invertebrate animals and algae living in an intertidal area. Community ecologists often restrict their studies to groups of species that all make their living in a similar way. Animal ecologists call such groups *guilds,* while plant ecologists use the term *life-form.* The field of community ecology concerns how the environment influences *community structure,* including the relative abundance and diversity of species, the subjects of chapter 16.

Most species are moderately abundant; few are very abundant or extremely rare. Frank Preston (1948) graphed the abundance of species in collections as distributions of species abundance, with each abundance interval twice the preceding one. Preston's graphs were approximately "bell-shaped" curves and are called "lognormal" distributions. Extensive analyses of ecological communities have now shown that the lognormal distribution is one of several distributions that give a reasonable match to the relative abundance of species. Regardless of distribution, however, the general pattern of species abundance encountered is one in which most species in communities are moderately abundant.

A combination of the number of species and their relative abundance defines species diversity. Two major factors define the diversity of a community: (1) the number of species in the community, which ecologists usually call *species richness,* and (2) the relative abundance of species, or *species evenness.* One of the most commonly applied indices of species diversity is the Shannon-Wiener index:

$$H' = -\sum_{i=1}^{s} p_i \log_e p_i$$

The relative abundance and diversity of species can also be portrayed using *rank-abundance curves.* Accurate estimates of species richness require carefully designed sampling programs.

Species diversity is higher in complex environments. Robert MacArthur (1958) discovered that five coexisting warbler species feed in different layers of forest vegetation and that the number of warbler species in North American forests increases with increasing forest stature. Various investigators have found that the diversity of forest birds increases with increased foliage height diversity. The niches of algae can be defined by their nutrient requirements. Heterogeneity in physical and chemical conditions across aquatic and terrestrial environments can account for a significant portion of the diversity among planktonic algae and terrestrial plants. Soil characteristics and depth to groundwater strongly influence the nature of local plant communities as shown in the Amazon River basin. Increased nutrient availability can reduce environmental complexity and lower algal and plant diversity.

Intermediate levels of disturbance promote higher diversity. Joseph Connell (1975, 1978) proposed that high diversity is a consequence of continually changing conditions, not of competitive accommodation at equilibrium. He proposed the *intermediate disturbance hypothesis,* which predicts that intermediate levels of disturbance foster higher levels of diversity. At intermediate levels of disturbance, a wide array of species can colonize open habitats, but there is not enough time for the most effective competitors to exclude the other species. Wayne Sousa (1979a), who studied the effects of disturbance on the diversity of sessile marine algae and invertebrates growing on intertidal boulders, found support for the intermediate disturbance hypothesis. Diversity in prairie vegetation also appears to be higher in areas receiving intermediate levels of disturbance. The effect of disturbance on diversity appears to depend on a trade-off between dispersal and competitive abilities.

The effects of human disturbance fall within the framework of the intermediate disturbance hypothesis. While less intensely used parts of the urban landscape can support high levels of bird diversity, intense urbanization reduces bird species richness and substantially alters bird species composition. For a number of other groups of organisms, however, ongoing studies are revealing a surprisingly diverse urban biota.

Key Terms

community 346
community structure 346
DNA barcoding 348
equilibrium 357

guild 346
intermediate disturbance hypothesis 357
life-form 346

lognormal distribution 347
rank-abundance curve 350
species diversity 348
species evenness 348

species rarefaction curve 348
species richness 348

Review Questions

1. What is the difference between a community and a population? What are some distinguishing properties of communities? What is a guild? Give examples. What is a plant life-form? Give examples.

2. Draw a "typical" lognormal distribution. Include properly labeled horizontal (x) and vertical (y) axes. You can use the lognormal distributions included in chapter 16 as models.

3. What are species richness and species evenness? How does each of these components of species diversity contribute to the value of the Shannon-Wiener diversity index (H')? How do species evenness and richness influence the form of rank-abundance curves?

4. Compare the "trophic" niches of warblers and diatoms as described by MacArthur (1958) and Tilman (1977). Why is it important that the ecologist be familiar with the niches of study organisms before exploring relationships between environmental complexity and species diversity?

5. Communities in different areas may be organized in different ways. For instance, C. Ralph (1985) found that in Patagonia in Argentina, as foliage height diversity increases, bird species diversity decreases. This result is exactly the opposite of the pattern observed by MacArthur (1958) and others reviewed in chapter 16. Design a study aimed at determining the environmental factors determining variation in bird species diversity across Ralph's Patagonian study sites.

6. According to the intermediate disturbance hypothesis, both low and high levels of disturbance can reduce species diversity. Explain possible mechanisms producing this relationship. Include trade-offs between competitive and dispersal abilities in your discussion.

7. The dams that have been built on many rivers often stabilize river flow by increasing flows below the dam during droughts and decreasing the amount of flooding during periods of high rainfall. Using the intermediate disturbance hypothesis, predict how stabilized flows would affect the diversity of river organisms below reservoirs.

8. Humans have been living in the tropical rain forests of the New World for at least 11,000 years. During this period, disturbance by humans has been a part of these tropical rain forests. Use the intermediate disturbance hypothesis to explain how recent disturbances threaten the biological diversity of these forests, while earlier disturbances apparently did not.

9. How would you design cities and their surrounding suburbs so that they have the potential to support higher levels of diversity than they do now?

heatherwest/RooM/Getty Images

Chapter

17

Species Interactions and Community Structure

The sea otter, *Enhydra lutris,* lives in nearshore environments from the Kuril Islands in the northwest Pacific Ocean to Alaska and Canada south to California. Despite living at relatively low population densities and being one of the smallest marine mammals, sea otters have great ecological influence on the communities in which they live. Because of that influence, which is exerted largely through their feeding on grazing benthic invertebrates, ecologists generally refer to sea otters as a *keystone species.*

Applications: Human Modification of Food Webs   380

Summary   381

Key Terms   382

Review Questions   382

CHAPTER CONCEPTS

17.1   A food web summarizes the feeding relations in a community.   367

  Concept 17.1 Review   368

17.2   Indirect interactions between species are fundamental to communities.   368

  Concept 17.2 Review   370

17.3   The feeding activities of a few keystone species may control the structure of communities.   371

  Concept 17.3 Review   377

17.4   Mutualists can act as keystone species.   378

  Concept 17.4 Review   379

LEARNING OUTCOMES

After studying this section you should be able to do the following:

17.1   Describe the major components of the Antarctic pelagic food web.

17.2   Interpret the feeding interactions indicated by a food web.

Some of the most easily documented examples of interactions within communities are feeding relationships. The feeding relationships in the ocean around Antarctica, one of the most productive marine environments on earth, provides a well-known example. Phytoplankton, especially diatoms, thrive in these frigid, windswept seas, where they are food for grazing zooplankton. One of the most important of these zooplankton is krill, shrimplike crustaceans. Krill are prey for a wide variety of larger plankton-feeding species, including crabeater seals, penguins, flying seabirds, and many species of fish and squid. The best known of the krill feeders are the baleen whales that once gathered in huge numbers to feed in Antarctic waters (fig. 17.1).

The krill-feeding fishes and squid are eaten by predaceous species, including emperor penguins, larger fish, and Weddell and

Figure 17.1   A food web in action: a feeding baleen whale.
John Tunney/Shutterstock.com

Ross seals. Leopard seals, a highly carnivorous species, feed on penguins and the smaller seals. Finally, the ultimate predators in this community are the killer whales, which eat seals, including leopard seals, and even attack and consume baleen whales. Huge populations of organisms live in the oceans surrounding Antarctica, all bound together in a tangle of feeding relationships.

How can we go beyond a confusing verbal description to a useful and easily understood summary of the feeding relationships within communities? One of the earliest approaches to the study of communities was to describe who eats whom. Since the beginning of the twentieth century, ecologists have meticulously described the feeding relationships in hundreds of communities. The resulting feeding relationships came to be called food webs. If we define a community as an association of interacting species, a moment's reflection will show that a **food web,** a summary of the feeding interactions within a community, is one of the most basic and revealing descriptions of community structure. A food web is, essentially, a community portrait based on feeding relationships (fig. 17.2). Other community portraits could be produced using competitive or mutualistic relationships.

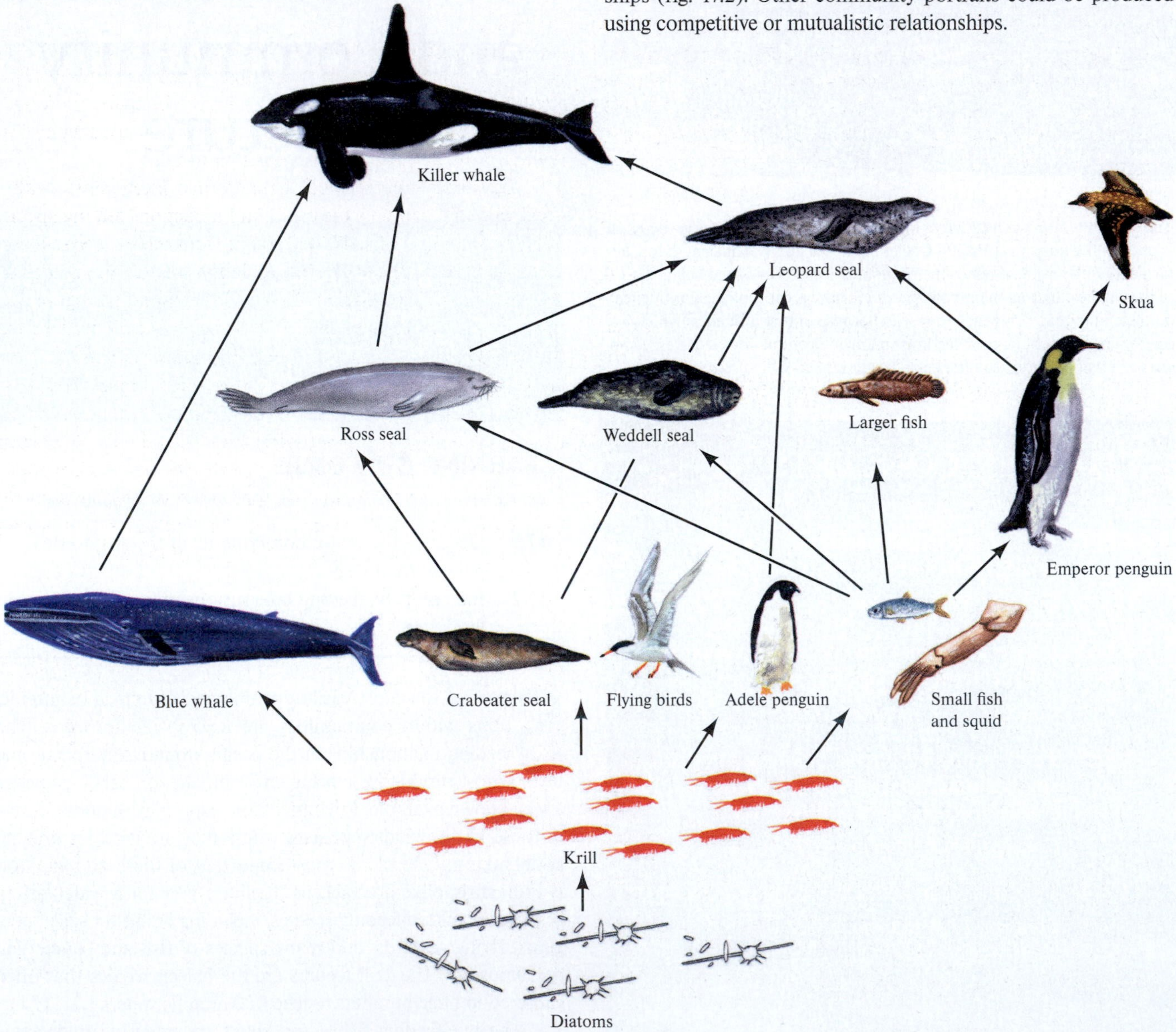

Killer whale

Leopard seal

Skua

Ross seal   Weddell seal   Larger fish

Emperor penguin

Blue whale   Crabeater seal   Flying birds   Adele penguin   Small fish and squid

Krill

Diatoms

Figure 17.2   The Antarctic pelagic food web.

17.1 Community Webs

LEARNING OUTCOMES

After studying this section you should be able to do the following:

17.3 Describe the influence of foundation species on communities.

17.4 Discuss the pros and cons of emphasizing strong interactions in food web studies.

17.5 Predict how the community food web associated with *Phragmites australis* would change if the insect-feeding bird *Cyanistes caeruleus* were removed from the ecosystem.

A food web summarizes the feeding relations in a community. The earliest work on food webs concentrated on simplified communities. One of the first of those food webs described the feeding relations on Bear Island in the high Arctic. Summerhayes and Elton (1923) studied the feeding relations there because they believed that the high Arctic, with few species, would be the best place to begin the study of food webs.

The work of Summerhayes and Elton revealed that even in these "impoverished faunas," feeding relations are complex and difficult to study, but they are much more manageable than the food webs of more diverse communities.

Strong Interactions and Food Web Structure

Robert Paine (1980) suggested that, in many cases, the feeding activities of a few species have a dominant influence on community structure. He called these influential trophic relations **strong interactions.** Paine also suggested that the defining criterion for a strong interaction is not necessarily quantity of energy flow but rather degree of influence on community structure. We will revisit this topic in the following section on keystone species, but for now, let's look at how recognizing interaction strength can affect depictions of food webs.

Paine's distinction between strong and weak interactions within food webs has been used to model the interactions within at least one terrestrial food web. Teja Tscharntke (1992) has worked intensively on a food web associated with the wetland reed *Phragmites australis* (fig. 17.3). This reed grows in large stands along the shores of rivers and other wetlands, where it is a **dominant, or foundation, species.** Foundation species are those that have substantial influences on community structure as a consequence of their high biomass—for example, abundant phragmites, an abundant tree in a forest, or a coral on a reef. Tscharntke's study site was along the Elbe River near Hamburg in northwestern Germany. Along the river, *Phragmites* is attacked by *Giraudiella inclusa,* a fly in the family Cecidomyidae, whose larvae develop within galls called "ricegrain" galls. At the study sites *Phragmites* is also attacked by *Archanara geminipuncta,* a moth in the family Noctuidae, whose larvae bore into the stems of *Phragmites.* Stem-boring by *A. geminipuncta* induces *Phragmites* to form side shoots, a response that provides additional sites for oviposition by the gall maker *G. inclusa.*

Tscharntke discovered that at least 14 species of parasitoid wasps attack *G. inclusa.* In winter, blue tits, *Cyanistes caeruleus* (fig. 17.4), move into stands of *Phragmites,* where they peck open the galls formed by *G. inclusa* and eat the larvae, causing mortality in this population as well as in its parasitoids.

Tscharntke represented these trophic interactions with a food web that captures the essential interactions among species in this community (fig. 17.5). Though Tscharntke's web still contains plenty of complexity, his depiction focuses the reader on the most important interactions in the community. Figure 17.5 suggests that feeding by blue tits strongly influences the parasitoids *Aprostocetus calamarius* and *Torymus arundinis* and their host, *G. inclusa,* in large gall clusters on main shoots. The other series of strong interactions involves the parasitoids *Aprostocetus gratus* and *Platygaster quadrifarius,* which attack the *G. inclusa* that inhabit small gall clusters in side shoots of *Phragmites.* These side shoots are in turn stimulated by the stem-boring larvae of the moth *A. geminipuncta.* Notice that blue tits only weakly influence populations on this side of the web.

Figure 17.3   Phragmites, the foundation of many food webs in wetland ecosystems.   Farlap/Alamy Stock Photo

Figure 17.4   The blue tit, *Cyanistes caeruleus,* is a Eurasian relative of the chickadees of North America that, like chickadees, gleans insects from vegetation.   Ingram Publishing/SuperStock

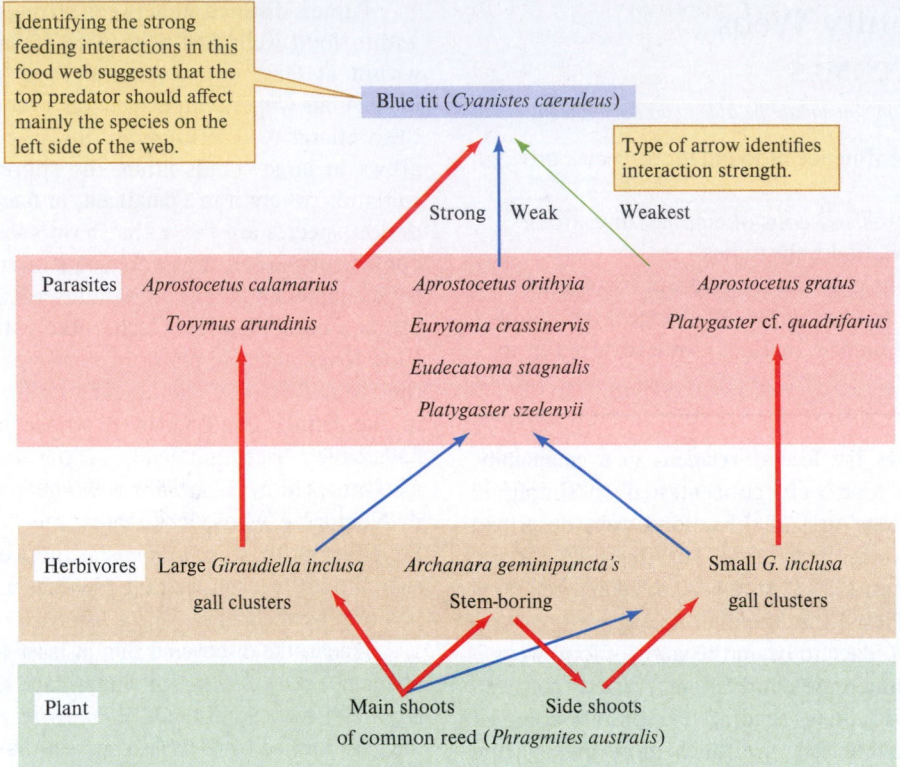

Figure 17.5   Food web associated with *Phragmites australis* (data from Tscharntke 1992).

By distinguishing between weak and strong interactions, Tscharntke produced an easily understood food web to represent the study community. Identifying strong interactions allows us to determine which species may have the most significant influences on community structure.

Concept 17.1 Review

1. What are the main advantages of including only strong linkages in a food web?
2. What was the primary way by which Tscharntke simplified the food web representing the interactions of blue tits and feeding on insects living on the wetland reed *Phragmites australis* (see fig. 17.5)?
3. In what other main way did Tscharntke simplify his study of trophic interactions in the wetland along the Elbe River?

17.2 Indirect Interactions

LEARNING OUTCOMES

After studying this section you should be able to do the following:

17.6   Distinguish between direct interactions and indirect interactions.

17.7   Define indirect commensalism and provide an example.

17.8   Explain apparent competition, and describe an experiment that could distinguish this type of indirect interaction from direct competition.

17.9   Predict how the results of the experiment summarized in figure 17.7 would have been changed, if the interaction between *Brassica nigra* and *Nassella pulchra* was actually direct competition, with small herbivorous mammals playing no role in the interaction.

Indirect interactions between species are fundamental to communities. Food webs emphasize *direct* trophic interactions between species. **Direct interactions** between two species, including competition, predation, herbivory, and mutualism (see chapters 13–15), involve positive or negative effects of one species on another without the involvement of an intermediary species. However, direct interspecific interactions can also result in ecologically significant *indirect* interactions between species. In **indirect interactions,** one species affects another through a third, intermediary species. Indirect interactions include trophic cascades, which we discuss in detail in chapter 18 (section 18.4), apparent competition, and indirect mutualism or commensalism.

Indirect Commensalism

As explained in Chapter 13, commensalism is an interaction between two species in which one species is benefited and the other is neither benefited nor harmed. **Indirect commensalism** occurs when the activities of one species indirectly—that is, through an intermediary species—benefit another species without itself being helped or harmed. Gregory Martinsen, Elizabeth Driebe, and Thomas Whitham of Northern Arizona University uncovered an indirect commensalism (fig. 17.6) in which beavers, *Castor canadensis,* indirectly benefit a herbivorous beetle species, *Chrysomela confluens,* through their effects on cottonwood trees, *Populus* spp. (Martinsen, Driebe, and Whitham 1998). As Martinsen, Driebe, and Whitham studied populations of *C. confluens* along the Weber River in northern Utah, they encountered higher densities of *C. confluens* in areas where beaver foraging activity was greatest. In these areas, the beetles were concentrated on cottonwood sprouts growing from the stumps of trees that had been felled by beavers. Searching for a mechanism to explain these higher beetle concentrations, Martinsen, Driebe, and Whitham found that, compared to leaves on undamaged cottonwood trees, the leaves on stump sprouts had higher concentrations of the defensive chemicals used by cottonwood trees to repel beavers and other herbivorous mammals. However, *C. confluens* is not repelled by the tree's defensive chemicals and instead uses them for its own defense. In addition, the

leaves on stump sprouts were also higher in nitrogen compared to leaves on uncut trees, making them a better food source for the beetles, which grew 20% larger and developed 10% faster when feeding on them. This is an example of a commensalism—beaver activity creates better conditions for the beetle population, without itself being affected. However, because the positive effect of beavers on the beetles is mediated through a third species, cottonwood trees, it is an *indirect* commensalism.

Apparent Competition

Ecologists generally infer interspecific competition, where interactions between the individuals of two species have a negative impact on their fitness. However, ecologists are discovering that in some cases these competition-like effects are the result of **apparent competition,** in which negative impacts are the result of two species sharing a predator or herbivore or by one species facilitating populations of a predator or herbivore of the second species. For example, if two prey species share a predator, increases in one of the prey populations may lead to increased numbers of the predator, which feed on and depress populations of the second prey species.

John Orrock, Martha Witter, and O. J. Reichman (2008) found that an assumed competitive interaction between an invasive exotic plant and a bunchgrass native to the grasslands of California is actually *apparent* competition. The invasive exotic plant, black mustard, *Brassica nigra,* has been long implicated in the competitive displacement of plants native to the California grasslands, including the native perennial bunchgrass, *Nassella pulchra.* However, Orrock, Witter, and Reichman hypothesized that *N. pulchra* populations might be limited more by small mammalian herbivores and granivores than by competition with *B. nigra.* They tested their hypothesis using a field experiment in which they established 14 study sites at varying distances from patches of *B. nigra.* At each study site, they applied three experimental treatments: (1) a fenced exclosure that excluded all small mammals, mainly mice, ground squirrels, and rabbits; (2) a control consisting of a fenced area that produced the disturbance associated with constructing an exclosure but allowed small mammal entry; and (3) an unfenced control. The researchers cultivated and broadcast *N. pulchra* seeds in each of their treatments during the growing season and then followed germination of seedlings and establishment of adult plants through spring, summer, and fall. They estimated small mammal activity using live traps during late summer, a time of peak mammal activity.

Orrock, Witter, and Reichman obtained results in support of their hypothesis (fig. 17.7). Small mammal activity was greatest near patches of *B. nigra* and declined away from patches of the invasive exotic plant. Small mammals apparently shelter in patches of *B. nigra* but do not feed on it. Instead, they feed on vegetation in the surroundings. As a consequence, the densities of both seedlings and adult *N. pulchra* were reduced in control plots, where small herbivorous mammals had access, near *B. nigra* patches. However, where small mammals were excluded, seedling and adult *N. pulchra* densities were high regardless of

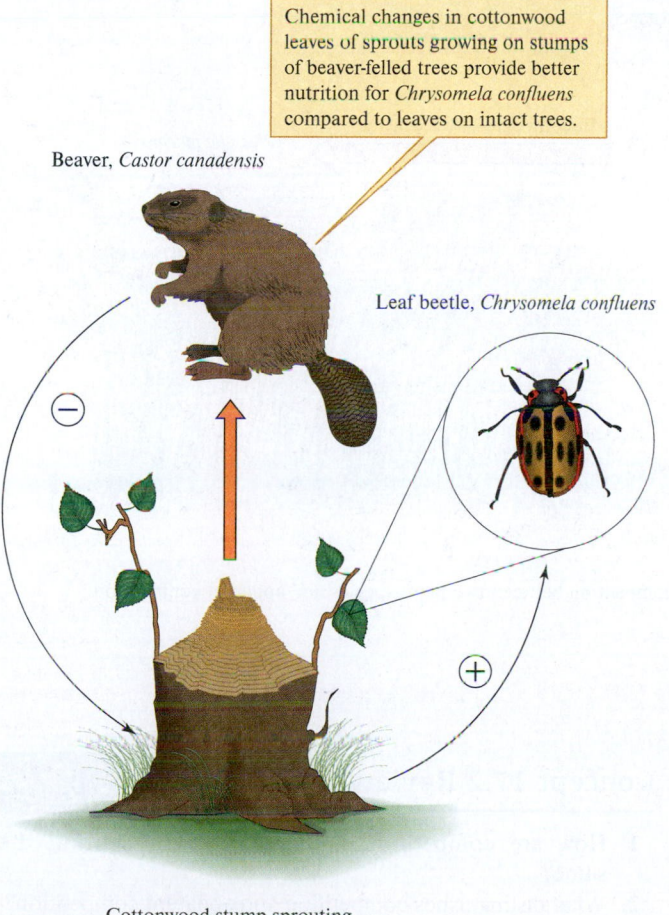

Chemical changes in cottonwood leaves of sprouts growing on stumps of beaver-felled trees provide better nutrition for *Chrysomela confluens* compared to leaves on intact trees.

Beaver, *Castor canadensis*

Leaf beetle, *Chrysomela confluens*

Cottonwood stump sprouting after tree felled by beaver.

Figure 17.6 An indirect commensalism. Beavers positively affect a leaf beetle population by felling cottonwood trees, inducing a leaf chemistry in the sprouts growing from the resulting stumps that repels beavers but is favorable to the leaf beetles (based on data from Martinsen, Driebe, and Whitham 1998).

Controls: mammal herbivore access

Herbivorous mammals shelter in *Brassica nigra* patches and forage in the surroundings, creating a gradient of decreasing herbivore pressure away from the patches.

Increasing bunchgrass density

Nassella pulchra

Brassica nigra

Treatment: mammal herbivore exclosure

In the absence of herbivorous mammals, the density of bunchgrass seedlings and adult plants does not vary with distance from *Brassica nigra* patches.

Uniform bunchgrass density

Nassella pulchra

Brassica nigra

Figure 17.7   A field experiment designed to distinguish between direct competition between two plant species and apparent competition (based on data from Orrock, Witter, and Reichman 2008).

their distance from *B. nigra* patches. Orrock, Witter, and Reichman propose that *B. nigra* patches provide areas of protective cover for the small mammals in their study area from which they forage across the surrounding landscape. The result is a gradient of decreasing herbivore pressure away from *B. nigra* patches, producing a gradient of increasing density in the *N. pulchra* population with distance from *B. nigra* patches. Orrock, Witter, and Reichman concluded that while the distribution patterns suggest competition, *B. nigra* actually suppresses *N. pulchra* indirectly by sheltering mammalian herbivores that feed on the native grass. In short, this is an example of apparent competition (fig. 17.8).

Concept 17.2 Review

1. How are competition and apparent competition the same?
2. What distinguishes competition and apparent competition?
3. How would the results of the experiment shown in figure 17.7 change if the effect of *Brassica nigra* on *Nassella pulchra* were the result of direct competition, with small herbivorous mammals playing no significant role?

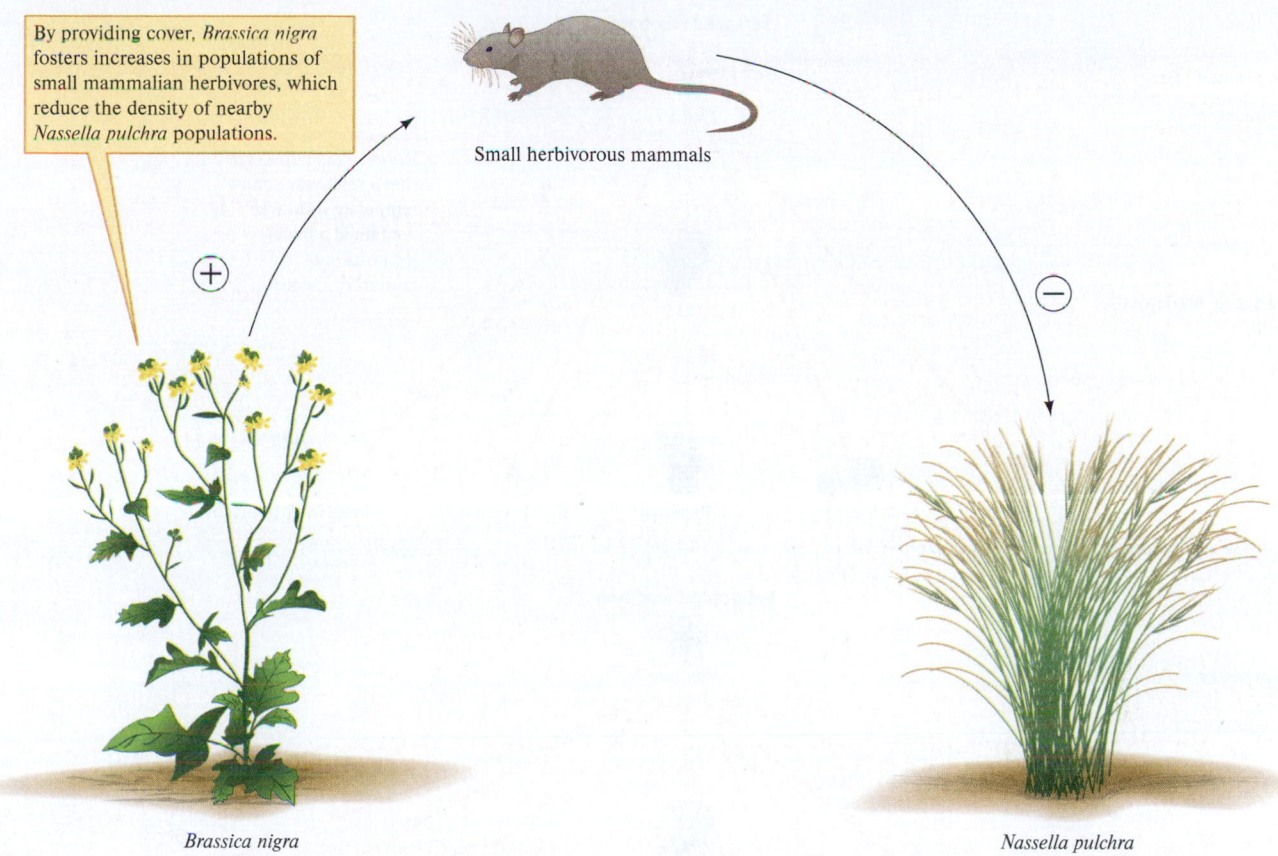

By providing cover, *Brassica nigra* fosters increases in populations of small mammalian herbivores, which reduce the density of nearby *Nassella pulchra* populations.

Small herbivorous mammals

Brassica nigra

Nassella pulchra

Figure 17.8   Apparent competition. *Brassica nigra* negatively affects *Nassella pulchra* through its positive effects on populations of small mammalian herbivores, which feed on the seeds, seedlings, and adults of *N. pulchra* (based on data from Orrock, Witter, and Reichman 2008).

17.3 Keystone Species

LEARNING OUTCOMES

After studying this section you should be able to do the following:

17.10 Define *keystone species*.

17.11 Distinguish between keystone species and dominant, or foundation, species.

17.12 Discuss Paine's experiments that led to his proposing the keystone species hypothesis.

17.13 Outline the various mechanisms by which keystone species influence the structure of communities.

The feeding activities of a few keystone species may control the structure of communities.   Robert Paine (1966, 1969) proposed that the feeding activities of a few species have inordinate influences on community structure. He called these **keystone species,** a concept that was later refined with further research. Paine's keystone species hypothesis emerged from a chain of reasoning. First, he proposed that predators might keep prey populations below their carrying capacity. Next, he reasoned the potential for competitive exclusion would be low in populations kept below carrying capacity. Finally, he concluded that if keystone species reduced the likelihood of competitive exclusion, their activities would increase the number of species that could coexist in communities. In other words, Paine predicted that some predators may increase species diversity.

Food Web Structure and Species Diversity

Paine began his studies by examining the relationship between overall species diversity within food webs and the proportion of the community represented by predators. He cited studies that demonstrated that as the number of species in marine zooplankton communities increases, the proportion that are predators also increases. For instance, the zooplankton community in the Atlantic Ocean over continental shelves includes 81 species, 16% of which are predators. In contrast, the zooplankton community of the Sargasso Sea contains 268 species, 39% of which are predators (Grice and Hart 1962). Paine set out to determine if similar patterns occur in marine intertidal communities.

Paine described a food web from the intertidal zone at Mukkaw Bay, Washington, which lies in the north temperate zone at 49° N. This food web is typical of the rocky shore community along the west coast of North America (fig. 17.9). The base of this food web consists of nine dominant intertidal invertebrates: two species of chitons, two species of limpets, a mussel, three species of acorn barnacles, and one species of gooseneck barnacle. Paine pointed out that the sea star *Pisaster* commonly consumes two other prey species in other areas, bringing the total food web diversity to 13 species. Ninety percent of the energy consumed by the middle-level predator, *Thais,* consists of barnacles. Meanwhile, the top predator, *Pisaster,* obtains 90% of its energy from a mixture of chitons (41%), mussels (37%), and barnacles (12%).

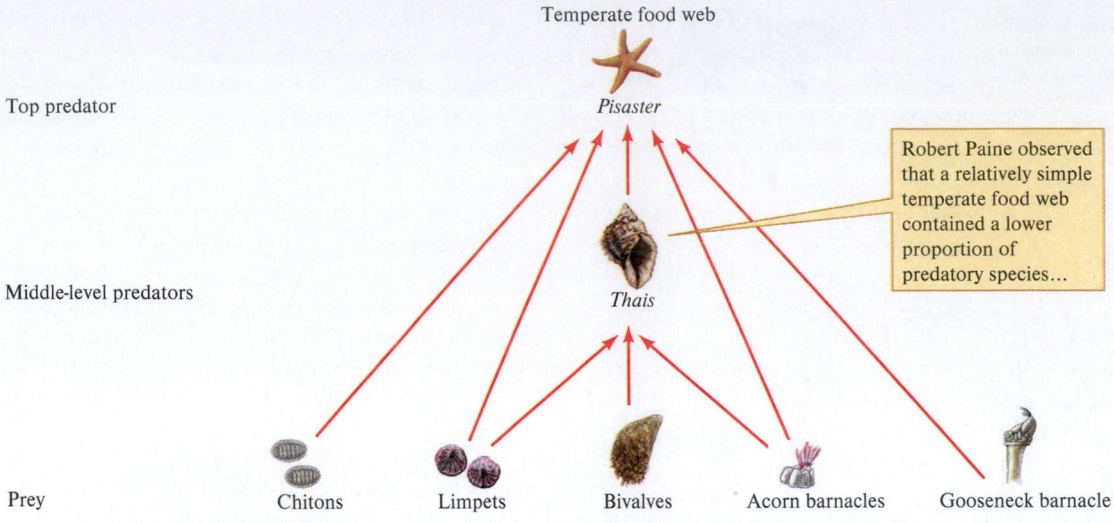

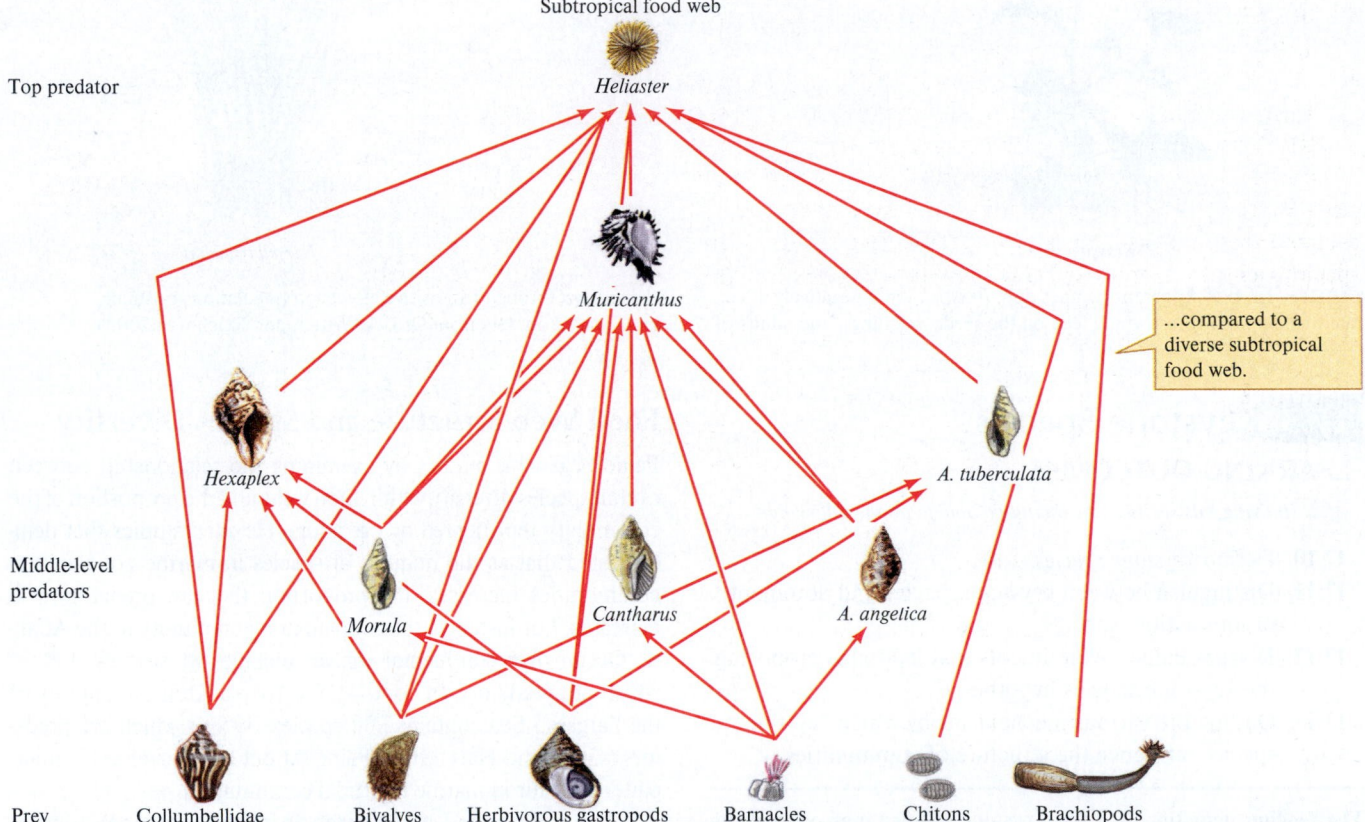

Figure 17.9   Roots of the keystone species hypothesis: Does a higher proportion of predators in diverse communities indicate that predators contribute to higher species diversity?

Paine also described a subtropical food web (31° N) from the northern Gulf of California, a much richer web that included 45 species. But like the food web at Mukkaw Bay, Washington, the subtropical web was topped by a single predator, the sea star *Heliaster kubiniji* (see fig. 17.9). However, six predators occupy middle levels in the subtropical web, compared to one middle-level predator at Mukkaw Bay. Because four of the five species in the snail family Columbellidae are also predaceous, the total number of predators in the subtropical web is 11. These predators feed on the 34 species that form the base of the food web.

Paine found that as the number of species in his intertidal food webs increased, the proportion of the web represented by predators also increased, a pattern similar to that in the zooplankton communities that had helped inspire Paine's hypotheses. As Paine went from Mukkaw Bay to the northern Gulf of California, overall web diversity increased from 13 species to 45 species, a 3.5-fold increase. However, at the same time, the number of predators in the two webs increased from 2 to 11, a 5.5-fold increase. According to Paine's predation hypothesis, this higher proportion of predators produces higher predation pressure on prey populations, which in turn promotes the higher diversity in the northern Gulf of California intertidal zone. Encouraged by these patterns, Paine designed a field experiment to evaluate his hypothesis that intertidal predators enhance species diversity.

Experimental Removal of Sea Stars

For his first experiment, Paine removed the top predator from the intertidal food web at Mukkaw Bay and monitored the response of the community. He chose two study sites in the middle intertidal zone that extended 8 m along the shore and 2 m vertically. One site was designated as a control and the other as an experimental site. He removed *Pisaster* from the experimental site and relocated them in another portion of the intertidal zone. Each week, Paine checked the experimental site for the presence of *Pisaster* and removed any that might have moved in since his last visit.

Paine followed the response of the intertidal community to sea star removal for 2 years. Over this interval, the diversity of intertidal invertebrates in the control plot remained constant at 15, while the diversity within the experimental plot declined from 15 to 8, a loss of 7 species. This reduction in species richness supported Paine's keystone species hypothesis. However, if this reduction was due to competitive exclusion, what was the resource over which species competed?

As we saw in chapters 11 and 13, the most common limiting resource in the rocky intertidal zone is space. Within 3 months of removing *Pisaster* from the experimental plot, the barnacle *Balanus glandula* occupied 60% to 80% of the available space. One year after Paine removed *Pisaster, B. glandula* was crowded out by mussels, *Mytilus californianus,* and gooseneck barnacles, *Pollicipes polymerus.* Benthic algal populations also declined because of a lack of space for attachment. The herbivorous chitons and limpets also left, due to a lack of space and a shortage of food. Sponges were also crowded out and a nudibranch that feeds on sponges also left. After 5 years, the *Pisaster* removal plot was dominated by two species: mussels and gooseneck barnacles.

This experiment showed that *Pisaster* is a keystone species. When Paine removed it from his study plot, the community collapsed. However, did this one experiment demonstrate the general importance of keystone species in nature? To demonstrate this, we need more experiments and observations across a wide variety of communities. Paine followed his work at Mukkaw Bay with a similar experiment in New Zealand.

The intertidal community along the west coast of New Zealand is similar to the intertidal community along the Pacific coast of North America. The top predator is a sea star, *Stichaster australis,* that feeds on a wide variety of invertebrates, including barnacles, chitons, limpets, and a mussel, *Perna canaliculus.* During 9 months following Paine's removal of sea stars, the number of species in the removal plot decreased from 20 to 14 and the coverage of the area by the mussel

increased from 24% to 68%. As in Mukkaw Bay, the removal of a predaceous sea star produced a decrease in species richness and a significant increase in the density of a major prey species. Again, the mechanism underlying disappearance of species from the experimental plot was competitive exclusion due to competition for space.

These results show that intertidal communities thousands of kilometers apart that do not share any species of algae or genera of invertebrates are influenced by similar biological processes (fig. 17.10). This is reassurance to ecologists seeking general ecological principles. Many other studies quickly followed the lead taken by Paine's pioneering work.

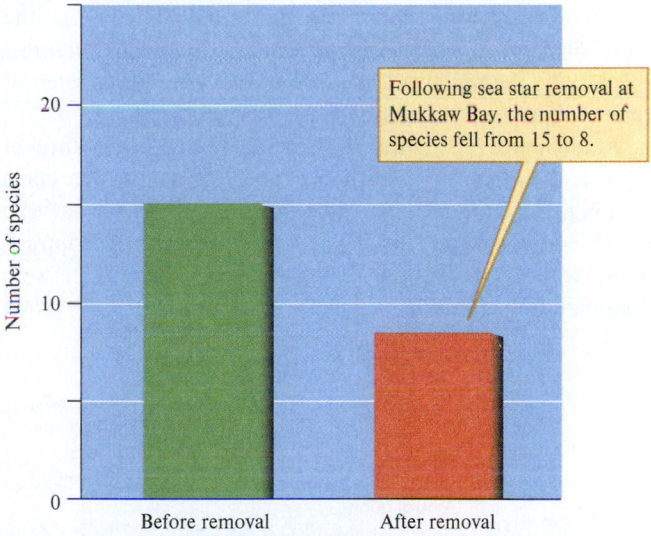

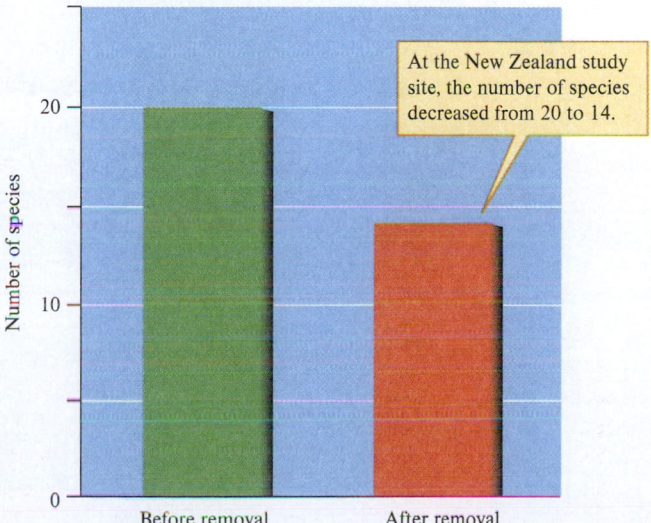

Figure 17.10   The effect of removing a top predator from two intertidal food webs (data from Paine 1966, 1971).

Snail Effects on Algal Diversity

Studies of how herbivorous intertidal snails influence the diversity of intertidal algae by Jane Lubchenco (1978) increased our understanding of the ecological details on which the influences of a keystone species can depend. Lubchenco observed that previous studies had indicated that herbivores sometimes increase plant diversity, sometimes decrease plant diversity, and sometimes seem to do both. She proposed that to resolve these apparently conflicting results it would be necessary to understand (1) the food preferences of herbivores, (2) the competitive relationships among plant species in the local community, and (3) how competitive relationships and feeding preferences vary across environments. Lubchenco used these criteria to guide her study of the influences of an intertidal snail, *Littorina littorea,* on the structure of an algal community.

Lubchenco studied the feeding preferences of *Littorina* in the laboratory. Her experiments indicated that algae fell into low, medium, or high preference categories. Generally, highly preferred algae were small, ephemeral, and tender like the green algae, *Enteromorpha* spp., while most tough, perennial species like the red alga *Chondrus crispus* were never eaten or eaten only if the snail was given no other choice.

Lubchenco also studied variation in the abundance of algae and *Littorina* in tide pools. She found that tide pools with high densities of *Enteromorpha,* one of the snail's favorite foods, contained low densities (4/m²) of snails. In contrast, pools with high densities of *Littorina* (233–267/m²) were dominated by *Chondrus,* a species for which the snail shows

low preference. Lubchenco reasoned that in the absence of *Littorina, Enteromorpha* competitively displaces *Chondrus.* She tested this idea by removing the *Littorina* from one of the pools in which they were present in high density and introducing them to a pool in which *Enteromorpha* was dominant. She monitored a third pool with a high density of the snails as a control.

The results of Lubchenco's removal experiment were clear (fig. 17.11). While the relative densities of *Chondrus, Enteromorpha,* and other ephemeral algae remained relatively constant in the control pool, the density of *Enteromorpha* declined with the introduction of *Littorina.* Meanwhile, *Enteromorpha* quickly increased in density and came to dominate the pool from which Lubchenco had removed the snails. In addition, as the *Enteromorpha* population in this pool increased, the population of *Chondrus* declined. Lubchenco began another addition and removal experiment in two other pools in the fall to check for seasonal effects on feeding and competitive relations. This second removal experiment produced results almost identical to the first. Where *Littorina* were added, the *Enteromorpha* population declined, whereas the *Chondrus* population increased. Where the snails were removed, the *Chondrus* population declined, while the *Enteromorpha* population increased.

What controls the local population density of *Littorina?* Apparently, the green crab, *Carcinus maenus,* which lives in the canopy of *Enteromorpha,* preys upon young snails and can prevent the juveniles from colonizing tide pools. Adult *Littorina* are much less vulnerable to *Carcinus* but rarely move

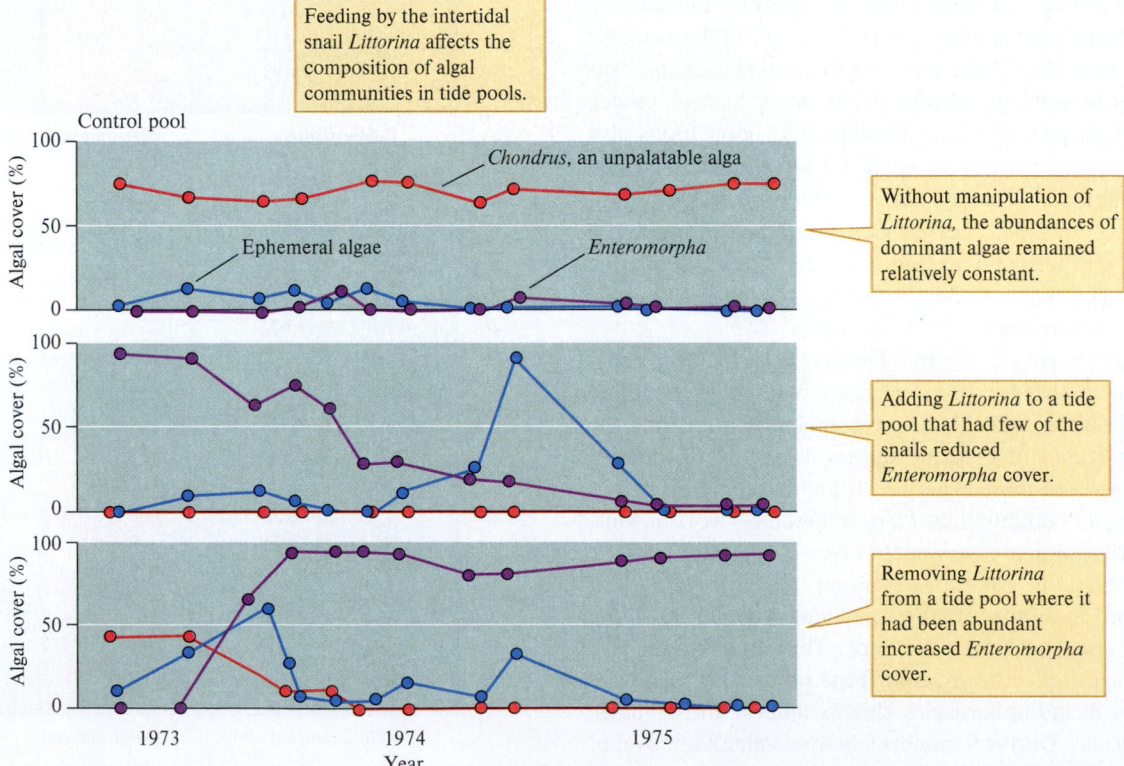

Figure 17.11   Effect of *Littorina littorea* on algal communities in tide pools (data from Lubchenco 1978).

to new tide pools. Populations of *Carcinus* are in turn controlled by seagulls. Here again, we begin to see the complexity of a local food web and the influences that trophic interactions within webs can have on community structure.

Therefore, within tide pools *Enteromorpha* can outcompete the other tide pool algae for space and *Enteromorpha* is the preferred food of *Littorina*. How might feeding by the snails affect the diversity of algae within tide pools? The relationship between the snails and the algal species they exploit is similar to the situation studied by Paine, where mussels were the competitively dominant species and one of the major foods of the sea star *Pisaster*.

Lubchenco examined the influence of *Littorina* on algal diversity by observing the number of algal species living in tide pools occupied by various densities of snails (fig. 17.12). As the density increased from low to medium, the number of algal species increased. Then, as the density increased further, from medium to high, the number of algal species declined.

How would you explain these results? At low density, the feeding activity by *Littorina* is not sufficient to prevent *Enteromorpha* from dominating a tide pool and crowding out some other species. At medium densities, the snail's feeding, which concentrates on the competitively dominant species, prevents competitive exclusion and so increases algal diversity. However, at high densities, the feeding requirements of the

population are so high that the snails eat their preferred algae as well as less-preferred species. Consequently, intense grazing by snails at high density reduces algal diversity.

What would happen if *Littorina* preferred to eat competitively inferior species of algae? This is precisely the circumstance that occurs on emergent substrata, rock surfaces that are not submerged in tide pools during low tide. On these emergent habitats, the competitively dominant algae are species in the genera *Fucus* and *Ascophyllum,* algae for which the snails show low preference. On emergent substrata, the snails continue to feed on ephemeral, tender algae such as *Enteromorpha,* largely ignoring *Fucus* and *Ascophyllum.* In this circumstance, Lubchenco found that algal diversity was highest when *Littorina* densities were low (see fig. 17.12).

Lubchenco's research improved our understanding of how trophic interactions can affect community structure. Her work demonstrated that the influence of consumers on the structure of food webs depends on their feeding preferences, the density of local consumer populations, and the relative competitive abilities of prey species. While Lubchenco moved the field well beyond the conceptual view held by ecologists when Paine first proposed the keystone species hypothesis, one basic element of the original hypothesis remained: Consumers can exert substantial control over food web structure; they can act as keystones.

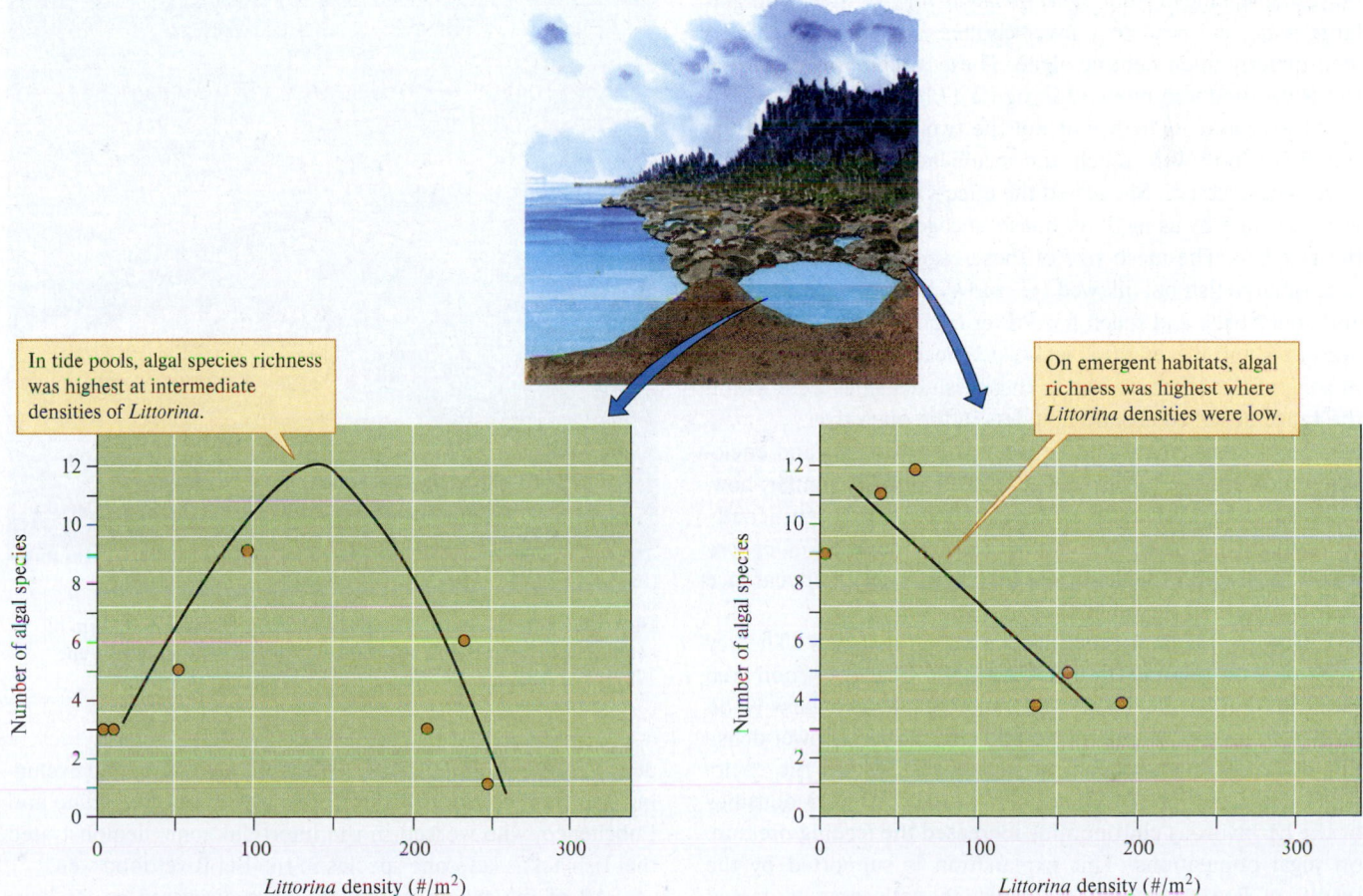

Figure 17.12   Effect of *Littorina littorea* on algal species richness in tide pools and emergent habitats (data from Lubchenco 1978).

Can predators act as keystone species in environments other than the intertidal zone? Here is an example from a riverine environment.

Fish as Keystone Species in River Food Webs

Mary Power (1990) tested the possibility that fish can significantly alter the structure of food webs in rivers. She conducted her research on the Eel River in northern California, where most precipitation falls from October to April, sometimes producing torrential winter flooding. During the summer, however, the flow of the Eel River averages less than 1 m³ per second.

In early summer, the boulders and bedrock of the Eel River are covered by a turf of the filamentous alga *Cladophora* (fig. 17.13*a*). However, the biomass of the algae declines by midsummer and what remains has a ropy, prostrate growth form and a "webbed" appearance (fig. 17.13*b*). These mats of *Cladophora* support dense populations of larval midges in the fly family Chironomidae. One chironomid, *Pseudochironomus richardsoni*, is particularly abundant. *Pseudochironomus* feeds on *Cladophora* and other algae and weaves the algae into retreats, altering their appearance in the process.

Chironomids are eaten by predatory insects and the young (known as *fry*) of two species of fish: a minnow called the California roach, *Hesperoleucas symmetricus*, and three-spined sticklebacks, *Gasterosteus aculeatus*. These small fish are eaten by young steelhead trout, *Oncorhynchus mykiss*. Steelhead and large roach eat predatory invertebrates, and large roach also feed directly upon benthic algae. These interactions form the Eel River food web pictured in figure 17.14.

Power asked whether or not the two top predators in the Eel River food web, roach and steelhead, significantly influence web structure. She tested the effects of these fish on food web structure by using 3 mm mesh to cage off 12 areas 6 m² in the riverbed. The mesh size of these cages prevented the passage of large fish but allowed free movement of aquatic insects and stickleback and roach fry. Power excluded fish from six of her cages and placed 20 juvenile steelhead and 40 large roach in each of the other six cages. These fish densities were within the range observed around boulders in the open river.

Significant differences between the exclosures and enclosures soon emerged. Algal densities were initially similar; however, enclosing fish over an area of streambed significantly reduced algal biomass (fig. 17.15). In addition, the *Cladophora* within cages with fish had the same ropy, webbed appearance as *Cladophora* in the open river.

How do predatory fish decrease algal densities? The key to answering this question lies with the Eel River food web (see fig. 17.14). Steelhead and large roach feed heavily on predatory insects, young roach, and sticklebacks. Lower densities of these young roach and sticklebacks within the enclosures decreased predation on chironomids. Higher densities of the herbivorous chironomids increased the feeding pressure on algal populations. This explanation is supported by the results of Power's experiment in which enclosures contained lower densities of predatory insects and fish fry and higher

(a)

(b)

Figure 17.13   Seasonal changes in biomass and growth form of benthic algae in the Eel River, California: (*a*) in early summer, June 1989; (*b*) in late summer, August 1989.   (*a, b*) Mary E. Power

densities of chironomids (fig. 17.16). By enclosing and excluding fish from sections of the Eel River, Power, like Paine and Lubchenco, who worked in the intertidal zone, demonstrated that fish act as keystone species in the Eel River food web.

All of the examples that we have discussed so far have been aquatic. Do terrestrial communities also contain keystone

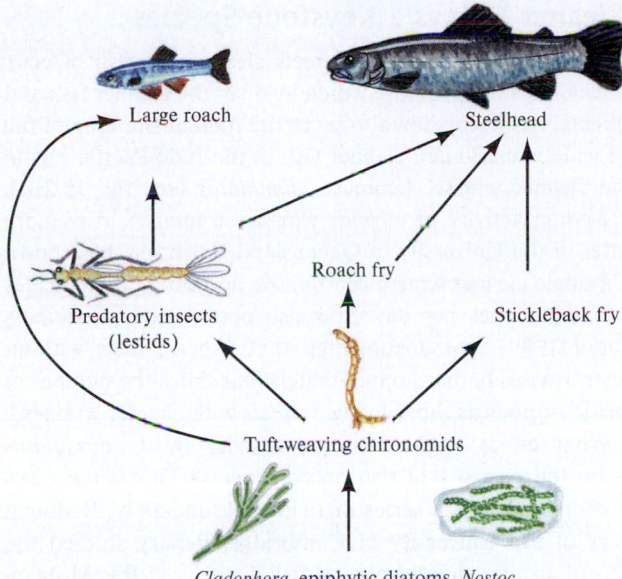

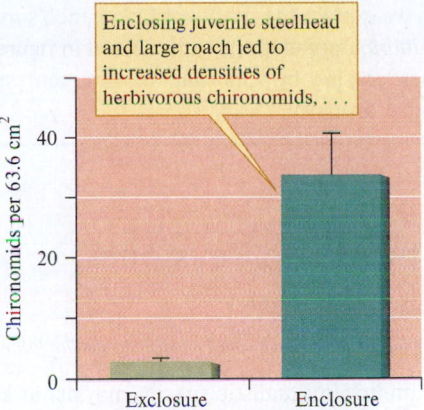

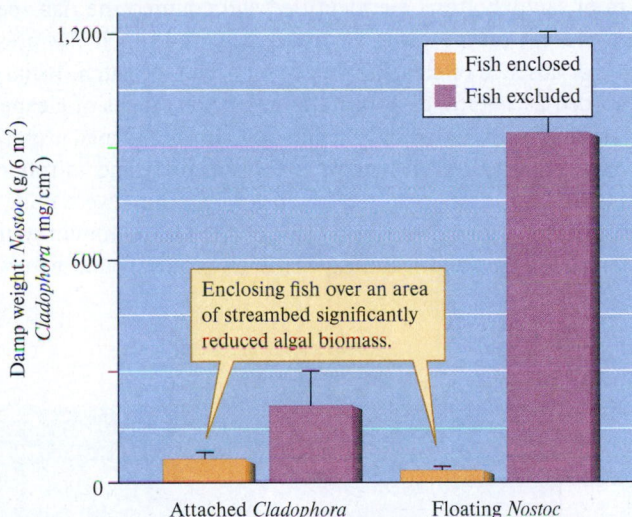

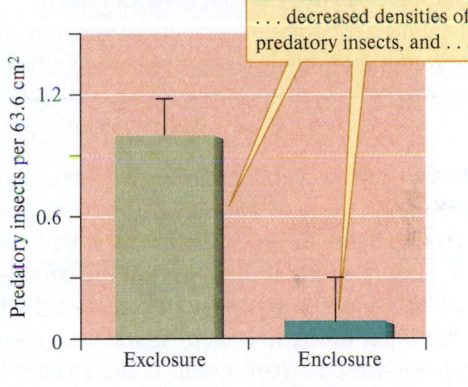

Figure 17.14   Food web associated with algal turf during the summer in the Eel River, California.

Figure 17.15   The influence of juvenile steelhead and California roach on benthic algal biomass in the Eel River (means, 1 standard error) (data from Power 1990).

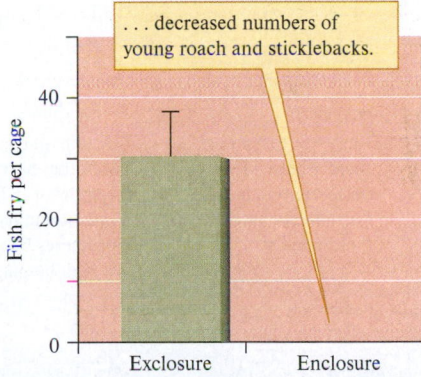

Figure 17.16   Effect of juvenile steelhead and California roach on numbers of insects and young (fry) roach and sticklebacks (means, 1 standard error) (data from Power 1990).

species? An increasing body of evidence indicates that they do, particularly in tropical forests (see fig. 17.22), where they create a "trophic cascade" (see section 17.2).

Many studies of food webs and keystone species have been done since Robert Paine's classic study of the intertidal food web at Mukkaw Bay, Washington. The studies have revealed a great deal of biological diversity, which has prompted biologists to ask what characterizes keystone species. This reflection is necessary to avoid the possibility that the term may become so inclusive that it becomes meaningless. The conclusions reached by a conference designed to address this question are summarized in figure 17.17 (Power et al. 1996). The conference participants were careful to distinguish between keystone and dominant, or foundation, species. In contrast to foundation

species, keystone species are those that, despite low biomass, exert strong effects on the structure of the communities they inhabit. As we shall see in the following discussion, those strong effects are not always positive, particularly where they involve invasive species.

Concept 17.3 Review

1. Paine discovered that intertidal invertebrate communities of higher diversity include a higher proportion of predator species. Did this pattern confirm Paine's predation hypothesis?

2. What was the major limitation of Paine's first removal experiment involving *Pisaster*?

3. How can we explain the results of Lubchenco's manipulation of *Littorina* populations summarized in figure 17.11?

4. Why does grazing by *Littorina* on emergent substrata reduce algal diversity?

17.4 Mutualistic Keystones

LEARNING OUTCOMES

After studying this section you should be able to do the following:

17.14 Explain how mutualistic species may act as keystone species.

17.15 Compare the research approaches used to document the roles of cleaner wrasses and ants as keystones in their respective communities.

Mutualists can act as keystone species. Returning to the classification of Power and colleagues shown in figure 17.17, the only requirements for keystone status is that the species in question have relatively low biomass in the community and that it has a high impact on community structure. Increasingly, ecologists are discovering that many mutualistic species meet these requirements. One such group is the cleaner fishes on coral reefs.

A Cleaner Fish as a Keystone Species

Many species of fish on coral reefs clean other fish of ectoparasites. This relationship, which involves the cleaner fish and its clients, has been shown to be a true mutualism. One of the most widely distributed cleaner fish in the Indo-Pacific region is the cleaner wrasse, *Labroides dimidiatus* (see fig. 15.21a). The feeding activity of cleaner wrasses is intense. Alexandra Grutter of the University of Queensland, Australia, has shown that a single cleaner wrasse can remove and eat 1,200 parasites from client fishes per day. She also performed experiments (Grutter 1999) that documented that fish on reefs without cleaner wrasses harbor approximately four times the number of parasitic isopods as those living on reefs with cleaner wrasses.

What effect might cleaning activity by *L. dimidiatus* have on the diversity of fish on coral reefs? This is the question addressed with a series of field experiments by Redouan Bshary of the University of Cambridge. Bshary studied the effects of cleaner wrasses on reef fish diversity at Ras Mohammed National Park, Egypt (Bshary 2003). The study area consists of a sandy bottom area approximately 400 m from shore dotted with reef patches in water depths from 2 to 6 m. Bshary chose 46 reef patches separated from other patches by at least 5 m of sandy bottom. He identified and counted the fish species present during dives on these reefs and noted the presence or absence of cleaner wrasses on each reef patch. Bshary recorded 29 natural disappearances or appearances of cleaner wrasses during his study. In addition, he performed experimental removals of cleaner wrasses from reefs and introductions of these cleaners to reef patches where there were none.

Bshary followed the responses of the fish community to natural disappearances and experimental removals and natural

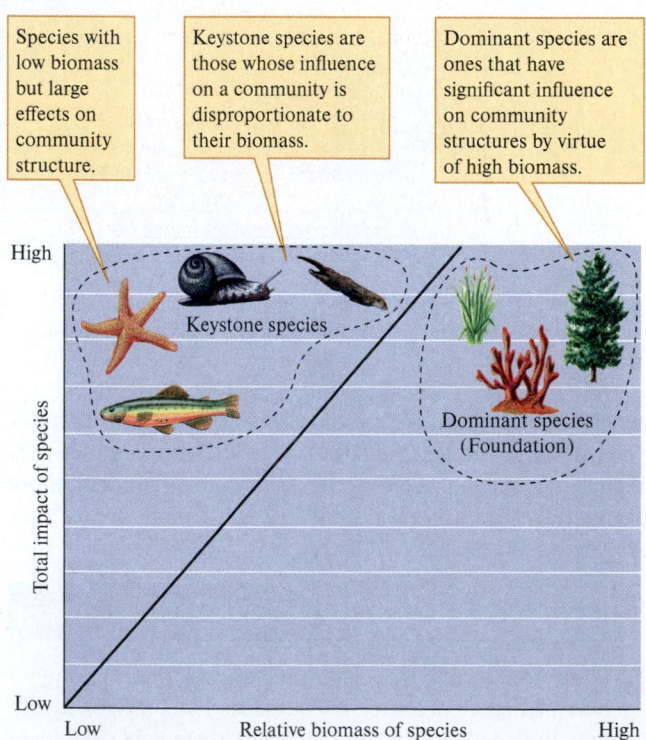

Figure 17.17   What is a keystone species (data from Power et al. 1996)?

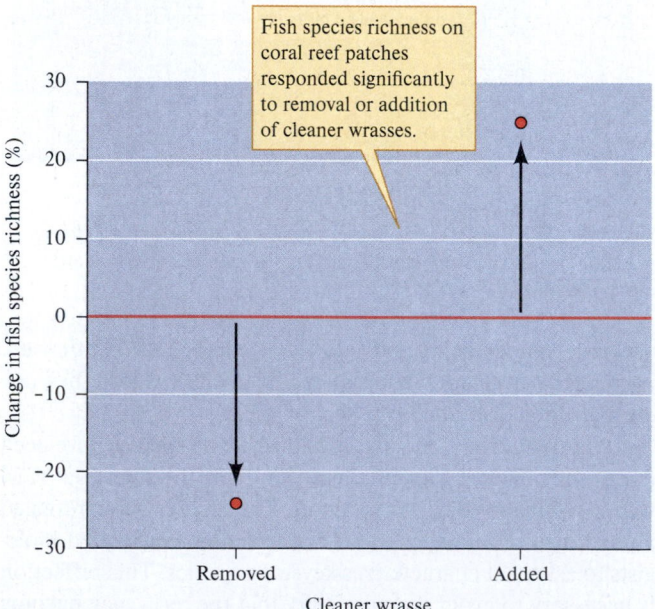

Figure 17.18   Results of experimental and natural removals or additions of cleaner wrasses, *Labroides dimidiatus,* to reef patches in the Red Sea (data from Bshary 2003).

colonization and experimental introductions. In doing so, he gained insights into the influence of these tiny mutualists on reef fish diversity. Figure 17.18 summarizes the responses of fish communities on reef patches 4 months following the natural or experimental addition or removal of cleaner wrasses. Bshary observed a median reduction in fish species richness of approximately 24% where cleaner wrasses disappeared or were removed. Where cleaner wrasses were added, either naturally or experimentally, he observed a median increase in fish species richness of 24%. Bshary's results indicate that the cleaner wrasse acts as a keystone species on the coral reefs of the Red Sea. Mutualists that act as keystone species have also been found on land.

Seed Dispersal Mutualists as Keystone Species

It appears that ants that disperse seeds have a significant influence on the structure of plant communities in the species-rich fynbos of South Africa. Caroline Christian (2001) observed that native ants disperse 30% of the seeds in the shrublands of the fynbos. The plants attract the services of these dispersers with food rewards on the seeds called elaiosomes. However, the Argentine ant, *Linepithema humile* (fig. 17.19), which does not disperse seeds, has invaded these shrublands. Christian documented how the invading Argentine ants have displaced (as they have in other regions) many of the native ant species in the fynbos. In addition, she discovered that the native ant species most impacted by Argentine ant invasion are those species most likely to disperse larger seeds.

Seed-dispersing ants are important to the persistence of fynbos plants because they bury seeds in sites where they are safe from seed-eating rodents and from fire. Fires are characteristic of Mediterranean shrublands such as the fynbos, and seeds are the only life stage of many fynbos plants

to survive fires. Consequently, ant dispersal is critical to the survival of many plant species. In a comparison of seedling recruitment following fire, Christian found substantial reductions in seedling recruitment by plants producing large seeds in areas invaded by Argentine ants (fig. 17.20). Meanwhile, small-seeded plants, whose dispersers are less affected by Argentine ants, showed no reduction in recruitment following fire. Christian's results, like Bshary's, reveal the influence of mutualists acting as keystone species within the communities they occupy. Other studies are revealing the importance of other mutualists, such as pollinators and mycorrhizal fungi, as keystone species.

Concept 17.4 Review

1. Bshary studied changes in fish species richness in response to both natural and experimental removals and additions of the cleaner fish *Labroides dimidiatus* (see fig. 17.18). Why did he not just focus on the response of fish species richness to natural additions and removals of the cleaner fish?
2. In many regions, native pollinator insects seem to be declining. Why is this a cause for concern among conservationists and ecologists?

Figure 17.19   The Argentine ant, *Linepithema humile,* has invaded and disrupted ant communities in many geographic regions. In the fynbos of South Africa, invading Argentine ants are displacing keystone ant species, which threatens the exceptional plant diversity of the fynbos.   Jesus Alberto Ramirez Viera/Getty Images

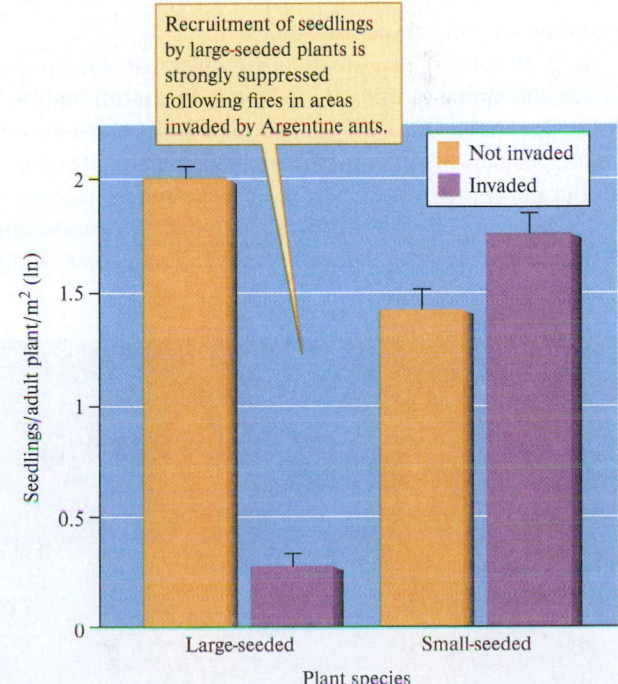

Recruitment of seedlings by large-seeded plants is strongly suppressed following fires in areas invaded by Argentine ants.

Figure 17.20   A comparison of recruitment of seedlings following fire in areas invaded by Argentine ants and areas not invaded shows the effects of the displacement of native seed-dispersing ants by Argentine ants (means, 1 standard error) (data from Christian 2001).

Applications

Human Modification of Food Webs

People have long manipulated food webs both as a consequence of their own feeding activities and by introducing species to or deleting species from existing webs. Consequently, either consciously or unwittingly, people have, themselves, acted as keystone species in communities.

Parasitoid Wasps: Apparent Competition and Biological Control

Earlier in this chapter, we learned about a vertebrate predator of parasitoids, the bird that ate these wasps; however, parasitoids can also be prey to other invertebrates. In fact, there are wasps that lay eggs inside parasitoid larvae, sometimes even while these larvae are still inside of their herbivore hosts, creating a Russian nesting doll of predation! These parasitoids of parasitoids are called **hyperparasitoids.**

Such threats to parasitoids are problematic for farmers who are attempting to using those insects to control herbivore threats (fig. 17.21). One such example are *Cotesia glomerata* and *C. rubecula,* parasitoid wasp species intentionally introduced from Europe to the United States to help control cabbage worm (*Pieris rapae*). Cabbage worm is an introduced, invasive species that eats many species of plants in the Brassicaceae family,

Figure 17.21 Parasitoids may be used for biological control to help protect crops, like the cabbage shown here, from herbivory by other insects such as the cabbage worm. Mikhail Kokhanchikov/Alamy Stock Photo

including cabbage crops. *C. glomerata and C. Rubecula* both are useful in helping to control this agricultural pest. The adult wasps lay eggs inside of the cabbage worm; when the eggs hatch into larvae, they feed on the host until they are ready to immerge and build cocoons. In the cocoons, they develop into the adult wasp stage, and the cycle can begin again. They may be attacked by hyperparasitoids either as larvae or while they are in their cocoons. In either case, the outcome is that it will be the hyperparasitoid that immerges from the cocoon instead of the host.

Dhaval K. Vyas and colleagues explored the food web dynamics of this system by sampling populations of *C. glomerata* and *C. rubecula* in Maryland, and populations of *C. glomerata* in Colorado (Vyas et al. 2020). Cocoons of both species were collected on cabbage plants and brought back to the lab to rear until any hyperparasitoids immerged. Vyas' group recorded eight different hyperparasitoid species. They then compared rates of hyperparasitoidism in the two wasp species to see if it helped explain how the two wasp species could coexist, given that they both depend on the same cabbage worm host.

In their native range in Europe, *C. glomerata* and *C. rubecula* can co-occur because they mostly parasitize different species of *Pieris* worms and thus do not compete with each other. However, in the Eastern United States, there is only *P. rapae.* Since *C. rubecula* is a specialist parasitoid on *P. rapae* in its native range, it is considered to be a better biological control agent than *C. glomerata.* Furthermore, *C. rubecula* produces one large larva per host, in contrast with *C. glomerata,* which produces many smaller larvae. The larger larvae of *C. rubecula* are usually able to kill the eggs or larvae of *C. glomerata* when both are laid in the same host (Geervliet et al. 2000). Taken together, this would suggest that *C. rubecula* would be expected to competitively exclude the other wasp species, and this does appear to have happened in some areas where both were introduced (Herlihy et al. 2012). However, *C. glomerata* has not been competitively excluded in some areas, including Maryland. How is this possible?

Vyas and colleagues found that although *C. glomerata* and *C. rubecula* were equally likely to be attacked by hyperparasitoids, *C. glomerata* had higher survivorship because this wasp species lays multiple cocoons in a single brood (fig. 17.22). Some larvae in the brood may die, while some of its siblings survive. In contrast, because *C. rubecula* produces a single larva at a time and thus a single cocoon, an attack by a hyperparasitoid generally resulted in 100% loss of that reproductive event. Indeed, it is hypothesized that the trait of producing multiple offspring in a single host likely evolved as a defense against predation (Mayhew 1998).

The result of this is apparent competition in which the *C. glomerata* may seem to be a good competitor against the *C. rubecula,* when in actuality it is predation by the hyperparasitoids, not interspecific competition, that is keeping *C. rubecula* in check. Thus, the hyperparasitoids are facilitating coexistence of the two parasitoid species in the Maryland communities, but this predation could even lead to the exclusion of *C. rubecula* over time. Given that *C. rubecula* is the better biological control agent for farmers, these ecological dynamics have important implications for management of crops.

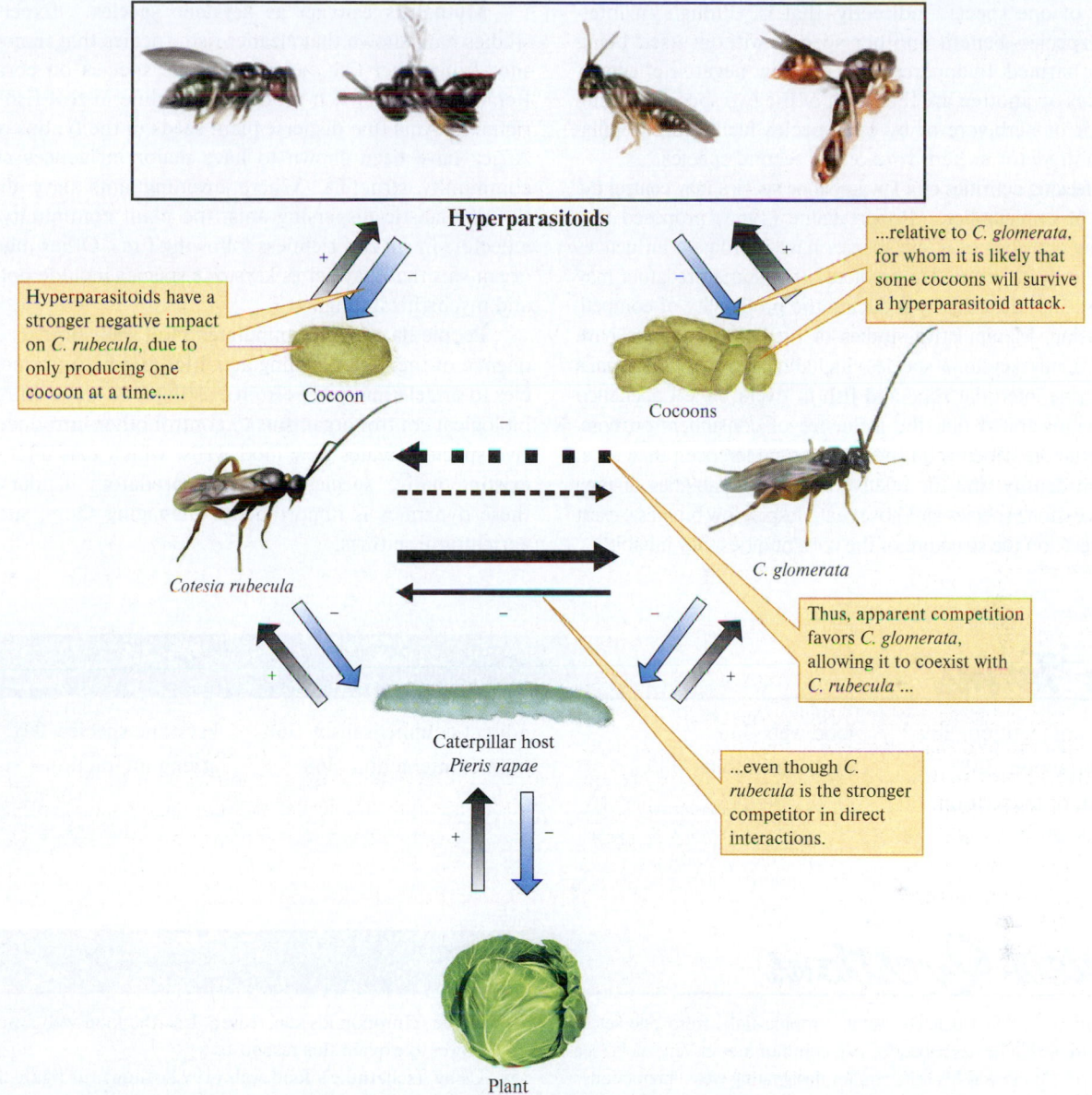

Hyperparasitoids

Hyperparasitoids have a stronger negative impact on *C. rubecula*, due to only producing one cocoon at a time......

...relative to *C. glomerata*, for whom it is likely that some cocoons will survive a hyperparasitoid attack.

Cocoon

Cocoons

Cotesia rubecula

C. glomerata

Thus, apparent competition favors *C. glomerata*, allowing it to coexist with *C. rubecula* ...

...even though *C. rubecula* is the stronger competitor in direct interactions.

Caterpillar host
Pieris rapae

Plant

Figure 17.22   In this food web diagram of four trophic levels, arrows indicate the costs and benefits of the interaction (i.e., predation or competition) to each organism's population (see table 13.1), with benefits (+) shown with black arrows and costs (−) shown in blue. Solid arrows show direct relationships, whereas dashed arrows are indirect. Four of the most common hyperparasitoids are shown here, with their two *Cotesia* hosts, Both *Cotesia glomerata* and *C. rubecula* parasitoid wasps compete for cabbageworm caterpillar hosts, but competition usually favors *C. rubecula*. In the fourth trophic level, hyperparasitoids use both *Cotesia* wasps as hosts, yet hyperparasitoids inflict more damage on *C. rubecula* than on *C. glomerata*. This differential effect of top-down pressures can create apparent competition that decreases the competitive advantage of *C. rubecula* over *C. glomerata* (adapted from Vyas et al. 2020).

Summary

A food web summarizes the feeding relations in a community. The earliest work on food webs concentrated on simplified communities in areas such as the Arctic islands. The level of food web complexity increased substantially, however, as researchers began to study complex communities. Studies of the food webs of tropical freshwater fish communities revealed highly complex networks of trophic interaction that persisted even in the face of various simplifications. A focus on strong interactions can simplify food web structure and identify those interactions responsible for most of the energy flow in communities.

Indirect interactions between species are fundamental to communities. Direct interspecific interactions can result in ecologically significant *indirect* interactions between species. In indirect interactions, one species affects another through a third, intermediary species. Indirect interactions include trophic cascades, apparent competition, and indirect mutualism or commensalism. Indirect commensalism occurs when the activities of one species indirectly—that is, through an intermediary species—benefit another species without itself being helped or harmed. In apparent competition, negative effects of one species on another are the result of the two species sharing a predator or herbivore or by one species facilitating populations of a predator or herbivore of the second species.

The feeding activities of a few keystone species may control the structure of communities. Robert Paine (1966) proposed that the feeding activities of a few species have inordinate influences on community structure. He predicted that some predators may increase species diversity by reducing the probability of competitive exclusion. Manipulative studies of predaceous species have identified many keystone species, including sea stars and snails in the marine intertidal zone and fish in rivers. Jane Lubchenco (1978) demonstrated that the influence of consumers on community structure depends on their feeding preferences, their local population density, and the relative competitive abilities of prey species. Keystone species are those that, despite low biomass, exert strong effects on the structure of the communities they inhabit.

Mutualists can act as keystone species. Experimental studies have shown that cleaner fish, species that remove parasites from other fish, act as keystone species on coral reefs. Removing cleaner fish produces a decline in reef fish species richness. Ants that disperse plant seeds in the fynbos of South Africa have been shown to have major influences on plant community structure. Where invading ants have displaced the mutualistic dispersing ants, the plant community suffers a decline in species richness following fires. Other mutualistic organisms that may act as keystone species include pollinators and mycorrhizal fungi.

People have long manipulated food webs both as a consequence of their own feeding activities and by introducing species to or deleting species from existing food webs. Introducing biological control organisms to control other introduced, invasive species creates new food webs, which can interact with existing native species, including predators. Understanding these dynamics is important for managing them, such as in agricultural settings.

Key Terms

apparent competition 369	food web 366	indirect commensalism 368	keystone species 371
direct interaction 368	hyperparasitoids 380	indirect interaction 368	strong interactions 367
dominant, or foundation, species 367			

Review Questions

1. Winemiller (1990) deleted "weak" trophic links from one set of food webs that he described for fish communities in Venezuela (see fig. 17.3). What was his criterion for designating weak interactions? Earlier, Paine (1980) suggested that ecologists could learn something by focusing on "strong" links in communities. How did Paine's criterion for determining a strong link differ from Winemiller's?

2. What is a keystone species? Paine (1966, 1969) experimented with two sea stars that act as keystone species in their intertidal communities along the west coast of North America and in New Zealand. Describe how the intertidal communities in these two areas are similar.

3. Explain how the experiments of Lubchenco (1978) showed that feeding preferences, population density, and competitive relations among food species all potentially contribute to the influences of "keystone" consumers on the structure of communities. What refinements did the work of Lubchenco add to the keystone species hypothesis?

4. When Power (1990) excluded predaceous steelhead and large roach from her river sites, the density of herbivorous insect larvae (chironomids) decreased. Use the food web described by Power to explain this response.

5. Using Tscharntke's food web (1992) shown in figure 17.5, predict which species would be most affected if you excluded the bird at the top of the web, *Cyanistes caeruleus.* What species would be affected less? Assume that *C. caeruleus* is a keystone species in this community.

6. Given our current understanding of the various ecological interactions in the cabbage/caterpillar/parasitoid/hyperparasitoid community characterized by Vyas et al. (shown in fig. 17.22), how would you advise cabbage farmers hoping to biologically control *Pieris* caterpillar damage in the long term?

7. All the keystone species work we have discussed in chapter 17 has concerned the influences of animals on the structure of communities. How could other groups of organisms, such as parasites and pathogens, act as keystone species?

Chapter 18

Primary and Secondary Production

Creatas Images/Jupiterimages

A world of green: a lush temperate rain forest in the Pacific Northwest state of Washington. The workings of the biosphere depend overwhelmingly on the production of energy-rich biomass by photosynthetic primary producers such as these green plants.

CHAPTER CONCEPTS

18.1 Terrestrial primary production is generally limited by temperature, moisture, and nutrients. 385
Concept 18.1 Review 387

18.2 Aquatic primary production is generally limited by nutrient availability. 387
Concept 18.2 Review 389

18.3 Primary producer diversity contributes to higher primary production. 390
Concept 18.3 Review 391

18.4 Consumers can influence rates of primary production in aquatic and terrestrial ecosystems through trophic cascades. 392
Concept 18.4 Review 396

18.5 Ecosystems with greater primary production generally support higher levels of secondary production. 396
Concept 18.5 Review 400

Applications: Using Stable Isotope Analysis to Study Feeding Habits 399
Summary 401
Key Terms 402
Review Questions 402

LEARNING OUTCOMES

After studying this section you should be able to do the following:

18.1 Distinguish between primary production and secondary production.
18.2 Define gross primary production, net primary production, secondary production, and trophic level.
18.3 Explain the significance of primary production to ecosystems.

The interactions between organisms and their environments are fueled by complex fluxes and transformations of energy. Sunlight shines down on the canopy of a forest—some is reflected, some is converted to heat energy, and some is absorbed by chlorophyll. Infrared radiation is

Figure 18.1   In most ecosystems, sunlight provides the ultimate source of energy to power all biological activity, such as the singing of this tree frog and the growth of the plant on which it sits.
Thomas Vinke/Getty Images

absorbed by the molecules in organisms, soil, and water, increasing their kinetic state and raising the temperature of the forest. Forest temperature affects the rate of biochemical reactions and transpiration by forest vegetation.

Forest plants use photosynthetically active radiation, or PAR (see chapters 5 and 7), to convert CO_2 into sugars and other forms of biomass, a process referred to as carbon fixation. The plants use some of the chemical energy in biomass to meet their own energy needs. Some fixed carbon goes directly into plant growth: to produce new leaves, to lengthen the tendrils of vines, to grow new root hairs, and so forth. Some biomass is stored as nonstructural carbohydrates, which act as energy stores in roots, seeds, or fruits.

A portion of the biomass produced by forest vegetation is consumed by herbivores, some is consumed by detritivores, and some ends up as soil organic matter. The energy contained in the biomass produced by forest vegetation powers bird flight through the forest canopy and fuels the muscle contractions of earthworms as they burrow through the forest soil. The forest vegetation is sunlight transformed, as are all the associated bacteria, fungi, and animals and all their activities (fig. 18.1).

We can view a forest as a system that absorbs, transforms, and stores energy. In this view, physical, chemical, and biological structures and processes are inseparable. When we look at a forest (or stream or coral reef) in this way, we view it as an ecosystem. As we saw in chapter 1, an ecosystem is a biological community plus all of the abiotic factors influencing that community. The term *ecosystem* and its definition were first proposed in 1935 by the British ecologist Arthur Tansley, who realized the importance of considering organisms and their environment as an integrated system. Tansley wrote: "Though the organisms may claim our primary interest, . . . we cannot separate them from their special environment, with which they form one physical system. It is the [eco]systems so formed which, from the point of view of the ecologist, are the basic units of nature on the face of the earth."

Ecosystem ecologists study the flows of energy, water, and nutrients in ecosystems and, as suggested by Tansley, pay as much attention to physical and chemical processes as they do to biological ones. Some fundamental areas of interest for ecosystem ecologists are primary production, secondary production, and nutrient cycling. We will discuss the first two topics in chapter 18 and nutrient cycling in chapter 19.

We saw in chapter 7 how the photosynthetic machinery of plants uses solar energy to synthesize sugars. In chapter 7, we considered photosynthesis from the perspective of the individual grass, tree, or cactus. Here we step back from the biochemical and physiological details of photosynthesis, and back even from the individual organism, to look at photosynthesis at the level of the whole ecosystem.

Primary production is the production of new organic matter, or biomass, by autotrophs in an ecosystem per unit area or volume during some period of time. In fact, not just primary production but all forms of ecological production are expressed per unit of area or volume (e.g., per m^2 or per m^3) and per unit of time (e.g., per hour or per year). Ecosystem ecologists distinguish between gross and net primary production. **Gross primary production** is the total primary production by all primary producers in the ecosystem. **Net primary production** is gross primary production minus respiration by primary producers; it is the amount of energy in the form of biomass available to the consumers in an ecosystem. **Secondary production** is the production of biomass by heterotrophic consumer organisms feeding on plants, animals, microbes, fungi, or detritus during some period of time, for example, per hour or per year. Secondary production, which is analogous to net primary production, includes consumer growth, reproduction, and, at the population level, mortality. Ecologists have measured primary production in a variety of ways but mainly as the rate of carbon uptake by primary producers or by the amount of biomass or oxygen produced.

We discussed feeding biology from a variety of perspectives in previous chapters. In chapter 7, we examined the biology of herbivores, detritivores, and carnivores. In chapter 14, we discussed the ecology of exploitative interactions, and in chapter 17, we used food webs as a means of representing the trophic structure of communities. Ecosystem ecologists are also concerned with trophic structure but have taken a different approach than population and community ecologists.

Ecosystem ecologists have simplified the trophic structure of ecosystems by arranging species into trophic levels based on the predominant source of their nutrition. A **trophic level** is a position in a food web and is determined by the number of transfers of energy from primary producers to that level. Primary producers occupy the first trophic level in ecosystems since they use inorganic forms of energy, principally light, to convert CO_2 into biomass. Herbivores and detritivores are often called primary consumers and occupy the second trophic level. Carnivores feeding on herbivores and detritivores are called secondary consumers and occupy the third trophic level. Predators that feed on carnivores occupy a fourth trophic level. Since each trophic level may contain several species, in some cases hundreds, an ecosystem perspective simplifies trophic structure.

Primary production, the conversion of inorganic forms of energy into organic forms, is a key ecosystem process. All consumer organisms, including humans, depend on primary production for their existence. Because of its importance and because rates of primary production vary substantially from one ecosystem to another, ecosystem ecologists study the factors controlling rates of primary production in ecosystems.

Patterns of natural variation in primary production provide clues to the environmental factors that control this key ecosystem process. Experiments test the importance of those controls. In chapter 18, we discuss the major patterns of variation in primary production in terrestrial and aquatic ecosystems and key experiments designed to determine the mechanisms producing those patterns. In the last sections of chapter 18, we consider secondary production and the use of stable isotope analysis to verify feeding habits of consumers.

18.1 Patterns of Terrestrial Primary Production

LEARNING OUTCOMES

After studying this section you should be able to do the following:

18.4 Describe the relationship between evapotranspiration and net primary production in terrestrial ecosystems.

18.5 Discuss evidence showing that nutrient availability can limit primary production in terrestrial ecosystems.

18.6 Design an experiment to test whether nutrient availability limits primary production in a particular terrestrial ecosystem.

Terrestrial primary production is generally limited by temperature, moisture, and nutrients. As we surveyed the major terrestrial biomes in chapter 2, you probably got a sense of the geographic variation in rates of primary production. Perhaps you also developed a feeling for the major environmental correlates with that variation. The variables most highly correlated with variation in terrestrial primary production are *temperature* and *moisture*. Highest rates of terrestrial primary production generally occur under warm, moist conditions.

Actual Evapotranspiration and Terrestrial Primary Production

Michael Rosenzweig (1968) estimated the influence of moisture and temperature on rates of primary production by plotting the relationship between annual net primary production and annual actual evapotranspiration. Annual **actual evapotranspiration (AET)** is the total amount of water that evaporates and transpires off a landscape during the course of a year and is measured in millimeters of water per year. The AET process is affected by both temperature and precipitation. The ecosystems showing the highest levels of primary production are those that are warm and receive large amounts of precipitation. Conversely, ecosystems show low

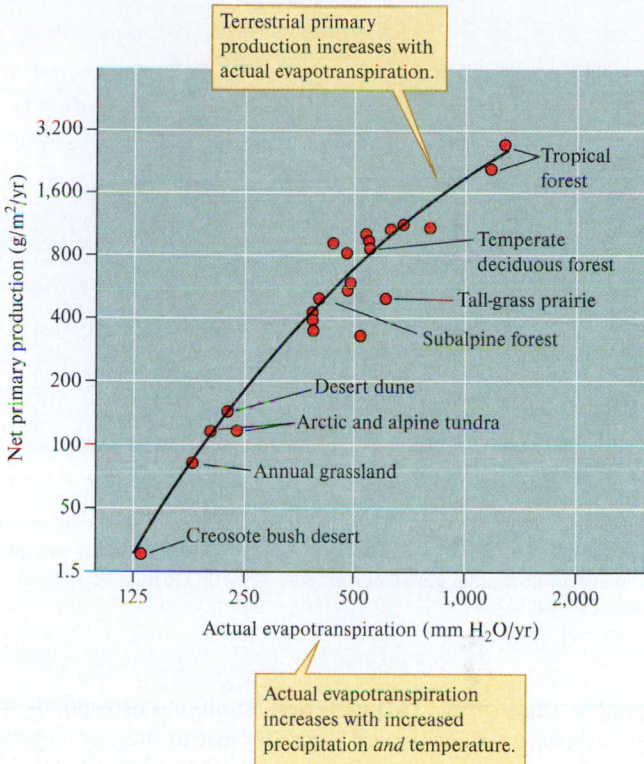

Figure 18.2   Relationship between actual evapotranspiration and net aboveground primary production in a series of terrestrial ecosystems (data from Rosenzweig 1968; Kaspari, O'Donnell, and Kercher 2000).

levels of AET because they receive little precipitation, or are very cold, or both. For instance, both hot deserts and tundra exhibit low levels of AET.

Figure 18.2 shows Rosenzweig's plot of the positive relationship between net primary production and AET, adjusted by later analyses by Michael Kaspari, Sean O'Donnell, and James Kercher (2000). Tropical forests show the highest levels of net primary production and AET. At the other end of the spectrum, hot, dry deserts and cold, dry tundra show the lowest levels. Intermediate levels occur in temperate forests, temperate grasslands, woodlands, and high-elevation forests. Figure 18.2 shows that AET accounts for a significant proportion of the variation in annual net primary production among terrestrial ecosystems.

Rosenzweig's analysis attempts to explain variation in primary production across the whole spectrum of terrestrial ecosystems. What controls variation in primary production within similar ecosystems? Osvaldo E. Sala and his colleagues (1988) at Colorado State University explored the factors controlling primary production in the central grassland region of the United States. Their study was based on data collected by the U.S. Department of Agriculture Soil Conservation Service at 9,498 sites. To make this large data set more manageable, the researchers grouped the sites into 100 representative study areas.

The study areas extended from Mississippi and Arkansas in the east, to New Mexico and Montana in the west, and from North Dakota to southern Texas. Primary production was highest in the eastern grassland study areas and lowest in the

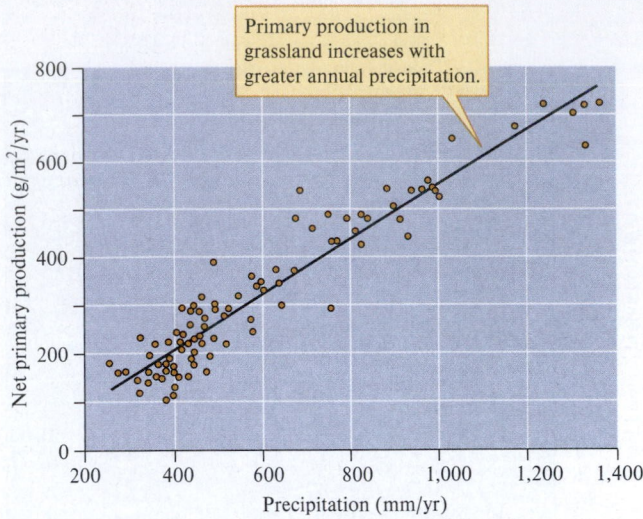

Primary production in grassland increases with greater annual precipitation.

Figure 18.3 Influence of annual precipitation on net aboveground primary production in grasslands of central North America (data from Sala et al. 1988).

western study areas. This east–west variation corresponds to the westward changes from tall-grass prairie to short-grass prairie that we reviewed in chapter 2. Sala and his colleagues found that this east–west variation in primary production among grassland ecosystems correlated significantly with the amount of rainfall (fig. 18.3).

Compare the plot by Sala and his colleagues (see fig. 18.3) with the one constructed by Rosenzweig (see fig. 18.2). How are they similar? How are they different? Both graphs have primary production plotted on the vertical axis as a dependent variable. However, while the Rosenzweig plot includes ecosystems ranging from tundra to tropical rain forest, the plot by Sala and his colleagues includes grasslands only. In addition, different variables are plotted on the horizontal axes of the two graphs. While Rosenzweig plotted actual evapotranspiration, which depends on temperature and precipitation, Sala and his colleagues plotted precipitation only. They found that including temperature in their analysis did not improve their ability to predict net primary production.

These researchers found strong correlations between AET or precipitation and rates of terrestrial primary production. However, their models did not completely explain the variation in primary production among the study ecosystems. For instance, in figure 18.2 ecosystems with annual AET levels of 500 to 600 mm of water showed annual rates of primary production ranging from 300 to 1,000 g per square meter. In figure 18.3, grassland ecosystems receiving 400 mm of annual precipitation had annual rates of primary production ranging from about 100 to 250 g per square meter. These differences in primary production challenge ecologists for an explanation.

Soil Fertility and Terrestrial Primary Production

Significant variation in terrestrial primary production can be explained by differences in soil fertility. Farmers have long known that adding fertilizers to soil can increase agricultural

production. However, it was not until the nineteenth century that scientists began to quantify the influence of specific nutrients, such as nitrogen (N) or phosphorus (P), on rates of primary production. Justus Liebig (1840) pointed out that nutrient supplies often limit plant growth. He also suggested that nutrient limitation to plant growth could be traced to a single limiting nutrient. This hypothetical control of primary production by a single nutrient was later called "Liebig's Law of the Minimum." We now know that Liebig's perspective was too simplistic. Usually several factors, including a number of nutrients, simultaneously affect levels of terrestrial primary production. However, his work led the way to a concept that remains true today: variation in soil fertility can significantly affect rates of terrestrial primary production.

Ecologists have increased primary production by adding nutrients to a wide variety of terrestrial ecosystems, including arctic tundra, alpine tundra, grasslands, deserts, and forests. For instance, Gaius Shaver and Stuart Chapin (1986) studied the potential for nutrient limitation in arctic tundra. They added commercial fertilizer containing nitrogen, phosphorus, and potassium to several tundra ecosystems in Alaska. They made a single application of fertilizer to half of their experimental plots and two applications to the remaining experimental plots.

Shaver and Chapin measured net primary production at their control and experimental sites 2 to 4 years after the first nutrient additions. Nutrient additions increased net primary production (by 23–300%) at all of the study sites. The response to fertilization was substantial and clear at most study sites. Four years after the initial application of fertilizer, net primary production on Kuparuk Ridge, for instance, was twice as high on the fertilized plots compared to the unfertilized control plots (fig. 18.4).

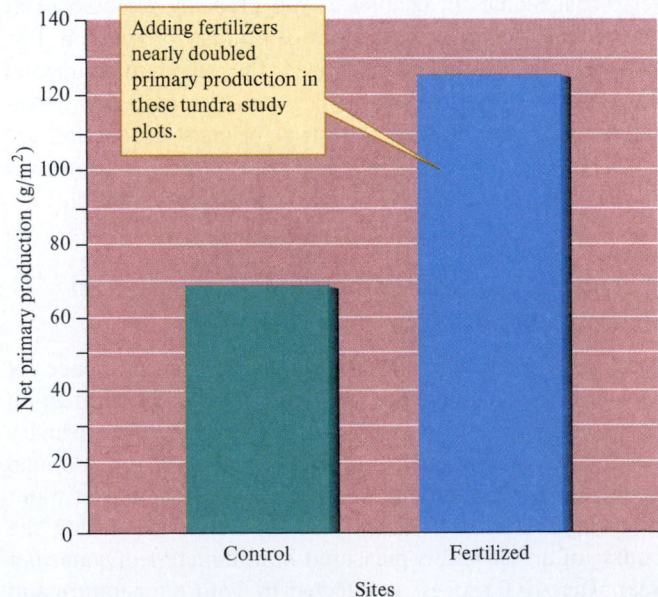

Adding fertilizers nearly doubled primary production in these tundra study plots.

Figure 18.4 Effect of addition of nitrogen, phosphorus, and potassium on net aboveground primary production in arctic tundra (data from Shaver and Chapin 1986).

Nutrient additions to alpine tundra indicate that the response of ecosystems to nutrient addition is affected by prior nutrient availability. William Bowman and his colleagues (1993) added nutrients to the alpine tundra on Niwot Ridge, Colorado. They conducted their experiment in adjacent dry alpine and wet alpine meadows at an elevation of 3,510 m. One of four treatments was applied in both the dry and wet alpine meadows: (1) control (no nutrient additions), (2) nitrogen added, (3) phosphorus added, and (4) nitrogen and phosphorus added. The researchers then measured soil nitrogen and phosphorus concentrations and annual net primary production in each study plot.

Initial concentrations of both nitrogen and phosphorus were higher in the wet meadow soils. Moreover, while fertilizing raised the concentrations of both nitrogen and phosphorus substantially in the dry meadow soils, fertilizing the wet meadow produced less change in soil nutrient concentrations.

Fertilizing produced greater increases in primary production in the dry meadow than in the wet meadow. Adding nitrogen to the dry meadow increased primary production by 63%. Adding nitrogen *and* phosphorus increased primary production by 178%. In contrast, the wet meadow showed relatively smaller but statistically significant responses to the additions of both nitrogen and phosphorus (fig. 18.5). Bowman and his colleagues suggest that these results show that nitrogen is the main nutrient limiting net primary production in the dry meadow and that nitrogen and phosphorus jointly limit net primary production in the wet meadow.

Experiments such as these have shown that despite the major influence of temperature and moisture on rates of primary production in terrestrial ecosystems, variation in nutrient availability can also have measurable influence. As we shall see in the next section, nutrient availability is the main factor limiting primary production in aquatic ecosystems.

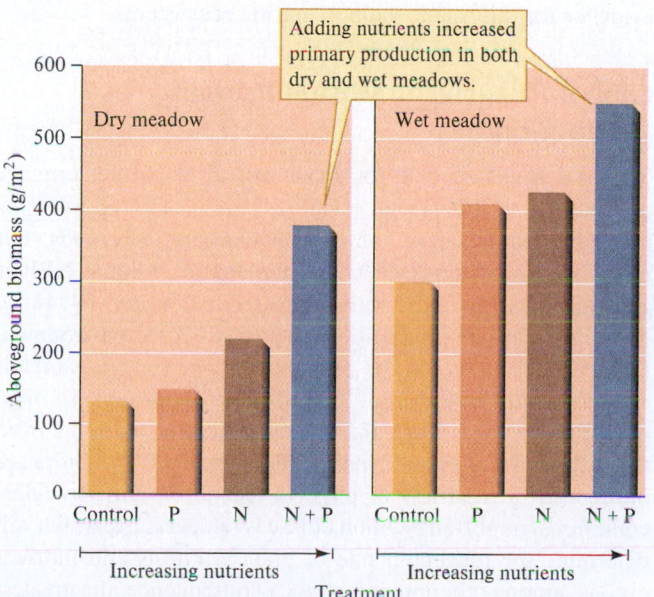

Figure 18.5   Effect of adding phosphorus (P) and/or nitrogen (N) on aboveground primary production in two environments in alpine tundra (data from Bowman et al. 1993).

Concept 18.1 Review

1. Why was precipitation alone, without temperature, sufficient to account for most of the variation in grassland net primary production across central North America (see fig. 18.3)?
2. How are the desert dune ecosystem and the arctic and alpine tundra ecosystems indicated in figure 18.2 the same?
3. How would actual evapotranspiration and net primary production in the desert dune ecosystem, which is a hot desert, and the arctic and alpine tundra ecosystems likely respond to a significant increase in precipitation?

18.2   Patterns of Aquatic Primary Production

LEARNING OUTCOMES

After studying this section you should be able to do the following:

18.7   Summarize the main factors limiting primary production in aquatic ecosystems.

18.8   Outline the types of evidence used to demonstrate nutrient limitation of primary production in aquatic ecosystems.

18.9   Compare nutrient limitation in freshwater and marine ecosystems.

Aquatic primary production is generally limited by nutrient availability.   Limnologists and oceanographers have measured rates of primary production and nutrient concentrations in many lakes and at many coastal and oceanic study sites. These studies have produced one of the best-documented patterns in the biosphere: the positive relationship between nutrient availability and rate of primary production in aquatic ecosystems.

Patterns and Models

A quantitative relationship between phosphorus, an essential plant nutrient, and phytoplankton biomass was first described for a series of lakes in Japan (Hogetsu and Ichimura 1954; Ichimura 1956; Sakamoto 1966). The ecologists studying this relationship found a remarkably good correspondence between total phosphorus and phytoplankton biomass. Later, Dillon and Rigler (1974) described a similar positive relationship between phosphorus and phytoplankton biomass for lake ecosystems throughout the Northern Hemisphere (fig. 18.6).

The data from Japan and North America strongly support the hypothesis that nutrients, particularly phosphorus, control phytoplankton biomass in lake ecosystems. However, what is the relationship between phytoplankton biomass and the rate of primary production? This relationship was explored by Val Smith (1979) for 49 lakes of the north temperate zone. The data from these lakes showed a strong positive correlation

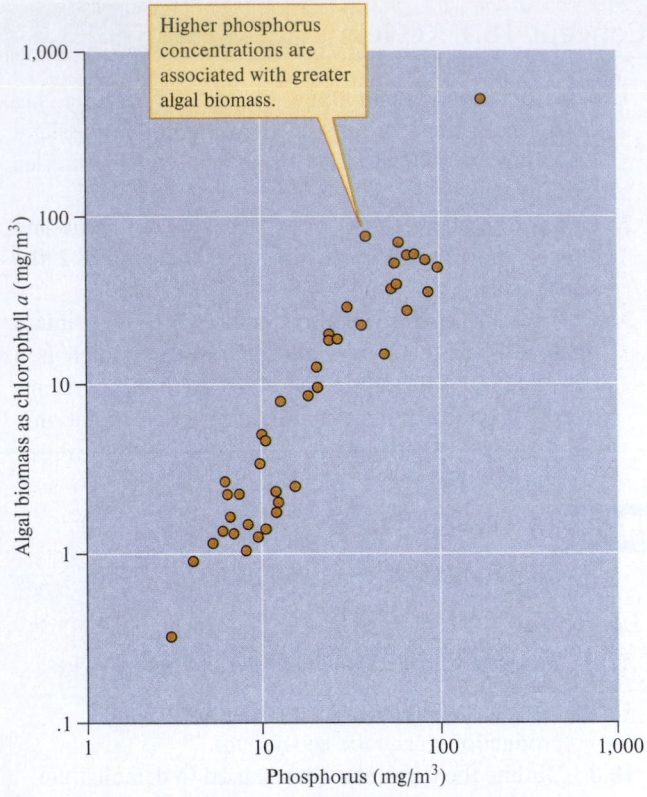

Figure 18.6   Relationship between phosphorus concentration and algal biomass in north temperate lakes (data from Dillon and Rigler 1974).

between chlorophyll concentrations and between total phosphorus concentration and photosynthetic rates. Aquatic ecologists have extended these correlational studies of the relationship between nutrient availability and primary production by manipulating nutrient availability in entire lake ecosystems.

Whole-Lake Experiments on Primary Production

The Experimental Lakes Area was founded in northwestern Ontario, Canada, in 1968, as a place in which aquatic ecologists could manipulate whole-lake ecosystems (Findlay and Kasian 1987; Mills and Schindler 1987). For instance, ecologists manipulated nutrient availability in a lake called Lake 226. They used a vinyl curtain to divide Lake 226 into two 8 ha basins, each containing about 500,000 m³ of water. Think about these numbers for a second. This was a huge experiment! Each subbasin of Lake 226 was fertilized from 1973 to 1980. The researchers added a mixture of carbon in the form of sucrose and nitrate to one basin, which increased phytoplankton biomass two to four times, and carbon, nitrate, and phosphate to the other basin, which increased phytoplankton biomass four to eight times. They stopped fertilizing the lakes after 1980 and then studied the recovery of the Lake 226 ecosystem from 1981 to 1983.

Both sides of Lake 226 responded significantly to nutrient additions. Prior to the manipulation, Lake 226 supported

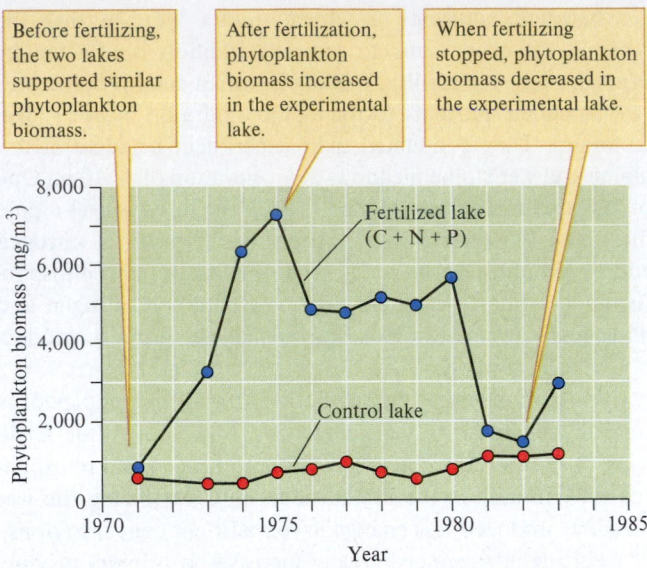

Figure 18.7   A whole-lake experiment shows the effect of nutrient additions (carbon (C) + nitrogen (N) + phosphorus (P)) on average phytoplankton biomass (data from Findlay and Kasian 1987).

about the same biomass of phytoplankton as two reference lakes (fig. 18.7). However, when experimenters began adding nutrients, the phytoplankton biomass in Lake 226 quickly surpassed that in the reference lakes. Phytoplankton biomass remained elevated in Lake 226 until the experimenters stopped adding fertilizer at the end of 1980. Then, from 1981 to 1983, the phytoplankton biomass in Lake 226 declined significantly.

In conclusion, both correlations—between phosphorus concentration and rate of primary production, and whole-lake experiments, involving nutrient additions—support the generalization that nutrient availability controls rates of primary production in freshwater ecosystems. Now, let's examine the evidence for this relationship in marine ecosystems.

Global Patterns of Marine Primary Production

Marine systems account for about half of global net primary productivity (NPP; Field et al. 1998). Nearly all of it is accomplished by algae, which includes large seaweeds, but most NPP is by single-celled phytoplankton. Global NPP is generally currently measured using satellite imagery (fig. 18.8). NPP in oceans is primarily a function of three environmental factors occurring in the photic zone where phytoplankton occur: (1) light availability, (2) availability of nutrients through upwelling (see fig. 3.7b), and (3) availability of nutrients from the breakdown of organic matter (Boyd et al. 2014). These are influenced by a variety of physical, chemical, and biological conditions. Importantly among these is temperature, which will determine the maximum rate of photosynthesis and nutrient cycling, among other processes. As a consequence, the greatest net primary productivity can be seen along continental margins, where there are upwellings of nutrients along the equator, and in shallow seas (Laufkötter et al. 2015).

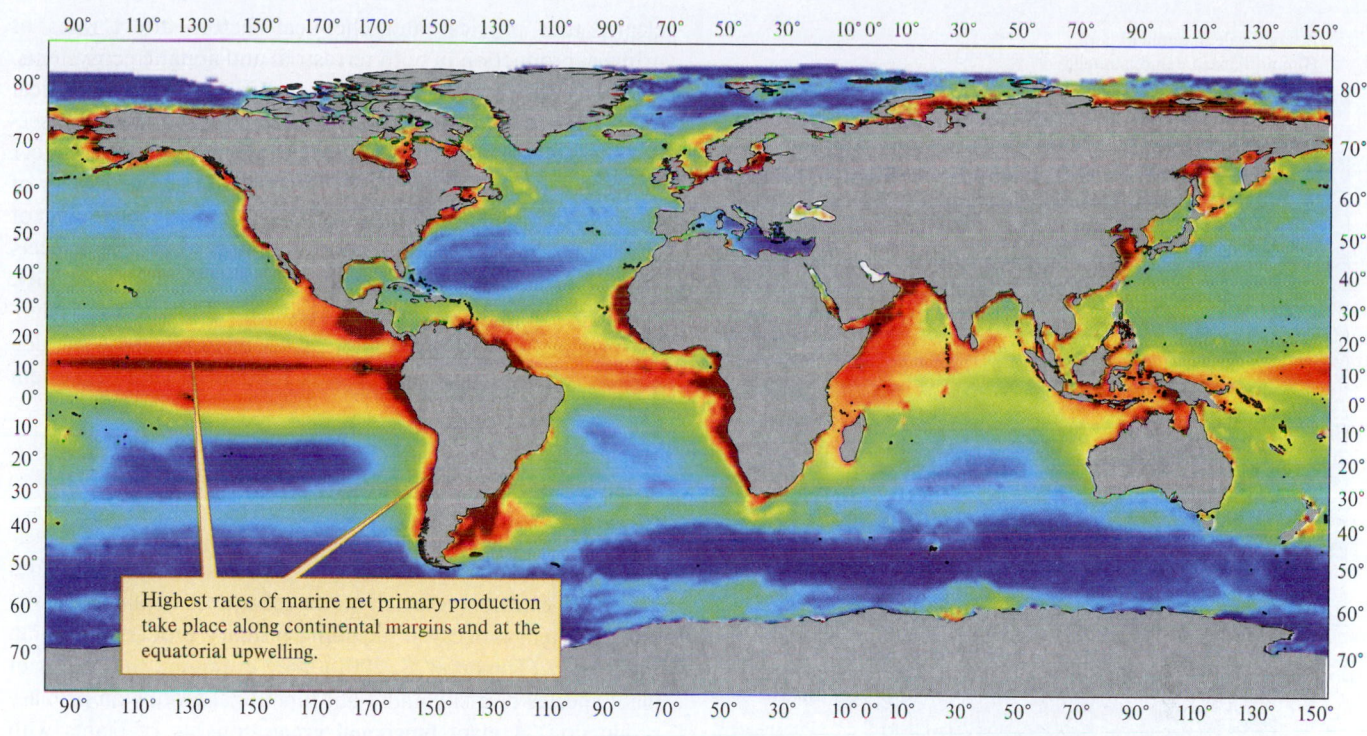

Highest rates of marine net primary production take place along continental margins and at the equatorial upwelling.

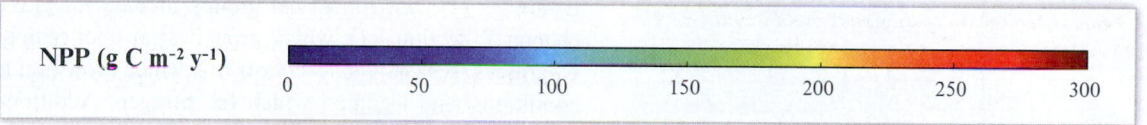

NPP (g C m⁻² y⁻¹)

$$\text{NPP (g C m}^{-2}\text{ y}^{-1})$$

0   50   100   150   200   250   300

Figure 18.8   Annual composite map of global areal of marine primary productivity rates, derived from Moderate Resolution Imaging Spectroradiometer (MODIS) Aqua satellite climatology from 2003 to 2012   (data from Boyd et al. 2014).

What is the experimental evidence for the role of nutrient availability for marine primary production? Edna Granéli and her colleagues (1990) have used nutrient enrichment to test whether nutrient availability limits primary production in the Baltic Sea.

In a test using a single algal species, Granéli added nutrients to filtered seawater from a series of study sites. She added nitrate to one experimental group, phosphates to another, and nothing to a third group of flasks (fig. 18.9). Notice that the flasks with additional nitrate showed increased chlorophyll *a* concentrations at all sites, while the flasks with additional phosphate had chlorophyll *a* concentrations very similar to the control flasks. What do these results indicate? They suggest that the rate of primary production in the Baltic Sea is limited by nutrients. However, in contrast to most freshwater lakes, the limiting nutrient appears to be nitrogen, not phosphorus.

Granéli did similar enrichment studies along a series of stations in the Kattegat, the Belt Sea, and the Skagerrak, where the salinity approaches that of the open ocean. However, in this second series of experiments, she used indigenous phytoplankton rather than a single standardized test species. Once again the concentrations of chlorophyll *a* were higher in the flasks to which nitrate had been added whereas the control

and phosphate treatment groups were virtually indistinguishable. Again, the results indicate nitrogen limitation along virtually the entire study area.

We can see from the imperfect relationships between nutrient concentration and phytoplankton biomass that other environmental factors are at play (figs. 18.6 and 18.9). In addition to temperature and light, CO_2 will also play a role, as will biotic processes including the intensity of predation on the zooplankton that feed on phytoplankton. As we shall discuss next, even primary producer diversity can affect productivity.

Concept 18.2 Review

1. Suppose that when you add nitrogen to one-half of a lake, you observe no change in phytoplankton biomass, but when you add phosphorus to the other half of the lake, phytoplankton biomass more than doubles. What is the most likely explanation of your results?

2. Suppose you fertilize a lake with nitrogen only, phosphorus only, and nitrogen plus phosphorus and observe no change in phytoplankton biomass. What is the most likely explanation of your results?

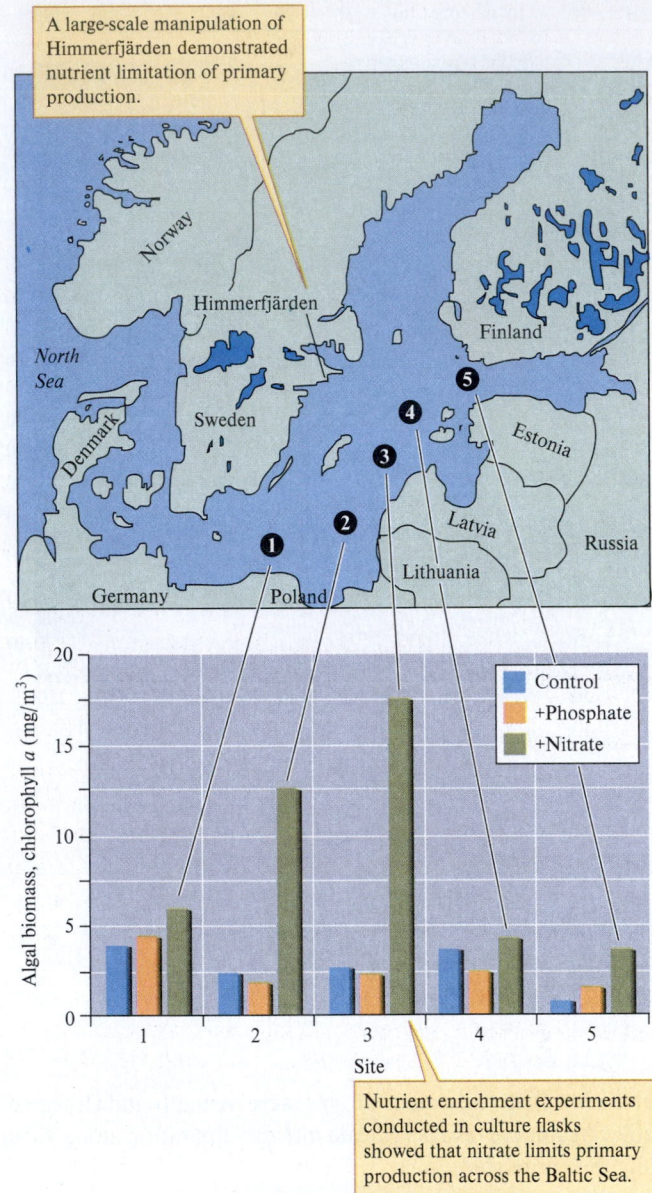

A large-scale manipulation of Himmerfjärden demonstrated nutrient limitation of primary production.

Nutrient enrichment experiments conducted in culture flasks showed that nitrate limits primary production across the Baltic Sea.

Figure 18.9 Nitrate control of primary production in the Baltic Sea (data from Granéli et al. 1990).

18.3 Primary Producer Diversity

LEARNING OUTCOMES

After studying this section you should be able to do the following:

18.10 List several components of overall biodiversity.

18.11 Distinguish between plant species diversity and plant functional diversity.

18.12 Discuss experimental evidence for a positive relationship between primary producer diversity and primary production in terrestrial and aquatic ecosystems.

Primary producer diversity contributes to higher primary production. Studies reviewed in sections 18.1 and 18.2 demonstrate

clearly that physical and chemical factors affect rates of primary production in both terrestrial and aquatic ecosystems. However, researchers have found that a number of biological factors can also influence rates of primary production. In section 18.3, we consider the contributions of primary producer diversity to rates of primary production. Our discussions of diversity thus far have focused on *species diversity*, the subject of chapter 16. We should note, however, that several other aspects of biological diversity, generally shortened to **biodiversity,** are of potential importance to ecological processes, including primary production. Other facets of biodiversity include genetic, physiological, anatomical, functional, and ecosystem diversity.

Terrestrial Plant Diversity and Primary Production

David Tilman of the University of Minnesota and colleagues have done extensive research on the influences of plant diversity on ecosystem processes, including primary production. In an early study (Tilman et al. 2001), they examined the effects of plant species diversity and plant functional groups on primary production. A **plant functional group** consists of plants with similar physiological and anatomical characteristics that influence their seasonality, resource requirements, and life histories. Examples of plant functional groups include C_3 grasses (see chapter 7, section 7.1), which grow best at cool temperatures, C_4 grasses (see chapter 7, section 7.1), which grow best in warm conditions; and legumes, which fix nitrogen. Additional functional groups included in the study were forbs and woody plants.

Tilman and colleagues estimated primary production on 168 grassland study plots, each measuring 9 m by 9 m, where the team varied the number of plant species and the plant functional groups present. The species planted in the study plots were chosen randomly from a pool of 18 species and varied from 1 species, which were monocultures, up to a maximum of 16. Plots were seeded with 10 g of seed per m^2 in May 1994 and 5 g per m^2 in May 1995, with the seed mass divided equally among the species planted. To maintain the experimental conditions, researchers regularly weeded study plots to remove any species not part of the experimental design.

Primary production was estimated once per year for 7 years, during the period of peak living plant biomass on study plots. Aboveground primary production in each plot was estimated by clipping the aboveground plant biomass in four 0.1 m by 3.0 m strips per plot. Belowground, root production was estimated from three 5 cm diameter by 30 cm deep soil cores taken in each clipped strip. Areas sampled were clipped and cored only once during the 7 years of the study. In the laboratory, plant biomass samples were sorted, to separate living from nonliving biomass, dried, and weighed.

The results of the study indicate strong influences of plant diversity on primary production. The number of plant species (i.e., plant species richness) in experimental plots was positively correlated with both aboveground and total primary production during all 7 years of the study. Figure 18.10 shows the positive relationship between plant species richness

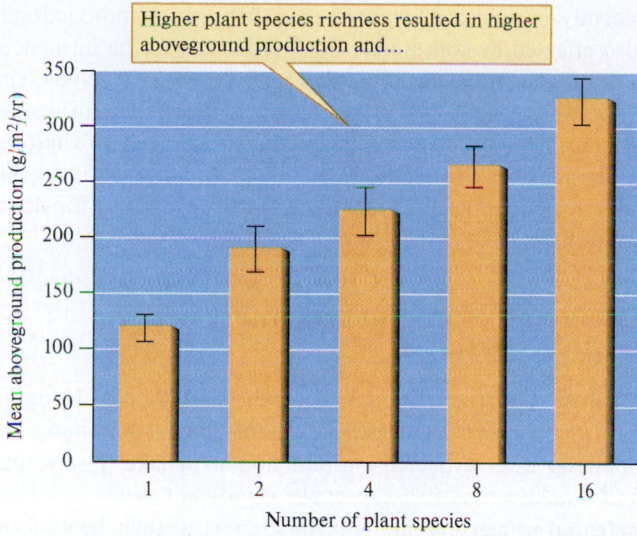

Higher plant species richness resulted in higher aboveground production and...

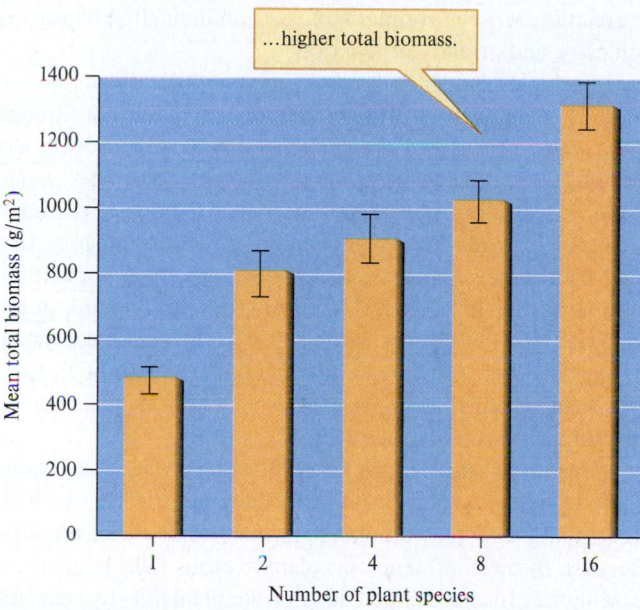

...higher total biomass.

Figure 18.10 Relationship between number of plant species in experimental plots and mean (± standard error) aboveground production and total biomass (data from Tilman et al. 2001).

and aboveground primary production and total biomass (aboveground plus belowground biomass) in 2000, the seventh year of the study. In that year, plots planted with 16 species averaged nearly three times the aboveground primary production and total biomass observed in plots planted with a single species (i.e., in monocultures). Functional group composition was also important. In particular, experimental plots planted with two particular plant functional groups, nitrogen-fixing legumes and C_4 grasses, supported higher levels of primary production on average. Follow-up analyses indicated that functional group composition and plant species richness were of approximately equal importance to primary production. The joint effects of species richness and functional groups strengthened during the course of the experiment, accounting for approximately one-third of variation in primary production across study plots at the beginning of the study and increasing to two-thirds of the variation by the end.

How do the effects of diversity on rates of primary production in terrestrial ecosystems compare to the effects of chemical and physical factors? In some cases, biological influences on primary production have been shown to be just as significant as those of temperature, moisture, and nutrients. For example, David Tilman, Peter Reich, and Forest Isbell found that increasing plant species richness in experimental plots from 4 to 16 species increased primary production as much as adding 54 kg of nitrogen fertilizer per hectare per year (Tilman, Reich, and Isbell 2012).

Algal Diversity and Aquatic Primary Production

Primary producer diversity has also been shown to increase primary production in experimental aquatic ecosystems. For example, Bradley Cardinale of the University of Michigan studied the influence of algal species richness on nutrient uptake and algal biomass in 150 small experimental streams (Cardinale 2011). The results of his study show a clear increase in nitrate uptake and biomass with increasing number of algal species. The positive effect of diversity was particularly strong in Cardinale's study in experiments that provided varied physical conditions analogous to those found in nature (fig. 18.11).

One of the greatest challenges to ecology is to determine the extent to which knowledge gained from small-scale experimental studies, such as those of Tilman and Cardinale, can be extrapolated to patterns and processes occurring on landscape and regional scales. A number of studies indicate that higher primary producer diversity is associated with higher primary production at these larger scales. For example, using data from 1,157 lakes in the United States, Emily Zimmerman and Bradley Cardinale (2014) found that, along with availability of nitrogen and phosphorus, algal diversity significantly correlated with lake primary production. On land, tree diversity has been linked to higher primary production across a sample of 12,000 Canadian and 54,000 Spanish forest study plots (Paquette and Messier 2011; Ruiz-Benito et al. 2014). Moreover, it has been found that plant diversity positively correlates with soil organic carbon in forests, shrublands, and grasslands (Chen et al. 2018). In the next section, we consider the effects of consumers on rates of primary production.

Concept 18.3 Review

1. What component of species diversity (see chapter 16, section 16.3) did Tilman's research group manipulate in their studies? What other components of species diversity could influence rates of primary production?
2. What does the effect of legumes on primary production in the Tilman experiment suggest about other factors limiting production on the experimental plots?
3. If plant species richness and functional group composition accounted for one-third to two-thirds of variation in primary production across study plots, what other factors likely accounted for the remainder of differences in primary production among plots?

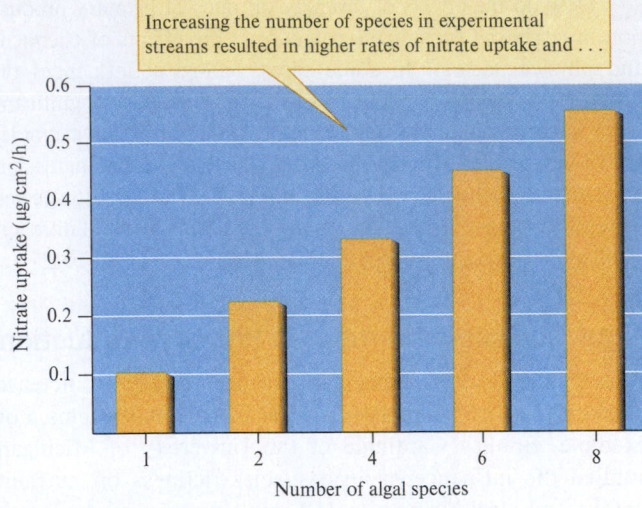

Increasing the number of species in experimental streams resulted in higher rates of nitrate uptake and . . .

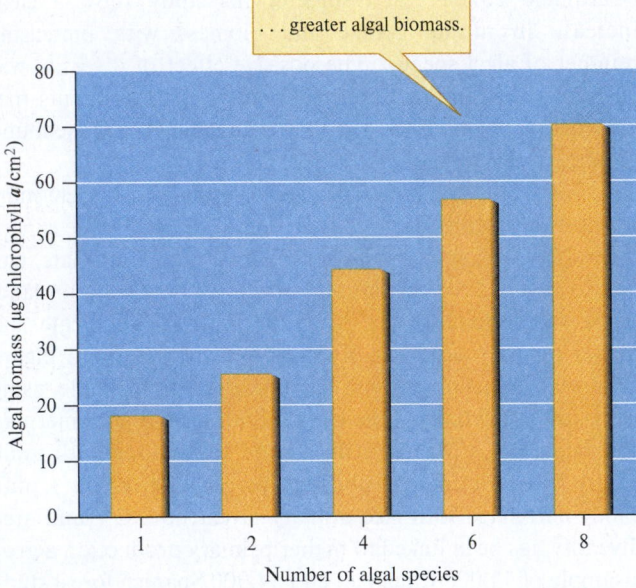

. . . greater algal biomass.

Figure 18.11 Relationship between number of algal species and mean nitrate uptake and mean algal biomass in experimental streams (data from Cardinale 2011).

18.4 Consumer Influences

LEARNING OUTCOMES

After studying this section you should be able to do the following:

18.13 Compare bottom-up and top-down controls on ecosystem processes.

18.14 Describe a trophic cascade.

18.15 Summarize evidence for trophic cascades in aquatic and terrestrial environments.

18.16 Explain the mechanisms producing a trophic cascade.

Consumers can influence rates of primary production in aquatic and terrestrial ecosystems through trophic cascades. In the first section of chapter 18, we emphasized the effects of physical and chemical factors on rates of primary production. More recently, ecologists have discovered that primary production is also affected by consumers. Ecologists refer to the influences of physical and chemical factors on ecosystems, such as temperature and nutrients, as **bottom-up controls.** The influences of consumers on ecosystems are known as **top-down controls.** In the previous two sections, we discussed bottom-up controls on rates of primary production. Here we discuss top-down control.

Piscivores, Planktivores, and Lake Primary Production

Stephen Carpenter, James Kitchell, and James Hodgson (1985) proposed that while nutrient inputs determine the potential rate of primary production in a lake, piscivorous and planktivorous fish can cause significant deviations from potential primary production. In support of their hypothesis, Carpenter and his colleagues (1991) cited a negative correlation between zooplankton size, an indication of grazing intensity, and primary production.

Carpenter and Kitchell (1988) proposed that the influences of consumers on lake primary production propagate through food webs. Since they visualized the effects of consumers coming from the top of food webs to the base, they called these effects on ecosystem properties *trophic cascades.* A **trophic cascade** involves effects of predators on prey that alter abundance, biomass, or productivity of a population, community, or trophic level across more than one link in the food web (fig. 18.12). Because predators' influence on ecosystem properties, such as primary production, occur through their effects on intermediary species, trophic cascades involve indirect interactions (see section 17.2).

Carpenter and Kitchell (1993) interpreted the trophic cascade in their study lakes as follows: Piscivores, such as largemouth bass, feed on planktivorous fish and invertebrates. Because of their influence on planktivorous fish, largemouth bass indirectly affect populations of zooplankton. By reducing populations of planktivorous fish, largemouth bass reduce feeding pressure on zooplankton populations. Large-bodied zooplankton, the preferred prey of size-selective fish (see chapter 7), soon dominate the zooplankton community. A dense population of large zooplankton reduces phytoplankton biomass and the rate of primary production. This interpretation of the trophic cascade is consistent with the negative correlation between zooplankton body size and primary production reported by Carpenter and his research team. This hypothesis is summarized in figure 18.13.

Carpenter and Kitchell tested their trophic cascade model by manipulating the fish communities in two lakes and using a third lake as a control. Figure 18.14 shows the overall design of their experiment. Two of the lakes contained substantial populations of largemouth bass. A third lake had no bass, due to occasional winterkill, but contained an abundance of planktivorous minnows. The researchers removed 90% of the largemouth bass from one experimental lake and put them into the other. They simultaneously removed 90% of the planktivorous minnows from the second lake

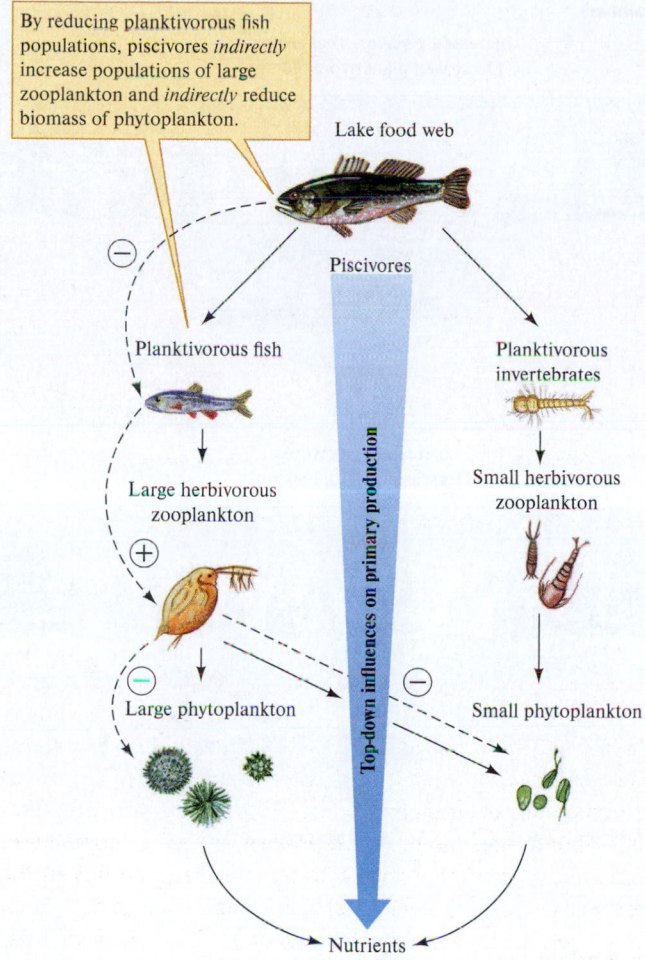

By reducing planktivorous fish populations, piscivores *indirectly* increase populations of large zooplankton and *indirectly* reduce biomass of phytoplankton.

Lake food web

Piscivores

Planktivorous fish

Planktivorous invertebrates

Large herbivorous zooplankton

Small herbivorous zooplankton

Large phytoplankton

Small phytoplankton

Top-down influences on primary production

Nutrients

Figure 18.12   The trophic cascade hypothesis, a result of "cascading" indirect interactions. Lines indicate both direct (solid) and indirect (dashed) relationships.

and introduced them to the first. They left a reference lake unmanipulated as a control.

The responses of the study lakes to the experimental manipulations support the trophic cascade hypothesis (see fig. 18.14). Reducing the planktivorous fish population led to reduced rates of primary production. In the absence of planktivorous minnows, the predaceous invertebrate *Chaoborus* became more numerous. *Chaoborus* fed heavily upon the smaller herbivorous zooplankton, and the herbivorous zooplankton assemblage shifted in dominance from small to large species. In the presence of abundant, large herbivorous zooplankton, phytoplankton biomass and rate of primary production declined.

Adding planktivorous minnows produced a complex ecological response. Increasing the planktivorous fish population led to increased rates of primary production. However, though the researchers increased the population of planktivorous fish in this experimental lake, they did so in an unintended way. Despite the best efforts of the researchers, a few bass remained. Therefore, by introducing a large number of minnows, they basically fed the remaining bass. An increased food supply combined with reduced population density induced a strong numerical response by the bass population (see chapter 10). The manipulation increased the reproductive rate of the remaining largemouth bass 50-fold, producing an abundance of young largemouth bass that feed voraciously on zooplankton.

The lake ecosystem responded to the increased biomass of planktivorous fish (young largemouth bass) as predicted at the outset of the experiment. The biomass of zooplankton decreased sharply, the average size of herbivorous zooplankton decreased, and phytoplankton biomass and primary production increased.

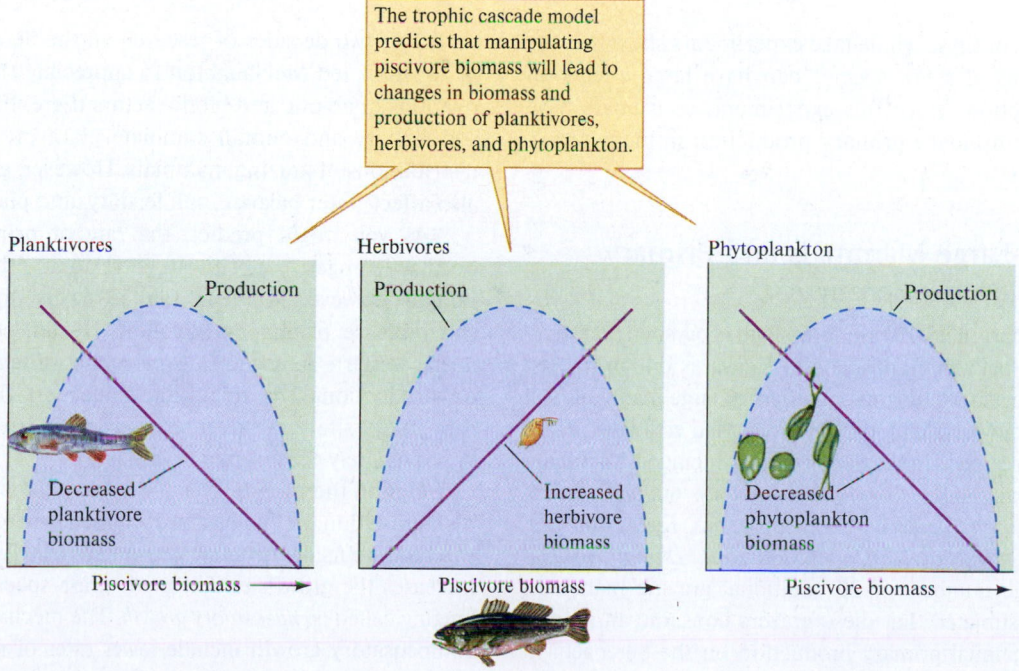

The trophic cascade model predicts that manipulating piscivore biomass will lead to changes in biomass and production of planktivores, herbivores, and phytoplankton.

Planktivores

Production

Decreased planktivore biomass

Piscivore biomass

Herbivores

Production

Increased herbivore biomass

Piscivore biomass

Phytoplankton

Production

Decreased phytoplankton biomass

Piscivore biomass

Figure 18.13   Predicted effects of piscivores on planktivore, herbivore, and phytoplankton biomass and production (data from Carpenter, Kitchell, and Hodgson 1985).

Experimental manipulations

Reduced piscivore (bass) biomass
Increased planktivore biomass

Increased piscivore (bass) biomass
Decreased planktivore biomass

Decreased herbivores
Increased phytoplankton

Responses

Increased herbivores
Decreased phytoplankton

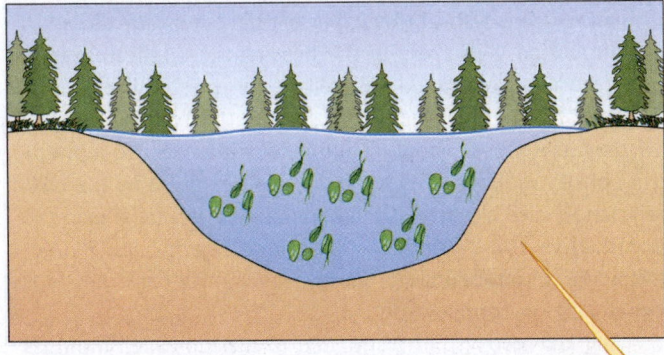
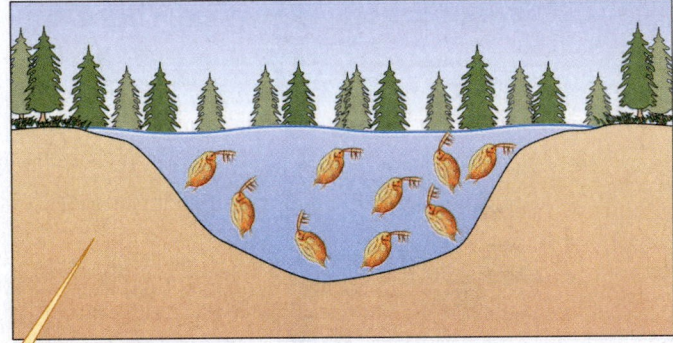

The responses of herbivores and phytoplankton to manipulations of piscivore and planktivore biomass support the trophic cascade model.

Figure 18.14   Experimental manipulations of ponds and responses.

The results of these whole-lake experiments show that the trophic activities of a few species can have large effects on primary production. Since this experiment, consumers have been shown to influence primary production in many terrestrial ecosystems.

Grazing by Large Mammals and Primary Production on the Serengeti

The Serengeti-Mara, a 25,000 km^2 grassland ecosystem that straddles the border between Tanzania and Kenya, is one of the last ecosystems on earth where great numbers of large mammals still roam freely. Sam McNaughton (1985) reported estimated densities of the major grazers in the Serengeti that included 1.4 million wildebeest, *Connochaetes taurinus albujubatus;* 600,000 Thomson's gazelle, *Gazella thomsonii;* 200,000 zebra, *Equus burchelli;* 52,000 buffalo, *Syncerus caffer;* 60,000 topi, *Damaliscus korrigum;* and large numbers of 20 additional grazing mammals. McNaughton estimated that these grazers consume an average of 66% of the annual primary production on the Serengeti. In light of this estimate, the potential for consumer influences on primary production seems very high.

Over two decades of research on the Serengeti ecosystem in Tanzania led McNaughton to appreciate the complex interrelations of abiotic and biotic factors there. For instance, both soil fertility and rainfall stimulate plant production and the distributions of grazing mammals. However, grazing mammals also affect water balance, soil fertility, and plant production.

As you might predict, the rate of primary production on the Serengeti is positively correlated with the quantity of rainfall. However, McNaughton (1976) also found that grazing can increase primary production. He fenced in some areas in the western Serengeti to explore the influence of herbivores on production. The migrating wildebeest that flooded into the study site grazed intensively for 4 days, consuming approximately 85% of plant biomass.

During the month after the wildebeest left the study area, biomass within the enclosures decreased, whereas the biomass of vegetation outside the enclosures increased (fig. 18.15). Grazing increases the growth rate of many grass species, a response to grazing called *compensatory growth*. The mechanisms underlying compensatory growth include lower rates of respiration due to lower plant biomass, reduced self-shading, and improved water balance due to reduced leaf area.

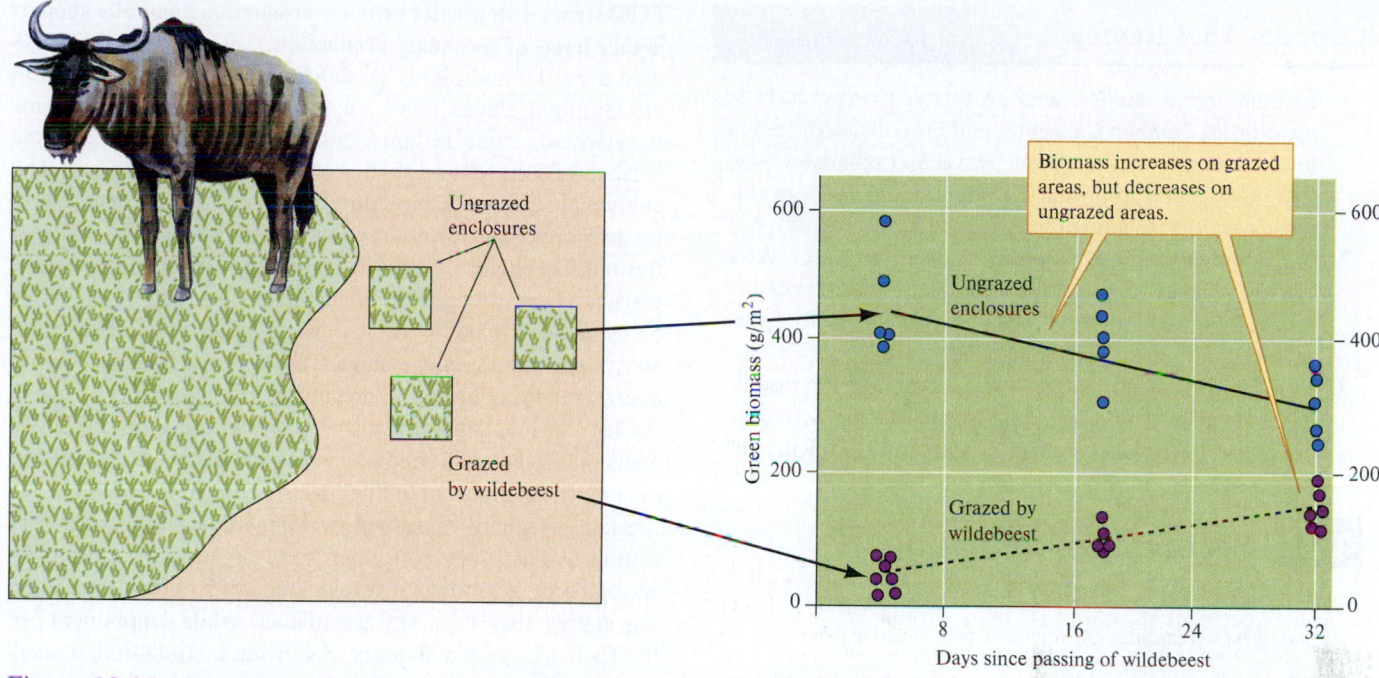

Figure 18.15   Growth response by grasses grazed by wildebeest (data from McNaughton 1976).

The compensatory growth observed by McNaughton was highest at intermediate grazing intensities (fig. 18.16). Apparently, light grazing is insufficient to produce compensatory growth and very heavy grazing reduces the capacity of the plant to recover. The large grazing mammals of the Serengeti have substantial influences on its rate of primary production. As McNaughton put it, "African ecosystems cannot be understood without close consideration of the large mammals. These animals interact with their habitats in complex and powerful patterns influencing ecosystems for long periods."

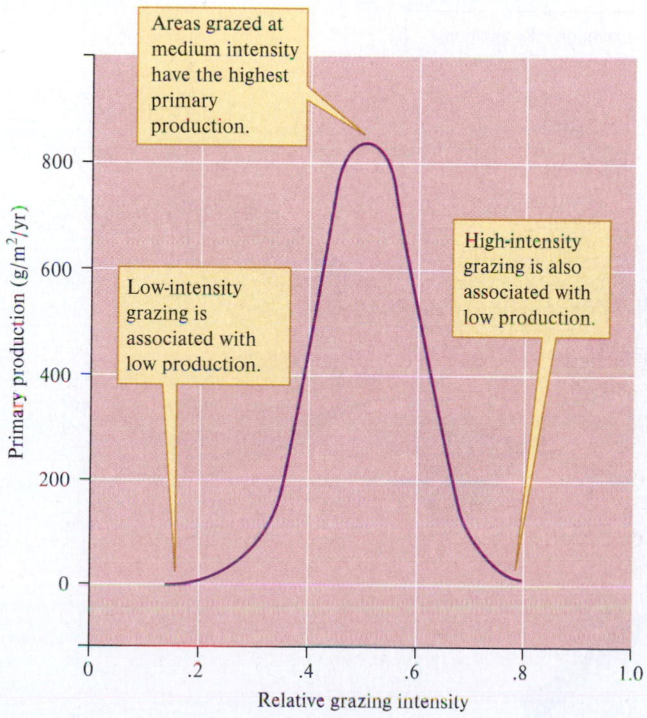

What McNaughton and his colleagues described is essentially the bottom portion of a potential trophic cascade in which the feeding activities of herbivores have a major influence on primary production. The full trophic cascade would include the 10 species of major carnivores on the Serengeti, including lions, hyenas, African wild dogs, leopards, and cheetahs. Consistent with the requirements for a trophic cascade, these predators appear to control the numbers of small and medium-sized mammalian herbivores (<150 kg) on the Serengeti (Sinclair, Mduma, and Brashares 2003). When predators were removed from a portion of the Serengeti for 8 years, the populations of five species of herbivores, for which there were data, increased several-fold. Populations of these herbivores decreased in density once the predators returned to the area. Unfortunately, there is no information on how primary production may have responded during the period of increased herbivore densities. We have already considered another system that may qualify as a terrestrial trophic cascade, involving wolves, elk, and riparian willow and aspen trees (Beschta et al. 2018; see section 14.4). However, the role of wolves in that system remains a focal point for debate, and more time will be required to see how primary producers ultimately respond to the return of wolves to the Yellowstone landscape (Fleming 2019).

In section 18.4, we've reviewed how consumers can influence rates of primary production. In the next section, we explore the connection between primary production and consumer production.

Figure 18.16   Grazing intensity and primary production of Serengeti grassland (data from McNaughton 1985).

Concept 18.4 Review

1. In their initial studies, leading to the trophic cascade hypothesis, Stephen Carpenter and his colleagues (1991) found a negative correlation between zooplankton size and phytoplankton primary production. What does this mean (see Investigating the Evidence 7 in Appendix A)?
2. Since increased phytoplankton biomass decreases water clarity in lakes, how should increased fishing pressure on the bass population in a lake ecosystem, such as that pictured in figure 18.12, affect lake clarity?
3. Why is it more difficult to obtain evidence for trophic cascades in terrestrial ecosystems such as the Serengeti, compared to the lakes studied by Carpenter and Kitchell?

18.5   Secondary Production

LEARNING OUTCOMES

After studying this section you should be able to do the following:

18.17 Define ecological efficiency and explain how this concept relates to Eltonian energy pyramids and secondary production.
18.18 Discuss the relative importance of top-down versus bottom-up controls on secondary production.
18.19 Discuss the influence of enrichment of primary production on secondary production by primary consumers and predators.
18.20 Summarize how studies of consumer responses to enrichment support ratio-dependent predator–prey models.

Ecosystems with greater primary production generally support higher levels of secondary production. We began chapter 18 with a partial and highly qualitative energy budget for a forest: Sunlight shines down on the canopy of a forest—some is reflected, some is converted to heat energy, and some is absorbed by chlorophyll and fuels photosynthesis. The energy budgets of ecosystems reveal that with each transfer or conversion of energy, some energy is lost. Some of the most ecologically significant of those energy losses take place as energy flows from one trophic level to another. While focusing on trophic levels can lead to better understanding of ecosystem processes, we should not forget that trophic level energy relations are an aggregate of processes occurring at the level of individual organisms, for example, as an herbivore feeds on a plant or a predator consumes its prey. As a model of energy relations of individuals, figure 18.17 summarizes the fate of energy consumed by carp studied in a laboratory setting. As with any consumer, food ingested by a carp has several fates. As food is digested, some of its constituents and the energy they bear are assimilated, while components of the food are egested as feces. Following assimilation, a small amount of energy is lost in the excretion of nitrogen waste products, while much more is lost in respiration as the individual carp meets its own energy needs. The remaining food and the energy it contains go into growth and reproduction. This remainder is *secondary production* at the level of the individual organism. Secondary production calculations at the population level must also incorporate mortality along with growth and reproduction.

One of the first ecologists to quantify the flux of energy through ecosystems was Raymond Lindeman.

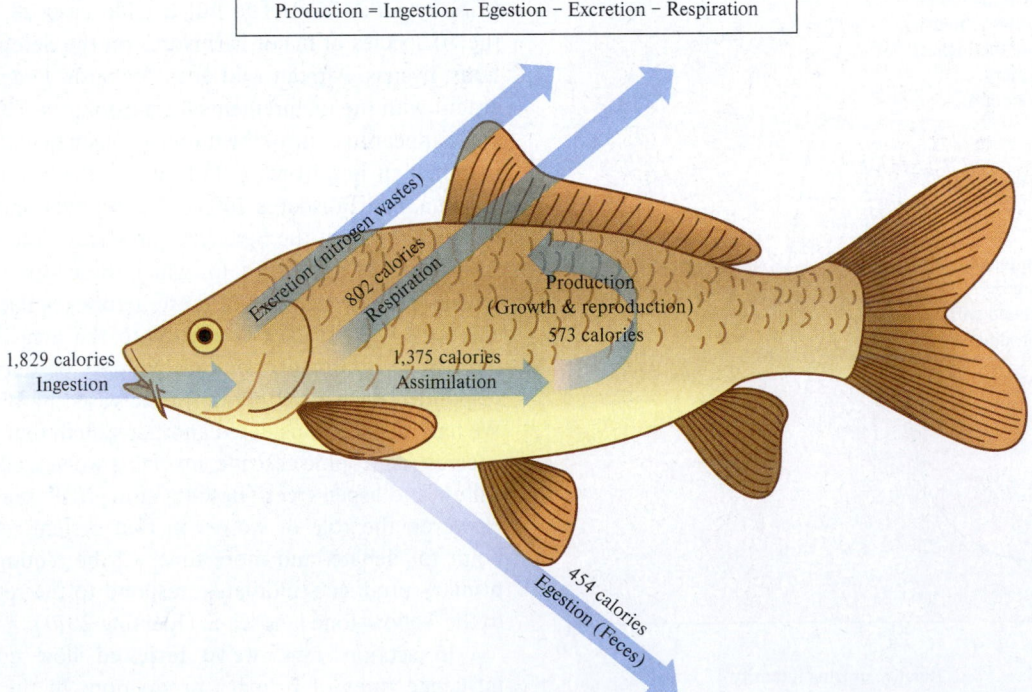

Production = Ingestion − Egestion − Excretion − Respiration

Excretion (nitrogen wastes)

802 calories
Respiration

Production
(Growth & reproduction)
573 calories

1,829 calories
Ingestion

1,375 calories
Assimilation

454 calories
Egestion (Feces)

Figure 18.17   Energy budget for a yearling carp. Production, the amount of biomass accruing as growth and reproduction, is determined by the rate of food ingestion minus energy losses with respiration, egestion of feces, and excretion of nitrogen wastes, for example, ammonium by fish or urea by mammals (data from Lindeman 1942).

A Trophic Dynamic View of Ecosystems

Raymond Lindeman (1942) received his PhD from the University of Minnesota in 1941, where his studies of the ecology of Cedar Bog Lake led him to a view of ecosystems far ahead of its time. Lindeman went from Minnesota to Yale University, where his association with G. E. Hutchinson from 1941 to 1942 led to the publication of a revolutionary paper with the provocative title "The Trophic-Dynamic Aspect of Ecology." In this paper, Lindeman articulated a view of ecosystems centered on flows of energy through ecosystems that remains influential to this day. Like Tansley before him, Lindeman pointed out the difficulty and artificiality of separating organisms from their environment and promoted an ecosystem view of nature. Lindeman concluded that the ecosystem concept is fundamental to the study of **trophic dynamics,** which he defined as the transfer of energy from one part of an ecosystem to another.

Lindeman suggested grouping organisms within an ecosystem into trophic levels: primary producers, primary consumers, secondary consumers, tertiary consumers, and so forth. In this scheme, each trophic level feeds on the one immediately below it. Energy enters the ecosystem as primary producers engage in photosynthesis using solar energy to convert CO_2 into biomass. Energy is transferred from one level to the next through consumption, only a portion of which is assimilated (some is defecated), and only a portion of that becomes production (some is respired), which results an increase in consumer mass. The percentage of biomass produced at a lower trophic level that is transferred to biomass produced at the next higher trophic level is called **ecological efficiency,** which varies from about 5% to 20%. As a result, the quantity of energy production decreases with each successive trophic level, which results in a pyramid-shaped distribution of energy among trophic levels. Lindeman called these trophic pyramids "Eltonian pyramids," since Charles Elton (1927) was the first to propose that the distribution of energy among trophic levels is shaped like a pyramid.

Figure 18.18 shows the distribution of annual production among trophic levels in Cedar Bog Lake and in Lake Mendota, Wisconsin. As predicted by Elton, the distribution of energy production across trophic levels in both lakes is shaped like a pyramid. This shape results from energy losses between trophic levels. Ecologists have proposed that these energy losses are also a major contributor to limiting the number of trophic levels in ecosystems. For instance, Lake Mendota supports four trophic levels, while Cedar Bog Lake includes only three.

Following Lindeman's pioneering work, many other ecologists have studied energy flow. Secondary production is key to understanding energy flow in ecosystems.

Top-down Versus Bottom-up Controls on Secondary Production

Just as we considered studies of the impacts of both lower and higher trophic levels on primary production, there is a large body of research investigating such controls on secondary production, and on herbivore populations in general. We have seen how predation pressure (i.e., top-down) can influence populations of their prey (e.g., Chapter 10), but bottom-up pressures should not be underestimated. Plants can vary in nutritional quality and abundance. They also can have a suite of defenses, both physical (e.g., thorns) and chemical (i.e., secondary plant compounds) that will affect availability as a resource to herbivores.

The relative strength of top-down (predator) versus bottom-up (resource) effects has important implications for understanding both trophic dynamics and evolution; for example, is it more important for an herbivorous insect to have traits that allow it to overcome plant defenses (bottom-up) or escape predation (top-down)? Research has shown that there can be energetic trade-offs between these (e.g., Murphy and Loewy 2015).

Certainly, which is more important may differ depending on which ecosystem or specific organism, but which is generally the stronger effect in nature? Because of the many plant defenses to prevent herbivory, it had long been assumed that bottom-up effects were more important. However, some researchers have argued that herbivores are more likely controlled not by resource availability but instead by top-down pressures; this is referred to as "the world is green" hypothesis (Hairston et al. 1960). That is, the very fact of the dominance of plants on our planet is evidence that herbivore populations are being kept in check by their predators. Which hypothesis is supported by the research?

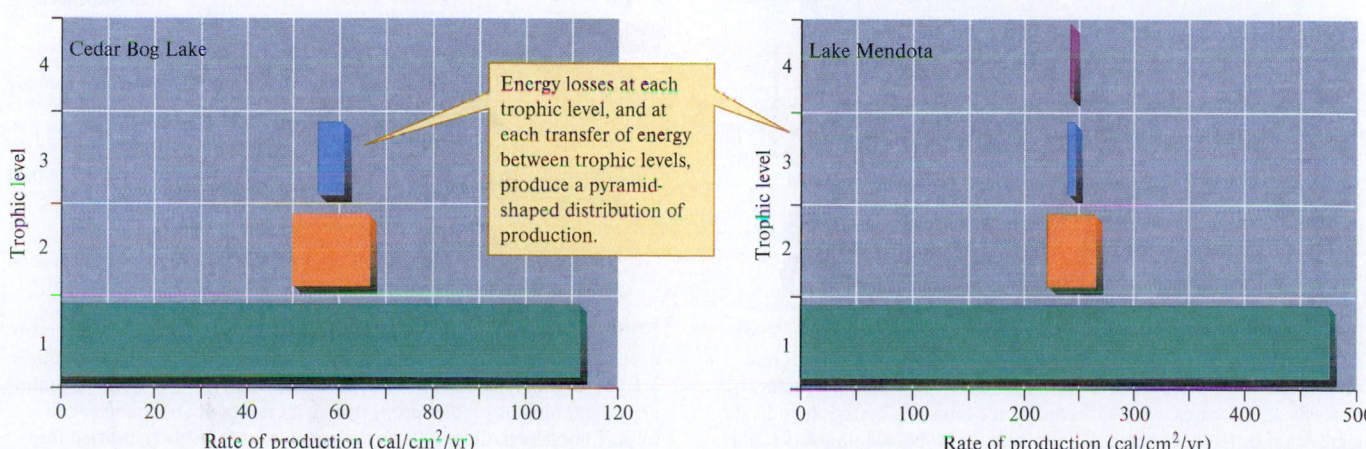

Figure 18.18   Annual production by trophic level in two lakes (data from Lindeman 1942).

To answer this question, Mayra Vidal and Shannon Murphy conducted a meta-analysis of the literature on top-down versus bottom-up effects for herbivorous insects (Vidal and Murphy 2018). Terrestrial insects are a good model organism to investigate trophic systems because of their tremendous diversity and the volume of studies that have been published on them. Vidal and Murphy reviewed the results of 1,617 publications on the trophic dynamics of insects; of these, they found 75 publications that had quantified both, that is, that had measured **tri-trophic interactions.** Most publications had multiple measures; thus, the final sample size of the meta-analysis was 365.

Vidal and Murphy discovered that while impacts from both trophic levels were important, top-down effects were significantly stronger than bottom-up effects in most cases, including for most types of herbivores (fig. 18.9). The only feeding guild for which top-down and bottom-up forces did not differ were mining insects; that is, those whose larvae feed by mining the middle tissues of a leaf. This may be true because their mode of feeding protects them from some types of predation. It is also important to note that they did not find any difference between studies that were done on natural versus experimental populations, further reflecting the consistency of their findings.

Linking Primary Production and Secondary Production

Given that top-down forces are generally stronger than bottom-up (at least for the largest group of herbivores on the planet), how much does resource availability actually matter? Is there a correlation between primary production and secondary production in an ecosystem? If primary production increases, will secondary production rise across trophic levels? Before addressing these questions, let's step back and consider that the relation between primary and secondary production does not occur in the abstract but rather results from the rates at which consumers feed on their food resources. These are precisely the interactions represented by alternative models reviewed in chapter 14,

where we compared the accuracy of functional response predictions made by prey-dependent versus ratio-dependent models of predator–prey interactions (see section 14.3). These two classes of predator–prey models also make contrasting predictions of how increased primary production, commonly referred to as "enrichment," will affect abundance (biomass or numbers) of consumers (Arditi and Ginzburg 1989). In an ecosystem with three trophic levels, both models predict a proportional increase in abundance of predators occupying the top trophic level (fig. 18.20). In contrast, while prey-dependent models predict no change in primary consumer abundance in response to enrichment, the ratio-dependent model predicts that primary consumer abundance will increase proportionally to the level of enrichment. In fact, the ratio-dependent model predicts proportional increases in abundance of consumers at all trophic levels in response to enrichment regardless of the number of trophic levels in the ecosystem. In contrast, prey-dependent models predict that an increase in abundance proportional to degree of enrichment will occur only at the top trophic level. For example, in the ecosystem with three trophic levels shown in figure 18.20, the prey-dependent models assume that feeding by top predators prevents an increase in primary consumer abundance. According to prey-dependent models, abundance at lower trophic levels may show nonlinear increases, show nonlinear decreases, or remain unchanged with enrichment.

One of the reasons for this may be the generally unseen role of microorganisms in facilitating and mitigating the relationships between plants and herbivores. Microorganisms act as symbiotic mutualists, pathogens, and parasites for both plants and herbivores, and as such can both increase and decrease availability of primary producers as a food source. An increasing number of studies show that the microbiome plays an important role in the relationship between insects and plants, creating its own trophic cascade (Ali et al. 2020). For example, viruses and bacterial pathogens can alter plant-insect interactions by changing the plant nutritional value or defenses (e.g., Patton et al. 2020). Meanwhile, bacteria in the gut of insects can help to break down plant defenses (Mason et al. 2019). Nonetheless experimental and observational studies show clearly that increased primary

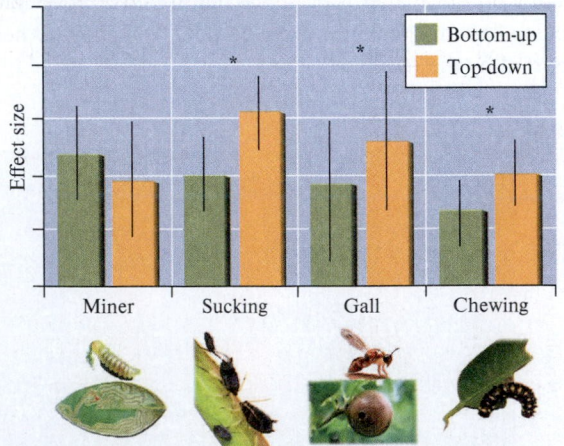

Figure 18.19   Bottom-up and top-down effects on four different feeding guilds of insects. Bar height is model estimate from a meta-analysis of 365 cases, with 95% confidence intervals. Asterisks indicate significant differences between bottom-up and top-down impacts (data from Vidal and Murphy 2018).

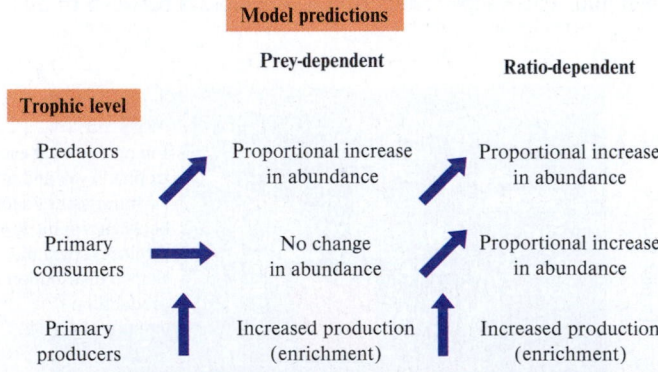

Figure 18.20   Contrasting predictions by prey-dependent and ratio-dependent models of functional response. While prey-dependent models predict varying responses to enrichment by different trophic levels, ratio-dependent models predict proportional increases in abundance at all trophic levels regardless of the number of trophic levels (modified from Arditi and Ginzburg 2012).

production is associated with greater abundance of consumers at all trophic levels, providing strong and growing support for the predictions of ratio-dependent predator–prey models (e.g., Bishop et al. 2006; Hanson and Peters 1984).

How does enrichment affect secondary production? There is clear evidence that increased primary production fosters greater secondary production in terrestrial, and aquatic ecosystems (e.g., McNaughton et al. 1989; Ware and Thomson 2005). As a sample from an extensive literature, figure 18.21 shows the positive correlation between annual precipitation and herbivore secondary production in African savanna ecosystems (Coe, Cumming, and Phillipson 1976). Because primary production in these savannas is tightly coupled with rainfall (Chamaille-Jammes, Fritz, and Murindagomo 2014; Zhu and Southworth 2013), the authors of this study could use annual precipitation in place of direct measurements of savanna primary production.

How does increased primary production affect production at levels above primary consumers? Figure 18.22 shows how availability of leaf litter in a detritus-based stream ecosystem influenced secondary production (Wallace et al. 1999). As on African savannas, secondary production by primary consumers, in this case, production of detritivores, rose with increased availability of food. In addition, there was a positive correlation between primary consumer production and predator production in this stream ecosystem. As Benke and Huryn (2010) suggested, the results of this study, and the many other studies of secondary production, have helped link ecosystem ecology to population and community ecology. In the Applications section, we review an invaluable tool to verifying the feeding habits of organisms, which are fundamental to energy flow studies.

Concept 18.5 Review

1. How can energy losses between trophic levels limit the number of trophic levels in an ecosystem?
2. Is it possible for two consumer populations to have equal biomass but differ in secondary production?
3. Would secondary production in a stream ecosystem dependent on inputs of detritus from a surround forest change, if detrital inputs to the stream—for example, leaves falling from nearby trees—were reduced? If so, in what ways?

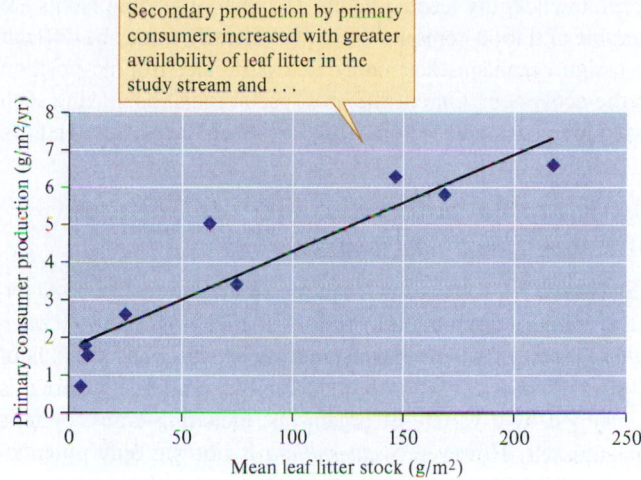

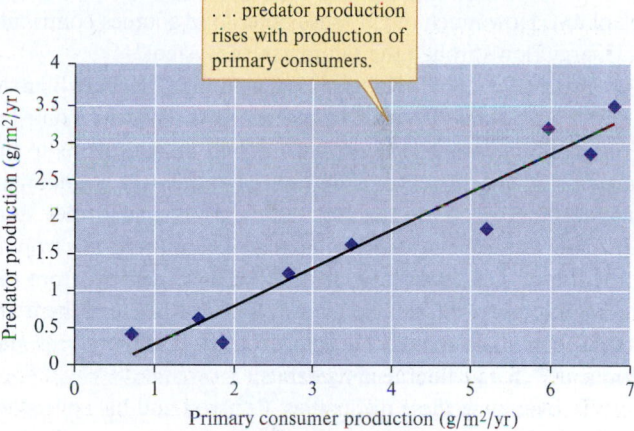

Figure 18.22 Correlations between leaf litter standing stock and secondary production by primary consumers and between primary consumer secondary production and secondary production by predators in a stream ecosystem (data from Wallace et al. 1999).

Applications

Using Stable Isotope Analysis to Study Feeding Habits

LEARNING OUTCOMES

18.21 Outline how stable isotope analysis can be used to determine the trophic level occupied by consumer organisms in an ecosystem.

18.22 Discuss how stable isotope analysis is revealing information about the trophic ecology of species that would be hidden to traditional sources of information on diets, such as stomach analysis.

One of the fundamental steps in studying trophic relations in ecosystems is to determine what organisms eat. While this task may sound easy, for many organisms it is not. If food items are easily identified and feeding habits are well studied and do not change significantly over time or from place to place, you may

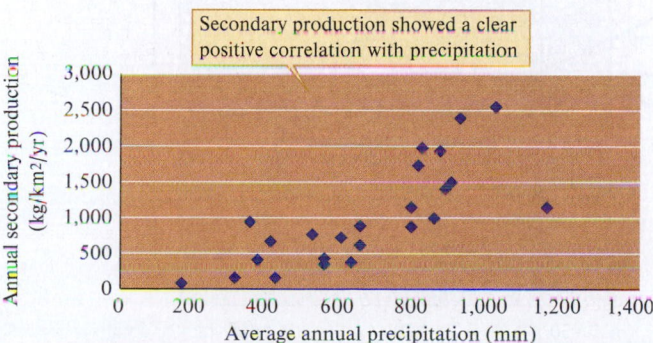

Figure 18.21 Relationship between average precipitation on African savanna ecosystems and annual secondary production by wild mammalian herbivores (data from Coe, Cumming, and Phillipson 1976).

accurately identify feeding habits. However, if feeding habits are variable or if food items are difficult to identify, it may be difficult to assign organisms accurately to a particular trophic position in the ecosystem. One of the most useful tools for making such assignments is stable isotope analysis (see chapter 6, section 6.3).

Using Stable Isotopes to Identify Sources of Energy in a Salt Marsh

The main energy source in a salt marsh in eastern North America is primary production by the salt marsh grass *Spartina alterniflora,* most of which is consumed as detritus. The detritus of *S. alterniflora* is carried into tidal creeks at high tide, where it is consumed by a variety of organisms, including crabs, oysters, and mussels. However, *S. alterniflora* is not the only potential source of food for these organisms. The waters of the salt marsh also contain organic matter from upland plants and carry phytoplankton. How much might these other food sources contribute to energy flow through the salt marsh ecosystem?

Bruce Peterson, Robert Howarth, and Robert Garritt (1985) used stable isotopes to determine the relative contributions of *S. alterniflora,* phytoplankton, and upland plants to the nutrition of the ribbed mussel, *Geukensia demissa,* a dominant filter-feeding species in the New England salt marsh they studied. The researchers pointed out that determining the trophic structure of salt marshes is difficult because detritus from different sources is difficult to identify visually, because there are several potential sources of detritus, and because organisms may frequently change their feeding habits.

To overcome these difficulties, Peterson and his colleagues used the ratios of stable isotopes of carbon, nitrogen, and sulfur. They used the stable isotopes of these three elements because

their ratios are different in phytoplankton, upland C_3 plants (see chapter 7), and *S. alterniflora,* a C_4 grass (fig. 18.23). Upland plants, with a $\delta^{13}C = -28.6 \ ^0/_{00}$, are the most depleted of ^{13}C, whereas *S. alterniflora,* with a $\delta^{13}C = -13.1 \ ^0/_{00}$, is the least depleted. However, ^{13}C ratios are similar in *S. alterniflora* and plankton, hence the value of also measuring ^{15}N and ^{34}S to distinguish them. *S. alterniflora,* with a $\delta^{34}S = -2.4 \ ^0/_{00}$, has the lowest relative concentration of ^{34}S, whereas plankton, with a $\delta^{34}S = +18.8 \ ^0/_{00}$, have the highest concentration of ^{34}S.

Using these differences in isotopic concentrations, the researchers were able to identify the relative contributions of potential food sources to the diet of the mussel (fig. 18.24). Their

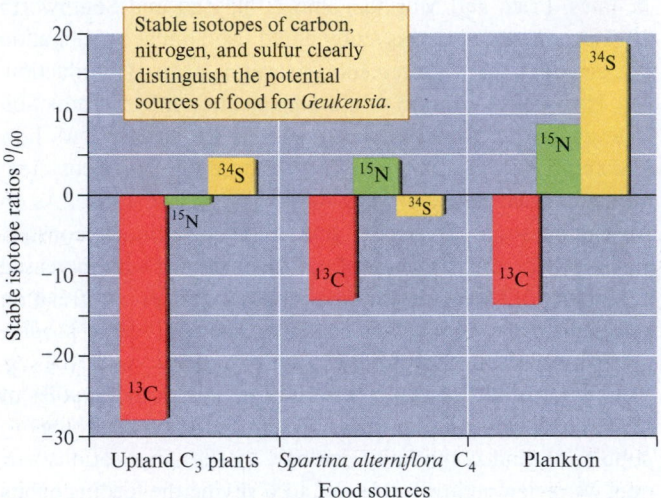

Figure 18.23   Isotopic content of potential food sources for the ribbed mussel, *Geukensia demissa,* in a New England salt marsh (data from Peterson, Howarth, and Garritt 1985).

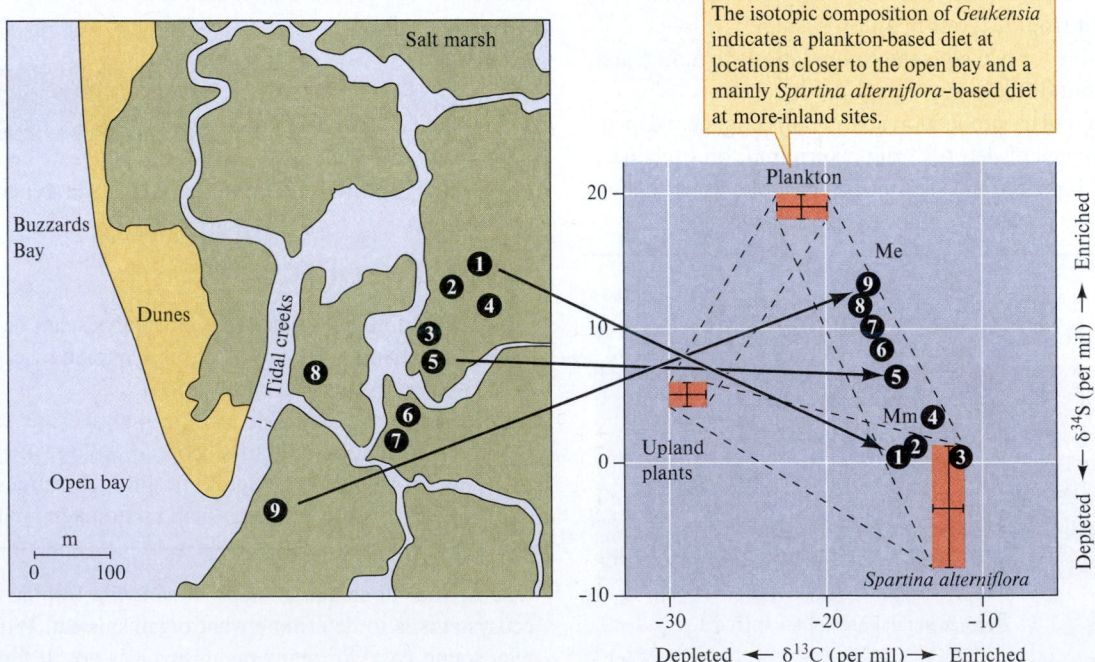

Figure 18.24   Variation in isotopic composition of ribbed mussels, *Geukensia demissa,* by distance inland in a New England salt marsh (data from Peterson, Howarth, and Garritt 1985).

analyses showed that *Geukensia* gets most of its energy from plankton and *S. alterniflora* but that the relative contributions of these two food sources depends on location. In the interior of the marsh, the mussel feeds mainly on *S. alterniflora,* but near the mouth of the marsh, it depends mainly on plankton. This is an example of how analyses of stable isotopes can provide us with a window to the otherwise hidden biology of species.

Stable isotope analyses continue to improve our understanding of energy flow through ecosystems. While energy flows through ecosystems in a one-way path, the elements, or nutrients, upon which organisms depend are recycled and may be used over and over again. The cycling of these nutrients is the subject of chapter 19.

Summary

We can view a forest, a stream, or an ocean as a system that absorbs, transforms, and stores energy. In this view, physical, chemical, and biological structures and processes are inseparable. When we look at natural systems in this way, we view them as ecosystems. An *ecosystem* is a biological community plus all of the abiotic factors influencing that community.

Primary production, the production of new biomass by autotrophs, is one of the most important ecosystem processes. Primary production is the amount of biomass produced per unit area or volume (e.g., per m^2 or per m^3) over some interval of time. Gross primary production is the total primary production by all primary producers in the ecosystem. Net primary production is gross primary production minus respiration by primary producers.

Terrestrial primary production is generally limited by temperature, moisture, and nutrients.   The variables most highly correlated with variation in terrestrial primary production are temperature and moisture. Highest rates of terrestrial primary production occur under warm, moist conditions. Temperature and moisture conditions can be combined in a single measure called annual actual evapotranspiration, or AET, which is the total amount of water that evaporates and transpires off a landscape during the course of a year. Annual AET is positively correlated with net primary production in terrestrial ecosystems. However, significant variation in terrestrial primary production results from differences in soil fertility.

Aquatic primary production is generally limited by nutrient availability.   One of the best-documented patterns in the biosphere is the positive relationship between nutrient availability and rate of primary production in aquatic ecosystems. Phosphorus concentration usually limits rates of primary production in freshwater ecosystems, whereas nitrogen concentration usually limits rates of marine primary production.

Primary producer diversity contributes to higher primary production.   Experimental field studies indicate strong influences of plant diversity and plant functional group composition on primary production. Primary producer diversity has also been shown to increase primary production in experimental aquatic ecosystems. The results of experimental studies of the influence of biodiversity on primary production are supported by patterns of variation in nature.

Consumers can influence rates of primary production in aquatic and terrestrial ecosystems through trophic cascades. Piscivorous fish can indirectly reduce rates of primary production in lakes by reducing the density of plankton-feeding fish, an effect called a trophic cascade. Reduced density of planktivorous fish can lead to increased density of herbivorous zooplankton, which can reduce the densities of phytoplankton and rates of primary production. Intensive grazing by large mammalian herbivores on the Serengeti increases annual net primary production by inducing compensatory growth in grasses.

Ecosystems with greater primary production generally support higher levels of secondary production.   Ecosystem ecologists have simplified the trophic structure of ecosystems by arranging species into trophic levels based on the predominant source of their nutrition. A trophic level is determined by the number of transfers of energy from primary producers to that level. As energy is transferred from one trophic level to another, energy is lost due to limited assimilation, respiration by consumers, and heat production. As a result of these losses, the quantity of energy in an ecosystem decreases with each successive trophic level, forming a pyramid-shaped distribution of energy among trophic levels. Estimates of changes in population density and average mass of consumers provide a basis for calculating secondary production between sampling periods. Both top-down and bottom-up forces can impact secondary production, but meta-analysis of tri-trophic interactions in insects suggests top-down effects may be more significant. A positive correlation between primary and secondary production has been demonstrated in terrestrial, freshwater, estuarine, and marine ecosystems. Patterns of consumer abundance in response to enrichment are consistent with predictions of ratio-dependent models of predator-prey interactions.

Stable isotope analysis can be used to trace the flow of energy through ecosystems. The ratios of different stable isotopes of important elements such as nitrogen and carbon are generally different in different parts of ecosystems. As a consequence, ecologists can use isotopic ratios to study the trophic structure and energy flow through ecosystems.

Key Terms

actual evapotranspiration
(AET)   385
biodiversity   390
bottom-up control   392

ecological efficiency   397
gross primary production   384
net primary production   384
plant functional group   390

primary production   390
secondary production   396
top-down control   392
tri-trophic interactions   398

trophic cascade   392
trophic dynamics   397
trophic level   384

Review Questions

1. Population, community, and ecosystem ecologists study structure and process. However, they focus on different natural characteristics. Contrast the important structures and processes in a forest from the perspectives of population, community, and ecosystem ecologists.

2. M. Huston (1994b) pointed out that the well-documented pattern of increasing annual primary production from the poles to the equator is strongly influenced by the longer growing season at low latitudes. The following data are from table 14.10 in Huston. The data cited by Huston are from Whittaker and Likens (1975).

Forest Type	Annual NPP (t/ha/yr)	Length of Growing Season (months)	Monthly NPP (t/ha/mo)
Boreal forest	8	3	2.7
Temperate forest	13	6	?
Tropical forest	20	12	?

Complete the missing data to compare the *monthly* production of boreal, temperate, and tropical forests. How does this short-term perspective of primary production in high-, middle-, and low-latitude forests compare to an annual perspective?

3. Many migratory birds spend approximately half the year in temperate forests during the warm breeding season and the other half of the year in tropical forest. Given the analyses you made in question 2, which forest appears to be more productive from the perspective of these migratory birds?

4. Field experiments demonstrate that variation in soil fertility influences terrestrial primary production. However, we cannot say that nutrients exert primary control. That role is still attributed to temperature and moisture. Why do ecologists still attribute the main control of terrestrial primary production to temperature and moisture? Consider the difference in primary production between arctic tundra and tropical forest (see fig. 18.2) and the extent to which nutrient additions (Shaver and Chapin 1986) changed primary production in tundra.

5. Compare the pictures of trophic structure that emerged from our discussions of food webs in chapter 17 with those in chapter 18. What are the strengths of each perspective? What are their limitations?

6. Suppose you are studying a community of small mammals that live on the boundary between a riverside forest and a semidesert grassland. One of your concerns is to discover the relative contributions of the grassland and the forest to the nutrition of small mammals living between the two ecosystems. Design a research program to find out. (Hint: The grassland is dominated by C_4 grasses and the forest by C_3 plants.)

7. In chapter 17, we examined the influences of keystone species on the structure of communities. In chapter 18, we reviewed trophic cascades. Discuss the similarities and differences between these two concepts. Compare the measurements and methods of ecologists studying keystone species versus those studying trophic cascades.

8. The studies of nutrient limitation of aquatic primary production that we reviewed focused almost entirely on lakes within the temperate zone. Suppose you are an ecologist interested in determining whether primary production in tropical lakes is subject to similar control by nutrient availability. Design a study to find out what controls rates of primary production in tropical lakes. Use all the sources of information at your disposal, including published research, surveys of natural variation, and large- and small-scale experiments.

Chapter 19

Nutrient Cycling and Retention

Photodisc Collection/Getty Images

A maple leaf falls into a lake. Decomposition of the leaf will begin quickly, releasing nutrients that will be available for uptake by primary producers in the lake, including the duckweed floating nearby.

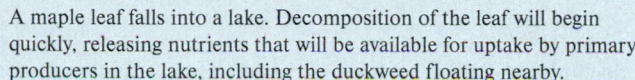

CHAPTER CONCEPTS

19.1 Nutrient cycles involve the storage of chemical elements in nutrient pools, or compartments, and the flux, or transfer, of nutrients between pools.   404

Concept 19.1 Review   408

19.2 Decomposition rate is influenced by temperature, moisture, and chemical composition of litter and the environment.   408

Concept 19.2 Review   412

19.3 Plants and animals can modify the distribution and cycling of nutrients in ecosystems.   412

Concept 19.3 Review   417

19.4 Disturbance generally increases nutrient loss from ecosystems.   417

Concept 19.4 Review   419

Applications: Altering Aquatic and Terrestrial
Ecosystems   419
Summary   421
Key Terms   422
Review Questions   422

LEARNING OUTCOMES

After studying this section you should be able to do the following:

19.1 Contrast the movement of energy and nutrients in ecosystems.

19.2 Summarize the ecological significance of nutrient cycling studies.

Exchange of nutrients between organisms and their environment is one of the essential aspects of ecosystem function. A diatom living in the surface waters of a lake absorbs an ion of phosphate from the surrounding water. It incorporates the phosphate into some of its DNA as it replicates its chromosomes during cell division. A few hours later, one of the diatom's daughter cells is eaten by a cladoceran, an algae-feeding member of the zooplankton. The cladoceran incorporates the

403

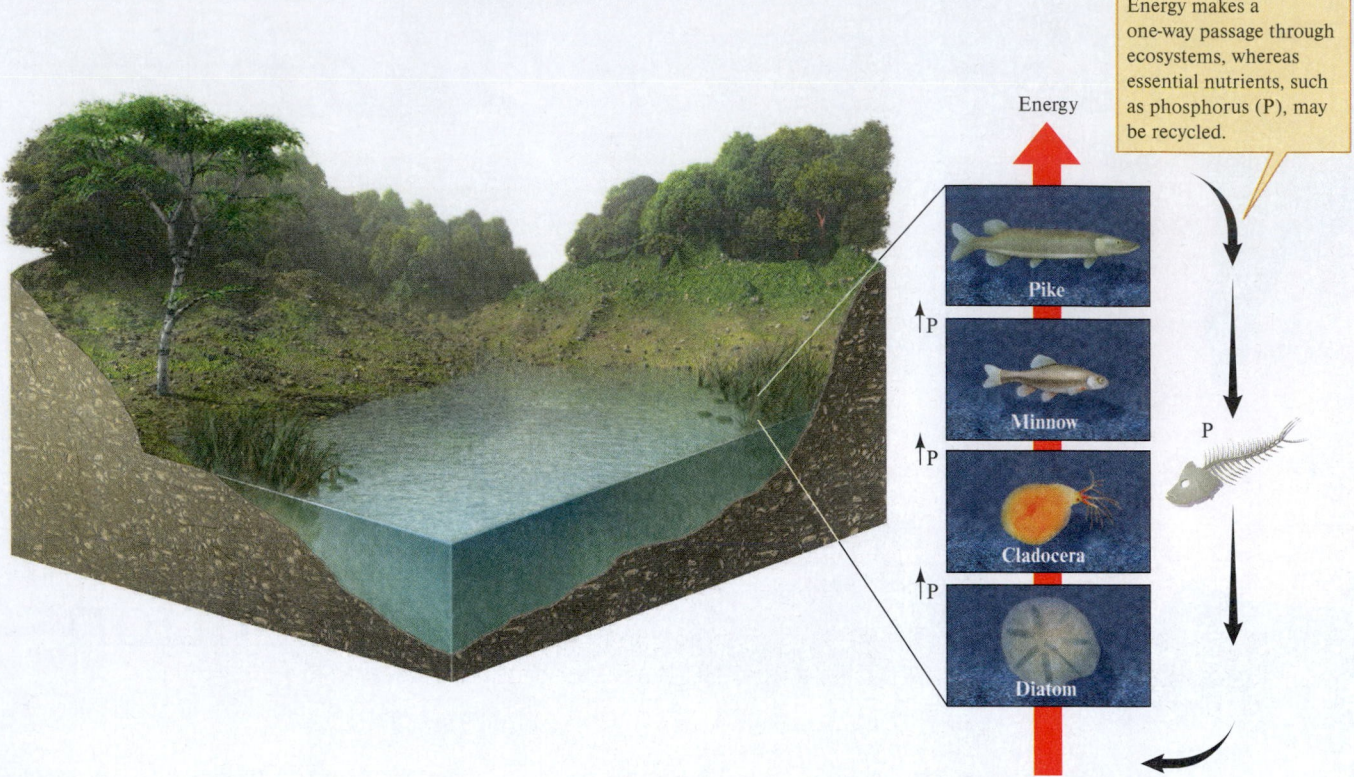

Energy makes a one-way passage through ecosystems, whereas essential nutrients, such as phosphorus (P), may be recycled.

Energy

Pike

Minnow

Cladocera

Diatom

P

Figure 19.1   Phosphorus cycle in a lake ecosystem.

phosphate into a molecule of ATP. The cladoceran lives 2 days more and then is eaten by a planktivorous minnow. Within the minnow, the phosphate is combined with a lipid to form a phospholipid molecule in the cell membrane of one of the minnow's neural cells. A few weeks later, the minnow is eaten by a northern pike and the phosphate is incorporated into part of the pike's skeleton. During the following winter, the pike dies and its tissues are attacked by bacteria and fungi that gradually decompose the pike, including its skeleton. During decomposition, the phosphate is dissolved in the surrounding water. The following spring the same ion of phosphate is taken up by another diatom, completing its cycle through the lake ecosystem (fig. 19.1).

In chapter 18, we saw that energy makes a one-way trip through ecosystems. In contrast, elements such as phosphorus (P), carbon (C), nitrogen (N), potassium (K), and iron (Fe) are recycled. Elements that are required for the development, maintenance, and reproduction of organisms are called *nutrients.* Ecologists refer to the use, transformation, movement, and reuse of nutrients in ecosystems as **nutrient cycling.** Because of the physiological importance of nutrients, their relative scarcity, and their influence on rates of primary production, nutrient cycling is one of the most significant ecosystem processes studied by ecologists.

Our understanding of nutrient cycles has developed under the guidance of ecosystem ecologists motivated by a passion for the movement of elements through the biosphere and who often worked in obscurity. For many of these scientists, few things are more stimulating than spending long hours debating about the main factors controlling the rate of mass loss

by decomposing logs, or how microbial uptake might affect the availability of nitrogen by plants growing under different climatic regimes. However, a passion for nutrient cycles is no longer limited to academic circles but now energizes people of all walks of life from around the world. This shift in interest is driven mainly by the looming consequences of our having dramatically altered some key nutrient cycles. The best known of these human-driven changes is the current imbalance in the carbon cycle that has resulted in a buildup of carbon dioxide in earth's atmosphere, accompanied by global warming (see chapter 23). However, other critical cycles are also out of balance and most scientists now agree that the fate of the biosphere, from the health of coral reefs to the state of the global economy, literally hangs on how well we understand and manage nutrient cycles, especially the carbon cycle. Three nutrient cycles play especially prominent roles: the *phosphorus cycle,* the *nitrogen cycle,* and the *carbon cycle.* In the next few pages, we review the major features of each of these cycles.

19.1 Nutrient Cycles

LEARNING OUTCOMES

After studying this section you should be able to do the following:

19.3   Describe nutrient sources and nutrient sinks.

19.4   Discuss the biological importance of phosphorus, nitrogen, and carbon.

19.5   Outline the major nutrient sources and sinks involved in the phosphorus, nitrogen, and carbon cycles.

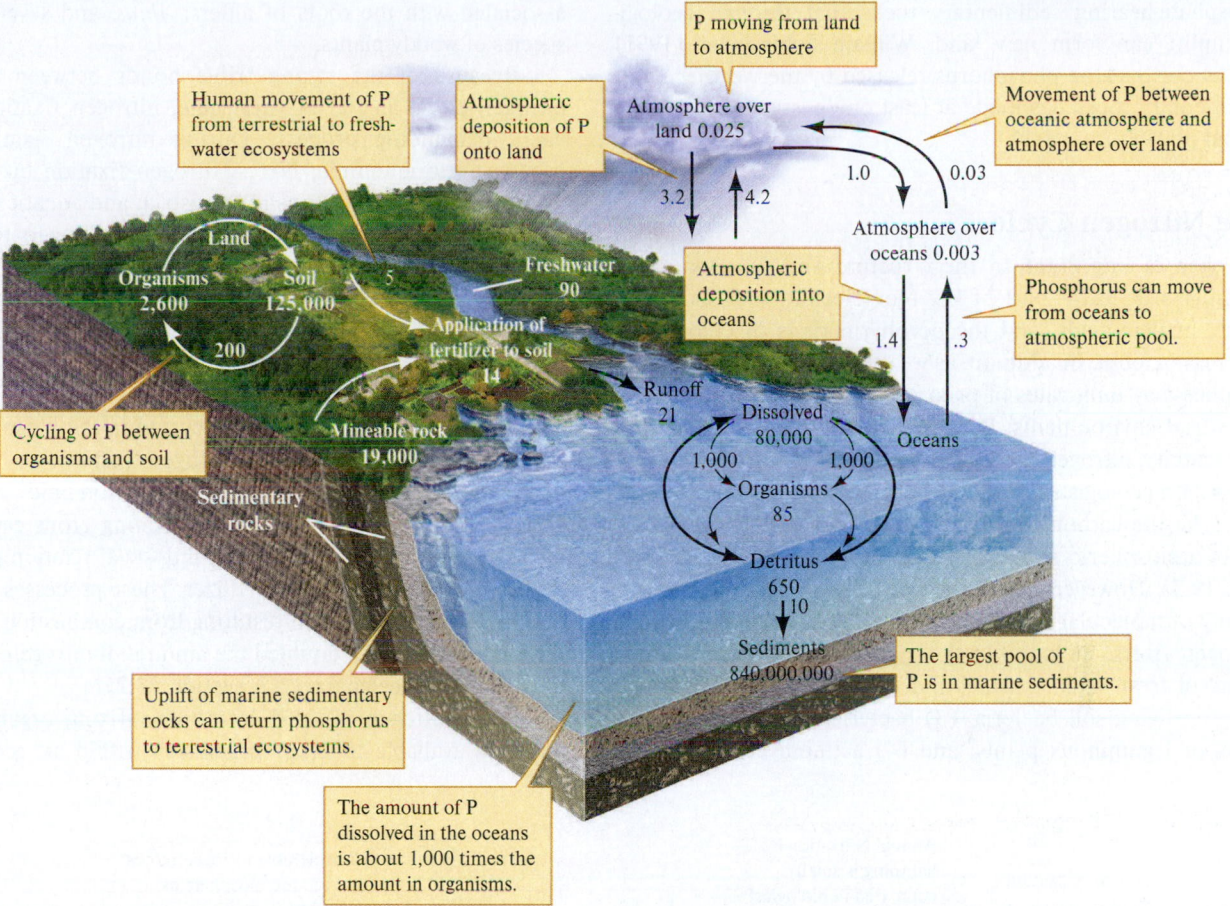

Figure 19.2   The phosphorus cycle. Numbers are 10^{12} g P or fluxes as 10^{12} g P per year (data from Schlesinger 1991, after Richey 1983; Meybeck 1982; Graham and Duce 1979).

Nutrient cycles involve the storage of chemical elements in nutrient pools, or compartments, and the flux, or transfer, of nutrients between pools.   While there are details that are unique to each nutrient cycle, there are some features that are common to all. One of those common features is that nutrients are stored in nutrient pools. A **nutrient pool** is the amount of a particular nutrient stored in a portion, or compartment, of an ecosystem. In addition, all nutrient cycles are, as the name implies, dynamic, with **nutrient flux** moving nutrients between the pools of an ecosystem. A major interest of ecologists studying nutrient cycling is to understand the factors controlling the distribution of nutrients among pools and the rates of nutrient flux among them. Theoretically, in a closed ecosystem, nutrients would cycle round and round as suggested by the phosphorus cycling shown in figure 19.1. However, ecosystems are not closed and nutrients may be lost from the ecosystem to a nutrient sink. A **nutrient sink** is a part of the biosphere where a particular nutrient is absorbed faster than it is released. For example, the phosphorus in the hypothetical lake in figure 19.1 could be lost to bottom sediments as an insoluble precipitate. Ecosystems may also gain nutrients from nutrient sources. A **nutrient source** is a portion of the biosphere where a particular nutrient is released faster than it is absorbed. For example, burning of fossil fuels acts as a source of carbon dioxide to the global ecosystem. We begin our discussion of the details of nutrient cycles with the phosphorus cycle.

The Phosphorus Cycle

Phosphorus is essential to the energetics, genetics, and structure of living systems. For instance, phosphorus forms part of the ATP, RNA, DNA, and phospholipid molecules. While of great biological importance, phosphorus is not very abundant in the biosphere. Consequently, phosphorus cycling has received a great deal of attention from ecosystem ecologists.

In contrast to carbon and nitrogen, the global phosphorus cycle does not include a substantial atmospheric pool (fig. 19.2). The largest quantities of phosphorus occur in mineral deposits and marine sediments. Sedimentary rocks that are especially rich in phosphorus are mined for fertilizer and applied to agricultural soils. Soil may contain substantial quantities of phosphorus. However, much of the phosphorus in soils occurs in chemical forms not directly available to plants.

Phosphorus is slowly released to terrestrial and aquatic ecosystems through the weathering of rocks. As phosphorus is released from mineral deposits, it is absorbed by plants and recycled within ecosystems. Mycorrhizae generally play a key role in the uptake of phosphorus by plants in terrestrial ecosystems. However, much phosphorus is washed into rivers and eventually finds its way to the oceans, where it remains in dissolved form until eventually finding its way to the ocean sediments. Ocean sediments are eventually transformed into phosphate-bearing sedimentary rocks that through geological uplift can form new land. William Schlesinger (1991) points out that the phosphorus released by the weathering of sedimentary rocks has made at least one passage through the global phosphorus cycle.

The Nitrogen Cycle

Nitrogen is important to the structure and functioning of organisms. It forms part of key biomolecules such as amino acids, nucleic acids, and the porphyrin rings of chlorophyll and hemoglobin. In addition, as we saw in chapter 18, nitrogen supplies may limit rates of primary production in marine and terrestrial environments. Because of its importance and relative scarcity, nitrogen has drawn a great deal of attention from ecosystem ecologists.

Like the carbon cycle, the nitrogen cycle also includes a major atmospheric pool in the form of molecular nitrogen, N_2 (fig. 19.3). However, few organisms can use this atmospheric supply of molecular nitrogen directly. These organisms, called nitrogen fixers, include (1) the cyanobacteria, or blue-green algae, of freshwater, marine, and soil environments; (2) certain free-living soil bacteria; (3) bacteria associated with the roots of leguminous plants; and (4) actinomycetes bacteria,

associated with the roots of alders, *Alnus,* and several other species of woody plants.

Because of the strong triple bonds between the two nitrogen atoms in the N_2 molecule, nitrogen fixation is an energy-demanding process. During nitrogen fixation, N_2 is reduced to ammonia, NH_3. Nitrogen fixation takes place under anaerobic conditions in terrestrial and aquatic environments, where nitrogen-fixing species oxidize sugars to obtain the required energy. Nitrogen fixation also occurs as a physical process associated with the high pressures and energy generated by lightning. Ecologists propose that all of the nitrogen cycling within ecosystems ultimately entered these cycles through nitrogen fixation by organisms or lightning. There is a relatively large pool of nitrogen cycled in the biosphere but only a small entryway through nitrogen fixation.

Humans have overcome this limitation and increased agricultural production globally by rotating crops capable of nitrogen fixation and through the industrial fixation of N_2 to produce ammonium (NH_4) fertilizer. These processes, in addition to sources of nitrogen resulting from combustion of fossil fuels, have more than doubled the amount of nitrogen entering the biosphere (see chapter 23, section 23.2).

Once nitrogen is fixed by nitrogen-fixing organisms, it becomes available to other organisms within an ecosystem.

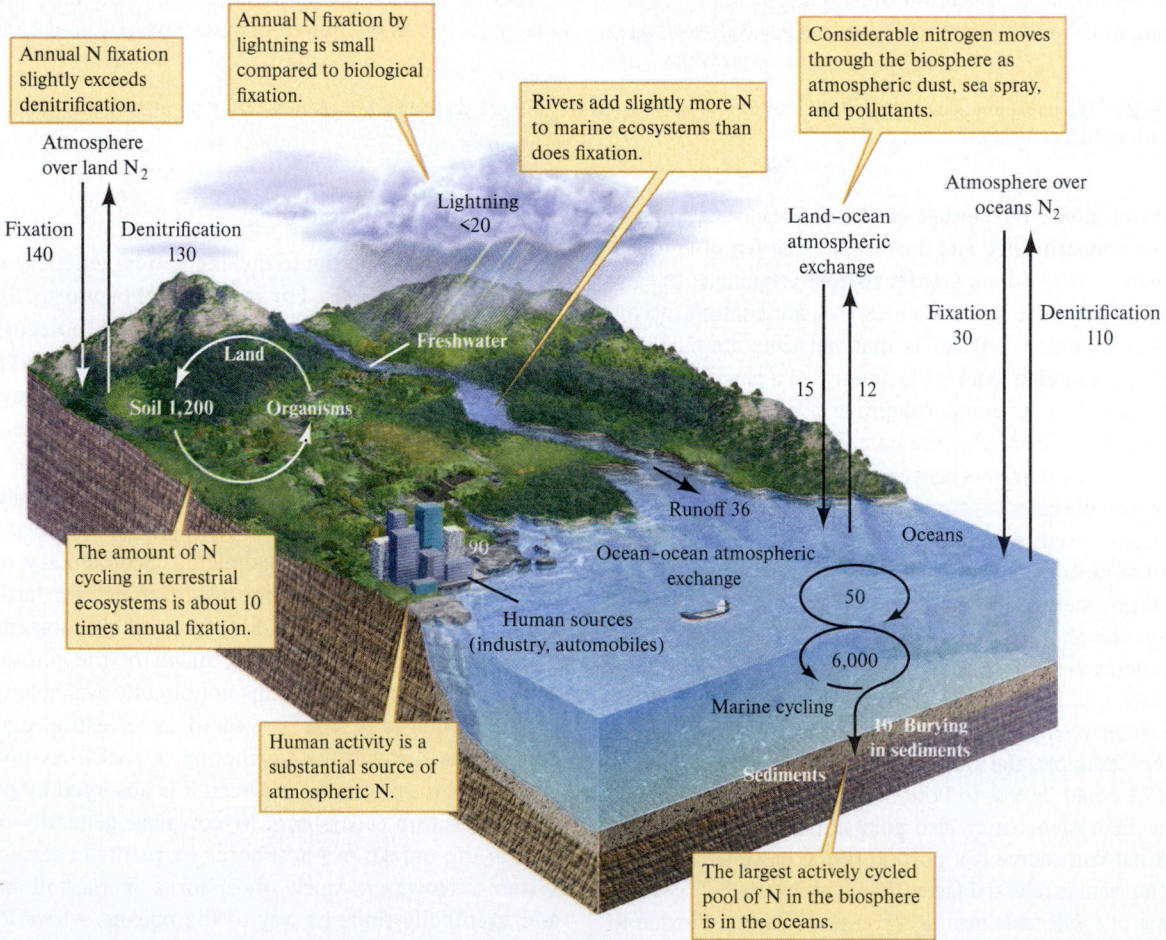

Figure 19.3   The nitrogen cycle. Numbers are storage or fluxes as 10^{12} g N per year (data from Schlesinger 1991, after Söderlund and Rosswall 1982).

Upon the death of an organism, the nitrogen in its tissues can be released by fungi and bacteria involved in the decomposition process. These fungi and bacteria release nitrogen as ammonium, NH_4^+, a process called ammonification. Ammonium can be converted to nitrate, NO_3^-, by other bacteria in a process called nitrification. Ammonium and nitrate can be used directly by bacteria, fungi, or plants. The nitrogen in dead organic matter can also be used directly by mycorrhizal fungi, which can be passed on to plants. The nitrogen in bacterial, fungal, and plant biomass may pass on to populations of animal consumers or back to the pool of dead organic matter, where it will be recycled again.

Nitrogen may exit the organic matter pool of an ecosystem through denitrification. Denitrification is an energy-yielding process that occurs under anaerobic conditions and converts nitrate to molecular nitrogen, N_2. The molecular nitrogen produced by denitrifying bacteria moves into the atmosphere and can reenter the organic matter pool only through nitrogen fixation. Ecologists estimate that the mean residence time of fixed nitrogen in the biosphere is about 625 years. They estimate that the mean residence time of phosphorus in the biosphere is on the order of thousands of years.

The Carbon Cycle

Carbon is an essential part of all organic molecules, and, carbon compounds such as carbon dioxide, CO_2, and methane,

CH_4, as constituents of the atmosphere, substantially influence global climate. This connection between atmospheric carbon and climate has drawn all nations of the planet into discussions of the ecology of carbon cycling.

Carbon moves between organisms and the atmosphere as a consequence of two reciprocal biological processes: photosynthesis and respiration (fig. 19.4). Photosynthesis removes CO_2 from the atmosphere, whereas respiration by primary producers and consumers, including decomposers, returns carbon to the atmosphere in the form of CO_2. In aquatic ecosystems, CO_2 must first dissolve in water before being used by aquatic primary producers. Once dissolved in water, CO_2 enters a chemical equilibrium with bicarbonate, HCO_3^-, and carbonate, CO_3^-. Carbonate may precipitate out of solution as calcium carbonate and may be buried in ocean sediments.

While some carbon cycles rapidly between organisms and the atmosphere, some remains sequestered in relatively unavailable forms for long periods of time. Carbon in soils, peat, fossil fuels, and carbonate rock would generally take a long time to return to the atmosphere. During modern times, however, fossil fuels have become a major source of atmospheric CO_2 as humans have tapped into fossil fuel supplies to provide energy for their economic systems.

The massive burning of fossil fuels, beginning with the industrial revolution, has increased concentrations of atmospheric CO_2 (see chapter 23). However, the rate of buildup has been significantly slower than scientists would predict given

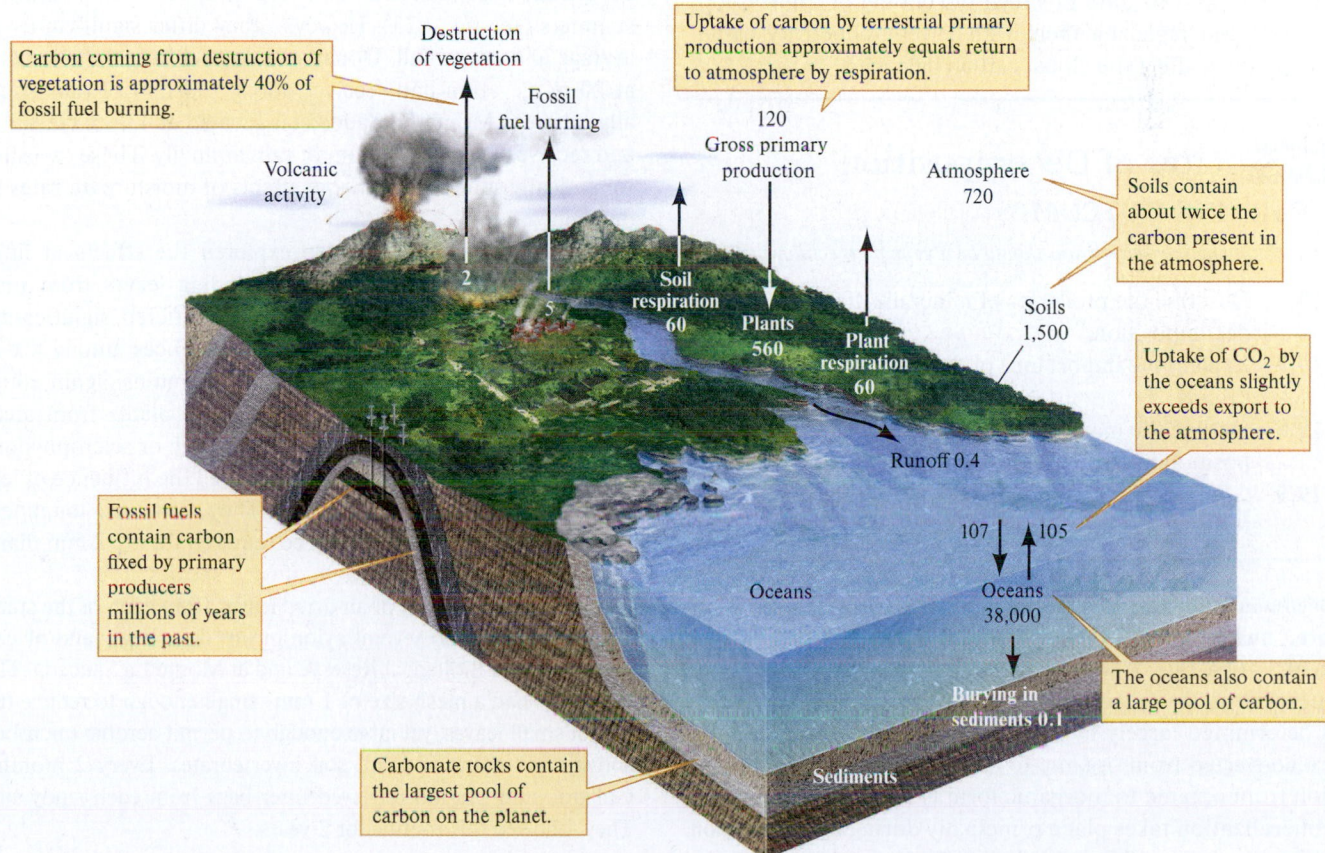

Figure 19.4   The carbon cycle. Numbers are storage as 10^{15} g or fluxes as 10^{15} g per year (data from Schlesinger 1991).

the rate of emissions to the atmosphere by fossil fuel burning minus known carbon sinks. As indicated in figure 19.4, the oceans are a known carbon sink, since the amount of carbon they absorb annually exceeds their emissions by 20^{15} g. Ecologists have assumed that the missing carbon sink was terrestrial, but where? The best current estimates are that the missing carbon sink is formed by a combination of northern and tropical forests (Stephens et al. 2007). However, much more research needs to be done to verify that these forests are the missing sinks, to estimate the relative rates of carbon uptake by the two forest types, and to understand how to sustain this critical uptake of atmospheric CO_2.

Ecosystem ecologists study the factors controlling the movement, storage, and conservation of nutrients within ecosystems. You can see broad outlines of these processes in figures 19.2–19.4. However, much remains to be learned, especially concerning the factors controlling rates of nutrient exchange within and between ecosystems. Nutrient exchange is substantially affected by the process of decomposition, the subject of the next section.

Concept 19.1 Review

1. Do the oceans act as a source or a sink for phosphorus (refer to fig. 19.2)?
2. What are the relative fluxes of nitrogen through fixation and denitrification on land and in the oceans (see fig. 19.3)?
3. What are two ways in which the cutting of tropical forests and replacing them with lower-productivity cattle pastures affect the global carbon balance?

19.2 Rates of Decomposition

LEARNING OUTCOMES

After studying this section you should be able to do the following:

19.6 Describe the processes of mineralization and decomposition.
19.7 Explain the importance of decomposition to nutrient cycling.
19.8 Outline the main factors controlling rates of decomposition in terrestrial ecosystems.
19.9 Compare the controls on decomposition rates in terrestrial and aquatic ecosystems.

Decomposition rate is influenced by temperature, moisture, and chemical composition of litter and the environment. The rate at which nutrients, such as nitrogen and phosphorus, are made available to the primary producers is determined largely by the rate at which nutrient supplies are converted from organic to inorganic forms. This conversion from organic to inorganic form is called **mineralization.** Mineralization takes place principally during **decomposition,** which is the breakdown of organic matter accompanied by the release of carbon dioxide. Consequently, ecosystem

ecologists consider decomposition as a key ecosystem process.

Decomposition of organic matter involves chemical and physical processes such as leaching and fragmentation of litter, and biological processes dominated by fungi and bacteria. The fine hyphae of fungi can penetrate litter and release external digestive enzymes that speed decomposition. Fragmentation and ingestion of litter by invertebrates also plays a key role during decomposition processes.

The rate of decomposition in ecosystems is significantly influenced by temperature, moisture, and the chemical composition of both plant litter and the environment. The chemical characteristics of plant litter that influence decomposition rates include nitrogen concentration, phosphorus concentration, carbon:nitrogen ratio, and lignin content. Ecologists have studied how several of these variables affect rates of leaf decomposition in Mediterranean ecosystems.

Decomposition in Two Mediterranean Woodland Ecosystems

Antonio Gallardo and José Merino (1993) studied how chemical and physical factors affect rates of decomposition of leaf litter in two Mediterranean woodland ecosystems. Their study sites were located at Doñana Biological Reserve and Monte La Sauceda in southwestern Spain. The mean annual temperature at the two sites differs by only 0.5°C: 16.7°C at Doñana versus 16.2°C at Monte La Sauceda. Both study sites experience Mediterranean climates with wet winters and dry summers (see fig. 2.23). However, they differ significantly in average annual rainfall. Doñana Biological Reserve is located at 20 m elevation and receives about 500 mm of rain annually, whereas Monte La Sauceda is located at 432 m elevation and receives about 1,600 mm of rain annually. These two sites were ideally suited to study the effects of moisture on rates of decomposition.

Gallardo and Merino also explored the effects of litter chemistry on decomposition by including leaves from nine species of native trees and shrubs that differed significantly in chemical composition. Chemical differences among leaves included differences in concentrations of tannins, lignin, nitrogen, and phosphorus. Many of the native plants from areas with a Mediterranean climate produce tough or sclerophyllous leaves. Gallardo and Merino also explored the influence of leaf toughness on decomposition rate. They estimated toughness by measuring the amount of force required for a 1.2 mm diameter rod to penetrate the leaves of each species.

Approximately 2 g of air-dried leaves from each of the study species was put into several nylon mesh "litter bags" and placed at the Doñana Biological Reserve and at Monte La Sauceda. The litter bags had a mesh size of 1 mm—small enough to reduce the loss of small leaves, yet large enough to permit aerobic microbial activity and entry of small soil invertebrates. Every 2 months, Gallardo and Merino retrieved litter bags from each study site. They followed this routine for 2 years.

In the laboratory, the researchers measured the mass of leaf tissue remaining in each of 6 to 10 replicate litter bags for

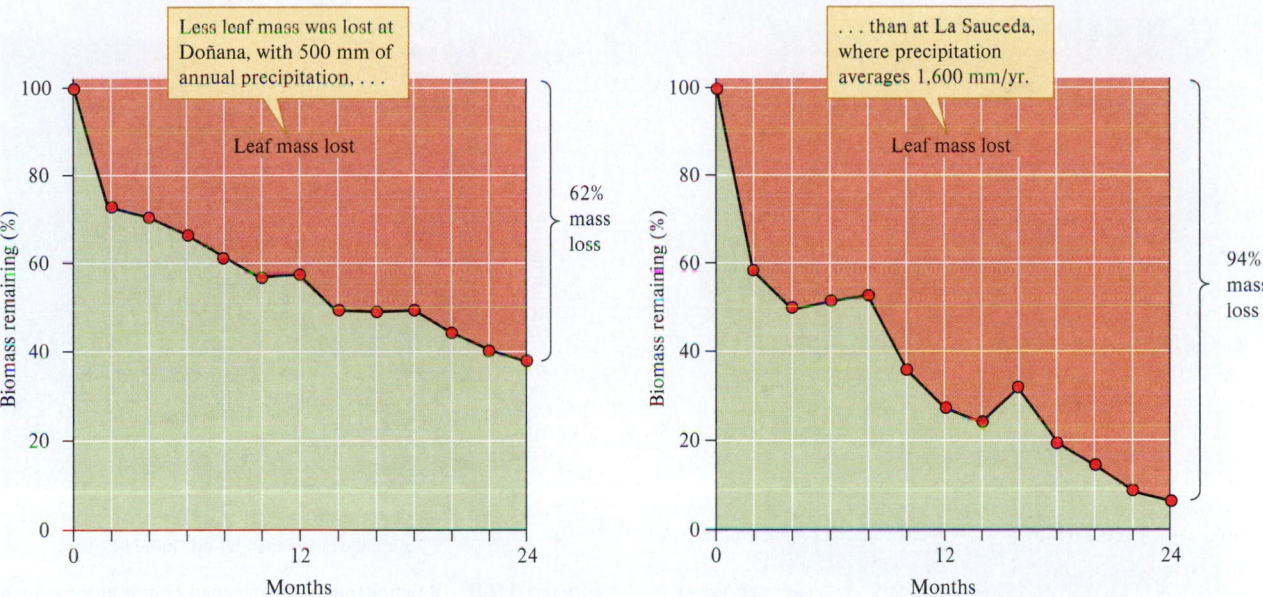

Figure 19.5   Decomposition of *Fraxinus angustifolia* leaves at drier and wetter sites (data from Gallardo and Merino 1993).

each leaf species. Figure 19.5 shows that the amount of leaf mass lost by ash leaves, *Fraxinus angustifolia*, was much higher at Monte La Sauceda. This higher decomposition rate probably reflects the higher precipitation at that site.

Though all types of leaves decomposed faster at Monte La Sauceda, differences in decomposition rates among leaf species were similar at the two sites. For instance, the leaves of ash, *Fraxinus*, showed the greatest mass loss at both study sites, while the oak, *Quercus lusitanica*, showed the lowest mass loss at both study sites. Differences in mass loss by the nine plant species studied reflected differences in the physical and chemical characteristics of their leaves. Gallardo and Merino found that the best predictor of mass loss at the Doñana site was the ratio of toughness to nitrogen content, toughness/%N, and that mass loss was a power function of this ratio:

$$\text{Mass} = 545[\text{toughness}/\%N]^{-0.38}$$

This is the equation for the line shown in figure 19.6. This equation indicates that tougher leaves with lower concentrations of nitrogen decomposed at a lower rate.

The greater mass losses at Monte La Sauceda demonstrate a positive influence of moisture on rates of decomposition, while differences in decomposition rates among leaf species show the influences of chemical composition on decomposition. Even toughness, which is a physical property, is a consequence of chemical composition, especially the concentration of lignin. As we will see in the next example, the ratio of lignin concentration to nitrogen content in leaves is also highly correlated with decomposition rates in temperate forest ecosystems.

Decomposition in Two Temperate Forest Ecosystems

Jerry Melillo, John Aber, and John Muratore (1982) used litter bags to study leaf decomposition of six species of trees

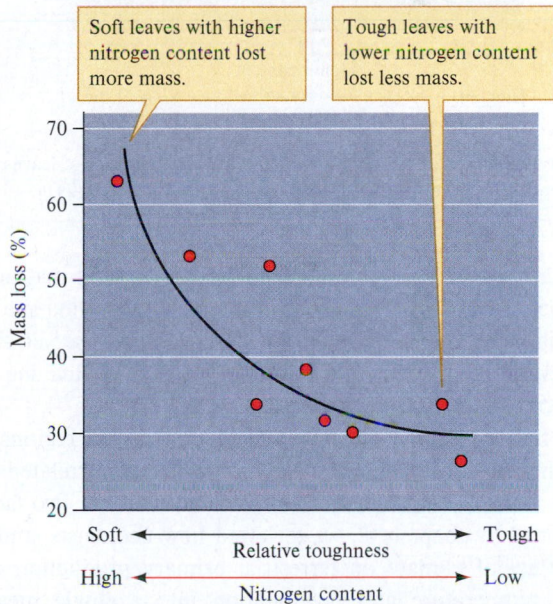

Figure 19.6   Influence of leaf toughness and nitrogen content on decomposition of the leaves of nine different shrub and tree species (data from Gallardo and Merino 1993).

in a temperate forest in New Hampshire. They also compared their results with decomposition of leaves from five tree species in a temperate forest in North Carolina.

In both the New Hampshire and North Carolina forests, the researchers found a negative correlation between the leaf mass remaining after 1 year of decomposition and the ratio of lignin to nitrogen concentrations in leaves, %lignin:%N. In other words, leaves with higher lignin:nitrogen ratios lost less mass during the year-long study. As you can see in figure 19.7, the amount of leaf mass remaining was lower at the North Carolina site than at the New Hampshire site. What factors were responsible for these higher rates of decomposition at the North

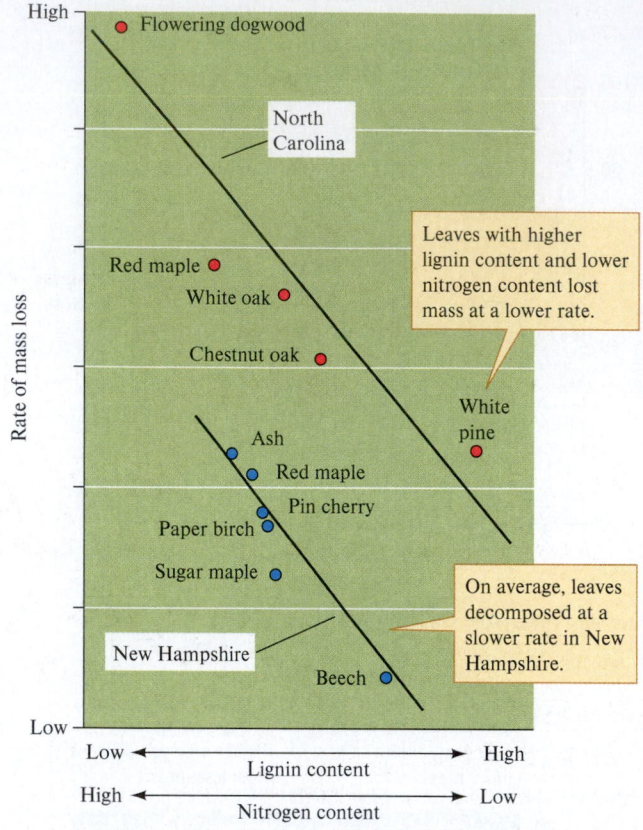

Figure 19.7   Influence of lignin and nitrogen content of leaves on decomposition (data from Melillo, Aber, and Muratore 1982).

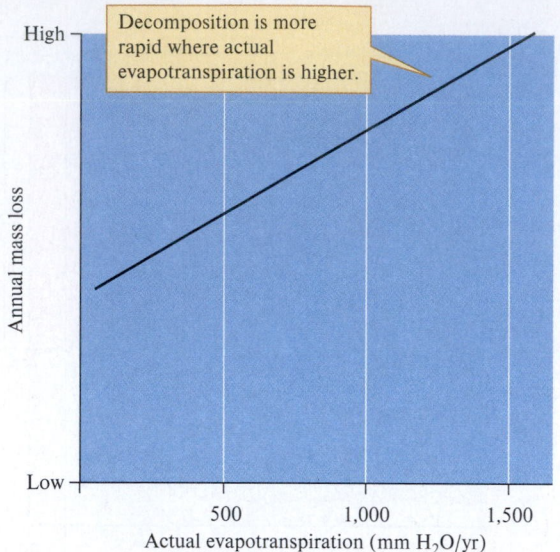

Figure 19.8   Relationship between actual evapotranspiration and decomposition (data from Meentemeyer 1978).

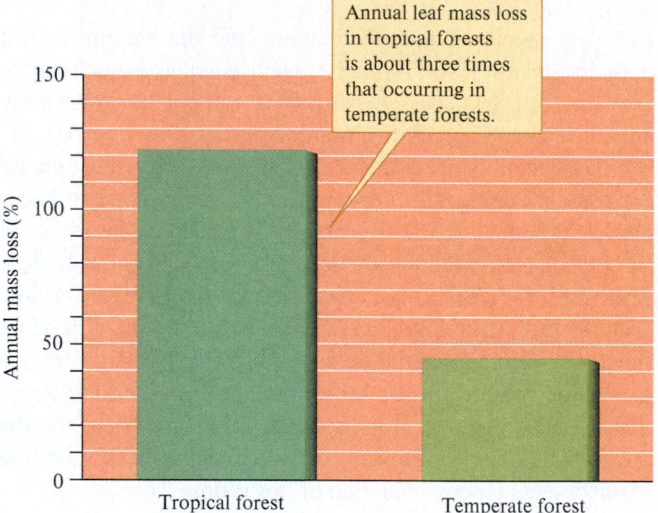

Figure 19.9   Decomposition in tropical and temperate forests (data from Anderson and Swift 1983).

Carolina site? Melillo and his colleagues suggested that higher nitrogen availability in the soils at the North Carolina site may contribute to the higher rates of decomposition observed there. However, higher temperatures at the North Carolina site may also contribute to higher decomposition rates.

Studies in both temperate and Mediterranean regions suggest that rates of decomposition are positively correlated with temperature and moisture. Can we combine these two factors into one? In chapter 18, we reviewed how ecologists studying the effect of climate on terrestrial primary production combined temperature and precipitation into a single measure called actual evapotranspiration, or AET. Vernon Meentemeyer (1978) analyzed the relationship between AET and decomposition and found a significant positive relationship (fig. 19.8).

If decomposition rates increase with increased evapotranspiration, how would you expect rates of decomposition in tropical and temperate ecosystems to compare? As you probably predicted, rates of decomposition are generally higher in tropical ecosystems. The average annual mass loss in tropical forests shown in figure 19.9 is 120%, or three times the average rate measured in temperate forests. These higher rates probably reflect the effects of higher AET in tropical forests and indicate complete decomposition in less than a year.

Soil nutrient content has also been shown to have a strong positive effect on rates of nutrient cycling in tropical forests. Three forest ecologists, Masaaki Takyu, Shin-Ichiro Aiba, and Kanehiro Kitayama, took advantage of natural variation in

nutrient content on different geological formations and different topographic situations to explore the factors influencing tropical rain forest functioning in Borneo (Takyu, Aiba, and Kitayama 2003). Takyu, Aiba, and Kitayama established research sites on three different rock types in two different locations in the landscape: on ridges and on lower slopes. Ridges tend to have soils with lower nutrient content compared to soils on lower slopes. The younger Quaternary sedimentary rock in their study area was approximately 30,000 to 40,000 years old. Soils developing on these rocks tended to have higher nutrient content compared to soils on the other rock types in the study, especially on lower slopes (fig. 19.10). Two older rock types were both approximately 40 million years old. One, a Tertiary sedimentary rock, supported soils that were considerably more fertile than the other, an ultrabasic rock, an igneous rock rich in iron and magnesium bearing minerals but with low silica content

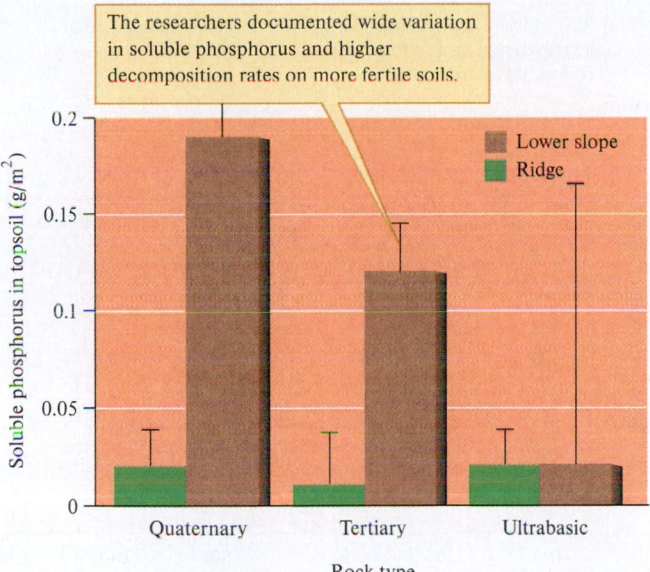

The researchers documented wide variation in soluble phosphorus and higher decomposition rates on more fertile soils.

Figure 19.10 Concentrations of soluble phosphorus in topsoils formed on three rock types and at two topographic positions in Borneo (means, 1 standard deviation) (data from Takyu, Aiba, and Kitayama 2003).

Takyu, Aiba, and Kitayama's study clearly demonstrated the influence of soil fertility on rates of decomposition and nutrient cycling. Because all study sites were at approximately the same elevation and all were on south-facing aspects, the researchers were able to isolate the influences of geological conditions, especially soil characteristics. They found higher rates of aboveground net primary production, higher rates of litter fall, and higher rates of decomposition on sites with higher concentrations of soluble phosphorus in topsoil, particularly on soils formed on the lower slopes of Quaternary and Tertiary rock formations. These results show that while climate may have a primary influence on decomposition rates, within climatic regions nutrient availability has an ecologically significant effect on decomposition and nutrient cycling rates.

In summary, decomposition in terrestrial ecosystems is influenced by moisture, temperature, soil fertility, and the chemical composition of litter, especially the concentrations of nitrogen and lignin. With the obvious exception of moisture, these factors also influence decomposition rates in aquatic ecosystems, which we examine next.

Decomposition in Aquatic Ecosystems

Jack Webster and Fred Benfield (1986) reviewed what was known about the decomposition of plant tissues in freshwater ecosystems. Among the most important variables that emerged from their analysis were leaf species, temperature, and nutrient concentrations in the aquatic ecosystem.

Webster and Benfield summarized the rates of leaf breakdown for 596 types of woody and nonwoody plants decaying in aquatic ecosystems and found that the average daily breakdown rate varied more than tenfold. As in terrestrial ecosystems, the chemical composition of litter significantly influences rates of decomposition in aquatic ecosystems.

Mark Gessner and Eric Chauvet (1994) studied the influence of litter chemistry on the rate of leaf decomposition. Their study site was a stream in the French Pyrenees. They included the leaves of several species of trees. The researchers found that leaves with a higher lignin content decomposed at a slower rate (fig. 19.11). What causes this? Gessner and Chauvet showed that a higher lignin content inhibits colonization of leaves by fungi, the main organisms responsible for decomposition of leaves in streams.

The nutrient content of stream water can also influence rates of decomposition. Keller Suberkropp and Eric Chauvet (1995) studied how water chemistry affects rates of leaf decomposition, using the leaves of yellow poplar, *Liriodendron tulipifera*. They placed the leaves in several temperate zone streams differing in water chemistry. The leaves in their litter bags decayed as an exponential function of time, following the relationship:

$$m_t = m_0 e^{-kt}$$

where:

m_t is the mass of leaves at time t

m_0 is the initial mass of leaves

e is the base of the natural logarithms

t is time in days

k is the daily rate of mass loss

We can use the constant k as an index of decay rate under particular environmental conditions. Suberkropp and Chauvet found that k varied significantly among their study streams. It turned out that leaves decayed faster—that is, had higher k—in streams with higher concentrations of nitrates (fig. 19.12). This result is consistent with the suggestion by Melillo's research

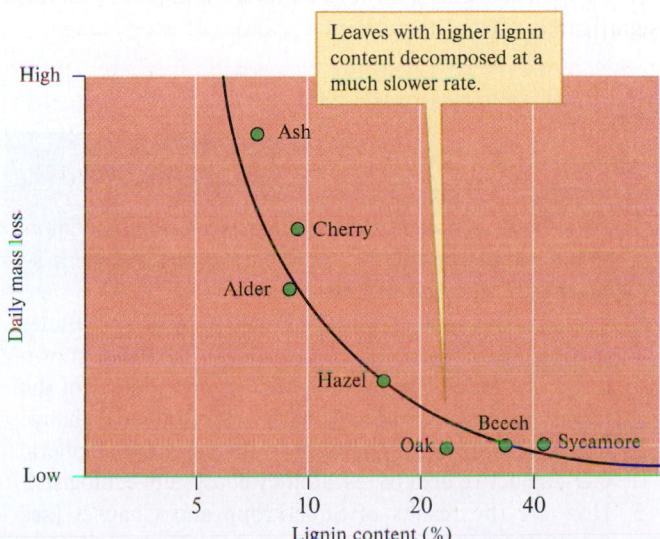

Leaves with higher lignin content decomposed at a much slower rate.

Figure 19.11 Lignin content of leaves and decomposition in an aquatic ecosystem (data from Gessner and Chauvet 1994).

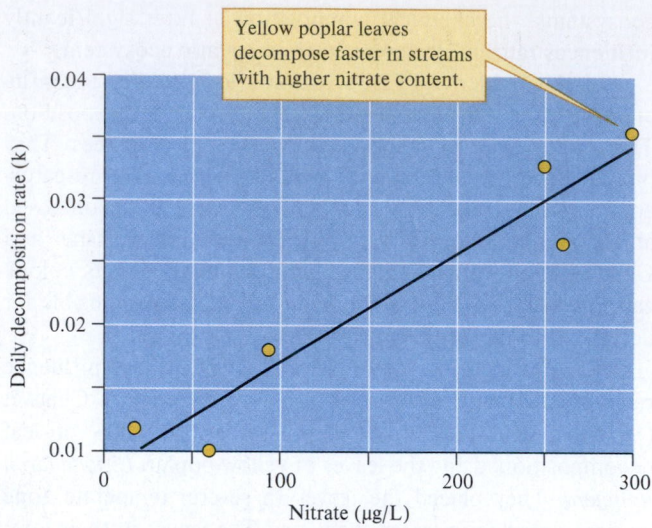

Figure 19.12   Stream nitrate and decomposition of *Liriodendron* leaves (data from Suberkropp and Chauvet 1995).

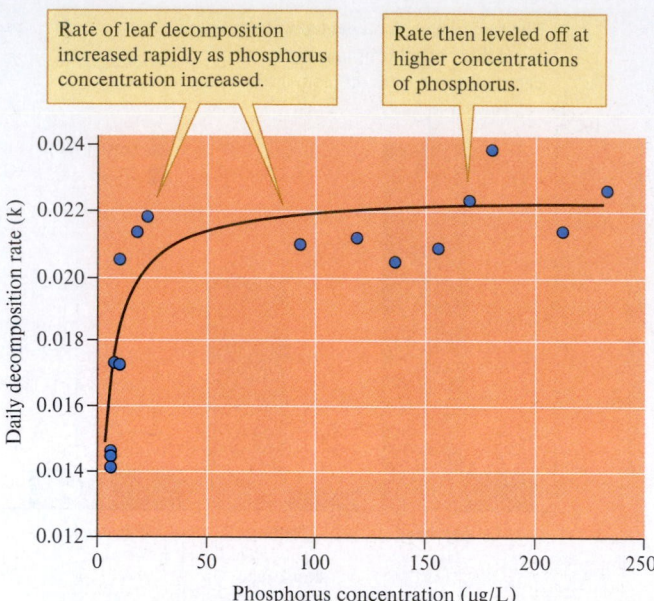

Figure 19.13   Phosphorus concentration of stream water and rate of decomposition of *Ficus insipida* leaves in tropical streams (courtesy of Amy Rosemond).

team that higher rates of decomposition at one of their study sites was due to higher availability of soil nitrogen.

Amy Rosemond and several colleagues of the University of Georgia conducted a similar study in streams draining a tropical forest in Costa Rica (Rosemond et al. 2002). They placed leaves of *Ficus insipida,* a tree that commonly grows along streams in Central America, at 16 stream sites that varied substantially in phosphorus concentration. Leaf decomposition rate increased markedly as phosphorus concentration increased to about 20 µg per liter, after which decomposition rate leveled off (fig. 19.13).

As in terrestrial ecosystems, litter chemistry and nutrient availability in the environment affect decomposition rates in aquatic ecosystems. The patterns discussed in this section emphasize the role played by the physical and chemical environment in the process of decomposition. As we shall see in the next section, however, animals and plants can also significantly affect the nutrient dynamics of ecosystems.

Concept 19.2 Review

1. Thousands of papers have been published on decomposition within ecosystems. Why have ecologists spent so much time studying decomposition?
2. The concentration of atmospheric CO_2, which continues to rise (see chapter 23, section 23.4), has been shown to be associated with an increased lignin content of the leaves of some plant species. Why is this potential change in leaf chemistry, in response to increased atmospheric CO_2, an active area of research by ecosystem ecologists?
3. How are the results of Suberkropp and Chauvet (see fig. 19.12) and Rosemond (see fig. 19.13) similar? How do their results differ?

19.3 Organisms and Nutrients

LEARNING OUTCOMES

After studying this section you should be able to do the following:

19.10 Describe the phenomenon of nutrient spiraling in stream ecosystems.
19.11 Predict how changes in velocity of nutrient movement (V) and time required for a nutrient atom to cycle (T) affect spiraling length (S).
19.12 Explain the relationship between nutrient spiraling length and stream retentiveness.
19.13 Discuss the influence of organisms on nutrients in aquatic and terrestrial ecosystems.
19.14 Explain the relationship between the N:P ratios of aquatic vertebrates and the ratio of N:P that they excrete.

Plants and animals can modify the distribution and cycling of nutrients in ecosystems.   How much do particular organisms affect ecosystem processes? With the threats of global change and mass extinction looming ever larger, this is one of the most important questions of our time. We do not yet know enough to offer a satisfying answer to this important question, but the information we do possess indicates that individual plant and animal species can substantially influence the distribution and dynamics of nutrients within ecosystems.

Nutrient Cycling in Streams and Lakes

Before we consider how stream animals influence the dynamics of nutrient turnover in streams, we have to consider some special features of this ecosystem. As we saw in chapter 3, the most distinctive feature of stream and river ecosystems is water flow. Jack Webster (1975) was the first to point out that because nutrients in streams are subject to downstream transport, there is little nutrient cycling in one place. Water currents move nutrients downstream. Webster suggested that rather than a stationary cycle, stream nutrient dynamics are better represented by a spiral. He coined the term **nutrient spiraling** to describe stream nutrient dynamics.

As an atom of a nutrient completes a cycle within a stream, it may pass through several ecosystem components such as an algal cell, an invertebrate, a fish, or a detrital fragment. Each of these ecosystem components may be displaced downstream by current and therefore contribute to nutrient spiraling. The length of stream required for an atom of a nutrient to complete a cycle is called the **spiraling length.** Spiraling length is related to the rate of nutrient cycling and average velocity of nutrient movement downstream. Denis Newbold and his colleagues (1983) represented spiraling length, S, as:

$$S = VT$$

where V is the average velocity at which a nutrient atom moves downstream and T is the average time for a nutrient atom to complete a cycle. If velocity, V, is low and the time to complete a nutrient cycle, T, is short, nutrient spiraling length is short. Where spiraling lengths are short, a particular nutrient atom may be used many times before it is washed out of a stream system.

The tendency of an ecosystem to retain nutrients is called **nutrient retentiveness.** In stream ecosystems, retentiveness is inversely related to spiraling length. Short spiraling lengths are equated with high retentiveness and long spiraling lengths with low retentiveness. Any factors that influence spiraling length affect nutrient retention by stream ecosystems.

Stream Invertebrates and Spiraling Length

Nancy Grimm (1988) showed that aquatic macroinvertebrates significantly increase the rate of nitrogen cycling in Sycamore Creek, Arizona. Streams in the arid American Southwest support high levels of macroinvertebrate biomass. Grimm estimated invertebrate population densities as high as 110,000 individuals per square meter and dry biomass as high as 9.62 g per square meter. More than 80% of macroinvertebrate biomass in Sycamore Creek was made up of species that feed on small organic particles, a feeding group that stream ecologists call *collector-gatherers.* The collector-gatherers of Sycamore Creek are dominated by two families of mayflies, Baetidae and Tricorythidae, and one family of Diptera, Chironomidae.

Grimm quantified the influence of macroinvertebrates on the nitrogen dynamics in the creek, where primary production is limited by nitrogen availability. She developed nitrogen budgets for stream invertebrates, mainly insect larvae and snails, by quantifying their rates of nitrogen ingestion, egestion (defecation), excretion, and accumulation during growth. By combining these rates with her estimates of macroinvertebrate biomass, Grimm was able to estimate the contribution of macroinvertebrates to the nutrient dynamics of the Sycamore Creek ecosystem.

Her measurements indicated that macroinvertebrates could play an important role in nutrient spiraling. To determine whether they play such a role, what information do we need? We need to know how much of the available nitrogen they ingest. If invertebrates ingest a large proportion of the nitrogen pool, then their influences on nitrogen spiraling may be substantial. Grimm measured the nutrient retention of Sycamore Creek as the daily difference between nitrogen inputs and outputs in her study area. Grimm set this rate of retention as 100% and then expressed her estimates of flux rates as a percentage of this total (fig. 19.14). Daily nitrogen ingestion rates by macroinvertebrates averaged about 131%. How can ingestion rates be greater than 100%? What this means is that the collector-gatherers in the study stream reingest nitrogen in their feces. This is a well-known habit of detritivores, many of which gain more nutritional value from their food by processing it more than once.

Grimm suggests that rapid recycling of nitrogen by macroinvertebrates may increase primary production in Sycamore Creek. Stream macroinvertebrates excreted and recycled 15% to 70% of the nitrogen pool as ammonia. By their high rates of feeding on the particulate nitrogen pool and their high rates of excretion of ammonia, the macroinvertebrates of the creek reduce the T in the equation for spiral length, $S = VT$. This effect coupled with the 10% of nitrogen tied up in macroinvertebrate biomass, which reduces V, reduces the nitrogen spiral length and increases the nutrient retentiveness of Sycamore Creek.

The Effect of Vertebrate Species on Nutrient Cycling in Aquatic Ecosystems

As we saw in chapter 7, organisms allocate nutrients in different ways. In general, adult animals of a given species maintain a relatively constant nutrient content of their body. Most herbivores and detritivores must overcome large differences between the low-nutrient content of their food and their own elemental requirements (see chapter 7, section 7.3). Remember that when an element is in high demand, it will be sequestered from food to build tissues, whereas other elements will be egested or excreted. Therefore, differences in the ratio of nitrogen:phosphorus (N:P) for different species could influence the ratio of N:P recycled into the environment.

Michael Vanni, Alexander Flecker, James Hood, and Jenifer Headworth (2002) sought to find out how animal species identity and N:P ratio affect nutrient cycling in the Rio Las Marías, a tropical stream in Venezuela. They measured the N:P ratio of 26 fish species and 2 amphibian species. The species in the study had diets of mainly algae and detritus. Excretion of nitrogen and phosphorus was quantified by confining individuals of each species in a plastic bag containing filtered stream water for about 1 hour and then analyzing the nutrient content of the water.

Vanni and his colleagues found a negative correlation between the excretion ratios of N:P and the N:P of each

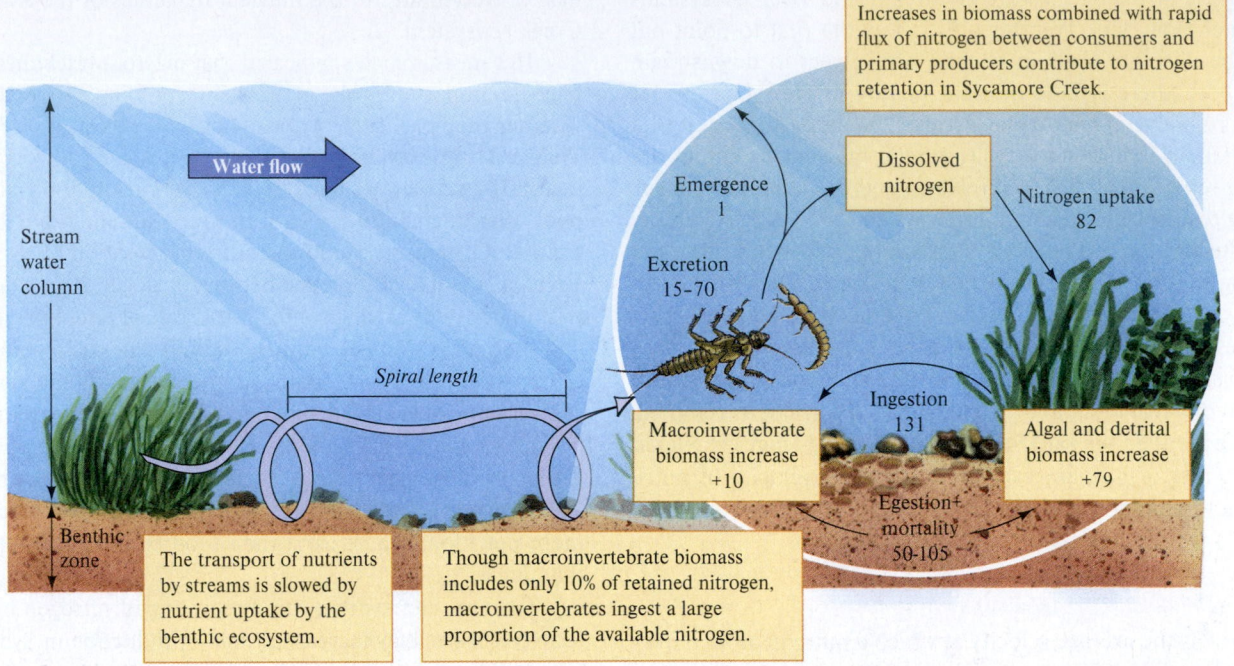

Increases in biomass combined with rapid flux of nitrogen between consumers and primary producers contribute to nitrogen retention in Sycamore Creek.

Water flow

Stream water column

Emergence 1

Dissolved nitrogen

Nitrogen uptake 82

Excretion 15–70

Spiral length

Ingestion 131

Macroinvertebrate biomass increase +10

Algal and detrital biomass increase +79

Egestion+ mortality 50–105

Benthic zone

The transport of nutrients by streams is slowed by nutrient uptake by the benthic ecosystem.

Though macroinvertebrate biomass includes only 10% of retained nitrogen, macroinvertebrates ingest a large proportion of the available nitrogen.

Figure 19.14   Nutrient spiraling in streams. Relative nitrogen fluxes are percentages of total nitrogen retained in Sycamore Creek, Arizona (data from Grimm 1988).

species (fig. 19.15). The N:P ratios of fish and amphibians varied from about 4 to 23. This variation in N:P ratio was mainly caused by differences in allocation of phosphorus for each species. For example, armored catfish species (Loricariidae) had the highest phosphorus content and lowest N:P. These fish species require large amounts of phosphorus relative to other fish and amphibians to make their bonelike armor. As a result, armored catfish had a low N:P ratio and excreted relatively

less phosphorus than fish and tadpole species with higher N:P ratios and lower phosphorus demands.

How important is the transport of nutrients by animals across ecosystem boundaries? The life cycle of anadromous Pacific salmon (*Oncorhynchus* spp.) is a dramatic example of how animals can affect nutrient cycling across ecosystem boundaries (see chapter 12 opener photo). *Oncorhynchus* spp. spend most of their lives at sea before they return to spawn and die in the streams and lakes where they were born. Annual spawning of *Oncorhynchus* spp. provides an important seasonal food resource for mammals and birds, and the marine-derived nutrients from salmon carcasses are utilized by multiple trophic levels in both freshwater and terrestrial ecosystems (fig. 19.16; Naiman et al. 2002).

As we shall see in the next section, consumers can also substantially affect nutrient cycling in terrestrial ecosystems.

Loricariids have a high P requirement, so they have low P excretion rates and higher N:P excretion ratios than . . .

. . . other fish and amphibians in the study that have a lower P requirement and low N:P excretion ratios

Excretion N:P ratio

Body N:P ratio

Figure 19.15   The relationship between the nutrient composition of vertebrate consumers and excretion ratios of nitrogen:phosphorus in a tropical stream (data from Vanni et al. 2002).

Figure 19.16   Salmon transport nutrients from marine to freshwater and terrestrial ecosystems during and after spawning.   wildnerdpix/123RF

Animals and Nutrient Cycling in Terrestrial Ecosystems

As we saw in chapter 16, burrowing animals, such as prairie dogs and pocket gophers, affect local plant diversity. These burrowers also alter the distribution and abundance of nitrogen within their ecosystems.

Pocket gophers can significantly affect their ecosystems because, as we discussed in chapter 16, their mounds may cover as much as 25% to 30% of the ground surface. This deposition represents a massive reorganization of soils and a substantial energy investment, since the cost of burrowing is 360 to 3,400 times that of aboveground movements. Estimates of the amount of soil deposited in mounds by gophers range from 10,000 to 85,000 kg per hectare per year.

Nancy Huntly and Richard Inouye (1988) found that pocket gophers altered the nitrogen cycle at the Cedar Creek Natural History Area in Minnesota by bringing nitrogen-poor subsoil to the surface (fig. 19.17). The result was greater horizontal heterogeneity in nitrogen availability and greater heterogeneity in light penetration. These effects on the nitrogen cycle in prairie ecosystems help explain some of the positive influences that pocket gophers have on plant diversity.

April Whicker and James Detling (1988) found that the feeding activities of prairie dogs also influence the distribution of nutrients within prairie ecosystems. This should not be surprising since these researchers estimate that prairie dogs consume or waste 60% to 80% of the net annual production from the grass-dominated areas around their colonies. One result of this heavy grazing is that aboveground biomass is reduced by 33% to 67% and the young grass tissue that remains is higher in nitrogen content (fig. 19.18). This higher nitrogen content may influence the behavior of bison, which spend a disproportionate amount of their time grazing near prairie dog colonies.

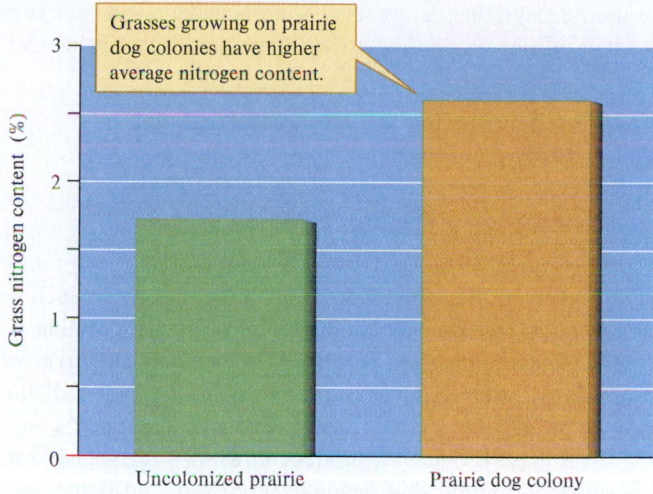

Figure 19.18 Early-season nitrogen content of grasses growing on uncolonized prairie and on a young prairie dog colony (data from Whicker and Detling 1988).

Bison and other large herbivorous mammals, such as moose and African buffalo, may also influence the cycling of nutrients within terrestrial ecosystems. Sam McNaughton and his colleagues (1988) report a positive relationship between grazing intensity and the rate of turnover of plant biomass in the Serengeti Plain of eastern Africa. Figure 19.19 suggests that increased grazing increases the rate of nutrient cycling.

McNaughton has built a model for nutrient cycling in grasslands that distinguishes between decomposition and grazing. He proposes that having a grazer pathway in an ecosystem speeds up the rate of nutrient cycling. Consequently, the large herbivores of the Serengeti are functionally similar to the collector-gatherer invertebrates of streams (see fig. 19.15). Both groups of organisms speed up the rate of nutrient cycling

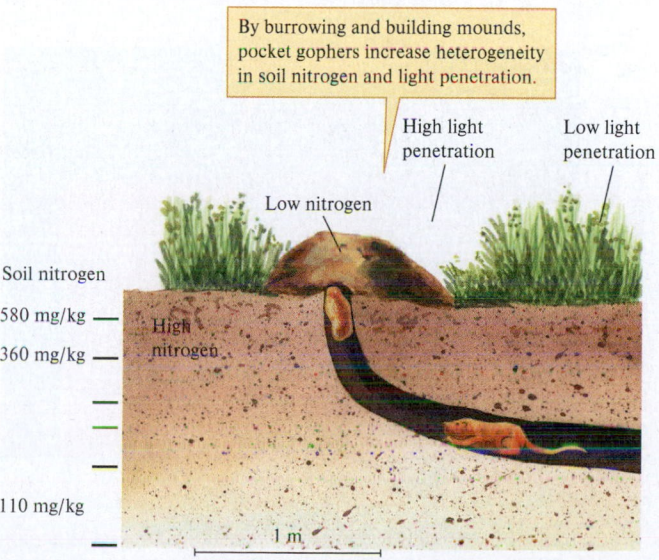

Figure 19.17 Pocket gophers and ecosystem structure (data from Huntly and Inouye 1988).

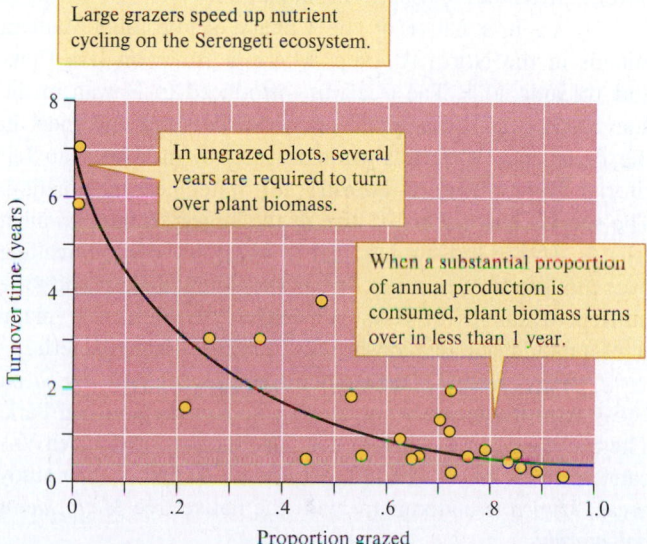

Figure 19.19 Effect of grazing on time required for turnover of plant biomass on the Serengeti ecosystem (data from McNaughton, Ruess, and Seagle 1988).

in their ecosystems. As we shall see, plants can also have substantial influences on the nutrient dynamics of ecosystems.

Plants and the Nutrient Dynamics of Ecosystems

Plants are not simply the passive recipients of influences from the physical environment or from animals and microbes. Plant species can influence ecosystem nutrient dynamics by a variety of mechanisms. Differences in plant species traits, such as nutrient uptake, allocation, and loss, affect nutrient cycling.

In general, plants in low-nutrient ecosystems tend to grow more slowly, which decreases their nutrient demand, and allocate more resources to roots to access nutrients in the soil. Plants in these ecosystems produce litter with high-lignin and low-nutrient content that decomposes slowly and deters herbivory. Ecosystems with greater nutrient availability generally support plants species that grow rapidly and that allocate more to shoots and leaves that contain nutrient levels that promote decomposition.

Studies mentioned earlier in this chapter showed that within an ecosystem, plant species exhibit wide variation in the lignin, carbon, nitrogen, and phosphorus content of their leaves, which affects rates of decomposition and mineralization of nutrients. In the following study, we will see how an introduced plant can modify nutrient levels in an ecosystem.

An Introduced Tree and Hawaiian Ecosystems

The Hawaiian Islands have been intensely invaded by both animals and plants. The native flora included approximately 1,200 species, of which over 90% were found nowhere else on earth. Humans have added approximately 4,600 exotic species to the Hawaiian flora. As you might expect, many of these exotic species significantly affect native populations, communities, and ecosystems. Peter Vitousek and Lawrence Walker (1989) found that an invading nitrogen-fixing tree, *Myrica faya*, is altering the nitrogen dynamics of ecosystems in Hawaii.

Myrica is a native of the Canary, Azore, and Madeira Islands in the North Atlantic, where it grows on lava flows and volcanic soils. The tree was introduced to Hawaii in the late 1800s as an ornamental or medicinal plant and then, in the 1920s and 1930s, was planted widely by the Hawaiian Territorial Department of Forestry for watershed reclamation. The species now occurs on five of the largest of the Hawaiian Islands. *Myrica* in Hawaii is highly invasive, quickly spreading over thousands of hectares. It has the potential to modify the nutrient dynamics of Hawaiian ecosystems because it maintains a mutualistic relationship with nitrogen-fixing bacteria.

Vitousek and Walker studied the influence of *Myrica* on ecosystem properties in the Hawaii Volcanoes National Park. Their study site was located near the summit of Kilauea Volcano at an elevation of 1,100 to 1,250 m. In one of their study areas, *Myrica* is codominant with the native tree *Metrosideros polymorpha*.

Vitousek and Walker estimated the rate of nitrogen fixation by *Myrica* in order to assess the rate at which the species adds nitrogen to the ecosystem. They put the contributions of

nitrogen fixation by the tree into context by also estimating the other nitrogen inputs to the ecosystem. They identified several indigenous sources of nitrogen fixation that included lichens, bacteria, and algae. Another major source of nitrogen to their study ecosystem was nitrogen in precipitation.

In the area where *Myrica* is codominant with the native tree *Metrosideros*, nitrogen fixation by *Myrica* is clearly the single largest source of nitrogen input to the ecosystem (fig. 19.20). As you might expect, the leaf tissues of *Myrica* contain a higher concentration of nitrogen than those of *Metrosideros* (fig. 19.21). The higher nitrogen content of the leaves is associated with a higher rate of decomposition and greater nitrogen release during decomposition. The result of these processes is increased nitrogen content in the soils associated with *Myrica*.

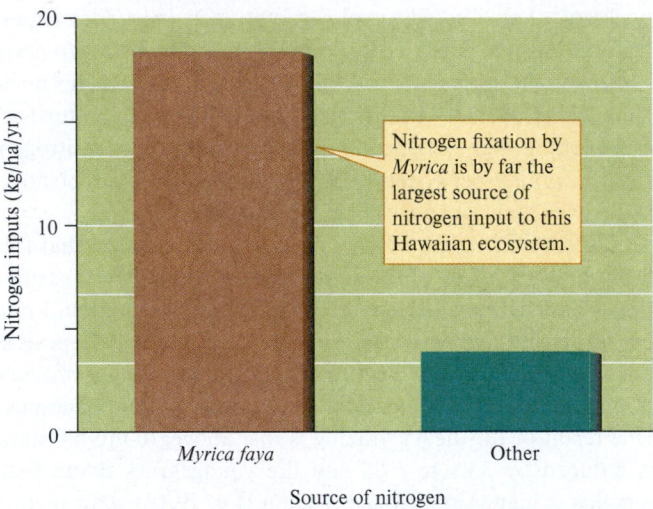

Figure 19.20   Nitrogen enrichment of Hawaiian ecosystems by an introduced tree, *Myrica faya* (data from Vitousek and Walker 1989).

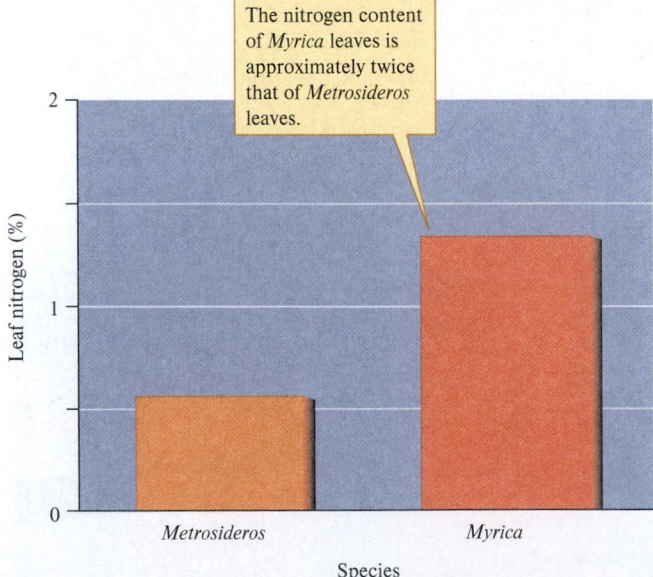

Figure 19.21   Leaf nitrogen content of a native tree, *Metrosideros polymorpha,* and an introduced tree, *Myrica faya* (data from Vitousek and Walker 1989).

In summary, invasive plants can substantially alter the nutrient dynamics of ecosystems. Vitousek and Walker suggest that the effects of invading species on ecosystem properties may offer the opportunity to trace the influences of individual species through the ecosystem and back to the population. Native species undoubtedly have similar effects. However, the effects of invading species are often more apparent, particularly where they are changing fundamental characteristics of entire ecosystems.

The introduction of exotic plants to ecosystems may be considered as a type of disturbance. In the next section, we review how disturbance increases rates of nutrient loss from ecosystems.

Concept 19.3 Review

1. The Great Plains of North America once supported bison herds numbering in the tens of millions. How did the near extermination of the bison likely affect nutrient cycling on the Great Plains?
2. How might nitrogen and phosphorus composition and excretion ratios differ for invertebrate consumers compared to vertebrate consumers?
3. Why might restoration of native plant communities to their original structure be difficult after exotic plants such as *Myrica,* in Hawaii, have occupied a site for a significant length of time?

19.4 Disturbance and Nutrients

LEARNING OUTCOMES

After studying this section you should be able to do the following:

19.15 Describe the effects of disturbance by deforestation on nutrient loss from the Hubbard Brook Experimental Forest.
19.16 Predict what would have happened to nutrient losses during the deforestation experiment at the Hubbard Brook Experimental Forest, if Likens and Bormann had not applied herbicides on their deforested basin over a period of 3 years.
19.17 Explain why the Bear Brook ecosystem exports more phosphorus than it receives as inputs in some years, while in other years, it exports less than it receives as inputs.

Disturbance generally increases nutrient loss from ecosystems. In the previous section, we saw how macroinvertebrates may increase nutrient retention by stream ecosystems. In this section, we consider evidence that disturbance affects nutrient retention.

Disturbance and Nutrient Loss from Forests

As Gene Likens and Herbert Bormann watched, work crews felled the trees covering an entire stream basin in the Hubbard Brook Experimental Forest of New Hampshire. The felling of these trees was a key part of an experiment that Likens and Bormann had designed to study how forests affect the loss of nutrients, such as nitrogen, from forested lands (Bormann and Likens 1994; Likens and Bormann 1995). They had studied two small stream valleys for 3 years before cutting the trees in one of the valleys. The undisturbed stream valley would act as a control against which to compare the response of the deforested stream valley (fig. 19.22).

Before they deforested the experimental basin, Likens and Bormann inventoried the distribution of nutrients. Those measurements indicated that over 90% of the nutrients in the ecosystem were tied up in soil organic matter. Most of the rest, 9.5%, was in vegetation. They estimated the rates at which some organisms fix atmospheric nitrogen and the rates at which weathering releases nutrients from the granite bedrock of the stream basins. They also measured the input of nutrients to the forest ecosystem from precipitation and nutrient outputs with stream water. The annual nutrient outputs in streamflow amounted to less than 1% of the amount contained within the forest ecosystems. After this preliminary work, Likens and Bormann cut the trees on their experimental stream basin. They then used herbicides to suppress regrowth of vegetation in their experimental basin and continued to apply herbicides for 3 years.

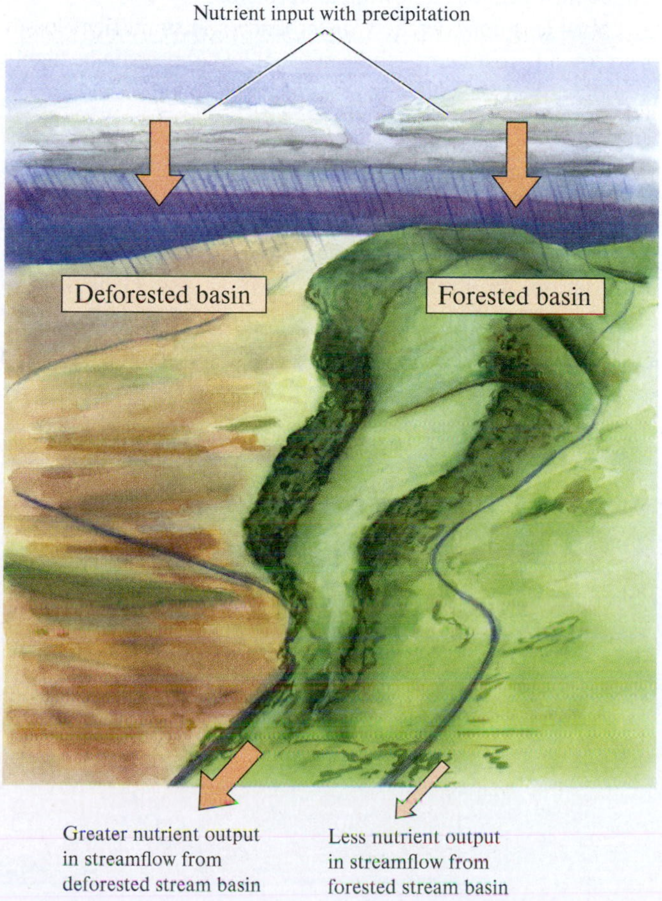

Figure 19.22   This whole stream basin manipulation demonstrated the influence of forest trees on nutrient budgets of northeastern hardwood forests.

The increased rates of nutrient loss following forest cutting were dramatic. The connection between forest cutting and increased nutrient output is shown clearly by plotting nutrient concentrations in the streams draining experimental and control stream basins. Figure 19.23 shows the highly significant increases in nitrate losses following deforestation, which were 40× to 50× higher than those in the deforested basin.

The development of vegetation following a disturbance affects many ecosystem processes, including nutrient cycling (see chapter 20). Monica Turner and her colleagues (2009) demonstrated biological influences on nutrient retention from forest ecosystems following the stand-replacing fires of 1988 in Yellowstone National Park, Wyoming. These forests are dominated by lodgepole pine (*Pinus contorta*) with a diverse understory plant community of understory plants that quickly regenerated following the fire (Turner et al. 2003). The studies of the Yellowstone fires suggest that the young, rapidly growing *P. contorta* forest and understory vegetation was a nitrogen sink in the ecosystem. Turner's work also indicates that not all ecosystems lose nutrients following a disturbance and that many factors affect nutrient retention over time.

What do these results suggest about the role of vegetation in preventing losses of nitrogen from forest ecosystems? Over the short term, at least, uptake by vegetation in ecosystems with high rates of plant growth should be able to rapidly reduce nitrogen loss following disturbance.

Now let's consider how disturbance affects nutrient losses from stream ecosystems, where nutrient loss appears to be highly episodic and associated with disturbance during flooding.

Flooding and Nutrient Export by Streams

How do the nutrient dynamics of stream ecosystems respond to variations in streamflow? Judy Meyer and Gene Likens (1979) examined the long-term dynamics of phosphorus in Bear Brook, a stream ecosystem in the Hubbard Brook Experimental Forest. They found that during periods of average flow, the ratio of annual phosphorus inputs to exports varied from 0.56 to 1.6 and that the balance depended on stream discharge. Meyer and Likens found that exports were highly episodic and associated with periods of high flow.

How did Meyer and Likens determine the phosphorus dynamics of Bear Brook ecosystem? They measured the geological and meteorological inputs and the geological exports of phosphorus in the stream, which they divided into three size fractions: (1) dissolved phosphorus, <0.45 μm; (2) phosphorus associated with fine particles, 0.45 μm to 1 mm; and (3) phosphorus associated with coarse particles, >1 mm.

Meyer and Likens inventoried the movement and storage of phosphorus in the Bear Brook ecosystem. They measured the inputs of dissolved phosphorus at 12 seeps (areas of groundwater input) along the length of the stream. They also measured meteorological inputs, which included precipitation and forest litter falling or blowing into the stream. The only significant export of phosphorus in the ecosystem was transport with streamflow. The researchers measured the amount of particulate matter transported by the Bear Brook ecosystem by collecting the organic matter deposited behind the weir (a small dam used to measure streamflow) on the stream and by collecting organic matter captured by nets set at seven sites. They estimated the amount of organic matter stored by removing all the organic matter in 42 randomly located 1 m² areas of stream bottom.

During 1974 to 1975, Meyer and Likens estimated an almost exact balance between inputs and exports of phosphorus, with approximately 1,250 mg of phosphorus per square meter of input and approximately 1,300 mg of phosphorus per square meter of output. Despite a balance between input and output, their data indicated significant transformation of phosphorus-size fractions. Phosphorus inputs to Bear Brook were almost evenly divided between dissolved (28%), fine particulate (37%), and coarse particulate (35%) fractions.

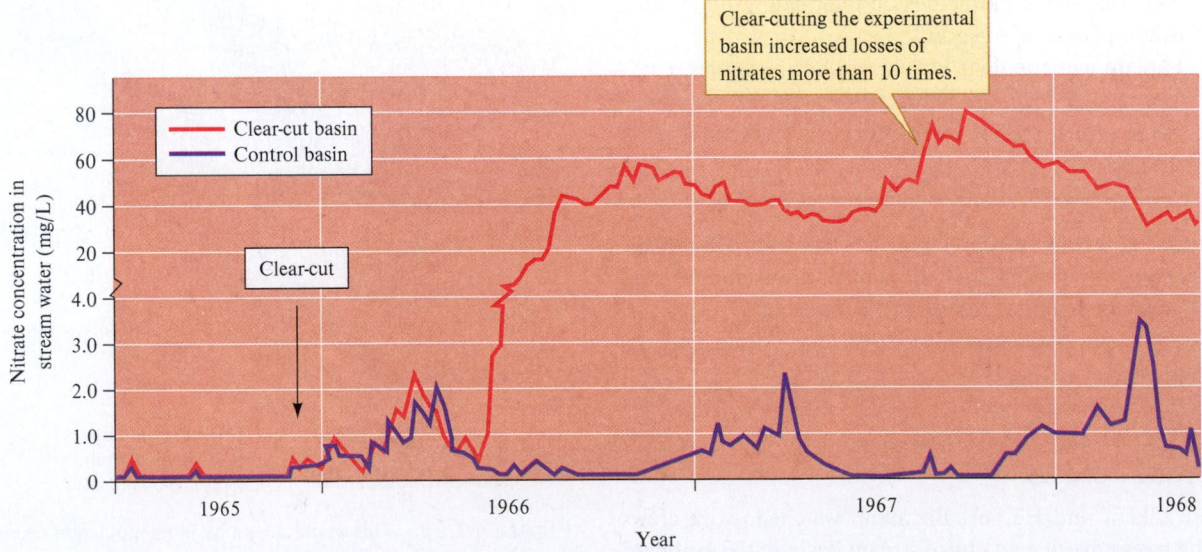

Figure 19.23   Deforestation and nitrate loss from a deciduous forest ecosystem (data from Likens et al. 1970).

However, 62% of exports were fine particulates. Clearly, physical and biological processes converted dissolved and coarse particulate forms of phosphorus into fine particulate forms.

Meyer and Likens used their estimates to reconstruct the long-term phosphorus dynamics of Bear Brook. During 1974 to 1975, the ratio of phosphorus export to input was almost exactly one (1.04). However, the Meyer and Likens model of phosphorus dynamics indicated considerable year-to-year variation in this ratio during the period from 1963 to 1975, which included the wettest and driest years in the 20-year precipitation record for the area. The predicted ratio of output to input over this interval was highly correlated with annual streamflow. The ratio ranged from 0.56 in the driest year to 1.6 in the wettest year (fig. 19.24). In other words, during the driest year only 56% of phosphorus inputs were exported, whereas during the wettest year exports amounted to 160% of inputs. In wet years the stream ecosystem's standing stocks of phosphorus were reduced by high levels of export.

The patterns of inputs and exports of phosphorus from Bear Brook were highly pulsed during Meyer and Likens's study (fig. 19.25). The researchers estimated that from 1974 to 1975, 48% of total annual input of phosphorus to Bear Brook entered during 10 days and that 67% of exports left the ecosystem during 10 days. The annual peak in phosphorus input was associated with autumn leaf fall, and an annual pulse of export was associated with spring snowmelt. Most phosphorus export, however, was irregular because it was driven by flooding caused by intense storms that may occur during any month of the year. If we consider floods as a source of disturbance, the behavior of stream ecosystems is consistent with the generalization that disturbance increases the loss of nutrients from ecosystems.

Aquatic ecologists study the nutrient dynamics of aquatic ecosystems like Bear Brook because, as we saw in chapter 18, nutrient availability is a key regulator of aquatic primary production. As we shall see in the Applications section, nutrient enrichment of ecosystems by human activity is a worldwide problem.

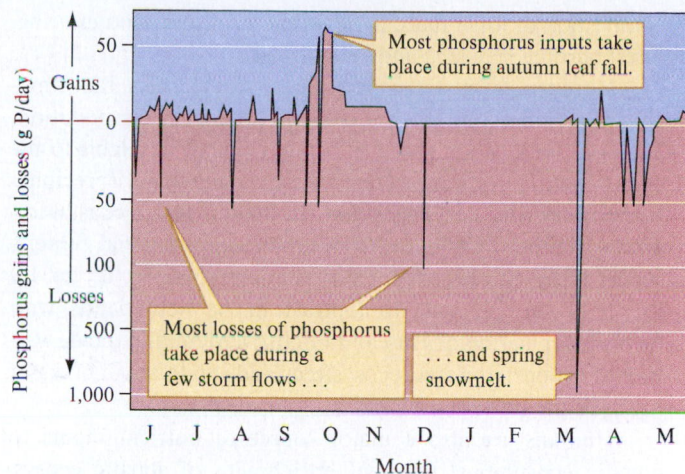

Figure 19.25 Daily gains and losses of phosphorus (P) by the Bear Brook ecosystem from 1974 to 1975 (data from Meyer and Likens 1979).

Concept 19.4 Review

1. What major conclusion can we draw from the pioneering experiment by Likens and Bormann?
2. What do the results of Likens and Bormann and those of Turner and her colleagues suggest about the role of vegetation in preventing losses of nitrogen in forest ecosystems?
3. Flood control on streams and rivers has often been cited as a potential threat to populations of aquatic animals and riparian trees that require flooding for reproduction. How might flow regulation also alter stream ecosystem nutrient dynamics?

Applications

Altering Aquatic and Terrestrial Ecosystems

LEARNING OUTCOMES

After studying this section you should be able to do the following:

19.18 Describe how human population density in river basins is related to the amount of nitrate carried in rivers.

19.19 Discuss how land cover in the Baltimore Ecosystem Study LTER is related to the export of nitrates in stream water.

19.20 Design a conceptual plan for reducing nitrate export from the Baltimore Ecosystem.

Human activity increasingly affects ecosystem nutrient cycles. Agriculture and forestry can remove nutrients from ecosystems. However, increasingly, human activity enriches ecosystems with nutrients, especially with nitrogen (see chapter 23) and phosphorus. Nitrogen enrichment comes from a variety of sources:

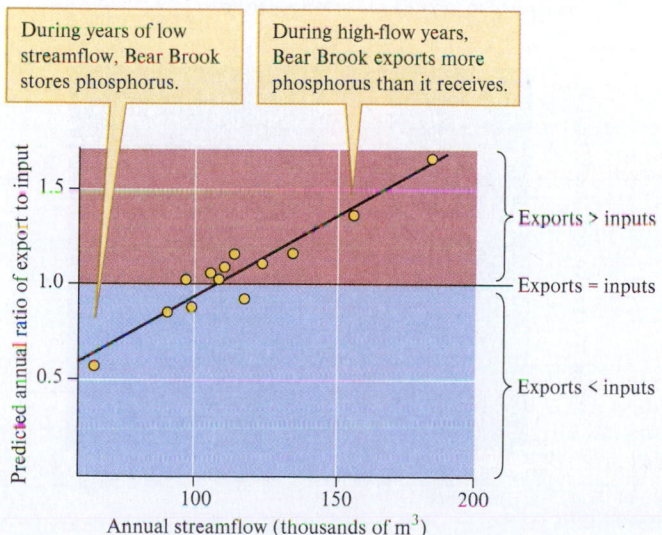

Figure 19.24 Annual streamflow and ratio of phosphorus export to input in Bear Brook, New Hampshire (data from Meyer and Likens 1979).

combustion of fossil fuels, agricultural fertilizers, land clearing, forest burning, industry, and animal waste.

Nitrogen from anthropogenic sources enters the atmosphere as emissions or particulates, producing air pollution. Nitrogen fertilization occurs when biologically available forms of nitrogen fall from the atmosphere as either **wet** (precipitation) or **dry** (dry fall) **deposition.** In the temperate coastal forests of southern Chile, far from urban and industrial centers, inputs from nitrogen deposition amount to about 0.1 to 1.0 kg per hectare per year. In contrast, in the Netherlands, with its high population density and intense agriculture, the deposition of nitrogen to forest ecosystems adds up to about 60 kg of nitrogen per hectare per year.

Humans are also a major source of nutrient inputs to aquatic ecosystems. Nutrient enrichment of aquatic ecosystems can result in water quality problems and **eutrophication,** a process generally resulting in increased primary production, anoxic conditions, and reduced biodiversity. Benjamin Peierls and his colleagues (1991) examined the relationship of human population density within river basins and nitrate concentration and export by 42 major rivers. These rivers, which deliver approximately 37% of the total freshwater flow to the oceans, support human population densities ranging from 1 to 1,000 individuals per square kilometer.

Peierls noted that while the concentration and export of nitrate by rivers is affected by complex biotic, abiotic, and anthropogenic factors, a single variable, human population density, explains most of the variation in nitrate concentration and export (fig. 19.26). The most probable sources of nitrate enrichment of river ecosystems are sewage disposal, atmospheric deposition, agriculture, and deforestation, all of which generally increase with increased human population density. The broad range of sources of nitrate is one of the main reasons it is difficult to control nitrogen pollution.

Cities are home to an increasing proportion of the human population. Worldwide, more people live in or near urban areas than in the more rural countryside. This concentration of people in cities is also associated with the concentration and transformation of energy, materials, and waste in a small area, which can have disproportionate impacts on nutrient cycling. Prompted by a need for better understanding of urban areas, ecological studies of nutrient pools and fluxes within whole watersheds, like those of Hubbard Brook Experimental Forest, are being applied to these human-dominated ecosystems.

One of the first questions asked by ecologists is whether urban ecosystems within a watershed are a source or a sink for nutrients. Peter Groffman and his colleagues (2004) compared the nitrogen budgets of forested, agricultural, and urbanized watersheds as part of the Baltimore Ecosystem Study LTER (see chapter 16, Applications) with surprising results. They constructed a budget by calculating inputs and outputs of nitrogen for each watershed type. Nitrate concentrations were measured as a major component of nitrogen outputs in streams draining each watershed. As expected, mean nitrate concentrations were very low in the forested watershed (fig. 19.27). However, nitrate levels in dense urban areas were lower than those in either the suburban or agricultural watersheds.

What caused higher nitrate concentrations in streams draining the agricultural and suburban watersheds? First, Groffman and his colleagues compared nitrogen inputs into the different watersheds. All of these watersheds receive similar amounts of atmospheric nitrogen deposition. The first major difference in inputs was in the application of fertilizers. Estimates of nitrogen inputs from fertilizers were highest for agricultural areas, but application rates of fertilizer to residential lawns in suburban areas were also substantial. They also found that centralized urban sanitary systems can leak

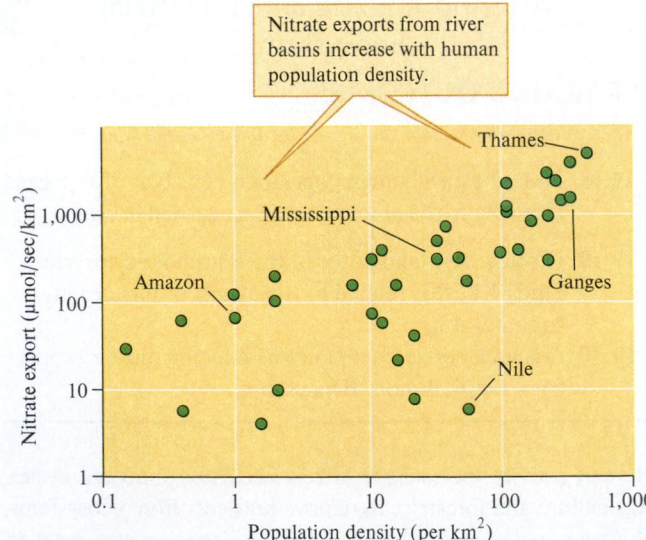

Figure 19.26   Human population density and nitrate export from river basins (data from Peierls et al. 1991).

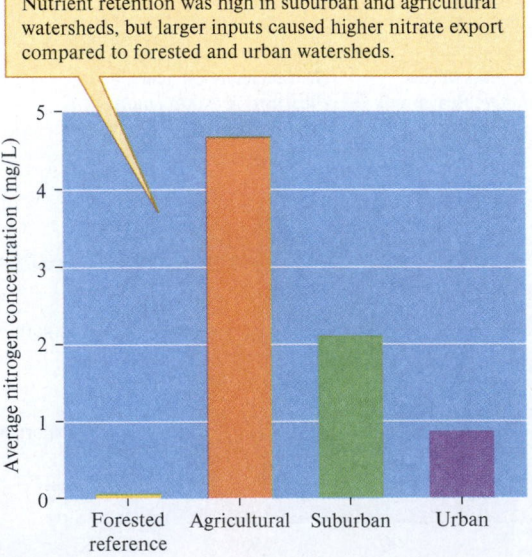

Figure 19.27   Nitrate export in streams draining watersheds with different land use in Baltimore County, Maryland (data from Pickett et al. 2008).

nitrogen, but are controllable. In contrast, residential septic systems in suburban areas are designed to discharge nitrate and can contribute to nitrate concentrations in streams.

The difference in nitrogen outputs from these watersheds could not be explained by inputs alone. Another surprising result of Groffman's study was that the suburban watershed retained an estimated 75% of nitrogen inputs, approaching retention values of forested watersheds (95%). Therefore, even though the suburban areas had higher nitrogen inputs, only a quarter of this nitrogen was exported downstream. Groffman and his colleagues attribute this high capacity for nitrogen retention in suburban areas to actively growing lawns, woodlots, and riparian areas that serve as nitrogen sinks. These permeable areas within the watershed have the potential to decrease nitrogen export to surface water through biological uptake and denitrification of nitrate. High nitrogen retention has also been observed for the large metropolitan area of Phoenix, Arizona (Baker et al. 2001). These results call into question a common assumption that human alteration of the environment limits ecological processes in urbanized areas.

If urban areas can retain a large proportion of nitrogen inputs, why do we observe high nitrate export to rivers worldwide dominated by human activity? Studies of nitrogen budgets in cities are very limited and, to date, incomplete. For instance, the nitrogen budget for Baltimore did not include the large flux of food into and sewage out of the study watersheds. Wastewater is piped out for treatment on much larger streams or bays where wastewater effluent can be safely discharged. Most cities around the globe are facing the challenge of minimizing nitrogen pollution without proper infrastructure and planning. In a review of urban ecological studies, Emily Bernhardt and her colleagues (2008) suggest that cities need to integrate nitrogen reduction strategies into their management plans to minimize impacts on water quality. However, while scientists can make recommendations based on their research, they do not make or implement policy. Cities are dynamic, complex, socio-ecological systems that require close cooperation of scientists, managers, and citizens to minimize nutrient pollution.

In summary, we know there is a direct connection between human activity and nutrient enrichment of ecosystems and that nutrient enrichment has a number of negative consequences. Application of an ecosystem approach to studies of nitrogen pools and fluxes within urban watersheds has advanced our understanding of nutrient cycling in urban systems, but much work remains at this frontier of ecology.

Summary

Nutrient cycles involve the storage of chemical elements in nutrient pools, or compartments, and the flux, or transfer, of nutrients between pools. The elements organisms require for development, maintenance, and reproduction are called *nutrients*. Ecologists refer to the use, transformation, movement, and reuse of nutrients in ecosystems as *nutrient cycling*. The *carbon, nitrogen,* and *phosphorus cycles* have played especially prominent roles in studies of nutrient cycling. The rate of CO_2 buildup in the atmosphere has been significantly slower than scientists would predict given the rate of emissions to the atmosphere by fossil fuel burning minus known carbon sinks. The best current estimates are that the missing carbon sink is formed by a combination of northern and tropical forests.

Decomposition rate is influenced by temperature, moisture, and chemical composition of litter and the environment. The rate of decomposition affects the rate at which nutrients, such as nitrogen and phosphorus, are made available to primary producers. Rates of decomposition in terrestrial ecosystems are higher under warm, moist conditions. The rate of decomposition in terrestrial ecosystems increases with nitrogen content and decreases with the lignin content of litter. The chemical composition of litter and the availability of nutrients in the surrounding environment also influence rates of decomposition in aquatic ecosystems.

Plants and animals can modify the distribution and cycling of nutrients in ecosystems. The dynamics of nutrients in streams are best represented by a spiral rather than a cycle. The length of stream required for an atom of a nutrient to complete a cycle is called the spiraling length. Stream macroinvertebrates can substantially reduce spiraling length of nutrients in stream ecosystems. Animals can also alter the distribution and rate of nutrient cycling in terrestrial ecosystems. The nitrogen and phosphorus composition of different species of fish was correlated with the ratio of N:P they excreted. Spawning salmon can increase nutrient availability into streams and lakes. Nitrogen-fixing plants increase the quantity and rates of nitrogen cycling in terrestrial ecosystems.

Disturbance generally increases nutrient loss from ecosystems. Vegetation exerts substantial control on nutrient retention by terrestrial ecosystems. Vegetative controls on nutrient loss from forest ecosystems are important after disturbance by fire. Nutrient loss by stream ecosystems is highly pulsed and associated with disturbance by flooding.

Nutrient enrichment by humans is altering aquatic and terrestrial ecosystems. Nitrate concentration and export by the earth's major rivers correlate directly with human population density. A study of an urban watershed showed that cities can have higher nutrient retention than previously expected. Ecosystem-scale studies can be used as a tool to improve management of urban areas and to address challenging water quality problems.

Key Terms

decomposition 408	nutrient cycling 404	nutrient retentiveness 413	nutrient spiraling 413
dry deposition 420	nutrient flux 405	nutrient sink 405	spiraling length 413
eutrophication 420	nutrient pool 405	nutrient source 405	wet deposition 420
mineralization 408			

Review Questions

1. Of all the naturally occurring elements in the biosphere, why have the cycles of carbon, nitrogen, and phosphorus been so intensively studied by ecologists? (Hint: Think about the kinds of organic molecules of which these elements are constituents. Also think back to our discussions, in chapter 18, of the influences of nitrogen and phosphorus on rates of primary production.)

2. Parmenter and Lamarra (1991) studied decomposition of fish and waterfowl carrion in a freshwater marsh. During the course of their studies, they found that the soft tissues of both fish and waterfowl decomposed faster than the most rapidly decomposing plant tissues. Explain the rapid decomposition of these animal carcasses.

3. Review figure 18.2, in which Rosenzweig (1968) plotted the relationship between actual evapotranspiration and net primary production. How do you think that decomposition rates change across the same ecosystems? Using what you learned in chapter 19, design an experiment to test your hypothesis.

4. Melillo, Aber, and Muratore (1982) suggested that soil fertility may influence the rate of decomposition in terrestrial ecosystems. Design an experiment to test this hypothesis. If you test for the effects of soil fertility, how will you control for the influences of temperature, moisture, and litter chemistry?

5. Many rivers around the world have been straightened and deepened to improve conditions for navigation. Side effects of these changes include increased average water velocity and decreased movement of water into shallow riverside environments such as eddies and marginal wetlands. What are the probable influences of these changes on nutrient spiraling length? Use the model of Newbold et al. (1983) in your discussion.

6. Likens and Bormann (1995) found that vegetation substantially influences the rate of nutrient loss from small stream catchments in the northern hardwood forest ecosystem. How do vegetative biomass and rates of primary production in these forests affect their capacity to regulate nutrient loss? How much do you think vegetation affects nutrient movements in desert ecosystems?

7. McNaughton, Ruess, and Seagle (1988) proposed that grazing by large mammals increases the rate of nitrogen cycling on the savannas of East Africa. Explain how passing through a large mammal could increase the rate of breakdown of plant biomass. How could the disappearance of the large mammals of East Africa affect nutrient cycling on the savanna?

8. The fynbos of South Africa is famous for the exceptional diversity of its plant community. Witkowski (1991) showed that invading *Acacia* are enriching the fynbos soil with nitrogen. How might enriching soil nitrogen affect nutrient cycling and primary production in this ecosystem? What mechanisms would likely produce your predicted changes?

9. Kauffman and his colleagues (1993) estimated that burning the tropical forest at their study site resulted in the loss of approximately 21 kg per hectare of phosphorus. This quantity is about 11% to 17% of the total pool of phosphorus. If total annual inputs of phosphorus to the ecosystem, mainly by rain and "dry fall," amount to about 0.2 kg per hectare per year (Murphy and Lugo 1986), how long would it take these inputs to make up for a single agricultural burn such as that created by Kauffman and his colleagues? Assuming a constant rate of loss, how many burns would it take to totally exhaust existing supplies?

Chapter 20

Succession and Stability

Design Pics Inc./Alamy Images

Maple trees and conifers along the edge of a bog in Maine. The accumulation of plant organic matter will gradually fill in this bog, which, as it dries and develops soil, will be colonized by grasses, shrubs, and trees. Eventually, a succession of physical and biological changes will transform this aquatic ecosystem into a forest.

CHAPTER CONCEPTS

20.1 Community changes during succession include increases in species diversity and changes in species composition.   425

Concept 20.1 Review   429

20.2 Ecosystem changes during succession include increases in biomass, primary production, respiration, and nutrient retention.   429

Concept 20.2 Review   432

20.3 Mechanisms that drive ecological succession include facilitation, tolerance, and inhibition.   432

Concept 20.3 Review   436

20.4 Community stability may be due to lack of disturbance or community resistance or resilience in the face of disturbance.   436

Concept 20.4 Review   440

Applications: Ecological Succession Informing Ecological Restoration   440

Summary   443

Key Terms   443

Review Questions   443

LEARNING OUTCOMES

After studying this section you should be able to do the following:

20.1 Define succession.
20.2 Distinguish between primary succession and secondary succession.
20.3 Describe the differences between pioneer community and climax community.

The first recorded visit to Glacier Bay gave no hint of its eventual contributions to our understanding of biological communities and ecosystems. In 1794, Captain George Vancouver visited the inlet to what is today called Glacier Bay, Alaska (fig. 20.1). He could not pass beyond the inlet to the bay, however, because his way was blocked by a mountain of ice (Vancouver and Vancouver 1798).

423

(a)

(b)

Figure 20.1 Glacier Bay National Park, Alaska, a laboratory for studying ecological succession. (*a*) The lower reaches of the Muir Glacier in August 1941. Note the bare terrain in the foreground of the photo. (*b*) The same scene 63 years later in 2004, at which point the Muir Glacier had retreated 8 km up the valley and out of the view to the left, leaving open water where the ice had been hundreds of meters thick. Notice that the once bare foreground is now covered by thick vegetation.
(*a*) USGS photo courtesy of the National Snow and Ice Data Center and Glacier Bay National Park and Perserve Archive; (*b*) Photo courtesy of Bruce Molnia, USGS

In 1879, John Muir explored the coast of Alaska, relying heavily on Vancouver's earlier descriptions. Muir (1915) commented in his journal that Vancouver's descriptions were excellent guides except for the area within Glacier Bay. Where Vancouver had met "mountains of ice," Muir found open water. He and his guides from the Hoona tribe paddled their canoe through Glacier Bay in rain and mist, feeling their way through uncharted territory. They eventually found the glaciers, which Muir estimated had retreated 30 to 40 km up the glacial valley since Vancouver's visit 85 years earlier.

Muir found no forests at the upper portions of the bay. He and his party had to build their campfires with the stumps and trunks of long-dead trees exposed by the retreating glaciers. Muir recognized that this "fossil wood" was a remnant of a forest that had been covered by advancing glaciers centuries earlier. He also saw that plants quickly colonized the areas uncovered by glaciers and that the oldest exposed areas, where Vancouver had met his mountains of ice, already supported forests.

Muir's observations in Glacier Bay were published in 1915 and read the same year by the ecologist William S. Cooper. Encouraged by Muir's descriptions, Cooper visited Glacier

Bay in 1916 in what was the beginning of a lifetime of study (Cooper 1923, 1931, 1939). Cooper saw Glacier Bay as the ideal laboratory for the study of ecological **succession,** the change in plant, animal, and microbial communities in an area following disturbance or the creation of new substrate. Glacier Bay was ideal for the study of succession because the history of glacial retreat could be accurately traced back to 1794 and perhaps further.

Because succession around Glacier Bay occurs on newly exposed geological substrates, not significantly modified by organisms, ecologists refer to this process as **primary succession.** Primary succession also occurs on newly formed volcanic surfaces such as lava flows. In areas where disturbance destroys a terrestrial community without destroying the soil, the subsequent succession is called **secondary succession.** For instance, secondary succession occurs after agricultural lands are abandoned or after a forest fire.

The first organisms to colonize in a successional sequence following a disturbance or the formation of a new geologic surface, form a **pioneer community.** The late successional community that can persist until disrupted by disturbance is called the **climax community.** The nature of the climax community depends upon environmental circumstances. The climax community around Glacier Bay is determined by the prevailing climate and local topography. On well-drained, steep slopes, the climax community is hemlock forest. In poorly drained soil on shallow slopes, the climax community is muskeg.

The nature of the climax community and the process of succession have been the focus of intense research and discussion for over a century. The debate between two ecologists was particularly prominent during the period when Cooper was studying succession at Glacier Bay. Frederic E. Clements (1916, 1936) compared ecological succession to the development of an organism and the climax community to a kind of superorganism. In Clements's view, succession was driven mainly by interactions between species that facilitated the progress of succession and that, given a set of environmental conditions, the species composition of the climax community was highly predictable. In addition, he proposed that essential species interactions sustained the climax community. In contrast, Henry A. Gleason (1926, 1939) advanced the idea that ecological communities, including the climax community, were not predicable associations of highly interdependent species but rather the result of a coincidence of independent species distributions along environmental gradients.

We have learned a great deal about ecological succession since the debates between Clements and Gleason, and the weight of the evidence now supports a view intermediate between their positions. For instance, we have reviewed in previous chapters how interactions between species can have far-reaching effects on the species diversity and composition of communities, which is consistent with the ideas of Clements. We have also learned that chance, which was prominent in Gleason's view, plays a significant role in the assembly of communities, particularly on islands and other isolated habitats (see Concept 22.2, chapter 22). In addition, the evidence indicates that a wide range of ecological processes drive succession (Pickett, Cadenasso, and Meiners 2009) and that the structure of climax communities is not as predictable as once thought.

Further, these late successional communities are not static but commonly change in response to disturbance, environmental change, and their own internal dynamics. Some of the most frequently observed patterns of change in community and ecosystem properties during succession and the mechanisms responsible for those changes are subjects covered in chapter 20. We also consider a companion topic: community and ecosystem stability.

20.1 Community Changes During Succession

LEARNING OUTCOMES

After studying this section you should be able to do the following:

20.4 Describe approaches to studying succession in different environments.

20.5 Compare patterns of community change during succession in various environments, for example, between stream communities and forest communities.

20.6 Explain the benefits and limitations of a chronosequence approach to understanding succession.

Community changes during succession include changes in species composition and species diversity. Some of the most detailed studies of ecological succession have focused on succession leading to a forest climax. Forest succession can manifest in different ways, resulting in different plant communities, depending on factors of chance and environment.

Primary Succession at Glacier Bay

William Cooper's studies of succession at Glacier Bay revealed an impressive successional change from a pioneer community of low herbaceous vegetation and fast-growing trees leading to mature forests (fig. 20.2). The plots established by Cooper were later sampled by his student, Donald Lawrence (Reiners et al. 1971), who maintained them until 1988. The sites remained unsampled until they were rediscovered in 2016 through painstaking work by Brian Buma, Sarah Bisbing,

John Krapek, and Glenn Wright (2017). The eight permanent plots established by Cooper in 1961 are now the longest-running primary succession study in the world.

The plots were initially chosen for similarity in physical features but differing substantially in age. In other words, they were established as a **chronosequence,** a series of communities or ecosystems representing a range of ages or times since disturbance. The eight study sites were below 100 m elevation and were on *glacial till*, an unstratified and unsorted material deposited by a glacier. Because they were established as permanent plots, they were also intended to be an opportunity to watch succession in action, that is, to test theories based on the chronosequence.

Both initial composition and changes in the vegetation of these permanent plots over time reveals a great deal about succession. Donald Lawrence, with his colleagues William Reiners and Ian Worley, used the chronosequence to develop hypotheses about the sequence of vegetation at Glacier Bay, concluding that the sites had gone through an orderly progression with only some variability (Reiners et al. 1971). This successional sequence supported Clements view that early species created conditions that supported the establishment of predictable later species. At Glacier Bay, primary succession began with moss and lichens which developed the soil, increasing in diversity over time until vascular herbaceous species established, such as *Dryas, Epilobium,* and *Equisetum* and the early successional *Salix* shrubs. Their chronosequence suggested that these gave way to other pioneer tree species such as *Populus* and *Alnus,* which were ultimately outcompeted by larger trees, such as *Picea* and *Tsuga* (fig. 20.2). Because older plots contained more species than younger plots, they also concluded that species richness generally increased over time, particularly of low shrubs and herbs.

However, later analysis of these same plots over time revealed a different story. By re-discovering the historic plots, Buma and colleagues were able to see if, in fact, the vegetation followed similar trajectories over time, leading to an inevitable convergence of a high-diversity *Picea* climax forest as Lawrence's group initially predicted (Buma et al. 2019). To test this hypothesis, Buma's group used the data from all

Figure 20.2 The classic story of succession at Glacier Bay, Alaska involves a series of stages, including (*a*) a pioneer stage dominated by herbaceous vegetation; (*b*) middle stages of fast-growing trees, such as these willows (*Salix*); and (*c*) a late stage of mature forest.
(*a, b*) Source: National Park Service; (*c*) Harald Sund/Getty Images

vegetation surveys done by Cooper and Lawrence, plus 2 years of new data—a total of 15 repeat observations per plot over the span of a century. They also used expanded plots and satellite imagery to assess the development of the forest at larger scales.

Buma and colleagues did not find that all plots were progressing to an inevitable end. There was change: in most plots, early abundance by a nitrogen-fixing, mat-forming *Dryas drummondii* (fig. 20.3) did give way to dominance by shrubs or trees. However, *Salix*-dominated plots did not ultimately become *Alnus* or *Picea*-dominated over time. In 2017, there were 3 apparently self-replacing (sustained) states: *Alnus*-dominated (3 plots), *Salix*-dominated (4 plots), and *Picea*-dominated (1 plot). The single *Picea* plot was the only one that was never colonized by *Salix;* the closed canopy, thick litter, and ability to vegetatively-reproduce allowed *Salix* to

competitively exclude the "later successional" *Alnus* and *Picea* in plots where it dominated early. Similarly, for plots where Alnus established early (the 1950's), the self-replacing nature of the resprouting species led to several successive generations and has, so far, excluded the other canopy species. Thus, species composition appeared to be more stochastic than deterministic, that is, based on what species happened to reach a location first as all are capable of vegetative reproduction, excluding other seeds from establishing.

Importantly, Buma also found that although there were some early increases in species richness, diversity generally either decreased or stayed the same, when considering the same plots over time, in contrast with what was seen in the chronosequence (fig. 20.4). That is, as individual trees became larger, there were less resources, particularly light, to support more species.

(a)

(b)

Figure 20.3   (*a*) Primary succession depends on lichens, mosses, and mat-forming herbs like this *Dryas drummondii* that can establish on little to no soil, shown here in flower. Dryas is especially helpful for the establishment of other plants because it is a nitrogen-fixer. (*b*) This photo was taken of Cooper's quadrat 2 in 1921 of Dryas in seed. By 2017, Dryas had been replaced in this and all other plots by later successional species.   (*a*) Mantonature/Getty Images; (*b*) Brian Buma

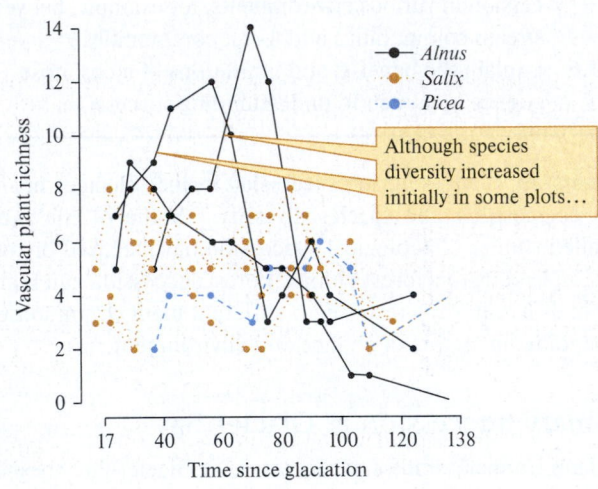

(a)

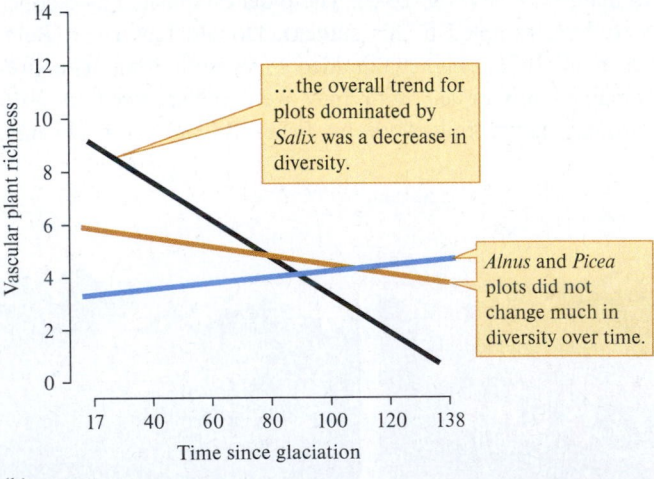

(b)

Figure 20.4   Change in vascular plant richness over time in eight long term plots of Glacier Bay, Alaska. They are categorized by which of three tree species was dominant in the final sampling period of 2017. (*a*) Change in species richness for each plot shows the dynamic nature of plant community composition. (*b*) The general trajectory of species richness over time for each community type.

Secondary Succession in Temperate Forests

The Piedmont Plateau of eastern North America includes some of the most convenient places to study secondary succession. The deciduous forests of this region were intensively cleared and cultivated beginning approximately three centuries ago. As fields were abandoned and new forest areas cleared, the region was progressively converted to a patchwork of abandoned fields of various ages interspersed with a few areas of undisturbed forest.

This situation provided Henry J. Oosting (1942) with study sites in virtually every stage of secondary succession for his now-classic studies of succession in the Piedmont Plateau of North Carolina. David Johnston and Eugene Odum (1956) described the pattern of succession on the Piedmont Plateau as follows. The first species to colonize and dominate abandoned fields are crabgrass, *Digitaria sanguinalis,* and horseweed, *Erigeron canadense.* During the second year of succession, the fields are dominated by either aster, *Aster pilosis,* or ragweed, *Ambrosia artemisiifolia.* A few years later the field is covered by broomsedge, *Andropogon virginicus,* with a scattering of shrubs and small trees. Pine seedlings may appear as early as the third year and may form a closed canopy in 10 to 15 years. Pine seedlings cannot grow in the shade of larger pines, but the seedlings of many deciduous trees can. Consequently, 40- to 50-year-old pine forests generally have a well-developed understory of young deciduous trees. These deciduous trees, especially oak, *Quercus,* and hickory, *Carya,* become the dominant trees by about 150 years, and the pines decline to a few scattered individuals. Because *Quercus* and *Carya* can reproduce in their own shade, the late successional oak-hickory forest is considered the climax stage.

Oosting's data show that the number of woody plant species increased during secondary succession on the Piedmont Plateau (fig. 20.5). The successional sequence began with a single species of woody plant invading soon after fields were abandoned and began to level off at 50 to 60 species after approximately 150 years.

How does animal diversity change across the same successional sequence? Johnston and Odum studied the birds living on thirteen 20-acre study sites ranging in successional age from 1 to 150 years and supporting vegetation ranging from grassland to mature oak-hickory forest. The increase in bird diversity across this successional sequence closely paralleled the increase in woody plant diversity observed by Oosting (fig. 20.6). During the grass-forb stage of succession, the bird community consisted of 2 species generally associated with grasslands. In the grass-shrub stage, the diversity of birds increased to 8 to 13 species. In 25- to 35-year-old pine forests, the diversity of birds was 10 to 12 species, increasing to 22 to 24 species in the later stages of succession.

In summary, over a period of approximately one and a half centuries, abandoned fields in eastern North America undergo successional changes in plant and bird communities that involve changes in species composition and increases in species diversity. Similar successional changes occur in the marine intertidal zone but on an even shorter timescale: approximately one and a half years instead of one and a half centuries.

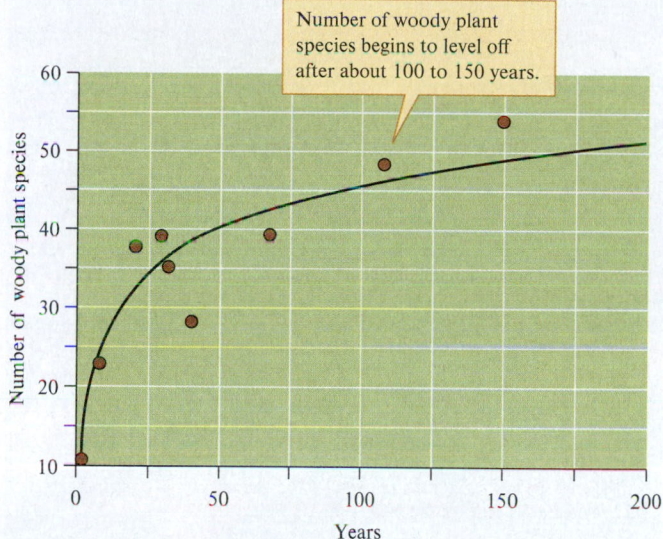

Figure 20.5   Change in woody plant species richness during secondary forest succession in eastern North America (data from Oosting 1942).

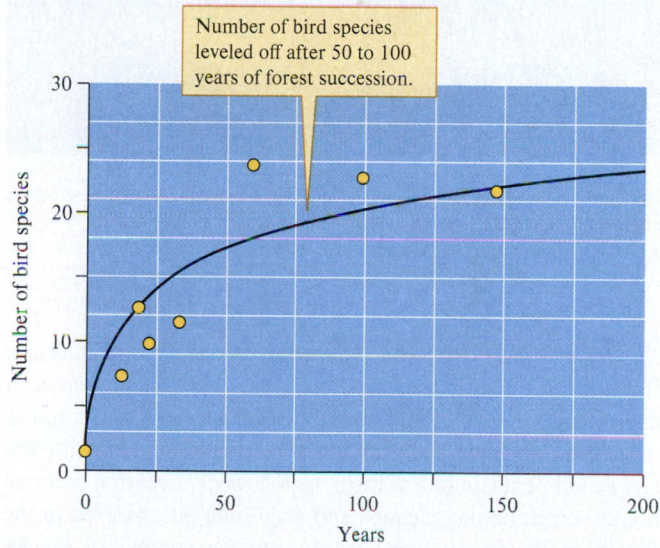

Figure 20.6   Change in number of breeding bird species during secondary forest succession (data from Johnston and Odum 1956).

Succession in Rocky Intertidal Communities

When we discussed the influence of disturbance on local species diversity in chapter 16, we saw how an intertidal boulder stripped of its cover of attached organisms was soon colonized by algae and barnacles (fig. 20.7). Looking back on that pattern of community change we can now see it as an example of ecological succession. Wayne Sousa (1979a, 1979b) showed that the first species to colonize open space on intertidal boulders were a green alga in the genus *Ulva* and the barnacle *Chthamalus fissus.* The next arrivals were several species of perennial red algae: *Gelidium coulteri, Gigartina leptorhynchos,*

Figure 20.7   Succession in the intertidal zone involves colonization and competition for limited space among species as different as attached marine algae, sea anemones, mussels, and barnacles.
Jim Zipp/Science Source

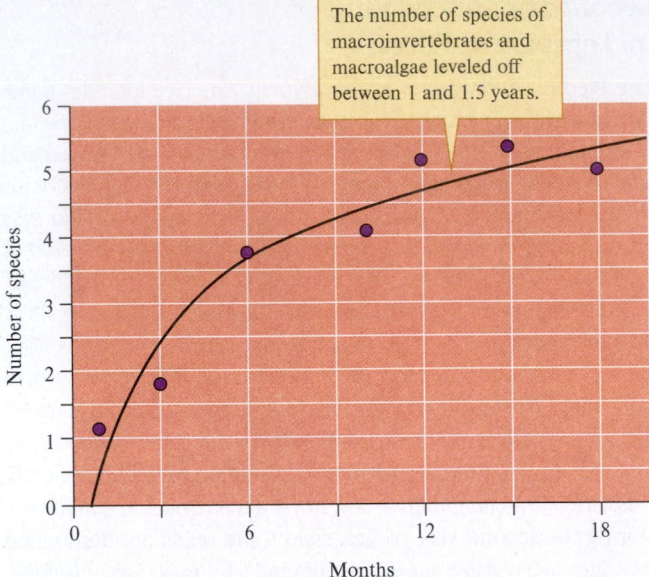

The number of species of macroinvertebrates and macroalgae leveled off between 1 and 1.5 years.

Figure 20.8   Succession in number of macroinvertebrate and macroalgal species on intertidal boulders (data from Sousa 1979a).

Rhodoglossum affine, and *Gigartina canaliculata.* Finally, if there was no disturbance for 2 to 3 years, *G. canaliculata* grew over the other species and dominated 60% to 90% of the space.

Sousa explored succession on intertidal boulders with several experiments. In one of them, he followed succession on small boulders that he had cleaned and stabilized. As observed in the Piedmont Plateau in forest succession, the number of species increased with time (fig. 20.8). Notice in the figure that the average number of species increased until about 1 to 1.5 years and then leveled off at about five species.

In both the intertidal and stream communities, where successional change is relatively rapid, researchers did not have to rely on a chronosequence or comparing historical data, but instead could document succession directly as it occurred on their study sites.

Succession in Stream Communities

Rapid succession has been well documented in Sycamore Creek, Arizona, which was studied for nearly two decades by Stuart Fisher and his colleagues (1982). Sycamore Creek, a tributary of the Verde River, lies approximately 32 km northeast of Phoenix, Arizona, where it drains approximately 500 km^2 of mountainous desert terrain. Evaporation nearly equals precipitation within the Sycamore Creek catchment, so

flows are generally low and often intermittent. However, the creek is subject to frequent flash floods with sufficient power to completely disrupt the community and initiate succession.

Fisher's research team reported on the successional events following one such flood. Intense floods occurred on Sycamore Creek on August 6, 12, and 16, of 1979, with peak flows of 7, 3, and 2 m^3 per second, respectively. Floods of this intensity mobilize the stones and sand of the stream, scouring some areas and depositing sediments in others. In the process, most stream organisms are destroyed. The three floods of August 1979 eliminated approximately 98% of algal and invertebrate biomass in Sycamore Creek.

In 63 days following these floods, Fisher and his colleagues observed rapid changes in both the diversity and composition of algae and invertebrates. Patterns among primary producers were especially clear. Two days after the floods, the majority of the stream bottom consisted of bare sand with some patches of diatoms. Five days after the flood, diatoms covered about half the streambed. Within 13 to 22 days, diatoms almost completely covered the stream bottom. Other algae, especially bluegreen algae and mats consisting of a mixture of the green alga *Cladophora* and blue-green algae, appeared in significant quantities by day 35. By day 63, the bottom of Sycamore Creek consisted of a patchwork of areas dominated by diatoms, blue-green algae, and mats of *Cladophora* and blue-green algae. The diversity of diatoms and other algae, as measured by H' (see chapter 16), leveled off after only 5 days and then began to decline after about 50 days (fig. 20.9).

Invertebrate diversity was strongly influenced by a single dominant species of crane fly, family Tipulidae, *Cryptolabis* sp. (fig. 20.10). Large numbers of *Cryptolabis* larvae in Sycamore Creek depressed H' diversity for all collections except for the collection on day 35, when most of the population emerged as adults. Throughout their collections over the 63-day period of the study, the researchers reported that they collected 38 to 43 species of aquatic invertebrates out of a total of 48 species

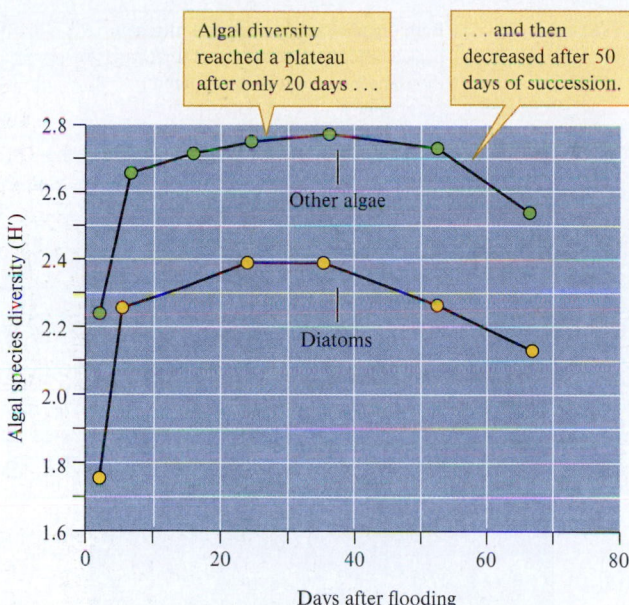

Figure 20.9   Algal species diversity during succession in Sycamore Creek, Arizona (data from Fisher et al. 1982).

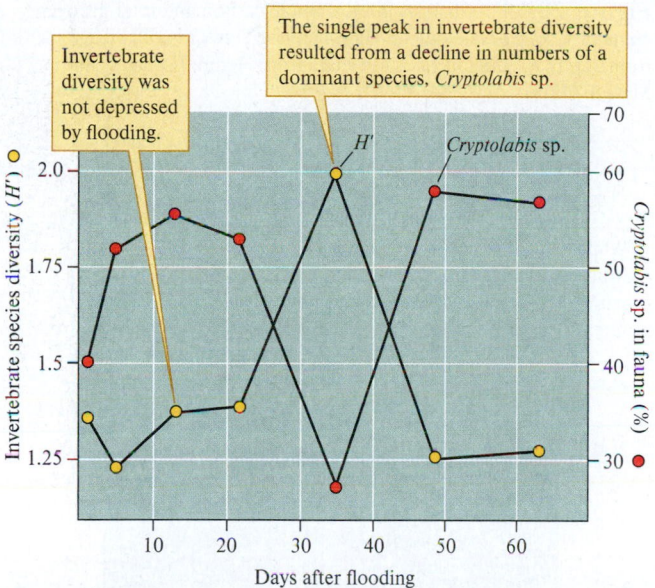

Figure 20.10   Invertebrate species diversity during succession in Sycamore Creek, Arizona (data from Fisher et al. 1982).

collected during their studies. In other words, most macroinvertebrate species survived the flood.

Where did these invertebrate species find refuge from the devastating floods of August 1979? The invertebrate community of Sycamore Creek is dominated by insects whose adults are terrestrial. During the floods of August, many adult insects were in the aerial stage and the flood passed under them. These aerial adults were the source of most invertebrate recolonization of the flood-devastated Sycamore Creek.

As we have seen, ecological succession involves changes in community structure—some predictable, and some not. These changes can also include ecosystem structure and function. These ecosystem changes with succession are the subject of the next section.

Concept 20.1 Review

1. How might the various ways that we observe succession affect what we understand about it?
2. What are the primary mechanisms producing the great differences in succession rates in forests versus rocky intertidal, communities?

20.2 Ecosystem Changes During Succession

LEARNING OUTCOMES

After studying this section you should be able to do the following:

20.7   Outline changes in soils during succession in the Hawaiian Islands.

20.8   Describe how Hawaiian Island ecosystems change over the course of millions of years.

20.9   Outline the patterns of ecosystem change in Sycamore Creek, Arizona, following disturbance by flooding.

20.10   Compare the pattern and timing of nitrogen retention and loss during succession in Sycamore Creek, Arizona, and from soils developing on Hawaiian Island lava flows.

Ecosystem changes during succession include increases in biomass, primary production, respiration, and nutrient retention. As succession changes the diversity and composition of communities, ecosystem properties change as well. In the previous section, we saw how plant and animal community structure changes during primary and secondary succession. In this section, we review evidence that many ecosystem properties also change during succession. For instance, many properties of soils, such as the nutrient and organic matter content, change during the course of succession.

Four Million Years of Ecosystem Change

The studies in Glacier Bay, Alaska that documented ecosystem change over the course of a century were impressive. However, 1794, when Captain George Vancouver encountered a wall of ice at the mouth of Glacier Bay, the island of Kauai in the Hawaiian Island chain supported forest ecosystems growing on soils that had developed on lava flows that were already over 4 million years old. The Hawaiian Islands have formed over a hot spot on the Pacific tectonic plate and have been transported on that plate to the northwest, forming a chain of islands that vary greatly in age (fig. 20.11). The youngest island in the group is the big island of Hawaii, which is currently growing over the hot spot. The big island is made up of volcanic rocks that vary from fresh lava flows to flows that are approximately 150,000 years old. Meanwhile, the islands to the northwest are sequentially older. As in Glacier Bay, teams of ecologists have probed the chronosequence represented by the Hawaiian Island chain for information on ecosystem development. However, in Hawaii the chronosequence spans not hundreds of years but millions.

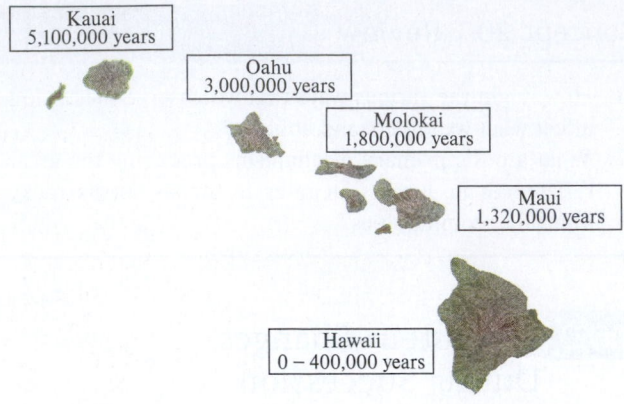

Figure 20.11   The Hawaiian archipelago creates a natural chronosequence for studying succession due to the varying ages of the islands.

Lars Hedin, Peter Vitousek, and Pamela Matson (2003) examined nutrient distributions and losses on a chronosequence of forest ecosystems on the islands of Hawaii, Molokai, and Kauai. The youngest ecosystems, which were on Hawaii, had developed on basaltic lava flows that were 300; 2,100; 20,000; and 150,000 years old. The study site on Molokai had developed on rocks that were 1.4 million years old, and the oldest study site, which was on Kauai, was 4.1 million years old. All sites currently have an average annual temperature of about 16°C and receive approximately 2,500 mm of precipitation annually. They also all support forest communities dominated by the native tree *Metrosideros polymorpha*.

Over the chronosequence represented by their six study sites, Hedin, Vitousek, and Matson encountered significant changes in a wide range of soil features. Earlier studies had demonstrated that primary production in the Hawaiian forest ecosystems is limited by nitrogen early in succession and by phosphorus later in succession. Organic matter, which is absent from fresh lava (fig. 20.12), increased in soils over the first 150,000 years of the chronosequence (fig. 20.13). However, in the Hawaiian chronosequence, organic matter was lower at the 1.4- and 4.1-million-year-old sites. Figure 20.13 also shows that changes in soil nitrogen content followed almost precisely the pattern exhibited by soil organic matter.

The pattern of change in the total phosphorus content of soils was remarkably different (fig. 20.14). The total amount

Figure 20.12   Pioneer species that can grow on little to no soil fix carbon and nitrogen from the atmosphere in their tissues, thus making these resources available in the soil when they decompose.   Westend61/SuperStock

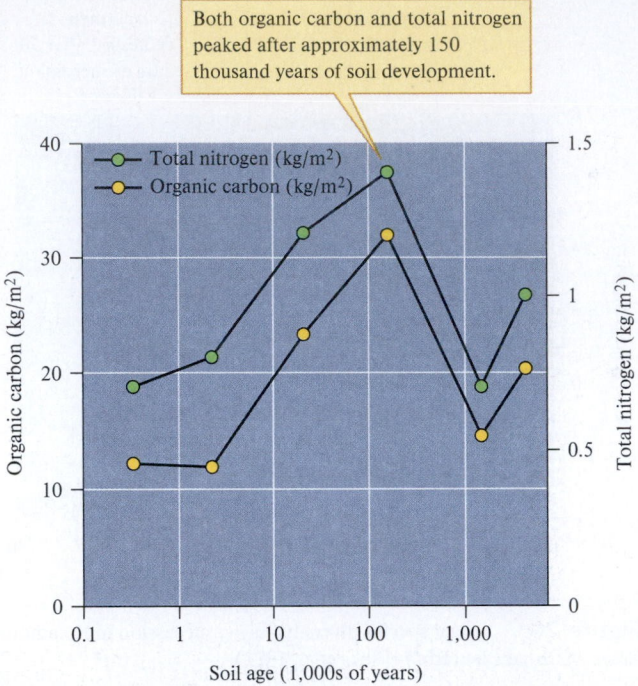

Figure 20.13   Changes in the organic carbon and total nitrogen content of soils developing on Hawaiian lava flows ranging in age from 300 to 4.1 million years old (data from Hedin, Vitousek, and Matson 2003).

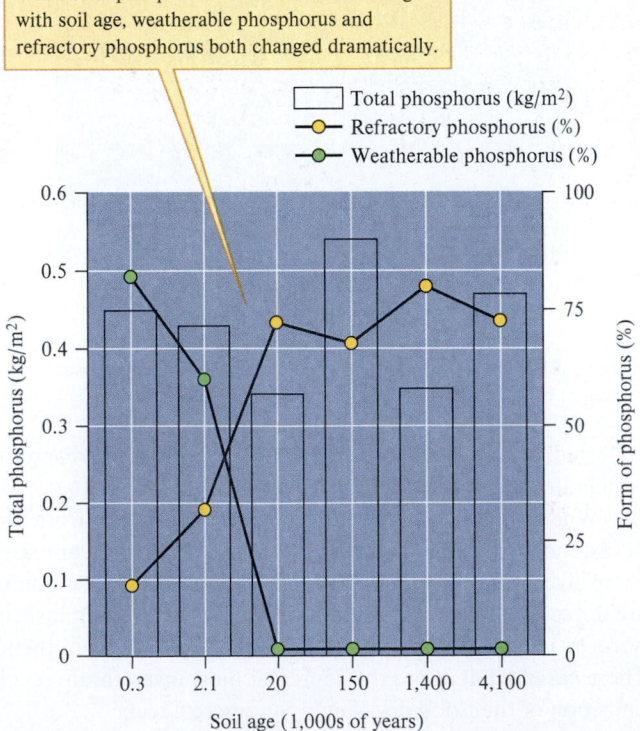

Figure 20.14   Changes in the total phosphorus and percentages of total phosphorus in weatherable and refractory (low availability) forms in soils developing on Hawaiian lava flows ranging in age from 300 to 4.1 million years old (data from Hedin, Vitousek, and Matson 2003).

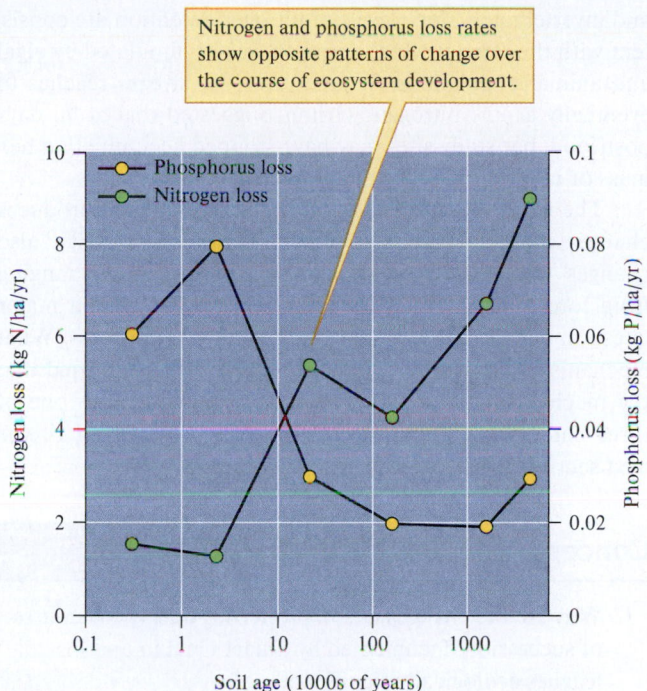

Figure 20.15   Nitrogen and phosphorus loss rates from soils developing on Hawaiian lava flows ranging in age from 300 to 4.1 million years old (data from Hedin, Vitousek, and Matson 2003).

of phosphorus in soils showed no obvious pattern of change with site age. However, the forms of phosphorus changed substantially over the chronosequence. Weatherable mineral phosphorus was largely depleted by 20,000 years. Meanwhile, the percentage of soil phosphorus in refractory forms, which are not readily available to plants, increased, varying from 68% to 80% of total phosphorus across ecosystems that had developed on lava flows 20,000 years old or older. On these older soils, primary production is limited by phosphorus availability.

Hedin, Vitousek, and Matson found changes in rates of nutrient loss across the chronosequence. Over the course of 4 million years of ecosystem development, these tropical forest ecosystems show progressively higher rates of nitrogen loss but decreased rates of phosphorus loss (fig. 20.15). In other words, for approximately 2,000 years these ecosystems are highly retentive of nitrogen, but as nitrogen content increases in their soils, they begin to lose nitrogen at a higher rate. Most losses are due to leaching to groundwater. In contrast, as phosphorus becomes progressively less available, and eventually limiting to primary production, in these ecosystems, they become more retentive of phosphorus. As we shall see in the next example, analogous changes in nutrient retention by a stream ecosystem occur over a period of just 90 days.

Succession and Stream Ecosystem Properties

Fisher's research group observed a pattern of rapid biomass accumulation followed by slower rates of increase during just 63 days of postflood succession in Sycamore Creek, Arizona. Algal biomass increased rapidly for the first 13 days following

disturbance and then increased more slowly from day 13 to day 63 (fig. 20.16). Sixty-three days after the flood, algal biomass showed clear signs of leveling off. The biomass of invertebrates, the chief animal group in Sycamore Creek, increased rapidly for 22 days following the flood and then, like the algal portion of the ecosystem, began to level off.

Ecosystem metabolic parameters showed even clearer signs of leveling off before the end of the 63-day study (fig. 20.17). Gross primary production (see chapter 18), measured as grams of O_2 produced per square meter per day, increased very rapidly until day 13, increased more slowly between days 13 and 48, and then leveled off between days 48 and 63. Total ecosystem

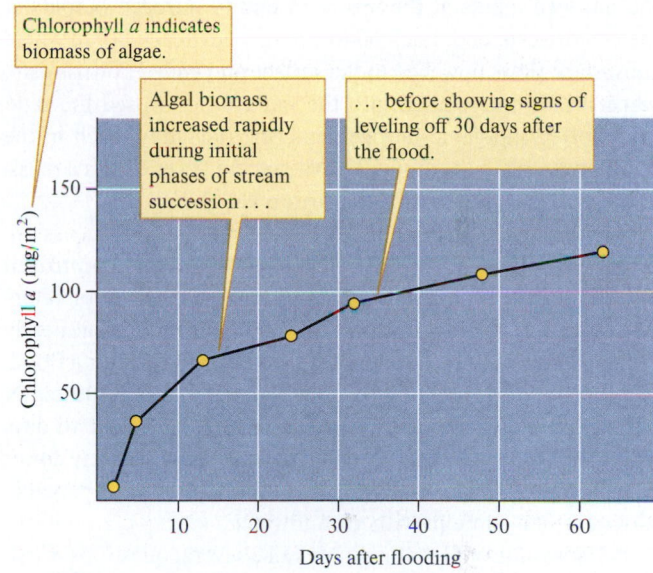

Figure 20.16   Changes in biomass during stream succession (data from Fisher et al. 1982).

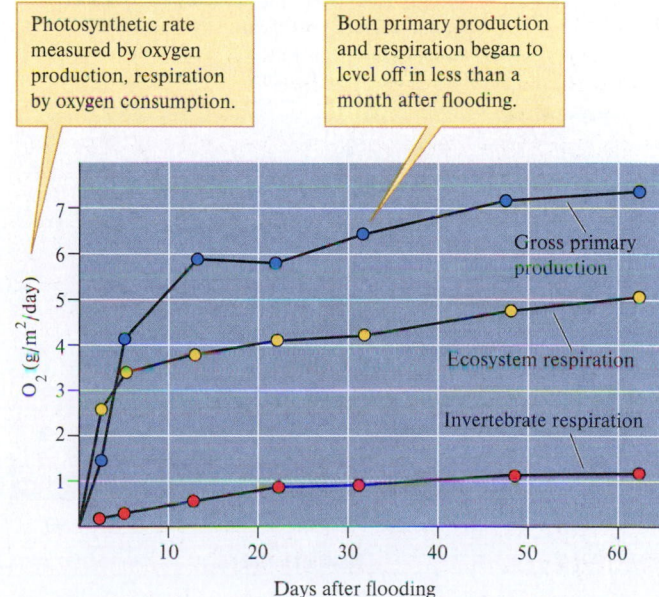

Figure 20.17   Ecosystem processes during succession in Sycamore Creek, Arizona (data from Fisher et al. 1982).

respiration, measured as oxygen consumption per square meter per day, increased quickly for only 5 days after the flood and then began to level off. Respiration by invertebrates, which at its maximum represented about 20% of total ecosystem respiration, leveled off by day 63.

Nancy Grimm (1987) studied nitrogen dynamics in Sycamore Creek following floods that occurred from 1981 to 1983. As in the earlier studies by Fisher and his colleagues (1982), Grimm found that during succession, algal biomass and whole ecosystem metabolism quickly reached a maximum and then leveled off, as did the quantity of nitrogen in the system.

In addition, however, Grimm examined patterns of nitrogen retention during stream succession. She estimated the nitrogen budget in each of her study reaches by comparing the nitrogen inputs at the upstream end to nitrogen outputs at the downstream end. Each 60 to 120 m study reach began where subsurface flows upwelled to the surface and ended downstream, where water disappeared into the sand. Grimm used the ratio of dissolved inorganic nitrogen entering the study reach in the upwelling zone to the amount leaving at the lower end as a measure of nitrogen retention by the stream ecosystem.

Figure 20.18 shows that in the early stages of succession, approximately equal amounts of dissolved inorganic nitrogen entered and left Grimm's study reaches. The level of retention increased rapidly during succession, leveling off at nearly 200 mg N per square meter per day, about 28 days after a flood. In other words, the study reach was accumulating 200 mg N per square meter per day. Then, between 28 days and 90 days after the flood, the study reach showed progressively lower retention until it eventually exported more dissolved inorganic nitrogen than came in with groundwater.

The results of Grimm's study raise several questions. First, what mechanisms underlie retention? Grimm attributes most retention by the Sycamore Creek ecosystem to uptake by algae

and invertebrates, since levels of nitrogen retention are consistent with the rates at which nitrogen was accumulated by algal and animal populations. What causes the stream reaches to eventually export nitrogen? Grimm suggested that at 90 days postflood, her study sites may have stopped accumulating biomass or may have even begun to lose biomass.

The major point here is that succession, which produces changes in species composition and species diversity, also changes the structure and function of ecosystems ranging from forests to streams. However, we are left with a major question concerning this important ecological process: What mechanisms drive succession? Ecologists have proposed that the mechanisms underlying succession may fall into one of three categories. Those mechanisms are the subject of the next section.

Concept 20.2 Review

1. Why are the changes in soil properties during the course of succession documented by Stuart Chapin and his colleagues ecologically significant?
2. What would equal levels of nitrogen input and output in the stream reaches (sections) studied by Nancy Grimm indicate?
3. What do the measurements made across the chronosequence on Hawaiian Island lava flows, which spanned millions of years, suggest about successional studies of ecosystem change covering months, years, or even centuries?

20.3 Mechanisms of Succession

LEARNING OUTCOMES

After studying this section you should be able to do the following:

20.11 Describe the mechanisms involved in facilitation, tolerance, and inhibition models of successional change.
20.12 Discuss experiments that revealed the existence of inhibition and facilitation during succession in marine intertidal communities.
20.13 Describe the roles of inhibition and facilitation during forest succession.
20.14 Explain how inhibition and facilitation can operate simultaneously during succession at Glacier Bay, Alaska.

Mechanisms that drive ecological succession include facilitation, tolerance, and inhibition. An early model for successional change proposed by Frederic E. Clements (1916) emphasized the role of facilitation as a driver of ecological succession. Later, Joseph Connell and Ralph Slatyer (1977) proposed three alternative models of succession: (1) facilitation, (2) tolerance, and (3) inhibition. This classic paper stimulated ecologists to think beyond facilitation and to also consider tolerance and inhibition as mechanisms underlying successional change (fig. 20.19).

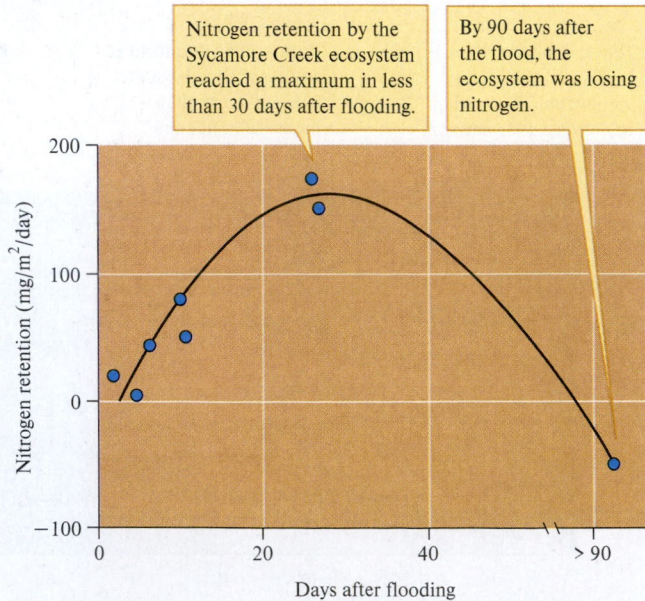

Figure 20.18 Nitrogen retention during stream succession (data from Grimm 1987).

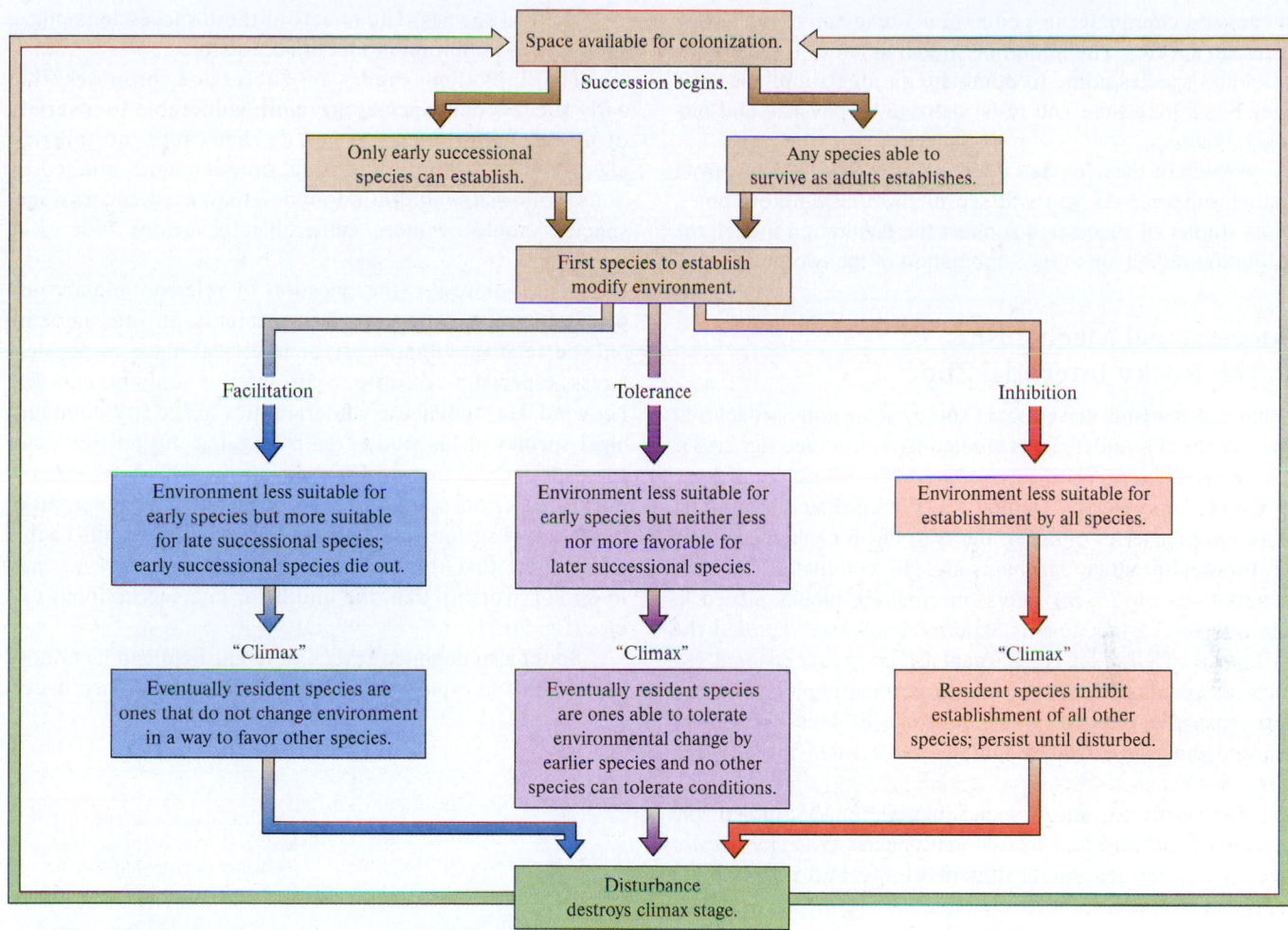

Figure 20.19   Alternative successional mechanisms (data from Connell and Slatyer 1977).

Facilitation

The **facilitation model** proposes that many species may attempt to colonize newly available space but only certain species, with particular characteristics, are able to establish themselves. These species, capable of colonizing new sites, are called pioneer species. According to the facilitation model, pioneer species modify the environment in such a way that it becomes less suitable for themselves and more suitable for species characteristic of later successional stages. In other words, these early successional species "facilitate" colonization by later successional species. Early successional species disappear as they make the environment less suitable for themselves and more suitable for other species. Replacement of early successional species by later successional species continues in this way until resident species no longer facilitate colonization by other species. This final stage in a chain of facilitations and replacements is the climax community.

Tolerance

According to the **tolerance model,** the initial stages of colonization are not limited to a few pioneer species. Juveniles of

species dominating at climax can be present from the earliest stages of succession, like the *Salix* seen in both pioneer and climax communities in Glacier Bay. Second, species colonizing early in succession do not facilitate colonization by species characteristic of later successional stages. They do not modify the environment in a way that makes it more suitable for later successional species. Later successional species are simply those tolerant of environmental conditions occurring later in succession. The climax community is established when the list of tolerant species has been exhausted.

Inhibition

Like the tolerance model, the **inhibition model** assumes that any species that can survive in an area as an adult can colonize the area during the early stages of succession. However, the inhibition model proposes that the early occupants of an area modify the environment in a way that makes the area less suitable for both early and late successional species. Simply, early arrivals inhibit colonization by later arrivals. Later successional species can invade an area only if space is opened up by disturbance of early colonists. In this case, succession culminates in a community made up of long-lived, resistant species. The inhibition model assumes that late successional species come to dominate an area simply because they live a long time and resist damage by physical and biological factors.

Which of these models does the weight of evidence from nature support? As you will see in the following examples, most studies of succession support the facilitation model, the inhibition model, or some combination of the two.

Successional Mechanisms in the Rocky Intertidal Zone

What mechanisms drive succession by algae and barnacles in the intertidal boulder fields studied by Sousa (see fig. 20.8)? The alternative mechanisms proposed by Sousa were those of Connell and Slatyer: facilitation, tolerance, and inhibition. Sousa used a series of experiments to test for the occurrence of these alternative mechanisms. He conducted his first experiments on 25 cm² plots on concrete blocks placed in the intertidal zone. In this experiment, Sousa explored the influence of *Ulva* on recruitment by later successional red algae by keeping *Ulva* out of four experimental plots and leaving four other control plots undisturbed. This experiment showed that *Ulva* strongly inhibits recruitment by red algae (fig. 20.20).

In a second set of experiments, Sousa studied the effects of the middle successional species *G. leptorhynchos* and *G. coulteri* on establishment of the late successional *G. canaliculata*. He selectively removed middle successional species from a set of four experimental plots while monitoring another set of four control plots. These experiments were conducted in 100 cm² areas on natural substrate, dominated by either *G. leptorhynchos* or *G. coulteri*. When Sousa removed these middle successional species, the experimental plots were quickly reinvaded by *Ulva* and eventually by significantly higher densities of *G. canaliculata*, the late

successional species. The effects of these successional algae support the inhibition model of succession.

The inhibition model of succession proposes that early successional species are more vulnerable to a variety of physical and biological factors that cause mortality. If algal succession in the intertidal boulder fields studied by Sousa follows the inhibition model, then early successional species should be more vulnerable to various sources of mortality.

Sousa addressed the question of relative vulnerability of algal species with several experiments. In one, he studied the relative vulnerability of intertidal algae to physical stress, especially exposure to air, intense sunlight, and drying wind. He studied the vulnerabilities of the five dominant algal species in his study area by tagging 30 individuals of each species and monitoring their survivorship for 2 months during a period when low tide occurred during the afternoon, when air temperatures are highest. The results of this study show that the early successional species, *Ulva*, had lower survivorship than the middle or late successional species (fig. 20.21).

Sousa also designed several different field and laboratory experiments to explore differential vulnerability to herbivores.

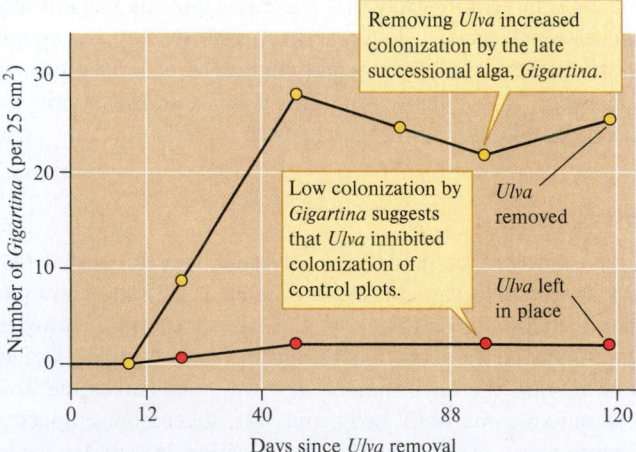

Figure 20.20   Evidence for inhibition of later successional species (data from Sousa 1979a).

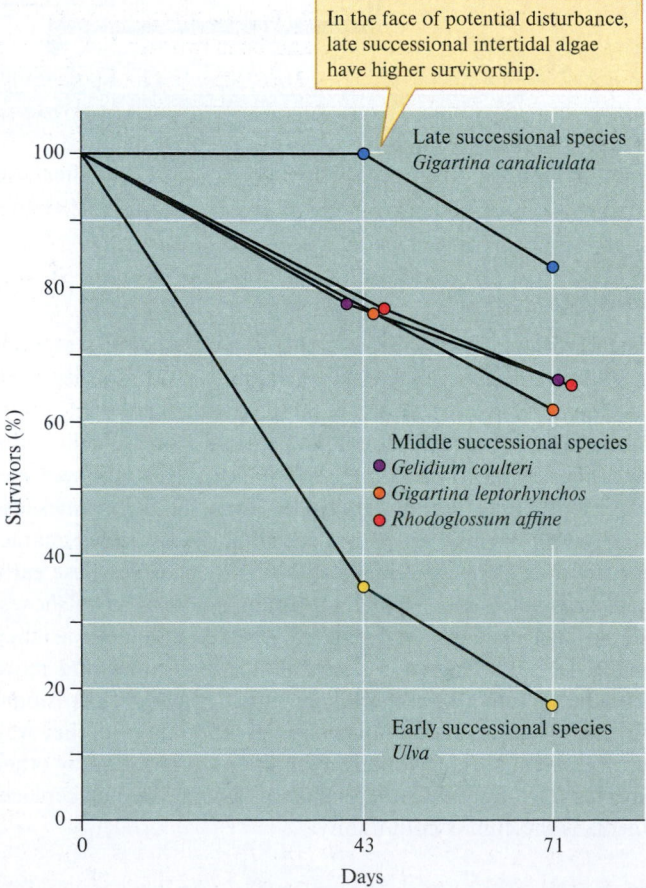

Figure 20.21   Survivorship of early, middle, and late successional species during a period of higher air temperatures at low tide (data from Sousa 1979b).

The results of all these experiments indicated that the early successional species *Ulva* is more vulnerable to herbivores than are later successional species. These results and those of the several other manipulations performed by Sousa support the inhibition model of succession.

Some studies of intertidal succession, however, have demonstrated facilitation. Teresa Turner (1983) pointed out that the bulk of intertidal studies had supported the inhibition model and that the few studies documenting facilitation had shown that facilitation was not obligate. However, she went on to report a case of obligate facilitation during intertidal succession.

Turner described the successional sequence at her Oregon study site as follows. High waves during winter storms create open space in the lower intertidal zone. In May, these open areas are colonized by *Ulva,* the same early colonist of open areas in Sousa's study area, over 1,000 km south of Turner's study site. *Ulva* is eventually replaced by several middle successional species, especially the red algae *Rhodomela larix, Cryptosiphonia woodii,* and *Odonthalia floccosa.* Through this middle stage, the pattern of succession appears much as that in the intertidal boulder field studied by Sousa. However, in the lower intertidal area studied by Turner, the dominant late successional species was not an alga but a flowering plant, the surfgrass *Phyllospadix scouleri.*

Turner proposed that recruitment of *Phyllospadix* by seeds depends on the presence of macroscopic algae. The seeds of *Phyllospadix* are large and bear two parallel, barbed projections. These projections hook and hold the seeds to attached algae. From this attached position, the seed germinates, first producing leaves and then roots by which the plant will anchor itself to the underlying rock. Once established, *Phyllospadix* spreads and consolidates space by vegetative growth.

Turner tested whether recruitment by *Phyllospadix* is facilitated by attached algae by clearing eight 0.25 m^2 plots of all attached algae. She then compared the number of new *Phyllospadix* seeds in these plots with the number in eight nearby control plots. The control plots remained undisturbed with their algal populations intact except that all *Phyllospadix* seeds were removed at the start of the study. Turner's control areas were dominated by the red alga *R. larix,* a species prominent in the middle successional stages in her study area and to which *Phyllospadix* seeds attach.

Turner set up and manipulated her study plots in September and then checked them the following March, after the period of seed dispersal. Over the fall and winter, a brown alga, *Phaeostrophion irregulare,* colonized the removal plots but the bladelike form of this species apparently does not allow attachment by *Phyllospadix* seeds. When Turner checked the removal and control plots, she found a total of 48 seeds, 46 on the control plots (all attached to *Rhodomela*) and 2 on the removal plots (fig. 20.22). Both seeds on the removal plots were attached to two isolated branches of *Rhodomela* that had sprouted from remnant holdfasts.

During 3 years, Turner systematically searched an area of about 200 m^2 for *Phyllospadix* seeds and found a total of 298.

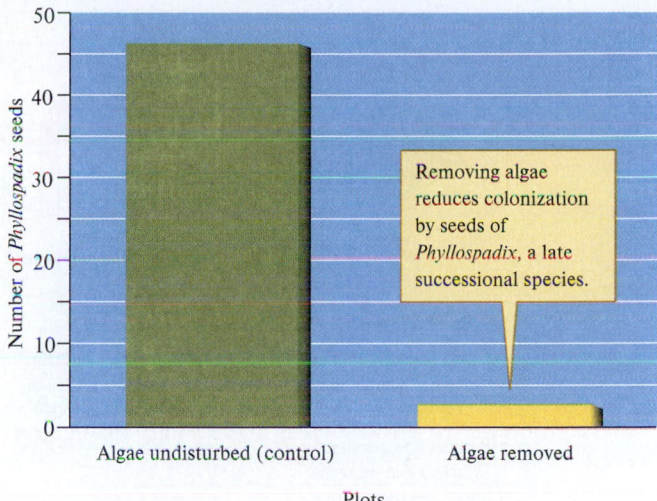

> Removing algae reduces colonization by seeds of *Phyllospadix,* a late successional species.

Figure 20.22   Evidence for facilitation of colonization by an intertidal plant, *Phyllospadix scouleri* (data from Turner 1983).

All were attached to algae. These data support the hypothesis that middle successional algae facilitate recruitment and establishment of *Phyllospadix* and that this facilitation is obligate. As a consequence of Turner's study and other studies, we can say that facilitation and inhibition occur during intertidal succession. Other research, which we review in the next example, has shown that facilitation and inhibition also occur during succession in abandoned agricultural fields.

Mechanisms in Old Field Succession

Catherine Keever (1950) studied succession on old fields of the Piedmont Plateau of North Carolina. She conducted some of the earliest experiments on the mechanisms regulating the early stages of succession in temperate forests. As we saw earlier in this chapter, the first species to colonize and dominate abandoned fields is generally crabgrass, *Digitaria sanguinalis. Digitaria* is usually followed by horseweed, *Erigeron canadense.* During the second year of succession, many fields are dominated by aster, *Aster pilosis.* A few years later the fields are covered by broomsedge, *Andropogon virginicus.* The objective of Keever's study was to determine the causes of this early pattern of species replacements.

Keever's experiments showed that *Erigeron* inhibits the growth of *Aster* and so the replacement of *Erigeron* by *Aster* follows the inhibition model. However, *Aster* stimulates the growth of its successor, *Andropogon,* so this second species replacement follows the facilitation model. Therefore, early succession on the Piedmont Plateau of eastern North America appears to involve a mixture of mechanisms (fig. 20.23).

In this and the previous two sections, we have discussed community and ecosystem changes and the mechanisms producing those changes. In the next section, we consider a companion topic: community and ecosystem stability.

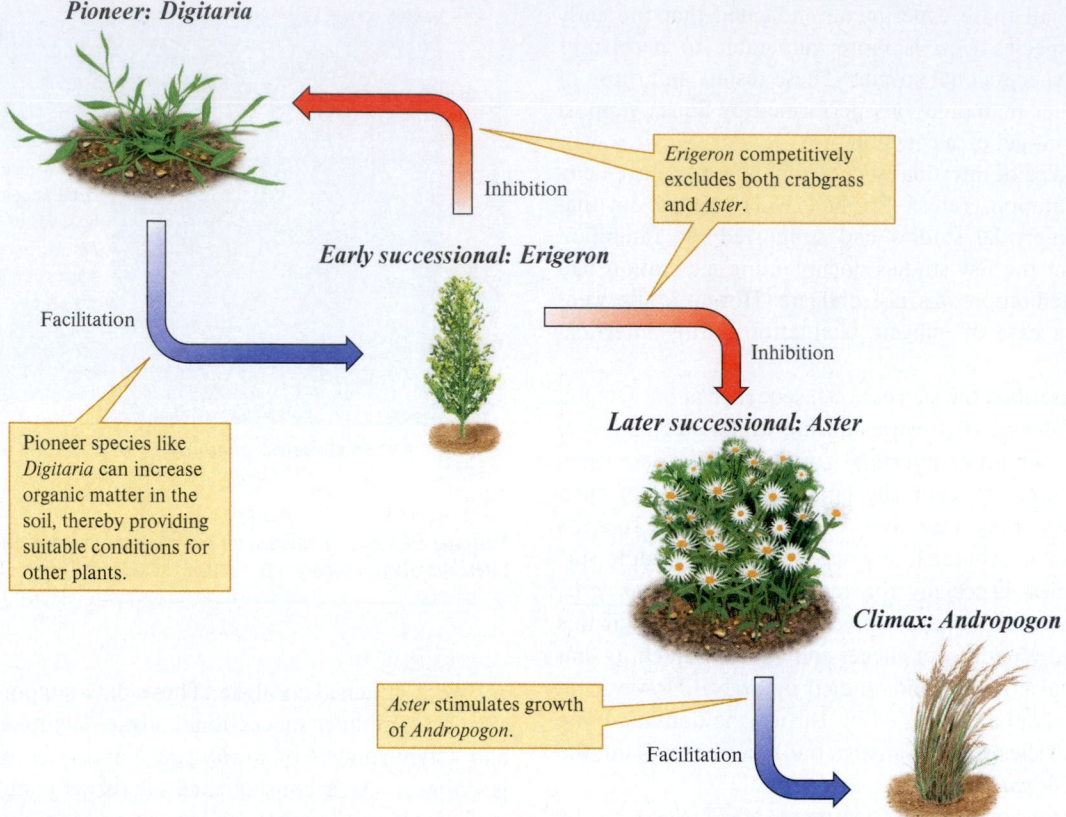

Pioneer: Digitaria

Inhibition

Erigeron competitively excludes both crabgrass and *Aster*.

Facilitation

Early successional: Erigeron

Inhibition

Pioneer species like *Digitaria* can increase organic matter in the soil, thereby providing suitable conditions for other plants.

Later successional: Aster

Climax: Andropogon

Aster stimulates growth of *Andropogon*.

Facilitation

Figure 20.23   Observations of succession on abandoned agricultural fields suggested both facilitation and inhibition mechanisms (based on Keever 1950).

Concept 20.3 Review

1. What is the role of disturbance in the Connell and Slatyer succession model (see fig. 20.19)?
2. Suppose *Gigartina* had colonized the plots where Sousa had removed *Ulva* and where he had left *Ulva* in place at the same rates (see fig. 20.20). This result would be consistent with which successional model?
3. What pattern of colonization by *Gigartina* in Sousa's *Ulva* removal experiment would have been consistent with the facilitation model?

20.4 Community and Ecosystem Stability

LEARNING OUTCOMES

After studying this section you should be able to do the following:

20.15 Define the concepts of stability, resistance, and resilience as applied to ecological communities.

20.16 Describe how altering the fertilizer mix in the Park Grass Experiment changed community composition.

20.17 Discuss how the Park Grass Experiment indicated relative stability in community composition over time.

20.18 Explain how level of taxonomic resolution (plant growth forms versus individual species) influences perception of community stability in the Park Grass Experiment.

20.19 Discuss how geologic features affect community resilience in Sycamore Creek.

Community stability may be due to lack of disturbance or community resistance or resilience in the face of disturbance. The simplest definition of **stability** is the absence of change. A community or ecosystem may be stable for a variety of reasons. One reason may be that there has been no disturbance. For instance, the benthic communities of the deep sea may remain stable over long periods of time because of constant physical conditions. The type of stability resulting from an absence of disturbance, if it exists, is not particularly interesting to ecologists.

Ecologists are more interested in how communities and ecosystems may remain stable even when exposed to potential disturbance. Consequently, ecologists generally define stability as the persistence of a community or ecosystem in the face of disturbance. Stability may result from two very different characteristics. **Resistance** is the ability of a community or ecosystem to maintain structure and/or function in the face of potential disturbance. However, stability may also result from the ability of a community to return to its original structure after a disturbance. The ability to bounce back after disturbance is called **resilience.** A resilient community or ecosystem may be completely disrupted by disturbance but quickly return to its former state.

Ecologists ask many questions about stability. Are some communities and ecosystems more resistant than others? What factors determine differences in resistance among communities and ecosystems? Are some ecosystems and communities more resilient than others? What factors determine the rate of recovery of community structure and ecosystem processes following disturbance? However, few studies have been conducted at scales appropriate to address these questions. One of the main problems faced by ecologists interested in community and ecosystem stability is the need to conduct detailed studies over a long period of time. There are a few studies that meet this requirement; one of them is the Park Grass Experiment.

Lessons from the Park Grass Experiment

The Park Grass Experiment is the prototype of all long-term experimental studies in ecology. It was started at the Rothamsted Experimental Station in Hertfordshire, England, between 1856 and 1872. The purpose of the experiment was to study the effects of several fertilizer treatments on the yield and structure of a hay meadow community. Because the Park Grass Experiment has continued without interruption for one and a half centuries, it provides one of the most valuable records of long-term community dynamics. That record provides some unique insights into the nature of community stability.

Jonathan Silvertown (1987) used data from the Park Grass Experiment to respond to the suggestion that existing studies do not conclusively demonstrate that any ecological community is stable. Silvertown pointed out that the Park Grass Experiment is one of the few studies of terrestrial communities that have been carried out in sufficient detail and over sufficient time to provide a test of stability.

The composition of the plant community at the Park Grass Experiment has been monitored since 1862. This record reveals at least one level of stability. Over this period, virtually no new species have colonized the meadow. Changes in the community have occurred as a consequence of increases or decreases in species already present in the meadow at the beginning of the experiment.

Silvertown used variation in community composition as a measure of stability. He represented composition as the proportion of the community consisting of grasses, legumes, or other species. The analysis of composition was restricted

to the period 1910 to 1948 to avoid the early period of the experiment when the meadow community was adjusting to the various fertilizer treatments. Figure 20.24 shows the relative proportions of grasses, legumes, and other plants on plots receiving three different treatments: plot 3, no fertilizer; plot 7, P, K, Na, and Mg; and plot 14, N, P, K, Na, and Mg. The differences in vegetation on the three plots were mostly produced by the different fertilizer treatments and developed early in the Park Grass Experiment.

The proportion of grasses, legumes, and other plants in the study plots varied from year to year, mainly in response to variation in precipitation. Despite this annual variation, figure 20.24 indicates that the proportions of three plant groups remained remarkably similar over the interval of the study. A quantitative analysis of trends in biomass revealed no significant changes in the biomass of the three plant groups in plots 3 and 7 and only a minor, but statistically significant, decrease in the biomass of grasses in plot 14. In other words, the data presented in figure 20.24 show remarkable stability in the proportion of grasses, legumes, and other species.

Does the stability of Silvertown's three major groups of plants in the Park Grass Experiment hold up if we examine community structure at the species level? It turns out that while the proportions of grasses, legumes, and other species remained fairly constant, populations of individual species changed substantially. Mike Dodd and his colleagues (1995) used census data from 1920 to 1979 to examine plant population trends. The result of their analysis showed that some species increased in abundance, some decreased, some showed no trend, while others increased and then decreased (fig. 20.25).

The contrasting results obtained by Silvertown and by Dodd's project suggest that whether a community or ecosystem appears stable may depend on how we view it. At a very coarse level of resolution, the Park Grass community has remained absolutely stable. It was a meadow community when the Park Grass Experiment began in 1856 and it remains so today. When Silvertown increased the resolution to distinguish between grasses, legumes, and other species, the community again appeared stable. However, when Dodd and his colleagues increased the resolution still further and examined trends in the abundances of individual species, the Park Grass community no longer appeared stable.

Are there stable natural communities? The answer to this question may depend on how you make your measurements. The ecologist interested in addressing any question concerning community stability is faced with several practical problems. Generally, an adequate study requires a great deal of time, which limits the possibility of replication. One solution to this problem is to study communities and ecosystems, such as Sycamore Creek, Arizona, that undergo more frequent disturbance and show relatively rapid recovery. These systems offer the opportunity to compare recoveries from multiple disturbances.

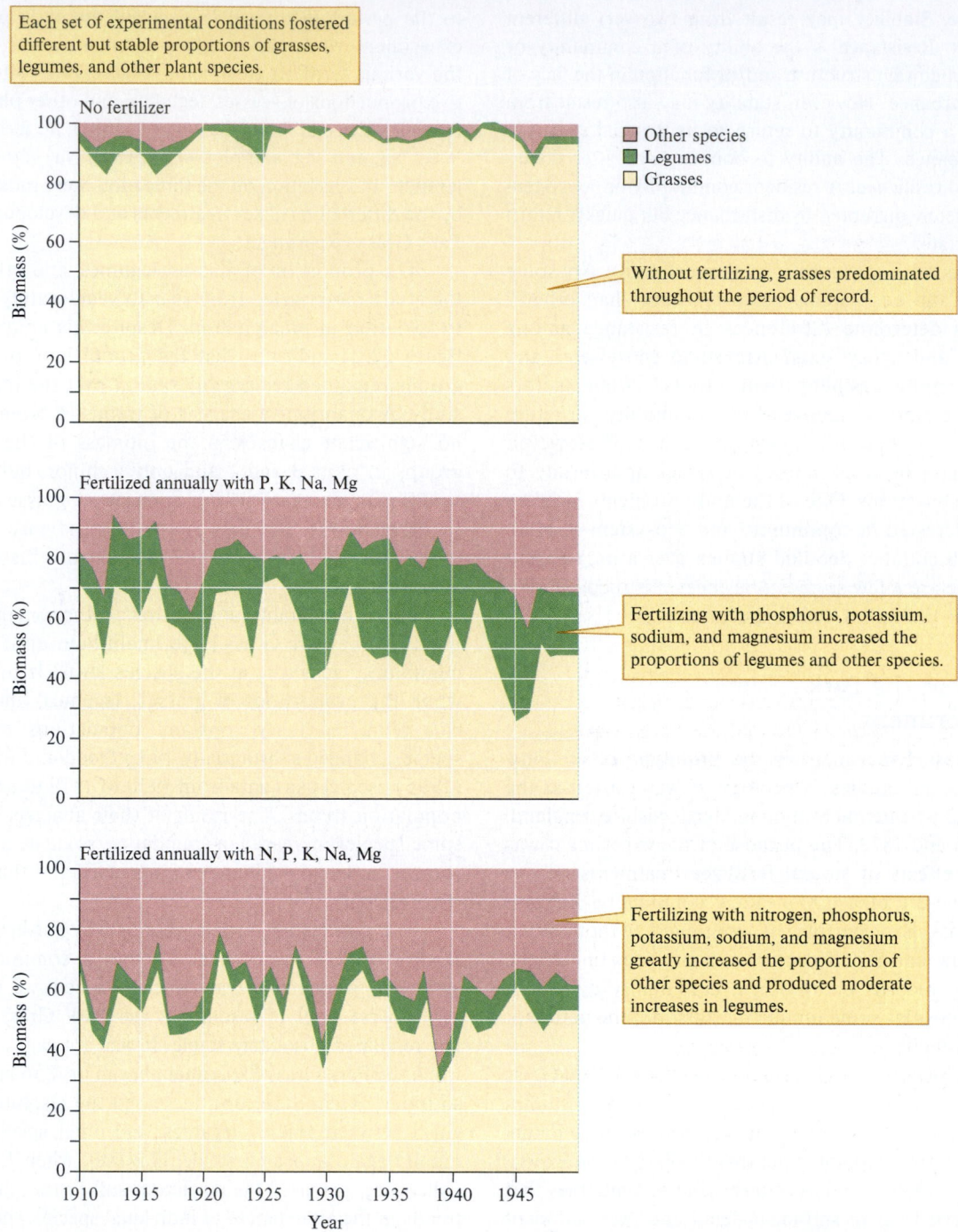

Each set of experimental conditions fostered different but stable proportions of grasses, legumes, and other plant species.

No fertilizer

■ Other species
■ Legumes
□ Grasses

Without fertilizing, grasses predominated throughout the period of record.

Fertilized annually with P, K, Na, Mg

Fertilizing with phosphorus, potassium, sodium, and magnesium increased the proportions of legumes and other species.

Fertilized annually with N, P, K, Na, Mg

Fertilizing with nitrogen, phosphorus, potassium, sodium, and magnesium greatly increased the proportions of other species and produced moderate increases in legumes.

Figure 20.24   Proportions of grasses, legumes, and other plant species under three experimental conditions (data from Silvertown 1987).

Replicate Disturbances and Desert Stream Stability

Numerous studies of disturbance and recovery in Sycamore Creek, Arizona, have produced a highly detailed picture of community, ecosystem, and population responses. This detailed picture suggests that ecologists have just begun to probe the subtleties of ecological stability. For instance, one study shows that resistance in the spatial structure of the Sycamore Creek ecosystem underlies spatial variation in ecosystem resilience.

Maury Valett and his colleagues (1994) studied the interactions between surface and subsurface waters in Sycamore Creek in order to study the influence of these linkages on ecosystem resilience. They tested the hypothesis that ecosystem resilience is higher where hydrologic linkages between the surface and subsurface water increase the supply of nitrogen. They proposed a controlling role for nitrogen because it is the nutrient that limits primary production in Sycamore Creek.

Valett and his colleagues intensively studied two stream sections at middle elevations in the 500 km^2 Sycamore

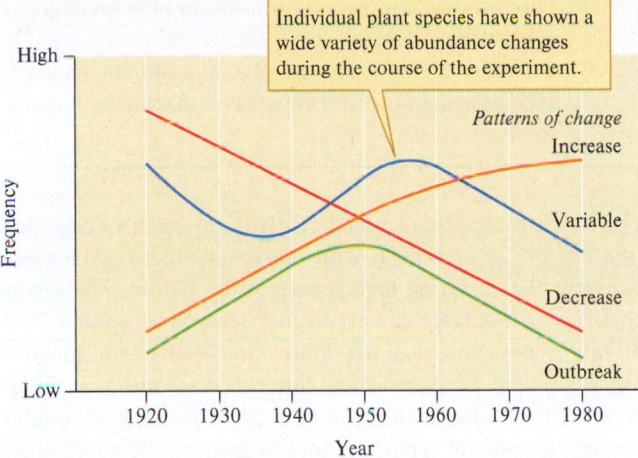

Figure 20.25   Patterns of species abundance during 60 years of the Park Grass Experiment (data from Dodd et al. 1995).

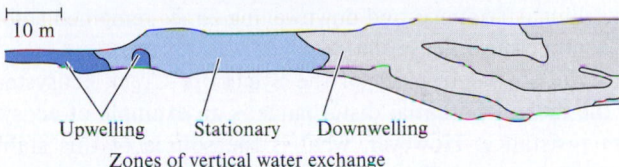

Figure 20.26   Patterns of upwelling and downwelling in a reach of Sycamore Creek, Arizona (data from Valett et al. 1994).

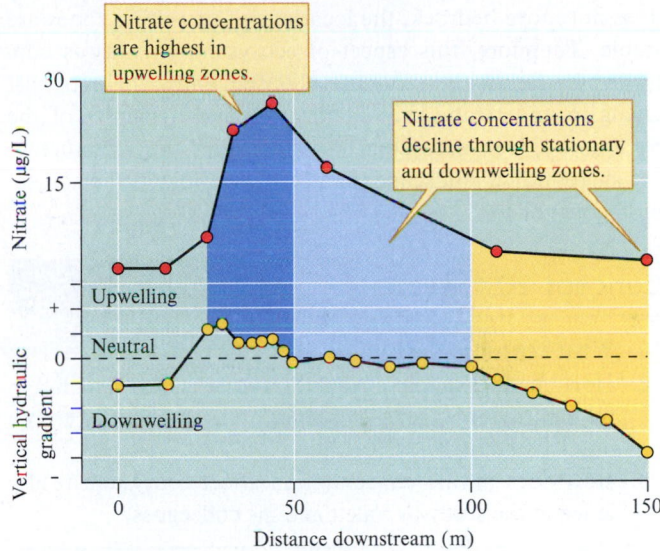

Figure 20.27   Relationship of nitrate to vertical hydraulic gradient in Sycamore Creek, Arizona (data from Valett et al. 1994).

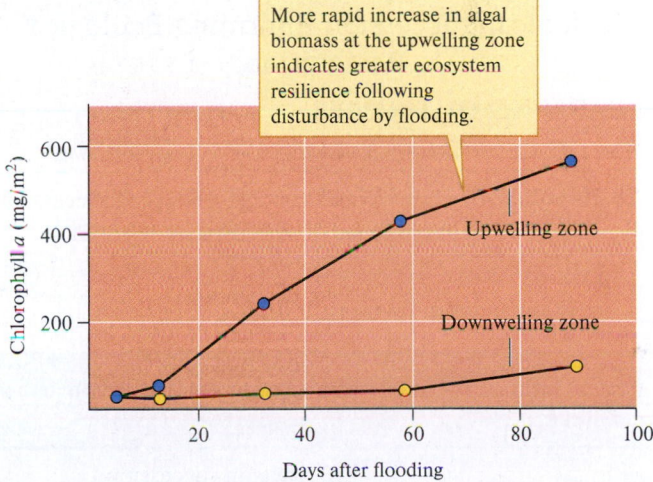

Figure 20.28   Changes in algal biomass, measured as chlorophyll *a*, following flooding at upwelling and downwelling zones (data from Valett et al. 1994).

Creek catchment. They measured the flow of water between the surface and subsurface along these reaches with devices called *piezometers.* Piezometers can be used to measure the vertical hydraulic gradient, which indicates the direction of flow between surface water and water flowing through the sediments of a streambed. Positive vertical hydraulic gradients indicate flow from the streambed to the surface in areas called upwelling zones. Negative vertical hydraulic gradients indicate flow from the surface to the streambed, which occurs in downwelling zones. Zero vertical hydraulic gradients indicate no net exchange between surface waters and water flowing within the sediments. Areas with zero vertical hydraulic gradients are called stationary zones.

Valett and his colleagues measured vertical hydraulic gradient along the lengths of both study sections, producing hydrologic maps for both. The upper end of each study reach was an upwelling zone. The middle reaches were stationary zones and the lower reaches were downwelling zones. Figure 20.26 shows the distributions of these zones across one of the study reaches.

The concentration of nitrate in surface water in the two study reaches varies directly with vertical hydraulic gradient (fig. 20.27). Upwelling zones, which are fed by nitrate-rich waters upwelling from the sediments, have the highest concentrations of nitrate. Nitrate concentrations gradually decline with distance downstream through the stationary and downwelling zones.

The higher concentrations of nitrate in the upper reaches of each study section are associated with higher algal production. Figure 20.28 indicates that algal biomass accumulates at a higher rate in upwelling zones compared to downwelling zones. Valett and his colleagues used rate of algal biomass accumulation as a measure of rate of recovery from disturbance. Because the rate at which algal biomass accumulates in upwelling zones is so much higher than in downwelling zones, they concluded that the rate of ecosystem recovery is higher in upwelling zones. This pattern supports their hypothesis that algal communities in upwelling zones are more resilient.

The team also found that while flash floods devastated the biotic community, the spatial arrangement of upwelling, stationary, and downwelling zones remained stable. In other words, this aspect of the spatial structure of the Sycamore Creek ecosystem is highly resistant to flash flooding. The location of upwelling, stationary, and downwelling zones remained stable after numerous intense floods.

The spatial stability of the Sycamore Creek ecosystem in the face of potential disturbance is an example of ecosystem resistance. However, what is the source of this stable spatial structure? This spatial stability can be explained by considering geomorphology, especially the distribution of bedrock. Subsurface water is forced to the surface in areas where bedrock lies close to the surface. Upwelling zones in Sycamore Creek are located in such areas, and since flooding does not move bedrock, the locations of upwelling zones are stable. Therefore, this aspect of ecosystem stability is controlled by landscape structure. Consequently, the ecologist trying to understand the organization and dynamics of the Sycamore Creek ecosystem must consider the structure of the surrounding landscape. Landscape ecology is the subject of chapter 21.

Concept 20.4 Review

1. What causes community resilience?
2. How might taxonomic resolution—that is, how precisely we identify organisms—influence an assessment of community stability?
3. How does nitrate concentration affect ecosystem resilience in the study by Valett and his colleagues?

Ecological Succession Informing Ecological Restoration

LEARNING OUTCOMES

After studying this section you should be able to do the following:

20.20 Summarize how various factors influencing succession can be managed to facilitate ecosystem restoration.

20.21 Discuss ways to improve restoration of tropical forest ecosystems.

20.22 Compare the effectiveness of abandonment versus recontouring for restoring areas impacted by logging roads.

The massive impacts of human activity on earth's ecosystems (see fig. 16.22) present us with pressing challenges. One is to conserve the surviving biodiversity of the planet. The second challenge is to restore damaged ecosystems to acceptable levels of biodiversity, physical structure, and ecosystem functioning, a process called **ecological restoration** (Gann et al. 2019). A specialized area within the science of ecology, **restoration ecology,** focuses on exploring ways to improve the effectiveness of ecological restoration by providing a conceptual framework to guide such work. The concepts derived from studies of ecological succession have much to contribute to ecological restoration (Walker, Walker, and Hobbs 2007). Some ecologists have even gone so far as to define ecological restoration as "essentially the manipulation of succession in order to achieve some predetermined goal" (Walker, Velázquez, and Shiels 2009).

Applying Succession Concepts to Restoration

As we saw earlier in the chapter, succession involves changes occurring over the course of decades to centuries, whereas ecological restoration is generally restricted to timescales of years to a few decades. Consequently, the practitioners of ecological restoration attempt to accelerate the process of succession. Successional studies have shown that a broad range of physical, chemical, and biological factors influence the process. Potentially influential factors include the abiotic characteristics of sites, differences in plant dispersal rates, ease of plant establishment, interaction between plant species, and herbivory. While some of these environmental factors facilitate successional change in community and ecosystem properties, others inhibit change. Table 20.1 lists a number of factors that

Table 20.1

Factors influencing successional change and restoration actions to manage their impact on the process (after del Moral, Walker, and Bakker 2007)		
Factor	**Influence on succession**	**Restoration actions**
Physical and chemical stressors	Potentially reduce plant establishment	Reduce physical stressors by, for example, adding organic matter and nutrients to restoration area
Plant dispersal rates	Can limit site colonization by poor dispersers	Actively add seeds to restoration site, attract animal seed dispersers
Plant establishment	Limited by availability of suitable "safe" sites	Create suitable safe sites
Interspecific plant interactions	May inhibit or facilitate establishment and growth	Control strongly inhibitory species; foster species that facilitate late successional species
Herbivory	Can restrict establishment of plants attractive to herbivores	Protect vulnerable plants until well established

affect the timing and course of succession along with actions that can be taken during restoration to counteract or reinforce their effects. Some results of applying these restoration principles are explored in the following case studies.

Restoring Tropical Forest

Tropical forests are increasingly the focus of intense economic activity ranging from clearing for agriculture to mining. For instance, the mining of bauxite, an ore of aluminum, occurs mainly in tropical and subtropical regions and can have major impacts on these landscapes (fig. 20.29). As a result, there is an increasing demand for restoration of these diverse ecosystems.

A research group in Brazil recently studied restoration of tropical forest impacted by bauxite mining in Central Amazonia (Dias et al. 2012). The type of forest the group focused on, the *igapó*, is particularly challenging because of its periodic flooding, which shortens the growing season and limits the use of fertilizers that can cause eutrophication of wetlands connected to the flooded forest.

The research group tested whether adding forest floor litter collected in undisturbed *igapó* and seeding could increase the effectiveness of restoration on bauxite mine tailings. Their research addresses two of the factors affecting succession listed in table 20.1: physical and chemical stressors and plant dispersal rates. Adding organic matter to the soil surface helps to reduce soil drying and eventually adds organic matter to the bauxite tailings. Seeding the site with preferred plant species augmented natural plant dispersal processes. In addition, seeding provided the research group with an opportunity to ensure that nitrogen-fixing plants would colonize the tailings and add nitrogen to the nutrient-poor bauxite tailings in a form with less chance of contributing to wetland eutrophication. The positive contributions of litter addition and seeding to promoting succession on bauxite tailings included

faster plant growth, greater leaf area, increased plant species richness, and higher seedling densities, a result presented in figure 20.30.

Encouraging dispersal of the many bird-dispersed seeds in tropical forests can also augment dispersal of plants onto tropical forest restoration sites. This was the approach taken by Aaron Shiels and Lawrence Walker in their studies of restoration of landslides on steep mountain slopes of the Luquillo Experimental Forest in Puerto Rico (Shiels and Walker 2003). They augmented the deposition of bird-dispersed seeds onto the landslides under study by placing perches, made of tree branches from the surrounding forests, on some of their study plots. Bird-dispersed seeds were found only below their introduced perches. Control study plots without perches received inputs of wind-dispersed seeds only. Consequently, perches increased the diversity of seeds on experimental plots (fig. 20.31).

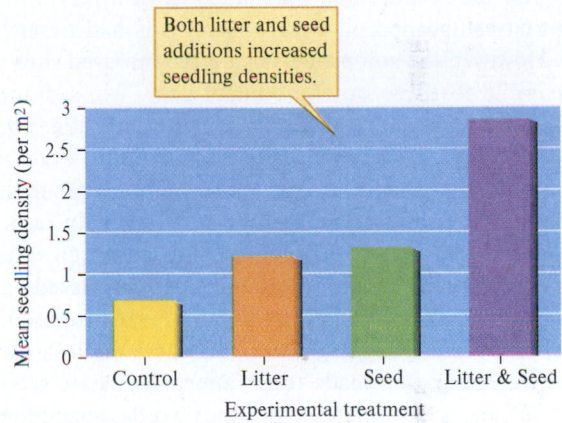

Figure 20.30   The effects of adding forest litter and seeds on mean densities of seedlings growing on bauxite mine tailings in Brazil (data from Dias et al. 2012).

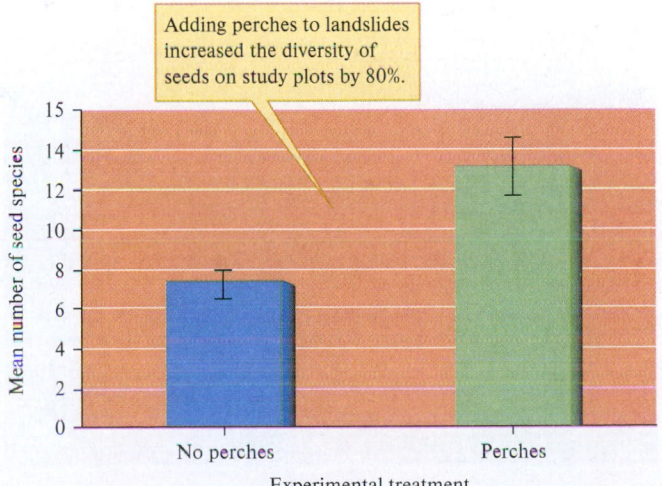

Figure 20.31   Effect of artificial perches on mean seed species richness (± one standard error) on landslides to facilitate reestablishment of bird-dispersed plant species in the Luquillo Experimental Forest, Puerto Rico (data from Shiels and Walker 2003).

Figure 20.29   Bauxite mines, such as this one near Perth, Australia, can remove vegetation from large areas and produce massive tailing piles. Ecological restoration is generally needed to stabilize exposed soils and tailings to prevent erosion and sedimentation damage to nearby ecosystems.   Howard Davies/Getty Images

Restoring Logging Road Beds

A dense network of roads is one of the common legacies of timber harvests (fig. 20.32). These logging roads have a number of environmental impacts, including heavy erosion and sedimentation of nearby watercourses. In a 2013 study of road decommissioning in northern Idaho, Rebecca Lloyd, Kathleen Lohse, and Ty Ferré compared the results of simply abandoning logging roads and allowing succession to proceed with recontouring roads and their surroundings to restore the original contours of the site. In their study, the team examined both vegetation establishment and changes in soil characteristics on three types of sites: control areas on which a road had never been built, abandoned roads, and recontoured roads.

The team found that recontouring accelerated restoration. After just 10 years, the amount of bare ground on recontoured sites had decreased to levels only observed on abandoned roads after 30 years (fig. 20.33). The vegetation on 10-year-old recontoured road sites was still very different from that established on sites where roads had never been built. However, the soil properties on recontoured sites were comparable to those on the control sites, while dramatically different from those on abandoned roads (fig. 20.34). In fact, the 10-year recontoured and control sites were statistically indistinguishable in regard to five properties measured in the O and A soil horizons of the study sites. In regard to soil properties at least, recontoured sites had been restored to a pre-disturbance condition in just 10 years. Based on the results of their study, Lloyd, Lohse, and Ferré proposed that simply abandoning logging roads sets them off in a successional trajectory that leads to an alternative state of structure and ecosystem function. In other words, abandonment will not lead to restoration to ecological conditions comparable to those prevailing before disturbance by road building. This is an excellent example of how restoration ecology can inform the process of ecological restoration practice.

Figure 20.32   Logging roads, such as this one in Oregon, which create a surface bare of vegetation and a topographic anomaly in the landscape, are a significant source of erosion and of sedimentation in nearby streams and rivers.   U.S. Fish & Wildlife Service/Steve Hillebrand

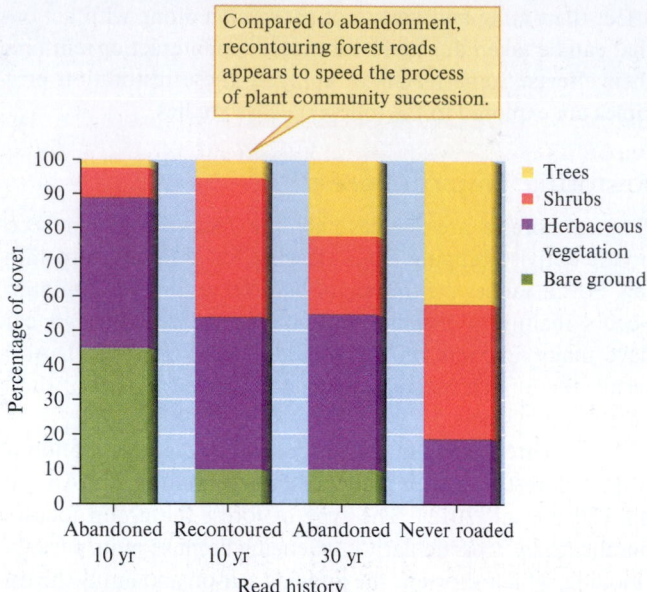

> Compared to abandonment, recontouring forest roads appears to speed the process of plant community succession.

Figure 20.33   Approaches to restoring vegetation to logging roads in northern Idaho: abandonment versus recontouring by eliminating the road-bed cut into a hillside and restoring the natural slope of the land (data from Lloyd, Lohse, and Ferré 2013).

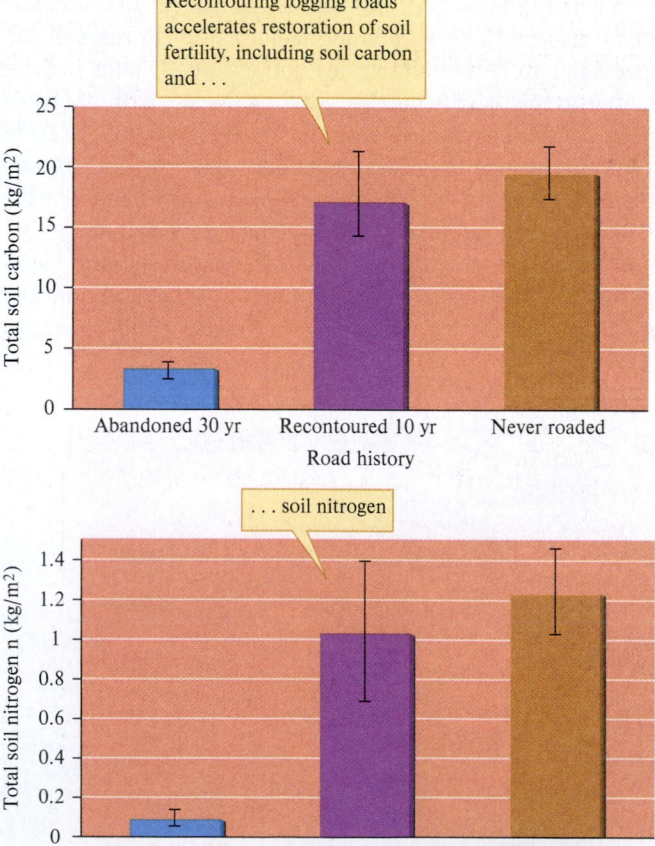

> Recontouring logging roads accelerates restoration of soil fertility, including soil carbon and . . .

> . . . soil nitrogen

Figure 20.34   Influence of recontouring on restoring carbon and nitrogen to soils disturbed by logging roads. Values are sample means plus or minus one standard error (data from Lloyd, Lohse, and Ferré 2013).

Summary

Succession is the change in plant, animal, and microbial communities in an area following disturbance or the creation of new substrate. *Primary succession* occurs on newly exposed geological substrates not significantly modified by organisms. *Secondary succession* occurs in areas where disturbance destroys a community without destroying the soil. Succession generally ends with a climax community whose populations remain stable until disrupted by disturbance.

Community changes during succession can include both increases and decreases in species diversity and changes in species composition. Studies that used permanent plots observed over a century at Glacier Bay suggested that stochastic processes determined climax species and that overall species diversity either declined or remained constant over time. Whereas, secondary forest succession on the Piedmont plateau and the intertidal zone showed patterns of increasing diversity in both flora and fauna. Secondary forest succession on the Piedmont Plateau takes about 150 years. Meanwhile, succession in the intertidal zone requires 1 to 3 years and succession within a desert stream occurs in less than 2 months.

Ecosystem changes during succession include increases in biomass, primary production, respiration, and nutrient retention. During ecosystem development on lava flows in Hawaii, organic matter and nitrogen content of soils increased over the first 150,000 years and then declined by 1.4 and 4.1 million years. Weatherable mineral phosphorus in soils was largely depleted on lava flows 20,000 years old or older. The percentage of soil phosphorus in refractory form made up the majority of phosphorus on lava flows 20,000 years old or older. Nitrogen losses from these ecosystems increased over time, while phosphorus losses decreased. Several ecosystem properties change predictably during succession in Sycamore Creek, Arizona, including biomass, primary production, respiration, and nitrogen retention.

Mechanisms that drive ecological succession include facilitation, tolerance, and inhibition. Glacier Bay shows evidence of the tolerance model, whereas succession in old fields and the intertidal zone suggest a combination of both facilitation and inhibition.

Community stability may be due to lack of disturbance or community resistance or resilience in the face of disturbance. Ecologists generally define stability as the persistence of a community or ecosystem in the face of disturbance. Resistance is the ability of a community or ecosystem to maintain structure and/or function in the face of potential disturbance. The ability to bounce back after disturbance is called resilience. A resilient community or ecosystem may be completely disrupted by disturbance but quickly returns to its former state. Studies of the Park Grass Experiment suggest that our perception of stability is affected by the scale of measurement. Studies in Sycamore Creek indicate that resilience is sometimes influenced by resource availability and that resistance may result from landscape-level phenomena.

The massive impacts of human activity on earth's ecosystems challenge us to restore them to acceptable levels of biodiversity, physical structure, and ecosystem functioning, a process called ecological restoration. Restoration is informed by a specialized area of ecological studies: restoration ecology. Ecological restoration draws substantially on the concepts of ecological succession to accelerate the pace of restoration.

Key Terms

chronosequence   425

climax community   424

ecological restoration   440

facilitation model   433

inhibition model   433

pioneer community   424

primary succession   424

resilience   437

resistance   437

restoration ecology   440

secondary succession   424

stability   436

succession   424

tolerance model   433

Review Questions

1. As we saw in figure 20.6, Johnston and Odum (1956) documented substantial change in the richness of bird species in a successional sequence going from the earliest stages in which the plant community was dominated by grasses and forbs to mature oak-hickory forests. Use MacArthur's (see chapter 16) studies (1958, 1961) of foliage height diversity and bird diversity to explain the patterns of diversity increase observed by Johnston and Odum.

2. Would you expect the number of species to remain indefinitely at the level shown in figure 20.8? Space on large, stable boulders in Sousa's study site is dominated by the algal *G. canaliculata* and support 2.3 to 3.5 species, not the 5 shown in figure 20.8. Explain. (Hint: How long did Sousa follow his study boulders?)

3. Early researchers studying succession in Glacier Bay created theories about deterministic plant communities and increasing diversity that were disproven by subsequent studies at the

same site. How and why could the earlier theories be rejected? Although they turned out not to be true in Glacier Bay, their succession theories do seem to be supported by research in other ecosystems. How is this possible? What does this tell us about the nature of scientific inquiry?

4. In most studies of forest succession such as that of Reiners and colleagues (1971) and Oosting (1942), researchers study succession along a chronosequence. This approach is called a "space for time substitution." What are some major assumptions of a space for time substitution? What advantages for studying succession are offered by systems like Sycamore Creek?

5. The rapid succession shown by the Sycamore Creek ecosystem is impressive. How might natural selection influence the life cycles of the organisms living in Sycamore Creek? Imagine a creek that floods about twice per century. How quickly would you expect the community and ecosystem to recover following one of these rare floods? Explain your answer in terms of natural selection by flooding on the life cycles of organisms.

6. Why is Glacier Bay an example of the tolerance model? Why might facilitation and inhibition be more commonly found in succession studies?

7. When Mount St. Helens in Washington erupted in 1980, it created a gradient in disturbance. In the pumice plains near the eruption, the devastation was almost total. The extent of disturbance was much less in the farthest reaches of the blast zone. Describe how the intensity of disturbance around Mount St. Helens may affect the rate of forest succession.

8. Species have come and gone in response to changing global climates during the history of the earth. Some of the mass extinctions of the past have resulted in the disappearance of over 90% of existing species. What do these biological changes suggest about the long-term stability of the species composition of climax communities?

9. Predict the characteristics of a frequently disturbed community/ecosystem versus a largely undisturbed community/ecosystem. What do your predictions suggest about a future biosphere increasingly disturbed by a growing human population?

Chapter

21

Landscape Ecology

Martial Colomb/Getty Images

A landscape in Provence, France. The structure of this landscape is apparent in the sharp contrasts in plant cover between the patches of well-tended lavender and nearby Mediterranean woodland. The landscape ecology of such a place would include sources of patch distribution and structure and the exchanges of materials, energy, and organisms across patch boundaries.

Applications: Landscape Approaches to
   Mitigating Urban Heat Islands 463

Summary 465

Key Terms 466

Review Questions 466

CHAPTER CONCEPTS

21.1 Landscape structure includes the size, shape, composition, number, and position of patches, or landscape elements, in a landscape. 447

Concept 21.1 Review 450

21.2 Landscape structure influences processes such as the flow of energy, materials, and species across a landscape. 450

Concept 21.2 Review 455

21.3 Landscapes are structured and change in response to geological processes, climate, activities of organisms, and fire. 455

Concept 21.3 Review 463

LEARNING OUTCOMES

After studying this section you should be able to do the following:

21.1 Define a landscape from the perspective of landscape ecology.

21.2 Discuss the differences between landscape ecology and other subdisciplines of ecology, such as community and ecosystem ecology.

In every region on earth and at every stage in history, human survival has required a basic understanding of landscapes. In contemporary ecology, a **landscape** is a heterogeneous area consisting of distinctive patches—which landscape ecologists refer to as **landscape elements**—organized into a mosaic-like pattern. The elements of a mountain landscape may include forests, meadows, bogs, and streams, while those in an urban landscape include parks, industrial districts, and residential areas.

While our distant ancestors did not articulate a formal definition of landscape, their lives and livelihoods clearly reflected their understanding of landscape structure and process. Hunters and gatherers were familiar with variation across the landscapes in which they lived. They learned where to find plants useful as food or medicines, and where to find game animals, including where the animals hid, fed, watered, and how they moved with the seasons. Later, pastoralists learned how to locate forage for livestock, how the most productive pastures changed with the seasons and between years of drought and years with ample rain, and where in the landscape predators and other dangers were likely to be encountered (fig. 21.1). Settled agriculturalists learned which areas were most suitable for planting row crops, which were best for orchards and vineyards, and how to work and shape the land to guide the movement of water and avoid soil losses (fig. 21.2). The establishment of cities required managing the movements of food, waste, and water between the urban center and the surrounding agricultural and wild lands (fig. 21.3).

With mounting environmental pressures from human populations, the need for understanding landscapes has grown, not diminished, and that growing need has created the modern science of landscape ecology. Jianguo Wu and Richard Hobbs (2007), of Arizona State University and University of Western, Australia, respectively, point out that the precise meaning of the term *landscape ecology*, first coined by the German geographer Carl Troll (1939), is still debated among landscape ecologists. However, drawing from its many definitions, Wu and Hobbs identify a thematic thread uniting the discipline and on that basis define **landscape ecology** as "the science and art of studying and influencing the relationship between spatial pattern and ecological processes across hierarchical levels of biological organization and different scales in space and time." Though most landscape ecologists have worked at larger spatial scales, the concepts of landscape ecology have been applied to spatial patterns and ecological processes ranging from those relevant to ground beetles moving across a few meters of grassland (Wiens, Schooley, and Weeks 1997) to very

Figure 21.2   Successful agriculturalists must have a basic understanding of landscape structure and process. These terraced rice fields are the result of human engineering of the landscape to retain water and prevent erosion.   KingWu/iStockphoto

Figure 21.3   With the development of cities, humans began to interact intensively with the landscape at larger scales. This 2,000-year-old Roman aqueduct in Segovia, Spain, once transported water 18 km to the center of the Roman town.   lovelypeace/123RF

large regional scales measured in thousands of km². In addition, while the concepts of landscape ecology were first developed in terrestrial settings (Forman 1995; Forman and Godron 1986; Turner, Gardner, and O'Neill 2001), they can be applied in aquatic environments as well.

Three facets of landscape ecology distinguish it from the other subdisciplines of ecology presented in this text. The first is that landscape ecology is generally highly interdisciplinary. Gunther Tress, Bärbel Tress, and Gary Fry (2005) point out that **interdisciplinary research** involves researchers from multiple disciplines working closely to produce an understanding that integrates across disciplines. Interdisciplinary research can include several scientific disciplines or extend beyond the boundaries of the natural sciences into the social sciences and humanities. The second characteristic distinguishing landscape ecology from other subdisciplines is that it has included humans and human influences on landscapes since its

Figure 21.1   Managing large bands of grazing animals requires detailed knowledge of local landscapes, especially the locations of good forage, water, and shelter.   NPS Photo by Jim Peaco

beginnings. As a consequence, landscape ecology often plays a central role where ecologists attempt to restore degraded landscapes. Third, and perhaps most central to the discipline, landscape ecology focuses on understanding the extent, origin, and ecological consequences of spatial heterogeneity across multiple spatial scales.

The full conceptual scope of landscape ecology cannot be covered in a single chapter (see Wu 2013). However, we will sample the discipline by reviewing some studies concerning core areas of landscape ecology. In earlier chapters, we discussed structure, process, and change within the context of populations, communities, and ecosystems. In this chapter, we revisit structure, process, and change within the context of landscapes.

21.1 Landscape Structure

LEARNING OUTCOMES

After studying this section you should be able to do the following:

21.3 Describe landscape structure, including the roles of patches and matrix in defining landscape structure.

21.4 Discuss how variation in number and shapes of ecosystem patches (e.g., forest fragments) contribute to landscape structure.

21.5 Explain how patch shape contributes to edge effect.

21.6 Explain how a unit of measure can influence the estimated perimeter of a patch (e.g., a patch of forest or an oceanic island).

21.7 Discuss the implications of Mandelbrot's ideas to the ecology of different organisms (e.g., eagles vs. barnacles).

Landscape structure includes the size, shape, composition, number, and position of patches, or landscape elements, in a landscape. Much of ecology focuses on studies of structure and process; landscape ecology is no exception. We are all familiar with the structure, or anatomy, of organisms. In chapter 9, we discussed the structure of populations, and in chapters 16 to 20 we considered the structure of communities and ecosystems. However, what constitutes landscape structure? **Landscape structure** consists mainly of the size, shape, composition, number, and position of patches, or landscape elements, within a landscape. As you look across a landscape you can usually recognize its constituent ecosystems as distinctive patches, which might consist of woods, fields, ponds, marshes, or towns. Landscape ecologists define a **patch** as a relatively homogeneous area that differs from its surroundings—for example, an area of forest surrounded by agricultural fields. The patches within a landscape form the mosaic that we call landscape structure. The background in this mosaic is called the **matrix,** the element within the landscape that is the most continuous, spatially.

Most questions in landscape ecology require that ecologists quantify landscape structure. The following examples show how this has been done on some landscapes and how some aspects of landscape structure are not obvious without quantification.

The Structure of Six Landscapes in Ohio

In 1981, G. Bowen and R. Burgess published a quantitative analysis of several Ohio landscapes. These landscapes consisted of forest patches surrounded by other types of ecosystems. Six of the 10 km by 10 km areas analyzed are shown in figure 21.4. If you look carefully at this figure you see that the landscapes, which are named after nearby towns, differ considerably in total forest cover, the number of forest patches, the average area of patches, and the shapes of patches. Some of the landscapes are well forested, and others are not. Some contain only small patches of forest, while others include some large patches. In some landscapes, the forest patches are long and narrow, whereas in others they are much wider. These general differences are clear enough, but we would find it difficult to give more precise descriptions unless we quantified our impressions.

First, let's consider total forest cover. Forest cover varies substantially among the six landscapes. The Concord landscape, with 2.7% forest cover, is the least forested. At the other extreme, forest patches cover 43.6% of the Washington landscape. Differences between these extremes are clear, but what about some of the less obvious differences? Compare the Monroe and Somerset landscapes (see fig. 21.4) and try to estimate which is more forested and by how much. Somerset may appear to have greater forest cover, but how much more? You may be surprised to discover that Somerset, with 22.7% forest cover, has twice the forest cover of the Monroe landscape, which includes just 11.8% forest cover (fig. 21.5). This substantial difference could mean the difference between persistence and local extinction for some forest species.

Now let's examine the size of forest patches in each of the landscapes. Again, the median area of forest patches differs significantly across the landscapes. The smallest median areas are in the Monroe landscape, 3.6 ha, and the Concord landscape, 4.1 ha. The Washington landscape has the largest median patch area.

Now, look back at figure 21.4 and try to estimate which of the landscapes contains the greatest number, or highest density, of forest patches. The Somerset landscape, with 244 forest patches, has the highest patch density, and the Monroe landscape, with 180 patches, has the next highest density of forest patches. Obviously, the Concord landscape has the lowest density of forest patches, with only 46. The Boston landscape, with 86 forest patches, contains the next lowest density of forest patches.

Now let's look at a more subtle feature of landscape structure: patch shape. Bowen and Burgess quantified patch shape by the ratio of patch perimeter to the perimeter (circumference) of a circle with an area equal to that of the patch. Their formula was:

$$S = \frac{P}{2\sqrt{\pi A}}$$

where:

S = patch shape

P = patch perimeter

A = patch area

Quantifying landscape structure may reveal relationships not apparent visually. Compare your impression of the landscapes shown here to quantitative representations of some attributes presented in figures 21.5 and 21.6.

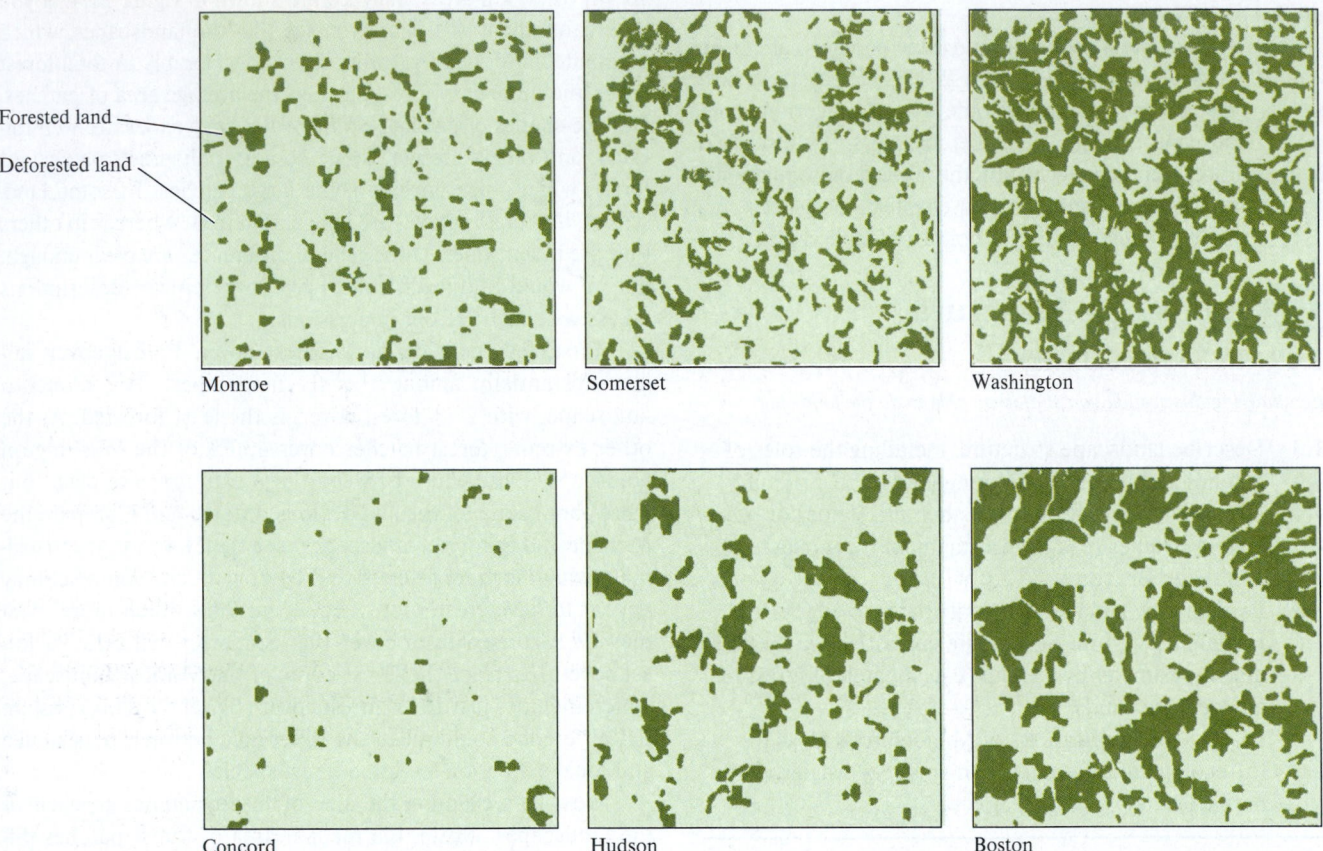

Forested land

Deforested land

Monroe   Somerset   Washington

Concord   Hudson   Boston

Figure 21.4   Forest fragments, shown as dark green, in six landscapes in Ohio (data from Bowen and Burgess 1981).

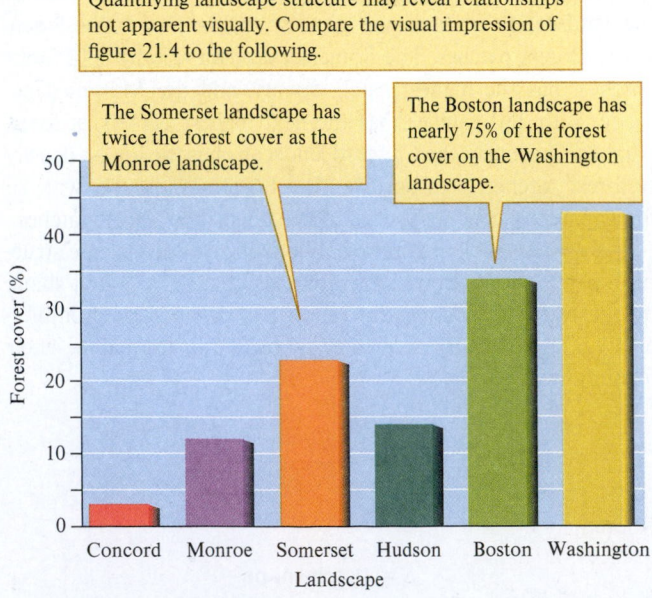

Quantifying landscape structure may reveal relationships not apparent visually. Compare the visual impression of figure 21.4 to the following.

The Somerset landscape has twice the forest cover as the Monroe landscape.

The Boston landscape has nearly 75% of the forest cover on the Washington landscape.

Figure 21.5   Percentage of forest cover in six landscapes in Ohio (data from Bowen and Burgess 1981).

How do you translate differences in the value of this index into shape? If S is about equal to 1, the patch is approximately circular. Increasing values of S indicate less circular patch shapes. High values of S generally indicate elongate patches and a long perimeter relative to area. In other words, such patches have more edge habitat relative to interior habitat.

The edges of habitat patches, such as forest patches surrounded by agricultural lands, do not have sharply defined boundaries. Instead, the edges of such habitats are **ecotones,** which are characterized by physical and biological transitions from one ecosystem type to another. Ecotones often support a mix of species from both ecosystems, for which they represent a transition, plus some species unique to the ecotone. Consequently, ecotones are often areas of distinctive ecological conditions and higher species richness compared to the ecosystems on either side of an ecotone—a phenomenon referred to as **edge effect.** The species associated with ecotones are often called "edge" species, while those associated with the interiors of ecosystems away from an ecotone are called "interior" species. We will revisit the ecological significance of edges in chapter 23 (see section 23.3).

Bowen and Burgess calculated the shapes, *S*, for the forest patches in each of their landscapes and then determined the median shape for each (fig. 21.6). The Concord landscape, with a median *S* of 1.16, contains the most circular patches of the six landscapes. The Washington landscape, with a median *S* of 1.6, contains the least circular patches. As we shall see in the next example, landscape ecologists have developed methods for representing landscape structure that go well beyond the classical methods used by Bowen and Burgess.

Historically, geometry, which means "earth measurement," could offer only rough approximations of complex landscape structure. Today, an area of mathematics called *fractal geometry* can be used to quantify the structure of complex natural shapes. Fractal geometry was developed by Benoit Mandelbrot (1982) to provide a method for describing the dimensions of natural objects as diverse as ferns, snowflakes, and patches in a landscape. Fractal geometry offers unique insights into the structure of nature.

The Fractal Geometry of Landscapes

During the development of fractal geometry, Mandelbrot asked a deceptively simple question: "How long is the coast of Great Britain?" This is analogous to estimating the perimeter of a patch in a landscape. Think about this question. At first, you might expect there to be only one, exact answer. For simple shapes with smooth outlines such as squares and circles, the assumption of a single answer is approximately correct. However, an estimate of the perimeter of a complex shape often depends on the size of the measuring device. In other words, if you measure the coastline of Great Britain, you will find that your measurement depends on the size of the ruler you use. If you were to step off the perimeter of Great Britain in 1 km lengths, which is like using a ruler 1 km long, you would get a

smaller estimate than if you made your measurements with a 100 m ruler. If you measured the coastline with a 10 cm ruler, you would get an even larger estimate of the perimeter. The reason a larger ruler gives a smaller estimate is that the large ruler misses many of the nooks and crannies along the coast. These smaller features show up in estimates made with smaller rulers.

Mandelbrot's answer to his question about the British coastline was "Coastline length depends on the scale at which it is measured!" We can see the ecological significance of this finding by considering some of its consequences to organisms. Bruce Milne (1993) measured the coastline of Admiralty Island off the coast of southeastern Alaska. He made his measurements from the perspective of two very different animal residents of the island, bald eagles and barnacles.

Milne considered how the measured length of Admiralty Island's coastline depends on the length of the measuring device. Figure 21.7 plots ruler length on the horizontal axis and estimated length of coastline on the vertical axis. The straight line that joins the dots slopes downward to the right. As Mandelbrot suggested, the estimated coastline length decreases as ruler length increases.

Now, what "ruler" are bald eagles and barnacles using? The distribution of eagle nests around Admiralty Island are about 0.782 km apart. This measurement of inter-nest distance gives us an estimate of the length of coastline required by a bald eagle territory on the island. In contrast, barnacles range from 1 to a few centimeters in basal diameter and they are

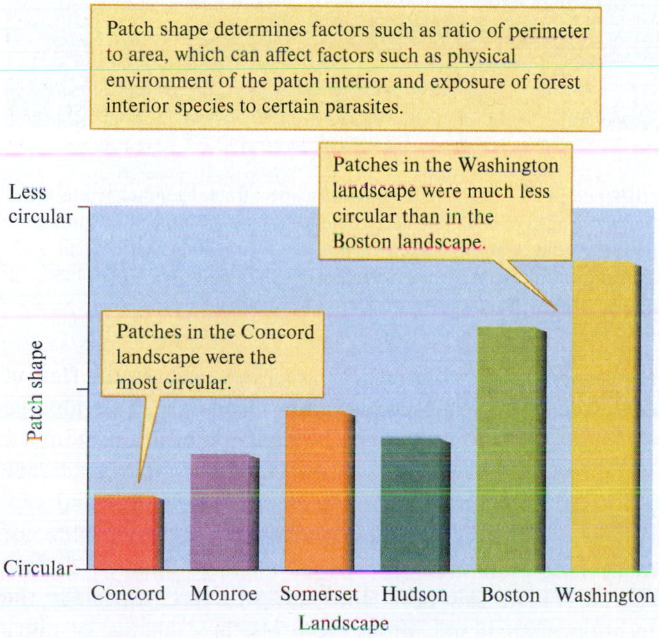

Patch shape determines factors such as ratio of perimeter to area, which can affect factors such as physical environment of the patch interior and exposure of forest interior species to certain parasites.

Patches in the Washington landscape were much less circular than in the Boston landscape.

Patches in the Concord landscape were the most circular.

Figure 21.6 Relative shapes of forest patches in six landscapes in Ohio (data from Bowen and Burgess 1981).

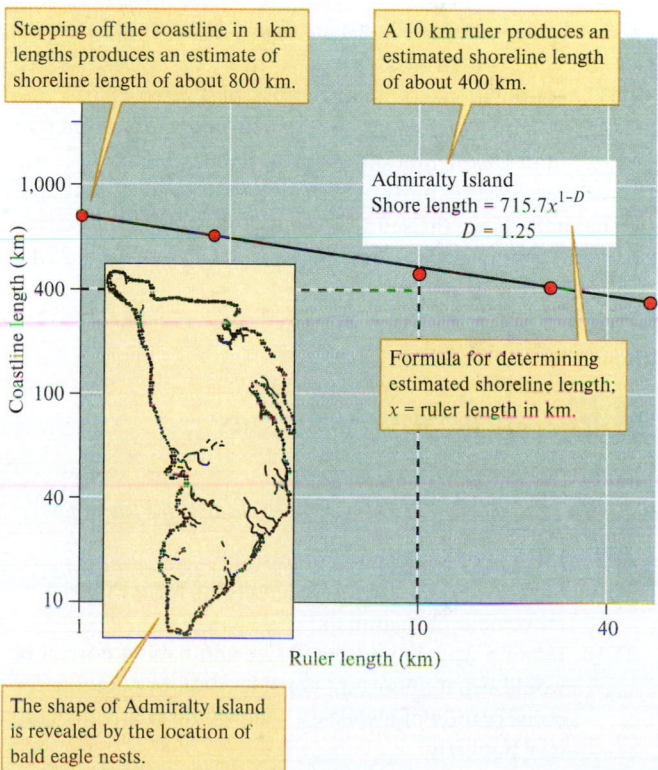

Stepping off the coastline in 1 km lengths produces an estimate of shoreline length of about 800 km.

A 10 km ruler produces an estimated shoreline length of about 400 km.

Admiralty Island
Shore length = $715.7x^{1-D}$
$D = 1.25$

Formula for determining estimated shoreline length; x = ruler length in km.

The shape of Admiralty Island is revealed by the location of bald eagle nests.

Figure 21.7 Relationship between ruler length and the measured length of the coastline of Admiralty Island, Alaska (data from Milne 1993).

sedentary. Barnacles need only a small area of solid surface to attach themselves and are often packed side by side along a rocky shore. Milne estimated that an individual barnacle requires about 2 cm (0.00002 km) of coastline.

Milne assumed that the eagles are, in effect, using a ruler 0.782 km long to step off the perimeter of the island and that barnacles use a ruler 0.00002 km long. Milne's analysis estimates that from the eagle's perspective, the perimeter of Admiralty Island is just a bit over 760 km. However, to a barnacle stepping off the coastline with its tiny ruler, the perimeter is over 11,000 km! Any of us would probably have assumed that the barnacle population "sees" a lot more of the spatial complexity around Admiralty Island. However, without Mandelbrot's fractal geometry, it would be difficult to predict that the difference in island perimeter for eagles and barnacles would be as great as 760 versus 11,000 km. At the conclusion of his analysis, Milne challenges us to imagine how long the coastline of Admiralty Island must be from the perspective of crude oil molecules. This is the length of coastline that determines the cost of a thorough cleanup after oil spills like the one in Venezuela in 2020 or the *Exxon Valdez* in 1989 (fig. 21.8).

As in other areas of science, describing aspects of landscape structure, such as the length of the coastline of Admiralty Island or the size, shape, and number of forest patches in Ohio landscapes, is not an end in itself. Landscape ecologists study landscape structure because it influences landscape processes and change. These are the next topics we will discuss.

Concept 21.1 Review

1. In the landscapes shown in figure 21.4, what is patch and what is matrix?
2. The populations of many forest bird species in eastern North America have declined following deforestation and fragmentation of forest habitat (see fig. 21.4), whereas many birds associated with open grassland habitats have thrived. From a landscape perspective, how could forest succession (chapter 20, section 20.1) change this situation?

21.2 Landscape Processes

LEARNING OUTCOMES

After studying this section you should be able to do the following:

21.8   List major landscape processes.
21.9   Discuss how landscape structure influences the movement of mammals.
21.10 Explain why habitat patch size and habitat corridors would be an important consideration in planning for conservation of a species, such as the Glanville fritillary butterfly.
21.11 Explain how understanding the chemistry of a lake may require looking beyond the boundary of the lake to consider the broader landscape context.

(a)

(b)

Figure 21.8   Perspective on landscapes: fractal geometry tells us that the length of coastline accessible to (*a*) the oil molecules spilling from the hold of an oil tanker, such as the *Exxon Valdez* (the larger ship shown in this photo) is much greater than that used by (*b*) bald eagles.   (*a*) AP Photo; (*b*) Shutterstock/Rocky Grimes

Landscape structure influences processes such as the flow of energy, materials, and species across a landscape.   Landscape ecologists study how the size, shape, composition, number, and position of ecosystems in the landscape affect **landscape processes**. Though less familiar than physiological and ecosystem processes, landscape processes are responsible for many important ecological phenomena. In chapter 20 (see fig. 20.27), we saw how landscape structure, especially the location of shallow bedrock, controls the exchange of nutrients between subsurface and surface waters and local rates of primary production in Sycamore Creek, Arizona. As we will see in the following examples, landscape structure affects other ecologically important processes such as the dispersal of organisms, extinction of local populations, and the flux of water between groundwater and lakes.

Landscape Structure and the Dispersal of Mammals

Landscape ecologists have proposed that landscape structure, especially the size, number, and isolation of habitat patches, can influence the movement of organisms between potentially suitable habitats. For instance, populations of desert bighorn sheep live in the isolated mountain ranges of the southwestern United States and northern Mexico, with individuals moving frequently among the ranges (fig. 21.9). The group of subpopulations of desert bighorn sheep living in an area such as the deserts of southern California constitute a metapopulation (see section 10.2). The rate of movement of individuals between such subpopulations can significantly affect the persistence of a species in a landscape.

Human activity often produces habitat fragmentation, which occurs where a road cuts through a forest, a housing development eliminates an area of shrubland, or tracts of tropical rain forest are cut to plant pastures. Because habitat fragmentation is increasing, ecologists study how landscape structure affects the movements of organisms, movements that might mean the difference between population persistence and local extinction.

James Diffendorfer, Michael Gaines, and Robert Holt (1995) studied how patch size affects the movements of three small mammal species: cotton rats, *Sigmodon hispidus,* prairie voles, *Microtus ochrogaster,* and deer mice, *Peromyscus maniculatus.* They divided a 12 ha prairie landscape in Kansas into eight 5,000 m^2 areas. The prairie vegetation was mowed to maintain three patterns of fragmentation (fig. 21.10). The least fragmented areas consisted of large, 50 m by 100 m patches. The areas with medium fragmentation each contained 6 medium 12 m by 24 m patches. The most fragmented landscapes contained 10 or 15 small 4 m by 8 m patches.

The researchers predicted that animals would move farther in the more fragmented landscapes consisting of small habitat patches. In fragmented landscapes, individuals must move farther to find mates, food, and cover. They also predicted that animals would stay longer in the more isolated patches within fragmented landscapes. Consequently, the proportion of animals moving would decrease with habitat fragmentation.

The rodent populations were monitored on the study site by trapping them with Sherman live traps and tagging newly caught individuals twice each month from August 1984 to May 1992. Over the course of their 8-year study, Diffendorfer, Gaines, and Holt amassed a data set consisting of 23,185 captures. They used these data to construct movement histories for individual animals in order to test their predictions. They expressed movements as *mean square distances,* a measurement that estimates the size of an individual's home range. A home range is the area that an animal occupies on a daily basis.

The behavior of two of the three study species supports the hypothesis that small mammals move farther in more fragmented landscapes. As predicted, *Peromyscus* and *Microtus,*

(a)

(b)

Figure 21.9 Fragmented landscapes: (*a*) the small mountain ranges of the southwestern United States and northern Mexico provide island-like habitats for populations of (*b*) desert bighorn sheep, which move frequently between the mountain ranges of the region. (*a*) Jim Lundgren/Pixtal/age fotostock; (*b*) NPS Photo by Mark Lellouch

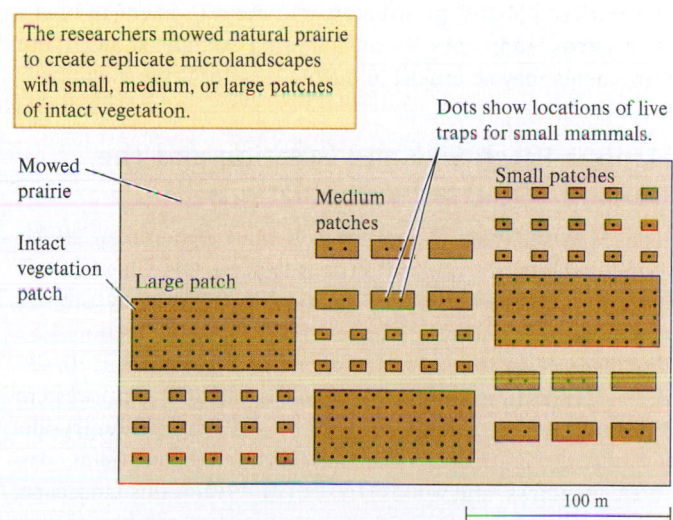

The researchers mowed natural prairie to create replicate microlandscapes with small, medium, or large patches of intact vegetation.

Dots show locations of live traps for small mammals.

Mowed prairie

Intact vegetation patch

Large patch

Medium patches

Small patches

100 m

Figure 21.10 Experimental landscape for the study of small mammal movements (data from Diffendorfer, Gaines, and Holt 1995).

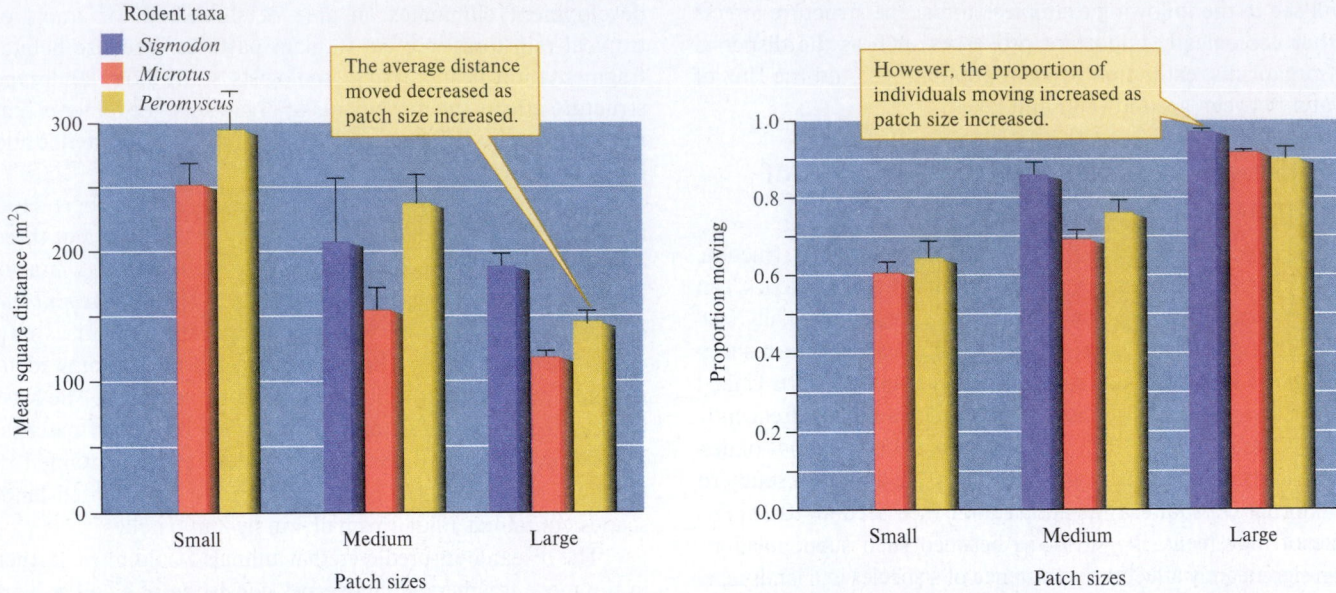

Figure 21.11 Influence of patch size on small mammal movements within experimental landscapes (means, 1 standard error) (data from Diffendorfer, Gaines, and Holt 1995).

living in small patches, moved farther than individuals living in medium or large patches (fig. 21.11). However, the movements of *Sigmodon* in medium and large patches did not differ significantly.

The proportion of *Sigmodon, Microtus,* and *Peromyscus* moving within the 5,000 m^2 experimental areas supported the hypothesis that animal movements decrease with habitat fragmentation (see fig. 21.11). A larger proportion of *Sigmodon* moved within large patch areas than moved within areas with medium patches. Because few *Sigmodon* were captured within small patch areas, their movements within these areas could not be analyzed. A larger proportion of *Microtus* and *Peromyscus* moved within large and medium patches than moved within small patches.

In summary, this experiment shows a predictable relationship between landscape structure and the movement of organisms across landscapes. As the following example shows, those movements may be crucial to maintaining local populations.

Habitat Patch Size and Isolation and the Density of Butterfly Populations

Ilkka Hanski, Mikko Kuussaari, and Marko Nieminen (1994) found that the local population density of the Glanville fritillary butterfly, *Melitaea cinxia,* is significantly affected by the size and isolation of habitat patches (see chapter 4, section 4.5). The researchers studied a metapopulation (see chapter 10, section 10.2) of these butterflies on Åland Island in southwestern Finland. Their study site consisted of 15.5 km^2 of countryside, a landscape consisting of small farms, cultivated fields, pastures, meadows, and woods (fig. 21.12). Within this landscape, habitat suitable for the butterfly consisted of patches of their larval food plant, *Plantago lanceolata,* which generally occurs in pastures and meadows.

Figure 21.12 Much of the landscape of southwestern Finland consists of a patchwork of pastures, meadows, and woods.
Kai Honkanen/ PhotoAlto

There were 50 patches of potential habitat within the study area. Forty-two of these patches were occupied by the butterflies in 1991. The patches ranged in area from 12 to 46,000 m^2 and supported populations ranging from 0 to 2,190 individuals. The habitat patches also varied in their degree of isolation from other habitat patches. The distance from habitat patches to the nearest occupied patch varied from 30 m to 1.6 km. However, Hanski and his colleagues found that, from a statistical perspective, the best index of isolation combined distances to neighboring habitat patches and the numbers of butterflies living on those patches.

Habitat patch area influenced both the size and density of the populations. Total population size within a patch increased with patch area. However, population density decreased as patch area increased (fig. 21.13). Thus, though large habitat patches supported larger numbers of individuals than smaller patches, population density was lower on large patches.

The team also found that more isolated patches supported lower densities of butterflies. Isolation influences local population density in these populations because local populations are partly maintained by immigration of *Melitaea* from other patches. For instance, during 1 week of sampling the butterflies in one patch, about 15% of the males and 30% of the females were recaptures from surrounding patches.

This experiment determined that area and isolation of patches strongly influence the size and density of *Melitaea* populations. One conclusion that we can draw from these patterns is that landscape structure is important for understanding the distribution and abundance of the butterflies. It turns out that landscape structure also affects the persistence of local populations. Between 1991 and 1992, Hanski and his colleagues recorded three extinctions of local populations and five colonizations of new habitat patches. All these extinctions and colonizations occurred in small patches with small populations.

The studies by Diffendorfer and colleagues and Hanski and colleagues show that the movement of organisms and the characteristics of local populations are significantly influenced by landscape structure.

Habitat Corridors and Movement of Organisms

One long-standing approach to reducing the negative impact of fragmentation and isolation on populations has been to connect habitat fragments with corridors of similar habitat. A growing number of experimental studies have filled many of the gaps in our understanding of the effects of habitat corridors on movement by organisms.

Nick Haddad and Kristen Baum (Haddad 1999; Haddad and Baum 1999) studied the influence of corridors on the movements of butterflies associated with early successional habitats. Their study site was the Savannah River Site, South Carolina, a National Environmental Research Park, where they created patches of open habitat in dense 40- to 50-year-old forests of pine. Patches of open habitat were squares, 128 m on a side, with an area of 1.64 ha, the approximate size of forestry clear-cuts in the surrounding region. With the help of the staff of the Savannah River Institute, Haddad and Baum created 27 openings: 8 that were isolated and 19 that were connected by corridors of open habitat to other patches. Within each patch all trees were removed and the slash (debris) was burned.

In a study of movements of butterflies between patches, Haddad (1999) focused his attention on two butterfly species: the common buckeye, *Junonia coenia,* and the variegated fritillary, *Euptoieta claudia.* Both of these species are specialized for life in early successional habitats and avoid pine forests. Haddad used mark and recapture techniques to study butterfly movements. He marked a total of 1,260 common buckeye butterflies and 189 variegated fritillaries. Haddad subsequently recaptured 239 common buckeye butterflies and 47 variegated fritillaries. His results showed clearly that corridors increased the frequency of movements by both butterfly species between patches. In a companion study, Haddad and Baum (1999) also documented higher densities of both butterfly species in open habitats connected by corridors (fig. 21.14).

However, the influences of corridors on movement of organisms between open habitats within the Savannah River Site extend far beyond butterflies. Research conducted by a team of 10 investigators (Tewksbury et al. 2002) also showed higher rates of movement by common buckeye and variegated fritillary butterflies between patches connected by corridors. They also discovered higher rates of pollination of plants growing in connected patches and higher rates of seed dispersal, mainly by birds. In summary, these studies and many others have now shown that habitat corridors can

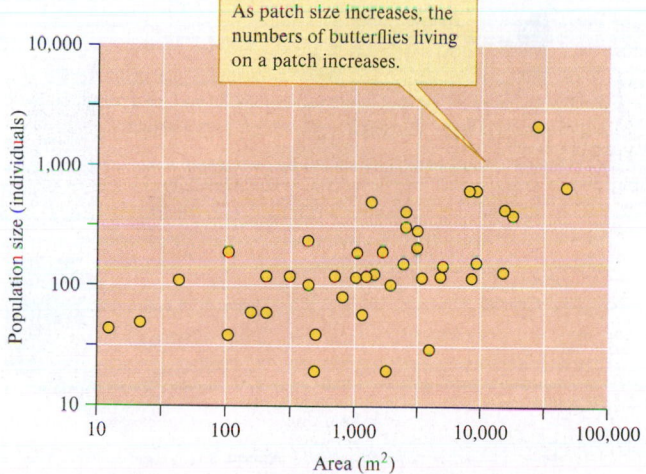

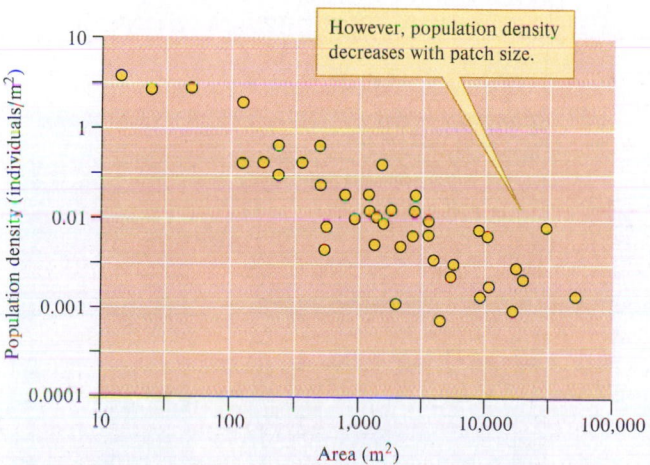

Figure 21.13   Relationship between habitat patch area and population size and density of the butterfly *Melitaea cinxia* in a landscape on Åland Island, Finland (data from Hanski, Kuussaari, and Nieminen 1994).

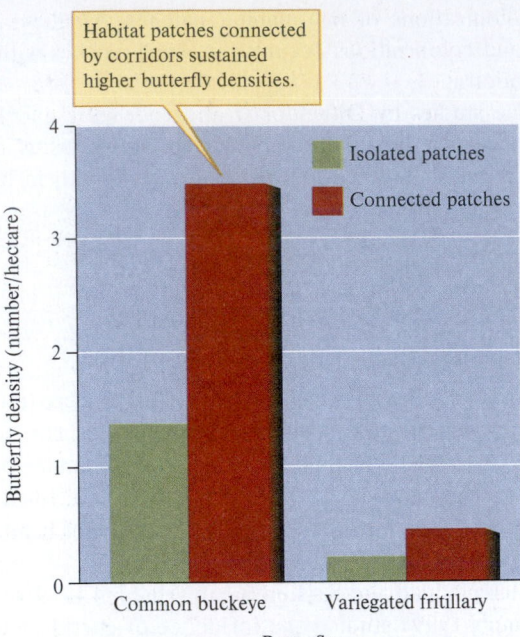

Habitat patches connected by corridors sustained higher butterfly densities.

Figure 21.14   Densities of common buckeye, *Junonia coenia,* and variegated fritillary, *Euptoieta claudia,* butterflies in early successional patches connected by corridors of open habitat or isolated by surrounding pine forest (data from Haddad and Baum 1999).

facilitate the movement of organisms between otherwise isolated habitat fragments.

As we will see in the next example, landscape structure can also influence the chemical and physical characteristics of ecosystems.

Landscape Position and Lake Chemistry

Katherine Webster and her colleagues (1996) at the Center for Limnology at the University of Wisconsin and the U.S. Geological Survey explored how the position of a lake in a landscape affects its chemical responses to drought. Drought can affect a wide range of lake ecosystem properties, including nutrient cycling and the concentrations of dissolved ions. However, two lakes may respond differently to drought. For instance, while drought increased the concentration of dissolved substances in Lake 239 at the Experimental Lakes Area in Ontario, Canada, it decreased them in Nevins Lake, Michigan.

Webster and her colleagues set out to determine whether the contrasting chemical responses of lakes to drought can be explained by the position of the lake in the landscape. They worked in northern Wisconsin, where they defined the landscape position of a lake as its location within a hydrologic flow system. The team quantified the position of a lake within a hydrologic flow system as the proportion of lake water supplied by groundwater.

The sources of water for a lake are precipitation, surface water, and groundwater flow. Different lakes receive different proportions of their water from these sources, and these proportions depend on a lake's position in the landscape. Figure 21.15 shows a series of lakes along a hydrologic flow

system in northern Wisconsin. Morgan Lake, which receives the bulk of its water from precipitation, occupies the upper end of this continuum. Lakes such as this one occupy high points in the hydrologic flow system and are called "hydrologically mounded" lakes. These lakes are sources of water for the rest of the hydrologic flow system. Crystal Lake and Sparkling Lake, which occupy intermediate positions within the hydrologic flow system and receive significant inflows of groundwater, are "groundwater flow through" lakes. Finally, at the lower end of the flow system are the "drainage" lakes that receive significant surface drainage as well as groundwater drainage. Webster and her colleagues estimated that Morgan Lake receives no groundwater inflow, while Trout Lake, at the

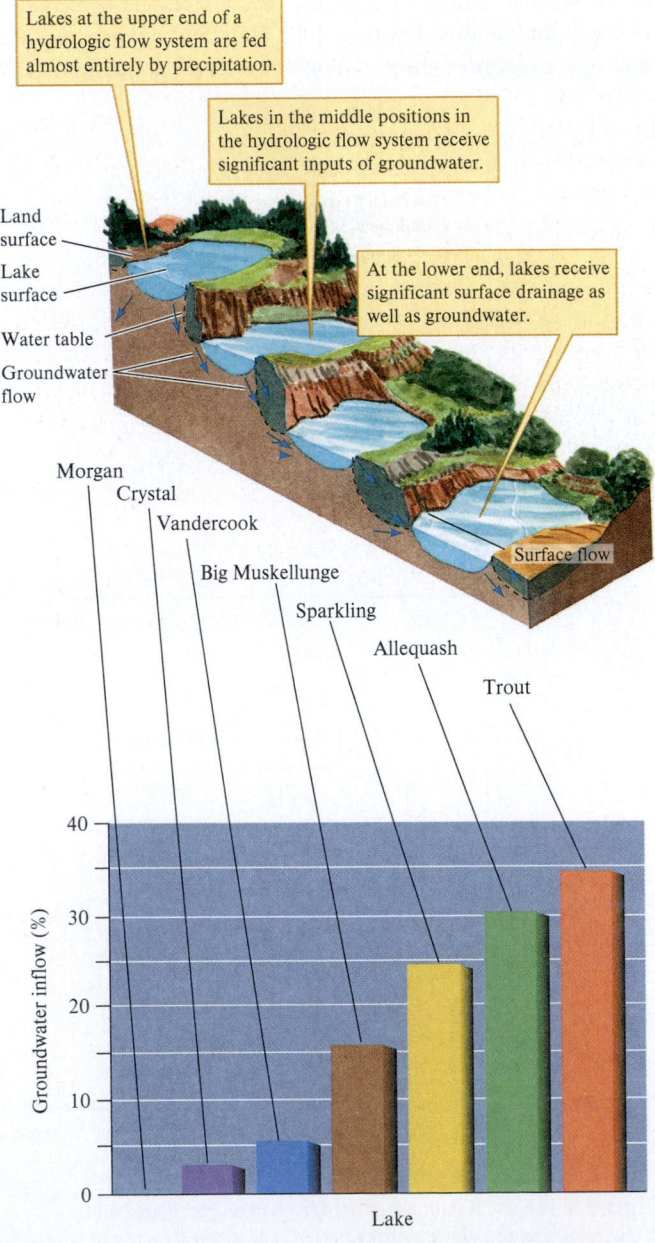

Lakes at the upper end of a hydrologic flow system are fed almost entirely by precipitation.

Lakes in the middle positions in the hydrologic flow system receive significant inputs of groundwater.

At the lower end, lakes receive significant surface drainage as well as groundwater.

Figure 21.15   Lake position in the landscape and proportion of water received as groundwater (data from Webster et al. 1996).

lower end of the hydrologic flow system, receives 35% of its inflow as groundwater.

The responses of these seven lakes to a drought were studied from 1986 to 1990. As you might expect, the levels of the lakes dropped during this 4-year drought. However, the amount of drop in lake level was related to a lake's position in the landscape (fig. 21.16). The level of Morgan Lake, at the upper end of the hydrologic flow system, dropped 0.7 m, while the levels of Vandercook, Big Muskellunge, Crystal, and Sparkling Lakes, in the middle of the hydrologic flow system, dropped 0.9 to 1.0 m. Meanwhile, the levels of Trout and Allequash Lakes, the two drainage lakes at the lower end of the hydrologic flow system, dropped very little.

Landscape position also significantly influenced a lake's chemical responses to the drought. The *concentrations* of dissolved ions such as calcium (Ca^{2+}) and magnesium (Mg^{2+}) increased in the majority of the lakes. However, the increase in ion concentration was highest at the upper and lower ends of the hydrologic flow system. Meanwhile, the combined *mass* of Ca^{2+} and Mg^{2+} increased in the three lakes at the lower end but did not change in Morgan Lake, at the upper end of the flow system, and either decreased or did not change in the lakes occupying the middle portions of the hydrologic flow system (see fig. 21.16).

The researchers concluded that the increased mass of Ca^{2+} and Mg^{2+} seen at the lower end of the hydrologic flow system was due to an increased proportion of inflows from groundwater and surface water, sources rich in Ca^{2+} and Mg^{2+}. The declines in mass of Ca^{2+} and Mg^{2+} in Big Muskellunge Lake are likely due to reduced inflow of ion-rich groundwater. The stability of Ca^{2+} and Mg^{2+} mass in Morgan Lake was attributed to its isolation from the groundwater flow system. Morgan Lake receives almost no groundwater even during wet periods. Regardless of the mechanisms, the chemical responses of these lakes to the drought were related to their positions in the landscape.

In the first section of this chapter, we reviewed the concept of landscape structure. In this section, we explored the connection between landscape structure and landscape processes. But what creates landscape structure? Landscape structure, like the structure of populations, communities, and ecosystems, changes in response to an interplay between dynamic processes. We explore the sources of landscape structure and change in the next section.

Concept 21.2 Review

1. What do the patterns shown in figure 21.11 suggest about the relative impact of fragmentation of prairie habitat on populations of *Sigmodon, Microtus,* and *Peromyscus*?
2. Habitat corridors are widely recommended for conservation of species whose populations are restricted to isolated patches of habitat. Why?
3. Are there any potential risks associated with increasing the exchange of individuals between habitat patches through the creation of habitat corridors?

21.3 Origins of Landscape Structure and Change

LEARNING OUTCOMES

After studying this section you should be able to do the following:

21.12 Discuss the role of geological processes in creating and maintaining landscape structure.
21.13 Describe the role of ecosystem engineers in landscapes.
21.14 Predict how landscape structure would be altered following the extinction of important ecosystem engineers.
21.15 Discuss the influences of fire and fire suppression on landscape structure in Mediterranean climates.

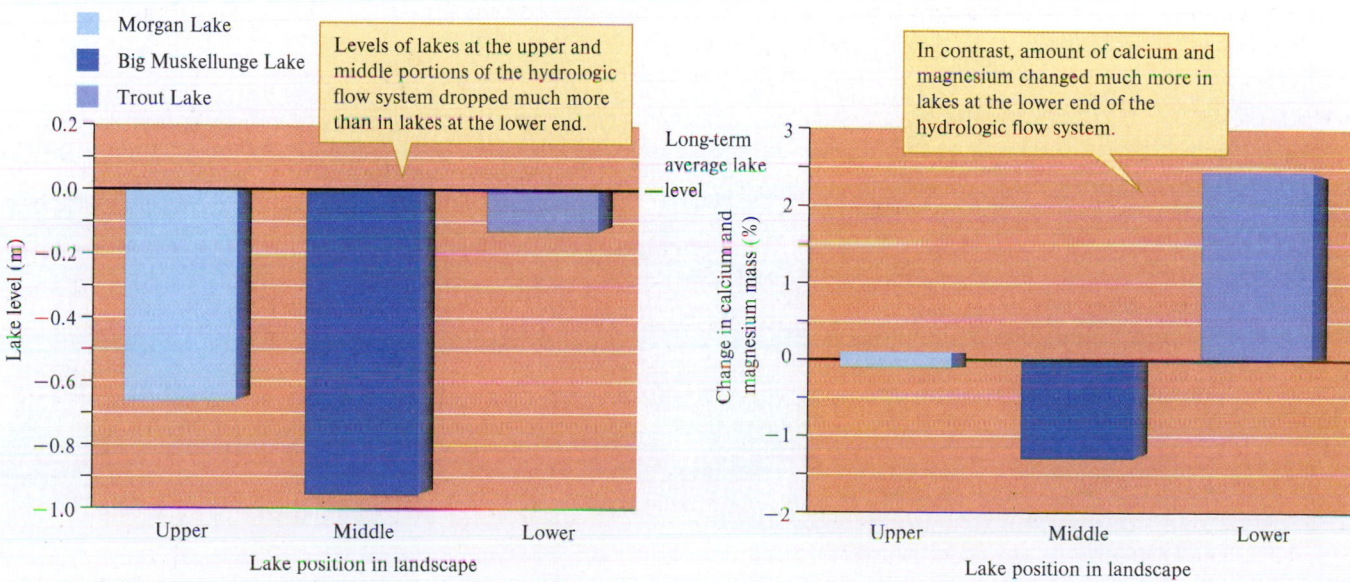

Figure 21.16 Lake position in a hydrologic flow system and response to a severe drought (data from Webster et al. 1996).

Landscapes are structured and change in response to geological processes, climate, activities of organisms, and fire. What creates the patchiness we see in landscapes? Many forces combine in numerous ways to produce the patchiness that we call landscape structure. In this section, we review examples of how geological processes, climate, organisms, and fire contribute to landscape structure.

Geological Processes, Climate, and Landscape Structure

The geological features produced by processes such as volcanism, sedimentation, and erosion provide a primary source of landscape structure. For instance, the alluvial deposits along a river valley provide growing conditions different from those on thin, well-drained soils on nearby hills. A volcanic cinder cone in the middle of a sandy plain offers different environmental conditions than the surrounding plain (fig. 21.17). Distinctive ecosystems may develop on each of these geological surfaces, creating patchiness in the landscape. In the following example, we shall see how distinctive soils contribute to vegetative patchiness in a Sonoran Desert landscape.

Soil and Vegetation Mosaics in the Sonoran Desert

The Sonoran Desert includes many long, narrow mountain ranges separated by basins or valleys. The mountains and basins in this region originated in movements of the earth's crust that began 12 to 15 million years ago. As the mountains were uplifted and the adjacent basins subsided, erosion removed materials from the mountain slopes. This eroded material was deposited in the surrounding basins, forming sloping plains, or *bajadas,* at the bases of the mountains. Sediment deposits in these basins may be over 3 km deep.

From a distance, the bajadas of the Sonoran Desert may appear to be uniform environments, especially against the backdrop of a rugged desert mountain (see fig. 16.1). However, Joseph McAuliffe (1994) has shown that bajadas in the Sonoran Desert near Tucson, Arizona, consist of a complex mosaic of distinctive **landforms.** His studies have shown that intermittent erosion and deposition operating over the past 2 million years have produced a complex landscape.

McAuliffe established study sites on the bajadas of three mountain ranges, where he studied soils and plant distributions. In all three study areas, he found a wide range of soil types and plant distributions that corresponded closely to soil age and structure.

Let's look at some of the patterns McAuliffe found on the bajada associated with the northern end of the Tucson Mountains. Going from left to right in figure 21.18, the first

Figure 21.17 Geologic features, such as the basaltic lava flow in the foreground and the volcanic cinder cones in the background, contribute to the structure of landscapes by adding geologic surfaces with distinctive physical and chemical properties.   John A. Karachewski

Figure 21.18   Soil ages on an outwash plain, or bajada, associated with the Tucson Mountains, Arizona; colors used only to show locations of different soils in landscape (data from McAuliffe 1994).

soils you see are of early Pleistocene age and are approximately 1.8 to 1.9 million years old. Going northward along the bajada, to the right in figure 21.18, the next soils in the sequence date from the middle to late Pleistocene and are hundreds of thousands of years old. These soils are followed by Holocene deposits that are less than 11,000 years old and are associated with an ephemeral desert water course called Wildhorse Wash. Near the Holocene soils, McAuliffe found soils that dated from the late Pleistocene. These soils were 25,000 to 75,000 years old.

In the space of a few kilometers, McAuliffe found patches of soil that were (1) almost 2 million years old, (2) hundreds of thousands of years old, (3) tens of thousands of years old, and (4) less than 11,000 years old. Because soil-building processes occur over long periods of time, these soils of vastly different ages also differ substantially in structure. Figure 21.19 shows McAuliffe's drawings of typical profiles of Holocene, middle to late Pleistocene, and early Pleistocene soils. The Holocene soils had low amounts of clay and calcium carbonate ($CaCO_3$) and poorly developed soil horizons. They also lacked a *caliche layer,* a hardpan soil horizon formed by precipitation of $CaCO_3$. Middle to late Pleistocene soils had a much higher clay content than Holocene soils, and early Pleistocene soils contained even more clay. These clay layers in the older soil profiles are called **argillic horizons.** Middle to late and early Pleistocene soils also contained more $CaCO_3$ and were underlain by a thick layer of caliche.

These differences in soil structure influence the distributions of perennial plants across the Tucson Mountains bajada (fig. 21.20). McAuliffe found that the relative abundances of two shrubs, *Larrea tridentata* and *Ambrosia deltoidea,* accounted for much of the variation in perennial plant distributions and that plant distributions map clearly onto soils of different ages. For instance, *Ambrosia* is most abundant on middle to late Pleistocene soils, whereas *Larrea* dominates on Holocene soils and on early Pleistocene soils. Other perennial plant species dominate on the eroding side slopes of early Pleistocene soils. Climate also influences landscape structure.

Climate and Landscape Structure

The soils studied by McAuliffe show evidence of the particular climate in which they formed. The soil mosaic along the bajada east of the Tucson Mountains consists of patches of material deposited during floods originating in these mountains from nearly 2 million years ago to less than 11,000 years ago. The deposits were laid down during times when the climate produced intense storms that caused flooding and erosion. Materials eroded from mountain slopes were deposited as alluvium on the surrounding bajadas.

These alluvial deposits were gradually changed, and these changes were dependent on climate. Two of the prominent features of the older soils studied by McAuliffe were the formation of a clay-rich argillic horizon and the formation of a $CaCO_3$-rich caliche layer. Both of these soil features are the

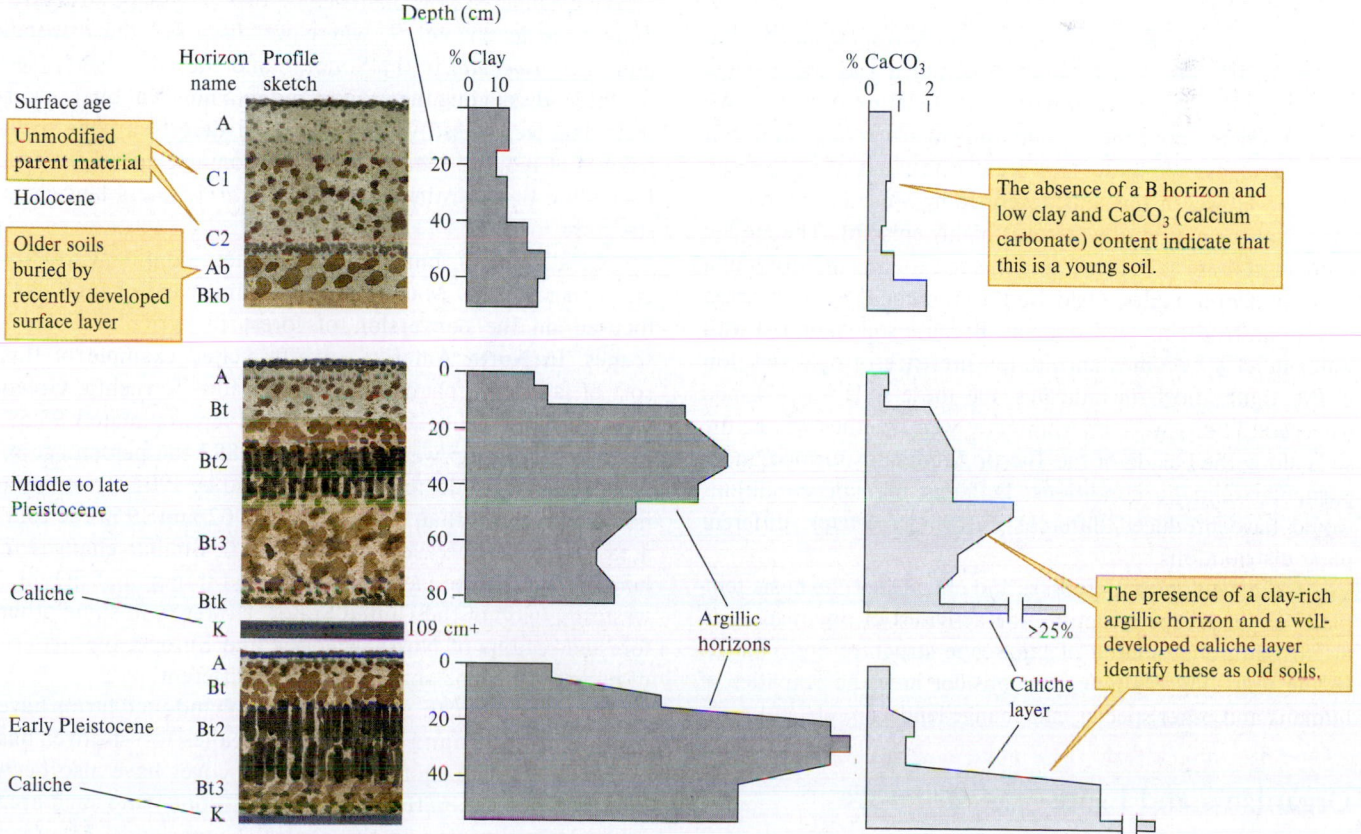

Figure 21.19   Structural features of young to old desert soils on the Tucson Mountains bajada (data from McAuliffe 1994).

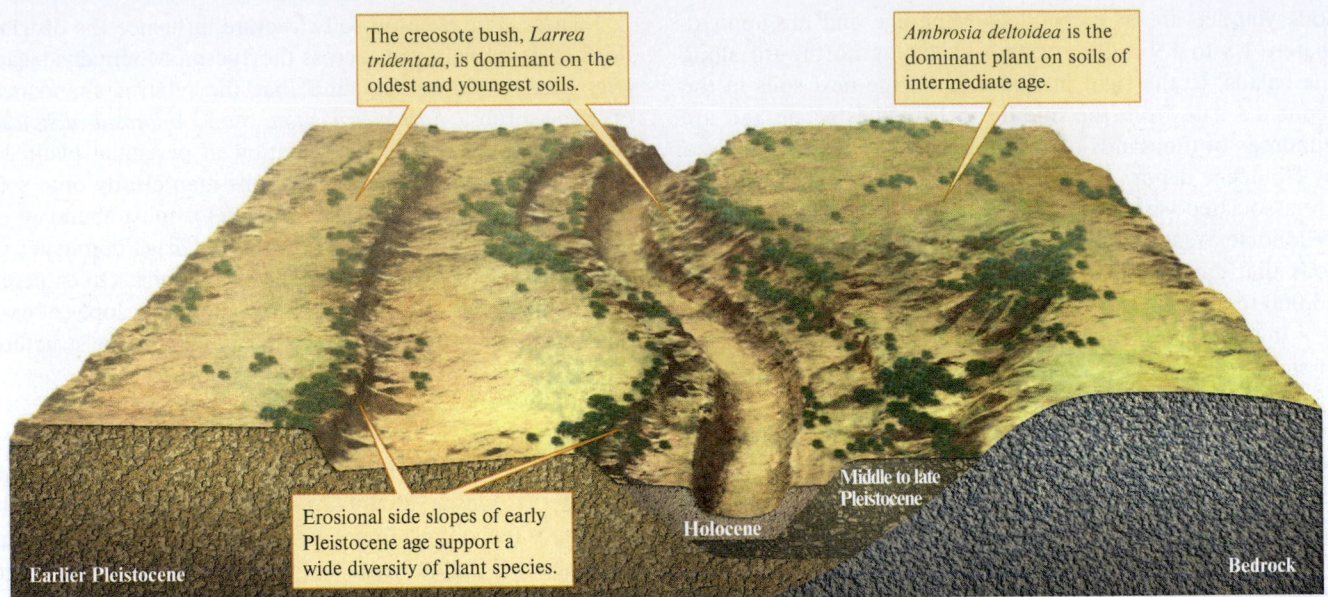

The creosote bush, *Larrea tridentata*, is dominant on the oldest and youngest soils.

Ambrosia deltoidea is the dominant plant on soils of intermediate age.

Middle to late Pleistocene

Holocene

Erosional side slopes of early Pleistocene age support a wide diversity of plant species.

Earlier Pleistocene

Bedrock

Figure 21.20   Association between vegetation and soils of different ages and structure on the Tucson Mountains bajada; colors used only to show locations of different soils in landscape (data from McAuliffe 1994).

result of water transport. Clay particles are transported as a colloidal suspension, whereas the $CaCO_3$ is transported in dissolved form. Consequently, the clays precipitate out of suspension higher in the soil profile than the $CaCO_3$. The result is the layering of an argillic horizon over a caliche layer as shown in figure 21.19.

Water, working on alluvial deposits, is responsible for the soil structure observed by McAuliffe, but it was water delivered to the landscape under particular climatic conditions. We can get a clue about those conditions by observing some soil characteristics. We know that argillic horizons are deposited by water. However, the soils described by McAuliffe also offer clues that the action of water was highly episodic. The argillic horizon in these soils is red, and this red color is the result of a buildup of iron oxides. Oxidation of iron could have occurred only in an oxidizing environment. Because soil saturated with water quickly becomes anoxic, the presence of oxidized iron in the argillic horizon indicates that these soils were formed when conditions were intermittently wet. In other words, the soils along the bajada of the Tucson Mountains formed under particular climatic conditions. Different climatic conditions would have produced different soils and, perhaps, different plant distributions.

While geological processes and climate set the basic template for landscape structure, the activities of organisms can be an additional source of landscape structure and change. In the following example, we consider how the activities of humans and other species can change landscape structure.

Organisms and Landscape Structure

Organisms of all sorts influence the structure of landscapes. Organisms that cause changes in the physical environment

sufficient to influence the structure of landscapes, ecosystems, or communities are often referred to as **ecosystem engineers** (Jones, Lawton, and Shachak 1994). While the following discussion focuses on the influences of animals, plants create much of the distinctive patchwork we call landscape structure. For an example of how plants can create landscape structure, think back to chapter 9, where we discussed the distribution of *L. tridentata* in the Sonoran and Mojave Deserts (see fig. 9.12). By shading the ground and retaining leaf litter under their canopies, these widely dispersed desert shrubs create patches of reduced temperature variation and higher fertility. By adding these distinctive patches, *Larrea* alters landscape structure.

Of all species, humans are the most dominant modifiers of landscapes. Many studies of landscape change have focused on the conversion of forest to agricultural landscapes. In North America, an often-cited example of this sort of landscape change is that of Cadiz Township, Green County, Wisconsin (fig. 21.21). In 1831, approximately 93.5% of Cadiz Township was forested. By 1882 the percentage of forested land had decreased to 27%, and by 1902 forest cover had fallen to less than 9%. Between 1902 and 1950 the total area of forest decreased again to 3.4%. Similar changes in landscape structure have been observed throughout the midwestern region of the United States. However, in some other forested regions of North America and Europe, the pattern of recent landscape change has been different.

In eastern North America, many abandoned farms have reverted to forest and in these landscapes forest cover has increased. Recent increases in forest cover have also been observed in some parts of northern Europe. One such area is the Veluwe region in the central Netherlands. Maureen Hulshoff (1995) reviewed the landscape changes that have

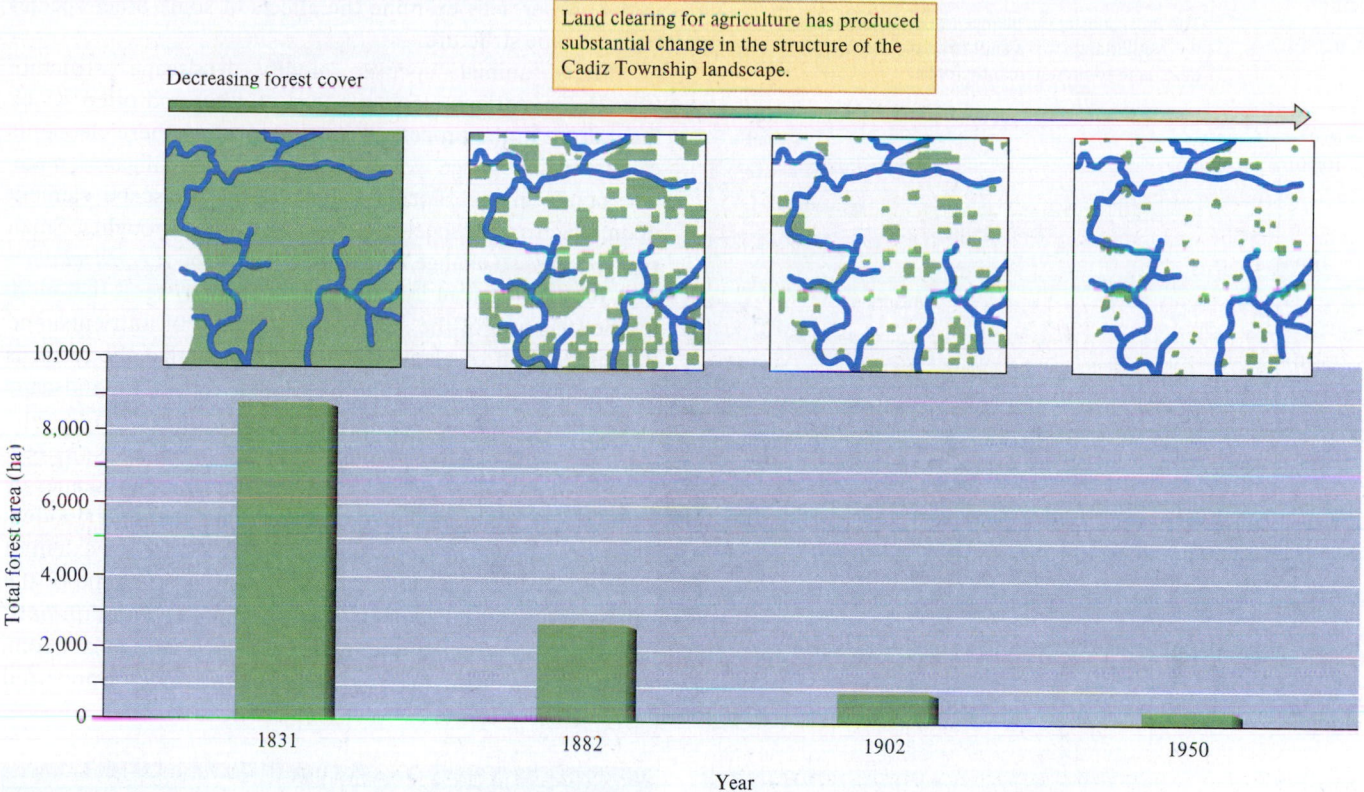

Land clearing for agriculture has produced substantial change in the structure of the Cadiz Township landscape.

Decreasing forest cover

Figure 21.21   Human activity reduced the forest cover in Cadiz Township, Wisconsin (data from Curtis 1956, maps after Curtis 1956).

occurred in the Veluwe region during the past 1,200 years. The Veluwe landscape was originally dominated by a mixed deciduous forest. Then, from A.D. 800 to 1100, people gradually occupied the area and cut the forest. Consequently, forests were gradually converted to heathlands, which are landscapes dominated by low shrubs and used for livestock foraging. Later, small areas of cropland were interspersed with the extensive heathlands. During the tenth and eleventh centuries, some areas were devegetated completely and converted to areas of drifting sand. The problem of drifting sand continued to increase until the end of the nineteenth century, when the Dutch government began planting pine plantations on the Veluwe landscape, a practice that continued into the twentieth century.

Figure 21.22 shows the changes in the composition of the Veluwe landscape from 1845 to 1982. The greatest change over this period was a shift in dominance from heathlands to forests. In 1845, heathlands made up 66% of the landscape, while forests constituted 17%. By 1982, coverage by heathlands had fallen to 12% of the landscape and forest coverage had risen to 64%. The figure also shows modest but ecologically significant changes in the other landscape elements. The area of drift sand reached a peak in 1898 and then dropped and held steady at 3% to 4% from 1957 to 1982. Urban areas established a significant presence beginning in 1957. Finally, coverage by agricultural areas has varied from 9% to 16% over the study interval, the least variation shown by any of the landscape elements.

As total coverage by forest and heathlands changed within the Veluwe landscape, the number and average area of forest and heath patches also changed. These changes indicate increasing fragmentation of heathlands and decreasing fragmentation of forests. For instance, between 1845 and 1982, the number of forest patches declined, while the average area of forest patches increased. During this period, the number of heath patches increased until 1957. Between 1957 and 1982, the number of heath patches decreased as some patches were eliminated. The average area of heath patches decreased rapidly between 1845 and 1931 and then remained approximately stable from 1931 to 1982.

During the period that Cadiz Township in Wisconsin was losing forest cover, this landscape element was increasing in the Veluwe district of the Netherlands. These two examples show how human activity has changed landscape structure. However, what forces drive human influences on landscapes? In both Cadiz Township and the Veluwe landscape, the driving forces were economic. A developing agricultural economy converted Cadiz Township from forest to farmland. The Veluwe landscape was converted from heathland to forest as the local sheep-raising economy collapsed in response to the introduction of synthetic fertilizers and inexpensive wool from Australia.

At the beginning of the twenty-first century, economically motivated human activity continues to change the structure of landscapes all over the globe. We examine current trends in land cover at the global scale in chapter 23. Before we do

The most substantial change in this landscape in the Netherlands was a shift from predominantly heathland to predominantly forest.

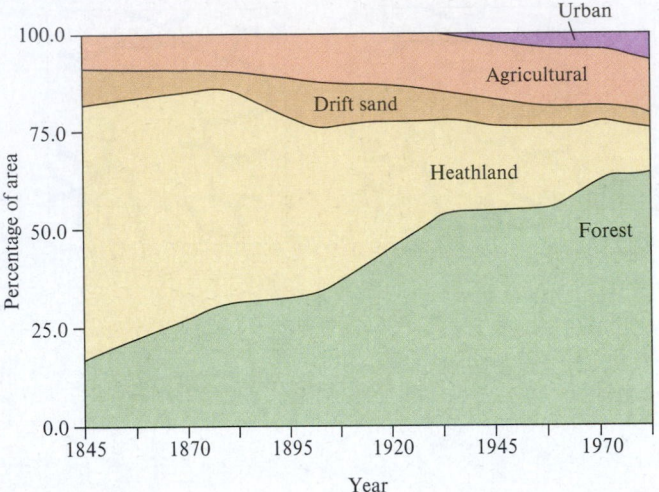

Figure 21.22   Change in a Dutch landscape (data from Hulshoff 1995).

that, however, let's examine the effects of some other species on landscape structure.

Many animal species modify landscape structure (fig. 21.23). African elephants feed on trees and often knock them down in the process. As a consequence, these elephants can gradually change woodland to grassland. Alligators maintain ponds in the Florida Everglades, a landscape element upon which many species depend to survive droughts. Small species can also change landscapes. Kangaroo rats, *Dipodomys* spp., of the American Southwest dig burrow systems that modify the structure of the soil, the distribution of nutrients, and the distribution of plants to such an extent that the result is recognizable from aerial photos. Similar effects on landscape structure are created by termites and ants.

One of the most adept modifiers of landscapes is the beaver, *Castor canadensis* (fig. 21.24). Beavers alter landscapes by cutting trees, building dams on stream channels, and flooding the surrounding landscape. Beaver dams increase the extent of wetlands in the landscape, alter the hydrologic regime of the catchment, and trap sediments, organic matter, and nutrients. The selective cutting of trees adds patchiness to the plant community and reduces the abundance of tree species preferred

(a)

(b)

(c)

(d)

Figure 21.23   Species with significant impacts on landscape structure. (*a*) African elephants control the extent of tree cover in some landscapes. (*b*) Alligators build and maintain ponds in wetland landscapes. (*c*) Feeding and burrowing by kangaroo rats introduce added patchiness into desert landscapes. (*d*) Termite mounds, such as the one on which this African antelope is standing, add distinctive landscape features.

(*a*) Shutterstock/Krishna Utkarsh Pandit; (*b*) Jason Edwards/National Geographic/Getty Images; (*c*) Alford W. Cooper/Science Source; (*d*) Shutterstock/martin613

Figure 21.24   Beavers are among nature's most active landscape engineers.   Dean Fikar 2013/iStockphoto

as food. These effects add several novel ecosystems to the landscape.

These influences of beavers on landscape structure once shaped the face of entire continents. At one time, beavers modified nearly all the temperate stream valleys in the Northern Hemisphere. The range of beavers in North America extended from arctic tundra to the Chihuahuan and Sonoran Deserts of northern Mexico, a range of approximately 15 million km^2. Before European colonization, the North American beaver population was 60 to 400 million. However, fur trappers eliminated beavers from much of their historical range and nearly drove them to extinction. With protection, North American beaver populations are recovering and large areas once again show the influence of beavers on landscape structure.

Robert Naiman and his colleagues have carefully documented the substantial effects of beavers on landscape structure (e.g., Naiman et al. 1994). Much of their work has focused on the effects of beavers on the 298 km^2 Kabetogama Peninsula in Voyageurs National Park, Minnesota. Following their near extermination, beavers reinvaded the Kabetogama Peninsula beginning about 1925. From 1927 to 1988, the number of beaver ponds on the peninsula increased from 64 to 834, a change in pond density from 0.2 to 3.0 per square kilometer. Over this 63-year period, the area of new ecosystems created by beavers, including beaver ponds, wet meadows, and moist meadows, increased from 200 ha, about 1% of the peninsula, to 2,661 ha, about 13% of the peninsula. Foraging by beavers altered another 12% to 15% of upland areas.

Beaver activity has changed the Kabetogama Peninsula from a landscape dominated by boreal forest to a complex mosaic of ecosystems. Figure 21.25 shows how beavers have changed a 45 km^2 catchment on the peninsula. These maps show that, between 1940 and 1986, beavers increased landscape complexity within this catchment. Similar changes have occurred over nearly the entire peninsula.

Naiman and his colleagues quantified the effects of beaver over 214 km^2, or 72%, of the Kabetogama Peninsula and found that between 1927 and 1988, beavers transformed most of the landscape. Within the study area, there are about 2,763 ha of

low-lying area that can be impounded by beavers. In 1927, the majority of the landscape, 2,563 ha, was dominated by forest. In 1927, moist meadow, wet meadow, and pond ecosystems covered only 200 ha. By 1988, moist meadows, wet meadows, and beaver ponds covered over 2,600 ha and boreal forest was limited to 102 ha.

The changes in landscape structure induced by beavers substantially alter landscape processes such as nutrient retention. Beaver activity between 1927 and 1988 increased the quantity of most major ions and nutrients in the areas affected by impoundments (fig. 21.26). The total quantity of nitrogen increased by 72%, while the amounts of phosphorus and potassium increased by 43% and 20%, respectively. The quantities of calcium, magnesium, iron, and sulfate stored in the landscape were increased by even greater amounts.

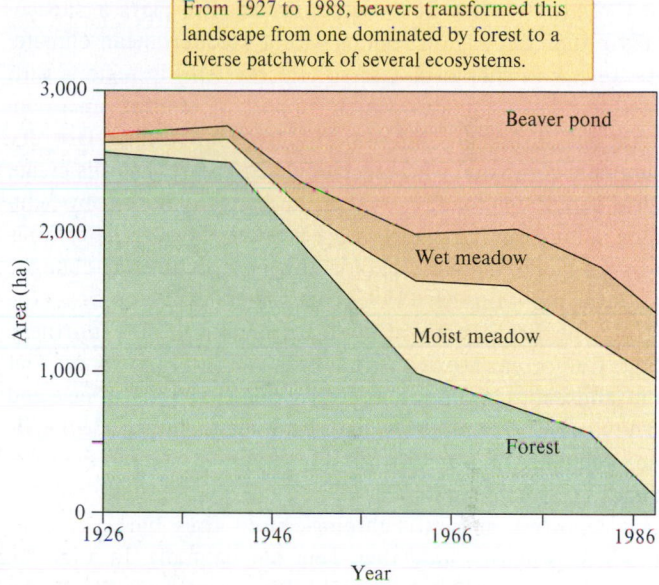

From 1927 to 1988, beavers transformed this landscape from one dominated by forest to a diverse patchwork of several ecosystems.

Figure 21.25   Beaver activity has changed landscape on the Kabetogama Peninsula, Minnesota (data from Naiman et al. 1994).

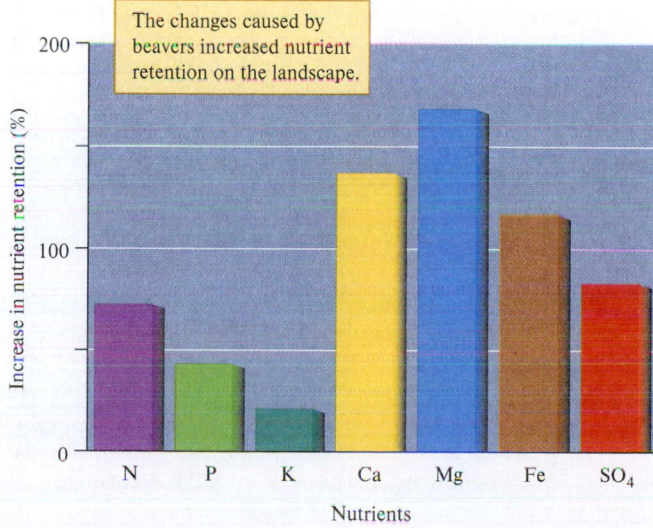

The changes caused by beavers increased nutrient retention on the landscape.

Figure 21.26   Nutrient retention on the Kabetogama Peninsula after alteration by beavers (data from Naiman et al. 1994).

Naiman and his colleagues offer three possible explanations for increased ion and nutrient storage in this landscape: (1) beaver ponds and their associated meadows may trap materials eroding from the surrounding landscape, (2) the rising waters of the beaver ponds may have captured nutrients formerly held in forest vegetation, and (3) the habitats created by beavers may have altered biogeochemical processes in a way that promotes nutrient retention. Whatever the precise mechanisms, beaver activity has substantially altered landscape structure and processes on the Kabetogama Peninsula.

Fire and the Structure of a Mediterranean Landscape

Fire contributes to the structure of landscapes ranging from tropical savanna to boreal forest. However, as the dramatic fires in California during 2020 demonstrated, fire plays a particularly prominent role in regions with a Mediterranean climate. As we saw in chapter 2, terrestrial ecosystems in regions with Mediterranean climates, which support Mediterranean woodlands and shrublands, are subject to frequent burning. Hot, dry summers combined with vegetation rich in essential oils create ideal conditions for fires, which can be easily ignited by lightning or humans. In regions with a Mediterranean climate, fire is responsible for a great deal of landscape structure and change.

Richard Minnich (1983) used satellite photos to reconstruct the fire history of southern California and northern Baja California, Mexico, from 1971 to 1980, and found that the landscapes of both areas consist of a patchwork of new and old burns. Though these regions experience similar Mediterranean climates and support similar natural vegetation, their fire histories diverged significantly in the early twentieth century. For centuries, natural lightning-caused fires burned, sometimes for months, until they went out naturally. In addition, Spanish and Anglo-American residents would set fire to the land routinely to improve grazing for cattle and sheep. Then,

early in the twentieth century, various government agencies in southern California began to suppress fires in order to protect property within an increasingly urbanized landscape.

Minnich proposed that the different fire histories of southern California and northern Baja California might produce landscapes of different structure. He suggested that fire suppression allowed more biomass to accumulate and set the stage for large, uncontrollable fires. His specific hypothesis was that the average area burned by wildfires would be greater in southern California.

Minnich tested his hypothesis using satellite images taken from 1972 to 1980 (fig. 21.27). He found that between 1972 and 1980 the total area burned in the two regions was fairly similar (fig. 21.28). However, the size of burns differed significantly between the two regions. The frequency of small burns below 1,000 ha was higher in northern Baja California, while large burns above 3,000 ha were more frequent in southern California. Consequently, median burn size in southern

Figure 21.27   Areas of Mediterranean shrubland in southern California periodically burn over large areas, destroying human habitations in the process.   FEMA Photo/Kevin Galvin

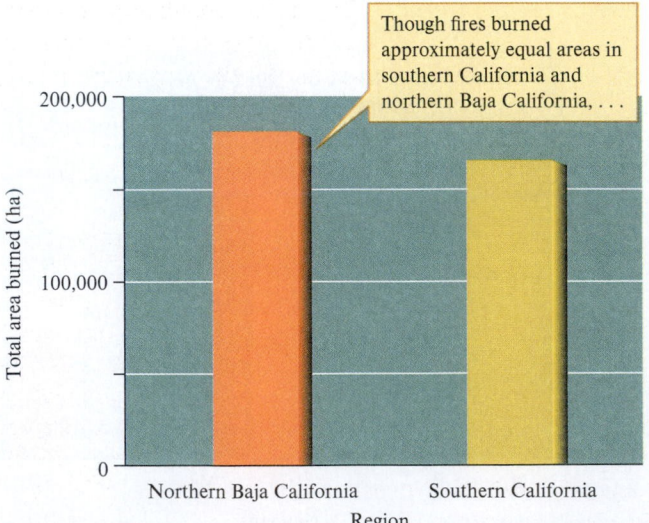

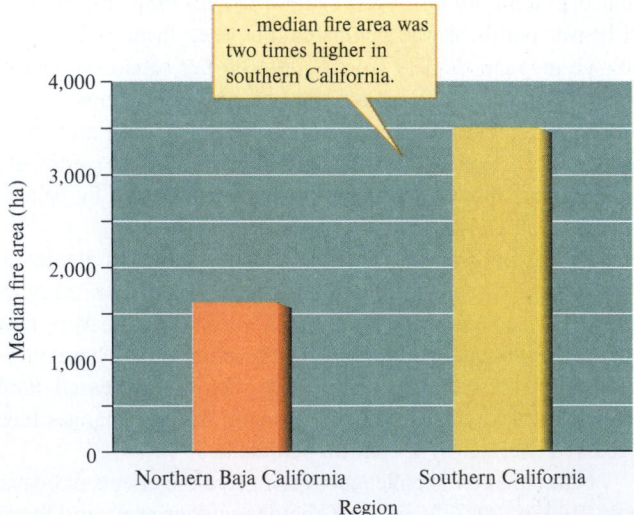

Figure 21.28   Characteristics of fires in the Mediterranean landscapes of southern California and northern Baja California from 1972 to 1980 (data from Minnich 1983).

California, 3,500 ha, was over twice that observed in northern Baja California, 1,600 ha (see fig. 21.28).

More recently, researchers from the College of Natural Resources, University of California, Berkeley and the U.S. Department of Agriculture tested this hypothesis by comparing 18,000 individual fire records from 1911 through 1924 (early suppression period) to those from 2002 through 2015 (Collins et al. 2019). Interestingly, they found no significant difference in the number of fires but there were large differences in the size of the fires. Fires that burned more than 12,000 ha accounted for only 0–6% of total burned area in the early suppression period but increased to 53–73% in the contemporary period. These changes were attributed not only to increased fuel loading as a result of fire suppression but also to climate change.

In this section, we have seen how geological processes, climate, the activities of organisms, and fire can contribute to landscape structure and change. Because human activity has often elevated the temperatures of cities compared to surrounding areas, there is growing interest in reducing heat storage in urban landscapes. That is the subject that we take up in the Applications section.

Concept 21.3 Review

1. How are "landscape engineers" similar to keystone species? How are they different?
2. Can a dominant species of tree in a forest or coral on a coral reef (see fig. 17.17) be an ecosystem engineer?
3. The patterns shown in figure 21.28 support Minnich's hypothesis that fire protection in southern California would produce a difference in median burn area. Why was analysis of pre- versus post-fire suppression also necessary to test this hypothesis? What are the other explanations for the results seen in both experiments?

Landscape Approaches to Mitigating Urban Heat Islands

LEARNING OUTCOMES

After studying this section you should be able to do the following:

21.16 Define heat wave and describe the threat that heat waves pose to human populations.
21.17 Summarize the history of human migration from rural areas to cities and the underlying reasons for this population shift.
21.18 Outline the mechanisms responsible for urban heat islands.
21.19 Review the ways to mitigate the growing urban heat island effect.

Hurricanes and tornadoes, which are capable of destroying entire towns, are among the most impressive displays of nature's raw power. Blizzards, often accompanied by extreme cold, can blanket entire regions with deep windblown snow, halting road and air travel and leaving communities without electrical power. However, the weather-related phenomenon responsible for the greatest loss of human life is none of these three sources of devastation. That distinction goes to **heat waves,** periods of abnormally hot and unusually humid weather, typically lasting for two or more days. A 1995 heat wave in Chicago was responsible for the deaths of an estimated 1,000 people. A 2010 heat wave in Russia killed 20,000 people. The deadliest weather in recent decades, however, was an extreme heat wave in western and southern Europe in 2003, when over 40,000 people died from heat stress (Bouchama 2004). The spatial distribution of deaths during these and other heat waves was not random but occurred disproportionately in urban centers (Dousset et al. 2011; Gabriel and Endlicher 2011; Luber and McGeehin 2008). Alarmingly, climate scientists project that frequency, intensity, and duration of heat waves will increase during the twenty-first century (Perkins-Kirkpatrick and Lewis 2020).

Urban ecology, which we introduced in chapter 1 (see fig. 1.2b), is an increasingly important area of ecological research, including studies of biodiversity in urban landscapes (chapter 16, Applications) and nutrient fluxes across the urban landscape (chapter 19, Applications). Urban ecology is also of fundamental importance, however, because cities are where most people now live. The global population has been steadily shifting from rural areas to urban centers. According to the United Nations, the proportion of the world's population living in cities increased from 30% in 1950 to over 50% today and is projected to increase to 68% in 2050 (United Nations, Department of Economic and Social Affairs, Population Division 2018). City dwellers already make up a much larger proportion of the population in some countries. For instance, the urban population of the United States makes up over 80% of the total and in Australia approximately 90% of the population is urban. Demographers predict that the movement from rural to urban environments will continue for the foreseeable future, particularly in Asia and Africa.

People are attracted to cities for many reasons. Chief among them are greater economic opportunities. Cities also provide access to a greater range of goods and services. In addition to the economic incentives for relocating, cities generally offer closer proximity to health care and schools for children compared to rural areas. There are, however, costs that come with migration to cities. Far too many rural migrants live in slums, where they join the ranks of the urban poor and are exposed to air and water pollution and higher crime rates. The physical environment of cities can also challenge rural migrants. One of the most noticeable differences they experience is the higher temperatures in urban centers compared to rural areas. This phenomenon, known as an **urban heat island,** results mainly from replacing the trees and other vegetation in a landscape with buildings, paved roads, and concrete walks. Vegetation moderates surface and air temperatures by reflecting some sunlight, shading the land surface, and evaporatively cooling surroundings as plants transpire water from their leaves. In contrast, the structures that generally dominate the urban landscape absorb solar energy, increasing surface and air temperatures. The energy used in cities for transportation and for air circulation and climate control of buildings generates large amounts of waste heat, increasing the urban heat island effect. Finally, tall buildings and narrow streets reduce air circulation and trap heat, further increasing the temperature of the urban heat island. Figure 21.29, which features two satellite images of Atlanta, Georgia, and surroundings, shows the dramatically warmer areas of the urban core and other heavily urbanized parts of the city compared to surrounding areas with substantial coverage of trees and other vegetation.

We discussed an example of an urban heat island in chapter 5 (see Applications), where the snail, *Arianta arbustorum*, has gone extinct in Basel, Switzerland, as the city gradually warmed during the twentieth century. Snails are not the only organisms to suffer the consequences of urban heat. As we've seen, people suffer in urban heat islands too, where they are subject to higher mortality rates during heat waves. Unfortunately, recent studies show that the public health challenges posed by urban heat islands are increasing. For instance, a study by Dana Habeeb, Jason Vargo, and Brian Stone of the Urban Climate Lab at the Georgia Institute of Technology revealed that the frequency, duration, and intensity of heat waves increased in 50 large metropolitan areas in the United States from 1961 to 2010 (Habeeb, Vargo, and Stone 2015). They found, in addition, that the average number of days that the populations of these cities were exposed to heat waves each year increased from 28.6 to 50.8 from the 1960s to the 2000s, an increase of 78% (fig. 21.30). Some of the trends in heat waves are the result of the global warming we'll discuss in chapter 23, but most of it resulted from the structure of the urban landscape itself. It turns out that the average rate of temperature increase in the majority of large U.S. metropolitan areas during the past half-century has been twice as fast as in nearby rural areas (Stone, Vargo, and Habeeb 2012). This is a direct result of the potentially deadly urban heat island effect, which will grow worse during the twenty-first century unless mitigated.

Encouragingly, there are many effective ways to mitigate the urban heat island effect that can be implemented locally. One means for reducing the urban heat island effect is to increase the vegetation cover in the urban landscape. Trees are particularly effective at reducing heat gain by reflecting sunlight, shading land surfaces, and cooling their surroundings as they transpire water through their leaves. Green roofs, another way to increase vegetative cover, cool the roof surface and the surrounding air by a combination of shading, evaporation, and transpiration (fig. 21.31). By insulating the rooftop, green roofs

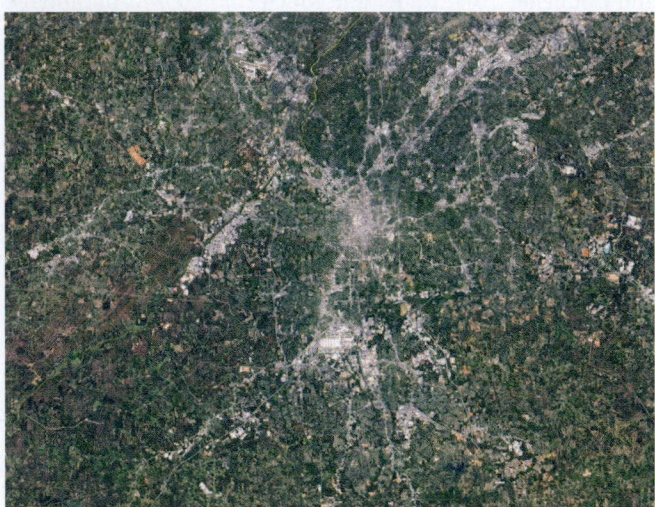

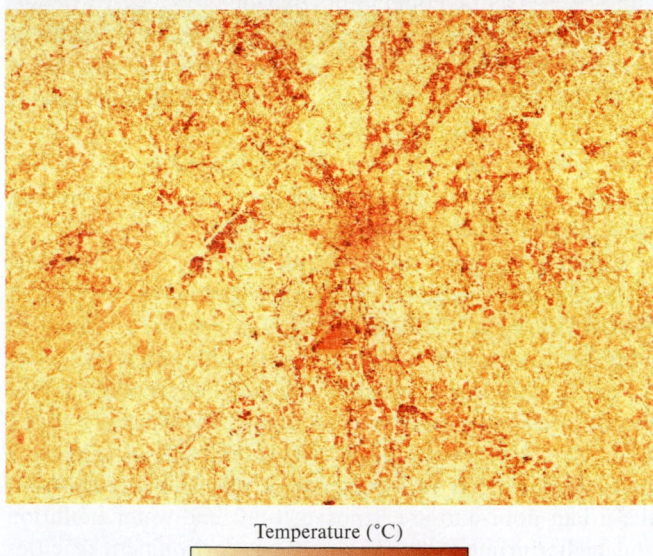

Temperature (°C)

18   24   30

Figure 21.29   An urban heat island: true color (upper) and thermal (low) images of Atlanta, Georgia, show how the most urbanized portions of the landscape have elevated surface temperatures compared to parts of the landscape with greater vegetation cover.   Source: Marit Jentoft-Nilsen, based on Landsat-7 data/NASA

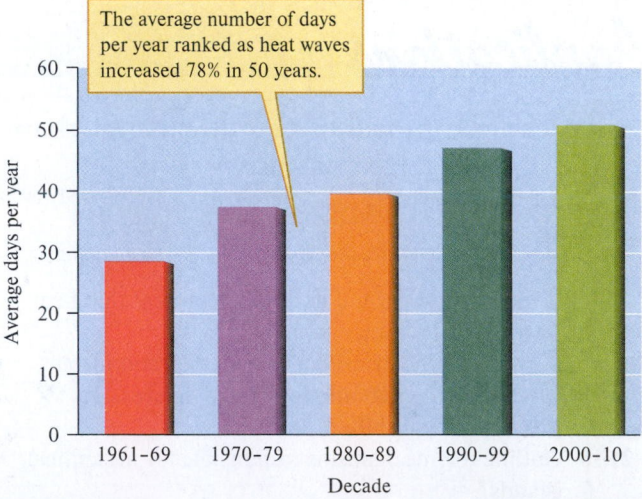

Figure 21.30   Average number of days per year in which the 50 largest metropolitan areas in the United State experienced heat waves, during the decades from1960 to 2010 (data from Habeeb, Vargo, and Stone 2015).

(a)   (b)

Figure 21.31   Reducing the contribution of roofs to an urban heat island: (*a*) a conventional commercial roof absorbs and stores a large amount of solar energy, which it radiates to its surroundings; (*b*) in contrast, a green roof insulates and shades the roof surface, which decreases the solar energy absorbed and radiated into the urban landscape. In addition, green roofs have a cooling effect through evaporation from planting beds and transpiration by vegetation.   (*a*) denboma/123RF; (*b*) Alison Hancock/Shutterstock

also reduce building cooling and heating requirements and, as a result, decrease the flow of waste heat from buildings into the urban landscape. Adopting more energy-efficient transportation systems can also reduce production of waste heat in cities. Finally, installing highly reflective roofing and paving materials can further reduce the amount of heat absorbed by an urban landscape. Several studies have now shown that combinations of these mitigation approaches have reduced the urban heat island effect by 1–7°C (Stone, Vargo, and Habeeb 2012). Levels of cooling similar to these will be essential in the decades ahead to combat the public health challenges posed by urban heat islands.

Summary

A landscape is a heterogeneous area composed of several ecosystems. The ecosystems making up a landscape generally form a mosaic of visually distinctive patches. These patches are called *landscape elements. Landscape ecology* is the study of landscape structure and processes.

Landscape structure includes the size, shape, composition, number, and position of patches, or landscape elements, in a landscape. Most questions in landscape ecology require that ecologists quantify landscape structure. Until recently, however, geometry, which means "earth measurement," could offer only rough approximations of complex landscape structure. Edge effects and ecotones increase on a landscape as habitats are fragmented. Today, an area of mathematics called fractal geometry can be used to quantify the structure of complex natural shapes. One of the findings of fractal geometry is that the length of the perimeter of complex shapes depends on the size of the device used to measure the perimeter. One implication of this result is that organisms of different sizes may use the environment in very different ways.

Landscape structure influences processes such as the flow of energy, materials, and species across a landscape. Landscape ecologists have proposed that landscape structure, especially the size, number, and isolation of habitat patches, can influence the movement of organisms between potentially suitable habitats. The group of subpopulations living on such habitat patches make up a metapopulation. Studies of the movements of small mammals in a prairie landscape show that a smaller proportion of individuals moves in more fragmented landscape but that the individuals that do move will move farther. The local population density of the Glanville fritillary butterfly, *Melitaea cinxia,* is lower on larger and on isolated habitat patches. Small populations of this butterfly and desert bighorn sheep are more vulnerable to local extinction. Habitat corridors have been shown to increase rates of movement among isolated habitat patches. The source of water for lakes in a Wisconsin lake district is determined by their positions in the landscape, which in turn determine their hydrologic and chemical responses to drought.

Landscapes are structured and change in response to geological processes, climate, activities of organisms, and fire. Geological features produced by processes such as volcanism, sedimentation, and erosion interact with climate to provide a primary source of landscape structure. In the Sonoran Desert, plant distributions map clearly onto soils of different ages and form a vegetative mosaic that closely matches soil mosaics. This mosaic will gradually shift as geological processes and climate gradually change the soil mosaic. While geological processes and climate set the basic template for landscape structure, the activities of organisms (ecosystem engineers), from plants to elephants, can be an additional source of landscape structure and change. Economically motivated human activity changes the structure of landscapes all over the globe. Beavers can quickly change landscape structure and processes over large regions. Fire contributes to the structure of landscapes ranging from tropical savanna to boreal forest. However, fire plays a particularly prominent role in regions with a Mediterranean climate.

Heat waves combined with the urban heat island effect are the weather-related phenomena responsible for the greatest loss of human life around the world. Urban heat islands result mainly from replacing the trees and other vegetation in a landscape with buildings, paved roads, and concrete walks. The urban heat island effect can be reduced by increasing vegetative cover in the urban landscape, adopting more energy-efficient building designs and transportation, and installing highly reflective roof and paving materials.

Key Terms

argillic horizon 457	heat wave 463	landscape 445	landscape structure 447
ecosystem engineers 458	interdisciplinary	landscape ecology 446	matrix 447
ecotone 448	research 446	landscape element 445	patch 447
edge effect 448	landform 456	landscape process 450	urban heat island 463

Review Questions

1. How does landscape ecology differ from ecosystem and community ecology? Contrast the questions that community, ecosystem and landscape ecologists would ask about a forest.

2. The green areas represent forest fragments surrounded by agriculture. Landscapes 1 and 2 contain the same total forest area. Which landscape, 1 or 2, will contain more forest interior species? Explain.

3. Consider the options for preserving patches of riverside forest shown in landscapes 3 and 4. Again, the two landscapes contain the same total area of forest but the patches in the two landscapes differ in shape. Which of the two would be most dominated by forest-edge species?

4. How do the sizes of patches in a landscape affect the movement of individuals among habitat patches and among portions of a

Landscape 1

Landscape 2

Landscape 3

Landscape 4

metapopulation? Consider the hypothetical landscapes shown in question 3. Which of the two landscapes would promote the highest rate of movement of individuals between forest patches?

5. Use fractal geometry and the niche concept (see chapters 9, 13, and 16) to explain why the canopy of a forest should accommodate more species of predaceous insects than insect-eating birds. Assume that the numbers of bird and predaceous insect species are limited by competition. (Hint: Milne's study [1993] of barnacles and bald eagles on Admiralty Island should provide a beginning for your argument.)

6. Analyses such as Milne's comparison (1993) of bald eagles and barnacles demonstrate that organisms of different sizes interact with the environment at very different spatial scales. With this in mind, consider the experiments of Diffendorfer and colleagues (1995) on the influence of habitat fragmentation on movement patterns of small mammals. Think about the size of their experimental study area (see fig. 21.10). How might a manipulation of this size have affected the movements of prairie birds? How would their manipulation have affected the movements of ground-dwelling beetles?

7. How do the activities of animals affect landscape heterogeneity? You might use either beaver or human activity as your model. What parallels can you think of between the influence of animal activity on landscape heterogeneity and the intermediate disturbance hypothesis (see chapter 16, section 16.4)? Which is concerned with the effect of disturbance on species diversity?

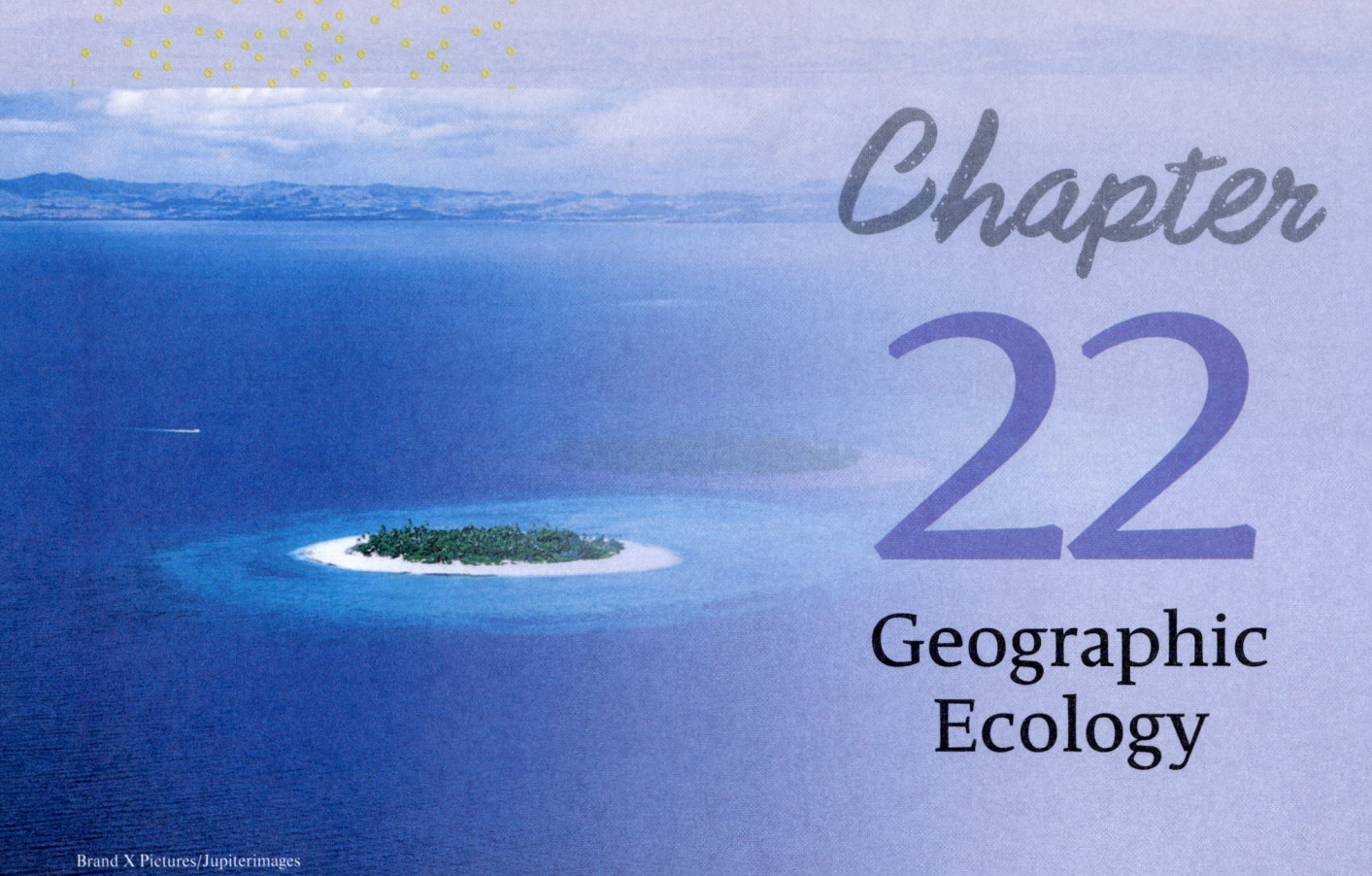

Brand X Pictures/Jupiterimages

Chapter
22
Geographic Ecology

Aerial view of Bounty Island, Fiji. The physical isolation of oceanic islands has long attracted the attention of scientists interested in understanding the factors that influence the diversity and composition of biological communities. Over time, studies of oceanic islands inspired research that extended, by analogy, to places far removed from their wave-swept shores.

CHAPTER CONCEPTS

22.1 On islands and habitat patches on continents, species richness increases with area and decreases with isolation. 470

Concept 22.1 Review 473

22.2 Species richness on islands can be modeled as a dynamic balance between immigration and extinction of species. 473

Concept 22.2 Review 478

22.3 Species richness generally increases from middle and high latitudes to the equator. 478

Concept 22.3 Review 482

22.4 Long-term historical and regional processes can significantly influence species richness and diversity. 482

Concept 22.4 Review 484

Applications: Global Positioning Systems, Remote Sensing, and Geographic Information Systems 485

Summary 487

Key Terms 488

Review Questions 488

LEARNING OUTCOMES

After studying this section you should be able to do the following:

22.1 Identify the central research focus of Humboldt's expedition to the Americas.
22.2 Outline some of the research topics that are part of geographic ecology.

Geographic ecology began on June 5, 1799, as Alexander von Humboldt and Aimé Bonpland sailed out of the port of Coruña in northwest Spain. Their small Spanish ship managed to slip past a British naval blockade and sail on, first to the Canary Islands and then to South America. Humboldt was a Prussian engineer and scientist and Bonpland was a French botanist. Humboldt came equipped with the finest scientific instruments of the time and was prepared to systematically survey the lands that he and Bonpland would visit. He wrote a letter to a friend a few hours before his ship left port outlining his purpose for the expedition: "I shall try to find out how the forces of nature interreact upon one another and *how the geographic environment influences plant and animal life* [emphasis added]."

Humboldt and Bonpland carried passports issued by the court of King Carlos IV of Spain, giving them permission to conduct scientific studies throughout the Spanish Empire, which then stretched from California to Texas in North America and south to the tip of South America. They had complete access to a vast area of the earth's surface that was essentially unexplored scientifically, and they put that access to productive use. Because their discoveries were so numerous and their explorations so thorough, Simón Bolívar, the liberator of most of Spanish America, referred to Humboldt as "the discoverer of the New World."

Humboldt's expedition was one of the most ambitious scientific explorations of the age (Helfrich 2004). During the course of their expedition, Humboldt and Bonpland traveled nearly 10,000 km through South and North America. They traveled on foot, by canoe, and on horseback, visiting latitudes ranging between 12°S and 52°N. They also climbed to nearly 5,900 m on the slopes of Chimborazo, the highest ascent by anyone in history up to that time (fig. 22.1).

The physical feats of their expedition, however, never took precedent over their scientific purpose. For instance, on their climb of Chimborazo, they faced the uncertain dangers of high altitude. Yet, as blood oozed from their lips and gums, Humboldt and Bonpland recorded the altitudinal distributions of plants and animals. Later, Humboldt organized their observations of climate and plant distributions into ingenious visual representations of plant geography. What he did not accomplish, he inspired others to do. One of those inspired to follow in Humboldt's footsteps was Charles Darwin. Darwin said that his reading of Humboldt's expedition to South America set the course of his whole life.

Robert H. MacArthur (1972) defined **geographic ecology** as the "search for patterns of plant and animal life that can be put on a map." MacArthur's map might include an archipelago of islands, a region, or a series of continents. Today we would add to MacArthur's definition that geographic ecology is the study of ecological structure and process at large geographic scales. Though geographic ecology began long before MacArthur, with explorers such as Humboldt, Darwin, and Wallace, MacArthur put an indelible quantitative stamp on the field when he and E. O. Wilson published their first models of island biogeography.

The development of geographic ecology continues as new generations of scientists equipped with a diversity of tools,

Figure 22.1   On the slopes of Chimborazo, a 6,310 m high volcanic peak in the Andes Mountains of Ecuador, Alexander von Humboldt and Aimé Bonpland meticulously recorded the altitudinal distributions of plants.   Norman Owen Tomalin/Bruce Coleman/Avalon

both ancient and modern, search for the elusive patterns that can be put on maps. There is a heightened sense of urgency, however, as the prospect of global climate change presses us to understand the forces controlling large-scale patterns of biological diversity and the geographic ranges of species. The breadth of geographic ecology is as vast as its subject. Consequently, we concentrate our discussions in this chapter on just a few aspects of the field: island biogeography, latitudinal patterns of species diversity, and the influences of large-scale regional and historical processes on biological diversity.

22.1 Area, Isolation, and Species Richness

LEARNING OUTCOMES

After studying this section you should be able to do the following:

22.3 Discuss the relationships of area and isolation to species richness on islands.

22.4 Interpret graphical evidence for the influences of island area and island isolation on species richness.

22.5 Explain why studies of species richness on oceanic islands can be applied to understanding patterns of species richness on habitat patches, such as mountains and lakes, on continents.

On islands and habitat patches on continents, species richness increases with area and decreases with isolation.

A quantitative relationship between area and number of species was first developed by Olof Arrhenius (1921), a pioneer in the area of geographic ecology. Arrhenius made his observations on islands near Stockholm, Sweden, where he worked within several plant communities that he identified by names such as *herb–Pinus wood and shore association.* He counted the number of species within areas of various sizes and then developed a mathematical description of the relationship between area sampled and number of plant species. However, Arrhenius worked at scales much smaller than the geographic focus of this chapter. To see the first quantitative work on geographic patterns, we have to move to a later time.

Island Area and Species Richness

Frank Preston (1962a) examined the relationship between number of species and the area of islands in the West Indies. As shown in figure 22.2*a,* the fewest bird species live on the smallest islands and the most on the largest islands, which are Cuba and Hispaniola. The relationship between island area and number of species is not just a property of bird assemblages. Sven Nilsson, Jan Bengtsson, and Stefan Ås (1988) explored patterns of species richness among woody plants, carabid beetles, and land snails on 17 islands in Lake Mälaren, Sweden. The islands ranged in area from 0.6 to 75 ha and all were forested. The researchers were careful to choose islands that showed few or no signs of human disturbance. One of the results of their study was that island area was the best single predictor of species richness in all three groups of organisms.

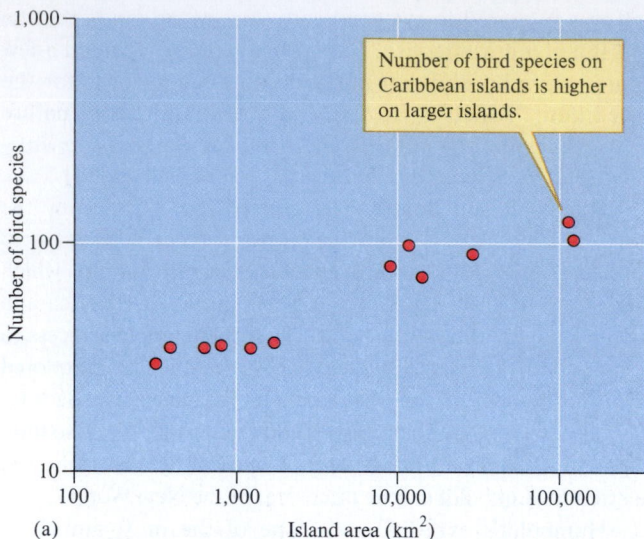

(a)

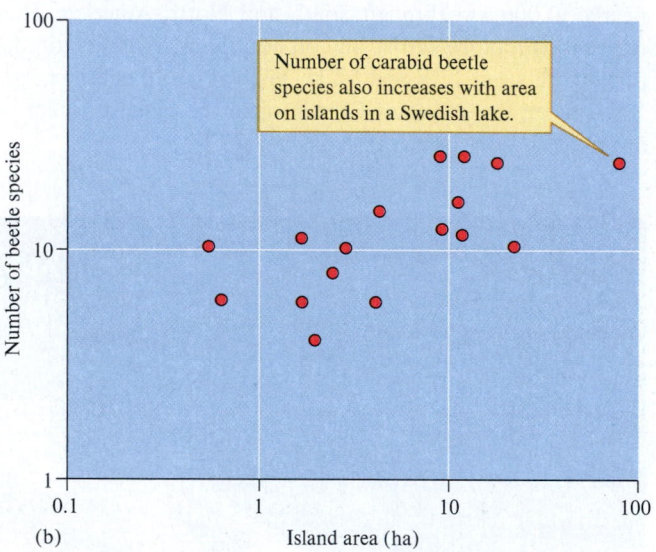

(b)

Figure 22.2   Relationship between island area and number of species (data from Preston 1962a; Nilsson, Bengtsson, and Ås 1988).

Figure 22.2*b* shows the relationship found between island area and number of carabid beetles.

When most of us think of islands, the picture that generally comes to mind is a small bit of land in the middle of an ocean. However, many habitats on continents are so isolated that they can be considered islands.

Habitat Patches on Continents: Mountain Islands

The many isolated mountain ranges that extend across the southwestern regions of North America are now continental islands. During the late Pleistocene, 11,000 to 15,000 years ago, forest and woodland habitats extended unbroken from the Rocky Mountains to the Sierra Nevada in California. Then, as the Pleistocene ended and the climate warmed, forest and alpine habitats contracted to the tops of the high mountains scattered across the American Southwest. As montane habitats retreated to higher elevations, woodland, shrubland, grassland, or desert scrub vegetation invaded the lower elevations. As a consequence of these changes, once-continuous forest and alpine vegetation was converted to a series of island-like habitat patches associated with mountains and therefore called *montane*.

As montane vegetation contracted to mountaintops, montane animals followed. Mark Lomolino, James H. Brown, and Russell Davis (1989) studied the diversity of montane mammals on isolated mountains in the American Southwest. They focused on the distributions of 26 species of nonflying forest mammals that occur on 27 montane islands. They chose mountain ranges that had been studied thoroughly enough so that their mammal faunas were well known. The list of species, which included shrews, ermine, squirrels, chipmunks, and voles, was limited to species that show a strong association with montane environments.

The team found that montane mammal richness was positively correlated with habitat area. As figure 22.3 shows, the area of the 27 montane islands ranged from less than 7 km² to over 10,000 km², while the number of montane mammals on

them ranged from 1 to 16. The strongly positive relationship between montane area and richness of montane mammal species in the American Southwest described by Lomolino and his colleagues has held up in the face of repeated testing and analysis (Lawlor 1998).

Lakes as Islands

Lakes can also be considered as habitat islands—aquatic environments isolated from other aquatic environments by land. However, lakes differ widely in their degree of isolation. Seepage lakes, which receive no surface drainage, are completely isolated, while drainage lakes, which have stream inlets and/or outlets, are less isolated (see chapter 21).

William Tonn and John Magnuson (1982) studied patterns of species composition and richness among fish inhabiting lakes in northern Wisconsin. They focused their research on 18 lakes in the Northern Highlands Lake District of Wisconsin and Michigan. The study was conducted in Vilas County, Wisconsin, which includes over 1,300 lakes (fig. 22.4). With so many lakes at their disposal, Tonn and Magnuson could match lakes carefully for a variety of characteristics. All 18 study lakes had similar bottom substrates and similar maximum depths. However, the lakes spanned a considerable range of surface area (2.4–89.8 ha). Ten of the lakes were drainage lakes or spring fed and eight lakes were seepage lakes. Eight lakes had a history of low oxygen content during winter.

Tonn and Magnuson collected a total of 23 species, 22 in summer and 18 in winter. If we combine their winter and summer collections on each lake and plot total species richness against area, there is a significant positive relationship (see fig. 22.4). Once again, we see that the number of species increases with the area of an insular environment. However, these researchers worked with a single lake district. Is there a relationship between lake area and diversity when lakes from several regions are included in the analysis?

Clyde Barbour and James H. Brown (1974) studied patterns of species richness across a worldwide sample of 70 lakes. The lakes in their sample ranged in area from 0.8 to 436,000 km², while the number of fish species ranged from

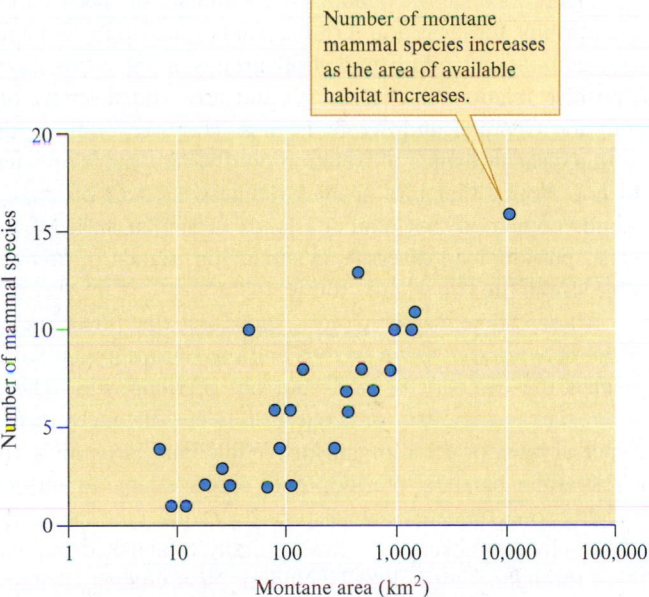

Figure 22.3 Area of montane habitat and number of montane mammal species on isolated mountain ranges in the American Southwest (data from Lomolino, Brown, and Davis 1989).

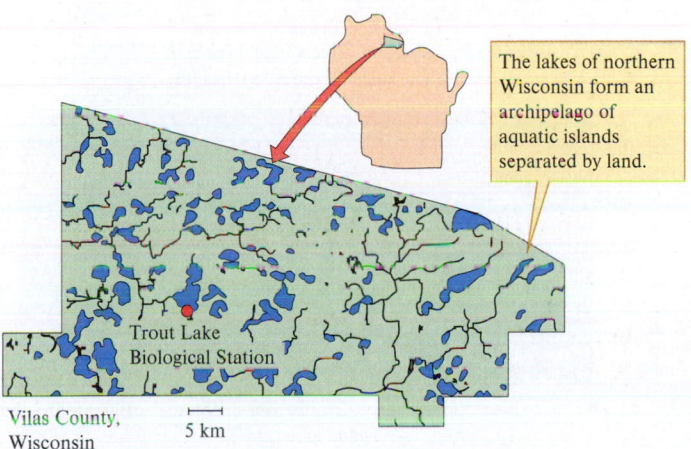

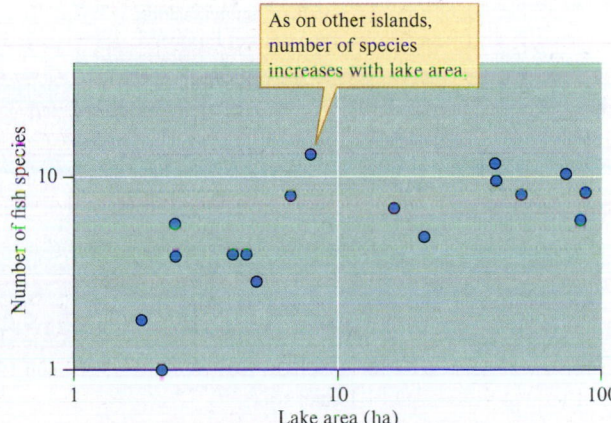

Figure 22.4 Lake area and number of fish species in lakes of northern Wisconsin (data from Tonn and Magnuson 1982).

5 to 245. Barbour and Brown also found a positive relationship between area and fish species richness.

Island Isolation and Species Richness

There is often a negative relationship between the isolation of an island and the number of species it supports. However, because organisms differ substantially in dispersal rates, an island that is very isolated for one group of organisms may be completely accessible to another group.

Marine Islands

Robert MacArthur and Edward O. Wilson (1963) found evidence that isolation reduces bird diversity on Pacific islands. In figure 22.5, islands less than 800 km from New Guinea, a potential source of colonists to the other islands, are represented by gold dots; islands greater than 3,200 km from New Guinea are represented by red dots; islands at intermediate distances from New Guinea are represented by blue dots. How does figure 22.5 indicate an effect of isolation on species richness? Compare the relative numbers of species on islands of approximately equal area but that lie at different distances from New Guinea. On average, those closest to New Guinea support a larger number of bird species than those at intermediate or far distances.

Comparative studies of diversity patterns on islands remind us that different organisms have markedly different dispersal abilities. Mark Williamson (1981) summarized the data for the relationship between island area and species richness for various groups of organisms inhabiting the Azore and Channel Islands. The Azore Islands lie approximately 1,600 km west of the Iberian Peninsula, while the Channel Islands are very near the coast of France. Though vastly different in distance from mainland areas, both island groups experience moist temperate climates and have biotas that are of European origin. Consequently, a comparison of their biotas should reveal the potential influence of isolation on diversity.

Figure 22.6 shows Williamson's summary of species area relationships for ferns and fernlike plants (pteridophytes) and land- and water-breeding birds. Both groups of organisms show a positive relationship between island area and diversity on both the Channel and Azore Islands. However, while birds show a clear influence of isolation on diversity, pteridophytes do not. Notice that bird species richness is lower on Azore Islands compared to Channel Islands of similar size. Meanwhile, pteridophyte diversity is similar on islands of comparable size in the two island groups.

These differences in pattern show that the 1,600 km of ocean between the Azore Islands and the European mainland reduces the diversity of birds but not pteridophytes. These differences in the effect of isolation reflect differences in the dispersal rates of these organisms. While land birds must fly across water barriers, pteridophytes produce large quantities of light spores that are easily dispersed by wind. One species of pteridophyte, bracken fern, has naturally established populations throughout the globe, including New Guinea, Britain, Hawaii, and New Mexico. When we consider the potential effects of isolation on diversity, we must also consider the dispersal capabilities of the study organisms.

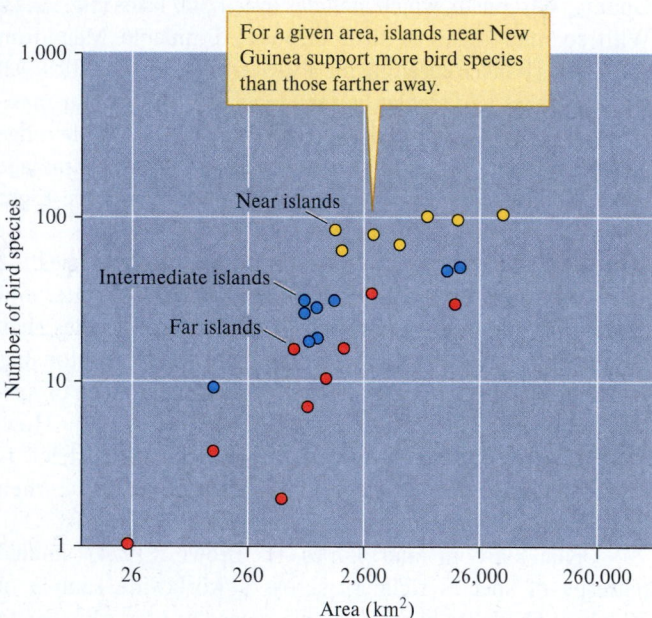

Figure 22.5   Distance from New Guinea and bird species richness on Pacific islands (data from MacArthur and Wilson 1963).

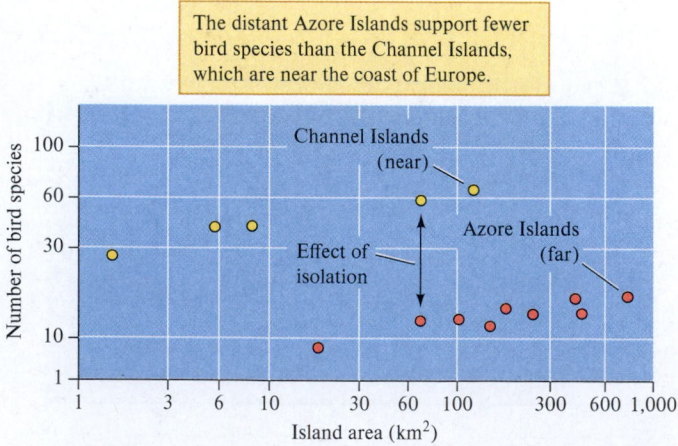

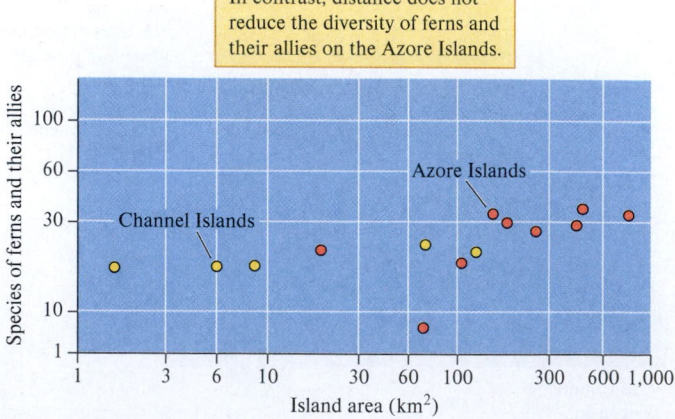

Figure 22.6   Influence of isolation on diversity of birds and ferns and their allies on the Channel and Azore islands (data from Williamson 1981).

Isolation and Habitat Islands on Continents

Has an isolation effect been observed across habitat islands on continents? Lomolino, Brown, and Davis (1989), whom we have already discussed, found a strong negative relationship between isolation and the number of montane mammal species living on mountaintops across the American Southwest. They measured isolation as the distance of a mountain range from potential sources of colonists. Colonists can emigrate from the southern Rocky Mountains, with 23 montane mammal species, or from the Mogollon Rim, with 16 montane mammal species. For mountains closer to the southern Rocky Mountains, isolation was measured as the distance from the mountain island to the nearest point of the Rocky Mountains. If the mountain was closer to the Mogollon Rim, the researchers used a composite measure of distance that weighted the distances to the Mogollon Rim and the southern Rocky Mountains by the diversities of their montane mammals. Figure 22.7 shows the results of their analysis, which shows a strong negative relationship between isolation (distance to species source areas) and species richness of montane mammals. The analysis clearly shows that both area and isolation affect the diversity of mammals on habitat patches on continents.

What does the negative effect of isolation on these mountain mammal populations indicate about local assemblages? The higher diversity of mountain islands near the southern Rocky Mountains and/or to the Mogollon Rim suggests that montane mammals continue to disperse across the intervening woodlands and grasslands. Without immigration, near and far mountains of equal area would support similar numbers of montane mammal species. This result suggests that the diversity of organisms on islands is maintained by a dynamic process that involves ongoing colonization. A dynamic rather than a static view of island diversity is the foundation of an equilibrium model of island biogeography that we discuss in the next section.

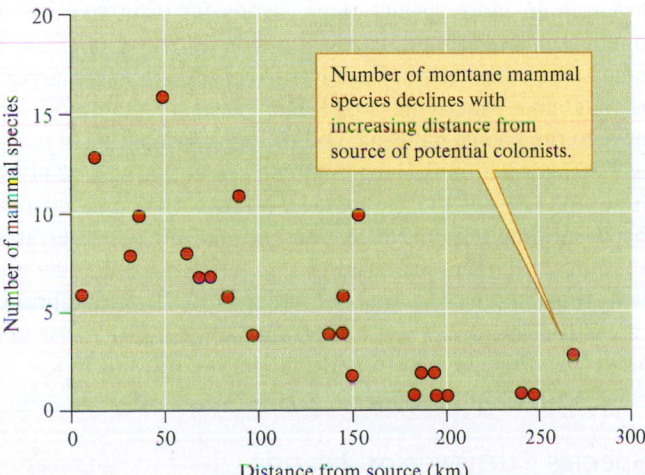

Figure 22.7   Distance from large montane areas and number of montane mammal species on isolated mountain ranges of the American Southwest (data from Lomolino, Brown, and Davis 1989).

Concept 22.1 Review

1. In chapter 21, we discussed the influences of habitat fragmentation from the perspective of populations (see figs. 21.11 and 21.13). Drawing from the information in this section, how do you think fragmentation will affect species richness—for instance, the number of bird species living in forest fragments?

2. In figure 22.7, the number of mammal species on the isolated mountain ranges varies greatly for a given distance from large montane areas, for instance, at a distance of 150 km. What is the likely source of much of this variation?

22.2   The Equilibrium Model of Island Biogeography

LEARNING OUTCOMES

After studying this section you should be able to do the following:

22.6   Summarize the equilibrium model of island biogeography.

22.7   Explain how the observation of species turnover on the California Channel Islands supports the equilibrium model of island biogeography.

22.8   Discuss experimental evidence in support of the equilibrium model of island biogeography.

Species richness on islands can be modeled as a dynamic balance between immigration and extinction of species. The examples we just reviewed show clear relationships between species richness and island area and isolation. When confronted with such a pattern, scientists look for explanatory mechanisms. What mechanisms might increase species richness on large islands and reduce richness on small and isolated islands? MacArthur and Wilson (1963, 1967) proposed a model that explained patterns of species diversity on islands as the result of a balance between rates of immigration and extinction (fig. 22.8). This model is called the **equilibrium model of island biogeography.**

Figure 22.8 shows that the model presents rates of immigration and extinction as a function of numbers of species on islands. How might rates of immigration and extinction be influenced by the numbers of species on an island? To answer this question, we need to understand what MacArthur and Wilson meant by rates of immigration and extinction. They defined the *rate of immigration* as the rate of arrival of *new* species on an island. *Rate of extinction* was the rate at which species went extinct on the island. MacArthur and Wilson reasoned that rates of immigration would be highest on a new island with no organisms, since every species that arrived at the island would be new. Then, as species began to accumulate on an island, the rate of immigration would decline, since fewer and fewer arrivals would be new species. They called the point at which the immigration line touches the

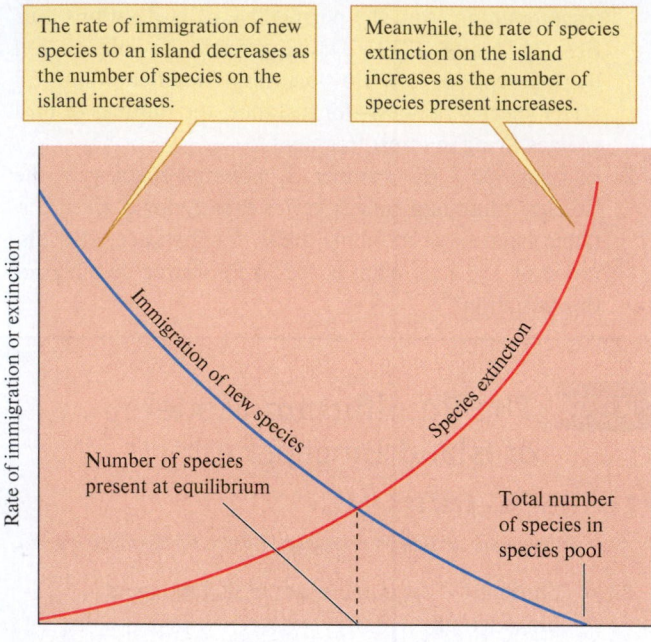

Figure 22.8 Equilibrium model of island biogeography (data from MacArthur and Wilson 1963).

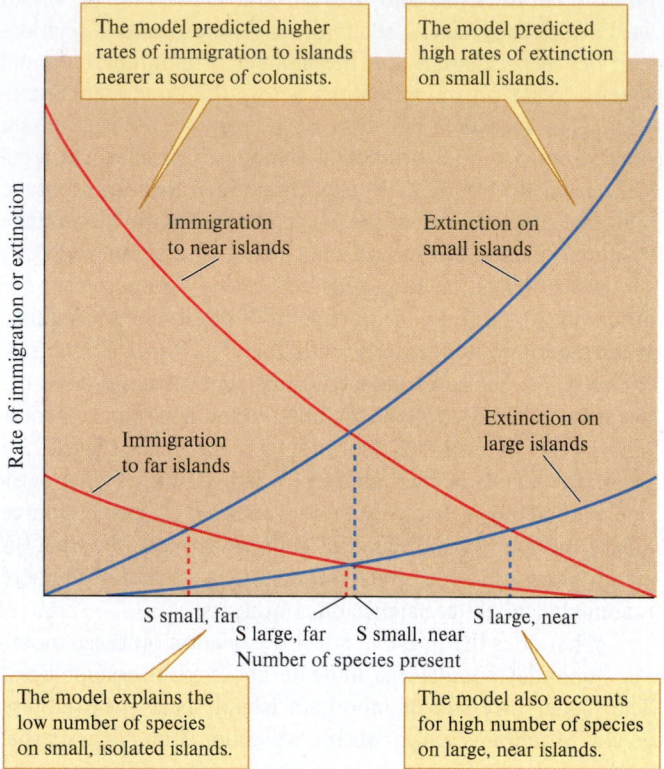

Figure 22.9 Island distance and area and rates of immigration and extinction (data from MacArthur and Wilson 1963).

horizontal axis P because it is the point representing the entire "pool" of species that might immigrate to the island.

How might numbers of species on an island affect the rate of extinction? MacArthur and Wilson predicted that the rate of extinction would rise with increasing numbers of species on an island for three reasons: (1) the presence of more species creates a larger pool of potential extinctions; (2) as the number of species on an island increases, the population size of each must diminish; and (3) as the number of species on an island increases, the potential for competitive interactions between species will increase.

Since the immigration line falls and the extinction line rises as number of species increases, the two lines must cross as shown in figure 22.8. What is the significance of the point where the two lines cross? The point where the two lines cross predicts the number of species that will occur on an island. Thus, the equilibrium model represents the diversity of species on islands as a result of a dynamic balance between immigration and extinction.

MacArthur and Wilson used the equilibrium model to predict how island size and isolation should affect rates of immigration and extinction. They proposed that the rate of immigration is mainly determined by an island's distance from a source of immigrants, for example, the distance of an oceanic island from a mainland. They proposed that rates of extinction on islands would be determined mainly by island size. These predictions are represented in figure 22.9. Notice that the figure predicts that large, near islands will support the greatest number of species, whereas

small, far islands will support the lowest number of species. The model predicts that small, near islands and large, far islands will support intermediate numbers of species.

The predictions of the equilibrium model of island biogeography are consistent with the patterns of island diversity reviewed in the previous section. Large islands hold more species than small islands, and islands near sources of immigrants hold more species than islands far from sources of immigrants. We should expect the equilibrium model to be consistent with known variation in species richness across islands, since MacArthur and Wilson designed their model to explain the known patterns. Did the equilibrium model make any new predictions? The main new predictions were (1) that island diversity is the outcome of a highly dynamic balance between immigration and extinction and (2) that the rates of immigration and extinction are determined mainly by the isolation and area of islands. In other words, the equilibrium model predicts that the species composition on islands is not static but changes over time. Ecologists call this change in species composition **species turnover.**

Species Turnover on Islands

Turnover of bird species was demonstrated on the California Channel Islands by Jared Diamond (1969). Diamond surveyed the birds of the nine California Channel Islands in 1968,

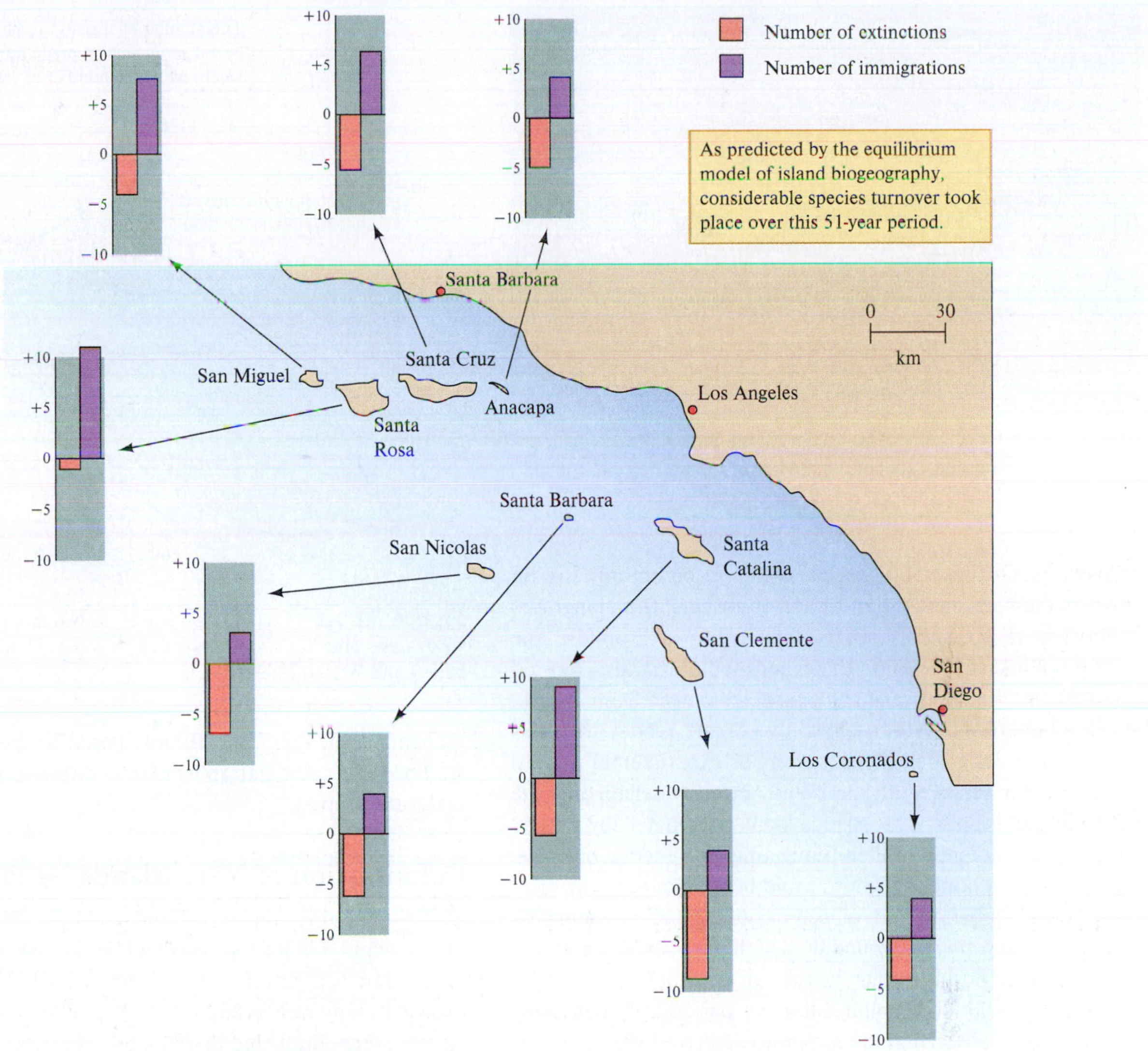

Figure 22.10   Extinction and immigration of bird species on the California Channel Islands between 1917 and 1968 (data from Diamond 1969).

approximately 50 years after an earlier survey by A. B. Howell. The islands range in area from less than 3 to 249 km² and lie 12 to 61 km from the coast of southern California (fig. 22.10). Howell had thoroughly censused all of the islands except for San Miguel and Santa Rosa Islands, where he had difficulty getting permission to do bird surveys. In his later study, Diamond had full access to all the islands and was able to survey all land and water birds.

The results of Diamond's study support the equilibrium model of island biogeography. The number of bird species inhabiting the California Channel Islands remained almost constant over the 50 years between the two censuses. However, this stability in numbers of species was the result of an approximately equal number of immigrations and extinctions on each of the islands (see fig. 22.10). Diamond's study is an excellent example of how theory can guide field ecology. He discovered the dynamics underlying the diversity of birds on the California Channel Islands because he went out to test the MacArthur-Wilson equilibrium model of island biogeography. Additional insights into this model have been provided by experiments.

Experimental Island Biogeography

As Diamond conducted his surveys of the California Channel Islands, Daniel Simberloff and Edward O. Wilson were engaged in experimental studies of mangrove islands in the Florida Keys (Simberloff and Wilson 1969; Wilson and Simberloff 1969). The Florida Keys support very large stands of mangroves, which are dominated by the red mangrove, *Rhizophora mangle*. Many of these stands occur as small islands that lie hundreds of meters from the nearest large patch of mangroves (fig. 22.11). Simberloff and Wilson chose eight of these small mangrove islands for their experimental study. Their study islands were roughly circular and varied from 11 to 18 m in diameter and 5 to 10 m in height. The distance of islands from large areas of mangroves that could act as a source of colonists varied from 2 to 1,188 m.

The main fauna inhabiting the small mangrove islands of the Florida Keys are arthropods, chiefly insects. Simberloff and Wilson estimated that of the approximately 4,000 species of insects in the Florida Keys, about 500 species inhabit

Figure 22.11   The mangrove islands in the Florida Keys, which number in the thousands, are convenient places to test the equilibrium model of island biogeography.   H.W. Kitchen/Science Source

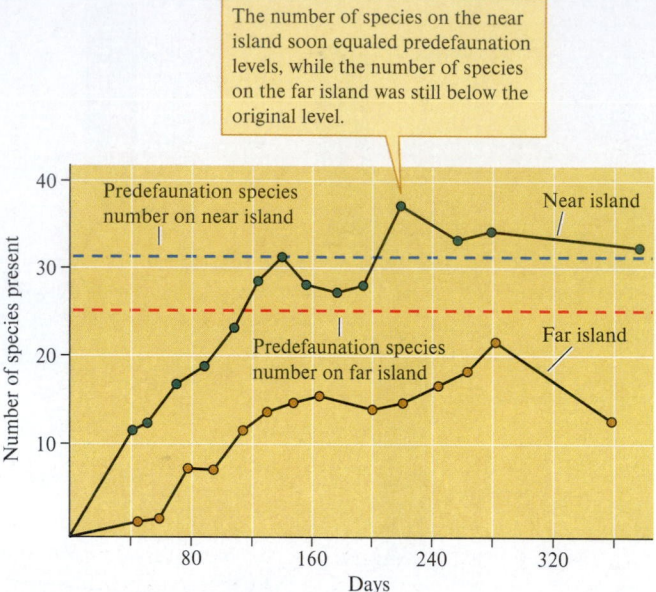

The number of species on the near island soon equaled predefaunation levels, while the number of species on the far island was still below the original level.

Figure 22.12   Colonization curves for two mangrove islands that were "near" and "far" from sources of potential colonists (data from Simberloff and Wilson 1969).

mangroves. Of these 500 species, about 75 commonly live on small mangrove islands. In addition to insects, the mangroves supported 15 species of spiders and other arthropods. The number of insect species on the experimental islands averaged 20 to 40 and the number of spider species ranged from 2 to 10.

Simberloff and Wilson chose two of the islands to act as controls and designated the six others as experimental islands. They carefully surveyed all the islands prior to defaunating the experimental islands. The islands were defaunated by enclosing them with a tent and then fumigating with methyl bromide. Fumigating was done at night to avoid heat damage to the mangrove trees. Simberloff and Wilson examined the trees immediately after fumigating and found that, with the possible exception of some wood-boring insect larvae, all arthropods had been killed. They followed recolonization by periodically censusing the arthropods on each island for approximately 1 year.

The number of species recorded on the two control islands was virtually identical at the beginning and end of the experiment. Though the number did not change significantly over the period of study, Simberloff and Wilson reported that species composition changed considerably. In other words, there had been species turnover on the control islands, a result consistent with the equilibrium model of island biogeography.

The equilibrium model was also supported by the recolonization studies of the experimental islands. Following defaunation, the number of arthropod species increased on all of the islands. All the islands, except the farthest island, eventually supported about the same number of species as they did prior to defaunation (fig. 22.12). Again, however, the composition of arthropods on the islands was substantially different, indicating species turnover. Species turnover is also indicated by the colonization histories of individual islands, which include many examples of species appearing and then disappearing from the community.

Island colonization can be followed either by removing the organisms from existing islands, as Simberloff and Wilson did when they defaunated their mangrove islands, or by creating new islands. Many new islands formed in a large lake in southern Sweden when the level of the lake was dropped at the end of the nineteenth century. Fortunately, some biologists

recognized the rare opportunity offered by the new islands and studied their colonization by plants. These studies have continued for a century.

Colonization of New Islands by Plants

The site of this long-term study is Lake Hjälmaren, which covers about 478 km^2 in Sweden (fig. 22.13). The level of Lake Hjälmaren was lowered 1.3 m between 1882 and 1886 and exposed many new islands. The first plant surveys of the new islands were conducted in 1886, and the islands were surveyed again in 1892, from 1903 to 1904, from 1927 to 1928, and from 1984 to 1985. Håkan Rydin and Sven-Olov Borgegård (1988) summarized the earlier surveys of these new islands and conducted their own surveys in 1985. The result was a unique long-term record of the colonization of 40 islands.

The study islands vary in area from 65 m^2 to over 25,000 m^2 and support a limited diversity of plants. Rydin and Borgegård estimated that approximately 700 species of plants occur around Lake Hjälmaren. Of these 700 plant species, the number recorded on individual islands during the first century of their existence varied from 0 to 127. As expected, this variation in species richness correlated positively with island area over the entire history of the islands and accounted for 44% to 85% of the variation in species richness among islands. Measures of island isolation accounted for 4% to 10% of the variation in plant species richness among islands through the 1903–04 census. Island isolation did not account for significant variation in species richness among islands in subsequent censuses.

Rydin and Borgegård used the censuses of 30 islands to estimate rates of plant immigration and extinction (see fig. 22.13). The historical record documents many immigrations and extinctions. There was a slight excess of immigrations over extinctions on small- and medium-sized islands during the

Plant species richness on the islands of Lake Hjälmaren, like arthropod richness on the mangrove islands studied by Simberloff and Wilson, appears to be maintained by a dynamic interplay between immigration and local extinction. Many studies support the basic predictions of the equilibrium model of island biogeography. However, many questions remain.

For instance, why do larger islands support more species? Is the greater species richness on large islands due to a direct effect of area, or do large islands support higher species richness because they include a greater diversity of habitats? Rydin and Borgegård found that measures of habitat diversity on the study islands accounted for only 1% to 2% of the variation in plant species richness. However, they point out that while some large islands with few habitats support low numbers of plant species, some small islands, with diverse habitats, support higher species richness than would be expected on the basis of area alone. The researchers point out that it is very difficult to separate the effects of habitat diversity from the effects of area. As we shall see in the next example, there is at least one experiment that came close to demonstrating that species richness on islands can be directly affected by area.

Manipulating Island Area

Daniel Simberloff (1976) tested the effect of island area on species richness experimentally. He surveyed the arthropods inhabiting nine mangrove islands that ranged in area from 262 to 1,263 m². The distance of these islands from large areas of mangrove forest ranged from 2 to 432 m. The islands were up to five times the size of the mangrove islands that he and Wilson had fumigated in their earlier study of recolonization, and therefore contained a larger number of arthropod species.

Simberloff kept one island as a control, while reducing the area of the eight other islands by 32% to 76%. Island area was reduced during low tide by removing whole sections of the islands. Workers cut mangroves off below the high tide level and loaded the cut trees and branches on a barge. They then moved the cut material away from the island, where they sank it into deeper water (green mangrove wood sinks). Simberloff reduced the area of four experimental islands twice and the area of the other four experimental islands once only.

The results of Simberloff's experiment show a positive relationship between area and species richness. In all cases where island area was reduced, species richness decreased (fig. 22.14). Meanwhile, species richness on the control island, which was not changed in area, increased slightly. Additional insights are offered by the contrasting histories of islands whose areas were reduced once and those whose areas were reduced twice. For instance, the area of Mud 2 island was reduced from 942 to 327 m² and the richness of its arthropod fauna fell from 79 to 62 species. The area of Mud 2 was not reduced further, and its arthropod richness remained almost constant. Meanwhile, the islands whose area was reduced twice lost species with each reduction in area. Simberloff's results showed that area itself, without increased habitat heterogeneity, has a positive influence on species richness.

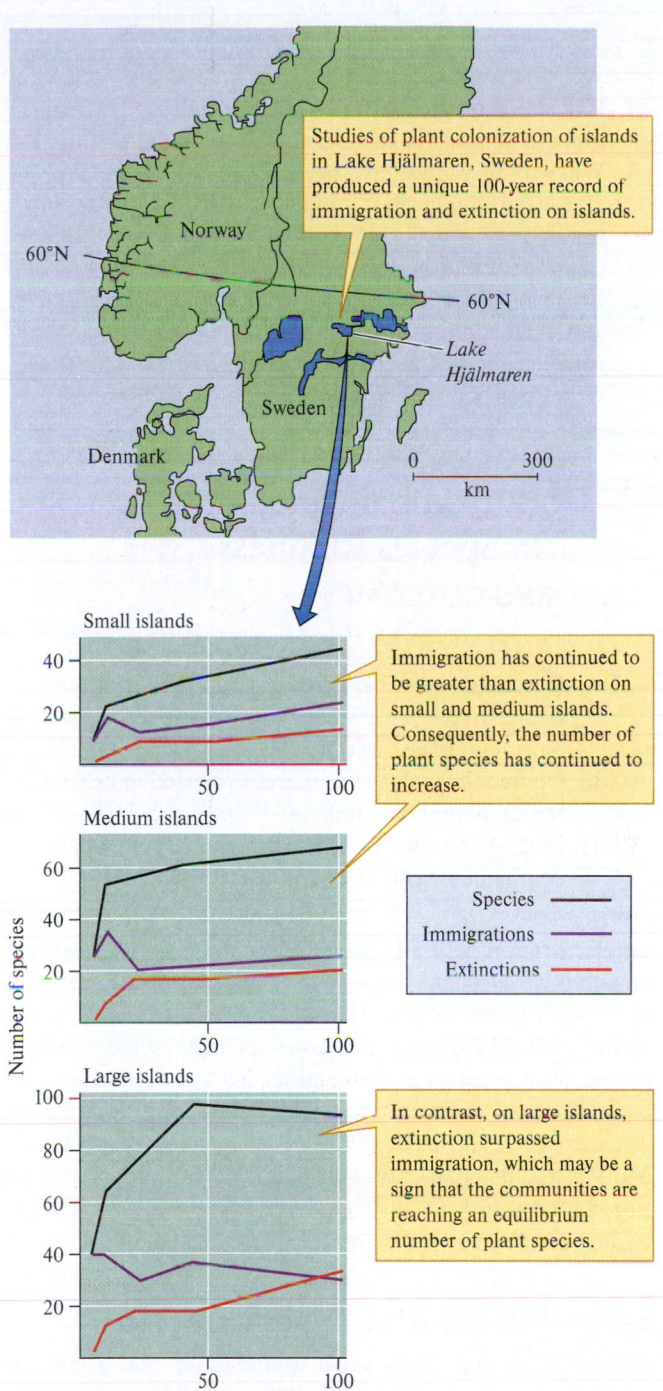

Figure 22.13 Species number, immigration, and extinction on 30 islands in Lake Hjälmaren, Sweden (data from Rydin and Borgegård 1988).

entire 100 years of record. What do these higher rates of immigration indicate? They show that small and medium islands continue to accumulate species. In contrast, large islands attained approximately equal rates of immigration and extinction sometime between 1928 and 1985. Over this period, approximately 30 plants became extinct on each large island and another 30 new species arrived. In other words, it appears that the number of species may have reached equilibrium on large islands.

The observed patterns of colonization were consistent with the predictions of the equilibrium model of island biogeography.

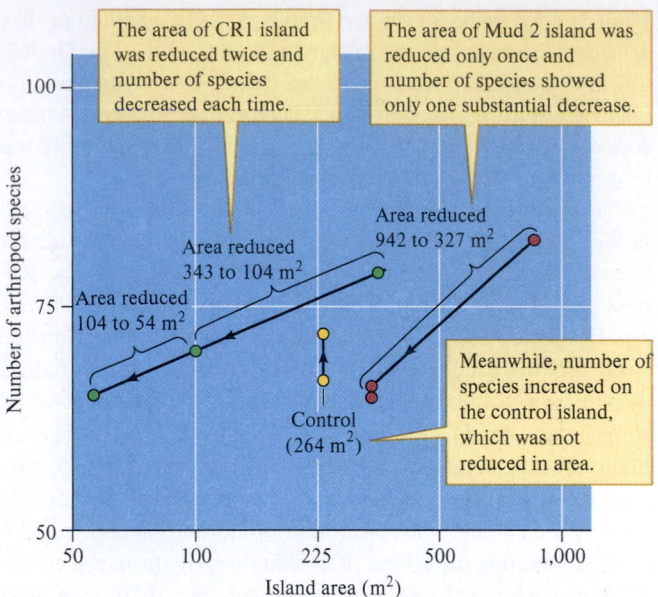

The area of CR1 island was reduced twice and number of species decreased each time.

The area of Mud 2 island was reduced only once and number of species showed only one substantial decrease.

Area reduced 942 to 327 m²

Area reduced 343 to 104 m²

Area reduced 104 to 54 m²

Meanwhile, number of species increased on the control island, which was not reduced in area.

Control (264 m²)

Figure 22.14 Effect of reducing mangrove island area on number of arthropod species (data from Simberloff 1976).

Island Biogeography Update

The equilibrium theory of island biogeography has had a major influence on the disciplines of biogeography and ecology. However, much has been discovered in the 50-plus years since MacArthur and Wilson proposed their theory. For instance, James Brown and Astrid Kodric-Brown (1977) showed how higher rates of immigration to near islands can reduce extinction rates. As a consequence, we now know that, contrary to the original MacArthur-Wilson model, island distance from sources of colonists can influence rates of extinction. Similarly, Mark Lomolino (1990) also extended the original model when he proposed the target hypothesis, demonstrating that island area can have a significant effect on rates of immigration to islands. Brown and Lomolino (2000) pointed out that we have also discovered that species richness is not in equilibrium on many islands. In addition, we now know that species richness on islands is affected by differences among species groups in their speciation, colonization, and extinction rates. And perhaps most significantly, area and isolation are only two of several environmental factors that affect species richness on islands. Brown and Lomolino suggest that we may be on the eve of another revolution in theories that will replace the MacArthur-Wilson model. If so, it will be the result of research largely inspired by their theory as well as by our fascination for the islands themselves.

Experiments on islands, such as those of Simberloff and Wilson, demonstrate the value of an experimental approach to answering ecological questions. However, there are important ecological patterns that occur over such large scales that experiments are virtually impossible. The ecologists who study these large-scale patterns must rely on other approaches. In the next section, we discuss one of these important large-scale patterns: latitudinal variation in species richness.

Concept 22.2 Review

1. Why are virtually all estimates of immigration and extinction rates on islands underestimates of the true rates?
2. What result would have been grounds for Diamond to reject the equilibrium model of island biogeography based on his studies of the California Channel Islands (see fig. 22.10)?
3. In the course of studies by Simberloff and Wilson (1969) and Simberloff (1976), several mangrove islands were defaunated and several were partially destroyed to reduce island area. Do such experiments raise ethical issues?

22.3   Latitudinal Gradients in Species Richness

LEARNING OUTCOMES

After studying this section you should be able to do the following:

22.9 Describe the typical latitudinal gradient in species richness seen in various groups of organisms, such as birds, butterflies, and trees.

22.10 Outline the major hypotheses proposed to explain typical latitudinal gradients in species richness.

22.11 Discuss the contribution of area to latitudinal and continental patterns of species richness.

Species richness generally increases from middle and high latitudes to the equator. Most groups of organisms are more species-rich in the tropics. This well-known increase in species richness toward the equator became apparent by the middle of the eighteenth century as taxonomists, led by Carolus Linnaeus, described tropical species sent back to Europe by explorers. These explorers and later naturalists, such as Humboldt, Darwin, and Wallace, described overwhelming biological diversity in the tropics. Today, two and a half centuries later, we are still trying to catalog this diversity and do not even know its full extent.

Latitudinal Gradient Hypotheses

Figures 22.15 and 22.16 show examples of how plant species richness (Reid and Miller 1989) and bird species richness (Dobzhansky 1950) decrease toward the poles. Despite some exceptions to the equatorial peak in species diversity (fig. 22.17), the pattern of increased numbers of species in the tropics is pervasive and dramatic. This pattern challenges ecologists for an explanation. Many mechanisms have been proposed to explain latitudinal gradients in species richness. James H. Brown (1988) grouped the hypotheses proposed to explain geographic gradients in species richness into six categories.

1. Time Since Perturbation

The *time since perturbation hypothesis* proposes that there are more species in the tropics because the tropics are older and they are disturbed less frequently. That is, more species occur in the tropics because (1) there has been more time for

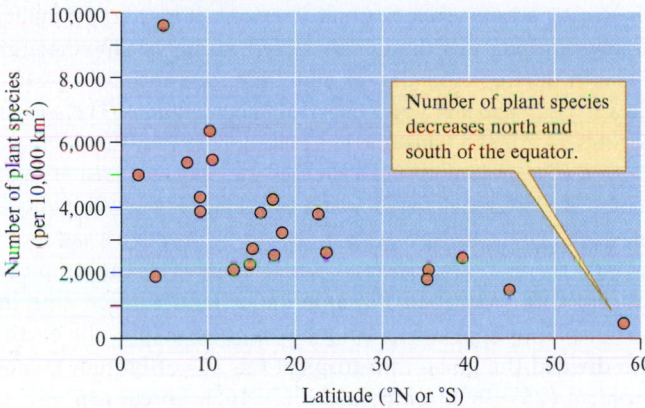

Figure 22.15   Variation in number of vascular plant species with latitude in the Western Hemisphere (data from Reid and Miller 1989).

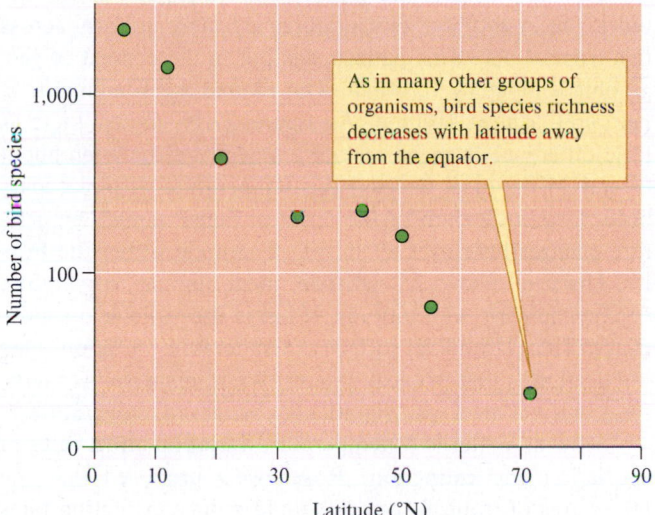

Figure 22.16   Latitudinal variation in number of bird species from Central to North America (data from Dobzhansky 1950).

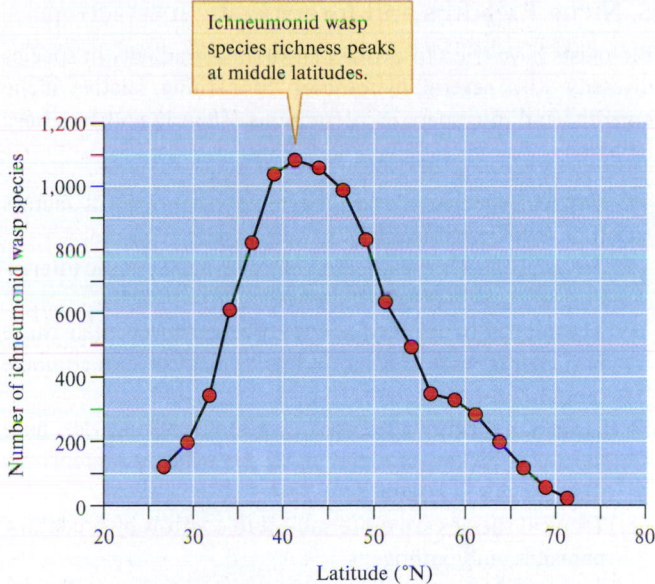

Figure 22.17   An exception to the general decline in species number with latitude: latitudinal variation in ichneumonid wasp species richness (data from Janzen 1981).

speciation and (2) less frequent perturbation reduces extinction rates. The proponents of this hypothesis assume that the tropics have remained relatively stable whereas middle and high latitudes have been repeatedly disrupted by the advance and retreat of glaciers. However, in chapter 16 we saw that intermediate levels of disturbance may increase local diversity. In counterpoint to this hypothesis, Joseph Connell (1978) proposed that the extraordinary diversity of tropical rain forests and coral reefs is maintained by frequent disturbance.

2. Productivity

The authors of the *productivity hypothesis* observe that two of the most diverse environments on earth, coral reefs and tropical rain forests, are also extraordinarily productive. This hypothesis proposes that high productivity contributes to high species richness. It assumes that with more energy to divide among organisms, specialized consumers will have larger populations. Since larger populations generally have lower probabilities of extinction than smaller populations, extinction rates should be lower in more productive environments. However, Brown points out that this hypothesis

must somehow explain the reduction in species diversity that accompanies nutrient enrichment and increased primary production (see chapter 16).

3. Environmental Heterogeneity

The *environmental heterogeneity hypothesis* proposes that the tropics contain more species because they are more heterogeneous than temperate regions. Daniel Janzen (1967c) and George Stevens (1989) pointed out that, compared to high-latitude species, most tropical species occur in far fewer environments along altitudinal and latitudinal gradients.

Michael Rosenzweig (1992), however, cautioned that we cannot consider habitat structure independently of the organisms living in a region. Species within more diverse communities tend to subdivide the environment more finely. Consequently, species diversity and habitat heterogeneity are not necessarily independent factors. For instance, when George Cox and Robert Ricklefs (1977) estimated the number of habitats used by birds in Panama versus four Caribbean islands, they found an inverse relationship between numbers of species and the number of habitats used by the birds. In other words, birds appeared to restrict their habitat use in the presence of more species.

4. Favorableness

The *favorableness hypothesis* proposes that the tropics provide a more favorable environment than do high latitudes. As we saw in chapter 2, the variation in temperature in high-latitude environments is much greater than in tropical environments. Biologists have proposed that the correspondence between low diversity and the physical variability of high latitudes is no accident. While many species are well adapted to physically harsh conditions, most species on earth are not. Biologists have proposed that physically extreme environments restrict the diversity of organisms.

5. Niche Breadths and Interspecific Interactions

Biologists have tried to explain latitudinal gradients in species diversity with several hypotheses concerning relative niche breadths and interspecific interactions. Their hypotheses have included:

1. Tropical species are limited more by biological factors than by physical factors.
2. Tropical species are affected more by interspecific interactions than by intraspecific interactions.
3. The niches of tropical species overlap more than those of higher-latitude species and so tropical species compete more intensively.
4. Tropical species are more specialized—that is, have narrower niches—and therefore compete less intensively than species at higher latitudes.
5. Tropical species are more subject to controls by predators, parasites, and pathogens.
6. Compared to temperate species, tropical species are involved in more mutualistic interactions.

Brown suggested that these hypotheses present the ecologist with a number of difficulties. Notice that some of these hypotheses are contradictory. In addition, they are difficult to test and do not address the primary differences between the tropics and higher latitudes. For instance, even if the niches of tropical species differ consistently from those of species at higher latitudes, we must determine the causes of those differences. Brown suggests that biological processes, such as competition and predation, must play a secondary role in determining species diversity gradients. He proposes that the ultimate causes of geographic gradients in species richness must be physical differences between the tropics and higher latitudes. The following hypothesis uses differences in rates of speciation and extinction to explain latitudinal patterns of diversity.

6. Differences in Speciation and Extinction Rates

Ultimately, the number of species in a particular area reflects the rate at which new species have been added to the species pool minus the rate at which they have disappeared. Species are added to species pools by either immigration or speciation. However, Rosenzweig (1992) proposed that when we consider the diversity of whole biogeographic provinces, immigration can be largely discounted and speciation will be the primary source of new species. Species are removed from species pools by extinction. Therefore, tropical species richness is greater than at higher latitudes because the tropics have experienced higher rates of speciation and/or lower rates of extinction. However, Brown would remind us here that we need to determine the physical mechanisms that produce differences in speciation and extinction rates in tropical versus higher latitudes.

Area and Latitudinal Gradients in Species Richness

John Terborgh (1973) and Michael Rosenzweig (1992) proposed that the greater species richness of the tropics can be explained by the greater area covered by tropical regions. It may not be immediately apparent that the tropics, which mainly occupy the area between the tropics of Cancer and Capricorn, include a greater area of both land and water than do higher latitudes. The reason for this is that the typical world map is based on the *Mercator Projection*, a projection that increases the apparent area at high latitudes. However, if you look at a world globe, you will immediately see that the tropical areas of the earth constitute a vast area.

Is there a greater land surface area in the tropics? Rosenzweig quantified the amount of land surface area in various latitudinal zones using a computer map of the earth. He divided the globe into tropical (± 26° of latitude), subtropical (26°–36°), temperate (36°–46°), boreal (46°–56°), and tundra (>56°). He then measured the area of land within these latitudinal zones and found that the area of land within the tropics far exceeds that of other areas (fig. 22.18).

Not only is there more land (and water) at tropical latitudes, but in addition, temperatures are more uniform across this tropical belt. This pattern was put in the context of geographic ecology by Terborgh, who plotted mean annual temperatures against latitude. As figure 22.19 shows, there is little difference in mean annual temperatures between about 0° and 25° latitude. Because this temperature pattern occurs both north and south of the equator, mean annual temperature changes little over about 50° of latitude within the tropics. However, above 25° latitude, mean annual temperature declines linearly with latitude. What is the biological significance of this latitudinal pattern of temperature variation? One implication is that tropical organisms can disperse over large areas and not meet with significant changes in temperature.

How do patterns of temperature variation affect rates of speciation and extinction? Rosenzweig proposed that the larger area of tropical regions should reduce extinction rates in two ways. First, large, physically similar areas will allow tropical species to be distributed over a larger area. Within these larger areas, there should be more refuges in which to

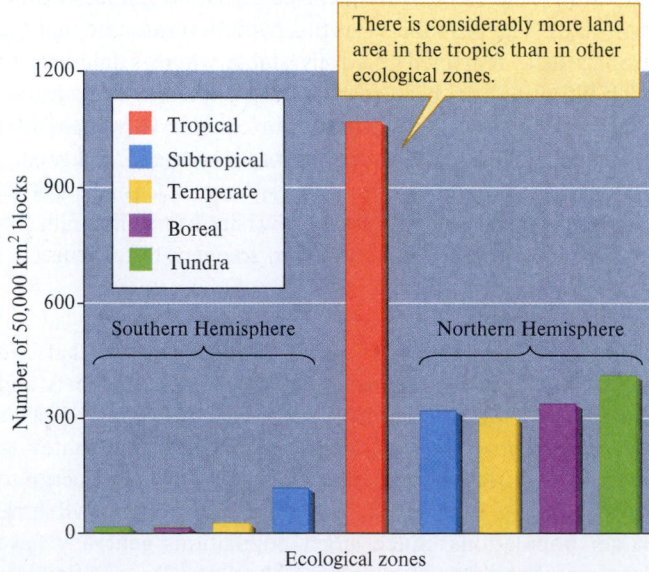

Figure 22.18 Land area in five latitudinal biomes (data from Rosenzweig 1992).

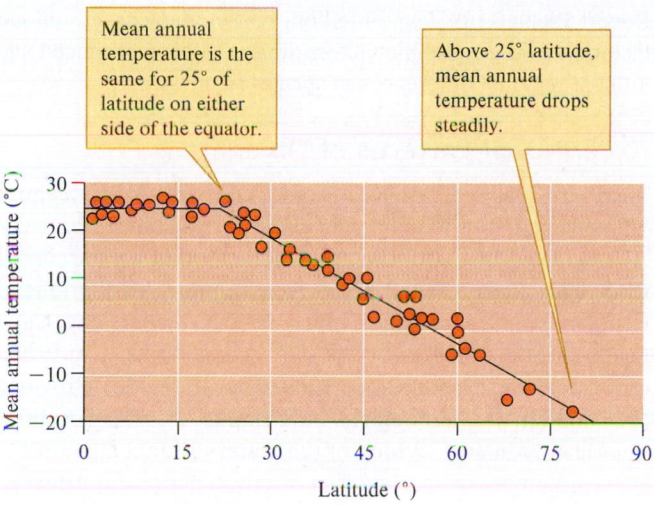

Figure 22.19   Mean annual temperature by latitude (data from Rosenzweig 1992; after Terborgh 1973).

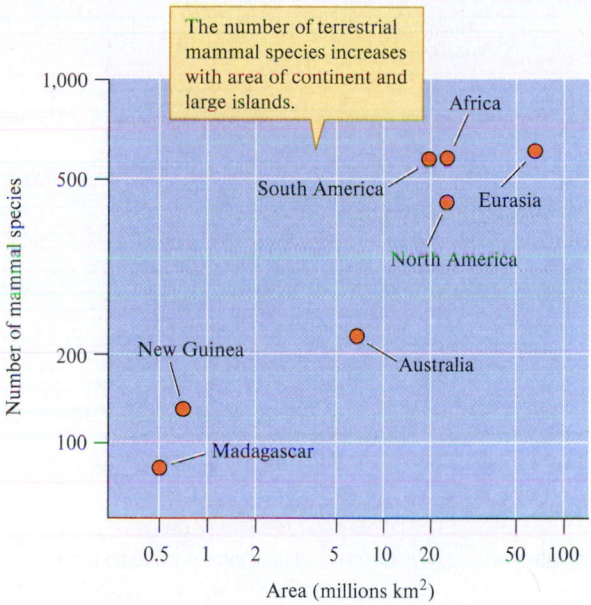

Figure 22.20   Relationship between area of continents and large islands and number of nonflying terrestrial mammals (data from Brown 1986).

survive environmental disturbances. Because of their larger range, tropical species should also have greater total population sizes. Larger populations are less likely to become extinct.

Rosenzweig also proposed that larger species ranges should increase rates of allopatric speciation. He reasoned that geographic barriers, such as mountain ranges or deep canyons, are more likely to form within large species ranges than within small species ranges. Therefore, since geographic isolation initiates allopatric speciation, speciation rates are likely to be higher in tropical regions.

Earlier in this chapter, we saw that species richness increases with island area. However, do larger continents also harbor more species? We explore this question in the following example.

Continental Area and Species Richness

Karl Flessa (1975, 1981) was the first to examine the relationship between continental area and species richness. He found a strong positive relationship between mammalian richness and the area of continents, large islands, and island groups. Flessa found a significant positive relationship whether his index of mammalian richness was total number of orders, families, or genera. James H. Brown (1986) performed an analysis similar to Flessa's; however, he restricted his analysis to the five major continents plus two large tropical islands, Madagascar and New Guinea. Brown also excluded flying mammals and analyzed patterns of mammalian diversity at the level of genera and species. Like Flessa, Brown found a strong positive relationship between mammalian richness and area (fig. 22.20). Madagascar, with the smallest land area, supports the lowest mammalian richness. Eurasia, with the largest land area, supports the greatest mammalian diversity. The continents of Australia, South America, North America, and Africa, with intermediate areas, contain intermediate levels of mammalian richness.

What do these analyses by Flessa and Brown have to do with higher tropical species richness? Rosenzweig proposed

that the greater area of the tropics (see fig. 22.18) is a primary cause of the higher diversity. If differences in area produce differences in species richness, then we should see a positive relationship between continental area and species richness. Flessa and Brown have shown such a relationship. Now, let's go back to the tropics.

Do tropical regions with different areas differ in species richness? If Rosenzweig's explanation for the greater tropical diversity is correct, tropical regions of different areas should support different levels of biological diversity. Rosenzweig examined patterns of diversity among fruit-eating mammals and plants in tropical rain forests ranging from Australia to Amazonia. The result was a strong positive relationship between area and diversity, a result that supports the area hypothesis (fig. 22.21). The smallest area of tropical rain forest, Australia, contains the smallest number of fruit-eating mammal and plant species. Amazonia, with the largest rain forest area, contains the greatest number of fruit-eating mammal and plant species.

In summary, many factors may contribute to higher tropical species richness, including (1) time since perturbation, (2) productivity, (3) environmental heterogeneity, (4) favorableness, (5) niche breadths and interspecific interactions, and (6) differences in speciation and extinction rates. However, several lines of evidence support the hypothesis that differences in surface area play a primary role in determining latitudinal gradients in species richness. Can we conclude that we fully understand the mechanisms underlying latitudinal gradients in diversity? Brown concluded that while we are close to understanding the mechanisms controlling variation in species richness across islands, "The distributions of species and higher taxa within continents are more complex and for the most part remain to be deciphered." As we shall see in the next section, some of that unexplained complexity is due to historical and regional differences between the continents.

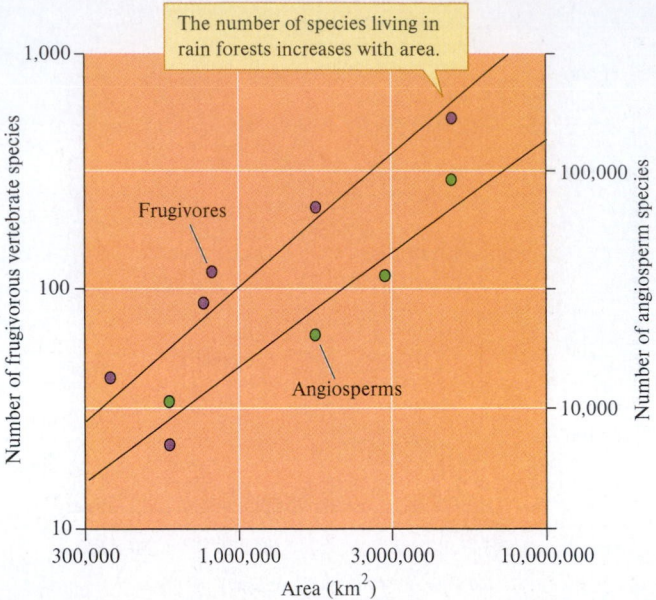

Figure 22.21 The relationship between rain forest area, from Australia to Amazonia, and numbers of flowering plant (angiosperm) species and number of fruit-eating (frugivorous) vertebrate species (data from Rosenzweig 1992).

Concept 22.3 Review

1. Why is there no one factor that seems to explain latitudinal gradients in species diversity?
2. What major pattern do patterns of island diversity and continental diversity have in common?

22.4   Historical and Regional Influences

LEARNING OUTCOMES

After studying this section you should be able to do the following:

22.12 Discuss evidence for historical influences on regional differences in species richness.
22.13 Explain how historical factors related to the processes of speciation and extinction can influence regional differences in species richness.

Long-term historical and regional processes can significantly influence species richness and diversity. Area and isolation explain much of the variation in species diversity and composition across islands. Area appears to account for much of the variation in biological diversity across continents. Additional variation in local diversity appears to be due to differences in habitat heterogeneity, disturbance, predation, and successional age of the local community, factors that we discussed in chapters 16, 17, and 20. However, as the following examples show, these factors are not adequate to explain many geographic differences in biological diversity and community organization.

Robert Ricklefs (1987) pointed out that in many cases, unique historical and geographic factors appear to have produced significant regional differences in species richness.

Exceptional Patterns of Diversity

There are major differences in species richness that cannot be explained by differences in area. For instance, consider the regions with Mediterranean climates that we discussed in chapter 2 (see fig. 2.23), which support Mediterranean woodlands and shrublands. Such regions include the Cape region of South Africa (90,000 km^2), southwestern Australia (320,000 km^2), and the California Floristic Province (324,000 km^2). These regions have similar climates but differ significantly in area. Which of these areas should contain the greatest number of species? The positive relationship between area and species richness that we have seen repeatedly earlier in this chapter, leads us to predict that southwestern Australia and the California Floristic Province, with more than three times the area of the Cape region of South Africa, will contain the greatest biological diversity. Southwestern Australia and the California Floristic Province have the same area and approximately the same number of species. However, as figure 22.22 shows, the Cape region, the smallest area, contains more than twice the number of plant species as the other two regions.

The failure of area to explain a significant regional diversity pattern is not unique to this example. For instance, Roger Latham and Robert Ricklefs (1993) reported a striking contrast in diversity of temperate zone trees that cannot be explained by an area effect. The temperate forest biome covers approximately equal areas in Europe (1.2 million km^2), eastern Asia (1.2 million km^2), and eastern North America (1.8 million km^2). The species area relationship would lead us to predict that

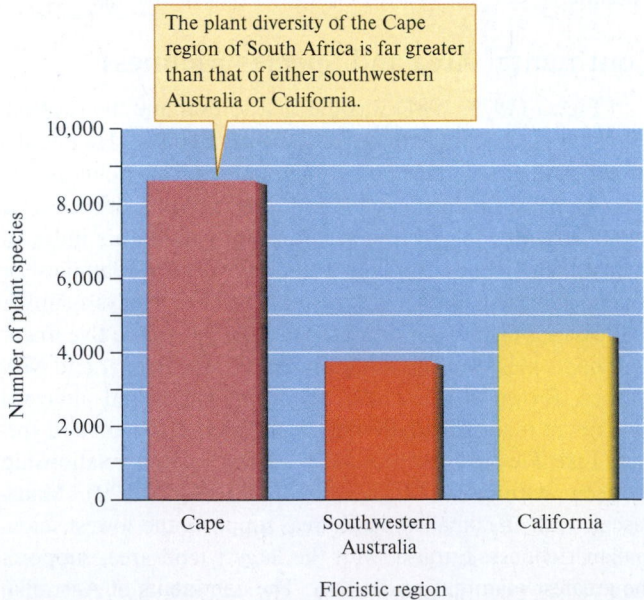

Figure 22.22 Number of plant species living in three regions with Mediterranean climates (data from Bond and Goldblatt 1984).

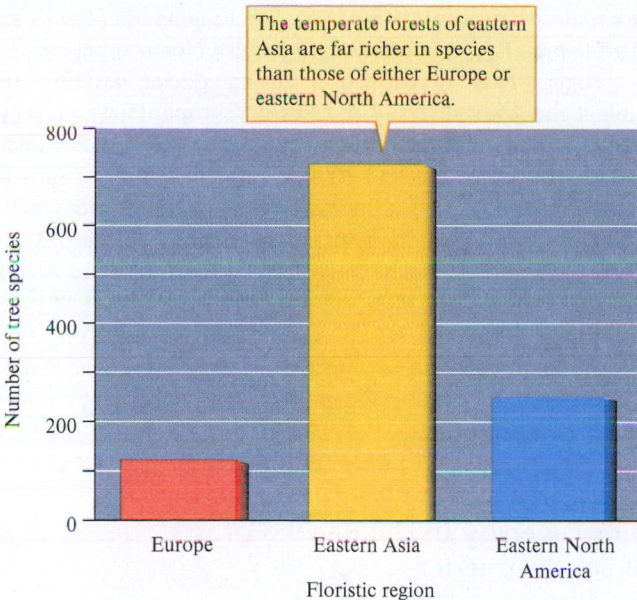

The temperate forests of eastern Asia are far richer in species than those of either Europe or eastern North America.

Figure 22.23 Number of tree species in three temperate forest regions (data from Latham and Ricklefs 1993).

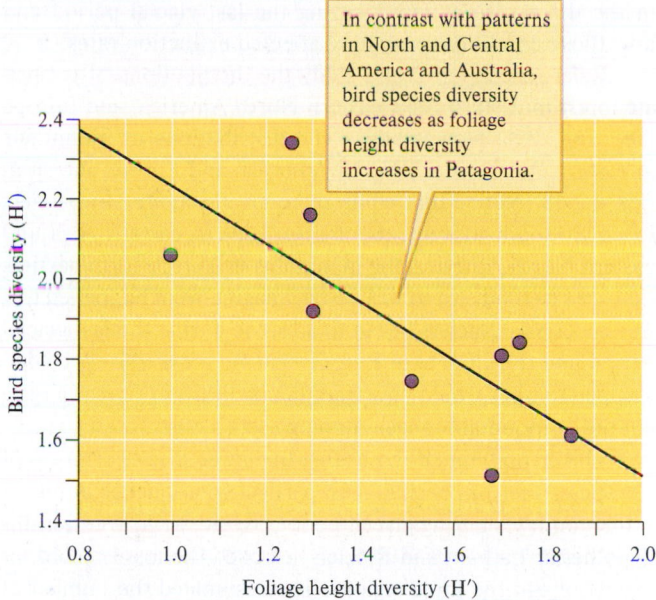

In contrast with patterns in North and Central America and Australia, bird species diversity decreases as foliage height diversity increases in Patagonia.

Figure 22.24 Foliage height diversity and bird species diversity in Patagonia (data from Ralph 1985).

these three regions would support approximately equal levels of biological diversity. However, figure 22.23 shows that eastern Asia contains nearly three times more tree species than eastern North America and nearly six times more species of trees than Europe.

Let's discuss another example, this time involving birds. In chapter 16 we reviewed several studies that showed a positive relationship between foliage height diversity and bird species diversity. These studies of bird communities showed that greater vertical heterogeneity in the plant community (foliage height diversity) is associated with greater bird species diversity in areas as widespread as northeastern North America, Puerto Rico, Panama, and Australia (e.g., see fig. 16.10). However, biologists have found that this positive relationship is not universal.

John Ralph (1985) studied the relationship between foliage height diversity and bird species diversity in northern Patagonia, Argentina. He chose study sites distributed along a 50 km moisture gradient in which annual precipitation ranged from 200 to 2,500 mm annually. This precipitation gradient produced a gradient in vegetative structure from grasslands and shrublands through beech forests. Contrary to what had been found in other regions, Ralph found a negative correlation between foliage height diversity and bird species diversity (fig. 22.24). He recorded higher diversity in shrub habitats than in beech forests with greater foliage height diversity.

Historical and Regional Explanations

How can we explain these exceptional patterns of biological diversity? What mechanisms produced these patterns that are contrary to generalizations discussed in this chapter and in earlier chapters? In each case, it appears that geography and history offer convincing explanations.

The Cape Floristic Region of South Africa

Pauline Bond and Peter Goldblatt (1984) attributed the unusual species richness of the Cape floristic region to several historic and geographic factors. Selection for a distinctively Mediterranean flora in southern Africa began during the late Tertiary period, about 26 million years ago. At that time, the climate became progressively cooler and drier, conditions that selected for succulence, fire resistance, and smaller, sclerophyllous leaves. The initial sites for evolution of the Cape flora were likely in south-central Africa, not in the Cape region itself. At that time, Africa lay farther south and the Cape region had a cool, moist climate and supported an evergreen forest.

As Africa drifted northward, the climate of southern Africa became more arid and the ancestors of today's Cape flora gradually migrated toward the Cape region. By the time Africa neared its present latitudinal position during the late Pliocene, about 3 million years ago, southern Africa was very arid and the Cape region had a Mediterranean climate. Bond and Goldblatt suggest that plant speciation within this region was promoted by the highly dissected landscape, the existence of a wide variety of soil types, and repeated expansion, contraction, and isolation of plant populations during the climatic fluctuations of the Pleistocene. They suggest that extinction rates were reduced by the existence of substantial refuge areas, even during times of peak aridity.

The Diversity of Temperate Trees

How did eastern Asia, eastern North America, and Europe, three temperate regions of approximately equal area and climate, end up with such different numbers of tree species? Latham and Ricklefs offer persuasive geographic and historical reasons. They propose that we need to consider what trees in the three regions faced during the last glacial period and how those conditions may have affected extinction rates.

Refer to chapter 2 and study the distributions of temperate forest in eastern Asia, eastern North America, and Europe (see fig. 2.29). Now examine the distributions of mountains in eastern Asia, eastern North America, and Europe shown in figure 2.38. Notice that while there are no mountain barriers to north–south movements of organisms in eastern Asia and eastern North America, the mountains in Europe form barriers that are oriented east to west. Now imagine what happened to a tree species as glaciers began to advance during the last glacial period and the climate of Europe became progressively colder. Temperate trees would have had their southward retreat largely cut off by mountain ranges running east to west.

This hypothesis proposes that the lower species richness of European trees has been at least partly a consequence of higher extinction rates during glacial periods. How would you test this hypothesis? Latham and Ricklefs searched the fossil record for extinctions in the three regions. They estimated the number of genera that have become extinct in the three regions during the last 30 to 40 million years. Their analysis showed that most of the plant genera that once lived in Europe have become extinct. A larger proportion of genera has become extinct in Europe than in either eastern Asia or eastern North America (fig. 22.25).

Now consider eastern North America. The only mountain range, the Appalachians, runs north to south. Consequently, in eastern North America, temperate trees had an avenue of retreat in the face of advancing glaciers and cooling climate. The movement of temperate tree populations in the face of climate change has been well documented by paleontologists such as Margaret Davis (see fig. 10.6). There are also no

mountain barriers in eastern Asia, where temperate trees can migrate even farther south than in eastern North America.

Higher rates of extinction during glacial periods can explain the lower diversity of trees in Europe. However, why does eastern North America include fewer tree species than eastern Asia? Latham and Ricklefs conclude that the fossil record and present-day distributions of temperate trees indicate that most temperate tree taxa originated in eastern Asia. These Asian taxa subsequently dispersed to Europe and North America. In addition, after the dispersal routes between eastern Asia and eastern North America were closed off, speciation continued in Asia, producing several endemic Asian genera. In other words, there are fewer tree species in eastern North America because most taxa originated in eastern Asia and never dispersed to North America.

Bird Diversity in the Beech Forests of South America

Why do the structurally complex beech forests of South America support lower bird diversity than the simpler shrub habitats? Dolph Schluter and Robert Ricklefs (1993) suggested that the lower bird diversity in beech forests may be due to the restricted geographic distribution of these temperate forests in South America. The shrub habitats where Ralph (1985) recorded the greatest diversity of birds occupy a much larger area within South America.

In addition to their small area, South American beech forests are also isolated from subtropical and tropical forests by vast areas of arid and semiarid vegetation. The biota of these forests includes endemic species of frogs, birds, and mammals, which suggests that South American beech forests have been isolated for a long time. Schluter and Ricklefs suggested that small area and isolation may account for the low diversity of birds in South American beech forests. They cite the relationship found by Ralph as "a convincing demonstration of how effects of local habitat on species diversity may be superseded by regional ones."

In summary, much geographic variation in species richness can be explained by historical and regional processes. The ecologist interested in understanding patterns of diversity at large spatial scales must consider processes occurring over similarly large scales and over long periods of time. As we shall see in chapter 23, a large-scale, long-term perspective is also essential for understanding global ecology.

Concept 22.4 Review

1. Why should history have such a strong influence on regional diversity patterns?
2. How does the combined evidence from studies of the flora of Mediterranean regions (fig. 22.22) and the diversity of trees in temperate forest regions increase confidence that historical differences can outweigh the potential influence of area on diversity?

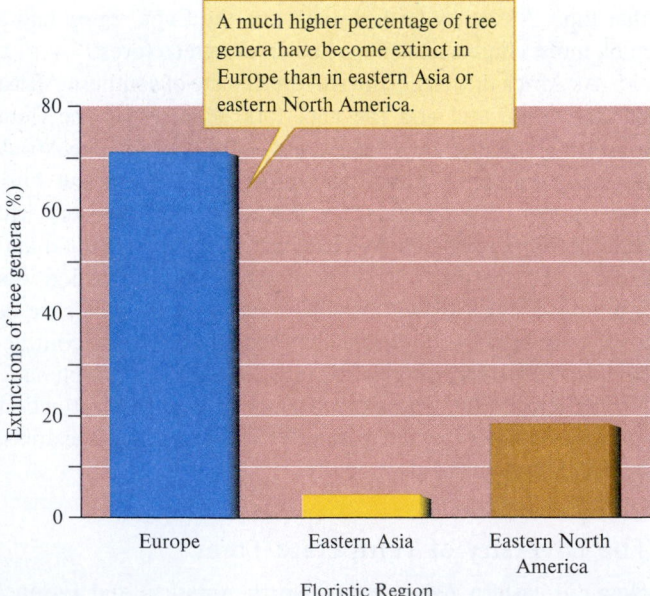

Figure 22.25 Extinctions of tree genera in Europe, eastern Asia, and eastern North America since the middle Tertiary period (data from Latham and Ricklefs 1993).

Applications

Global Positioning Systems, Remote Sensing, and Geographic Information Systems

LEARNING OUTCOMES

After studying this section you should be able to do the following:

22.14 Discuss the importance of knowing location in studies of geographic ecology.

22.15 Explain how remote sensing has revolutionized the study of large-scale ecological processes.

In 1972, Robert MacArthur defined geographic ecology as the study of patterns you can put on a map. Spatial distributions that can be put on maps are still the center of geographic ecology, but the nature of "maps" has changed tremendously. Modern tools have revolutionized the field. Today, geographic ecologists generally record their data on geographic information systems, which are computer-based systems that store, analyze, and display geographic information. In addition, the geographic ecologists of today also have access to more information of greater accuracy because of remote sensing and global positioning systems.

Global Positioning Systems

What is the location? This is one of the most basic questions the geographer can ask. Scientists, engineers, navigators, and explorers have spent centuries devising methods to measure elevation, latitude, and longitude. Recent technological advances have improved the accuracy of these measurements.

Alexander von Humboldt, the founder of geographic ecology, would appreciate these recent technological advances. As he explored South and North America, he carefully determined the latitude, longitude, and elevation of important geographic features. For instance, Humboldt was particularly interested in verifying the existence and location of a waterway called the Casiquiare Canal. The Casiquiare reportedly connected the Orinoco River with the Rio Negro, which flows into the Amazon. A connection between two major river systems would make the Casiquiare unique, but its existence was widely doubted.

Humboldt halted his expedition at the junction of the Casiquiare and the Rio Negro so that he could record the latitude and longitude. Biting insects tormented the explorers as they waited for nightfall. Luckily, that night the clouds parted and Humboldt could see the stars well enough to take sightings and determine their position. At other times, he was not so lucky. He once waited for nearly a month for the weather to clear sufficiently to make his sightings on the stars. Today, equipped with a global positioning system, Humboldt could have determined the latitude and longitude of the junction of the Casiquiare and the Rio Negro anytime he wished, regardless of weather.

A **global positioning system (GPS)** determines locations on the earth's surface, including latitude, longitude, and altitude, using satellites as reference points. These satellites, which orbit the earth at a height of about 21,000 km, continuously transmit their position and the time. The satellites keep track of time with an extremely accurate atomic clock that loses or gains 1 second in about 30,000 years. A global positioning system receives the signals broadcast by these satellites. Because the system also includes an extremely accurate clock, the time required for the satellite signal to reach the receiver can be used as a measure of the distance between the two. With measurements of the distance to four satellites, a global positioning system can determine the latitude, longitude, and altitude of any point on earth with great accuracy (fig. 22.26).

While navigation satellites and global positioning systems can accurately locate places on the ground, other satellites provide a wealth of other information about those localities. These "remote sensing" satellites transmit pictures of the earth that are extremely valuable to ecologists.

Remote Sensing

Remote sensing refers to gathering information about an object without direct contact with it, mainly by gathering and processing electromagnetic radiation emitted or reflected by the object. Using this definition, the original remote sensor was the eye. However, we generally associate remote sensing with technology that extends the senses, technology ranging from binoculars and cameras to satellite-mounted sensors.

Remote sensing satellites are generally fitted with electro-optical sensors that scan several bands of the electromagnetic

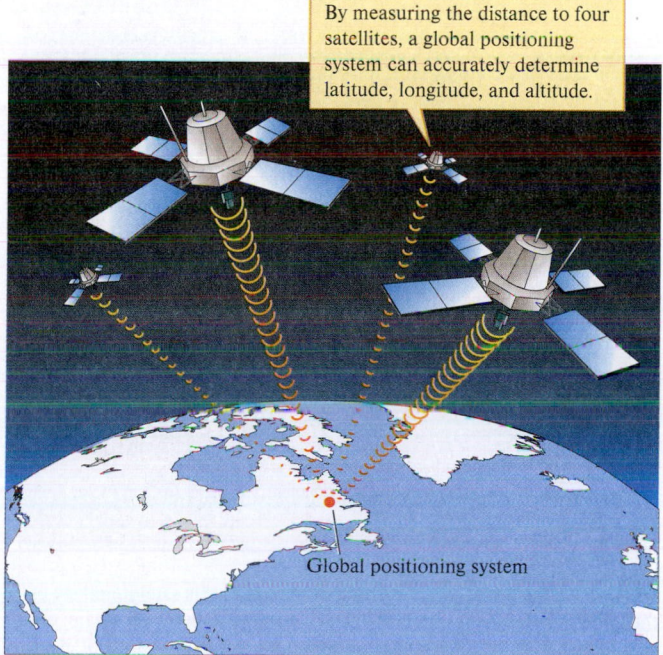

By measuring the distance to four satellites, a global positioning system can accurately determine latitude, longitude, and altitude.

Global positioning system

Figure 22.26   Global positioning systems determine latitude, longitude, and altitude by measuring the distance from several satellites.

spectrum. These sensors convert electromagnetic radiation into electrical signals that are in turn converted to digital values by a computer. These digital values can be used to construct an image. The earliest of the *Landsat* satellites monitored four bands of electromagnetic radiation, two bands of visible light (0.5–0.6 μm and 0.6–0.7 μm), and two bands in the near infrared (0.7–0.8 μm and 0.8–1.1 μm). From this beginning, satellite imaging systems have gotten progressively more sophisticated in terms of both the number of wavelengths scanned and the spatial resolution.

Satellite-based remote sensing has produced detailed images of essentially every square meter of the earth's surface. These images provide very useful information to ecologists, especially for landscape and geographic ecology. Ecologists have used remote sensing to monitor the biomass of vegetation using indices of "greenness." In the arid American Southwest, vegetative biomass indicates the positions of moist mountain areas.

Mountain Islands in the American Southwest

Norman Roller and John Colwell (1986) reviewed how satellites gathering information at a coarse level of resolution could be useful for conducting ecological surveys, particularly when those surveys concern large geographic areas. Earlier in this chapter, we discussed one such survey conducted by Lomolino and his colleagues on the mammals of montane "islands" in the American Southwest. In their study, these researchers relied principally on a 1982 map of the vegetation of the area. They could have also used satellite imagery from the same region to delineate their study areas.

In the satellite image shown in figure 22.27, red, orange and yellow patches in New Mexico and Arizona correspond

to the forested montane islands studied by Lomolino and his colleagues. However, unlike a static map, satellite estimates of plant biomass can be made frequently and offer the possibility of detecting temporal variation. This is particularly important in a region like the American Southwest that experiences high year-to-year variability in precipitation and plant production.

Marine Primary Production from Space

Mary Jane Perry (1986) demonstrated the use of remote sensing to study marine primary production. In chapter 21, we concentrated on patchiness in terrestrial environments. However, marine environments are also highly patchy, especially in regard to primary production (fig. 22.28). Perry pointed out that we know little about interannual variation in marine primary production, particularly in the open ocean.

The main reasons we know little about the dynamics of production in the open ocean are that (1) the open ocean ecosystem is so vast, covering approximately 332 million km^2; (2) there are limited numbers of oceanographic ships and other sea-based sampling devices; and (3) open ocean studies are expensive. Perry proposed that satellite remote sensing of ocean color is the best tool at our disposal for studying regional and global marine primary production. Figure 22.28 shows a remote sensing image taken along the west coast of North America. The colors indicate concentrations of chlorophyll *a*, a measure of phytoplankton biomass. The image

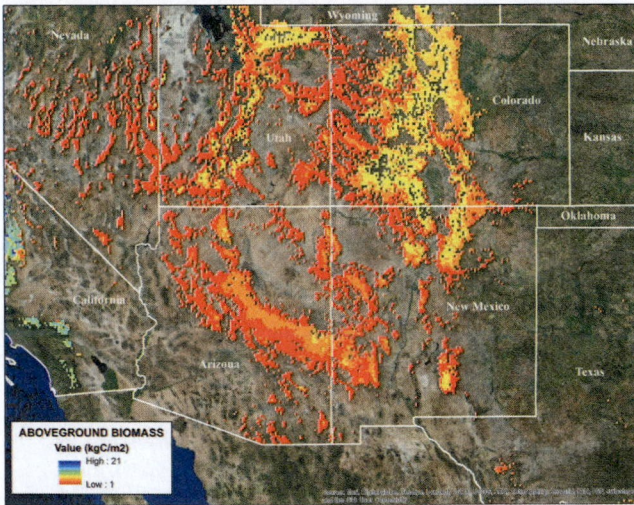

Figure 22.27 Higher vegetative biomass in this satellite image of the American Southwest occurs mainly on mountain ranges, many of which are distributed across the arid region as isolated islands of elevated production (Collatz et al. 2014). Source: Collatz, G.J., C. Williams, B. Ghimire, S. Goward, and J. Masek. 2014. *NACP CMS: Forest Biomass and Productivity, 1-degree and 5-km, Conterminous US, 2005.* Data set. Available on-line [http://daac.ornl.gov] from Oak Ridge National Laboratory Distributed Active Archive Center, Oak Ridge, Tennessee, USA. DOI: 10.3334/ORNLDAAC/1221

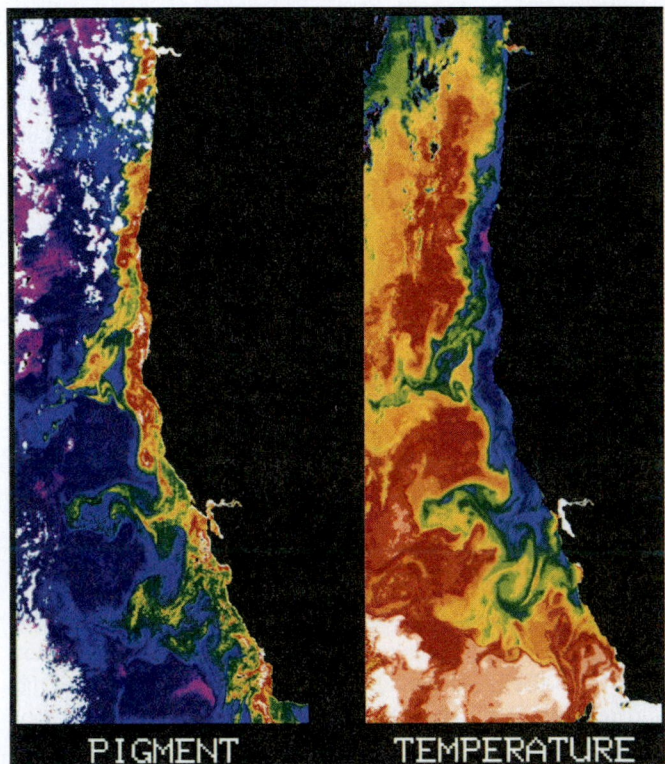

Figure 22.28 Concentrations of chlorophyll *a* and temperature along the west coast of North America determined from satellite imagery. In the image on the left, yellow and red indicate areas of high phytoplankton biomass. These coincide with cooler sea surface temperatures associated with upwelling, which are shown in the right image by violet, purple, and blue colors (Perry 1986).   Science and Society/Superstock

indicates a high degree of patchiness; higher concentrations of phytoplankton occur in a narrow coastal band. The figure also shows that the coastal waters supporting high phytoplankton biomass are also colder than offshore areas. What process is suggested by the combination of cold water and high phytoplankton biomass? The combination suggests coastal upwelling, which brings nutrient-rich deep water to the surface.

Perry reports that though oceanographers had long known that phytoplankton populations are patchy, they had no idea of the complexity of that patchiness. Some phytoplankton patchiness occurs far out at sea, and there is also a high degree of temporal variation in patch location. Patches of high production that form along coastlines can move offshore at rates of 2 to 7 km per day. It would be impossible for a ship moving at less than 30 km per hour to capture such spatially and temporally complex patterns of production. However, the satellite that took the image shown in figure 22.28 can scan a path 1,600 km wide with a resolution of about 1 km^2 and resample the same area at 5- to 6-day intervals.

This example and the previous one show that satellite-based remote sensing can gather large amounts of data over large areas. This ability solves some of the sampling problems associated with studying large-scale ecological phenomena. However, these large quantities of data create another problem. Ecologists need a system for storing, sorting, analyzing, and displaying these large quantities of geographic information. This is the problem addressed by geographic information systems.

Geographic Information Systems

In the days of Humboldt, geographers often had too little data. Today, with new tools for gathering great quantities of information, geographers and geographic ecologists can be overwhelmed by data. **Geographic information systems,** computer-based systems for storing, sorting, analyzing, and displaying geographic data, are designed to handle large quantities of data. Sometimes geographic information systems are confused with computerized mapmaking. While these systems can produce maps, they do much more. Much of population ecology is concerned with understanding the factors controlling the distribution and abundance of organisms. However, the geographic context of populations often has been lost. Geographic information systems preserve this geographic information. Because they preserve geographic context, the systems provide ecologists with a valuable tool for exploring large-scale population responses to climate change.

As we shall see in chapter 23, rapid global change challenges the field of ecology to address large-scale environmental questions. As ecologists address these compelling questions, geographic information systems, global positioning systems, and remote sensing will be increasingly valued parts of their tool kit.

Summary

Geographic ecology focuses on large-scale patterns of distribution and diversity of organisms, such as island biogeography, latitudinal patterns of species diversity, and the influences of large-scale regional and historical processes on biological diversity.

On islands and habitat patches on continents, species richness increases with area and decreases with isolation. Larger oceanic islands support more species of most groups of organisms than small islands. Isolated oceanic islands generally contain fewer species than islands near mainland areas. Many habitats on continents are so isolated that they can be considered as islands. Species richness on habitat islands, such as mountain islands in the American Southwest, increases with area and decreases with isolation. Lakes can also be considered as habitat islands. They are aquatic environments isolated from other aquatic environments by land. Fish species richness generally increases with lake area. Species richness is usually negatively correlated with island isolation. However, because organisms differ substantially in dispersal rates, an island that is very isolated for one group of organisms may be completely accessible to another group.

Species richness on islands can be modeled as a dynamic balance between immigration and extinction of species. The equilibrium model of island biogeography proposes that the difference between rates of immigration and extinction determines the species richness on islands. The equilibrium model of island biogeography assumes that rates of species immigration to islands are mainly determined by distance from sources of immigrants. The model assumes that rates of extinction on islands are determined mainly by island size. The predictions of the equilibrium model of island biogeography are supported by observations of species turnover on the islands and by colonization studies of mangrove islands in Florida and new islands in Lake Hjälmaren, Sweden.

Species richness generally increases from middle and high latitudes to the equator. Most groups of organisms are more species-rich in the tropics. Many factors may contribute to higher tropical species richness, including (1) time since perturbation, (2) productivity, (3) environmental heterogeneity, (4) favorableness, (5) niche breadths and interspecific interactions, and (6) differences in speciation and extinction rates. Several lines of evidence support the hypothesis that differences in surface area play a primary role in determining latitudinal gradients in species richness.

Long-term historical and regional processes can significantly influence species richness and diversity. Much geographic variation in species richness can be explained by historical and regional processes. Some exceptional situations that seem to have resulted from unique historical and regional processes include the exceptional species richness of the Cape floristic region of South Africa, the high species richness of temperate trees in eastern Asia, and the low bird diversity in beech forests of South America. The ecologist interested in understanding large-scale patterns of species richness must consider processes occurring over similarly large scales and over long periods of time.

Global positioning systems, remote sensing, and geographic information systems are important tools for effective geographic ecology. A global positioning system determines locations on the earth's surface, including latitude, longitude, and altitude, using satellites as reference points. Remote sensing satellites are generally fitted with electro-optical sensors that scan several bands of the electromagnetic spectrum. These sensors convert electromagnetic radiation into electrical signals that are in turn converted to digital values by a computer. These digital values can be used to construct an image. Geographic information systems are computer-based systems that store, analyze, and display geographic information. Global positioning systems, remote sensing, and geographic information systems are increasingly valuable parts of the ecologist's tool kit. Ecologists are using these new tools to study large-scale, dynamic ecological phenomena such as interannual variation in regional terrestrial primary production, dynamics of marine primary production, and potential population responses to climate change.

Key Terms

equilibrium model of island
  biogeography   473
geographic ecology   469

geographic information system   487
global positioning system (GPS)   485

remote sensing   485
species turnover   474

Review Questions

1. The following data (corrected from Preston 1962a) give the area and number of bird species on islands in the West Indies:

Island	Area (km²)	Log₁₀ Area	# Species	Log # Species
Cuba	110,900	5.045	124	2.093
Hispaniola	76,250	4.882	106	2.025
Jamaica	11,000	4.041	99	1.996
Puerto Rico	8,875	3.948	79	1.898
Bahamas	13,950	4.145	74	1.869
Virgin Islands	500	2.699	35	1.544
Guadalupe	1,700	3.230	37	1.568
Dominica	750	2.875	36	1.556
St. Lucia	620	2.792	35	1.544
St. Vincent	390	2.591	35	1.544
Granada	340	2.531	29	1.462

The numbers are expressed in two ways: as simple measurements and counts and as the logarithms of area and numbers of species. Use these data to plot your own species–area relationship. Plot area on the horizontal axis and number of species on the vertical axis. First plot the simple measurements of area and species number on one graph, and then plot the logarithms of area and species number on another graph. Which gives you the tighter relationship between area and species richness?

2. Refer to figure 22.5, which MacArthur and Wilson (1963) used to show how isolation affects species richness on islands. Find a detailed map of the Pacific Ocean and locate New Guinea. Next locate as many of the "near," "intermediate," and "far" islands on the map as you can. This will give you a better sense of the distances represented by the islands. How do the numbers of species on near, intermediate, and far islands support the hypothesis that island isolation tends to reduce species richness?

3. We discussed how Diamond (1969) documented immigrations and extinctions on the California Channel Islands by comparing his censuses of the birds of the islands with the birds recorded over 50 years earlier. Disregarding the numbers for San Miguel and Santa Rosa Islands, which were not well censused in 1917, Diamond showed that an average of approximately six bird species became extinct on the California Channel Islands between 1917 and 1968. During the same period, an average of approximately five new bird species immigrated to the islands. Diamond suggested that his estimates of immigration and extinction were likely underestimates of the actual rates. Explain why his comparative study produced underestimates of rates of immigration and extinction.

4. Diamond's estimates (1969) of numbers of species immigrating and numbers that became extinct (six versus five) were virtually identical. Is this near equality in numbers of extinction and immigration consistent with the equilibrium model of island biogeography? Explain.

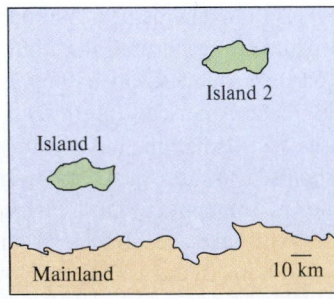

5. Suppose you are about to study the bird communities on the islands shown above, which are identical in area but lie at different distances from the mainland. According to the equilibrium model of island biogeography, which of the islands should experience higher rates of immigration? What does the equilibrium model of island biogeography predict concerning relative rates of extinction on the two islands?

6. Now, suppose you are going to study the bird communities on the islands shown below, which lie equal distances from the

mainland but differ in area. According to the equilibrium model of island biogeography, what should be the relative rates of immigration to the two islands? On which island should the rate of extinction be lower? Explain.

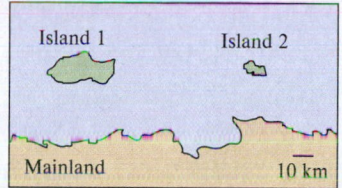

7. Review the major hypotheses proposed to explain the higher species richness of tropical regions compared to temperate and high-latitude regions. How are each of these hypotheses related to relative rates of speciation and extinction in tropical regions and temperate and high-latitude regions?

8. Explain how speciation and extinction rates might be affected by the area of continents. What evidence is there to support your explanation? What does the influence of area on rates of extinction and speciation have to do with higher species richness in tropical regions compared to temperate and high-latitude regions?

Purestock/SuperStock

Chapter 23

Global Ecology

View of North and South America from 35,000 km above earth. The necessity to address major environmental challenges, and having the tools to do so, have combined to make global ecology one of the most active and widely discussed areas of science.

CHAPTER CONCEPTS

23.1 The El Niño Southern Oscillation, a large-scale atmospheric and oceanic phenomenon, influences ecological systems on a global scale. 493
Concept 23.1 Review 499

23.2 Human activity has greatly increased the quantity of fixed nitrogen cycling through the biosphere. 499
Concept 23.2 Review 500

23.3 Rapid changes in global patterns of land use threaten biological diversity. 500
Concept 23.3 Review 504

23.4 Human activity is changing the composition of the atmosphere. 504
Concept 23.4 Review 508

Applications: Impacts of Global Climate Change 508
Summary 511
Key Terms 512
Review Questions 512

LEARNING OUTCOMES

After studying this section you should be able to do the following:

23.1 Discuss the significance of the *Apollo 8* photo of earth to perception of the planet by people around the world.

23.2 Outline how human activities, through agriculture and industry, are impacting the global environment.

23.3 Explain how the greenhouse effect influences earth's surface temperature.

23.4 List some of the impacts of global climate change on biotic systems and how we can determine the global scale of these impacts.

During the final days of December 1968, the *Apollo 8* mission to the moon transmitted images of the earth rising above the moon's horizon. For the first time, in

Figure 23.1   Oasis in space: earthrise over the moon's horizon.
NASA

color, we could see how earth appears from our nearest neighbor in the solar system (fig. 23.1). The human response to the sight of the earth framed by a bleak lunar landscape is captured by the words of the *Apollo 8* astronauts: "The earth from here is a grand oasis in the . . . vastness of space."

That image, of earth as a shining blue ball against the blackness of space, instantaneously changed the perspective that most people held of the planet and made it easier to think of the earth as a single ecological system.

At the beginning of the twenty-first century, it is important that we keep that perspective alive. The rapid pace of global change challenges ecologists to study ecological phenomena at a global scale. Global ecology is the scientific study of the relationship between biotic systems and environmental processes at a planetary scale. It is multidisciplinary in nature, with climatology, oceanography, geography, chemistry, and other fields intersecting with biology to shed light on processes that can only be fully understood when considering the earth as an ecosystem. These processes include research on responses to global climate, biogeochemical cycling, land use changes, movement of species, and global impacts of pollution. The Department of Global Ecology, founded in 2002 on the campus of Stanford University, states, "The goal of this research is understanding the ways these interactions shape the behavior of the earth system, including its responses to future changes." To this point, a strong focus within global ecology is the effect of human activities, as we have had more dramatic impacts on the global ecosystem than any other species, as well as being the first to be able to mitigate these impacts. In his influential paper on global ecology, Peter Vitousek (1994) identified connections between human activities and loss of biodiversity, with particular emphasis on changes in the nitrogen cycle, land use, and atmospheric CO_2 (fig. 23.2).

The reason why activities in one part of the globe, such as agriculture and industry, have impacts on a global scale is that

the atmosphere is not static. The atmosphere and the oceans are in continuous motion as a consequence of the uneven heating of the earth's surface. In chapters 2 and 3, we reviewed the major patterns of atmospheric and oceanic circulation (see figs. 2.4, 2.5, 3.3, and 3.7b). These circulatory systems link the various regions of the globe into a single physical system by moving heat energy and materials from one part of the biosphere to another. Because of global circulation, pollutants produced in one part of the globe eventually reach all parts of the globe.

Because many global-scale phenomena are mediated through the atmosphere, let's look briefly at the structure and origins of earth's atmospheric system.

The Atmospheric Envelope and the Greenhouse Earth

The earth is wrapped in an atmospheric envelope that makes the biosphere a hospitable place for life as we know it. Clean, dry air at the earth's surface is approximately 78.08% nitrogen, 20.94% oxygen, 0.93% argon, 0.03% carbon dioxide, and less than 0.00005% ozone. Air also contains variable concentrations of water vapor and trace quantities of helium, hydrogen, krypton, methane, and neon. The concentrations of these gases change with altitude. The highest concentrations

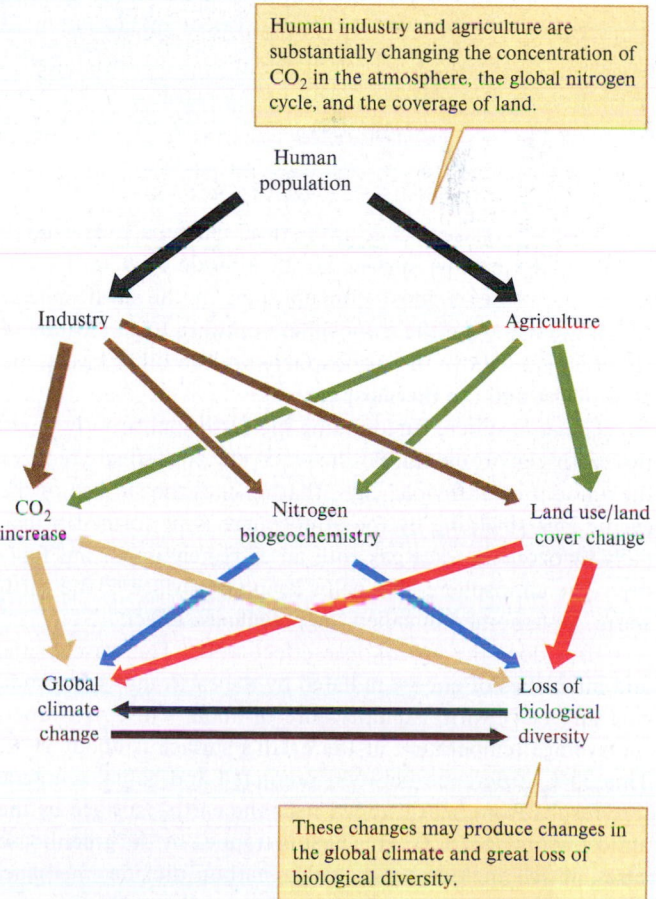

Figure 23.2   Some causes and potential consequences of global environmental change (data from Vitousek 1994).

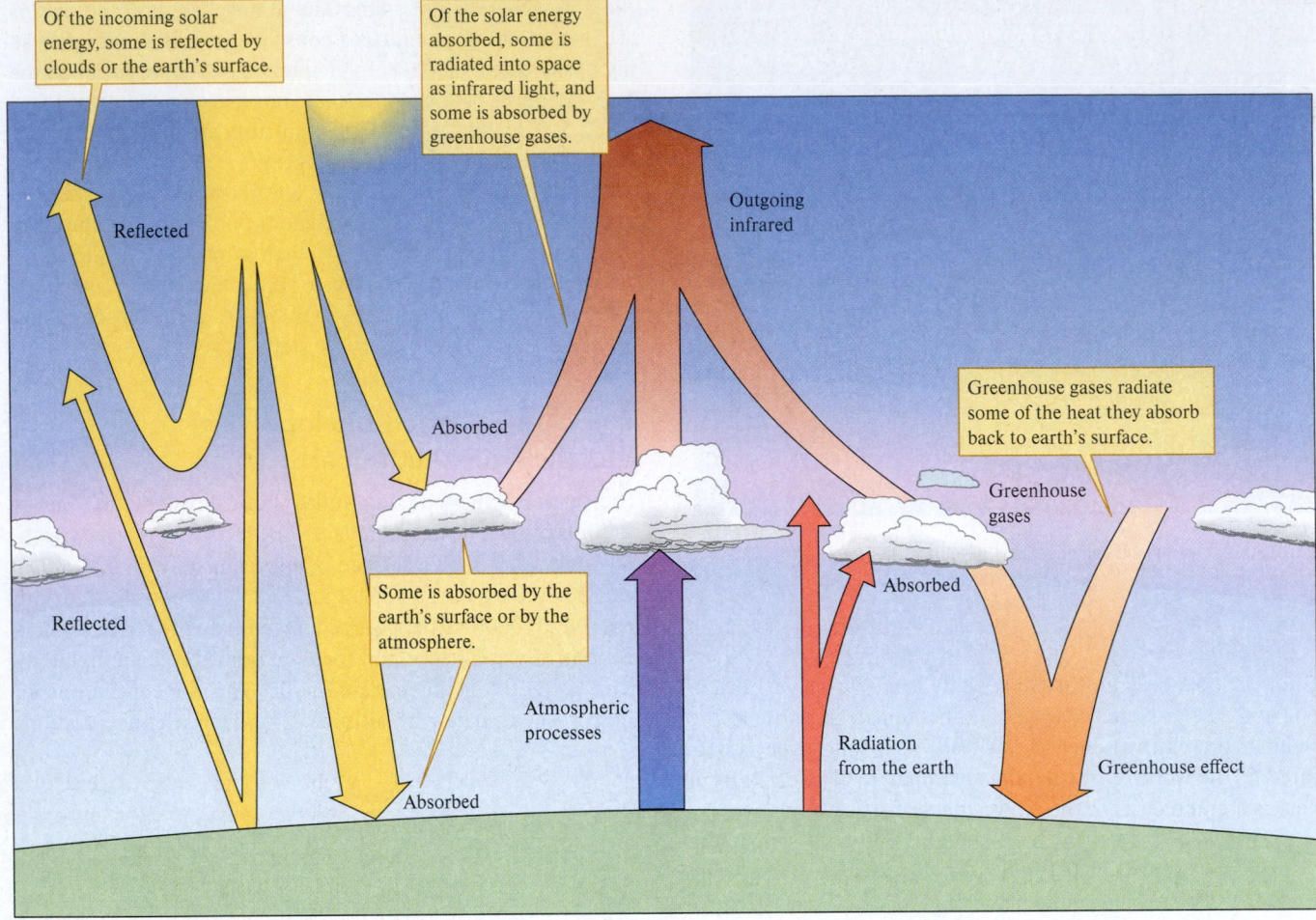

Of the incoming solar energy, some is reflected by clouds or the earth's surface.

Of the solar energy absorbed, some is radiated into space as infrared light, and some is absorbed by greenhouse gases.

Reflected

Outgoing infrared

Absorbed

Greenhouse gases radiate some of the heat they absorb back to earth's surface.

Greenhouse gases

Reflected

Some is absorbed by the earth's surface or by the atmosphere.

Absorbed

Atmospheric processes

Radiation from the earth

Greenhouse effect

Absorbed

Figure 23.3   The greenhouse effect: heat trapping by earth's atmosphere.

of atmospheric gases occur in the **troposphere,** a layer extending from the earth's surface to an altitude of 9 to 16 km. However, ozone is most concentrated in the **stratosphere,** which extends from the troposphere outward to an altitude of about 50 km. Above the troposphere are two other layers, the **mesosphere** and the **thermosphere.**

The atmosphere surrounding the earth significantly modifies earth's environment. For instance, the atmosphere reduces the amount of ultraviolet light that reaches the surface of the earth. This shielding by the atmosphere is performed principally by ozone, a trace gas with an extremely important function. The atmosphere also helps keep the surface of the earth warm, a phenomenon called the **greenhouse effect.**

How does the greenhouse effect work? The wavelengths and intensities of energy radiated by the earth into space indicate an object with a temperature of about −18°C. However, the average temperature at the earth's surface is about 15°C. This 33°C difference between predicted and actual temperature results from heat trapped near the earth's surface by the atmosphere (fig. 23.3). This heat is trapped by the greenhouse gases, which include water vapor, carbon dioxide, methane, ozone, nitrous oxide, and chlorofluorocarbons. Notice that several of these greenhouse gases are products of biological activity. Like the glass of a greenhouse, these gases absorb

infrared radiation emitted by a solar-heated earth and reemit most of that energy back to the earth.

Let's look briefly at a budget of solar energy for the earth. About 30% of the solar energy shining on earth is reflected back into space by clouds, by particles in the atmosphere, or by the surface of the earth. Approximately 70% of the solar energy shining on the earth is absorbed either by the atmosphere or by the earth's surface. This energy is reemitted as infrared radiation. Some of the infrared radiation from the atmosphere is radiated into space, and some is radiated toward the surface of the earth. Most of the infrared radiation from the earth's surface is absorbed by greenhouse gases in the atmosphere and radiated back to the earth's surface. By radiating infrared radiation back to the earth's surface, greenhouse gases trap heat energy and raise the earth's surface temperature.

How do we know that ecosystems are responding to changes in climate? We have discussed evidence of climate change in different organisms as shown by range and elevation shifts (chapter 10, sections 10.1 and Applications) and phenological changes (chapter 12, Applications). Because organisms will also respond to local and nonclimatic factors, to determine whether there is evidence of an actual "climate fingerprint," it is important that such studies consider patterns across many taxa and geographic regions and over long time periods. For example, in the study of phenological shifts by plants in Europe, Menzel and colleagues (2006) included over 100,000 time series records (see chapter 12, Applications). In one of the most widely cited works on the impact of climate change, Camille Parmesan and Gary Yohe (2003) compiled and evaluated the results of hundreds of studies on biological responses to climate. Among their compelling findings of a climate fingerprint, Parmesan and Yohe observed that among 893 taxonomic groups investigated, 434 had observed range and abundance shifts, with 80% of these shifts consistent with climate change predictions. They performed a meta-analysis on studies of 99 species of birds, butterflies, and alpine herbs for which data were available and found a mean change of 6.1 (+/−2.4) km per decade toward the poles for those species showing shifts in latitudinal distributions. Or, since temperature changes in 1 kilometer of latitude are equivalent to those occurring in 1 meter of altitude, those species for which altitude was measured shifted an average of 6.1 (+/−2.4) m upward per decade. For 172 species of herbs, shrubs, trees, birds, butterflies, and amphibians, meta-analysis also found a mean shift toward an earlier spring by 2.3 days (95% CI 1.7–3.2) per decade. For the remainder of studies on phenological shifts, which included 677 different species, 87% had observed changes in timing of blooming, hatching, and other temperature-cued events that were consistent with climate models. Overall, Parmesan and Yohe reported highly significant and nonrandom patterns of change in accordance with climate records across the published record covering more than 1,700 taxonomic groups. Although compilations of existing research do not give good coverage of the entire globe or all types of ecosystems, they provide compelling evidence that global climate change is affecting species on our planet.

Our discussion begins with a large-scale atmosphere-ocean system that has global effects on ecological systems. From this general discussion of climatic systems, we review some key human influences on the biosphere. Vitousek (1994) discussed three aspects of global change caused by human activity: (1) changes in the nitrogen cycle, (2) changes in landscapes, and (3) changes in atmospheric CO_2. Because these environmental changes influence global climate and biological diversity (see fig. 23.2), it is important to understand their causes and interactions.

23.1   A Global System

LEARNING OUTCOMES

After studying this section you should be able to do the following:

23.5   Discuss the history of how the Southern Oscillation was discovered.

23.6   Explain the relationship between the Southern Oscillation and El Niño and La Niña.

23.7   Outline oceanic and climatic conditions during El Niños and La Niñas.

23.8   Discuss the influences of the El Niño Southern Oscillation on populations and ecosystems around the world.

The El Niño Southern Oscillation, a large-scale atmospheric and oceanic phenomenon, influences ecological systems on a global scale.   Large-scale atmospheric and oceanic systems exert global-scale influences on ecological systems. One of the most thoroughly studied of these large-scale systems is the *El Niño Southern Oscillation.* The name **El Niño** originated when this climatic system seemed limited to the west coast of South America. During an El Niño, a warm current appears off the west coast of Peru, generally during the Christmas season (El Niño refers to the Christ child). The term **Southern Oscillation** refers to an oscillation in atmospheric pressure that extends across the Pacific Ocean. Before we discuss the behavior and effects of the El Niño Southern Oscillation, let's review the historical origins of our present-day knowledge of the system. This knowledge, which we take for granted today, took most of the twentieth century to gather.

The Historical Thread

In 1904, Gilbert Walker, a British mathematician, was appointed Director General of Observatories in India. Walker arrived in India shortly after a disastrous famine from 1899 to 1900 caused by crop failures during a drought. This tragic event led him to search for a way to predict the rainfall associated with the Asian monsoons. Walker (1924) eventually found a correspondence between barometric pressure across the Pacific Ocean and the amount of rain falling during the monsoons. He found that reduced barometric pressure in the eastern Pacific was accompanied by increased barometric pressure in the western Pacific. In a similar fashion, when the barometric pressure fell in the western Pacific, it rose in the eastern Pacific. Walker called this oscillation in barometric pressure the Southern Oscillation.

Today, meteorologists monitor the state of the Southern Oscillation with the Southern Oscillation Index. The value of the index is determined by the difference in barometric pressure between Tahiti and Darwin, Australia (fig. 23.4). Walker noticed that low values of the Southern Oscillation Index were associated with drought in Australia, Indonesia, India, and parts of Africa. Walker also suggested that winter temperatures in Canada were somehow connected to the Southern Oscillation. His studies led him to a global perspective on climate, a perspective well ahead of his time. Walker was highly criticized for suggesting a climatic link between such widely separated regions. However, he did not waver from his view. He assured his critics that the climatic connection between regions would eventually be explained when the proper measurements could be made.

The connection between Walker's Southern Oscillation and patterns of ocean temperature during El Niños was eventually described by Jacob Bjerknes (1966, 1969), who was able to connect the Southern Oscillation with El Niño because of a fortuitous coincidence. A strong El Niño from 1957 to 1958 coincided with the International Geophysical Year, during which oceanographic vessels made simultaneous observations across the Pacific and Indian Oceans. For the first time, scientists had extensive oceanographic data during a strong El Niño.

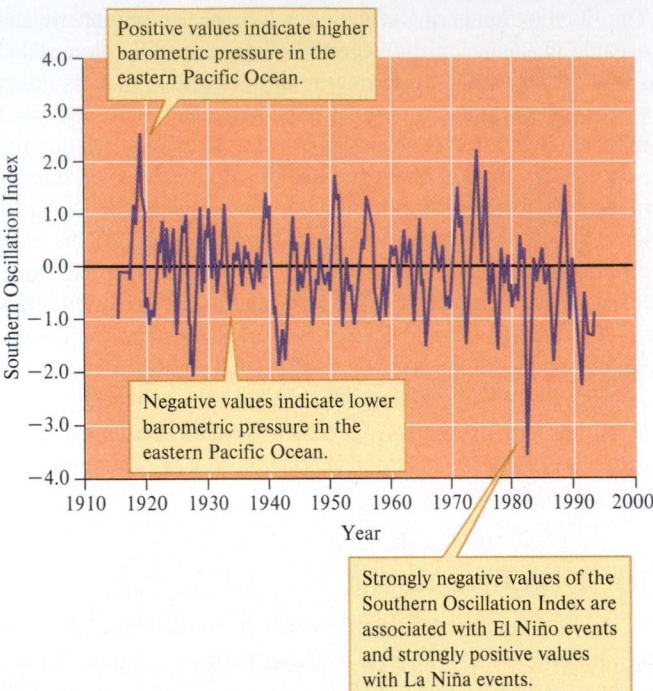

Figure 23.4   The Southern Oscillation Index shows the difference in barometric pressures between Tahiti and Darwin, Australia.

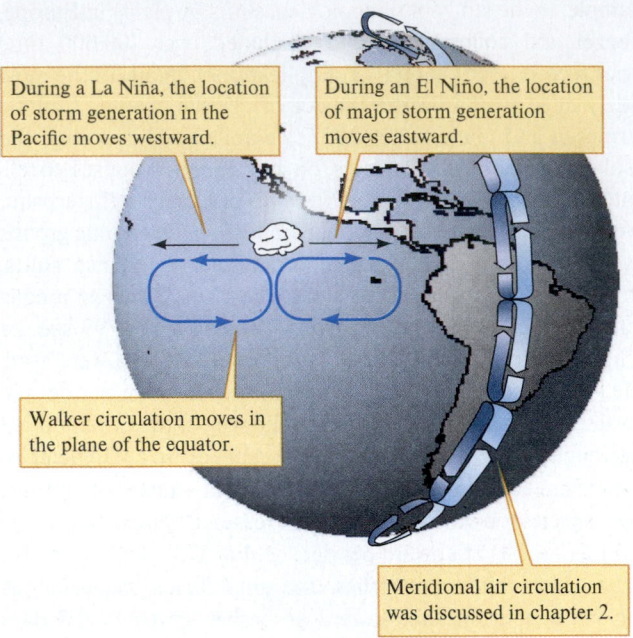

Figure 23.5   Walker circulation, El Niño, and La Niña.

Those data showed that the warm waters associated with El Niño were not limited to the west coast of South America but extended far out into the Pacific Ocean.

Bjerknes proposed that the gradient in sea surface temperature across the central Pacific Ocean produces a large-scale atmospheric circulation system that moves in the plane of the equator, as shown in figure 23.5. Air over the warmer western Pacific rises, flows eastward in the upper atmosphere, and then sinks over the eastern Pacific. This air mass then flows westward along with the southeast trade winds, gradually warming and gathering moisture. This westward-flowing air eventually joins the rising air in the western Pacific. As this warm and moist air rises, it forms rain clouds. Bjerknes called this atmospheric system **Walker circulation** after Sir Gilbert Walker.

Bjerknes, like Walker before him, possessed a global perspective before it became commonplace. Eugene Rasmusson (1985) referred to Bjerknes's model, which coupled oceanic and atmospheric circulation, as a "grand hypothesis." This hypothesis stimulated decades of research and has led to a greatly enhanced understanding of the El Niño Southern Oscillation.

El Niño and La Niña

The El Niño Southern Oscillation is a highly dynamic, large-scale weather system that involves variation in sea surface temperature and barometric pressure across the Pacific and Indian Oceans. We discuss the El Niño Southern Oscillation (ENSO) here because recent discoveries show that this system drives a great deal of climatic variability around the globe. This system affects the climate of North America, South America, Australia, southern Asia, Africa, and parts of southern Europe (fig. 23.6). This

climatic variability has substantial influences on the distribution of organisms, structure of communities, and ecosystem processes.

During the mature phase of an El Niño, the sea surface in the eastern tropical Pacific Ocean is much warmer than average and the barometric pressure over the eastern Pacific is lower than average. The combination of warm sea surface temperatures and low barometric pressure promotes the formation of storms over the eastern Pacific Ocean. These storms bring increased precipitation to much of North and South America. During an El Niño, the sea surface in the western Pacific is cooler than average and the barometric pressure is higher than average. These conditions produce drought over much of the western Pacific region, including Australia.

A period of lower sea surface temperature and higher-than-average barometric pressure in the eastern tropical Pacific of the ENSO is called a **La Niña.** La Niña brings drought to much of North and South America. During a La Niña, a pool of warm seawater moves far into the western Pacific. This warm water combined with lower barometric pressures in the western Pacific Ocean generates many storms. Consequently, La Niña brings higher-than-average precipitation to the western Pacific. It appears that La Niña and El Niño represent opposite extremes in the El Niño Southern Oscillation cycle.

While often associated with the tropics, the influence of the El Niño Southern Oscillation extends well into temperate regions. For instance, an El Niño is associated with higher-than-average precipitation over much of the western and southeastern United States and adjacent regions of Mexico. La Niña events are consistently associated with drought over most of the same region. The El Niño Southern Oscillation also affects temperatures over large geographic areas. During El Niño, event much of the northern United States, Canada, and Alaska are much warmer than average. During a La Niña,

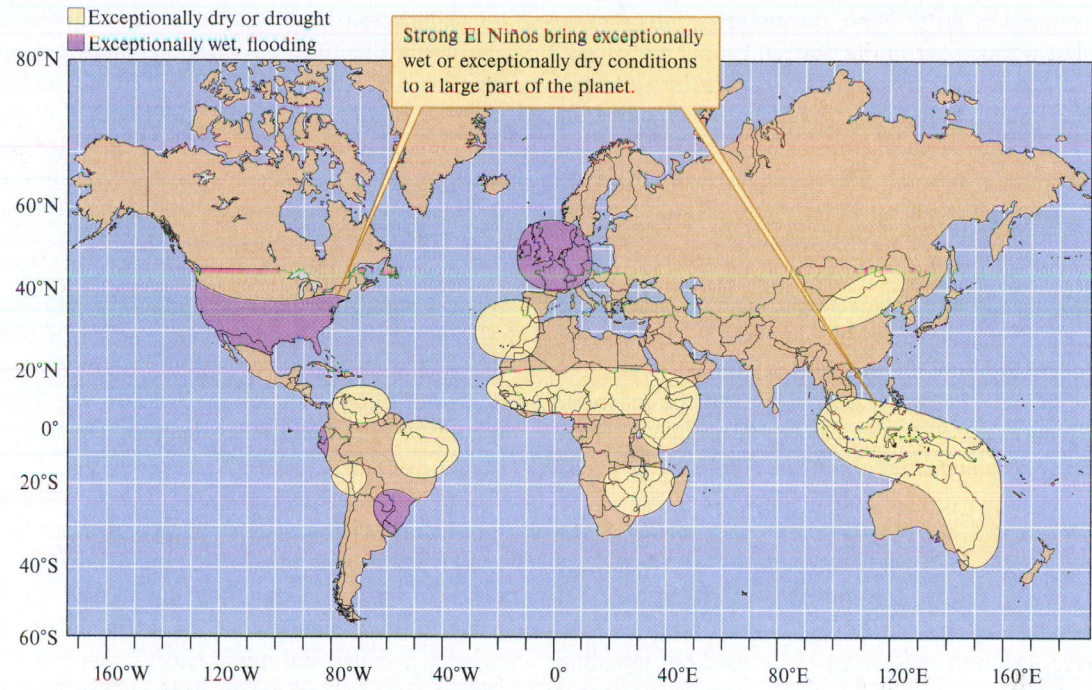

Figure 23.6   Effects of the exceptionally strong El Niño of 1982 to 1983 on patterns of global precipitation (data from Diaz and Kiladis 1992).

these regions are colder than average. As you might expect, this global climate system affects ecological systems around the globe.

El Niño Southern Oscillation and Marine Populations

Some of the most dramatic ecological responses to the El Niño Southern Oscillation occur in marine populations along the west coast of South America. Long before the recent discovery of the global extent of its effects, El Niño was known to produce declines in coastal populations of anchovies and sardines and the seabirds that feed on them. How does El Niño induce these population declines? They are produced by changes in the pattern of sea surface temperatures and coastal circulation. Figure 23.7*b* shows sea surface temperatures off the west coast of South America during average conditions. Notice that under average conditions, coastal waters are relatively cool along most of the west coast of South America and that a tongue of cool water extends westward toward the open Pacific Ocean. This cool water is brought to the surface by upwelling. Upwelling along the coast is driven by the southeast trade winds, while the offshore upwelling is driven by the east winds of the Walker circulation.

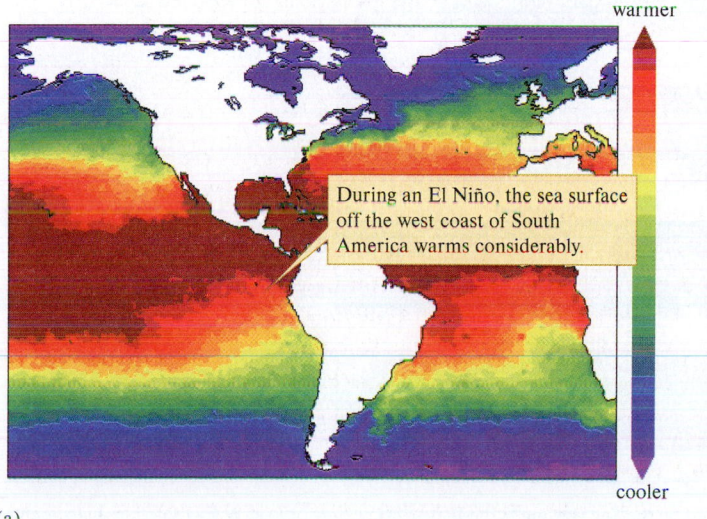

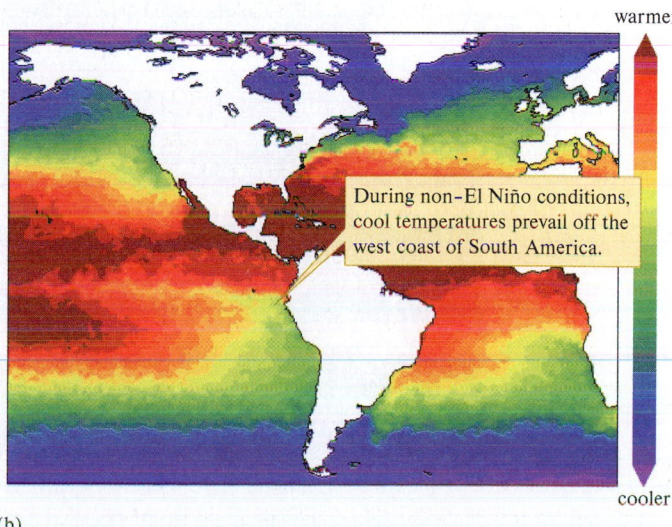

(a)   (b)

Figure 23.7   Sea surface temperature during (a) El Niño and (b) non-El Niño conditions.

With the onset of an El Niño, the easterly winds slacken and the pool of warm water in the western Pacific moves eastward. Eventually this pool of warm water reaches the west coast of South America and then moves north and south along the coast (fig. 23.7a). During the mature phase of an El Niño, the warm surface water along the west coast of South America shuts off upwelling. Consequently, the supply of nutrients that upwelling usually delivers to surface waters is also shut off. A lower nutrient supply reduces primary production by phytoplankton. This decline in primary production reduces the supply of food available to consumers in the coastal food web and is followed by declines in populations of fish and their predators.

Recently, more attention has been paid to the increased primary production associated with events that follow El Niño. Once upwelling resumed after the 1998, 2003 and 2005 El Niño events, major phytoplankton blooms were observed in the Pacific Ocean, with the most intense bloom occurring during the strong La Niña of 1998 (Ryan et al. 2006). Figure 23.8 shows how satellite observations of chlorophyll can track primary production at global scales. It has been proposed that it is the increase in nutrient availability caused by upwelling, rather than temperature itself that is responsible for increase in primary production, which is consistent with demonstrations of nutrient limitation of marine primary production reviewed in chapter 18 (section 18.2).

Changes in the rate and distribution of primary production, such as those shown in figure 23.8, induced reproductive failure, migration, and widespread death among seabird populations in the Galápagos Islands and along the west coast of South America during the 1982–83 El Niño. Many seabirds abandoned their nests with the onset of this El Niño and migrated either north or south along the coast of South America. Virtually no birds reproduced and most of the migrating birds starved. The adult populations of three seabird species on the coast of Peru declined from 6.01 million to 330,000 between March 1982 and May 1983, a decline of approximately 95%.

The 1982–83 El Niño also had a major impact on the fur seal population, mainly through reductions in food supply (fig. 23.9). Fur seals *Arctocephalus australis,* primarily feed on the anchoveta, *Engraulis ringens. Engraulis* normally lives at depths of 0 to 40 m. However, during the 1982–83 El Niño, it moved away from fur seal colonies to cooler water at depths of up to 100 m.

In response, both on the mainland and on the Galápagos, female fur seals increased their foraging time. Since females are away from their young while foraging, the pups in both populations did not get enough food and all died. On the Galápagos, nearly 100% of mature male fur seals died, while the mortality of adult females and nonterritorial males was approximately 30%. A large fur seal colony at Punta San Juan, Peru, declined from 6,300 to 4,200 individuals.

As the previous examples show, El Niño has well-documented effects on marine populations along the coast of South America. However, as we shall see in the following examples, El Niño can have ecological effects that extend far from the west coast of South America.

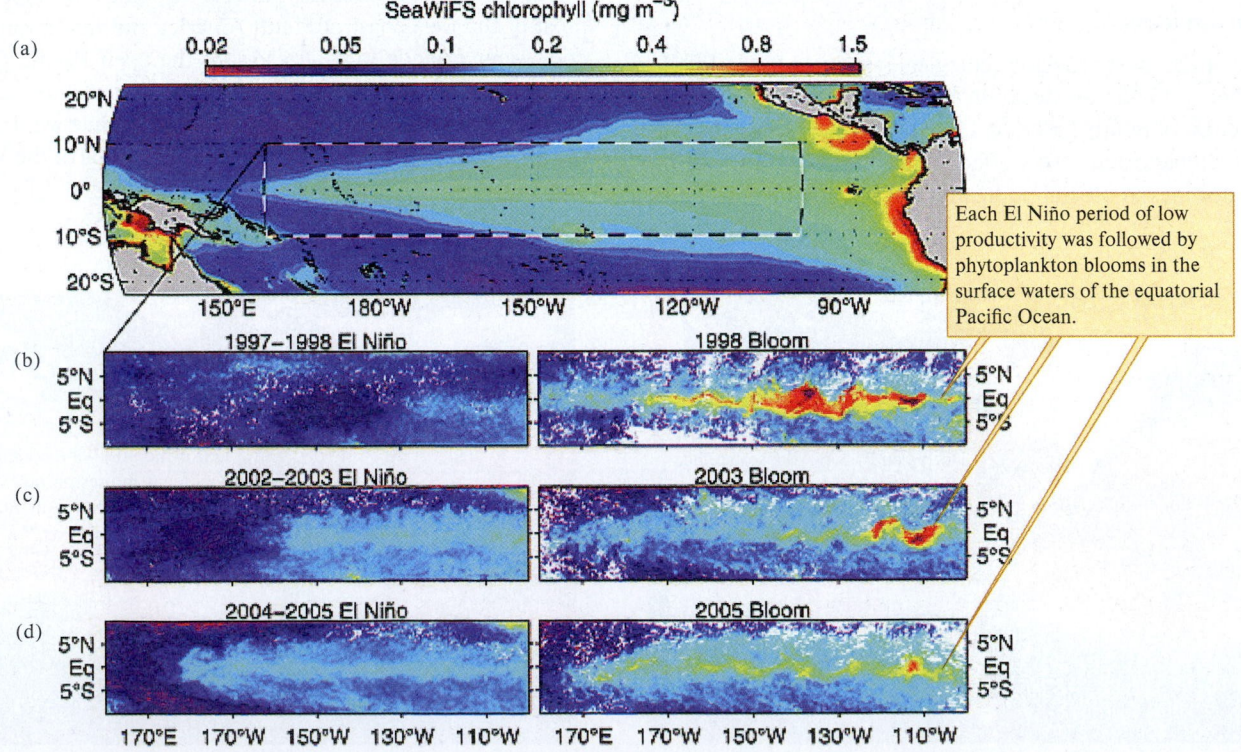

Each El Niño period of low productivity was followed by phytoplankton blooms in the surface waters of the equatorial Pacific Ocean.

Figure 23.8 Chlorophyl concentrations in the Pacific Ocean as detected by the Sea-viewing Wide Field-of-view Sensor (SeaWiFS) satellite instrument. Increases in chlorophyl, an indicator of phytoplankton growth, followed each of three El Niño events (data from Ryan et al. 2006). Source: SeaWiFS/ NASA Goddard Space Flight Center

Figure 23.9   Southern fur seals, *Arctocephalus australis,* resting on a rocky island. Populations of southern fur seals off the Pacific coast of South America and around the Galápagos Islands are decimated when the coming of El Niño reduces marine primary production.
Image Source/Getty Images

El Niño and the Great Salt Lake

El Niño can influence weather in continental areas far from the central Pacific Ocean (see fig. 23.6). The exceptionally strong El Niño of 1982–83 was the source of many moisture-bearing storms that penetrated deep into the interior of North America. These storms substantially increased precipitation within the basin of the Great Salt Lake. Another wet period, within this lake basin, soon followed with the El Niño of 1986–87. The effects of increased precipitation on the Great Salt Lake ecosystem were dramatic. Between 1983 and 1987, the level of the lake rose 3.7 m and its salinity decreased from over 100 to 50 g per liter.

Wayne Wurtsbaugh and Therese Smith Berry of Utah State University documented the ecological responses to these physical changes (Wurtsbaugh 1992; Wurtsbaugh and Smith Berry 1990). Prior to the 1982–83 El Niño, high salinity limited the zooplankton community of the Great Salt Lake to a few salt-tolerant species, especially the brine shrimp, *Artemia franciscana.* How high was the salinity of the lake? For comparison, the salinity of the open ocean is about 34 to 35 g per liter. At 100 g per liter, the lake's salinity was nearly three times that of seawater. However, by 1985–87, the lake's salinity dropped to approximately 50 g per liter, only 50% higher than seawater, and the open lake was invaded by the predaceous insect *Trichocorixa verticalis.*

Wurtsbaugh and Smith Berry found that the colonization by the predatory insects induced a trophic cascade (see chapter 18) in the pelagic zone of the Great Salt Lake (fig. 23.10). Predation reduced the population of brine shrimp from approximately 12,000 to 74 per cubic meter. Though other grazing zooplankton moved into the lake when its salinity dropped, the overall grazing rate was still greatly reduced. As predicted by the trophic cascade model, phytoplankton biomass increased significantly. This increase in phytoplankton biomass was accompanied by reduced water transparency and greatly reduced nutrient concentrations in lake water.

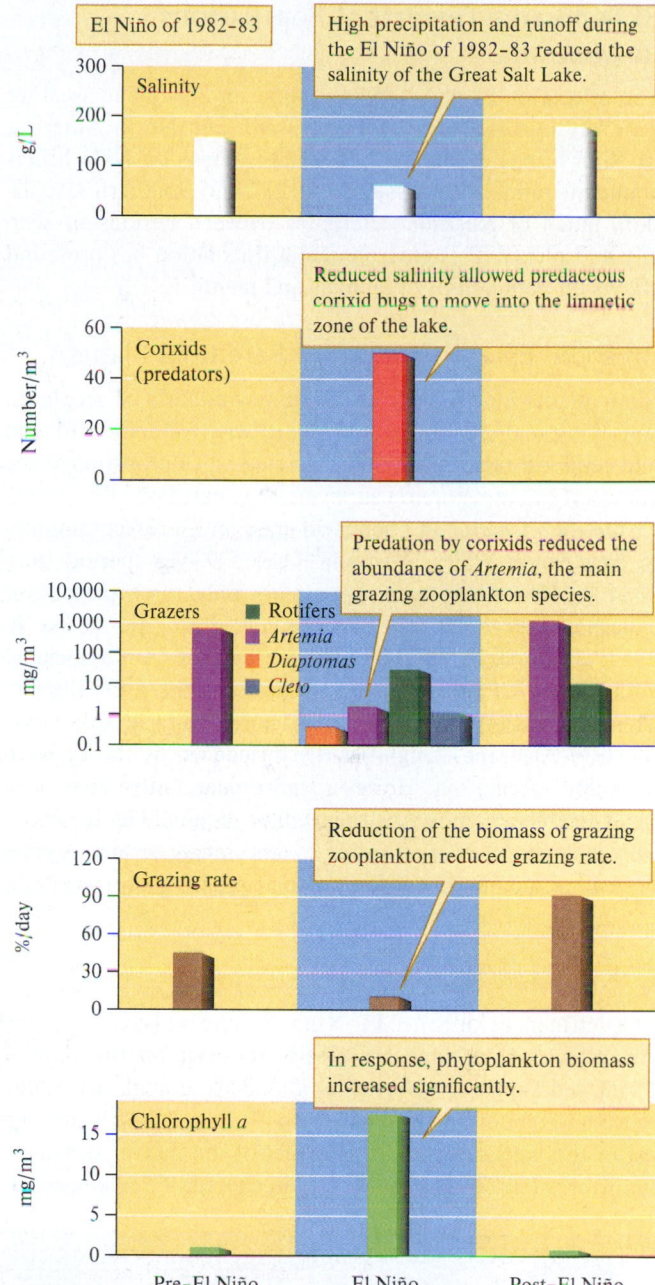

Figure 23.10   The El Niño of 1982–83 created conditions for a trophic cascade (data from Wurtsbaugh 1992).

Then from 1987 to 1990, the level of the lake fell 2.8 m and the salinity returned to over 100 g per liter. With this increase in salinity, nearly all the changes observed in the Great Salt Lake ecosystem were reversed. *Trichocorixa* was eliminated from the open lake, brine shrimp populations increased, and phytoplankton biomass declined. This example shows how a large-scale climate system can control local community and ecosystem structure and processes. It also suggests that ecological responses to global climate change are complicated by biological interactions and phenomena such as trophic cascades and keystone species effects.

El Niño and Terrestrial Populations in Australia

The effects of El Niño and La Niña on Australian weather generally mirror their effects on South and North America. El Niño brings drought to Australia whereas La Niña brings abundant rainfall. Because of the El Niño Southern Oscillation, much of Australia alternates between periods of scarcity and plenty. This environmental fluctuation has profound effects on populations of animals and plants.

Episodic Establishment by Perennial Plants

Many plants infrequently establish new cohorts of seedlings. Episodic plant establishment is particularly common in arid and semiarid regions. Graham Harrington (1991) studied the effects of soil moisture on survival of the narrow-leaf hopbush, *Dodonaea attenuata,* in a semiarid grass and shrub community in New South Wales, Australia. Over a 97-year period from 1884 to 1981, there were apparently only three periods of widespread establishment of this plant: during the 1890s, in 1952, and in 1974. All three of these periods were associated with the La Niña side of the El Niño Southern Oscillation. This association suggests that plant community structure over much of Australia is significantly influenced by the El Niño Southern Oscillation. However, more quantitative studies of plant establishment in Australia will be required before we can fully understand its relationship to this climate system. A great deal more quantitative information is available on the effects of climate on kangaroo populations.

El Niño and Kangaroo Populations

The influences of the El Niño Southern Oscillation on Australian populations are clearly reflected in the biology of the red kangaroo, *Macropus rufus.* This animal, which can reach a weight of 93 kg and a length of about 2.5 m, is the largest of the kangaroos and the largest of the native Australian herbivores (fig. 23.11). As we saw in chapter 9, red kangaroos

Figure 23.11 Populations of red kangaroos, *Macropus rufus,* are substantially influenced by El Niño and La Niña.   Carolina Garcia Aranda/ Getty Images

occupy nearly the entire arid and semiarid interior of Australia (see fig. 9.2). The range occupied by red kangaroos is a region where occasional moist periods are interspersed with severe droughts.

The reproductive biology of the red kangaroo is not tied to any fixed seasonal cycle but instead responds rapidly to changing environmental conditions, particularly increased rainfall and vegetative production. Neville Nicholls (1992) described the reproductive response of red kangaroos to a wet period as follows. During wet years, when food is plentiful, female red kangaroos will simultaneously have a juvenile, or "joey," following her, a younger offspring in her pouch nursing, and a quiescent embryo that will enter the pouch as soon as the current occupant leaves. As soon as this replacement occurs, the mother mates again, producing another embryo that will develop to the point where it is ready to enter the pouch. These relay tactics enable female red kangaroos to produce independent offspring at 240-day intervals.

Under marginal conditions, females continue to reproduce. However, most young die soon after they leave the pouch. If food becomes even scarcer, females will stop lactating and the young die in the embryo stage. Red kangaroos stop breeding only in response to severe, prolonged droughts. During droughts, these kangaroos wander widely in search of food. Females range over areas greater than 18 km^2 and males over areas of about 36 km^2. During severe droughts, females, which may reach sexual maturity as young as 15 to 20 months old, may not breed until they are over 3 years old.

When abundant rains finally come, the female hormonal system responds rapidly. The kangaroos breed quickly and young enter the pouch within 60 days of the onset of significant rainfall. Nicholls suggests that this strategy ensures a short interval between the return of good conditions and recruitment of young into the population. By reproducing large numbers of young when conditions are favorable, the animals increase the size of the adult population that will face each drought induced by El Niño.

Stuart Cairns and Gordon Grigg (1993) quantified how rainfall in southern Australia affected populations of red kangaroos (fig. 23.12). The study population reached a peak of 2,175,000 in 1981 and then declined to 745,000 by 1984. This decline was a response to drought associated with the 1982–83 El Niño. The population rebounded in 1985, held approximately steady through 1986, but then declined somewhat in 1987, during another El Niño. Then, in the last of the records analyzed by Cairns and Grigg, the population began to grow rapidly in response to the abundant rains and plant production induced by the 1988–89 La Niña. This record suggests a tight coupling of *M. rufus* populations to the El Niño Southern Oscillation. This coupling is a mirror image of the influences of El Niño on Galápagos finch populations (see fig. 11.17).

In many situations, biologists studying local populations must consider the influences of large-scale systems such as the El Niño Southern Oscillation. ENSO is only one such system; the Northern Atlantic Oscillation (NAO) is another, which

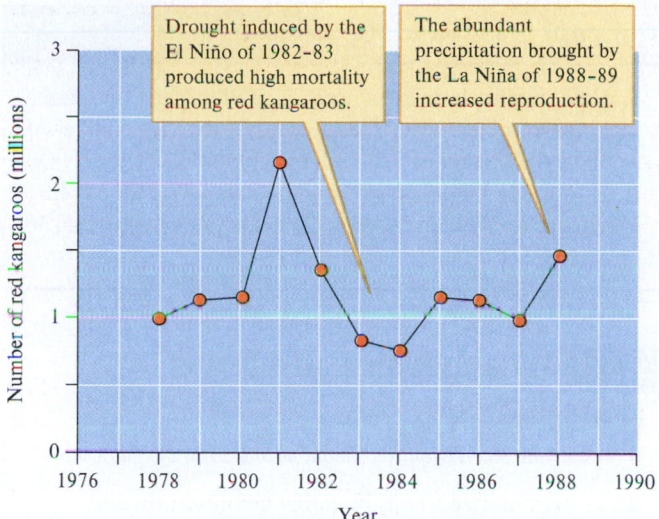

Drought induced by the El Niño of 1982–83 produced high mortality among red kangaroos.

The abundant precipitation brought by the La Niña of 1988–89 increased reproduction.

Figure 23.12   El Niño, La Niña, and population dynamics of the red kangaroo (data from Cairns and Grigg 1993).

dominates climate patterns in the northern Atlantic region. While ENSO and NAO are normal patterns of climate fluctuation on our planet, other global processes affecting climate and the movement of nutrients and organisms on our planet are not. Increasingly, ecologists must consider the effects of human modifications of the global environment. Because of the increasing size of the human population and the intensity of human activity, we are rapidly modifying global nutrient cycles, the face of the land, and even the composition of the atmosphere. The effects of the growing human population on the global environment are the subject of the next three sections.

Concept 23.1 Review

1. How are El Niño and La Niña related to the concepts of top-down versus bottom-up control of populations and ecosystems (chapter 18, section 18.4)?
2. How does the example of El Niño and the Great Salt Lake confound the concepts of top-down and bottom-up control?
3. The example of El Niño and the Great Salt Lake might lead us to what general conclusion concerning the concepts of top-down and bottom-up control?

23.2 Human Activity and the Global Nitrogen Cycle

LEARNING OUTCOMES

After studying this section you should be able to do the following:

23.9   Summarize human activities that alter the global nitrogen cycle.

23.10   Describe some ways that changes to global nitrogen cycles impact terrestrial and aquatic ecosystems.

Human activity has greatly increased the quantity of fixed nitrogen cycling through the biosphere.   When we reviewed the nitrogen cycle in chapter 19, we saw that nitrogen enters the cycle through the process of nitrogen fixation. This is the process by which N_2 gas is converted into reactive nitrogen (N_r), which includes ammonia (NH_3), nitrogen dioxide (NO_2), nitrous oxide (N_2O) and other forms. For millions of years, the only organisms that could fix nitrogen were nitrogen-fixing bacteria. Then, as humans developed intensive agricultural and industrial processes that fix nitrogen, we began to manipulate the nitrogen cycle on a massive scale.

The nitrogen fixed in terrestrial environments by free-living nitrogen-fixing bacteria and nitrogen-fixing plants totals between 40 and 127 teragrams (Tg) of nitrogen (N) per year (1 Tg = 10^{12} g). Nitrogen fixation in marine environments adds 121 to 177 Tg N per year; fixation by lightning adds about 5 Tg N per year. These estimates of nonhuman sources of fixed nitrogen total approximately 166 to 302 Tg N per year (Battye et al. 2017).

Human additions to the nitrogen cycle have increased five-fold over the past 60 years (Battye et al. 2017; fig. 23.13). The greatest source of this increase is the industrial production of fertilizers, primarily NH_3, both through mining and synthetic creation. Fertilizers currently contribute 110 Tg N per year. Another significant source is the emission of NOx (including NO_2) from the combustion of fossil fuels, including coal-fired electrical generation as well as internal combustion engines; the total is 38 Tg N per year. These two sources account for 78% of anthropogenic sources of fixed nitrogen.

Shailesh Kharol and colleagues (2018) from the United States and Canada used satellite imagery to model the geographic distribution of nitrogen deposition across those countries. They found large "hot spots" of NH_3 and NO_2 deposition in the United States, corresponding to urban/industrial centers and agricultural centers, respectively (fig. 23.14).

Humans also fix nitrogen by planting crops. At some point, agriculturists learned that rotating legumes such as alfalfa and soybeans with grains such as oats and maize could increase crop yields. We now know that those increased grain yields are due mainly to nitrogen additions to the soil by the legumes. Soybean production increased worldwide by 300% between 1910 and 1960 (Shurtlef and Aoyagi 2004). It has been estimated that crops currently fix about 43 Tg N per year. Taken with the other sources mentioned above, this means that human production of reactive nitrogen is now approximately 190 Tg N per year—much higher than natural production in terrestrial systems and nearly as much as all natural sources combined (fig. 23.13).

Anthropogenic nitrogen is problematic in a number of ways. Forms of reactive nitrogen contribute to acid rain, the reduction of ozone, eutrophication, and gaseous forms can have a climate change potential approximately 250 times that of CO_2 (IPCC, 2013). Reactive nitrogen can poison groundwater and change community dynamics of plants by favoring invasive species. Some argue that anthropogenic nitrogen is as serious a global crisis as increased CO_2 (Battye et al. 2017).

By creating environmental conditions favorable to some species and unfavorable to others, large-scale nitrogen

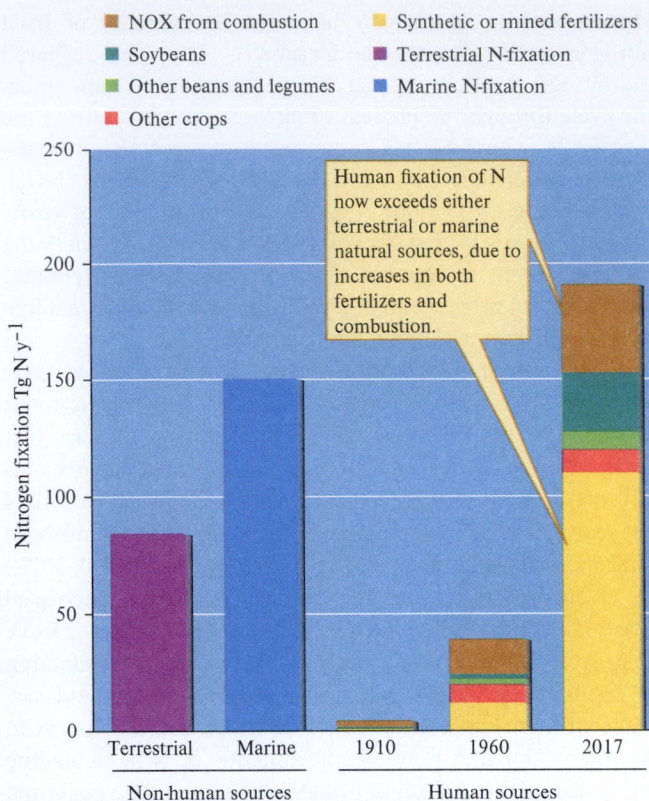

Figure 23.13 Non-human and human sources of fixed nitrogen. Human sources are shown for three time periods. Natural sources also include a small amount from lightning (not shown here) (data from Battye et al. 2017).

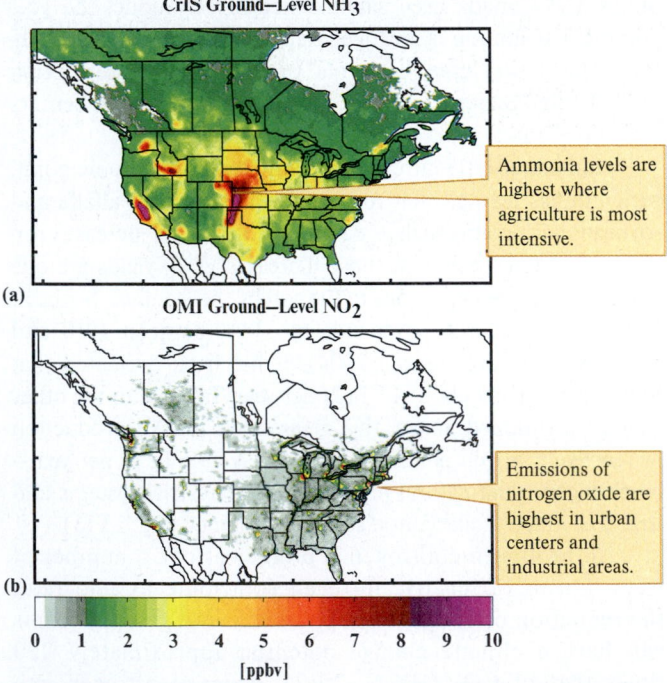

Figure 23.14 Deposits of NH_3 and NO_2 over the United States and Canada, as determined by satellite imagery (data from Kharol et al. 2018).

enrichment threatens the health of ecosystems and biological diversity. As we shall see in the next section, however, changes in land cover pose a more direct threat to biological diversity.

Concept 23.2 Review

1. How might human-induced alterations to the global nitrogen cycle impact aquatic ecosystems (see chapters 3, 18, and 19, sections 3.2, 18.1, and 19.1)?
2. How might human-induced alterations to the global nitrogen cycle affect terrestrial ecosystems (see chapters 15, 16, and 19, sections 15.1, 16.3, and 19.1)?

23.3 Changes in Land Cover

LEARNING OUTCOMES

After studying this section you should be able to do the following:

23.11 Summarize trends in global deforestation and describe how remote sensing has been used to document those trends.
23.12 Describe the negative consequences of global deforestation.
23.13 Explain how edge effects compound the impacts of tropical deforestation.

Rapid changes in global patterns of land use threaten biological diversity. Humans have changed the face of the earth. Human activities, mainly agriculture and urbanization, have significantly altered one-third to one-half of the ice-free land surface of the earth. Marshes have been drained and filled to build urban areas or airports. Tropical forests have been cut and converted to pasture. The courses of rivers have been changed. The Aral Sea in central Asia has been so starved for water that it is nearly dry. Vitousek suggested that changes in land cover may be the greatest single threat to biological diversity (see fig. 23.2). Let's review some of the changes in land cover and the mechanisms that make landscape changes such a powerful threat to biological diversity. A widely cited example of land cover change is the cutting of forests.

Deforestation

According to the Food and Agriculture Organization of the United Nations (FAO) State of the World's Forests 2020 report, forests covered 30.8% of global land area in 2020. This is a decrease from 32.5% in 1990, representing the loss of 420 million ha in three decades (FAO 2020). This is alarming because most of the world's biodiversity exists in forests, providing habitats for 80% of amphibian species, 75% of bird species, and 68% of mammal species. Tropical forests alone account for 60% of all vascular plant species. These organisms and their ecosystems provide a wide variety of ecosystem services for humans, including food, pollination of crops, firewood, medicines, and homes.

More than half of the world's forests are found in only five countries: Brazil, Canada, China, Russian Federation, and the United States; China representing the largest portion (20%) and Brazil next most (12%) (FAO 2020). Historically, the greatest

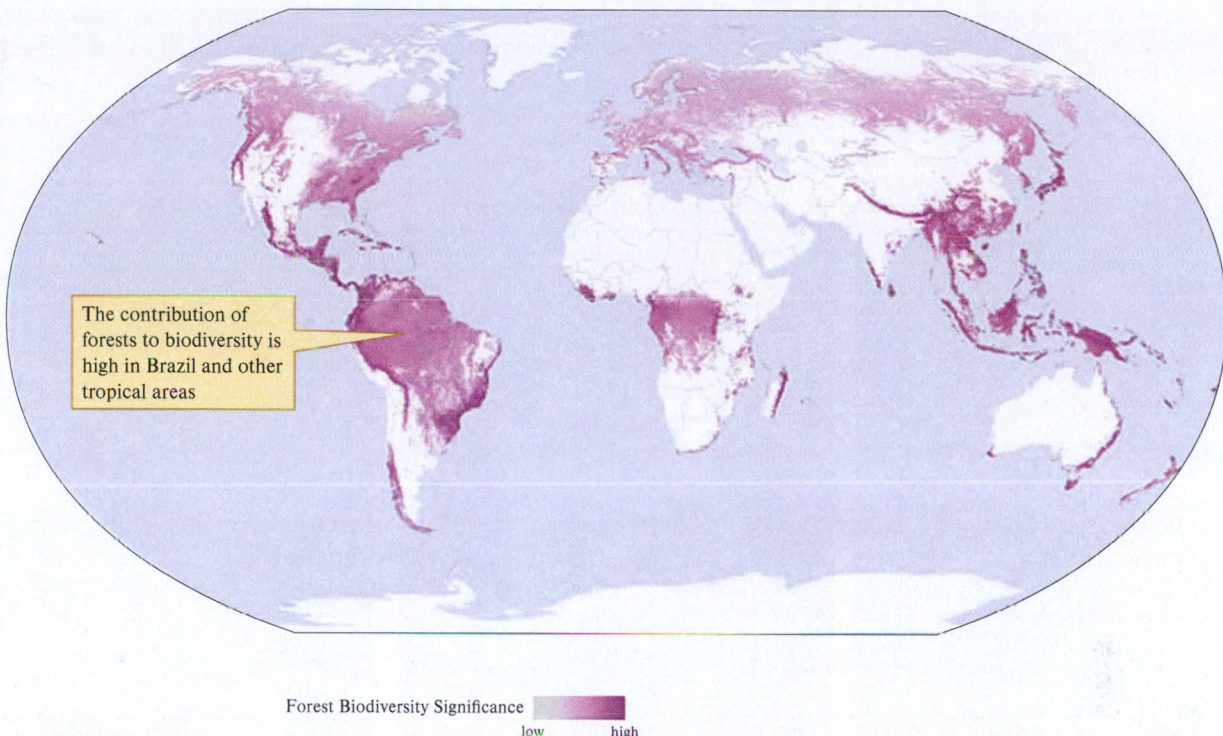

The contribution of forests to biodiversity is high in Brazil and other tropical areas

Forest Biodiversity Significance
low   high

Figure 23.15   Significance of forest biodiversity across the globe, as measured by the regional contribution to mammal, bird, amphibian and conifer species. (data from Hill et al. 2019).

forest losses have occurred in South America, particularly Brazil. Brazil is particularly important because of its contribution to global biodiversity (fig. 23.15). As just one indication, Brazil has more tree species than any other country, over 9,000, with approximately half as endemic species (Beech et al. 2017). Loss of forest in the Brazilian Amazon presents not only an extinction risk to many species but also threats to the nearly 900,000 indigenous peoples who live there (https://www.iwgia.org).

While deforestation in Brazil have often been understood to be very high, actual estimates of those rates varied widely. To address the need to directly measure change in tree cover, we began using satellite imagery. In the early 1990's, Skole and Tucker set out to provide an accurate estimate of deforestation rates in the Amazon Basin using remote sensing by *Landsat* satellites (Skole and Tucker 1993). The images they used, *Landsat* Thematic Mapper photos, provide high-resolution information. As you can see in figure 23.16, Thematic Mapper photos clearly show areas of deforestation, regrowth on deforested plots, and areas of isolated forest. Skole and Tucker entered these high-resolution images into a geographic information system (see chapter 22, Applications), which they used to create computerized maps of deforestation within the Amazon Basin.

Let's look at the maps in figure 23.17 and see what they tell us. First, notice that large areas—cerrado—colored light brown in the map, were not forested. These areas, concentrated in the southeastern Amazon Basin, have a semiarid climate and support scrubby vegetation. Some areas, shown in light violet, were covered by clouds and could not be analyzed. Skole and Tucker used the 1978 image to estimate the amount of deforestation that had occurred

prior to 1978. They then compared the 1978 and 1988 photos to determine the amount of deforestation during that decade. They divided the Amazon Basin into 16 km by 16 km squares for their analysis. One of those areas is enlarged in insets on figure 23.17. These insets show the amount of deforestation in 1978 (fig. 23.17*a*) and in 1988 (fig. 23.17*b*). Notice that the deforested area in the inset increased significantly between 1978 and 1988.

Skole and Tucker used their analyses of 16 km by 16 km areas to estimate the percentage of the land surface that had been deforested across the entire Amazon Basin. On their maps, white indicates completely forested areas, while various colors indicate increasing degrees of deforestation. At one end of their spectrum, gray indicates 0.25% to 5% deforested; at the other end, red indicates 90% to 100% deforested. Notice that the color of the area covered by the inset is purple on the 1978 map and green on the 1988 map. Change in color on the map of the entire basin from 1978 to 1988 reflects the increase in deforested area during that period.

How much of the Brazilian Amazon has been deforested? Skole and Tucker estimated that the total area deforested by 1988 was 230,000 km^2, at an annual rate of 15,000 km^2. That same year, the Brazilian government founded the Amazon Deforestation Monitoring Project (PRODES; www.obt.inpe.br/prodes), which also used GIS to measure changes in tree cover. According to its numbers, peak deforestation occurred in 2004 when deforestation was taking place at a rate of over 27,000 km^2/year. Implementation of laws to decrease deforestation were made that successfully decreased deforestation, and Alexandra Tyukavina and colleagues (2017) reported that by 2013 the rate of deforestation in the Brazilian Amazon was

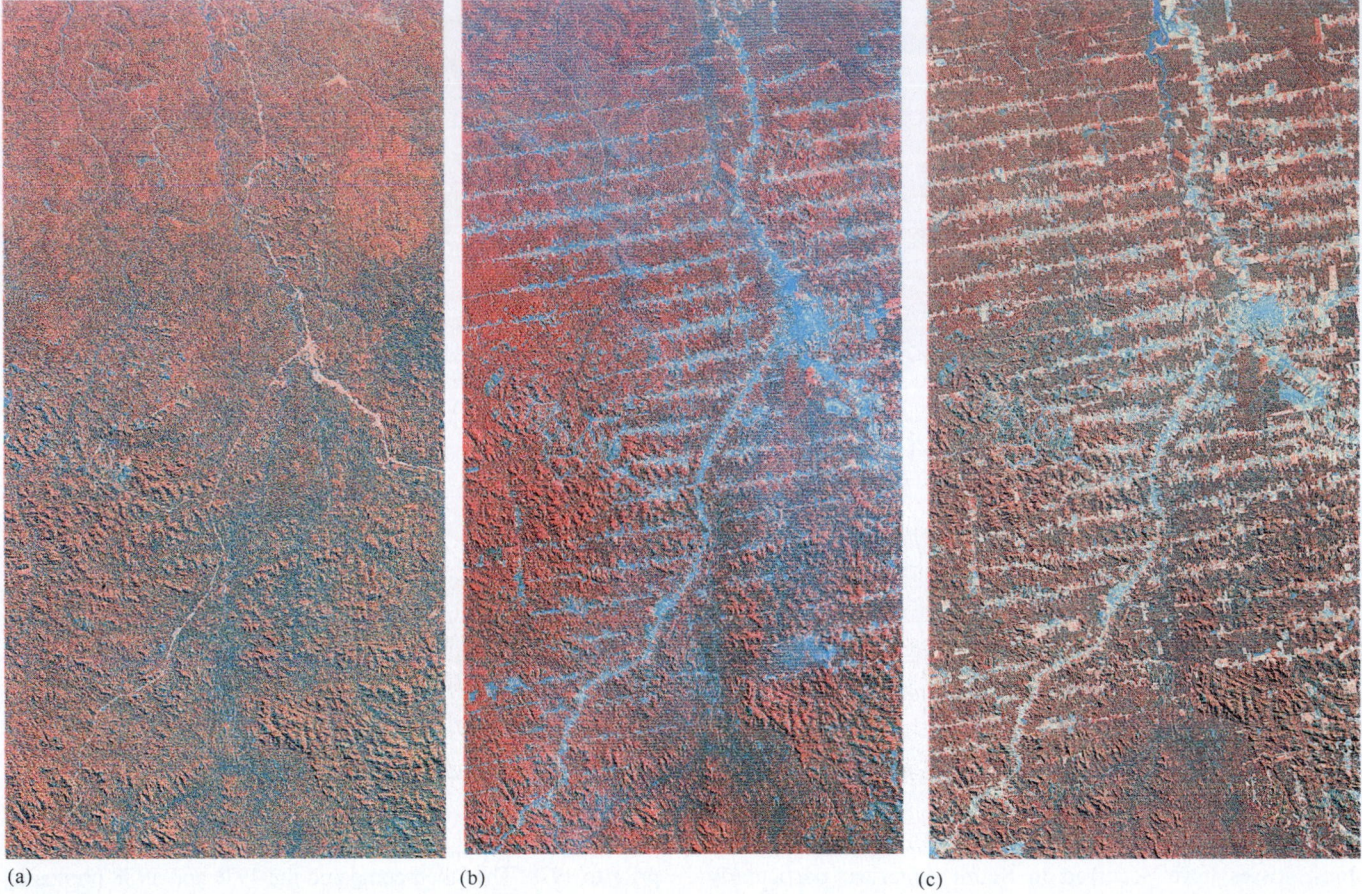

(a)

(b)

(c)

Figure 23.16   Information on tropical deforestation from satellite images: deforestation in Rondônia State, Brazil (light areas), in (*a*) 1975, (*b*) 1986, and (*c*) 1992.   (*a*) Landsat images courtesy of Eros Data Center/USGS; (*b*) Landsat images courtesy of Eros Data Center/USGS; (*c*) InterNetwork Media/Getty Images;

approximately 6,000 km^2/yr. However, deforestation in Brazil has increased again in recent years to pre-1990's rates (Escobar 2020). This has been due to a change in administration which relaxed deforestation regulations, combined with devastating fires linked to climate change (Seymour and Harris 2020).

Remote sensing and GIS are now being used not only to identify general deforestation patterns but even to distinguish between types degradation. The use of geographic information systems in ecology is growing in importance and sophistication, and our ability to measure large-scale processes such as deforestation is expected to continue to improve.

Edge Effects and Forest Fragmentation

The area of forest removed does not give a complete picture of the ecological effects of deforestation. When a tract of forest is cut, the adjacent forest is affected by changes in the physical environment along its edges, by reduced habitat area, and by isolation. Let's look at the nature of these "edge effects" in Amazonian forest fragments.

In 1979, Brazil's National Institute of Amazonian Research and the World Wildlife Fund began a long-term study of tropical forest fragmentation. This research project took advantage of a Brazilian law that requires that 50% of land developed in the Amazon Basin remain forested. The researchers worked with ranchers to leave forested tracts in particular areas to facilitate research on the ecological influences of forest fragment size and isolation. The fragments studied were 1, 10, 100, and 200 ha (fig. 23.18). These were compared to areas of 1, 10, 100, and 1,000 ha in undisturbed forest.

When a small fragment of forest is isolated by cutting the surrounding forest, its edge is exposed to greater amounts of solar radiation and wind. Wind and sun combine to change the physical environment within forest fragments. The physical environment along forest edges is hotter and drier and the intensity of solar radiation is higher. These physical changes, in turn, affect the structure of the forest community. Tree mortality is higher along the edges of forest fragments, and the forest overstory decreases while the thickness of the understory vegetation increases. Fragmentation also decreases the diversity of many animal groups, including monkeys, birds, bees, and carrion and dung beetles. Some of these reductions in animal populations may have significant impacts on key ecological processes such as pollination and decomposition.

Because edge effects, isolation, and reduced habitat area negatively affect biological diversity within tropical forest fragments, Skole and Tucker extended their analysis of deforestation in the Amazon Basin to include these effects. They assumed that edge effects extend for 1 km from the forest edge. Edge effects more than doubled the area—from 230,000 to 588,000 km^2—of Amazonian forest affected by deforestation.

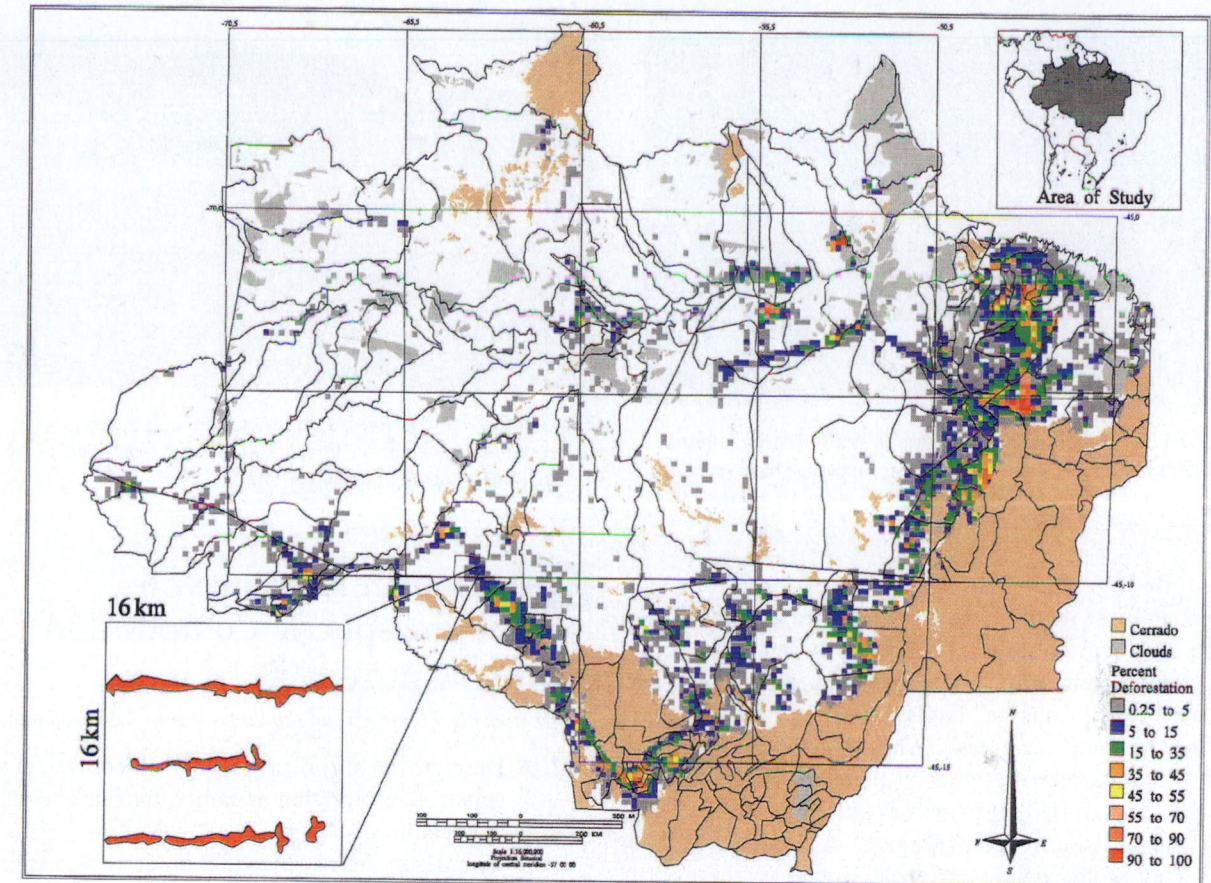

(a)

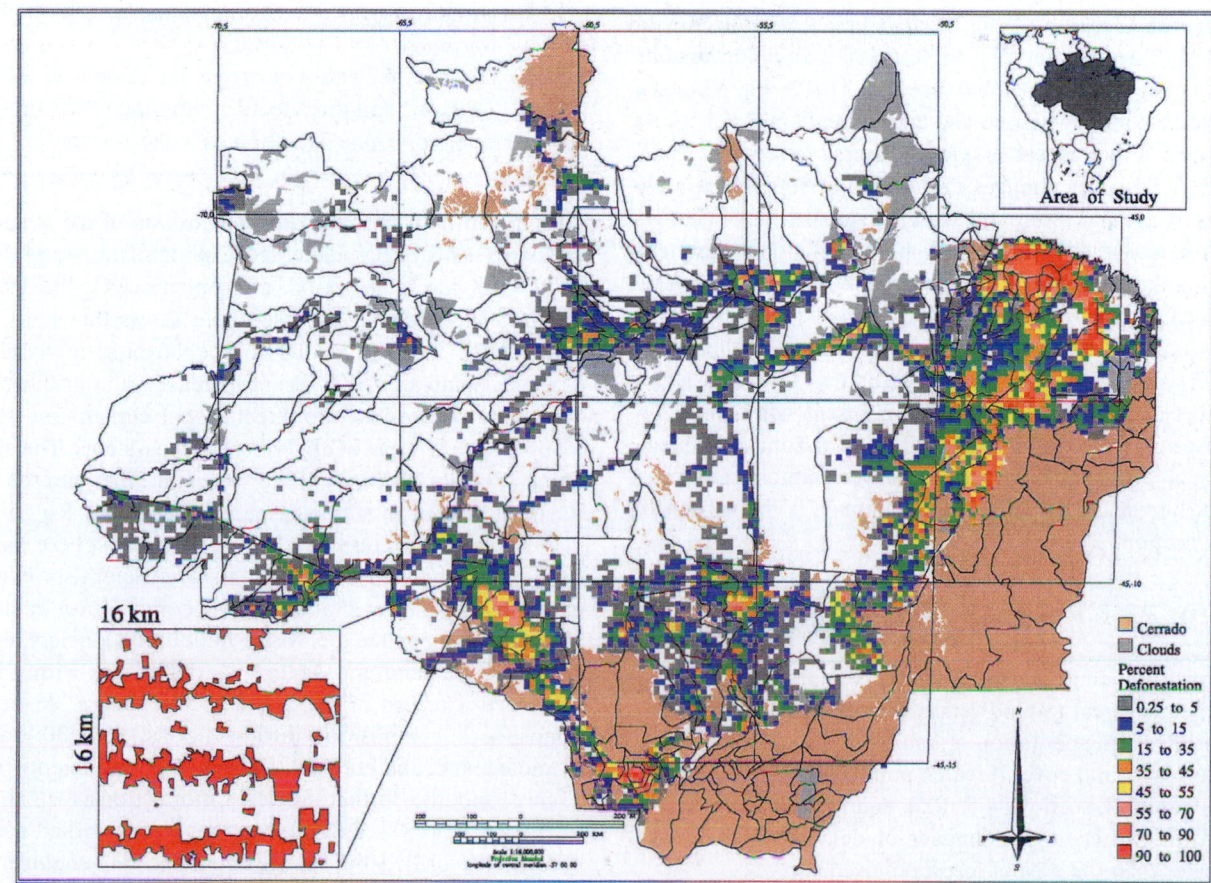

(b)

Figure 23.17   Deforestation in Amazonia between (*a*) 1978 and (*b*) 1988 (data from Skole and Tucker 1993).

Figure 23.18 Deforestation in the Amazon River Basin. Forest fragments left by clear-cutting the surrounding landscape have very different physical environments than intact forest.   Richard O. Bierregaard, Jr./ University of North Carolina, Biology Department

Figure 23.19 Deforestation is not limited to the tropics: deforestation in a temperate forest of the Pacific Northwest of the United States.   steve estvanik/Shutterstock

Although most people have focused on tropical deforestation, massive deforestation has occurred in temperate and boreal regions; the majority of old-growth temperate forests in northwestern North American has been cut (fig. 23.19). It has been estimated that human activity has transformed approximately half the ice-free land cover of the earth (Riggio et al. 2020), with some saying that there is as little as 5% completely free from human modification (Kennedy et al. 2019). In the process, many of the major terrestrial biomes of the earth (see chapter 2) have been highly fragmented. Others, such as tropical dry forest, have been nearly eliminated by conversion to agriculture. Because of the negative effect of reduced area on diversity (see chapter 22), these massive land conversions present a major threat to global diversity. That is why Vitousek suggested that human-caused changes to land cover constitute the greatest direct threat to global diversity (see fig. 23.2). However, land cover changes also have the potential to contribute, directly and indirectly, to rapid global climate change. One of the ways that deforestation may affect global climate is by altering the influence of forests on atmospheric concentrations of CO_2; the TREES-3 project estimated that, based on their estimates, deforestation results in annual additions of 602–1,237 million metric tons of carbon to the atmosphere (Achard et al. 2014). Global forests represent significant CO_2 sinks, especially in tropical and subtropical zones (Requena Suarez et al. 2019). The connection between atmospheric composition, human activity, and global climate is the subject of the next section.

Concept 23.3 Review

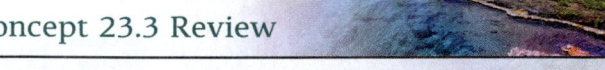

1. Why is reducing forest area through deforestation a fundamental threat to biodiversity (see chapter 22)?
2. How does fragmentation of habitat as a result of human changes in land cover threaten populations (see chapters 4, 10, and 21, sections 4.5, 10.2, and 21.2)?
3. Why is the ecological impact of deforestation always greater than the area of forest removed?

23.4 Human Influence on Atmospheric Composition

LEARNING OUTCOMES

After studying this section you should be able to do the following:

23.14 Describe the historical relationship between atmospheric concentration of carbon dioxide and global temperatures.
23.15 Summarize the evidence that modern increases in atmospheric concentration of carbon dioxide are related to the burning of fossil fuels.
23.16 Outline projected environmental effects of global warming.
23.17 Discuss the history of ozone depletion over the Antarctic and the effects of banning CFCs on the extent of ozone depletion over the Antarctic.

Human activity is changing the composition of the atmosphere. Industrial activity has increased exponentially since the year 1800. Over the same period, atmospheric CO_2 has increased steadily. The evidence discussed here shows that most of this atmospheric increase is due to the burning of fossil fuels. Vitousek pointed out that recent increases in atmospheric CO_2 concentration are likely to affect global climate and will certainly affect the biota of all terrestrial ecosystems. The effect of human activity on atmospheric CO_2 and other gases is one of the most thoroughly studied aspects of global ecology.

The concentration of CO_2 in the atmosphere has been dynamic over much of earth's history. Scientists have very carefully reconstructed atmospheric composition by studying air bubbles trapped in ice. As ice built up on glaciers in places such as Greenland and Antarctica, air spaces within the ice preserved a record of the ancient atmosphere. A record of atmospheric composition during the last 160,000 years was extracted and analyzed by a joint team of scientists from France and the former Soviet Union (Lorius et al. 1985; Barnola et al. 1987). This international team studied a 2,083 m core of ice drilled by Soviet scientists and engineers near the Antarctic station of Vostok. Vostok, located in eastern

Antarctica at a latitude of over 78° S, has a mean annual temperature of –55°C, ideal conditions for preserving samples of the atmosphere in ice. The Vostok research station sits on the high Antarctic plateau, where the ice is about 3,700 m thick. The amazing physical feat of extracting such a long ice core in such difficult physical circumstances is equaled by the dramatic climatic record contained within the Vostok ice core.

To extract air trapped within ice, scientists place sections of an ice core into a chamber and create a vacuum, removing traces of the current atmosphere in the process. The ice, still under vacuum, is then crushed and the air it contains is released into the chamber. Sampling devices then measure the CO_2 concentration of the air released from the ice. The Barnola team made 66 measurements of CO_2 along the length of the Vostok ice core. The scientists made measurements of CO_2 every 25 m along the length of the ice core from about 850 m depth to the bottom of the core. These lower sections of the core correspond to ages from 50,000 to 160,000 years. Because there were many fractures in the core above 850 m depth, the upper portion of the core was generally sampled at intervals greater than 25 m.

Another team of scientists drilling at a second location (Lüthi et al. 2008) extended the atmospheric CO_2 record, determined from Antarctic ice cores, to 800,000 years before the present (fig. 23.20). Overall, the record shows that CO_2 concentrations have oscillated between low concentrations of approximately 170 to 300 parts per million (ppm). About

750,000 years ago, the atmospheric concentration of CO_2 was less than 180 ppm. This early period in the ice core corresponds to an ice age. Then, about 740,000 years ago, the atmospheric concentration of CO_2 began to rise abruptly. This rise in CO_2 corresponds to a warmer interglacial period. High levels of CO_2 persisted until about 680,000 years ago. This 800,000-year record now provides measurements of atmospheric CO_2 concentrations spanning eight glacial (ice age) and interglacial periods.

Notice that the fluctuations in CO_2 within the ice core correspond to variation in temperature (see fig. 23.20). The periods of low CO_2 correspond to the low temperatures experienced during ice ages, whereas the periods of high CO_2 correspond to warmer, interglacial periods (Jouzel et al. 2007).

The most recent measurements in the Antarctic ice cores are about 2,000 years old. How has atmospheric CO_2 varied during the most recent 2,000 years? W. Post and colleagues (1990) assembled atmospheric CO_2 records from a number of sources to estimate atmospheric concentrations during the last 1,000 years (fig. 23.21). The first 700 years of the record come from the South Pole ice core, which was analyzed by Ulrich Siegenthaler and colleagues (1988) of the University of Bern, Switzerland. This record shows that the concentration of CO_2 remained relatively constant for approximately 800 years. Another study at the University of Bern provided a CO_2 record for the most recent 200 years (Friedli et al. 1986). This part of the CO_2 record comes from the Siple ice core, from Siple

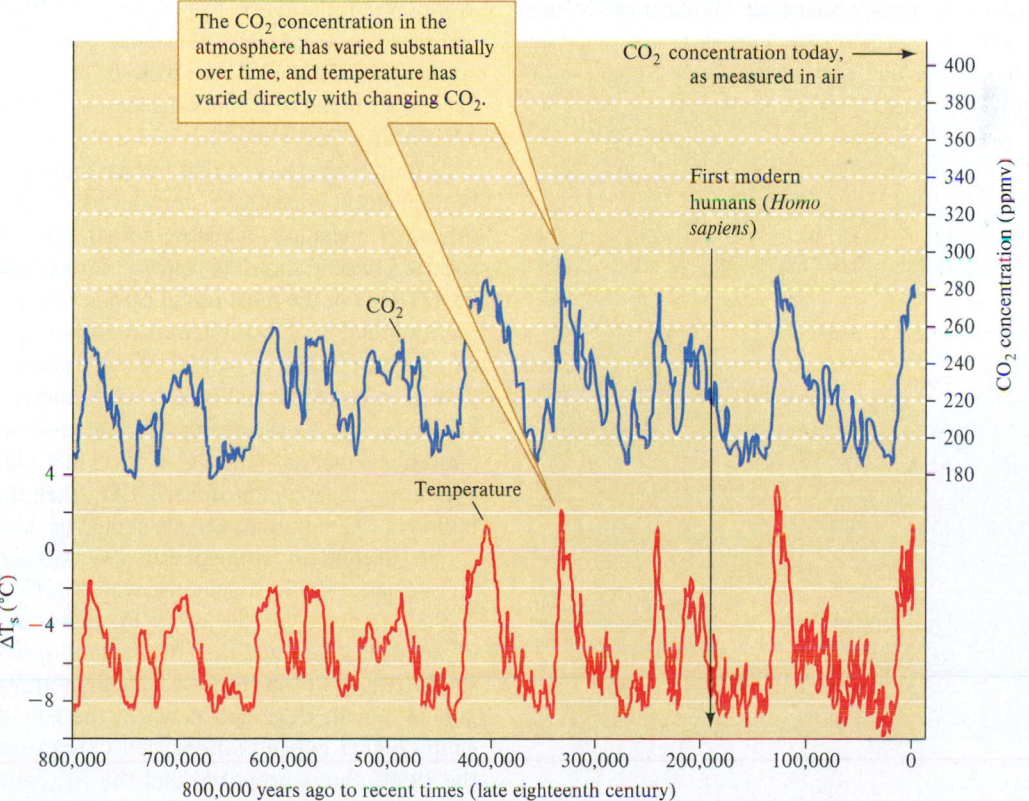

Figure 23.20   Variation in temperature and CO_2 concentrations over the last 800,000 years, as determined by ice core records from Vostok, Antarctica. CO_2 levels are currently the highest they have been in human history. (CO_2 data from Lüthi et al. 2008; temperature data from Jouzel et al. 2007. Current CO2 levels from NOAA 2021)

Station at about 75° S latitude. While the Siple ice core does not allow us to look as far back in time as the deep core records, it provides a very detailed estimate of recent concentrations of atmospheric CO_2. H. Friedli and colleagues dated the beginning of the Siple record at about A.D. 1744. At that time, about two and a half centuries ago, the atmospheric concentration of CO_2 was about 277 ppm. This estimated concentration is almost identical to those made by the Siegenthaler team for the same time period using the South Pole ice core. Therefore, both the South Pole and Siple ice cores indicate that the CO_2 concentration in the middle 1700s was approximately the same as at the end of the Vostok record, about 2,000 years earlier.

The Siple record showed that CO_2 increased exponentially from 1744 to 1953. The Friedli team estimated the 1953 concentration of CO_2 at 315 ppm. However, the trace in CO_2 concentrations shown in figure 23.21 extends beyond 1953 and above 315 ppm. Where do these later measurements come from? These later CO_2 concentrations are direct measurements made on Mauna Loa, Hawaii, by Charles Keeling and his associates over a period of about 40 years (Keeling and Whorf 1994).

Keeling's measurements complement the ice core data from the Vostok, South Pole, and Siple stations in two ways. First, they extend the record into the present. Second, they help validate the measurements of CO_2 made from the ice cores. How do Keeling's measurements lend credence to the ice core data? Look carefully at the plot of CO_2 concentrations shown in figure 23.21. Notice that two of the measurements made from the Siple ice core overlap the period when Keeling and his team made measurements at Mauna Loa. Notice

also that the two estimates made independently by Keeling at Mauna Loa and by Friedli and his colleagues from the Siple ice core are almost identical.

The data in figure 23.21 indicate that during the nineteenth and twentieth centuries the concentration of atmospheric CO_2 increased dramatically. Measurements at Mauna Loa continue to this day, and there is no indication of an end to the increase. Since the first publication of this textbook, we have had to extend the y-axis of this graph by 50% to accommodate the rising curve; in summer of 2020 Mauna Loa recorded a record high value of 417 ul/L (Scripps CO_2 Program).

This period of increase coincides with the Industrial Revolution. However, what evidence is there that human activity caused this observed increase? Vitousek provided evidence by pointing out that the annual increase in atmospheric carbon in the form of CO_2 is about 3,500 Tg (1 Tg = 10^{12} g), while the annual burning of fossil fuels releases about 5,600 Tg carbon as CO_2. Therefore, fossil fuel burning alone produces more than enough CO_2 to account for recent increases in atmospheric concentrations. Furthermore, the increase in atmospheric concentrations of CO_2 from the burning of fossil fuels corresponds to increasing temperatures over the same period (fig. 23.22).

If we look carefully at the pattern of CO_2 increase between 1880 and 2016, we find additional evidence for a human influence. Figure 23.22 shows three interruptions in the otherwise steady increase in the burning of fossil fuels. Those periods correspond to three major disruptions of global economic activity: World War I, the Great Depression, and World War II. At the end of each of these major global upheavals, the increase in atmospheric CO_2 resumed. These patterns provide circumstantial evidence that humans are responsible for the modern increase in atmospheric CO_2. However, there is also direct evidence.

Additional evidence that human industrial activity is at the heart of recent increases in atmospheric CO_2 and therefore likely driving temperature increases comes from analyses of atmospheric concentrations of various carbon isotopes (see chapter 18). One of the most useful carbon isotopes for determining the contribution of fossil fuels to atmospheric CO_2 is radioactive ^{14}C. Because ^{14}C has a half-life of 5,730 years, fossil fuels, which have been buried for millions of years, contain very little of this carbon isotope. Consequently, burning fossil fuel adds CO_2 to an atmosphere that has little ^{14}C. If fossil fuel additions are a major source of increased atmospheric CO_2, then the relative concentration of ^{14}C in the atmosphere should be declining.

A decline in atmospheric ^{14}C was first described by Hans Suess (1955), a scientist with the U.S. Geological Survey. Suess made his discovery by analyzing the ^{14}C content of wood. He analyzed the ^{14}C content of wood laid down by single trees at various times during their growth. He found that annual growth rings laid down in the late 1800s had significantly higher concentrations of ^{14}C than those laid down in the 1950s. Suess proposed that the ^{14}C content in wood was being progressively reduced because burning of fossil fuels was reducing the atmospheric concentration of ^{14}C. Because of his pioneering work, reduced atmospheric ^{14}C as a consequence of fossil fuel burning is called the **Suess effect.**

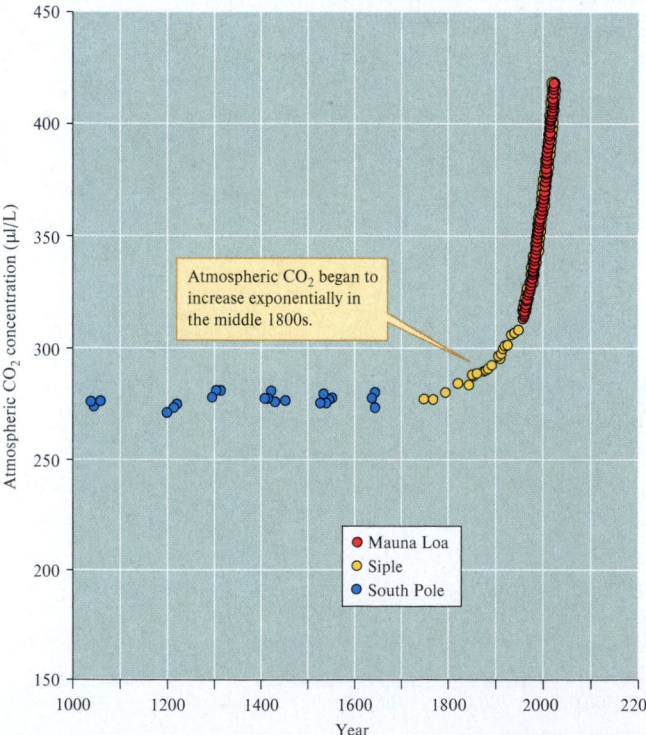

Within the figure: Atmospheric CO_2 began to increase exponentially in the middle 1800s.

Legend: ● Mauna Loa ○ Siple ● South Pole

Figure 23.21   A 1,000-year atmospheric CO_2 record. Note that there are many more measurements from Mauna Loa, where data are collected almost monthly. (data from Post et al. 1990; Neftel et al. 1994; Keeling et al. 2001; Scripps CO_2 Program 2020).

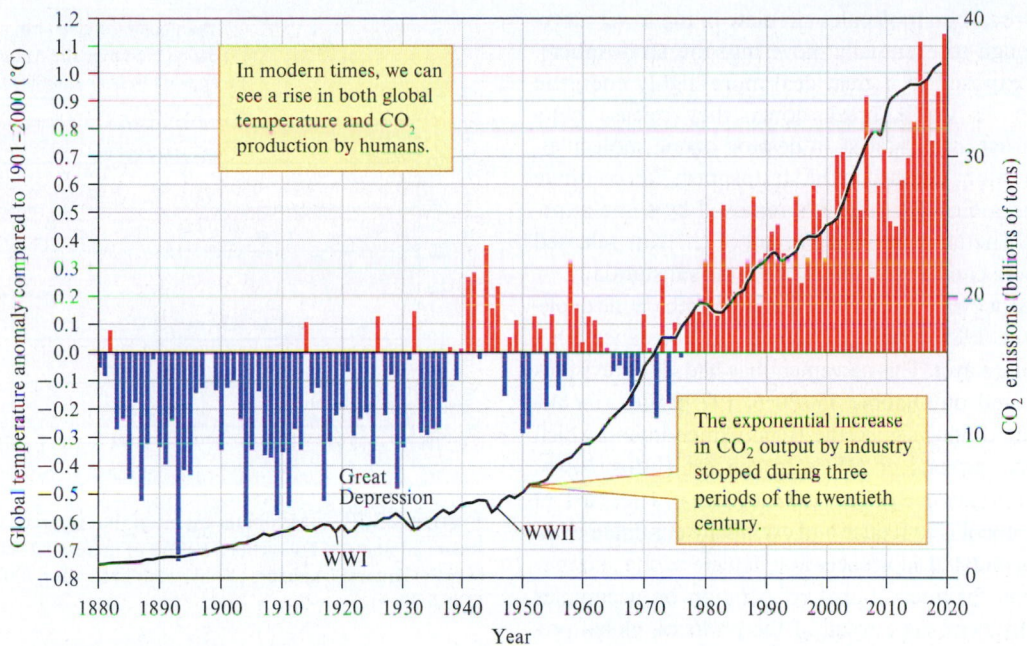

Figure 23.22   Increases in CO_2 from fossil fuel burning and deviations in mean annual land and ocean temperatures during the modern era. Blue bars indicate temperatures lower than long-term average; red bars indicate temperatures higher than long-term average (data from Bacastow and Keeling 1974; Knorr 2009; and https://www.ncdc.noaa.gov/sotc/global/202101).

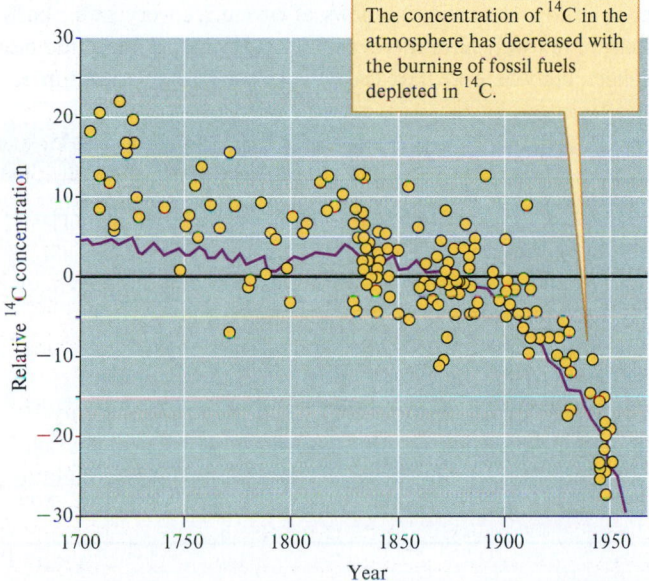

Figure 23.23   The Suess effect (data from Bacastow and Keeling 1974).

Robert Bacastow and Charles Keeling (1974) compiled ^{14}C data from several studies of ^{14}C in trees and plotted the date when the wood was formed against the relative ^{14}C content of the wood. As figure 23.23 shows, the concentration of ^{14}C was fairly stable from A.D. 1700 until about 1850. After 1850, ^{14}C concentrations in wood declined significantly. The line shows the predictions of ^{14}C made by a model built by Bacastow and Keeling. Their model made these predictions based on global patterns of fossil fuel burning and estimated rates of exchange of carbon between the ocean, the earth's biota, and the atmosphere.

Depletion and Recovery of the Ozone Layer

In 1985, scientists of the British Antarctic Survey discovered a major reduction in the amount of ozone, O_3, in the stratosphere over the Antarctic. Stratospheric ozone absorbs potentially harmful ultraviolet light, particularly UV-B light, or radiation. Because high-energy, UV-B radiation is capable of destroying biological molecules and damaging living tissue, the ozone layer is critical for the well-being of life on earth. The British scientific team also analyzed historical measurements of ozone, which demonstrated clearly that the total amount of ozone over the Antarctic had been declining since the 1970s. Depletion of earth's ozone layer was not the first sign of human influence on the environment. However, it was a clear and dramatic indication that human impact on the environment had achieved truly global proportions.

Though the ozone hole was centered over the Antarctic far from most human population centers, its discovery generated widespread concern. Perhaps the greatest fear was that the breakdown of the ozone layer over the Antarctic might be a prelude to breakdown of the protective ozone layer over the entire earth, endangering humans as well as crops, wild plants, and animals. Other scientists had warned that the ozone layer was threatened by human activities. However, it was the discovery of the ozone hole that aroused world concern and stimulated international action. Attention was quickly focused on stopping the production of chlorofluorocarbons, or CFCs, organic chemicals containing carbon, chlorine, and fluorine that were widely used as refrigerants. Because CFCs are very stable molecules, their concentrations in the atmosphere gradually increased after their introduction in the 1930s. By the 1970s, the concentrations of CFCs had been increased sufficiently that they could be detected everywhere.

Chlorofluorocarbon molecules circulate in the lower atmosphere long enough to eventually move into the stratosphere, where they are exposed to a great deal more highly energetic ultraviolet light. As CFCs break down, they release chlorine, which can act as a catalyst to destroy ozone molecules. A single chlorine atom released in the stratosphere can continue to destroy ozone molecules until it is removed by some atmospheric process. Therefore, a small amount of chlorine released in the stratosphere can deplete the ozone layer substantially.

World concern over the dangers associated with ozone depletion prompted the 1987 Montreal Protocol on Substances that Deplete the Ozone Layer. This agreement has had eight revisions since that time, and on October 15, 2016, 197 parties (including the European Union) signed the Kigali Amendment, which added hydrofluorocarbons or HFCs to the list (https://www.state.gov/e/oes/eqt/chemicalpollution/83007.htm). The goal of the Montreal Protocol is to reduce and eventually eliminate emissions of human-generated substances that deplete ozone and may represent a model for international cooperation on a complex environmental problem. As a result of the protocol, global production of CFCs was reduced from over 1 million tons annually to less than 50,000 tons by the end of 2003.

How has the ozone hole over the Antarctic changed since its discovery in 1985? It grew reaching its maximum area, so far, in 2000, when it covered nearly 29.9 million km^2 (NASA 2017). In 2002, the Antarctic ozone hole closed quickly, suggesting that the ozone layer was recovering. However, the third-largest ozone hole was recorded in 2003, when it reached 28.4 million km^2 (fig. 23.24). Encouragingly, the year 2003 also saw the first reported evidence that the ozone layer is recovering, when several scientists reported (Newchurch et al. 2003) evidence for a slowdown in stratospheric ozone loss from 1997 to 2003. By 2012, the maximum extent of the Antarctic ozone hole had decreased to 21.1 million km^2 and was showing clear signs of stabilizing (fig. 23.25). It appears that cooperation by the international community in the banning of CFCs has begun to reverse

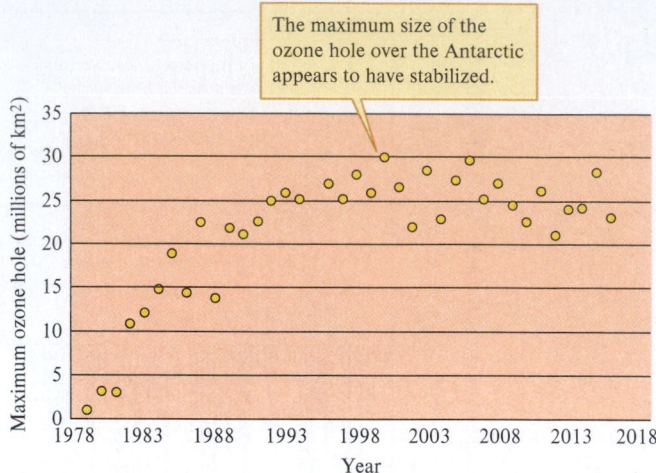

Figure 23.25   Maximum extent of the Antarctic ozone hole in square kilometers. Following a period of increasing size, the ozone hole over the Antarctic appears to have stabilized after 2000 (data from NASA 2017).

the process of ozone depletion. Atmospheric scientists predict that recovery of the stratospheric ozone layer will likely take at least another half century (Kuttippurath et al. 2013; Kramarova et al. 2014). However, the news of ozone recovery says clearly that we not only have the capacity to seriously damage the biosphere, but where we have the will, we can also act to restore it.

Concept 23.4 Review

1. What can we conclude from the evidence summarized by figures 23.20 to 23.23?
2. What aspects of global warming are widely supported by available evidence?
3. Are there uncertainties remaining regarding global warming?
4. Why may the history of CFCs in the atmosphere in the years following the Montreal Protocol offer encouragement as humanity strives to reverse the modern buildup of atmospheric CO_2?

Impacts of Global Climate Change

LEARNING OUTCOMES

After studying this section you should be able to do the following:

23.18   Summarize evidence that the earth has warmed over the past century.
23.19   Outline the threats of global climate change to biodiversity.
23.20   Describe how predicted climate change would impact human populations and infrastructure.

Figure 23.24   The ozone hole in September 2003. Even as scientists verified that the rate of ozone depletion was declining, the second-largest ozone hole ever discovered developed over Antarctica.   SVS/TOMS/NASA

How will human-caused changes in atmospheric composition and associated climate change affect earth's ecosystems and human populations? In 1957 Roger Revelle and Hans Suess wrote: "Human beings are now carrying out a large-scale geophysical experiment of a kind that could not have happened in the past nor be reproduced in the future. Within a few centuries we are returning to the atmosphere and oceans the concentrated organic carbon stored in sedimentary rocks over hundreds of millions of years. This experiment, if adequately studied and documented, may yield a far-reaching insight into the processes determining weather and climate."

Half a century after Revelle and Suess made their prophetic statement, the results of the unprecedented experiment are being recorded around the planet. During the past century, the average global temperature has risen approximately 0.9°C. Record high temperatures are being observed in many areas of the globe, with the hottest temperature on earth reliably measured in modern times, 54°C, occurring in 2013 in Death Valley, California, and in 2016 in Mitribah, Kuwait (Samenow 2017). There is broad agreement among earth scientists that increased atmospheric concentration of greenhouse gases, especially CO_2, is a major contributor to global warming (National Academies of Sciences, Engineering and Medicine 2016). This consensus follows from an understanding of the basic physics of the greenhouse effect (see fig. 23.3). Climate models predict that without reductions in emissions of CO_2 and other greenhouse gases, global temperatures will increase an additional 2.6° to 4.8°C over the next century (U.S. National Academy of Sciences and The Royal Society 2014). Evidence that the earth system continues to warm includes temperature records from long-term ground and satellite measurements, melting of glaciers and arctic sea ice, reductions in Northern Hemisphere spring snow cover, increasing ocean heat content, and continuing sea level rise. Based on observations of responses to existing warming and modeling studies, scientists predict a wide range of environmental and economic consequences of rapid global warming (IPCC 2018).

Some of the most dramatic examples of the impact of climate change were the historic fires of 2020, seen across the world, including Indonesia, Siberia, the western United States and Australia. These wildfires disrupted forest ecosystems and caused extensive damage to human infrastructure. We know that they are due to climate change in part because the area burned by wildfires has increased over time (fig. 23.26a); the area that burned in California in 2020 was double the previous record (fig. 23.26b). The fires across Australia burned an unprecedented 18.4 million ha; an area the size of Syria. These fires were the clear consequence of 2019 being the hottest and driest year on record (fig. 23.27).

Shifts in Biodiversity and Widespread Extinction of Species

Earth's biota is responding to climate change. We reviewed how plant and animal species are shifting their distributions earlier in this chapter and in chapter 10 (see section 10.1 and Applications) and how the phenologies of species are changing in

chapter 12 (see Applications). Chapter 1 also included references to evidence of evolutionary adaptation of species to past and present climate change (see section 1.2). Since earth's temperature has varied substantially during its history (see fig. 23.20), why are scientists concerned about how continued warming may threaten biodiversity? Major reasons for the concern are the magnitude and speed of contemporary warming. The earth has warmed 4° to 5°C since the last ice age; however, that warming took place over approximately 7,000 years. Current climate models are predicting a similar magnitude of warming by the end of this century, a rate of warming over 70 times faster. This pace of change and estimates of contemporary rates of species extinction have led scientists to suggest we are in the middle of a mass extinction comparable to past mass extinctions, such as the one that included the demise of the dinosaurs (Robertson et al. 2004; Barnosky et al. 2011). A meta-analysis of 131 published works found that 7.9% (95% CI: 6.2% to 8.8%) extant species are predicted to go extinct from climate change, with endemic species facing closer to 14% (Urban 2015). This represents hundreds of thousands of species expected to be lost if the current pace of global change is not slowed.

Climate change is in the process of altering the dynamics of entire ecosystems. Climate models indicate that regions now supporting tropical rain forest will be subject to drying with projected climate change sufficient to threaten the existence of these especially diverse ecosystems. Global warming has already contributed to forest dieback in temperate zones around the world as a result of increased incidence of disease and insect attack (Anderegg, Kane, and Anderegg 2013). Forest dieback accompanied by increased temperatures and drought is one of the mechanisms that causes greater frequency and extent of wildfires around the world.

Threats to biodiversity are not limited to the terrestrial environment. Global warming and ocean acidification threaten marine biodiversity, including coral reef ecosystems. Ocean warming challenges coral reefs because, as we saw in chapter 3 (see section 3.2), reef-building corals have narrow temperature tolerances. In addition, the increasing atmospheric concentration of CO_2 pushes more of the gas into solution in the oceans where it is lowering the pH of seawater. This ocean acidification, which inhibits formation of shells and coral skeletons, poses a grave threat to coral reefs, the pinnacle of marine diversity and productivity. Acidification coupled with elevated temperatures may lead to the disappearance of coral reefs by the end of the century (Pandolfi et al. 2011). The loss of coral reefs, which are habitat for many commercially important fish populations, will have a major negative impact on the food supplies of hundreds of millions of people. This is only one of many negative impacts of climate change on human populations.

Human Impacts of Climate Change

The list of how climate change can potentially affect human populations is very long. The extensive damage caused by the 2020 fires is only one example. The heat waves and drought associated with these fires are also predicted to increase the frequency of crop failures and to disrupt food supplies. In addition, heat waves are a source of direct physiological stress and mortality in human

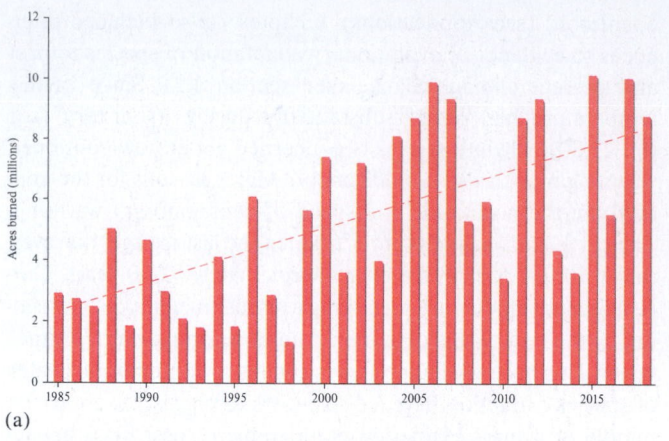

(a)

(b)

Figure 23.26   (*a*) Acres of land burned by wildfires in the United States since 1985 show a positive trend over time (as shown by the dotted line). (*b*) Wildfires in California in 2020 exceeded 4 million acres, more than double the previous record (nbcnews.com). (data from Union of Concerned Scientists https://www.ucsusa.org/resources/climate-change-and-wildfires).   (b) NASA Earth Observatory image by Joshua Stevens, using Landsat data from the U.S. Geological Survey, and MODIS data from NASA EOSDIS/LANCE and GIBS/Worldview

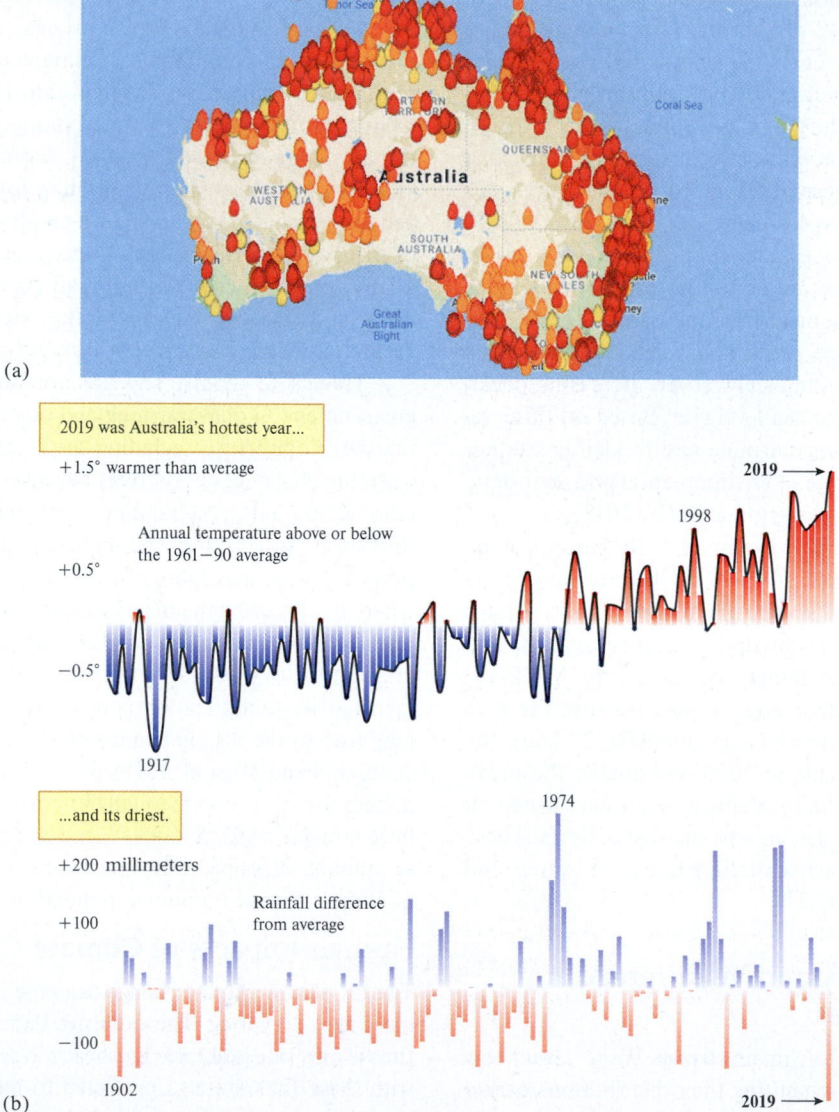

(a)

(b)

Figure 23.27   (*a*) The conditions that precluded the historic 2020 fires in Australia (map produced by Government of Western Australia from satellite data MyFireWatch), (*b*) were precluded by historically hot and dry conditions. (data from Australia Government Bureau of Meteorology).

populations. For instance, a heat wave in the summer of 2003 contributed to the deaths of over 70,000 people in Europe (Robine et al. 2008), and another in 2015 left over 2,300 people dead in India (Whiteman 2015). As we learned in chapter 21, both global climate change and more local heat island effects contribute to these tragedies.

While some areas are experiencing increased drought, the warmer atmosphere can also hold more water vapor, thus causing intense precipitation events in some places.

The consequences of such events can be catastrophic to infrastructure, including bridges, roadways, and drainage systems, and to human safety. For instance, in 1999, two storms that dropped over 110 cm of rain in Vargas State, Venezuela, produced thousands of landslides in the mountainous terrain, flooding, and debris flows. Some of the boulders carried by the Venezuelan flooding were more than 10 m in diameter. The massive debris flows triggered by the storm buried entire coastal towns, causing approximately $2 billion in property damage (fig. 23.28). In addition, the storm and its aftermath killed an estimated 19,000 people. Intense rainfall and flooding can also increase exposure of people to waterborne diseases such as cholera and polio, especially in developing countries.

Shifting of the distribution of earth's tropical biota northward and to higher elevations carries potential health consequences. Environmental scientists have been particularly concerned with the potential spread of insect-transmitted tropical diseases such as malaria and dengue fever to cooler climates. Recent events appear to justify those concerns. For example, locally acquired cases of dengue fever were recorded in Texas in 2004–05 and in southern Florida in 2009–13 (Eisen and Moore 2013; USGS 2013). Meanwhile, malaria-transmitting mosquitoes are shifting their distributions to higher elevations in the mountains of Africa.

Sea level rise will impact coastal populations. Melting of glaciers and polar ice is already producing sea level rise that will eventually flood critical coastal areas, if not stopped. Such flooding would create a human and economic disaster. As we saw in chapter 11, human populations are concentrated in coastal regions (see fig. 11.23, p. 254). Much of economic activity and critical infrastructure is also concentrated along coasts. For example, in 2011 coastal regions in the United States alone contributed approximately $6.6 trillion to the national economy

Figure 23.28   U.S. and Venezuelan geologists surveying a boulder field deposited by a debris flow originating in the Sierra de Avila Mountains (in the background) following heavy rainfall over a two-day period in December 1999. The debris flow completely obliterated single- story homes in its path killing an estimated 19,000 people.   M.C. Larsen/USGS

(National Research Council 2013). Unfortunately, much of the infrastructure that this economic activity requires is vulnerable to coastal flooding. Similar concentrations of economic activity and infrastructure occur in Asia and Europe. The direct human impact is also potentially serious. Extensive flooding of low-lying countries, such as Bangladesh, has the potential to create hundreds of millions of climate refugees.

It would be easy to give up in the face of what predicted climate change threatens. However, there is much that can be done to mitigate its effects and stave off the worst-case scenarios. Ironically, when the time for action has arrived, many voices from stakeholders in the status quo counsel us to proceed slowly. Encouragingly, the potential damage of climate change is rallying individuals and governments everywhere to work toward reducing climate-related threats to the biosphere—ultimately, our only life-support system. In our quest to reduce and hopefully reverse human damage to the earth, ecological knowledge will play a key role. As a student of ecology, you have built a foundation for being an informed participant in the discussion of how to address one of the most daunting challenges of our age.

Summary

This chapter focuses on global-scale processes and phenomena, including large-scale weather systems and global change induced by humans. We are the only species that exerts global-scale influences on the environment.

The earth is wrapped in an atmospheric envelope that makes the biosphere a hospitable place for life as we know it. The earth's atmosphere reduces the amount of ultraviolet light reaching the surface. The atmosphere also helps to keep the surface of the earth warm through the *greenhouse effect.*

The surface of the earth is kept warmer than it would be by the greenhouse gases, including water vapor, methane, ozone, nitrous oxide, chlorofluorocarbons, and carbon dioxide.

The El Niño Southern Oscillation, a large-scale atmospheric and oceanic phenomenon, influences ecological systems on a global scale.   The El Niño Southern Oscillation is a highly dynamic, large-scale weather system that involves variation in sea surface temperature and barometric pressure across the Pacific and Indian Oceans. During the mature phase of an El Niño, the sea surface in the eastern tropical Pacific Ocean is much warmer than average and the barometric pressure over the eastern Pacific is lower than average. El Niño brings increased precipitation to much of North and parts of South America and drought to the western Pacific. Periods of lower sea surface temperature and higher-than-average barometric pressures in the eastern tropical Pacific have been named La Niña events. La Niña brings drought to much of North and South America and higher-than-average precipitation to the western Pacific. The variation in weather caused by the El Niño Southern Oscillation has dramatic effects on marine and terrestrial populations around the world.

Human activity has greatly increased the quantity of fixed nitrogen cycling through the biosphere. For millions of years, the only organisms that could fix nitrogen were nitrogen-fixing bacteria. The total amount of nitrogen fixed by these historical sources is approximately 130 Tg N per year. The nitrogen now fixed as a consequence of human activity is more than all nonhuman sources of fixed nitrogen combined. Large-scale nitrogen enrichment may threaten biological diversity by creating environmental conditions favorable to some species at the expense of others.

Rapid changes in global patterns of land use threaten biological diversity. Human activities, mainly agriculture and urbanization, have significantly altered one-third to one-half of the ice-free land surface of the earth. A widely cited example of land cover change is tropical deforestation. From 1978 to 1988, the rate of deforestation in the Amazon Basin of Brazil averaged about 15,000 km^2 per year. By 1988, the total area deforested within the Amazon Basin was 230,000 km^2. By adding in edge effects and the effects of isolation, the area of Amazonian forest affected by deforestation increases from 230,000 km^2 to 588,000 km^2. By 2013, the rate of deforestation

in the Brazilian Amazon was reduced to approximately 6,000 km^2 per year but has increased again in recent years due to a change in administration and thus environmental law enforcement. Massive deforestation has also occurred outside of the tropics. Because of the negative effect of reduced habitat area on diversity, these massive land conversions present a major threat to global biological diversity.

Human activity is changing the composition of the atmosphere. Analyses of air trapped in ice show that the concentration of CO_2 in the atmosphere has varied widely during the last 800,000 years and closely parallels variation in global temperatures. High levels of atmospheric CO_2 have corresponded to higher global temperatures. The buildup of atmospheric CO_2 during the past two centuries has reached levels of atmospheric CO_2 not equaled in the past 800,000 years. The present level of CO_2 in the atmosphere is strongly influenced by burning of fossil fuels. Increases in atmospheric CO_2 concentration are associated with increased global temperatures and a variety of environmental impacts. Encouragingly, the response of the global community to ozone depletion by CFCs and evidence of recent ozone recovery suggest that we have the potential to address and remedy global-scale environmental problems.

Based on observations of responses to existing warming and modeling studies, scientists predict a wide range of environmental and human consequences as a result of rapid global warming. The pace of climate change and estimates of contemporary rates of species extinction have led scientists to suggest that we are in the middle of a mass extinction comparable to past mass extinctions. Environmental scientists predict serious impacts of climate change on human populations, economic systems, and infrastructure.

Key Terms

El Niño 493	mesosphere 492	Suess effect 506	Walker circulation 494
greenhouse effect 492	Southern Oscillation 493	thermosphere 492	
La Niña 494	stratosphere 492	troposphere 492	

Review Questions

1. Ecologists are now challenged to study global ecology. The apparent role played by humans in changing the global environment makes it imperative that we understand the workings of the earth as a global system. What are some of the main differences between global ecology and, for instance, the study of interspecific competition (see chapter 13) or forest succession (see chapter 20)? How will these differences affect the design of studies at the global scale?

2. Geologists, atmospheric scientists, and oceanographers have been conducting global-scale studies for some time. What role will information from these disciplines play in the study of global ecology? Why will global ecological studies generally be pursued by interdisciplinary teams? How can ecologists play a useful role in global studies?

3. What changes in sea surface temperatures and atmospheric pressures over the Pacific Ocean accompany El Niño? What physical changes accompany La Niña? How do El Niño and La Niña affect precipitation in North America, South America, and Australia?

4. Review evidence that the El Niño Southern Oscillation significantly influences populations around the globe. Considering our discussions in chapters 18 and 19 of physical controls on rates of terrestrial primary production and decomposition, how does the El Niño Southern Oscillation likely affect these ecosystem processes in Australia or the American Southwest? How would you test your ideas?

5. In chapter 23, we briefly discussed how humans have more than doubled the quantity of fixed nitrogen cycling through the biosphere. In chapter 15, we reviewed studies by Nancy Johnson (1993) on the effects of fertilization on the mutualistic relationship between mycorrhizal fungi and grasses. The increases in fixed nitrogen cycling through the biosphere, particularly that portion deposited by rain, are analogous to a global-scale fertilization experiment. Reasoning from the results of Johnson's study, how should increased fixed nitrogen supplies affect the relationship between mycorrhizal fungi and their plant partners? How would you test your ideas?

6. As we saw in chapters 18 and 19, nitrogen availability seems to control the rates of several ecosystem processes. How should nitrogen enrichment affect rates of primary production and decomposition in terrestrial, freshwater, and marine environments? How could you test your ideas? What role might geographic comparisons play in your studies?

7. Ecologists predict that global diversity is threatened by land use change and by the reductions in habitat area and the fragmentation that accompany land use change. Vitousek (1994) suggested that land use change may be the greatest current threat to biological diversity (see fig. 23.2). What role do studies of diversity on islands and species area relationships on continents (see chapter 22) play in these predictions?

8. Skole and Tucker (1993) and others documented the rate and extent of recent deforestation in the Amazon Basin in Brazil. However, scientists have documented agricultural activity in New World tropical forests beginning at least 6,000 years ago. What makes present-day deforestation in the Amazon Basin different from this historical activity? What does the long history of agriculture in the Amazon Basin suggest about the potential for coexistence of agriculture and biological diversity there?

9. Review the long-term atmospheric CO_2 record as revealed by studies of air trapped in ice cores. What is the evidence that burning of fossil fuels is responsible for recent increases in atmospheric CO_2 concentrations?

10. What evidence is there that variation in atmospheric CO_2 concentration is linked to variation in global temperatures? How might rapid changes in global temperatures lead to the extinction of large numbers of species? How might changes in global temperatures affect agriculture around the world?

Appendix A

Investigating the Evidence

The Scientific Method–Questions and Hypotheses

LEARNING OUTCOMES

After studying this section you should be able to do the following:

1.1 Distinguish between questions and hypotheses in the scientific process.

1.2 Discuss the scientific method, emphasizing hypothesis testing.

Ecologists explore the relationships between organisms and environment using the methods of science. The series of Investigating the Evidence discussions introduce various aspects of the scientific method and its application to ecology. While each box describes only a small part of science, taken together, they represent a substantial introduction to the philosophy, techniques, and practice of ecological science.

Let us begin the discussion with the most basic element, the nature of science. The word *science* comes from a Latin word meaning "to know." Broadly speaking, science is a way of obtaining knowledge about the natural world using certain formal procedures. Those procedures, which make up what we call "the scientific method," are outlined in figure 1. Despite a great diversity of approaches to doing science, sound scientific studies have many methodological characteristics in common. The most universal and critical aspects of the scientific method are: asking questions and forming testable hypotheses.

Questions and Hypotheses

What do scientists do? Simply put, scientists ask and attempt to find answers to questions about the natural world. Questions are the guiding lights of the scientific process. Without them, exploration of nature lacks focus and yields little understanding of the world. Let's consider a question asked by an ecologist discussed in chapter 1. The main question asked by Robert MacArthur in his studies of warblers (section 1.2)

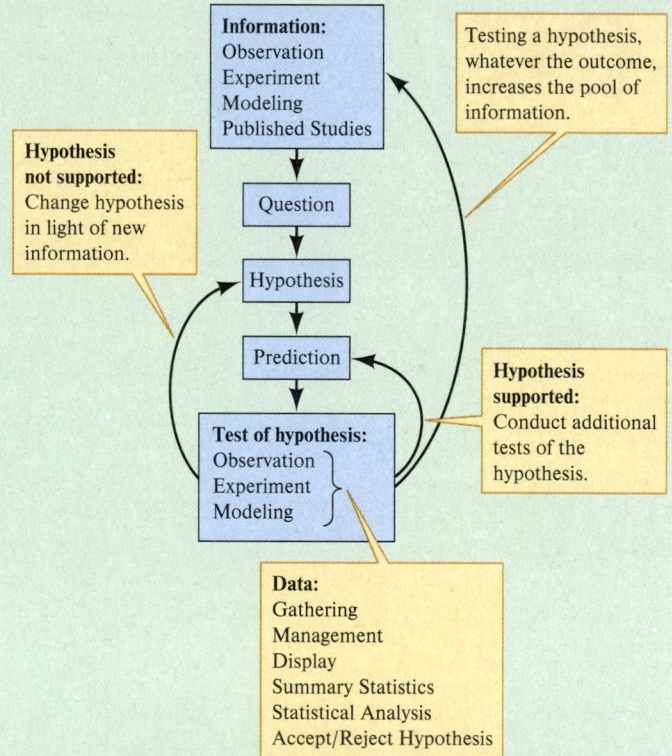

Figure 1 Graphic summary of the scientific method. The scientific method centers on the use of information to propose and test hypotheses through observation, experiment, and modeling.

was something like the following: "How can several species of insect-eating warblers live in the same forest without one species eventually excluding the others through competition?" While this focus on questions may seem obvious, one of the most common questions asked of scientists at seminars and professional meetings is "What is your question?"

If scientists are in the business of asking questions about nature, where does a hypothesis enter the process?

A hypothesis is a possible answer to a question. MacArthur's main hypothesis (possible answer to his question) was "Several warbler species are able to coexist because each species feeds on insects living in different zones within trees."

Once a scientist or team of scientists proposes a hypothesis (or multiple alternative hypotheses), the next step in the scientific method is to determine its validity by testing predictions that follow from the hypothesis. Three fundamental ways to test hypotheses are through observation, experiments, and modeling. A model is a tool that helps describe and/or predict patterns that we observe; it can be mathematical as in the case of a population growth curve, or it can be conceptual such as the flow diagram in figure 1. By definition, models simplify reality; for example, no population will perfectly follow a mathematical formula for growth. However, having generalizations like these allows us to test hypotheses such as whether a population of birds is growing or shrinking and by how much. Conceptual models, such as a diagram that shows the relationships between all of the factors that could affect population growth, are often used to generate hypotheses. Observation and/or experimentation can then provide the data to build mathematical models to quantify these relationships.

CRITIQUING THE EVIDENCE 1

1. How does the development of new research tools, such as aerial drones and stable isotope analysis, affect the process of science as outlined by figure 1 of this Investigating the Evidence discussion?

Information
Hypothesis
Predictions
✓ **Testing**

Investigating the Evidence 2

Determining the Sample Mean

LEARNING OUTCOMES

After studying this section you should be able to do the following:

2.1   Calculate the mean of a sample drawn from a population.

2.2   Explain how a sample mean represents a statistical estimate.

One of the most common and important steps in the processing of data is the production of summary statistics. First, what is a statistic? A statistic is a number that is used by scientists to estimate a measurable characteristic of an entire population. Population characteristics of interest to an ecologist might include features such as average mass, growth rate, or air temperature. In order to determine the exact average value of such population characteristics, the ecologist would have to measure every individual in the population. Clearly, the opportunity to measure or test all the individuals in a population for any characteristic is extremely rare. For instance, an ecologist studying reproductive rate in a population of birds would be unlikely to locate and study all the nests in the population. As a consequence, ecologists generally estimate reproductive rates for birds, or other characteristics of any population, using samples drawn randomly from the population. An ecologist working with a population of rare plants, for example, might locate 11 seedlings and calculate the average height of these 11 individual plants. This average calculated from the sample of 11 seedlings would be the **sample mean.** The sample mean is a statistical estimate of the true population mean.

The sample mean is one of the most common and useful summary statistics. It is a statistic that we use extensively, for example, as we discussed mean temperature or mean precipitation for biomes around the world (see fig 2.6). How is the sample mean calculated? Consider the following sample of seedling height:

Seedling number	1	2	3	4	5	6	7	8	9	10	11
Height in cm	3	6	8	7	2	4	9	4	5	7	8

What was the average height of seedlings in the population at the time of the study? Since we did not locate all the seedlings, we cannot know the true population mean, or parameter. However, our sample of 11 seedlings allows us to calculate a sample mean as follows:

$$\text{Sum of measurements} = \Sigma X$$
$$\Sigma X = 3+6+8+7+2+4+9+4+5+7+8$$
$$\Sigma X = 63$$

We calculate the sample mean by dividing the sum of measurements by the number of seedlings measured:

$$\text{Sample mean} = \overline{X}$$
$$n = \text{sample size, or } 11$$
$$\overline{X} = \frac{\Sigma X}{n}$$
$$\overline{X} = \frac{63}{11}$$
$$\overline{X} = 5.7 \text{ cm}$$

Again, 5.7 cm, the sample mean, is the ecologist's estimate of the true mean height of seedlings in the entire population at the time of the study.

(continued)

CRITIQUING THE EVIDENCE 2

1. If you measured the heights of 100 seedlings randomly drawn from the hypothetical population, instead of the 11 measured in the example, would the sample mean be likely to be exactly 5.7 cm?

2. Would the mean height of a sample of 100 seedlings likely be closer to the true population mean than the mean of a sample of 11?

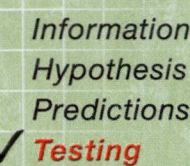

Information
Hypothesis
Predictions
✓ *Testing*

Investigating the Evidence 3

Determining the Sample Median

LEARNING OUTCOMES

After studying this section you should be able to do the following:

3.1   Calculate the median of a sample drawn from a population.

3.2   Explain the conditions under which it is appropriate to use a median, rather than a mean, to represent the typical member of a population.

In Investigating the Evidence 2, we determined the sample mean. However, while the sample mean is one of the most common and useful of summary statistics, it is not the most appropriate statistic for some situations. One of the assumptions underlying the use of the sample mean is that the observations used to calculate it are drawn from a population with a normal, or bell-shaped, distribution. However, where the distribution of values within a population deviates substantially from a normal distribution, it may be better to use another estimator of the population "average." An alternative statistic is the **sample median.** The sample median is the middle value in a sample. Let's determine the median for the sample we used to determine a sample mean in Investigating the Evidence 2. Here is the table summarizing that sample:

Table 1   Heights of 11 seedlings of a rare plant species

Seedling number	1	2	3	4	5	6	7	8	9	10	11
Height in cm	3	6	8	7	2	4	9	4	5	7	8

To determine the sample median, it's convenient to reorder the observations in a sample from lowest to highest:

Table 2   Heights of 11 seedlings of a rare plant species with observations ordered from lowest to highest

Samples: low to high	1	2	3	4	5	6	7	8	9	10	11
Height in cm	2	3	4	4	5	6	7	7	8	8	9

Because there was an odd number of observations in this sample, 11, there is a middle value in the series of observations, with 5 observations with higher value and 5 with lower value. That middle value occurs at sample rank number 6 and the height of the seedling with this rank happens to be 6 cm. Therefore, the sample median is 6 cm. Notice that this value is very similar to the sample mean we calculated in Investigating the Evidence 2, which was 5.7 cm. In this case, the sample mean and sample median give similar estimates of the average within the population.

However, where a population contains a few very large or very small values, the sample median may give a better estimate of the average within the population. Consider the following sample of the abundance of mayfly nymphs living within 0.1 m^2 areas of stream bottom. The sample was taken from a high mountain stream of the southern Rocky Mountains and the mayfly species was *Baetis bicaudatus:*

Table 3   Number of *Baetis bicaudatus* nymphs in 0.1 m^2 benthic samples

Samples: low to high	1	2	3	4	5	6	7	8	9	10	11	12
Number of nymphs	2	2	2	3	3	4	5	6	6	8	10	126

In this case, because one of the samples contained 126 nymphs, the sample median and sample mean give very different estimates of the average number of *B. bicaudatus* living within a 0.1 m^2 area of the study stream. *Because this sample has an even number of observations, the sample median is determined as the average of the two middle observations.* That is:

$$\text{Sample median} = \frac{4 + 5}{2} = 4.5 \text{ B. bicaudatus per 0.1 m}^2.$$

The sample mean is $\overline{X} = \frac{\Sigma X}{n} = \frac{177}{12} = 14.8$ *B. bicaudatus* per 0.1 m^2. This estimate of the population mean is more than three times the estimate provided by the sample median. In this case it is clear that the sample median, which again is the middle value in an ordered sample of observations, more closely estimates the number of *B. bicaudatus* that you are likely to encounter within 0.1 m^2 of benthic habitat in this mountain stream.

CRITIQUING THE EVIDENCE 3

1. Suppose you find that an organism is fairly evenly distributed across most of a habitat but is found at 10 times typical numbers in a few hotspots. In such a situation, which

would be a better description of the typical population density—the mean or the median?

2. Why are comparisons of per capita income in different countries generally reported as differences in median income, rather than mean income?

At this time, there is no universally accepted mathematical symbol for sample median.

Information
Hypothesis
Predictions
✓ Testing

Investigating the Evidence 4

Variation in Data

LEARNING OUTCOMES

After studying this section you should be able to do the following:

4.1 Describe the range of a sample.

4.2 Calculate a sample variance and standard deviation.

4.3 Discuss why, compared to the sample range, the standard deviation generally gives a better representation of variation in a sample.

In Investigating the Evidence 2, we calculated the sample mean and in Investigating the Evidence 3, we determined the sample median. The mean and median are different ways of representing the middle, or typical, within a sample of a population. Another important question we can ask is, how much *variation* is there around the average? This is important for several reasons. For example, two or more samples may have the same mean but have quite different amounts of variation among the observations within each sample. Knowing the variation within samples, as well as their means or medians, is critical to comparing them statistically (see Investigating the Evidence 17 and 18).

Suppose you are studying a population of whitefish living in Lake Lucerne, Switzerland (see chapter 4, fig. 4.5) and, as part of the study, you need to estimate the variation in length of a small sample of young whitefish taken from the lake. Here are your measurements:

Specimen	1	2	3	4	5	6	7	8	9	10
Total length (mm)	60	62	56	53	53	59	62	41	58	58

The simplest index of variation would be the **range**, which is the difference between the largest and the smallest observation. In this case the range would be:

$$\text{Range} = 62 - 41 = 21 \text{ mm}$$

The range does not represent variation in samples well, since very different sets of observations can have the same range. A better representation of the variation in a sample is one that factors in all the observations relative to the sample mean. One such index that is most commonly used is called the sample **variance**. The variance is calculated as follows:

First we calculate the sample mean as we did in Investigating the Evidence 2.

$$\overline{X} = 56.2 \text{ mm (For practice you could calculate this sample mean using the data given above.)}$$

The variance is calculated by squaring the differences between the sample mean and each of the observations, adding them up to produce the "sum of squares," and dividing by the sample size minus 1. Let's do this in steps. First let's calculate the sum of squares.

$$\text{Sum of squares} = \Sigma(X - \overline{X})^2$$

Using the fish length measurements given in the table above:

$$\Sigma(X - \overline{X})^2 = (60 - 56.2)^2 + (62 - 56.2)^2$$
$$+ (56 - 56.2)^2 + (53 - 56.2)^2$$
$$+ (53 - 56.2)^2 + (59 - 56.2)^2$$
$$+ (62 - 56.2)^2 + (41 - 56.2)^2$$
$$+ (58 - 56.2)^2 + (58 - 56.2)^2$$

Taking the differences gives:

$$\Sigma(X - \overline{X})^2 = (3.8)^2 + (5.8)^2 + (-0.2)^2 + (-3.2)^2$$
$$+ (-3.2)^2 + (2.8)^2 + (5.8)^2$$
$$+ (-15.2)^2 + (1.8)^2 + (1.8)^2$$

Squaring the differences yields:

$$\Sigma(X - \overline{X})^2 = 14.44 + 33.64 + 0.4 + 10.24$$
$$+ 10.24 + 7.84 + 33.64$$
$$+ 231.04 + 3.24 + 3.24$$

(continued)

Adding the squared differences gives the sum of squares:

$$\text{Sum of squares} = \Sigma(X - \bar{X})^2 = 347.6 \text{ mm}^2$$

The sample variance is calculated by dividing the sum of squares by the sample size minus 1. The sample size in this case is 10 measurements.

$$\text{Sample variance} = s^2 = \frac{\Sigma(X - \bar{X})^2}{n - 1}$$

Now putting in the values,

$$s^2 = \frac{347.6 \text{ mm}^2}{9}$$

and dividing,

$$s^2 = 38.6 \text{ mm}^2$$

This, then, is the sample variance. However, notice that the units of the sample variance are square mm, not mm. Because the sample variance is expressed in squares of the original units, generally we take the square root of the variance to calculate a measure of variation called the sample **standard deviation.**

$$\text{Standard deviation} = s = \sqrt{s^2}$$

Calculating the standard deviation for our data:

$$s = \sqrt{38.6 \text{ mm}^2} = 6.2 \text{ mm}$$

While it took a little effort to calculate it, the standard deviation, 6.2 mm, provides us with a standardized index of the variation in length of the fish used in your study. Fortunately, most electronic calculators make these calculations automatically, once the data are entered. The sample standard deviation along with the sample mean enable us to make statistical comparisons of samples.

CRITIQUING THE EVIDENCE 4

1. Why do the standard deviation and variance generally represent variation in a sample better than the range?
2. Can samples drawn from two different populations have approximately the same mean (e.g., body weight) yet have different variances? Why?

Information
Hypothesis
Predictions
✓ Testing

Investigating the Evidence 5

Laboratory Experiments

LEARNING OUTCOMES

After studying this section you should be able to do the following:

5.1   Describe the basic design of a laboratory experiment.
5.2   Discuss the relative strengths and weaknesses of laboratory experiments and field observations in ecological studies.

One of the most powerful ways to test a hypothesis is through an experiment. Experiments used by ecologists generally fall into one of two categories—field experiments and laboratory experiments. Field and laboratory experiments generally provide complementary information or evidence, and differ somewhat in their design. Here we discuss the design of laboratory experiments.

In a laboratory experiment, the researcher attempts to keep all factors relatively constant except one. The one factor that is not kept constant is the one of interest to the experimenter, and it is the one that the experimenter varies across experimental conditions. Let's draw an example of a laboratory experiment discussed in chapter 5. Based upon published studies, Michael Angilletta (2001) concluded that geographically separated populations of the eastern fence lizard, *Sceloporus undulatus,* may differ physiologically or behaviorally.

Angilletta designed a laboratory experiment to test the hypothesis that populations of *S. undulatus* from regions with significantly different climates differ in how temperature affects their rates of metabolizable energy intake. The results of that experiment are summarized by figure 5.10. What we want to consider here is the design of the experiment that produced those results. What factors do you think Angilletta may have attempted to control in this experiment? First, he used similar numbers of lizards from two populations. He tested 20 lizards from both populations at 33°C, 13 from New Jersey at 30° and 36°C, and 14 from South Carolina at 30° and 36°C. A second factor that Angilletta controlled was lizard size. Lizards from both populations used in the experiments had an average body mass of approximately 5.4 g. Since males and females may differ physiologically, Angilletta included approximately equal numbers of males and females in his experiments. He also was careful to expose all the lizards to the same quality of light and to the same numbers of hours of light and darkness, and he maintained them in the same kinds of experimental enclosures. Angilletta also fed all the lizards in his experiment the same type of food: live crickets. The list could go on but these are the major factors controlled in this experiment.

Now, what factors did Angilletta vary in that experiment? For each study population, New Jersey or South Carolina, he varied a single factor: temperature. In the experiment, Angilletta maintained lizards from New Jersey and South Carolina at three temperatures: 30°, 33°, and 36°C and estimated their rates of metabolizable energy intake at these three temperatures. Angilletta's experiment revealed that lizards from both populations have a maximum metabolizable energy intake at 33°C. This result suggests, contrary to the study's hypothesis, that the optimal temperature for feeding does not differ for the two populations. However, the experiment also showed that at 33°C *S. undulatus* from South Carolina have a higher metabolizable

energy intake compared to lizards from New Jersey. This result provides evidence of the geographic differences that Angilletta thought might exist across the range of *S. undulatus*. The power of this experiment to reveal the influence of temperature on lizard performance resulted from the ability of the researcher to control all significant factors but the one of interest. In this case the main factor of interest was temperature.

CRITIQUING THE EVIDENCE 5

1. What is the greatest strength of laboratory experiments in ecological research?
2. Why do ecologists generally supplement information resulting from laboratory experiments with field observations or experiments?

Information
Hypothesis
Predictions
✓ *Testing*

Investigating the Evidence 6

Sample Size

LEARNING OUTCOMES

After studying this section you should be able to do the following:

6.1 Define the term *sample size*.

6.2 Design a study, including the number of samples to be taken, to document the impact of future floods on the number of benthic insect species living in Tesuque Creek.

The number of observations included in a sample, that is, sample size, has an important influence on the level of confidence we place on conclusions based on that sample. Let's examine a simple example of how sample size affects our estimate of some ecological feature. Consider an ecologist interested in how disturbance by flash flooding may affect the number of benthic insect species living in a stream. The stream is Tesuque Creek at about 3,000 m elevation in the mountains above Santa Fe, New Mexico. A flash flood, which completely disrupted one fork of Tesuque Creek, left a second, similar-sized fork undisturbed. Nine months after the flood, samples were taken to determine if there was a difference in the number of species of mayflies (Order Ephemeroptera), stoneflies (O. Plecoptera), caddisflies (O. Trichoptera), and beetles (O. Coleoptera) living in similar-sized reaches of the two forks. Samples of the benthic community were taken at 5 m intervals with a Surber sampler, which has a 0.1 m² metal frame, or quadrat, and an attached net. As a stream ecologist disturbs the bottom material within the quadrat of a Surber sampler, the net trailing in the current catches benthic organisms that are dislodged. In the study of Tesuque Creek, the number of benthic insect species captured in each 0.1 m² sample ranged from one to six in the disturbed fork and from two to eight in the undisturbed fork. However,

our question concerns the total number of species in each fork and the number of benthic samples required to make a good estimate of that number of species.

Figure 1 plots the data in a way that provides an answer to both questions. The Surber samples are plotted in the exact

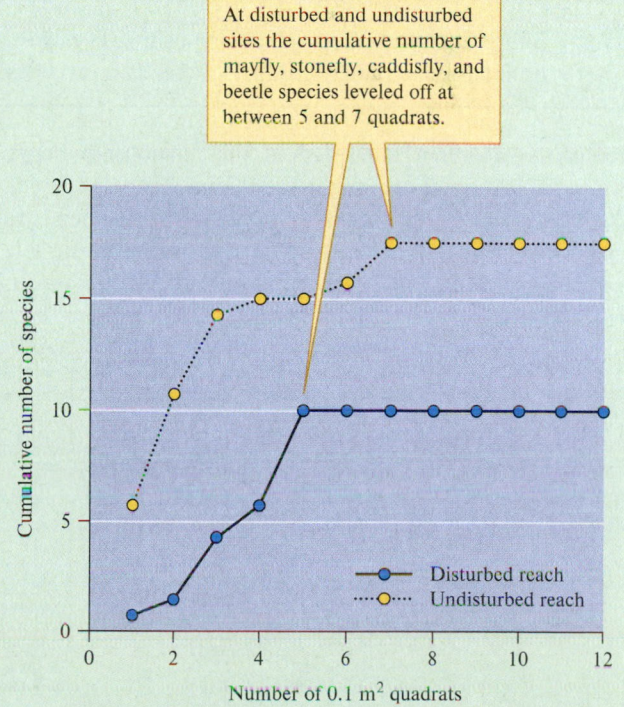

At disturbed and undisturbed sites the cumulative number of mayfly, stonefly, caddisfly, and beetle species leveled off at between 5 and 7 quadrats.

Figure 1   The cumulative number of species increased with the number of quadrats studied in both disturbed and undisturbed streams, eventually leveling off at a sample size of five to seven quadrats.

(continued)

order they were taken, beginning with the first that was taken at the downstream end of each study reach and ending with the twelfth sample taken 55 m upstream from the first. As shown in figure 1, each of the first few samples adds to the cumulative number of species collected at each site, which rises steeply at first and then levels off at a maximum number of species in each study reach. The cumulative number of species stopped increasing at a sample size of seven quadrats in the undisturbed study reach and at five quadrats in the disturbed study reach.

How many samples should a researcher take? In the case of the benthic community just examined, seven replicate counts from 0.1 m^2 quadrats appears to be sufficient to estimate the number of benthic mayfly, stonefly, caddisfly, and beetle species living in a short reach of a small, high-elevation stream in the Rocky Mountains. (We will revisit this study, which concerns estimating the number of species in a community, in Investigating the Evidence 16.) In contrast, to make generalizations about global

patterns of rooting among plants, Schenk and Jackson (2002) reported on 475 root profiles at 209 locations (see chapter 6, section 6.2). The number of samples necessary depends on the amount of variability in the system under study and the spatial and temporal scope of the study. However, whether the scope of a project is large or small, sample size is one of the most important components of study design.

CRITIQUING THE EVIDENCE 6

1. When designing an ecological study, it is important to take a sufficient number of samples to test the hypothesis under study. Why might a researcher try to collect a sufficient number of samples to test the hypothesis but not more?
2. Judging from the data displayed in figure 1, how did disturbance by flash flooding affect the number of mayfly, stonefly, caddisfly, and beetle species living in the disturbed reach?

Information
Hypothesis
Predictions
✓ Testing

Investigating the Evidence 7

Scatter Plots and the Relationship Between Variables

LEARNING OUTCOME

After studying this section you should be able to do the following:

7.1 Interpret scatter plots indicating positive, negative, or uncorrelated relationships between two variables.

Ecologists are often interested in the relationship between two variables, which we might call X and Y. For example, in chapter 7 we reviewed a study of how the size of pumas, variable X, is related to the size of prey that they take, variable Y (see chapter 7, fig. 7.19). The scatter plot shown in figure 7.19

is one of an infinite number of possible relationships between two variables.

Let's consider just three of the possible relationships, which are shown in figure 1. The most basic scatter plot is one in which there is no relationship between X and Y. This situation is represented by figure 1a, where there is no correlation between values of X and values of Y. As a result, the scatter plot forms a more or less circular pattern. In contrast the pattern shown in figure 1b, which represents the situation where as X increases, Y decreases, follows an approximately linear pattern that slopes downward to the right. The relationship between

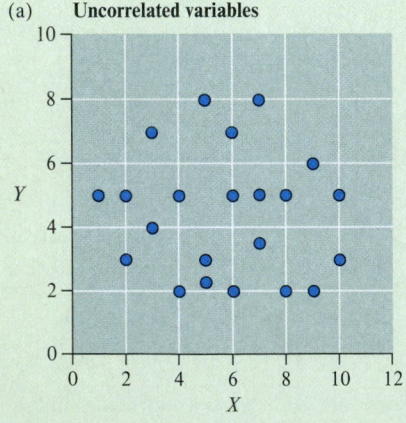
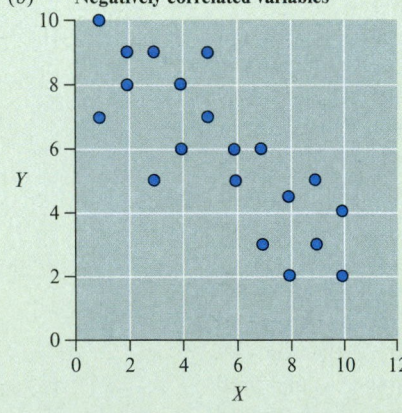
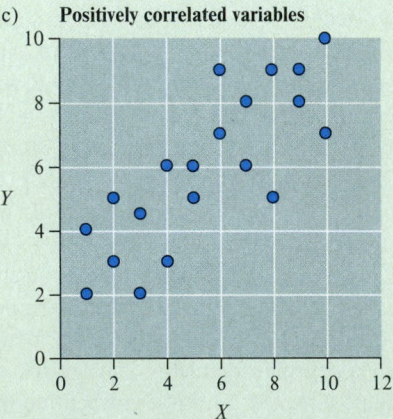

Figure 1   A scatter plot is a useful tool for exploring relationships between any two variables X and Y.

rainfall and green woodhoopoe reproduction we examined in chapter 8 shows this pattern (see chapter 8, fig. 8.19). This type of relationship between X and Y is called a negative correlation. The opposite pattern, shown by figure 1c, is called a positive correlation. When two variables are positively correlated, increases in X are associated with increases in Y. For instance, increased body size in populations of pumas is correlated with increased size of the prey they eat (see chapter 7, fig. 7.19).

CRITIQUING THE EVIDENCE 7

1. Suppose that during a field study you find a positive correlation between some variable X and another variable Y—for example, between higher concentrations of the stable isotope ^{13}C in the tissues of migrating redstarts and later arrival dates on their northern breeding grounds (see chapter 1, section 1.2). Does this positive correlation demonstrate that an increase in X directly causes an increase in Y?

Information
Hypothesis
Predictions
✓ Testing

Investigating the Evidence 8

Estimating Heritability Using Regression Analysis

LEARNING OUTCOMES

After studying this section you should be able to do the following:

8.1 Describe regression analysis.

8.2 Appraise the relative degrees to which traits are heritable based on regression coefficients.

As we have seen, the extent to which phenotypic variation in a trait is determined by genetic variation affects its potential to evolve by natural selection. In other words, the potential for a trait to evolve is affected by the trait's heritability. How can we estimate the heritability of a particular trait? One common method is through regression analysis. Regression analysis is a statistical technique used to explore the extent to which one factor, called the **independent variable** (usually symbolized as X), determines the value of another variable, which we call the **dependent variable** (usually represented by the symbol Y). In regression analysis, we construct X–Y plots as we did when we explored scatter plots and correlation (Investigating the Evidence 7). However, regression analysis is used to determine the equation for a line, called a **regression line,** that best fits the relationship between X and Y. When the relationship between X and Y follows a straight line, such as those in chapter 8, fig. 8.21, the regression equation is:

$$Y = bX + a$$

In this equation, a is the point at which the line crosses the Y axis, which is called the Y intercept, and b, which is the slope of the line, is the **regression coefficient.**

Let's use a natural system to learn more about regression analysis and its use in heritability studies. In heritability

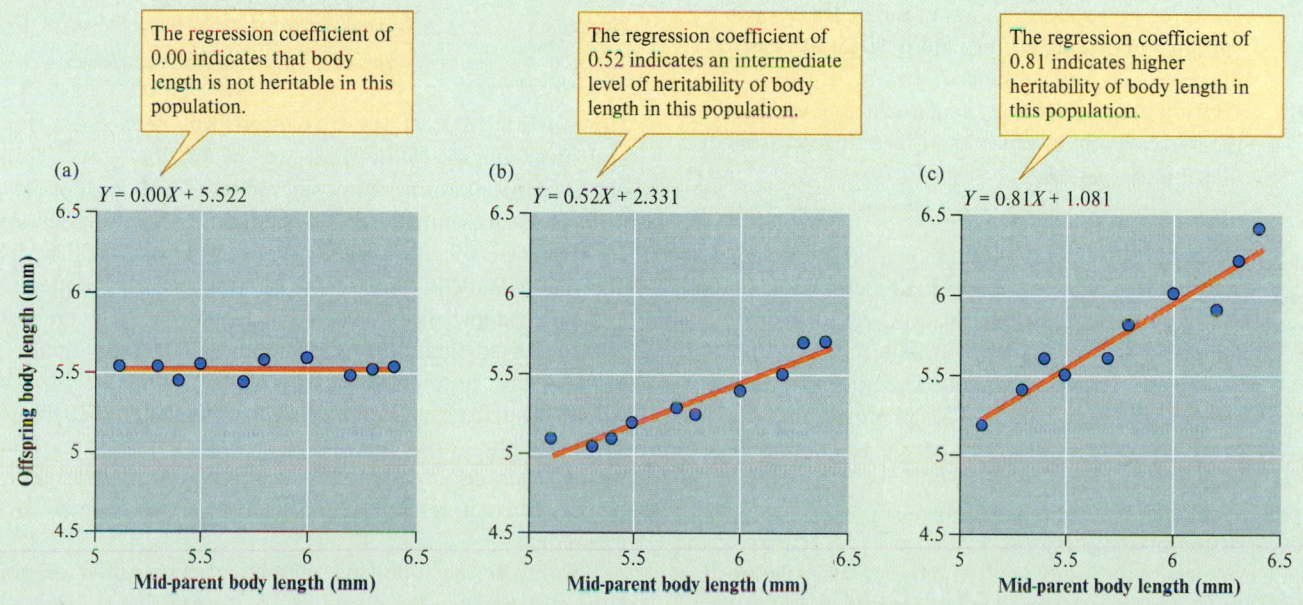

Figure 1   Regression analyses indicating degree of heritability of body length in three hypothetical populations of waterlily leaf beetles.

(continued)

studies, we are interested in the extent to which the characteristics of parents determine the characteristics of offspring. For instance, a team of Dutch scientists studying waterlily leaf beetles (Pappers et al. 2002) explored the heritability of body size, measured as body length, in different populations of the beetle. As we saw in the studies of Randy Thornhill, male body size can have a very significant influence on mating success (see chapter 8, fig. 8.13). To determine the heritability of body length, they conducted regression analyses using the body length of parents as the independent variable, and body length of the offspring as the dependent variable. Because both parents contribute to the genotype of the offspring, the value used for parental body length is the "mid-parent body length," which is the average of the two parents' body lengths. Let's consider the relationships between length of parents and offspring, and use regression analysis to estimate heritability of body length in some hypothetical populations of waterlily leaf beetles.

Consider the three scatter plots shown in figure 1 and the lines drawn through the scatter of points. Again, these are much like the scatter plots we examined in chapter 7 but with regression lines drawn through each. The regression coefficient in each of the graphs indicates the level of heritability in the three hypothetical populations. In population (a), the regression coefficient of 0.00 indicates that there is no relationship between parental body length and the body length of offspring. This result is apparent from just the scatter plot, which shows that parents of any length, large or small, can have small or large offspring. In this population it appears that the variation in body length among the offspring is determined entirely by environmental effects. In contrast, body length has a heritability of 0.52 in population (b) and 0.81 in population (c). What do these values indicate? With a heritability of 0.52, we can conclude that about half of the variation in body length in population (b) results from genetic effects, and about half from environmental effects, such as food quality, temperature, and so forth. The regression coefficient of 0.81 in population (c) indicates that more of the variation in body length in the offspring in that population is the result of genetic effects.

CRITIQUING THE EVIDENCE 8

1. What are the evolutionary implications of the patterns shown in figure 1?

Information
Hypothesis
Predictions
✓ Testing

Investigating the Evidence 9

Clumped, Random, and Regular Distributions

LEARNING OUTCOMES

After studying this section you should be able to do the following:

9.1   Describe how relative values of mean density and variance in density correspond to random, regular, and clumped distributions.

9.1   Given a table of values, calculate density variance to mean ratios and interpret their meaning in terms of distribution patterns.

Imagine sampling a population of plants or animals to determine the distribution of individuals across the habitat. One of the most basic questions that you could ask is "How are individuals in the population distributed across the study area?" How might they be distributed? The three basic patterns that we've discussed in this section are clumped, random, and regular distributions. The first step toward testing statistically between these three types of distributions is to sample the population to estimate the mean (Investigating the Evidence 2) and variance (Investigating the Evidence 4) in density of the population across the study area. The theoretical relationships between variance in density and mean density are as follows:

Distributions	Relation of variance to mean
Clumped	Variance > Mean, or Variance/Mean > 1
Random	Variance = Mean, or Variance/Mean = 1
Regular	Variance < Mean, or Variance/Mean < 1

How do we connect these relationships between variance and mean density with what we see on the ground? In a clumped distribution, many sample plots will contain few or no individuals while some will contain a large number. As a consequence, the variance among sample plots will be high and the variance in density will be greater than the mean. In contrast, sample plots of a population with a regular distribution will all include a similar number of individuals. As a result, the variance in density across samples will be low when taken from a population with a regular distribution; therefore, the variance will be less than the mean. Meanwhile, in a randomly distributed population, the variance in density across the habitat will be approximately equal to the mean density.

Consider the following samples of three different populations of herbaceous plants growing on a desert landscape.

Each sample is the number counted in a randomly located $1 m^2$ area at the study site.

Sample	a	b	c	d	e	f	g	h	i	j	k
Number of Species A	6	2	6	5	2	5	7	3	5	9	8
Number of Species B	5	6	6	5	5	5	5	6	4	5	5
Number of Species C	20	1	2	1	15	3	1	1	10	2	2

The distribution of individuals among the samples of species A, B, and C is quite different. For instance, each of the samples contained approximately the same number of individuals of species B. In contrast, the numbers of species C varied widely among samples. Meanwhile, counts of species A showed a level of variation somewhere in between variations in species B and C. The samples of species A, B, and C may give the impression of random, regular, and clumped distributions. We can quantify our visual impressions by calculating the sample means and sample variances for the densities of species A, B, and C:

Statistic	Species A	Species B	Species C
Sample mean, $\overline{X}$	5.27	5.18	5.27
Sample variance, s^2	5.22	0.36	44.42
Ratio of variance to mean, $\dfrac{s^2}{\overline{X}}$	0.99	0.07	8.43

While the mean density calculated from the samples was very similar for all three species, their variance in density among samples was quite different. As a consequence, the ratios of sample variance to sample means, $\dfrac{s^2}{\overline{X}}$, were also different. While $\dfrac{s^2}{\overline{X}}$ for species A was nearly 1, this ratio was much less than 1 for species B, and much greater than 1 for species C. These results show how the $\dfrac{s^2}{\overline{X}}$ ratio can quantify the visual impression of random, regular, and clumped distributions that we formed when we inspected the samples of the three species. Can we conclude from the $\dfrac{s^2}{\overline{X}}$ ratios we calculated that species A has a random distribution, that species B has a regular distribution, and that C has a clumped distribution? While it is likely that they do, in science we need to attach probabilities to such conclusions. To do that, we need to consider these samples of species A, B, and C from a statistical perspective. We will look at the statistics of these samples in Investigating the Evidence 10.

CRITIQUING THE EVIDENCE 9

1. According to the results of Phillips and MacMahon, what is the approximate value of the ratio of variance in shrub density to mean shrub density (variance/mean) for young, medium-age, and older creosote bushes (see chapter 9, fig. 9.13)?

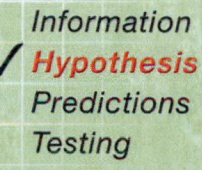

Information
✓ *Hypothesis*
Predictions
Testing

Investigating the Evidence 10

Hypotheses and Statistical Significance

LEARNING OUTCOME

After studying this section you should be able to do the following:

10.1   Describe the concept of "statistical significance."

In Investigating the Evidence 1, we reviewed the roles of questions and hypotheses in the process of science. Briefly, we considered how scientists use information to formulate questions about the natural world and convert their questions to hypotheses. A hypothesis, we said, is a possible answer to a question.

Let's use the distributions we considered in Investigating the Evidence 9 to examine the nature of scientific hypotheses in more detail. In that discussion we examined samples from three populations of plants, from which we calculated the following statistics:

Statistic	Species A	Species B	Species C
Mean density, $\overline{X}$	5.27	5.18	5.27
Variance in density, s^2	5.22	0.36	44.42
Ratio of variance to mean, $\dfrac{s^2}{\overline{X}}$	0.99	0.07	8.43

Recall that in a random distribution the ratio of the variance to the mean equals 1, that is, $\dfrac{s^2}{\overline{X}} = 1.0$.

As we have stated repeatedly, the center of scientific investigation is the hypothesis. In the case of these three populations, an appropriate hypothesis would be that in each case, $\dfrac{s^2}{\overline{X}}$ *does not* differ *significantly* from 1. Such a hypothesis,

(continued)

which proposes no significant difference between an observation and a statistical expectation or between populations, is called a *null hypothesis*. How well does this hypothesis match the results of our population counts? While $\frac{s^2}{\overline{X}}$ for species A comes close to 1.0 $\left(\frac{s^2}{\overline{X}} = 0.99\right)$, the calculated value of $\frac{s^2}{\overline{X}}$ does not exactly equal 1 for any of the populations. However, we need to remember that the numbers listed in the table are statistical *estimates* of the true, or actual, variance to mean ratio for each of the populations. It is unlikely that any of the $\frac{s^2}{\overline{X}}$ values, which were calculated from a sample of observations, would exactly match the true variance to mean ratio in any of the three study populations. Because of our limited sample size, we expect to see some difference between the statistical estimate and the theoretical expectation of $\frac{s^2}{\overline{X}} = 1.0$, even in a population known to have a random distribution.

The critical point here is to have a basis for judging whether an observed $\frac{s^2}{\overline{X}}$ differs *sufficiently* from the theoretical expectation of $\frac{s^2}{\overline{X}} = 1.0$ to be *statistically significant*. A statistically significant difference is one that is unlikely to occur by chance. By tradition, the level of significance used in most scientific investigations is $P < 0.05$, or less than 1 chance in 20.

Let's go back to our populations of plant species A, B, and C. How can we tell if any of the three $\frac{s^2}{\overline{X}}$ values in the table differs significantly from 1? That is, how can we determine

if the probability of obtaining each value by chance in a population that actually has a random distribution, is less than 0.05? This is generally done by comparing an observed value with tables of theoretically derived values. For now, we can use our judgment to make some predictions. Consider species A. The probability that we could observe a mean value of $\frac{s^2}{\overline{X}} = 0.99$ by chance in a population with a random distribution is likely to be much greater than 0.05. As a consequence, we are likely to accept the hypothesis that species A has a random distribution. In contrast, the values of $\frac{s^2}{\overline{X}}$ observed for species B and C (0.07 and 8.43) differ so much from 1.0 that the probability of obtaining these results by chance from a population that is actually randomly distributed is likely to be low. If it is less than 0.05, we will reject the hypothesis that these populations are randomly distributed, and tentatively accept the alternative hypotheses of clumped or regular distributions. We will gradually work up to evaluating whether the $\frac{s^2}{\overline{X}}$ ratios listed in the table differ significantly from $\frac{s^2}{\overline{X}} = 1.0$ in Investigating the Evidence 11 and 12.

CRITIQUING THE EVIDENCE 10

1. When using a significance level of 0.05, how often will we reject hypotheses that are actually correct? For instance, how often will we reject the hypothesis that the individuals in a study population are randomly distributed when, in fact, they really are randomly distributed?

Information
Hypothesis
Predictions
✓ Testing

Investigating the Evidence 11

Frequency of Alternative Phenotypes in a Population

LEARNING OUTCOMES

After studying this section you should be able to do the following:

11.1 Explain the chi-square goodness of fit test.
11.2 Interpret the statistical significance of results of a chi-square goodness of fit test using a statistical table.

Ecologists often ask questions about observed frequencies of individuals in a population relative to some theoretical or expected frequencies. For example, an ecologist studying the nesting habits of Darwin's finches may be interested in the frequency at which finches nest in alternative nest sites relative to the availability of the alternative nest sites in the environment. Another ecologist studying the habitat association of a plant may be interested in its relative frequencies in sandy, loamy, or clay soils. An ecologist studying the mating behavior of some species may want to determine whether alternative male phenotypes (males with different physical and/or behavioral characteristics) occur at different frequencies in the population.

A common method to test hypotheses concerning the relationship between observed and hypothesized frequencies is the **chi-square** (χ^2) "goodness of fit" test. This test is used to judge how well an observed distribution of frequencies matches one "expected" from a particular hypothesis. Let's explore this test using the frequency of alternative male phenotypes in a population of side-blotched lizards, *Uta stansburiana*. Barry Sinervo (Sinervo and Lively 1996) found that males in populations of side-blotched lizards of the coastal range of central California include three male phenotypes: very aggressiv orange-throat males, moderately aggressive blue-throat males, and sneaker yellow-throat males. Sinervo and his colleagues also found that these three male phenotypes vary in their frequencies over time. Let's consider the following hypothetical table of data from a population and test the hypothesis that the three male phenotypes are present in equal frequencies in the population.

Male phenotype	Observed frequency (O)	Expected frequency (E)
Orange-throat	12	17
Blue-throat	30	17
Yellow-throat	9	17

Before proceeding with our test, let's reflect on what we are doing. We would like to know if there are differences in the frequencies of male phenotypes in this population of side-blotched lizards. Therefore, we went out and obtained a sample of 51 males from the population. This sample included 12 orange-throat males, 30 blue-throat males, and 9 yellow-throat males. Our sample is an estimate of the actual frequencies of the three phenotypes in the larger population that we are studying. Because our hypothesis is that there are no differences in frequencies among the male phenotypes, the "expected" frequencies, in the third column of our table, are equal $\left(\frac{51}{3} = 17\right)$.

Let's use the chi-square (χ^2) "goodness of fit" test to determine how well our observed frequency distribution matches the expected frequency distribution. The value of χ^2 is calculated as follows:

$$\chi^2 = \Sigma \frac{(O - E)^2}{E}$$

In this equation, O is the observed frequency of a particular phenotype and E is the expected frequency. If we enter the values from the table, we get the following:

$$\chi^2 = \frac{(12 - 17)^2}{17} + \frac{(30 - 17)^2}{17} + \frac{(9 - 17)^2}{17}$$
$$\chi^2 = \frac{(-5)^2}{17} + \frac{(13)^2}{17} + \frac{(-8)^2}{17}$$
$$\chi^2 = \frac{25}{17} + \frac{169}{17} + \frac{64}{17}$$
$$\chi^2 = 1.47 + 9.94 + 3.76$$
$$\chi^2 = 15.17$$

The next step in the chi-square test is to determine whether the difference between observed and expected values, as indexed by our calculated χ^2, is "significant." We determine significance by comparing our calculated value of χ^2 with a table of χ^2 values, which are included in most statistics textbooks. In order to find the appropriate, or critical, value of χ^2 from such a table, we need to know two things. First, we need to choose a level of significance, which, as we saw in Investigating the Evidence 10, is generally $P < 0.05$. Second, we need to know the "degrees of freedom," which, in this case, is the number of male phenotypes (3), or n minus 1.

$$\text{degrees of freedom} = n - 1$$
$$= 3 - 1$$
$$= 2$$

What does "degrees of freedom" mean? It is the number of values we can pick freely without being constrained by other values within a set. In the case of the frequency of three male phenotypes, given a particular sample size, once we know the frequencies of two of the phenotypes, we automatically know the third. For instance, with a sample of 51 lizards, once we determine that there are 12 orange-throat males and 30 blue-throat males in our sample, the frequency of yellow-throat males is constrained to be 9. Therefore, in this sample of three male phenotypes, there are two degrees of freedom.

In a table of critical values of chi-square (Appendix B, table B.3), you will find that for a significance level of $P = 0.05$ and 2 degrees of freedom, the critical χ^2 value is 5.991. Since our calculated value of χ^2 (15.17) is greater than 5.991, we reject the hypothesis that the three male phenotypes are present in equal frequencies in our study population. In other words, we have evidence of significant differences in frequency among the three phenotypes. Moreover, we can attach a probability statement to this conclusion, which is $P < 0.05$.

CRITIQUING THE EVIDENCE 11

1. Would the results of the goodness of fit analysis described here have been significant if we had chosen a significance level of $P = 0.01$?
2. Suppose we study another population of side-blotched lizards in which there are five male phenotypes. If we did a "goodness of fit" analysis to test for an equal distribution of males among phenotypes in this population, what would be the degrees of freedom for our test and what would be the critical value of χ^2 for $P = 0.05$?

Information
Hypothesis
Predictions
✓ **Testing**

Investigating the Evidence 12

A Statistical Test for Distribution Pattern

LEARNING OUTCOMES

After studying this section you should be able to do the following:

12.1 Calculate chi-square given mean density and variance in density estimated by a sample of a population.

12.2 Given chi-square values and degrees of freedom from a variance/mean ratio test of distributions, determine whether a population is more likely to have a regular, random, or clumped distribution. (Use a graph showing the distribution of critical chi-square values.)

Suppose you are studying the life history of three species of herbaceous plants in a desert landscape. As part of that study, you are interested in determining the pattern of distribution of individuals in each population. Your hypothesis states that the individuals in each population are randomly distributed across the landscape. Your alternative hypotheses propose that individuals are either clumped or uniformly distributed. In Investigating the Evidence 9, we reviewed a study that suggested very different patterns of distribution in three plant populations. In that example, the sample means and sample variances of the three hypothetical populations of plant species A, B, and C were as follows:

Statistic	Species A	Species B	Species C
Sample mean, $\overline{X}$	5.27	5.18	5.27
Sample variance, s^2	5.22	0.36	44.42
Ratio of sample variance to sample mean, $s^2/\overline{X}$	0.99	0.07	8.43

Also in Investigating the Evidence 9, we reviewed how in a randomly distributed population, variance/mean = 1, while in a regularly distributed population, variance/mean < 1, and in a clumped population, variance/mean > 1. Given these relationships, the $s^2/\overline{X}$ ratios in the above table suggest that species A has a random distribution, species B has a regular distribution, and species C has a clumped distribution. However, since the values in the table are statistical *estimates* of the true variance/mean ratios in the study populations, we need to do a statistical test to determine the significance of our results.

The first step in our test is to establish a hypothesis. In each case, our hypothesis states that the variance/mean ratio of the population equals 1. We next need to determine a significance level, which we have seen is generally $P < 0.05$. We can use a chi-square test to determine whether a sample variance/sample mean ratio is significantly different from 1 as follows:

$$\chi^2 = \frac{s^2(n-1)}{\overline{X}}$$

Here, $n-1$ is the degrees of freedom, which is the sample size minus 1. In the case of our plant study, the sample size, n, is the number of sample quadrats studied for each population, which was 11. Therefore, the degrees of freedom is 10.

For our sample of the species A population, the calculation is:

$$\chi^2 = \frac{5.22 \times 10}{5.27}$$

$$\chi^2 = 9.905$$

For our sample of the species B population, the calculation is:

$$\chi^2 = \frac{0.36 \times 10}{5.18}$$

$$\chi^2 = 0.695$$

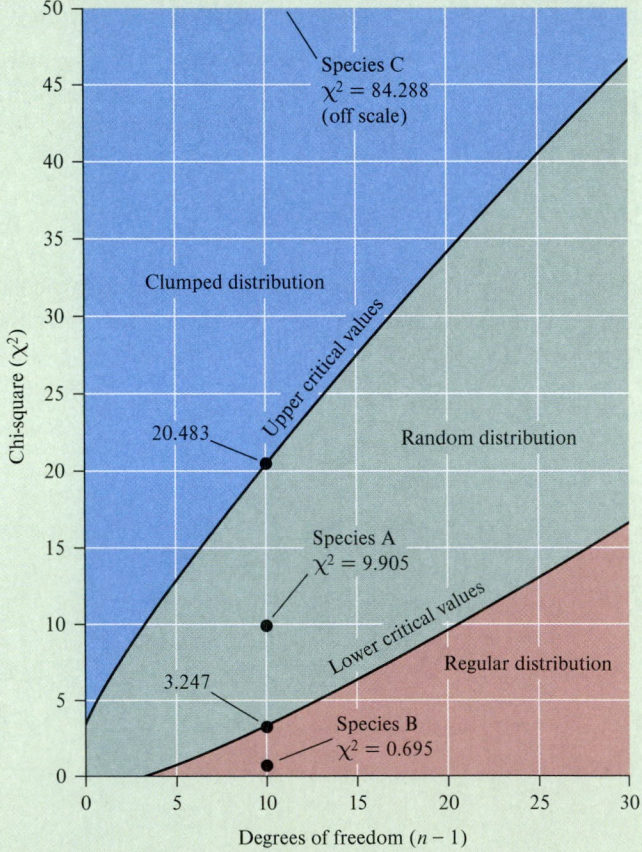

Figure 1   Figure for determining critical values of chi-square for variance/mean ratio test of random, regular, or clumped distributions.

For our sample of the species C population, the calculation is:

$$\chi^2 = \frac{44.42 \times 10}{5.27}$$
$$\chi^2 = 84.288$$

How do we determine if these values of chi-square are statistically significant at $P < 0.05$? Here, we need to consider whether the $s^2/\overline{X}$ ratios are significantly *greater* than 1, *or* significantly *less* than 1. Therefore, in contrast to the situation that we analyzed in chapter 11, we will compare our chi-square values to two critical values, one small and one large.

The situation we are considering is pictured in figure 1. With degrees of freedom of 10 ($11 - 1 = 10$), the critical values of chi-square are $\chi^2 = 3.247$ and $\chi^2 = 20.483$. As shown in figure 1, these values of chi-square fall on the lines that form the boundaries of the area shaded green. For values of chi-square within the green area, we accept the hypothesis that the variance/mean ratio in the population equals 1, and that the population has a random distribution. Values of chi-square in the blue zone indicate a clumped distribution, while values in the red zone indicate a regular distribution. Returning to the populations of species A, B, and C, we accept the hypothesis that the variance/mean ratio of species A ($\chi^2 = 9.905$) does not differ significantly from 1, and therefore, we conclude that it has a random distribution. Meanwhile, because the value of chi-square for species B ($\chi^2 = 0.695$) is less than the critical value of 3.247, we reject the hypothesis that the variance/mean ratio for species B equals 1, and conclude that it has a regular distribution. Using similar logic, we conclude that species C ($\chi^2 = 84.288$) has a clumped distribution.

CRITIQUING THE EVIDENCE 12

1. Do the results of the chi-square test for species B show beyond a doubt that its population has a regular distribution?
2. How could we improve our confidence in the conclusion that species B has a regular distribution?

Information
Hypothesis
Predictions
✓ Testing

Investigating the Evidence 13

Field Experiments

LEARNING OUTCOMES

After studying this section you should be able to do the following:

13.1 Describe the difference between a typical laboratory experiment and a field experiment.
13.2 Explain the importance of the several elements of a sound field experiment.

Field experiments have played a key role in the assessment of the importance of competitive interactions in nature. Joseph Connell (1974) and Nelson Hairston, Sr. (1989), two of the pioneers in the use of field experiments in ecology, outlined their proper design and execution. Connell points out that one of the most substantial differences between laboratory and field experiments is that in the laboratory setting, the investigator controls all important factors but one, the factor of interest. In contrast, in field experiments, all factors are allowed to vary naturally (the investigator generally has no choice) while the factor of interest is controlled, or manipulated, by the investigator.

Both laboratory and field experiments have played an important role in ecology, but it is the field experiment that provides the key to unlock the secrets of complex interactions in nature. Why is it that field experiments are more useful in this regard? Connell pointed out that compared to laboratory experiments, the results of field experiments can be more directly applied to understanding relationships in nature because "interactions with other organisms, and the natural variation in the abiotic environment, are included in the experiment." The best field experiments are those that are executed with the least disturbance to the natural community. The utility of field experiments, however, depends on several design features.

Knowledge of Initial Conditions

To test for change in response to experimental manipulation, you have to know what conditions were like before the manipulation. Departures from initial conditions indicate a response to the experimental treatments. For instance, in his experiments on competition between barnacles, Connell first estimated the initial population density of one of the species in all his study plots (see fig. 13.19). Brown and his colleagues (Brown and Munger 1985; Heske, Brown, and Mistry 1994) were also careful to measure the population densities of all rodent species in their study plots several times before excluding large granivorous rodents from their experimental plots.

Controls

As in laboratory experiments, field experiments must include controls. Without controls it would be impossible to determine whether or not an experimental treatment has had an effect. Tansley (1917) created controls for his experiments on competition by growing each of his potential competitors by themselves

(continued)

in acidic and basic soils. What was the control for the experiment on competition among desert rodents? Brown's research team created controls by surrounding study plots with their mouse-proof fence but then cutting holes 6.5 cm in diameter in the fences to allow large granivorous rodents to move freely into and out of the plots.

Replication

Field experiments, where possible, should include replication. Why? Ecological systems and environmental conditions are variable, both in time and space. Replication is intended to capture this variation. The question posed by the experimenter is whether an experimental effect is apparent *despite* variation. Ecologists use statistics to make such a judgment. Without replication, you would never know if the results could be repeated in either time or space.

What is considered acceptable study design has changed over the decades, reflecting increasing familiarity and concern for statistical analysis. In Tansley's experiments on how competition may restrict the distribution of *Galium* species to particular soils, replication was totally lacking. In Tansley's experiment, each condition (soil type) was represented

just one time. Connell's later experiments with barnacles included some replication, but it was still limited at each tidal level. However, since there was a great deal of consistency in response across tidal levels, we can accept that competition acts as a significant force limiting barnacle distributions within the intertidal zone. In contrast to these earlier experiments, the more recent experiments by Brown on competition among desert rodents were replicated sufficiently for statistical analysis and repeated a second time.

The reviews by Connell and Hairston provide a guide to field experimentation as it has been conducted in the past few decades. However, as we see in section VI, the need for experimentation at large scales is forcing ecologists to further expand their concept of experimental design.

CRITIQUING THE EVIDENCE 13

1. Why did Brown's research team (see section 13.4) create controls by completely fencing study plots and then cutting holes in their sides to allow free passage of rodents into and out of the plot? Why not just compare the density of small rodents in the large granivore removal plots with their densities in the surrounding desert?

Information
Hypothesis
Predictions
✓ *Testing*

Investigating the Evidence 14

Standard Error of the Mean

LEARNING OUTCOMES

After studying this section you should be able to do the following:

14.1 Calculate a standard error of the mean.
14.2 Explain why larger samples of a population generally produce a more precise estimate of the population mean.

When we introduced the sample mean in Investigating the Evidence 2, we pointed out how it is an estimate of the actual, or true, population mean. A second sample from a population would probably have a different sample mean and a third sample would have yet another. How close is a given sample mean to the true population mean? The answer to this question will depend on two factors: the variation within the population and the number of observations or measurements in our sample from the population. Here, we begin to build a way of representing the precision of a given estimate of a population mean. Our first step is to calculate a statistic called the standard error of the mean, usually called the **standard error,** $s_{\bar{X}}$. The standard

error is calculated from the sample variance (see Investigating the Evidence 4) and the sample size as follows:

$$s_{\bar{X}} = \sqrt{\frac{s^2}{n}}$$

$$s_{\bar{X}} = \frac{s}{\sqrt{n}}$$

where

s^2 = sample variance
s = sample standard deviation
n = sample size

For a concrete example, let's use the body length measurements for a sample of loach minnows, *Tiaroga cobitis,* from a tributary of the San Francisco River in southwestern New Mexico. Suppose you are comparing the body sizes of loach minnows in populations exposed to predation by flathead catfish, an introduced species, to populations not exposed to predation by this introduced fish. To do so, you need to estimate body sizes in several populations.

Your sample from the San Francisco River was:

Specimen	1	2	3	4	5	6	7	8	9	10
Total length (mm)	60	62	56	53	53	59	62	41	58	58

The mean of this sample (see Investigating the Evidence 2) is:

$$\overline{X} = 56.2 \text{ mm}$$

Moreover, its standard deviation is:

$$s = 6.2 \text{ mm}$$

Since the number of fish in our sample is 10, the standard error for this sample of loach minnows is:

$$s_{\overline{X}} = \frac{6.2 \text{ mm}}{\sqrt{10}}$$

$$s_{\overline{X}} = \frac{6.2 \text{ mm}}{3.16}$$

$$s_{\overline{X}} = 1.96 \text{ mm}$$

Now, let's consider the hypothetical situation in which we obtained a sample of loach minnows from another study site on the Gila River. This second sample happened to yield the same sample mean and the same standard deviation. However, instead of 10 loach minnows, this second sample included 50 loach minnows. The standard error calculated from this sample is:

$$s_{\overline{X}} = \frac{6.2 \text{ mm}}{\sqrt{50}}$$

$$s_{\overline{X}} = \frac{6.2 \text{ mm}}{7.07}$$

$$s_{\overline{X}} = 0.88 \text{ mm}$$

Notice that because there were more minnows in this second sample, the size of the standard error is considerably smaller. In other words, our second sample mean is a more precise estimate of the true population mean. This is shown in the form of a graph in figure 1. In figure 1a the points indicate the sample means for our two samples and the vertical bars, above and below the points, are plus and minus one standard error. The same statistics are plotted in figure 1b as a bar graph and only the upper standard error bar is shown, which is a common way to plot such data. In either case, the smaller standard error around the sample mean for the Gila River indicates that our estimate of the mean length of loach minnows in the population is more precise there than that for the population in the San Francisco River. In Investigating the Evidence 15, we will use the standard error to derive a more quantitative expression of precision called the confidence interval.

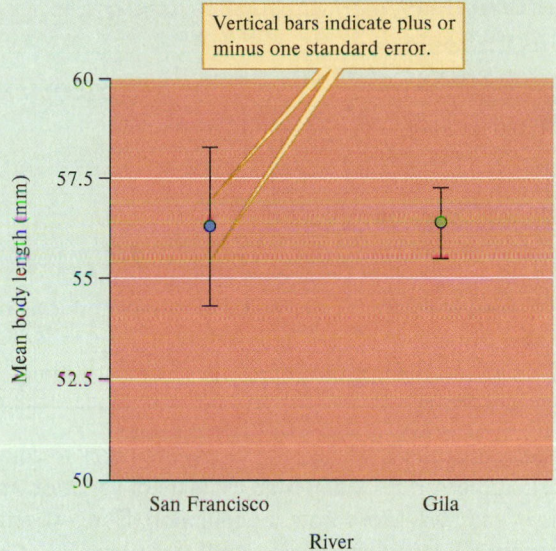

(a)

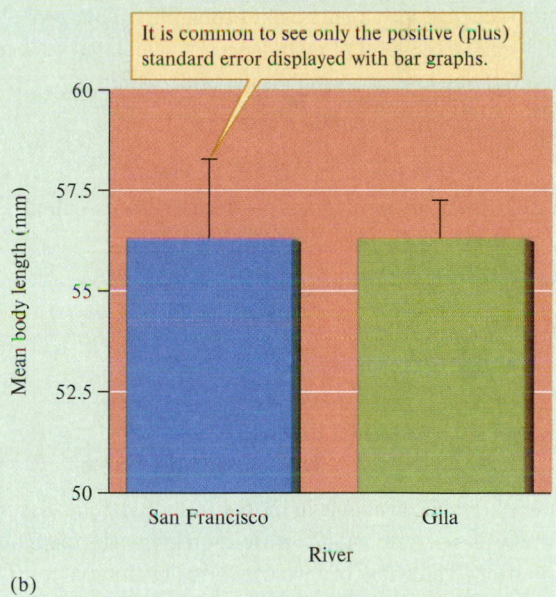

(b)

Figure 1   Average body length of loach minnows and standard errors calculated from samples collected in the San Francisco River ($n = 10$) and the Gila River ($n = 50$). Smaller standard error for the sample from the Gila River is the result of a larger sample size.

CRITIQUING THE EVIDENCE 14

1. When sampling a population to estimate a population mean, why, from a statistical perspective, is it generally better to have a larger sample size?
2. When might it be preferable to take smaller, rather than larger, samples of a population?

Investigating the Evidence 15

Confidence Intervals

LEARNING OUTCOMES

After studying this section you should be able to do the following:

15.1 Distinguish between significance level and level of confidence.

15.2 Explain the meaning of a 95% confidence interval.

In Investigating the Evidence 14, we reviewed how to calculate the standard error, $s_{\bar{x}}$, which is an estimate of variation among means of samples drawn from a population. Here, we will use the standard error to calculate a confidence interval. A **confidence interval** is a range of values within which the true population mean occurs with a particular probability. That probability, which is called the **level of confidence,** is calculated as 1 minus the significance level, α, which is generally 0.05:

$$\text{Level of confidence} = 1 - \alpha$$
$$\text{Level of confidence} = 1 - 0.05 = 0.95$$

Using this level of confidence produces what is called a 95% confidence interval that is calculated as follows:

$$\text{Confidence interval for } \mu = \bar{X} \pm s_{\bar{x}} \times t$$

where

$$\mu = \text{true population mean}$$
$$\bar{X} = \text{sample mean}$$
$$s_{\bar{x}} = \text{standard error}$$
$$t = \text{value from a Student's } t \text{ table}$$

A Student's t table, available in most statistics textbooks, summarizes the values of a statistical distribution known as the Student's t distribution. The value of t we use for calculating a confidence interval is determined by the degrees of freedom ($n - 1$) and the significance level, which in this case is $\alpha = 0.05$.

Let's calculate a 95% confidence interval using the body length measurements for the sample of loach minnows, *Tiaroga cobitis,* that we used to calculate a mean and standard error in Investigating the Evidence 14:

$$\bar{X} = 56.2 \text{ mm}$$
$$s_{\bar{x}} = 1.96 \text{ mm}$$

This sample of body lengths included measurements of 10 fish ($n = 10$), and so the degrees of freedom for this sample ($n - 1$) is 9. Using a significance level of 0.05 and degrees of freedom of 9, we find that the critical value of t from a Student's t table is 2.26 (table B.1 of Appendix B). Therefore, the 95% confidence interval calculated from this sample is:

$$\text{Confidence interval for } \mu = \bar{X} \pm s_{\bar{x}} \times t$$
$$= 56.2 \text{ mm} \pm 1.96 \text{ mm} \times 2.26$$
$$= 56.2 \text{ mm} \pm 4.43 \text{ mm}$$

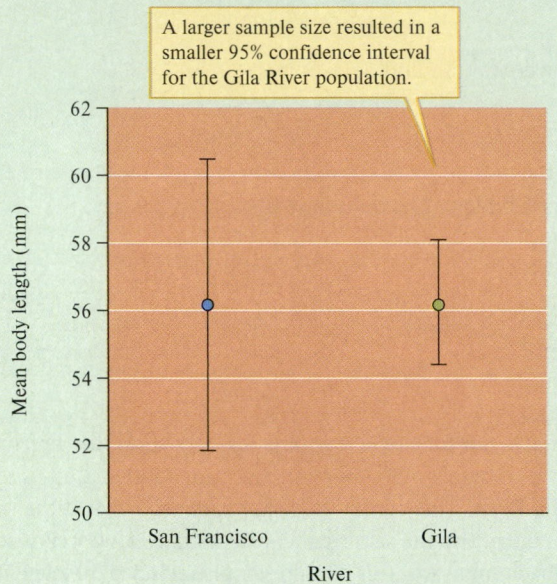

A larger sample size resulted in a smaller 95% confidence interval for the Gila River population.

Figure 1 Average body length of loach minnows and 95% confidence intervals calculated from samples collected in the San Francisco River ($n = 10$) and Gila River ($n = 50$).

With this confidence interval, we can say that there is a 95% probability that the true mean body length in this population of loach minnows is somewhere between 60.63 mm (56.2 mm + 4.43 mm) and 51.77 mm (56.2 mm − 4.43 mm).

This is shown graphically in figure 1, along with the mean and 95% confidence interval for the sample of loach minnows from the Gila River that we first considered in Investigating the Evidence 14. Notice that the 95% confidence interval for the Gila River sample is much smaller. This smaller confidence interval is the result of the larger sample size from the Gila River ($n = 50$), which produced a smaller standard error ($s_{\bar{x}} = 0.88$) and a smaller critical t value (2.01), since the degrees of freedom is 49. As a consequence of having a larger sample, our estimate of the true population mean has been narrowed to a much smaller range for the Gila River population of loach minnows.

CRITIQUING THE EVIDENCE 15

1. What is the 95% confidence interval for the Gila River sample of loach minnows?

2. What value of t would you use from table B.1 for calculating a 95% confidence interval, if your sample size was 18 and your significance level was $\alpha = 0.05$?

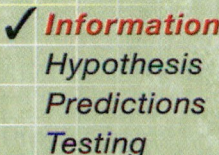

Investigating the Evidence 16

Estimating the Number of Species in Communities

LEARNING OUTCOMES

After studying this section you should be able to do the following:

16.1 Explain the difficulties involved in trying to estimate the total number of species in a community.

16.2 Discuss ways to reduce the effort necessary for making a comparison of the relative species richness of communities.

How many species are there? This is one of the most fundamental questions that an ecologist can ask about a community. With increasing threats to biological diversity, species richness is also one of the most important community attributes we might measure. For instance, estimates of species richness are critical for determining areas suitable for conservation, for diagnosing the impact of environmental change on a community, and for identifying critical habitat for rare or threatened species. However, determining the number of species in a community is not a simple undertaking. Sound estimates of species richness for most taxa require a carefully designed, standardized sampling program. Here we will review some of the basic factors that an ecologist needs to consider when designing such a sampling program to gather information about species richness within and among communities.

Sampling Effort

The number of species recorded in a sample of a community increases with higher sampling effort. We reviewed a highly simplified example of this in Investigating the Evidence 6, where we considered how numbers of quadrats influenced estimates of species richness in the benthic community of a small Rocky Mountain stream. In that example, a relatively small sample size was required. However, often far more effort is required. For example, Petri Martikainen and Jari Kouki (2003) estimated that to verify the presence or absence of threatened beetle species in the boreal forests of Finland required a sample of over 400 beetle species. They also suggested that a sample of over 100,000 individual beetles may be required to assess just 10 forest areas in Finland for their suitability to serve as conservation areas for threatened beetle species. To reduce the sampling effort required to estimate species richness, community ecologists and conservationists often focus on groups of organisms that are reliable indicators of species richness.

Indicator Taxa

Because of the great cost and time of making thorough inventories of species diversity, ecologists have proposed a wide variety of taxa as indicators of overall biological diversity. Indicator taxa have generally been well-known and conspicuous groups of organisms such as birds and butterflies. However, it appears that indicator taxa need to be chosen with caution. For example, John Lawton of Imperial College in the United Kingdom and

12 colleagues (Lawton et al. 1998) attempted to characterize biological diversity along a disturbance gradient in the tropical forest of Cameroon, Africa, using indicator taxa. In addition to birds and butterflies, Lawton and his colleagues sampled flying beetles, beetles living in the forest canopy, canopy ants, leaf-litter ants, termites, and soil nematodes. They sampled these taxa from 1992 to 1994 and spent several more years sorting and cataloging the approximately 2,000 species collected. This work required approximately 10,000 scientist hours. Unfortunately, the conclusion at the end of this massive study was that no one group serves as a reliable indicator of species richness for other taxonomic groups. Lawton and his colleagues estimated that their survey included from one-tenth to one-hundredth the total number of species in their study site. Citing their own experience, they concluded that characterizing the full biological diversity of just 1 hectare of tropical forest would require from 100,000 to 1 million scientist hours. As a consequence of these time constraints, ecologists will likely continue to focus their studies of diversity on smaller groups of taxa. However, even with a restricted taxonomic focus, it is important to standardize sampling across study communities.

Standardized Sampling

Standardizing sampling effort and technique is generally necessary to provide a valid basis for comparing species richness across communities. For example, Frode Ødegaard of the Norwegian Institute for Nature Research took great care to standardize sampling as he compared species richness among plant-feeding beetles living in a tropical dry forest and in a tropical rain forest in Panama (Ødegaard 2006). Ødegaard sampled both forests from a canopy crane that provided access to similar areas of forest (~0.8 ha). He standardized the amount of time he spent sampling each tree or vine, and he used the same sampling techniques in both forests. Ødegaard also sampled the beetles on approximately the same number of plant species—50 species in the dry forest and 52 in the rain forest. His efforts resulted in the capture of very similar numbers of individual beetles in the two forests: 35,479 in dry forest versus 30,352 in rain forest. However, his collections in rain forest included 37% more beetle species than in dry forest: 1,603 species in rain forest versus 1,165 in dry forest. Because Ødegaard took care to standardize sampling, we can conclude that the species richness of plant-feeding beetles was probably higher at his rain forest study site. If his sampling efforts were uneven, we could not reach such a conclusion.

CRITIQUING THE EVIDENCE 16

1. A complete list of species has not been determined for any area of substantial habitat anywhere on earth. Why not?
2. Why do most surveys of species diversity focus on restricted groups of relatively well-known organisms such as plants, mammals, and butterflies?

Investigating the Evidence 17

Using Confidence Intervals to Compare Populations

LEARNING OUTCOMES

After studying this section you should be able to do the following:

17.1 Calculate a confidence interval.

17.2 Determine whether differences between two populations are statistically significant, based on comparisons of confidence intervals.

In Investigating the Evidence 15, we reviewed how to calculate confidence intervals for the true population mean as:

Confidence interval for $\mu = \overline{X} \pm s_{\overline{X}} \times t$

Here, we will use the confidence intervals calculated from samples of two populations to create a visual comparison of the populations.

Suppose you are studying the recovery of the food web of a mountain stream from the effects of a flood. Part of your study involves estimating the biomass of each of the consumers in the food web. One of the species you are studying is *Neothremma alicia,* a small species of caddisfly, an insect in the order Trichoptera, that spends its larval stage grazing on diatoms living on the tops of stones in swift mountain streams. In your study site there are two forks in a stream, one that flooded 2 months before you took your sample, and one that did not flood. Otherwise, the two forks of the stream are similar. The following are the dry weights (milligrams) of *N. alicia* that you collected in 0.1 m² quadrats in each study stream:

| Quadrat Number | | | | | | | | |
1	2	3	4	5	6	7	8	9
Flooded (mg)								
4.83	3.00	3.63	1.20	2.97	1.17	1.95	0.98	1.46
Unflooded (mg)								
7.08	5.18	5.97	3.64	5.14	3.05	4.23	3.14	3.71

Using these data, we can calculate the means and standard errors for each sample, with the following results:

Flooded stream ($f = flooded$):

$$\overline{X}_f = 2.354 \text{ mg}/0.1 \text{ m}^2$$

$$S_{\overline{X}_f} = \frac{s_f}{\sqrt{n}} = \frac{1.329 \text{ mg}/0.1 \text{ m}^2}{3} = 0.443 \text{ mg}/0.1 \text{ m}^2$$

Unflooded stream ($u = unflooded$):

$$\overline{X}_u = 4.571 \text{ mg}/0.1 \text{ m}^2$$

$$S_{\overline{X}_u} = \frac{s_u}{\sqrt{n}} = \frac{1.371 \text{ mg}/0.1 \text{ m}^2}{3} = 0.457 \text{ mg}/0.1 \text{ m}^2$$

Now, using our degrees of freedom ($n - 1 = 8$) and a level of confidence of 0.95, we find that our critical value of Student's *t* is 2.31. Using this *t* value, we can calculate a confidence interval for each population:

Confidence interval for $\mu = \overline{X} \pm s_{\overline{X}} \times t$

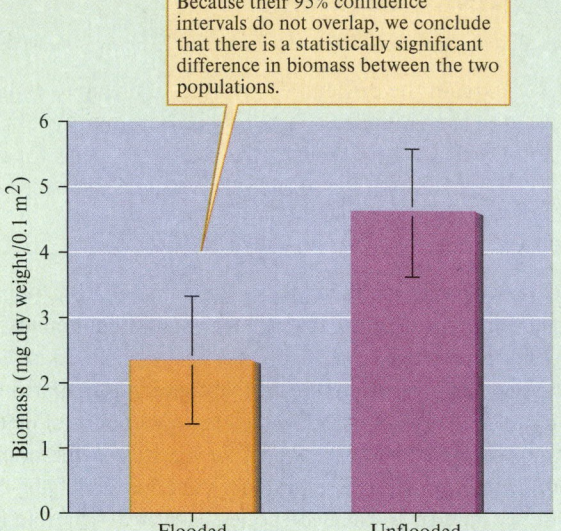

Because their 95% confidence intervals do not overlap, we conclude that there is a statistically significant difference in biomass between the two populations.

Figure 1 Means and 95% confidence intervals of biomass of *Neothremma alicia* (class Insecta, order Trichoptera) in a recently flooded and an unflooded stream in the Rocky Mountains.

Flooded stream:

$$\mu_f = 2.354 \text{ mg}/0.1 \text{ m}^2 \pm 0.443 \text{ mg}/0.1 \text{ m}^2 \times 2.31$$

$$\mu_f = 2.354 \text{ mg}/0.1 \text{ m}^2 \pm 1.023 \text{ mg}/0.1 \text{ m}^2$$

Unflooded stream:

$$\mu_u = 4.571 \text{ mg}/0.1 \text{ m}^2 \pm 0.457 \text{ mg}/0.1 \text{ m}^2 \times 2.31$$

$$\mu_u = 4.571 \text{ mg}/0.1 \text{ m}^2 \pm 1.056 \text{ mg}/0.1 \text{ m}^2$$

These sample means and confidence intervals are plotted in figure 1. Recall from Investigating the Evidence 15 that the true population means for each of the study populations has a 95% chance of falling somewhere within the 95% confidence intervals. Now notice that the 95% confidence intervals for the two samples do not overlap. This indicates that there is less than a 5% chance that the two samples were drawn from populations of *N. alicia* with the same mean biomass per unit area. In other words, we have a basis for saying that there is a statistically significant difference in biomass of *N. alicia* in the two study streams. In Investigating the Evidence 18, we will take a somewhat different approach to making a statistical comparison of these two populations.

CRITIQUING THE EVIDENCE 17

1. Why do larger sample sizes improve the ability of a researcher to detect statistical differences between populations?

2. How would increasing the level of confidence from 0.95 to 0.99 affect the range of values included in the confidence intervals for the abundance of *Neothremma alicia* in the flooded and unflooded study streams?

Information
Hypothesis
Predictions
✓ Testing

Investigating the Evidence 18

Comparing Two Populations with the t-Test

LEARNING OUTCOMES

After studying this section you should be able to do the following:

18.1 Compare two populations statistically using a t-test.
18.2 Given a value of t, degrees of freedom for a particular t-test, and a table of critical t values, judge whether the means of some measurements made for two populations are different statistically.

In Investigating the Evidence 17, we used confidence intervals to compare the biomasses of two populations of the diatom-feeding caddisfly, *Neothremma alicia*. That comparison indicated that the population living in a stream that had flooded recently had a lower biomass per unit area. Here, we will use the same samples of the two populations to test for significant differences using a **t-test**, a method for statistical comparison of two samples. The t-test involves calculating the statistic t and comparing the value with a table of critical values of t. The hypothesis is that the populations from which the samples were drawn have the same mean, and the alternative hypothesis is that the population means are different. If the calculated t statistic is *less than* the critical value of t, the hypothesis that the populations are not different is accepted. If the calculated t statistic is *greater than* the critical value of t, this hypothesis is rejected.

As an example, let's use the t-test to compare our two samples of *N. alicia*:

Quadrat Number								
1	2	3	4	5	6	7	8	9
Flooded (mg)								
4.83	3.00	3.63	1.20	2.97	1.17	1.95	0.98	1.46
Unflooded (mg)								
7.08	5.18	5.97	3.64	5.14	3.05	4.23	3.14	3.71

The statistic t for this comparison is calculated as:

$$t = \frac{|\bar{X}_f - \bar{X}_u|}{s_{\bar{X}_f - \bar{X}_u}}$$

In this equation:
$\bar{X}_f$ = mean of sample from flooded stream
$\quad = 2.354$ mg/0.1 m^2
$\bar{X}_u$ = mean of sample from unflooded stream
$\quad = 4.571$ mg/0.1 m^2
$s_{\bar{X}_f - \bar{X}_u}$ = standard error of the difference between means, which is calculated as:

$$s_{\bar{X}_f - \bar{X}_u} = \sqrt{\frac{s_p^2}{n_f} + \frac{s_p^2}{n_u}}$$

Here:

n_f = number of quadrats in sample from flooded stream
n_u = number of quadrats in sample from unflooded stream
s_p^2 = pooled estimate of the variance

The pooled estimate of the variance for our two samples is calculated as:

$$s_p^2 = \frac{SS_f + SS_u}{DF_f + DF_u}$$

In this equation:

SS_f = sum of squares for sample from flooded stream
SS_u = sum of squares for sample from unflooded stream
DF_f = degrees of freedom for sample from flooded stream
DF_u = degrees of freedom for the sample from the unflooded stream

We calculated the sum of squares for a sample in Investigating the Evidence 4 as:

$$\text{Sum of squares} = \Sigma(X - \bar{X})^2$$

Using this equation, we can calculate the sums of squares for the two populations, with these results:

$$SS_f = 14.138 \ (\text{mg/0.1 m}^2)^2$$
$$SS_u = 15.031 \ (\text{mg/0.1 m}^2)^2$$

and degrees of freedom, which we considered in Investigating the Evidence 11, is $n - 1 = 8$ for both samples. Using $DF = 8$ for both populations, we can calculate the pooled estimate of the variance:

$$s_p^2 = \frac{14.138 \ (\text{mg/0.1 m}^2)^2 + 15.031 \ (\text{mg/0.1 m}^2)^2}{8 + 8}$$

$$s_p^2 = 1.823 \ (\text{mg/0.1 m}^2)^2$$

Using this value, we can now calculate the standard error of the difference between means:

$$s_{\bar{X}_f - \bar{X}_u} = \sqrt{\frac{s_p^2}{n_f} + \frac{s_p^2}{n_u}}$$

$$s_{\bar{X}_f - \bar{X}_u} = \sqrt{\frac{1.823(\text{mg/0.1 m}^2)^2}{9} + \frac{1.823(\text{mg/0.1 m}^2)^2}{9}}$$

$$s_{\bar{X}_f - \bar{X}_u} = \sqrt{0.203(\text{mg/0.1 m}^2)^2 + 0.203(\text{mg/0.1 m}^2)^2}$$

$$s_{\bar{X}_f - \bar{X}_u} = 0.637 \ \text{mg/0.1 m}^2$$

(continued)

Now, we have all the values we need to calculate t:

$$t = \frac{|\bar{X}_f - \bar{X}_u|}{s_{\bar{X}_f - \bar{X}_u}}$$

$$t = \frac{|2.354 \text{ mg/0.1 m}^2 - 4.571 \text{ mg/0.1 m}^2|}{0.637 \text{ mg/0.1 m}^2}$$

$$t = 3.480$$

At this point in our t-test, we need to compare the calculated t with the appropriate critical value. Again, we need to know two factors: the desired level of significance, which will be $P < 0.05$, and the degrees of freedom. In this case we use the pooled degrees of freedom:

$$DF_{pooled} = DF_f + DF_u = 8 + 8 = 16$$

The critical value of Student's t for $P < 0.05$ and $DF = 16$, is 2.12. Since our calculated value of t, 3.480, is greater than this critical value, the probability that the population means are the same is *less than* 0.05. Therefore, we reject the hypothesis that the mean biomass of *N. alicia* per unit area is the same in the two streams and accept the alternative hypothesis that the mean biomass of this caddisfly differs in the two streams.

CRITIQUING THE EVIDENCE 18

1. How would the outcome of our statistical comparison of the two *Neothremma alicia* populations have been affected if we had chosen a level of significance of $P < 0.01$ (see table B.1 of Appendix B)?

Information
Hypothesis
Predictions
✓ *Testing*

Investigating the Evidence 19

Assumptions for Statistical Tests

LEARNING OUTCOMES

After studying this section you should be able to do the following:

19.1 Summarize the importance of the normal distribution to the t-test and to estimating 95% confidence intervals.

19.2 List some characteristics of ecological systems that do not generally show a normal distribution.

In Investigating the Evidence 18, we compared samples from two populations using the t-test to judge whether there was a statistically significant difference between the populations. While the t-test is one of the most valuable tools for comparisons of pairs of samples, like any tool there are situations where it is appropriate to use a t-test and others where it is not. The t-test is based on a number of assumptions, as are other statistical tests.

One requirement of the t-test is that the populations being compared have equal variances. Another assumption of the t-test is that each of the samples is drawn from a population with a **normal distribution.** We first considered this assumption in Investigating the Evidence 3, when we discussed the assumptions underlying calculating the sample mean as a way of estimating the average, or typical, in a population. As we saw, the sample mean was appropriate for one population (a sample of seedling heights) but not appropriate for another (a sample of stream invertebrate densities) that we considered. A normal distribution is also assumed for calculating 95% confidence intervals (Investigating the Evidence 15) and for regression analysis (Investigating the Evidence 8).

Let's consider the assumption of a normal distribution in a bit more detail. A normal distribution has a particular shape. As shown in figure 1, a normal distribution is bell-shaped and proportioned in such a way that predictable percentages of the observations, or measurements, will fall within one, two, or three standard deviations of the mean (see Investigating the Evidence 4). If the characteristic of interest is not normally distributed, then we cannot be certain, for instance, that a 95% confidence interval will be accurate or that two sample means compared using a t-test are statistically different. Fortunately, many of the kinds of measurements made by ecologists, such as weights of individuals, body lengths or lengths of appendages, running speeds, or rates of photosynthesis, have normal distributions. In addition, the fit of measurements to a normal distribution does not have to be exact. For example, the t-test will produce reliable results if the distribution of measurements is fairly symmetrical around the mean. In addition, the t-test can give reliable results with some differences in variances, as long as the sizes of samples being compared are similar.

However, there are some important attributes of ecological systems that are not distributed normally. These include population densities (numbers per unit area) of plants or animals, proportions of different species in a community, percentage of time that an animal spends in different activities, and exponential rates of litter decay. One way to analyze such data is to use statistical methods that do not assume a normal distribution. We entered this area in Investigating the Evidence 3 where we discussed the sample median. We will explore this area further in Investigating the Evidence 20 to 22.

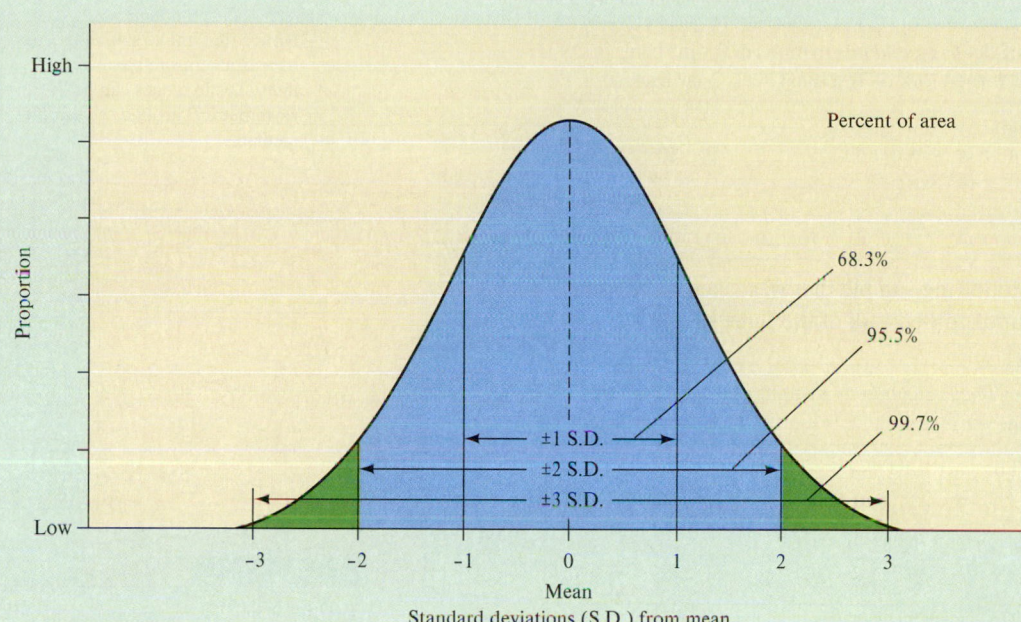

Figure 1   Example of a normal distribution, showing the percentages of observations included within one, two, or three standard deviations of the mean.

1. Suppose you sample two populations for a characteristic that has a normal distribution in both populations but is much more variable in one of the populations compared to the other. In general, would it be appropriate to test for statistical differences in the characteristic in the two populations using a *t*-test?

Information
Hypothesis
Predictions
✓ *Testing*

Investigating the Evidence 20

Variation Around the Median

LEARNING OUTCOMES

After studying this section you should be able to do the following:

20.1   Define the interquartile range for a set of samples from which a median has been determined.

20.2   Calculate an interquartile range for a set of samples of a population.

20.3   Describe the circumstance in which it is appropriate to use a median and interquartile range to represent a population sample instead of a mean and variance.

The question we consider now is how to represent variation in samples drawn from populations in which measurements or observations do not have normal distributions. When analyzing normally distributed measurements, depending on our purpose, we can estimate and represent variation using the range, variance, standard deviation, standard error, or 95% confidence interval. However, most of these indices of variation are not appropriate for non-normal distributions.

To help us consider how to represent variation when analyzing non-normal distributions, let's return to a sample of mayfly nymphs that we considered in Investigating the Evidence 3 (table 1). Suppose you are studying the recovery of this population following disturbance by a flash flood. The sample was taken from the south fork of Tesuque Creek, New Mexico, a high mountain stream of the southern Rocky Mountains. This fork had flooded 1 year before the sample was taken.

(continued)

Table 1   Number of *Baetis bicaudatus* nymphs in 0.1 m² benthic samples from the disturbed fork of Tesuque Creek, New Mexico

Quadrats: low to high											
1	2	3	4	5	6	7	8	9	10	11	12
Number of nymphs											
2	2	2	3	3	4	5	6	6	8	10	126

Now consider the following sample that was taken on the same date, but from an undisturbed fork of the same stream.

Table 2   Number of *Baetis bicaudatus* nymphs in 0.1 m² benthic samples from the undisturbed fork of Tesuque Creek, New Mexico

Quadrats: low to high											
1	2	3	4	5	6	7	8	9	10	11	12
Number of nymphs											
12	30	32	35	37	38	42	48	52	58	71	79

In Investigating the Evidence 3, we determined the median density of *B. bicaudatus* in the disturbed fork (table 1) as:

Sample median = 4 + 5 = 4.5 *B. bicaudatus* per 0.1 m² quadrat

The median density of *B. bicaudatus* in the undisturbed fork (table 2) is:

Sample median = 38 + 42

= 40 *B. bicaudatus* per 0.1 m² quadrat

The median indicates that the density of *B. bicaudatus* is 10 times higher in the undisturbed fork. Now, how can we represent the variation around these medians? One common method to represent variation in cases such as these is to divide the samples into four equal parts, called quartiles, and use the range of measurements between the upper bound of the lowest quartile and the lower bound of the highest quartile. This representation of variation in a sample is called the **interquartile range.** In table 3, the data in tables 1 and 2 have been divided into quartiles with different colors:

Table 3   Number of *Baetis bicaudatus* nymphs in 0.1 m² benthic samples from the undisturbed and disturbed forks of Tesuque Creek, New Mexico. The first, second, third, and fourth quartiles are shaded orange, yellow, green, and blue, respectively.

Quadrats: low to high											
1	2	3	4	5	6	7	8	9	10	11	12
Number of nymphs, disturbed fork											
2	2	2	3	3	4	5	6	6	8	10	126
Number of nymphs, undisturbed fork											
12	30	32	35	37	38	42	48	52	58	71	79
Quartiles **1st**		2nd			**3rd**			4th			

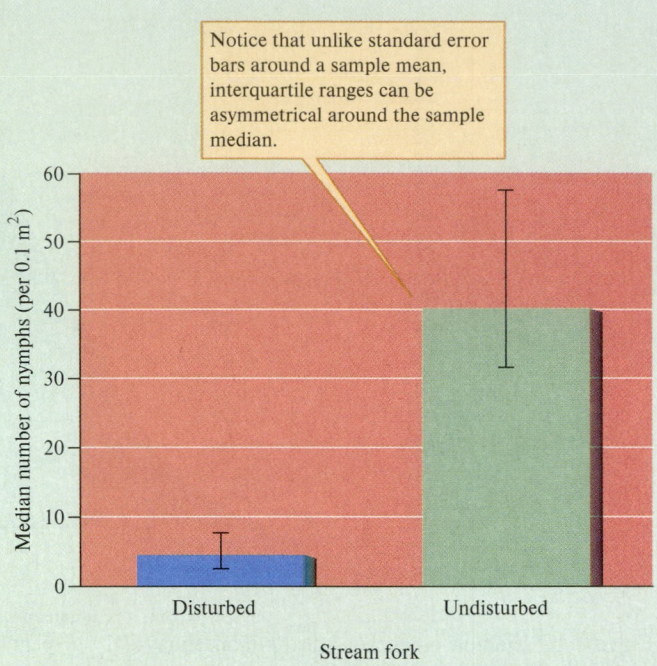

Notice that unlike standard error bars around a sample mean, interquartile ranges can be asymmetrical around the sample median.

Figure 1   Medians and interquartile ranges of mayfly nymphs, *Baetis bicaudatus,* in 0.1 m² quadrats in disturbed and undisturbed forks of Tesuque Creek, New Mexico.

Notice that the interquartile range for the undisturbed fork is 32 to 58; for the disturbed fork, the interquartile range is 2 to 8. Notice that 50% of the quadrat counts in each sample fall within this range. The medians and interquartile ranges for each of the populations are plotted in figure 1, which shows that they do not overlap. However, is there a statistically significant difference in density in the two stream forks? To answer that question, we will need a method for comparing samples that does not assume a normal distribution. We will make that comparison in Investigating the Evidence 21.

CRITIQUING THE EVIDENCE 20

1. Why can an interquartile range around a median, even when sample sizes are large, be asymmetrical?
2. Why are the standard error bars around the mean (see Investigating the Evidence 14) always symmetrical?

Information
Hypothesis
Predictions
✓ Testing

Investigating the Evidence 21

Comparison of Two Samples Using a Rank Sum Test

LEARNING OUTCOMES

After studying this section you should be able to do the following:

21.1 Compare two sample medians using the Mann-Whitney test.

21.2 Evaluate the statistical significance of Mann-Whitney U values, using sample sizes and a table of critical values for the Mann-Whitney test statistic.

Suppose you are studying the exchange of organic matter between forests and streams and the landscape you are studying is a mosaic of patches of two forest types: deciduous and coniferous. Part of your study involves determining whether there is a difference in the amount of detritus in streams draining patches of deciduous forest versus those draining coniferous forest. In an initial phase of the study, you take random measurements of the amounts of detritus (g dry weight per m^2) in two streams: one draining a deciduous forest patch and one draining a coniferous forest patch:

Measurements	1	2	3	4	5	6	7
Deciduous forest	40.6	34.2	366.5	26.9	23.1	42.8	51.1
Coniferous forest	161.1	123.5	182.3	216.6	110.9	121.2	542.4

Your hypothesis is that there is no difference in the amounts of detritus that these two streams contain. However, it turns out that the distribution of detritus within the streams is not normal, and so a sample mean will not accurately reflect the typical amount of detritus per square meter. In addition, a *t*-test is not appropriate for making a statistical comparison of detritus standing stock in the two ecosystems. The alternative is to use a statistical test that does not assume a normal distribution and compares medians *not* means. One such procedure is the Mann-Whitney test, which uses ranks of measurements or observations made in two populations, rather than the measurements themselves to make a statistical comparison. Here are the same data ordered (ranked) from smallest to largest:

We can now calculate the Mann-Whitney statistic U for the two streams. Let's begin with the stream draining the deciduous forest:

$$U_d = (n_d)(n_c) + \left[\frac{(n_d)(n_d + 1)}{2} \right] - T_d$$

$$U_d = (7)(7) + \left[\frac{(7)(7 + 1)}{2} \right] - 34$$

$$U_d = 49 + 28 - 34$$

$$U_d = 43$$

The Mann-Whitney statistic for the coniferous stream can be calculated in the same way as:

$$U_c = (n_d)(n_c) + \left[\frac{(n_c)(n_c + 1)}{2} \right] - T_c$$

Or more simply as:

$$U_c = (n_d)(n_c) - U_d$$
$$U_c = (7)(7) - 43$$
$$U_c = 6$$

At this point in the Mann-Whitney procedure, the larger of the two U values is compared to a table of critical values (Appendix table B.2). The applicable critical values are determined by significance level, generally $P < 0.05$, and the sample sizes, n_1 and n_2, which in this case are $n_d = 7$ and $n_c = 7$. Examining table B.2, we find that the critical value of the Mann-Whitney test statistic for our comparison is 41. Since $U_d = 43$ is greater than 41, we reject the hypothesis that the two streams contain the same standing stock of detritus and accept the alternative hypothesis that the standing stocks of detritus in these two particular streams are different.

Measurements (deciduous patch)	Ranks	Measurements (coniferous patch)	Ranks
23.1	1	110.9	7
26.9	2	121.2	8
34.2	3	123.5	9
40.6	4	161.1	10
42.8	5	182.3	11
51.1	6	216.6	12
355.6	13	542.4	14
$n_d = 7$ (measurements)	$T_d = \Sigma$ ranks $= 34$	$n_c = 7$ (measurements)	$T_c = \Sigma$ ranks $= 71$

(continued)

CRITIQUING THE EVIDENCE 21

1. Can we conclude from this study that streams draining deciduous versus coniferous forest patches contain different amounts of detritus?

Information
Hypothesis
Predictions
✓ *Testing*

Investigating the Evidence 22

Sample Size Revisited

LEARNING OUTCOMES

After studying this section you should be able to do the following:

22.1 Discuss sample size as it relates to large-scale geographic studies.

22.2 Contrast the challenges related to sample size faced by researchers doing large-scale ecological research versus those doing laboratory-based research on ecological questions (e.g., research on competitive interactions among microorganisms).

In Investigating the Evidence 6, we considered the number of samples necessary to obtain a reasonably precise estimate of the number of species in two simple communities. In Investigating the Evidence 16, we reconsidered the same question in relation to very complex communities, concluding that, in some situations, the sampling efforts required for precise estimates of the number of species must be intense. In general, the sample size necessary to detect statistically significant differences, or effects, increases with the variation of the system under study. Here we consider another question, "What determines sample size?" Another way of putting this question is "What is a replicate observation or measurement?"

For small-scale studies, the answers to these questions are clear. For instance, in laboratory studies of the running performance of an animal species, or the photosynthetic rate of a plant species, the number of individuals measured would determine sample size. In an experimental field study of the effects of nitrogen availability on plant diversity, the number of field plots in which the investigator manipulated soil nitrogen would determine sample size. However, the answers to these questions may not be as obvious, as ecologists begin to address larger-scale ecological problems. For example, in Investigating the Evidence 21, we compared the standing stocks of detritus in two streams one that drained a deciduous forest and one that drained a coniferous forest, and found significant differences between the two streams. In that comparison, the number of

measurements of detritus standing stock in each stream, which was 7, determined the sample size.

However, based on the comparison of the two study streams, can we conclude that streams draining conifer forests, *in general,* contain higher standing stocks of detritus compared to deciduous forest streams? We cannot reach such a general conclusion. Why not? The basic reason is that the study outlined in Investigating the Evidence 21 included only one stream draining each type of forest. In other words, relative to the general relationship between type of forest and amount of detritus in associated streams, our sample size was 1. Even if we made 100 measurements of detritus in the two study streams and, as a consequence, obtained very precise estimates of the amount of detritus that each held, the sample size relative to the broader question would still be one stream of each type.

How do we increase sample size for such a study? To do so we would need to locate and study several streams associated with deciduous forests and coniferous forests. Ideally, we would sample beyond a particular landscape and include streams in several landscapes throughout a region. The number of *different* streams of each type that we sampled within the region would determine the sample size. To make statements beyond the regional scale, we would need to sample several regions within a continent. The requirements of ecological research at very large scales soon taxes the limited resources of any single investigator or team of investigators. As a consequence, ecologists studying at large spatial scales increasingly turn to computer-based systems for data gathering and analysis, a topic discussed in the Applications section of chapter 22. Another way to increase our ability to make inferences is to utilize information gathered and published by other research teams. Doing so puts local and regional studies in broader contexts. In Investigating the Evidence 23, we discuss some approaches to searching this literature.

CRITIQUING THE EVIDENCE 22

1. How might whole-earth scale, global ecology be affected by sample size considerations?

Information
Hypothesis
Predictions
✓ **Testing**

Investigating the Evidence 23

Discovering What's Been Discovered

LEARNING OUTCOMES

After studying this section you should be able to do the following:

23.1 Discuss the essential role of literature search in scientific research.

23.2 Summarize tactics for conducting searches of the scientific literature.

Throughout this series of discussions of investigating the evidence, we have emphasized one main source of evidence—original research. While original research is the foundation on which science rests, our emphasis has neglected one of the most valuable sources of information, the published literature. A researcher in any discipline needs to keep up with developments in his or her areas of interest and in related areas. In addition, some researchers may use published literature to weigh the evidence for or against some hypothesis or theory. In the section in chapter 13 titled "Evidence for Competition in Nature" (section 13.2), we reviewed two studies that took such an approach. Students of ecology may use published literature to learn more about a particular subject, to read additional papers by researchers whose work interests them, or to do literature surveys in support of their own independent research. As pointed out in the preface of this book, however, the explosive pace of scientific discovery makes staying current very difficult. Fortunately, there are now many databases and searching tools that can help.

The databases available for searching ecological literature are far too many to review here. Therefore, we'll focus on three contemporary ones: *Google Scholar, Proquest Natural Science Collection,* and *Web of Science,* which are widely available in university libraries and include many journals of significance to ecology. Some of their characteristics are listed in the accompanying table.

The important point here is that these databases provide access to millions of published papers often covering many decades of research. Of course, few people would want to spend the time laboriously sorting through all those articles. Fortunately, each of these databases includes a powerful search tool that will help you locate articles of interest. Let's consider some basic tips on how to use these tools to conduct an effective search.

You should generally begin your search by summarizing your subject or research interest. Next, divide your subject into major concepts or key terms. Be sure to think of alternative terms for the same subject—for example, beetles or Coleoptera, daisy or Asteraceae, competition or interference. Next determine the time period in which you are interested. For instance, do you want only the most current literature on your subject, or do you want all literature available in the database?

Once you have your terms listed and have selected a time period, try a search using one or more terms. If you get too many references, too few references, or unwanted references, you can use Boolean Logical Operators to adjust your search. The main Boolean Logical Operators are **and, or,** and **not.** The operator **or** will *broaden* your search and will generally yield more references. For instance, the search specified by "daisy **or** Asteraceae" will retrieve references containing *either* daisy *or* Asteraceae. In contrast the operator **and** will *narrow* your search. The search specified by "daisy **or** Asteraceae **and** desert" will retrieve references containing *either* daisy *or* Asteraceae, but restrict the list of references to those concerned with these flowers in *desert areas*. The search specified by "daisy **or** Asteraceae **and** alpine" would yield literature on these flowers in alpine zones. If you want to exclude certain types of references from your search, you may choose to use the operator **not.** For example, the search "daisy **or** Asteraceae **not** sunflower" will exclude references that include the term sunflower.

Another useful tool for refining searches is the wild card. A wild card is used to locate references including a particular word or term with alternative endings. For example, you may encounter references to the insect order to which beetles

Selected electronic databases that cover the ecological literature

Database	URL	Advantages	Limitations
Google Scholar	scholar.google.com	Free to use, search results include 80–90% of all publications in English. Simple interface	A search engine only; full text not always available without a subscription to one of the below services
ProQuest Natural Science Collection	www.proquest.com	Full text access to three databases covering biological, environmental, agricultural and related fields, managed by an editorial team	Requires library subscription
Web of Science	webofknowledge.com	A citation indexing service with many tools in addition to full text access, including citation analysis with graphical representation	Requires library subscription

(continued)

belong as Coleoptera, coleopteran, or coleopterans. In the three databases listed below, an asterisk, *, is generally used as a wild card. In all of these databases, the search term *coleoptera** would locate references that included the terms Coleoptera, coleopteran, or coleopterans. Similarly, the search term *dais** would retrieve references to both daisy and daisies.

This review is intended to suggest only general guidelines to searching literature. There are many other databases besides the ones listed here, and the creators of all of them work very hard to improve the functioning of their products. As a consequence, the operating details of the various databases are highly dynamic. Therefore, you should periodically review the tips and instructions provided with any database that you might use. The main point of this discussion is to open a door to the rich world of ecological literature, to the world of discovery. Exploring that world can quickly extend your knowledge of the discipline of ecology far beyond the introduction provided by this textbook.

CRITIQUING THE EVIDENCE 23

1. When and why is it often necessary to narrow a search of the research literature?
2. When and why may it be important to broaden a literature search?

Statistical Tables

Table B.1 **Critical Values of Student's** *t*

DF	$\alpha = 0.10$	$\alpha = 0.05$	$\alpha = 0.02$	$\alpha = 0.01$	DF	$\alpha = 0.10$	$\alpha = 0.05$	$\alpha = 0.02$	$\alpha = 0.01$
1	6.31	12.71	31.82	63.66	22	1.72	2.07	2.51	2.82
2	2.92	4.31	6.96	9.92	24	1.71	2.06	2.49	2.80
3	2.35	3.18	4.54	5.84	26	1.71	2.06	2.48	2.78
4	2.13	2.78	3.75	4.60	28	1.70	2.05	2.47	2.76
5	2.01	2.57	3.36	4.03	30	1.70	2.04	2.46	2.75
6	1.94	2.45	3.14	3.71	35	1.69	2.03	2.44	2.72
7	1.89	2.36	3.00	3.50	40	1.68	2.02	2.42	2.70
8	1.86	2.31	2.90	3.36	45	1.68	2.01	2.41	2.69
9	1.83	2.26	2.82	3.25	50	1.68	2.01	2.40	2.68
10	1.81	2.23	2.76	3.17	60	1.67	2.00	2.39	2.66
11	1.80	2.20	2.72	3.11	70	1.67	1.99	2.38	2.65
12	1.78	2.18	2.68	3.06	80	1.66	1.99	2.37	2.64
13	1.77	2.16	2.65	3.01	90	1.66	1.99	2.37	2.63
14	1.76	2.14	2.62	3.00	100	1.66	1.98	2.36	2.63
15	1.75	2.13	2.60	2.95	120	1.66	1.98	2.36	2.62
16	1.75	2.12	2.58	2.92	150	1.66	1.98	2.35	2.61
17	1.74	2.11	2.57	2.90	200	1.65	1.97	2.35	2.61
18	1.73	2.10	2.55	2.88	300	1.65	1.97	2.34	2.59
19	1.73	2.09	2.54	2.86	500	1.65	1.96	2.33	2.59
20	1.72	2.09	2.53	2.85	∞	1.65	1.96	2.33	2.58

The above values were computed as described by Zar (1996: App 18). More extensive tables of Student's *t* are found in Rohlf and Sokal (1995:7) and Zar (1996: App 18–19).

Table B.2   Critical Values of the Mann-Whitney Test Statistic

$\alpha = 0.10$

n_1	$n_2 = 2$	3	4	5	6	7	8	9	10	11	12	13	14	15	16	17	18	19	20
2				10	12	14	15	17	19	21	22	24	25	27	29	31	32	34	36
3		9	12	14	16	19	21	23	26	28	31	33	35	38	40	42	45	47	49
4		12	15	18	21	24	27	30	33	36	39	42	45	48	50	53	56	59	62
5	10	14	18	21	25	29	32	36	39	43	47	50	54	57	61	65	68	72	75
6	12	16	21	25	29	34	38	42	46	50	55	59	63	67	71	76	80	84	88
7	14	19	24	29	34	38	43	48	53	58	63	67	72	77	82	86	91	96	101
8	15	21	27	32	38	43	49	54	60	65	70	76	81	87	92	97	103	108	113
9	17	23	30	36	42	48	54	60	66	72	78	84	90	96	102	108	114	120	126
10	19	26	33	39	46	53	60	66	73	79	86	93	99	106	112	119	125	132	138
11	21	28	36	43	50	58	65	72	79	87	94	101	108	115	122	130	137	144	151
12	22	31	39	47	55	63	70	78	86	94	102	109	117	125	132	140	148	156	163
13	24	33	42	50	59	67	76	84	93	101	109	118	126	134	143	151	159	167	176
14	25	35	45	54	63	72	81	90	99	108	117	126	135	144	153	161	170	179	188
15	27	38	48	57	67	77	87	96	106	115	125	134	144	153	163	172	182	191	200
16	29	40	50	61	71	82	92	102	112	122	132	143	153	163	173	183	193	203	213
17	31	42	53	65	76	86	97	108	119	130	140	151	161	172	183	193	204	214	225
18	32	45	56	68	80	91	103	114	125	137	148	159	170	182	193	204	215	226	237
19	34	47	59	72	84	96	108	120	132	144	156	167	179	191	203	214	226	238	250
20	36	49	62	75	88	101	113	126	138	151	163	176	188	200	213	225	237	250	262

$\alpha = 0.05$

n_1	$n_2 = 2$	3	4	5	6	7	8	9	10	11	12	13	14	15	16	17	18	19	20
2							16	18	20	22	23	25	27	29	31	32	34	36	38
3				15	17	20	22	25	27	30	32	35	37	40	42	45	47	50	52
4			16	19	22	25	28	32	35	38	41	44	47	50	53	57	60	63	66
5		15	19	23	27	30	34	38	42	46	49	53	57	61	65	68	72	76	80
6		17	22	27	31	36	40	44	49	53	58	62	67	71	75	80	84	89	93
7		20	25	30	36	41	46	51	56	61	66	71	76	81	86	91	96	101	106
8	16	22	28	34	40	46	51	57	63	69	74	80	86	91	97	102	108	113	119
9	18	25	32	38	44	51	57	64	70	76	82	89	95	101	107	114	120	126	132
10	20	27	35	42	49	56	63	70	77	84	91	97	104	111	118	125	132	138	145
11	22	30	38	46	53	61	69	76	84	91	99	106	114	121	129	136	143	151	158
12	23	32	41	49	58	66	74	82	91	99	107	115	123	131	139	147	155	163	171
13	25	35	44	53	62	71	80	89	97	106	115	124	132	141	149	158	167	175	184
14	27	37	47	57	67	76	86	95	104	114	123	132	141	151	160	169	178	188	197
15	29	40	50	61	71	81	91	101	111	121	131	141	151	161	170	180	190	200	210
16	31	42	53	65	75	86	97	107	118	129	139	149	160	170	181	191	202	212	222
17	32	45	57	68	80	91	102	114	125	136	147	158	169	180	191	202	213	224	235
18	34	47	60	72	84	96	108	120	132	143	155	167	178	190	202	213	225	236	248
19	36	50	63	76	89	101	114	126	138	151	163	175	188	200	212	224	236	248	261
20	38	52	66	80	93	106	119	132	145	158	171	184	197	210	222	235	248	261	273

continued

Table B.2   Critical Values of the Mann-Whitney Test Statistic—Concluded

$\alpha = 0.01$

n_1	$n_2 = 2$	3	4	5	6	7	8	9	10	11	12	13	14	15	16	17	18	19	20
2																			38
3								27	30	33	35	38	41	43	46	49	52	54	57
4					24	28	31	35	38	42	45	49	52	55	59	62	66	69	72
5				25	29	34	38	42	46	50	54	58	63	67	71	75	79	83	87
6			24	29	34	39	44	49	54	59	63	68	73	78	83	87	92	97	102
7			28	34	39	45	50	56	61	67	72	78	83	89	94	100	105	111	116
8			31	38	44	50	57	63	69	75	81	87	94	100	106	112	118	124	130
9		27	35	42	49	56	63	70	77	83	90	97	104	111	117	124	131	138	144
10		30	38	46	54	61	69	77	84	92	99	106	114	121	129	136	143	151	158
11		33	42	50	59	67	75	83	92	100	108	116	124	132	140	148	156	164	172
12		35	45	54	63	72	81	90	99	108	117	125	134	143	151	160	169	177	186
13		38	49	58	68	78	87	97	106	116	125	135	144	153	163	172	181	190	200
14		41	52	63	73	83	94	104	114	124	134	144	154	164	174	184	194	203	213
15		43	55	67	78	89	100	111	121	132	143	153	164	174	185	195	206	216	227
16		46	59	71	83	94	106	117	129	140	151	163	174	185	196	207	218	230	241
17		49	62	75	87	100	112	124	136	148	160	172	184	195	207	219	231	242	254
18		52	66	79	92	105	118	131	143	156	169	181	194	206	218	231	243	255	268
19		54	69	83	97	111	124	138	151	164	177	190	203	216	230	242	255	268	281
20	38	57	72	87	102	116	130	144	158	172	186	200	213	227	241	254	268	281	295

The values in the above table are derived, with permission of the publisher, from the extensive tables of Milton (1964, *Journal of the American Statistical Association* 59:925–34). See Zar (1996:App 86–97) for some sample sizes and significance levels not included above.

Table B.3   Critical Values of Chi-Square

DF	$\alpha = 0.10$	$\alpha = 0.05$	$\alpha = 0.025$	$\alpha = 0.01$	DF	$\alpha = 0.10$	$\alpha = 0.05$	$\alpha = 0.025$	$\alpha = 0.01$
1	2.706	3.841	5.024	6.635	21	29.615	32.671	35.479	38.932
2	4.605	5.991	7.378	9.210	22	30.813	33.924	36.781	40.289
3	6.251	7.815	9.348	11.345	23	32.007	35.172	38.076	41.638
4	7.779	9.488	11.143	13.277	24	33.196	36.415	39.364	42.980
5	9.236	11.070	12.833	15.086	25	34.382	37.652	40.646	44.314
6	10.645	12.592	14.449	16.812	26	35.563	38.885	41.923	45.642
7	12.017	14.067	16.013	18.475	27	36.741	40.113	43.195	46.963
8	13.362	15.507	17.535	20.090	28	37.916	41.337	44.461	48.278
9	14.684	16.919	19.023	21.666	29	33.711	39.087	42.557	45.722
10	15.987	18.307	20.483	23.209	30	40.256	43.773	46.979	50.892
11	17.275	19.675	21.920	24.725	31	41.422	44.985	48.232	52.191
12	18.549	21.026	23.337	26.217	32	42.585	46.194	49.480	53.486
13	19.812	22.362	24.736	27.688	33	43.745	47.400	50.725	54.776
14	21.064	23.685	26.119	29.141	34	44.903	48.602	51.966	56.061
15	22.307	24.996	27.488	30.578	35	46.059	49.802	53.203	57.302
16	23.542	26.296	28.845	32.000	36	47.212	50.998	54.437	58.619
17	24.769	27.587	30.191	33.409	37	48.363	52.192	55.668	59.893
18	25.989	28.869	31.526	34.805	38	49.513	53.384	56.896	61.162
19	27.204	30.144	32.852	36.191	39	50.660	54.572	58.120	62.428
20	28.412	31.410	34.170	37.566	40	51.805	55.758	59.342	63.691

The above values were computed as described by Zar (1996:App 16). More extensive tables of chi-square are found in Rohlf and Sokal (1995:24–25) and Zar (1996:App 13–16). Values of chi-square for degrees of freedom (n) greater than 40 may be approximated very accurately (Zar 1984:482), as follows:

$$\chi_{\alpha,v} = v \left(1 - 2/9v + c\sqrt{2/9v}\right)^3,$$

where the appropriate value of c is

$\alpha =$	0.10	0.05	0.025	0.01
$c =$	1.28155	1.64485	1.95996	2.32635

Abbreviations Used in This Text

°C	degrees Centigrade	°N	degrees north	mg H_2O/m^3	milligrams of water per cubic meter
0/00	grams of salt per kilogram of water	CPOM	coarse particulate organic matter	g H_2O/m^3	grams of water per cubic meter
λ	geometric rate of increase	FPOM	fine particulate organic matter	W_d	water taken by drinking
kPa	kilopascal			W_f	water taken in with food
MPa	megapascal	PDSI	Palmer Drought Severity Index	W_a	water absorbed from the air
N	newton				
Pa	pascal	MEI	metabolizable energy intake	W_e	water lost by evaporation
cm	centimeter			W_s	water lost with various secretions and excretions
cm^2	square centimeter	$\overline{X}$	sample mean		
cm^3	cubic centimeter	n	sample size		
g	gram	$\sum X$	sum of measurements or observations	W_r	water taken from soil by roots
kg	kilogram				
mg	milligram	Ψ	water potential of a solution	W_t	water lost by transpiration
mg/L	milligrams per liter			W_i	internal water
μg/L	micrograms per liter	$\Psi_{solutes}$	reduction in water potential due to dissolved substances	PEP	phosphoenolpyruvate
Tg	terragrams			PGA	phosphoglyceric acid
ha	hectare			RuBP	ribulose bisphosphate
km	kilometer	Ψ_{matric}	reduction in water potential due to matric forces within plant cells	NH_4^+	ammonium
km^2	square kilometer			NO_2^-	nitrite
km^3	cubic kilometer			NO_3^-	nitrate
m	meter	$\Psi_{pressure}$	reduction in water potential due to negative pressure created by water evaporating from leaves	Fe	elemental iron
m^2	square meter			Fe^{2+}	ferrous iron
m^3	cubic meter			CO	carbon monoxide
μm	micrometer			P_{max}	maximum rate of photosynthesis
mm	millimeter	Ψ_{soil}	water potential of soil		
nm	nanometer	Ψ_{plant}	water within plant cells	I_{sa}	irradiance required to saturate photosynthesis
μmol	micromole	δX	± the relative concentration of the heavier isotope, for example: D, ^{13}C, ^{15}N, or ^{34}S in $^0/_{00}$		
L	liter			N_{e1}	number of prey 1 encountered per unit of time
ml	milliliter				
μl	microliter			E_1	energy gained by feeding on an individual prey 1 minus the costs of handling
Ca^{2+}	calcium ion	H_s	total heat stored in the body of an organism		
Cl^-	chloride ion				
Fe	iron	H_m	heat gained from metabolism	C_s	cost of searching for the prey
K	potassium				
Mg	magnesium	H_{cd}	heat gained or lost through conduction	H_1	time required for "handling" an individual of prey 1
Mg^{2+}	magnesium ion				
N	nitrogen	H_{cv}	heat lost or gained by convection		
Na	sodium				
Na^+	sodium ion	H_r	heat gained or lost through electromagnetic radiation	S^2	sample variance
P	phosphorus			RAPD	randomly amplified polymorphic DNA
AET	actual evapotranspiration	H_e	heat lost through evaporation		
PAR	photosynthetically active radiation			h^2	heritability of a trait
		mg H_2O/L	milligrams of water per liter	V_G	genetic variance
°S	degrees south			V_P	phenotypic variance

V_E	variance in phenotype due to environmental effects on the phenotype	N_0	initial number of individuals	t	value of the statistic t, either calculated or determined from a Student's t table
V_T	total variation	e	*base of natural logarithms*		
V_{GE}	variation due to gene-by-environment interactions	χ^2	chi-square	H'	the value of the Shannon-Wiener diversity index
		O	observed frequency of a particular phenotype		
V_R	unexplained residual variation	E	expected frequency	p_i	the proportion of the ith species
		K	carrying capacity		
X	independent variable	GSI	gonadosomatic index	$\log_e$	the natural logarithm
Y	dependent variable	N_1 and N_2	population sizes of species 1 and 2	kcal	kilocalories
a	Y intercept			R_{sample}	the isotopic ratio in the sample, for example, $^{13}C{:}^{12}C$ or $^5N{:}^{14}N$
b	regression coefficient; slope of the line	K_1 and K_2	carrying capacities of species 1 and 2		
PCR	polymerase chain reaction	r_1 and r_2	intrinsic rates of increase for species 1 and 2	$R_{standard}$	the isotopic ratio in the standard, for example, $^{13}C{:}^{12}C$ or $^{15}N{:}^{14}N$
R_0	net reproductive rate				
T	generation time	$s_{\bar{x}}$	standard error	m_t	mass of leaves at time t
r	per capita rate of increase	s	sample standard deviation	m_0	initial mass of leaves
I_{max}	maximum per capita rate of increase; intrinsic rate of increase	N_h	number of hosts	e	base of the natural logarithms
		N_p	number of parasites or predators		
N	population size	α	significance level	k	daily rate of mass loss
N_t	number of individuals at time t	μ	true population mean	U	Mann-Whitney statistic
				CFC	chlorofluorocarbons

Global Biomes

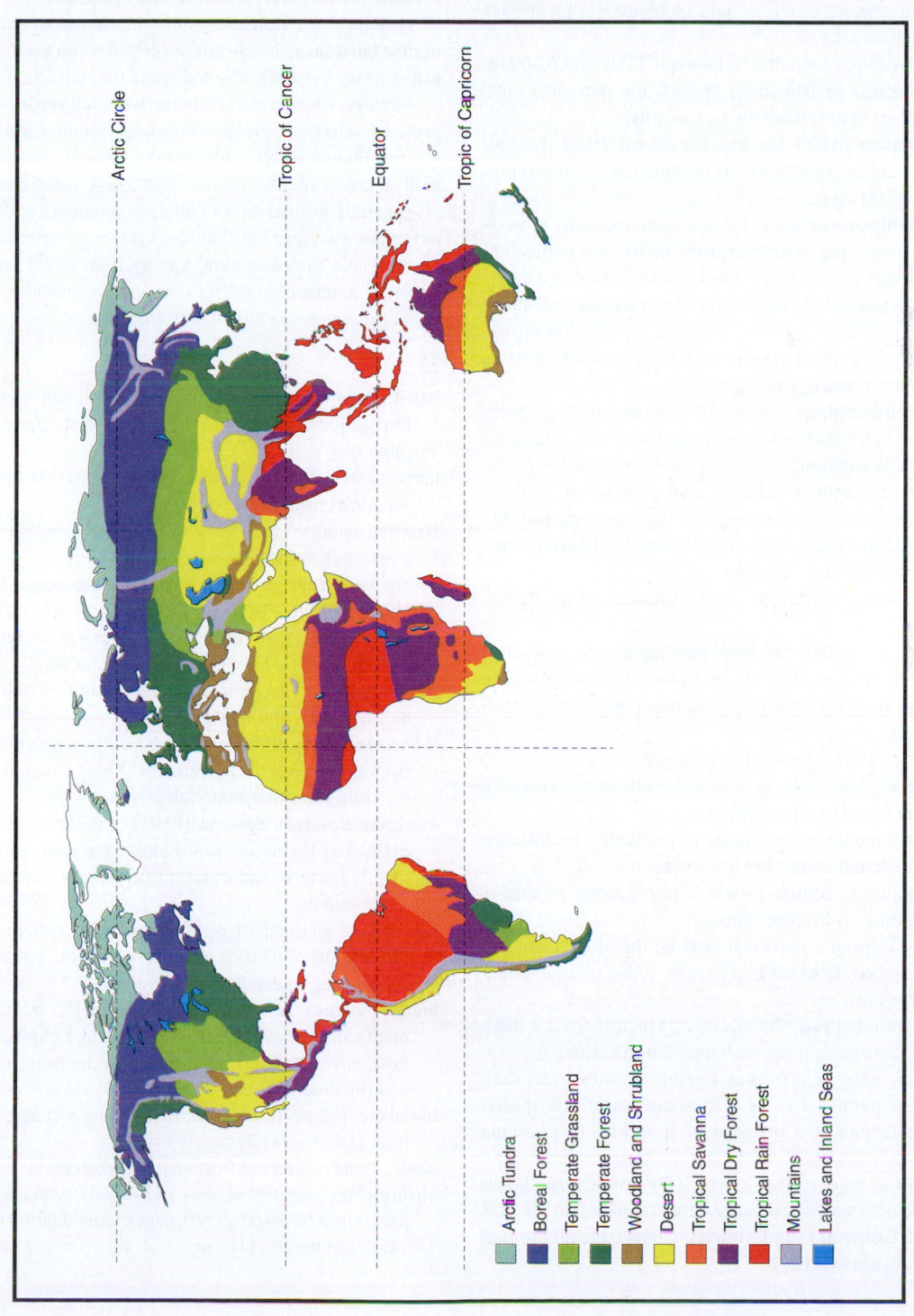

Arctic Circle

Tropic of Cancer

Equator

Tropic of Capricorn

- Arctic Tundra
- Boreal Forest
- Temperate Grassland
- Temperate Forest
- Woodland and Shrubland
- Desert
- Tropical Savanna
- Tropical Dry Forest
- Tropical Rain Forest
- Mountains
- Lakes and Inland Seas

Design elements: Cover Image: Anna A. Sher

Glossary

A

abiotic nonliving; physical or chemical rather than biological.

abundance the total number of individuals, or biomass, of a species present in a specified area.

abyssal zone a zone of the ocean depths between 4,000 and 6,000 m.

acclimation physiological adjustment to change in a particular environmental factor, such as temperature or salinity.

actual evapotranspiration (AET) the amount of water lost from an ecosystem to the atmosphere due to a combination of evaporation and transpiration by plants.

adaptation an evolutionary process that changes anatomy, physiology, or behavior, resulting in an increased ability of a population to live in a particular environment. The term is also applied to the anatomical, physiological, or behavioral characteristics produced by this process.

adhesion-adapted a term applied to seeds with hooks, spines, or barbs that disperse by attaching to passing animals.

aeroecology the interdisciplinary study of ecology of the earth's atmospheric boundary layer and the myriad airborne organisms that depend on this environment for their existence. Relevant disciplines include atmospheric science, animal behavior, ecology, evolution, earth science, computational biology, and engineering.

age distribution the distribution of individuals among age groups in a population; often called age structure.

agriculture the growing of crops and livestock for human consumption.

A horizon a biologically active soil layer consisting of a mixture of mineral materials, such as clay, silt, and sand, as well as organic material, derived from the overlying O horizon; generally characterized by leaching.

allele one of the alternative forms of the same gene.

allele frequencies the proportions in which the different forms of a gene (alleles) are found in a population.

allelopathy the phenomenon of an organism producing a substance that is harmful or lethal to another organism.

allopatric describes the condition in which populations or species have nonoverlapping geographic ranges.

allozyme alternative form of a particular enzyme that differs structurally but not functionally from other allozymes coded for by different alleles at the same locus.

amensalism interaction between organisms in which there is a negative impact on one organism but no impact on the other.

apparent competition negative effects as a result of two species sharing a predator or herbivore or as a consequence of one species facilitating populations of a predator or herbivore of a second species.

arbuscular mycorrhizae mycorrhizae in which the mycorrhizal fungus produces arbuscules (sites of exchange between plant and fungus), hyphae (fungal filaments), and vesicles (fungal energy storage organs within root cortex cells).

arbuscule a bush-shaped organ on an endomycorrhizal fungus that acts as a site of material exchange between the fungus and its host plant.

archaea prokaryotes distinguished from bacteria on the basis of structural, physiological, and other biological features.

argillic horizon a subsoil characterized by an accumulation of clays.

aril a fleshy covering of some seeds that attracts birds and other vertebrates, which act as dispersers of such seeds.

artificial selection selective breeding techniques used to develop or maintain desirable traits in domesticated plants and animals.

atoll a circle of low islands and coral reefs encircling a lagoon, generally formed on a submerged mountain called a seamount.

autotroph an organism that can synthesize organic molecules using inorganic molecules and energy from either sunlight (photosynthetic autotrophs) or from inorganic molecules, such as hydrogen sulfide (chemosynthetic autotrophs).

B

balanced growth cell growth in which all cell constituents, such as nitrogen, carbon, and DNA, increase at approximately the same rate.

barrier reef a long, ridge-like reef that parallels the mainland and is separated from it by a deep lagoon.

Batesian mimicry evolution of a non-noxious species to resemble a poisonous or inedible species.

bathypelagic zone a zone within the deep ocean that extends from about 1,000 to 4,000 m.

behavioral ecology study of the relationships between organisms and environment that are mediated by behavior.

benthic an adjective referring to the bottom of bodies of waters such as seas, lakes, and streams.

B horizon a subsoil in which materials leached from above, generally from the A horizon, accumulate. May be rich in clay, organic matter, iron, and other materials.

biochemical oxygen demand (BOD) a measure of organic pollution defined as the amount of dissolved oxygen required by microbes, mainly bacteria and fungi, to decompose the organic matter in a water sample.

biodiversity an inclusive term that encompasses all aspects of biological diversity, including genetic, behavioral, physiological, anatomical, species, and ecosystem diversity.

biome biomes are distinguished primarily by their predominant plants and are associated with particular climates. They consist of distinctive plant formations such as the tropical rain forest biome and the desert biome.

biosphere the portions of earth that support life; also refers to the total global ecosystem.

biotic living or derived from living organisms.

birthrate the number of new individuals produced in a population generally expressed as births per individual or per thousand individuals in the population.

boreal forests northern forests that occupy the area south of arctic tundra. Though dominated by coniferous trees, they also contain aspen and birch. Also called taiga.

bottom-up control control of a community or ecosystem by physical or chemical factors such as temperature or nutrient availability.

bundle sheath structure that surrounds the leaf veins of C_4 plants, made up of cells, where four-carbon acids produced during carbon fixation are broken down to three-carbon acids and CO_2.

C

caliche a calcium carbonate–rich hardpan soil horizon; the extent of caliche formation can be used to determine the age of desert soils.

CAM (crassulacean acid metabolism) photosynthesis a photosynthetic pathway largely limited to succulent plants in arid and semiarid environments, in which carbon fixation takes place at night, when lower temperatures reduce the rate of water loss during CO_2 uptake. The resulting four-carbon acids are stored until daylight, when they are broken down into pyruvate and CO_2.

carnivore an organism that consumes flesh; approximately synonymous with *predator*.

carrying capacity (K) the maximum population of a species that a particular ecosystem can sustain.

caste a group of individuals that are physically distinctive and engage in specialized behavior within a social unit, such as a colony.

C_4 photosynthesis in C_4 photosynthesis, CO_2 is fixed in mesophyll cells by combining it with phosphoenol pyruvate, or PEP, to produce a four-carbon acid. Plants using C_4 photosynthesis are generally more drought tolerant than plants employing C_3 photosynthesis.

character displacement changes in the physical characteristics of a species' population as a consequence of natural selection for reduced interspecific competition.

chemosynthetic autotroph refers to autotrophs that use inorganic molecules as a source of carbon and energy.

chi-square (χ^2) a statistic used to measure how much a sample distribution differs from a theoretical distribution.

C horizon a soil layer composed of largely unaltered parent material, little affected by biological activity.

chronosequence a series of communities or ecosystems representing a range of ages or times since disturbance.

climate the atmospheric conditions prevailing in an area over a long period.

climate change a shift in prevailing atmospheric conditions at a regional or global scale, typically in relation to increased levels of atmospheric carbon from the burning of fossil fuels and other anthropogenic activity.

climate diagram a standardized form of representing average patterns of variation in temperature and precipitation that identifies several ecologically important climatic factors such as relatively moist periods and periods of drought.

climax community the late successional community that can persist until disrupted by disturbance.

clumped distribution a pattern of distribution in a population in which individuals have a much higher probability of being found in some areas than in others; in other words, individuals are aggregated rather than dispersed.

clutch size the number of eggs laid by a bird, reptile, amphibian, or fish. The term is also sometimes applied to the number of seeds produced by a plant.

cohesion a property of water whereby molecules stick together, making the transport of water through plants possible.

cohort a group of individuals of the same age.

cohort life table a life table based on individuals born (or beginning life in some other way) at the same time.

colonization cycle the situation in which stream populations are maintained through a dynamic interplay between downstream drift and upstream dispersal.

combined response the combined effect of functional and numerical responses by consumers on prey populations; determined by multiplying the number of prey eaten per predator times the number of predators per unit area, giving the number of prey eaten per unit area. Combined response is generally expressed as a percentage of the total number of prey.

commensalism an interaction between two species in which one species is benefited and the other is neither benefited nor harmed.

common garden experiment an experiment in which individuals from two or more populations are transplanted and grown in the same, or "common," environment, generally used to study the contributions of environment versus genetic factors to differences between populations.

community an association of interacting species living in a particular area; also often defined as all of the organisms living in a particular area.

community structure attributes of a community such as the number of species or the distribution of individuals among species within the community.

comparative method a method for reconstructing evolutionary processes and mechanisms that involves comparisons of different species or populations in a way that attempts to isolate a particular variable or characteristic of interest, while randomizing the influence of confounding, or confusing, variables on the variable of interest across the species or populations in the study.

competition coefficient a coefficient expressing the magnitude of the negative effect of individuals of one species on individuals of a second species.

competitive exclusion principle the principle that two species with identical niches cannot coexist indefinitely.

competitive plants according to Grime (1977), competitive plants occupy environments where disturbance intensity is low and the intensity of stress is also low.

conduction the movement of heat between objects in direct physical contact.

confidence interval a range of values within which the true population mean occurs with a particular probability called the level of confidence.

consumer an organism that obtains its energy by feeding on other organisms; a heterotroph.

convection the process of heat flow or transfer to a moving fluid, such as wind or flowing water.

Coriolis effect a phenomenon caused by the rotation of the earth, which produces a deflection of winds and water currents to the right of their direction of travel in the Northern Hemisphere and to the left of their direction of travel in the Southern Hemisphere.

C_3 photosynthesis the photosynthetic pathway used by most plants and all algae, in which the product of the initial reaction is phosphoglyceric acid, or PGA, a three-carbon acid.

cyanobacteria a phylum of bacteria that photosynthesizes, formerly called "blue green algae."

D

decomposition the breakdown of organic matter accompanied by the release of carbon dioxide and other inorganic compounds; a key process in nutrient cycling.

density the number of individuals in a population per unit area.

density-dependent factor biotic factors in the environment, such as disease and competition, are often called density-dependent factors because their effects on populations may be related to, or depend on, local population density.

density-independent factor abiotic factors in the environment, such as floods and extreme temperature, are often called density-independent factors because their effects on populations may be independent of population density.

dependent variable the variable traditionally plotted on the vertical, or "Y," axis of a scatter plot.

desert an arid biome occupying approximately 20% of the land surface of the earth in which water loss due to evaporation and transpiration by plants exceeds precipitation during most of the year.

detritivore an organism that feeds on nonliving organic matter, usually on the remains of plants.

diffusion transport of material due to the random movement of particles; net movement is from areas of high concentration to areas of low concentration.

direct interaction negative or positive interaction between two species, including competition, predation, herbivory, and mutualism, that occurs without the involvement of an intermediary species.

directional selection a form of natural selection that favors an extreme phenotype over other phenotypes.

disruptive selection a form of natural selection that favors two or more extreme phenotypes over the average phenotype in a population.

distribution the geographic range of an organism or the spatial arrangement of individuals in a local population.

disturbance Grime (1977) defined disturbance from the perspective of plants as any process that limits plants by destroying plant biomass. Sousa (1984) also defined disturbance from an organismic perspective as any discrete, punctuated killing, displacement, or damaging of one or more individuals (or colonies) that directly or indirectly creates an opportunity for new individuals (or colonies) to become established. White and Pickett (1985) defined disturbance more broadly as any relatively discrete event that disrupts ecosystem, community, or population structure and changes resources, substrate availability, or the physical environment.

DNA deoxyribonucleic acid; the long molecule that contains genetic coding in living organisms.

DNA barcoding a genetic analysis tool that involves sequencing the same small region of DNA across many individuals.

dominant species or foundation species organisms, such as abundant, forest tree species or reef coral species, that substantially influence community structure as a consequence of their abundance.

drift the active or passive downstream movement of stream organisms.

drought an extended period of dry weather during which precipitation is reduced sufficiently to damage crops, impair the functioning of natural ecosystems, or cause water shortages for human populations.

dry deposition fallout of particulate material from the air by means other than with precipitation in rain or snow.

E

ecological efficiency the percentage of energy in the biomass at a lower trophic level that is transferred to the biomass at the next higher trophic level; ecological efficiency varies from approximately 5% to 20%.

ecological restoration a process for restoring damaged ecosystems to an acceptable level of diversity, physical structure, and functioning, using methods that accelerate an ecosystem's recovery from damage or even from total destruction. See *restoration ecology.*

ecological stoichiometry the study of the balance of multiple chemical elements in ecological interactions, for example, in trophic interactions.

ecology the study of the relationships between organisms and the environment.

ecosystem a biological community plus all of the abiotic factors influencing that community.

ecosystem engineer an organism that causes changes in the physical environment sufficient to influence the structure of landscapes, ecosystems, or communities.

ecotone a spatial transition from one type of ecosystem to another; for instance, the transition from a woodland to a grassland.

ecotype locally adapted and genetically distinctive population within a species.

ectomycorrhizae an association between a fungus and plant roots in which the fungus forms a mantle around roots and a netlike structure around root cells.

ectotherm an organism that relies mainly on external sources of energy for regulating body temperature.

edge effect the distinctive ecological conditions and higher species richness that often occur at an ecotone; ecotones generally support a mix of species from both ecosystems on either side of the transition between ecosystems as well as some species unique to the ecotone.

elaiosome a structure on the surface of some seeds generally containing oils attractive to ants, which act as dispersers of such seeds.

El Niño a large-scale coupled oceanic-atmospheric system that has major effects on climate worldwide. During an El Niño, the sea surface temperature in the eastern Pacific Ocean is higher than average and barometric pressure is lower.

endemic existing in only one place on earth, to describe a species.

endotherm an organism that relies mainly on internal sources of energy for regulating body temperature.

environmental enrichment increasing the complexity of the environment of captive animals to foster behaviors characteristic of the species in the wild; may be critical to the survival of captive-bred animals reintroduced to the natural environment.

epilimnion the warm, well-lighted surface layer of lakes.

epipelagic zone the warm, well-lighted surface layer of the oceans.

epiphyte a plant, such as an orchid, that grows on the surface of another plant but is not parasitic.

equilibrium a state of balance in a system in which opposing factors cancel each other.

equilibrium model of island biogeography a model proposing that the number of species on islands is the result of a dynamic balance between rates of immigration and extinction.

estivation a dormant state that some animals enter during the summer; involves a reduction of metabolic rate.

estuary the lowermost part of a river, which is under the influence of the tides and is a mixture of seawater and freshwater.

eusociality highly specialized sociality generally including (1) individuals of more than one generation living together, (2) cooperative care of young, and (3) division of individuals into sterile, or nonreproductive, and reproductive castes.

eutrophic a term applied to lakes, and sometimes to other ecosystems, with high nutrient content and high biological production.

eutrophication nutrient enrichment of an ecosystem, generally resulting in increased primary production and reduced biodiversity. In lakes, eutrophication leads to seasonal algal blooms, reduced water clarity, and, often, periodic fish mortality as a consequence of oxygen depletion.

evaporation the process by which a liquid changes from liquid phase to a gas, as in the change from liquid water to water vapor.

evolution a process that changes populations of organisms over time. Since evolution ultimately involves changes in the frequency of heritable traits in a population, we can define evolution more precisely as a change in gene frequencies in a population.

ex situ literally, "out of place," referring to management or conservation taking place outside of a species natural habitat, such as in a zoo or garden.

experiments a scientific approach to test a hypothesis that typically involves manipulating variables of interest.

exploitation an interaction between species that enhances the fitness of the exploiting individual—the predator, the pathogen, etc.—while reducing the fitness of the exploited individual—the prey, host, etc.

exponential population growth population growth that produces a J-shaped pattern of population increase. In exponential population growth, the change in numbers with time is the product of the per capita rate of increase, r, and population size, N.

extrafloral nectary nectar-secreting glands found on structures other than flowers, such as leaves.

F

facilitation model according to the facilitation model, pioneer species modify the environment in such a way that it becomes less suitable for themselves and more suitable for species characteristic of later successional stages.

facultative mutualism a mutualistic relationship between two species that is not required for the survival of the two species.

fecundity the number of eggs or seeds produced by an organism.

fecundity schedule a table of birthrates for females of different ages in a population.

female sex that produces larger, more energetically costly gametes (eggs or ova).

fitness the number of offspring contributed by an individual relative to the number of offspring produced by other members of the population. Ultimately defined as the relative genetic contribution of individuals to future generations.

flood pulse concept a theory of river ecology identifying periodic flooding as an essential organizer of river ecosystem structure and functioning.

food web a summary of the feeding relationships within an ecological community.

forbs herbaceous plants other than graminoids.

foundation species See *dominant species*.

freshwater water containing less than 0.5 g of dissolved salts per liter, < 0.5 0/00.

freshwater wetlands swamps and marshes, for example, that occupy low-lying areas within landscapes and are generally inundated with water for some part of each year.

fringing reef a coral reef that forms near the shore of an island or continent.

functional response an increase in animal feeding rate, which eventually levels off, that occurs in response to an increase in food availability.

fundamental niche the physical conditions under which a species might live, in the absence of interactions with other species.

functional traits those morphological, physiological, and phenological characteristics of an organism that affect fitness through their influence on growth, reproduction, or survival, often also considered as those characteristics that define an organism's ecological role.

G

genes units of heredity; a segment of DNA composed of sequences of nucleotides.

genetically modified organisms (GMOs) organisms whose genetic makeup has been modified through the process of genetic engineering, generally through the deletion of genes or insertion of novel genes, often from different organisms.

genetic drift change in gene frequencies in a population due to chance or random events.

genetic engineering alteration of the genetic makeup of an organism through the introduction or deletion of genes; for example, the introduction of bacterial genes into crop plants to give them more resistance to insect pests.

geographic ecology the study of ecological structure and process at large geographic scales; sometimes defined as the study of ecological patterns that can be put on a map.

geographic information system a computer-based system that stores, analyzes, and displays geographic information, generally in the form of maps.

geometric population growth population growth in which generations do not overlap and in which successive generations differ in size by a constant ratio.

geometric rate of increase (λ) the ratio of the population size at two points in time: $\lambda = N_{t+1}/N_t$, where N_{t+1} is the size of the population at some future time and N_t is the size of the population at some earlier time.

germination the sprouting of seeds.

global positioning system (GPS) a device that determines locations on the earth's surface, including latitude, longitude, and altitude, using radio signals from satellites as references.

gonadosomatic index (GSI) an index of reproductive effort calculated as ovary weight divided by body weight and adjusted for the number of batches of offspring produced per year.

graminoids grasses and grasslike plants, such as sedges and rushes.

granivore an animal that feeds chiefly on seeds.

greenhouse effect warming of the earth's atmosphere and surface as a result of heat trapped near the earth's surface by gases in the atmosphere, especially water vapor, carbon dioxide, methane, ozone, nitrous oxide, and chlorofluorocarbons.

gross primary production the total biomass produced by the primary producers in an ecosystem per unit area or volume—for example, per m^2 or per m^3 over some interval of time.

growth form See *life-form*.

guild a group of organisms that make their living in a similar way; for example, the seed-eating animals in a desert, the fruit-eating birds in a tropical rain forest, or the filter-feeding invertebrates in a stream.

gyre a large-scale, circular oceanic current that moves to the right in the Northern Hemisphere and to the left in the Southern Hemisphere.

H

hadal zone the deepest parts of the oceans, below about 6,000 m.

Hamilton's rule the conditions under which helping kin should be favored by natural selection: $R_gB - C > 0$, where R_g is the genetic relatedness of the helper and the recipient of the help, B is the reproductive benefit gained by the recipient, and C is the reproductive cost to the helper of giving aid.

haplodiploidy sex inheritance in which males are haploid and females are diploid.

Hardy-Weinberg principle a principle that in a population mating at random in the absence of evolutionary forces, allele frequencies will remain constant.

heat the kinetic energy resulting from molecular motion in a mass of a substance, also referred to as heat energy. Heat can be transferred by conduction, convection, or radiation.

heat wave a period of abnormally hot and unusually humid weather, typically lasting for two or more days.

herbivore a heterotrophic organism that eats plants.

heritability the proportion of total phenotypic variation in a trait attributable to genetic variation; determines the potential for evolutionary change in a trait.

hermaphrodite an individual capable of producing both sperm or pollen and eggs or ova.

heterotroph an organism that uses organic molecules both as a source of both carbon and energy.

hibernation a dormant state, involving reduced metabolic rate, that occurs in some animals during the winter.

homeotherm an organism that uses metabolic energy to maintain a relatively constant body temperature; such organisms are often called warm-blooded.

hydrologic cycle the sun-driven cycle of water through the biosphere through evaporation, transpiration, condensation, precipitation, and runoff.

hydrophilic moisture attracting.

hydrophobic moisture repelling.

hyperosmotic a term describing organisms with body fluids with a lower concentration of water and higher solute concentration than the external environment.

hyphae long, thin filaments that form the basic structural unit of fungi.

hypolimnion the deepest layer of a lake below the epilimnion and thermocline.

hypoosmotic a term describing organisms with body fluids with a higher concentration of water and lower solute concentration than the external environment.

hyporheic zone a zone below the benthic zone of a stream; a zone of transition between surface water flow and groundwater.

I

in situ literally "in place"; management or conservation taking place in the natural habitat of a species.

inbreeding mating between close relatives. Inbreeding tends to increase levels of homozygosity in populations and often results in offspring with lower survival and reproductive rates.

inclusive fitness overall fitness, which is determined by the survival and reproduction of an individual, plus the survival and reproduction of genetic relatives of the individual.

independent variable the variable traditionally plotted on the horizontal, or "X," axis of a scatter plot.

indirect commensalism an interaction in which one species benefits another species indirectly, through an intermediary species, without itself being helped or harmed.

indirect interaction negative or positive interaction between two species, including trophic cascades, apparent competition, and indirect mutualism or commensalism, that is mediated through a third species.

inhibition model a model of succession that proposes that early occupants of an area modify the environment in a way that makes the area less suitable for both early and late successional species.

insectivore a heterotrophic organism that eats insects.

interdisciplinary research investigations that involve researchers from multiple disciplines working closely to produce an understanding that integrates across disciplines; may include several scientific disciplines or extend beyond the boundaries of the natural sciences into the social sciences and humanities.

interference competition a form of competition involving direct antagonistic interactions between individuals, including territoriality, chemical poisoning, and overgrowing, which result in direct harm or reducing the competitor's access to resources, such as food or light.

intermediate disturbance hypothesis a proposal that high diversity is maintained by changing environmental conditions and that highest levels of diversity will occur at intermediate levels of disturbance.

interquartile range a range of measurements that includes the middle 50% of the measurements or observations in a sample, bounded by the lowest value of the highest 25% of measurements and the highest value of the lowest 25% of measurements.

intersexual selection sexual selection occurring when members of one sex choose mates from among the members of the opposite sex on the basis of some anatomical or behavioral trait, generally leading to the elaboration of that trait.

interspecific competition competition between individuals of different species.

intertidal zone the zone between the highest and lowest tides along marine shores. See *littoral zone.*

intrasexual selection sexual selection in which individuals of one sex compete among themselves for mates.

intraspecific competition competition between individuals of the same species.

intrinsic rate of increase the maximum per capita rate of population increase; may be approached by a population with a stable age distribution growing under ideal environmental conditions.

irradiance the level of light intensity, often measured as photon flux density.

I_{sat} the irradiance required to saturate the photosynthetic capacity of a photosynthetic organism.

Island biogeography a field of study within biogeography that seeks to explain patterns of diversity in isolated natural communities.

isoclines of zero population growth lines, in the graphical representation of the Lotka-Volterra competition model, where population growth of the species in competition is zero.

isosmotic a term describing organisms with body fluids containing the same concentration of water and solutes as the external environment.

iteroparity reproduction that involves production of an organism's offspring in two or more events, generally spaced out over the lifetime of the organism.

K

keystone species species that, despite low biomass, exert strong effects on the structure of the communities they inhabit.

kin selection selection in which individuals increase their inclusive fitness by helping increase the survival and reproduction of relatives (kin) that are not offspring.

K selection a term referring to the carrying capacity of the logistic growth equation; a form of natural selection that favors more efficient utilization of resources such as food and nutrients. K selection is predicted to be strongest in those situations where a population lives as densities near carrying capacity much of the time.

L

landform any distinctive feature of the earth's surface.

landscape a heterogeneous area consisting of distinctive patches, or landscape elements, organized into a mosaic-like pattern.

landscape ecology the study of the relationship between spatial pattern and ecological processes over a range of scales.

landscape elements the ecosystems in a landscape, which generally form a mosaic of visually distinctive patches.

landscape process the exchange of materials, energy, or organisms among the patches, or elements, that make up a landscape.

landscape structure the size, shape, composition, number, and position of different patches, or landscape elements, within a landscape.

La Niña the opposite of an El Niño. During a La Niña, the sea surface temperature in the eastern Pacific Ocean is lower than average and barometric pressure is higher.

large-scale phenomena phenomena of a geographic scale rather than a local scale.

lentic concerning still water ecosystems, generally refers to freshwater lakes or ponds or the organisms living in these environments.

level of confidence 1 minus the significance level, α, which is generally 0.05, for example, level of confidence $= 1 - 0.05 = 0.95$.

life-form the life-form of a plant is a combination of its structure and its growth dynamics. Plant life-forms include trees, vines, annual plants, sclerophyllous vegetation, grasses, and forbs.

life history the adaptations of an organism that influence aspects of its biology, such as the number of offspring it produces, its survival, and its size and age at reproductive maturity.

life table a table of age-specific survival and death, or mortality, rates in a population.

light compensation point when oxygen is produced at the same rate as it is used by photosynthesizing organisms.

lifetime reproductive success the total number of offspring produced by an individual over the course of a lifetime.

limnetic zone the open lake beyond the littoral zone.

limiting resource an environmental factor used by an organism for which an increase in abundance will correspond to an increase in growth of the organism or its population size.

lithosol soils very low in organic matter and composed of rock fragments.

littoral zone the shallowest waters along a lake or ocean shore; where rooted aquatic plants may grow in lakes.

loci plural of **locus,** the location along the length of a particular chromosome where a gene is located.

logistic equation $dN/dt = r_{max}N(K - N/K)$.

logistic population growth a pattern of growth that produces a sigmoidal, or S-shaped, population growth curve; population size levels off at carrying capacity (K).

lognormal distribution a distribution of relative abundances of species produced by plotting the abundance of species as a frequency distribution in which each abundance interval is twice the preceding one, 1, 2, 4, 8, etc.; the result follows an approximately normal distribution.

lotic concerning running water ecosystems, generally refers to freshwater creeks and rivers or the organisms living in these environments.

M

macroclimate the prevailing weather conditions of a region over a long period of time.

male sex that produces smaller, less costly gametes (sperm or pollen).

mangrove forest a forest of subtropical and tropical marine shores dominated by salt-tolerant woody plants, such as *Rhizophora* and *Avicennia.*

matric force a force resulting from water's tendency to adhere to the walls of containers such as cell walls or the soil particles lining a soil pore.

matrix the landscape element within a landscape mosaic that is the most continuous spatially, for example, the forest that surrounds small, isolated patches of meadow.

Mediterranean woodland and shrubland a biome associated with mild, moist winter conditions and usually with dry summers between about 30° and 40° latitude. The vegetation of this biome is usually characterized by small, tough (sclerophyllous) leaves and adaptations to periodic fire. It is found around the Mediterranean Sea and in western North America, Chile, southern Australia, and southern Africa. It is known by many local names such as chaparral, garigue, maquis, and fynbos.

meristematic tissue tissue made up of the actively dividing cells responsible for plant growth.

mesopelagic zone a middle-depth zone of the oceans, extending from about 200 to 1,000 m.

mesosphere a layer in the earth's atmosphere, extending from 64 to 80 km above the earth's surface; temperatures drop steeply with altitude in this atmospheric layer.

meta-analysis a statistical tool to test the effect size of a phenomenon using many different studies, typically used as a way of summarizing a large body of published literature.

metabolic heat energy released within an organism during the process of cellular respiration.

metabolic water water released during oxidation of organic molecules.

metalimnion a depth zone between the epilimnion and hypolimnion characterized by rapid decreases in temperature and increases in water density with depth. Often used synonymously with the term *thermocline.*

metapopulation a group of spatially separated subpopulations connected by active exchange of individuals among subpopulations.

microclimates smaller areas that differ in temperature and/or moisture than the prevailing climate; caused by factors such as a distinctive substrate, location, or aspect.

microsatellite DNA sequence of tandemly repetitive DNA, in which a few base pairs, for example, three base pairs, are repeated up to 100 times.

mineralization the breakdown of organic matter from organic to inorganic form during decomposition.

modeling a tool used to represent a concept or system such as to make it easier to understand, and/or to make predictions or test hypotheses about it. Can be quantitative or conceptual.

Müllerian mimicry comimicry among several species of noxious organisms.

mutualism interactions between individuals of different species that benefit both partners.

mycorrhizae a mutualistic association between fungi and the roots of plants.

N

natal territory territory where an individual was born.

natural experiment an empirical study that takes advantage of natural phenomena that are outside of the control of an investigator to create levels for comparison, such as before versus after a natural disturbance.

natural history the study of how organisms in a particular area are influenced by factors such as climate, soils, predators, competitors, and evolutionary history, involving field observations rather than carefully controlled experimentation or statistical analyses of patterns.

natural selection differential reproduction and survival of individuals in a population due to environmental influences on the population; proposed by Charles Darwin as the primary mechanism driving evolution.

negative phototaxis movement of an organism away from light.

neritic zone a coastal zone of the oceans, extending to the margin of a continental shelf, where the ocean is about 200 m deep.

net photosynthesis total CO_2 uptake (or energy fixed) during photosynthesis minus the CO_2 produced (or energy consumed) by the plant's, or alga's, own respiration.

net primary production net primary production is gross primary production minus respiration by primary producers; it is the amount of energy in the form of biomass produced per unit area or volume—for example, per m^2 or per m^3—over some interval of time that is available to the consumers in an ecosystem.

net reproductive rate (R_0) the average lifetime number of female offspring produced by an individual female in a population.

neutralism when there is no net benefit or loss to either organism involved in an interaction.

niche the environmental factors that influence the growth, survival, and reproduction of a species.

niche partitioning an evolutionary process whereby organisms decrease direct competition by differentiating in their use of resources by altering which, how, or when resources are used.

normal distribution a bell-shaped distribution, proportioned so that predictable proportions of observations or measurements fall within one, two, or three standard deviations of the mean.

numerical response change in the density of a predator population in response to increased prey density.

nutrient chemical substance required for the development, maintenance, and reproduction of organisms.

nutrient cycling the use, transformation, movement, and reuse of nutrients in ecosystems.

nutrient flux the movement of nutrients between nutrient pools in an ecosystem.

nutrient pool the amount of a particular nutrient stored in a portion, or compartment, of an ecosystem.

nutrient retentiveness the tendency of an ecosystem to retain nutrients.

nutrient sink a part of the biosphere where a particular nutrient is absorbed faster than it is released.

nutrient source a portion of the biosphere where a particular nutrient is released faster than it is absorbed.

nutrient spiraling a representation of nutrient dynamics in streams, which, because of downstream displacement of organisms and materials, are better represented by a spiral than a cycle.

O

obligate mutualism a mutualistic relationship in which species are so dependent upon the relationship that they cannot live in its absence.

oceanic zone the open ocean beyond the continental shelf with water depths generally greater than 200 m.

O (organic) horizon the most superficial soil layer containing substantial amounts of organic matter, including whole leaves, twigs, other plant parts, and highly fragmented organic matter.

oligotrophic a term generally referring to lakes of low nutrient content, abundant oxygen, and low primary production.

omnivore a heterotrophic organism that eats a wide range of food items, usually including both animal and plant matter.

optimal foraging theory theory that attempts to model how organisms feed as an optimizing process, a process that maximizes or minimizes some quantity, such as energy intake or predation risk.

optimization a process that maximizes or minimizes some quantity.

organic compound a chemical compound, or molecule, that contains carbon bound to other elements such as hydrogen and nitrogen; examples include carbohydrates, proteins, fats, and alcohols.

osmoregulation regulation of internal salt concentration and water using a variety of anatomical structures and physiological processes; an energy-consuming process.

osmosis diffusion of water down its concentration gradient.

P

Palmer Drought Severity Index (PDSI) an index of drought that uses temperature and precipitation to represent moisture conditions in a region relative to long-term average temperature and precipitation within the region.

parasite an organism that lives in or on another organism, called the host, deriving benefits from it; parasites typically reduce the fitness of the host, but do not generally kill it.

parasitism the symbiotic relationship between a parasite and its host.

parasitoid an insect whose larva consumes its host and kills it in the process; parasitoids are functionally equivalent to predators.

patch a relatively homogeneous area in a landscape that differs from its surroundings, for example, an area of forest surrounded by agricultural fields.

pathogen any organism that induces disease, a debilitating condition, in their hosts; common pathogens include viruses, bacteria, and protozoans.

pelagic a term referring to marine life zones or organisms above the bottom; for instance, tuna are *pelagic* fish that live in the *epipelagic* zone of the oceans.

per capita rate of increase usually symbolized as *r*, equals per capita birthrate minus per capita death rate: $r = b - d$.

phenology the study of the relationship between climate and the timing of ecological events such as the date of arrival of migratory birds on their wintering grounds, the timing of spring plankton blooms, and the onset and ending of leaf fall in a deciduous forest.

phenotype the observable characteristics of an organism that result from interactions between the genetic makeup of the individual and its environment.

phenotypic plasticity variation among individuals in form and function as a result of environmental influences.

pheromone a chemical substance secreted by some animals for communication with other members of their species.

philopatry a term, which means literally "love of place," used to describe the tendency of some organisms to remain in the same area throughout their lives.

photon flux density the number of photons of light striking a square meter surface each second.

photorespiration a metabolic process occurring mainly in C_3 plants under hot, dry conditions that reduces the efficiency of photosynthesis; initiated by the enzyme rubisco as it combines oxygen with RuBP; occurs in the light, consumes oxygen and energy, and releases CO_2.

photosynthesis process in which the photosynthetic pigments of plants, algae, or bacteria absorb light and transfer their energy to electrons; the energy carried by these electrons is used to synthesize ATP and NADPH, which in turn serve as donors of electrons and energy for the synthesis of sugars.

photosynthetically active radiation (PAR) wavelengths of light between 400 and 700 nm that photosynthetic organisms use as a source of energy.

photosynthetic autotroph an organism that uses carbon dioxide as a carbon source and light as an energy source to synthesize organic compounds.

phreatic zone the region below the hyporheic zone of a stream; contains groundwater.

phytoplankton microscopic photosynthetic organisms that drift with the currents in the open sea or in lakes.

pioneer community the first organisms to colonize in a successional sequence following a disturbance or the formation of a new geologic surface.

piscivore a predator that eats fish.

pistil female organ of a flower.

plant functional group a group of plants with similar physiological and anatomical characteristics that influence their seasonality, resource requirements, and life histories—for example, C3 grasses, which grow best at cool temperatures; C4 grasses, which grow best in warm conditions; and legumes, which fix nitrogen.

P_{max} maximum rate of photosynthesis for a particular species of plant growing under ideal physical conditions.

poikilotherm organisms whose body temperature varies directly with environmental temperatures; commonly called cold-blooded.

polygenic traits traits determined by the effects of many genes.

polymorphic locus a locus, or gene, that occurs as more than one allele, each of which synthesizes a different allozyme.

population a group of individuals of a single species inhabiting a specific area.

population dynamics an area of population ecology concerned with the factors influencing the expansion, decline, or maintenance of populations, including rates of births, deaths, immigration, and emigration.

population genetics study of the genetics of populations.

positive phototaxis movement of an organism toward light.

predator a heterotrophic organism that kills and eats other organisms for food; usually an animal that hunts and kills other animals for food.

predator satiation a defensive tactic in which prey reduce their individual probability of being eaten by occurring at very high densities; predators can capture and eat only so many preys and so become satiated when prey are at very high densities.

prey-dependent functional response a functional response in which rate of feeding by a predator is a function of prey population size only.

primary producers organisms that convert inorganic forms of energy, such as that from the sun, into forms that are usable by the next level of a food chain. For example, plants are primary producers in most terrestrial ecosystems.

primary production the production of organic matter, or biomass, by photosynthetic and chemosynthetic autotrophs in an ecosystem per unit area or volume—for example, per m^2 or per m^3—during some period of time, for example, per hour or per year; in most ecosystems, the most significant autotrophs are photosynthetic.

primary succession succession on newly exposed geological substrates, not significantly modified by organisms, for instance, on newly formed volcanic lava or on substrate exposed during the retreat of a glacier.

principle of allocation the principle that if an organism allocates energy to one function, such as growth or reproduction, it reduces the amount of energy available to other functions, such as defense.

priority effect a situation in which a species establishing itself in greater numbers in a place before the establishment of potential competitors wins in competition with species arriving later.

prokaryotes organisms with cells that have no membrane-bound nucleus or organelles. The prokaryotes include the bacteria and the archaea.

psychrophilic organisms that live and thrive at temperatures below 20°C.

pubescence soft or fine hairs, such as those found on leaves or parts of insects.

Q

quantitative genetics the mathematical treatment of continuously varying traits and how they respond to natural selection.

R

radiation the transfer of heat through electromagnetic radiation, mainly infrared light.

rain shadow effect the phenomenon of an area receiving lower precipitation because wind-bearing moist air is blocked by mountains or hills, causing the release of moisture before reaching the leeward side, thus producing an arid "shadow."

random distribution a distribution in which individuals within a population have an equal chance of living anywhere within an area.

range the difference between the largest and smallest values in a set of measurements or observations.

rank-abundance curve a curve that portrays the number of species in a community and their relative abundance; constructed by plotting the relative abundance of species against their rank in abundance.

ratio-dependent functional response a functional response in which the rate of feeding by a predator is a function of the ratio of prey population size to predator population size, that is, the number of prey per predator.

realized niche the actual niche of a species whose distribution is restricted by biotic interactions such as competition, predation, disease, and parasitism.

regression coefficient the slope of a regression line.

regression line the line that best fits the relationship between two variables, X and Y.

regular distribution a distribution of individuals in a population in which individuals are uniformly spaced.

relative humidity the percent water content of air relative to a potential maximum; relative humidity = water vapor density/saturation water vapor density × 100.

remote sensing gathering information about an object without direct contact with it, mainly by gathering and processing electromagnetic radiation emitted or reflected by the object; such measurements are typically made from remote sensing satellites.

reproductive effort the allocation of energy, time, and other resources to the production and care of offspring, generally involving reduced allocation to other needs such as maintenance and growth.

resilience the capacity to recover structure and function after disturbance; a highly resilient community or ecosystem may be completely disrupted by disturbance but quickly returns to its former state.

resistance the capacity of a community or ecosystem to maintain structure and/or function in the face of potential disturbance.

resource (or exploitative) competition intraspecific or interspecific competition for limited resources, generally not involving direct antagonistic interactions between individuals.

resource limitation limitation of population growth by resource availability.

restoration ecology an area of ecological research focused on exploring ways to improve the effectiveness of ecological restoration of damaged ecosystems.

rhizome rootlike, enlarged, underground stem of some plants from which new plants can develop.

rhodopsin light-absorbing pigments found in the eyes of animals and in bacteria and archaea.

riparian vegetation vegetation growth along rivers or streams.

riparian zone the transition between the aquatic environment of a river or stream and the upland terrestrial environment, generally subject to periodic flooding and elevated groundwater table.

river continuum concept a model that predicts change in physical structure, dominant organisms, and ecosystem processes along the length of temperate rivers.

river ecosystem synthesis a theory of river ecology, predicting that flow conditions and geologic setting may be of greater significance in determining ecological characteristics than the position of a river section along the course of a river system.

r selection a term referring to the per capita rate of increase; a form of natural selection favoring higher population growth rate. r selection is predicted to be strongest in disturbed habitats.

rubisco the enzyme that catalyzes the initial reaction in photosynthesis that combines CO_2 and ribulose bisphosphate.

ruderals plants or animals that live in highly disturbed habitats and that may depend on disturbance to persist in the face of potential competition from other species.

S

salinity the salt content of water.

salt marsh a marine shore ecosystem dominated by herbaceous vegetation, found mainly along sandy shores from temperate to high latitudes.

saltwater water containing more than 0.5 g of dissolved salts per liter, > 0.5 0/00.

sample mean the average of a sample of measurements or observations; an estimate of the true population mean.

sample median the middle value in a series of measurements or observations, chosen so that there are equal numbers of measurements in the series that are larger than the median and smaller than the median.

saturation water vapor pressure the pressure exerted by the water vapor in air that is saturated with water vapor.

scatterhoarded a term applied to seeds gathered by mammals and stored in scattered caches or hoards.

secondary producers organisms that feed on primary producers and are themselves an energy source for organisms in the next level of a food chain.

secondary production the production of biomass by heterotrophic organisms feeding on plants, animals, microbes, fungi, or detritus during some period of time, for example, per hour or per year. Secondary production at the level of an individual organism includes only consumer growth and reproduction. At the population level, it also includes mortality.

secondary succession succession where disturbance has destroyed a terrestrial community without destroying the soil; for instance, forest succession following a forest fire or logging.

selection coefficient(s) the relative selection costs or benefits (decreased or increased fitness) associated with a particular biological trait.

selective pressure the extent to which an organism's environment influences the fitness of that organism; the force that drives natural selection.

self-incompatibility incapacity of a plant to fertilize itself; such plants must receive pollen from another plant in order to develop seeds.

self-thinning reduction in population density as a stand of plant increases in biomass, due to intraspecific competition.

self-thinning rule (−3/2) a rule resulting from the observation that plotting the average weight of individual plants in a stand against density often produces a line with an average slope of approximately −3/2.

semelparity reproduction that involves production of all an organism's offspring in one event, generally over a short period of time.

sexual selection results from differences in reproductive rates among individuals as a result of differences in mating success due to intrasexual selection, intersexual selection, or a mixture of the two forms of sexual selection.

sigmoidal population growth curve an S-shaped pattern of population growth, with population size leveling off at the carrying capacity of the environment.

size-selective predation prey selection by predators based on prey size.

small-scale phenomena phenomena that take place on a local scale.

sociality group living generally involving some degree of cooperation between individuals.

sociobiology a branch of biology concerned with the study of social relations.

solifluction the slow movement of tundra soils down slopes as a result of annual freezing and thawing of surface soil and the actions of water and gravity.

Southern Oscillation an oscillation in atmospheric pressure that extends across the Pacific Ocean.

spate sudden flooding in a stream.

species diversity a measure of diversity that increases with species evenness and species richness.

species evenness the relative abundance of species in a community or collection.

species rarefaction curve a type of model that defines the number of samples likely needed to record all or almost all species at a location of interest.

species richness the number of species in a community or collection.

species turnover changes in species composition on islands resulting from some species becoming extinct and others immigrating.

spiraling length the length of stream required for an atom of a nutrient to complete a cycle from release into the water column to reentry into the benthic ecosystem.

stability the persistence of a community or ecosystem in the face of disturbance, usually as a consequence of a combination of resistance and resilience.

stabilizing selection a form of natural selection that acts against extreme phenotypes; can act to impede changes in populations.

stable age distribution a population in which the proportion of individuals in each age class is constant, as a result of age-specific survival, l_x, and age-specific reproduction, m_x, remaining constant over time.

stable isotope analysis analysis of the relative concentrations of stable isotopes, such as ^{13}C and ^{12}C, in materials; used in ecology to study the flow of energy and materials through ecosystems.

stamen male organ of a flower.

standard deviation the square root of the variance.

standard error an estimate of variation among means of samples drawn from a population.

static life table a life table constructed by recording the age at death of a large number of individuals; the table is called static because the method involves a snapshot of survival within a population during a short interval of time.

stratosphere a layer of earth's atmosphere that extends from about 16 km to an altitude of about 50 km.

stream order a numerical classification of streams whereby they occur in a stream drainage network. In this classification, headwater streams are first-order streams, joining of two first-order streams forms a second-order stream, joining of two second-order streams forms a third-order stream, and so forth.

stress in general, stress consists of any strong negative environmental conditions that induce physiological responses in an organism or alter the structure or functioning of an ecosystem. Grime (1977) defined stress in relation to plants as external constraints that limit the rate of dry matter production, or growth, of all or part of the vegetation.

stress-tolerant plants plants that live under conditions of high stress but low disturbance.

strong interactions feeding activities of a few species that have a dominant influence on community structure.

subpopulation a portion of a larger population, with which it sustains a connection through immigration and emigration.

succession the gradual change in plant and animal communities in an area following disturbance or the creation of new substrate.

Suess effect reduced concentration of ^{14}C in the atmosphere as a consequence of fossil fuel burning.

survivorship curve a graphical summary of patterns of survival in a population.

sympatric describes the condition in which populations or species have overlapping geographic ranges.

T

taiga northern forests that occupy the area south of arctic tundra. Though dominated by coniferous trees they also contain aspen and birch. Also called *boreal forest*.

temperate forest deciduous or coniferous forests generally found between 40° and 50° latitude, where annual precipitation averages anywhere from about 650 mm to over 3,000 mm; this biome receives more winter precipitation than temperate grasslands receive.

temperate grassland grasslands growing in middle latitudes that receive between 300 and 1,000 mm of annual precipitation, with maximum precipitation usually falling during the summer months.

temperature a measure of the average kinetic energy, or energy of motion, of the molecules, in a mass of a substance, for example, a mass of air or of water.

thermal neutral zone the range of environmental temperatures over which the metabolic rate of a homeothermic animal does not change.

thermocline a depth zone in a lake or ocean through which temperature changes rapidly with depth, generally about 1°C per meter of depth.

thermohaline circulation (THC) a global pattern of water movement in the oceans that is driven by differences in water density, where density is determined primarily by temperature (thermo) and salinity (haline).

thermophilic a term applied to organisms that tolerate or require high-temperature environments.

thermosphere the outer layer of the earth's atmosphere beginning approximately 80 km above the earth's surface.

tolerance model a model of succession in which initial stages of colonization are not limited to a few pioneer species, juveniles of species dominating at climax can be present from the earliest stages of succession, and species colonizing early in succession do not facilitate colonization by species characteristic of later successional stages. Later successional species are simply those tolerant of environmental conditions early in succession.

top-down control the control or influence of consumers on ecosystem processes.

torpor a state of low metabolic rate and lowered body temperature.

tri-trophic interactions trophic dynamics that involve three levels, such as one that includes a primary producer, the herbivores that consume it, and the predators of the herbivores.

trophic (feeding) biology the study of the feeding biology of organisms.

trophic cascade effects of predators on prey that alter abundance, biomass, or productivity of a population, community, or trophic level across more than one link in the food web.

trophic dynamics the transfer of energy from one part of an ecosystem to another.

trophic level trophic position in an ecosystem; for instance, primary producer, primary consumer, secondary consumer, tertiary consumer, and so forth.

tropical dry forest a broadleaf deciduous forest growing in tropical regions having pronounced wet and dry seasons; trees drop their leaves during the dry season.

tropical rain forest a broadleaf evergreen forest growing in tropical regions where conditions are warm and wet year-round.

tropical savanna a tropical grassland dotted with scattered trees; characterized by pronounced wet and dry seasons and periodic fires.

troposphere a layer of the atmosphere extending from the earth's surface to an altitude of 9 to 16 km.

t-test a statistical test used to compare pairs of samples of measurements with approximately normal distributions and equal variances.

tundra a northern biome dominated by mosses, lichens, and dwarf willows, receiving low to moderate precipitation and having a very short growing season.

type I survivorship curve a pattern of survivorship in which there are high rates of survival among young and middle-aged individuals followed by high rates of mortality among the aged.

type II survivorship curve a pattern of survivorship characterized by constant rates of survival throughout life.

type III survivorship curve a pattern of survivorship in which a period of extremely high rates of mortality among the young is followed by a relatively high rate of survival.

U

upwelling movement of deeper ocean water to the surface; occurs most commonly along the west coasts of continents and around Antarctica.

urban ecology the study of urban areas as complex, dynamic ecological systems, influenced by interconnected, biological, physical, and social components.

urban heat island a metropolitan area that is significantly warmer than nearby rural areas; heating results mainly from replacing the trees and other vegetation in a landscape with buildings, paved roads, and concrete walks and from waste energy from human activity in the urban landscape.

V

vapor pressure deficit the difference between the actual water vapor pressure and the saturation water vapor pressure at a particular temperature.

variance a measure of variation in a population or a sample from a population.

vesicle storage organ in vesicular-arbuscular mycorrhizal fungi.

W

Walker circulation a large-scale atmospheric circulation system that moves in the plane of the equator.

water potential the capacity of water to do work, which is determined by its free energy content; water flows from positions of higher to lower free energy. Increasing solute concentration decreases water potential.

water vapor pressure the atmospheric pressure exerted by the water vapor in air; increases as the water vapor in air increases.

weather the state of the atmosphere (e.g., temperature and precipitation) of a particular place at a particular time. Not to be confused with climate.

wet deposition washing of materials from the atmosphere onto the earth surface with rain or snow.

Z

zonation of species pattern of separation of species into distinctive vertical habitats or zones.

zooplankton animals that drift in the surface waters of the oceans or lakes; most zooplankton are microscopic.

References

Abdollahinejad, B., H. Pasalari, A. J. Jafari, A. Esrafili, and M. Farzadkia, 2020. Bioremediation of diesel and gasoline-contaminated soil by co-vermicomposting amended with activated sludge: Diesel and gasoline degradation and kinetics. *Environmental Pollution*, 114584.

Achard, F., H. D. Eva, H.-J. Stibig, P. Mayaux, J. Gallego, T. Richards, and J.-P. Malingreau. 2002. Determination of deforestation rates of the world's humid tropical forests. *Science* 297:999–1002.

Ackert, L. T. 2006. The role of microbes in agriculture: Sergei Vinogradskii's discovery and investigation of chemosynthesis, 1880–1910. *Journal of the History of Biology* 39(2):373–406.

Adams, P. A., and J. E. Heath. 1964. Temperature regulation in the sphinx moth, *Celerio lineata*. *Nature* 201:20–22.

Adler, P. B., D. Smull, K. H. Beard, R. T. Choi, T. Furniss, A. Kulmatiski, and K. E. Veblen. 2018. Competition and coexistence in plant communities: intraspecific competition is stronger than interspecific competition. *Ecology Letters* 21(9):1319–29.

Agrawal, D. C. 2010. Photosynthetic solar constant. *Latin American Journal of Physics Education* 4:46–50.

Agrawal, A. A. 2020. A scale-dependent framework for trade-offs, syndromes, and specialization in organismal biology. *Ecology* 101(2):e02924.

Ali, J. G., C. K. Casteel, K. E. Mauck, and O Trase. 2020. Chemical ecology of multitrophic microbial interactions: plants, insects, microbes and the metabolites that connect them. *Journal of Chemical Ecology* 46(8):645–48.

Allen, B., M. A. Nowak, and E. O. Wilson. 2013. Limitations of inclusive fitness. *Proceedings of the National Academy of Sciences of the United States* 110:20135–39.

Allen, E. B., and M. F. Allen. 1986. Water relations of xeric grasses in the field: interactions of mycorrhizae and competition. *New Phytologist* 104:559–71.

Anderegg, W. R. L., J. M. Kane, and L. D. L. Anderegg. 2013. Consequences of widespread tree mortality triggered by drought and temperature stress. *Nature Climate Change* 3:30–36.

Anderson, J. M., and M. J. Swift. 1983. Decomposition in tropical forests. In S. L. Sutton, T. C. Whitmore, and A. C. Chadwick,

eds. *Tropical Rain Forest: Ecology and Management*. Oxford: Blackwell Scientific Publications.

Angilletta, M. J., Jr. 2001. Thermal and physiological constraints on energy assimilation in a widespread lizard (*Sceloporus undulatus*). *Ecology* 82:3044–56.

Aono, Y., and S. Saito. 2010. Clarifying springtime temperature reconstructions of the medieval period by gap-filling the cherry blossom phenological data series at Kyoto, Japan. *International Journal of Biometeorology* 54:211–9.

Arakawa, H. 1956. Climatic change as revealed by the blooming dates of the cherry blossoms at Kyoto. *Journal of Meteorology* 13:599–600.

Arditi, R., and L. R. Ginzburg. 1989. Coupling in predator-prey dynamics: ratio-dependence. *Journal of Theoretical Biology* 139:311–26.

Arditi, R., and L. R. Ginzburg. 2012. *How Species Interact: Altering the Standard View on Trophic Ecology.* Oxford: Oxford University Press.

Arrhenius, O. 1921. Species and area. *Journal of Ecology* 9:95–9.

Arsenault, S. V., K. M. Glastad, and B. G. Hunt. 2019. Leveraging technological innovations to investigate evolutionary transitions in insect sociality. *Current Opinion in Insect Science* 34:27–32.

Bacastow, R., and C. D. Keeling. 1974. Atmospheric carbon dioxide and radiocarbon in the natural carbon cycle: II. Changes from A.D. 1700 to 2070 as deduced from a geochemical model. In G. M. Woodwell and E. V. Pecan, eds. *Carbon and the Biosphere*. BHNL/CONF 720510. Springfield, Va.: National Technical Information Service.

Baker, L. A., D. Hope, Y. Xu, J. Edmonds, and L. Lauver. 2001. Nitrogen balance for the central Arizona–Phoenix (CAP) ecosystem. *Ecosystems* 4:582–602.

Baker, M. C., L. R. Mewaldt, and R. M. Stewart. 1981. Demography of white-crowned sparrows (*Zonotrichia leucophrys nuttalli*). *Ecology* 62:636–44.

Baldridge, E., D. J. Harris, X. Xiao, and E. P. White. 2016. An extensive comparison of species-abundance distribution models. *PEERJ* 4:e2823.

Baldwin, J., and P. W. Hochachka. 1970. Functional significance of isoenzymes in thermal acclimation: acetylcholinesterase from trout brain. *Biochemical Journal* 116:883–87.

Banerjee, R., J. Bhattacharya, and P. Majumdar. 2020.

Exponential-growth prediction bias and compliance with safety measures in the times of COVID-19. *arXiv preprint arXiv:2005.01273*

Baptista, L. F., P. W. Trail, and H. M. Horblit. 1997. Family Columbidae (doves and pigeons). In J. del Hoyo, A. Elliott, and J. Sargatal, eds. *Handbook of the birds of the World.* Vol. 4. *Sand Grouse to Cuckoos*. Barcelona: Lynx Edicions.

Barbour, C. D., and J. H. Brown. 1974. Fish species diversity in lakes. *American Naturalist* 108:473–89.

Barnes, R. S. K., and R. N. Hughes. 1988. *An Introduction to Marine Ecology,* 2nd edition. Oxford: Blackwell Scientific Publications.

Barnola, J. M., D. Raynaud, Y. S. Korotkevich, and C. Lorius. 1987. Vostok ice core provides 160,000-year record of atmospheric CO_2. *Nature* 329:408–14.

Barnosky, A. D., N. Matzke, S. Tomiya, G. O. U. Wogan, B. Swartz, T. Quental, C. Marshall, J. L. McGuire, E. L. Lindsey, K. C. Maguire, B. Mersey, and E. A. Ferrer. 2011. Has the Earth's sixth mass extinction already arrived? *Nature* 471:51–7.

Bässler, C., T. Hothorn, R. Brandl, and J. Müller. 2013. Insects overshoot the expected upslope shift caused by climate warming. *PLoS ONE* 8:e65842.

Baur, B., and A. Baur. 1993. Climatic warming due to thermal radiation from an urban area as possible cause for the local extinction of a land snail. *Journal of Applied Ecology* 30:333–40.

Baur, B., and A. Baur. 2013. Snails keep the pace: shift in upper elevation limit on mountain slopes as a response to climate warming. *Canadian Journal of Zoology* 91:596–9.

Beech, E., M. Rivers, S. Oldfield, and P. Smith. 2017. GlobalTreeSearch: the first complete global database of tree species and country distributions. *Journal of Sustainable Forestry* 36(5):454–89.

Béjà, O., L. Aravind, E. V. Koonin, M. T. Suzuki, A. Hadd, L. P. Nguyen, S. B. Jovanovich, C. M. Gates, R. A. Feldman, J. L. Spudich, E. N. Spudich, and E. F. Delong. 2000. Bacterial rhodopsin: evidence for a new type of phototrophy in the sea. *Science* 289:1902–6.

Béjà, O., E. N. Spudich, J. L. Spudich, M. Leclerc, and E. F. Delong. 2001. Proteorhodopsin phototrophy in the ocean. *Nature* 411:786–9.

Béjà, O., M. T. Suzuki, J. F. Heidelberg, W. C. Nelson, C. M. Preston, T. Hamada, J. A. Eisen, C. M. Fraser, and E. F. Delong. 2002. Unsuspected diversity among marine aerobic anoxygenic phototrophs. *Nature* 415:630–3.

Bellec, L., M. A. Cambon-Bonavita, L. Durand, J. Aube, N. Gayet, R. Sandulli, and D. Zeppilli. 2020. Microbial communities of the shallow-water hydrothermal vent near Naples, italy, and chemosynthetic symbionts associated with a free-living marine nematode. *Frontiers in Microbiology*, 11:2023.

Benke, A. C. 2010. Secondary production. *Nature Education Knowledge* 1(10):23. www.nature.com/scitable/knowledge/library/secondary-production-13234142.

Benke, A. C. 2011. Secondary production, quantitative food webs, and trophic position. *Nature Education Knowledge* 2(2):2. www.nature.com/scitable/knowledge/library/secondary-production-quantitative-food-webs-and-trophic-17653963.

Benke, A. C., and A. D. Huryn. 2010. Benthic invertebrate production—facilitating answers to ecological riddles in freshwater ecosystems. *Journal of the North American Benthological Society* 29:264–85.

Bennett, A. F., and R. E. Lenski. 2007. An experimental test of evolutionary trade-offs during temperature adaptation. *Proceedings of the National Academy of Sciences of the United States* 104(Suppl. 1): 8649–54.

Bennett, K. D. 1983. Postglacial population expansion of forest trees in Norfolk, UK. *Nature* 303:164–67.

Bernhardt, E. S., L. E. Band, C. J. Walsh, and P. E. Berke. 2008. Understanding, managing, and minimizing urban impacts on surface water nitrogen loading. *Annals of the New York Academy of Sciences* 1134:61–96.

Berry, J., and O. Björkman. 1980. Photosynthetic response and adaptation to temperature in higher plants. *Annual Review of Plant Physiology* 31:491–543.

Bertschy, K. A., and M. G. Fox. 1999. The influence of age-specific survivorship on pumpkinseed sunfish life histories. *Ecology* 80:2299–313.

Beschta, R. L., and W. J. Ripple. 2013. Are wolves saving Yellowstone's aspen? A landscape-level test of a behaviorally mediated trophic cascade: comment. *Ecology* 94:1420–5.

Beschta, R. L., L. E. Painter, and W. J. Ripple. 2018. Trophic cascades at multiple spatial scales shape recovery of young aspen in Yellowstone. *Forest Ecology and Management*, 413:62–9.

Bethel, W. M., and J. C. Holmes. 1977. Increased vulnerability of amphipods to predation due to altered behavior induced by larval parasites. *Canadian Journal Zoology* 55:110–5.

Bi, K., T. Linderoth, S. Singhal, D. Vanderpool, J. L. Patton, R. Nielsen, and J. M. Good. 2019. Temporal genomic contrasts reveal rapid evolutionary responses in an alpine mammal during recent climate change. *PLoS Genetics* 15(5):e1008119.

Birch, L. C. 1948. The intrinsic rate of natural increase of an insect population. *Journal of Animal Ecology* 17:15–26.

Birnie-Gauvin, K., M. H. Larsen, J. Nielsen, and K. Aarestrup. 2017. 30 years of data reveal dramatic increase in abundance of brown trout following the removal of a small hydrodam. *Journal of Environmental Management* 204:467–71.

Bishop, M. J., B. P. Kelaher, M. P. Lincoln Smith, P. H. York, and D. J. Booth. 2006. Ratio-dependent response of a temperate Australian estuarine system to sustained nitrogen loading. *Oecologia* 149:701–8.

Bjerknes, J. 1966. A possible response of the atmospheric Hadley circulation to equatorial anomalies of ocean temperature. *Tellus* 18:820–9.

Bjerknes, J. 1969. Atmospheric teleconnections from the equatorial Pacific. *Monthly Weather Review* 97:163–72.

Blair, R. 2004. The effects of urban sprawl on birds at multiple levels of biological organization. *Ecology and Society* 9:2 [online].

Blanco, J. F., and F. N. Scatena. 2005. Floods, habitat hydraulics, and upstream migration of *Neritina virginea* (Gastropoda: Neritidae) in northeastern Puerto Rico. *Caribbean Journal of Science* 41:55–74.

Blanco, J. F., and F. N. Scatena. 2006. Hierarchical contribution of river-ocean connectivity, water chemistry, hydraulics, and substrate to the distribution of diadromous snails in Puerto Rican streams. *Journal of the North American Benthological Society* 25:82–98.

Blanco, J. F., and F. N. Scatena. 2007. The spatial arrangement of *Neritina virginea* (Gastropoda: Neritidae) during upstream migration in a split-channel reach. *River Research and Applications* 23:235–45.

Bloom, A. J., F. S. Chapin III, and H. A. Mooney. 1985. Resource limitation in plants—an economic analogy. *Annual Review of Ecology and Systematics* 16:363–92.

Boag, P. T., and P. R. Grant. 1978. Heritability of external morphology in Darwin's finches. *Nature* 274:793–4.

Boag, P. T., and P. R. Grant. 1984a. Darwin's finches (*Geospiza*) on Isla Daphne Major, Galápagos: breeding and feeding ecology in a climatically variable environment. *Ecological Monographs* 54:463–89.

Boag, P. T., and P. R. Grant. 1984b. The classical case of character release: Darwin's finches (*Geospiza*) on Isla Daphne Major, Galápagos. *Biological Journal of the Linnean Society* 22:243–87.

Bobbink, R., and J. H. Willems. 1987. Increasing dominance of *Brachypodium pinnatum* (L.) Beauv. in chalk grasslands: a threat to a species-rich ecosystem. *Biological Conservation* 40:301–14.

Bobbink, R., and J. H. Willems. 1991. Effect of different cutting regimes on the performance of *Brachypodium pinnatum* in Dutch chalk grassland. *Biological Conservation* 56:1–21.

Boden, T. A., G. Marland, and R. J. Andres. 2017. Global, regional, and national fossil-fuel CO_2 emissions. Carbon Dioxide Information Analysis Center, Oak Ridge National Laboratory, U.S. Department of Energy, Oak Ridge, Tenn., U.S.A. doi 10.3334/CDIAC/00001_V2017.

Bollinger, G. 1909. *Zur Gastropoden-fauna von Basel und Umgebung.* Ph.D. dissertation, University of Basel, Switzerland.

Bolser, R. C., and M. E. Hay. 1996. Are tropical plants better defended? Palatability and defenses of temperate vs. tropical seaweeds. *Ecology* 77:2269–86.

Bond, P., and P. Goldblatt. 1984. Plants of the Cape Flora. *Journal of South African Botany,* Supplementary Volume No. 13.

Bormann, F. H., and G. E. Likens. 1994. *Pattern and Process in a Forested Ecosystem.* New York: Springer-Verlag.

Borrego, N. 2020. Socially tolerant lions (*Panthera leo*) solve a novel cooperative problem. *Animal Cognition* 23(2): 327–36.

Bouchama, A. 2004. The 2003 European heat wave. *Intensive Care Medicine* 30(1):1–3.

Bowen, G. W., and R. L. Burgess. 1981. A quantitative analysis of forest island pattern in selected Ohio landscapes. ORNL/TM 7759. Oak Ridge National Laboratory, Oak Ridge, Tenn.

Bowman, W. D., T. A. Theodose, J. C. Schardt, and R. T. Conant. 1993. Constraints of nutrient availability on primary production in two alpine tundra communities. *Ecology* 74:2085–97.

Boyd, P. W., S. Sundby, and H.-O. Pörtner. 2014. Cross-chapter box on net primary production in the ocean. In Field, C. B., V. R. Barros, D. J. Dokken, K. J. Mach, M. D. Mastrandrea, T. E. Bilir, M. Chatterjee, K. L. Ebi, Y. O. Estrada, R. C. Genova, B. Girma, E. S. Kissel, A. N. Levy, S. MacCracken, P. R. Mastrandrea, and L. L. White, eds. *Climate Change 2014: Impacts, Adaptation, and Vulnerability. Part A: Global and Sectoral Aspects. Contribution of Working Group II to the Fifth Assessment Report of the Intergovernmental Panel on Climate Change,* 133–6. Cambridge, England/New York: Cambridge University Press.

Boyles, J. G., P. M. Cryan, G. F. McCracken, and T. H. Kunz. 2011. Economic importance of bats in agriculture. *Science* 332:41–2.

Braithwaite, V. A., and A. G. V. Salvanes. 2005. Environmental variability in the early rearing environment generates behaviourally flexible cod: implications for rehabilitating wild populations. *Proceedings of the Royal Society B* 272:1107–13.

Brenchley, W. E. 1958. *The Park Grass Plots at Rothamsted.* Harpende: Rothamsted Experimental Station.

Brisson, J., and J. F. Reynolds. 1994. The effects of neighbors on root distribution in a creosote bush (*Larrea tridentata*) population. *Ecology* 75:1693–702.

Brock, T. D. 1978. *Thermophilic Microorganisms and Life at High Temperatures.* New York: Springer-Verlag.

Brouwer, R. 1983. Functional equilibrium: sense or nonsense? *Netherlands Journal of Agricultural Science* 31:335–48.

Brown, J. H. 1984. On the relationship between abundance and distribution of species. *American Naturalist* 130:255–79.

Brown, J. H. 1986. Two decades of interaction between the MacArthur-Wilson model and the complexities of mammalian distributions. *Biological Journal of the Linnean Society* 28:231–51.

Brown, J. H. 1988. Species diversity. In A. A. Meyers and P. S. Giller, eds. *Analytical Biogeography* London: Chapman and Hall.

Brown, J. H., and A. Kodric-Brown. 1977. Turnover rates in insular biogeography: effects of immigration on extinction. *Ecology* 58:445–9.

Brown, J. H., and M. V. Lomolino. 2000. Concluding remarks: historical perspective and the future of island biogeography theory. *Global Ecology and Biogeography* 9:87–92.

Brown, J. H., D. W. Mehlman, and G. C. Stevens. 1995. Spatial variation in abundance. *Ecology* 76:2028–43.

Brown, J. H., and J. C. Munger. 1985. Experimental manipulation of a desert rodent community: food addition and species removal. *Ecology* 66:1545–63.

Bshary, R. 2003. The cleaner wrasse, *Labroides dimidiatus,* is a key organism for reef fish diversity at Ras Mohammed National Park, Egypt. *Journal of Animal Ecology* 72:169–72.

Buma, B., S. Bisbing, J. Krapek, and G. Wright. 2017. A foundation of ecology rediscovered: 100 years of succession on the William S. Cooper plots in Glacier Bay, Alaska. *Ecology* 98(6):1513–23.

Buma, B., S. M. Bisbing, G. Wiles, and A. L. Bidlack. 2019. 100 yr of primary succession highlights stochasticity and competition driving community establishment and stability. *Ecology* 100(12):e02885.

Byers, J. E. 2000. Competition between two estuarine snails: implications for invasions of exotic species. *Ecology* 81:1225–39.

Byers, J. E., and L. Goldwasser. 2001. Exposing the mechanism and timing of impact of nonindigenous species on native species. *Ecology* 82:1330–43.

Cairns, S. C., and G. C. Grigg. 1993. Population dynamics of red kangaroos (*Macropus rufus*) in relation to rainfall in the south Australian pastoral zone. *Journal of Applied Ecology* 30:444–58.

Calder, W. A. 1994. When do hummingbirds use torpor in nature? *Physiological Zoology* 67:1051–76.

Calow, P., and G. E. Petts. 1992. *The Rivers Handbook.* London: Blackwell Scientific Publications.

Canington, S. L. 2018. *Gorilla beringei* (Primates: Hominidae). *Mammalian Species* 50(967):119–33.

Cardinale, B. J. 2011. Biodiversity improves water quality through niche partitioning. *Nature* 472:86–U113.

Carpenter, F. L., M. A. Hixon, C. A. Beuchat, R. W. Russell, and D. C. Patton. 1993. Biphasic mass gain in migrant hummingbirds: body composition changes, torpor, and ecological significance. *Ecology* 74:1173–82.

Carpenter, S. R., T. M. Frost, J. F. Kitchell, T. K. Kratz, D. W. Schindler, J. Shearer, W. G. Sprules, M. J. Vanni, and A. P. Zimmerman. 1991. Patterns of primary production and herbivory in 25 North American lake ecosystems. In J. Cole, G. Lovett, and S. F. Findlay, eds. *Comparative Analyses of Ecosystems: Patterns, Mechanisms, and Theories.* New York: Springer-Verlag.

Carpenter, S. R., and J. F. Kitchell. 1988. Consumer control of lake productivity. *BioScience* 38:764–9.

Carpenter, S. R., and J. F. Kitchell. 1993. *The Trophic Cascade in Lakes.* Cambridge, England: Cambridge University Press.

Carpenter, S. R., J. F. Kitchell, and J. R. Hodgson. 1985. Cascading trophic interactions and lake productivity. *BioScience* 35:634–9.

Carroll, S. P., and C. Boyd. 1992. Host race radiation in the soapberry bug: natural history with the history. *Evolution* 46:1052–69.

Carroll, S. P., H. Dingle, T. R. Famula, and C. W. Fox. 2001. Genetic architecture of adaptive differentiation in evolving host races of the soapberry bug, *Jadera haematoloma. Genetica* 112:257–72.

Carroll, S. P., H. Dingle, and S. P. Klassen. 1997. Genetic differentiation of fitness-associated traits among rapidly evolving populations of the soapberry bug. *Evolution* 51:1182–8.

Carroll, S. P., S. P. Klassen, and H. Dingle. 1998. Rapidly evolving adaptations to host ecology and nutrition in the soapberry bug. *Evolutionary Ecology* 12:955–68.

Carruthers, R. I., T. S. Larkin, H. Firstencel, and Z. Feng. 1992. Influence of thermal ecology on the mycosis of a rangeland grasshopper. *Ecology* 73:190–204.

Caughley, G. 1977. *Analysis of Vertebrate Populations.* New York: John Wiley & Sons.

Caughley, G., J. Short, G. C. Grigg, and H. Nix. 1987. Kangaroos and climate: an analysis of distribution. *Journal of Animal Ecology* 56:751–61.

Census of Marine Life. 2016. www.coml.org/

Chak, S. T. C., J. E. Duffy, K. M. Hultgren, and D. R. Rubenstein. 2017. Evolutionary transitions towards eusociality in snapping shrimps. *Nature Ecology & Evolution* 1(4):1–7.

Chamaille-Jammes, S., H. Fritz, and F. Murindagomo. 2014. Spatial patterns of the NDVI-rainfall relationship at the seasonal and interannual time scales in an African savanna. *International Journal of Remote Sensing* 27:5185–200.

Chapman, R. N. 1928. The quantitative analysis of environmental factors. *Ecology* 9:111–2.

Chapman, V. J. 1977. *Wet Coastal Ecosystems.* Amsterdam: Elsevier Scientific Publishing.

Charnov, E. L. 1973. *Optimal Foraging: Some Theoretical Explorations* Ph.D. Dissertation, University of Washington, Seattle.

Charnov, E. L. 2002. Reproductive effort, offspring size and benefit-cost ratios in the classification of life histories. *Evolutionary Ecology Research* 4:749–58.

Charnov, E. L., R. Warne, and M. Moses. 2007. Lifetime reproductive effort. *American Naturalist* 170: E129–42.

Chatterjee, S., and S. Sharma. 2019. Microplastics in our oceans and marine health. Special issue, *Field Actions Science Reports. The Journal of Field Actions* 19:54–61.

Chaudhari, S., V. K. Varanasi, S. Nakka, P. C. BhoMwik, C. R. Thompson, D. E. Peterson, and M. Jugulam. 2020. Evolution of target and non-target based multiple herbicide resistance in a single Palmer amaranth (*Amaranthus palmeri*) population from Kansas. *Weed Technology* 34(3):447–53.

Chen, J. L., Y. C. Lin, H. Y. Fu, and C. S. Yang. 2019. The blue-green sensory rhodopsin srm from *Haloarcula marismortui* attenuates both phototactic responses mediated by sensory rhodopsin I and II in *Halobacterium salinarum*. *Scientific reports*, 9(1), 1–11.

Chen, S., W. Wang, W. Xu, Y. Wang, H. Wan, D. Chen, and D. Zhou. 2018. Plant diversity enhances productivity and soil carbon storage. Proceedings of the *National Academy of Sciences* 115(16):4027–32.

Chen, I.-C., J. K. Hill, R. Ohlemüller, D. B. Roy, and C. D. Thomas. 2011. Rapid range shifts of species associated with high levels of climate warming. *Science* 333:1024–26.

Chiariello, N. R., C. B. Field, and H. A. Mooney. 1987. Midday wilting in a tropical pioneer tree. *Functional Ecology* 1:3–11.

Christian, C. E. 2001. Consequences of a biological invasion reveal the importance of mutualism for plant communities. *Nature* 413:635–39.

Clark, T. J., and A. D. Luis. 2020. Nonlinear population dynamics are ubiquitous in animals. *Nature Ecology & Evolution* 4(1):75–81.

Clausen, J., D. D. Keck, and W. M. Hiesey. 1940. *Experimental Studies on the Nature of Species. I. The Effect of Varied Environments on Western North American Plants.* Washington, D.C.: Carnegie Institution of Washington, Publication no. 520.

Clements, F. E. 1916. *Plant Succession: An Analysis of the Development of Vegetation.* Washington, D.C.: Carnegie Institution of Washington, Publication no. 242.

Clements, F. E. 1936. Nature and structure of the climax. *Journal of Ecology* 24:252–84.

Cleveland, C. J., M. Betke, P. Federico, J. D. Frank, T. G. Hallam, J. Horn, J. D. López Jr., G. F. McCracken, R. A. Medellín, A. Moreno-Valdez, C. G. Sansone, J. K. Westbrook, and T. H. Kunz. 2006. Economic value of the pest control service provided by Brazilian free-tailed bats in south-central Texas. *Frontiers in Ecology and Environment* 4:238–43.

Coe, M. J., D. H. Cumming, and J. Phillipson. 1976. Biomass and production of large African herbivores in relation to rainfall and primary production. *Oecologia* 22:341–54.

Coleridge, S.T. 1798. "The Rime of the Ancient Mariner." In William Wordsworth and Samuel Taylor Coleridge, *Lyrical Ballads: With a Few Other Poems.* London: J. & A. Arch.

Coley, P. D., and T. M. Aide. 1991. Comparison of herbivory and plant defenses in temperate and tropical broad-leaved forests. In P. W. Price et al., eds. *Plant-Animal Interactions: Evolutionary Ecology in Tropical and Temperate Regions.* New York: John Wiley & Sons.

Collatz, G. J., C. Williams, B. Ghimire, S. Goward, and J. Masek. 2014. *NACP CMS: Forest Biomass and Productivity, 1-degree and 5-km, Conterminous US, 2005.* Data set. Available online [http://daac.ornl.gov] from Oak Ridge National Laboratory Distributed Active Archive Center, Oak Ridge, Tennessee, USA. http://dx.doi.org/10.3334/ORNLDAAC/1221.

COML. 2010. Census of marine life. www.coml.org/.

Condit, R., B. J. Le Boeuf, P. A. Morris, and M. Sylvan. 2007. Estimating population size in asynchronous aggregations: a Bayesian approach and test with elephant seal censuses. *Marine Mammal Science* 23:834–5.

Connell, J. H. 1961a. The effects of competition, predation by *Thais lapillus* and other factors on natural populations of the barnacle, *Balanus balanoides. Ecological Monographs* 31:61–104.

Connell, J. H. 1961b. The influence of interspecific competition and other factors on the distribution of the barnacle *Chthamalus stellatus. Ecology* 42:710–23.

Connell, J. H. 1974. Ecology: field experiments in marine ecology. In R. N. Mariscal, ed. *Experimental Marine Biology.* New York: Academic Press.

Connell, J. H. 1975. Some mechanisms producing structure in natural communities: a model and evidence from field experiments. In M. L. Cody and J. Diamond, eds. *Ecology and Evolution of Communities.* Cambridge, Mass.: Harvard University Press.

Connell, J. H. 1978. Diversity in tropical rain forests and coral reefs. *Science* 199:1302–10.

Connell, J. H. 1983. On the prevalence and relative importance of interspecific competition: evidence from field experiments. *American Naturalist* 122:661–96.

Connell, J. H., and R. O. Slatyer. 1977. Mechanisms of succession in natural communities and their role in community stability and organization. *The American Naturalist* 111:1119–44.

Cook, L. M., B. S. Grant, I. J. Saccheri, and J. Mallet. 2012. Selective bird predation on the peppered moth: the last experiment of Michael Majerus. *Biology Letters* 8:609–12.

Cooper, P. D. 1982. Water balance and osmoregulation in a free-ranging tenebrionid beetle, *Onymacris unguicularis,* of the Namib Desert. *Journal of Insect Physiology* 28:737–42.

Cooper, W. S. 1923. The recent ecological history of Glacier Bay, Alaska. *Ecology* 4:93–128, 223–46, 355–65.

Cooper, W. S. 1931. A third expedition to Glacier Bay, Alaska. *Ecology* 12:61–95.

Cooper, W. S. 1939. A fourth expedition to Glacier Bay, Alaska. *Ecology* 20:130–55.

Coppock, D. L., J. K. Detling, J. E. Ellis, and M. I. Dyer. 1983. Plant herbivore interactions in a North American mixed-grass prairie. l. effects of black-tailed prairie dogs on intraseasonal aboveground plant biomass and nutrient dynamics and plant species diversity. *Oecologia* 56:1–9.

Costello, M. J., M. Coll, R. Danovaro, P. Halpin, H. Ojaveer, and P. Miloslavich. 2010. A census of marine biodiversity knowledge, resources, and future challenges. *PLoS ONE* 5:e12110.

Coupland, R. T., and R. E. Johnson. 1965. Rooting characteristics of native grassland species in Saskatchewan. *Journal of Ecology* 53:475–507.

Cox, G. W., and R. E. Ricklefs. 1977. Species diversity, ecological release, and community structure in Caribbean landbird faunas. *Oikos* 29:60–6.

Culp, J. M., and G. J. Scrimgeour. 1993. Size dependent diel foraging periodicity of a mayfly grazer in streams with and without fish. *Oikos* 68:242–50.

Curtis, J. T. 1956. The modification of mid-latitude grasslands and forests by man. In W. L. Thomas, Jr., ed. *Man's Role in Changing the Face of the Earth.* Chicago: University of Chicago Press.

Damuth, J. 1981. Population density and body size in mammals. *Nature* 290:699–700.

Darwin, C. 1842a. *Journal of Researches into the Geology and Natural History of the Various Countries Visited During the Voyage of H.M.S. 'Beagle' Under the Command of Captain FitzRoy, N. N. from 1832-1836.* London: Henry Colborn.

Darwin, C. 1842b. *The Structure and Distribution of Coral Reefs.* London: Smith, Elder and Company. Reprinted by the University of California Press, Berkeley, 1962.

Darwin, C. 1859. *The Origin of Species by Means of Natural Selection, or the Preservation of Favored Races in the Struggle for Life.* New York: Modern Library.

Darwin, C. 1862. On the two forms, or dimorphic condition, in the species of *Primula,* and on their remarkable sexual relations. In P. H. Barrett, ed. *The Collected Papers of Charles Darwin.* Chicago: University of Chicago Press.

Darwin, C. 1868. *The Variation of Animals and Plants Under Domestication.* London: John Murray.

Darwin, C. 1871. *The Descent of Man, and Selection in Relation to Sex.* London: John Murray.

Darwin, C., and F. Darwin. 1896. *The Life and Letters of Charles Darwin, Including an Autobiographical Chapter.* New York: D. Appleton and Company.

Daston, L. 2011. The empire of observation. In L. Daston and E. Lunbeck eds. *Histories of Scientific Observation,* 81–113. Chicago: The University of Chicago Press.

Davis, M. B. 1981. Quaternary history and the stability of forest communities. In D. C. West, H. H. Shugart, and D. B. Botkin, eds. *Forest Succession: Concepts and Application.* New York: Springer-Verlag.

Davis, M. B. 1983. Quaternary history of deciduous forests of eastern North America and Europe. *Annals of the Missouri Botanical Garden* 70:550–63.

Davis, M. B. 1989. Retrospective studies. In G. E. Likens, ed. *Long-Term Studies in Ecology.* New York: Springer-Verlag.

Davis, M. B., and R. G. Shaw. 2001. Range shifts and adaptive responses to Quaternary climate change. *Science* 292:673–9.

Davis, M. B., R. G. Shaw, and J. R. Etterson. 2005. Evolutionary responses to climate change. *Ecology* 86:1704–14.

Dawson, T. E., S. Mambelli, A. H. Plamboeck, P. H. Templer, and K. P. Tu. 2002. Stable isotopes in plant ecology. *Annual Review of Ecology and Systematics* 33:507–99.

Dayan, T., and D. Simberloff. 2005. Ecological and community-wide character displacement: the next generation. *Ecology Letters* 8:875–94.

Deegan, L. A., D. S. Johnson, R. S. Warren, B. J. Peterson, J. W. Fleeger, S. Fagherazzi, and W. M. Wollheim. 2012. Coastal eutrophication as a driver of salt marsh loss. *Nature* 490(7420):388–92.

Deevey, E. S. 1947. Life tables for natural populations of animals. *Quarterly Review of Biology* 22:283–314.

de Leon, L. F., E. Bermingham, J. Podos, and A. P. Hendry. 2010. Divergence with gene flow as facilitated by ecological differences: within-island variation in Darwin's finches. *Philosophical Transactions of the Royal Society B: Biological Sciences* 365:1041–52.

De León, L. F., A. Cornejo, R. G. Gavilán, and C. Aguilar. 2020. Hidden biodiversity in Neotropical streams: DNA barcoding uncovers high endemicity of freshwater macroinvertebrates at small spatial scales. *bioRxiv.*

del Moral, R., L. R. Walker, and J. P. Bakker. 2007. Insights gained from succession for the restoration of landscape structure and function. In R. del Moral, L. R. Walker, and J. P. Bakker, eds. *Linking Restoration and Ecological Succession.* New York: Springer.

Denno, R. F., and G. K. Roderick. 1992. Density-related dispersal in planthoppers: effects of interspecific crowding. *Ecology* 73:1323–34.

Diamond, J. M. 1969. Avifaunal equilibria and species turnover rates on the Channel Islands of California. *Proceedings of the National Academy of Sciences* 64:57–63.

Dias, A. T. C., R. L. Bozelli, R. M. Darigo, F. de A. Esteves, H. F. dos Santos, M. P. Figueiredo-Barros, M. F. Q. S. Nunes, F. Roland, L. R. Zamith, and F. R. Scarano. 2012. Rehabilitation of a bauxite tailing substrate in central Amazonia: the effect of litter and seed addition on flood-prone forest restoration. *Restoration Ecology* 20:483–9.

Diaz, H. F., and G. N. Kiladis. 1992. Atmospheric teleconnections associated with the extreme phases of the Southern Oscillation. In H. F. Diaz and V. Markgraf, eds. *El Niño Historical and Paleoclimatic Aspects of the Southern Oscillation.* Cambridge, England: Cambridge University Press.

Dick, G. J. 2019. The microbiomes of deep-sea hydrothermal vents: distributed globally, shaped locally. *Nature Reviews Microbiology* 17(5):271–83.

Diffendorfer, J. E., M. S. Gaines, and R. D. Holt. 1995. Habitat fragmentation and movements of three small mammals (*Sigmodon, Microtus,* and *Peromyscus*). *Ecology* 76:827–39.

Dillon, P. J., and F. H. Rigler. 1974. The phosphorus-chlorophyll relationship in lakes. *Limnology and Oceanography* 19:767–73.

Dobrynin, P., S. Liu, G. Tamazian, Z. Xiong, A. A. Yurchenko, K. Krasheninnikova, S. Kliver, A. Schmidt-Kuntzel, K. P. Koepfli, W. Johnson, L. F. K. Kuderna, R. Garcia-Perez, M. de Manuel, R. Godinez, A. Komissarov, A. Makunin, V. Brukhin, W. L. Qiu, L. Zhou, F. Li, J. Yi, C. Driscoll, A. Antunes, T. K. Oleksyk, E. Eizirik, P. Perelman, M. Roelke, D. Wildt, M. Diekhans, T. Marques-Bonet, L. Marker, J. B. JunWang, J. Wang, G. J. Zhang, and S. J. O'Brian. 2015. Genomic legacy of the African cheetah, *Acinonyx jubatus. Genome Biology* 16:277.

Dobzhansky, T. 1937. *Genetics and the Origin of Species.* New York: Columbia University Press.

Dobzhansky, T. 1950. Evolution in the tropics. *American Scientist* 38:209–21.

Dodd, M., J. Silvertown, K. McConway, J. Potts, and M. Crawley. 1995. Community stability: a 60-year record of trends and outbreaks in the occurrence of species in the Park Grass Experiment. *Journal of Ecology* 83:277–85.

Dongmei, H., and J. C. Matthew. 2018. Delineating multiple salinization processes in a coastal plain aquifer, northern China: hydrochemical and isotopic evidence. *Hydrology and Earth System Sciences* 22:3473–91.

Dorado-Liñán, I., M. Valbuena-Carabaña, I. Cañellas, L. Gil, and G. Gea-Izquierdo. 2020. Climate change synchronizes growth and iWUE across species in a temperate-submediterranean mixed oak forest. *Frontiers in Plant Science* 11:706.

Douglas, M. R., and P. C. Brunner. 2002. Biodiversity of central alpine *Coregonus* (Salmoniformes): impact of one-hundred years of management. *Ecological Applications* 12:154–72.

Dousset, B., F. Gourmelon, K. Laaidi, A. Zeghnoun, E. Giraudet, P. Bretin, E. Maurid, and S. Vandentorren. 2011. Satellite monitoring of summer heat waves in the Paris metropolitan area. *International Journal of Climatology* 31:313–23.

Edney, E. B. 1953. The temperature of wood lice in the sun. *Journal of Experimental Biology* 30:331–49.

Edwards, B. P., and A. C. Smith. 2020. bbsBayes: An R package for hierarchical Bayesian analysis of North American Breeding Bird Survey data. *bioRxiv.*

Ehleringer, J. R. 1980. Leaf morphology and reflectance in relation to water and temperature stress. In N. C. Turner and P. J. Kramer, eds. *Adaptations of Plants to Water and High Temperature Stress.* New York: Wiley-Interscience.

Ehleringer, J. R., O. Björkman, and H. A. Mooney. 1976. Leaf pubescence: effects on absorptance and photosynthesis in a desert shrub. *Science.* 192:376–77.

Ehleringer, J. R., T. E. Cerling, and B. R. Helliker. 1997. C_4 photosynthesis, atmospheric CO_2, and climate. *Oecologia* 112:285–99.

Ehleringer, J. R., and C. Clark. 1988. Evolution and adaptation in *Encelia* (Asteraceae). In L. D. Gottlieb and S. K. Jain, eds. *Plant Evolutionary Biology.* London: Chapman and Hall.

Eisen, L., and C. G. Moore. 2013. *Aedes* (*Stegomyia*) *aegypti* in the continental United States: a vector at the cool margin of its geographic range. *Journal of Medical Entomology* 50:467–78.

Elser, J. J., W. F. Fagan, R. F. Denno, D. R. Dobberfuhl, A. Folarin, A. Huberty, S. Interlandi, S. S. Kilham, E. McCauley, K. L. Schulz, E. H. Siemann, and R. W. Sterner. 2000. Nutritional constraints in terrestrial and freshwater food webs. *Nature* 408:578–80.

Elton, C. 1924. Periodic fluctuations in the numbers of animals: their causes and effects. *British Journal of Experimental Biology* 2:119–63.

Elton, C. 1927. *Animal Ecology.* London: Sidgewick & Jackson.

Endler, J. A. 1980. Natural selection on color patterns in *Poecilia reticulata. Evolution* 34:76–91.

Endler, J. A. 1995. Multiple-trait coevolution and environmental gradients in guppies. *Trends in Ecology & Evolution* 10:22–9.

Epps, M. J., and A. E. Arnold. 2019. Interaction networks of macrofungi and mycophagous beetles reflect diurnal variation and the size and spatial arrangement of resources. *Fungal Ecology* 37:48–56.

Evaristo, J., and J. J. McDonnell. 2017. Prevalence and magnitude of groundwater use by vegetation: a global stable isotope meta-analysis. *Scientific Reports* 7(1):1–12.

Escobar, H. 2020. Deforestation in the Brazilian Amazon is still rising sharply. *Science* 369(6504):613.

FAO. 1972. *Atlas of the Living Resources of the Sea.* 3rd ed. Rome: FAO.

FAO and UNEP. 2020. The State of the World's Forests 2020. Forests, biodiversity and people. Rome. https://doi.org/10.4060/ca8642en

Feggestad, A. J., P. M. Jacobs, X. D. Miao, and J. A. Mason. 2004. Stable carbon isotope record of Holocene environmental change in the central Great Plains. *Physical Geography* 25:170–90.

Feldman, G., D. Clark, and D. Halpern. 1984. Satellite color observations of the phytoplankton distribution in the eastern equatorial Pacific during the 1982–1983 El Niño. *Science* 226:1069–71.

Field, C. B., M. J. Behrenfeld, J. T. Randerson, and P. Falkowski. 1998. Primary production of the biosphere: integrating terrestrial and oceanic components. *Science* 281(5374):237–40.

Field, J., and H. Toyoizumi. 2020. The evolution of eusociality: no risk-return tradeoff but the ecology matters. *Ecology Letters* 23(3):518–26.

Fernández-Pascual, E., E. Mattana, and H. W. Pritchard. 2019. Seeds of future past: climate change and the thermal memory of plant reproductive traits. *Biological Reviews* 94(2):439–56.

Fietz, J., F. Tataruch, K. H. Dausmann, and J. U. Ganzhorn. 2003. White adipose tissue composition in the free-ranging fat-tailed dwarf lemur (*Cheirogaleus medius;* Primates), a tropical hibernator. *Journal of Comparative Physiology B: Biochemical Systemic and Environmental Physiology* 173:1–10.

Findlay, D. L., and S. E. M. Kasian. 1987. Phytoplankton community responses to nutrient addition in Lake 226, Experimental Lakes Area, northwestern Ontario. *Canadian Journal of Fisheries and Aquatic Sciences* 44(Suppl. 1):35–46.

Fisher, S. G., L. J. Gray, N. B. Grimm, and D. E. Busch. 1982. Temporal succession in a desert stream ecosystem following flash flooding. *Ecological Monographs* 52:93–110.

Fitter, A., and R. K. M. Hay. 1987. *Environmental Physiology of Plants* London: Academic Press.

Fleming, P. J. S. 2019. They might be right, but give no strong evidence that "trophic cascades shape recovery of young aspen in Yellowstone National Park": A fundamental critique of methods. *Forest Ecology and Management* 454:117283.

Flessa, K. W. 1975. Area, continental drift and mammalian diversity. *Paleobiology* 1:189–94.

Flessa, K. W. 1981. The regulation of mammalian faunal similarity among continents. *Journal of Biogeography* 8:427–38.

Forel, F. A. 1892. *Le Léman: Monograhie limnologique.* Tome I, Géographie, Hydrographie, Géologic, Climatologic, Hydrologie. Lausanne, F. Rouge. Reprinted Genève, Slatkine Reprints, 1969.

Forman, R. T. T. 1995. *Land Mosaics: the Ecology of Landscapes and*

Regions. Cambridge, England: Cambridge University Press.

Forman, R. T. T., and M. Godron. 1986. *Landscape Ecology.* New York: Wiley.

Foster, J. D., A. G. Ellis, L. C. Foxcroft, S. P. Carroll, and J. Le Roux. 2019. The potential evolutionary impact of invasive balloon vines on native soapberry bugs in South Africa. *NeoBiota,* 49;19.

Frank, P. W., C. D. Boll, and R. W. Kelly. 1957. Vital statistics of laboratory cultures of *Daphnia pulex* De Geer as related to density. *Physiological Zoology* 30:287–305.

Frazer, N. B., J. W. Gibbons, and J. L. Greene. 1991. Life history and demography of the common mud turtle *Kinosternon subrubrum* in South Carolina, USA. *Ecology* 72:2218–31.

Friedli, H., H. Lötscher, H. Oeschger, U. Siegenthaler, and B. Stauffer. 1986. Ice core record of the $^{13}C/^{12}C$ ratio of atmospheric CO_2 in the past two centuries. *Nature* 324:237–38.

Friedmann, H. 1955. The honeyguides. *Bulletin of the United States National Museum* 208:1–292.

Gabriel, K. M. A., and W. R. Endlicher. 2011. Urban and rural mortality rates during heat waves in Berlin and Brandenburg, Germany. *Environmental Pollution* 159:2044–50.

Gadagkar, R. 2010. Sociobiology in turmoil again. *Current Science* 99:1036–41.

Galen, C. 1989. Measuring pollinator-mediated selection on morphometric floral traits: bumblebees and the alpine sky pilot, *Polemonium viscosum. Evolution* 43(4):882–90.

Gallardo, A., and J. Merino. 1993. Leaf decomposition in two Mediterranean ecosystems of southwest Spain: influence of substrate quality. *Ecology* 74:152–61.

García-Herrera, R., J. Día, R. M. Trigo, J. Luterbacher, and E. M. Fischer. 2010. A review of the European summer heat wave of 2003. *Critical Reviews in Environmental Science and Technology* 40(4):267–306.

Garner, G., I. A. Malcolm, J. P. Sadler, and D. M. Hannah. 2017. The role of riparian vegetation density, channel orientation and water velocity in determining river temperature dynamics. *Journal of Hydrology* 553:471–85.

Gause, G. F. 1934. *The Struggle for Existence.* Baltimore: Williams & Wilkins. Reprinted by Hafner Publishing Company, New York, 1969.

Gause, G. F. 1935. Experimental demonstration of Volterra's periodic oscillation in the numbers of animals. *Journal of Experimental Biology* 12:44–8.

Gauslaa, Y. 1984. Heat resistance and energy budget in different Scandinavian plants. *Holarctic Ecology* 7:1–78.

Geervliet, J. B. F., M. S. W. Verdel, H. Snellen, J. Schaub, M. Dicke, and L. E. M. Vet. 2000. Coexistence and niche segregation by field populations of the parasitoids *Cotesia glomerata* and *C. rubecula* in the Netherlands: predicting field performance from laboratory data. *Oecologia* 124:55–63.

Gelin, M. L., L. C. Branch, D. H. Thornton, A. J. Novaro, M. J. Gould, and A. Caragiulo. 2017. Response of pumas (*Puma concolor*) to migration of their primary prey in Patagonia. *PLoS One* 12(12):e0188877.

Gessner, M. O., and E. Chauvet. 1994. Importance of stream microfungi in controlling breakdown rates of leaf litter. *Ecology* 75:1807–17.

Gibbs, H. L., and P. R. Grant. 1987. Ecological consequences of an exceptionally strong El Niño event on Darwin's finches. *Ecology* 68:1735–46.

Gibbs, R. J. 1970. Mechanisms controlling world water chemistry. *Science* 170:1088–90.

Gleason, H. A. 1926. The individualistic concept of the plant association. *Torrey Botanical Club Bulletin* 53:7–26.

Gleason, H. A. 1939. The individualistic concept of the plant association. *American Midland Naturalist* 21:92–110.

Glynn, P. W. 1983. Crustacean symbionts and the defense of corals: coevolution of the reef? In M. H. Nitecki, ed. *Coevolution.* Chicago: University of Chicago Press.

Graham, W. F., and R. A. Duce. 1979. Atmospheric pathways of the phosphorus cycle. *Geochimica et Cosmochimica* 43:1195–1208.

Grande, J. M., P. M. Orozco-Valor, M. S. Liébana, and J. H. Sarasola. 2018. Birds of prey in agricultural landscapes: the role of agriculture expansion and intensification. In *Birds of Prey,* 197–28. Springer, Cham.

Granéli, E., K. Wallström, U. Larsson, W. Granéli, and R. Elmgren. 1990. Nutrient limitation of primary production in the Baltic Sea area. *Ambio* 19:142–51.

Grant, B. S., A. D. Cook, C. A. Clarke, and D. F. Owen. 1998. Geographic and temporal variation in the incidence of melanism in peppered moth populations in America and Britain. *Journal of Heredity* 89(5):465–71.

Grant, B. R., and P. R. Grant. 1989. *Evolutionary Dynamics of a Natural Population.* Chicago: University of Chicago Press.

Grant, P. R. 1986. *Ecology and Evolution of Darwin's Finches.* Princeton, N.J.: Princeton University Press.

Grassle, J. F. 1991. Deep-sea benthic biodiversity. *BioScience* 41:464–9.

Green, S. J., E. R. Dilley, C. E. Benkwitt, A. C. Davis, K. E. Ingeman, T. L. Kindinger, and M. A. Hixon. 2019. Trait-mediated foraging drives patterns of selective predation by native and invasive coral-reef fishes. *Ecosphere* 10(6):e02752.

Grice, G. D., and A. D. Hart. 1962. The abundance, seasonal occurrence and distribution of the epizooplankton between New York and Bermuda. *Ecological Monographs* 32:287–309.

Grime, J. P. 1977. Evidence for the existence of three primary strategies in plants and its relevance to ecological and evolutionary theory. *American Naturalist* 111:1169–94.

Grime, J. P. 1979. *Plant Strategies and Vegetation Processes.* New York: John Wiley & Sons.

Grimm, N. B. 1987. Nitrogen dynamics during succession in a desert stream. *Ecology* 68:1157–70.

Grimm, N. B. 1988. Role of macroinvertebrates in nitrogen dynamics of a desert stream. *Ecology* 69:1884–93.

Grimm, N. B., S. H. Faeth, N. E. Golubiewski, C. L. Redman, J. Wu, X. Bai, and J. M. Briggs. 2008. Global change and the ecology of cities. *Science* 319:756–60.

Grinnell, J. 1917. The niche-relationships of the *California thrasher. Auk* 34:427–33.

Grinnell, J. 1924. Geography and evolution. *Ecology* 5:225–9.

Groffman, P. M., N. L. Law, K. T. Belt, L. E. Band, and G. T. Fisher. 2004. Nitrogen fluxes and retention in urban watershed ecosystems. *Ecosystems* 7:393–403.

Grosholz, E. D. 1992. Interactions of intraspecific, interspecific, and apparent competition with host-pathogen population dynamics. *Ecology* 73:507–14.

Gross, J. E., L. A. Shipley, N. T. Hobbs, D. E. Spalinger, and B. A. Wunder. 1993. Functional response of herbivores in food-concentrated patches: tests of a mechanistic model. *Ecology* 74:778–91.

Grutter, A. S. 1999. Cleaner fish really do clean. *Nature* 398:672–73.

Gurevitch, J., L. L. Morrow, A. Wallace, and J. S. Walsh. 1992. A meta-analysis of competition in field experiments. *The American Naturalist* 140(4):539–72.

Gunderson, D. R. 1997. Trade-off between reproductive effort and adult survival in oviparous and viviparous fishes. *Canadian Journal of Fisheries and Aquatic Sciences* 54:990–98.

Habeeb, D., J. Vargo, and B. Stone Jr. 2015. Rising heat wave trends in large US cities. *Natural Hazards* 76:1651–65.

Haddad, N. M. 1999. Corridor and distance effects on interpatch movements: a landscape experiment with butterflies. *Ecological Applications* 9:612–22.

Haddad, N. M., and K. A. Baum. 1999. An experimental test of corridor effects on butterfly densities. *Ecological Applications* 9:623–33.

Hadley, N. F., A. Savill, and T. D. Schultz. 1992. Coloration and its thermal consequences in the New Zealand tiger beetle *Neocicindela perhispida. Journal of Thermal Biology* 17:55–61.

Hadley, N. F., and T. D. Schultz. 1987. Water loss in three species of tiger beetles (*Cicindela*): correlations with epicuticular hydrocarbons. *Journal of Insect Physiology* 33:677–82.

Hairston, N.G., F. E. Smith, and L. B. Slobodkin. 1960. Community structure, population control, and competition. *American Naturalist.* 94:421–5.

Hairston, N. G., Sr. 1989. *Ecological Experiments: Purpose, Design, and Execution.* Cambridge, England: Cambridge University Press.

Hall, J. B., and M. D. Swaine. 1976. Classification and ecology of closed-canopy forest in Ghana. *Journal of Ecology* 64:913–51.

Hamilton, W. D. 1964. The genetical evolution of social behavior, I and II. *Journal of Theoretical Biology* 7:1–52.

Hamilton, W. J., III, and M. K. Seely. 1976. Fog basking by the Namib Desert beetle, *Onymacris unguicularis. Nature* 262:284–85.

Hanski, I. 1982. Dynamics of regional distribution: the core and satellite hypothesis. *Oikos* 38:210–21.

Hanski, I., M. Kuussaari, and M. Nieminen. 1994. Metapopulation structure and migration in the butterfly *Melitaea cinxia. Ecology* 75:747–62.

Hanson, J. M., and R. H. Peters. 1984. Empirical prediction of crustacean zooplankton biomass and profundal macrobenthos in lakes. *Canadian Journal of Fisheries and Aquatic Sciences* 41:439–45.

Hardie, K. 1985. The effect of removal of extraradical hyphae on water uptake by vesicular-arbuscular mycorrhizal plants. *New Phytologist* 101:677–84.

Hardy, G. H. 1908. Mendelian proportions in a mixed population. *Science* 28:49–50.

Harnik, P. G., C. Simpson, and J. L. Payne. 2012. Long-term differences in extinction risk among the seven forms of rarity. *Proceedings of the Royal Society B: Biological Sciences* 279(1749):4969–76.

Harper, J. L., P. H. Lovell, and K. G. Moore. 1970. The shapes and sizes of seeds. *Annual Review of Ecology and Systematics* 1:327–56.

Harrington, G. N. 1991. Effects of soil moisture on shrub seedling survival in a semi-arid grassland. *Ecology* 72:1138–49.

Haukioja, E., K. Kapiainen, P. Niemelä, and J. Tuomi. 1983. Plant availability hypothesis and other explanations of herbivore cycles: complementary or exclusive alternatives? *Oikos* 40:419–32.

Havens, K. E. 1994. A preliminary characterization of the Lake Okeechobee (Florida, USA) food web. *Bulletin of the North American Benthological Society* 11:97.

Hawn, A. T., A. N. Radford, and M A. du Plessis. 2007. Delayed breeding affects lifetime reproductive success differently in male and female green woodhoopoes. *Current Biology* 17:844–9.

Heap, I. 2007. International survey of herbicide-resistant weeds. www.weedscience.org.

Hedin, L. O., P. M. Vitousek, and P. A. Matson. 2003. Nutrient losses over four million years of tropical forest development. *Ecology* 84:2231–55.

Hegazy, A. K. 1990. Population ecology and implications for conservation of *Cleome droserifolia:* a threatened xerophyte. *Journal of Arid Environments* 19:269–82.

Helfrich, G. 2004. *Humboldt's Cosmos: Alexander von Humboldt and the Latin American Journey That Changed the Way We See the World.* New York: Gotham.

Hendry, A. P., S. K. Huber, L. F. de León, A. Herrel, and J. Podos. 2009. Disruptive selection in a bimodal population of Darwin's finches. *Philosophical transactions of the Royal Society B-Biological Sciences* 276:753–9.

Hengeveld, R. 1988. Mechanisms of biological invasions. *Journal of Biogeography* 15:819–28.

Hengeveld, R. 1989. *Dynamics of Biological Invasions.* New York: Chapman and Hall.

Herlihy, M. V., R. G. Van Driesche, M. R. Abney, J. Brodeur, A. B. Bryant, R. A. Casagrande, D. A. Delaney, T. E. Elkner, S. J. Fleischer, R. L. Groves, D. S. Gruner, J. P. Harmon, G. E. Heimpel, K. Hemady, T. P. Kuhar, C. M. Maund, A. M. Shelton, and Z. Seaman. 2012. Distribution of *Cotesia rubecula* (Hymenoptera: Braconidae) and its displacement of *Cotesia glomerata* in eastern North America. *Florida Entomologist* 95:461–7.

Heske, E. J., J. H. Brown, and S. Mistry. 1994. Long-term experimental study of a Chihuahuan Desert rodent community: 13 years of competition. *Ecology* 75:438–45.

Hickling, R. L., D. B. Roy, J. K. Hill, R. Fox, and C. D. Thomas. 2006. The distributions of a wide range of taxonomic groups are expanding polewards. *Global Change Biology* 12:450–5.

Hogetsu, K., and S. Ichimura. 1954. Studies on the biological production of Lake Suwa. 6. The ecological studies in the production of phytoplankton. *Japanese Journal of Botany* 14:280–303.

Hölldobler, B., and E. O. Wilson. 1990. *The Ants.* Cambridge, Mass.: The Belknap Press of Harvard University Press.

Holland, J. N., and D. L. DeAngelis. 2009. Consumer-resource theory predicts dynamic transitions between outcomes of interspecific interactions. *Ecology Letters* 12:1357–66.

Holling, C. S. 1959a. The components of predation as revealed by a study of small mammal predation of the European pine sawfly. *The Canadian Entomologist* 91:293–320.

Holling, C. S. 1959b. Some characteristics of simple types of predation and parasitism. *Canadian Entomologist* 91:385–98.

Holmes, R. T., J. C. Schultz, and P. Nothnagle. 1979. Bird predation on forest insects: an exclosure experiment. *Science* 206:462–63.

Horn, J. W., and T. H. Kunz. 2008. Analyzing NEXRAD doppler radar images to assess nightly dispersal patterns and population trends in Brazilian free-tailed bats (*Tadarida brasiliensis*). *Integrative and Comparative Biology* 48:24–39.

Horton, K. G., F. A. La Sorte, D. Sheldon, T. Y. Lin, K. Winner, G. Bernstein, and A. Farnsworth. 2020. Phenology of nocturnal avian migration has shifted at the continental scale. *Nature Climate Change* 10(1):63–8.

Houde, A. E. 1997. *Sex, Color, and Mate Choice in Guppies.* Princeton, N.J.: Princeton University Press.

Howe, W. H., and F. L. Knopf. 1991. On the imminent decline of the Rio Grande cottonwoods in central New Mexico. *Southwestern Naturalist* 36:218–24.

Huang, H. T., and P. Yang. 1987. The ancient cultured citrus ant. *BioScience* 37:665–7.

Hubbell, S. P., and L. K. Johnson. 1977. Competition and nest spacing in a tropical stingless bee community. *Ecology* 58:949–63.

Hudelson, K. E., D. C. Muir, P. E. Drevnick, G. Köck, D. Iqaluk, X. Wang, and A. T. Fisk. 2019. Temporal trends, lake-to-lake variation, and climate effects on Arctic char (*Salvelinus alpinus*) mercury concentrations from six High Arctic lakes in Nunavut, Canada. *Science of The Total Environment* 678:801–12.

Hudson, P. J., A. P. Dobson, and D. Newborn. 1992. Do parasites make prey vulnerable to predation? *Journal of Animal Ecology* 61:681–92.

Hudson, M. A. R., C. M. Francis, K. J. Campbell, C. M. Downes, A. C. Smith, and K. L. Pardieck. 2017. The role of the North American Breeding Bird Survey in conservation. *The Condor: Ornithological Applications* 119(3):526–45.

Huey, R. B., T. Garland Jr., and M. Turelli. 2019. Revisiting a key innovation in evolutionary biology: Felsenstein's "phylogenies and the comparative method". *The American Naturalist* 193(6):755–72.

Huffaker, C. B. 1958. Experimental studies on predation: dispersion factors and predator-prey oscillations. *Hilgardia* 27:343–83.

Hulshoff, R. M. 1995. Landscape indices describing a Dutch landscape. *Landscape Ecology* 10:101–11.

Huntly, N., and R. Inouye. 1988. Pocket gophers in ecosystems: patterns and mechanisms. *BioScience* 38:786–93.

Huston, M. 1980. Soil nutrients and tree species richness in Costa Rican forests. *Journal of Biogeography* 7:147–57.

Huston, M. 1994a. Biological diversity, soils, and economics. *Science* 262:1676–9.

Huston, M. 1994b. *Biological Diversity.* New York: Cambridge University Press.

Hutchinson, G. E. 1957. Concluding remarks. *Cold Spring Symposia on Quantitative Biology* 22:415–27.

Hutchinson, G. E. 1959. Homage to Santa Rosalia or why are there so many kinds of animals? *American Naturalist* 93:145–59.

Hutchinson, G. E. 1961. The paradox of the plankton. *American Naturalist* 95:137–45.

Ichimura, S. 1956. On the standing crop and productive structure of phytoplankton community in some lakes of central Japan. *Japanese Botany Magazine Tokyo* 69:7–16.

Inbar, M., A. Eshel, and D. Wool. 1995. Interspecific competition among phloem feeding insects mediated by induced host plant sinks. *Ecology* 76:1506–15.

Inouye, D. W. 2008. Effects of climate change on phenology, frost damage, and floral abundance of montane wildflowers. *Ecology* 89:353–62.

Inouye, D. W., and O. R. Taylor Jr. 1979. A temperate region plant-ant-seed predator system: consequences of extrafloral nectar secretion by *Helianthella quinquenervis*. *Ecology* 60:1–7.

Iriarte, J. A., W. L. Franklin, W. E. Johnson, and K. H. Redford. 1990. Biogeographic variation of food habits and body size of the American puma. *Oecologia* 85:185–90.

Isack, H. A., and H.-U. Reyer. 1989. Honeyguides and honey gatherers: interspecific communication in a symbiotic relationship. *Science* 243:1343–6.

IUCN. 2007. Red list of threatened species. www.iucnredlist.org/.

Jakobsson, A., and O. Eriksson. 2000. A comparative study of seed number, seed size, seedling size and recruitment in grassland plants. *Oikos* 88:494–502.

Janzen, D. H. 1966. Coevolution of mutualism between ants and acacias in Central America. *Evolution* 20:249–75.

Janzen, D. H. 1967a. Fire, vegetation structure, and the ant X acacia interaction in Central America. *Ecology* 48:26–35.

Janzen, D. H. 1967b. Interaction of the bull's-horn acacia (*Acacia cornigera* L.) with an ant inhabitant (*Pseudomyrmex ferruginea* F. Smith) in eastern Mexico. *The University of Kansas Science Bulletin* 47:315–558.

Janzen, D. H. 1967c. Why mountain passes are higher in the tropics. *American Naturalist* 101:233–49.

Janzen, D. H. 1981. The peak in North American ichneumonid species richness lies between 38° and 42°. *Ecology* 62:532–37.

Janzen, D. H. 1985. Natural history of mutualisms. In D. H. Boucher, ed. *The Biology of Mutualism: Ecology and Evolution.* London: Croom Helm.

Jarvis, J. U. M. 1981. Eusociality in a mammal: cooperative breeding in naked mole-rat colonies. *Science* 212:571–73.

Jenny, H. 1980. *The Soil Resource.* New York: Springer-Verlag.

Johnson, N. C. 1993. Can fertilization of soil select less mutualistic mycorrhizae. *Ecological Applications* 3:749–57.

Johnson, N. C., J. H. Graham, and F. A. Smith. 1997. Functioning of mycorrhizal associations along the mutualism–parasitism continuum. *The New Phytologist* 135:575–85.

Johnson, N. C., J. D. Hoeksema, J. D. Bever, V. B. Chaudhary, C. Gehring, J. Klironomos, R. Koide, R. M. Miller, J. Moore, P. Moutoglis, M. Schwartz, S. Simard, W. Swenson, J. Umbanhowar, G. Wilson, and C. Zabinski. 2006. From Lilliput to Brobdingnag: extending models of mycorrhizal function across scales. *BioScience* 56:889–900.

Johnson, N. C., D. L. Rowland, L. Corkidi, L. M. Egerton-Warburton, and E. B. Allen. 2003. Nitrogen enrichment alters mycorrhizal allocation at five mesic to semiarid grasslands. *Ecology* 84:1895–908.

Johnston, D. W., and E. P. Odum. 1956. Breeding bird populations in relation to plant succession on the Piedmont of Georgia. *Ecology* 37:50–62.

Jones, C. G., J. H. Lawton, and M. Shachak. 1994. Organisms as ecosystem engineers. *Oikos* 69:373–86.

Jordan, C. F. 1985. Soils of the Amazon rain forest. In G. T. Prance and T. E. Lovejoy, eds. *Amazonia.* Oxford: Pergamon Press.

Jost, C., G. Devulder, J. A. Vucetich, R. O. Peterson, and R. Arditi. 2005. The wolves of Isle Royale display scale-invariant satiation and ratio-dependent predation on moose. *Journal of Animal Ecology* 74:809–16.

Jouzel, J., V. Masson-Delmotte, O. Cattani, G. Dreyfus, S. Falourd, G. Hoffmann, B. Minster, J. Nouet, J. M. Barnola, J. Chappellaz, H. Fischer, J. C. Gallet, S. Johnson, M. Leuenberger, L. Loulergue, D. Luethi, H. Oerter, F. Parrenin, G. Raisbeck, D. Raynaud, A. Schilt, J. Schwander, E. Selmo, R. Souchez, R. Spahni, B. Stauffer, J. P. Steffensen, B. Stenni, T. F. Stocker, J. L. Tison, M. Werner, and E. W. Wolff. 2007. Orbital and millennial Antarctic climate variability over the past 800,000 years. *Science* 317:793–7.

Kairiukstis, L. A. 1967. In J. L. Tselniker (ed.) *Svetovoi rezhim fotosintez i produktivnost lesa.* (*Light Regime, Photosynthesis and Forest Productivity.*) Nauka, Moscow.

Kalka, M. B., A. R. Smith, and E. K. V. Kalko. 2008. Bats limit arthropods and herbivory in a tropical forest. *Science* 320:71.

Kallio, P., and L. Kärenlampi. 1975. Photosynthesis in mosses and lichens. In J. P. Cooper, ed. *Photosynthesis and Productivity in Different Environments.* Cambridge, England: Cambridge University Press.

Kane, J. M., and T. E. Kolb. 2010. Importance of resin ducts in reducing ponderosa pine mortality from bark beetle attack. *Oecologia* 164:601–9.

Karr, J. R., and D. R. Dudley. 1981. Ecological perspective on water quality goals. *Environmental Management* 5:55–68.

Kaspari, M., S. O'Donnell, and J. R. Kercher. 2000. Energy, density, and constraints to species richness: ant assemblages along a productivity gradient. *American Naturalist* 155:280–93.

Katz, C. H. 1985. A nonequilibrium marine predator-prey interaction. *Ecology* 66:1426–38.

Kauffman, J. B., R. L. Sanford Jr., D. L. Cummings, I. H. Salcedo, and E. V. S. B. Sampaio. 1993. Biomass and nutrient dynamics associated with slash fires in neotropical dry forests. *Ecology* 74:140–51.

Kauffman, M. J., J. F. Brodie, and E. S. Jules. 2010. Are wolves saving Yellowstone's aspen? A landscape-level test of a behaviorally mediated trophic cascade. *Ecology* 91:2742–55.

Kauffman, M. J., J. F. Brodie, and E. S. Jules. 2013. Are wolves saving Yellowstone's aspen? A landscape-level test of a behaviorally mediated trophic cascade: reply. *Ecology* 94:1425–31.

Keeler, K. H. 1981. A model of selection for facultative nonsymbiotic mutualism. *American Naturalist* 118:488–98.

Keeler, K. H. 1985. Benefit models of mutualism. In D. H. Boucher, ed. *The Biology of Mutualism: Ecology and Evolution.* London: Croom Helm.

Keeling, C. D., and T. P. Whorf. 1994. Atmospheric CO_2 records from sites in the SIO air sampling network. In T. A. Boden, D. P. Kaiser, R. J. Sepanski, and F. W. Stoss, eds. *Trends '93: A Compendium of Data on Global Change.* ORNL/ CDIAC-65. Oak Ridge, Tenn.: Carbon Dioxide Information Analysis Center, Oak Ridge National Laboratory.

C. D. Keeling, S. C. Piper, R. B. Bacastow, M. Wahlen, T. P. Whorf, M. Heimann, and H. A. Meijer. 2001. Exchanges of atmospheric CO_2 and $^{13}CO_2$ with the terrestrial biosphere and oceans from 1978 to 2000. I. Global aspects. *SIO Reference Series,* No. 01-06. San Diego: Scripps Institution of Oceanography, 88 pages.

Keever, C. 1950. Causes of succession on old fields of the Piedmont, North Carolina. *Ecological Monographs* 20:230–50.

Keith, L. B. 1963. *Wildlife's Ten-Year Cycle.* Madison, Wis.: University of Wisconsin Press.

Keith, L. B. 1983. Role of food in hare population cycles. *Oikos* 40:385–95.

Keith, L. B., J. R. Cary, O. J. Rongstad, and M. C. Brittingham. 1984. Demography and ecology of a declining snowshoe hare population. *Wildlife Monographs* 90:1–43.

Kelly, A. E., and M. L. Goulden. 2008. Rapid shifts in plant distribution with recent climate change. *Proceedings of the National Academy of Sciences of the United States* 105:11823–26.

Kempton, R. A. 1979. The structure of species abundance and measurement of diversity. *Biometrics* 35:307–21.

Kettlewell, H. B. D. 1959. Darwin's missing evidence. *Scientific American* 200:48–53.

Kevan, P. G. 1975. Sun-tracking solar furnaces in high arctic flowers: significance for pollination and insects. *Science* 189:723–6.

Kidron, G. J., E. Barzilay, and E. Sachs. 2000. Microclimate control upon sand microbiotic crusts, western Negev Desert, Israel. *Geomorphology* 36:1–18.

Killingbeck, K. T., and W. G. Whitford. 1996. High foliar nitrogen in desert shrubs: an important ecosystem trait or defective desert doctrine? *Ecology* 77:1728–37.

Klemmedson, J. O. 1975. Nitrogen and carbon regimes in an ecosystem of young dense ponderosa pine in Arizona. *Forest Science* 21: 163–8.

Knorr, W. 2009. Is the airborne fraction of anthropogenic CO_2 emissions increasing? *Geophysical Research Letters* 36.

Knutson, R. M. 1974. Heat production and temperature regulation in eastern skunk cabbage. *Science* 186:746–7.

Knutson, R. M. 1979. Plants in heat. *Natural History* 88:42–7.

Kodric-Brown, A. 1993. Female choice of multiple male criteria in guppies: interacting effects of dominance, coloration and courtship. *Behavioral Ecology and Sociobiology* 32:415–20.

Kolber, Z. S., C. L. Van Dover, R. A. Niederman, and P. G. Falkowski. 2000. Bacterial photosynthesis in surface waters of the open ocean. *Nature* 407:177–9.

Kontiainen, P., J. E. Brommer, P. Karell, and H. Pietiäinen. 2007. Heritability, plasticity and canalization of Ural owl egg size in a cyclic environment. *Journal of Evolutionary Biology* 21:88–96.

Korpimäki, E. 1988. Factors promoting polygyny in European birds of prey—a hypothesis. *Oecologia* 77:278–85.

Korpimäki, E., and K. Norrdahl. 1991. Numerical and functional responses of kestrels, short-eared owls, and long-eared owls to vole densities. *Ecology* 72:814–26.

Kramarova, N. A., E. R. Nash, P. A. Newman, P. K. Bhartia, R. D. McPeters, D. F. Rault, C. J. Seftor, P. Q. Xu, and G. J. Labow. 2014. Measuring the Antarctic ozone hole with the new Ozone Mapping and Profiler Suite (OMPS). *Atmospheric Chemistry and Physics* 14:2353–61.

Krebs, C. J., R. Boonstra, S. Boutin, and A. R. E. Sinclair. 2001. What drives the 10-year cycle of snowshoe hares? *BioScience* 51:25–36.

Krebs, C. J., S. Boutin, R. Boonstra, A. R. E. Sinclair, J. N. M. Smith, M. R. T. Dale, K. Martin, and R. Turkington. 1995. Impact of food and predation on the snowshoe hare cycle. *Science* 269:1112–5.

Kuhlbrodt, T., A. Griesel, M. Montoya, A. Levermann, M. Hofmann, and S. Rahmstorf. 2007. On the driving processes of the Atlantic meridional overturning circulation. *Reviews of Geophysics* 45:RG2001, doi:10.1029/2004RG000166

Kuhlbrodt, T., S. Rahmstorf, K. Zickfeld, F. B. Vikebø, S. Sundby, M. Hofmann, P. M. Link, A. Bondeau, W. Cramer, and C. Jaeger. 2009. An integrated assessment of changes in the thermohaline circulation. *Climatic Change* 96:489–537.

Kunte, K., W. Zhang, A. Tenger-Trolander, D. H. Palmer, A. Martin, R. D. Reed, S. P. Mullen, and M. R. Kronforst. 2014. Doublesex is a mimicry supergene. *Nature* 507:229–32.

Kunz, T. H., S. A. Gauthreaux Jr., N. I. Hristov, J. W. Horn, G. Jones, E. K. V. Kalko, R. P. Larkin, G. F. McCracken, S. M. Swartz, R. B. Srygley, R. Dudley, J. K. Westbrook, and M. Wikelski. 2008. Aeroecology: probing and modeling the aerosphere. *Integrative and Comparative Biology* 48:1–11.

Kuttippurath, J., F. Lefèvre, J.-P. Pommereau, H. K. Roscoe, F. Goutail, A. Pazmiño, and J. D. Shanklin. 2013. Antarctic ozone loss in 1979–2010: first sign of ozone recovery. *Atmospheric Chemistry and Physics* 13:1625–35.

Lack, D. 1947. *Darwin's Finches.* Cambridge, England: Cambridge University Press.

Lamberti, G. A., and V. H. Resh. 1983. Stream periphyton and insect herbivores: an experimental study of grazing by a caddisfly population. *Ecology* 64:1124–35.

Larcher, W. 1995. *Physiological Plant Ecology.* 3rd ed. Berlin: Springer.

Latham, R. E., and R. E. Ricklefs. 1993. Continental comparisons of temperate-zone tree species diversity. In R. E. Ricklefs and D. Schluter, eds. *Species Diversity in Ecological Communities.* Chicago: University of Chicago Press.

Latombe, G., D. Fortin, and L. Parrott. 2014. Spatio-temporal dynamics in the response of woodland caribou and moose to the passage of grey wolf. *Journal of Animal Ecology* 83:185–98.

Laundré, J. W. 2010. Behavioral response races, predator-prey shell games, ecology of fear, and patch use of pumas and their ungulate prey. *Ecology* 91:2995–3007.

Lavorel, S., and E. Garnier. 2002. Predicting changes in community composition and ecosystem functioning from plant traits: revisiting the Holy Grail. *Functional Ecology* 16:545–56.

Lawlor, T. E. 1998. Biogeography of Great Basin mammals: paradigm lost? *Journal of Mammalogy* 79:1111–30.

Lawton, J. H., D. E. Bignell, B. Bolton, G. F. Bloemers, P. Eggleton, P. M. Hammond, M. Hodda, R. D. Holt, T. B. Larsen, N. A. Mawdsley, N. E. Stork, D. S. Srivastava, and A. D. Watt. 1998. Biodiversity inventories, indicator taxa and effects of habitat modification in tropical forest. *Nature* 391:72–6.

Le Boeuf, B. J., and R. M. Laws. 1994. *Elephant Seals: Population Ecology, Behavior, and Physiology.* Berkeley: University of California Press.

Lebo, M. E., J. E. Reuter, C. R. Goldman, C. L. Rhodes, N. Vucinich, and D. Mosely. 1993. Spatial variations in nutrient and particulate matter concentrations in Pyramid Lake, Nevada, USA, during a dry period. *Canadian Journal of Fisheries and Aquatic Science* 50:1045–54.

Leonard, P. M., and D. J. Orth. 1986. Application and testing of an index of biotic integrity in small, coolwater streams. *Transactions of the American Fisheries Society* 115:401–14.

Levang-Brilz, N., and M. E. Biondini. 2002. Growth rate, root development and nutrient uptake of 55 plant species from the Great Plains Grasslands, USA. *Plant Ecology* 165:117–44.

Leverich, W. J., and D. A. Levin. 1979. Age-specific survivorship and reproduction in *Phlox drummondii. American Naturalist* 113:881–903.

Levins, R. 1968. *Evolution in Changing Environments.* Princeton, N.J.: Princeton University Press.

Liebig, J. 1840. *Chemistry in Its Application to Agriculture and Physiology.* London: Taylor and Walton.

Ligon, D. 1999. *The Evolution of Avian Breeding Systems.* Oxford: Oxford University Press.

Ligon, J. D., and S. H. Ligon. 1978. Communal breeding in green woodhoopoes as a case for reciprocity. *Nature* (London) 276:496–8.

Ligon, J. D., and S. H. Ligon. 1982. The cooperative breeding behavior of the green wood hoopoe. *Scientific American* 247:126–34.

Ligon, J. D., and S. H. Ligon. 1989. Green woodhoopoe. In I. Newton, ed. *Lifetime Reproduction in Birds.* London: Academic Press Ltd.

Ligon, J. D., and S. H. Ligon. 1991. Green woodhoopoe: life history traits and sociality. In P. B. Stacey and W. D. Koenig, eds. *Cooperative Breeding in Birds: Long-Term Studies of Behavior and Ecology.* Cambridge, England: Cambridge University Press.

Likens, G. E., and F. H. Bormann. 1995. *Biogeochemistry of a Forested Ecosystem.* 2nd ed. New York: Springer-Verlag.

Likens, G. E., F. H. Bormann, N. M. Johnson, D. W. Fisher, and R. S. Pierce. 1970. Effects of forest cutting and herbicide treatment on nutrient budgets in the Hubbard Brook watershed-ecosystem. *Ecological Monographs* 40:23–47.

Lilleskov, E. A., T. J. Fahey, T. R. Horton, and G. M. Lovett. 2002. Belowground ectomycorrhizal fungal community change over a nitrogen deposition gradient in Alaska. *Ecology* 83:104-15.

Lindeman, R. L. 1942. The trophic-dynamic aspect of ecology. *Ecology* 23:399-418.

Lindström, E. R., H. Andrén, P. Angelstam, G. Cederlund, B. Hörnfeldt, L. Jäderberg, P. A. Lemnell, B. Martinsson, K. Sköld, and J. E. Swenson. 1994. Disease reveals the predator: sarcoptic mange, red fox predation, and prey populations. *Ecology* 75:1042-9.

Linklater, W. L. 2004. Wanted for conservation research: behavioral ecologists with a broader perspective. *BioScience* 54:352-60.

Lloyd, R. A., K. A. Lohse, and T. P. A. Ferré. 2013. Influence of road reclamation techniques on forest ecosystem recovery. *Frontiers in Ecology and the Environment* 11:75-81.

Lomolino, M. V. 1990. The target hypothesis—the influence of island area on immigration rates of non-volant mammals. *Oikos* 57:297-300.

Lomolino, M. V., J. H. Brown, and R. Davis. 1989. Island biogeography of montane forest mammals in the American Southwest. *Ecology* 70:180-94.

Long, S. P., and C. F. Mason. 1983. *Saltmarsh Ecology.* Glasgow: Blackie.

Lorius, C., J. Jouzel, C. Ritz, L. Merlivat, N. I. Barkov, Y. S. Korotkevich, and V. M. Kotlyakov. 1985. A 150,000-year climatic record from antarctic ice. *Nature* 316:591-96.

Lotka, A. J. 1925. *Elements of Physical Biology.* Baltimore, Md.: Williams and Wilkins.

Lotka, A. J. 1932a. Contribution to the mathematical theory of capture. I. conditions for capture. *Proceedings of the National Academy of Sciences* 18:172-200.

Lotka, A. J. 1932b. The growth of mixed populations: two species competing for a common food supply. *Journal of the Washington Academy of Sciences* 22:461-9.

Lubchenco, J. 1978. Plant species diversity in a marine intertidal community: importance of herbivore food preference and algal competitive abilities. *American Naturalist* 112:23-39.

Luber, G., and M. McGeehin. 2008. Climate change and extreme heat events. *American Journal of Preventative Medicine* 35:429-35.

Lüthi, D., M. Le Floch, B. Bereiter, T. Blunier, J.-M. Barnola, U. Siegenthaler, D. Raynaud, J. Jouzel, H. Fischer, K. Kawamura, and T. F. Stocker. 2008. High-resolution carbon dioxide concentration record 650,000-800,000 years before present. *Nature* 453:379-82.

MacArthur, R. H. 1972. *Geographical Ecology.* New York: Harper & Row.

MacArthur, R. H., and J. W. MacArthur. 1961. On bird species diversity. *Ecology* 42:594-8.

MacArthur, R. H., and E. R. Pianka. 1966. On optimal use of a patchy environment. *American Naturalist* 100:603-9.

MacArthur, R. H., and E. O. Wilson. 1963. An equilibrium theory of insular zoogeography. *Evolution* 17:373-87.

MacArthur, R. H., and E. O. Wilson. 1967. *The Theory of Island Biogeography.* Princeton, N.J.: Princeton University Press.

MacLulich, D. A. 1937. Fluctuation in the numbers of the varying hare (*Lepus americanus*). *University of Toronto Studies in Biology Series No. 43.*

Mandelbrot, B. 1982. *The Fractal Geometry of Nature.* New York: W. H. Freeman.

Margulis, L., M. Chapman, R. Guerrero, and J. Hall. 2006. The last eukaryotic common ancestor (LECA): acquisition of cytoskeletal motility from aerotolerant spirochetes in the Proterozoic Eon. *Proceedings of the National Academy of Sciences of the United States of America* 103:13080-5.

Margulis, L., and R. Fester. 1991. *Symbiosis as a Source of Evolutionary Innovation: Speciation and Morphogenesis.* Cambridge, Mass.: MIT Press.

Marquis, R. J., and C. J. Whelan. 1994. Insectivorous birds increase growth of white oak through consumption of leaf-chewing insects. *Ecology* 75:2007-14.

Marshall, D. L. 1990. Non-random mating in a wild radish, *Raphanus sativus. Plant Species Biology* 5:143-56.

Marshall, D. L., and M. W. Folsom. 1991. Mate choice in plants: an anatomical to population perspective. *Annual Review of Ecology and Systematics* 22:37-63.

Marshall, D. L., and O. S. Fuller. 1994. Does nonrandom mating among wild radish plants occur in the field as well as in the greenhouse? *American Journal of Botany* 81:439-45.

Martikainen, P., and J. Kouki. 2003. Sampling the rarest: threatened beetles in boreal forest biodiversity inventories. *Biodiversity and Conservation* 12:1815-31.

Martinsen, G. D., E. M. Driebe, and T. G. Whitham. 1998. Indirect interactions mediated by changing plant chemistry: beaver browsing benefits beetles. *Ecology* 79:192-200.

Mathews, F., M. Orros, G. McLaren, M. Gelling, and R. Foster. 2005. Keeping fit on the ark: assessing the suitability of captive-bred animals for release. *Biological Conservation* 121:569-77.

Matter, S. F., M. Ezzeddine, E. Duermit, J. Mashburn, R. Hamilton, T. Lucas, and J. Roland. 2009. Interactions between habitat quality and connectivity affect immigration but not abundance or population growth of the butterfly, *Parnassius smintheus. Oikos* 118:1461-70.

May, R. M. 1989. Honeyguides and humans. *Nature* 338:707-8.

McAuliffe, J. R. 1994. Landscape evolution, soil formation, and ecological patterns and processes in Sonoran Desert bajadas. *Ecological Monographs* 64:111-48.

McKinney, M. L. 2002. Urbanization, biodiversity, and conservation. *BioScience* 52:883-90.

McLachlan, A., and A. Dorvlo. 2005. Global patterns in sandy beach macrobenthic communities. *Journal of Coastal Research* 21:674-87.

McNaughton, S. J. 1976. Serengeti migratory wildebeest: facilitation of energy flow by grazing. *Science* 191:92-4.

McNaughton, S. J. 1985. Ecology of a grazing ecosystem: the Serengeti. *Ecological Monographs* 55:259-94.

McNaughton, S. J., M. Oesterheld, D. A. Frank, and K. J. Williams. 1989. Ecosystem level patterns of primary productivity and herbivory in terrestrial habitats. *Science* 341:142-4.

McNaughton, S. J., R. W. Ruess, and S. W. Seagle. 1988. Large mammals and process dynamics in African ecosystems. *BioScience* 38:794-800.

Meehl, G. A., and C. Tabaldi. 2004. More intense, more frequent, and longer lasting heat waves in the 21st century. *Science* 305:994-7.

Meentemeyer, V. 1978. An approach to the biometeorology of decomposer organisms. *International Journal of Biometeorology* 22:94-102.

Meinhardt, K., and C. Gehring. 2013. Tamarix and soil ecology. In A. Sher and M. F. Quigley, eds. *Tamarix: A Case Study of Ecological Change in the American West.* Oxford: Oxford University Press.

Melillo, J. M., J. D. Aber, and J. F. Muratore. 1982. Nitrogen and lignin control of hardwood leaf litter decomposition dynamics. *Ecology* 63:621-6.

Mendel, G. 1866. Versuche über Pflanzen-Hybriden (Experiments in plant hybridization). *Verhandlungen des Naturforschenden Vereines, Abhandlungen, Brünn* 4:3-47.

Menzel, A., T. H. Sparks, N. Estrella, E. Koch, A. Aasaa, R. Ahas, K. Alm-Kübler, P. Bissolli, O. Braslavavska, A. Briede, F. M. Chmielewski, Z. Crepinsek, Y. Curnel, Å. Dahl, C. Defila, A. Donnelly, Y. Filella, K. Jatczak, F. Måge, A. Mestre, Ø. Nordli, J. Peñuelas, P. Pirinen, V. Remišová, H. Scheifinger, M Striz, A. Susnik, A. J. H. van Vliet, F.-E. Wielgolaski, S. Zach, and A. Zust. 2006. European phenological response to climate change matches the warming pattern. *Global Change Biology* 12:1969-76.

Mertz, D. B. 1972. The *Tribolium* model and the mathematics of population growth. *Annual Review of Ecology and Systematics* 3:51-106.

Messier, F. 1994. Ungulate population models with predation: a case study with the North American moose. *Ecology* 75:478-88.

Meybeck, M. 1982. Carbon, nitrogen, and phosphorus transport by world rivers. *American Journal of Science* 282:401-50.

Meyer, J. L., and G. E. Likens. 1979. Transport and transformation of phosphorus in a forest stream ecosystem. *Ecology* 60:1255-69.

Miller, R. B. 1923. First report on a forestry survey of Illinois. *Illinois Natural History Bulletin* 14:291-377.

Miller-Rushing, A. J., and R. B. Primack. 2008. Global warming and flowering times inThoreau's Concord: a community perspective. *Ecology* 89:332-41.

Mills, E. L., J. H. Leach, J. T. Carlton, and C. L. Secor. 1994. Exotic species and the integrity of the Great Lakes. *BioScience* 44:666-76.

Mills, K. H., and D. W. Schindler. 1987. Preface. *Canadian Journal of Fisheries and Aquatic Sciences* 44 (Suppl. 1):3-5.

Milne, B. T. 1993. Pattern analysis for landscape evaluation and characterization. In M. E. Jensen and P. S. Bourgeron, eds. *Ecosystem Management: Principles and Applications.* Gen. Tech. Report PNW-GTR-318. Portland, Ore.: U.S. Department of Agriculture Forest Service, Pacific Northwest Research Station.

Milton, R. C. 1964. An extended table of critical values for the Mann-Whitney (Wilcoxon) two-sample statistical. *Journal of the American Statistical Association* 59:925-34.

Minnich, R. A. 1983. Fire mosaics in southern California and northern Baja California. *Science* 219:1287-94.

Mitter, C., R. W. Poole, and M. Matthews. 1993. Biosystematics of the Heliothinae (Lepidoptera: Noctuidae). *Annual Review of Entomology* 38:207-25.

Moles, A. T., D. D. Ackerly, C. O. Webb, J. C. Tweddle, J. B. Dickie, A. J. Pitman, and M. Westoby. 2005a. Factors that shape seed mass evolution. *Proceedings of the National Academy of Sciences of the United States of America* 102:10540-44.

Moles, A. T., D. D. Ackerly, C. O. Webb, J. C. Tweddle, J. B. Dickie, and M. Westoby. 2005b. A brief history of seed size. *Science* 307:576-80.

Molles, M. C., Jr. 1978. Fish species diversity on model and natural reef patches: experimental insular biogeography. *Ecological Monographs* 48:289-305.

Mooney, H. A. 1972. The carbon balance of plants. *Annual Review of Ecology and Systematics* 3:139-45.

Moore, J. 1983. Responses of an avian predator and its isopod prey to an acanthocephalan parasite. *Ecology* 64:1000-15.

Moore, J. 1984a. Altered behavioral responses in intermediate hosts—an acanthocephalan parasite strategy. *American Naturalist* 123:572-77.

Moore, J. 1984b. Parasites that change the behavior of their host. *Scientific American* 250:108-15.

Moran, P. A. P. 1949. The statistical analysis of the sunspot and lynx cycles. *Journal of Animal Ecology* 18:115-6.

Morita, R. Y. 1975. Psychrophilic bacteria. *Bacteriological Reviews* 39:144–67.

Mosser, J. L., A. G. Mosser, and T. D. Brock. 1974. Population ecology of *Sulfolobus acidocaldarius*. I. temperature strains. *Archives for Microbiology* 97:169–79.

Muir, J. 1915. *Travels in Alaska*. Boston: Houghton Mifflin.

Müller, K. 1954. Investigations on the organic drift in north Swedish streams. *Reports of the Institute of Freshwater Research of Drottningholm* 35:133–48.

Müller, K. 1974. Stream drift as a chronobiological phenomenon in running water ecosystems. *Annual Review of Ecology and Systematics* 5:309–23.

Munger, J. C., and J. H. Brown. 1981. Competition in desert rodents: an experiment with semipermeable exclosures. *Science* 211:510–2.

Murie, A. 1944. The wolves of Mount McKinley. *Fauna of the National Parks of the U.S., Fauna Series No. 5*. Washington, D.C.: U.S. Department of the Interior, National Park Service.

Murphy, S. M., S. M. Leahy, L. S. Williams, and J. T. Lill. 2010. Stinging spines protect slug caterpillars (Limacodidae) from multiple generalist predators. *Behavioral Ecology* 21:153–60.

Murphy, S. M., T. M. Stoepler, K. Grenis, and J. T. Lill. 2014. Host ontogeny determines parasitoid use of a forest caterpillar. *Entomologia Experimentalis et Applicata* 150:217–25.

Muscatine, L., and C. F. D'Elia. 1978. The uptake, retention, and release of ammonium by reef corals. *Limnology and Oceanography* 23:725–34.

Naiman, R. J., R. E. Bilby, D. E. Schindler, and J. M. Helfield. 2002. Pacific salmon, nutrients, and the dynamics of freshwater and riparian ecosystems. *Ecosystems* 5:399–417.

Naiman, R. J., G. Pinay, C. A. Johnston, and J. Pastor. 1994. Beaver influences on the long-term biogeochemical characteristics of boreal forest drainage networks. *Ecology* 75:905–21.

NASA. 2017. http://ozonewatch.gsfc.nasa.gov/meteorology/annual_data.html.

NASA's Goddard Institute for Space Studies (GISS). 2017. https://climate.nasa.gov/vital-signs/global-temperature.

National Academies of Sciences, Engineering, and Medicine. 2016. *Attribution of Extreme Weather Events in the Context of Climate Change*. Washington, D.C.: The National Academies Press.

National Research Council. 2013. *Abrupt Impacts of Climate Change: Anticipating Surprises*. Washington, D.C.: The National Academies Press.

Neftel, A., H. Friedli, E. Moor, H. Lötscher, H. Oeschger, U. Siegenthaler, and B. Stauffer. 1994. Historical CO_2 record from the Siple Station ice core. In *Trends: A Compendium of Data on Global Change, 1994*. Carbon Dioxide Information Analysis Center, Oak Ridge National Laboratory, U.S. Department of Energy, Oak Ridge, Tenn., U.S.A.

Newbold, J. D., J. W. Elwood, R. V. O'Neill, and A. L. Sheldon. 1983. Phosphorus dynamics in a woodland stream ecosystem: a study of nutrient spiraling. *Ecology* 64:1249–65.

Newchurch, M. J., E.-S. Yang, D. M. Cunnold, G. C. Reinsel, J. M. Zawodny, and J. M. Russell III. 2003. Evidence for slowdown in stratospheric ozone loss: first stage of ozone recovery. *Journal of Geophysical Research* 108(D16), 4507, doi:10.1029/2003JD003471, 2003 (published online).

Nicholls, N. 1992. Historical El Niño/Southern Oscillation variability in the Australasian region. In H. F. Diaz and V. Markgraf, eds. *El Niño Historical and Paleoclimatic Aspects of the Southern Oscillation*. Cambridge, England: Cambridge University Press.

Nilsson, S. G., J. Bengtsson, and S. Ås. 1988. Habitat diversity or area *per se*? Species richness of woody plants, carabid beetles and land snails on islands. *Journal of Animal Ecology* 57:685–704.

NOAA. 2020. National Oceanic & Atmospheric Administration, U.S. Department of Commerce. cpc.ncep.noaa.gov

NOAA. 2015. National Oceanic & Atmospheric Administration, U.S. Department of Commerce. www.cpc.ncep.noaa.gov/products/analysis_monitoring/regional_monitoring/palmer/2015/11-28-2015.gif.

NOAA. 2017. National Oceanic & Atmospheric Administration, U.S. Department of Commerce. www.ncdc.noaa.gov/cag/time-series/global/globe/land_ocean/ytd/12/1880-2017?trend=true&trend_base=10&firsttrendyear=1880&lasttrendyear=2017.

Nobel, P. S. 1977. Internal leaf area and cellular CO_2 resistance: photosynthetic implications of variations with growth conditions and plant species. *Physiologia Plantarum* 40:137–44.

Nowak, M. A., C. E. Tarnita, and E. O. Wilson. 2010. The evolution of eusociality. *Nature* 437:1291–8.

Ødegaard, F. 2006. Host specificity, alpha- and beta-diversity of phytophagous beetles in two tropical forests in Panama. *Biodiversity and Conservation* 15:83–105.

O'Donoghue, M., S. Boutin, C. J. Krebs, and E. J. Hofer. 1997. Numerical responses of coyotes and lynx to the snowshoe hare cycle. *Oikos* 80:150–62.

O'Donoghue, M., S. Boutin, C. J. Krebs, G. Zuleta, D. L. Murray, and E. J. Hofer. 1998. Functional responses of coyotes and lynx to the snowshoe hare cycle. *Ecology* 79:1193–208.

Ohrtman, M., and K. D. Lair. 2013. Tamarisk and Salinity: an overview. In A. Sher and M. F. Quigley, eds. *Tamarix: A Case Study of Ecological Change in the American West*. Oxford: Oxford University Press.

Oosting, H. J. 1942. An ecological analysis of the plant communities of Piedmont, North Carolina. *The American Midland Naturalist* 28:1–126.

Orrock, J. L., M. S. Witter, and O. J. Reichman. 2008. Apparent competition with an exotic plant reduces native plant establishment. *Ecology* 89:1168–74.

Packer, C., and A. E. Pusey. 1982. Cooperation and competition within coalitions of male lions: kin selection or game-theory? *Nature* 296:740–2.

Packer, C., and A. E. Pusey. 1983. Cooperation and competition in lions: reply. *Nature* 302:356.

Packer, C., and A. E. Pusey. 1997. Divided we fall: cooperation among lions. *Scientific American* 276:52–9.

Packer, C., D. A. Gilbert, A. E. Pusey, and S. J. O'Brien. 1991. A molecular genetic analysis of kinship and cooperation in African lions. *Nature* 351:562–5.

Paine, R. T. 1966. Food web complexity and species diversity. *American Naturalist* 100:65–75.

Paine, R. T. 1969. A note on trophic complexity and community stability. *American Naturalist* 103:91–3.

Paine, R. T. 1971. A short-term experimental investigation of resource partitioning in a New Zealand rocky intertidal habitat. *Ecology* 52:1096–106.

Paine, R. T. 1980. Food webs: linkage, interaction strength and community infrastructure. *Journal of Animal Ecology* 49:667–85.

Pandolfi, J. M., S. R. Connolly, D. J. Marshall, and A. L. Cohen. 2011. Projecting coral reef futures under global warming and ocean acidification. *Science* 333:418–22.

Pappers, S. M., G. van der Velde, N. J. Ouborg, and J. M. van Groenendael. 2002. Genetically based polymorphisms in morphology and life history associated with putative host races of the water lily leaf beetle *Galerucella nymphaeae*. *Evolution* 56:1610–21.

Paquette, A., and C. Messier. 2011. The effect of biodiversity on tree productivity: from temperate to boreal forests. *Global Ecology and Biogeography* 20:170–80.

Park, T. 1948. Experimental studies of interspecific competition. I. competition between populations of flour beetles *Tribolium confusum* Duval and *Tribolium castaneum* Herbst. *Ecological Monographs* 18:267–307.

Park, T. 1954. Experimental studies of interspecific competition. II. temperature, humidity and competition in two species of *Tribolium*. *Physiological Zoology* 27:177–238.

Park, T., D. B. Mertz, W. Grodzinski, and T. Prus. 1965. Cannibalistic predation in populations of flour beetles. *Physiological Zoology* 38:289–321.

Park, Y.-M. 1990. Effects of drought on two grass species with different distribution around coastal sand dunes. *Functional Ecology* 4:735–41.

Parmenter, R. R., and V. A. Lamarra. 1991. Nutrient cycling in a freshwater marsh: the decomposition of fish and waterfowl carrion. *Limnology and Oceanography* 36:976–87.

Parmenter, R. R., C. A. Parmenter, and C. D. Cheney. 1989. Factors influencing microhabitat partitioning among coexisting species of arid-land darkling beetles (Tenebrionidae): behavioral responses to vegetation architecture. *The Southwestern Naturalist* 34:319–29.

Parmesan, C., and G. Yohe. 2003. A globally coherent fingerprint of climate change impacts across natural systems. *Nature* 421:37–42.

Parolo, G., and G. Rossi. 2008. Upward migration of vascular plants following a climate warming trend in the Alps. *Basic and Applied Ecology* 9:100–107.

Pascoal, S., T. Cezard, A. Eik-Nes, K. Gharbi, J. Majewska, E. Payne, M. G. Richey, M. Zuk, and N. W. Bailey. 2014. Rapid convergent evolution in wild crickets. *Current Biology* 24:1369–74.

Pearcy, R. W. 1977. Acclimation of photosynthetic and respiratory carbon dioxide exchange to growth temperature in *Atriplex lentiformis* (Torr.) Wats. *Plant Physiology* 59:795–9.

Pearcy, R. W., and A. T. Harrison. 1974. Comparative photosynthetic and respiratory gas exchange characteristics of *Atriplex lentiformis* (Torr.) Wats. in coastal and desert habitats. *Ecology* 55:1104–11.

Peierls, B. L., N. F. Caraco, M. L. Pace, and J. J. Cole. 1991. Human influence on river nitrogen. *Nature* 350:386–7.

Perry, M. J. 1986. Assessing marine primary production from space. *BioScience* 36:461–7.

Peters, R. H., and K. Wassenberg. 1983. The effect of body size on animal abundance. *Oecologia* 60:89–96.

Peterson, B. J., R. W. Howarth, and R. H. Garritt. 1985. Multiple stable isotopes used to trace the flow of organic matter in estuarine food webs. *Science* 227:1361–3.

Phillips, D. L., and J. A. MacMahon. 1981. Competition and spacing patterns in desert shrubs. *Journal of Ecology* 69:97–115.

Pianka, E. R. 1970. On *r*- and *K*-selection. *American Naturalist* 104:592–7.

Pianka, E. R. 1972. *r*- and *K*-selection or *b* and *d* selection. *American Naturalist* 106:581–8.

Pickett, S. T. A., M. L. Cadenasso, J. M. Grove, P. M. Groffman, L. E. Band, C. G. Boone, W. R. Burch Jr., C. S. B. Grimmond, J. Hom, J. C. Jenkins, N. L. Law, C. H. Nilon, R. V. Pouyat, K. Szlavecz, P. S. Warren, and M. A. Wilson. 2008.

Beyond urban legends: an emerging framework of urban ecology, as illustrated by the Baltimore Ecosystem Study. *BioScience* 58:139–50.

Pickett, S. T. A., M. L. Cadenasso, and S. J. Meiners. 2009. Ever since Clements: from succession to vegetation dynamics and understanding to intervention. *Applied Vegetation Science* 12:9–21.

Podos, J. 2010. Acoustic discrimination of sympatric morphs in Darwin's finches: a behavioural mechanism for assortative mating? *Philosophical Transactions of the Royal Society B: Biological Sciences* 365:1031–9.

Post, W. M., T.-H. Peng, W. R. Emanuel, A. W. King, V. H. Dale, and D. L. DeAngelis. 1990. The global carbon cycle. *American Scientist* 78:310–26.

Power, M. E. 1990. Effects of fish on river food webs. *Science* 250:811–4.

Power, M. E., D. Tilman, J. A. Estes, B. A. Menge, W. J. Bond, L. S. Mills, G. Daily, J. C. Castilla, J. Lubchenco, and R. T. Paine. 1996. Challenges in the quest for keystones. *BioScience* 46:609–20.

Preston, F. W. 1948. The commonness, and rarity, of species. *Ecology* 29:254–83.

Preston, F. W. 1962a. The canonical distribution of commonness and rarity: part I. *Ecology* 43:185–215.

Preston, F. W. 1962b. The canonical distribution of commonness and rarity: part II. *Ecology* 43:410–32.

Rabinowitz, D. 1981. Seven forms of rarity. In H. Synge, ed. *The Biological Aspects of Rare Plant Conservation*. New York: John Wiley & Sons.

Rahmstorf, S. 2002. Ocean circulation and climate during the past 120,000 years. *Nature* 419:207–214.

Raine, N. E., P. Willmer, and G. N. Stone. 2002. Spatial structuring and floral avoidance behaviour prevent ant-pollinator conflict in a Mexican ant-acacia. *Ecology* 83:3086–96.

Ralph, C. J. 1985. Habitat association patterns of forest and steppe birds of northern Patagonia, Argentina. *The Condor* 87:471–83.

Rasmusson, E. M. 1985. El Niño and variations in climate. *American Scientist* 73:168–77.

Ratnieks, F. L., K. R. Foster, and T. Wenseleers. 2011. Darwin's special difficulty: the evolution of "neuter insects" and current theory. *Behavioral Ecology and Sociobiology* 65:481–92.

Reichard, J. D., S. I. Prajapati, S. N. Austad, C. Keller, and T. H. Kunz. 2010. Thermal windows on Brazilian free-tailed bats facilitate thermoregulation during prolonged flight. *Integrative and Comparative Biology* 50:358–70.

Reid, W. V., and K. R. Miller. 1989. *Keeping Options Alive: The Scientific Basis for Conserving Biodiversity*. Washington, D.C.: World Resources Institute.

Reiners, W. A., I. A. Worley, and D. B. Lawrence. 1971. Plant diversity in a chronosequence at Glacier Bay, Alaska. *Ecology* 52:55–69.

Requena Suarez, D., Rozendaal, D. M., de Sy, V., Phillips, O. L., Alvarez-Dávila, E., Anderson-Teixeira, K., Araujo-Murakami, A., Arroyo, L., Baker, T. R., Bongers, F., Brienen, R. J. W., Carter, S., Cook-Patton, S. C., Feldpausch, T. R., Griscom, B. W., Harris, N., Hérault, B., Honorio Coronado, E. N., Leavitt, S. M., Lewis, S. L., Marimon, B. S., Monteagudo Mendoza, A., N'dja, J. K., N'Guessan, A. E., Poorter, L., Qie, L., Rutishauser, E., Sist, P., Sonké, B., Sullivan, M. J., Vilanova, E., Wang, M. M., Martius, C., and Herold, M. 2019. Estimating aboveground net biomass change for tropical and subtropical forests: Refinement of IPCC default rates using forest plot data. *Global Change Biology* 25(11): 3609–24.

Revelle, R., and H. E. Suess. 1957. Carbon dioxide exchange between atmosphere and ocean and the question of an increase of atmospheric CO_2 during the past decades. *Tellus* 9:18–27.

Ricciardi, A., and F. G. Whoriskey. 2004. Exotic species replacement: shifting dominance of dreissenid mussels in the Soulanges Canal, upper St. Lawrence River, Canada. *Journal of the North American Benthological Society* 23:507–14.

Richey, J. E. 1983. The phosphorus cycle. In B. Bolin and R. B. Cook, eds. *The Major Biogeochemical Cycles and Their Interaction*. New York: John Wiley & Sons.

Ricklefs, R. E. 1987. Community diversity: relative roles of local and regional processes. *Science* 235:167–71.

Ripple, W. J., and R. L. Beschta. 2004. Wolves and the ecology of fear: can predation risk structure ecosystems? *BioScience* 54:755–66.

Ripple, W. J., and R. L. Beschta. 2007. Restoring Yellowstone's aspen with wolves. *Biological Conservation* 138:514–9.

Risch, S. J., and C. R. Carroll. 1982. Effect of a keystone predaceous ant, *Solenopsis geminata*, on arthropods in a tropical agroecosystem. *Ecology* 63:1979–83.

Robertson, D. S., M. C. McKenna, O. B. Toon, S. Hope, and J. A. Lillegraven. 2004. Survival in the first hours of the Cenozoic. *Geological Society of America Bulletin* 116:760–8.

Robertson, G. P., M. A. Huston, F. C. Evans, and J. M. Tiedje. 1988. Spatial variability in a successional plant community: patterns of nitrogen availability. *Ecology* 69:1517–24.

Robine, J. M., S. L. K. Cheung, S. Le Roy, H. Van Oyen, C. Griffiths, J. P. Michel, and F. R. Herrmann. 2008. Death toll exceeded 70,000 in Europe during the summer of 2003. *Comptes Rendus Biologies* 33:171–8.

Rohlf, F. J., and R. R. Sokal. 1995. *Statistical Tables*. 3rd ed. San Francisco: W. H. Freeman and Co.

Roland, J., N. Keyghobadi, and S. Fownes. 2000. Alpine Parnassius butterfly dispersal: effects of landscape and population size. *Ecology* 81:1642–53.

Roller, N. E. G., and J. E. Colwell. 1986. Coarse-resolution satellite data for ecological surveys. *BioScience* 36:468–75.

Root, T. 1988. *Atlas of Wintering North American Birds*. Chicago: University of Chicago Press.

Rosemond, A. D., C. M. Pringle, A. Ramirez, M. J. Paul, and J. L. Meyer. 2002. Landscape variation in phosphorus concentration and effects on detritus-based tropical streams. *Limnology and Oceanography* 47:278–89.

Rosenzweig, M. L. 1968. Net primary productivity of terrestrial environments: predictions from climatological data. *American Naturalist* 102:67–84.

Rosenzweig, M. L. 1992. Species diversity gradients: we know more and less than we thought. *Journal of Mammalogy* 73:715–30.

Rowland, D., A. A. Sher, and D. L. Marshall. 2004. Cottonwood population response to salinity. *Canadian Journal of Forest Research* 34:1458–66.

Roy, B. A. 1993. Floral mimicry by a plant pathogen. *Nature* 362:56–8.

Ruiz-Benito, P., L. Gómez-Aparicio, A. Paquette, C. Messier, J. Kattge, and M. A. Zavala. 2014. Diversity increases carbon storage and tree productivity in Spanish forests. *Global Ecology and Biogeography* 23:311–22.

Ryan, J. P., Ueki, I., Chao, Y., Zhang, H., Polito, P. S., and Chavez, F. P., 2006. Western Pacific modulation of large phytoplankton blooms in the central and eastern equatorial Pacific. *Journal of Geophysical Research: Biogeosciences* 111:G2.

Rydin, H., and S. O. Borgegård. 1988. Plant species richness on islands over a century of primary succession: Lake Hjälmaren. *Ecology* 69:916–27.

Saccheri, I., M. Kuussaari, M. Kankare, P. Vikman, W. Fortelius, and I. Hanski. 1998. Inbreeding and extinction in a butterfly metapopulation. *Nature* 392:491–4.

Sage, R. F. 1999. Why C_4 photosynthesis? In R. F. Sage and R. K. Monson, eds. C_4 *Plant Biology*. San Diego, Calif.: Academic Press.

Sakamoto, M. 1966. Primary production by phytoplankton community in some Japanese lakes and its dependence on lake depth. *Archive für Hydrobiologie* 62:1–28.

Sala, O. E., W. J. Parton, L. A. Joyce, and W. K. Laurenroth. 1988. Primary production of the central grassland regions of the United States. *Ecology* 69:40–5.

Sala, O. E., Chapin, F.S., J. J. Armesto, E. Berlow, J. Bloomfield, R. Dirzo, E. Huber-Sanwald, L. F. Huenneke, R. B. Jackson, A. Kinzig, and R. Leemans 2000.

Global biodiversity scenarios for the year 2100. *Science*, 287:1770–4.

Samenow, J. 2017. Iranian city soars to record 129 degrees: Near hottest on Earth in modern measurements. *The Washington Post*. www.washingtonpost.com/news/capital-weather-gang/wp/2017/06/29/iran-city-soars-to-record-of-129-degrees-near-hottest-ever-reliably-measured-on-earth/?utm_term=.55a3478c6feb.

Schenk, H. J., and R. B. Jackson. 2002. The global biogeography of roots. *Ecological Monographs* 72:311–28.

Schlesinger, W. H. 1991. *Biogeochemistry: An Analysis of Global Change*. New York: Academic Press.

Schluter, D., T. D. Price, and P. R. Grant. 1985. Ecological character displacement in Darwin's finches. *Science* 227:1056–59.

Schluter, D., and R. E. Ricklefs. 1993. Species diversity: an introduction to the problem. In R. E. Ricklefs and D. Schluter, eds. *Species Diversity in Ecological Communities*. Chicago: University of Chicago Press.

Schmidt-Nielsen, K. 1964. *Desert Animals: Physiological Problems of Heat and Water*. Oxford: Clarendon Press.

Schmidt-Nielsen, K. 1969. The neglected interface. The biology of water as a liquid-gas system. *Quarterly Review of Biophysics*. 2:283–304.

Schmidt-Nielsen, K. 1983. *Animal Physiology: Adaptation and Environment*. 3rd ed. Cambridge, England: Cambridge University Press.

Schneider, D. W., and J. Lyons. 1993. Dynamics of upstream migration in two species of tropical freshwater snails. *Journal of the North American Benthological Society* 12:3–16.

Schoener, T. W. 1983. Field experiments on interspecific competition. *American Naturalist* 122:240–85.

Schoener, T. W. 1985. Some comments on Connell's and my reviews of field experiments on interspecific competition. *American Naturalist* 125:730–40.

Schoener, T. W. 2009. I.1 ecological niche. In S.A. Levin, ed. *The Princeton Guide to Ecology*. Princeton: Princeton University Press.

Scholander, P. F., R. Hock, V. Walters, F. Johnson, and L. Irving. 1950. Heat regulation in some arctic and tropical mammals and birds. *Biological Bulletin* 99:237–58.

Schultz, T. D., and N. F. Hadley. 1987. Microhabitat segregation and physiological differences in co-occurring tiger beetle species, *Cicindela oregona* and *Cicindela tranquebarica*. *Oecologia*, 73:363–70.

Schultz, T. D., M. C. Quinlan, and N. F. Hadley. 1992. Preferred body temperature, metabolic physiology, and water balance of adult *Cicindela longilabris*: a comparison of populations from boreal habitats and climatic refugia. *Physiological Zoology* 65:226–42.

Schumacher, H. 1976. *Korallenriff*. Munich: BLV Verlagsgellschaft mbH.

Scripps CO$_2$ Program. 2020. Atmospheric CO$_2$ Data. https://scrippsco2.ucsd.edu/data/atmospheric_co2/primary_mlo_co2_record.html. Accessed November 27, 2020.

Seiwa, K., and K. Kikuzawa. 1991. Phenology of tree seedlings in relation to seed size. *Canadian Journal of Botany* 69:532–8.

Serrano, D., and J. L. Tella. 2003. Dispersal within a spatially structured population of lesser kestrels: the role of spatial isolation and conspecific attraction. *Journal of Animal Ecology* 72:400–10.

Setälä, H., and V. Huhta. 1991. Soil fauna increase *Betula pendula* growth: laboratory experiments with coniferous forest floor. *Ecology* 72:665–71.

Shaver, G. R., and F. S. Chapin III. 1986. Effect of fertilizer on production and biomass of tussock tundra, Alaska, U.S.A. *Arctic and Alpine Research* 18:261–8.

Sher, A. A., D. L. Marshall, and S. A. Gilbert. 2000. Competition between native *Populus deltoides* and invasive *Tamarix ramosissima* and the implications of reestablishing flooding disturbance. *Conservation Biology* 14:1744–54.

Sherman, P. W., J. U. M. Jarvis, and S. H. Braude. 1992. Naked mole rats. *Scientific American* 257:72–8.

Shiels, A. B., and L. R. Walker. 2003. Bird perches increase forest seeds on Puerto Rican landslides. *Restoration Ecology* 11:457–65.

Shine, R., and E. L. Charnov. 1992. Patterns of survival, growth, and maturation in snakes and lizards. *American Naturalist* 139:1257–69.

Shochat, E., S. B. Lerman, J. M. Anderies, P. S. Warren, S. H. Faeth, and C. H. Nilon. 2010. Invasion, competition, and biodiversity loss in urban ecosystems. *BioScience* 60:199–208.

Siegenthaler, U., H. Friedli, H. Loetscher, E. Moor, A. Neftel, H. Oeschger, and B. Stauffer. 1988. Stable-isotope ratios and concentrations of CO$_2$ in air from polar ice cores. *Annals of Glaciology* 10:151–6.

Silvertown, J. 1987. Ecological stability: a test case. *American Naturalist* 130:807–10.

Simberloff, D., and W. Boeklin. 1981. Santa Rosalia reconsidered: size ratios and competition. *Evolution* 35:1206–28.

Simberloff, D. S. 1976. Experimental zoogeography of islands: effects of island size. *Ecology* 57:629–48.

Simberloff, D. S., and E. O. Wilson. 1969. Experimental zoogeography of islands: the colonization of empty islands. *Ecology* 50:278–96.

Sinclair, A. R. E., S. Mduma, and J. S. Brashares. 2003. Patterns of predation in a diverse predator-prey system. *Nature* 425:228–90.

Sinervo, B., and C. M. Lively. 1996. The rock-paper-scissors game and the evolution of alternative male strategies. *Nature* 380:240–3.

Skole, D., and C. Tucker. 1993. Tropical deforestation and habitat fragmentation in the Amazon: satellite data from 1978 to 1988. *Science* 260:1905–10.

Smil, V. 1990. Nitrogen and phosphorus. In B. L. Turner II, W. C. Clark, R. W. Kates, J. F. Richards, J. T. Mathews, and W. B. Meyer, eds. *The Earth as Transformed by Human Action.* Cambridge, England: Cambridge University Press.

Smith, R. L. 1996. *Ecology and Field Biology.* New York: HarperCollins.

Smith, V. H. 1979. Nutrient dependence of primary productivity in lakes. *Limnology and Oceanography* 24:1051–64.

Söderlund, R., and T. Rosswall. 1982. The nitrogen cycles. In O. Hutzinger, ed. *The Handbook of Environmental Chemistry,* vol. 1, part B, *The Natural Environment and the Biogeochemical Cycles.* New York: Springer-Verlag.

Sousa, W. P. 1979a. Disturbance in marine intertidal boulder fields: the nonequilibrium maintenance of species diversity. *Ecology* 60:1225–39.

Sousa, W. P. 1979b. Experimental investigations of disturbance and ecological succession in a rocky intertidal algal community. *Ecological Monographs* 49:227–54.

Sousa, W. P. 1984. The role of disturbance in natural communities. *Annual Review of Ecology and Systematics* 15:353–91.

Stephens, B. B., K. R. Gurney, P. P. Tans, C. Sweeney, W. Peters, L. Bruhwiler, P. Ciais, M. Ramonet, P. Bousquet, T. Nakazawa, S. Aoki, T. Machida, G. Inoue, N. Vinnichenko, J. Lloyd, A. Jordan, M. Heimann, O. Shibistova, R. L. Langenfelds, L. P. Steele, R. J. Francey, and A. S. Denning. 2007. Weak northern and strong tropical land carbon uptake from vertical profiles of atmospheric CO$_2$. *Science* 316:1732–5.

Stevens E. D., J. W. Kanwisher, and F. G. Carey. 2000. Muscle temperature in free-swimming giant Atlantic bluefin tuna (*Thunnus thynnus* L.). *Journal of Thermal Biology* 25:419–23.

Stevens, G. C. 1989. The latitudinal gradient in geographical range: how so many species coexist in the tropics. *American Naturalist* 133:240–56.

Stevens, O. A. 1932. The number and weight of seeds produced by weeds. *American Journal of Botany* 19:784–94.

Stimson, J. 1990. Stimulation of fat-body production in the polyps of the coral *Pocillopora damicornis* by the presence of mutualistic crabs of the genus *Trapezia.* *Marine Biology* 106:211–8.

Stoepler, T. M., J. T. Lill, and S. M. Murphy. 2011. Cascading effects of host size and host plant species on parasitoid resource allocation. *Ecological Entomology* 36:724–35.

Stone, B., J. Vargo, and D. Habeeb. 2012. Managing climate change in cities: will climate action plans work? *Landscape and Urban Planning* 107:263–71.

Stork, N. E. 2007. Australian tropical forest canopy crane: new tools for new frontiers. *Austral Ecology* 32:4–9.

Strassmann, J. 2001. The rarity of multiple mating by females in the social Hymenoptera. *Insectes Sociaux* 48:1–13.

Strong, D. R., L. A. Szyska, and D. Simberloff. 1981. Tests of community-wide character displacement against null hypotheses. *Evolution* 33:897–913.

Suberkropp, K., and E. Chauvet. 1995. Regulation of leaf breakdown by fungi in streams: influences of water chemistry. *Ecology* 76:1433–45.

Suess, H. E. 1955. Radiocarbon concentration in modern wood. *Science* 122:415–7.

Sugihara, G. 1980. Minimal community structure: an explanation of species abundance patterns. *American Naturalist* 116:770–87.

Summerhayes, V. S., and C. S. Elton. 1923. Contribution to the ecology of Spitsbergen and Bear Island. *Journal of Ecology* 11:214–86.

Takyu, M., S.-I. Aiba, and K. Kitayama. 2003. Changes in biomass, productivity and decomposition along topographical gradients under different geological conditions in tropical lower montane forests on Mount Kinabalu, Borneo. *Oecologia* 134:397–404.

Tan, C. C. 1946. Mosaic dominance in the inheritance of color patterns in the lady-bird beetle, *Harmonia axyridis.* *Genetics* 31:195–210.

Tan, C. C., and J. C. Li. 1934. Inheritance of the elytral color patterns of the lady-bird beetle, *Harmonia axyridis* Pallas. *American Naturalist* 68:252–65.

Tansley, A. G. 1917. On competition between *Galium saxatile* L. (*G. hercynicum* Weig.) and *Galium sylvestre* Poll. (*G. asperum* Schreb.) on different types of soil. *Journal of Ecology* 5:173–9.

Tansley, A. G. 1935. The use and abuse of vegetational concepts and terms. *Ecology* 16:284–307.

Taper, M. L., and T. J. Case. 1992. Coevolution among competitors. *Oxford Series in Evolutionary Biology.*

Terborgh, J. 1973. On the notion of favorableness in plant ecology. *American Naturalist* 107:481–501.

Terborgh, J. 1988. The big things that run the world: a sequel to E. O. Wilson. *Conservation Biology* 2:402–3.

Terborgh, J., K. Feeley, M. Silman, P. Nuñez, and B. Balukjian. 2006. Vegetation dynamics of predator-free land-bridge islands. *Journal of Ecology* 94:253–63.

Terborgh, J., L. Lopez, P. Nuñez, V. M. Rao, G. Shahabuddin, G.

Orihuela, M. Riveros, R. Ascanio, G. H. Adler, T. D. Lambert, and L. Balbas. 2001. Ecological meltdown in predator-free forest fragments. *Science* 294:1923–6.

Terra, L. S. W. Unpublished Light Trap Data. Vila do Conde, Portugal: Estação Aquícola.

Tewksbury, J. J., D. J. Levey, N. M. Haddad, S. Sargent, J. L. Orrock, A. Weldon, B. J. Danielson, J. Brinderhoff, E. I. Damschen, and P. Townsend. 2002. Corridors affect plants, animals and the interactions in fragmented landscapes. *Proceedings of the National Academy of Sciences of the United States of America* 99:12923–6.

Thibault, K. M., and J. H. Brown. 2008. Impact of an extreme climatic event on community assembly. *Proceedings of the National Academy of Sciences of the United States of America* 105:3410–5.

Thomson, D. A., and C. E. Lehner. 1976. Resilience of a rocky intertidal fish community in a physically unstable environment. *Journal of Experimental Marine Biology and Ecology* 22:1–29.

Thompson, R. S., and K. H. Anderson. 2000. Biomes of western North America at 18,000, 6000, and 0 ^{14}C yr B.P. reconstructed from pollen and packrat midden data *Journal of Biogeography,* 27:555–84.

Thornhill, R. 1981. Panorpa (Mecoptera: Panorpidae) scorpionflies: systems for understanding resource-defense polygyny and alternative male reproductive efforts. *Annual Review of Ecology and Systematics* 12:355–86.

Thornhill, R., and J. Alcock. 1983. *The Evolution of Insect Mating Systems.* Cambridge, Mass.: Harvard University Press.

Thorp, J. H., M. C. Thoms, and M. D. Delong. 2006. The riverine ecosystem synthesis: biocomplexity in river networks across space and time. *River Research and Applications* 22:123–47.

Thorp, J. H., M. C. Thoms, and M. D. Delong. 2008. *The River Ecosystem Synthesis: Towards Conceptual Cohesiveness in River Science.* Amsterdam: Academic Press.

Tilman, D. 1977. Resource competition between planktonic algae: an experimental and theoretical approach. *Ecology* 58:338–48.

Tilman, D. 1994. Competition and biodiversity in spatially structured habitats. *Ecology* 75:2–16.

Tilman, D., P. B. Reich, and F. Isbell. 2012. Biodiversity impacts ecosystem productivity as much as resources, disturbance, or herbivory. *Proceedings of the National Academy of Sciences of the United States* 109:10394–7.

Tilman, D., R. B. Reich, J. Knops, D. Wedin, T. Mielke, and Clarence Lehman. 2001. Diversity and productivity in a long-term grassland experiment. *Science* 294:843–5.

Tinbergen, N. 1963. On methods and aims of ethology. *Zeitschrift für Tierpsychologie* 20:410–33.

Tinghitella, R. M. 2008. Rapid evolutionary change in a sexual signal: genetic control of the mutation "flatwing" that renders male field crickets (*Teleogryllus oceanicus*) mute. *Heredity* 100:261–7.

Tinghitella, R. M., and M. Zuk. 2009. Asymmetric mating preferences accommodated the rapid evolutionary loss of a sexual signal. *Evolution* 63:2087–98.

Tinghitella, R. M., M. Zuk, M. Beveridge, and L. W. Simmons. 2011. Island hopping introduces Polynesian field crickets to novel environments, genetic bottlenecks and rapid evolution. *Journal of Evolutionary Biology* 24:1199–211.

Todd, A. W., and L. B. Keith. 1983. Coyote demography during a snowshoe hare decline in Alberta. *Journal of Wildlife Management* 47:394–404.

Tonn, W. M., and J. J. Magnuson. 1982. Patterns in the species composition and richness of fish assemblages in northern Wisconsin lakes. *Ecology* 63:1149–66.

Toolson, E. C. 1987. Water profligacy as an adaptation to hot deserts: water loss rates and evaporative cooling in the Sonoran Desert cicada, *Diceroprocta apache* (Homoptera, Cicadidae). *Physiological Zoology* 60:379–85.

Toolson, E. C., and N. F. Hadley. 1987. Energy-dependent facilitation of transcuticular water flux contributes to evaporative cooling in the Sonoran Desert cicada, *Diceroprocta apache* (Homoptera, Cicadidae). *Journal of Experimental Biology* 131:439–44.

Tosi, J., and R. F. Voertman. 1964. Some environmental factors in the economic development of the tropics. *Economic Geography* 40:189–205.

Toumey, J. W., and R. Kienholz. 1931. Trenched plots under forest canopies. *Yale University School of Forestry Bulletin* 30:1–31.

Tracy, R. L., and G. E. Walsberg. 2000. Prevalence of cutaneous evaporation in Merriam's kangaroo rat and its adaptive variation at the subspecific level. *Journal of Experimental Biology* 203:773–81.

Tracy, R. L., and G. E. Walsberg. 2001. Intraspecific variation in water loss in a desert rodent, *Dipodomys merriami*. *Ecology* 82:1130–7.

Tracy, R. L., and G. E. Walsberg. 2002. Kangaroo rats revisited: re-evaluating a classic case of desert survival. *Oecologia* 133:449–57.

Trappe, J. M. 2005. A. B. Frank and mycorrhizae: the challenge to evolutionary and ecologic theory. *Mycorrhiza* 15:277–81.

Tress, G., B. Tress, and G. Fry. 2005. Clarifying integrative research concepts in landscape ecology. *Landscape Ecology* 20:479–93.

Troll, C. 1939. Luftbildplan und okologische bodenforschung.

Zeitschraft der Gesellschaft fur Erdkunde Zu Berlin, pp. 241–98.

Tscharntke, T. 1992. Cascade effects among four trophic levels: bird predation on galls affects density-dependent parasitism. *Ecology* 73:1689–98.

Turner, M. G., R. H. Gardner, and R. V. O'Neill. 2001. *Landscape Ecology in Theory and Practice: Pattern and Process.* New York: Springer-Verlag.

Turner, M. G., W. H. Romme, and D. B. Tinker. 2003. Surprises and lessons from the 1988 Yellowstone fires. *Frontiers in Ecology and the Environment* 1:351–8.

Turner, M. G., E. A. H. Smithwick, D. B. Tinker, and W. H. Romme. 2009. Variation in foliar nitrogen and aboveground net primary production in young postfire lodgepole pine. *Canadian Journal of Forest Research* 39:1024–35.

Turner, T. 1983. Facilitation as a successional mechanism in a rocky intertidal community. *The American Naturalist* 121:729–38.

Turner, T. F., and J. C. Trexler. 1998. Ecological and historical associations of gene flow in darters (Teleostei: Percidae). *Evolution* 52:1781–801.

Tyukavina, A., M. C. Hansen, P. V. Potapov, S. V. Stehman, K. Smith-Rodriguez, C. Okpa, and R. Aguilar. 2017. Types and rates of forest disturbance in Brazilian Legal Amazon, 2000–2013. *Science Advances* 3:e1601047.

Ulrich, W., M. Ollik, and K. I. Ugland. 2010. A meta-analysis of species-abundance distributions. *Oikos* 119:1149–55.

United Nations, Department of Economic Affairs, Population Division. 2014. World Urbanization Prospects: The 2014 Revision, Highlights (ST/ESA/SER.A/352).

United Nations Population Information Network. www.un.org/popin/.

Urban, M. C. 2015. Accelerating extinction risk from climate change. *Science* 348:571–3.

U.S. Bureau of the Census, International Data Base. www.census.gov/pub/ipc/www/idbnew.html.

USDA Agricultural Research Service. 2011. www.ars.usda.gov/Research/docs.htm?docid=11059&page=6.

USGS. 2013. http://diseasemaps.usgs.gov/2013/del_us_human.html.

USGS, USFWS. 2005. The cranes: status survey and conservation action plan, whooping crane (*Grus americana*). www.npsc.nbs.gov.

U.S. National Academy of Sciences and The Royal Society. 2014. *Climate Change Evidence and Causes.* http://nas-sites.org/americasclimatechoices/events/a-discussion-on-climate-change-evidence-and-causes/.

Utida, S. 1957. Cyclic fluctuations of population density intrinsic to the host-parasite system. *Ecology* 38:442–9.

Valett, H. M., S. G. Fisher, N. B. Grimm, and P. Camill. 1994. Vertical hydrologic exchange and ecological

stability of a desert stream ecosystem. *Ecology* 75:548–60.

Van Bael, S. A., J. D. Brawn, and S. K. Robinson. 2008. Birds defend trees from herbivores in a Neotropical forest canopy. *Proceedings of the National Academy of Sciences of the United States* 100:8304–7.

Vancouver, G., and J. D. Vancouver. 1798. *A Voyage of Discovery to the North Pacific Ocean.* London: G. G., and J. Robinson.

Vanni, M. J., A. S. Flecker, J. M. Hood, and J. L. Headworth. 2002. Stoichiometry of nutrient recycling by vertebrates in a tropical stream: linking species identity and ecosystem processes. *Ecology Letters* 5:285–93.

Vannote, R. L., G. W. Minshall, K. W. Cummins, J. R. Sedell, and C. E. Cushing. 1980. The river continuum. *Canadian Journal of Fisheries and Aquatic Sciences* 37:130–7.

Verhulst, P. F., and A. Quetelet. 1838. Notice sur la loi que la population suit dans son accroisissement. *Corresponce in Mathematics and Physics* 10:113–21.

Violle, C., M. L. Navas, D. Vile, E. Kazakou, C. Fortunel, I. Hummel, and E. Garnier. 2007. Let the concept of trait be functional! *Oikos* 116:882–92.

Vitale, J., and W. H. Schlesinger. 2011. Historical analysis of the spring arrival of migratory birds to Dutchess County, New York: A 123-year record. *Northeastern Naturalist* 18:335–46.

Vitousek, P. M. 1994. Beyond global warming: ecology and global change. *Ecology* 75:1861–76.

Vitousek, P. M., and L. R. Walker. 1989. Biological invasion by *Myrica faya* in Hawaii: plant demography, nitrogen fixation, ecosystem effects. *Ecological Monographs* 59:247–65.

Volterra, V. 1926. Variations and fluctuations of the number of individuals in animal species living together. Reprinted 1931. In R. Chapman. *Animal Ecology.* New York: McGraw-Hill.

Vucetich, J. A., and R. O. Peterson. 2012. *The population biology of Isle Royale wolves and moose: an overview.* www.isleroyalewolf.org.

Vucetich, J. A., R. O. Peterson, and C. L. Schaefer. 2002. The effect of prey and predator densities on wolf predation. *Ecology* 83:3003–13.

Walker, G. T. 1924. Correlation in seasonal variations of weather, no. 9: a further study of world weather. *Memoirs of the Indian Meteorology Society* 24:275–332.

Walker, J. C. G. 1986. *Earth History: The Several Ages of the Earth.* Boston: Jones and Bartlett Publishers.

Walker, L. R., E. Velázquez, and A. Shiels. 2009. Applying lessons from ecological succession to the restoration of landslides. *Plant and Soil* 324:157–68.

Walker, L. R., J. Walker, and R. J. Hobbs. 2007. *Linking Restoration and Ecological Succession.* New York: Springer.

Wallace, J. B., S. L. Eggert, J. L. Meyer, and J. R. Webster. 1999. Effects of resource limitation on a detrital-based ecosystem. *Ecological Monographs* 69:409–42.

Walter, H. 1985. *Vegetation of the Earth.* 3rd ed. New York: Springer-Verlag.

Ware, D. M., and R. E. Thomson. 2005. Bottom-up ecosystem trophic dynamics determine fish production in the Northeast Pacific. *Science* 308:1280–4.

Waters, T. F. 1977. Secondary production in inland waters. *Advances in Ecological Research* 10:91–164.

Watwood, M. E., and C. N. Dahm. 1992. Effects of aquifer environmental factors on biodegradation of organic contaminants. In *Proceedings of the International Topical Meeting on Nuclear and Hazardous Waste Management Spectrum '92.* La Grange Park, Ill.: American Nuclear Society.

Webster, J. R. 1975. Analysis of potassium and calcium dynamics in stream ecosystems on three southern Appalachian watersheds of contrasting vegetation. Ph.D. thesis, University of Georgia, Athens.

Webster, J. R., and E. F. Benfield. 1986. Vascular plant breakdown in freshwater ecosystems. *Annual Review of Ecology and Systematics* 17:567–94.

Webster, K. E., T. K. Kratz, C. J. Bowser, J. J. Magnuson, and W. J. Rose. 1996. The influence of landscape position on lake chemical responses to drought in northern Wisconsin. *Limnology and Oceanography* 41:977–84.

Weinberg, W. 1908. On the demonstration of heredity in man. In S. H. Boyer, ed. *Papers on Human Genetics.* 1963. Englewood Cliffs, N.J.: Prentice Hall.

Werner, E. E., and G. G. Mittelbach. 1981. Optimal foraging: field tests of diet choice and habitat switching. *American Zoologist* 21:813–29.

West, P. M., and C. Packer. 2002. Sexual selection, temperature, and the lion's mane. *Science* 297:1339–49.

West, S. A., and A. Gardner. 2013. Adaptation and inclusive fitness. *Current Biology* 23:R577–R584.

Westoby, M. 1984. The self-thinning rule. *Advances in Ecological Research* 14:167–255.

Westoby, M., M. Leishman, and J. Lord. 1996. Comparative ecology of seed size and dispersal. *Philosophical Transactions of the Royal Society of London Series B* 351:1309–18.

Wetzel, R. G. 1975. *Limnology.* Philadelphia: W. B. Saunders.

Whicker, A. D., and J. K. Detling. 1988. Ecological consequences of prairie dog disturbances. *BioScience* 38:778–85.

White, C. S., and J. T. Markwiese. 1994. Assessment of the potential for *in situ* bioremediation of cyanide and nitrate contamination at a heap leach mine in central New Mexico. *Journal of Soil Contamination* 3:271–83.

White, J. 1985. The thinning rule and its application to mixtures of plant populations. In J. White, ed. *Studies in Plant Demography.* New York: Academic Press.

White, J., and J. L. Harper. 1970. Correlated changes in plant size and number in plant populations. *Journal of Ecology* 58:467–85.

White, P. S., and S. T. A. Pickett. 1985. Natural disturbance and patch dynamics: an introduction. In S. T. A. Pickett and P. S. White, eds. *The Ecology of Natural Disturbance and Patch Dynamics.* New York: Academic Press.

Whiteman, H. 2015. India heat wave kills 2,330 people as millions wait for rain. CNN World. www.cnn.com/2015/06/01/asia/india-heat-wave-deaths/.

Whittaker, R. H. 1956. Vegetation of the Great Smoky Mountains. *Ecological Monographs* 26:1–80.

Whittaker, R. H. 1965. Dominance and diversity in land plant communities. *Science* 147:250–60.

Whittaker, R. H. 1975. *Communities and Ecosystems.* New York: Macmillan.

Whittaker, R. H., and G. E. Likens. 1973. The primary production of the biosphere. *Human Ecology* 1:299–369.

Whittaker, R. H., and G. E. Likens. 1975. The biosphere and man. In *Primary Productivity of the Biosphere.* New York: Springer-Verlag.

Whittaker, R. H., and W. A. Niering. 1965. Vegetation of the Santa Catalina Mountains, Arizona: a gradient analysis of the south slope. *Ecology* 46:429–52.

Wiebe, H. H., R. W. Brown, T. W. Daniel, and E. Campbell. 1970. Water potential measurement in trees. *BioScience* 20:225–26.

Wiens, J. A., R. L. Schooley, and R. D. Weeks. 1997. Patchy landscapes and animal movements: do beetles percolate? *Oikos* 78:257–64.

Williams, M. 1990. Forests. In B. L. Turner II, W. C. Clark, R. W. Kates, J. F. Richards, J. T. Mathews, and W. B. Meyer, eds. *The Earth as Transformed by Human Action.* Cambridge, England: Cambridge University Press.

Williams-Guillén K., I. Perfecto, and J. Vandermeer. 2008. Bats limit insects in a neotropical agroforestry system. *Science* 320:70.

Williamson, M. 1981. *Island Populations.* Oxford: Oxford University Press.

Willmer, P. G., and G. N. Stone. 1997. Ant deterrence in *Acacia* flowers: how aggressive ant-guards assist seed-set in *Acacia* flowers. *Nature* 388:165–7.

Wilson, E. O. 1980. Caste and division of labor in leaf-cutter ants (Hymenoptera: Formicidae: *Atta*), I: The overall pattern in *A. sexdens. Behavioral Ecology and Sociobiology* 7:143–56.

Wilson, E. O., and D. S. Simberloff. 1969. Experimental zoogeography of islands: defaunation and monitoring techniques. *Ecology* 50:267–78.

Winemiller, K. O. 1990. Spatial and temporal variation in tropical fish trophic networks. *Ecological Monographs* 60:331–67.

Winemiller, K. O. 1992. Life history strategies and the effectiveness of sexual selection. *Oikos* 63:318–27.

Winemiller, K. O. 1995. Fish ecology. pp. 49–65. In Vol. 2 *Encyclopedia of Environmental Biology.* New York: Academic Press.

Winemiller, K. O., and K. A. Rose. 1992. Patterns of life-history diversification in North American fishes: implications for population regulation. *Canadian Journal of Fisheries and Aquatic Sciences* 49:2196–218.

Winston, M. L. 1992. Biology and management of Africanized bees. *Annual Review of Entomology* 37:173–93.

Witkowski, E. T. F. 1991. Effects of invasive alien acacias on nutrient cycling in the coastal lowlands of the Cape fynbos. *Journal of Applied Ecology* 28:1–15.

Wu, J. 2013. Key concepts and research topics in landscape ecology revisited: 30 years after the Allerton Park workshop. *Landscape Ecology* 28:1–11.

Wu, J., and R. Hobbs. 2007. Landscape ecology: The-state-of-the-science. In J. Wu and R. Hobbs, eds *Key Topics in Landscape Ecology.* Cambridge, England: Cambridge University Press.

Wurtsbaugh, W. A. 1992. Food-web modification by an invertebrate predator in the Great Salt Lake (USA). *Oecologia* 89:168–75.

Wurtsbaugh, W. A., and T. Smith Berry. 1990. Cascading effects of decreased salinity on the plankton, chemistry, and physics of the Great Salt Lake (Utah). *Canadian Journal of Fisheries and Aquatic Science* 47:100–109.

WWF. 2006. *Living Planet Report 2006.* Gland, Switzerland: WWF—World Wide Fund for Nature.

Yang, M. M., C. Mitter, and D. R. Miller. 2001. First incidence of inquilinism in gall-forming psyllids, with a description of the new inquiline species (Insecta, Hemiptera, Psylloidea, Psyllidae, Spondyliaspidinae). *Zoologica Scripta* 30:97–113.

Yoda, K., T. Kira, H. Ogawa, and K. Hozumi. 1963. Intraspecific competition among higher plants. XI. self-thinning in overcrowded pure stands under cultivated and natural conditions. *Journal of Biology Osaka City University* 14:107–29.

Zar, J. H. 1984. *Biostatistical Analysis.* 2nd ed. Englewood Cliffs, N.J.: Prentice Hall.

Zar, J. H. 1996. *Biostatistical Analysis.* 3rd ed. Upper Saddle River, N.J.: Prentice Hall.

Zhu, L., and J. Southworth. 2013. Disentangling the relationships between net primary production and precipitation in southern Africa savannas using satellite observations from 1982 to 2010. *Remote Sensing* 5:3803–25.

Zickfeld, K., A. Levermann, M. G. Morgan, T. Kuhlbrodt, S. Rahmstorf, and D. W. Keith. 2007. Expert judgements on the response of the Atlantic meridional overturning circulation to climate change. *Climatic Change* 82:235–65.

Zimmerman, E. K., and B. J. Cardinale. 2014. Is the relationship between algal diversity and biomass in North American lakes consistent with biodiversity experiments? *Oikos* 123:267–78.

Zuk, M., and G. R. Kolluru. 1998. Exploitation of sexual signals by predators and parasitoids. *Quarterly Review of Biology* 73:415–38.

Zuk, M., J. T. Rotenberry, and R. M. Tinghitella. 2006. Silent night: adaptive disappearance of a sexual signal in a parasitized population of field crickets. *Biology Letters* 2:521–24.

A

abiotic factors, 78, 100, 198, 245, 248, 252–53, 297, 384, 401
abundance, 198
abyssal zone, 50
adaptations, 2, 8–10, 29, 80–82, 99–102, 105–6, 111, 127, 134, 136–38, 164, 213, 255, 274, 547, 552
  behavioral, 46
  herbivore, 90
  elaborate, 117
  evolutionary, 279, 509
acclimation, 109
actual evapotranspiration (AET), 385
adaptation, 80
adhesion-adapted, 259
Admiralty Island, 449–50, 467
Africa, 21–25, 38–39, 65, 157, 216, 252, 341, 356, 463, 481–83, 493–94, 511, 531
  East, 5, 24, 38, 69, 71
  North, 140, 160
  Northwest, 38
  South, 59, 91, 186, 191, 379, 382, 483, 487
  southern, 28, 38, 127, 185, 483, 552
  southwestern, 24–26, 49, 134
  West, 20
age distribution, 224
agriculture, 22, 27, 29, 31–32, 63, 78, 96–97, 100, 328, 341, 385, 419, 420, 441, 459, 463, 491, 500, 504, 512–13, 547
A horizon, 19
Åland Island, 452–53
Alaska, 31, 33, 35–38, 49–50, 117, 141, 197, 222, 224, 240, 249, 254, 306, 356–57, 365, 386, 423–26, 432, 449, 494
algae, 7, 36, 46, 67, 72, 151, 159, 217, 220–21, 300–2, 336–37, 343–44, 346–47, 351–56, 358, 363, 373–76, 388, 403, 406, 413, 416, 427–28, 431–32, 434, 435, 548, 553
  artificial, 159
  benthic, 56, 163, 376
  blue-green algae, 45, 406, 428, 548
  dominant, 374–75
  ephemeral, 374
  epiphytic, 54
  green, 45–46, 374
  macroscopic, 435
  middle successional, 435
  mutualistic, 344
  photosynthetic, 153
  planktonic, 163, 353, 363
  powdered, 159
  preferred, 374–75
  red, 46, 427, 434
  species, 373–75, 391–92, 434
  tender, 375
  tide pool, 375
  unicellular, 336
algal biomass, 300–2, 376, 391, 431–32, 439
  benthic, 302, 377
  increased, 302
  reduced, 376–77
algal diversity, 374–75, 378, 391, 429
alleles, 80–81, 84–87, 93–94, 99–100
  alternative, 84
  counting, 85
  dominant, 80
  frequencies, 86
  new, 86
  pairing, 86
  total, 86
allelopathy, 278
Allequash Lake, 455
allopatric, 293
allozyme, 256
alligators, 63, 460
alpine plants, 39, 81–82, 112–13
Alps, 38, 81, 83, 84, 99, 123, 234
alternif lora, 281, 400
Amazon forest, 354
amensalism, 278
American Southwest, 137, 199, 413, 460, 470–71, 473, 486–87

ammonium, 155, 156, 336–37, 396, 406–7
  excrete, 336
amphibians, 113, 212, 270, 414, 493
ancestors, 91, 106, 137, 191, 197, 256, 344, 483
  marine, 63
  common, 79
  wild, 96, 100
  distant, 446
Andrews Forest, 31–32
animal populations, 88, 189, 213, 218, 272, 281, 305, 432, 502
animal species, 21, 79, 88, 105, 195, 293, 412–13, 460, 509
Anolis lizards, 295
Antarctica, 28, 51, 53, 109, 365–66, 504–5, 508
ants, 183–84, 189–90, 192, 194, 258, 328–36, 341, 344, 378–79, 382
  driver, 185
  fewer, 334
  guarding, 333
  high densities of, 334, 344
  influence of, 332, 335
  medium-sized, 191
  mutualistic, 340
  mutualistic dispersing, 382
  prevented, 334
  protective, 333
  removed, 332
  resident, 331–32
  stinging, 278
  worker, 192
apparent competition, 369
aphids, 141, 281, 334
aquatic ecosystems, 45, 75, 385, 387, 390–91, 399, 401, 405, 407, 408, 411–13, 419–21, 499
  experimental, 391, 401
aquatic invertebrates, 208–9, 214, 361, 428
aquatic organisms, 46, 65, 71, 73, 83, 125, 128, 130–31, 141, 143, 148
Arabian Sea, 50

arbuscular mycorrhizae, 327
arbuscule, 327
archaea, 150
arctic tundra, 12, 36, 102, 136, 386, 402, 461
Arctic Tundra Boreal Forest Temperate Grassland, 12
Argentina, 30, 189, 216, 306, 364, 483
argillic horizons, 457–58, 464
aril, 259
Arianta arbustorum, 123–24, 233, 464
Aripo River, 176–77, 195
Arizona, 17, 26, 114, 136–37, 141, 189, 196, 199, 207–8, 216, 291–92, 295, 346, 361, 369, 413–14, 421, 428–29, 431, 437–39, 443, 446, 454, 456, 486
arthropods, 53, 140–41, 178–80, 262, 300–1, 303–4, 318, 321, 323, 475–77
  dead, 178–80
  herbivorous, 301
artificial selection, 96
Asia, 20–21, 24–25, 30–34, 38–39, 43, 49, 64, 80–81, 84–85, 90, 104, 160, 197, 210, 248–49, 252, 356, 463, 481–84, 487, 493–94, 500, 511
aspen, 34, 305–6, 333–35
Asterionella, 353, 357
  growth rate of, 353
  high ratios, 353
Atlantic Ocean, 47, 65, 249, 371
Atlas Mountains, 38
atmosphere, 3, 15, 35, 42, 47–48, 58, 62–63, 75, 103, 128–32, 147, 151–52, 404–8, 420–21, 430, 490–94, 499, 504–9, 511–12
  ancient, 504
  earth's, 404, 492, 511
  lower, 508
  oceanic, 405
  upper, 494
  warmer, 509

atmospheric CO_2, 2, 152–53, 407–8, 412, 491, 493, 504–6, 508, 512
atoll, 54
Australia, 11, 21–22, 24–25, 28, 38, 93–94, 198–99, 202, 213, 249, 258–59, 312, 352, 378, 441, 446, 459, 463, 482–83, 494, 509–12
  eastern, 199
  northeastern, 20
  northern, 24
  semiarid southeastern, 66
  southern, 28, 31, 199, 498
  western, 199
autotrophs, 149–50, 154–56, 170, 336, 384, 401
Avila Mountains, 511
Azores Islands, 472

B

Bacastow, 507
bacteria, 13, 19, 20, 67–68, 149–51, 153, 155–56, 168–71, 197, 300, 302, 326, 333, 339, 398, 404, 406–7, 499
  actinomycetes, 406
  benzene-degrading, 169
  chemoautotrophic, 155–56
  chemosynthetic, 170
  cold-loving, 110
  denitrifying, 407
  heterotrophic, 170
  nitrifying, 155–56, 169
  prokaryotic, 150
  sulfur-oxidizing, 155
  symbiotic, 50
Bajada of Tucson Mountains, 456
balanced growth, 336
Baltic Sea, 33, 49–51, 389–90
barnacles, 47, 59, 200–13, 217, 243, 290, 315, 358, 371–73, 428, 434, 449–50, 467, 528
  acorn, 29, 315, 371
  gooseneck, 371–73
  individual, 290, 450
barometric pressure, 493–94, 511–12
barrier reef, 54
basins, 50–51, 57, 61, 65, 69, 75–76, 388, 417–20, 456
  basin control, 418
  deforested, 417–18, 448, 501, 512
  largest ocean, 50
bass, 74, 234, 339, 392–94
Bässler, 234
Batesian mimicry, 160
bathypelagic zone, 50

bats, 46, 115, 255, 299–301, 303–4, 321–24
  economic benefits of, 321
  emerging, 322
  excluded, 321
  female, 321
  foliage gleaning, 303
  followed, 322
  insectivorous, 323–24
  observed, 322
  pest control by, 299, 321–22
Bavarian Forest National Park, 234
bays, 57, 59, 296–97, 421
beaches, 59, 104, 120, 126, 243
  black, 104
  white, 104
Bear Brook, 417–19
Bear Island, 367
beavers, 317, 369, 460–62, 466
  effects of, 461
  repels, 369
beech, 8–9, 268, 410–11, 483–84, 487, 501
bees, 6–7, 25, 80, 118, 160, 181–82, 184, 189, 191–92, 202–3, 215–19, 235, 319, 326, 333–34, 341–44, 394–95, 502
  mimic, 160
  open, 341
  stinging, 160
  stingless, 202–3, 213
  worker, 215
beetles, 84–87, 120–21, 126, 136–37, 139, 178, 192, 235, 285, 288, 309, 312, 320, 347, 361, 369
  active, 121
  adult, 320
  black, 120
  carabid, 470
  dung, 502
  ground-dwelling, 467
  ladybird, 87
  spotted cucumber, 159
  tenebrionid, 148
behavioral ecology, 173
benthic, 50
Bering Sea, 49–50
B horizon, 19
Big Sulphur Creek, 301
biochemical oxygen demand (BOD), 67
biodiversity, 390
biomass, 267, 275, 280–81, 288, 297, 300–2, 306, 345, 376, 384, 390–92, 394, 397–98, 401, 407, 415, 423, 429, 430–32, 437, 441–43
  aboveground, 387, 415
  adult, 271

belowground, 391
  dry, 413
  energy-rich, 383
  equal, 399
  high, 367, 375
  highest, 313
  increased, 393
  insect-devouring, 321
  invertebrate, 428
  leaf, 158
  living, 157, 336
  low, 279, 377–78, 382
  macroinvertebrate, 413–14
  new, 401
  nonliving, 390
  phytoplankton, 387–89, 392–93
  shoot, 167–68, 331
  vegetative, 352, 422, 484
biomes, 11–19, 25, 27, 36–38, 40–43, 45, 49, 51, 53, 67, 198, 385, 504, 515
  isolated, 37
  largest, 30
  shrubland/woodland, 27, 198
  temperate grassland, 30, 40
  tropical savanna, 43
  varied, 198
biosphere, 2–3
biotic, 78
birch, 34–36, 167–68, 268, 306, 356, 410
bird diversity, 361, 363, 427, 443, 472, 484, 487
birds, 4, 8, 25, 63, 80, 91–92, 99, 111, 115, 120–21, 160–61, 184–86, 206–14, 217, 219, 223, 227, 230, 234–35, 246–47, 255, 259, 269–75, 283, 300, 302–4, 312, 323, 341, 347, 342, 361–63, 472, 474–75, 479, 484, 488, 496
  altricial, 270–71, 275
  aquatic, 117
  banded adult, 223
  breeding, 205
  counted, 206
  dry forest, 22
  excluded, 303
  female, 175
  fewer native local, 362
  fruit-eating, 346
  humming, 21, 123, 125–26, 326
  infected, 318
  insect-eating, 467
  larger, 284
  male, 175
  marine, 47
  marked, 220

  migratory/migrating, 402, 496
  nesting, 35
  resident, 36
  smaller, 283
  water-breeding, 472
  wild, 318, 343
  young, 217, 247
bird species, 91, 207, 212, 226, 234, 274, 347, 351–53, 361–64, 427, 443, 450, 472–75, 478–79, 483, 488, 500
  breeding, 427
  fewer, 472
  migratory, 36
  native woodland, 362
  new, 488
  richness, 361–63, 472, 478–79
  woodland, 362
birthrates, 230, 233, 235, 238, 243–45, 249, 250, 253
bison, 29, 31, 34, 359, 415, 417
Bjerknes's model, 494
Black Seas, 50, 52, 73
bluegills, 166–68
Bolinas Lagoon, 296
bollworms, 322
boreal forest(s), 32–35, 38, 43, 51, 108, 125, 151, 213, 268, 305–7, 402, 461–62, 466
  cool, 108
  undisturbed, 307
  taiga, 32
bottom-up control, 392
Bounty Island, 468
Brassica nigra, 368–71
Brazil, 20, 22, 47, 61, 216, 441, 500, 441, 500
  South-central, 24
  Southern, 30
breeding season, 206, 223, 256, 402
breeding territories, 185
British Isles, 31, 235, 241–42, 289
Brown, James H., 206, 211, 291–93, 471, 478, 481
buffalo, 187, 339, 394
  African, 415
  east African, 225
bumblebees, 6, 118, 326, 334
bundle sheath, 152
burrows, 102, 104, 124–25, 140–41, 160, 312, 359
  mammal, 104
butterflies, 80, 87, 125, 160, 178, 181, 197, 221–23, 235, 273, 319, 326, 361, 452–54, 478, 493
  common buckeye, 453–54
  female, 197

C

cacti, 12–13, 27, 246–48
  pear, 246
California, 17, 27–28, 30–31, 49–50, 70, 81–82, 108–9, 111, 150, 182, 197, 200, 212, 217, 296, 301, 351, 365, 369, 372–73, 469–70, 473–75, 478, 482, 488
  central, 196–97, 244
  coastal, 296
  northern, 197, 199, 361, 376
  separated, 362
  southern, 197, 199, 451, 462–63
caliche, 27
camel, 115, 139–40, 147, 326
CAM photosynthesis, 150–51, 153, 170
Canada, 34–35, 36, 72, 102, 206, 222–23, 235, 241–42, 249, 251, 264, 295, 306, 316, 365, 388, 454, 493–94, 499–500
  Central, 33
  North-central, 69
  Northern, 35, 240
  Northeastern, 57
  southern, 30, 197
  western, 135
Canary Islands, 468
cannibalism, 162, 282, 320
Cape flora, 483
Cape Floristic Region, 483, 487
Cape May, 4
Caribbean islands, 274, 295, 470, 479
Caribbean Sea, 49–50
caribou, 34–36
carnivore, 156
carp, 67, 74, 396
carrying capacity (*K*), 243
caterpillars, 197, 222, 246–47, 282, 284–85, 322, 326
catfish, 414
cattle, 22, 25, 29, 332, 339, 354, 462
caste, 190
Cedar Bog Lake, 397
Cedar Creek Natural History Area, 415
Central Alps, 83–84
Central America, 20–21, 216, 252, 330, 352, 412, 483
Central Gulf, 351–52
Channel Islands, 472–75
character displacement, 293
Chauvet, Eric, 411
cheetahs, 96, 395
chemosynthesis, 53, 153–54, 156, 171

chemosynthetic autotroph, 150
Chihuahuan Desert, 104, 204, 291–92
China, 11, 30–31, 49, 50, 57, 84, 97, 248, 251, 500
C horizon, 19
chronosequence, 425
Chthamalus, 201–2, 290–91, 427
  adult, 201–2, 290
  distribution of, 290
  larval, 290
cicadas, 140–41, 148, 313
  periodical, 313–14
cities, 32, 63, 67, 70, 94, 239, 361, 420–21, 446, 463–65
citizen scientist, 6
clay, 17–19, 42, 132, 261, 354–55, 457–58
climate change, 4, 7–8, 10, 19, 40, 42, 124, 202, 217–18, 233–36, 271, 463, 484–93, 497, 499, 502, 508–9, 512
  associated, 509
  contemporary, 272
  ecological consequences of, 7, 272
  global, 470, 490–91, 493, 497, 504, 508, 511
  historical, 219
  predicted, 508, 511
  projected, 509
  rapid, 210, 221
  rapid global, 504
  recent, 274
climate diagram, 16
climax community, 424
clouds, 15, 42, 47, 485, 492, 494, 501
clumped distribution, 202
clutch size, 232
CO_2, 1, 108, 134, 149–56, 163, 170–71, 382, 395, 405, 489, 502–4, 506–7, 509
  measurements of, 502, 504
CO_2 concentrations, 152, 171, 489, 505–6, 513
  atmospheric, 513
  internal, 152
coarse particulate organic matter. *See* CPOM
coast, 21, 25, 31, 36, 49–50, 54–55, 58, 63, 81, 111–12, 158, 197, 199–201, 206, 227, 229, 241, 244, 279, 290, 295–96, 371, 373, 424, 449, 472, 475, 486, 493–97
Coconut Island, 338
cod, 53, 194

cohesion, 131
cohort, 224
cohort life table, 224
colonies, 173, 184, 190–95, 202–5, 213, 216, 223, 243, 332, 338, 359, 415, 496
  cave, 322
  collected, 338
  competing, 193
  distributed, 213
  leaf-cutter, 190
  maternity, 321
  new, 203, 216, 332
  rival, 202–3
  scattered, 223
  sexdens, 190
  single, 190
  small, 223
  unprotected, 337
  wild, 216
colonization cycle, 220
Colorado, 6, 39, 40, 64, 199, 319, 380, 385, 387
coloration, 160, 173, 175–76, 212
  aposematic, 171
  brighter, 175
  bright warning, 284
  male, 175
Columbia River, 67
combined response, 312
commensalism, 278, 326
common garden experiment, 81
community, 346
community structure(s), 345–47, 351, 363, 365–67, 373, 375, 377–79, 381–82, 429, 437, 498
comparative method, 189
competition coefficients, 286
competitive exclusion, 202, 282–83, 285, 288–90, 297–98, 352–53, 357–58, 371, 373, 375, 382
competitive interactions, 202, 204, 279, 289, 297, 320, 474
competitive plant, 268
conceptual models, 7
conduction, 110
Congo River, 65
conifer forest, 18
  dense, 103
  mature, 34
  mixed, 207
conservation, 2, 83, 117, 127, 133, 136–39, 172, 193–95, 212, 214, 216, 321, 360–61, 379, 385, 408, 450, 455
consumer, 13
convection, 110

Cook Islands Atiu, 94
Cooper, William, 425
coral reefs, 49, 54–57, 73, 75, 162, 277, 336, 378–79, 382, 404, 479, 509
corals, 54, 56–57, 76, 277, 335–38, 344, 509
  crabs inhabiting, 338
  hard, 224
  protecting, 337
  reef-building, 56, 75, 326, 335
  unprotected, 337
  young, 56, 76
Coral Sea, 49–50
Coriolis effect, 14–16, 42, 48–49
corn, 21, 98
Costa Rica, 4, 202, 216, 221, 355, 412
cottonwood trees, 369
  felling, 369
  female, 228, 276
  undamaged, 369
coyotes, 306
C3 photosynthesis, 151
C4 photosynthesis, 152
CPOM (coarse particulate organic matter), 67–68
crabs, 53, 154, 160, 336–38, 344, 400
  commercial, 338
  corals harboring, 337
  mutualistic, 337–38
  protective, 338
  removed, 337–38
creeks, 61, 400
crickets, 93–95, 100, 107, 179–81
  female, 94–96
  large, 179
  male, 95
  males guarding, 180
crop(s), 40, 97–98, 152, 321, 324, 339, 380, 406, 499–500, 507
  agricultural, 100
  cabbage, 380
  cotton, 322
  nitrogen-fixing, 497
  planting row, 446
crustaceans, 52–53, 160, 166, 317, 335–38, 365
currents, 47–49, 51–52, 56, 58, 61–62, 66, 70, 220, 311, 413
  deep-water, 51–52
  ocean(ic), 51, 56, 150, 217
  saltwater, 62
Cuyahoga River, 72
Cyanistes caeruleus, 367–68, 382
  insect-feeding bird, 367
cyanobacteria, 45–46, 52, 76, 339, 406

D

Dahm, 169
daisies, 258, 334
dams, 229, 364, 460
damselfish, 173, 279–80
  cocoa, 280
  individual, 279
  threespot, 279
Daphne Major Island, 246, 248, 284
darter species, 74, 255–57, 261, 274
  larger, 256–57, 274
  smaller, 261
Darwin, Charles, 54, 79, 100, 245, 469
death rates, 198, 233, 237–38, 242–45, 248–50, 253–53
Death Valley, 109, 111–12, 509
deciduous forest(s), 18, 30–32, 119, 123, 260–61, 284, 352, 385, 427
  mature, 352
  mixed, 459
  virgin, 32
decomposition, 408
defenses, 90, 157–60, 162, 170, 284, 306, 340, 397
  alternative, 340, 344
  behavioral, 160
  chemical, 170, 284
  physical, 157–58, 170
deforestation, 35, 212, 360, 417–18, 420, 450, 500–4, 512
  amount of, 501
  massive, 512
  rate of, 501, 512
density, 198
density-dependent factor, 245
density-independent factor, 245
deserts, 11–15, 17, 19–20, 24–28, 30, 40, 42, 51, 66, 103–4, 134–36, 139, 159, 163–64, 167, 204, 386, 451, 458, 461
  biological, 59
  central Asian, 104
  coastal, 134
  driest, 127, 200
  dry, 385
  encountered, 51
  hot, 25, 26, 111, 129, 163–64, 345, 385, 387
  hottest, 108, 200
  plants, 25, 27, 105, 111–12, 138, 347
  rodents, 291, 297
  southern, 213
  surrounding, 207
  terrestrial, 53
  tropical rain forest, 15
  warm, 136

detritivores, 156, 159, 170–71, 326, 384, 399, 413
Detroit River, 72
diffusion, 130
direct interaction, 368
directional selection, 87
disruptive selection, 88
disease, 74, 93, 105, 197–98, 237, 241, 244–45, 252, 295, 299–301, 303–5, 307, 309, 311, 313, 315, 317, 319, 321, 323–24, 357, 509, 511
distribution, 198
disturbance, 267
dolphin, 117
Doñana Biological Reserve, 408
DNA barcoding, 348
dominant, or foundation, species, 367
drift, 220
drought, 24–30, 40–42, 135, 138–39, 141, 229, 233, 246–48, 259, 268, 283–84, 304, 446, 454–55, 460, 465, 493–95, 498–99, 509, 511–12
  extended, 66
  frequent, 31
  prolonged, 498
  seasonal, 24, 32, 42
  simulated, 136
  year-round, 26, 129
Dryas, 102, 112, 425–26
dry deposition, 420
dry season(s), 16, 21–22, 66, 122, 186, 242, 246–47, 330

E

eagles, 10, 271, 447, 449–50, 467
earthworms, 19, 384
Ebro River, 223–24
E. coli, 106
ecological efficiency, 397
ecological restoration, 440
ecological stoichiometry, 157
ecological succession, 423–25, 427, 429, 432, 435, 440, 443
ecologists, 1–5, 7–10, 19, 40, 45, 50, 57, 74, 82–83, 90, 107, 111, 127, 145–46, 151–52, 154, 156–58, 161, 164–68, 170–71, 174–75, 184–85, 193, 197–98, 202, 211, 213, 219–21, 223, 241, 245, 262, 267–69, 272, 274, 278–81, 283, 285, 290–91, 300, 303–6,

326, 330, 338, 340, 344, 366, 369, 346, 348, 352, 355–57, 363, 373, 375, 378, 384–88, 392, 396–97, 401–2, 404–8, 410, 413, 420–21, 424, 429, 432–38, 440, 443, 445, 447, 449–51, 465, 474, 478, 485–88, 491, 499, 512–13
  animal, 346, 363
  aquatic, 388, 419
  behavioral, 189, 191, 195
  desert, 204
  early, 204
  plant, 209, 258, 363
  population, 227–28, 230, 233, 236, 243, 250, 255
ecology, 1–3, 10, 440, 445–47, 449, 451, 453, 455, 457, 459, 461, 463, 465–67, 468–71, 473, 475, 477, 479–81, 483, 484, 485–89, 490–513
  geographic, 3, 10, 468–71, 473, 475, 477, 479–81, 483, 485–89
  global, 10, 484, 490–513
  landscape, 3, 440, 445–47, 449, 451, 453, 455, 457, 459, 461, 463, 465–67
ecosystem, 2–3, 8, 11–12, 124, 146, 154, 336, 356, 361, 384–85, 387–88, 390, 392–94, 396–99, 401–8, 411–21, 425, 429, 431–32, 435–40, 442–45, 447–48, 450, 454–55, 458, 486, 494, 497, 500
  aquatic, 46, 48, 73
  engineers, 458
  forest, 409, 417–21, 429–31, 440, 509
  geographic, 483, 485–86, 488
  global, 405, 491
  population, 2, 238, 268–69, 275, 487
  river, 66–68, 76, 413, 420
  stream, 67, 76, 399, 412–13, 417–19, 431–32
  wildlife, 303
ecosystem processes, 66, 356, 390, 392, 396, 401, 404, 412, 418, 431, 437, 450, 494, 512–13
ecosystem structures, 415, 429, 497
ecotone, 448
ecotype, 82
ectomycorrhizae, 327
ectotherm, 111
edge effect, 448

Eel River, 376–77
elaiosome, 259
Elbe River, 367–68
elephants, 25, 312, 326, 460, 466
elk, 218–19, 313, 395
Ellenton Bay, 233
Ellesmere Island, 102
El Niño, 247–48, 490, 493–99, 511–12
emigration, 92, 99, 198, 215–17, 221–23, 228, 248, 251, 311
endangered species, 9, 96, 197, 213, 216, 233, 272, 360
  managing, 9
  preserving, 213
  saving, 213
endemic, 211
endotherm, 111
energy, 13, 45–46, 66–68, 105–8, 110–11, 115, 117–18, 120–23, 128, 130, 143–44, 146, 148–51, 153–57, 159, 161–63, 165–71, 188, 198, 200, 255–56, 260–63, 268–69, 271, 275, 278–79, 297, 300, 319, 321, 326, 340, 344, 367, 371, 384–85, 396–97, 399–400
  budget, 198, 254, 262, 269, 275, 330, 340, 344, 396
  chemical, 108, 156, 384
  electromagnetic, 151
  kinetic, 101–3
  limited, 105, 165, 254–55, 262, 274
  metabolic, 111, 119, 125
  metabolizable, 107–8, 113–14
  solar, 14, 42, 44–45, 47, 51, 75, 102, 150, 384, 397, 464–65, 492
  stores, 384, 401
  wave, 58
energy content, 107, 111, 131–32, 151, 166, 107, 340
  free, 131, 132
  higher free, 132
  total, 151, 170
England, 66, 160, 356, 437
ENSO. *See* El Niño Southern Oscillation
environmental enrichment, 193
epilimnion, 69
epipelagic zone, 50
epiphytes, 4, 20, 34, 153
erosion, 19, 22, 38, 212, 346, 441–42, 446, 456–58, 465
estuaries, 59, 61–64, 73, 76
equilibrium, 357
equilibrium model of island biogeography, 473

estivation, 121
estuary, 59
Euptoieta claudia, 453–54
Eurasia, 30, 43, 223, 279, 481
Europe, 30, 38, 72, 160, 217–19, 239, 242, 249, 252, 273, 380, 463, 478, 482–84, 511
   Central, 80, 123
   eastern, 30, 217
   forested, 30
   northern, 234
   northwestern, 48, 123
   south-central, 33
   southern, 494
eusociality, 184
eutrophication, 71, 420
Eutrophic lake, 71–72
evaporation, 111
evolution, 2, 6–10, 13, 40, 45, 79, 84, 86–91, 95–100, 105–6, 128, 152, 155, 174, 176, 178, 183–86, 189, 191–93, 245, 264, 283, 327, 338–40, 344, 483
   defined, 84
   explored, 89
   rapid, 8, 100
   social, 184, 195
evolutionary change, 7, 79, 84, 86–87, 89, 91–92, 96–97, 99, 176
experiments, 5–6, 283, 288, 298, 310–11, 332–33, 388
   Gause's experiments, 283, 288, 298, 310–11
   Janzen's experiments, 332–33
   Whole-Lake Experiments on Primary Production, 388
exploitation, 40, 59, 242, 252, 299–301, 304, 310, 317, 320, 323–24, 341
exponential population growth, 240
ex situ, 6
extinction, 9, 92, 96, 122–25, 210, 212–13, 216, 288, 310–11, 412, 447, 451, 455, 468, 473–84, 487–89, 501, 509, 512–13
   potential, 474
   produced, 123, 315
   rates, 478–81, 483–84, 487, 489
extrafloral nectaries, 133, 334, 340–41, 344

F

facilitation model, 433
facultative mutualism, 326

farmers, 31, 97, 321–24, 354, 380, 386
fats, 115, 134, 150
feces, 134–35, 138, 318, 396, 413
fecundity, 230, 232–33, 235–36, 253, 255, 256, 266, 268–70, 275
   low, 269
   schedule, 230
feeding activities, 326, 365, 367, 371, 380, 382, 395, 415
feeding biology, 73–74, 283–84
feeding habits, 168, 291, 383, 385, 399–400
feeding rate(s), 163–64, 312, 314–17, 322
   animal's, 164
   increased, 312
   individual, 322
   plotted, 315
feeding relations, 365, 367, 381
females, 89–90, 94–96, 100, 172–75, 177–79, 181, 185–88, 191–92, 194–95, 197, 222–23, 230, 232, 243–44, 250, 257, 262, 453, 496, 498
   adult, 96, 187, 496
   discriminating, 95
   encroaching, 187
   first-breeding, 223
   individual, 188, 230, 235
   less-choosy, 95
   local, 173
   mature, 183, 264
   older, 186
   selective, 174
   yellow, 173
   young, 186
ferns, 4, 449, 472
fertility, 25, 28, 30–31, 33, 42, 102, 328, 351, 356, 386, 394, 401, 411, 458
fertilization, 182, 328, 330, 356, 386, 388, 420, 513
fertilizers, 386, 420, 441, 499–500
   artificial, 344
   inorganic, 31, 328
   nitrogen-potassium-phosphorus, 307
   synthetic, 459
finches, 79, 91–92, 229, 245–48, 283–84, 293–95
   large ground, 283
   medium ground, 91–92, 99, 246–48, 283–84, 293
fine particulate organic matter. *See* FPOM
Finland, 89, 95–96, 99, 108, 220–21, 304, 390, 452–53

fires, 24, 25, 33, 40, 379, 418, 462–63, 502, 509
   natural lightning-caused, 462
   uncontrollable, 462
fish, 67–68, 71–76, 83, 143–44, 160, 172–73, 193, 206, 220, 227, 255–56, 262–66, 268–71, 279–80, 365–66, 376–79, 392–93, 413–14, 471–72, 496, 509
   assigned, 84
   benthic, 74
   bioluminescent, 51, 326
   butterfly, 54
   captive-bred, 193
   client, 378
   common aquarium, 242
   deep-sea, 111
   enclosed, 377
   endothermic, 117–18, 120
   excluded, 376
   freshwater, 67, 72, 131, 144–45, 256, 381
   freshwater bony, 143, 144–45, 148
   headwater, 67, 83
   homeothermic, 117
   identical, 173
   individual, 83
   krill-feeding, 365
   large, 376
   larger, 57, 365–66
   larval, 264
   managing, 197, 213
   marine bony, 143–44, 148
   medium-sized, 166
   migratory, 68
   moving, 83
   piscivorous, 401
   plankton-feeding, 67, 401
   predaceous, 173, 175, 307
   predatory, 376
   reef, 160, 217, 351–52, 378, 382
   schooling, 173
   sedentary reef, 217
   size-selective, 392
   small, 337, 376
   species, 47, 67, 72, 83, 262–64, 357, 378–79, 382, 413–14, 471–72
   streamlined benthic, 256
   tolerant, 72, 212
   valued commercial, 194
fish communities, 73–74, 351–52, 379, 381–82, 392
Fisher, Stuart, 428
fitness, 87
fledglings, 90, 185, 187, 218
flocks, 25, 185–86, 188, 195
flooding, 40, 66, 130, 272, 355, 376, 418–19, 421, 429,

439–41, 457, 460, 495, 511
   coastal, 511
   periodic, 76
   producing torrential winter, 376
   seasonal, 355
flood pulse concept, 66
floras, 102, 258–59, 261, 482
Florida, 90–91, 206, 216–17, 317, 352, 460, 475, 487, 511
Florida Keys, 475–76
Floscularia conifera, 226
flowers, 6, 80, 82, 97, 102, 112, 121–22, 133, 181–82, 202, 247–48, 319, 320, 325, 333–35
   acacia's, 333
   arctic, 102
   cactus, 247
   colored, 80
   distinct, 80
   distributed, 202
   garden pea, 80
   mature, 246
   opened, 182, 247
   perfect, 174
   scarlet beebalm, 326
   tapping, 182
flies, 36, 95, 160, 197, 284–85, 334, 339, 342–43
food webs, 13, 365–68, 371–73, 375–77, 380–82, 384, 392, 402
   coastal, 496
   intertidal, 372–73, 377
   local, 375
   long manipulated, 380
   subtropical, 372
foraging, 122, 165–68, 192, 198, 256, 322, 331, 461, 496
forb, 258
forests, 4, 31–32, 228, 301–3, 348, 351–56, 385–86, 391, 408–10, 417–18, 420–22, 424, 445, 453, 459, 484, 500–4
   adjacent, 502
   ancient, 32
   clearing, 29
   coniferous, 31–32, 135, 222
   contorta, 418
   cottonwood, 228
   dense, 165
   distinctive, 355
   dry, 21–22, 24, 42–43, 59, 121–23, 202–3, 321, 504
   evergreen, 483
   fir, 38
   flooded, 441
   hardwood, 417
   hemlock, 424

forests—*Cont.*
  high-elevation, 385
  hypothetical, 348
  intact, 504
  intervening, 222
  isolated, 501
  losing, 459
  low-latitude, 402
  low-stature, 355
  mature conifer, 34
  mature oak-hickory, 427, 443
  mixed, 354–55
  northern, 34
  old, 453
  old-growth, 32
  once-continuous, 471
  open, 222
  protecting, 360
  redwood, 255
  riparian, 260
  riverside, 402, 466
  salt marshes and mangrove, 59
  submarine, 54
  supported, 424, 429
  surrounding, 32, 66, 77, 357,
    441, 502
  temperate coastal, 420
  total, 447, 459, 466
  undisturbed, 502
  virgin, 212
fossil fuels, 361, 405–7, 420, 490,
    504–7
  burning of, 405, 407, 504,
    506–7, 512–13
foxes, 36, 300, 303–6, 323–24
FPOM (fine particulate organic
    matter), 67–68
free-tailed bats, 115, 321–24
freshwater, 45
freshwater ecosystems, 73, 388,
    401, 405, 411
freshwater wetland, 59
fringing reef, 54
frogs, 160, 171, 272, 484
frosts, 28, 334–36
fruits, 34, 90–91, 97, 122,
    260, 384
functional response, 164
functional traits, 13
fundamental niche, 198
fungi, 13, 32, 36, 67, 115, 150,
    155, 157, 170, 189, 278,
    319, 325–30, 335, 339,
    404, 407–8

G

Galápagos Islands, 79–80, 91,
    229, 235, 245, 246, 248,
    496–97
galls, 278–79, 367–68
Gause's experiments, 283, 288,
    298, 310–11

gene, 80
gene flow, 48, 51, 255–57, 274
genera, 63, 84, 110, 189, 256,
    330, 373, 481, 484
generations, 79–80, 86–89, 95,
    106, 173, 177–78, 184,
    194, 197, 231, 233,
    238–40, 272–73, 306,
    309–10
  overlapping, 231, 238–40, 252
genetically modified organisms
    (GMOs), 97
genetic differences, 81–84,
    89–91, 110, 257, 261
genetic diversity, 80, 86, 92–94,
    96, 99–100
genetic engineering, 96
genetic drift, 78, 84, 86, 92–95,
    99–100
genetics, 80–81, 84, 89, 93,
    99–100, 405
genetic variation, 81–84, 86, 89,
    92, 99, 100, 176
genotype frequencies, 84–86,
    100
Genovesa Island, 229, 247–48
geographic ecology, 3, 10,
    468–71, 473, 475, 477,
    479–81, 483, 485–89
  defined, 469, 485
  effective, 488
geographic information
    system, 487
geometric population
    growth, 238
geometric rate of increase
    (λ), 230
Georgia, 412, 464
germination, 259
Ghana, 356
Glacier Bay, 423–26, 429, 432,
    433, 443–44
glaciers, 47, 75, 83, 219, 424–25,
    479, 484, 504, 509, 511
Glanville fritillary, 95–96, 100,
    222, 224, 450, 452, 465
global ecology, 10, 484, 490–513
global positioning system
    (GPS), 485
gonadosomatic index (GSI), 262
GPS. *See* global positioning
    system
graminoid, 258
granivore, 291
Grant, Peter, 91, 229, 245–46,
    283
Grant, Rosemary, 229, 245–46
grasshoppers, 31, 114–15, 126
grasslands, 17–18, 29, 30, 31,
    260, 352, 356, 359,
    361, 370, 385, 386, 402,
    427, 470

chalk, 360
  inhabits subalpine, 114
  open, 31
  primary production in, 386
  semiarid, 135, 136
  semidesert, 402
  seminatural, 259
  subtropical, 152
gray whales, 10, 196–97
grazing, 29, 301, 324, 354, 359,
    378, 394, 415, 422, 462
Great Basin Desert of North
    America, 25
Great Britain, 241–42, 356, 449
Great Lakes, 72–73, 77
Great Salt Lake, 65, 71, 130,
    497, 499
Great Smoky Mountains, 208
  forested, 66
greenhouse effects, 139, 176–77,
    182–83, 194–95, 259,
    261, 298, 492
Greenland, 36, 504
Greenland Sea, 49
grey kangaroo, 198–99
  western, 198–99
Grimm, Nancy, 361, 413–14, 432
Grimsö Wildlife Research
    Station, 303
gross primary production, 384
growing season, 26, 30–31, 34,
    42, 284, 290, 321, 369,
    402, 441
growth form, 258
growth rates, 89, 242, 248, 282,
    296–97, 308, 332, 338,
    353, 394
guild, 346
Gulf Stream, 48–49, 51
guppies, 175–78, 189, 242, 270
  adult, 175
  female, 175, 177–78
guppy coloration, 195
  male, 177
gut, beetle's, 320
gyres, 48, 51, 76

H

habitats, 4, 73, 136–37, 180,
    184, 185, 199, 202, 448,
    451–53, 455, 462, 465,
    470–71, 477, 479
  breeding, 184–85
  cold-water, 83
  common, 185
  complex, 60
  critical, 10
  disturbed, 165, 255, 266–67,
    275
  drier, 136
  drier dune, 135
  driest, 22

drought-prone, 135
dry, 136–37
  destruction, 125, 214, 241, 360
  fragmentation, 451–52, 467,
    473
  early successional, 453
  edge, 448
  emergent, 375
  experimental, 301
  fragmented, 93
  human, 63
  hunting, 361
  interior, 448
  intertidal, 57
  isolated, 424
  littoral, 264
  local, 484
  natural, 73, 288
  nesting, 34
  new, 266
  open grassland, 450
  potential, 452
  replacing wildlife, 63
  shady, 163, 312
  shrub, 483–84
  sunny, 170
  undisturbed, 73
  vegetated, 123
  warmed, 233
  wetter, 22
  woodland, 470
hadal zone, 50
Hamilton's rule, 184
haplodiploidy, 191
Hardy-Weinberg principle, 86
hare populations, 300, 305, 309
  foxes control, 300
  lynx and snowshoe, 305, 310
  lynx control snowshoe, 310
Hawaii, 50, 93–95, 338, 416–17,
    429–30, 443, 472, 506
heat, 102
heat waves, 463–64, 466, 511
hemlock, 207, 208, 218–19, 221,
    233, 258, 280
herbaceous plants, 29, 30, 158,
    258, 280, 297–98, 319
  perennial, 35–36
herbicides, 97–99, 417
herbivores, 156–59, 162, 164,
    170–71, 300–1, 323, 326,
    330–31, 368–69, 374,
    380, 382, 384, 393–96,
    397–99, 434–35
  dominant, 301
  large, 30, 208, 332, 415
  medium-sized mammalian,
    395
  potential, 158, 344
  roving, 31
  sheltering mammalian,
    370–71

small mammalian, 369, 371
specialized, 159
wild mammalian, 399
heritability, 89
hermaphrodite, 174
heterotroph, 150
hibernation, 120–23, 125–26
homeotherm, 111
honey badgers, 341, 344
honeybees, 160, 182, 192, 215–16
hot springs, 53, 109–10, 155
Hubbard Brook Experimental
   Forest, 417–18, 420
Hudson River, 273
human activity, 25, 27, 29, 419,
   421, 440, 443, 459, 463,
   490–91, 493, 499–500,
   504, 506–7, 512
human influences, 21–22, 25,
   27, 29–31, 32, 34, 37, 40,
   53, 56, 59, 63, 68, 504,
   506–7
human populations, 27, 31, 57,
   59, 68, 72, 76, 248–50,
   252–53, 255, 360–61,
   491, 499, 502, 508–9,
   511–12
   global, 251
   growing, 25, 444, 499
   sustainable global, 252
hummingbirds, 21, 118, 121–23,
   125, 325–26
   abundant, 126
   broad-tailed, 121–22
   female ruby-throated, 326
   rufus, 121
   weighing, 121
   wild, 121
Hungary, 250–51
hunters, 37, 162, 212, 304, 446
hydrologic cycle, 47
hydrophilic, 134
hydrophobic, 134
hyperosmotic organisms, 128–29,
   130–31, 143–44, 147–48
hyperparasitoids, 380
hyphae, 327
hypolimnion, 69
hypoosmotic organisms, 128–29,
   130–31, 143–44, 147–48
hyporheic zone, 65

I

IBI. See Index of Biological
   Integrity
ice age, 34, 83, 217–18, 240,
   505, 509
Illinois, 228
immigration, 86, 99, 198,
   215–17, 221, 228, 248,
   251, 310–11, 316, 453,
   473–78, 480, 487–89

documented, 488
periodic, 311
surpassed, 477
inbreeding, 93
inclusive fitness, 184
Index of Biological Integrity
   (IBI), 73–76
India, 21–22, 248, 493, 510
Indiana Dunes National
   Lakeshore, 258
Indian Oceans, 47, 50, 54, 249,
   493–94, 511
indirect commensalism, 368
indirect interaction, 368
Indus River, 24
inhibition model, 433
Inland Seas, 12
Inouye, David, 334–36, 340,
   344, 415
insects, 20–22, 118, 120–22, 157,
   160, 178, 181–82,
   278–79, 281, 284,
   312–13, 332–33, 380,
   475–76
   adult, 429
   aquatic, 160, 166, 220, 376
   dead, 181
   eating, 246
   eusocial, 189
   flower-visiting, 319
   gleans, 367
   predatory, 376, 497
   singing, 313
insectivore, 74
in situ, 4, 6
interdisciplinary research, 446
interference competition, 279
intermediate disturbance
   hypothesis, 357
intersexual selection, 174–75,
   177–78, 187, 194
interspecific competition, 279
intertidal zone, 49, 50, 54,
   57–59, 61, 75–76, 200–2,
   243, 290–91, 358,
   371–74, 376–77, 427
intrasexual selection, 174
intraspecific competition, 279
intrinsic rate of increase, 244
invertebrates, 47, 51, 54, 60,
   63, 67, 72, 143, 144,
   301, 358–59, 363, 373,
   428–29, 431–32
   filter-feeding, 346
   predatory, 376
irradiance, 163
Isat, 163
islands, 54, 56, 79, 93–95,
   99–100, 197, 212,
   245–47, 293–95, 316,
   429–30, 449–50, 468–78,
   481–82, 486–89

aquatic, 471
biological, 42
continental, 470
distant outlying, 93
experimental, 476–77
farthest, 476
green, 207
individual, 476
inhabit, 92, 99
intermediate, 472
isolated, 246, 473–74, 486
largest, 470
large tropical, 481
mangrove, 475–78, 487
medium-sized, 476
mountain, 470, 473, 486–87
neighboring, 94, 473,
   476, 487
rates of extinction on,
   474, 485
remote, 93
rocky, 497
small, 474–75, 477, 487
smallest, 470
species richness on, 468, 473,
   477–78, 487–88
tropical Bahama, 158
urban heat, 101, 123, 233,
   463–66
urbanized landscapes
   thermal, 123
volcanic, 79
islands of Hawaii, 429–30
Isle Royale, 160, 316–17
isoclines of zero population
   growth, 286
isopods, 25, 148, 282, 318
isosmotic, 130
iteroparity, 267

J

jaguars, 298
Janzen's experiments, 332–33
Japan, 31, 49–50, 101, 135, 260,
   272, 309, 387

K

Kabetogama Peninsula, 461–62
Kaneohe Bay, 338
kangaroos, 25, 198, 498
Kansas, 30, 40–41, 451
Keeler's model, 338, 340, 344
Keeling, Charles, 506–7
Keller's theory, 341
Kenya, 185, 190, 341, 394
keystone species, 365, 367, 371,
   373, 375–82, 402, 463
kin selection, 184
Klamath River, 312
Kronforst, Marcus, 80–81
K selection, 266
Kuril Islands, 365

L

Labroides dimidiatus, 378–79
lagoons, 54, 56, 172–73
Lake, Lawrence, 167
Lake, Morgan, 454–55
Lake Baikal, 69–70
Lake Erie, 72–73
Lake Hjälmaren, 476–77, 487
Lake Mendota, 397
lakes, 8–9, 45, 47, 69–72,
   75–77, 241, 264–66,
   353–54, 387–89, 391–93,
   396–97, 450–51, 454–55,
   471–72, 497
   bottom of, 71
   clear, 65
   clearest, 70
   deepest, 69
   desert, 71
   drainage, 454–55, 471
   experimental, 77, 388, 392–93
   fish inhabiting, 471
   groundwater flow through,
      454
   healthy, 72
   high-latitude, 238
   hydrologically mounded, 454
   hypersaline, 129
   hypothetical, 405
   large, 69, 166, 476
   larger, 69
   match, 471
   oligotrophic, 72
   oligotrophic and eutrophic,
      71–72
   open, 69, 497
   productive, 70
   reference, 388, 393
   rift, 69
   saline, 130
   seepage, 471
   shallow, 69
   small, 265, 295
   temperate zone, 70, 72
Lake Tanganyika, 69–71
landform, 456
landscape ecology, 3, 440,
   445–47, 449, 451, 453,
   455, 457, 459, 461, 463,
   465–67
landscape element, 445
landscape process, 450
landscapes, 3, 9–10, 21, 24–26,
   29–30, 59, 61, 64, 69–70,
   103, 202–4, 360–61,
   441–42, 445–46, 458–67
   agricultural, 458
   bleak lunar, 491
   changed, 461
   complex, 62, 456
   degraded, 447

landscapes—Cont.
    dissected, 483
    dry forest, 122
    fragmented, 451, 465
    frozen, 119
    hypothetical, 467
    local, 101–2, 124, 446
    mountain, 445
    mountain forest, 359
    northern, 219
    open, 35
    rain forest, 20
    salt marsh, 59–60
    semiarid, 104
    southern African, 127
    structure, 440, 445, 447,
        449–58, 460–63, 465–66
    surrounding, 65, 70, 113, 124,
        241, 321, 359, 369, 440,
        460, 462, 502
    tropical savanna, 24
    understanding, 446
    undulating, 61
    urban, 363, 445, 463–66
    urbanized, 462
    weathered, 129
    well-watered, 95
    wetland, 460
landscape structure, 447
land snails, 101, 123–25, 233,
    361, 470
    common, 123
    studied, 123
La Niña, 494
large-scale phenomena, 202
Larrea tridentata, 204, 345–46,
    457–58
larva, 255, 300, 398
lava f lows, 416, 424, 429, 431, 443
leaf beetles, 369
leaf-cutter ants, 189–92, 195
lentic, 45
legumes, 258, 390–91, 437–39,
    499
lesser kestrels, 221, 223–24, 235
Leverich, 226, 231, 240
lichens, 34–35, 36, 163, 339,
    416, 425–26
life history, 255
life table, 224
life-form, 346
lifetime reproductive success,
    186
light, 35–36, 44–46, 49–51,
    75–76, 107–8, 112, 114,
    146, 149–52, 167, 170,
    268, 297–98, 317–18,
    330–32
    blue, 45
    colored, 501
    compensation point, 46
    dim, 164

incoming, 139
    red, 151
    shed, 191, 220, 310, 491
    visible, 104, 111–12, 114, 151,
        486
Ligon, David, 184
Likens, Gene, 402, 417–19
limiting resource, 45
limnetic zone, 69
lions, 186, 187
lipids, 46, 136, 338, 404
lithosol, 25
littoral zone, 50
Littorina, 374–76, 378
liverworts, 357
livestock species, 341
lizards, 13, 79, 89, 107–8,
        113–14, 262–64, 269–71,
        274, 352
    eastern fence, 107–8,
        113–14, 264
    populations of, 108, 263
loci, 256
locus, 80
Los Hermanos Islands, 293, 295
logistic equation, 244
logistic population growth, 242
lognormal distribution, 347
lotic, 45
Lotka, 285–86, 297–98, 307–8,
        311–12, 323
Louisiana crayfish, 171
lowlands, 27, 40, 81–82, 330, 366
LTER (Long Term Ecological
        Research), 204, 361
Lutra lutra, 160
lynx, 34, 305–6, 309–10
Lyons, John, 221

M

Mabul Island, 44
MacArthur, Robert, 4, 10, 166,
        266, 268–69, 352–53,
        363, 445, 469, 472–74,
        478, 488
macroclimate, 101–2, 124–25,
        141, 200
macroinvertebrates, 413, 417, 428
Madagascar, 122, 481
Madeira Islands, 416
Maine, 199, 352, 423
Major, Daphne, 246–47, 283,
        293–95
males, 50, 88, 92, 94–95,
        173–75, 177–81, 183,
        185–88, 191–92, 194–95,
        197, 222, 232, 250, 264
    adult, 187
    bluehead wrasse, 173
    bright, 175
    colorful, 175–76, 178
    courting, 174

dominant, 175
evolution of ornamentation in,
        172, 174, 194
first-breeding, 223
flatwing, 95
fourth-ranked, 188
infanticidal, 187–88
large, 88, 180–81
larger, 180–81, 194
largest, 180
less-colorful, 194
more-colorful, 194
multiple, 192
non-singing, 95
nonterritorial, 496
repelling attacking, 187
singing, 94
single, 187
smallest, 180
territory-holding, 173
third-ranked, 188
top-ranked, 188
unrelated, 187–88
winged, 190
winged reproductive, 332
yellow, 173
young, 32
younger, 162
mammals, 2, 116–19, 161,
        209–11, 214, 217, 235,
        259, 262, 270–71,
        346, 359, 361, 394–95,
        450–51
    burrowing, 19, 359
    comparable-sized, 210
    eusocial, 189
    excluded flying, 481
    fruit-eating, 481
    grazing, 152, 157, 392
    large, 161, 209, 219, 227,
        394–95, 423
    large-footed, 198
    large native, 36
    large North American, 382
    large sea, 47
    largest, 45
    nonflying forest, 471
    nonflying terrestrial, 481
    rain forest, 116
    small, 211, 219, 369–71, 402,
        451, 465, 467
    smallest marine, 365
mangrove forests, 59, 61–64,
        76, 477
    intact, 64
    replanting, 64
marine environments, 50, 52–53,
        57, 76, 338, 486, 499, 513
    nearshore, 54
    productive, 365
marine invertebrates, 131, 143,
        148, 227, 262, 313, 358

marine organisms, 47–48,
        50–51, 75
marshes, 59, 62, 161, 400,
        447, 500
    coastal, 197
    estuaries and salt, 63–64
Maryland, 361, 420
Massachusetts, 272
mass extinctions, 412, 444,
        509, 512
matings, 86, 92–93, 173, 178–79,
        181–83, 192–93, 194, 197
    controlled, 84
    selective, 86
    success, 174, 178–81, 194
matric forces, 131
matrix, 447
Maumee River, 212
Mauna Loa, 50, 506
mayflies, 255, 324, 413
meadows, 95–96, 123, 221,
        213, 221, 224, 260, 437,
        445, 452
    dry, 95–96, 221, 385–87
    wet, 385, 461
    wet mountain, 334
Mediterranean climates, 28, 408,
        455, 462, 466, 482–83
Mediterranean woodland and
        shrubland, 27
MEI. See metabolizable energy
        intake
Melitaea cinxia, 95–96, 100, 221,
        452–53, 465
meristematic tissue, 319
mesopelagic zone, 50
mesosphere, 492
meta-analysis, 146
metabolic heat, 110
metabolic water, 134
metabolism, 110–11, 154,
        157, 336
metabolizable energy intake
        (MEI), 107–8, 113–14
metalimnion, 69
metapopulation, 215, 221–24,
        235–36, 451–52,
        465, 467
Mexico, 20–22, 30, 50, 57, 104,
        137, 206, 213, 216, 237,
        296, 321, 324, 334, 461
mice, 31, 80, 219, 292, 300, 367
microbes, 29, 53, 67–68, 109–10,
        384, 416
    heat-loving, 110
microbial activity, 109–10,
        125, 169
microclimates, 11, 17, 42, 102–5,
        120, 124–25, 129, 135,
        139, 141, 200
microorganisms, 220–21, 279,
        288, 326

microplastics, 54
microsatellite DNA 83
migration, 22, 197, 254, 272, 463, 496
  annual, 197
  birds, 272–73, 496
  human, 462
  long, 149
mineralization, 408
Minnesota, 390, 397, 415, 461
minnows, 376, 392–93, 404
Mississippi, 64, 90, 206, 249, 385, 420
Missouri, 64, 256
mites, 19, 20, 303, 311, 367
  complex array herbivorous, 311
  predatory, 311
modeling, 3, 6–8, 10, 285, 338, 509, 512
  Lotka and Volterra's modeling exercise, 285
  studies, 509, 512
Mojave Deserts, 204, 458
moles, 151, 190, 259
monarch butterflies, 197, 326
Mongolia, 26, 249
montane habitats, 470–71
Montreal Protocol, 508
moon, 57–58, 75, 490
moose, 34, 161, 164, 165, 300, 316, 415
mortality, 178, 201, 204–5, 224–27, 231, 235, 245, 247, 262–63, 265, 275–76, 280–81, 283, 434, 464
  adult, 262–64
  adult fish, 263
  barnacle, 202
  higher, 92, 99, 175, 186, 202, 204, 262
  increased, 247, 303
mosses, 35, 36, 103, 110, 111, 165, 357, 426
moths, 118, 161, 178, 223, 273, 307, 321, 347–48, 367
  cotton bollworm, 321–22
  female, 322
  female bollworm, 322
  free-flying, 119
  heated, 118
  peppered, 161
  sample of, 347
mountains, 12, 17–18, 37–40, 42–43, 103, 198–99, 207–8, 451, 456–57, 470–71, 473, 481, 484, 486, 509–11
  collected, 316
  distributions of, 17, 484
  high, 40, 200, 212, 468

high equatorial, 39
isolated, 38, 451, 470–72, 473
low, 38
male, 174
midlatitude, 39
narrow, 456
polar, 38
slopes, 33, 202, 234, 260, 456–57
Mount Desert Island, 352
Mount Kilimanjaro, 37, 39
Mount St. Helens, 444
Muir Glacier, 424
Mukkaw Bay, 371–73, 377
Müllerian mimicry, 160
Murphy, Shannon, 161, 284–86, 398
muskrats, 303, 317
mussels, 53, 57, 59, 371–73, 400, 428
  ribbed, 400
  zebra, 73
mutualism, 278–79, 295, 297–98, 320, 325–32, 336–42, 343–45, 346, 368, 378, 381
  ancient, 344
  ant-acacia, 336
  ant-protection, 330
  benefits of, 326, 336, 343
  complex, 338
  coral-centered, 343
  coral-crustacean, 336
  enhancing, 341
  evolution of, 338
  examples of, 326, 333
  facultative, 325, 338, 344
  facultative plant protection, 338
  honeyguide-human, 344
  human, 341
  indirect, 368, 381
  key, 330, 343
  plant-animal, 326
  plant-centered, 327, 344
  temperate ant-plant, 334
  tropical swollen thorn acacia-ant, 334
mycorrhizae, 21, 42, 325, 326–29, 338, 341, 343–45, 405, 547
  arbuscular, 326–27, 343
  plants form, 326
mycorrhizal fungi, 325–31, 335–36, 343–44, 351, 379, 382, 407, 513

N

naked mole rats, 189–92, 195
Namib Desert, 26, 134, 148
natural experiment, 6
natal territory, 185

natural history, 11
natural selection, 78–81, 83–85, 87–93, 95–100, 102, 111–12, 126, 158, 160–62, 165, 170–71, 175–76, 194–95, 262–63, 265
  extrafloral, 334
  meal of, 325
  nectar, 121–22, 133, 202, 246, 331, 333–35, 340, 344
  process of, 81, 86–87, 97
  sources of, 172, 174–75, 194, 278
  theory of evolution by, 78–80, 88, 96–99
negative phototaxis, 317
Negev Desert, 103
neritic zone, 50
nests, 184, 186, 189–91, 203–4, 229, 232, 264, 330, 332, 340–44, 380–81, 449, 496
  ant, 332, 338
  new, 382
  occupied, 204
  parent's, 217
  prospective, 203
  pumpkinseed, 264
  sexdens, 189
  young birds, 217
Netherlands, 420, 458–60
net photosynthesis, 163
net primary production, 384
net reproductive rate (R_0), 230
Nevada, 354
New England, 31, 400
New Guinea, 472, 481, 488, 502
New Hampshire, 280, 409–10, 417, 419
New Jersey, 107–8, 114, 249
New Mexico, 104, 203, 216, 222, 228, 229, 385, 472, 486
  central, 104, 228
New River, 74
New York, 32, 274
New Zealand, 30, 31, 104, 120, 126, 218, 373, 382
niche, 198
niche partitioning, 284
Nile, 64, 420
nitrogen, 72–73, 155–59, 167, 169–70, 327, 336, 353, 384–91, 400–1, 404–9, 413–16, 418–22, 430–32, 438–41, 499
  amounts of, 414, 500
  bacteria release, 407
  dissolved inorganic, 432
  export, 432
  fix, 390, 498, 512
  fixed, 406, 497–99, 512

molecular, 339, 406–7
  preventing losses of, 418–20
  retained, 414
  role of vegetation in preventing losses of, 418–20
  source of, 169, 404, 416
  stream reingest, 413
  total, 413, 428
Norris, Ryan, 496
Norway, 36, 273, 304, 390, 477
numerical response, 219
nutrients, 6, 20–22, 72, 149, 156–57, 165–67, 268, 279–81, 325–31, 343–44, 351–53, 383–89, 401–2, 412–13, 414–22
  absorbing, 336
  abundant, 163, 167, 170, 241
  captured, 462
  cycling, 404
  distribution of, 203, 348, 403, 415, 417, 458
  enhanced, 343
  extract, 27
  flux, 405
  gain, 405
  harvest, 326
  immobile, 327
  inorganic, 31, 71, 327–28
  large quantities of, 70, 336
  leach, 20
  lower, 328
  marine-derived, 414
  mineral, 19
  moving, 405
  organisms recycle, 32
  pool, 405
  releasing, 403
  renew, 62
  replenishes, 70
  retentiveness, 413
  scavenge, 20
  single, 386
  sink, 405
  source, 405
  spiralling, 413
  supplemental, 329
  supply of, 167, 496
  transport of, 413

O

obligate mutualism, 326
observation, 3–6, 10
Oceania, 94, 249, 252
oceanic islands, 37, 248, 447, 468, 470, 474
  isolated, 248, 487
  scattered, 259
  submerged, 54
oceanic zone, 50

oceans, 44–45, 47–54, 57, 62–64, 69–71, 75–76, 129–30, 270, 273, 365–66, 401, 405–8, 468, 470, 507–9
  clearest, 50
  deepest, 50
  open, 29, 48, 51, 53, 59, 62, 70–71, 75, 173, 389, 486, 497
  world's, 69
offspring, 79–80, 87, 173, 175, 177, 181, 184–88, 191, 194, 230–31, 254–56, 263, 267–71, 274–75, 308
  dependent, 187
  female, 231–32, 235
  female's, 181
  fewer, 162
  heterozygous, 85
  independent, 498
  low numbers of, 268–69
  produced, 208
  producing, 175, 184
  provisioned, 275
  size of, 253–56, 260, 266, 270, 274
  small, 266, 269, 275
  well-cared-for, 256
Ohio, 256, 347, 361
O (organic) horizon, 17
oligotrophic, 71
omnivore, 74
optimal foraging theory, 165
optimization, 166
orchids, 4, 21, 150, 258, 278, 326
Oregon, 31–32, 70, 435, 442
organic compounds, 150
organic matter, 17–21, 25, 30–32, 33, 36, 38, 53, 63–64, 66–68, 407, 417–18, 429–30, 436, 441, 443
  dead, 21, 407
  decomposed, 19
  fallen, 17, 42
  new, 382
  nonliving, 156
  soluble, 18
organisms, 1–4, 101–2, 104–7, 109–11, 123–33, 146–51, 155–59, 217–21, 252–57, 260–62, 274–78, 398–400, 403–8, 451–53, 472–73
  attached, 427
  beach, 104
  consumer, 13, 108, 384, 399
  deep-sea, 53
  desert, 25, 127
  dominant, 49
  dominant photosynthesizing, 45
  ectothermic, 115

elements, 421
endothermic, 119
faces, 128
free-ranging nocturnal, 3
group, 150
heterotrophic, 149, 156, 170
homeothermic, 111
hyperosmotic, 131
hypoosmotic, 131
individual, 1–2, 9–10, 124, 129, 145, 155, 183, 384, 396
intertidal, 58
isosmotic, 313
larger, 46, 316, 323
live-bearing, 230
microscopic, 52
modified, 97
mountain, 38
mutualistic, 382
nitrogen-fixing, 406
non-haplodiploid, 192
noxious, 160
oceanic, 52
photosynthesizing, 13, 45
photosynthetic, 75, 151, 163
population growth by, 240, 252
real, 253
respiring, 45
river, 66–67, 363
shape-shifting, 173
single-celled, 45
social, 192, 213
soft-bodied, 46
soil, 17
stream, 76, 220, 235, 428
suggested grouping, 397
super, 105
territorial, 213
tide pool, 59
tropical, 480
well-studied, 235, 271
wild, 97
Orinoco River, 485
osmoregulation, 143
osmosis, 130
otters, 160, 171
owls, 34, 89, 185, 217, 219–20, 305, 307
ozone, 491–92, 507, 508

P

Pacific Ocean, 50, 65, 93, 196, 221, 249, 337, 488, 493–94, 496, 512
  central, 494, 497
  eastern North, 10
  eastern tropical, 494, 512
  northwest, 365
  open, 495
  western, 50, 494

Paine, Robert, 367, 371–73, 375–77, 382
Palmer Drought Severity Index, 11, 40–41, 42
parasites, 96, 199, 216, 299–300, 304, 310, 317, 320, 322, 344, 368, 378, 382, 449, 480
  external, 303
  intestinal, 318
  intracellular, 320
  pathogenic, 303, 324
  protozoan, 320, 324
parasitism, 278
parasitoids, 95, 191, 284–85, 298, 300, 307, 309, 323–24, 367
patch, 447
pathogens, 115, 199, 216, 299, 300, 304, 317, 319, 322–24, 382, 480
Pecos River, 67
pelagic, 50
penguins, 117, 365–66
per capita rate of increase (r), 233
perches, 121, 342–43, 441–42
peregrine falcons, 212
Persian Gulf, 49–50
Peru, 493, 496, 502
phenology, 272, 284
phenotype, 81
phenotypic plasticity, 82
pheromone, 203
philopatry, 185
photon flux density, 151
photorespiration, 152
photosynthesis, 2, 13, 45, 46–48, 53, 102, 108–9, 125, 139, 151–54, 163, 165, 170–71, 326, 384
  high rates of, 63, 153, 184
  light-dependent reactions of, 151–52
  optimal temperature for, 108–9
  rate of, 108–10, 163
  saturate, 163
  stimulate, 48, 155
photosynthetic autotroph, 150
photosynthetically active radiation (PAR), 151
phreatic zone, 65
phytoplankton, 52, 67–68, 72, 76, 353–54, 365, 387–89, 392–94, 396, 400–1, 486–87, 496–97
Piedmont Plateau, 427–28, 435, 443, 427
pigmentation, 83–84, 89, 113–14
pines, 34, 158, 197, 208, 280, 427, 453
pioneer community, 424

piscivore, 74
pistil, 181
plant functional group, 390
plants, 11–13, 79–83, 89–91, 111–13, 131–38, 146–54, 156–59, 163–65, 167–68, 181–83, 258–62, 278–82, 323–35, 340–44, 345–47
  adjacent, 222, 334
  adult, 370
  animal-pollinated, 326
  annual, 165, 224, 226, 230–31, 242, 346
  arctic and alpine, 112–13
  aromatic, 27
  chemical defenses of, 157–58, 170
  coffee, 321, 324
  colonizing, 328
  communities, 9, 17, 41, 346, 350–53, 358, 378, 380, 382, 424, 436–37, 444, 458, 470, 483
  cotton, 321–23
  covered, 303
  crop, 97, 151, 321
  cushion, 112–13
  distribution of, 200, 208, 458–60
  domesticated, 96, 100
  dominant, 11, 458
  endothermic, 119–20
  exotic, 416
  experimental, 167, 301–3
  fernlike, 472
  fertilizer, 354
  fire-resistant, 29
  food, 149, 247, 305–6
  forest, 384
  forest understory, 164
  fynbos, 379
  graminoid, 258–59
  green, 46, 383
  groups of, 152, 425, 437
  high affinity C_4, 152
  hummingbird-pollinated, 325
  hydroelectric, 131
  inconspicuous, 10
  individual, 135, 143, 147, 278–83, 357, 412
  invasive, 416
  invasive exotic, 369
  juvenile, 227
  land, 326
  larval food, 452
  leguminous, 406
  living, 21, 330
  lowland, 81–82
  maternal, 182, 260
  mating behavior of, 181–82
  medicinal, 40, 416
  mid-elevation, 82

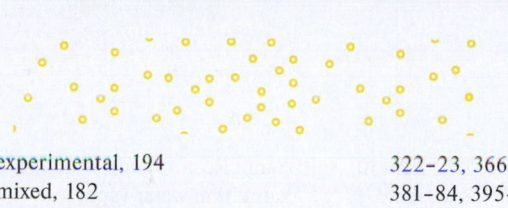

mustard, 319
mutant, 341
mutualistic, 343
mycorrhizal, 328–31
native, 91, 408
new, 99, 260
nitrogen-fixing, 421, 441
nonmutualistic, 339
nonmycorrhizal, 326, 328
ocotillo, 138
parent, 211
perennial, 227, 267, 457, 498
rain forest, 129, 139
refuge, 323
riparian, 68
rosette, 112
salt marsh, 290
seed, 182
selected, 272
shade, 163
small, 258
small-seeded, 379
soapberry, 92
stress-tolerant, 267, 275
succulent, 13, 153
surviving, 281
temperature regulation by,
    111–12
tobacco, 159
tomato, 118
tropical, 21, 158
umbrella-shaped, 139
understory, 418
upland, 400
vascular, 117
vulnerable, 440
water acquisition by, 134–35
water uptake by, 127, 145, 148
weedy, 267
wild, 97, 507
wind-pollinated, 326
woody, 207, 258, 260, 300,
    390, 406, 470
young, 330
plant species, 167–68, 217,
    258–59, 272–73, 346–47,
    355, 359–60, 379,
    390–91, 416, 427, 438,
    476–78, 481–82
bird-dispersed, 441
individual, 136, 437
perennial, 457
preferred, 441
single, 209, 345
vascular, 479
woody, 427
plant tissues, 132, 157–59, 170,
    326, 333, 339–40, 411
P_{max}, 163
poikilotherm, 111
pollination, 182, 327, 333–34,
    341, 343, 453, 502

experimental, 194
    mixed, 182
pollution, 9, 57, 67, 75, 169, 170,
    253, 351, 491
    control nitrogen, 420
    environmental, 75, 169
polygenic traits, 87
polymorphic locus, 256
ponds, 35–36, 45, 47, 64, 166,
    176–77, 264, 317, 394,
    395, 447, 460
    beaver, 460–62
    coastal, 351
population, 197
    dynamics, 216
    genetics, 84
population density, 89, 206,
    208–13, 214, 217, 245,
    249, 252–53, 280–82,
    296–97, 306, 312, 382,
    401, 452–53
    high, 29, 57, 207, 210, 213–14,
        245, 249, 296, 301,
        305, 420
    human, 22, 248–49, 252,
        419–21
    local, 215, 217, 220–21, 230,
        235, 374, 382, 452–53, 465
    low, 211, 213, 241, 244–45,
        314, 362
    natural, 180
    reduced, 393
Portugal, 29, 160
positive phototaxis, 317
Potentilla glandulosa, 81–82
prairies, 30, 298, 358–60, 415
    short-grass, 30, 386
    species-rich North
        American, 30
    tall-grass, 30, 385–86
    uncolonized, 415
precipitation, 12, 14–18, 20,
    30–33, 36, 38–40, 42,
    47–48, 51, 198–200,
    385–87, 416–18, 419,
    428, 454
    abundant, 15, 498
    annual, 12, 16, 26, 137, 200,
        385–86, 399, 408, 483
    global, 493
    high, 43, 497
    higher-than-average, 494, 512
    increased, 385, 494, 497, 509–11
    low, 43
    maximum, 30
    moderate, 42
    monthly, 16
    plotted, 386
    unpredictable, 42
predators, 160–61, 166, 175–77,
    185–87, 194–95, 219–20,
    299–301, 303–15, 317,

322–23, 366, 371–73,
    381–84, 395–96, 398–99
    active, 50
    aquatic, 312, 316
    attacking, 337
    attendant, 31
    avian, 305
    coral, 343
    dangerous, 191
    effective, 150, 171, 175, 178
    effects of, 299, 382, 392
    efficient, 162
    excluded mammalian, 307
    satiation, 313
    size-selective, 323
    terrestrial vertebrate, 304
    top, 367, 371–73, 376, 397–98
    ultimate, 366
    visual, 175–76
    wide variety of, 176, 334–36
prey, 50, 160–62, 164–67, 170,
    173, 299–300, 303–5,
    307–16, 323–24, 337,
    339, 365, 369, 392, 396
    abundant, 167
    animal, 35, 159
    arthropod, 141, 179–80
    density, 164, 219, 236, 312,
        314–15
    different-sized, 161
    individual, 312
    large, 161–62
    live, 194, 309
    noxious, 171
    per capita availability of, 314–15
    potential, 166–67, 168
    potential insect, 163
    preferred, 392
    processing, 166
    ratio of, 299, 314–16
    selected, 167
    shelter terrestrial, 312
    structure of, 299–300, 323
    tropical regions, 162
    well-defended, 161
prey-dependent functional
    response, 314
prey populations, 160–62, 291,
    303–5, 307, 309–11, 323,
    369, 371, 373
    changing, 89
    effects of predators on, 313
    owl's, 89
    prey species, 89, 160, 166,
        304, 313, 316, 323–25,
        369, 371, 375, 382
    dominant large, 165
    lynx's staple, 305
    new, 168
    single, 166, 316
primary producers, 13
primary production, 390

primary succession, 424
principle of allocation, 105, 165
priority effect, 288
prokaryote, 150
protozoan parasite Adelina
    tribolii, 324
psychrophilic, 109
pubescence, 102
Puerto Rico, 441, 483, 488
Pyramid Lake, 354

Q

quantitative genetics, 87
quantitative models, 7
Queensland, 375
Quercus alba, 228

R

radiation, 111, 113, 120, 125,
    490, 507
    active, 150–51, 170, 200, 384
    electromagnetic, 110–11,
        486, 488
    heat gain by, 111–13
    infrared, 383, 492
radish, wild, 181–83, 189, 194
rainfall, 12, 14, 24–25, 103, 138,
    185–86, 200, 245–47,
    252, 386, 394, 399,
    493, 498
    abundant, 22, 498
    high, 246–47, 364
    seasonal, 43
rain forests, 5, 12, 19–22, 34,
    139–39, 149, 268, 356,
    378, 380, 482
    central African, 21
    lush temperate, 383
    nearest, 22
    total, 501
rain shadow effect, 17
random distribution, 202
rank-abundance curve, 350
ratio-dependent functional
    response, 315
rats, 137, 189–90
    kangaroo, 134–35, 137, 148,
        291–92, 460
recolonization, 477
    followed, 476
    invertebrate, 428
red foxes, 298, 303–6
    population of, 303
red kangaroos, 198–99, 202,
    498–99
    female, 498
Red Sea, 49–51, 55, 379
realized niche, 198
redwoods, 254–55
reefs, 51, 173, 279, 337, 357, 367, 378
    barrier, 54, 56, 75–76
    fringing, 54, 56, 75–76, 326

refuges, 37, 83, 102, 160, 214, 218, 299, 310–13, 323–24, 429, 480
  earthly, 37
  establishing protected, 214
  ice age, 218
  physical, 312
  spatial, 312, 324
reindeer, 34, 36, 102
regular distribution, 202
relative humidity, 129
remote sensing, 485
repellents, 159–60, 333
reproduction, 87–89, 173–74, 184, 186, 191–92, 198, 219–20, 228–29, 233, 254, 261–63, 267–72, 275–75, 283–83, 396
  asexual, 174, 230
  continuous, 231
  deferring, 262
  delay, 186, 262
  delayed, 186
  delaying, 186, 262
  early, 267
  episodic, 236
  female, 174
  higher, 88, 307
  increased, 499
  lower, 88, 186
  male, 174
  natural, 230
  organisms defer, 254, 262, 275
  pulsed, 230–31
  reduced, 91, 233
  repeated, 267
  seasonal, 314
  sexual, 173–74, 194, 319
reproductive effort, 262
reptiles, 107, 113, 230, 262, 270, 275, 377
reservoirs, 47, 66, 68, 75–76, 311, 364
  largest, 47–48
resilience, 437
resistance, 437
resource (or exploitative) competition, 280
resource limitation, 279
respiration, 71, 102, 140, 163, 336, 394, 396, 401, 407, 423, 429, 431–31, 443
  cellular, 110, 134
  gross primary production minus, 384, 401
  invertebrate, 431
restoration, 417, 440–41, 442–43
  accelerated, 442
  ecology, 440
  studied, 441
rhodopsin, 150
Ricklefs, Robert, 483–84

Rio Grande, 64, 206, 228–30, 276
Rio Negro, 67, 485
riparian vegetation, 105
riparian zone, 65
river basins, 64–65, 197, 419–20
river continuum concept, 67
river ecosystem synthesis, 67
river systems, 67–68, 261, 485
river valleys, 27, 206, 208, 249, 456
RM endothermy, 117
roots
  actual, 205
  high, 330
  higher, 167
  increased, 168
  influenced, 330
  large, 119
  lower, 168, 330
  penetrating, 19, 135–35
  pine, 280
  prop, 59–60
  reduced, 329
  shallow, 34, 135, 139
  strong, 47
Rosemond, Amy, 412
Rosenzweig, Michael, 385–86, 422, 480–82
r selection, 266
rubisco, 151
ruderal, 267
Russia, 30, 33–34, 36, 273, 463
Rwanda, 249–50

S

saguaro cactus, 139–39, 148, 196, 207
Sahara Desert, 24
salinity, 44–45, 49, 51, 56, 59, 62, 67, 71, 75–76, 129–30, 279, 497
  high, 128, 497
  increasing, 63
  lake's, 497
  oceanic, 71
  reduced, 59, 497
  stable, 56
  total, 51
  world average, 71
salmon, 83, 267, 273
salt marsh, 59
salts, 27, 51–52, 61, 128, 130, 144–45, 148, 279
  absorbing, 144
  amount of, 51, 67
  inorganic, 130
  internal, 143
saltwater, 45
Santa Catalina Mountains, 207
Santa Cruz Island, 92, 99, 294–95

Santa Rosa Islands, 475, 488
saturation water vapor pressure, 129
savannas, 12, 17, 22–24, 59, 199, 399, 422
scatterhoarded, 259
Sceloporus undulatus, 107–8, 113–14, 264
scorpionflies, 178–80, 189, 194–95
  female, 178
  male, 178–81
scorpions, 11, 13, 140–41, 148, 165
Scotland, 201, 290
seabirds, 54, 173, 495
seals, 117, 366
  female fur, 496
  northern elephant, 237, 243–44
  southern fur, 497
sea otters, 117, 362
seas, 47, 49–51, 53–54, 56–59, 62–65, 69, 76, 117, 151, 155, 159, 202, 220, 237–38, 386
  high, 217
  high-latitude, 75
  open, 31, 52, 54, 105
  shallow, 50, 389
  smaller, 50
  tan desert, 207
  windswept, 365
sea stars, 47, 56, 59, 337, 373–75, 382
  predatory, 337
  predatory crown-of-thorns, 56
seaweeds, 54, 59, 75, 159, 194
secondary producers, 13
secondary production, 396
secondary succession, 424
sedimentation, 442, 456, 465
selection
  coefficient, 340
  environmental, 95
  intrasexual, 174–75, 178, 181, 187, 194–95
  kin, 183–84, 188–89, 191–92, 195
  population genetics and natural, 78–79, 81, 83, 85, 87, 89, 91, 93, 95, 97–99, 162
  strong, 90, 314, 334
selective pressure, 13
self-incompatibility, 181
self-thinning, 209, 280–80, 297–98
semelparity, 267
sexual selection, 174–76, 181, 189, 194–95
Shannon-Wiener index, 348–50, 352, 363

sharks, 143–43, 148, 230, 256, 270, 307
  dogfish, 263
sheep, 175, 224–27, 462
  desert bighorn, 451, 465
Shetland Islands, 160
shrimp, 192, 262, 337, 344
  brine, 497
  pistol, 336–38
shrubs, 21, 28–30, 32, 34, 103–4, 108–9, 200, 204–5, 306, 345–47, 359, 361, 408–9, 423, 427
  abundant, 5
  azalea, 255
  excavated, 205
  fruit-bearing, 33
  large, 205
  low, 103, 425, 459
  low-growing, 36
  scattered, 137
  short-lived perennial, 163, 170
  small, 204, 346
  tall, 103
  understory, 306
  young, 204
Siberia, 33, 34, 69–70, 197
sigmoidal population growth curve, 243
singing, 94–95, 384
size-selective predation, 161
small-scale phenomena, 202
snails, 123, 164, 220–21, 233, 296, 374–75, 382, 413, 464
  invasive marine, 293–96
  migrating, 221
snakes, 107, 160, 191–92, 206, 262–63, 275, 278, 300
snowshoe hares, 34, 302–7, 323–22
  population cycles in, 306–7
soapberry bugs, 90–91
  populations of, 90–92, 99
sociality, 184
Society Islands, 94
sociobiology, 173
soils, 17, 19–20, 22, 24–28, 30–31, 102–4, 128–35, 167–71, 327–30, 354–57, 405–7, 410–11, 415–17, 429–30, 456–58
  acidic, 32, 289–90, 295, 353
  bare, 103, 228
  basic, 289–90, 353
  brown, 30
  clay, 132
  collected, 328
  contaminate, 169
  contaminated, 169
  distinctive, 456
  drained, 424

dry, 136
experimental, 167
exposed, 438
fertile, 22, 30–32, 42, 168, 326, 411
fertilizing, 329
increased, 437
infertile, 42, 168
mountain, 38
nitrogen-poor, 167
nutrient-rich, 167
old, 457
older, 431, 457
overlying, 104
poorly developed, 42
salty, 279
shallow, 201
studied, 456
supported, 410
surrounding, 132
tropical, 67
tundra, 36
unsterilized, 335
unwatered, 135
volcanic, 416
well-drained, 456
young, 457
solar energy, 14, 42, 44–45, 47, 75, 102, 384, 397, 464, 465, 492
incoming, 492
solifluction, 36
Sonoran Desert, 26, 127, 133, 138–41, 196, 207, 346, 456, 461
Sousa, Wayne, 355–56, 425–26, 431–33, 435, 442
South Africa, 29, 59, 186, 190, 376, 380, 420, 480, 485
South America, 20, 23–25, 38–39, 49, 55, 161, 189, 216, 466–67, 479, 482, 485, 488, 491–95, 508–9
northern, 330
southern, 79
South Carolina, 107–8, 114, 233, 395–96, 453
South Dakota, 359–60
southern Arizona, 207, 216, 346
Southern Hemisphere, 14–16, 28, 30, 31, 42, 48, 50–51, 103, 480
Southern Oscillation Index, 493–94
soybeans, 97–100, 151, 499
spate, 220
sparse desert vegetation, 26
species abundance, 345, 347, 350, 353, 355, 358, 359, 362–63, 439
species composition, 39, 43, 63, 73–74, 354, 423–24, 427,

432, 443–44, 471, 474, 476
species diversity, 345, 348–53, 356–58, 360, 363–64, 371–73, 390–91, 421–23, 427, 443–43, 467, 470, 478–80, 482, 484
algal, 428
calculating, 350
communities bird, 353
foliage height diversity and bird, 351–53, 483
high bird, 353
higher, 348–50, 360, 372
increased, 443
invertebrate, 428
local, 427
low bird, 353
lower, 350
reduced, 360
species evenness, 348
species extinction, 474, 509, 511
species rarefaction curve, 348
species richness, 348–51, 361, 363–64, 373, 382, 391, 425–27, 468, 470–74, 476–78, 480–84, 487–88
algal, 375, 391
equal levels of, 348
exceptional, 487
high, 479, 487
higher tropical, 481, 487
highest, 362
ichneumonid wasp, 479
increased, 425
increased plant, 359, 441
increasing plant, 391
latitudinal gradients in, 478, 480–81, 487
lower, 484
reef fish, 382
seed, 441
total, 471
species turnover, 474
sperm, 81, 86, 173–74, 192
spiraling length, 413
sphinx moths, 118, 141
sponges, 45, 173, 337, 373
spruce, 9, 34, 38, 213, 305–6
stability, 436
stabilizing selection, 87
stable age distribution, 230
stable isotope analysis, 145
starlings, 210, 317–19, 362
stamen, 181
static life table, 224
storms, 21, 59, 212, 247, 293, 494, 497, 509
stratosphere, 492
stream order, 65
streams, 36–36, 64–67, 74–77, 105, 220–21, 254–56,

324, 346–47, 355, 399, 411–15, 418, 420–21, 431–32
arid-land, 66
clear, 35, 107, 221
clearest, 65
clear mountain, 175
first-order, 65
higher-order, 65
insect inhabits, 300
larger, 421
lower-order, 65
low predation, 177
medium-sized, 67, 212
mountain, 351
salty, 59
second-order, 65
smaller, 176
steady, 332
swift, 47
temperate zone, 411
tributary, 74, 254
stress, 267
stress-tolerant plant, 268
strong interactions, 367
subpopulation, 221
succession, 424
Suess effect, 506, 512
sugars, 108, 150–53, 243, 326, 331, 384
summer, 14, 24, 27–28, 30, 33, 35–36, 70–72, 102–3, 112, 121, 123, 125, 145–48, 199, 376–77
dry, 28, 42, 408, 462
early, 181, 376
late, 319, 369, 376
short, 36
soggy, 42
sun, 11, 13–14, 16, 41–42, 45, 47, 51, 57, 102, 111–14, 120–21, 124, 139–40, 163, 165
bright equatorial, 79
intense desert, 139
midday, 139, 139
sunfish, 256, 270
pumpkinseed, 262, 264–65, 275
sunflowers, 102, 326, 334
sunlight, 45, 47, 51, 61, 66, 70, 102, 111–13, 139, 150–51, 153, 163, 237, 384, 464
direct, 103, 139
intense, 27, 153, 434
surface temperatures, 48, 57, 123
bat's, 116
cooler sea, 486
earth's, 490, 492
elevated, 464
highest average, 48, 51
lower sea, 494, 512

minimum sea, 61
warm sea, 494
surface waters, 48, 51, 54, 62, 65, 70–71, 75, 354, 389, 403, 421, 438, 450, 454–55, 496
warm, 496
well-lighted, 109, 317
survival, 87–89, 91–92, 188, 193, 198, 216–17, 224–28, 230, 254–55, 262, 264, 274, 282, 291
adult pumpkinseed, 265, 275
differential, 78–79, 87, 99
equal, 99
estimating patterns of, 224
higher rates of, 87, 194, 226–27, 307
high pest, 323
history of, 215, 228, 235
human, 445
improved, 184
isopod, 282
larval, 96, 100
limiting, 282
lizard, 262
low, 228, 235, 323
lower rates of, 87–88
pattern of, 215, 223–28, 235
survivorship curve, 225
Sweden, 33–34, 227, 249–50, 260, 300, 303–4, 323, 390, 470, 476–77, 487
Swiss National Park, 234
Switzerland, 83, 123–25, 233–34, 464, 505
Sycamore Creek, 413–14, 428–29, 431–32, 436, 438–39, 443–44, 450
flood-devastated, 429
sympatric, 293

T

temperate forests, 12, 15, 17–18, 20, 29, 31–33, 42, 51, 158–59, 167, 402, 409–10, 427, 435, 483–84
coniferous, 31
cool, 18
northern, 199
old-growth, 32
temperate grassland, 29
temperature, 101
Tennessee, 66, 208, 486
terrariums, 179–80
terrestrial biomes, 11–13, 16, 19, 42, 45, 49, 67, 385
terrestrial ecosystems, 327, 330, 383, 385–86, 392, 394, 396, 401, 403, 405–6, 408, 411–12, 414–15, 419, 421

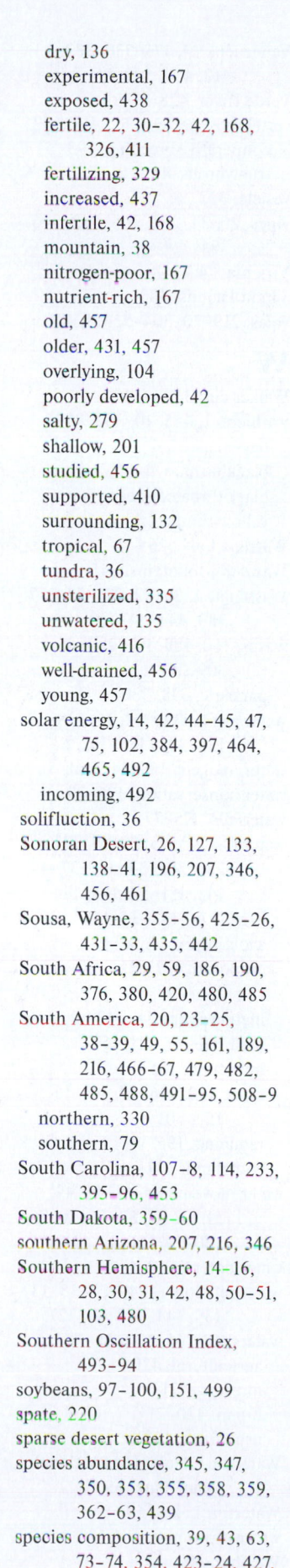

terrestrial ecosystems—*Cont.*
  rate of decomposition in, 421–22
  rates of primary production in, 387, 391
terrestrial isopods, 280, 282, 297, 317–18
terrestrial mutualists, 325, 336, 343
terrestrial organisms, 44, 46–47, 49, 130, 133, 143
terrestrial plants, 127, 132–33, 136, 147, 163, 257, 335, 352–53, 363
Texas, 9, 240, 322, 469, 511
  south-central, 321
  southwestern, 321, 323–24
Texas Gulf Coast, 241
thermal neutral zone, 115
thermocline, 47
thermohaline circulation, 51
thermophilic, 110
thermosphere, 492
—3/2 self-thinning rule, 281
tides, 50, 57–59, 61–63, 75, 200
tiger beetles, 120–21, 136, 199, 213
  black, 120
  overheating, 139
  predatory, 120
tigers, 63, 210–12
Tilman, David, 167–68, 280, 298, 353, 364, 390–91
Togiak River, 64
tolerance model, 433
Tomales Bay, 296
top-down control, 392
torpor, 121
traits, 13, 79–82, 86–89, 91–92, 96, 99, 174–75, 194
  adaptive, 95
  behavioral, 79, 175
  biological, 45, 189
  competitive, 181
  divergent, 75
  functional, 13, 41–42
  genetically based, 94
  heritability of, 89
  heritable, 84
  life history, 256
  phenotypic, 89
  polygenic, 87, 100
  rare, 80
  studied reproductive, 264
trees, 4, 20–22, 24, 30–31, 34–34, 202–3, 228–30, 255, 260–61, 277–79, 416–17, 460, 463–64, 483–84
  abundant, 367
  alder, 255
  aspen, 313, 395
  beaver-felled, 369

birch, 34, 168
canopy, 4–5
cherry, 272
chestnut, 8
conifer, 34, 103
coniferous, 38, 150
dominant, 34, 427
fast-growing, 425
flat-podded golden rain, 90
giant Eucalyptus, 31
intact, 369
invading nitrogen-fixing, 416
largest, 32, 210
long-dead, 424
mangrove, 61, 63, 476
maple, 221, 423
mature, 209, 213
medium-sized, 346
mesquite, 140–41
native, 279, 408, 416
older, 228–29
palo verde, 346
pine, 157, 197
potential nest, 203, 205
redwood, 254–55
riparian, 419
roost, 197
round-podded golden rain, 90
saltcedar, 279
scattered, 22–24
shade-tolerant understory, 32
single, 506
single rain forest, 20
small, 299–300, 427
soapberry, 90
temperate zone, 482
uncut, 369
walnut, 278
young, 24, 32, 228–29
Tri-trophic interactions, 398
tropical forests, 22, 32, 153, 300–1, 303, 376, 402, 408, 410, 412, 421–22, 441, 500
  lowland, 303, 321, 323
  world's, 498
tropical rain forests, 4, 19–22, 25, 32, 34, 40, 42–43, 53, 56, 123, 125, 364, 377, 479, 481
trophic (feeding) biology, 150
trophic cascade, 392
trophic dynamics, 397
trophic level, 384
tropical dry forest, 21
tropical rain forest, 19
tropical savanna, 22
troposphere, 492
trout, 67, 72, 74, 83, 107, 125, 220, 255, 454
  rainbow, 107, 125
Trout Lake, 454–55
Truckee River, 354

Tucson Mountains, 456–58
tunas, 117–18
tundra, 12, 18–19, 33, 35–36, 42–43, 51, 111, 385–87, 402, 480
Turkey, 212, 217–18
turtles, 136, 224, 232
  common mud, 227, 231–33
type I survivorship curve, 227
type II survivorship curve, 227
type III survivorship curve, 227

U
Ugland, 348
ultraviolet light, 151, 171, 492, 507–8, 511
upwelling, 48, 51–52, 76, 388, 439–40, 486, 495–96
  coastal, 487
  nitrate-rich waters, 439
upwelling zones, 432, 439–40
ural owls, 89–90
  female, 89
  population of, 90, 99
urban ecology, 3, 361, 362, 463
urban heat island, 463
urchins, 76, 159

V
valleys
  deforested stream, 417
  small stream, 417
  temperate stream, 461
  undisturbed stream, 417
Vancouver, 423–24
vapor pressure deficit, 129
vegetation, 19, 30–32, 38, 102–5, 123–24, 137, 167, 352, 354–55, 359–60, 407, 417–19, 421–22, 441–42, 462–66
  clean salt marsh, 63
  clipping, 359
  competing, 228
  dense, 28
  desert scrub, 470
  forest floor, 280
  grassland, 30
  intact, 449
  low, 42, 151
  mangrove, 62
  natural, 21, 24, 123, 462
  natural woodland, 362
  restoring, 441, 442
  riparian, 65, 68, 105, 125, 260, 313
  semiarid, 484, 498, 511
  support scrubby, 501
  thick, 424, 426, 505
  understory, 418, 502
  woody, 23, 29, 30, 36, 207, 258–60, 346, 411, 427, 470

Venezuela, 24, 176, 330, 382, 413, 450, 509, 511
Verde River, 428
vertebrate species, 262, 413, 482
  cooperative-breeding, 193
  frugivorous, 482
vesicle, 327
vines, 20, 34, 258, 260, 332, 346, 384
Virginia, 74
Virgin Islands, 488
voles, 219–20, 304–5, 451, 471

W
Walker circulation, 494
warblers, 1, 4–5, 10, 352, 364
  bay-breasted, 4
  blackburnian, 4
  black-throated green, 4
  yellow-rumped, 1, 4
Warrens Lake, 265
Wasatch Mountains, 345
Washington, 30, 32, 371–73, 377, 383, 444, 447–49
wasps, 160, 189, 191, 284–85, 309, 367
  parasitic, 378, 380
water availability, 12, 16, 128–29, 132–33, 135, 136, 355
water budgets, 139, 141, 148
water conservation, 136–39
waterfalls, 175–77
water loss, 13, 25, 27, 29, 127–28, 130, 133–36, 137–44, 148, 151–53, 165, 184
  avenue of, 133–34, 143
  avoiding excessive, 184
  evaporative, 27, 129–30, 136–42, 148
  high evaporative, 142
  high rates of, 141, 151
  minimum, 165
  rates, 130, 136–39, 141, 151, 153, 201
  reducing, 136
  withstand, 139
water movements, 44–45, 49, 51, 56–57, 62, 66, 70, 75, 130–31, 141–42, 148
water potential, 131
water relations, 124, 127–35, 137, 139, 144, 146–48, 327
watersheds, 65, 420–21
  agricultural, 420
  forested, 421
  urban, 420–21
  urbanized, 420
Waterton-Glacier International Peace Park, 69
Waterton Lakes, 69
water vapor pressure, 129
Weather, 40

Weber River, 369
Weinberg, 84–86
wet deposition, 420
wetlands, 59, 61–64, 66, 95, 367, 441, 460
whales, 10, 45, 53, 117, 197, 224, 240, 365–66
  baleen, 365–66
whitefish, 72, 83, 99
White Mountains, 114
Whole-Lake Experiments on Primary Production, 388
wildebeest, 25, 394–95
wildfires, 462, 509

wildflowers, 30, 319, 345
willows, 34–35, 113, 306, 313, 425
Wind Cave National Park, 359–60
winter, 14, 25–31, 33–36, 42, 51, 54, 70–71, 101, 107, 109, 120–21, 149, 197, 205–7, 216, 223, 281, 300, 404, 471, 493
  cold, 42, 101
  long, 33, 35
  warmer, 509
  wet, 408

Wisconsin, 199, 397, 454, 458–59, 471
wolves, 31, 34, 36, 164–65, 184, 300, 313, 316, 395
  gray, 313
  preying, 315
worm, 155, 300
giant tube, 53, 155
spiny-headed, 317, 324

X

xylem, 132, 146–47

Y

yeast, 243, 310
Yellowstone National Park, 69, 110, 313, 418

Z

zebras, 2–3, 25, 187, 394
zonation of species, 59
zooplankton, 52–53, 68, 76, 165, 167, 210, 336, 365, 371–72, 389, 392–93, 401, 403, 497